Springer Tracts in Modern Physics

Volume 288

Springer Tracts in Modern Physics provides comprehensive and critical reviews of topics of current interest in physics. The following fields are emphasized:

- Particle and Nuclear Physics
- Condensed Matter Physics
- Light Matter Interaction
- Atomic and Molecular Physics

Suitable reviews of other fields can also be accepted. The Editors encourage prospective authors to correspond with them in advance of submitting a manuscript. For reviews of topics belonging to the above mentioned fields, they should address the responsible Editor as listed in "Contact the Editors".

Joachim Stöhr

The Nature of X-Rays and Their Interactions with Matter

Springer

Joachim Stöhr
SLAC National Accelerator Laboratory
Menlo Park, CA, USA

ISSN 0081-3869 ISSN 1615-0430 (electronic)
Springer Tracts in Modern Physics
ISBN 978-3-031-20743-3 ISBN 978-3-031-20744-0 (eBook)
https://doi.org/10.1007/978-3-031-20744-0

This Springer imprint is published by the registered company Springer Nature Switzerland AG
The registered company address is: Gewerbestrasse 11, 6330 Cham, Switzerland

To Linda,
My ski buddy, my best friend, my muse,
my love, my wife.

Preface

Starting in the mid-1970s, I have had the extraordinary luck of experiencing and participating in two revolutions in x-ray science. It is the wonderful developments I witnessed during these nearly fifty years and my role in the development of x-ray techniques and the formulation of scientific cases for advanced synchrotron radiation (SR) facilities and x-ray free electron lasers (XFELs) that have led me to take on the endeavor of writing the present book. The other impetus was the recognition that the very process would allow me to learn what I craved to understand on a deeper level or for the first time.

The period 1970–2000 saw the development of SR sources and science, based on the utilization of electron storage rings for the production of x-rays and the development of modern x-ray techniques. Storage ring-based sources offer high photon flux over a broad spectral range and by use of monochromators provided tunable high-intensity x-rays for the first time, soon outdating previous x-ray sources based on Bremsstrahlung and characteristic line emission. The early development and use of SR continued in the 1980s–90s through the extension to undulator sources and optimization of dedicated electron storage rings. Today, the development of SR sources continues with the goal of building "ultimate" storage rings that produce diffraction limited x-ray beams into the multi-keV range.

In the period 2005–2010, a second revolution occurred with the advent of XFELs based on the development of high brightness electron sources in conjunction with linear accelerators. In electron storage rings, the size of the continually circulating electron bunches is limited by an equilibrium between the forces of the focusing magnets that overcome the Coulombic repulsion between the electrons and the destabilizing radiation losses. Linear accelerators offer the advantage that the electron beam is not subjected to radiation losses, and exquisitely prepared electron bunches can therefore be used to create radiation in a long undulator, with the electron bunch then being discarded. While the repetition rate of linacs is considerably lower than that of storage rings, each XFEL pulse is in principle sufficient to obtain a spectrum or diffraction pattern.

Over the years, my particular interest and efforts have been the development of *soft x-ray techniques* and their applications. My first book *NEXAFS Spectroscopy*, published in 1992 [1], dealt with the use of x-rays for problems in surface science and chemistry. It was influenced by what I learned at EXXON Research and Engineering Company about catalytic processes and about thin polymer films at the IBM Almaden Research Center. My second book, *Magnetism: From Fundamentals to Nanoscale Dynamics*, co-authored with H. C. Siegmann and published in 2006 [2] was largely the result of what I learned from my colleagues at IBM Almaden, from the literature and especially from Hans Siegmann. By inviting Hans to Stanford in 2000 after he had to retire from ETH, I helped him to actively continue what he loved most, doing physics. He taught me the mysteries and magic of magnetism and in the process allowed me to tap into his deep knowledge of science.

The present book was written during the last ten years or so, after finishing management periods as Director of the Stanford Synchrotron Radiation Lightsource (SSRL) and the Linac Coherent Light Source (LCLS) at SLAC/Stanford. I finally felt that I had time (partly through retirement) to dig deeper into scientific questions I had wondered about, and naturally explored topics associated with my own research. I therefore could have added the words "soft" to x-rays and "resonant" to interactions in the title of the book, but the present broader title is intended to pull in the entire x-ray community. I have tried to connect classical, quantum, and quantum electrodynamics (QED) concepts. Their unified formulation required repetition of conventional semi-classical treatments that may be found in other books, but what I consider novel is the introduction of QED and quantum optics concepts to x-ray science. Whatever area of x-ray science you come from, I hope that you will find the book pedagogical and interesting and sections of it novel and fascinating.

I have found that some of the most difficult questions to answer are naive childlike questions since they often go to the core of things. The impetus for writing this book was partly my curiosity of understanding the nature of "light", especially the concepts of coherence, interference, and diffraction, and how to describe x-ray interactions with matter. I completely underestimated the difficulty of doing so and repeatedly wondered how I could finish this book without understanding the very nature of "light" itself to my own satisfaction. I was consoled by Feynman's words *"Even the experts don't understand it as well as they would like"*. In physics, "understanding" typically means describing observations through mathematical formulations that through certain assumptions and symmetries reduce the infinite complexity of nature. In the process, I often wondered about alternative formulations, say by extraterrestrials.

Writing this book was not a "brain dump" of my knowledge. In many cases, I first had to read the literature to learn the material, either from scratch or beyond a cursory understanding. Most of my time was spent on mathematical derivations of different phenomena—a lonesome endeavor—which resulted in long detailed drafts of chapters. Mind you, the joyous periods, when a solution emerged or a section had been finished, were interleaved with periods of agonizing over things. Even the search for a minus sign, a factor of 2, π or i can be frustrating, time consuming, and drive you crazy! On the other hand, I often thought about the privilege of being a scientist. Like artists, true scientists want to explore and create, but we are fortunate to get paid to pursue our passion and do not fall into the category of "starving artists".

I very much like how Steve Jobs described the creative process at a Stanford commencement speech: "When you first start off trying to solve a problem, the first solutions you come up with are very complex, and most people stop there. But if you keep going, and live with the problem and peel more layers of the onion off, you can often times arrive at some very elegant and simple solutions." I often found myself working out a problem with full mathematical rigor, only to isolate its physical essence after stripping away or simplifying much of the math. I have strived to interpret the abstract mathematical formulations by simple mental pictures. It is these pictures that remain even after you have forgotten the details.

I seriously started to work on this book after stepping down from management in 2013. Getting older, I wanted to pursue my passions rather than to deal with managerial responsibilities, which increasingly appeared to be "golden handcuffs". My favorite days as a manager were those when my calendar allowed me to devote time to learning and writing and especially the "group meetings" with my students and collaborators. After "retiring", I spent the winters at beautiful Red Mountain Resort in British Columbia, Canada, where periods of writing were interleaved with my other passion, deep powder downhill skiing, and classical and skating cross-country skiing with my wife Linda.

One of the most significant realizations for me in writing this book was that the understanding of light (or the world around us) resembles an infinite series of steps that appears bottomless. I therefore had to draw the line at a certain level of my own understanding. In the end, I also had to make choices of what to include in the final version of the book. My detailed write-ups and derivations amounted to about 1500 pages in the format of this book, and I had to omit topics and leave out details of derivations. I did not formulate homework problems for students, hoping that the book provides plenty of material to entice the reader's scientific curiosity. I personally have learned more from digging into curiosity-inspired problems than homework assignments.

The final material chosen for the book naturally reflects my personal taste and expertise developed through my 50-year journey from synchrotron radiation to XFEL developments and research. As a scientist, whose first goal was opening up the soft x-ray region, previously plagued by a gap from about 250–3000 eV due to carbon contaminated optics at the low end and Be window transmission at the high end, I have admired and benefitted from the pioneering contributions of Burton Henke, largely based on laboratory-based x-ray sources, which culminated in the modern compilation of x-ray scattering factors and cross sections. They are now widely available in updated form on the web[1] through the efforts of Eric Gullikson. My learning process was also aided by key books: D. Attwood's and A. Sakdinawat's *X-Rays and Extreme Ultraviolet Radiation: Principles and Applications*, J. Als-Nielsen and D. McMorrow's *Elements of Modern X-Ray Physics*, J. J. Sakurai's *Modern Quantum Mechanics*, J. W. Goodman's books *Fourier Optics* and *Statistical Optics*, E. Hecht's *Optics*, R. Loudon's *The Quantum Theory of Light* and M. O.

[1] https://henke.lbl.gov/optical_constants/.

Scully and M. S. Zubairy's *Quantum Optics*. I also spent considerable time learning quantum optics from other books and papers, referenced in the text.

Of course, I learned from many people in person like my students, of whom Bill Schlotter and Diling Zhu became true x-ray "jocks". I especially befitted from my more recent postdocs and collaborators, in particular, Ives Acremann, Hermann Dürr, Stefan Eisebitt, Jan Lüning, Hendrik Ohldag, Alexander Reid, and Andreas Scherz. Andreas sent me in a new direction when he showed me data that he and my student Benny Wu had measured at LCLS in 2012. Rather than turning the problem over to theoreticians, as is often done by experimentalists like me, I wanted to understand the experimental results on my terms. So I followed Andreas' and my intuition about their meaning. This required learning quantum optics and formulating the phenomenon of diffraction in a photon picture.

During this arduous journey, filled with periods of inspiration and self-doubts, I realized the existing gap between the conventional concepts used in the x-ray community and the modern QED formalism and language of quantum optics. I hope that the present book contributes to overcoming this barrier, especially because the two communities increasingly mingle in XFEL-based science. Most of all, I hope that the book inspires young scientists to push beyond my understanding.

Over the years, I benefitted from discussions with (in alphabetical order) Massimo Altarelli, John Arthur, David Attwood, Uwe Bergmann, Martin Beye, Henry Chapman, Ryan Coffee, Georgi Dakovski, Hubert Ebert, Alexander Föhlisch, Kelly Gaffney, Frank De Groot, Giacomo Ghiringhelli, Janos Hajdu, Jerry Hastings, Olav Hellwig, Zhirong Huang, Ludvig Kjellsson, Jianbin Liu, Bob Nagler, Josef Nordgren, Nils Mårtensson, Martin Magnuson, Shaul Mukamel, Z. Y. (Jeff) Ou, Bruce Patterson, Lars Pettersson, Claudio Pellegrini, John Rehr, Andrei Rogalev, Jan-Erik Rubensson, Evgeny Saldin, Robin Santra, Bob Schoenlein, Sebastian Stepanow, Gerrit van der Laan, Michel Van Veenendaal, Ivan Vartanyants, and Linda Young.

Special thanks to Frithjof Nolting and Pietro Gambardella for sponsoring my sabbatical at PSI/ETH and Pietro for allowing me to use his beautiful experimental data, to Faris Gel'mukhanov for sharing his insights and detailed derivations of RIXS phenomena, to Joseph Goodman who patiently helped me with various optics problems over the years, and finally to my good friend Anders Nilsson for many exchanges on science, philosophy, and spirituality.

Possible corrections and updates of the book will be posted on my website [3].

Emerald Hills, CA, USA Joachim Stöhr

References

1. J. Stöhr, *NEXAFS Spectroscopy* (Springer, Heidelberg, 1992)
2. J. Stöhr, H.C. Siegmann, *Magnetism: From Fundamentals to Nanoscale Dynamics* (Springer, Heidelberg, 2006)
3. https://stohr.sites.stanford.edu/

Contents

About the Author

Joachim Stöhr received his Ph.D. from TU Munich, Germany, and, after spending time at Exxon and IBM Research Labs, joined Stanford University as a professor of Photon Science in 2000. He was the director of the Stanford Synchrotron Radiation Lightsource (2005–09) and the founding director of the Linac Coherent Light Source (2009–13). He has written two prior books, *NEXAFS Spectroscopy* (Springer, 1992) and *Magnetism: From Fundamentals to Nanoscale Dynamics* (Springer, 2006) with H. C. Siegmann. In 2011, he received the Davisson-Germer Prize in Surface Physics from the American Physical Society. He has been a professor emeritus of Photon Science since 2017. For more details see: https://stohr.sites.stanford.edu/

Chapter 1
Introduction and Overview

Mehr Licht! (More light!)
Last words of Johann Wolfgang von Goethe (1749–1832)

1.1 About the Present Book

We begin with a brief outline of the motivations and objectives of the book.

1.1.1 Motivations

The primary background and impetus for the present book is the development of increasingly bright sources of x-rays over the last fifty years and their use to obtain novel information about the electronic, spin, and atomic structure of matter. This development has paralleled or was interwoven with the development of optical light sources as illustrated in Fig. 1.1. While the invention of the light bulb, the discovery of x-rays, and the invention of the laser are widely recognized as revolutions, the more recent developments of new x-ray sources based on large accelerators may arguably also be associated with revolutionary developments.

All five scientific revolutions shown in Fig. 1.1 involve paradigm shifts in the utilization of "light". Within the context of the present book, we define "light" to cover the spectral range from optical to x-ray frequencies illustrated in Fig. 1.2.

Light is distinguished from longer wavelength electromagnetic (EM) radiation by its quantum detection process through the photoelectric effect. The true photon nature of what we call EM radiation, however, is not restricted to the so-defined wavelength range. It is a universal feature of EM radiation despite the power of Maxwell's classical theory of electromagnetic waves. At longer wavelengths it just becomes increasingly difficult to detect photons since typical detectors at room tem-

J. Stöhr, *The Nature of X-Rays and Their Interactions with Matter*, Springer Tracts in Modern Physics 288, https://doi.org/10.1007/978-3-031-20744-0_1

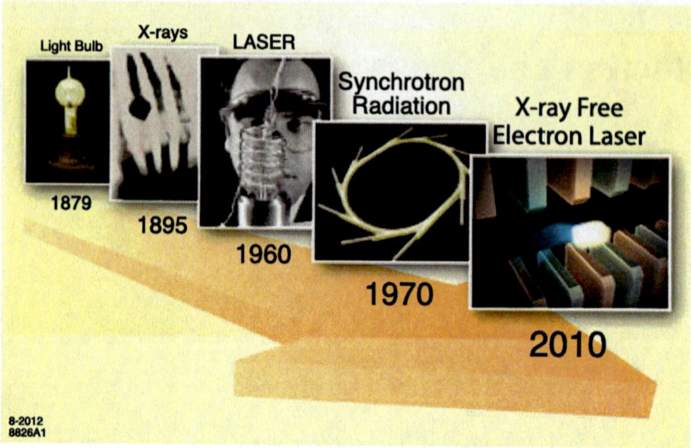

Fig. 1.1 Revolutions in the development of sources of light with time, indicated by approximate years

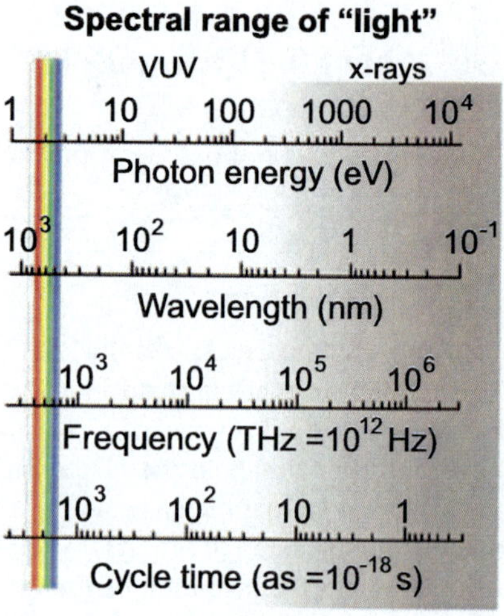

Fig. 1.2 Definition of "light" within this book, defined through its photoelectric detection process

perature have a thermal background noise corresponding to about 25 meV. Today, even micowave photons may be detected by use of photon counters based on Josephson junctions [1].

While the first three developments shown in Fig. 1.1 are revolutionary because of their practical impact on our everyday life, the development of the laser and the last two events are revolutionary because of the 10–15 orders of magnitude increase in photon brightness or photon density. Such increases are not just evolutionary but, using Thomas Kuhn's definition [2], can rightfully be referred to as revolutionary.

To put this development in a quantitative perspective, we show in Fig. 1.3 the number of photons produced in a birth volume of size $V = \lambda^3$ by different types of

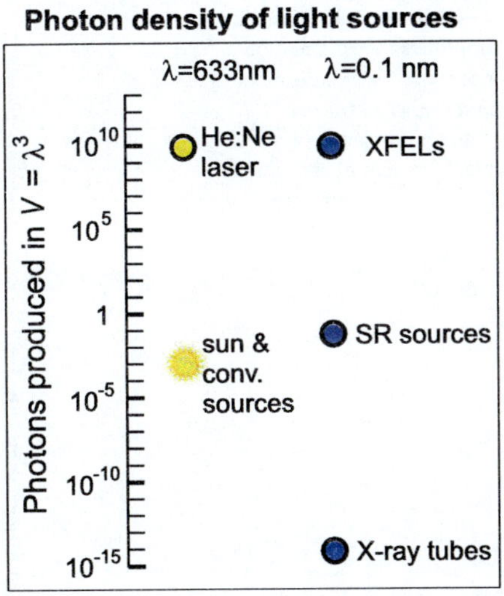

Fig. 1.3 Number of photons generated in a birth volume of order λ^3, the so-called photon degeneracy parameter, for different sources [4]

sources. We shall see that this number corresponds to the so-called *photon degeneracy parameter* or the number of first-order coherent photons [3, 4]. It is also proportional to the *peak brightness* of a source which is the conventional measure of its quality [5–7]. In this book we will use the terms brightness and brilliance interchangeably.

The photon energy dependence or spectral distribution of the peak brightness of conventional lasers, storage rings or "synchrotrons", and XFELs is illustrated in Fig. 1.4 [8]. Note that today the peak brightness of conventional lasers still falls off steeply beyond about 100 eV where accelerator-based sources take over.

1.1.2 Objectives

The goal of the present book is to present a modern description of the *nature* of x-rays and their *interactions* with the atomic and electronic structure of matter. Building on the pioneering quantum description of radiation by Dirac [9] and Heitler [10], and modern books on x-ray science [5, 6, 11], we here extend today's usual "semi-classical" treatment of x-rays and their interactions, based on a mixture of wave and photon concepts. The semi-classical treatment is reviewed and directly linked to that in modern quantum optics which is based on the fundamental description of light by quantum electrodynamics (QED). The quantum optics formalism, pioneered by Glauber [12–14] in the 1960s, has been widely used in the laser community [15–19], and it becomes increasingly important when the photon degeneracy parameter, or number of photons per coherence volume, is much larger than 1 as illustrated in Fig. 1.3. This is the case for XFELs where the emitted radiation contains on average a large number of photons per coherence volume. The formulation of x-ray/matter interactions through quantum optics readily allows for the simultaneous presence of many photons that may cooperate in a non-linear fashion, leading to fascinating new phenomena.

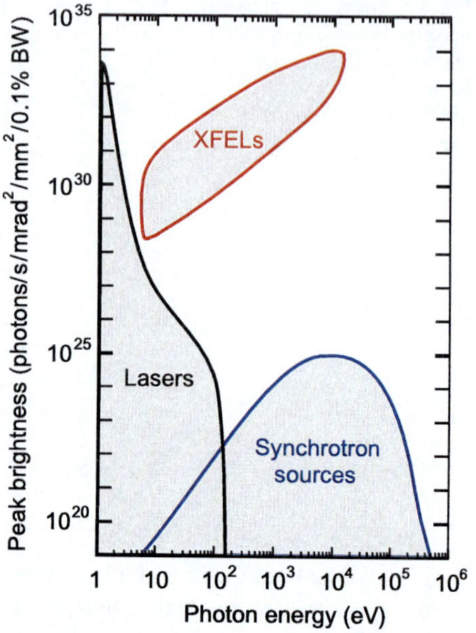

Fig. 1.4 Peak brilliance or brightness versus photon energy of conventional lasers and higher harmonic generation sources (lasers, black envelope), synchrotron sources (blue envelope), and x-ray free electron lasers (red envelope). Figure adapted from [8]

Our book complements and goes beyond other modern monographs on x-ray science that utilize a "semi-classical" description, such as D. Attwood's and A. Sakdinawat's *Soft X-Rays and Extreme Ultraviolet Radiation* [5] and J. Als-Nielsen's and D. McMorrow's *Elements of Modern X-Ray Physics* [6]. Mind you, the semi-classical description has served us well over the last 100 years and because of its deep roots will continue to guide our thinking in the interpretation of experiments, especially those performed with synchrotron radiation. But it is time to go beyond it.

XFEL experiments utilize ultrashort and intense pulses that give rise to many -photon or non-linear effects. Many XFEL experiments today utilize femtosecond optical pulses as time references and to trigger the dynamics of electronic, spin, and structural phenomena. While x-ray science in the twentieth century has largely explored time-averaged phenomena, the new century will see increasing emphasis on dynamical phenomena which underlie the function of systems. X-ray scientists have only begun to appreciate the concepts and language of quantum optics and to apply them for the interpretation of XFEL experiments [4].[1] The present book is envisioned to help bridge the existing gap in language and concepts used in the optical and x-ray laser communities which are increasing mingling.

[1] It is interesting that part of the optical laser community continues to use the classical field-based formalism, reflected by the terms "strong-field" phenomena and "higher order susceptibilities". The persistent use of this formalism may be traced to its long history and the fact that early laser experiments were also interpreted in the field-based language of nuclear magnetic resonance pioneered by Bloch, Purcell, and Rabi.

The *semi-classical* description of light explains diffraction through the Huygens-Fresnel principle of wave interference but avoids the specification of how the diffracted intensity is detected. The crucial detection step was addressed in Dirac's famous and bold statement that photons only interfere with themselves and not with each other [9]. Hence the conventional diffraction pattern consists of a binary assembly of 0 and 1 photon counts, corresponding to the destructive and constructive interference of a single photon wavefunction with itself. The various formulations of quantum mechanics [20] support the puzzling wave-particle duality. It is supplemented by Bohr's complementarity principle introduced in 1935 [21] which states that in a given experiment one may only observe either the wave or photon aspect of light but not both.

The acceptance of the wave-particle duality has allowed us to this day to use either the photon or wave concept to describe x-ray experiments. We typically use the *wave* concept to formulate interference and diffraction and the quantum concept of *photons* to describe atomic transitions and their spectroscopic observations. A key part of the present book is to provide a deeper understanding of when the wave and photon descriptions are equivalent and when the equivalence breaks down, i.e. when the semi-classical framework needs to be abandoned. This is elucidated here through the formulation of diffraction in a photon picture, facilitated by use of the concepts of modern quantum optics [14–16].

The validity of the semiclassical theory is based on the *linearity* of conventional quantum theory, which is only of *first order* within the more complete framework of QED [22]. Within QED, today accepted as the fundamental theory of light and matter that is valid to infinite order, light consists of elementary particles—photons. This is clearly stated by Feynman in his reader-friendly 1985 book "QED: The Strange Theory of Light and Matter" [23]. In the first order, these photons are independent and do not interact (Dirac's statement), while with increasing order correlations between photons become allowed. The incompleteness of the wave or first order QED formulation is most beautifully demonstrated when the quantum formulation of diffraction is extended to the second order. This is a key part of our book and in the process we establish an important and remarkable new paradigm that within QED, diffraction images are direct signatures of different quantum states of light.

The formulation of diffraction in a photon picture was famously introduced by Feynman in his Lectures on Physics [24], where he remarked that photon interference "*has in it the heart of quantum mechanics. In reality it contains the only mystery.*" The tricky aspect of the photon picture was similarly expressed in Einstein's 1951 statement, "*All the fifty years of conscious brooding have brought me no closer to the answer to the question, 'What are light quanta?' Nowadays every fool thinks he knows the answer, but he is deluding himself.*" Feynman unified the photon with the wave description by assigning probability amplitudes to photons and electrons which closely resemble wave fields.

The probability amplitude formulation becomes cumbersome with increasing order in QED [4], and for this reason we shall here utilize the more elegant albeit more abstract formulation of light and its interactions based on photon creation and annihilation operators. This formulation pioneered by Dirac [9] in the first order was extended to higher order in QED by Glauber [14]. We will show that the semi-classical description of light is equivalent to a *first order* approximation within QED, but that it breaks down in second order QED.

Clearly, choices had to be made in covering different x-ray phenomena and techniques. The present book emphasizes the processes associated with "x-rays in/x-rays out" in the most important forms of x-ray absorption, emission, scattering, and diffraction. We also describe the evolution of light-matter interactions with increasing energy from infrared to hard x-ray photons which naturally directs the attention to x-ray absorption edges or *resonances* which break up the otherwise smooth photon energy response. Due to their large cross section, they are also best suited to investigate the appearance of non-linear effects that arise in XFEL studies. Resonance phenomena form the basis of modern x-ray absorption techniques such as NEXAFS[2] [29] or XANES[3] and the important resonant elastic (REXS) and inelastic (RIXS) x-ray scattering techniques. Resonances also play an important role today in protein crystallography which is increasingly dominated by the use of Multiwavelength Anomalous Diffraction or MAD, where the photon energy is tuned around heavy atom absorption resonances [31, 32]. They furthermore exhibit strong polarization dependent effects which allow the separation of charge and spin phenomena in spectroscopy, microscopy, and diffractive imaging [11, 29, 33, 34].

We do *not* discuss extended x-ray absorption fine structure (EXAFS), which is reviewed in [25–27], details of conventional non-resonant x-ray crystallography which are well covered in the literature [6, 35, 36], and photoemission spectroscopy whose different aspects are treated in other books [37–39].

1.2 The Nature of Light: From Light Rays to QED

1.2.1 Early Concepts of the Nature of Light

Over the centuries, light has been the primary tool for seeing and understanding the world and universe around us. The early belief was that light was created by a deity, like the Egyptian Sun God Ra or the biblical God who said "Let there be light". Early theories of light were put forward by the Greek philosophers, and the concepts of light and vision remained closely linked until about 1600 when Johannes Kepler initiated a more physics-based approach to light itself [40].

While the concept of light as the opposite of darkness is fundamentally simple, the understanding of the true nature of light constitutes one of the most debated problems in the history of science. To quote Roy Glauber, who received the 2005 Nobel Prize in Physics for the development of modern quantum optics, "*Few problems of physics*

[2] The name NEXAFS for "near edge x-ray absorption fine structure" was created to distinguish it from EXAFS [25–27] and SEXAFS [28].

[3] The name XANES for "x-ray absorption near edge structure" was coined by A. Bianconi in 1980 [30].

Fig. 1.5 Light rays falling into Grand Central Station in New York City

have received more attention in the past than those posed by the dual wave-particle properties of light" [14].

The concept of "light rays" was discussed already around 300 BC by Euclid [40]. This simple concept, based on light propagation on straight trajectories, is nicely illustrated by the photo in Fig. 1.5, showing beams of light falling through windows into Grand Central Station in New York City.

This ray picture was prominently supported by Isaac Newton (1643–1727), who in 1704 put forward the view that light is corpuscular in the book *Opticks: A Treatise of the Reflections, Refractions, Inflections and Colours of Light*. Newton's influence in the scientific community largely allowed the corpuscular light theory to prevail during the eighteenth century, despite challenges by Francesco Grimaldi (1618–63), Robert Hooke (1635–1703), and especially Christiaan Huygens (1629–1695) whose theory of light was published in 1690 in his book *Traité de la lumiére* (Treatise on Light). These scientists pointed out that the light path can deviate from rectilinear behavior, i.e. it can "diffract" around corners.

Today we typically credit Huygens with introducing spherical wavefronts and Thomas Young (1773–1829) [41] and Augustin J. Fresnel (1788–1827) [42] for their use to explain diffraction around 1800.[4] Fresnel also suggested the transverse nature of light and established the concepts of linear, circular, and elliptical polarization [45] which today are used to explain the magneto-optical effect discovered by Faraday in 1846 [46] and Pasteur's discovery of chirality in 1850 [47].

[4] It was known earlier that the resolution of microscopes suffered from "diffraction effects" [43]. Credit for the formulation of the diffraction limit is typically given to Ernst Abbe (1840–1905) who in 1873 stated the minimum resolvable distance to be $d_{\min} = \lambda/(2 \sin \alpha)$, where α is the aperture's half angle of the microscope's objective [44].

The central part of the wave description of light that became consolidated in the first quarter of the nineteenth century is today expressed by the magical *Huygens-Fresnel principle*, which states that every point in a light path may be considered a source of a spherical wave.[5] Based on this principle, today's diffraction theory was developed by Gustav R. Kirchhoff (1824–1887), John William Strutt (Lord Rayleigh) (1842–1919), and Arnold Sommerfeld (1868–1951) [48]. The notion of light waves was famously supported by James Clerk Maxwell's (1831–1879) unified description of light and electromagnetic (EM) waves proposed in 1861 [49, 50]. The validity of this concept on the macroscopic length scale was proven by experiments of Heinrich Hertz around 1888 [51].

The nineteeth century concluded with two seminal developments. Around 1879, Thomas Edison (1847–1931) invented a reliable long-lasting electric light bulb.[6] Today it is hard to imagine the modern industrialized world without artificial light sources, and for this reason we have listed Edison's invention as the first modern revolution in "light" sources in Fig. 1.1.

The second revolution shown Fig. 1.1 was the discovery of x-rays in 1895 by Wilhelm C. Röntgen (1845–1923) [52]. Their use has allowed us to see the invisible, from medical imaging to the atomic structure of matter. Because x-ray science constitutes the central topic of this book, we shall discuss its detailed development separately, after finishing our historical journey into the nature of light.

By 1900, all observations about the behavior of light, like the phenomena of attenuation (absorption), refraction, and diffraction, could be explained by the notion that light is an electromagnetic wave.

1.2.2 The Quantum Era and Wave-Particle Duality

What appeared to be a complete description of light was challenged through a paper, published on December 14, 1900, by Max Planck (1858–1947), which accounted for the spectrum of light emitted by thermal sources [53, 54]. Planck's empirical formula described the so-called blackbody spectrum, i.e. the frequency dependent intensity distribution of light emitted by a hot gas in a closed ("black") box through a small opening. Planck's derivation which included a mysterious constant $h = 6.55 \times 10^{-27}$ [erg $\times$ sec], he termed "Naturconstante", ushered in the quantum age.

The granular nature of light emerged more directly in 1905 through Einstein's explanation of the photoelectric effect [55], observed earlier in 1887 by Heinrich Hertz (1857–1894) [56]. Einstein introduced the idea that light is granular and that

[5] It corresponds to the mathematical expansion of a plane wave as a linear combination of spherical waves.

[6] Earlier efforts, including Alessandro Volta's demonstration of a glowing wire in 1800, resulted in incandescent bulbs with extremely short lives, high production expense, and the need for high electric currents made them commercially unattractive. Edison filed for an electric lamp patent (U.S. 223,898) on November 4, 1879 which was granted on January 27, 1880. It was based on "a carbon filament or strip, coiled and connected to platina contact wires".

such light "quanta" can transfer their energy to electrons and emit them from solids. In 1916, Millikan [57] experimentally verified Einstein's all-important photoelectric equation through a detailed study of the photoelectric effect, measuring Planck's constant to within 0.5%.

In 1909 Einstein published a less known paper [58], where he derived a fluctuation formula for Planck's blackbody radiation which contained a wave and a particle term. He suggested an inherent *wave-particle duality* in the nature of light that combines two seemingly different concepts. He concluded [59], "*It is, therefore, my opinion that the next stage of the development of theoretical physics will bring us a theory of light which can be regarded as a kind of fusion of the wave theory and the emission theory.*"

The final breakthrough on the origin of Planck's law came in 1916/17 through Einstein's famous rate equation model that is based on how an atomic gas can achieve equilibrium at a given temperature through photon absorption and emission [60, 61]. In the work he introduced two types of light emission processes: spontaneous and stimulated emission.[7]

Historically, *stimulated* emission of visible light was long considered less important and it was more readily understandable since stimulated absorption and emission are completely symmetric processes for a strong classical field. It became of central importance in the development of nuclear magnetic resonance (NMR) in the 1940s and of the maser (microwave amplification by stimulated emission of radiation) and laser (light amplification by stimulated emission of radiation) in the period 1954–1960. In practice, the emission of conventional optical sources such as light bulbs and x-ray tubes is dominated by *spontaneous* emission which accounts for their chaotic properties. Its introduction was the truly revolutionary part of Einstein's paper since it could not be readily understood classically. Einstein was led to its introduction through the existence of spontaneous decays in radioactive materials. Its fundamental origin was explained by Dirac in 1927 [62, 63] as a consequence of the quantum mechanical zero-point field whose fluctuations trigger spontaneous decays.[8]

The complete formulation of a quantum theory of light and matter remained an elusive goal until the mid-1920s. The term "quantum mechanics" was introduced by Max Born (1882–1970) in 1924 [67] and the term for a quantum of light, the "photon", was coined in 1926 by Gilbert N. Lewis (1875–1946) [68]. The quantum revolution occurred over the three year period 1925–1928. Although one may today distinguish nine different formulations of quantum mechanics as reviewed by Styer et

---

[7] Einstein was quite excited about his work as evidenced by a letter to his friend Michele Besso on August 11, 1916: "Es ist mir ein prächtiges Licht über die Absorption und Emission der Strahlung aufgegangen - es wird Dich interessieren. Eine verblüffend einfache Ableitung der Planck' schen Formel, ich möchte sagen *die* Ableitung. Alles ganz quantisch." In translation "A splendid light has dawned on me about the absorption and emission of radiation—it will be of interest to you. A stunningly simple derivation of Planck's formula, I might say *the* derivation. Everything completely quantized."

[8] It appears [64] that Planck first conjectured the concept of a residual energy of the radiation field [65], but it was a fuzzy concept at the time. The term "Nullpunktsenergie" or zero-point energy was phrased by A. Einstein and O. Stern in 1912 [66].

al. [20], this period produced the best known three versions of non-relativisitic QM: *Matrix mechanics* by Werner Heisenberg (1901–1976), Born and Pascual Jordan (1902–1980) [69–71],[9] *wave mechanics* by Louis de Broglie (1892–1987) [73][10] and Erwin Schrödinger (1887–1961) [76–80], and so-called *transformation theory* by Paul Dirac (1902–1984) [62, 63, 81, 82].

In particular, Dirac formally quantized the EM field and showed that one must treat its ground state as a dynamical system which contains the previously conjectured zero-point energy. Its associated fluctuations provided the mysterious driving force for Einstein's spontaneous emission. In a sense, photon emission could now be perceived as amplification of the zero-point energy.

1.2.3 New Insight in the Period 1930–1950

The period from the 1930s to the 1950s saw several important developments in the description of light.

In 1935 A. Einstein, B. Podolsky, and N. Rosen [83], published a famous paper, often referred to as the "EPR" paper, that outlined the "paradoxical" properties of certain two-particle states. In the same year, Schrödinger [84] coined the term "entangled states" to emphasize the fact that the properties of two quantum particles may be inextricably coupled. At the heart of the EPR paper was the authors' conviction that it was unacceptable that for entangled states, quantum mechanics allowed the existence of a correlation between measurements made between spatially separated particles, referred to by Einstein as "spooky action at a distance". At the time, the success of quantum mechanics, however, led most physicists to pay little attention to the EPR paper. But that changed with the development of quantum optics in the 1960s as will be discussed in this book.

Another important development in the mid-1930s was the work by Pieter H. van Cittert (1889–1959) and Frits Zernike (1888–1966) resulting in the *van Cittert–Zernike (VCZ) theorem* [85], which allows the description of spatially incoherent light

[9] Born and Jordan did not share Heisenberg's 1932 Nobel Prize which he actually received in 1933. In 1933 Heisenberg wrote to Born, "The fact that I am to receive the Nobel Prize alone, for work done in Göttingen in collaboration—you, Jordan, and I—depresses me and I hardly know what to write to you. I am, of course, glad that our common efforts are now appreciated, and I enjoy the recollection of the beautiful time of collaboration. I also believe that all good physicists know how great was your and Jordan's contribution to the structure of quantum mechanics—and this remains unchanged by a wrong decision from outside. Yet I myself can do nothing but thank you again for all the fine collaboration and feel a little ashamed." [72].

[10] De Broglie first suggested the existence of a wave-particle duality not only for light but particles. He conjectured that particles are accompanied by "phase waves" [74]. The work published in more detail in 1925 [73] earned him the 1929 Nobel Prize in Physics *"for his discovery of the wave nature of electrons"*. De Broglie's hypothesis was proven in 1927 by the Davisson and Germer experiment of electron diffraction by crystals [75].

emitted by thermal sources and stars.[11] It laid the foundation for optical coherence theory and statistical optics, developed in more detail in the 1960s by Emil Wolf and Leonard Mandel [86, 87]. The VCZ theorem is arguably one of the most important theorems in modern optics since it characterizes the *spatial* coherence properties of a source. Its analogue, describing the *temporal* coherence, is the Wiener–Khintchine theorem [88], which was also formulated in the 1930s.

In 1947 Dennis Gabor conceived the idea of holography and, by employing conventional filtered-light sources, developed the basic technique. Holography did not become commercially feasible until the advent of the laser. Another advance occurred in the 1950s through Alfred Kastler's (1902–1984) development of optical pumping that earned him the 1966 Nobel Prize in Physics [89]. Following earlier work by A. Rubinowicz in 1918 [90] on the polarization dependence of electromagnetic transitions, Kastler's work led to the assignment of a photon angular momentum or spin of $+\hbar$ or $-\hbar$ for right and left circularly polarized light, which today is widely utilized for the study of magnetic materials [34].

1.2.4 Beyond Quantum Mechanics: Quantum Electrodynamics

In 1928 Dirac discovered an equation describing the motion of electrons in an external EM field [81, 82] which incorporated both the requirements of quantum theory and the theory of special relativity. It naturally led to the emergence of the electronic spin, whose interesting history has been discussed by Stöhr and Siegmann [34].

Dirac's theory provided the foundation of *quantum electrodynamics* (QED), although it became soon apparent that it predicted an infinite mass of the electron [91]. Also, since each of the infinite number of oscillators in the quantum mechanical vacuum state possessed a zero-point energy, it implied that the EM vacuum had an infinite energy density and that the electron sea had an infinite negative charge density. These infinities arose from evaluating non-convergent integrals whenever corrections were attempted to Dirac's lowest order quantum formulations. The incompleteness of Dirac's theory became apparent by experiments in 1947/48 through the measurements of the Lamb shift of the energy levels of hydrogen by W. E. Lamb and R. C. Retherford [92] and the anomalous magnetic moment of the electron by H. M. Foley and P. Kusch [93].

The difficulties were resolved between 1948 and 50 through the diagrammatic formulations of QED by Richard P. Feynman (1918–1988), expressed by the famous Feynman diagrams, and the independently developed operator methods by Julian S. Schwinger (1918–1994) in the USA and Sin-Itiro Tomonaga (1906–1979) in Japan. All three shared the 1965 Nobel Prize in Physics since their theories were shown to be equivalent by Freeman Dyson (1923–2020) in 1949 [94]. While a complete theory of

[11] Zernike received the Nobel Prize in Physics in 1953 for the development of the phase contrast microscope.

light in the form of quantum electrodynamics (QED) now existed, its application to practical problems in optics remained a challenge. The devil remained in the details!

Feynman also elucidated the equivalence of the disparate wave and particle pictures in first order QED. Using his own space-time probability amplitude formulation of QM [95], he explained Young's double-slit experiment in his *Lectures on Physics* [24] by use of the particle concept for photons or electrons which, in first order QED, have the same probability amplitudes [23]. This leads to the simple picture where one may approximate a wave by summing over all probabilistic paths or probability amplitudes of a large number of independent photons. Absorption can then be viewed either as the attenuation of the incident wave or the loss of some of the photons. Diffraction can be pictured either as the scattering and interference of the incident wave or fractions of the incident photons being scattered into directions out of the beam. From a quantum perspective, Young's double-slit experiment proceeds in single-photon processes whose probabilistic repetition leads to the buildup of the diffraction pattern [4].

1.2.5 Development of the Laser and Quantum Optics Around 1960

The ten year period around 1960 was of seminal importance for our understanding of light and consisted of several scientific breakthroughs. The best known ones are the development of the maser and laser in the late 1950s through work by Nikolay Basov, Gordon Gould, Aleksandr Prokhorov, Arthur Schawlow, and Charles Townes. While the credit for the "invention of the laser" remains controversial, the first laser was operated by Theodore Maiman on May 16, 1960. It constitutes the third revolution in Fig. 1.1.

The invention of the laser, in particular, overcame the difficulty to produce strong coherent radiation in the optical wavelength region. It followed in the wake of the development of RADAR sources, short for RAdio Detection And Ranging,[12] during the Second World War. The high intensity coherent microwave radiation produced by such sources led to the development of NMR in the 1940s.

The late 1950s resulted in another seminal experimental and theoretical discovery of the properties of light, the Hanbury Brown–Twiss (HBT) effect [96–98]. It triggered the development of *statistical optics*[13] in the 1960s by Emil Wolf and Leonard Mandel [86, 101] and *quantum optics* by Roy Glauber [12–14]. The ensuing dispute between the two camps has been chronicled by Bromberg [102].

The experiments conducted by Robert Hanbury Brown and Richard Twiss utilized thermal light either produced by a laboratory source or the star Sirius and observed

[12] The term RADAR was coined in 1939 by the United States Signal Corps.

[13] Statistical optics is based on the statistical description of light, which extended the original work of Max Born [99], who together with Emil Wolf wrote their famous textbook, first published in 1959 [100].

a distance- or time-dependent correlation between the arrival of two photons. It was found to peak when the two photons of wavelength around 500 nm arrived at the same time and point and decreased when either their temporal separation was increased at the same point or their spatial separation was increased at the same time [4]. Remarkably, the interpretation of their results was possible within both statistical optics, where it was attributed to statistical intensity fluctuations [103] and through quantum optics, where it was interpreted as the interference of the probability amplitudes of the two photons [14, 104].

As discussed in this book, the HBT effect is a second order correlation effect in QED which for chaotic light happens to be also describable through statistical optics. We will, however, show that the quantum optics approach is more fundamental since it can account for effects, such as diffraction of entangled photons, which are outside the purview of statistical optics.

In 1964 John Bell published a seminal paper [105], establishing *Bell's inequalities* that allowed addressing the incompleteness of quantum mechanics suggested by the Einstein-Podolsky-Rosen paradox. Extensive laser experiments starting in the 1970s confirmed the violation of Bell's inequalities in support of quantum mechanics. In the process, the extraordinary properties of entangled photons became the basis for the entire field of quantum information science [106]. This development has been honored by the 2022 Nobel Prize in Physics awarded to Alain Aspect, John F. Clauser, and Anton Zeilinger "for experiments with entangled photons, establishing the violation of Bell inequalities and pioneering quantum information science".

More generally, optical laser experiments have revealed the extraordinary power of the use of various aspects of light to explore the quantum world. The development is continuing to this day, owing to the enormous number of possible (quantum) states of light. There is still plenty of room for the discovery of new weird and wonderful quantum properties through optical and x-ray experiments. While the advent of conventional lasers has greatly advanced our understanding of the photon nature of light [4], the question arises whether XFELs can unveil additional aspects of "light" itself? As evidenced by the increase of Compton scattering with photon energy, the particle aspects become more pronounced for x-rays, and the intensity of XFEL beams is approaching levels where the small probability of photon-photon scattering through higher order coupling in QED can be studied [107].

At the end of the historical journey through the understanding of light, we summarize in Fig. 1.6 key milestones associated with our discussion above. These milestones in the understanding of optical light are discussed in more detail in [40, 108, 109]. We shall extend this picture to x-rays in the following sections.

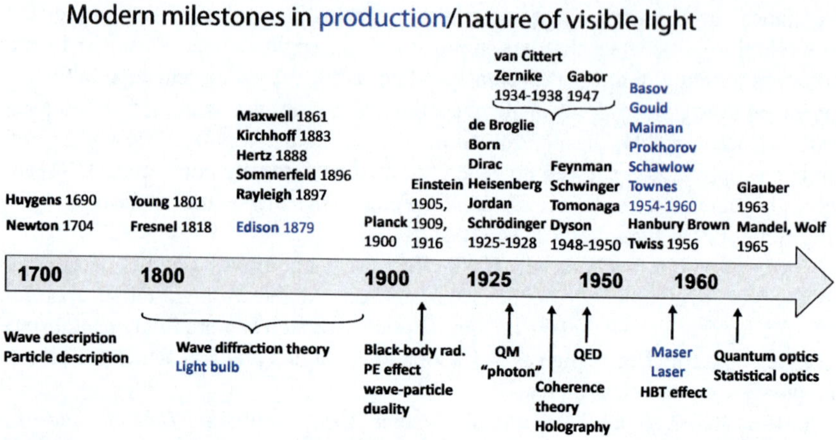

Fig. 1.6 Important milestones from around 1700 to about 1965 associated with the production of visible light (blue) and theoretical and experimental advances in understating the nature of light (black)

1.3 The X-Ray Revolution

1.3.1 The Discovery and Early Utilization of X-Rays

The second modern revolution in the production of what we have defined as "light" was the discovery of x-rays in 1895 by Wilhelm C. Röntgen (1845–1923) [52], as indicated in Fig. 1.1. Because of the central role of x-ray science in this book, we shall discuss its development now that we have completed the historical journey associated with visible light. In our discussion we can of course benefit from the previous discussion of visible light which is only distinguished from x-rays by the photon energy or wavelength as shown in Fig. 1.2.

Röntgen's work was immediately accepted as a revolutionary discovery by both scientists and society at large. Only six years later he was awarded the first Nobel Prize in Physics "*in recognition of the extraordinary services he has rendered by the discovery of the remarkable rays subsequently named after him.*" Röntgen referred to the radiation discovered by him as "X-Strahlen" (x-rays), with "X" reflecting their unknown character and "rays" because of the shadow images they formed of bone and other dense objects. In German and many other languages, the "X" is replaced by "Röntgen" in honor of their inventor. In further experiments, Röntgen did not succeed in observing their refraction by prisms, their focusing by lenses or their deflection by magnetic fields. He concluded that x-rays were not susceptible to conventional refraction and reflection and were distinct from Lenard's cathode rays, which were shown to be electrons by J. J. Thomson in 1897.

The early developments in x-ray science are partly chronicled by numerous Nobel Prizes, summarized in Fig. 1.7 [110]. The early Nobel Prizes were awarded in Physics

Nobel Prizes based on the Utilization of X-Rays

Physics

1901: WILHELM RÖNTGEN

1914: MAX VON LAUE

1915: SIR WILLIAM HENRY BRAGG and SIR WILLIAM LAWRENCE BRAGG

1917: CHARLES BARKLA

1924: KARL MANNE SIEGBAHN

1927: ARTHUR COMPTON

1981: KAI SIEGBAHN

Medicine

1946: HERMANN JOSEPH MULLER

1962: FRANCIS CRICK, JAMES WATSON and MAURICE WILKINS

1979: ALAN M. CORMACK and SIR GODFREY N. HOUNSFIELD

Chemistry

1936: PETER DEBYE

1962: MAX PERUTZ and SIR JOHN KENDREW

1964: DOROTHY HODGKIN

1976: WILLIAM LIPSCOMB

1985: HERBERT HAUPTMAN and JEROME KARLE

1988: JOHANN DEISENHOFER, ROBERT HUBER and HARTMUT MICHEL

1997: PAUL D. BOYER and JOHN E. WALKER

2003: PETER AGRE, and RODERICK MACKINNON

2006: ROGER KORNBERG

2009: VENKATRAMAN RAMAKRISHNAN, THOMAS STEITZ and ADA YONATH

2012: ROBERT LEFKOWITZ and BRIAN KOBILKA

Fig. 1.7 List of Nobel Prizes associated with the nature and applications of x-rays

for contributions to the understanding of x-rays and their interactions with matter. Later Nobel Prizes in Chemistry and Medicine are typically associated with pioneering structure determinations of macromolecules such as proteins that are used to imply their function.

While the use of x-rays for medical imaging was demonstrated by Röntgen in the process of their discovery, their interactions with matter in the form of atoms, molecules, and solids began to be investigated by others at the dawn of the twentieth century. The work focused on three fundamental types of the x-ray responses: absorption, emission, and scattering. They are also the key processes discussed in this book.

X-ray absorption and emission spectroscopy was pioneered by the English physicist Charles G. Barkla (1877–1944), who in 1905 found that x-rays are polarized perpendicular to the direction of propagation [111] and thus must be transverse waves.[14] He also showed that each element contained characteristic x-ray absorption thresholds [112] and first distinguished the primary x-ray emission lines, labelling them the K and L series [113]. The work, for which he won the 1917 Nobel Prize in Physics, is summarized in his 1918 paper [114].

The characteristic change of the photon energies of the emitted K lines was studied in more detail in 1913/1914 by Henry G. J. Moseley (1887–1915) in England, leading

[14] Interestingly, the Braggs were Barkla's vociferous opponents since they originally thought that γ- and x-rays were a neutral pair of particles, an electron coupled to a positively charged particle.

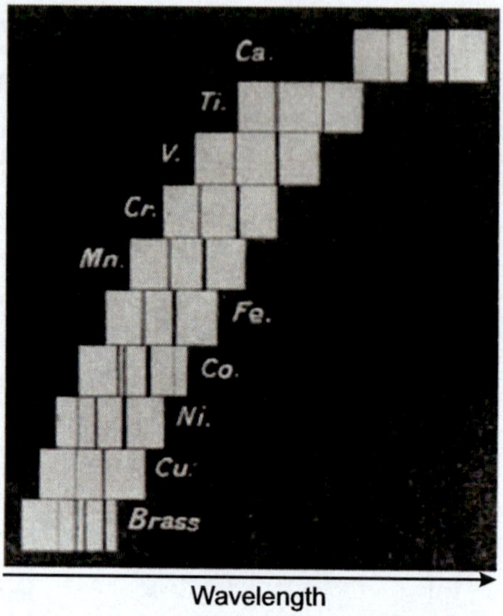

Fig. 1.8 Moseley's photographic recording of K_α and K_β emission lines for different 3D elements and brass (an alloy of Cu and Zn) [116]

to Moseley's law [115–117], which states that the K emission energies scale with the square of the atomic numbers. It also provided a practical means of using x-ray spectroscopy to determine atomic composition of samples and their order in Mendeleev's periodic table. Moseley's observation of the dependence of the atomic emission lines for different 3D transition metals is shown in Fig. 1.8.

Moseley's untimely death in 1915 during the battle of Gallipoli at age 27 likely cost him the Nobel Prize. This is evidenced by the fact that two Nobel Prizes in Physics for x-ray spectroscopy were later given to Barkla in 1917 and to Swedish physicist Manne Siegbahn (1886–1978) in 1924. Manne Siegbahn began his studies of x-ray spectroscopy in 1914 and initially used the same type of spectrometer as Moseley. He then developed improved spectrometers to make very accurate measurements of the x-ray wavelengths emitted by different atoms [118, 119], resulting in an almost complete understanding of the electron shell [120].

Barkla's early studies of the x-ray *absorption* fine structure of matter were continued in more detail in 1918 by Wilhelm Stenström and Hugo Fricke in Manne Siegbahn's laboratory, and an early theoretical interpretation was put forward by Walter Kossel in 1920. Significant progress in its interpretation was however only made after the development of quantum theory when in 1931 Ralf de L. Kronig (1904–1995) published a seminal paper [121]. It may be considered the founding theoretical paper of modern x-ray absorption spectroscopy. The interesting history of x-ray absorption spectroscopy has been reviewed by Christian Brouder under the pseudonym R. Stumm von Bordwehr[15] [122] and by Lytle [123].

[15] General Stumm von Bordwehr is a comical figure, "the man without qualities", in a novel by Robert Musil.

The early spectroscopic x-ray absorption and emission studies were complemented by studies of the diffractive properties of x-rays. On June 8, 1912, Paul Knipping and Walter Friedrich, who were assistants of Max von Laue (1879–1960), first reported the diffraction of x-rays by crystals, and the results were published in 1913 by W. Friedrich, P. Knipping and M. Laue [124], with Laue writing the theoretical and the two others the experimental part of the paper. Only a year later, Laue received the 1914 Nobel Prize in Physics *"for his discovery of the diffraction of x-rays by crystals."*

The work in Germany was immediately followed up in England by William Henry Bragg (1862–1942) and William Lawrence Bragg (1890–1971). Bragg's law $n\lambda = 2d \sin\theta$ was first presented on November 11, 1912, to the Cambridge Philosophical Society and published in 1913 [125]. In the following year, Paul Peter Ewald (1888–1985) devised a graphic method of solving the Bragg equation for crystals, now known as Ewald's sphere, where the scattering vector, corresponding to the momentum transfer, is equal to a reciprocal lattice vector [126]. The Braggs were awarded the 1915 Nobel Prize in Physics. The work of von Laue, the Braggs, and Ewald gave birth to the field of x-ray crystallography that has revolutionized our understanding of the atomic structure of all forms of matter.

The particle-like nature of light suggested by Einstein in 1905 was supported in 1922 by an experiment carried out by Arthur H. Compton (1892–1962). He studied the scattering of x-rays from electrons in a carbon target and found that the scattered x-rays had a longer wavelength than the incident ones, and that the shift in wavelength increased with scattering angle [127]. Compton explained and modeled the data by assuming a particle (photon) nature for light and applying conservation of energy and momentum to the collision between the photon and the electron.

Historically, Einstein's photoemission and Compton's scattering effects were widely accepted as proof that light cannot be explained purely as a wave phenomenon.[16] Compton's measurement and explanation earned him a share of the Nobel Prize in physics in 1927.

1.3.2 Development of Synchrotron Radiation Sources

The first x-ray tubes emitted relatively weak intensities consisting of atomic emission lines and a broad bremsstrahlung background. After J. J. Thomson's (1856–1940) discovery of the fact that "cathode rays" were negatively charged "electrons" in 1897, Emil Wiechert (1861–1928) [130] and Alfred-Marie Liénard (1869–1958) independently worked out what is now called the Liénard-Wiechert potentials of a moving charge [131, 132]. In particular, Liénard's paper of 1898 already contained a

---

[16] It was later realized that both effects can be explained by assuming classical electromagnetic waves with quantum properties associated only with the atomic and electronic building blocks of matter [128]. Today, the experiment performed in 1986 by Grangier, Roger, and Aspect [129] is often considered as additional proof for the existence of photons.

Fig. 1.9 First synchrotron light emitted by the 70 MeV synchrotron at general electric. The light is visible as a bright spot below the center of the picture on the left [134]

basic theory of synchrotron radiation. It did not require use of the theory of relativity because Maxwell's equations are invariant under transformations between moving systems. In 1912 the theory of synchrotron radiation was worked out in more detail by the mathematician G. A. Schott [133]. The radiation could not be observed, however, until circular electron accelerators became available. Synchrotron radiation was first observed in 1947 at General Electric Research Laboratory on a 70 MeV synchrotron [134], as shown in Fig. 1.9.

In the 1960s the use of synchrotron radiation for the study of matter began at synchrotrons in Washington, D.C (USA), Tokyo (Japan), Hamburg (Germany), and Frascati (Italy). In accelerator-based sources the kinetic energy of electrons is increased from tens of keV in conventional x-ray tubes to energies of the order of a GeV, surpassing the relativistic limit set by the rest mass of the electron, 511 keV. The relativistic reduction of the Coulomb repulsion between the electrons then gives rise to the emission of radiation into a narrow forward cone when the electrons are bent by a magnetic field.

In 1968, first experiments done on an electron *storage ring* located at the University of Wisconsin in Madison, named Tantalus I, revealed the superiority of storage rings over synchrotrons. In a storage ring, the electron bunches circulate at a fixed energy at a high repetition rate of about 500 MHz for periods up to many hours, providing a stable broadband radiation spectrum of high intensity and stability. Despite the nearly exclusive use of storage rings for the production of x-rays today, the historical name "synchrotron radiation" (SR) has been maintained.

Today, one distinguishes three stages of further development after the start of the "synchrotron revolution" around 1970 [5–7]. The *first* generation of storage ring sources was primarily operated for high energy physics and the by-product of synchrotron radiation was used parasitically. The *second* generation utilized storage rings that were operated specifically for the production of x-rays. The *third* generation of sources was specifically designed for optimizing the radiation quality or *spectral brightness* of the source and included the use of undulators and wigglers.[17] One may roughly associate the first generation with the years <1980 with third generation sources taking over after about 1990.

The first and second generation of synchrotron radiation (SR) experiments mainly benefitted from the large number of electrons $N_e \sim 10^{10}$ that could be concentrated in individual bunches due to relativistic effects and the high pulse repetition rate $R \sim 500$ MHz that could be achieved by the use of storage rings. The average flux, Φ, or number of photons per unit time per unit area, emitted by such sources, simply scales as $\Phi \propto R\, N_e$. The reason for this scaling is that the fields of the individual electrons in the centimeter sized bunches do not act as a single macro-charge but rather add incoherently.

In third generation synchrotron sources the electron bunch size was optimized for x-ray emission by arrangement of electron optical elements around the ring, and undulators were used to enhance the brightness by periodically wiggling the electron bunch. This increased the average brightness relative to a single arc bending magnet source by a factor $n_u^2 \sim 10^4$, where $n_u \sim 100$ is the number of undulator periods. In the process, the broad bandwidth bending magnet spectrum becomes concentrated into a narrow bandwidth around a selected photon energy. The factor of 10^4 enhancement in average brightness became the driving force for the construction of numerous third generation light sources around the world. It is continued today by the quest to build "ultimate" storage rings where the emission is limited only by the quantum mechanical space-momentum uncertainty relation [7, 137].

1.3.3 The Advent of X-Ray Free Electron Lasers

The final revolution in x-ray science in Fig. 1.1, caused by the construction and use of XFELs, happened at the dawn of the twenty-first century. The foundation for this development was laid in the 1970s by John Madey and coworkers at Stanford University, USA [138–140]. The concept was based on creating a radiation cavity by use of mirrors that reflected the radiation back and forth across a straight section in a storage ring, similar to a conventional laser cavity.

In the 1980s Madey's concepts were extended and the mirrors which limited the spectral range were omitted. The novel concept which allowed the generation of radiation up into the hard x-ray regime was based on creating a self-ordering effect in a single pass of a small electron bunch through a long undulator, a process called *self-amplified spontaneous emission* (SASE), owing to the fact that it resembles the

[17] The term brightness and its significance were introduced for the description of SR by K.-J. Kim in 1986 [135] following an earlier paper that focused on x-ray coherence [136].

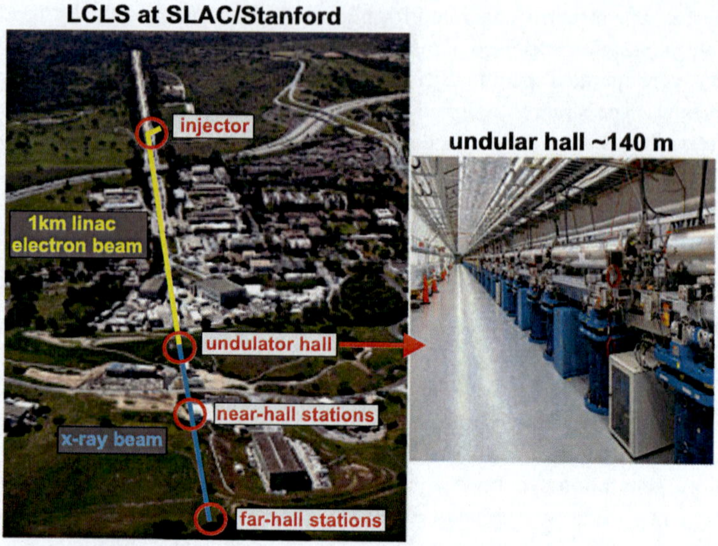

Fig. 1.10 Schematic of the Linac Coherent Light Source (LCLS) XFEL superimposed on an arial photograph of SLAC National Accelerator Laboratory at Stanford University, California. On the right is a picture of the long undulator hall where the x-rays are produced

amplification of spontanaeous emission (ASE) in optical lasers. The process was first discussed by the Russian scientists A. Kondratenko and E. Saldin [141, 142]. Shortly thereafter, the theory was solidified by the Italian scientists R. Bonifacio, C. Pellegrini, and L. Narducci [143], who clarified the fundamental physics and presented a simple model based on a single FEL parameter that depends on the electron beam density, its energy, the undulator period, and its magnetic field. The detailed historical development of FELs has been reviewed by Pellegrini [144], and its modern theory is given in [7, 145].

The first FEL facility built for user operation was the SASE VUV/soft x-ray facility FLASH (Free electron LASer Hamburg) at DESY in Hamburg, Germany, which began operation in 2005. It utilized a superconducting linac, covering the vacuum ultraviolet spectral range $\hbar\omega < 100\,\text{eV}$.

In 1992 Claudio Pellegrini presented a concrete plan of building a SASE-based "hard x-ray" FEL (XFEL) using the existing 3-km-long linac at SLAC/Stanford equipped with a high-quality photoinjector [146]. This later led to the Linac Coherent Light Source (LCLS) which utilized the last 1 km of an existing linac, providing an electron beam energy 4–14 GeV that covered the spectral range of about 300 eV–10 keV with 120 pulses/s. LCLS started its user program in 2009 [147]. It is currently being upgraded (LCLS-II) by adding a superconducting accelerator structure to the first km of the linac which provides 10^6 pulses/s. An areal view of the large-scale facility which began operation in 2009 and a photograph of the undulator hall where the photons are created are shown in Fig. 1.10.

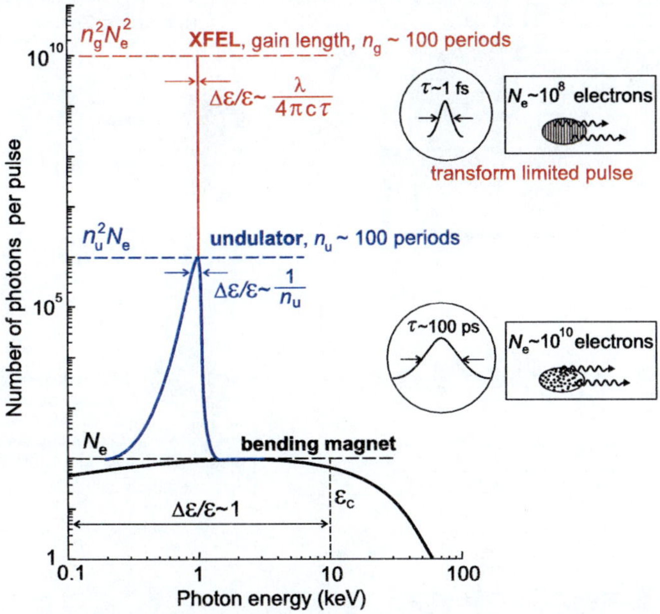

Fig. 1.11 Comparison of the spectral distribution and photons per pulse for a bending magnet and undulator source in a storage ring with an ideal XFEL, emitting diffraction, and transform-limited pulses. We assumed $N_e = 10^{10}$ electrons per bunch for the storage ring, corresponding to a charge of 1.6 nC, a beam energy $\mathcal{E}_e = 3$ GeV, and an undulator length of $n_u \simeq 100$ periods. For the XFEL we assumed the same beam energy, a gain length of $n_g \simeq n_u \simeq 100$ periods, and that the $N_e = 10^8$ electrons per bunch are ordered in sheets to produce coherent pulses of about 1 fs duration

During the 1990s the XFEL concept was further developed by accelerator scientists. The revolutionary evolution of the emitted x-ray spectrum from bending magnets to undulators in storage rings to a spatially and temporally coherent XFEL is illustrated in Fig. 1.11. In the figure we have assumed the production of a single coherent x-ray pulse of about 1 fs duration, which is a typical length of a single spike in a SASE pulse [4, 148].

Linacs have the advantage over storage rings that a better spatial and temporal compression of electron bunches can be achieved. The bunches are used only once before they are discarded. For example, an electron bunch compressed to a length of $\ell = 3\mu$m produces an x-ray pulse of the same length corresponding to $\tau = \ell/c = 10$ fs (1 fs=10^{-15} s). More importantly, when the electron bunches are sent through of a long undulator, regions of the bunches self-order in the field of the spontaneously emitted synchrotron radiation into sheets separated by the wavelength. The in-phase emission of the ordered electron sheets then results in a coherent radiation enhancement, which scales with the square of the number of electrons N_e^2 in the ordered regions. For short bunch lengths of about 1 fs that may contain $N_e \simeq 10^8$ electrons, the number of photons per pulse is about 10^{10} as shown in Fig. 1.11.

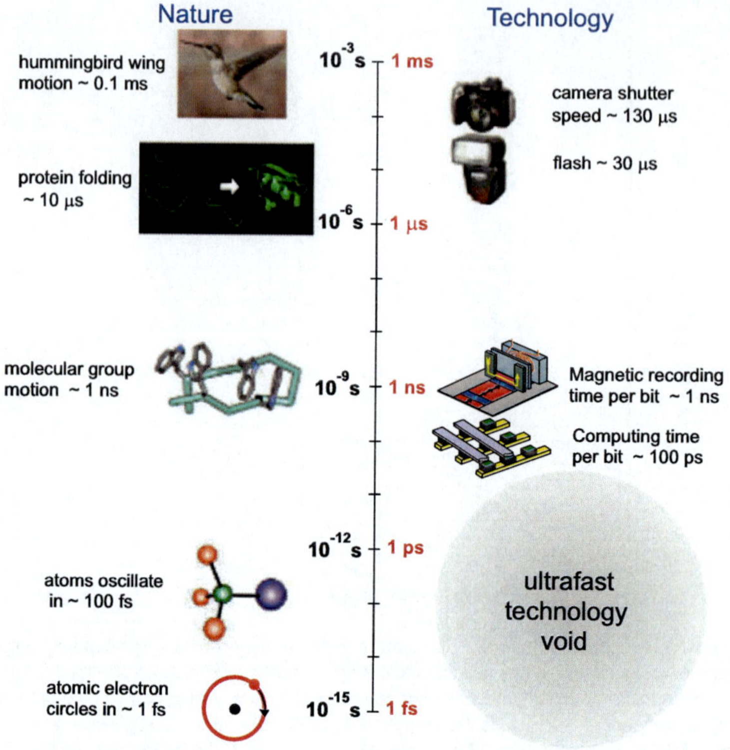

Fig. 1.12 Timescales involved in processes in nature and technology. One may obtain an approximate time duration Δt of a process by considering the energy $\Delta\mathcal{E}$ involved by use of the relation $\Delta t\,\Delta\mathcal{E} \simeq \hbar$, where $\hbar = 0.66\,\text{fs eV}$

To take advantage of the extraordinary XFEL capabilities, x-ray scientists began to develop a convincing *scientific case* in various scientific disciplines, required for funding XFEL construction. The early scientific case was largely based on the use of high intensity ultrashort pulses which for the first time opened the door for recording movies of atomic motions [149]. The femtosecond XFEL pulses are faster than the timescale of atomic motion which proceeds with the speed of sound of about 1 nanometer per picosecond ($1\,\text{ps} = 10^{-12}$ s). They may even capture charge and spin dynamics during the creation and dissociation of chemical bonds. The timescales of various motions in nature and technology are illustrated in Fig. 1.12. They follow the rule of thumb "the smaller the faster", expected from the concept of inertia (Newton's first law).

In 2010, the FERMI FEL in Trieste, Italy, which constituted a *seeded version* of a single-pass high-gain FEL, became operational. Seeded operation provides improved control of the FEL spectrum, pulse duration, and synchronization with conventional laser sources for pump and probe applications. An initial seed pulse, provided by a conventional, high peak power, pulsed laser operating at 260 nm is temporally

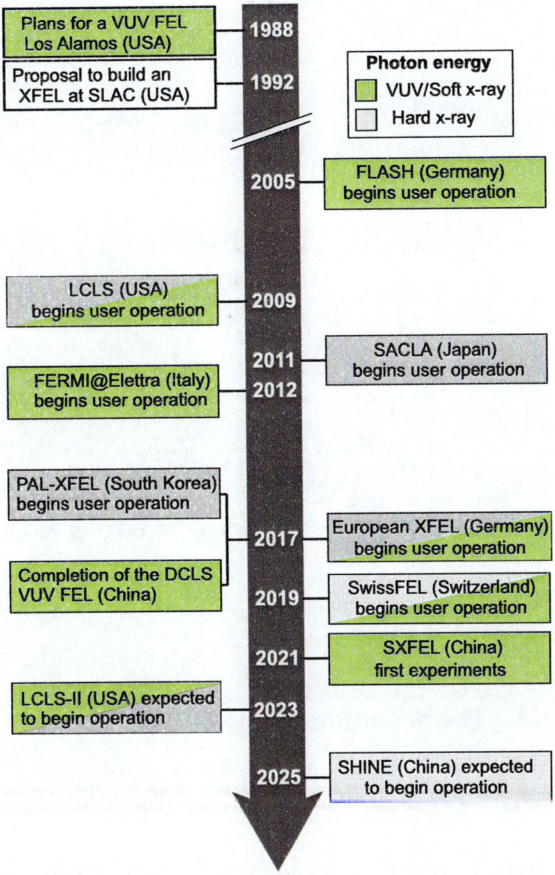

Fig. 1.13 Timeline of developments of free electron lasers covering the vacuum ultra violet (VUV) to hard x-ray spectral range. Colors indicate presently covered spectral range

synchronized with and superimposed on electron bunches from a linac, causing a spatial density modulation at the seed wavelength that also contains higher harmonics. One of the higher harmonics in the vacuum ultraviolet (originally up to 20 nm, later extended to about 4 nm) is then amplified by an undulator to a high peak power level. The temporal duration of the FEL radiation is approximately that of the seed laser (about 100 fs).

The upconversion process used in FERMI becomes increasingly difficult with increasing photon energy, and SASE XFELs therefore typically use a variant called self-seeding. A broadband SASE photon pulse is first sent through a monochromator and is then again superimposed on the original electron bunch for amplification [150, 151].

Figure 1.13 summarizes the historical development of free electron lasers in the vacuum ultra violet (VUV) to hard x-ray regions around the world. The presently covered spectral range is indicated in color, which in some cases has been extended after the original construction.

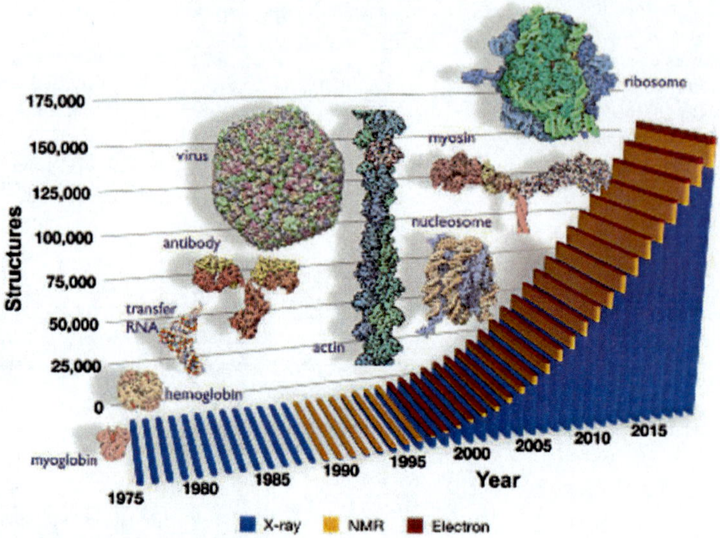

Fig. 1.14 Growth of protein structure determinations per year by x-ray crystallography, NMR, and electron microscopy deposited into protein data bank. Figure courtesy of Anton Barty and Henry Chapman, DESY

1.4 The Scientific Power of X-Rays

1.4.1 From Optical to X-Ray Response of Matter

Today, x-rays are well known for medical imaging and for their ability to reveal the atomic structure of matter owing to their short wavelength. X-ray diffraction from the atomic building blocks of matter has been responsible for some of the great breakthroughs during the twentieth century: the understanding of the atomic structure of materials and the fundamental worker molecules of life, proteins. The remarkable increase of x-ray structure determinations of proteins and the dominant use of x-rays are illustrated in Fig. 1.14.

The process responsible for diffraction is elastic scattering, corresponding to wiggling of the total atomic electron clouds followed by re-emission into random directions. It is referred to as *Rayleigh scattering* in the optical and *Thomson scattering* in the x-ray regions. X-ray diffraction in the form of Bragg diffraction results from coherent addition or constructive interference of Thomson scattering into specific directions. Today, the determination of *de novo* crystal structures of biological macromolecules is naturally dominated by the utilization of synchrotron and XFEL sources due to their high intensity. Importantly, however, it is the tunability of the x-ray energy to absorption edges of heavy embedded atoms, like the Se K-edge at 12.6 keV, which underlies the majority of protein structure determinations [32]. The

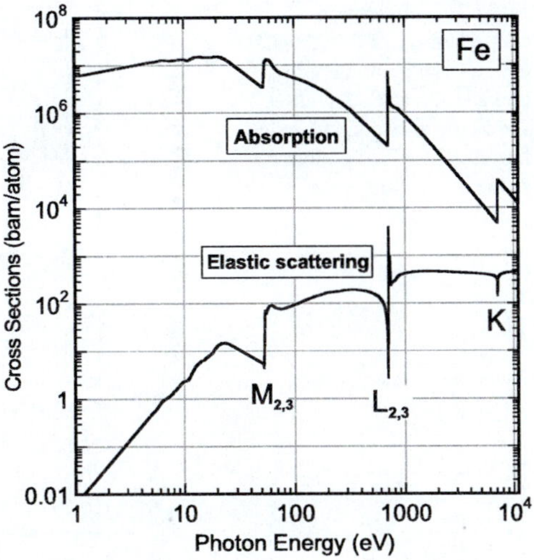

Fig. 1.15 Interaction cross sections of photons with Fe atoms as a function of photon energy extending from the optical to the x-ray region (1 barn = 10^{-10} nm^2)

so-called **S***ingle-wavelength* **A***nomalous* **D***iffraction* (SAD) and **M***ulti-wavelength* **A***nomalous* **D***iffraction* (MAD) methods facilitate the solution of the "phase problem" by exploiting the "anomalous" changes in the scattering phase near the absorption edge [31, 32, 152].

The changes of the scattering and absorption cross sections from the visible to the x-ray range, and the dramatic resonance effects at absorption "edges", are illustrated in Fig. 1.15 for Fe atoms in the elementary metal. The x-ray absorption cross section is composed of contributions from the individual atomic shells. In the optical range it is determined by transitions within the outermost valence shell, and with increasing photon energy, the additional excitation of electrons from inner shells, leading to the emission of photoelectrons, gives rise to absorption steps or "edges". At photon energies above about 100 eV the absorption cross section decreases, accounting for deeper penetration of x-rays into materials, whereas the non-resonant Thomson scattering cross section levels off and remains approximately constant.

With increasing photon energy one can liberate electrons from solids by overcoming the surface potential or workfunction, utilized in ultraviolet photoelectron spectroscopy (UPS), or at higher photon energies eject core electrons in x-ray photo electron spectroscopy (XPS) [37]. The excitation of inner shell electrons by x-rays leads to another fundamental difference between the absorption and emission of optical and x-ray photons. The created core holes may be filled by outer shell electrons, with the electronic decay resulting in the emission of photons as well as Auger electrons. The Auger decay channel actually dominates for the $1s$ or K-shell excitation of atoms lighter than Mn (Z = 25) and for nearly all atoms for $2p$ or L-shell excitation.

More details of characteristic changes of absorption from the infrared to the x-ray region are shown in Fig. 1.16 for liquid water [153–155], represented by the polarization averaged linear absorption coefficient μ_x. The lowest infrared to optical region

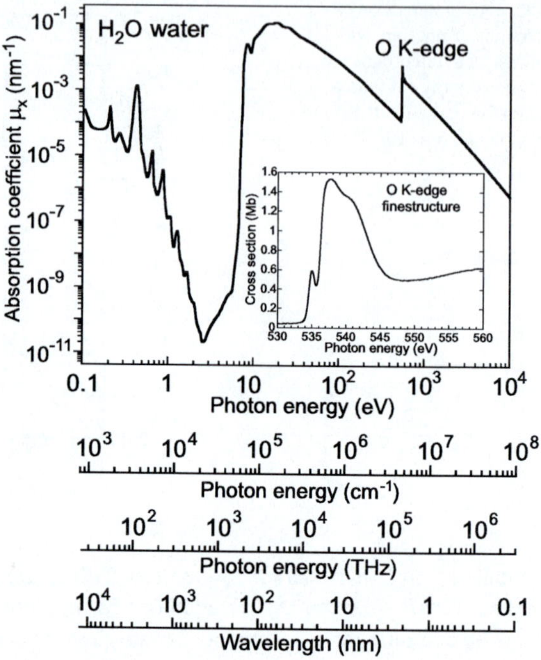

Fig. 1.16 Absorption coefficient μ_x of liquid water over a large photon energy range [153, 154]. The inset shows the enlarged O K-edge fine structure [155], plotted as the absorption cross section per atom σ_{abs} ($1\,\mathrm{Mb} = 10^{-4}\,\mathrm{nm}^2$)

reveals detailed structure arising from rotational, vibrational, and electronic excitations. The x-ray response above 100 eV is considerably simpler and dominated by the O K-edge absorption increase near 540 eV. Closer inspection of the K-edge reveals the fine structure shown enlarged in the inset below [155]. It arises from transitions of O 1s core electrons to unoccupied molecular states of the H_2O molecules [29] with modifications due to H-H interactions in water [155].

Another example is given in Fig. 1.17, where the absorption response of Co metal and the insulator CoO are compared. The data have been synthesized over a large energy range from various sources [154, 156–161]. In the optical region, absorption in the insulator CoO is suppressed by the existence of a band gap of $\simeq 2.5\,\mathrm{eV}$, while Co metal shows a rather smooth response. Above 100 eV, the x-ray response differs at the positions of the oxygen and cobalt absorption resonances, as indicated. The fine structure near the Co L_3 absorption threshold, shown in the inset, reflects the difference in the local bonding configurations in the two cases [33, 34]. While the characteristic atomic absorption "edges" lead to atomic specificity, the fine structure near these "edges", referred to as near edge x-ray absorption fine structure (NEXAFS) or x-ray absorption near edge structure (XANES), provides information on chemical bonding or coordination of the absorbing atom.

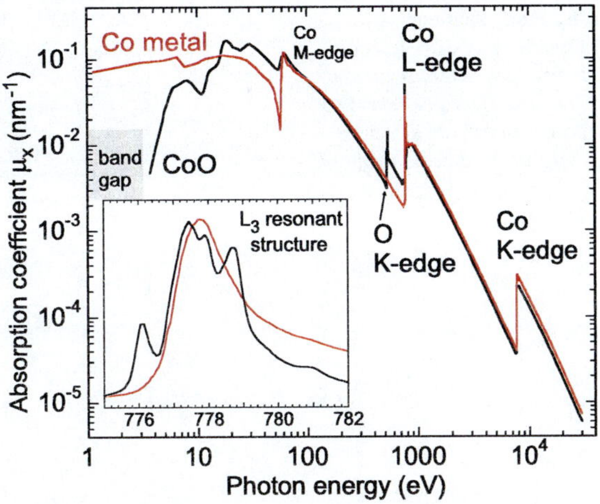

Fig. 1.17 Frequency dependence of the x-ray absorption coefficient μ_x for Co metal [154, 157, 158, 160] (red) and CoO [154, 156, 159, 161] (black) over a wide photon energy range. Note in particular the strong difference of absorption in the optical region due to the 2.5 eV band gap in CoO

1.4.2 The Importance of X-Ray Resonances

The importance of x-ray resonances, which occur at atomic absorption thresholds ("edges"), is based on the following key properties: (1) their elemental or atomic specificity, (2) their enhanced cross sections, (3) their chemical or bonding sensitivity, and (4) their polarization or orientational dependence. Owing to the short x-ray wavelength these properties may be utilized to obtain nanoscale structural resolution through spectro-microscopy. While the elemental specificity and enhanced cross sections of the resonances are apparent from Fig. 1.17, we shall briefly demonstrate the chemical and polarization effects.

1.4.2.1 Chemical or Bonding Specificity

The chemical information regarding *local bonding* is beautifully illustrated in Fig. 1.18 for different covalent bonds of selected C atoms with its C, N, and O neighbors [162, 163].

In the shown molecular complexes, central C atoms form bonds with different neighbor atoms through the formation of molecular orbitals. One typically distinguishes π and σ molecular orbitals which are respectively oriented perpendicular and parallel to the internuclear axis [29]. In x-ray absorption, one observes the transitions of C 1s core electrons into the *empty* parts of these orbitals, commonly denoted as π^* and σ^* orbitals. The different empty orbitals are then directly revealed by different resonances in the x-ray absorption spectra.

The splitting and shifting of resonances arise from both chemical core level shifts as observed in x-ray photoelectron spectroscopy (XPS) as well as from the splitting of the empty valence orbitals [29]. The change in energy position and intensity

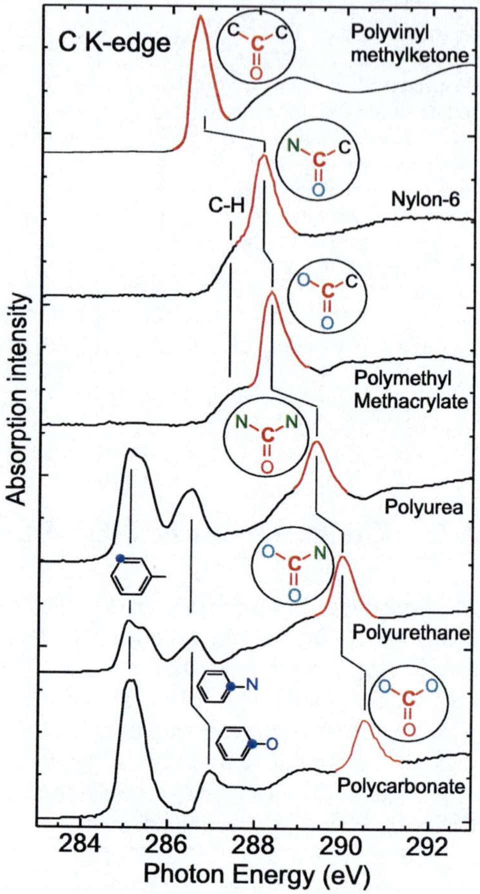

Fig. 1.18 Near edge resonances of carbon atoms that are covalently bonded to C, N, and O atoms, revealing the sensitivity of resonance intensities and peak positions to chemical environments [162, 163]

of different resonances may be used to study changes in molecular bonding upon chemical transformations in various forms, for example in molecule-surface reactions [29, 164] or even in transient reaction intermediates on ultrafast time-scales [165, 166].

1.4.2.2 Polarization Dependent Spectroscopy

The importance of the *polarization dependence* of the resonances in x-ray absorption spectra is illustrated in Fig. 1.19. One typically distinguishes the limiting cases of linear and circular polarization.

In linearly polarized light, the electric field vector which defines the polarization direction oscillates in magnitude along an axis. When interacting with a sample, the polarization vector acts as a search light for empty valence orbitals oriented along the

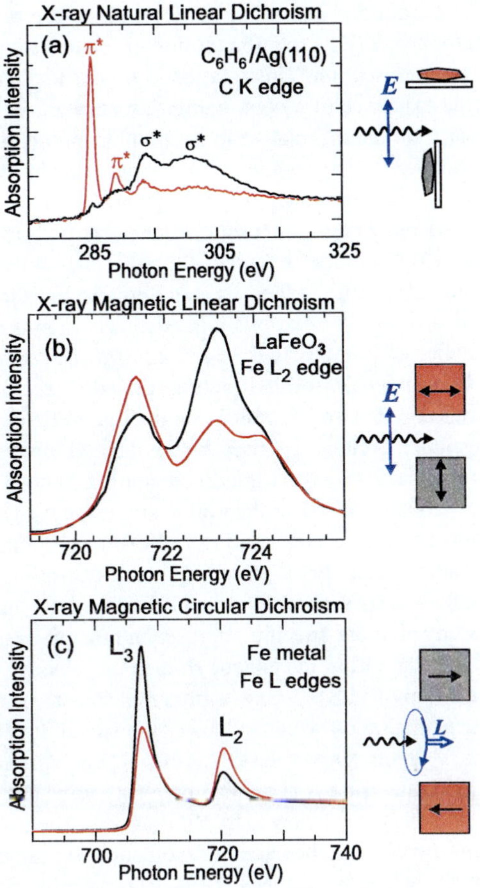

Fig. 1.19 **a** X-ray natural linear dichroism illustrated by the polarization dependence of the C K-edge absorption spectrum of benzene (C_6H_6) bonded to a Ag(110) surface, revealing that the molecule lies down on the surface [167]. **b** X-ray magnetic linear dichroism of the Fe L_2 resonances for a $LaFeO_3$ film due to the preferential orientation of the antiferromagnetic axis [168]. **c** X-ray magnetic circular dichroism for a ferromagnetic Fe film, revealing the direction and size of the magnetic moments [160]

axis. Parallel alignment maximizes the transition intensity, so that the polarization dependence directly determines the valence orbital orientation in the sample [29, 164]. The so-called *x-ray natural linear dichroism* (XNLD) effect is illustrated in Fig. 1.19a for benzene molecules bonded to a Ag(110) surface [167]. Polarized x-ray absorption spectroscopy may even detect the existence of a minute orientational order in structurally highly disordered systems [169–172].

The related effect of *x-ray magnetic linear dichroism* (XMLD) exists only in the presence of a magnetic alignment axis in a sample [34, 173–175]. It arises from the spin-orbit coupling which leads to a distortion of the charge density relative to the preferred spin orientation axis. The charge asymmetry then causes a difference in the x-ray absorption intensity when the polarization vector is parallel and perpendicular to the magnetic axis. XMLD is often used today for the determination of the orientation of the antiferromagnetic axis in thin films as illustrated in Fig. 1.19b [168].

Circular polarization corresponds to the *temporal* clockwise (right) or counter-clockwise (left) rotation motion of the electric field at a *fixed point* along the propagation direction. This rotation in *time* leads to a lack of time reversal symmetry and the existence of a *photon angular momentum*. The orientation of this photon angular momentum relative to an angular momentum or magnetic moment in a sample then leads to a prominent *x-ray magnetic circular dichroism* (XMCD) effect [33, 34, 176–179], as shown in Fig. 1.19c [160].

If one takes a snapshot in *time*, the polarization vector of the light wave shows a helical motion in *space* along the propagation direction creating a lack of inversion symmetry. When the bonding around the absorbing atom also lacks a center of inversion or exhibits a handedness in space, the relative alignments of the two helicities give rise to the *x-ray natural circular dichroism* (XNCD) effect [180–187]. In a molecular orbital picture a handedness of the charge density arises from the mixing of *p* and *d* orbitals by the ligand field and this "twist" or handedness in the valence orbitals is sensed by the handed electric field. This effect is typically small and does not exist in the dipole approximation.

Today, soft x-ray absorption studies may also be performed with table-top higher harmonic laser sources [188, 189]. In particular, polarization and time-dependent studies using the M-edges of the 3*d* transition metals Fe, Co, and Ni, which occur below 100 eV as shown for Co in Fig. 1.17, have emerged as a convenient way to study element-specific femtosecond magnetization dynamics [190, 191].

Polarization dependent studies are also of importance in x-ray emission spectroscopy (XES). Today, x-ray emission studies are typically carried out after resonant excitation which enhances the signal. It is then referred to as *resonant inelastic x-ray scattering* or RIXS. While RIXS is in general a complicated process [192–197], the process is quite simple for atoms and molecules bonded to surfaces [164, 198, 199]. As reviewed by Nilsson and Pettersson [199], RIXS has provided detailed information on the nature of molecular bonding to surfaces which plays a key role in heterogeneous catalysis [199–201]. As an example we show in Fig. 1.20 C K-shell absorption and emission spectra of oriented benzene C_6H_6 molecules on Cu(110) and Ni(100) surfaces [202, 203].

For benzene, the π (perpendicular to molecular plane) and σ (in plane) orbitals can be separately observed through the polarization dependence of the transitions in absorption and emission. Resonances in the x-ray absorption spectra shown on the right of Fig. 1.20a correspond to transitions of a 1*s* core electron into the lowest unoccupied molecular orbitals of benzene which are *anti bonding* in nature. The polarization dependence is due to the alignment of the incident linear polarization (electric field) vector relative to the spatial direction of molecular orbitals [29], as previously shown in Fig. 1.19a.

In contrast, XES or RIXS spectra reflect transitions from filled valence states, typically responsible for *bonding*, into the core hole created in the x-ray absorption step. In RIXS, illustrated on the left of Fig. 1.20, a polarization dependence arises from the transverse nature of the emitted light with the polarization vector lying in a plane perpendicular to the photon emission direction [198]. For emission normal to the surface, shown in blue, only σ orbitals are observed. At glancing emission from

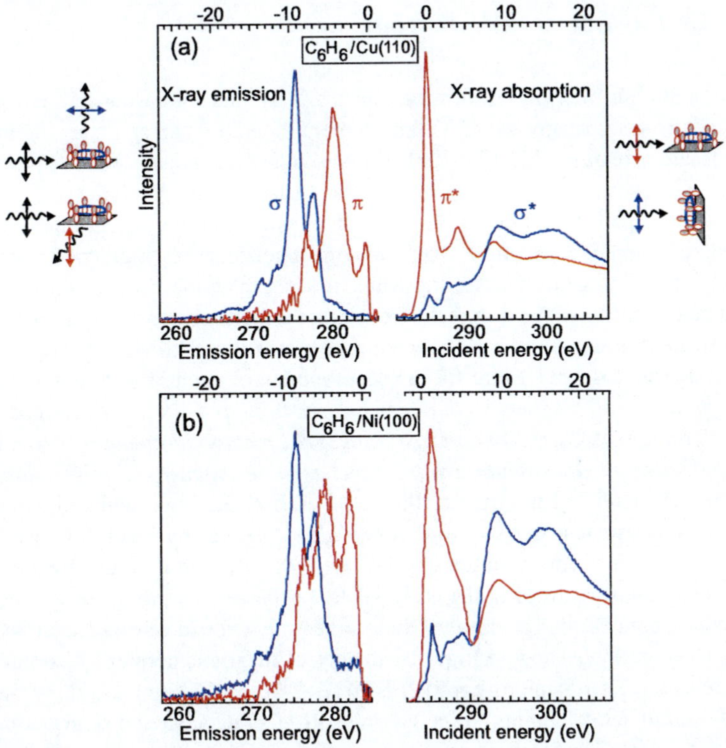

Fig. 1.20 **a** Illustration of the complementarity of polarization dependent C $1s \leftrightarrow 2p$ x-ray absorption and emission spectra for benzene (C_6H_6) molecules bonded in a lying down configuration to a Cu(110) surface [202, 203]. In x-ray absorption (right) the empty valence orbitals aligned parallel to the electric field polarization vector are revealed by the spectra, as shown on the right. In x-ray emission (left), the probed orbitals are determined by the emission direction. In the shown red spectrum the parallel component has been eliminated. **b** Complementary spectra for C_6H_6 bonded to a Ni(100) surface [202], revealing increased surface chemical bonding through changes of resonances in position and intensity

the surface, both perpendicular (π) and parallel (σ) bonds to the surface are observed. In the shown red emission spectrum on the left of Fig. 1.20, the contributions from σ orbitals, given by the blue spectrum, have been subtracted from the original spectrum.

The case of C_6H_6 on Cu(110) shown in Fig. 1.20a reflects a relatively weak chemisorption bond reflected by the minor modification of the molecular benzene orbitals. This is in contrast to the case of C_6H_6 on Ni(100) shown in Fig. 1.20b, where both XNLD and RIXS spectra directly reveal the distortion of the molecular orbitals upon chemisorption through changes of the spectral features in position and intensity [202].

1.4.3 X-Ray Spectro-Microscopy

The combination of x-ray absorption spectroscopy and microscopy is utilized in x-ray spectro-microscopy which yields element, bonding, and spin specific images of nanoscale structures [11, 204–206]. X-ray imaging techniques are based on either real space images or the inversion of reciprocal space diffraction images.

As an example of the novel utilization of x-rays we briefly discuss their impact on modern magnetism research. Originally, the determination of magnetic structure was the domain of neutron scattering which first revealed the atomic spin structures of bulk single crystals. The progress of x-ray research may be put into perspective by the 1994 press release by the Nobel committee awarding the Prize in Physics to B. N. Brockhouse and C. G. Shull for the development of neutron techniques, stating: "*Neutrons are small magnets ... (that) can be used to study the relative orientations of the small atomic magnets. (In contrast)...the x-ray method has been powerless, and in this field of application neutron diffraction has since assumed an entirely dominant position.*" Ironically, at the time of the press release, the importance of x-rays for modern magnetism research had already been demonstrated by the discovery of x-ray magnetic dichroism effects in the late 1980s [174, 176] and their use for the study of magnetic nanostructures in the early 1990s [207]. By now, the use of x-rays has taken over from neutrons because of their larger spin dependent cross sections [34].

As an example we show in Fig. 1.21 images of magnetic domains recorded with photoemission electron microscopy (PEEM) [204, 208]). Figure 1.21a shows images of the *ferromagnetic* domains ("bits") written on a CoPtCr coated disk in a computer hard drive [207]. The images were obtained with circularly polarized soft x-rays tuned to the Co L_3 resonance. In (b) we show the naturally occuring *antiferromagnetic* domains in an epitaxially grown $LaFeO_3$ film, recorded with linearly polarized x-rays tuned to the Fe L_2 resonance [209]. In (c) and (d) the correlation between the ferromagnetic domains in a thin Co film and the antiferromagnetic domains in a NiO single crystal underneath is demonstrated. In the figure the original assignment of the orientation of the antiferromagnetic axis [210, 211] has been changed by 90° based on more recent work [212–214].

The functioning of a magnetic memory cell is illustrated in Fig. 1.22 by time dependent nanoscale images recorded with scanning transmission x-ray microscopy (STXM).

The experimental geometry is schematically shown in Fig. 1.22a [215], and the images recorded at the Advanced Light Source (ALS) in Berkeley in (b) [216]. They reveal the remarkable switching of the magnetic orientation of the blue layer in (a), whose direction represents either a 0 or 1 bit. It evolves by motion of a vortex core through the bit instead of a simple rotation of the entire magnetic bit. Not shown is that a current pulse sent in the opposite direction switches the magnetic bit back into the original direction [216, 217].

The images reveal the extraordinary capabilities of modern storage rings which, in contrast to XFELs, provide high intensity stability required for state-of-the-art measurements of small signals. Elemental specificity is used to pick out the signal of

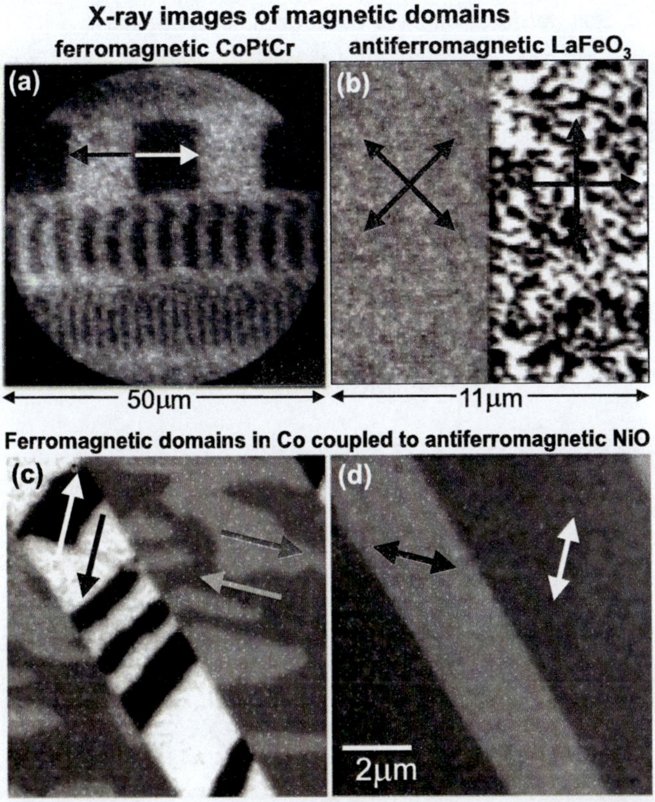

Fig. 1.21 PEEM images of magnetic domains recorded with polarized x-rays. **a** Images of magnetic bits in a CoPtCr magnetic recording medium recorded with circularly polarized x-rays tuned to the Co L_3 resonance [207]. The arrows indicate the ferromagnetic alignment directions. **b** Images of antiferromagnetic domains in $LaFeO_3$ recorded with horizontal linear polarization alignment and the photon energy tuned to the Fe L_2 resonance [209]. The domains are only visible when there is an unequal projection of the polarization axis onto the two antiferromagnetic axes indicated by double arrows. **c** Ferromagnetic domains in a Co film following the antiferromagnetic domains of the NiO single crystal template underneath, shown in (**d**). The assignment of the orientation of the antiferromagnetic domains [210, 211] has been changed by 90° based on later work [212, 213]. The ferromagnetic images were taken with circular polarized x-rays tuned to the Co L_3 resonance, the antiferromagnetic images of the same region utilized linear polarized x-rays tuned to the Ni L_2 resonance

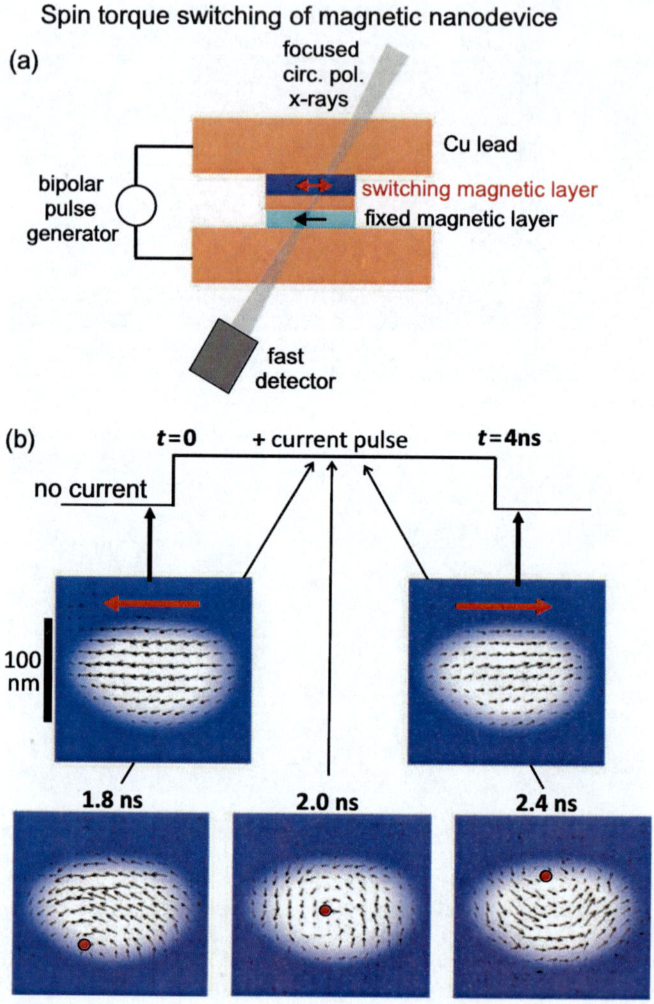

Fig. 1.22 STXM images of the mechanism underlying the spin torque switching of a magnetic layer Co layer (blue) in a memory device. A schematic of the experimental arrangement is shown in **a** [215] and time-dependent magnetic images are shown in **b** [216]

a 2-nm-thick Co layer in an oval 180×110 nm pillar buried under a total of 250-nm-thick metal (mostly Cu). Magnetic contrast is achieved by use of circularly polarized x-rays, spatial resolution of 30 nm through x-ray zone-plate focusing, and temporal resolution through the 70 ps long pulses emitted by the x-ray storage ring. The x-ray pulses are synchronized with ultrafast electronics with 30 ps rise time.

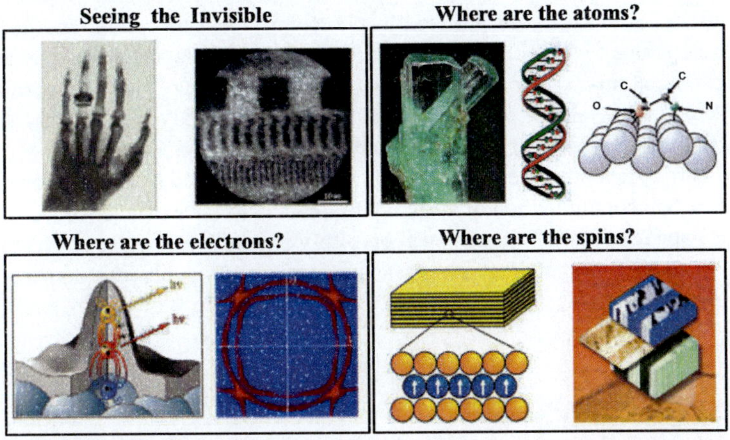

Fig. 1.23 Illustration why x-rays are so useful. Top left: X-rays allow imaging the invisible, illustrated by a medical application and by an image showing the bit structure of a magnetic recording disk. Top right: X-Ray crystallography reveals where the atoms are in crystals, macromolecules, or chemisorbed molecules. Bottom left: X-ray absorption and emission (shown on left) spectroscopy can reveal the bonding orbitals on individual atoms and angle resolved photoemission spectroscopy reveals the electron states at the Fermi surface of a high temperature superconductor (right). Bottom right: Magnetic linear and circular dichroism studies reveal the spin structure in ferromagnetically coupled multilayers (left) and near ferromagnetic/antiferromagnetic interfaces (right)

1.4.4 Summary of Key Capabilities of Synchrotron Radiation

We may summarize the capabilities of synchrotron radiation x-rays for the study of properties of matter during the first one hundred years as shown in Fig. 1.23. During the twentieth century, x-ray science primarily concentrated on the investigation of the static or equilibrium properties of matter.

The first well-known application of x-rays shown in Fig. 1.23 is a result of their penetrating nature, allowing one to see below the surface of objects, utilized in medical x-ray imaging. This capability became a world wide sensation immediately after Röntgen's first experiments, and unlike many other scientific discoveries, its importance was immediately apparent even to the lay person and played a role in Röntgen receiving the Nobel Prize only six years after his discovery.

The second well-known use of x-rays is a result of the wavelength being of the same spatial dimension as the atomic building blocks of matter. This has made x-ray diffraction the most important technique for the unraveling of the atomic structure of matter, ranging from solids to organic matter. As an example, by 2018 of the about 145 thousand solved macromolecular structures entered into the worldwide protein data bank about 130 thousand were solved by use of x-ray diffraction [218] with the overwhelming majority using synchrotron radiation.

While the first two applications utilize "hard" x-rays with energies of 10 keV or higher, "soft" x-rays with energies in the 100–1000 eV range also have important applications. Soft x-rays are utilized in photoemission, absorption, and emission spectroscopy due to their sensitivity to the electronic and magnetic structure associated with the atomic valence electrons that form the bonding "glue" between atoms [29, 34, 37]. The valence electrons are either probed directly by photoemission or indirectly through resonant transitions between core and valence electrons. In contrast to hard x-rays, the use of soft x-rays was long impeded by their absorption in air and carbon contamination of x-ray optics, leading to the existence of a soft x-ray spectral gap [219, 220]. This was overcome only around 1980 by use of synchrotron sources in conjunction with ultra-high vacuum optics and sample environments [220, 221].

1.5 Science with XFELs

We conclude this chapter by highlighting key new capabilities of XFELs in different areas of science. We note that, today, scientific teams typically utilize all existing XFEL facilities (see Fig. 1.13) where "beamtime" can be secured through proposals. The examples illustrated below are therefore a testament of the spirit of international scientific collaborations.

1.5.1 Snapshots of the Atomic Structure of Matter: Probe Before Destruction

As first suggested by Neutze et al. [222], the combined high intensity and short duration of XFEL pulses may be used to record snapshots of the *atomic structure* of a sample before it is destroyed by the high intensity pulse. The essence of the "probe before destruction" concept is based on the fact that the speed of light is about a million times faster than the speed of sound. In practice, a picture of the equilibrium atomic structure may therefore be captured by a femtosecond pulse before there is significant change of the atomic positions on the 1 picosecond timescale. The large deposited energy per pulse initially heats the electronic system on the femtosecond timescale before the energy is transferred to the lattice, leading to large thermally induced atomic motion that results in bond breaking, melting, and ultimately sample explosion [222].

This temporal evolution leads to the counterintuitive ability of XFEL pulses to actually avoid structural and crystallographic damage which may occur at longer exposure times required for synchrotron radiation [223], with the caveat that the sample needs to be replaced after each shot. This "probe before destruction" concept only applies, however, to the positions of the atomic cores (or nuclei) but not to the valence electronic structure since charge and spin densities can change on the timescale of order of 1–10 fs (see below).

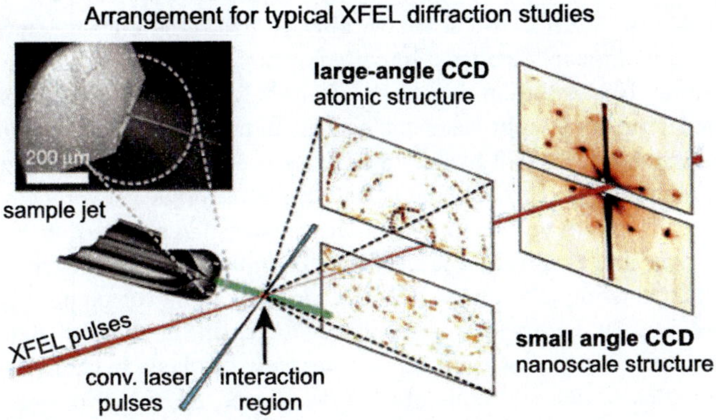

Fig. 1.24 Typical experimental arrangement for XFEL diffraction experiments, adapted from Chapman et al. [226]. A jet-based sample delivery system provides either a continuous-stream or pulsed beam. A sample response may be initiated by a short pulse conventional laser that also defines a time-zero reference for time-resolved XFEL diffraction patterns. A position sensitive CCD detector records the wide angle diffraction pattern, while the small angle pattern transmitted through a central opening is recorded with a second CCD detector

The requirement of shot-to-shot sample replacement led to the development of new sample delivery systems [224, 225], which are incorporated into a typical experimental arrangement for XFEL diffraction studies, as schematically illustrated in Fig. 1.24. Various kinds of samples can be delivered by a liquid jet as shown in Fig. 1.24, such as macromolecules and small protein crystals surrounded by water or different liquids, while atomic clusters may be delivered by gas dynamic virtual nozzles [224, 225]. For solid samples, pulse-synchronized sample rastering techniques may be used [148, 227] since the required motion is quite small due to the focused micron-size XFEL beam.

Using the experimental geometry in Fig. 1.24, Chapman et al. [226] in 2011 successfully demonstrated the "probe before destruction" concept with atomic resolution. Instead of single protein molecules envisioned by Neutze et al. [222], they used small protein crystals which diffract significantly stronger. The diffraction patterns could also be directly compared to those recorded with synchrotron radiation. Another experiment in the same year by Seibert et al. [228] probed the shape and internal structure of single viruses on the nanometer length scale. These experiments established XFELs as new powerful tools for structure determinations of macromolecules in their natural room temperature environment, based on the counterintuitive trick to beat radiation damage by use of ultra-intense and ultra-fast x-ray pulses. The required size of protein crystals, whose growth is often an impediment, may also be reduced with the ultimate goal of single molecule imaging [222]. The XFEL-based development of protein crystallography has been reviewed by Brändén and Neutze [229] and Barends et al. [230].

One area, where XFEL studies have already had a significant impact, is the understanding of arguably the most important chemical process on Earth, oxygenic photosynthesis. It is based on the action of an enzyme known as photosystem II, that uses solar energy to split water molecules, and in the process creating oxygen molecules plus electrons and hydrogen radicals needed to convert carbon dioxide into the organic fuel glucose, $C_6H_{12}O_6$, and different forms of hydrocarbons [231, 232]. About 2.5 billion years ago, this process resulted in our O_2-rich atmosphere, the ozone layer and eventually the evolution of multicellular lifeforms. Today, its detailed understanding is envisioned to provide a blueprint for future technologies based on using solar energy to satisfy our ever-increasing energy demands.

Over the last twenty years, remarkable progress has been made in understanding this complex process in terms of the structure of the central oxygen evolving center, Mn_4CaO_5. As reviewed by Barber [231], an early picture was obtained with synchrotron radiation-based diffraction [233–235] and x-ray spectroscopic studies [236, 237], which has been increasingly refined through XFEL studies, as reviewed by Cox, Pantazis, and Lubitz [232] and Bergmann et al. [238]. Uncertainties associated with potential x-ray damage of synchrotron based studies were only overcome through XFEL studies, which revealed the atomic structure of the central Mn_4CaO_5-cluster and its function during the so-called Kok cycle [239] shown in Fig. 1.25 [238]. Combined diffraction and x-ray emission studies revealed the key insertion of a new oxygen atom (open red circle) in the Mn cluster during the second flash 2F, followed by initial O–O bond formation coupled to Mn reduction after the third flash 3F as indicated in the figure [240].

1.5.2 *Creation and Characterization of Transient States of Matter*

Another broad theme of XFEL applications is *time dependent* snapshots of fleeting atomic arrangements by diffractive imaging and of the associated electronic structure by x-ray absorption and emission spectroscopy. Below we briefly discuss the use of XFELs to detect and characterize *transient states* of matter in the form of chemical transition states in molecules and transient collective states in liquids.

1.5.2.1 Femtochemistry

The field of femtochemistry explores the very act underlying chemistry, the making and breaking of chemical bonds. It requires techniques that can image on the atomic length scale and the intrinsic femtosecond timescale over which the valence electron density (molecular orbitals) can change. The field was pioneered by Ahmed Zewail, who in 1999 was awarded the Nobel Prize in Chemistry "for his studies of the transition states of chemical reactions using femtosecond spectroscopy" [241]. The

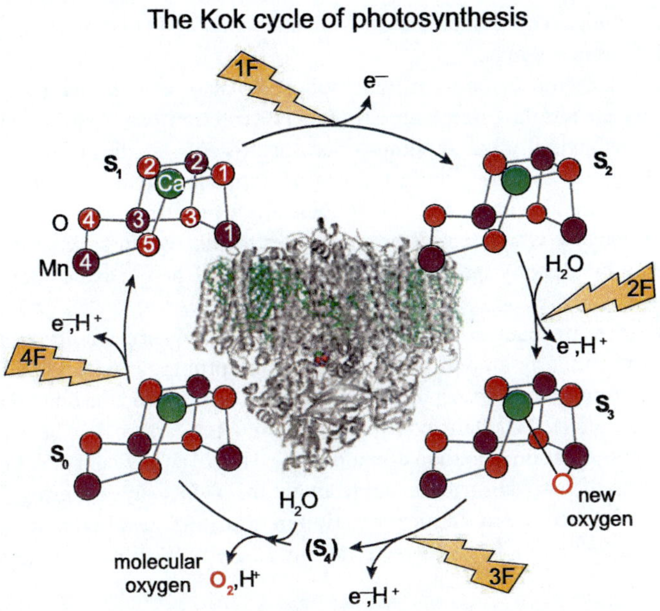

Fig. 1.25 Kok cycle of photosynthesis [239]. The overall structure of the protein (center) is surrounded by the four-step catalytic cycle (Kok cycle), revealed by light flashes 1F–4F. For each of the stable states S_0, S_1, S_2, and S_3, the x-ray diffraction structure of the catalytic Mn_4CaO_5 cluster (O in red, Mn in purple, and Ca in green) obtained from XFEL measurements is also shown. Figure adapted from Bergmann et al. [238]

early work was based on combining the femtosecond time resolution of laser pulses with the atomic length scale resolution of electron diffraction. XFELs naturally have both, and this ability has given rise to the new field of ultrafast x-ray science. It has been argued that this has led to an ultrafast x-ray spectroscopic revolution in chemical dynamics [242].

The often quoted quest of femtochemistry is a *molecular movie* with femtosecond frames and atomic resolution that shows the breaking or formation of chemical bonds. While this goal will most likely be reached for simple molecules, the real problem of practical importance is much more complicated. It involves the understanding of catalytic pathways in multi-component systems that can inspire the development of new energy-efficient technologies. In practice, many chemical transformations are catalytically enhanced and proceed in multi-component systems from equilibrium structures and states, defined by energy wells in the potential energy surface, through *transition states* that exist at energy barriers along a reaction path. These transition states are hard to capture or observe not only because of their short lifetimes but because they may involve only tiny fractions of the equilibrium atomic populations. We are only beginning to tackle some simple problems in the vast area of chemical and biological activity. This is facilitated by the elemental and bonding state specificity of resonantly tuned x-rays.

In practice, the observation of transition states also requires that the process is triggered on an ultrafast timescale so that its evolution can then be followed in time. This has naturally led to pump-probe spectroscopy studies of *photochemical* reactions triggered by optical laser pulses down to tens of femtoseconds durations. Optical pulses are also used in XFEL studies as a trigger, followed by femtosecond x-ray probe pulses, which can determine either the change in atomic positions by diffraction or bonding by spectroscopy as reviewed by Chergui and Collet [243] and Wernet [244].

By now, there are many examples of utilizing the combined capability of triggering a chemical reaction by an optical pump pulse and probing it with an XFEL pulse tuned to an atom-specific absorption resonance. By this method the bond breaking of carbon monoxide (CO), catalyzed by a ruthenium (Ru) surface, has been observed, followed by the creation of carbon dioxide (CO_2) [165, 166]. Other studies have used a combination of x-ray absorption and resonant inelastic x-ray scattering (RIXS) to map out the detailed photochemical reaction in transition metal complexes such as $Fe(CO)_5$ and $[Fe^{III}(CN)_6]^{3-}$ as reviewed by Jay, Kunnus, Wernet, and Gaffney [245].

1.5.2.2 Transient Density Fluctuations

It is well known that nanoscale ordering phenomena associated with charge, spin, and orbital order exist in correlated materials [246]. In many cases these states have sufficiently long lifetimes to be imaged by diffraction or real space microscopy. It has recently been observed, however, that charge order in $La_{2-x}Ba_xCuO_4$ exhibits fluctuations on picosecond timescales accessible with XFELs [247]. Since picosecond temporal fluctuations correspond to meV energy changes that are difficult to resolve by inelastic x-ray scattering in the energy domain, time-dependent x-ray scattering measurements with XFELs appear to provide important new capabilities.

Similar to correlated materials, liquids where small molecular units interact loosely through hydrogen bonds are also expected to exhibit nanoscale order associated with changes in orientational order, conformation, and atomic density within the first few coordination shells of the individual molecules. Such systems may appear globally homogeneous, and the local structural motives may be difficult to detect owing to their dynamic fluctuations. Judging from Fig. 1.12, the laws of inertia would seem to allow fluctuation of molecular groups on sub-nanosecond timescales.

Such locally different structural motives are today considered as an explanation of the unusual thermodynamic properties of *liquid* water. Water is the most important liquid for our existence and is the only common substance that appears under everyday conditions in all three states of aggregation—gas, liquid, and solid. It is essential for many energy-related chemical transformations and environmental processes. From a scientific point of view, water is a liquid with strange thermodynamic properties; its maximum density occurs at 4° C and increases upon melting (ice floats

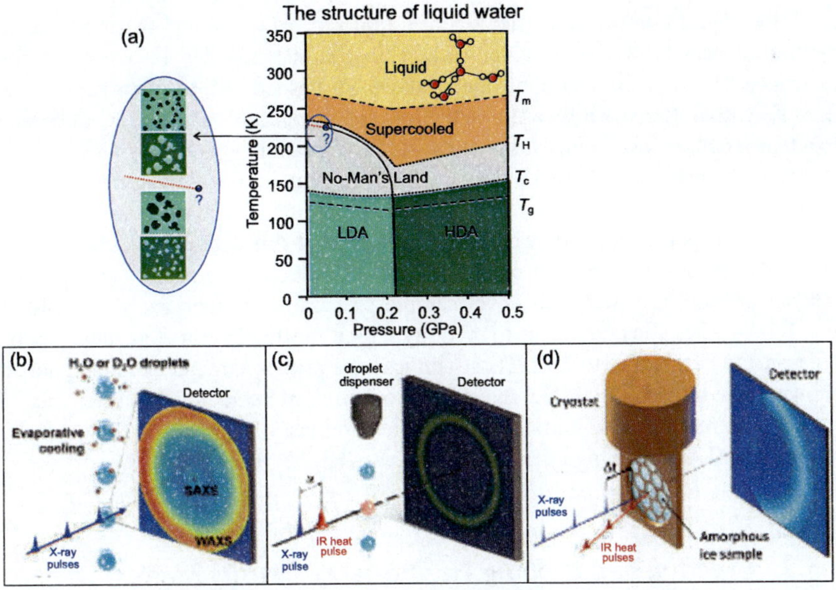

Fig. 1.26 **a** Schematic phase diagram of water connecting the liquid, supercooled, and glassy states labeled low density amorphous (LDA) and high density amorphous (HDA). The gray-colored area depicts the "no-man's land" which is conjectured to contain an illusive critical point, indicated in blue. The left inset shows two proposed fluctuating phases, separated by the dashed red Widom line, that consist of different mixtures of LDA and HDA, based on the latest XFEL diffraction studies [251, 252]. **b, c** Schematic diagrams of preparing supercooled water in the no-man's land region. **b** A continuous series of $\sim 14\,\mu$m water droplets cooled through evaporation as a function of distance between the dispenser and the measurement point. **c** Heating of the supercooled water droplets with infrared pulses before the x-ray pulses. **d** Ultrafast heating of amorphous ice samples with infrared pulses. Figure courtesy of Anders Nilsson [252]

on top), its viscosity decreases under pressure, and it has a high surface tension, to name a few.

The structure of water was already studied by x-ray diffraction in the 1930s, and in 1933 Bernal and Fowler [248] deduced an essentially tetrahedral structure in which each water molecule has four nearest-neighbor molecules. They concluded that modifications of the diffraction patterns with temperature from below 4° C (ice) to 340° C indicated subtle structural changes that evolved continuously into each other. Today, the complexity of possible networks of H_2O molecules is evidenced by the fact that there are 17 experimentally confirmed polymorphs of ice, with snowflakes corresponding to one of them [249].

Over the years, the liquid phases of water have attracted particular attention with experimental neutron and x-ray studies complemented by molecular dynamics computations as reviewed by Nilsson and Pettersson [250]. The pressure-temperature phase diagram emerging from these studies is shown in Fig. 1.26a.

Without going into details, the key to a uniform structural and thermodynamic picture appears to be the existence of a critical point, indicated in blue, with the extension of a so-called Widom line,[18] shown in dashed red, to a temperature of 228° K at 1 bar. Remarkably, water in the form of small droplets can be super cooled to a temperature close to this value without crystallization.

Over the last ten years or so, much effort has gone into exploring the so-called no-man's land indicated in gray in Fig. 1.26a, which is believed to contain the indicated critical point (blue circle). The combination of diffraction experiments with sub-100 fs pulses, where each pulse yields a diffraction pattern that can be separately analyzed, together with novel sample preparation techniques, illustrated in Fig. 1.26b–d has allowed access to the gray region for the first time, as reviewed by Nilsson and Perakis [251, 252]. These studies have proposed two fluctuating phases, shown in the inset in Fig. 1.26a, that consist of different fluctuating mixtures related to two pure macroscopic liquid phases of the two glassy states labeled low density amorphous (LDA) and high density amorphous (HDA).

1.5.3 Creation and Probing High Energy Density Matter

Remarkably, x-rays interact more strongly with atomic core than valence electrons, as illustrated for Ne atoms in Fig. 1.27. The figure shows that at x-ray energies that exceed the binding energies of the innermost $1s$ electrons, the interaction cross section with different atomic shells decreases from the inside out. This allows the creation of "hollow atoms" with x-rays, as beautifully demonstrated in early XFEL experiments [255, 256]. In this context it is important that the cross sections shown in Fig. 1.27 are *single photon* absorption or photoemission cross sections. Once an electron has been excited from a given core shell, the excitation probability of additional electrons from the same core shell is exceedingly small [256].

The stripping of electrons from atomic shells also occurs in solids. After a femtosecond pulse excitation, a solid is turned into an *electronically* hot state. However, atomic motion and bond breaking will only occur on picosecond or longer timescales because of the greater mass of atoms or ions than electrons, ultimately ending in the explosion of the sample. The properties of such created hot plasma-like states are of interest in the field termed *high energy density* (HED) science [257]. This relatively new subfield of physics has strong connections with plasma physics, nuclear physics, and astrophysics [258, 259]. HED matter may be created with multi-terawatt lasers and more recently with XFELs. The short XFEL pulses can both create and probe the HED state, and their complete penetration provides true "bulk" information. This is a significant advantage over optical pulses, which after creation of the HED state cannot penetrate it. For the combined reasons of being ultrashort and penetrating, XFEL pulses have become powerful new tools for HED science.

---

[18] The Widom line, named in honor of Benjamin Widom, defines a line in the pressure-temperature plane where the correlation length has its maximum.

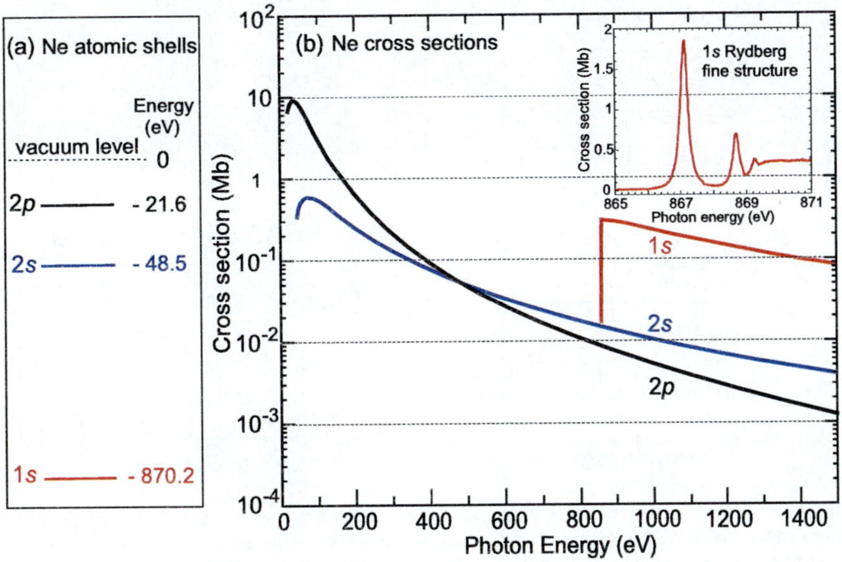

Fig. 1.27 Photon excitation cross for different shells of Ne atoms. The atomic shell structure and binding energies are shown in (**a**), and the respective ionization cross sections are shown in (**b**) [253]. The inset shows the fine structure due to Rydberg resonances just below the $1s$ (K-shell) ionization threshold at 870.2 eV [254]

An early example that directly reveals the evolution of the simple metal Al from room temperature to the HED regime is shown in Fig. 1.28. Al has two core absorption edges, an L-edge ($2p$ excitation) at 73 eV and a K-edge ($1s$ excitation) at 1560 eV. The shown work [260, 261] utilized $\hbar\omega$ =93 eV XUV radiation from the FLASH facility.[19]

Figure 1.28a shows the change in transmission through an Al film as a function of fluence for incident 15 fs pulses of $\hbar\omega = 93$ eV or 20 eV above the lowest energy $L_{2,3}$ absorption threshold. With increasing fluence, the sample is seen to become nearly transparent. This is caused by increasingly stripping single electrons from the $2p$ core shell until at the highest fluence all illuminated atoms have a $2p$ core hole. Then further absorption becomes prohibited since the excitation probability of a second electron out of the $2p$ core shell of the same atom is negligibly small as shown by early XFEL experiments [255, 256].[20] Also, the $2p$ core shell excitation dominates over direct excitation of valence electrons (see Fig. 1.27). The changes

[19] Later work also explored the K-edge response [262] but will not be discussed here.

[20] The excitation of more than one-electron from a core shell of atoms only occurs after all valence electrons are removed, as first observed in He-like Ne^{8+} [256].

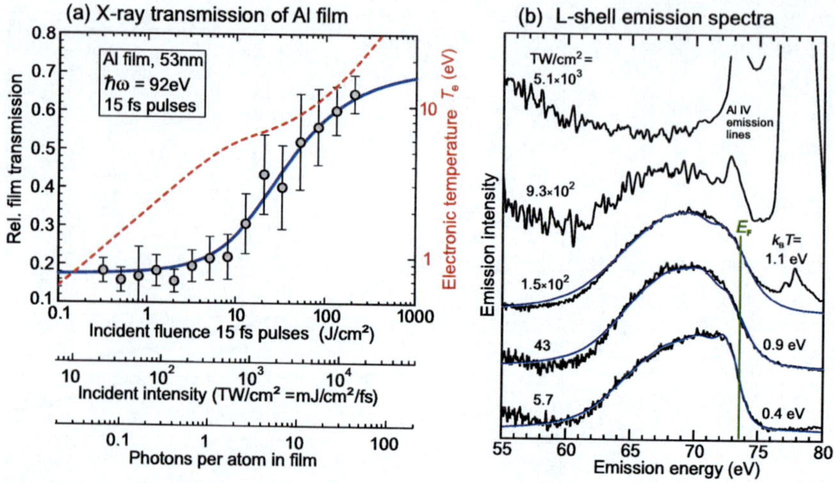

Fig. 1.28 Response of Al metal as a function of incident intensity of incident 93 eV pulses, about 20 eV above the lowest energy $L_{2,3}$ core shell. **a** Change in *transmission* [260] as a function of fluence of 15 fs pulses. The highest fluence for the used 15 fs pulses of 200 J/cm^2 corresponds to an intensity of 13 J/cm^2/fs (1.3 × 10^4 TW/cm^2). The circles represent the experimental data points, and the solid line is a theoretical prediction. The dashed curve refers to the electron temperature in eV (right scale) of the valence electrons at the end of the 15 fs FEL pulse, but before the L-shell holes are filled at about 40 fs after creation. **b** Change of the emission spectrum as a function of intensity (fluence/pulse length) for 35 fs excitation pulses, revealing the transition from a transiently hot solid to the HED regime [261]. The blue curves are simulations and shown in green is the position of the Fermi level E_F where changes in electronic temperature manifest themselves

in transmission reflect the saturation of the single $2p$ core electron excitations per atom.[21]

The complementary emission spectrum shown in Fig. 1.28b shows that intensities below about 1.5 × 10^2 TW/cm^2, causing an electronic temperature in the sample of up to $k_B T \simeq 1$ eV, modify the band occupation in Al near the Fermi level, E_F, reflected by the change in slope of the Fermi-Dirac distribution around E_F. At the highest incident intensities, the emission spectrum is dominated by intense plasma line emission that obscures any possible survival of valence band structure.

1.5.4 *Non-linear X-Ray Interactions with Matter*

The field of non-linear (NL) optics was born with the advent of the laser, and its beginning is often associated with the discovery of second harmonic generation in 1961 by Franken et al. [264]. It is typically defined classically as the case where the

[21] Since the created core holes, which create Auger electrons, decay on a longer time (30 fs [263]) than the pulse length, this process is not seen in the shown transmission measurement.

material response not longer depends linearly on the action of the electric field of light [265]. In the language of x-ray science and quantum optics, used in this book, NL optical processes may occur by the presence of more than a single photon within the atomic coherence volume of the atomic building blocks of matter.

There is a remarkable exception to the above definition of NL x-ray science, based on the requirement of more than a single incident photon at a time. In 1971, Eisenberger and McCall [266] conducted an experiment where radiation from a 17 keV Mo x-ray tube source was scattered by a beryllium crystal. After about two weeks of accumulation of coincidence events in the scattered x-rays, they observed the splitting of single incident photons into two entangled photons, referred to as spontaneous parametric down conversion. This tour de force experiment, conducted about 40 years before the advent of XFELs, may rightfully be considered the first NL x-ray experiment.

The key to understanding NL interactions of XFEL x-rays with matter is the *photon degeneracy parameter* whose increase with the advent of optical lasers and XFELs is shown in Fig. 1.3. It exceeds unity only for such sources. This means that all synchrotron radiation experiments are basically determined by interaction processes with matter that occur one-photon-at-a-time since the probability that two or more photons are simultaneously present is negligibly small. XFELs therefore offer the opportunity for the first time to study processes that are driven by two or more photons. This case was first treated quantum mechanically by Maria Göppert-Mayer in 1931 [267], who derived the expressions for two-photon absorption and photoemission and stimulated Rayleigh/Raman scattering, long before such processes could be observed.[22]

Over the last 15 years or so, different non-linear x-ray interactions with matter have been observed and non-linearity thresholds for XFEL radiation have been discussed in [268–272]. The observation of x-ray transparency in non-resonant absorption by Nagler et al. in 2009, discussed above, represents the first XFEL experiment revealing a NL response [260, 261]. It was followed by the demonstration of the creation of multi-electron excitations and the creation of hollow Ne atoms in 2010 [255, 256]. These studies were later extended with atomic and molecular [273, 274] and solid samples [259, 262, 275–277].

Another milestone is the study by Rohringer et al. in 2012 [278], demonstrating amplified spontaneous emission in Ne gas (see below). It was followed by the demonstration of stimulated x-ray emission [271, 278–282] and stimulated resonant inelastic scattering [283–286]. Other NL studies involved non-linear mixing of optical light and x-rays [287], second harmonic generation [288], two-photon absorption [272, 289, 290], four wave mixing [291], and anomalous Compton scattering [292].

Below we give two brief examples of XFEL induced non-linear processes.

[22] I would like to thank Shaul Mukamel for pointing this out to me.

1.5.4.1 Amplified Spontaneous Emission

A unique capability of XFELs is their ability to create a population inversion in atoms, similar to the first step in an optical laser. This requires an intense short pulse, or a spike within a pulse, whose intensity is sufficient to excite a significant fraction of atoms in a sample before they can decay again. For x-rays, this process involves kicking out core electrons from atoms, resulting in an excited atomic state that has a core hole. The process has to happen on a timescale shorter than the core hole decay time (a few femtoseconds).

Following such a pump pulse, x-rays will first be spontaneously emitted during the decays of electrons from outer shells into the core hole. In his original introduction of spontaneous and stimulated emission, Einstein already considered the emission directions of spontaneous and stimulated radiation based on arguments of momentum conservation [60]. Spontaneous emission is in random directions, while stimulated emission is directional, with the photon driving the stimulated decay cloning itself in every aspect. By using a gas cell shaped as a cylinder, this cloning process will preferentially occur in the long direction of the cylinder, and consecutive stimulation events will lead to a directional photon avalanche, as in a laser cavity. The enhancement is referred to as *amplified spontaneous emission* (ASE) [293].

The x-ray ASE effect, first demonstrated by Rohringer et al. [278], utilized the geometry illustrated in Fig. 1.29. XFEL SASE pulses of 960 ± 4 eV photon energy, $\simeq 50$ fs duration, and an intensity up to 2×10^{17} W/cm^2 (200 J/fs/cm^2) were used to induce a population inversion of the Ne Kα transition (849 eV). The Ne gas was contained in an elongated cell along the beam direction which allowed the spontaneous Kα x-rays created in the front part of the cell to be amplified through stimulated decays along the incident beam direction. The 960 eV pump and 849 eV stimulated beams were then separated by a grating spectrometer of about 2 eV resolution and imaged by a CCD detector. A strong increase of about four orders of magnitude of the spontaneously emitted Kα radiation by the ASE process was observed upon doubling the incident pulse energy from 0.12 to 0.24 mJ.

1.5.4.2 Stimulated Forward Scattering and Loss of Diffraction

Rather than letting the stimulation process be triggered by uncontrolled spontaneous decays as in ASE, one may utilize a high intensity XFEL pulse whose temporal coherence length is longer than the core hole decay time. Such a pulse, created, e.g. by sending a SASE pulse through a monochromator, can then control both the excitation as well as the core hole decay process. The pulse necessarily has a narrow energy width and may be tuned to an atomic absorption resonance. When such a pulse is sent through a sample in the form of a thin film, with its energy tuned to an atomic resonance, the random spontaneous scattering observed at synchrotron sources will be replaced with increasing XFEL intensities by *stimulated emission* into the forward direction of the incident photons.

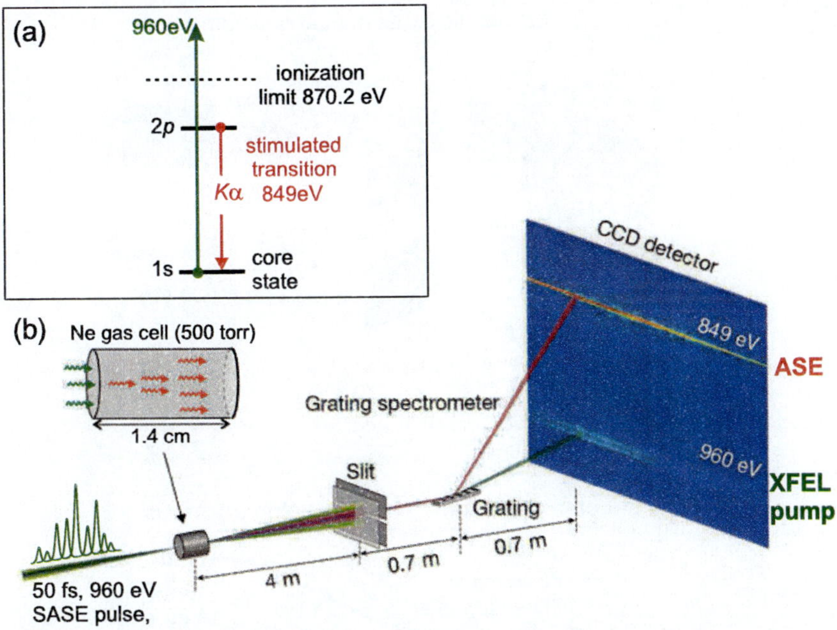

Fig. 1.29 **a** Energy levels and transitions of interest in Ne atoms. **b** Experimental scheme used to demonstrate amplified spontaneous emission (ASE) in a Ne gas cell [278]. The coherent spikes in the incident SASE pulse of about 50 fs and 960 eV nominal energy were of sufficient intensity to create K-shell holes in Ne atoms, overcoming their natural 2.4 fs decay lifetime. The spontaneously emitted Kα line of 849 eV by atoms at the front end of the cell is then amplified along the cell in the ASE process and separated from the incident 960 eV photons by a grating spectrometer to become visible on the CCD detector

If the thin film sample now contains nanoscale domains in either composition, thickness, bond orientation, or magnetization directions, the spontaneous diffraction pattern, observed at finite momentum transfer with a synchrotron source, will vanish with increasing XFEL intensity since all photons are scattered through stimulated decays in the forward direction, only. At sufficiently high intensity, absorption and stimulated emission will become the same and the film becomes transparent. Such *self-induced transparency* was first observed in the optical regime in 1967 [294, 295]. In the x-ray region, this effect was predicted theoretically by Stöhr and Scherz [296] and the disappearance of the diffraction pattern of magnetic domains in Co/Pd multilayer sample was demonstrated experimentally in 2016 by Wu et al. [297]. They used 50 fs SASE XFEL pulses, monochromatized to a bandwidth of 0.2 eV, and tuned to the Co L_3 resonance (see inset of Fig. 1.17).

In Fig. 1.30a we show the arrangement used in a later experiment by Chen et al. [298] with SASE XFEL pulses as short as 5 fs. They simultaneously monitored the interplay between the centrally transmitted and out-of-beam diffracted intensity by a similar Co/Pd film with nanoscale magnetic stripe domains.

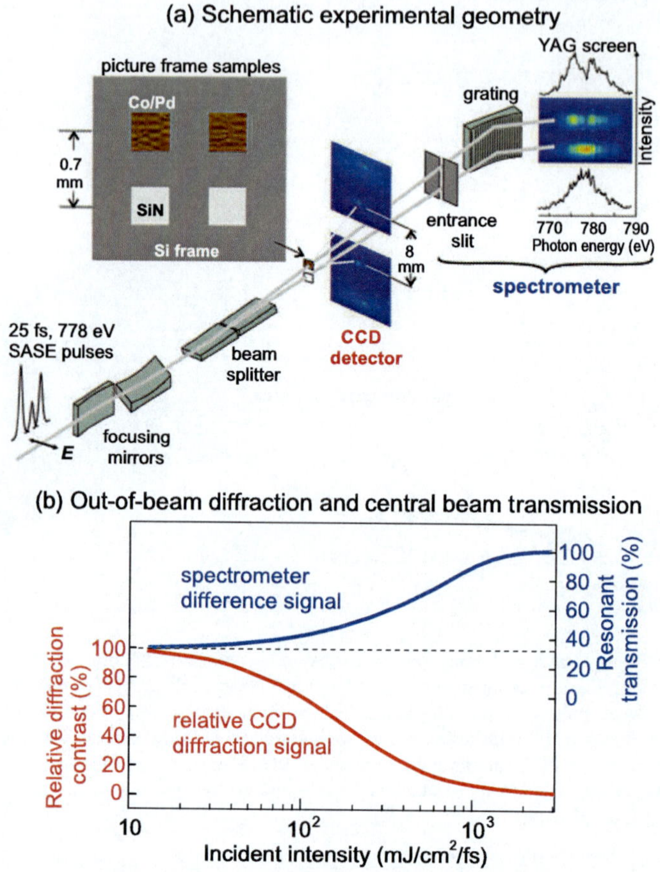

Fig. 1.30 Experimental geometry for simultaneous pulse-by-pulse measurements of the transmitted and diffracted response of a Co/Pd thin film [298]. Incident SASE pulses of ~25 fs length are focused onto the sample plane and split by the sharp edge of a mirror, with one half propagating through a Co/Pd/SiN sample in a picture frame and the other through a pure SiN reference film for normalization purposes. The horizontal magnetic stripe domains in Co/Pd produce strong first and third order Bragg diffraction peaks on a pnCCD detector. The spatially separated undiffracted beams are allowed to propagate into a downstream spectrometer. A grating disperses the two offset beams onto a yttrium aluminum garnet (YAG) fluorescence screen, yielding separate single-shot sample and reference spectra around the Co L_3 resonance. The shown first and third order diffraction images on the CCD detector and spectra on the YAG screen are real data averaged over many pulses

Figure 1.30b shows the statistical average of the first order diffracted intensity (red curve), corresponding to the small intensity spots on the CCD separated by 8 mm in (a), and the normalized sample transmission signal (blue) corresponding of the peak value of the normalized Co L_3 absorption signal, represented in (a) by the size of the central dip in the YAG screen intensity. The two signals show complementary behavior with increasing incident intensity, reflecting the increase in sample transmission at the cost of the loss of the out-of-beam diffracted intensity. At the highest intensity the sample has become transparent and stimulated forward scattering has replaced conventional spontaneous out-of-beam diffraction.

1.5.4.3 A Historical Perspective

It is interesting to put the last experiment into perspective of the historical evolution of stronger sources of electromagnetic radiation. As discussed in Sect. 1.2.5, NMR was facilitated by the development of stronger RADAR microwave sources in the 1940s, which in the late 1950s was followed by the advent of the maser/laser, and another 50 years later by the development of XFELs. The key breakthroughs with these sources were the creation of intense coherent radiation with degeneracy parameters far exceeding unity.

In each case, the coherent control of transitions between two quantum states became possible for the first time. Coherent control means that one may picture the concerted action of many photons, generated within a source or coherence volume of order λ^3, as a strong classical wave that can cause both up and down transitions in a two-level system. The cramming of photons into the coherence volume becomes increasing difficult with decreasing λ, accounting for the timeline of the historical evolution. Another way of putting this into perspective is shown in Fig. 1.31, revealing the coherent control of quantum states with increasing energy separation, from μeV in NMR to eV for lasers to keV for XFELs.

The figure also illustrates the electromagnetic aspect of radiation. In NMR one utilizes the *magnetic* component of a strong coherent microwave field to drive transitions between nuclear states split by a static magnetic field as illustrated in Fig. 1.31a. The process involves a nuclear *spin flip* and is described by the famous *Bloch equations* [299]. For the laser case, shown in Fig. 1.31b, the two levels are *electronic* valence states, with the transition driven by the *electric* component of the optical field. At optical frequencies, the magnetic field component that acts on spins becomes unimportant since the electronic spins can no longer follow the rapid oscillatory motion, and the process is described by the *optical Bloch equations* [15, 300]. With the advent of XFELs, multi-photon or strong field effects allow the control of atomic core-to-valence transitions in the keV range as shown in Fig. 1.31c.

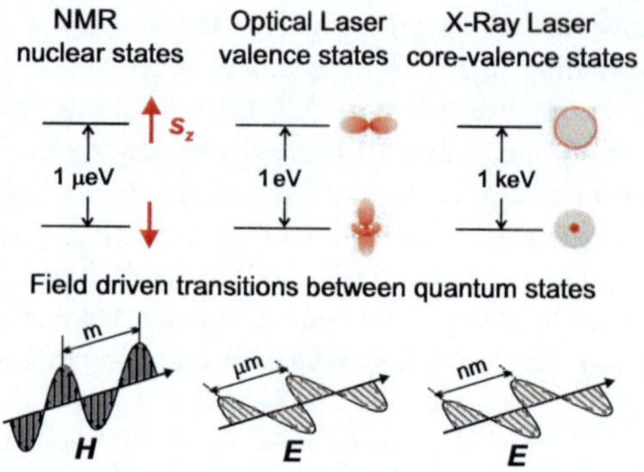

Fig. 1.31 Historical milestones due to the development of improved sources of electromagnetic radiation. The development of radar in the 1940s enabled control of spin transitions through the **H**-field in nuclear magnetic resonance. This development was extended through the invention of the maser and laser into the eV range, corresponding to **E**-field driven transitions between electronic valence states. The advent of XFELs has allowed the control of **E**-field driven core-to-valence transitions in atoms in the keV range

References

1. Y.F. Chen, D. Hover, S. Sendelbach, L. Maurer, S.T. Merkel, E.J. Pritchett, F.K. Wilhelm, R. McDermott: Phys. Rev. Lett. **107**, 217 401 (2011)
2. T. Kuhn, *The Structure of Scientific Revolutions* (University of Chicago Press, Chicago, 1962)
3. B. Lengeler, Naturwissenschaften **88**, 249 (2001)
4. J. Stöhr, Adv. Opt. Photonics **11**, 215 (2019)
5. D. Attwood, A. Sakdinawat, *X-Rays and Extreme Ultraviolet Radiation: Principles and Applications*, 2nd edn. (Cambridge University Press, Cambridge, 2017)
6. J. Als-Nielsen, D. McMorrow, *Elements of Modern X-Ray Physics*, 2nd edn. (Wiley, New York, 2011)
7. K.J. Kim, Z. Huang, R. Lindberg, *Synchrotron Radiation and Free-Electron Lasers* (Cambridge University Press, Cambridge, 2017)
8. J. Ullrich, A. Rudenko, R. Moshammer, Ann. Rev. Phys. Chem. **63**, 635 (2012)
9. P.A.M. Dirac, *Quantum Mechanics*, 4th edn. (Oxford University Press, Oxford, 1958)
10. W. Heitler, *The Quantum Theory of Radiation*, 3rd edn. (Clarendon Press, Oxford, 1954)
11. C. Jacobsen, *X-Ray Microscopy* (Cambridge University Press, Cambridge, 2020)
12. R.J. Glauber, Phys. Rev. **130**, 2529 (1963)
13. R.J. Glauber, Phys. Rev. **131**, 2766 (1963)
14. R. Glauber, Optical coherence and photon statistics, in *Quantum Optics and Electronics*, ed. by A.B.C. deWitt, C. Cohen-Tannoudji (Gordon and Breach, New York, 1965)
15. R. Loudon, *The Quantum Theory of Light*, 3rd edn. (Clarendon Press, Oxford, 2000)
16. M.O. Scully, M.S. Zubairy, *Quantum Optics* (Cambridge University Press, Cambridge, 1997)
17. Z.Y.J. Ou, *Multi-Photon Quantum Interference* (Springer, Heidelberg- New York, 2007)
18. Y. Shih, *An Introduction to Quantum Optics: Photon and Biphoton Physics* (CRC Press, Boca Raton, Florida, 2011)
19. Z.Y.J. Ou, *Quantum Optics for Experimentalists* (World Scientific, Singapore, 2017)
20. D.F. Styer et al., Am. J. Phys. **70**, 288 (2002)
21. N. Bohr, Phys. Rev. **48**, 696 (1935)
22. X.F. Pang, Y.P. Feng, *Quantum Mechanics in Nonlinear Systems* (World Scientific, New Jersey, 2005)

23. R.P. Feynman, *QED: The Strange Theory of Light and Matter* (Princeton University Press, Princeton NJ, 1985)
24. R.P. Feynman, R.B. Leighton, M. Sands, *The Feynman Lectures on Physics*, vol. III (Addison-Wesley, Reading, 1965)
25. D.C. Koningsberger, E.R. Prins, *X-Ray Absorption: Principles, Applications, Techniques of EXAFS, SEXAFS and XANES* (Wiley and Sons, New York, 1988)
26. J.J. Rehr, R.C. Albers, Rev. Mod. Phys. **72**, 621 (2000)
27. G. Bunker, *Introduction to XAFS: A Practical Guide to X-ray Absorption Fine Structure Spectroscopy* (Cambridge University Press, Cambridge, 2010)
28. J. Stöhr, "SEXAFS": everything you always wanted to know about SEXAFS but were afraid to ask, in *X-Ray Absorption: Principles, Applications, Techniques of EXAFS, SEXAFS and XANES*, ed. by D.C. Koningsberger, R. Prins (Wiley and Sons, New York, 1988), p. 443
29. J. Stöhr, *NEXAFS Spectroscopy* (Springer, Heidelberg, 1992)
30. A. Bianconi, Appl. Surf. Sci. **6**, 392 (1980)
31. W.A. Hendrickson, Science **254**, 51 (1991)
32. W.A. Hendrickson, Q. Rev. Biophysics **47**, 49 (2014)
33. F. de Groot, A. Kotani, *Core Level Spectroscopy of Solids* (CRC Press, Boca Raton, USA, 2008)
34. J. Stöhr, H.C. Siegmann, *Magnetism: From Fundamentals to Nanoscale Dynamics* (Springer, Heidelberg, 2006)
35. B.E. Warren, *X-Ray Diffraction* (Dover Publications, New York, 1999)
36. C. Hummond, *The Basics of Crystallography and Diffraction*, 4th *Edition* (Oxford University Press, Oxford, 2015)
37. S. Hüfner, *Photoelectron Spectroscopy: Principles and Applications* (Spinger, Berlin, 2003)
38. E.S. Hüfner, *Very High Resolution Photoelectron Spectroscopy* (Spinger, Berlin, 2007)
39. E.J. Woicik, *Hard X-Ray Photoelectron Spectroscopy (HARPES)* (Spinger, Berlin, 2016)
40. O. Darrigol, *A History of Optics: From Greek Antiquity to the Nineteenth Century* (Oxford University Press, Oxford, 2012)
41. T. Young, *A Course of Lectures on Natural Philosophy and the Mechanical Arts* (Printed for J. Johnson by W. Savage, London, 1807)
42. A. Fresnel, Mémoires de l'Académie Royale des Sciences de l'Institut de France p. 339 (1818)
43. M.A. Lauterbach: Optical Nanoscopy p. 1 (2012)
44. E. Abbe, Archiv Mikrosk. Anatomie **9**, 413 (1873)
45. A. Fresnel, Œuvres **1**, 731 (1822)
46. M. Faraday, Phil. Trans. Roy. Soc. London **1**, 136 (1846)
47. L. Pasteur, Annales de chimie et de physique **28**, 56 (1850)
48. J.W. Goodman, *Introduction to Fourier Optics*, 4th edn. (W. H. Freeman, New York, 2017)
49. J.C. Maxwell, Phil. Mag. **90**, 11 (1861)
50. J.C. Maxwell: *A Treatise on Electricity and Magnetism, Vol, I and II* (Clarendon Press, Oxford, 1873)
51. H.R. Hertz, Ann. Physik **270**, 551 (1888)
52. W. C. Röntgen submitted his first communication, *Über eine neue Art von Strahlen*, to the Proceedings of the Würzburg Physico-Medical Society on December 28, 1895
53. M. Planck, Verh. Deut. Phys. Ges. **2**, 202 (1900)
54. M. Planck, Verh. Deut. Phys. Ges. **2**, 237 (1900)
55. A. Einstein, Ann. Physik **17**, 132 (1905)
56. H.R. Hertz, Ann. Physik **267**, 983 (1887)
57. R.A. Millikan, Phys. Rev. **7**, 355 (1916)
58. A. Einstein, Phys. Z. **10**, 185 (1909)
59. A. Einstein, Phys. Z. **10**, 817 (1909)
60. A. Einstein, Verh. Deut. Phys. Ges. **18**, 318 (1916)
61. A. Einstein, Phys. Z. **18**, 121 (1917)
62. P.A.M. Dirac, Proc. Roy. Soc. A **114**, 243 (1927)
63. P.A.M. Dirac, Proc. Roy. Soc. London A **114**, 710 (1927)

64. J. Mehra, H. Rechenberg, Found. Phys. **29**, 91 (1999)
65. M. Planck, Verh. Deut. Phys. Ges. **13**, 138 (1911)
66. A. Einstein, O. Stern, Ann. Physik **40**, 551 (1913)
67. M. Born, Z. Physik **26**, 379 (1924)
68. G.N. Lewis, Nature **118**, 874 (1926)
69. W. Heisenberg, Z. Phys. **33**, 879 (1925)
70. M. Born, P. Jordan, Z. Physik **34**, 858 (1925)
71. M. Born, W. Heisenberg, P. Jordan, Z. Physik **35**, 557 (1926)
72. M. Born, *My Life: Recollections of a Nobel Laureate* (Taylor and Francis, New York, 1978)
73. L. de Broglie, Ann. Phys. (Paris) **3**, 22 (1925)
74. L. de Broglie, Nature **112**, 540 (1923)
75. C. Davisson, L.H. Germer, Phys. Rev. **303**, 705 (1927)
76. E. Schrödinger, Ann. Phys. **79**, 361 (1926)
77. E. Schrödinger, Ann. Phys. **79**, 489 (1926)
78. E. Schrödinger, Ann. Phys. **79**, 734 (1926)
79. E. Schrödinger, Ann. Phys. **80**, 437 (1926)
80. E. Schrödinger, Ann. Phys. **81**, 109 (1926)
81. P.A.M. Dirac, Proc. Roy. Soc. A **117**, 610 (1928)
82. P.A.M. Dirac, Proc. Roy. Soc. A **118**, 351 (1928)
83. A. Einstein, B. Podolsky, N. Rosen, Phys. Rev. **47**, 777 (1935)
84. E. Schrödinger, Naturwiss. **23**, 807 (1935)
85. P. H. van Cittert, Physica **1**, 201 (1934); F. Zernike: Physica **5**, 785 (1938)
86. L. Mandel, E. Wolf, *Optical Coherence and Quantum Optics* (Cambridge University Press, Cambridge, 1995)
87. E. Wolf, *Introduction to the Theory of Coherence and Polarization of Light* (Cambridge University Press, Cambridge, 2007)
88. N. Wiener, Acta Math. **55**, 117 (1930); A. Khintchine: Mathematische Annalen **109**, 604 (1934)
89. The history of the development of magnetic spectroscopy is nicely reviewed in Alfred Kastler's Nobel lecture available on the Nobel website http://www.nobel.se/physics
90. A. Rabinowicz, Z. Physik **19**, 442 (1918)
91. For review see: L. M. Brown, Edt., *Renormalization: From Lorentz to Landau (and Beyond)*, (Springer, New York, 1993)
92. J.W.E. Lamb, R.C. Retherford, Phys. Rev. **72**, 241 (1947)
93. H. M. Foley and P. Kusch, Phys. Rev. **73**, 412 (1948); P. Kusch and H. M. Foley Phys. Rev. **74**, 250 (1948)
94. F. Dyson, Phys. Rev. **75**, 486 (1949)
95. R.P. Feynman, Rev. Mod. Phys. **20**, 367 (1948)
96. R.H. Brown, R.Q. Twiss, Nature **178**, 1046 (1956)
97. R.H. Brown, R.Q. Twiss: Proc. Roy. Soc. (London) A **242**, 300 (1957)
98. R.H. Brown, R.Q. Twiss: Proc. Roy. Soc. (London) A **243**, 291 (1958)
99. In 1933 Max Born published a comprehensive optics textbook entitled *Optik: Ein Lehrbuch der Elektromagnetischen Lichttheorie* published by Springer Verlag, whose extension co-authered by Emil Wolf 1959 marked the first edition of "Born and Wolf". The latter utilized Born's statistical interpretation of quantum mechanics, outlined in his 1954 Nobel lecture *The Statistical Interpretation of Quantum Mechanics*, available as a PDF file at the official website of the Nobel Prize
100. M. Born, E. Wolf, *Principles of Optics: Electromagnetic Theory of Propagation, Interference, and Diffraction of Light* (Cambridge University Press, Cambridge, 1999)
101. L. Mandel, E. Wolf, Rev. Mod. Phys. **37**, 231 (1965)
102. J.L. Bromberg, Arch. Hist. Exact Sci. **70**, 243 (2016)
103. E.M. Purcell, Nature **178**, 1449 (1956)
104. U. Fano, Am. J. Phys. **29**, 539 (1961)
105. J.S. Bell, Physics **1**, 195 (1964)

106. A. Zeilinger, Rev. Mod. Phys. **71**, S288 (1999)
107. A.D. Piazza, C. Müller, K.Z. Hatsagortsyan, C.H. Keitel, Rev. Mod. Phys. **84**, 1177 (2012)
108. E. Wolf, Opt. News **5**, 24 (1979)
109. A. Duncan, M. Janssen, Stud. Hist. Phil. Mod. Phys. **39**, 634 (2008)
110. http://www.nobel.se/
111. C.G. Barkla, Phil. Trans. Roy. Soc. A **204**, 467 (1905)
112. C.G. Barkla, C.L. Sandler, Phil. Mag. **17**, 739 (1909)
113. C.G. Barkla, J. Nicol, Proc. Phys. Soc. London **24**, 9 (1911)
114. C.G. Barkla, Phil. Trans. Roy. Soc. **A217**, 315 (1918)
115. H.G.J. Moseley, C.G. Darwin, Phil. Mag. **26**, 210 (1913)
116. H.G.J. Moseley, Phil. Mag. **26**, 1024 (1913)
117. H.G.J. Moseley, Phil. Mag. **27**, 703 (1914)
118. M. Siegbahn, Phil. Mag. **37**, 601 (1919)
119. M. Siegbahn, A.B. Leide, Phil. Mag. **38**, 647 (1919)
120. M. Siegbahn, *Spektroskopie der Röntgenstrahlen* (Julius Springer Verlag, Berlin, 1924)
121. R. de L. Kronig: Z. Phys. **70**, 317 (1931)
122. R.S. von Bordwehr, Ann. Phys. France **14**, 377 (1989)
123. F.W. Lytle, J. Synchr. Rad. **6**, 123 (1999)
124. Experimental x-ray diffraction patterns were first reported on June 8, 1912 by W. Friedrich and P. Knipping, Sitz. Ber. Bayer. Akad. Wiss., 311-322 (1912). In 1913 the results were published by W. Friedrich, P. Knipping and M. Laue, Ann. Phys. **346**, 971 (1913), with Laue writing the theoretical and the two others the experimental part of the paper
125. W.H. Bragg, W.L. Bragg, Proc. R. Soc. London A **88**, 428 (1913)
126. P.P. Ewald, Z. Phys. **14**, 465 (1913)
127. A.H. Compton, Phys. Rev. **22**, 409 (1923)
128. G. Greenstein, A.G. Zajonc, *The Quantum Challenge: Modern Research on the Foundations of Quantum Mechanics*, 2nd edn. (Jones and Bartlett, Sudbury, MA, 2006)
129. P. Grangier, G. Roger, A. Aspect, Europhys. Lett. **1**, 173 (1986)
130. The credit for the discovery of the electron is typically given to J. J. Thomson who received the Nobel Prize for it. However, the same results were obtained independently by Emil Wiechert at about the same time. Wiechert's many contributions to physics and geophysics are discussed at: http://verplant.org/history-geophysics/Wiechert.htm
131. H. Wiedemann, *Synchrotron Radiation* (Springer, Berlin, 2003)
132. A. Hofmann, *The Physics of Synchrotron Radiation* (Cambridge University Press, Cambridge, 2004)
133. G. A. Schott, *Electromagnetic Radiation and the Mechanical Reactions Arising from It*. Adams Prize Essay (Cambridge University Press, 1912)
134. J.P. Blewett, Nucl. Instrum. Methods A **266**, 1 (1988)
135. K.J. Kim, Nucl. Instrum. Meth. Phys. Res. A **246**, 71 (1986)
136. D. Attwood, K. Halbach, K.J. Kim, Science **228**, 1265 (1985)
137. M. Bei, M. Borland, Y. Cai, P. Elleaume, R. Gerig, K. Harkay, L. Emery, A. Hutton, R. Hettel, R. Nagaoka, D. Robin, C. Steier, Nucl. Instum. Methods Phys. Res. A **622**, 518 (2010)
138. J.M.J. Madey, J. Appl. Phys. **42**, 1906 (1971)
139. L.R. Elias, W.M. Fairbank, J.M.J. Madey, H.A. Schwettman, T.I. Smith, Phys. Rev. Lett. **36**, 717 (1976)
140. D.A.G. Deacon, L.R. Elias, J.M.J. Madey, G.J. Ramian, H.A. Schwettman, T.I. Smith, Phys. Rev. Lett. **38**, 892 (1977)
141. A.M. Kondratenko, E.L. Saldin, Part. Accel. **10**, 207 (1980)
142. Y.S. Derbenev, A.M. Kondratenko, E.L. Saldin, Nucl. Instrum. Meth. A **193**, 415 (1982)
143. R. Bonifacio, C. Pellegrini, L.M. Narducci, Opt. Commun. **50**, 373 (1984)
144. C. Pellegrini, Eur. Phys. J. H **37**, 659 (2012)
145. C. Pellegrini, A. Marinelli, S. Reiche, Rev. Mod. Phys. **88**, 015 006 (2016)
146. C. Pellegrini, *A 4 to 0.1 nm FEL Based on the SLAC Linac*, in ed. by M. Cornacchia, H. Winick. Proceedings Workshop on 4th Generation Light Sources. SSRL/SLAC Rep. 92/02, pp. 364–375

147. P. Emma et al., Nat. Photonics **4**, 641 (2010)
148. C. Bostedt, S. Boutet, D.M. Fritz, Z. Huang, H.J. Lee, H.T. Lemke, A. Robert, W.F. Schlotter, J.J. Turner, G.J. Williams: Rev. Mod. Phys. **88**, 015 007 (2016)
149. G. K. Shenoy and J. Stöhr, (eds.), *LCLS-The First Experiments*, SLAC publication SLAC-R-611 (2000)
150. J. Amann et al., Nat. Photonics **6**, 693 (2012)
151. D. Ratner et al.: Phys. Rev. Lett. **114**, 054 801 (2015)
152. W.A. Hendrickson, J.R. Horton, D.M. LeMaster, EMBO J. **9**, 665 (1991)
153. J.E. Bertie, Z. Lan, Appl. Spectrosc. **50**, 1047 (1996)
154. B.L. Henke, E.M. Gullikson, J.C. Davis, At. Data Nucl. Data Tables **54**, 181 (1993)
155. J.A. Sellberg et al., Nature **510**, 381 (2014)
156. R.J. Powell, W.E. Spicer, Phys. Rev. B **2**, 2182 (1970)
157. M.A. Ordal, L.L. Long, R.J. Bell, S.E. Bell, R.R. Bell, J.R.W. Alexander, C.A. Ward, Appl. Opt. **22**, 1099 (1983)
158. M.A. Ordal, R.J. Bell, J.R.W. Alexander, L.L. Long, M.R. Querry, Appl. Opt. **24**, 4493 (1985)
159. F.M.F. de Groot, M. Abbate, J. van Elp, G.A. Sawatzky, Y.J. Ma, C.T. Chen, F. Sette, J. Phys.: Condens. Matter **5**, 2277 (1993)
160. R. Nakajima, J. Stöhr, Y. Idzerda, Phys. Rev. B **59**, 6421 (1999)
161. T.J. Regan, H. Ohldag, C. Stamm, F. Nolting, J. Lüning, J. Stöhr, R.L. White: Phys. Rev. B **64**, 214 422 (2001)
162. S.G. Urquhart, H. Ade, J. Phys. Chem. B **106**, 8531 (2002)
163. B. Watts, S. Swaraj, D. Nordlund, J. Lüning, H. Ade: J. Chem. Phys. **134**, 024 702 (2011)
164. A. Nilsson, J. Elec. Spectrosc. Relat. Phenom. **126**, 3 (2002)
165. M. Dell Angela et al.: Science **339**, 1302 (2013)
166. H. Öström et al., Science **347**, 978 (2015)
167. J.L. Solomon, R.J. Madix, J. Stöhr, Surf. Sci. **255**, 12 (1991)
168. J. Lüning, F. Nolting, A. Scholl, H. Ohldag, J. Seo, J. Fompeyrine, J.P. Locquet, J. Stöhr, Phys. Rev. B **67**, 214 433 (2003)
169. K. Weiss, C. Wöll, E. Böhm, B. Fiebranz, G. Forstmann, B. Peng, V. Scheumann, D. Johannsmann, Macromolecules **31**, 1930 (1998)
170. J. Stöhr, M.G. Samant, A. Cossy-Favre, J. Díaz, Y. Momoi, S. Odahara, T. Nagata, Macromolecules **31**, 1942 (1998)
171. J. Stöhr, M.G. Samant, J. Elec. Spectrosc. Relat. Phenom. **98**, 189 (1999)
172. J. Stöhr, M.G. Samant, J. Lüning, A.C. Callegari, P. Chaudhari, J.A. Doyle, J.A. Lacey, S.A. Lien, S. Purushothaman, J.L. Speidell, Science **292**, 2299 (2001)
173. B.T. Thole, G. van der Laan, G.A. Sawatzky, Phys. Rev. Lett. **55**, 2086 (1985)
174. G. van der Laan, B.T. Thole, G.A. Sawatzky, J.B. Goedkoop, J.C. Fuggle, J.M. Esteva, R. Karnatak, J.P. Remeika, H.A. Dabkowska, Phys. Rev. B **34**, 6529 (1986)
175. J. Kuneš, P.M. Oppeneer, Phys. Rev. B **67**, 024 431 (2003)
176. G. Schütz, W. Wagner, W. Wilhelm, P. Kienle, R. Zeller, R. Frahm, G. Materlik, Phys. Rev. Lett. **58**, 737 (1987)
177. J. Stöhr, H. König, Phys. Rev. Lett. **75**, 3748 (1995)
178. D. Weller, J. Stöhr, R. Nakajima, A. Carl, M.G. Samant, C. Chappert, R. Mégy, P. Beauvillain, P. Veillet, G. Held, Phys. Rev. Lett. **75**, 3752 (1995)
179. J. Stöhr, J. Magn. Magn. Mater. **200**, 470 (1999)
180. J. Goulon, C. Goulon-Ginet, A. Rogalev, V. Gotte, C. Malgrange, C. Brouder, J. Chem. Phys. **108**, 6394 (1998)
181. L. Alagna, T. Prosperi, S. Turchini, J. Goulon, A. Rogalev, C. Goulon-Ginet, C.R. Natoli, R.D. Peacock, B. Stewart, Phys. Rev. Lett. **80**, 4799 (1998)
182. C.R. Natoli, C. Brouder, P. Sainctavit, J. Goulon, C. Goulon-Ginet, A. Rogalev, Eur. Phys. J. B **4**, 1 (1998)
183. O. Plashkevych, V. Carravetta, O. Vahtras, H. Ågren, Chem. Phys. **232**, 49 (1998)
184. S. Turchini, N. Zema, S. Zennaro, L. Alagna, B. Stewart, R.D. Peacock, T. Prosperi, J. Am. Chem. Soc. **126**, 4532 (2004)

185. S. Villaume, P. Norman, Chirality **21**, E13 (2009)
186. Y. Izumi, M. Tanabe, A. Imazu, A. Mimoto, M. Tanaka, A. Agui, T. Muro, K. Nakagawa, J. Am. Chem. Soc. **138**, 074 305 (2013)
187. O. Takahashi, M. Kimoto, L.G.M. Pettersson, Chem. Phys. **450–451**, 109 (2015)
188. Y. Pertot et al., Science **355**, 264 (2017)
189. D. Popmintchev et al., Phys. Rev. Lett. **120**, 093 002 (2018)
190. F. Siegrist et al., Nature **571**, 240 (2019)
191. M. Hofherr et al., Sci. Adv. **6**, 8717 (2020)
192. F. Gel'mukhanov, H. Ågren, Phys. Rep. **312**, 87 (1999)
193. A. Kotani, S. Shin, Rev. Mod. Phys. **73**, 203 (2001)
194. L.J.P. Ament, M. van Veenendaal, T.P. Devereaux, J.P. Hill, J. van den Brink, Rev. Mod. Phys. **83**, 705 (2011)
195. J.P. Rueff, A. Shukla, Rev. Mod. Phys. **82**, 847 (2010)
196. M. van Veenendaal, *Theory of Scattering and Absorption of X-Rays* (Cambridge University Press, Cambridge, 2015)
197. F. Gel'mukhanov, M. Odelius, S.P. Polyutov, A. Föhlisch, V. Kimberg, Rev. Mod. Phys. **93**, 035 001 (2021)
198. A. Nilsson et al., Appl. Phys. A **65**, 147 (1997)
199. A. Nilsson, L.G.M. Pettersson, Surf. Sci. Rep. **55**, 49 (2004)
200. F. Zaera, Prog. Surface Sci. **69**, 1 (2001)
201. A. Nilsson, L.G.M. Pettersson, B. Hammer, T. Bligaard, C.H. Christenen, J.K. Nørskov, Catalysis Lett. **100**, 111 (2005)
202. M. Weinelt, N. Wassdahl, T. Wiell, O. Karis, J. Hasselström, P. Bennich, A. Nilsson, J. Stöhr, M. Samant, Phys. Rev. B **58**, 7351 (1998)
203. L. Triguero, A. Föhlisch, P. Väterlein, J. Hasselström, M. Weinelt, L.G.M. Pettersson, Y. Luo, H. Ågren, A. Nilsson, J. Am. Chem. Soc. **122**, 12 310 (2000)
204. S. Anders, H.A. Padmore, R.M. Duarte, T. Renner, T. Stammler, A. Scholl, M.R. Scheinfein, J. Stöhr, L. Séve, B. Sinkovic, Rev. Sci. Instrum. **70**, 3973 (1999)
205. A.P. Hitchcock, J. Elec. Spectrosc. Relat. Phenom. **200**, 49 (2015)
206. J. Feng, A. Scholl, Photoemission electron microscopy, in *Springer Handbook of Microscopy*, ed. by P.W. Hawkes, J.C.H. Spence (Springer Handbooks, New York, 2019)
207. J. Stöhr, Y. Wu, B.D. Hermsmeier, M.G. Samant, G.R. Harp, S. Koranda, D. Dunham, B.P. Tonner, Science **259**, 658 (1993)
208. J. Stöhr, S. Anders, IBM. J. Res. Develop. **44**, 535 (2000)
209. A. Scholl, J. Stöhr, J. Lüning, J.W. Seo, J. Fompeyrine, H. Siegwart, J.P. Locquet, F. Nolting, S. Anders, E.E. Fullerton, M.R. Scheinfein, H.A. Padmore, Science **287**, 1014 (2000)
210. H. Ohldag, A. Scholl, F. Nolting, S. Anders, F.U. Hillebrecht, J. Stöhr, Phys. Rev. Lett. **86**, 2878 (2001)
211. H. Ohldag, T.J. Regan, J. Stöhr, A. Scholl, F. Nolting, J. Lüning, C. Stamm, S. Anders, R.L. White, Phys. Rev. Lett. **87**, 247 201 (2001)
212. H. Ohldag, G. van der Laan, E. Arenholz, Phys. Rev. B **79**, 052 403 (2009)
213. G. van der Laan, N.D. Telling, A. Potenza, S.S. Dhesi, E. Arenholz, Phys. Rev. B **83**, 064 409 (2011)
214. E. Arenholz, G. van der Laan, R.V. Chopdekar, Y. Suzuki, Phys. Rev. Lett. **98**, 197 201 (2007)
215. Y. Acremann, J.P. Strachan, V. Chembrolu, S.D. Andrews, T. Tyliszczak, J.A. Katine, M.J. Carey, B.M. Clemens, H. Siegmann, J. Stöhr, Phys. Rev. Lett. **96**, 217 202 (2006)
216. J.P. Strachan, V. Chembrolu, Y. Acremann, X.W. Yu, A.A. Tulapurkar, , T. Tyliszczak, J.A. Katine, M.J. Carey, M.R. Scheinfein, H. Siegmann, J. Stöhr, Phys. Rev. Lett. **100**, 247 201 (2008)
217. V. Chembrolu, J.P. Strachan, X.W. Yu, A.A. Tulapurkar, T. Tyliszczak, J.A. Katine, M.J. Carey, J. Stöhr, Y. Acremann, Phys. Rev. B **80**, 024 417 (2009)
218. See https://www.rcsb.org
219. See http://www-ssrl.slac.stanford.edu/stohr and https://stohr.sites.stanford.edu/
220. J. Stöhr, D. Denley, P. Perfetti, Phys. Rev. B **18**, 4132 (1978)

221. Z. Hussain, E. Umbach, D.A. Shirley, J. Stöhr, J. Feldhaus, Nucl. Inst. Methods **195**, 115 (1982)
222. R. Neutze, R. Wouts, D. van der Spoel, E. Weckert, J. Hajdu, Nature **406**, 752 (2000)
223. J.C.H. Spence, Sruct. Dyn. **4**, 044 027 (2017)
224. D.P. DePonte, R.B. Doak, M. Hunter, Z. Liu, U. Weierstall, J.C.H. Spence, Micron **40**, 507 (2009)
225. L.M.G. Chavas, L. Gumprecht, H.N. Chapman, Struct. Dyn. **2**, 041 709 (2015)
226. H.N. Chapman, P. Fromme, A. Barty, T.A. White, R.A. Kirian et al., Nature **470**, 73 (2011)
227. D.A. Sherrell, A.J. Foster, L. Hudson, B. Nutter, J. O'Hea, S. Nelson, O. Paré-Labrosse, S. Oghbaey, R.J.D. Miller, R.L. Owena, J. Synchrotron Rad. **22**, 1372 (2015)
228. M.M. Seibert et al., Nature **470**, 78 (2011)
229. G. Brändén, R. Neutze, Science **373**, 6558 (2021)
230. T.R.M. Barends, B. Stauch, V. Cherezov, I. Schlichting, Nat. Rev. Methods Primers **2**, 59 (2022)
231. J. Barber, Quart. Rev. Biophysics **49**, e14 (2016)
232. N. Cox, D.A. Pantazis, W. Lubitz, Ann. Rev. Biochem. **89**, 795 (2020)
233. A. Zouni, H.T. Witt, J. Kern, P. Fromme, N. Krauss et al., Nature **409**, 739 (2001)
234. K.N. Ferreira, T.M. Iverson, K. Maghlaoui, J. Barber, S. Iwata, Science **303**, 1831 (2004)
235. Y. Umena, K. Kawakami, J.R. Shen, N. Kamiya, Nature **473**, 55 (2011)
236. J. Yano, J. Kern, K.D. Irrgang, M.J. Latimer, U. Bergmann et al., Proc. Nat. Acad. Sci. **102**, 12 047 (2005)
237. C. Glöckner, J. Kern, M. Broser, A. Zouni, V. Yachandra, J. Yano, J. Biol. Chem. **288**, 22 607 (2013)
238. U. Bergmann, J. Kern, R.W. Schoenlein, P. Wernet, V.K. Yachandra, J. Yano, Nat. Rev. Phys. **3** (2021)
239. B. Kok, B. Forbush, M. McGloin, Photochem. Photobiol. **11**, 457 (1970)
240. J. Kern et al., Nature **563**, 421 (2018)
241. A.H. Zewail, Angew. Chem. Int. Ed. **39**, 2586 (2000)
242. P. Kraus, M. rch, S. K. Cushing, D. M. Neumark, S. R. Leone, Nat. Rev. Chem. **2**, 82 (2018)
243. M. Chergui, E. Collet, Chem. Rev. **117**, 11 025 (2017)
244. P. Wernet, Phil. Trans. R. Soc. A **377**, 20170 464 (2019)
245. R.M. Jay, K. Kunnus, P. Wernet, K.J. Gaffney, Annu. Rev. Phys. Chem. **73**, 187 (2022)
246. Y. Tokura, Phys. Today **56**, 50 (2003)
247. M. Mitrano, S. Lee, A.A. Husain, L. Delacretaz, M. Zhu et al., Sci. Adv. **5**, eaax3346 (2019)
248. J.D. Bernal, R.H. Fowler, J. Chem. Phys. **1**, 515 (1933)
249. C.G. Salzmann, J. Chem. Phys. **150**, 060 901 (2019)
250. A. Nilsson, L.G.M. Pettersson, Nat. Comm. **6**, 8998 (2015)
251. A. Nilsson, F. Perakis, X-ray studies of water, in *Synchrotron Light Sources and Free-Electron Lasers: Accelerator Physics, Instrumentation and Science Applications*, ed. by E.J. Jaeschke, S. Khan, J.R. Schneider, J.B. Hastings (Springer International Publishing, New York, 2020)
252. A. Nilsson, Journal of Non-Crystalline Solids, X **14**, 100095 (2022)
253. J.J. Yeh, I. Lindau, At. Data Nucl. Data Tables **32**, 1 (1985)
254. M. Coreno, L. Avaldi, R. Camilloni, K.C. Prince, M. de Simone, J. Karvonen, R. Colle, S. Simonucci, Phys. Rev. A **59**, 2494 (1999)
255. L. Young et al., Nature **466**, 56 (2010)
256. G. Doumy et al., Phys. Rev. Lett. **106**, 083 002 (2011)
257. R.P. Drake, Introduction to high-energy-density physics, in *High-Energy-Density Physics*, ed. by R.P. Drake (Springer, Berlin, 2018)
258. R.P. Drake, Phys. Plasmas **16**, 055 501 (2009)
259. S.H. Glenzer, L.B. Fletcher, E. Galtier, B. Nagler, R. Alonso-Mori et al., J. Phys. B: At. Mol. Opt. Phys. **49**, 092 001 (2016)
260. B. Nagler et al., Nat. Phys. **5**, 693 (2009)
261. S.M. Vinko, U. Zastrau, S. Mazevet, J. Andreasson, S. Bajt, T. Burian, J. Chalupsky, H.N. Chapman, J. Cihelka, D. Doria et al., Phys. Rev. Lett. **104**, 225 001 (2010)

262. D.S. Rackstraw, O. Ciricosta, S.M. Vinko, B. Barbrel, T. Burian et al., Phys. Rev. Lett. **114**, 015 003 (2015)
263. C.O. Almbladh, A.L. Morales, G. Grossmann, Phys. Rev. B **39**, 3489 (1989)
264. P.A. Franken, A.E. Hill, C.W. Peters, G. Weinreich, Phys. Rev. Lett. **7**, 118 (1961)
265. R.W. Boyd, *Nonlinear Opt.* (Academic Press, Cambridge, MA, 2008)
266. P. Eisenberger, S.L. McCall, Phys. Rev. Lett. **26**, 684 (1971)
267. M. Göppert-Mayer, Ann. Phys. (Leipzig) **401**, 273 (1931)
268. T. Pfeifer, C. Spielmann, G. Gerber, Rep. Prog. Phys. **69**, 443 (2006)
269. S.P. Hau-Riege, *High-Intensity X-Ray-Interaction with Matter* (Wiley-VCH, Weinheim, Germany, 2011)
270. T. Wang et al., Phys. Rev. Lett. **108**, 267 403 (2012)
271. S. Schreck, M. Beye, A. Föhlisch, J. Mod. Opt. **62**, S34 (2015)
272. J. Szlachetko et al., Sci. Rep. **6**, 33 292 (2016)
273. A. Rudenko, L. Inhester, K. Hanasaki, X. Li, S.J. Robatjazi, B. Erk, R. Boll, K. Toyota, Y. Hao, O. Vendrell et al., Nature **546**, 129 (2017)
274. L. Young, K. Ueda, M. Gühr, P.H. Bucksbaum, M. Simon, S. Mukamel, N. Rohringer et al., J. Phys. B: Atomic, Mol. Opt. Phys. **51**, 032 003 (2018)
275. S.M. Vinko, O. Ciricosta, B.I. Cho, K. Engelhorn, H.K. Chung, C.R.D. Brown, T. Burian, J. Chalupský, R.W. Falcone, C. Graves et al., Nature **482**, 59 (2012)
276. H. Yoneda, Y. Inubushi, M. Yabashi, T. Katayama, T. Ishikawa, H. Ohashi, H. Yumoto, K. Yamauchi, H. Mimura, H. Kitamura, Nat. Comm. **5**, 5080 (2014)
277. O.S. Humphries, R.S. Marjoribanks, Q.Y. van den Berg, E.C. Galtier, M.F. Kasim, H.J. Lee, A.J.F. Miscampbell, B. Nagler, R. Royle, J.S. Wark, S.M. Vinko, Phys. Rev. Lett. **125**, 195 001 (2020)
278. N. Rohringer et al., Nature **481**, 488 (2012)
279. M. Beye, S. Schreck, F. Sorgenfrei, C. Trabant, N. Pontius, C. Schüßler-Langeheine, W. Wurth, A. Föhlisch, Nature **501**, 191 (2013)
280. H. Yoneda, Y. Inubushi, K. Nagamine, Y. Michine, H. Ohashi, H. Yumoto, K. Yamauchi, H. Mimura et al., Nature **524**, 446 (2015)
281. T. Kroll et al., Phys. Rev. Lett. **120**, 133 203 (2018)
282. T. Kroll, C. Weninger, F.D. Fuller, M.W. Guetg, A. Benediktovitch, Y. Zhang, A. Marinelli, R. Alonso-Mori, A. Aquila, M. Liang et al., Phys. Rev. Lett. **125**, 037 404 (2020)
283. C. Weninger et al., Phys. Rev. Lett. **111**, 233 902 (2013)
284. V. Kimberg, A. Sanchez-Gonzalez, L. Mercadier, C. Weninger, A. Lutman, D. Ratner, R. Coffee, M. Bucher, M. Mucke, M. Agåker et al., Faraday Discuss. **194**, 305 (2016)
285. N. Rohringer, Philos. Trans. R. Soc. A **377**, 20170 471 (2019)
286. D.J. Higley, Z. Chen, M. Beye, M. Hantschmann, A.H. Reid, V. Mehta, O. Hellwig, G.L. Dakovski et al., Commun. Phys. **5**, 83 (2022)
287. T.E. Glover, D.M. Fritz, M. Cammarata, T.K. Allison, S. Coh, J.M. Feldkamp et al., Nature **488**, 603 (2012)
288. S. Shwartz, M. Fuchs, J.B. Hastings, Y. Inubushi, T. Ishikawa, T. Katayama, D.A. Reis, T. Sato, K. Tono, M. Yabashi, S. Yudovich, S.E. Harris, Phys. Rev. Lett. **112**, 163 901 (2014)
289. K. Tamasaku, E. Shigemasa, Y. Inubushi, T. Katayama, K. Sawada, H. Yumoto, H. Ohashi, H. Mimura, M. Yabashi, K. Yamauchi et al., Nat. Photonics **8**, 313 (2014)
290. S. Ghimire, M. Fuchs, J.B. Hastings, S.C. Herrmann, Y. Inubushi, J. Pines, S. Shwartz, M. Yabashi, D.A. Reis, Phys. Rev. A **94**, 043 418 (2016)
291. F. Bencivenga, R. Cucini, F. Capotondi, A. Battistoni, R. Mincigrucci, E. Giangrisostomi, A. Gessini, M. Manfredda, I.P. Nikolov, E. Pedersol et al., Nature **520**, 205 (2015)
292. M. Fuchs, M. Trigo, J. Chen, S. Ghimire, S. Shwartz, M. Kozina, M. Jiang, T. Henighan, C. Bray, G. Ndabashimiye et al., Nat. Phys. **11**, 964 (2015)
293. O. Svelto, *Principles of Lasers*, 5th edn. (Springer, Heidelberg, 2010)
294. S. L. McCall, E.L. Hahn, Phys. Rev. Lett. **18**, 908 (1967)
295. S. L. McCall, E.L. Hahn, Phys. Rev. **183**, 457 (1969)
296. J. Stöhr, A. Scherz, Phys. Rev. Lett. **115**, 107 402 (2015)

297. B. Wu, T. Wang, C.E. Graves, D. Zhu, W.F. Schlotter, J.J. Turner, O. Hellwig, Z. Chen, H.A. Dürr, A. Scherz, J. Stöhr, Phys. Rev. Lett. **117**, 027 401 (2016)
298. Z. Chen, D.J. Higley, M. Beye, M. Hantschmann, V. Mehta, O. Hellwig et al., Phys. Rev. Lett. **121**, 137 403 (2018)
299. F. Bloch, Phys. Rev. **70**, 460 (1946)
300. R.P. Feynman, F.L. Vernon, R.W. Hellwarth, J. Appl. Phys. **28**, 49 (1957)

Part I
Production of X-Rays and Their Description

Chapter 2
Production of X-Rays: From Virtual to Real Photons

2.1 Introduction

In this chapter we examine the fundamentals of accelerator-based x-ray sources and how the quality of the emitted radiation by many electrons contained in an electron bunch can be described through the equivalent concepts of *brightness*[1] and *coherence* of a source. In addition of the conventional description of the generation of electromagnetic (EM) radiation, given in books [1–5] and reader-friendly reviews [6–9], we also introduce a different treatment. It is based on the conversion of a virtual photon cloud attached to a charge into real photons when the charge experiences a change in its velocity.

This method, called the *Weizsäcker-Williams* (WW) formalism [10, 11], describes how the Coulomb fields attached to a relativistic electron bunch are converted into freely propagating electromagnetic radiation. The WW concept developed in the mid-1930s foreshadowed the formal development of QED in the late 1940s. The central idea is that the Coulomb or "velocity" fields of a charge in steady motion [12] may be thought of as a cloud of virtual photons attached to and moving with the charge. The energy stored in the Coulomb field is equated to that of the virtual photon cloud which may be partly converted into true photons or "radiation" when the constant velocity of the charge is changed. This happens, for example, when electrons are slowed down by a target, resulting in bremsstrahlung radiation or accelerated by a magnetic field, leading to synchrotron radiation.

The WW formalism conveniently allows an early introduction of the concept of *virtual photons* which are the carriers of the electromagnetic force and in QED constitute the quantum mechanical zero-point field. It also yields a simple geometrical picture of coherence based on the concept of a coherence cone, in which the EM fields of the individual electrons in a bunch add up coherently. The chapter also

[1] The term "brightness" is typically used in the USA, while the equivalent term "brilliance" is used preferentially in Europe.

J. Stöhr, *The Nature of X-Rays and Their Interactions with Matter*, Springer Tracts in Modern Physics 288, https://doi.org/10.1007/978-3-031-20744-0_2

discusses the characteristics of synchrotron radiation emitted by bending magnets and undulators and compares them to the radiation properties of XFELs.

The concepts used today to describe the EM fields associated with a charged particle that moves with an arbitrary velocity, including the effects of relativity, date back to around 1900 when Liénard [13] and independently Wiechert [14] derived a mathematical formulation now referred to as the *Liénard–Wiechert field equation* [2, 12, 15]. At the time, the electron had just been discovered by J. J. Thomson in 1897 and the difference between Lenard's cathode rays (electrons) and Röntgen's x-rays was known. The work of Liénard and Wiechert was extended in 1912 by Schott [16], who already conjectured the existence of what we now call synchrotron radiation.

EM radiation is produced by accelerating charges and the higher the energy of the moving charge the higher the energy of the radiation that can be produced. In the late 1960s and early 1970s, the advantages of using electron accelerators for the production of EM radiation in the form of x-rays became clear [17]. In accelerator-based sources the kinetic energy of electrons is increased from tens of keV in conventional x-ray tubes to a few GeV, surpassing the relativistic limit set by the rest mass energy of the electron, $m_e c^2 = 511\,\text{keV}$. The relativistic reduction of the Coulomb repulsion between the electrons allows the generation of electron bunches of high charge densities and the emission of radiation into a narrow forward cone when the electrons are radially accelerated in a magnetic field.

Over time, various generations of synchrotron radiation (SR) sources have improved the quality of the emitted x-rays. The quality of a source is quantitatively described by its *brightness*, which cannot be improved but only spoiled by the manipulation of x-rays through optics. The fundamental importance of brightness and its direct link to the non-trivial concept of what we call "coherence" will be discussed in detail in Chap. 4. It will then become clear that "coherence" can be more accurately defined through increasing orders of perturbations within quantum electrodynamics. At this point it suffices to use the word "coherence" in the conventional sense defined through wave interference, which corresponds to *first order* coherence in QED [9].

Owing to the energy width of the emitted spectrum, SR sources are never intrinsically *temporally* coherent, also called *transform limited*, and can only be made so by use of a monochromator. The quality of a SR source is given by its *average brightness*, defined by arbitrarily assuming a spectral bandwidth of $\Delta\lambda/\lambda = 0.1$. Today's SR sources are also not laterally coherent at higher photon energies. The goal of so-called ultimate storage rings is to achieve lateral coherence or *diffraction limited* performance over the photon energy range up to tens of keV [6, 8]. Figure 2.1 illustrates the limitations in lateral coherence of third generation storage rings relative to a completely coherent "ultimate" ring.

The difficulty of achieving a diffraction limited source lies in the effect of the emitted radiation on the electron bunch. In *storage rings*, an equilibrium is established between energy and momentum transfer between the electron and photon systems, which limits the achievable size and angular emission cone of the electron bunch. This is overcome in a *linear accelerator* (linac), where the radiation losses are sufficiently small that it is possible to prepare an exquisitely small electron bunch before it is made to radiate in an undulator. This ability led to the development of x-ray free electron

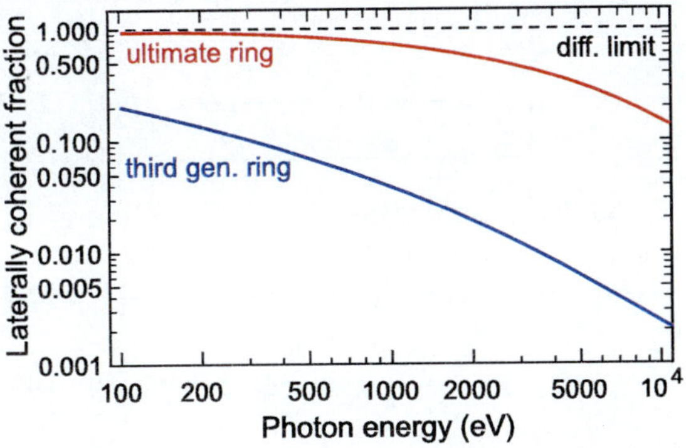

Fig. 2.1 Illustration of the laterally coherent fraction as a function of photon energy, for a hypothetical diffraction limited ring, and that available at a characteristic third generation source and for an envisioned "ultimate storage ring" [3, 6, 8]

lasers (XFELs). In the most common self-amplified spontaneous emission (SASE) XFELs, a well-prepared bunch of micrometer dimensions is sent through a long (tens of meters) undulator before it is discarded. Since each bunch is freshly prepared, the most brilliant radiation comes at an expense of a repetition rate ($\sim 10^2 - 10^5$ Hz) which is much smaller than the 500 MHz rate of a storage ring.

As a compromise between storage rings and XFELs, energy recovery linacs (ERLs) have been proposed [18, 19]. Such sources integrate a linac acceleration structure into a long straight section of a race track shaped magnet lattice resembling that of a storage ring. After an individual high-quality electron bunch is prepared in the linac section, it is allowed to circle once around the race track where it can serve many user stations. During its single cycle, the electron bunch maintains its properties since it requires thousands of cycles for a freshly prepared bunch to reach an equilibrium state. At the end of its cycle the bunch passes through the same long linac section and is decelerated to recover part of its energy before it is discarded in a beam dump. Energy recovery helps to reduce the operational costs of the facility. Energy recovery linacs have high repetition rates and therefore high average spectral brightness, but lower peak brightness than an XFEL.

2.2 Relativistic Concepts and Electron Bunch Compression

Before we derive the fields associated with relativistic electrons in accelerators, it is useful to briefly review some key relativistic concepts.

In the description of electron beams we use the well-known length and time transformations between inertial systems that move relative to each other with a constant velocity $\mathbf{v}$ [12, 20]. We define the *rest frame* of an object as the coordinate system in which it is at rest. The length measured in the rest frame is called the *proper length*, and the time measured by a clock in the rest frame is called the *proper time*. Following convention, we define the following relationships involving the ratio of v and the speed of light c,

$$\boxed{\beta \equiv \frac{v}{c} \quad \gamma \equiv \frac{1}{\sqrt{1-\beta^2}} \quad \beta \equiv \sqrt{1-\frac{1}{\gamma^2}} \simeq 1-\frac{1}{2\gamma^2}} \,. \tag{2.1}$$

With these conventions the relativistic energy $\mathcal{E}_e$ of the electron beam in the laboratory system, which is the sum of the electron kinetic energy $\mathcal{E}_{kin} = m_e v^2/2$ and the rest energy of the electron $m_e c^2 = 0.511$ MeV, is given by[2]

$$\gamma = \frac{\mathcal{E}_e}{m_e c^2}, \quad \text{or with } \mathcal{E}_e \text{ in [GeV]} : \; \boxed{\gamma = 1.96 \times 10^3 \, \mathcal{E}_e} \tag{2.2}$$

At a moderate accelerator energy of ten times the rest mass energy or 5 MeV, we have $\gamma \simeq 10$ and the particle travels already at $v = 0.995\,c$. As the particle energy is further increased, the particle velocity remains relatively constant at about the speed of light. In a linear accelerator, the electron velocity thus remains nearly constant once its energy is well above the *electron rest mass* energy of 0.511 MeV. Since energies of tens of MeV are considered small for typical high energy physics accelerators, one may consider the electrons in uniform motion through most of their path.[3]

If we compare the proper length l_0 in the rest frame to the length l along v, in a system moving with v relative to the rest frame, the lengths are related by the well-known Lorentz relationship

$$l = \frac{l_0}{\gamma}. \tag{2.3}$$

If l_0 and l are measured perpendicular to v, we have $v_\perp = \beta c = 0$ and $l_0 = l$. This shows that the length is always the largest in the rest frame, which is the well-known *Lorentz contraction* observed for a moving system. For an electron beam of 5 GeV we have $\gamma = 10^4$ and an electron bunch that appears to be $l = 1$ mm long in the laboratory frame, has a proper length $l_0 = 10$ m in the rest frame of the electron bunch. This bunch has a temporal length of $\Delta t = l/c \simeq 3$ ps in the laboratory frame.

The time behaves exactly opposite. If we measure a time t_0 in the rest frame, the time t in the moving frame is

$$t = t_0\,\gamma. \tag{2.4}$$

The proper time measured in the rest frame is therefore always the shortest, and we observe a *time dilation* for a moving system.

---

[2] The electron mass m_e in SI units is given by $m_e = 9.109 \times 10^{-31}$ VAs^3 m^{-2} $= 9.109 \times 10^{-31}$ kg.

[3] Further increases in energy originate from the increasing relativistic mass m_e^* of the particle $m_e^* = \gamma\, m_e$. In Einstein's theory of relativity, there is a crucial difference between particles with and without mass: All massless particles must travel at the speed of light, whereas massive particles can never attain this ultimate speed.

It is interesting and important that relativistic effects for the Coulomb forces between electrons exist both in the direction of motion and perpendicular to it. When two electrons are separated by a distance d_0 in their rest frame along the direction of motion $\mathbf{v}$ relative to an observer, the distance $d_{\|}$ seen by the observer is strongly contracted but the force is unchanged,

$$d_{\|} = \frac{d_0}{\gamma}\,, \qquad \mathbf{F}^{\|} = \mathbf{F}_0. \tag{2.5}$$

Thus through relativity one can produce for the observer an apparent small separation between the electrons without changing the electrostatic repulsion.

When two electrons are separated by a distance d_0 in their rest frame perpendicular to the direction of motion $\mathbf{v}$ relative to an observer, the distance $d_{\perp}$ seen by the observer is unchanged but the force is reduced according to

$$d_{\perp} = d_0\,, \qquad \mathbf{F}^{\perp} = \frac{\mathbf{F}_0}{\gamma}. \tag{2.6}$$

This remarkable result is due to the fact that in the perpendicular direction the force has both an "electric", $\mathbf{F}_E^{\perp}$, and "magnetic", $\mathbf{F}_B^{\perp}$ component with opposite signs, and therefore the magnetic contribution weakens the Coulomb repulsion. Thus the bunch can also be compressed in the perpendicular direction.

In practice, remarkable bunch compression results have been achieved at the SLAC National Accelerator Laboratory using its 3-km-long linear accelerator (linac) before it was converted for LCLS. The linac was capable of generating ultra-relativistic electron beams with energies up to 50 GeV. At an energy of 28 GeV, a Gaussian rms bunch length as short as $\sigma_z = 21\,\mu\text{m}$ in the laboratory frame was achieved while maintaining $N_e = 1.5 \times 10^{10}$ electrons in the bunch [21, 22]. Such bunches pass by a point in the laboratory in a time span $2\sigma_t = 140 \times 10^{-15}\,\text{s}$ or 140 femtoseconds (fs) and therefore create a peak current of $I \simeq N_e\, e/2\sigma_t \simeq 1.5 \times 10^4\,\text{A}$. Because the bunch can also be compressed in the perpendicular direction, high current densities can also be obtained. In fact, sub micron beam diameters have been achieved. Denoting the lateral cross section (area) as a, the current density is given by $j = N_e\, e/(2\sigma_t\, a)$, and values of $j \simeq 10^{16}\,\text{A/m}^2$ have been reached [23]. The manipulation of the shape and size of relativistic electron bunches has become a science in itself and so has their accurate characterization [24].

2.3 The Fields of a Moving Charge: Liénard–Wiechert Equations

The calculation of the fields associated with a single charged particle in arbitrary motion, including the effects of relativity, was first accomplished by Alfred-Marie Liénard (1869–1958) in 1898 [13] and independently by Emil Wiechert (1861–

1928) in 1900 [14]. In their honor, the mathematical formulation of the fields is called the *Liénard–Wiechert field equation*. We shall not derive the Liénard–Wiechert expression here but will simply state it and use it. For a detailed discussion the reader is referred to other texts [2, 12, 15]. The book of Hofmann [2] provides a particularly nice derivation of the field equations and their application to different situations. The tricky part in the calculation of the fields is the treatment of the relativistic effects referred to as "retardation effects".

Retardation effects arise from the finite velocity of light. Let us for a moment assume that the velocity of light was infinite. We could then describe all fields in a "snapshot picture" that instantaneously correlates time and position. For example, the **E**-field measured at a given time could be directly expressed as a function of the position of the charged particle at the same time, as in electrostatics. However, since the speed of light is finite, there is a time difference between the emission and observation of the light. As illustrated in Fig. 2.2, it takes the light a finite time $\Delta t = r^*/c$ to travel the distance r^*.

We can then no longer simply write down an expression for the field $\mathbf{E}(t)$, measured at an instantaneous time t, in terms of the particle properties (e.g. position, velocity, and acceleration) at that time since the relevant particle properties that determine $E(t)$ occurred at an earlier or retarded time $t^* = t - r^*/c$.

In order to calculate the field at an observer at time t we must know the position $\mathbf{r}(t^*) = \mathbf{r}^*$, the velocity $\mathbf{v}(t^*) = \mathbf{v}^*$, and the acceleration $\mathbf{a}(t^*) = \mathbf{a}^*$ of the charge q at the retarded time t^*. In our notation, the relation between t and t^* is given by the following two general equations [2]

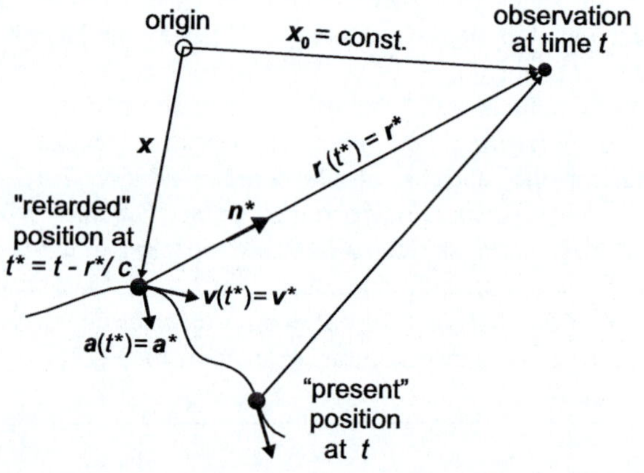

Fig. 2.2 Illustration of the retarded time concept. A charge moves on an arbitrary trajectory. The fields at the observer at a time t are determined by the position $\mathbf{r}(t^*) = \mathbf{r}^*$, the velocity $\mathbf{v}(t^*) = \mathbf{v}^* = d\mathbf{x}/dt^* = -d\mathbf{r}^*/dt^*$, and the acceleration $\mathbf{a}(t^*) = \mathbf{a}^*$ at the retarded time $t^* = t - r^*/c$. The observer is located at a fixed position $\mathbf{x}_0$ relative to the origin of the coordinate system

$$\boxed{t = t^* + \frac{r^*}{c}\,, \qquad \mathrm{d}t = \left(1 - \frac{\mathbf{n}^* \cdot \mathbf{v}^*}{c}\right) \mathrm{d}t^*} \tag{2.7}$$

These simple equations, however, may have very complicated solutions for certain particle motions!

Liénard and Wiechert succeeded in deriving equations that express the electric and magnetic fields of a *point charge* in arbitrary motion. Below we give these equations using the notation of Fig. 2.2, where at the time of emission the unit vector $\mathbf{n}^* = \mathbf{r}^*/r^*$ points from the charge to the observer [2, 12, 15].

The Liénard–Wiechert field $\mathbf{E}(t)$ of a point charge q detected by an observer at a time t is determined by the distance r^*, the velocity v^*, and acceleration a^* of the charge at the emission or retarded time $t^* = t - r^*/c$. Defining $\boldsymbol{\beta}^* = \mathbf{v}^*/c$ we have

$$\mathbf{E}(t) = \frac{q}{4\pi\epsilon_0} \underbrace{\frac{1-(\boldsymbol{\beta}^*)^2}{r^{*2}\,(1-\mathbf{n}^*\!\cdot\boldsymbol{\beta}^*)^3}\left[\mathbf{n}^* - \boldsymbol{\beta}^*\right]}_{\text{velocity field}}$$
$$+ \frac{q}{4\pi\epsilon_0} \underbrace{\frac{1}{c^2\, r^*\,(1-\mathbf{n}^*\!\cdot\boldsymbol{\beta}^*)^3}\left\{\mathbf{n}^* \times \left(\left[\mathbf{n}^* - \boldsymbol{\beta}^*\right] \times \mathbf{a}^*\right)\right\}}_{\text{acceleration field}}. \tag{2.8}$$

We have indicated all retarded quantities by an asterisk.

Our sign convention is that of Fig. 2.2, where $\mathbf{n}^* \cdot \mathbf{v}^* \geq 0$ is the velocity component of the charge toward the observer. Equation (2.8) is valid for any motion of a single charged particle, but it holds only if the particle is a point charge. The reason is that only then can the particle's instantaneous motion and position be described in terms of a single retarded time.

In (2.8) we have explicitly identified the two contributions to the electric field of the charge. Fields of a charge that moves with a *constant* velocity are called *velocity fields* or *Coulomb fields*. They are also referred to as *near fields* because they fall off fast with the distance to the observer as $1/r^{*2}$. *Velocity fields* correspond to $a^* = 0$ and are attached to and move with the charge. In the limit $v^* = a^* = 0$ we obtain from (2.8)

$$\mathbf{E}(t) = \mathbf{E}^{\text{rest}} = \frac{q}{4\pi\epsilon_0 r^{*2}}\,\mathbf{n}^* = \frac{q}{4\pi\epsilon_0 r^2}\,\mathbf{n}, \tag{2.9}$$

which is just the Coulomb field of a point charge, and by omission of the asterisk we have indicated that retardation effects are absent.

In contrast, the fields produced when a charged particle is accelerated are called *acceleration fields* but are best known as *electro magnetic radiation* which separates from the charge. Acceleration fields decrease more gradually with $1/r^*$ and owing to their dominance at large distance are also called *far fields*. We see from the second term in (2.8) that the far field **E** is perpendicular to $\mathbf{n}^*$ pointing from the charge at the retarded time t^* toward the observer.

For both terms in (2.8) the **E**- and **B**-fields are perpendicular and are linked by the following relation,

$$\boxed{\mathbf{B}(t) = \frac{1}{c}\left[\mathbf{n}^* \times \mathbf{E}(t)\right]}\,. \qquad (2.10)$$

2.4 Fields of a Charge in Uniform Motion: Velocity Fields

In the following we consider the strength, angular distribution, and time dependence of velocity fields surrounding moving charges. We start with the case of a moving single electron and will then consider the fields of many electrons moving as a concentrated bunch.

2.4.1 *Spatial Dependence of Velocity Fields of a Single Electron*

The limit of a stationary charge expressed by 2.9 is illustrated on the left side of Fig. 2.3. In the rest frame of the electron, the radial electric field by definition terminates on the negatively charged electron ($q = -e$), and there is no magnetic field so that $\mathbf{B} = 0$.

We now consider the case of an electron traveling down a linear accelerator with close to the speed of light. In practice, electrons in a linac rapidly reach a relativistic velocity and their further acceleration is then quite small and can be neglected to a good approximation. We assume that an electron sits at the origin of a frame K' which moves away with a velocity $+\mathbf{v}$ from our observer frame K. We can then transform the fields from the electron rest frame K' back into the observer frame K. We start in the frame K' and denote all quantities in that frame by primes. The fields in the primed electron rest frame are expressed in terms of the proper distance r'. If we denote the unit position vectors in the two frames as $\mathbf{r_0}$ and $\mathbf{r}'_0$, we have

$$\mathbf{E}' = \frac{q}{4\pi\epsilon_0\,(r')^2}\,\mathbf{r}'_0 \qquad \mathbf{B}' = 0. \qquad (2.11)$$

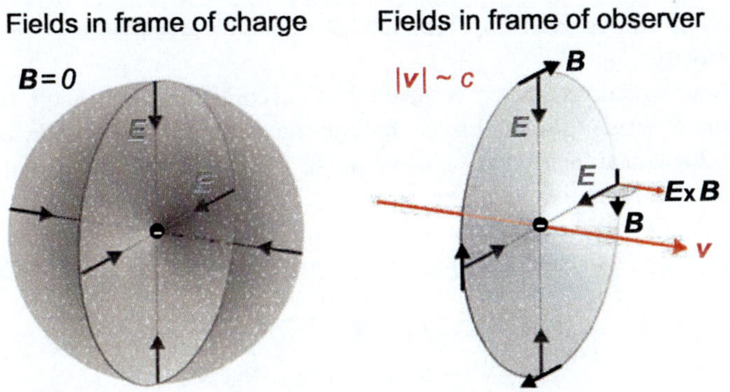

Fig. 2.3 Example of the electric and magnetic fields of a *single electron* in different reference frames. *Left*: Electric field in the frame of the electron with charge $q = -e$, which is the well-known radial Coulomb field with the electric field vector pointing toward the negative electron. There is no magnetic field. *Right*: The same electron is assumed to move with a highly relativistic (constant) velocity $|\mathbf{v}| \approx c$ relative to the stationary observer, and the fields are observed from the stationary frame. The electric field is now nearly contained in a plane perpendicular to $\mathbf{v}$ since the component $\mathbf{E}_\perp \perp \mathbf{v}$ is enhanced by a factor γ according to (2.15) and the component parallel to the field is reduced by a factor $1/\gamma^2$ according to (2.14) over the field in the rest frame of the electron. The moving charge also gives rise to a field $\mathbf{B}$ which lies in the plane perpendicular to $\mathbf{v}$ and at any point has the direction and magnitude $\mathbf{B} = (\mathbf{v} \times \mathbf{E})/c^2$. Both the electric, and magnetic fields move with the electron, and hence there is no "radiation"

The fields of the moving charge in the observer frame K are given by

$$\mathbf{E}_\parallel = \mathbf{E}'_\parallel \qquad \mathbf{B}_\parallel = \mathbf{B}'_\parallel = 0. \tag{2.12}$$

and

$$\mathbf{E}_\perp = \gamma \mathbf{E}'_\perp \qquad \mathbf{B}_\perp = \gamma \, \frac{\mathbf{v} \times \mathbf{E}'}{c^2}. \tag{2.13}$$

We now need to express the field $\mathbf{E}'$ in (2.11) in terms of coordinates r in the frame K since the distance in the two frames is different due to relativistic effects. We have the following relationships

$$\begin{aligned} \mathbf{r}_0, \mathbf{r}'_0 \parallel \mathbf{v}: &\qquad \mathbf{r}' = \gamma\, \mathbf{r} \\ \mathbf{r}_0, \mathbf{r}'_0 \perp \mathbf{v}: &\qquad \mathbf{r}' = \mathbf{r}. \end{aligned}$$

We obtain

$$E_\parallel = \frac{q}{4\pi\epsilon_0\, \gamma^2 r^2} = \frac{E'}{\gamma^2} \tag{2.14}$$

$$E_\perp = \frac{q\gamma}{4\pi\epsilon_0\, r^2} = \gamma\, E'. \tag{2.15}$$

These are our desired results for the fields of a charge moving with a constant relativistic velocity.

The above equations directly compare the fields of the moving charge $E_{\parallel}$ and $E_{\perp}$ seen by the observer to the fields E' that the same charge would produce if it was stationary located at the same distance r in the frame of the observer. We see from (2.14) and (2.15) that the field is reduced by a factor $1/\gamma^2 = 1 - \beta^2$ parallel to the propagation direction[4] and enhanced by a factor γ perpendicular to $\mathbf{v}$.

We can write (2.14) and (2.15) in a single general expression [2, 25, 26]

$$\mathbf{E}(\mathbf{r}) = \left[\frac{1-\beta^2}{(1-\beta^2\sin^2\theta)^{3/2}}\right] \underbrace{\frac{q}{4\pi\epsilon_0 r^2}\,\mathbf{r}_0}_{\mathbf{E}'(\mathbf{r})}\ , \tag{2.16}$$

where θ is the angle between the direction of observation $\mathbf{r}$ (unit vector $\mathbf{r}_0$)[5] and the direction of motion $\mathbf{v}$, as shown in Fig. 2.4. The direction of the $\mathbf{E}$-field is seen to be along $\mathbf{r}$, and the field *magnitude* is the largest in the plane of the charge perpendicular to $\mathbf{v}$.

Similar to (2.16), the corresponding $\mathbf{B}$-field expression is derived by use of (2.12) and (2.13) and is given by

$$\mathbf{B}(\mathbf{r}) = \left[\frac{1-\beta^2}{(1-\beta^2\sin^2\theta)^{3/2}}\right] \underbrace{\frac{q}{4\pi\epsilon_0 r^2 c^2}\,[\mathbf{v}\times\mathbf{r}_0]}_{\mathbf{B}'(\mathbf{r})}\ . \tag{2.17}$$

The field lines are circular about $\mathbf{v}$, as shown in Fig. 2.3 for the case of a single electron. The field *magnitude* is largest perpendicular to $\mathbf{v}$ in the plane that contains the charge and vanishes for $\mathbf{v} \parallel \mathbf{r}_0$ or $\theta = 0$.

The fields given by (2.16) and (2.17) are shown on the right side of Fig. 2.3 for the case of an electron moving with nearly the speed of light. The dramatic relativistic effect leads to fields that are squished along the propagation direction and are contained in an angular field cone of *half* width obtained from (2.14) and (2.15) as $E_{\perp}/E_{\parallel} = 1/\gamma$. The angular distribution of the fields is shown in Fig. 2.4 for selected cases of β. As γ increases the field pattern becomes increasingly compressed along the beam direction.

Figures 2.3 and 2.4 illustrate that the fields stay attached to and move with the charge. They do not "radiate" away from or separate from the charge.

---

[4] This seems to contradict (2.12), stating that $\mathbf{E}_{\parallel} = \mathbf{E}'_{\parallel}$, but one needs to realize that this assumes that the field $\mathbf{E}'_{\parallel}$ is expressed in terms of coordinates $\mathbf{r}'$ in the frame K'. The relativistic correction comes from expressing the field in terms of coordinates $\mathbf{r}$ in the frame K.

[5] Note that in (2.16) the distance r is *not* the distance r^* at the retarded time t^*. Rather, in the actual evaluation of the retarded field expression, r^* has been expressed in terms of the distance r at the observation time t.

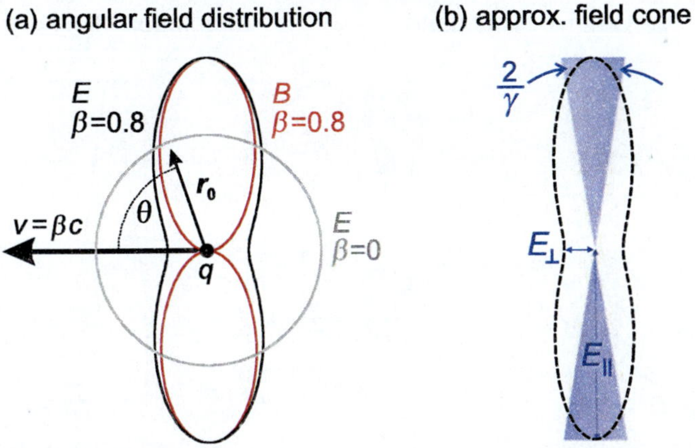

Fig. 2.4 **a** Angular patterns of the **E**- and **B**-fields according to (2.16) and (2.17) for a particle of charge q, traveling with a speed $v = \beta c$, for different values of β. **b** Approximate angular full width of the cone perpendicular to **v** that contains the fields, obtained from (2.14) and (2.15) as $2/\gamma$

2.4.2 *Temporal Dependence of Velocity Fields of a Single Electron*

The fields given by (2.16) and (2.17) express the size and direction of the fields attached to a single charge in uniform motion. A stationary observer will experience a field pulse when the charge moves by. The duration of the field pulse may be expressed in terms of the time interval Δt over which the field seen by the observer is large, or equivalently, as illustrated in Fig. 2.5a, by introducing the concept of an *observation cone*. The observation cone is simply the relativistic field cone of FWHM $2/\gamma$, drawn from the observer position. In the field cone approximation the field amplitude is constant (flat-top) when the point charge is located within the observation cone.

More accurately, the temporal field pulse has a shape that is determined by the component $\mathbf{E} \simeq \mathbf{E}_\perp$ given by

$$E(R,t) = \frac{q\,\gamma}{4\pi\epsilon_0\,R^2}\left(1+t^{*2}\right)^{-3/2} \tag{2.18}$$

where

$$t^* = \frac{\gamma\,v}{R}\,t. \tag{2.19}$$

It is plotted in Fig. 2.5b, and the pulse has a FWHM value

$$\left(1+t^{*2}\right)^{-3/2} = \frac{1}{2} \;\Rightarrow\; t^* = 0.8. \tag{2.20}$$

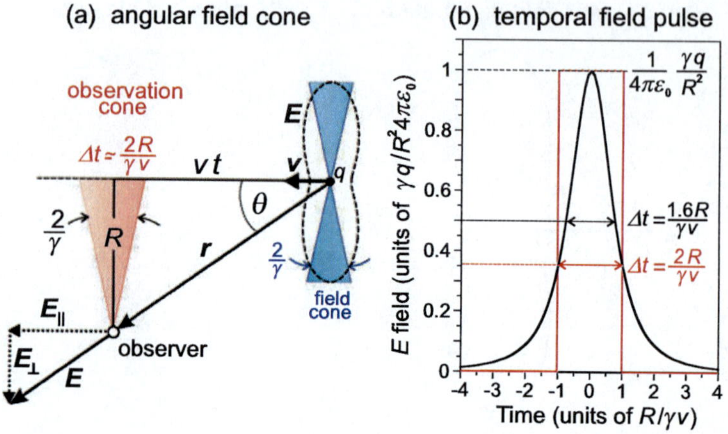

Fig. 2.5 **a** Geometry of an incident charge q of velocity **v** that passes by an observer at a perpendicular distance R. Attached to the charge we show the relativistic angular patterns of the total **E**-field and in blue the approximate field cone of width $2/\gamma$, according to Fig. 2.4. The red cone shown at the observer position with the same angular width is referred to as the *observation cone*. The observer sees a flat-top field amplitude only when the moving charge is located within this observation cone. **b** Field pulse magnitude created by the charge at the site of the observer according to (2.18). The FWHM of the actual pulse shown in black is given by $\Delta t = 1.6R/(\gamma\, v)$ and is equivalent to the shown red top flat-top approximation of width $\Delta t = 2R/(\gamma\, v)$

When the integral over the actual temporal shape shown in black in Fig. 2.5b is replaced by a flat-top shape shown in red, we find by comparison of the integrated field distribution $\int_{-\infty}^{\infty}(1+t^{*2})^{-3/2}\,\mathrm{d}t^* = 2$ an equivalent flat-top width $\Delta t = 2R/\gamma\, v$, as expected. The peak value is $E_{\max}(t) = \gamma\, q/(4\pi\epsilon_0 R^2)$, as shown.

The corresponding B-field is derived from (2.17) and has a peak value of $B_{\max} = E_{\max} v/c^2$. The fields become huge and the pulse duration extremely short for distances R of order nanometers. To obtain large fields at larger distances, say micrometers to millimeters from the charge trajectory, it is necessary to pack a large number ($\sim 10^{10}$) of electrons into a small volume by use of a linear accelerator. A sample can then experience these large fields when the beam is either sent directly through it [21, 27] or grazes by its surface.

2.4.3 *The Fields of a Relativistic Gaussian Electron Bunch*

The nature of the fields associated with more than one charge may be distinguished in terms of their coherent and incoherent superposition. This is facilitated by the relativistic field cone concept shown in Fig. 2.6 for two charges, separated by a distance d along the direction of motion in the frame of the observer.

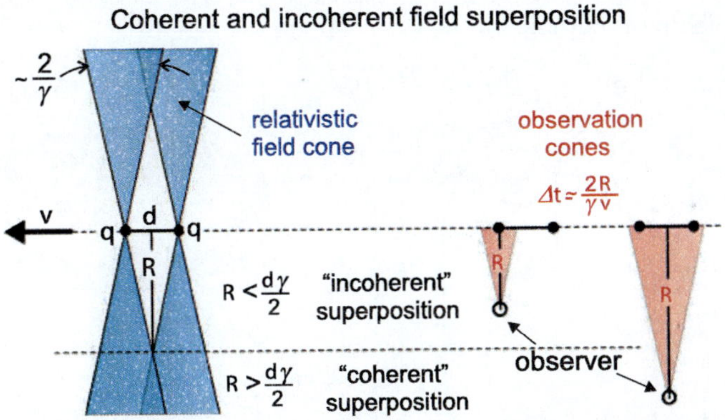

Fig. 2.6 Illustration of the field pulses due to two relativistic charges q, separated by a distance d along the direction of motion in the frame of the observer and moving in the same direction with a constant speed $v = \beta c$. The field cones (blue) are relativistically contracted and for $v \sim c$ have an opening angle $2/\gamma$ (see Fig. 2.5). An observer located at a vertical distance $R < d\gamma/2$ sees two separate field pulses due to the fact that the two charges never fall into the same observation cone (red) and the field superposition may be called "incoherent". An observer at distance $R > d\gamma/2$ only sees one pulse, corresponding to a "coherent" field superposition. The pulse length seen by the observer is $\Delta t \simeq 2R/(\gamma v)$ as indicated

The relativistic blue field cones of the two charges, introduced in Fig. 2.5, begin to overlap at a distance $R = d\gamma/2$. An observer stationed at $R < d\gamma/2$ sees two separate field pulses, while an observer at $R > d\gamma/2$ sees a single field pulse as shown by the red observation cone on the right. In the latter case, the fields produced by the two charges appear to originate from a single charge $2q$.

The simplifying concept of relativistic field cones associated with individual charges and the complementary concept of observation cones allows us to derive the fields associated with a *Gaussian distribution* of many charges as illustrated in Fig. 2.7 for three cases of observer positions.

2.4.3.1 Coherent Field Superposition

In Fig. 2.7a, the observer is located far away, at a distance $R \gg \sigma_z\gamma = v\,\sigma_t\gamma$. For typical values $\sigma_z = 50\,\mu\text{m}$ and $\gamma = 10^4$ the shown case describes observer distances of $R \gg 0.5\,\text{m}$. The concept illustrated in Fig. 2.6 shows that for the case of a Gaussian bunch traveling with speed $v \sim c$, the field cones produced by all charges in the bunch overlap at the observer position and therefore appear to originate from a single macro-charge $N_e e$. The maximum total field of the bunch $E(R)$ is then N_e-times that of the individual single charges given by (2.18).

When the total bunch falls into the observation cone, as shown in Fig. 2.7a the field seen by the observer is that of a macro-charge $N_e q$ given by

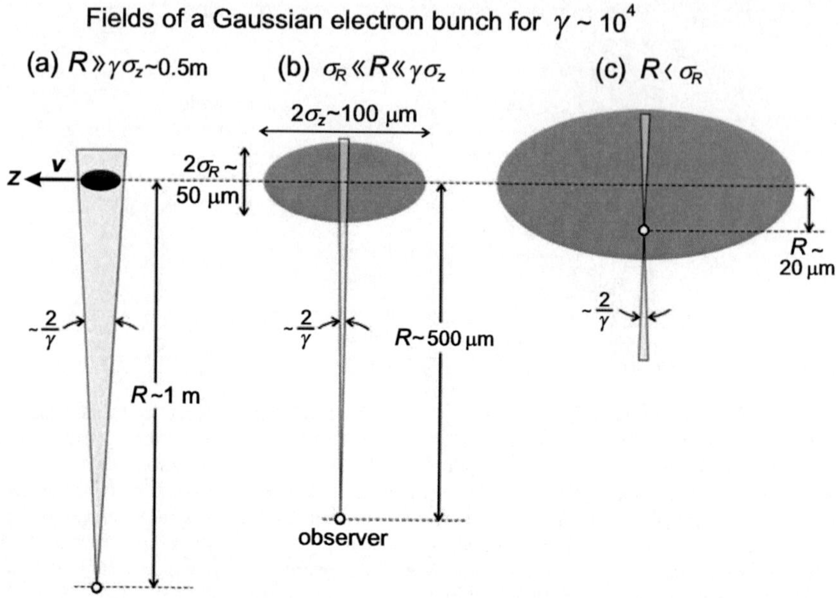

Fig. 2.7 Illustration of the observation or coherent field cones for a Gaussian electron bunch traveling along z at nearly the speed of light with $\gamma \simeq 10^4$ and typical rms widths $\sigma_z \simeq 50\,\mu$m along z, so that $\gamma\sigma_z = 0.5$ m, and $\sigma_R \simeq 25\,\mu$m perpendicular to z. In all cases the coherent fields at the observer arise only from charges in the observation cones. We consider only the maximum fields at an instant of time when the center of the moving bunch has its shortest distance R from the observer. **a** All fields superimpose coherently at the sight of the distant observer. **b** Only a slice of the bunch contributes to the fields at a given time and with time the field strength follows the longitudinal bunch shape $\exp(-z^2/2\sigma_z^2)$. **c** Now the fields of charges in both observation cones perpendicular to z contribute to the field, and the field is zero for $R = 0$ and increases with R until it saturates near the edge of the bunch

$$E(R) = \sum_{i=1}^{N_e} \underbrace{\frac{-e\gamma}{4\pi\epsilon_0 R^2}}_{E_e \text{ of one electron}} = N_e E_e \quad \text{for} \quad R \gg \sigma_z\gamma \tag{2.21}$$

The energy flow associated with the moving velocity fields is given by the magnitude of the Poynting vector, and the intensity I of dimension [energy/time/unit area] is given by

$$I = \epsilon_0 c\, E^2(R) = \epsilon_0 c N_e^2 E_e^2. \tag{2.22}$$

The intensity scales with the square of the number of charges and therefore their field contributions add *coherently*.

Next we consider the case in Fig. 2.7b, where $\sigma_R \ll R \ll \sigma_z\gamma$. This important case applies to an observer that is well outside the perpendicular bunch radius $\sigma_R \simeq 25\,\mu$m but still much closer than $\sigma_z\gamma \sim 1$ m. In this case we need to consider the Gaussian shape (see Appendix A.2.2) of the charge distribution $\rho(R,t)$. It is described in the longitudinal time coordinate t by an rms width σ_t and in the transverse coordinate R by an rms width σ_R according to

$$\rho(R,t) = \frac{-N_e e}{(2\pi)^{3/2}\,\sigma_t\,\sigma_R^2}\exp\left(-\frac{R^2}{2\sigma_R^2}\right)\exp\left(-\frac{t^2}{2\sigma_t^2}\right), \tag{2.23}$$

where $\sigma_t = \sigma_z/v \simeq \sigma_z/c$. The normalization is chosen so that the integral of the charge density $\rho(R,t)$ over time t and the transverse distance R is equal to the total charge $q = -N_e e$ of the electron bunch. The fields for $\sigma_R \ll R \ll \sigma_z\gamma$ are obtained as [27]

$$E(R,t) = -\frac{N_e e}{(2\pi)^{3/2}\,\epsilon_0\,c\,\sigma_t\,R}\left(1-\exp\left[-\frac{R^2}{2\sigma_R^2}\right]\right)\exp\left[-\frac{t^2}{2\sigma_t^2}\right]. \tag{2.24}$$

We also have $B(R,t) = E(R,t)/c$, which may be expressed in terms of the permeability of vacuum by use of $\mu_0 = 1/(\epsilon_0 c^2)$. As the bunch moves by the observer, the field strength will follow the longitudinal Gaussian charge distribution so that the amplitude of the temporal field pulse will follow the charge that is in the observation cone per unit time, i.e. the instantaneous current.

Remarkably, the fields fall off with $1/R$ rather than with $1/R^2$ for a point charge (see (2.16) and (2.17)) so that the B-field of the bunch has the same characteristic $1/R$ dependence as that around a current carrying wire.[6] Also, the fields do not explicitly depend on the electron beam energy $\mathcal{E}_e$ and are completely determined by three parameters: the total charge of the bunch $N_e e$, the longitudinal or temporal width $\sigma_t = \sigma_z/v \simeq \sigma_z/c$, and the lateral width or beam radius σ_R.

The field given by (2.24) is still related to the total number of charges N_e because of the well-defined Gaussian shape of the bunch. With time the number of contributing charges follows the Gaussian distribution given by (2.24), and the maximum field amplitude is

$$E_{\max} = -\frac{eN_e}{(2\pi)^{3/2}\epsilon_0\,\sigma_z R}. \tag{2.26}$$

The intensity associated with (2.24) scales with N_e^2. At a given time, the fields seen by the observer are a *coherent* superposition of those from the charges in the observation cone. The observation cone is thus equivalent to the coherence cone of the fields.

---

[6] In fact, (2.24) may be derived from Ampère's law for the **B**-field surrounding a current carrying wire

$$\oint_L \mathbf{B}\cdot d\mathbf{l} = 2\pi\,R\,B = \mu_0 I = \mu_0 d(N_e e)/dt. \tag{2.25}$$

This simplification is a consequence of relativity. For the derivation see https://stohr.sites.stanford.edu/published-unpublished-research under "Homework problems associated with the Magnetism book".

The case in Fig. 2.7c where the observer is located within the beam diameter, i.e. at a distance $R \ll \sigma_R$ from the beam trajectory, is also given by (2.24) and $B(R,t) = E(R,t)/c$. If the observer is positioned on the bunch trajectory $R = 0$, then all fields cancel by symmetry. When the observer shifts off the beam trajectory toward the edge of the bunch, the fields will steadily increase because of the increasing difference of the number of charges in the two opposite coherence cones perpendicular to the beam trajectory. The peak field occurs at $R = 1.58\,\sigma_R$ (see Fig. 2.8).

2.4.3.2 Incoherent Field Superposition

For a stationary observer located inside the incident beam radius, we also need to consider the fields due to individual electrons that pass by at very small distances. For example, if we choose the observer site at the center of an atom, then individual electrons may pass by within the atomic radius and it is very unlikely that more than one beam electron will be present within the atomic radius at a given time. Hence the atomic observer will only see consecutive field pulses from individual electrons. The field pulses are described by (2.18) and plotted in Fig. 2.5b. The fields are huge, and the pulse duration is extremely short for distances R of order of the Bohr radius $a_0 = 0.0529\,\text{nm}$. They may in fact lead to liberation of core electrons, called field ionization or impact ionization.

It is common to characterize the field strength at atomic sites relative to the distance b of the passing electron from the nucleus, where b is called the *impact parameter*. These fields will only arise from individual electrons in the beam since the probability of two electrons being simultaneously within an atomic distance is negligible. If we time-average over the entire pulse, several electrons, say n, may pass by the atomic observer at different times and the intensity will be

$$I = \epsilon_0 c \underbrace{\sum_i^n E_i^2}_{\text{incoherent}} = \epsilon_0 c n E^2 \quad \text{for} \quad b \approx a_0. \tag{2.27}$$

In this case the fields at the atomic sites add *incoherently*, as indicated. We will see that this distinction between coherent and incoherent fields leads to the virtual photon spectrum associated with the velocity or Coulomb fields of a Gaussian bunch shown in Fig. 2.11.

2.4.4 Generation of Huge Field Pulses: THz Fields and Radiation

The dependence of the fields on R given by (2.24) with $t = 0$ is shown in Fig. 2.8 for practical beam parameters that can be achieved in a linac [21].

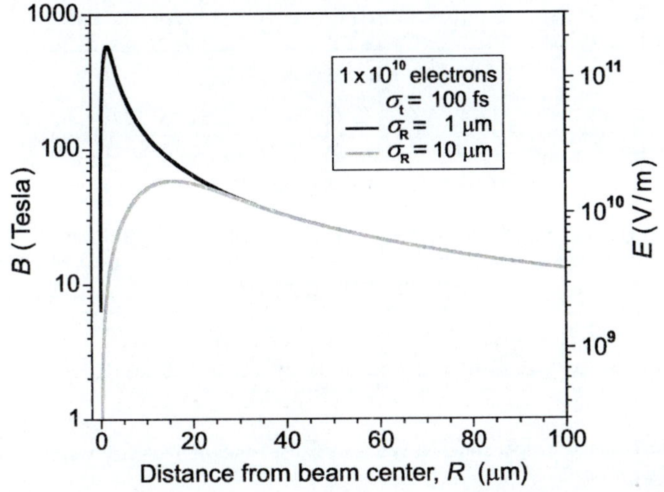

Fig. 2.8 Amplitude of the magnetic induction $\mathbf{B}(R)$ in [Tesla] and the electric field $\mathbf{E}(R)$ in [V/m] as a function of the perpendicular distance R from the beam center, according to (2.24) with $t = 0$. We have assumed 1×10^{10} electrons per pulse and a Gaussian bunch profile in all directions, with standard deviations $\sigma_R = \sigma_x = \sigma_y = 1\,\mu\text{m}$ (black) or $10\,\mu\text{m}$ (gray) perpendicular to the beam and $\sigma_z = 30\,\mu\text{m}$ corresponding to a temporal pulse width of $\sigma_t = 100\,\text{fs}$. The peak value of the field occurs at $R = 1.58\,\sigma_R$ from the beam center

For *low energy* (tens of keV) electron beams, the absence of relativistic effects limits the achievable peak current densities. Since the bunch length in the rest frame of the electrons is not relativistically increased for low energies ($l_0 = l$ in (2.3)), the strong Coulomb repulsion limits the number of electrons per bunch or pulse. Nevertheless, owing to the larger electron than x-ray interaction cross sections pulses of tens of femtosecond length can still be used for the study of ultrafast dynamics in many systems [28, 29].

2.4.5 *Frequency Spectrum of Gaussian Electron Bunches*

The beam parameters used for Fig. 2.8 correspond to electron bunches that have been compressed by state-of-the-art techniques [21]. More typical linac pulses have Gaussian temporal widths of a few picoseconds [30]. Since a temporal length of 1 ps = 10^{-12} s corresponds to a *cycle* frequency, measured in units [Hz] = [1/s], of 10^{12} Hz = 1 terahertz (THz), such pulses have an associated spectrum in the far-infrared or THz regime. Radiation of 1 THz cycle frequency has a photon energy of $\hbar\omega = 4.14\,\text{meV}$

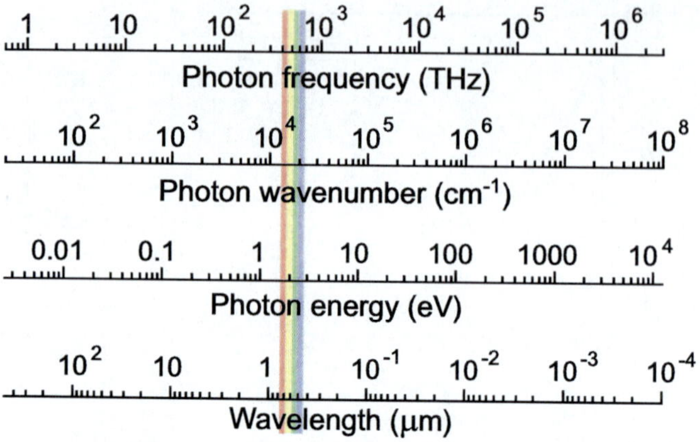

Fig. 2.9 Electromagnetic spectrum in the commonly used units of frequency, wavenumber, photon energy, and wavelength

and a wavelength of 300 μm.[7] For convenience we show in Fig. 2.9 an overview of the electromagnetic spectrum from the THz to the x-ray region in various commonly used units.

In practice, one typically links *field* pulses in the time and frequency domains by a Fourier transform. For a Gaussian lineshape the Fourier transform is again a Gaussian. One needs to distinguish the widths of *field* pulses that are Fourier transformed from the width of the typically measured *intensity* pulses, which correspond to squared fields. Pulse shapes and Fourier transforms are discussed in Appendices A.2 and A.4.1.

Figure 2.10a compares the shape of the temporal pulse due to the velocity fields attached to a relativistic electron bunch with that of a "half-cycle" freely propagating THz pulse. The velocity field pulse is a *unipolar* pulse while the freely propagating THz pulse also has a negative component required for wave propagation. The latter is however weak and extends for a long time. The Fourier transforms of the two pulses are shown in Fig. 2.10b. The frequency spectrum of the e-beam pulse peaks at zero frequency, while that of the "half-cycle" pulse looks similar but has zero field amplitude at zero frequency.

Usually, freely propagating "half-cycle" THz pulses are produced by laser excitation of matter [31–34]. Coherent THz radiation is also emitted by a bending magnet of a storage ring, referred to as coherent synchrotron radiation [35–38]. Higher THz frequencies may be achieved by insertion of a thin metal foil (e.g. Be) directly into a linac beam with electron bunch lengths as short as 100 fs. The foil strips off THz fields which can then freely propagate as coherent transition radiation that may be focused.

[7] Care has to be exercised in distinguishing the *cycle frequency* ν (units Hz) which is the inverse of the time that it takes a system to evolve through a complete cycle of $360° = 2\pi$ from the larger *angular frequency* $\omega = 2\pi\nu$, corresponding to the shorter time it takes a system to evolve through part of a cycle defined as $1\,\text{rad} = 360°/2\pi = 57.3°$.

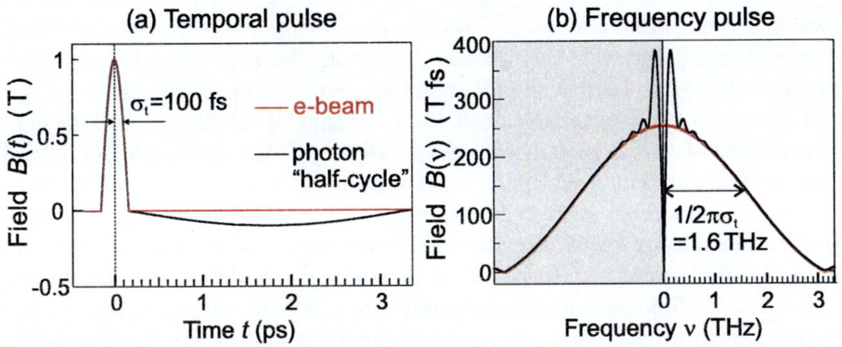

Fig. 2.10 **a** Comparison of temporal shapes of a velocity field pulse produced by a relativistic electron beam (rms width 100 fs) shown in red and the corresponding freely propagating "half-cycle" THz pulse (black) [31]. The latter also has a negative weak cycle extending to longer times, required for wave propagation. **b** Frequency spectra of the two pulses in (**a**). Only positive frequencies are allowed as indicated by shading. The frequency pulses are nearly identical except that the e-beam pulse has a finite DC component at zero frequency, while the photon pulse has no DC component

For example, using 70 fs LCLS electron bunches, a focused THz field strength of 4.4 GV/m has been achieved [39]. This is comparable to the velocity field of an electron beam at a distance of 100 μm, as shown in Fig. 2.8 [21].

2.5 Weizsäcker-Williams Method: Virtual Photon Spectrum

The right side of Fig. 2.3 shows that for a single charge moving with constant relativistic velocity, both the **E**- and **B**-fields are attached to and move with the charge. The fields do not separate from the charge and hence do not "radiate" away from it [1, 12]. For an ultra-relativistic charge, the electric and magnetic fields are perpendicular to the direction of charge motion **v**, so that $\mathbf{E} \times \mathbf{B} \propto \mathbf{v}$, with $v \simeq c$. The two fields are also perpendicular to each other, and they obey the relationship $|\mathbf{E}| = c|\mathbf{B}|$, just like an electromagnetic wave.[8]

When such an electron beam traverses a sample or passes by a surface at a small distance of order of microns, the velocity fields act on the sample just like an EM wave. The close resemblance between relativistic velocity fields and EM waves was first noticed by Fermi in 1924 [40] and in the mid-1930s was developed independently by Weizsäcker [10] and Williams [11] into a powerful method, now called the *Weizsäcker-Williams* (WW) method of virtual photons [12, 41, 42].[9] By use of this formalism one can calculate the scattering and absorption of the virtual photons,

---

[8] Strictly, $|\mathbf{E}| = c|\mathbf{B}|$ is only true for photons, but for highly relativistic electrons it is a good approximation.

[9] I would like to thank Max Zolotorev (LBNL) for discussions of the WW method.

taken to be real photons, by atoms. This was convenient in the early days [40] since photon cross sections were better known. Today they are conveniently tabulated by Henke et al. [43], Hubbell et al. [44] or can be downloaded from the web [45].

The central idea is that the velocity fields accompanying the charge may be thought of as a cloud of virtual photons carried by the charge. This cloud may be liberated through interactions of the charge. For example, in materials the fields and electrons may move at different speeds so that the fields are "stripped off". The photon cloud may also be exchanged between particles. For example, the electromagnetic repulsion between two electrons may be viewed as an exchange or transfer of virtual photons between them. Similarly, when an electron interacts with a nucleon, it is not the electron which actually "probes" the nucleon but the mediating virtual photons.[10] The WW method already provided a glimpse of virtual photons which through the development of QED in the late 1940s emerged as the carriers of the electromagnetic force.

Prominent examples of virtual photon exchange are [12, 42, 46, 47]

- Virtual photon conversion into synchrotron radiation upon interaction with a magnetic field.
- Virtual photon conversion into excited atomic electrons via photoelectric excitation, known as collisional ionization of atoms.
- Virtual photon conversion into bremsstrahlung radiation by scattering on the nuclear Coulomb field.
- Virtual photon conversion into transition radiation upon deceleration at a dielectric boundary.
- Virtual photon conversion into Čerenkov radiation when the particle outruns the attached fields.
- Virtual photon conversion into real photons during positron-electron annihilation.

> Each relativistic charged particle carries with it a *virtual photon spectrum*. Through interaction, the particle may radiate an EM spectrum that closely follows the virtual spectrum.

2.5.1 Virtual Photon Spectrum of a Gaussian Electron Bunch

In practice we are interested in the virtual photon spectrum of a Gaussian bunch of electrons, which corresponds to the velocity fields discussed in Sect. 2.4.3. We have seen that the distance and time-dependent velocity fields of a Gaussian bunch $E(t, R)$ always contain a *coherent* component whose intensity scales with the square

[10] The larger the energy and momentum transfer to the mediating virtual photons, the shorter their wavelength. Since only very short wavelength photons can "see" the constituents of the nucleon, the quarks, larger and larger accelerators are needed to see more and more details.

of both the number of charges and the fields of the individual electrons, $I \propto N_e^2 E_e^2$. Inside the beam radius there is an additional *incoherent* intensity which scales only linearly with the number of electrons, $I \propto N_e E_e^2$.

Classically, the EM energy is stored in equal parts in the **E**- and **B**-fields, and the energy density is given by [12],

$$\frac{d\mathcal{E}}{dV} = \frac{1}{2}\left(\epsilon_0 |E|^2 + \frac{|B|^2}{\mu_0}\right) = \epsilon_0 |E|^2 = \frac{1}{\mu_0}|B|^2 \tag{2.28}$$

On the right we have indicated that the energy may be expressed in terms of the E or B fields alone since they are related according to $|E| = c|B| = |B|/\sqrt{\epsilon_0\mu_0}$. In the following we shall consider the EM energy as expressed by the E-field expression.

2.5.2 *Coherent Virtual Spectrum: THz Photons*

The coherent part of the virtual photon intensity arises from THz fields as discussed qualitatively in Sect. 2.4.3.1. The Weizsäcker-Williams method first converts the time-dependent coherent fields $E(t)$ into frequency dependent fields $E(\omega)$ through a Fourier transform. For a Gaussian bunch, the coherent temporal E-fields have the form (2.24) or

$$E(R,t) = E_0(R)\, \exp\left[-\frac{t^2}{2\sigma_t{}^2}\right], \tag{2.29}$$

where

$$E_0(R) = -\frac{N_e e}{(2\pi)^{3/2}\,\epsilon_0\, c\, \sigma_t\, R}\left(1 - \exp\left[-\frac{R^2}{2\sigma_R^2}\right]\right). \tag{2.30}$$

The frequency spectrum is given by the Fourier transform (see Appendix A.4)

$$F(R,\omega) = \int_{-\infty}^{\infty} E(R,t)\, e^{-i\omega t}\, dt = \underbrace{\sqrt{2\pi}\,\sigma_t\, E_0(R)\, \exp\left[-\frac{\omega^2}{2(1/\sigma_t)^2}\right]}_{E(R,\omega)}, \tag{2.31}$$

where $E(R,t)$ and $E(R,\omega)$ have the same dimension. The transform is again a Gaussian with an rms width $\sigma_\omega = 1/\sigma_t$ and an amplitude $E_0(R)\sqrt{2\pi}\,\sigma_t$. When the square of the absolute values of the fields is integrated over time and frequency, respectively, the corresponding energies are conserved according to Parseval's theorem (see Appendix A.4), so that

$$\int_{-\infty}^{+\infty} |E(R,t)|^2\, dt = \sigma_t^2 \int_{-\infty}^{+\infty} |E(R,\omega)|^2\, d\omega = |E_0(R)|^2 \sqrt{\pi}\sigma_t. \tag{2.32}$$

Using this result, we can express the energy contained in the EM field through an integration of (2.28) that involves either the temporal fields $\epsilon_0|E(R,t)|^2$ and the volume element $\mathrm{d}V = \mathrm{d}A\, c\mathrm{d}t$ or the frequency fields $\epsilon_0|E(R,\omega)|^2$ and the element $\mathrm{d}V = \mathrm{d}A\, c\sigma_t^2\mathrm{d}\omega$, where $\mathrm{d}A$ is an area element in the plane perpendicular to the direction of charge motion. Hence in the frequency domain we have

$$\int_A \int_{-\infty}^{+\infty} \frac{\mathrm{d}\mathcal{E}}{\mathrm{d}V}\, \mathrm{d}A\, c\sigma_t^2 \mathrm{d}\omega = \int_A \int_{-\infty}^{+\infty} \epsilon_0 |E(R,\omega)|^2\, \mathrm{d}A\, c\sigma_t^2 \mathrm{d}\omega$$
$$= \int_A \int_{-\infty}^{+\infty} \frac{\mathrm{d}N_{\mathrm{ph}}\, \hbar\omega}{\mathrm{d}V}\, \mathrm{d}A\, c\sigma_t^2 \mathrm{d}\omega. \tag{2.33}$$

In the last line we have furthermore equated the EM energy density to that contained in N_{ph} virtual photons of energy $\hbar\omega$ in $\mathrm{d}V$.

By rewriting $|E_0(R)|^2$ defined in (2.30) in terms of the fine structure constant $\alpha_{\mathrm{f}} = e^2/(4\pi\epsilon_0\hbar c) \simeq 1/137$ we obtain,

$$|E_0(R)|^2 = \frac{N_{\mathrm{e}}^2\, \hbar\, \alpha_{\mathrm{f}}}{2\pi^2\, \epsilon_0\, c\, \sigma_t^2\, R^2} \left(1 - \exp\left[-\frac{R^2}{2\sigma_R^2}\right]\right)^2. \tag{2.34}$$

By use of this relation, comparison of the last two expressions in (2.33) yields

$$\frac{\mathrm{d}^2 N_{\mathrm{ph}}}{\mathrm{d}A\, \mathrm{d}\omega/\omega} = \frac{N_{\mathrm{e}}^2\, \alpha_{\mathrm{f}}}{2\pi^2\, \sigma_R^2\, (r^*)^2} \left[1 - e^{-(r^*)^2/2}\right]^2 e^{-\omega^2\sigma_t^2}, \tag{2.35}$$

where we have defined $r^* = R/\sigma_R$.

When (2.35) is integrated over the plane perpendicular to the moving charge (element $\mathrm{d}A$), it does not converge similar to the energy of a point charge. We may, however, calculate the energy contained in a finite circular area of radius R by integrating (2.35) in cylindrical coordinates according to

$$2\pi \int_0^R \frac{\mathrm{d}N_{\mathrm{ph}}}{\mathrm{d}A \mathrm{d}\omega/\omega}\, r\, \mathrm{d}r = \frac{N_{\mathrm{e}}^2\, \alpha_{\mathrm{f}}}{\pi}\, e^{-\omega^2\sigma_t^2} \underbrace{\int_0^{R/\sigma_R} \frac{1}{r'} \left[1 - e^{-(r')^2/2}\right]^2 \mathrm{d}r'}_{\mathcal{I}\left(\frac{R}{\sigma_R}\right)}, \tag{2.36}$$

where $r' = r/\sigma_R$. We obtain for the number of virtual THz photons per unit bandpass in the area πR^2,

$$\boxed{\text{Coherent virtual photons}: \quad \frac{\mathrm{d}N_{\mathrm{ph}}}{\mathrm{d}\omega/\omega} = \frac{N_{\mathrm{e}}^2\, \alpha_{\mathrm{f}}}{\pi}\, e^{-\omega^2\sigma_t^2}\, \mathcal{I}\left(\frac{R}{\sigma_R}\right)} \tag{2.37}$$

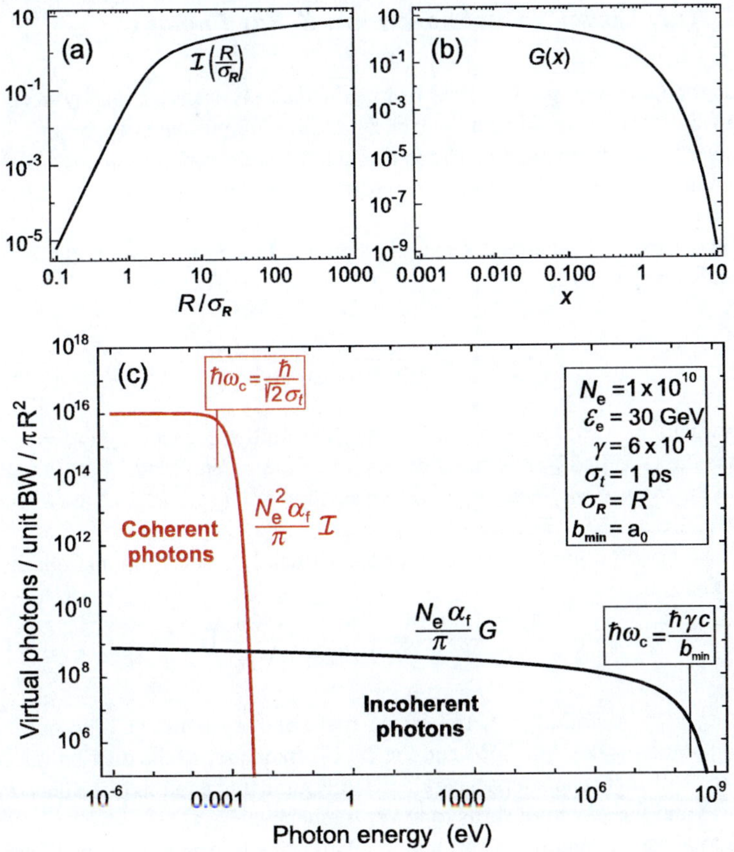

Fig. 2.11 **a** Function $\mathcal{I}(R/\sigma_R)$ defined in (2.36). **b** Function $G(x)$ defined in (2.43). **c** Virtual photon spectrum of a 30 GeV Gaussian electron beam ($\gamma = 6 \times 10^4$) of $N_e = 10^{10}$ electrons with temporal rms length σ_t and lateral rms width σ_R. The number of virtual photons is assumed to be contained within the lateral beam radius $R = \sigma_R$. The coherent spectrum is given by (2.37), and the cut off energy is 4.65×10^{-4} eV. The incoherent spectrum is given by (2.42), where we have assumed a minimum impact parameter $b_{\min} = a_0 = 0.0529$ nm, leading to a cut off energy of 2.23962×10^8 eV

The integral $\mathcal{I}(R/\sigma_R)$ is plotted in Fig. 2.11a, and for a circular area with radius $R = \sigma_R$ we have $\mathcal{I} = 4.55 \times 10^{-2}$. The Gaussian spectrum for this case is plotted in Fig. 2.11c as a red curve for the listed beam parameters. The spectrum has a cut off energy

$$\hbar\omega_c = \frac{\hbar}{\sqrt{2}\,\sigma_t} \tag{2.38}$$

which is also indicated in the figure. For the assumed rms pulse length of $\sigma_t = 1$ ps the cut off energy is 4.65×10^{-4} eV (0.11 THz).

2.5.3 Incoherent Virtual Spectrum: X-Ray Photons

Incoherent field superposition arises from *individual* electrons passing by an observer at ultrashort distances as discussed in Sect. 2.4.3.2, approximated by an "impact parameter", usually denoted b. The virtual photon spectrum as a function of energy is again obtained from the Fourier transform of the temporal field pulse of a *single electron* given by (2.18). The Fourier transform operation is valid because the field of a single electron is "coherent". Using the notation $b = R$, the field in the frequency domain is given by [12, 41]

$$E(b,\omega) = -\frac{e}{(2\pi)^{3/2}\epsilon_0\, c\, \sigma_t\, b}\, f\left(\frac{b\omega}{\gamma c}\right). \tag{2.39}$$

The prefactor is the same as in (2.30), and the unitless function $f(x) = x\, K_1(x)$ contains the modified Bessel function $K_1(x)$ of the second kind. $f(x)$ is unity for small x, i.e. for $\omega < \gamma c/b$, and goes to zero asymptotically as xe^{-x}, i.e. for $\omega > \gamma c/b$ [12]. We also have $\int_0^\infty f(x)\,\mathrm{d}x = \pi/2$.

We can now follow the derivation of the virtual THz spectrum and obtain for a *single* electron

$$\frac{\mathrm{d}^2 N_{\mathrm{ph}}(b,\omega)}{\mathrm{d}A\, \frac{\mathrm{d}\omega}{\omega}} = \frac{\alpha_{\mathrm{f}}}{2\pi^2\, b^2}\, f^2\left(\frac{b\omega}{\gamma c}\right), \tag{2.40}$$

where $\alpha_{\mathrm{f}} = e^2/(4\pi\epsilon_0\hbar c) \simeq 1/137$ is the fine structure constant. The number of pseudo photons scales with $1/b^2$ and the cut off frequency of the distribution scales with $1/b$. With increasing impact parameter, the number of photons therefore strongly decreases and the spectrum shifts to lower frequency.

The total pseudo photon spectrum of a relativistic electron is given by integration of (2.40) over impact parameter b

$$2\pi \int_{\infty}^{b_{\mathrm{min}}} \frac{\mathrm{d}^2 N_{\mathrm{ph}}(\omega)}{\mathrm{d}A\, \frac{\mathrm{d}\omega}{\omega}}\, b\,\mathrm{d}b = \frac{\alpha_{\mathrm{f}}}{\pi} \int_{\infty}^{b_{\mathrm{min}}} \frac{1}{b}\, f^2\left(\frac{b\omega}{\gamma c}\right) \mathrm{d}b. \tag{2.41}$$

Each electron therefore contributes independently and the total incoherent spectrum scales with the number of electrons in the bunch, N_{e}. The total incoherent *virtual photon spectrum* is obtained as

$$\boxed{\text{Incoherent virtual photons}: \quad \frac{\mathrm{d}N_{\mathrm{ph}}(\omega)}{\frac{\mathrm{d}\omega}{\omega}} = \frac{N_{\mathrm{e}}\,\alpha_{\mathrm{f}}}{\pi}\, G\left(\frac{b_{\mathrm{min}}\omega}{\gamma c}\right)}. \tag{2.42}$$

The function

$$G(x) = \frac{1}{2}\left\{[x\, K_0(x)]^2 + 2x K_0(x) K_1(x) - [x\, K_1(x)]^2\right\} \tag{2.43}$$

where $K_0(x)$ and $K_1(x)$ are modified Bessel functions of the second kind, is plotted in Fig. 2.11b. The cut off frequency ω_c expressed as a photon energy is given by

$$\hbar\omega_c \simeq \frac{\hbar\gamma c}{b_{\min}}. \tag{2.44}$$

It is seen to shift to higher energy with increasing γ and $1/b_{\min}$, as expected. The incoherent spectrum is plotted in Fig. 2.11c as a black curve, and the cut off energy is indicated.

In laboratory x-ray generators, non-relativistic electron beams are used to create bremsstrahlung x-rays, and the spectrum is well known to extend to the electron energy of the incident beam [48]. If we assume that the highest virtual photon energy is also determined by the beam energy $\mathcal{E}_e = \gamma m_e c^2$, we obtain $b_{\min} = \hbar/m_e c = \lambda_C/(2\pi) = 3.87 \times 10^{-4}$ nm, where $\lambda_C = 2.43 \times 10^{-3}$ nm is the *Compton wavelength*. For the incoherent spectrum shown in Fig. 2.11 we have assumed $b_{\min} \simeq a_0$, where $a_0 = 0.0529$ nm is the Bohr radius.

The large coherent intensity is due to the fact that the wavelength exceeds the longitudinal beam size of 300 μm (1 ps), and thus the virtual photons associated with the different electrons are in phase and add coherently. As a consequence, the number of coherent virtual photons is considerably larger than the corresponding number of incoherent photons in the same beam area. As shown by the integral $\mathcal{I}$ plotted in Fig. 2.11a, the coherent photon cloud is largely contained in an area approximately 10 times the beam size. In contrast, the incoherent intensity falls off as $1/R^2$ with the distance R from the electron positions and in practice is negligible outside the beam area. The cut off energy of the coherent spectrum increases for shorter electron bunch lengths σ_t according to $1/(\sqrt{2}\sigma_t)$, and it does not explicitly depend on the beam energy $\mathcal{E}_e$. In contrast, the cut off of the incoherent spectrum is largely determined by $\mathcal{E}_e$ (or γ) and the spectrum does not depend on the electron bunch length σ_t.

In the following section we shall discuss how the cloud of virtual photons can be spun off the charge and turned into true photons.

2.6 Acceleration Fields: Creation of EM Radiation

As discussed in Sect. 2.3, the Liénard–Wiechert fields associated with a moving charge come in two categories, *velocity fields* that remain attached to the charge and *acceleration fields* which separate from the charge as electromagnetic radiation.[11]

Today, one of the most important applications of the Liénard–Wiechert equation (2.8) is the description of "synchrotron radiation", which has evolved from a "waste product" and nuisance in high energy physics to its deliberate optimization

[11] The electric fields and radiation patterns associated with a charge in arbitrary motion can be visualized by the real-time radiation simulator written by Shintake [49].

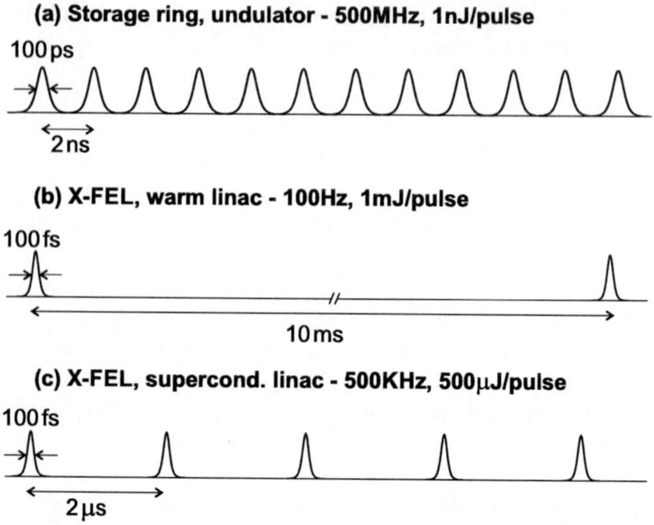

Fig. 2.12 Typical electron bunch lengths and repetition rates in storage rings and XFELs. The shown electron bunch structure is also that of the emitted x-ray pulses. The listed energy is that contained in the x-ray pulses

for research around the world. In practice, one does no longer use a synchrotron[12] but an electron storage ring which provides better beam stability. Synchrotron radiation is created by electrons or positrons of constant energy (a few GeV) and velocity that circle around the storage ring. The electrons are propelled forward by a microwave radiofrequency (rf) field that consists of "rf-buckets" that rotate around the ring. When the buckets are filled with many electrons, we speak of an "electron bunch". Typical rf frequencies are 500 Mz, corresponding to a temporal separation between buckets or bunches of 2 ns, as illustrated in Fig. 2.12.

In the stationary frame of an observer, *storage ring* bunches have a typical length of 30 mm, corresponding to a pulse length of 100 ps, a lateral cross section of about 100 μm, and one bunch contains of order 1 nC of charge (1.6 nC=10^{10} electrons). The total energy per x-ray pulse is about 1 nJ/pulse (see Fig. 2.12). The bunches are kept on the desired horizontal orbit by vertical magnetic fields produced at locations around the ring by dipole electromagnets. Other magnets, such as quadrupoles, are used to focus the bunches and sextupoles to correct for aberrations.

Bunches in *XFELs* are produced in a laser photocathode gun and are then accelerated and compressed in several linear accelerator sections. 100 fs long bunches may contain a similar number of electrons (10^{10}) as in storage rings with the number proportionately reduced for shorter bunches. After being sent through a long undulator consisting of permanent magnets that wiggle the bunch to produce radiation,

[12] In a synchrotron, the charged particles spiral out from a center, rather than circle at a constant radius.

the bunch is discarded in a beam dump. The length of the emitted photon pulses is approximately the same (within a factor of 2) as the length of the electron bunches. Differences may arise from the fact that the SASE process may only involve a fraction of the bunch, as discussed below. The pulse repetition rate depends on the type of linac used. As illustrated in Fig. 2.12, so-called warm linacs have typical rep rates of 100 Hz and superconducting linacs about 500 kHz. We shall see below that the characteristics of synchrotron and XFEL radiation are also significantly different.

In the following we shall discuss the creation of EM radiation in storage rings and XFELs and its properties.

2.6.1 Distortion of Field Lines: Radiation

The principle of inertia states that matter continues in its existing state of rest or uniform motion in a straight line, unless that state is changed by an external force. We therefore refer to reference frames that are in a state of constant, rectilinear motion with respect to one another as *inertial* frames. In inertial frames, the electric field lines emanating out from a charged particle to infinity are also straight. With increasing velocity, the electric field lines in the frame of a stationary observer become squished along the propagation direction, as shown in Fig. 2.3, but they are still straight and radial, as illustrated in Fig. 2.13a.

When a particle is accelerated, it moves relative to an inertial system and a distortion of the electric field lines occurs as shown in Fig. 2.13b. Here we have assumed that a particle is accelerated from a time t_0 to the time t_1 with an acceleration $\mathbf{a}$. We picture the event in the inertial frame of the particle associated with its uniform motion ($v = \text{const.}$) prior to t_0. The fact that the particle has accelerated from t_0 to t_1 is known only within a limited area—the *event horizon*—since the "signal of acceleration" travels away from the particle source at the finite speed of light c. Therefore, field lines outside the shaded radius $c(t_1 - t_0)$ around the position of the particle at time t_0 are the same as before. They still point to the original location of the particle at t_0. Within the event horizon, the field lines emerging from the particle are bent and join the old straight field lines at the event horizon. The bent field lines within the event horizon have a non-radial component along the direction of acceleration $\mathbf{a}$. It is this field distortion, with an $\mathbf{E}$ component parallel to $\mathbf{a}$ and traveling away from the particle with the speed of light, that we call *radiation*.

> *Electromagnetic radiation* is the *distortion of the electric field* $\mathbf{E}$ that is created parallel to the acceleration direction $\mathbf{a}$ of a charged particle. It moves away from the particle with the speed of light and the associated fields fall off with the distance r from the particle as $1/r$.

The calculation of the emitted acceleration fields, i.e. the synchrotron radiation spectrum, starts from considering the fields due to the motion of a single charge, which is described by the Liénard–Wiechert equation (2.8) [2]. The effect of many

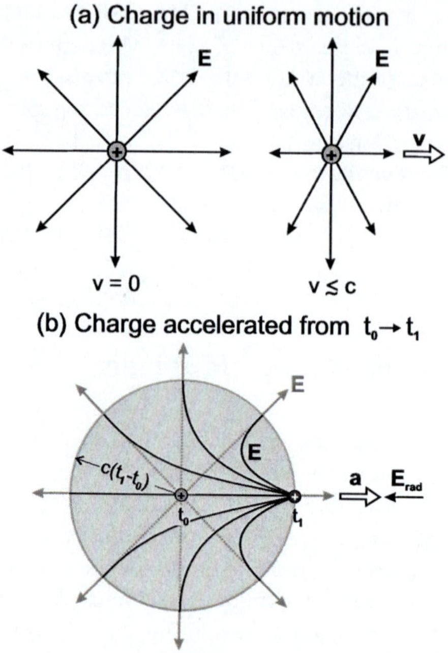

Fig. 2.13 **a** Electric fields **E** seen by an observer at rest that originate from a charge that is at rest or in uniform motion $v = const.$ (also see Fig. 2.3). In **b** we picture a charge in an inertial reference system ($v = const.$) that is accelerated from t_0 to t_1 along a direction **a**. The acceleration causes a distortion of the field lines within an "event horizon" shown in gray shading, given by the distance $c(t_1 - t_0)$ around the charge position at time t_0. Outside the event horizon the field lines still resemble those at time t_0. Continuity of the field lines at the event horizon determines the form of the bent field lines inside the horizon (adapted from Wiedemann [1]). The direction of the radiated field $\mathbf{E}_{\text{rad}}$ is also indicated

charges, constituting an electron bunch, is only considered later and leads to enhanced radiation and its picosecond pulse structure. For the calculation of the complete angular and time (or frequency) dependence of the fields we refer the reader to the books by Hofmann and Wiedemann [1, 2]. In the following we shall touch on the most important aspects.

2.6.2 The Angular Spectrum of a Single Accelerated Charge

In the calculation of the *angular radiation pattern* we can fortunately ignore the time dependence of the fields and therefore do not have to worry about retardation effects. We simply evaluate the fields of a point charge at the emission (retarded) time and give them at a fixed time later, without their time dependence. Obviously, our approach will not give information on the emitted frequency spectrum of the emitted radiation. We can then drop the asterisk superscripts in (2.8). By evaluating the cross products we obtain the radiation field **E** created by an electron ($q = -e$) observed in the direction $\mathbf{n} = \mathbf{r}/r$ as

$$\mathbf{E}(r) = -\frac{e}{4\pi\epsilon_0}\frac{[\mathbf{n}-\boldsymbol{\beta}](\mathbf{n}\cdot\mathbf{a}) - \mathbf{a}(1-\mathbf{n}\cdot\boldsymbol{\beta})}{c^2 r\,(1-\mathbf{n}\cdot\boldsymbol{\beta})^3}\,, \tag{2.45}$$

where r is the distance between the electron and the observer located in the direction of the unit vector **n**. Since $\mathbf{E}\cdot\mathbf{n}=0$, the E-field is always perpendicular to **n**.

The radiation pattern is given by the radiated x-ray intensity which is given by the Poynting vector

$$\mathbf{S} = \mathbf{E}\times\mathbf{H} = \frac{1}{\mu_0}\mathbf{E}\times\mathbf{B} \tag{2.46}$$

and because of (2.10), **S** can be written in terms of the square of the electric field

$$\mathbf{S} = \sqrt{\frac{\epsilon_0}{\mu_0}}\,|\mathbf{E}|^2\mathbf{n} = c\epsilon_0\,|\mathbf{E}|^2\mathbf{n}\,. \tag{2.47}$$

By means of (2.45) we obtain for our case of $\mathbf{v}\perp\mathbf{a}$,

$$\mathbf{S} = \frac{e^2}{16\pi^2\epsilon_0}\frac{\mathbf{a}^2(1-\mathbf{n}\cdot\boldsymbol{\beta})^2 - (\mathbf{n}\cdot\mathbf{a})^2(1-\boldsymbol{\beta}^2)}{c^3 r^2\,(1-\mathbf{n}\cdot\boldsymbol{\beta})^6}\,\mathbf{n}. \tag{2.48}$$

The Poynting vector gives the *emitted intensity* or energy/(time×area) into the direction **n**. The angular dependence of the *observed power* with dimension [energy/(time×solid angle)] is given by [2]

$$\frac{d\mathcal{P}}{d\Omega} = r^2|\mathbf{S}|\,(1-\mathbf{n}\cdot\boldsymbol{\beta}) = \frac{e^2}{16\pi^2\epsilon_0 c^3}\frac{\mathbf{a}^2(1-\mathbf{n}\cdot\boldsymbol{\beta})^2 - (\mathbf{n}\cdot\mathbf{a})^2(1-\boldsymbol{\beta}^2)}{(1-\mathbf{n}\cdot\boldsymbol{\beta})^5}\,. \tag{2.49}$$

Here the factor $1-\mathbf{n}\cdot\boldsymbol{\beta} = \mathrm{d}t/\mathrm{d}t^*$ accounts for the different time increments at the retarded time of emission t^* and the observation time t. The solid angle $\mathrm{d}\Omega$ has units [steradian].[13]

Following Fig. 2.14a, we can write the *power* emitted as a function of the angle θ between the observation direction **n** and the velocity **v** and the angle ϕ of **n** with respect to the orbit plane as

$$\frac{d\mathcal{P}}{d\Omega} = \frac{e^2a^2}{16\pi^2\epsilon_0 c^3}\frac{(1-\beta\cos\theta)^2 - (1-\beta^2)\sin^2\theta\,\cos^2\phi}{(1-\beta\cos\theta)^5}\,. \tag{2.50}$$

[13] Since the solid angle unit [steradian] and the plane angle unit [radian] are defined as ratios of two areas or lengths, they are actually dimensionless. The names are introduced to distinguish between dimensionless quantities of different nature.

In the ultrarelativistic limit we can simplify (2.50) by use of $\beta \simeq 1 - 1/(2\gamma^2)$, $\beta^2 = 1 - 1/\gamma^2$ and approximating the trigonometric functions to second order in the argument, i.e. $\cos\theta \simeq 1 - \theta^2/2$ and $\sin\theta \simeq \theta$, and use of $2\cos^2\phi - 1 = \cos(2\phi)$. We obtain

$$\frac{\mathrm{d}\mathcal{P}}{\mathrm{d}\Omega} = \underbrace{\frac{e^2 a^2 \gamma^6}{2\pi^2 \epsilon_0 c^3}}_{\mathcal{P}_0} \frac{1 + \gamma^4\theta^4 - 2\gamma^2\theta^2\cos(2\phi)}{\left(1 + \gamma^2\theta^2\right)^5}. \tag{2.51}$$

Integration of (2.51) over solid angle gives the total observed power

$$\mathcal{P}_{\mathrm{tot}} = \int_0^{2\pi}\int_0^{\infty} \frac{\mathrm{d}\mathcal{P}}{\mathrm{d}\Omega}\,\theta\,\mathrm{d}\theta\,\mathrm{d}\phi = \frac{\pi}{3\gamma^2}\mathcal{P}_0 = \frac{e^2 a^2 \gamma^4}{6\pi\epsilon_0 c^3}. \tag{2.52}$$

The integrated power can also be written in terms of the radius ρ of the electron orbit which is related to the acceleration a caused by the magnetic "field" B of the bending magnet (see (2.61)) according to

$$\rho = \frac{m_e c\,\beta\,\gamma}{e\,B}, \quad \text{and} \quad a = \frac{\beta^2 c^2}{\rho}. \tag{2.53}$$

We obtain

$$\mathcal{P}_{\mathrm{tot}} = \frac{e^2}{4\pi\epsilon_0}\frac{2c\beta^4\gamma^4}{3\rho^2} = \underbrace{\frac{e^2}{4\pi\epsilon_0 m_e c^2}}_{r_0}\frac{2c^3\beta^2\gamma^2 e^2 B^2}{3m_e c^2}, \tag{2.54}$$

where $r_0 = 2.818 \times 10^{-15}$ m is the classical electron radius.

The normalized power distribution $(\mathrm{d}\mathcal{P}/\mathrm{d}\Omega)/\mathcal{P}_0$ is plotted in Fig. 2.14b for three values of ϕ. The case $\phi = 45°$ shown in red is also the ϕ-averaged (cylindrically symmetric) distribution around $\theta = 0$ into which a single electron emits coherently. The red distribution has the same integrated power $\mathcal{P}_0\pi/(3\gamma^2)$ as the flat-top distribution shown in dashed red, so that we can define an effective 2D solid angle that contains the total power as

$$\mathrm{d}\Omega = \mathrm{d}\Omega_{\mathrm{coh}} = \frac{\pi}{3\gamma^2} \simeq \frac{1}{\gamma^2}. \tag{2.55}$$

The radiation pattern observed in the frame of the electron ($v = \beta = 0$) obtained from (2.50) is the well-known *dipole radiation pattern*. It has a node along $\mathbf{a}$ ($\theta = 90°$) and is constant for $\theta = 0$. The synchrotron radiation and dipole patterns are compared in Fig. 2.15.

We can now compare the angular and temporal field patterns associated with a single electron that moves with a constant relativistic velocity as discussed in Sect. 2.4.1 and one that is radially accelerated, as shown in Fig. 2.16. In the absence

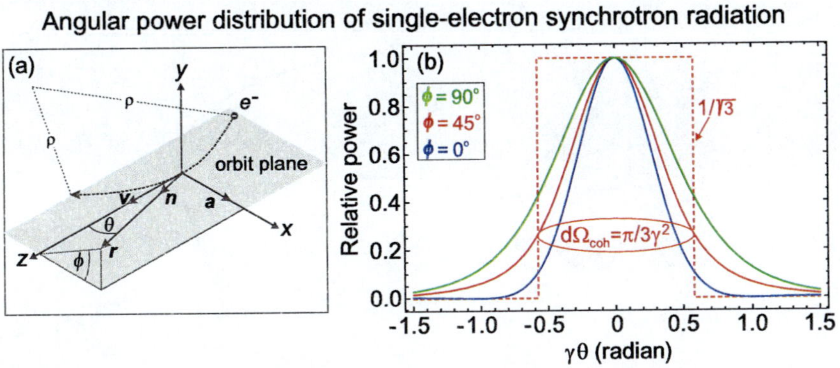

Fig. 2.14 **a** Geometry of a single electron on an orbit of radius ρ emitting synchrotron radiation into the observer direction $\mathbf{n} = \mathbf{r}/r$, at an angle θ from the tangential velocity direction $\mathbf{v}\|z$ and an azimuthal angle ϕ from the acceleration direction $\mathbf{a}\|x$. **b** Angular pattern of the observed power $(d\mathcal{P}/d\Omega)/\mathcal{P}_0$, according to (2.51), as a function of $\gamma\theta$ for three values of ϕ. The symmetric case $\phi = 45°$ shown in red corresponds to a 2D solid angle $d\Omega = \pi/(3\gamma^2)$ as indicated, and the integrated power under the solid red curve is the same as that under the flat-top dashed red curve

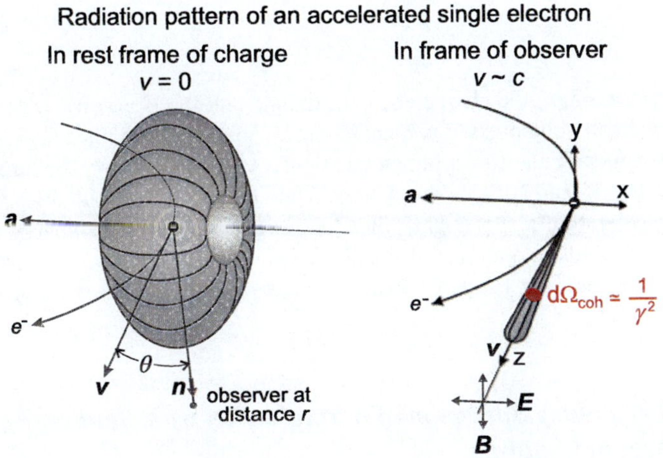

Fig. 2.15 Illustration of the radiation pattern emitted by a single electron moving on a circular orbit with a tangential velocity $\mathbf{v}$ and an acceleration $\mathbf{a}$ pointing toward the center of the orbit. On the left the radiation pattern is pictured in the rest frame of the charge ($v = 0$). It has the shape of the donut centered about the axis defined by the acceleration direction $\mathbf{a}$. On the right we show the characteristic pattern of synchrotron radiation when $v \sim c$. The solid angle containing most of the radiation is about $1/\gamma^2$ as discussed in the text

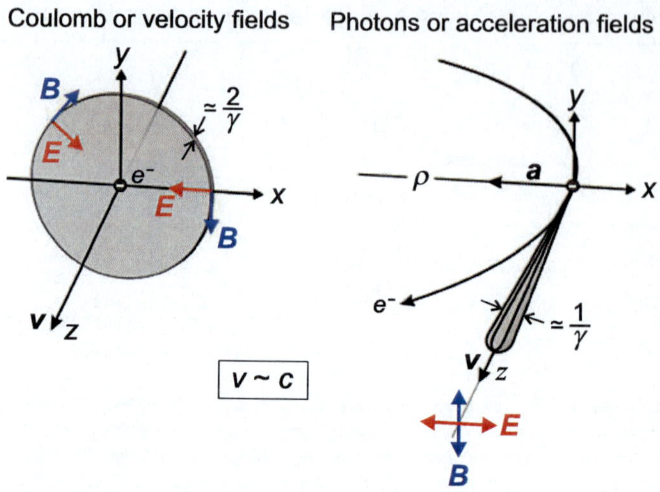

Fig. 2.16 Angular patterns containing the field intensities (fields squared) associated with a point charge that moves with a constant velocity $v \simeq c$, either on a straight line (left) or on an circle of radius ρ (right)

of an acceleration, **a**, the charge does not radiate and the **E** and **B** Coulomb fields are relativistically compressed perpendicular to **v** and the field intensity is confined to a sheet perpendicular to **v** with a total angular width of $\simeq 2/\gamma$. The directions of the concentric **E**- and **B**-fields are shown. When the electron is radially accelerated with an acceleration $\mathbf{a} \perp \mathbf{v}$ (right), the charge radiates part of its energy stored in the Coulomb fields into a cone of opening angle $\simeq 1/\gamma$ about **v**, as shown. In the orbit plane the emitted **E**- and **B**-fields are linearly polarized in the x–y plane, as indicated.

2.6.3 *Frequency Spectrum of a Single-Electron Bending Magnet Source*

We assume that an observer is located at the end of a "beam line" which is built out in a tangential direction of the storage ring. We first consider the case of a *single* charge and leave the discussion of the effect of a bunch of electrons for later. As the electron travels around the ring it will only radiate down the beam line into the direction of the observer when its radiation cone sweeps by. We shall assume that the direction of the radiation to the observer is well defined by an aperture as shown schematically in Fig. 2.17.

The duration of the x-ray pulse Δt, or "flash" of x-ray intensity, seen by the observer in Fig. 2.17, can be calculated from considering how the radiation cone of FWHM $\simeq 1/\gamma$ sweeps across the aperture in front of the observer. This is not simply

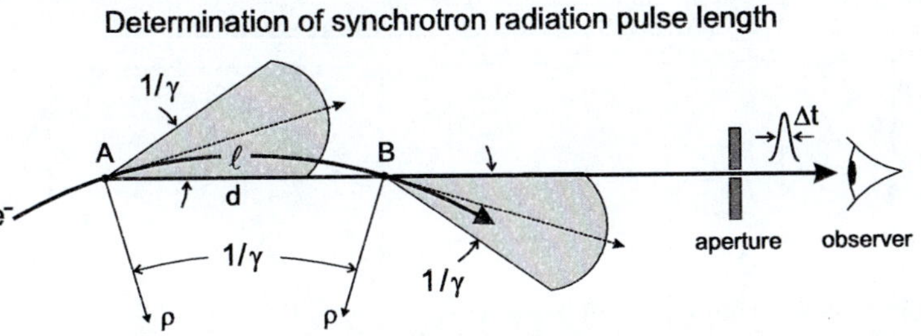

Fig. 2.17 Illustration of the x-ray pulse emitted by a single electron moving on an arc. The x-ray pulse seen by an observer through a small aperture has a temporal width Δt that determines the bandwidth $\Delta\omega$ of the synchrotron radiation spectrum as discussed in the text. The length of the arc from A to B is ℓ, that of the straight line is d, and ρ is the bending radius

the time it takes the electron to travel along the arc of length ℓ from point A to point B, but rather the difference of the time $t_e = \ell/v$ it takes the electron to move across the arc of length ℓ minus the time it takes the light to cover the straight line of length d from point A to point B. We assume that the magnetic field of the bending magnet is strong and take into account the small difference in the lengths of the arc ℓ and straight line d, and obtain

$$\Delta t = \frac{\ell}{v} - \frac{d}{c} = \frac{\rho}{\beta c \gamma} - \frac{\rho \sin\left(\frac{1}{\gamma}\right)}{c}. \tag{2.56}$$

The electron radiates in the direction of the observer over an arc length $\ell = \rho/\gamma$. For example, for a 3 GeV storage ring with $\gamma \simeq 5.9 \times 10^3$ and an arc of radius of ρ= 25m, we obtain $\ell \simeq 4$ mm. We can now evaluate (2.56) for small angles $1/\gamma$ by using the expansion $\sin(1/\gamma) = 1/\gamma - 1/6\gamma^3$ and obtain

$$\Delta t = \frac{2\rho}{3c\gamma^3}. \tag{2.57}$$

Using the same values for γ and ρ we find $\Delta t \simeq 1$ as, so that the radiation flash is extremely short. This time is comparable to the minimum cycle time $t_0 = \lambda/c$ of a propagating electromagnetic wave of x-ray wavelength λ. For example, for 1 keV ($\lambda = 1.2$ nm) x-rays we have $t_0 = 0.4$ as. The time Δt is much shorter than the time $\ell/c \simeq 13$ ps it takes an electron of speed $v \simeq c$ to travel across the $\ell \simeq 4$ mm arc in Fig. 2.17. It is also much shorter than the duration of a light pulse from an entire electron bunch, which is about ~100 ps. As discussed in Sect. 2.4.5 the total pulse length of the electron bunch determines the coherent part of the synchrotron spectrum as shown in Fig. 2.11.

In the frequency domain, the temporal flash of about $\Delta t = 1$ as contains Fourier components whose cut off frequency is determined by the Fourier relationship

$$\Delta t \, \Delta\omega = A, \tag{2.58}$$

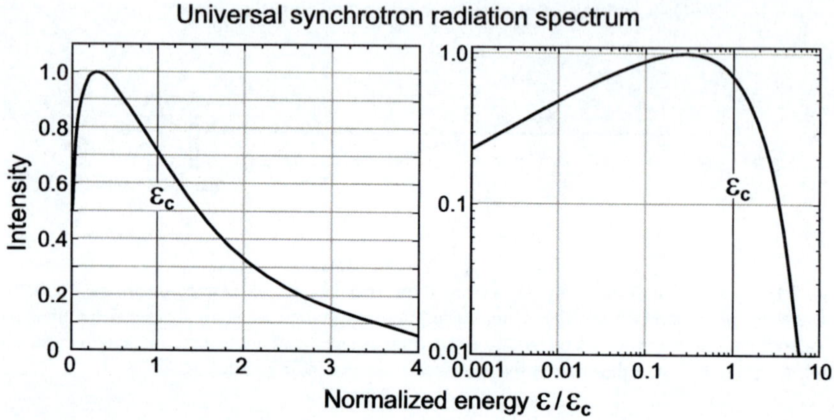

Fig. 2.18 Universal synchrotron radiation spectrum emitted from a bending magnet in an electron storage ring. We have normalized the energy axis to the critical energy $\mathcal{E}_C$ and normalized the peak intensity to 1. On the left, the spectrum is plotted on linear scales, on the right on log–log scales. The critical energy divides the total spectrum into two parts of equal area under the curve

where A depends on the shape of the temporal and frequency distribution functions. Taking $A = 1$ we obtain by use of (2.57) the so-called *critical energy* $\mathcal{E}_C$ of the emitted spectrum, given by

$$\mathcal{E}_C = \hbar \Delta\omega = \frac{\hbar}{\Delta t} = \frac{3\hbar c\, \gamma^3}{2\rho}. \tag{2.59}$$

When the emitted intensity is plotted as a function of energy normalized to $\mathcal{E}_C$, one obtains a universal synchrotron radiation spectrum [1, 2], which is shown in Fig. 2.18 on linear and logarithmic scales.

The critical energy divides the total emitted energy, the number of emitted photons times the photon energy, into two equal parts.

We can rewrite the critical energy $\mathcal{E}_C$ in terms of the magnetic field strength that causes the radial acceleration of the electrons **a** in Fig. 2.15. To do so, we use the expression for the classical *cyclotron frequency* of a non-relativistic electron given by

$$\omega_0 = \frac{e}{m_e} B. \tag{2.60}$$

For $B = 1\,\mathrm{T}$, we have $\omega_0 = 1.76 \times 10^{11}\,\mathrm{rad\,s^{-1}}$, corresponding to a cycle frequency $\nu_0 = \omega_0/2\pi \simeq 28\,\mathrm{GHz}$ and an energy $\hbar\omega_0 \simeq 0.12\,\mathrm{meV}$. For a relativistic electron, the mass in the laboratory frame is enhanced by a factor of γ given by (2.2), and accounting for this enhancement, we obtain the electron bending radius as

$$\rho = \frac{\gamma\, m_e\, v}{e\, B} = \frac{\gamma\, v}{\omega_0}. \tag{2.61}$$

Table 2.1 Typical parameters for a bending magnet source on a state-of-the-art third generation storage ring

Energy (GeV)	γ	B [T]	$\hbar\omega_0$ (μeV)	ρ (m)	ℓ (mm)	$\mathcal{E}_C$ (keV)	Δt (as)
3.0	5880	0.4	47	25	4	2.4	1

Listed are the electron beam energy, the relativistic parameter γ defined in (2.2), the magnetic induction B of a bending magnet, the associated cyclotron energy $\hbar\omega_0$ given by (2.60), the bending magnet radius ρ, the arc length $\ell = \rho/\gamma$, the critical energy $\mathcal{E}_C$ given by (2.62), and the duration of the single electron x-ray flash Δt where 1 as$=10^{-18}$ s

Inserting into (2.58) and using $v \simeq c$, we obtain the critical energy as

$$\boxed{\mathcal{E}_C = \frac{3}{2}\gamma^2\,\hbar\omega_0}. \tag{2.62}$$

We have summarized typical parameters for a bending magnet source on a third generation storage ring in Table 2.1.

As illustrated in Fig. 2.18, the photon energy range is determined by the critical energy given by (2.62) which through (2.58) is determined by the radiation flash $\Delta t \sim 10^{-18}$ s emitted by a single electron.

We have found in (2.62) that the synchrotron spectrum is related to the relativistically corrected fundamental cyclotron frequency ω_0 of the moving charge about the perpendicular bending magnet field. The relativistic correction is a reference frame effect which my be derived in a two-step process. First, the electron is accelerated in a perpendicular magnetic field due to the Lorentz force $\mathbf{F} = m_e\mathbf{a} = -e(\mathbf{v} \times \mathbf{B})$. This acceleration causes the emission of radiation whose frequency is simply the cyclotron frequency (2.60), $\omega_0 = eB/m_e$, where the field is that experienced by the electron in its rest frame. For the electron, the bending magnet field B moves toward it with relativistic speed $\mathbf{v}$ and this requires a correction of the field amplitude by a factor of γ so that in the electron rest frame we have

$$\omega_{el} = \gamma \frac{e}{m_e} B\,. \tag{2.63}$$

Due to the experienced Lorentz force acceleration, the electron radiates an EM wave with the frequency ω_{el} in its rest frame, as illustrated in Fig. 2.13. When observed by a stationary observer, the emitted radiation is subject to a Doppler shift. When the radiating electron moves toward an observer in the laboratory with speed $v = c\beta$, the frequency in the laboratory frame $\boldsymbol{\omega}$ is enhanced according to the relativistic Doppler formula [4]

$$\omega = \omega_{el}\,\gamma\,(1+\beta). \tag{2.64}$$

For a relativistic electron we have $\beta \simeq 1$, and the observer sees an EM wave with frequency

$$\omega = 2\gamma^2 \frac{e}{m_e} B = 2\gamma^2 \omega_0. \tag{2.65}$$

This frequency is approximately the same as the spectral band width given by (2.62). We can therefore make the following statement.

> The emitted frequency of *synchrotron radiation* is determined by the electron cyclotron frequency in the bending magnet field, enhanced by a factor γ due to the relativistic electron motion, and by another factor γ arising from the Doppler shift of the emitted radiation.

The two ways of looking at the origin of the spectrum have the common link that in both pictures the radiation is determined by electron motion across a short arc of length $\ell \simeq \rho/\gamma$, illustrated in Fig. 2.17. In one picture we consider the time associated with emission of radiation from this arc, in the other we calculate the electron precession frequency or cyclotron frequency associated with electron motion across this arc.

2.6.4 *Frequency Spectrum of a Single-Electron Undulator Source*

From the above discussion it is straightforward that one can narrow the spectral bandwidth of the radiation spectrum by increasing the time that the electron radiates into the direction of the observer. However, if one simply lowered the beam energy $\mathcal{E}_e$ and γ, one would lose the x-ray part of the spectrum. In order to produce a narrow frequency spectrum in the x-ray range, one needs to keep the electron beam energy at a few GeV and use the concept of an undulator. Then the radiation time is simply increased by a weakly undulating electron path that keeps the radiation cone pointing into the observer direction.

We assume that the undulator consists of n_U oscillation periods of length λ_U (in practice ~5 cm) for a total undulator length of $L = n_U\lambda_U$, as shown in Fig. 2.19. In practice, the electron bunch has a finite energy spread and angular divergence of the electron trajectories so that the undulator spectrum saturates at a length L of a few meters, and we shall assume this case corresponding to $n_U \sim 100$. We also assume that the weak magnetic field deflection of the electron beam is smaller than the natural opening angle $1/\gamma$ so that we can assume that both electrons and photons travel straight as shown.

Similar to the bending magnet case, we calculate the duration of the radiation flash emitted by an electron moving by the undulator period length λ_U, but because

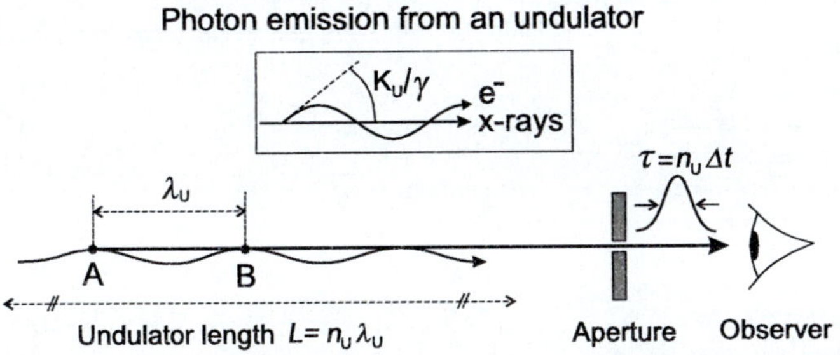

Fig. 2.19 Illustration of radiation from a weak field undulator with period λ_U, where the deflection angle associated with the electron beam trajectory is smaller than the natural opening angle $1/\gamma$. For each period a delay Δt builds up between the faster photons and the electrons which causes a total delay or pulse length $n_U \Delta t$ at the position of the observer. This causes a frequency bandwidth reduction by a factor n_U or an enhancement of the intensity emitted per unit bandwidth. Another way to visualize the enhanced peak intensity is by interference of the fields emitted from the n_U undulator periods

the field is weak we neglect the path difference between the arc and the straight line between points A and B.

$$\Delta t = \frac{\lambda_U}{v} - \frac{\lambda_U}{c} = \frac{\lambda_U}{c}\left(\frac{1-\beta}{\beta}\right) \simeq \frac{\lambda_U}{c}\left(\frac{1}{2\gamma^2}\right) \tag{2.66}$$

Each oscillation period causes the same delay so that the total pulse length seen by the observer is n_U times longer so that we obtain for the total undulator length $L = n_U \lambda_U$

$$\Delta t \simeq \frac{L}{2c\gamma^2} \tag{2.67}$$

The effective pulse length is longer than for the bending magnet case, as shown in Fig. 2.20, where we have assumed that the field is weak so that we can in both cases neglect the difference between the electron arcs and the straight light path.

Due to the longer traveling time of the electron in an undulator than in a bending magnet source, the effective bandwidth is reduced. If we take the bending magnet bandwidth to be the critical energy as shown in Fig. 2.18, we find a typical value $\mathcal{E}_C \simeq 2.5\,\text{keV}$ for a 3 GeV storage ring from Table 2.1. In contrast, for an undular of length L we obtain from (2.67) a bandwidth of

$$\Delta(\hbar\omega) = \frac{\hbar}{\Delta t} \simeq \frac{2c\gamma^2\hbar}{L}. \tag{2.68}$$

Assuming $L = 2\,\text{m}$ we find $\Delta\mathcal{E} \simeq 7\,\text{eV}$ for a typical undulator bandwidth.

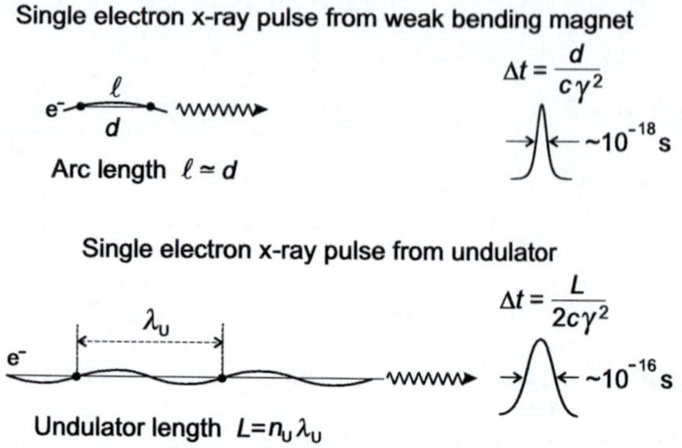

Fig. 2.20 Illustration of the temporal duration of x-ray pulses emitted by single electrons in a weak bending magnet and undulator where we can neglect the difference between the curvy electron and straight light paths. The time for the undulator case is longer by a factor $L/2d$ and the bandwidth is reduced by this factor

2.6.4.1 Interference Picture

Another way to visualize this results is that the wave emitted from point A can constructively interfere with that emitted from point B if the two differ by one wavelength or multiples n thereof. We obtain by use of (2.66)

$$n\lambda = c\Delta t = \frac{\lambda_U}{2\gamma^2}\,. \tag{2.69}$$

Thus at such wavelengths we will have an n_U-fold enhancement of the intensity. The undulator spectrum therefore exhibits a structure of peaked harmonics $n = 1, 2, \ldots$, resulting from interference of the fields emitted during the n_U radiation periods. For example, a 3 GeV storage ring has a $\gamma \simeq 6 \times 10^3$ and an undulator of period length 5 cm would therefore produce radiation of $\lambda \simeq 0.7$ nm or $\hbar\omega \simeq 1.8$ keV.

The form of (2.69) suggests a simple interpretation of the measured wavelength. In the frame of the electron the undulator fields look to the electron as an oncoming EM wave of wavelength λ_U/γ, shortened due to the relativistic contraction. This wave jiggles the electrons which in turn re-emit radiation of the same wavelength. When transformed into the laboratory frame of the observer, the radiation is Doppler shifted to an even shorter wavelength by a factor $1/2\gamma$.

2.6.4.2 Energy Conversion from Electrons to Photons

As illustrated in Fig. 2.11 we can picture the energy contained in relativistic electrons in terms of their attached virtual photon spectrum. When electrons are accelerated by passing through a perpendicular magnetic field, they radiate part of their energy as synchrotron radiation. The shape of the synchrotron spectrum resembles the virtual photon spectrum. Typically, the coherent part of the synchrotron spectrum which falls into the infrared to THz region is not shown, but it has been extensively utilized [36–38, 50]. The incoherent part resembles the synchrotron spectrum from a bending magnet shown in Fig. 2.18. In the x-ray region, the number of emitted photons scales with the number of electrons in the bunch like the incoherent virtual photon spectrum given by (2.42).

For undulators in high brightness storage rings, only about one photon is created per 10^4 electrons on passage through the device. For $N_e = 10^{10}$ electrons we obtain a maximum number of 10^6 photons per bunch for each x-ray pulse seen by an observer. The overall energy loss of the bunch per passage is quite small. Assuming an electron energy of 3 GeV and a typical undulator first harmonic photon energy of 1 keV, the average relative energy change is the amount of energy in the photon pulse relative to the total energy in the electron bunch, $\Delta\mathcal{E}_e/\mathcal{E}_e \sim (10^6 \times 10^3)/(10^{10} \times 3 \times 10^9) \sim 3 \times 10^{-11}$.

2.6.5 The Angular Distribution of Undulator Radiation

A more detailed calculation shows that (2.69) is valid only in the limit of weak undulator field **B** or small undulator period λ_U. The more general expression for the radiation wavelength includes additional terms depending on these parameters, combined in the dimensionless undulator parameter $K_U = (e/2\pi m_e c)\, B\, \lambda_U$, and containing the observation angle θ_{obs} from the forward direction. For a planar undulator the emitted first order wavelength is given by

$$\lambda = \frac{\lambda_U}{2\gamma^2}\left(1 + \frac{1}{2}K_U^2 + \gamma^2\theta_{obs}^2\right) . \tag{2.70}$$

Expressing λ_U in [mm] and the electron beam energy $\mathcal{E}_e$ in [GeV], we have for the first harmonic wavelength in [nm]

$$\lambda = 0.13\,\frac{\lambda_U}{\mathcal{E}_e^2}\left(1 + \frac{1}{2}K_U^2 + \gamma^2\theta_{obs}^2\right) . \tag{2.71}$$

Undulators are considerably brighter than bending magnet sources because of a dual effect. The peak intensity is enhanced at harmonic wavelengths up to a factor n_U, and the emission cone associated with these harmonics is reduced $\propto 1/n_U$, as illustrated in Fig. 2.21.

This is derived by rewriting (2.70) in terms of the on-axis wavelength λ_0 as

$$\lambda = \lambda_0\left(1 + \frac{\gamma^2\theta_{obs}^2}{1 + \frac{1}{2}K_U^2}\right) = \lambda_0\left(1 + \frac{\Delta\lambda}{\lambda_0}\right) , \tag{2.72}$$

where

$$\lambda_0 = \frac{\lambda_U}{2\gamma^2}\left(1 + \frac{1}{2}K_U^2\right) . \tag{2.73}$$

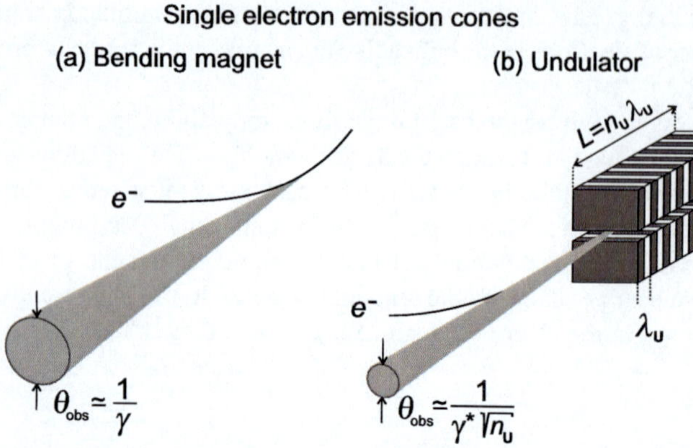

Fig. 2.21 Intensity emission cones of a single relativistic electron radiated by **a** a bending magnet source and **b** an undulator source of n_{U} magnetic periods of length λ_{U}. The 1D emission cone angles contain the intensity distributions to the first node as in Fig. 2.16 (twice the FWHM value θ_{obs} given by (2.75))

By use of the fact that the FWHM bandwidth of the the first undulator harmonic scales inversely with the number of undulator periods n_{U}, we obtain

$$\frac{\Delta\lambda}{\lambda_0} = \frac{1}{n_{\mathrm{U}}} = \frac{\gamma^2\theta_{\mathrm{obs}}^2}{1+\frac{1}{2}K_{\mathrm{U}}^2}. \tag{2.74}$$

Solving for θ_{obs} we obtain the FWHM of the 1D emission cone angle around the optical axis

$$\theta_{\mathrm{obs}} \simeq \frac{1}{\gamma^*\sqrt{n_{\mathrm{U}}}}, \quad \text{where} \quad \gamma^* = \frac{\gamma}{\sqrt{1+\frac{1}{2}K_{\mathrm{U}}^2}}. \tag{2.75}$$

The emission cone formed by two equal orthogonal planar FWHM angles θ (units [radian]) is linked to the solid angle of emission $\mathrm{d}\Omega$ (units [steradian]) by $\mathrm{d}\Omega = 4\pi\sin^2(\theta/4) = 2\pi[1-\cos(\theta/2)]$ which for small angles becomes $\mathrm{d}\Omega \simeq \pi\theta^2/4$. The solid angle of undulator radiation emitted by a single electron becomes

$$\mathrm{d}\Omega_{\mathrm{coh}} = \frac{\pi}{4\gamma^{*2}n_{\mathrm{U}}} \simeq \frac{1}{\gamma^{*2}n_{\mathrm{U}}}. \tag{2.76}$$

In the following sections we will discuss why the derivations given above for a single radiating electron can be readily extended to the case of an electron bunch.

2.7 The Synchrotron Radiation Spectrum of Electron Bunches

The case of a single electron forms the foundation for both the spectra and angular patterns of synchrotron radiation. In practice, electrons in electron storage rings come in bunches containing up to $N_e \sim 10^{10}$ electrons, corresponding to 1.6 nC of total charge. We now consider the spectra and angular patterns for this case.

2.7.1 *The Spectrum of an Electron Bunch and Its Temporal Coherence*

The frequency spectrum of both bending magnet and undulator sources is determined by that of the individual electrons. According to Fig. 2.20, the photon wavetrains emitted by individual electrons for both bending magnets and undulator sources, which are of duration $\Delta t \sim 10^{-18}$ s and $\Delta t \sim 10^{-16}$ s, respectively, are considerably shorter than the $\simeq 100$ ps electron bunch length so that the emitted x-ray fields consist of temporally *incoherent* superpositions of the temporally coherent fields emitted by the individual electrons as illustrated in Fig. 2.22.

The global temporal coherence properties of synchrotron radiation are similar to those of bremsstrahlung emitted by an x-ray tube or by optical light emitted by a conventional thermal source [51, 52]. Synchrotron radiation can only be made temporally coherent by use of a monochromator, which picks out an energy interval that corresponds to a temporally coherent fraction. For example, a grating monochromator picks out a Gaussian-shaped energy window, and the rms width $\sigma_\mathcal{E}$ of the intensity distribution corresponds to an rms Gaussian coherence time σ_t according to the relation,

$$\boxed{\text{Gaussian transform limit:}\ \sigma_t\,\sigma_\mathcal{E} = \frac{\hbar}{2} \quad \text{or} \quad \sigma_z\,\frac{\sigma_\mathcal{E}}{\hbar\omega} = \frac{\lambda}{4\pi}}. \tag{2.77}$$

This correlation defines the so-called *transform limit*, where on the right side we have given the expression in terms of the spatial rms pulse length $\sigma_z = c\sigma_t$ and the wavelength.

2.7.2 *The Emitted Radiation Cone and Lateral Coherence*

Similar to the temporal coherence being determined by the radiation emitted by single electrons, the spatial coherence is also benchmarked by that of a single electron. The best one can do in terms of spatial coherence is determined by the radiation cone emitted by a single electron illustrated in Fig. 2.21.

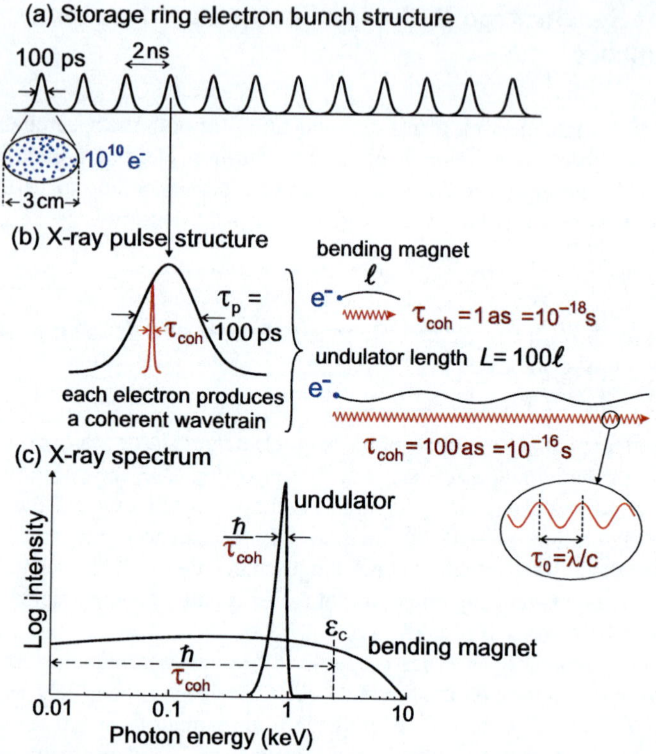

Fig. 2.22 **a** Electron bunches in a storage ring have a length of about 3 cm and consist of about 10^{10} electrons. The temporal duration of the bunch and the x-ray pulse length are about the same ($\sim$100 ps). **b** The wavetrains emitted by the individual electrons are coherent with coherence time $\tau_{coh} \sim 1$ as for a bending magnet source and $\tau_{coh} \sim 100$ as for an undulator source. There is no correlation between the positions of the individual electrons on the scale of the radiation wavelength, and the fields superimpose at random in the emitted pulse which is therefore spatially and temporally chaotic from a global perspective. **c** The width of the emitted spectrum corresponds to the transform limited width $\hbar/\tau_{coh}$, determined by the unresolved spikes of the temporally coherent radiation of the individual electrons

In order to understand the relation between the emission cone and lateral coherence we have a borrow the fundamental concept of the diffraction limit, discussed in more detail in Sect. 4.2.2. Lateral coherence is defined by the product of the lateral source area, A_s, and the associated solid angle into which the radiation is emitted, dΩ. The minimum product allowed by the position-momentum uncertainty principle is called the *diffraction limit* and for a lateral Gaussian distribution is given by (see Sect. 4.2.3)

$$\boxed{\text{Gaussian diffraction limit:}\;\; A_s\, d\Omega \geq A_{coh}\, d\Omega_{coh} = \frac{\lambda^2}{4}}. \tag{2.78}$$

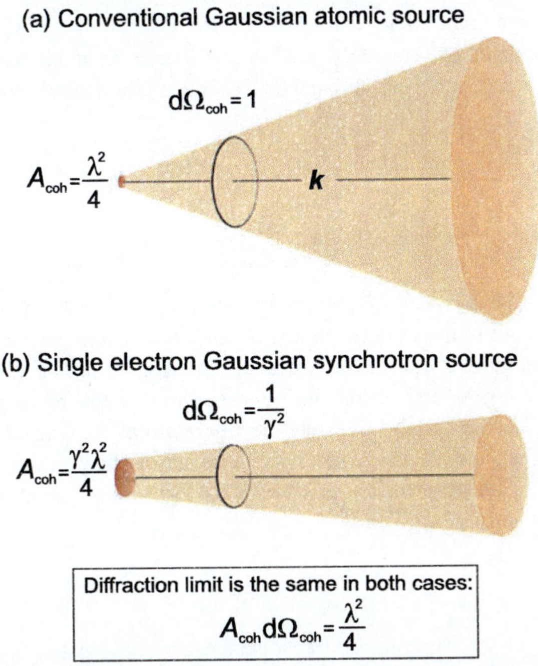

Fig. 2.23 Interplay between source coherence area and coherent emission angle for two types of sources with coherent Gaussian distributions that are laterally coherent, expressed by the diffraction limit $A_{coh}d\Omega_{coh} = \lambda^2/4$. **a** Coherent emission from the coherence area $A_{coh} = \lambda^2/4$ into the associated solid angle $d\Omega_{coh} = 1$ according to (2.78). **b** Coherent radiation emitted by individual electrons in a globally chaotic bending magnet synchrotron source ($n_u = 1$, $\gamma^* = \gamma$). Now the small relativistic solid emission angle $d\Omega_{coh} = 1/\gamma^2$ defines the single electron coherence area $A_{coh} = \gamma^2\lambda^2/4$ through the diffraction limit stated by (2.78) and at the bottom of the figure

In general, the minimum product, which defines the maximum spatial (or average) brightness, depends on the shape of the source intensity distribution as discussed in Sect. 4.2.6. For a given lateral shape, the uncertainty product is *conserved* and it has the value $\lambda^2/4$ for a Gaussian distribution and λ^2 for a square flat-top distribution. If we assume a Gaussian distribution, we have the situation illustrated in Fig. 2.23a.

Hence for a bremsstrahlung source to be diffraction limited one would have to make the spot size as small as $\lambda^2/4$, as illustrated in Fig. 2.23a. This cannot be done in practice. In contrast, for a synchrotron source this can in principle be achieved due to the relativistic effect, as shown in Fig. 2.23b. Now the solid emission angle for a *single electron* beam is $d\Omega_{coh} = 1/\gamma^2$ according to Fig. 2.21, so that the effective source size is given by

$$A_{coh} = \frac{\gamma^2\lambda^2}{4}. \tag{2.79}$$

Owing to the large value of γ^2 ($\simeq 3.5 \times 10^7$ for a 3 GeV ring) the single electron source coherence area is fortunately quite large. In order to make the radiation of the entire electron bunch coherent, one needs to squeeze the lateral bunch size into an area less than A_{coh}.

2.7.3 *Toward Ultimate Storage Rings*

Since the field radiated by a single electron is laterally coherent, it defines the diffraction limited product of the effective lateral source dimension and the angular emission angle, as stated for a Gaussian electron distribution by (2.78). While the relation between brightness, the diffraction limit and coherence will be discussed in more detail in Sect. 4.2.6, we here briefly state the key relations for Gaussian distributions in bending magnet and undulator sources in terms of the product of the rms lateral source size σ_r and the rms angular width σ_θ. The fundamental parameter is the so-called source *emittance* given by

$$\epsilon = \sigma_r \, \sigma_\theta . \tag{2.80}$$

This expression is a reformulation of Heisenberg's space-momentum uncertainty principle [53]

$$\sigma_x \, \sigma_p \geq \frac{\hbar}{2} . \tag{2.81}$$

Substituting $\sigma_x = \sigma_r$ and $\sigma_p = \hbar \sigma_q = \hbar k \sigma_\theta$ we obtain with $k = 2\pi/\lambda$ the *diffraction limited emittance* as

$$\boxed{\epsilon = \sigma_r \, \sigma_\theta = \frac{\lambda}{4\pi}} . \tag{2.82}$$

This important expression allows us to obtain the effective rms width and emission angle for a diffraction limited source.

2.7.3.1 Example: Diffraction Limited Undulator Source

The emission cone angle for an undulator given by (2.75) can be rewritten with the definition $L = \lambda_{\mathrm{U}} n_{\mathrm{U}}$ (see Fig. 2.21) and the on-axis wavelength λ_0 given by (2.73) as

$$\theta_{\mathrm{obs}} = \sqrt{\frac{1 + \frac{1}{2} K_{\mathrm{U}}^2}{\gamma^2 n_{\mathrm{U}}}} = \sqrt{\frac{2\lambda_0}{\lambda_{\mathrm{U}} n_{\mathrm{U}}}} = \sqrt{\frac{2\lambda_0}{L}} . \tag{2.83}$$

For a Gaussian distribution we obtain approximately

$$\sigma_\theta \simeq \frac{\theta_{\text{obs}}}{2} = \sqrt{\frac{\lambda_0}{2L}} \tag{2.84}$$

and by use of (2.82) we find the effective source rms width to be

$$\sigma_r = \frac{\sqrt{2\lambda_0 L}}{4\pi}. \tag{2.85}$$

These expressions agree with those proposed by Kim in 1987 [54] and are used in the modern x-ray literature discussing ultimate storage rings [6, 55].

For a typical undulator length of $L = 2$ m in a 3 GeV storage ring, we obtain for σ_r and λ_0 measured in units of [nm] and σ_θ in μrad,

$$\sigma_r \simeq 5000\sqrt{\lambda_0}, \quad \text{and} \quad \sigma_\theta \simeq 16\sqrt{\lambda_0}\ \mu\text{rad}. \tag{2.86}$$

In practice, the emission cone is of the order of a few microradians, corresponding to a beam diameter of less than 1 mm if one observes it 100 m away from the undulator source. Relative to bending magnet radiation the brightness in a typical undulator of about 100 periods is enhanced by a factor of $n_{\text{U}}^2 \simeq 10^4$.

Figure 2.24 puts the performance of third generation rings into perspective toward the goal of a diffraction limited source in "ultimate storage rings" [3, 6]. The achievement of the diffraction limit becomes increasingly difficult with increasing photon energies. In the third generation storage rings, for example, only about 20% of the emitted radiation is coherent at 100 eV, while the coherent fraction drops to about 0.2% at 10 keV.

2.7.4 Polarization of Synchrotron Radiation

While the x-ray intensity is governed by the transfer of energy from the electron to the photon system, the x-ray polarization depends on the transfer of angular momentum. For "light", the polarization is defined through the **E**-vector since we will later see in Sect. 7.3.3 that the magnetic **B**- or **H**-fields play essentially no role in the interaction of light with matter. Although we have not yet formally defined the polarization of light, we will nevertheless illustrate the basic concepts based on the formal definitions given later in Sect. 3.2.6.

For linearly polarized light, the **E**-vector oscillates in time along an axis perpendicular to the x-ray propagation direction. For circular polarization, **E** rotates in *time* either clockwise or anticlockwise about the propagation direction and different labels are found in the literature to describe the rotation sense of circular light. In this book we label the two circular cases as illustrated in Fig. 2.25 (also see Fig. 3.4) [27].

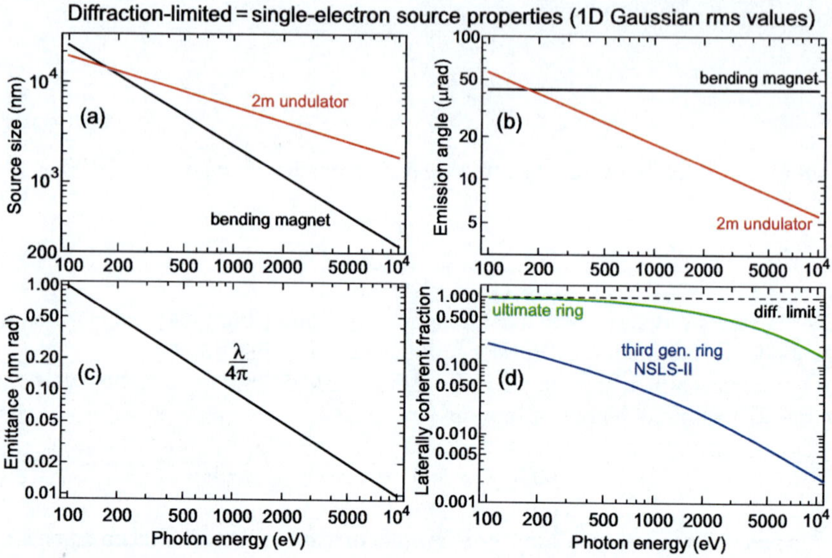

Fig. 2.24 **a** Diffraction limited source size and **b** emission angles corresponding to a radiating single electron in a storage ring bending magnet or a 2-m-long undulator. We have assumed a statistical variation of the lateral electron position described by a Gaussian rms distribution. **c** Diffraction limited emittance under the same assumption. **d** Coherent fraction of a state-of-the-art third generation storage ring (NSLS-II) and that could be achieved in an ultimate storage ring (also see Fig. 2.1) [3, 6, 8]

We start with the angular momentum **L** of a circulating electron in a storage ring. Its direction is defined by the circular motion of electrons according to the right hand rule in magnetism, as shown in the figure [27]. In the radiation process, both energy and angular momentum are transferred from the electron to the x-rays. If radiation is selected by an aperture in the plane of the electron orbit, as illustrated in the middle of the figure, the x-ray propagation direction z is perpendicular to **L** and the angular momentum component L_z is zero. The emitted radiation is *linearly polarized*.

Radiation that is emitted at a finite angle above or below the orbit plane will have a finite angular momentum projection L_z along the x-ray propagation direction z. Note that angular momentum is defined as a rotation in *time* at a given position along the propagation direction. For example, *below* the orbit plane, the projection L_z is along z and the circularly polarized wave has an angular momentum $L_z = +\hbar$. This may also be called "right circular" by use of the right hand rule. If the right thumb points into the propagation direction z, then the **E**-vector is seen to rotate in time in the direction of the fingers. The opposite rotation sense of **E** is encountered for radiation emitted *above* the orbit plane, as shown on the left of Fig. 2.25. Now the

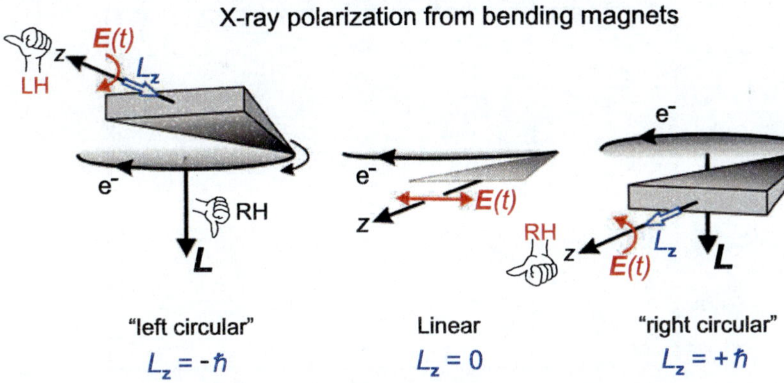

Fig. 2.25 Origin of polarized synchrotron radiation from a bending magnet source. Electrons moving on a circular orbit have an angular momentum **L** of the shown direction defined by the convention in magnetism [27]. The radiating electron transfers both energy and angular momentum to the emitted photons. For radiation emitted in the plane of the electron orbit, as shown in the middle, the x-ray propagation direction z is perpendicular to **L**. The emitted x-rays have $L_z = 0$ and are *linearly polarized*. For radiation emitted below or above the orbit plane, there is a finite projection of **L** along z, so that the emitted photons have angular momenta $L_z = \pm\hbar$, as shown. The radiation is circularly polarized. We illustrate the relationship between L_z and the temporal rotation of the **E**-vector by means of right and left hands with the thumb pointing into the propagation direction of light and the fingers defining the temporal rotation sense

wave has an angular momentum projection $L_z = -\hbar$ and the rotation sense follows the left hand rule and is "left circular".[14]

The polarization of bending magnet and wiggler sources thus depends on what part of the vertical emission angle is selected by a horizontal aperture. If no aperture is used, the integrated radiation is elliptically polarized and consists of a major linear polarization component $\mathbf{E}^{\parallel}$ in the horizontal plane of the ring, and a smaller vertical linear component $\mathbf{E}^{\perp}$. Since the circularly polarized components above and below the orbit plane have opposite angular momenta and the same intensity, there is no net circular polarization. For a bending magnet source it can be obtained by selecting radiation only above or below the orbit plane. The degree of circular radiation can be enhanced at the cost of intensity by moving a horizontal aperture away from the beam center.

The polarization also depends on the storage ring energy since the vertical opening angle of the emitted radiation decreases with increasing electron beam energy. Hence at higher beam energy the linear polarization in the beam center becomes more dominant and the circular polarization contained in the fraction above and below the orbit plane decreases. This is illustrated in Fig. 2.26 where we have plotted the dependence of the polarization on the storage ring energy $\mathcal{E}$ normalized to its critical

---

[14] Note that in the literature "right" and "left" circular polarization is not uniquely defined because the rotation sense of the **E**-vector is opposite at a given position in time and at a given time in space (see Sect. 3.2.6).

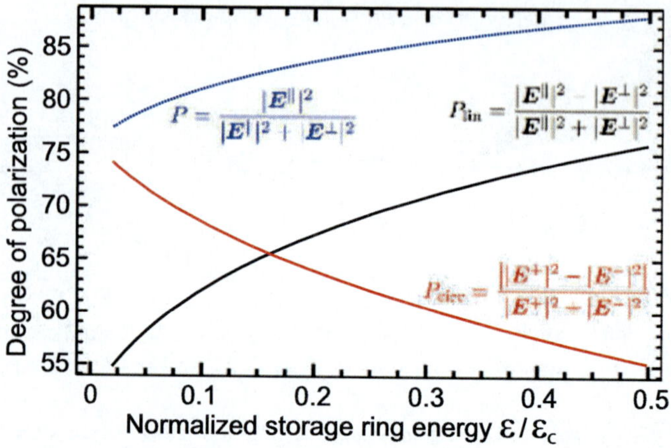

Fig. 2.26 Dependence of the linear and circular x-ray polarization emitted by a bending magnet for a storage ring energy $\mathcal{E}$ normalized to its critical energy $\mathcal{E}_C$. We have integrated the polarization from the beam center over the entire half emission angle. The linear component $\mathbf{E}^{\parallel}$ lies in the storage ring plane and $\mathbf{E}^{\perp}$ is the smaller linear component perpendicular to the orbit plane that causes the entire radiation to be elliptically polarized. The black and dotted blue curves correspond to two different definitions of the degree of linear polarization, as indicated (see Sect. 3.2.7)

energy $\mathcal{E}_C$. For the circular case, we have assumed that only the half beam below or above the ring plane is accepted and we have integrated over the entire half opening angle. For the linear polarization case, the black and dotted blue curves correspond to two different definitions of the degree of linear polarization as discussed in Sect. 3.2.7.

Today, purely linearly or circularly polarized radiation may be obtained by use of undulators in storage rings and XFELs, which in many cases allow complete polarization control [2–4, 56].

2.8 X-Ray Free Electron Lasers

In a storage ring, electron bunches are stored for hours and the bunch size assumes an equilibrium value that is determined by various opposing forces. The uncorrelated temporal emission by the electrons leads to an incoherent superposition of the radiation emitted by the individual electrons. The intensity then scales linearly with the number of electrons or charge N_e and not quadratically, N_e^2, as expected for a coherent superposition of the individual fields.

The question arises whether one can obtain a coherent superposition of the fields of all charges in the bunch which would yield a peak brightness enhancement by a factor $N_e \sim 10^{10}$ over a synchrotron source. According to Fig. 2.24a, this requires a diffraction limited *lateral* beam area with a Gaussian rms width of about $2\,\mu$m at a photon energy of 10 keV. The required longitudinal beam size is obtained from the *transform limit*, expressed for Gaussian distributions by (2.77). To achieve a bandwidth $\sigma_{\mathcal{E}}/(\hbar\omega)$ of 10^{-4} at 10 keV ($\lambda = 0.12$ nm) would require a longitudinal electron bunch length of $\sigma_z \simeq 0.1\,\mu$m. These dimensions cannot be achieved in a storage ring, which in the early 1980s led to the suggestion of using a linac [57, 58], where the radiation losses are sufficiently small that it is possible to squeeze the electrons into a tiny electron bunch before it is made to radiate in an undulator. The history of the x-ray free electron laser has been documented by two of its inventors Saldin [59] and Pellegrini [60], and the detailed principles have been reviewed in the book of Kim et al. [3].

2.8.1 The Radiation Emitted by SASE XFELs

Owing to the small size of electron bunches that can be produced in linear accelerators, XFEL radiation is essentially *laterally* coherent [51, 61, 62]. The *temporal* coherence is also different from synchrotron radiation since each x-ray pulse has a discernible substructure consisting of coherent spikes of about 1 fs length which are visible on the typical 10–100 fs scale of the total bunch length.

In a SASE-FEL, electrons spontaneously emit x-ray radiation that propagates slightly faster and can therefore interact with electrons located toward the front end of the bunch. The distance that the emitted photon moves ahead of the emitting electron is determined by the difference in speed of the electron $v = c\sqrt{\gamma^2 - 1}/\gamma$ and the light, c. When the electron travels a distance l_0, the photon moves ahead by the distance

$$\Delta l \simeq \frac{l_0}{2\gamma^2} = l_0 \frac{\lambda}{\lambda_U}. \tag{2.87}$$

Thus over one undulator period λ_U, the electrons slip back by exactly one radiation wavelength. If we assume a long undulator with 1000 periods, e.g. a 30-m-long undulator with $\lambda_U = 3$ cm (used at LCLS), the photons only advance by 1000 wavelengths. For $\lambda = 0.1$ nm, the light advances only 100 nm which is much less than a typical bunch length. This shows that it is in general difficult to induce complete order across the entire length of the electron bunch through the emitted light. More typically, charge ordering within the electron bunch through the SASE process occurs only over distances that are less than the bunch length l_{bun}.

The photon-electron interaction in the bunch modulates the electron energy sinusoidally with the radiation wavelength. Under the influence of the undulator magnetic field, the trajectory of electrons with increased modulated energy is bent less and they assemble in a plane ahead of those electrons with decreased modulated energy. The length of the electron trajectory changes, and the electrons within one radiation wavelength tend to form sheets perpendicular to the direction of propagation. The resulting order grows exponentially and with it the amplitude of the electromagnetic field. The exponential growth constant is called the *gain length* ℓ_{gain} and is of the

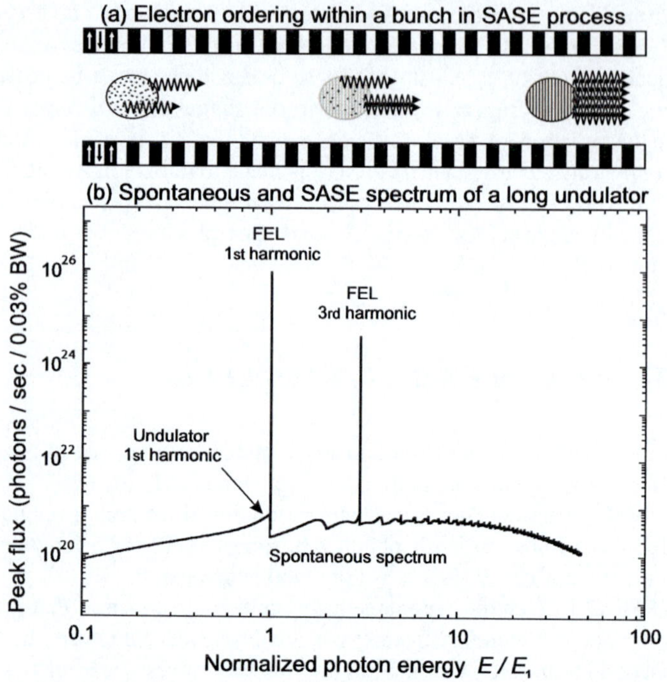

Fig. 2.27 **a** Schematic of the electron ordering effect within the electron bunch produced by the SASE process. We have assumed perfect ordering within the electron bunch, so that the electron sheets separated by a wavelength extend through the entire bunch. **b** Calculated characteristic spectrum with the parameters for LCLS with the photon energy plotted on a universal scale, normalized to the photon energy of the first harmonic. It falls at 920 eV for a linac beam energy of 4.8 GeV and at 8.3 keV for a beam energy of 14.35 GeV. The spontaneous spectrum of the 120 m undulator shows small peaks indicating the undulator harmonics. At the position of the first and third harmonics, strong and narrow peaks are observed with an intensity increase of four to five orders of magnitude. These peaks arise from gain through SASE associated with lasing. Spectrum courtesy of Roman Tachyn

order of a few meters. For a given electron bunch quality, the exponential growth saturates when the electrons have become optimally ordered in sheets.

The ordering effect is schematically illustrated in Fig. 2.27a, and a characteristic XFEL spectrum is shown in (b). The figure nicely illustrates the coherent enhancement of the characteristic narrow XFEL harmonic peaks superimposed on a relatively strong continuous background of synchrotron radiation. The crucial parameter that determines the length of charge ordered regions within the bunch is the *cooperation length* l_C. It is defined as the distance that the photons move ahead of the emitting electron over one SASE *gain length*, ℓ_{gain} [63],

$$l_C \simeq \ell_{gain} \frac{\lambda}{\lambda_U} \,. \tag{2.88}$$

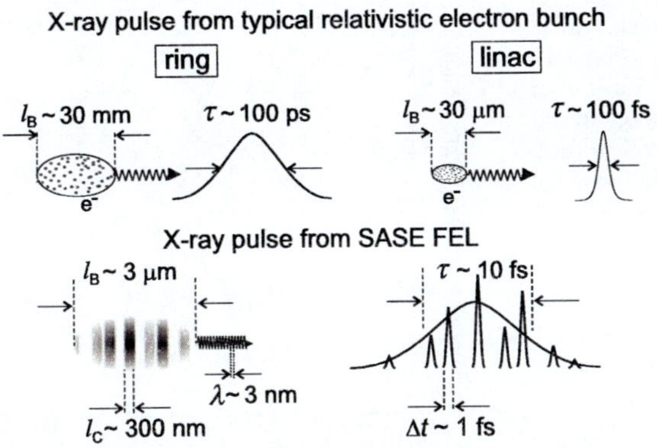

Fig. 2.28 When conventional electron bunches traverse an undulator, there is no correlation between the positions of the individual electrons on the scale of the radiation wavelength and their radiated fields superimpose at random. The emitted spontaneous radiation is longitudinally incoherent, and the x-ray pulse length is simply determined by the electron bunch length. For a SASE FEL, the interaction of the emitted light with the preceding electrons causes a substructure in the electron bunch, consisting of regions perpendicular to the direction of motion as shown on the bottom left. The regions extend over about 100 wavelengths and all electrons within the regions, ordered in sheets separated by the wavelength, emit coherent radiation. The resulting x-ray pulse, corresponding to the harmonic peaks in Fig. 2.27b, contains the same substructure as the electron bunch with coherent intensity spikes and a total x-ray pulse length that is determined by the electron bunch length

For example, at LCLS with $\lambda_U = 3 \times 10^{-2}$ m, the gain length has been measured to be $\ell_{gain} \simeq 3$ m, so that the cooperation length is $l_C \simeq 100\,\lambda$. Each cooperation length region produces a coherent spike in the emitted x-ray pulse, as schematically illustrated in Fig. 2.28. There is no correlation between the coherent spikes so that the pulses are *partially coherent* in the time domain. As a rule of thumb, a pulse of M fs duration contains about M coherent spikes. The temporally coherent fraction is approximately given by $1/M$ and the global pulse is *partially coherent*. The intensity in each spike fluctuates according to a Γ-function and becomes a Gaussian for a large number of spikes and a Poisson distribution for a single spike [64].

The width of the spikes determines the longitudinal coherence time τ_{coh} or length ℓ_{coh}. It may be pictured as multiples of the oscillation time τ_0 or wavelength related by $\tau_0 = \lambda/c$. One extreme is a classical EM wave whose wavelength is perfectly defined. The wave has no beginning and no end so that $\lambda/\Delta\lambda = \infty$. The other extreme is $\lambda/\Delta\lambda = 1$ or $\tau_{coh} = \tau_0$. Now the uncertainty in wavelength is as large as the wavelength itself. In practice, the coherence length has a maximum value that is equal to the macroscopic bunch length, as assumed in Fig. 2.27b.

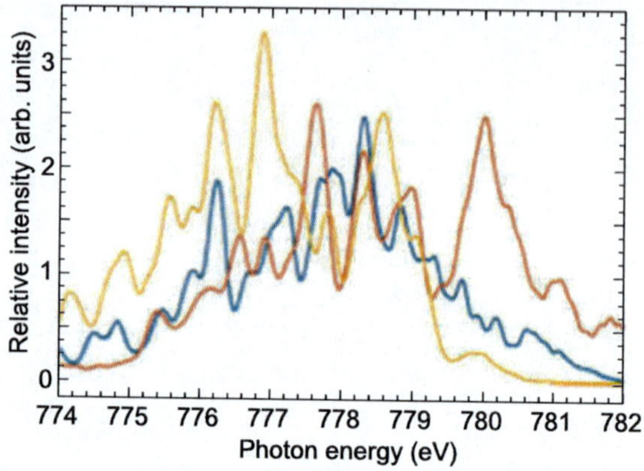

Fig. 2.29 Measured SASE pulse in the energy domain for a nominally 25 fs electron bunch with an energy spread of 0.45% of the central linac beam energy, corresponding to a photon energy spread of ~ 3.5 eV at the central photon energy of 778 eV (Co L_3 resonance) [65]. Different colors indicate the statistical fluctuations of individual ("single shot") pulses

2.8.2 Realistic Pulse Structure in SASE XFELs

The measured energy spectrum of a SASE pulse of nominally $\tau_p = 25$ fs duration with a central photon energy of 778 eV (Co L_3 resonance) and a photon energy spread of ~ 3.5 eV is shown in Fig. 2.29. The incident SASE radiation was dispersed by a grating monochromator set to 0.2 eV FWHM energy resolution onto a YAG screen where it was imaged with an optical camera [65].

The three colored intensity distributions correspond to single shots, revealing the variation of the structure and intensity of the spikes from pulse to pulse, with an average over many pulses assuming a Gaussian distribution. It has only recently become possible to measure the temporal pulse structure with attosecond resolution via angular streaking of Ne 1s photoelectrons produced by the x-rays [76] (see Fig. 4.4).

Simulations of the shot-to-shot variations of the internal SASE pulse structure in both the energy and time domain are shown by comparison of two single-shot pulses in Fig. 2.30 using similar parameters as those in Fig. 2.29. The figure also demonstrates the change of the pulse substructure upon monochromatization.

In the simulations, the energy spectrum of the SASE pulse was first calculated [65, 66] and then Fourier transformed to the time domain. We assumed an electron bunch length with a flat-top distribution of 25 fs length and an energy spread of 0.45% of the central linac beam energy (~ 3.5 eV). The effect of the monochromator is applied in the energy domain through a Gaussian window which in our case had a FWHM of 0.2 eV. The filtered energy spectrum is then transformed into the time domain to obtain the time structure of the pulse behind the monochromator. Since the monochromator acts in the energy domain, the temporal spike structure after the monochromator is not simply a convolution of the original one, but the temporal width of the coherent spikes is considerably increased to ~ 10 fs.

Most of the long undulator in a SASE XFEL is used to generate the internal order in the electron bunch. The emitted spectrum is mainly determined by only the final undulator section, reflecting the order in the electron bunch. This allows the use of

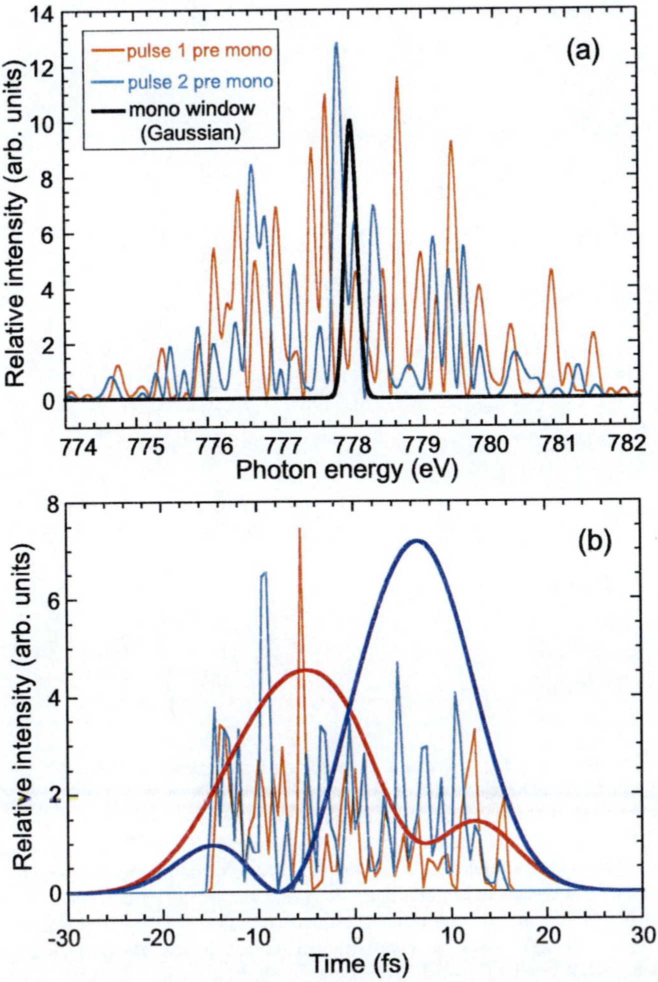

Fig. 2.30 **a** Simulated SASE pulses in the energy domain, generated by a 25 fs flat-top electron bunch with an energy spread of 0.45% of the central linac beam energy, corresponding to a photon energy spread of ~ 3.5 eV at the central photon energy of 778 eV (Co L_3 resonance). Superimposed in black is a Gaussian of 0.2 eV FWHM, which models the transmission window of a grating monochromator. **b** Temporal structure of the pulses in (**a**) shown in the same colors. Shown as a thick red line is the time structure of the orange pre-mono SASE pulse after it has been transmitted through the black Gaussian energy window in (**a**). The thick blue line corresponds to the temporal structure of the light blue SASE pulse after transmission through the monochromator window

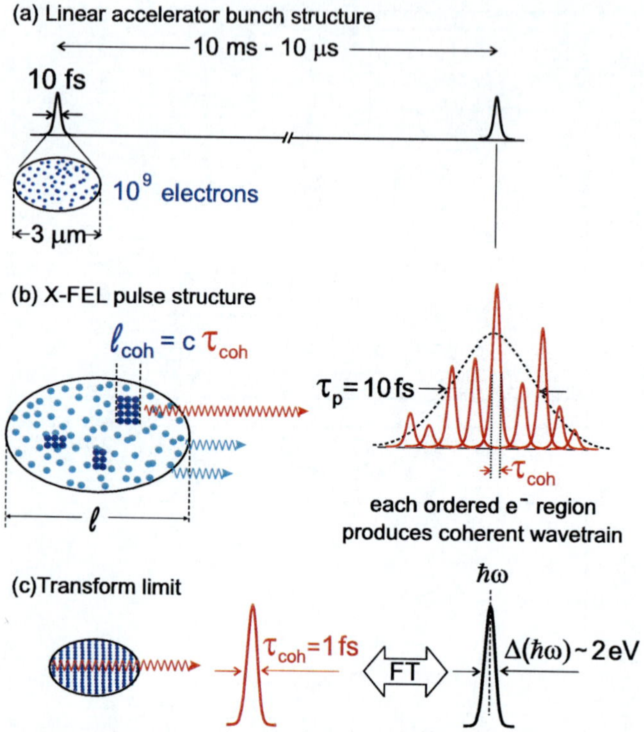

Fig. 2.31 **a** Electron bunches in a linear accelerator typically consist of $10^9 - 10^{10}$ electrons and through compression may have a length of $\sim 3\,\mu$m or ~ 10 fs. This comes at a price of lower rep rate than for storage rings. **b** The temporal duration of the bunch and the x-ray pulse length are about the same ($\tau_p \sim 10$ fs). The wavetrains emitted by small regions of electrons ordered with a periodicity of the wavelength form coherent spikes with $\tau_{coh} \sim 1$ fs. There is no correlation between the individual spikes, and the total single pulse is globally spatially and temporally partially coherent. (c) A very short electron bunch of ~ 300 nm may produce a single spike at $\lambda \sim 1$ nm which is nearly transform limited in the time–energy domain. For Gaussian distributions the FWHM values in the energy and time domain are related by $\Gamma_\mathcal{E}\Gamma_t = 1.825$ eV fs (see Table 4.2)

shorter undulators of length comparable to the gain length to produce variable states of polarization, demonstrated at LCLS through the production of nearly perfectly circularly polarized photons [56, 67].

2.8.3 *From SASE to Transform Limited Pulses*

Typically the number of coherent spikes in the SASE pulse is reduced proportionally to its total length, with the width of the coherent spikes remaining about the same. Pulses as short as a few hundred nanometers can be prepared by lowering the number of electrons resulting in only one or two spikes at wavelengths around 1 nm. The technique of self-seeding, where a SASE pulse is sent through a monochromator and then superimposed onto the electron bunch, may be used to more reliably produce single spike pulses [68]. The evolution of the x-ray pulses toward the single spike limit is schematically illustrated in Fig. 2.31.

While the structure and intensity of the SASE spikes vary from pulse to pulse, their average over many pulses assumes a Gaussian distribution, as shown by the dashed black curve in Fig. 2.31b. The temporal average renders XFEL sources, despite their name, to be statistically chaotic [69]. By use of ultrashort electron bunches of low charge or passing a bunch through a narrow aperture [70, 71], it is possible to produce individual near transform-limited pulses of about 1 fs duration as shown in Fig. 2.31c. Note, however, that consecutive pulses have no phase relation and the statistical average of such pulses is still chaotic.

At photon energies below about 100 eV it is possible to use high-harmonic pulses of conventional high power lasers to directly seed the SASE process, resulting in transform-limited pulses [72]. It has been shown that the so-generated pulses indeed behave statistically like a laser [73]. In the multi-keV x-ray range it has been proposed to use an energy recovery linac in conjunction with a FEL oscillator to create higher order coherent x-rays [74]. More recently, a related scheme has been proposed based on the use of an FEL oscillator in the extreme ultraviolet ($\sim$ 13 nm) that is upshifted in energy by a multi-stage cascade scheme to 0.6–0.2 nm (2–5 keV) [75].

References

1. H. Wiedemann, *Synchrotron Radiation* (Springer, Berlin, 2003)
2. A. Hofmann, *The Physics of Synchrotron Radiation* (Cambridge University Press, Cambridge, 2004)
3. K.J. Kim, Z. Huang, R. Lindberg, *Synchrotron Radiation and Free-Electron Lasers* (Cambridge University Press, Cambridge, 2017)
4. D. Attwood, A. Sakdinawat, *X-Rays and Extreme Ultraviolet Radiation: Principles and Applications*, 2nd edn. (Cambridge University Press, Cambridge, 2017)
5. J. Als-Nielsen, D. McMorrow, *Elements of Modern X-Ray Physics*, 2nd edn. (Wiley, New York, 2011)
6. M. Bei, M. Borland, Y. Cai, P. Elleaume, R. Gerig, K. Harkay, L. Emery, A. Hutton, R. Hettel, R. Nagaoka, D. Robin, C. Steier, Nucl. Instum. Methods Phys. Res. A **622**, 518 (2010)
7. B.W.J. McNeil, N.R. Thompson, Nat. Photonics **4**, 814 (2010)
8. M. Borland: J. Phys.: Conf. Ser. **425**, 042 016 (2013)
9. J. Stöhr, Adv. Optics and Photonics **11**, 215 (2019)
10. C.F. von Weizsäcker, Z. Phys. **88**, 612 (1934)
11. E. J. Williams, Kgl. Danske Videnskab. Selskab Mat.-fys. Medd. **13**, No. 4 (1935)
12. J.D. Jackson, *Classical Electrodynamics* (Wiley, New York, 1999)
13. A. Liénard, L'Eclairage Electrique **16**, 5 (1898)
14. E. Wiechert: Archives Neerlandaises p. 546 (1900)
15. W. Heitler, *The Quantum Theory of Radiation*, 3rd edn. (Clarendon Press, Oxford, 1954)
16. G. A. Schott, *Electromagnetic radiation and the mechanical reactions arising from it*, Adams Prize Essay (Cambridge University Press, 1912)
17. J.P. Blewett, Nucl. Instrum. Methods A **266**, 1 (1988)

18. M. Tigner, Nuovo Cimento **37**, 1228 (1965)
19. S.M. Gruner, D. Bilderback, I. Bazarov, K. Finkelstein, G. Krafft, L. Merminga, H. Padamsee, Q. Shen, C. Sinclair, M. Tigner, Rev. Sci. Instrum. **73**, 1402 (2002)
20. L.D. Landau, E.M. Lifshitz, *The Classical Theory of Fields* (Butterworth-Heinemann, Oxford, 2002)
21. S.J. Gamble, M.H. Burkhardt, A. Kashuba, R. Allenspach, S.S.P. Parkin, H.C. Siegmann, J. Stöhr: Phys. Rev. Lett. **102**, 217 201 (2009)
22. Sara J. Gamble: Ph.D. dissertation, Stanford University, Stanford, CA (2009), available on website http://www-ssrl.slac.stanford.edu/stohr
23. V. Balakin et al., Phys. Rev. Lett. **74**, 2479 (1995)
24. A. L. Cavalieri et al.: Phys. Rev. Lett. **94**, 114 801 (2005)
25. E.M. Purcell, *Electricity and Magnetism* (McGraw-Hill, New York, 1986)
26. D.J. Griffith, *Introduction to Electrodynamics* (Prentice-Hall, New Jersey, 1999)
27. J. Stöhr, H.C. Siegmann, *Magnetism: From Fundamentals to Nanoscale Dynamics* (Springer, Heidelberg, 2006)
28. R. Srinivasan, V.A. Lobastov, C.Y. Ruan, A.H. Zewail, Helv. Chim. Acta **86**, 1763 (2003)
29. K. Sokolowski-Tinten, X. Shen, Q. Zheng, T. Chase, R. Coffee, M. Jerman, R.K. Li, M. Ligges, I. Makasyuk, M. Mo, A.H. Reid, B. Rethfeld, T. Vecchione, S.P. Weathersby, H.A. Dürr, X.J. Wang: Struct. Dynamics **4**, 054 501 (2017)
30. C. Stamm, I. Tudosa, H.C. Siegmann, J. Stöhr, A.Y. Dobin, G. Woltersdorf, B. Heinrich, A. Vaterlaus: Phys. Rev. Lett. **94**, 197 603 (2005)
31. J. G. Zeibel: Ph.D. Thesis *Manipulating Wavepacket Dynamics with Half-Cycle Pulses*, University of Virginia, 2003
32. J. Shan, T.F. Heinz: "Terahertz radiation from semiconductors", in *Ultrafast Dynamical Processes in Semiconductors*, ed. by K.T. Tsen (Springer, Berlin, 2004)
33. S.G. Ganichev, W. Prettl, *Intense Terahertz Excitation of Semiconductors* (Oxford University Press, Oxford, 2006)
34. M. Tonouchi, Nature Photonics **1**, 97 (2007)
35. J.S. Nodvick, D.S. Saxon, Phys. Rev. **96**, 180 (1954)
36. G.P. Williams, C.J. Hirschmugl, E.M. Kneedler, P.Z. Takacs, M. Shleifer, Y.J. Chabal, F.M. Hoffmann, Phys. Rev. Lett. **62**, 261 (1989)
37. K. Ishi, Y. Shibata, T. Takahashi, H. Mishiro, T. Ohsaka, M. Ikezawa, Y. Kondo, T. Nakazato, S. Urasawa, N. Niimura, R. Kato, Y. Shibasaki, M. Oyamada, Phys. Rev. A **43**, 5597 (1991)
38. G.L. Carr, M.C. Martin, W.R. McKinney, K. Jordan, G.R. Neil, G.P. Williams, Nature **420**, 153 (2002)
39. Z. Wu, A.S. Fisher, J. Goodfellow, M. Fuchs, D. Daranciang, M. Hogan, H. Loos, A. Lindenberg: Rev. Sci. Instrum. **84**, 022 701 (2013)
40. E. Fermi, Z. Phys. **29**, 315 (1924)
41. W.K.H. Panofsky, M. Phillips, *Classical Electricity and Magnetism* (Dover, New York, 2005)
42. M. S. Zolotorev and K. T. McDonald *Classical Radiation Processes in the Weizsäcker-Williams Approximation*, https://arxiv.org/abs/physics/0003096
43. B.L. Henke, E.M. Gullikson, J.C. Davis, At. Data Nucl. Data Tables **54**, 181 (1993)
44. J.H. Hubbell, P.N. Trehan, N. Singh, B. Chand, D. Mehta, M.L. Garg, R.R. Garg, S. Singh, S. Puri, J. Phys. Ref. Data **23**, 339 (1994)
45. https://henke.lbl.gov/optical_constants/
46. M.L. Ter-Mikaelian, *High-Energy Electromagnetic Processes in Condensed Media* (John Wiley & Sons Inc, New York, 1972)
47. V. Ginzburg, V. Tsytovich, *Transition Radiation and Transition Scattering* (Institute of Physics Pub, Bristol, 1991)
48. H.P. Klug, L.E. Alexander, *X-Ray Diffraction Procedures* (Wiley, New York, 1974)
49. T. Shintake, Nucl. Inst. Meth. A **507**, 89 (2003)
50. M. Abo-Bakr, J. Feikes, K. Holldack, P. Kuske, W.B. Peatman, U. Schade, G. Wüstefeld, H.W. Hübers: Phys. Rev. Lett. **90**, 094 801 (2003)
51. I.A. Vartanyants, A. Singer: New J. Phys. **12**, 035 004 (2010)

52. I.A. Vartanyants, A. Singer: "Coherence properties of third-generation synchrotron sources and free-electron lasers", in *Synchrotron Light Sources and Free-Electron Lasers. Accelerator Physics, Instrumentation and Science Applications*, ed. by E. Jaeschke, S. Khan, J. Schneider, J. Hastings (Springer, Amsterdam, 2016), p. 821
53. W. Heisenberg, Z. Phys. **43**, 172 (1927)
54. K.-J. Kim: Proceedings of PAC 1987, p. 194
55. G. Geloni, V. Kocharyan, E. Saldin, J. Synch. Rad. **22**, 288 (2015)
56. A.A. Lutman et al., Nature Photonics **10**, 468 (2016)
57. A.M. Kondratenko, E.L. Saldin, Part. Accel. **10**, 207 (1980)
58. R. Bonifacio, C. Pellegrini, L.M. Narducci, Opt. Commun. **50**, 373 (1984)
59. E.L. Saldin, E.A. Schneidmiller, M.V. Yurkov: New J. Phys. **12**, 035 010 (2010)
60. C. Pellegrini, Eur. Phys. J. H **37**, 659 (2012)
61. I.A. Vartanyants et al.: Phys. Rev. Lett **107**, 144 801 (2011)
62. A. Singer et al.: Opt. Express **20**, 17 485 (2012)
63. R. Bonifacio, L.D. Salvo, P. Pierini, N. Piovella, C. Pellegrini, Phys. Rev. Lett. **73**, 70 (1994)
64. C. Pellegrini, A. Marinelli, S. Reiche: Rev. Mod. Phys. **88**, 015 006 (2016)
65. Zhao Chen, *X-Ray Studies of Magnetism in the 3d Transition Metals: From the Nanoscale to the Ultrafast to the Nonlinear Limit*, Ph. D. thesis, Department of Physics, Stanford University (2017)
66. T. Pfeifer, Y. Jiang, S. Düsterer, R. Moshammer, J. Ullrich, Opt. Lett. **35**, 3441 (2010)
67. D.J. Higley, K. Hirsch, G.L. Dakovski, E. Jal, E. Yuan, T. Liu, A.A. Lutman, J.P. MacArthur, E. Arenholz, Z. Chen, G. Coslovich, P. Denes, P.W. Granitzka, P. Hart, M.C. Hoffmann, J. Joseph, L.L. Guyader, A. Mitra, S. Moeller, H. Ohldag, M. Seaberg, P. Shafer, J. Stöhr, A. Tsukamoto, H.D. Nuhn, A.H. Reid, H.A. Dürr, W.F. Schlotter: Rev. Sci. Instrum. **87**, 033 110 (2016)
68. D. Ratner et al.: Phys. Rev. Lett. **114**, 054 801 (2015)
69. O.Y. Gorobtsov et al., Sci. Rep. **8**, 2219 (2018)
70. P. Emma, K. Bane, M. Cornacchia, Z. Huang, H. Schlarb, G. Stupakov, D. Walz: Phys. Rev. Lett. **92**, 074 801 (2004)
71. C. Bostedt, S. Boutet, D.M. Fritz, Z. Huang, H.J. Lee, H.T. Lemke, A. Robert, W.F. Schlotter, J.J. Turner, G.J. Williams: Rev. Mod. Phys. **88**, 015 007 (2016)
72. E. Allaria et al., Nat. Photonics **6**, 699 (2012)
73. O.Y. Gorobtsov et al., Nature Comm. **9**, 4498 (2018)
74. K.J. Kim, Y. Shvyd'ko, S. Reiche: Phys. Rev. Lett. **100**, 244 802 (2008)
75. V. Petrillo, M. Opromolla, A. Bacci, F. Broggi, I. Drebot, G. Ghiringhelli, E. Puppin, M.R. Conti, A.R. Rossi, M. Ruijter, S. Samsam, A. Tagliaferri, G. Rossi, L. Serafini: New J. Phys. **22**, 073 058 (2020)
76. N. Hartmann, G. Hartmann, R. Heider, M.S. Wagner, M. Ilchen et al., Nature Phot. **12**, 215 (2018)

Chapter 3
From Electromagnetic Waves to Photons

3.1 Introduction and Overview

In this chapter we review the description of EM radiation in a wave and quantum picture. The quantum picture has its origin in Dirac's quantization of the EM field in 1927 [1, 2], which is derived. The two pictures are equivalent in conventional quantum mechanics, which supports a wave-particle ambiguity. In the past, this has led to the widespread use of a "semi-classical" description of light that combines aspects of both descriptions. The fundamentals of the wave and particle concepts reviewed here serve as the basis for later chapters.

The wave description of light is discussed in terms of Maxwell's equations and the wave equation of the EM field. We then define various properties of EM waves such as their energy, momentum, intensity, flux, and fluence and in the process discuss the issue of normalizing the field to a volume in space. The description of light polarization is outlined with emphasis on linearly or circularly polarized light.

The quantum treatment of light is presented starting with the quantization of the EM field. The important concept of the zero-point field, which does not exist in a classical picture, is discussed and how it may be neglected in certain cases through a renormalization procedure. The basic quantum states of light, known as number or Fock states, are then introduced. They are used to define the so-called photon degeneracy parameter which has become of key importance with the advent of XFELs. The notion of different *modes* of light, defined through the propagation direction or wavevector, **k**, and the polarization, p, is introduced, and their fundamental relationship with the coherence volume is discussed. The chapter concludes by relating the field-based classical properties of light to the quantum properties.

J. Stöhr, *The Nature of X-Rays and Their Interactions with Matter*, Springer Tracts in Modern Physics 288, https://doi.org/10.1007/978-3-031-20744-0_3

3.2 Classical Description of Light

3.2.1 Maxwell's Equations and Their Symmetries

The properties of electric and magnetic fields are summarized in Maxwell's equations. In his book *Treatise on Electricity and Magnetism*, Maxwell summarized his theory in terms of 20 equations. As reviewed by Hunt [3], the form of today's four Maxwell equations is due to Oliver Heaviside who in 1884–1885 eliminated the vector and scalar potentials and expressed the electromagnetic relations purely in term of the electric, **E**, and magnetic, **H**, fields.

In the SI system of units, Maxwell's equations are

$$\nabla \cdot \mathbf{D} = \rho, \tag{3.1}$$

$$\nabla \cdot \mathbf{B} = 0, \tag{3.2}$$

$$\nabla \times \mathbf{E} = -\dot{\mathbf{B}}, \tag{3.3}$$

$$\nabla \times \mathbf{H} = \mathbf{j} + \dot{\mathbf{D}}. \tag{3.4}$$

The SI units of the different quantities are electric field E [V/m], magnetic field H [A/m], displacement $D = \epsilon\, \epsilon_0\, E$ [A s m^{-2}], magnetic induction $B = \mu\, \mu_0\, H$ [V s m^{-2} ≡ T], charge density ρ [A s m^{-3} ≡ C m^{-3}], and current density $\mathbf{j}$ [A m^{-2} = C s^{-1} m^{-2}]. The dotted quantities denote temporal derivatives.

Equations (3.1) and (3.2) determine the static electric and magnetic fields, respectively. In (3.3) we recognize Faraday's induction law. Ampère's law is contained in (3.4), but we see now that it is valid only for stationary electric currents where the dielectric field **D** does not change with time, i.e. $\dot{\mathbf{D}} = \partial \mathbf{D}/\partial t = 0$. In general, this contribution is non-zero.

For an assembly of point-like particles of charges q_i with velocities $\mathbf{v}_i$, located at positions $\mathbf{r}_i$, we can write the charge density in (3.1) by use of Dirac's delta function (see Appendix A.3) as

$$\rho(\mathbf{r}, t) = \sum_i q_i\, \delta(\mathbf{r} - \mathbf{r}_i(t)) \tag{3.5}$$

and the current density in (3.4) as

$$\mathbf{j}(\mathbf{r}, t) = \sum_i q_i\, \mathbf{v}_i\, \delta(\mathbf{r} - \mathbf{r}_i(t)). \tag{3.6}$$

Together with the Lorentz force equation,

$$\mathbf{F}_i = m_i \frac{d\mathbf{v}_i}{dt} = q_i \left[\mathbf{E}(\mathbf{r}_i(t), t) + \mathbf{v}_i \times \mathbf{B}(\mathbf{r}_i(t), t)\right] \tag{3.7}$$

we have a complete set of coupled equations that yields all results of classical electrodynamics.

In vacuum, Maxwell's equations are invariant under the parity transformation as discussed in [4], while this invariance typically breaks down in materials. The two main effects responsible for the breakdown are *transfer of energy and angular momentum* from the EM wave to the sample. The detailed understanding of the transfer of angular momentum from an atomic to a macroscopic length scale remains an important research topic in the area of modern magnetism [4], in particular the all optical switching of the magnetization [5].

3.2.2 The Electromagnetic Wave Equation

The most important feature of Maxwell's theory is that it predicts electromagnetic (EM) waves traveling at the speed of light c. It turns out that $c^2 = 1/\mu_0\epsilon_0$, hence the speed of light in vacuum, c, can be determined from experiments that measure the force between condenser plates or current carrying wires, experiments that do not seem to be connected to light at all. The equation that describes EM waves is derived from (3.4). In vacuum, we have $\mathbf{j} = 0$ since there are no charges, and with $\mathbf{D} = \epsilon_0\mathbf{E}$ and $\mathbf{B} = \mu_0\mathbf{H}$ we obtain

$$\epsilon_0\mu_0\dot{\mathbf{E}} - \nabla \times \mathbf{B} = 0. \tag{3.8}$$

Differentiating with respect to time t yields $\epsilon_0\mu_0\ddot{\mathbf{E}} - \nabla \times \dot{\mathbf{B}} = 0$ and with (3.3) we obtain

$$\epsilon_0\mu_0\ddot{\mathbf{E}} + \nabla \times (\nabla \times \mathbf{E}) = 0. \tag{3.9}$$

This is already the wave equation. We are left with only one field, namely $\mathbf{E}$, and the second derivative with respect to time is connected to the derivative with respect to location. By means of the identity $\nabla \times (\nabla \times \mathbf{E}) = \nabla(\nabla \cdot \mathbf{E}) - \nabla^2\mathbf{E}$ and since $\nabla \cdot \mathbf{E} = 0$ in vacuum we obtain the wave equation for the electric field vector of an EM wave traveling at the speed $c = 1/\sqrt{\mu_0\epsilon_0}$, in its familiar form.

In vacuum, and to a good approximation in air, *the wave equation for the electric field* is given by

$$\frac{1}{c^2}\ddot{\mathbf{E}} - \nabla^2\mathbf{E} = 0\,. \tag{3.10}$$

Any electric field $E(\mathbf{r}, t)$ that satisfies (3.10) is called an electromagnetic (EM)-wave. The equation shows that the *temporal variation* of $\mathbf{E}$ is directly linked with the *spatial variation*. The equation is invariant to time reversal as expected, and consequently, the optical path can be inverted as long as there is no damping of the wave.

One solution of (3.10) is $E(\mathbf{r}, t) = \text{const.}\, \mathrm{e}^{\pm \mathrm{i}(\mathbf{kr}-\omega t)}$. It describes a wave propagating in the direction of the wave vector $\mathbf{k}$ and has positive phase velocity. Another solution describing a wave that propagates in the opposite direction $-\mathbf{k}$ is $E(\mathbf{r}, t) = \text{const.}\, \mathrm{e}^{\pm \mathrm{i}(\mathbf{kr}+\omega t)}$. It has a negative phase velocity. The wave equation can also be written in terms of the magnetic induction $\mathbf{B}$ by simply substituting $\mathbf{E} \rightarrow \mathbf{B}$ in (3.10).

Within this book we shall make a choice of phase which follows that of Jackson [6] and is also commonly used in the X-ray literature [7, 8]. If we assume that the wave is in a pure linear or circular polarization state, discussed in more detail in Sect. 3.2.6, we define as follows.

We describe the *electric and magnetic fields of an EM wave* traveling in vacuum in the direction of the wavevector $\mathbf{k} = (\omega/c)\mathbf{k}_0$ as

$$\mathbf{E}(\mathbf{r}, t) = E_0 \left[\boldsymbol{\epsilon}_p \, e^{\mathrm{i}(\mathbf{k}\cdot\mathbf{r}-\omega t)} + \boldsymbol{\epsilon}_p^* \, e^{-\mathrm{i}(\mathbf{k}\cdot\mathbf{r}-\omega t)} \right], \tag{3.11}$$

and

$$\mathbf{B}(\mathbf{r}, t) = \frac{1}{c} \left[\mathbf{k}_0 \times \mathbf{E}(\mathbf{r}, t) \right]. \tag{3.12}$$

Here $\boldsymbol{\epsilon}_p$ is a unit polarization vector, which is real for linear polarized waves and complex for circularly polarized waves.

The electric field of a classical linearly polarized EM then has the real form,

$$\mathbf{E}(\mathbf{r}, t) = \mathbf{E}_0 \left[e^{\mathrm{i}(\mathbf{k}\cdot\mathbf{r}-\omega t)} + e^{-\mathrm{i}(\mathbf{k}\cdot\mathbf{r}-\omega t)} \right] = 2\, \mathbf{E}_0 \cos(\mathbf{k} \cdot \mathbf{r} - \omega t)\,. \tag{3.13}$$

Because of the factor of 2 on the right side, the electric field amplitude is sometimes defined as $2E_0$ in the literature. The real electric and magnetic fields for linearly polarized x-rays in space and time are shown in Fig. 3.1. It is remarkable that the wave equation describes EM waves with wave lengths $\lambda = 2\pi/k$ that span the enormous range from km to the dimensions of the atomic nucleus.

3.2.3 Energy, Momentum, Intensity, Flux, and Fluence

In the discussion of various x-ray phenomena in this book we will use different concepts and quantities to describe the properties of an incident x-ray beam, like its total energy, intensity, power or flux. It is therefore important to define these

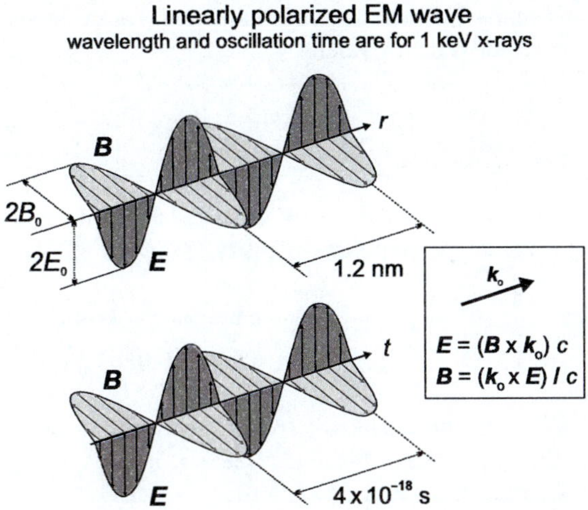

Fig. 3.1 Amplitude and phase of a linearly polarized electromagnetic wave according to (3.13). The electric field vector **E** characterizes the polarization direction. It is in phase with the magnetic field vector **B** and they are both perpendicular to the x-ray propagation, labeled by the unit wave vector $\mathbf{k_0}$

quantities and how they are linked. We do so here by using classical concepts based on the amplitude of the EM fields, but we shall see below that the same concepts can be used in a quantum description, where the fields are just replaced by the expectation value of the corresponding quantum mechanical operators.

The magnitude and direction of energy flow of an EM wave are given by the Poynting vector

$$\mathbf{S}(\mathbf{r}, t) = \mathbf{E}(\mathbf{r}, t) \times \mathbf{H}(\mathbf{r}, t), \tag{3.14}$$

whose dimension VA/m^2 shows that it represents the radiated power per unit area or the energy per unit time per unit area. It points in the direction of light propagation. With (3.11) and the **B**-field (3.12) expressed in terms of **H** through $\mathbf{B} = \mu_0 \mathbf{H}$ and use of $c = 1/\sqrt{\epsilon_0 \mu_0}$ we obtain

$$\mathbf{S}(\mathbf{r}, t) = \epsilon_0 \, c \, |\mathbf{E}(\mathbf{r}, t)|^2 \, \mathbf{k_0} = \mu_0 \, c \, |\mathbf{H}(\mathbf{r}, t)|^2 \, \mathbf{k_0}. \tag{3.15}$$

The amplitude of an electromagnetic wave propagating in a well-defined direction oscillates with time and position as expressed by (3.11) and (3.12).

Since x-ray detectors are too slow to resolve the oscillation time of the field, the measured response is always an intensity given by the cycle-averaged Poynting vector

$$I = \langle |\mathbf{S}(\mathbf{r}, t)| \rangle = \epsilon_0 \, c \, \langle |\mathbf{E}(\mathbf{r}, t)|^2 \rangle \tag{3.16}$$

with the cycle average given by

$$\langle |\mathbf{E}(\mathbf{r}, t)|^2 \rangle = 4|\mathbf{E}_0|^2 \, \big\langle \cos^2(\mathbf{k} \cdot \mathbf{r} - \omega t) \big\rangle = 2|\mathbf{E}_0|^2 \tag{3.17}$$

and similarly for the magnetic induction. The cycle-averaged magnitude of the Poynting vector is then given by

$$\langle|\mathbf{S}(\mathbf{r},t)|\rangle = 2\,\epsilon_0\,c\,|\mathbf{E}_0|^2 = 2\,\mu_0\,c\,|\mathbf{H}_0|^2. \tag{3.18}$$

The electric field $|\mathbf{E}|$ in [V m^{-1}] and magnetic field $|\mathbf{H}|$ in [A m^{-1}] are obtained from the incident power density $\langle\mathbf{S}\rangle$ in [W m^{-2}] = [V A m^{-2}] as

$$|\mathbf{E}_0| = 13.72\sqrt{\langle|\mathbf{S}|\rangle}, \tag{3.19}$$

$$|\mathbf{H}_0| = 0.0364\sqrt{\langle|\mathbf{S}|\rangle}. \tag{3.20}$$

Note that the prefactors are different from those in [4] owing to the factor 2 in the definitions of the fields (see Fig. 3.1).

Some important quantities associated with an EM wave include the *total energy* which is given by [6]

$$\mathcal{E} = \frac{1}{2}\int_V \left(\epsilon_0|\mathbf{E}|^2 + \frac{1}{\mu_0}|\mathbf{B}|^2\right)\mathrm{d}V = \epsilon_0\int_V |\mathbf{E}|^2\,\mathrm{d}V. \tag{3.21}$$

By confining the EM fields to a box of volume V and averaging over a wave period we obtain [6]

$$\boxed{\text{Energy:}\quad \langle\mathcal{E}\rangle = \epsilon_0 V\,\langle|\mathbf{E}(\mathbf{r},t)|^2\rangle = 2\epsilon_0 V|\mathbf{E}_0|^2.} \tag{3.22}$$

When the EM field is quantized, the energy given by (3.21) simply becomes the number of photons times the photon energy $\hbar\omega$. The *power* is the rate of energy flow. Writing the volume $V = A\,d$, where A is the lateral area to which the fields are confined and d the longitudinal dimension, we have with $d = c\,t$

$$\boxed{\text{Power:}\quad \mathcal{P}_0 = \frac{\langle\mathcal{E}\rangle}{t} = 2\epsilon_0\,c\,A\,|\mathbf{E}_0|^2\,,} \tag{3.23}$$

and it has the dimension [energy/time]. Using the same box normalization, the classical expression of the *momentum* of an EM wave is [6]

$$\boxed{\text{Momentum:}\quad \langle\mathbf{P}\rangle = \frac{\langle\mathcal{E}\rangle}{c} = \frac{2\,\epsilon_0 V\,|\mathbf{E}_0|^2}{c}.} \tag{3.24}$$

The momentum has the dimension [energy × (time/length)]. The momentum of light leads to *radiation pressure*. From a classical point of view, this arises from the Lorentz force due to the magnetic field of the EM wave $\mathbf{F} = q(\mathbf{v}\times\mathbf{B})$ that acts on charges that are set in motion by the Lorentz force due to the electric field $\mathbf{F} = q\mathbf{E}$.

Hence the momentum is in the direction of the Poynting vector. Another important property of an EM wave is its *intensity*, I, also called the *irradiance*, which is given by the Poynting vector

$$\boxed{\text{Intensity:} \quad I = \langle \mathbf{S} \rangle = 2\epsilon_0 \, c \, |\mathbf{E}_0|^2,} \tag{3.25}$$

and it has the dimension [power/unit area] or [energy/(time × unit area)] (see (3.23). The *photon flux*, Φ_0, is defined as the number of photons per second per unit area and is given by the magnitude of the average Poynting vector divided by the energy of a photon

$$\boxed{\text{Flux:} \quad \Phi_0 = \frac{\langle \mathbf{S} \rangle}{\hbar\omega} = \frac{2\,\epsilon_0 \, c \, |\mathbf{E}_0|^2}{\hbar\omega}.} \tag{3.26}$$

Finally, the *fluence* is defined as the electromagnetic energy per unit area A and is given by

$$\boxed{\text{Fluence:} \quad \mathcal{F} = \frac{\langle \mathcal{E} \rangle}{A} = \frac{2\,\epsilon_0 \, V \, |\mathbf{E}_0|^2}{A}.} \tag{3.27}$$

3.2.4 *Field Strength of X-Ray Beams*

In our above discussion we have relied on the classical concept of the strength of the EM fields given for the **E**-field component in SI units of [V/m] and for the **H**-field component in [A/m]. In principle, the rate of x-ray interactions with matter is indeed determined by the strength of the fields since they act on the charges and spins.[1] However, in practice, x-ray scientists do not think in terms of field strengths but rather define the flux or intensity in terms of the quantum concept of photons. Let us therefore calculate the size of the electric and magnetic fields in a typical x-ray beam available at third generation light sources and at XFELs.

We first calculate the average power density $\langle |\mathbf{S}| \rangle$ and fields for an x-ray beam at a typical storage ring. We use the *average* photon flux Φ_0 defined by (3.26), i.e. the number of photons per second per unit area. The cycle-averaged Poynting vector is then obtained by multiplying Φ_0 by the photon energy in [eV] and the conversion factor $1\text{ eV} = 1.6 \times 10^{-19}\,\text{W s}$. Assuming a photon flux $\Phi_0 = 10^{13}$ photons/(s mm^2) at a photon energy of 1000 eV and a band width of 1 eV, we obtain a power density $\langle |\mathbf{S}| \rangle \simeq 1.602 \times 10^3\ \text{W/m}^2$. This gives rather weak *average* fields $|\mathbf{E}_0| = 550$ V/m and $|\mathbf{H}_0| \simeq 1.5\ \text{A/m} = 1.8 \times 10^{-2}$ Oe.

[1] We shall later see that at optical and higher frequencies, one may in fact neglect the direct interaction of the magnetic field with the spins, so that one can calculate the interaction solely by considering the effect of the electric field on the electrons, with access to the spin arising only through the intra-atomic spin-orbit coupling (see Sect. 9.2.2).

Next we calculate the *peak* power density and fields during a single pulse. This value will be larger since we exclude the considerable dead time between pulses. We assume a typical 100 ps long pulse with 10^6 photons per pulse per eV and the same 1 mm^2 beam spot. This gives a power density per pulse of $\langle|\mathbf{S}|\rangle = 1.602 \times 10^6$ W/m^2 or $|\mathbf{E}_0| = 1.7 \times 10^4$ V/m and $|\mathbf{H}_0| = 46$ A/m = 0.58 Oe, which is about the size of the earth's magnetic field.

A XFEL is a pulsed source, and the relevant quantities are the peak fields. At a photon energy of 1000 eV an XFEL delivers about $\sim 10^{12}$ photons in an ultrashort flash of 10 fs, yielding a peak power of about 1×10^{10} W. This power is contained in the unfocused beam with a cross-sectional area of about $100 \times 100 \mu$m^2. In order to look at the extreme power density limit for such a beam, let us assume that we can focus the beam to a spot size of 1 μm^2 with a loss of a factor 2 through the optical system. This would yield a peak power density $\langle|\mathbf{S}|\rangle = 5 \times 10^{21}$ W/m^2. The fields are enormous in this case. We obtain $|\mathbf{E}_0| \simeq 1 \times 10^{12}$ V/m or on the scale of an atom 100 V/Å. The magnetic field is $|\mathbf{H}_0| = 2.6 \times 10^9$ A/m = 3.2×10^7 Oe = 3.2×10^3 T which rivals the molecular or exchange field in a ferromagnet.

The peak fields emitted by a typical synchrotron source are of order $|\mathbf{E}_0| \simeq 2 \times 10^4$ V/m and $|\mathbf{H}_0| \simeq 0.6$ Oe.

For a focused XFEL beam, the peak fields are $|\mathbf{E}_0| \simeq 1 \times 10^{12}$ V/m and $|\mathbf{H}_0| \simeq 3 \times 10^7$ Oe.

3.2.5 *Field Normalization Volume*

In practice, x-ray scientists describe beams in terms of the photon flux Φ_0 or number of photons n_0 that flow through a certain cross-sectional area of a sample per unit time. The flux is readily converted to intensity by multiplying by the photon energy. With the advent of x-ray lasers, one typically states the beam intensity instead of the flux, adopting the intensity units [J/(cm^2 s)] commonly used in the laser community.

The conversion of the classical field-based description to the quantum mechanical photon description requires dealing with the issue of the volume V that contains the fields and the photons with the EM energy (3.22) linked by

$$\mathcal{E} = 2\epsilon_0 |E_0|^2\, V = n_0 \hbar\omega. \tag{3.28}$$

In practice, the volume is typically taken to be the illuminated sample volume which is readily obtained from the measured beam diameter and the sample thickness.[2] We

[2] For x-rays, the natural choice of the volume V would actually be the atomic volume, since this is where the action is. While the atomic cross section is stated per atom, the flux is almost never stated per atom.

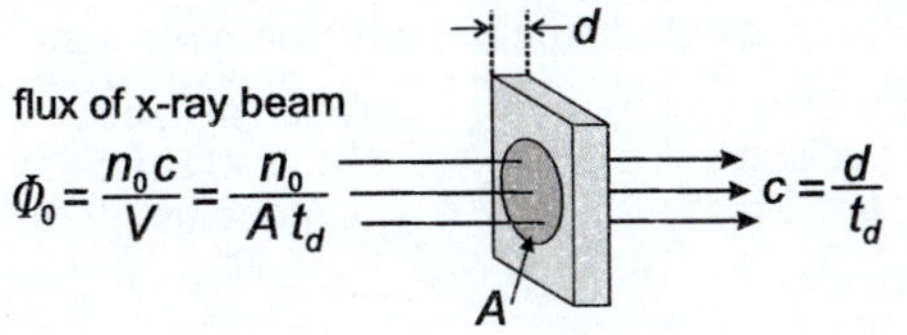

Fig. 3.2 Illustration of the x-ray flux, Φ_0, associated with n_0 photons through a sample volume $V = Ad$. One can then replace the thickness of the sample by the transit time t_d. By normalizing to the same time interval, one can state the flux as the number of photons flowing through the sample area A per chosen (i.e. unit) time interval

can then visualize the x-ray flux as illustrated in Fig. 3.2, as the number of photons flowing through a sample area A in a specified time. Classical and quantum concepts are then seamlessly linked according to

$$\langle \mathbf{S} \rangle = 2\epsilon_0 c\,|E_0|^2 = \frac{n_0\,c}{V}\hbar\omega = \Phi_0\,\hbar\omega. \tag{3.29}$$

We will come back to the tricky question of normalization volume V in Sect. 3.5 in conjunction with the quantization of the EM field and the definition of the coherence volume of a beam.

3.2.6 *The Description of Polarized EM Waves*

Our description of the polarization of light goes back to the work of Fresnel who around 1820 suggested the transverse nature of light and introduced the distinction between linearly and circularly polarized light [30]. Today, the polarization of "light" is defined through the motion of the *electric field* vector **E**. This convention is based on the fact that at optical and x-ray frequencies it is the **E** component of the EM field that determines the coupling to both the charges and spins of matter as discussed in Sect. 7.3.3. In the following we shall take a look at the mathematical description of light polarization. Since this has been extensively covered in textbooks from an optical light [9] and synchrotron radiation [10] perspective, we can here limit our discussion to important aspects that bear on the utilization of polarized x-rays, discussed in Sect. 7.9.3 and Chap. 11 below.

Polarized EM waves are described by (3.11) with different expressions for the unit polarization vector $\boldsymbol{\epsilon}_p$. The definition of *linear polarization* is simplified by the fact that the **E**-vector oscillates in time and space along an axis perpendicular to the propagation direction **k** as shown in Fig. 3.1. For linear polarization, $\boldsymbol{\epsilon}_p$ is *real* and if we choose the propagation direction **k** into the z direction, the vector **E** will lie in the $x - y$ plane. The two unit vectors $\boldsymbol{\epsilon}_x$ and $\boldsymbol{\epsilon}_y$ then define two orthogonal polarization directions. Since the **E**-field oscillates in time through the phase term $e^{-i\omega t}$, the directions $\pm\boldsymbol{\epsilon}_p$ with $p = x$ or y are equivalent.

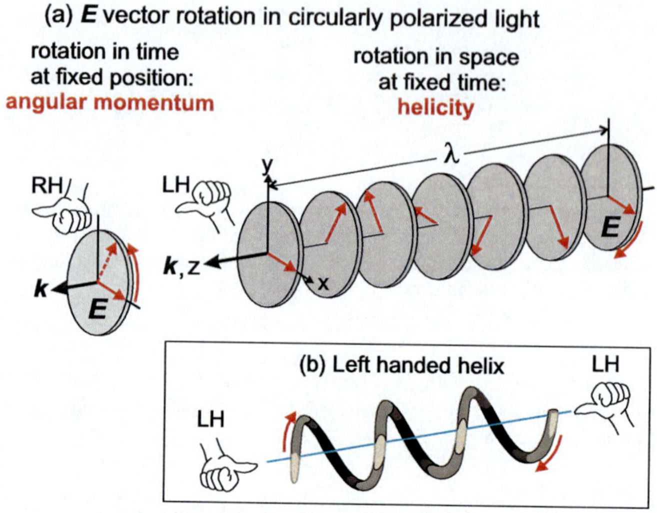

Fig. 3.3 **a** Illustration of the different rotation senses in space and time of the **E**-vector in circularly polarized light, which are associated with the concepts of *angular momentum* and *helicity*. **b** Rotation sense in a left circular helix which also reflects the spatial **E**-vector rotation sense in (**a**). The handedness of the helix is preserved upon rotation in space. It is the same if one looks at its handedness along z or $-z$, as illustrated by the direction of the thumb and fingers of a left hand

The description of a *circularly polarized* wave is more complicated since there is an ambiguity associated with the rotation sense of a circularly polarized wave in space and time, as illustrated in Fig. 3.3. The rotation sense is seen to be opposite in space and time, as reflected by the opposite signs of the spatial and temporal phases in $e^{i(\mathbf{k}\cdot\mathbf{r}-\omega t)}$. This has led to different definitions of circular polarization in the literature.

The two definitions of "right" or "left" handedness are based on using the thumb of the right and left hands, respectively, to define a direction in space, with the fingers defining the rotation sense relative to that direction. Using this concept, the light propagation direction is defined by the thumb, and the rotation is defined by that of the **E**-vector. Problem is that the rotation sense of **E** is opposite in time and space. In optics, the **E** rotation as left and right is envisioned in space [9], while in particle physics it is envisioned in time.

In this book we define the *photon angular momentum* to reflect the rotation of the **E**-vector in *time* at a given point around the propagation wavevector **k** as shown in Fig. 3.3a. The opposite rotation of **E** in *space* then follows a helix as illustrated more clearly in Fig. 3.3b. Circular light incorporates both properties of angular momentum (time) and helicity (space), and we can state as follows.

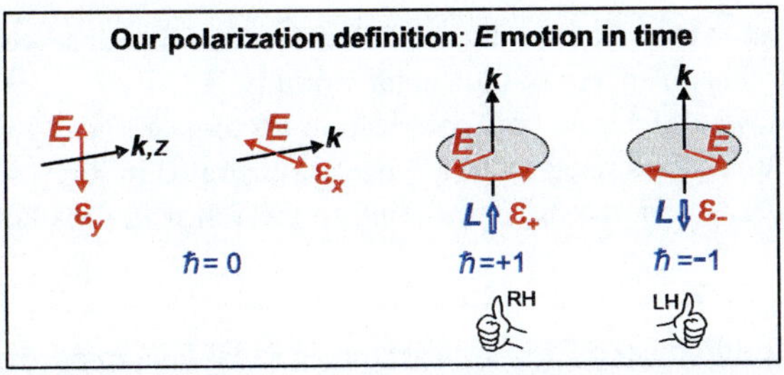

Fig. 3.4 Illustration of our definition of the two basis states of linearly and circularly polarized light, defined as the motion of the polarization vector **E** in *time*. For *linearly polarized* light, there are two orthogonal axes perpendicular to **k** along which **E** can oscillate in time. For *circularly polarized* light, the two basis states $\boldsymbol{\epsilon}_+$ and $\boldsymbol{\epsilon}_-$ are defined by the rotation sense of **E** in *time*, illustrated by right and left hands with the thumb pointing into the propagation direction **k** and the finger defining the rotation sense (also see Fig. 2.25). The unit vector $\boldsymbol{\epsilon}_+$ corresponds to an angular momentum of light $L_z = +\hbar$, and $\boldsymbol{\epsilon}_-$ to $L_z = -\hbar$, while linearly polarized light has $L_z = 0$

> *Circularly polarized light* incorporates the temporal property of angular momentum and spatial property of helicity. The existence of an angular momentum reflects the breaking of *time reversal symmetry*, while the existence of helicity embodies the lack of *inversion symmetry*, or the violation of parity.

Figure 3.3 illustrates the following interesting point. The *temporal* rotation of **E** at a fixed point occurs within a thin spatial slice shown as a gray disk. The thickness of this disk is much smaller than the wavelength λ. The motion of **E** within the disk for either linear and circular light is equivalent to the famous dipole approximation, which approximates the spatial motion by $e^{i\mathbf{k}\cdot\mathbf{r}} \simeq 1 + i\mathbf{k}\cdot\mathbf{r} \simeq 1$. This approximation is valid for the description of the Faraday effect which is related to the modern x-ray *magnetic* circular dichroism (XMCD) effect. Both only depend on the angular momentum of light, i.e. its temporal rotation, so that the effects can indeed be described within the dipole approximation (see Sect. 11.5).

In contrast, the phenomenon of optical activity of chiral systems requires a description beyond the dipole approximation. In this case the **k** dependence cannot be neglected since it accounts for the spatial twist of light. To the lowest order, the light twist or chirality can be accounted for by the approximation $e^{i\mathbf{k}\cdot\mathbf{r}} \simeq 1 + i\mathbf{k}\cdot\mathbf{r}$, which is used to calculate the optical activity and related x-ray *natural* circular dichroism (XNCD) effects (see Sect. 11.4).

Our discussion readily gives a simple explanation for the existence of the two main effects associated with circularly polarized x-rays. XMCD arises from the difference in parallel versus antiparallel coupling between the angular momentum of light (or "photon spin") and that of the electronic system, i.e. its magnetic moment. This was first revealed by the magneto-optical effect, discovered by Faraday's in 1846 [11].

In contrast, the XNCD effect is due to the difference in parallel versus antiparallel coupling between the spatial helicity of light and the chirality of the electronic charge, i.e. a spatial twist in the valence orbitals. This effect was first observed by Pasteur in 1850, leading to his discovery of molecular chirality [12].

In this book, we define the unit polarization vectors of linearly polarized light, $\boldsymbol{\epsilon}_x, \boldsymbol{\epsilon}_y$, and circularly polarized light, $\boldsymbol{\epsilon}_+, \boldsymbol{\epsilon}_-$, as illustrated in Fig. 3.4.

The unit polarization vectors correspond to the following mathematical definitions.

The two unit orthogonal polarization vectors in a real *linear polarized* basis and complex *circularly polarized* basis are given by

$$\boldsymbol{\epsilon}_x = \frac{1}{\sqrt{2}}(\boldsymbol{\epsilon}_- - \boldsymbol{\epsilon}_+), \quad \boldsymbol{\epsilon}_y = \frac{\mathrm{i}}{\sqrt{2}}(\boldsymbol{\epsilon}_- + \boldsymbol{\epsilon}_+),$$
$$\boldsymbol{\epsilon}_+ = -\frac{1}{\sqrt{2}}\left(\boldsymbol{\epsilon}_x + \mathrm{i}\boldsymbol{\epsilon}_y\right), \quad \boldsymbol{\epsilon}_- = \frac{1}{\sqrt{2}}\left(\boldsymbol{\epsilon}_x - \mathrm{i}\boldsymbol{\epsilon}_y\right) \tag{3.30}$$

Note that the complex nature of the collinear vectors $\boldsymbol{\epsilon}_+$ and $\boldsymbol{\epsilon}_-$ allows them to be orthogonal.

Our choice of signs is that used to describe light polarization in quantum mechanics through the first order spherical tensors (see Table A.2 in Appendix A.5).[3]

The most general form of light may be expressed as a linear combination of either the two linear or circular basis states in (3.30) according to [4, 9]

$$\mathbf{E}(z,t) = \frac{1}{\sqrt{2}}\left(\boldsymbol{\epsilon}_x\, E_x\, e^{i(kz-\omega t)+i\phi_x} + \boldsymbol{\epsilon}_y\, E_y\, e^{i(kz-\omega t)+i\phi_y}\right) \tag{3.31}$$

or

$$\mathbf{E}(z,t) = \frac{1}{\sqrt{2}}\left(\boldsymbol{\epsilon}_+\, E_+\, e^{i(kz-\omega t)+i\phi_+} + \boldsymbol{\epsilon}_-\, E_-\, e^{i(kz-\omega t)+i\phi_-}\right). \tag{3.32}$$

The linear combination coefficients are complex numbers with different amplitudes E_i and phases ϕ_i. This general form describes the following common forms of light.

- If the two amplitudes are the same, $E_i = E_j$ and the phases ϕ_i and ϕ_j are random, the wave is called *naturally* polarized, sometimes referred to as "unpolarized". The endpoints of the randomly oriented **E**-vector lie on a circle.
- If one of the two amplitudes E_x or E_y is zero, the wave is called *linearly* polarized and the endpoint of **E** oscillates along a line in space.
- If one of the two amplitudes E_+ or E_- is zero, the wave is called *circularly* polarized with angular momentum $+\hbar$ or $-\hbar$, and the endpoint of **E** traces out a circle.
- If the two amplitudes are different $E_i \neq E_j$ and the phases ϕ_i and ϕ_j are linked according to $\phi_i - \phi_j = const$, the wave is called *elliptically* polarized since the

[3] The minus sign in the definition of $\boldsymbol{\epsilon}_+$ is sometimes replaced by a plus sign. This is unimportant and corresponds to an arbitrary phase shift of π.

endpoint of **E** traces out an ellipse [9]. However, there are certain values $\phi_i - \phi_j = n\pi$ with $n = 0, 1, 2$, where the wave is linearly polarized along specific directions in the $(x - y)$ plane, and other values $\phi_i - \phi_j = m\pi/4$ with $m = 1, 2, 3, 5, 6, 7$ where the elliptical wave has an angular momentum and helicity [9].

Owing to the transverse nature of light, the **E**- and **B**-vectors are always in the plane perpendicular to the propagation direction. For this reason the term "unpolarized" light is a misnomer. For naturally polarized light one can always calculate the measured intensities as an incoherent sum of the two linear or circular fields, i.e. by simply adding intensities for the two polarizations.

3.2.7 *The Degree of X-Ray Polarization*

Synchrotron radiation emitted by bending magnets is elliptically polarized as discussed in Sect. 2.7.4, and it can be described in terms of its linear or circular components. The total emitted radiation is elliptically polarized and contains a dominant linear component in the plane of the storage ring, $\mathbf{E}_{\|}$, and the smaller field component perpendicular to the ring plane, $\mathbf{E}_{\perp}$. The entire emitted radiation contains no net circular polarization, but the balance between the two circular components may be broken by use of an aperture to select only parts of the beam above or below the orbit plane of the storage ring.

In general, the *degrees of polarization* are defined in terms of polarization dependent intensities

$$I_\alpha = \epsilon_0 c\, |\mathbf{E}_\alpha|^2, \tag{3.33}$$

where $\mathbf{E}_\alpha = \boldsymbol{\epsilon}_\alpha E_\alpha$ are the **E**-vector components in (3.31) and (3.32). In our previous linear polarization notation in Fig. 2.26 we have $\mathbf{E}_x = \mathbf{E}_{\|}$ and $\mathbf{E}_y = \mathbf{E}_{\perp}$ and the *degree of linear polarization* is typically defined as

$$P_{\text{lin}} = \frac{|\mathbf{E}_{\|}|^2 - |\mathbf{E}_{\perp}|^2}{|\mathbf{E}_{\|}|^2 + |\mathbf{E}_{\perp}|^2} \tag{3.34}$$

For the description of polarization dependent spectroscopic measurements it is more convenient to use the following related definition [13] (see Sect. 7.10.2)

$$P = \frac{|\mathbf{E}_{\|}|^2}{|\mathbf{E}_{\|}|^2 + |\mathbf{E}_{\perp}|^2}. \tag{3.35}$$

We have $0 \leq P_{\text{lin}} \leq 1$ and $0.5 \leq P \leq 1$, with typical values shown in Fig. 2.26.

The *degree of circularly polarized radiation* is defined in analogy to (3.34) as

$$P_{\mathrm{circ}} = \frac{\left||\mathbf{E}_+|^2 - |\mathbf{E}_-|^2\right|}{|\mathbf{E}_+|^2 + |\mathbf{E}_-|^2}. \tag{3.36}$$

With undulator sources have become common today to obtain essentially 100% linearly or circularly polarized radiation for both storage rings and XFELs.

3.3 Quantum Theoretical Description of X-Rays

Classical electrodynamics describes electromagnetic radiation as a wave of oscillating electric and magnetic fields. A complementary particle-like picture has also been discussed for centuries going back to Newton. This picture finds its modern realization in the now common concept of the "photon", a name coined by Gilbert Lewis in 1926 [14].

The concept of the quantized nature of the radiation field was first proposed by Einstein in 1905 [15] as part of what is now called the "old quantum theory". Einstein's visionary paper, which won him the Nobel Prize, is today mostly remembered in conjunction with the photoelectric effect, but its importance is much broader. He proposed that the production of electrons by ultraviolet light can be understood by assuming that the energy of light is not continuous but consists of "a finite number of energy quanta, localized in space, which move without being divided and which can be absorbed or emitted only as a whole". Ironically, in later years Einstein did not like the name "photon", although he clearly had this concept in mind.

The rigorous quantization of the EM field was carried out by Dirac in 1927 [1], who in the same paper also derived the quantum mechanical expression for x-ray absorption and emission, a few years later called by Fermi the "golden rule" [16]. Dirac's field quantization allowed first-principles calculations of the intensity of electronic transitions in x-ray absorption and x-ray emission, which was not possible within the old quantum theory. We shall introduce it now.

3.3.1 *The Quantized Electromagnetic Field: Birth of the Photon*

In the previous Sect. 3.2 we described the EM wave by its electric field **E** and magnetic induction **B**, which we have called a "field" because of its simple relationship in free space with the true magnetic field $\mathbf{H} = \mathbf{B}/\mu_0$. The fields provide a good intuitive description of the EM wave which has proven valuable in our classical considerations of electromagnetism. There is, however, an equivalent way to describe the EM wave. It is by means of the scalar potential Φ and the vector potential **A** which are linked to the **E**- and **B**-fields by the equations

$$\mathbf{E}(\mathbf{r}, t) = -\nabla\Phi(\mathbf{r}, t) - \frac{\partial \mathbf{A}(\mathbf{r}, t)}{\partial t} \tag{3.37}$$

and

$$\mathbf{B}(\mathbf{r}, t) = \nabla \times \mathbf{A}(\mathbf{r}, t) . \tag{3.38}$$

Since gauge transformations do not alter the **E**- and **B**-fields, we shall work in the so-called *Coulomb or transverse gauge* by choosing $\nabla \cdot \mathbf{A} = 0$ and $\Phi = 0$. This corresponds to the assumption that the fields are transverse without a longitudinal component. We obtain simplified relations of considerable importance. Note, that $\Phi = 0$ holds for EM waves in vacuum where no additional potentials due to charges exist.

The *vector potential* **A** of the EM waves defines the **E**- and **B**-fields and contains the complete information on field through the relations

$$\mathbf{E}(\mathbf{r}, t) = -\frac{\partial \mathbf{A}(\mathbf{r}, t)}{\partial t} \tag{3.39}$$

$$\mathbf{B}(\mathbf{r}, t) = \nabla \times \mathbf{A}(\mathbf{r}, t). \tag{3.40}$$

We also have according to (3.12)

$$\mathbf{B}(\mathbf{r}, t) = \frac{1}{\omega}\,[\mathbf{k} \times \mathbf{E}(\mathbf{r}, t)] = \mathrm{i}\,[\mathbf{k} \times \mathbf{A}(\mathbf{r}, t)] . \tag{3.41}$$

When we write an expression for the energy density of the EM field in terms of **A**, we can, by analogy with the harmonic oscillator, derive an important new concept, the quantization of the electromagnetic field.

In Sect. 3.2.3 we have seen that the *energy density* $\mathrm{d}\mathcal{E}/\mathrm{d}V$ of an EM wave can be expressed as a function of $|\mathbf{E}|^2$ and $|\mathbf{B}|^2$. By use of the link of the fields with the vector potential, we can write

$$\frac{\mathrm{d}\mathcal{E}}{\mathrm{d}V} = \frac{1}{2}\left(\epsilon_0|\mathbf{E}|^2 + \frac{1}{\mu_0}|\mathbf{B}|^2\right) = \frac{1}{2}\left(\epsilon_0\left|-\frac{\partial \mathbf{A}}{\partial t}\right|^2 + \frac{1}{\mu_0}|\mathbf{k} \times \mathbf{A}|^2\right) . \tag{3.42}$$

Let us confine the fields to some region of space described by a large cubic box with volume $L^3 = V$, known as the *quantization cavity*. It will drop out of all expressions for observable quantities. We can rewrite the right side of (3.42) in terms of a Hamiltonian $\mathcal{H}_{\mathrm{rad}} = \mathcal{E}$, and by use of $|\mathbf{k}| = \omega/c$ and $c^2 = 1/\epsilon_0\mu_0$ we get

$$\mathcal{H}_{\mathrm{rad}} = \frac{\epsilon_0 V}{2}|\dot{\mathbf{A}}|^2 + \frac{\epsilon_0 V\omega^2}{2}|\mathbf{A}|^2 . \tag{3.43}$$

This resembles the form of the Hamiltonian of a one-dimensional harmonic oscillator with mass m whose energy $\mathcal{E}_{\rm osc}$ depends on position x according to

$$\mathcal{E}_{\rm osc} = \frac{m}{2}\dot{x}^2 + \frac{m\,\omega^2}{2}x^2 \,. \tag{3.44}$$

We now utilize the well-established formalism for the harmonic oscillator, which is found to consist of quantized energy levels, as discussed in many textbooks. One rewrites (3.44) in the quantized Hamiltonian form

$$\mathcal{H}_{\rm osc} = \hbar\omega\left(a^\dagger a + \frac{1}{2}\right), \tag{3.45}$$

with the solutions given by

$$x = \sqrt{\frac{\hbar}{2m\omega}}\left(a^\dagger e^{i\omega t} + a e^{-i\omega t}\right), \tag{3.46}$$

where a and $a^\dagger$ are *ladder* operators that obey the commutation relation $[a, a^\dagger] = 1$ and act on eigenfunctions $|n\rangle$ according to

$$a^\dagger|n\rangle = \sqrt{n+1}\,|n+1\rangle \tag{3.47}$$

$$a|n\rangle = \sqrt{n}\,|n-1\rangle, \quad n \geq 1 \tag{3.48}$$

$$a|0\rangle = |0\rangle. \tag{3.49}$$

The states $|n\rangle$ are called the *number states* of the harmonic oscillator. We will see below that they become the basic states of the quantum theory of light.

The fields and vector potential of the oscillating EM field have the form $\exp[\pm \mathrm{i}(\omega t - \mathbf{k}\cdot\mathbf{r})]$, and by direct comparison with (3.46) we write $\mathbf{A}$ in such a form. Introducing the unit polarization vector $\boldsymbol{\epsilon}$ and using the correspondence of (3.43) and (3.44) or $m \leftrightarrow \epsilon_0 V$ and $x \leftrightarrow A$, we directly obtain from (3.46)

$$\mathbf{A}^p(\mathbf{r}, t) = \boldsymbol{\epsilon}_p\sqrt{\frac{\hbar}{2\epsilon_0 V\omega}}\left[a_{\mathbf{k}}\, e^{\mathrm{i}(\mathbf{k}\cdot\mathbf{r}-\omega t)} + a_{\mathbf{k}}^\dagger\, e^{-\mathrm{i}(\mathbf{k}\cdot\mathbf{r}-\omega t)}\right]. \tag{3.50}$$

The vector potential (and similarly the field) is that for a polarization state indicated by the label p and a wavevector state $\mathbf{k}$, which together define a so-called *mode* of the EM field. The quantized vector potential is contained in the volume V which is called the *quantization volume* or quantization cavity. At this stage, the volume is ill defined, and we only require it to be large enough to contain the EM field. We shall develop a better understanding of the meaning of V in Sect. 3.5. The Hamiltonian for the radiation field in a given polarization mode p is given in analogy to (3.45) as

$$\mathcal{H}_{\text{rad}} = \hbar\omega_{\mathbf{k}} \left(a_{\mathbf{k}}^{\dagger} a_{\mathbf{k}} + \frac{1}{2} \right). \tag{3.51}$$

We have now quantized the EM field. Its energy is quantized in units of $\hbar\omega_{\mathbf{k}}$, the energy we associate with a single photon. Remarkably, the Hamiltonian (3.51) contains two terms. The first is what we today call the photon energy, and the second term corresponds to the "zero-point energy". This term ("Nullpunktsenergie" in German) was phrased by A. Einstein and O. Stern in 1912 [31] who, in analyzing the specific heat of hydrogen gas, followed an earlier suggestion by Planck [17] about the existence of a residual energy that all oscillators have at absolute zero. It reflects the fact that the energy of a quantized system, in contrast to a classical system, can never reach the bottom of a potential well. This would require that the position coordinate and its momentum simultaneously have the precise values zero, in violation of the uncertainty principle.

In analogy of the eigenstates $|n\rangle$ of the harmonic oscillator, one defines photon states with similar properties, called *number states*, as discussed in more detail in Sect. 3.4.2. When the operator $a_{\mathbf{k}}^{\dagger}$ acts on a photon state, it adds a photon in the same state, and it is therefore called a *creation operator*, while $a_{\mathbf{k}}$ destroys a photon and is called an *annihilation operator*.

In quantum optics, (3.51) is written by summing over the two polarization basis states p and over all wavevectors $\mathbf{k}$

$$\mathcal{H}_{\text{rad}} = \sum_{p\mathbf{k}} \hbar\omega_k \left(a_{p\mathbf{k}}^{\dagger} a_{p\mathbf{k}} + \frac{1}{2} \right). \tag{3.52}$$

This equation is the basis for the concept of photons labeled by their *modes* $p\mathbf{k}$, the polarization p, the wavevector $\mathbf{k}$, and its associated photon energy $\hbar\omega_k = \hbar c|\mathbf{k}|$.

The quantized EM field contains photons in different *modes*, labeled by their polarization p and their wavevector $\mathbf{k}$. There are two orthogonal polarization modes, either the two perpendicular linear polarization modes or the two circular modes with opposite angular momenta. The wavevector magnitude $|\mathbf{k}|$ also defines the photon energy according to $\hbar\omega_k = \hbar c|\mathbf{k}|$.

The modes uniquely characterize the quantized field in terms of the propagation direction, $\mathbf{k}$, the direction of the $\mathbf{E}$-vector in the plane perpendicular to $\mathbf{k}$, and the photon energy $\hbar c|\mathbf{k}|$.

The quantized field also contains a "background" whose energy, called zero-point energy, is given by the term $\hbar\omega_k/2$ in (3.52). Real photons correspond to elementary excitations of the zero-point field, reflecting the creation of light out of darkness.

3.3.2 Zero-Point Energy and Virtual Photons

According to (3.52), the lowest possible energy is one-half quantum per mode of the field, expressed by $\hbar\omega_k/2$. It is the energy of the quantum mechanical ground state which by the uncertainty principle is greater than the minimum of a classical potential well. Even at absolute zero temperature, quantum mechanical systems in their ground state undergo fluctuations, and one obtains from (3.50)

$$\langle 0|\mathbf{A}^2|0\rangle = \frac{\hbar}{2\epsilon_0 V\omega}. \tag{3.53}$$

Today the original concept of the zero-point (ZP) energy, whose interesting history has been reviewed in [32], has been extended to cover all fields, which in the standard model of particle physics includes all gauge fields and the Higgs field [18]. One therefore also uses the term "ZP field" for the lowest energy state of a particular field. In quantum field theory the ground state of the fields is associated with the alternative name "vacuum energy". In cosmology, the vacuum energy is one possible explanation for the cosmological constant. A problem of quantum field theory is that the field contributions from all points in space results in infinite zero-point energy and renormalization is required.

From an EM radiation point of view, the ZP energy is caused by *virtual photons* with random directions and polarization states. Summing over the two possible polarization basis states (either the two orthogonal linear of circular polarization modes) there is one virtual photon per wavevector, $\mathbf{k}$, or energy, $\hbar\omega_k$, mode. In the treatment of photon interactions with matter, one typically ignores these virtual ZP photons and the interaction Hamiltonian (see 3.54 below) is renormalized to exclude them.

Upon closer inspection, however, the properties of an atom are slightly modified by its interaction with the vacuum field, leading to effects such as (i) spontaneous emission, (ii) the Lamb shift, i.e. the differential shift of the $2S_{1/2}$ and $2P_{1/2}$ levels in hydrogen due to the electric part of the ZP field, (iii) the slight deviation of the magnetic moment of the free electron from the value $2\mu_B$ due to the magnetic part of the ZP field, and (iv) the Casimir effect, leading to a force between two electrically neutral metal plates.

3.3.3 The Renormalized Hamiltonian of the Radiation Field

Given the caveats about the zero-point energy discussed above, we can describe the interaction of photons with matter by shifting the zero of energy to coincide with that of the vacuum state. This is done by subtracting $\frac{1}{2}\sum_{p,\mathbf{k}} \hbar\omega_{\mathbf{k}}$ from (3.52) [19, 20]. In the following we will denote the photon creation and annihilation operators in bold to more clearly indicate their operator nature.

The normalized Hamiltonian of the radiation field is

$$\mathcal{H}_{\text{rad}} = \sum_{p\mathbf{k}} \hbar\omega_{\mathbf{k}}\, \mathbf{a}^{\dagger}_{p\mathbf{k}}\, \mathbf{a}_{p\mathbf{k}}. \tag{3.54}$$

It is the basis for the concept of a photon with polarization p, wavevector $\mathbf{k}$, and energy $\hbar\omega_{\mathbf{k}}$.

In general, we also rewrite (3.50) by summing over the two polarization basis states p, defined by complex polarization vectors $\boldsymbol{\epsilon}_p$, and over all wavevectors $\mathbf{k}$.

The operator form of the quantized vector potential is

$$\mathbf{A}(\mathbf{r}, t) = \sum_{p\mathbf{k}} \sqrt{\frac{\hbar}{2\epsilon_0 V \omega_{\mathbf{k}}}} \left[\boldsymbol{\epsilon}_p\, \mathbf{a}_{p\mathbf{k}}\, e^{\mathrm{i}(\mathbf{k}\cdot\mathbf{r}-\omega_{\mathbf{k}}t)} + \boldsymbol{\epsilon}^{*}_{p}\, \mathbf{a}^{\dagger}_{p\mathbf{k}}\, e^{-\mathrm{i}(\mathbf{k}\cdot\mathbf{r}-\omega_{\mathbf{k}}t)}\right]. \tag{3.55}$$

V is known as the *quantization volume*, and we will discuss it more in Sect. 3.5. The vector potential is real, so that $\mathbf{A}(\mathbf{r}, t) = \mathbf{A}^{*}(\mathbf{r}, t)$. It is made up of *photon modes* labeled by the double index $\mathbf{k}p$. Its expectation value has units [Vs/m]. Note that photons are Bosons and that each mode $p\mathbf{k}$ can contain more than one photon.

We now need to establish some practical rules that will allow us to use the field quantization concept when calculating x-ray absorption and scattering processes. The key to such calculations is the evaluation of operator matrix elements that involve the creation and annihilation operators in the operator $\mathbf{A}$ in (3.55). Since $\mathbf{A}$ is linear in the creation and annihilation operators, it can be written as a sum of operators that induce absorption and emission of a photon.

The vector potential can be written as a sum of operators that induce absorption and emission of a photon.

$$\mathbf{A}(\mathbf{r}, t) = \mathbf{A}^{\text{ab}}(\mathbf{r}, t) + \mathbf{A}^{\text{em}}(\mathbf{r}, t) \tag{3.56}$$

with

$$\mathbf{A}^{\text{ab}}(\mathbf{r}, t) = \mathbf{A}^{+}(\mathbf{r}, t) = \sum_{p\mathbf{k}} \sqrt{\frac{\hbar}{2\epsilon_0 V \omega_{\mathbf{k}}}}\, \boldsymbol{\epsilon}_p\, \mathbf{a}_{p\mathbf{k}}\, e^{\mathrm{i}(\mathbf{k}\cdot\mathbf{r}-\omega_{\mathbf{k}}t)} \tag{3.57}$$

$$\mathbf{A}^{\text{em}}(\mathbf{r}, t) = \mathbf{A}^{-}(\mathbf{r}, t) = \sum_{p\mathbf{k}} \sqrt{\frac{\hbar}{2\epsilon_0 V \omega_{\mathbf{k}}}}\, \boldsymbol{\epsilon}^{*}_{p}\, \mathbf{a}^{\dagger}_{p\mathbf{k}}\, e^{-\mathrm{i}(\mathbf{k}\cdot\mathbf{r}-\omega_{\mathbf{k}}t)} \tag{3.58}$$

where $\mathbf{A}^+(\mathbf{r}, t)$ destroys and $\mathbf{A}^-(\mathbf{r}, t)$ creates photons.

X-ray absorption, corresponding to the destruction of a photon, and x-ray emission, corresponding to the creation of a photon, therefore involve matrix elements of the respective operators (3.57) and (3.58), only. In contrast, a non-resonant elastic x-ray scattering process involves destruction or removal of a photon in one state and creation or addition of a photon in another state with a different wave vector and polarization. This process therefore involves matrix elements of both operators, in particular, an operator of the form $\mathbf{A}^{ab}\mathbf{A}^{em}$. By learning the rules how these operators act on wave functions that describe photons we will be in a position to treat the interactions between photons and matter.

In quantum optics, different kinds of photon beams are distinguished through the properties of their quantized electric field rather than the vector potential [21–23, 33]. The two formulations are linked through the classical expression $\mathbf{E} = -\partial\mathbf{A}/\partial t$, and one uses a similar nomenclature for the description of the classical and quantum fields. We start from the expression of the classical field given by (3.13) and assume linear polarized light so that the classical field is given by the real expression

$$\mathbf{E}(\mathbf{r}, t) = \mathbf{E}_0 \left[e^{\mathrm{i}(\mathbf{k}\cdot\mathbf{r}-\omega t)} + e^{-\mathrm{i}(\mathbf{k}\cdot\mathbf{r}-\omega t)}\right] = 2\,\mathbf{E}_0 \cos(\mathbf{k}\cdot\mathbf{r} - \omega t). \tag{3.59}$$

To obtain an expression for the quantum field operator, we differentiate the quantized vector potential (3.55) with respect to time and obtain

$$\mathbf{E}(\mathbf{r}, t) = -\frac{\partial \mathbf{A}}{\partial t} = \mathrm{i}\sum_{p\mathbf{k}} E_0\left[\boldsymbol{\epsilon}_p\, \mathbf{a}_{p\mathbf{k}}\, e^{\mathrm{i}(\mathbf{k}\cdot\mathbf{r}-\omega_{\mathbf{k}} t)} - \boldsymbol{\epsilon}_p^*\, \mathbf{a}_{p\mathbf{k}}^\dagger\, e^{-\mathrm{i}(\mathbf{k}\cdot\mathbf{r}-\omega_{\mathbf{k}} t)}\right]. \tag{3.60}$$

It is convenient to get rid of the prefactor of i in (3.60) which is of no significance. We eliminate it by the phase substitution $(\mathbf{k}\cdot\mathbf{r} - \omega_{\mathbf{k}} t) \to (\mathbf{k}\cdot\mathbf{r} - \omega_{\mathbf{k}} t) - \pi/2$ which yields the phase notation of Scully and Zubairy [34]. From now on we will use the following quantum operator form for the electric field.

In analogy to the classical field given by (3.11), the quantum mechanical operator for the EM field is given by

$$\mathbf{E}(\mathbf{r}, t) = \underbrace{\sum_{p\mathbf{k}} E_0\, \boldsymbol{\epsilon}_p\, \mathbf{a}_{p\mathbf{k}}\, e^{\mathrm{i}(\mathbf{k}\cdot\mathbf{r}-\omega_{\mathbf{k}} t)}}_{\mathbf{E}^+(\mathbf{r},t)} + \underbrace{\sum_{p\mathbf{k}} E_0\, \boldsymbol{\epsilon}_p^*\, \mathbf{a}_{p\mathbf{k}}^\dagger\, e^{-\mathrm{i}(\mathbf{k}\cdot\mathbf{r}-\omega_{\mathbf{k}} t)}}_{\mathbf{E}^-(\mathbf{r},t)}, \tag{3.61}$$

where $\mathbf{E}^+(\mathbf{r}, t)$ destroys and $\mathbf{E}^-(\mathbf{r}, t)$ creates photons. The field amplitude associated with one photon per mode is given by

$$E_0 = \sqrt{\frac{\hbar\omega_{\mathbf{k}}}{2\epsilon_0 V}}. \tag{3.62}$$

We have now established an important link between the amplitudes of the electric field in the classical expression (3.59) and the quantum mechanical one (3.61) written in operator form. For a single mode the sum is omitted.

3.3.4 The Basic Quantum States of Light

In the quantization of the EM field we used the states of the harmonic oscillator, and, in practice, similar states, called *number states*, are typically used to calculate the basic quantum-optical properties of light. In the following we shall formally introduce these number states. Other states that represent different facets of the complex nature of light will be discussed later in Sect. 16.4.

3.3.5 Single Mode Number States

The simplest and most convenient states for calculations are the so-called *single-mode number states* or *Fock states*. As discussed earlier, it is sufficient to label the Hamiltonian and the vector potential of the quantized EM field by subscripts $p\mathbf{k}$, denoting the *mode* of the field, since the photon energy is specified by the magnitude of the wavevector through the relation $\hbar\omega = \hbar c|k|$.

The wavefunction of a single-mode number state is denoted $|n_{p\mathbf{k}}\rangle$, and $n_{p\mathbf{k}}$ is called the occupation number of the mode $p\mathbf{k}$. For example, the state $|0_{p\mathbf{k}}\rangle$ is the vacuum state containing no real photons while the state $|1_{p\mathbf{k}}\rangle$ contains 1 photon in the mode $p\mathbf{k}$, and so on.

The annihilation operators $\mathbf{a}_{p\mathbf{k}}$ and creation operators $\mathbf{a}^{\dagger}_{p\mathbf{k}}$ act on these states according to

$$\begin{aligned} \mathbf{a}^{\dagger}_{p\mathbf{k}}|n_{p\mathbf{k}}\rangle &= \sqrt{n_{p\mathbf{k}}+1}\,|n_{p\mathbf{k}}+1\rangle \\ \mathbf{a}_{p\mathbf{k}}|n_{p\mathbf{k}}\rangle &= \sqrt{n_{p\mathbf{k}}}\,|n_{p\mathbf{k}}-1\rangle \\ \mathbf{a}_{p\mathbf{k}}|0\rangle &= |0\rangle, \end{aligned} \tag{3.63}$$

where $|0\rangle$ is the empty or vacuum state. The photon annihilation operator $\mathbf{a}_{p\mathbf{k}}$ and creation operator $\mathbf{a}^{\dagger}_{p\mathbf{k}}$ *do not commute*,

$$\left[\mathbf{a}_{p\mathbf{k}}, \mathbf{a}^{\dagger}_{p'\mathbf{k}'}\right] = \mathbf{a}_{p\mathbf{k}}\mathbf{a}^{\dagger}_{p'\mathbf{k}'} - \mathbf{a}^{\dagger}_{p'\mathbf{k}'}\mathbf{a}_{p\mathbf{k}} = \delta_{p,p'}\,\delta_{\mathbf{k},\mathbf{k}'}. \tag{3.64}$$

In contrast, consecutive creation or annihilation operators do commute, e.g. $[\mathbf{a}_{p\mathbf{k}}, \mathbf{a}_{p'\mathbf{k}'}]=[\mathbf{a}^{\dagger}_{p\mathbf{k}}, \mathbf{a}^{\dagger}_{p'\mathbf{k}'}] = 0$ [24, 35].

Any number state of mode $p\mathbf{k}$ can be created by repeated use of the creation operator $\mathbf{a}^{\dagger}_{p\mathbf{k}}$, starting with the vacuum state $|0\rangle$, according to

$$|n_{p\mathbf{k}}\rangle = \frac{[\mathbf{a}^{\dagger}_{p\mathbf{k}}]^{n_{p\mathbf{k}}}}{\sqrt{n_{p\mathbf{k}}!}}|0\rangle \tag{3.65}$$

3.3.5.1 Photon Degeneracy Parameter

For a given term in (3.52), we can define an operator

$$\mathcal{N}_{p\mathbf{k}} = \mathbf{a}^{\dagger}_{p\mathbf{k}}\,\mathbf{a}_{p\mathbf{k}}\,, \tag{3.66}$$

which by use of (3.63) has the following properties

$$\begin{aligned}\langle n_{p\mathbf{k}}|\mathcal{N}_{p\mathbf{k}}|n_{p\mathbf{k}}\rangle &= \langle n_{p\mathbf{k}}|\mathbf{a}^{\dagger}_{p\mathbf{k}}\,\mathbf{a}_{p\mathbf{k}}|n_{p\mathbf{k}}\rangle = n_{p\mathbf{k}}\\ \langle n_{p\mathbf{k}}|\mathcal{N}^{\dagger}_{p\mathbf{k}}|n_{p\mathbf{k}}\rangle &= \langle n_{p\mathbf{k}}|\mathbf{a}_{p\mathbf{k}}\,\mathbf{a}^{\dagger}_{p\mathbf{k}}|n_{p\mathbf{k}}\rangle = n_{p\mathbf{k}} + 1\,.\end{aligned} \tag{3.67}$$

For obvious reasons $\mathcal{N}_{p\mathbf{k}}$ is called a *number operator* and its expectation value is called the *photon degeneracy parameter*. We see that for a single mode number state the degeneracy parameter is just the number of photons $n_{p\mathbf{k}}$.

3.4 The Properties of Single Mode Quantum States

Following Dirac's original field quantization, resulting in the basic number states of light, in the mid-1960s Glauber developed a classification of different quantum states of light within the broader framework of QED, which contains an infinite number of quantum states that may be created as linear combinations, statistical assemblies, or products of number states [21, 22]. In the following we will introduce two particular important quantum states, called *coherent states* and *chaotic states* which are the quantum analogues of what we semi-classically refer to as coherent and chaotic light. We will further restrict our discussion here to the two types of quantum states associated with a single propagation direction or wavevector, referred to in quantum optics as *single mode quantum states*. Our discussion here forms the basis for their extension to two-mode states used in Chap. 16 to formulate diffraction in a quantum picture.

All quantum states of light may be created from the zero-point (ZP) quantum vacuum, representing the state of darkness, through different forms of excitations. In practice, the excitation energy is supplied by electrons, either through quantum jumps to lower atomic states or the kinetic energy of electrons in accelerators. The quantum creation and destruction of light are described by the action of the electric field operator (3.61), written in terms of creation and annihilation operators.

We have seen that the basic quantum operators are the creation and annihilation operators $\mathbf{a}_{p\mathbf{k}}$ and $\mathbf{a}^{\dagger}_{p\mathbf{k}}$ of the vector potential. For a *single mode* $p\mathbf{k}$ we can write the electric field operator (3.61) as

$$\mathbf{E}(\mathbf{r},t) = \underbrace{E_0\, \boldsymbol{\epsilon}^*_p\, \mathbf{a}^{\dagger}_{p\mathbf{k}}\, e^{-\mathrm{i}(\mathbf{k}\cdot\mathbf{r}-\omega_{\mathbf{k}}t)}}_{\mathbf{E}^-(\mathbf{r},t)} + \underbrace{E_0\, \boldsymbol{\epsilon}_p\, \mathbf{a}_{p\mathbf{k}}\, e^{\mathrm{i}(\mathbf{k}\cdot\mathbf{r}-\omega_{\mathbf{k}}t)}}_{\mathbf{E}^+(\mathbf{r},t)}, \tag{3.68}$$

where E_0 is given by (3.62) and the creation and annihilation operators obey the commutation relations (3.64).

3.4.1 *Definition of Key Properties*

The properties of the single mode quantum states are evaluated by calculating expectation values of various kinds of operators [21, 22]. If we use $|\phi\rangle$ as a general state label of a quantum state, we can define the following fundamental properties of interest.

The photon degeneracy parameter n_{coh} gives the average number of *coherent* photons or the number of photons per mode, $p\mathbf{k}$, and is defined as

$$\text{Photon degeneracy parameter:} \quad n_{\mathrm{coh}} = \langle\phi|\mathcal{N}_{p\mathbf{k}}|\phi\rangle = \langle\phi|\mathbf{a}^{\dagger}_{p\mathbf{k}}\, \mathbf{a}_{p\mathbf{k}}|\phi\rangle. \tag{3.69}$$

Since *photons are Bosons*, there can be an unrestricted number of photons per mode. Photons that belong to the same mode are coherent.

The *coherent signal* is defined in terms of the expectation value of the electric field operator $\mathbf{E}$ as

$$\text{Coherent signal:} \quad S = \langle\phi|\mathbf{E}|\phi\rangle. \tag{3.70}$$

The fluctuations of the electric field are characterized by the variance of the field,

$$\text{Field variance:} \quad \langle\phi|(\Delta E)^2|\phi\rangle = \langle\phi|\mathbf{E}^2|\phi\rangle - \langle\phi|\mathbf{E}|\phi\rangle^2, \tag{3.71}$$

and the associated standard deviation is often referred to as the *uncertainty of the field*

$$\text{Field uncertainty:} \quad \Delta E = \sqrt{\langle\phi|(\Delta E)^2|\phi\rangle}. \tag{3.72}$$

We start by considering the properties of single mode number states and for brevity drop the mode index and use the short form $|n\rangle = |n_{p\mathbf{k}}\rangle$ for single mode number states.

3.4.2 Properties of Single Mode Number States

The degeneracy parameter given by (3.69) is obtained as

$$n_{\rm coh} = \langle n|\mathbf{a}^\dagger\,\mathbf{a}|n\rangle = n. \tag{3.73}$$

By use of the operator relations (3.63) we see that the coherent signal (3.70) or expectation value of the field operator (3.61) vanishes

$$S = \langle \mathbf{E}\rangle = \langle n|\mathbf{E}|n\rangle = 0. \tag{3.74}$$

This means that the number states do not carry information that can be detected by a homodyne detection scheme. The noise or the variance of the field defined by (3.71) is obtained as

$$\left\langle(\Delta E)^2\right\rangle = \langle n|\mathbf{E}^2|n\rangle = 2\,E_0^2\left(n+\frac{1}{2}\right) = \frac{E_1^2}{2}\left(n+\frac{1}{2}\right), \tag{3.75}$$

where we have used the notation $E_1 = 2E_0$ (see (3.13)). The field uncertainty is

$$\Delta E = \sqrt{2}\,E_0\sqrt{n+\frac{1}{2}} = \frac{E_1}{\sqrt{2}}\sqrt{n+\frac{1}{2}}. \tag{3.76}$$

In particular, we see from (3.75) that the zero-point field corresponding to $n = 0$ has an uncertainty of $\Delta E_{\rm ZP} = E_0 = E_1/2$.

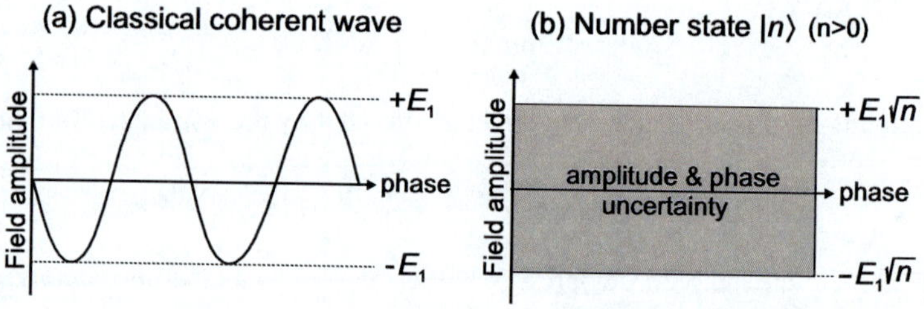

Fig. 3.5 **a** Graphic representations of the field of a classical wave and a number state $|n\rangle$. **a** Phase dependence of the amplitude of a classical wave, e.g. as a function of time at a fixed observation point. **b** Phase dependence of the electric field variation for a single mode number state $|n\rangle$. At any phase value, the amplitude can take a continuous range of values with an uncertainty $\Delta E = E_1\sqrt{n/2}$ given by (3.76), where we have ignored the constant contribution from the zero-point field

The well-defined field of a classical EM wave and that of number states $n > 1$ are schematically compared in Fig. 3.5. The number state field oscillates like a classical field with a frequency ω, but the phase is not defined so that the oscillations are smeared as indicated in gray shading.

Figure 3.5 reveals that the field variation of a single mode number state is in stark contrast with the stable amplitude and phase of a classical wave. The number state has a large amplitude uncertainty and a completely blurred or random phase, indicated by gray shading. Note that the uncertainty of the zero-point field, given by $\Delta E_{ZP} = E_0 = E_1/2$, is smaller than that shown in Fig. 3.5.

3.4.3 Properties of Single Mode Coherent States

For a given mode $\mathbf{pk}$, the most general solution of the time-dependent Schrödinger equation for the Hamiltonian (3.51) is a linear combination of number states [21]

$$|\Phi\rangle_k = c_0|0\rangle_k + c_1|1\rangle_k + c_2|2\rangle_k + \dots . \tag{3.77}$$

By choosing appropriate coefficients c_n we can construct various types of states. One specific linear combination in (3.77), the so-called *coherent state*, is of particular

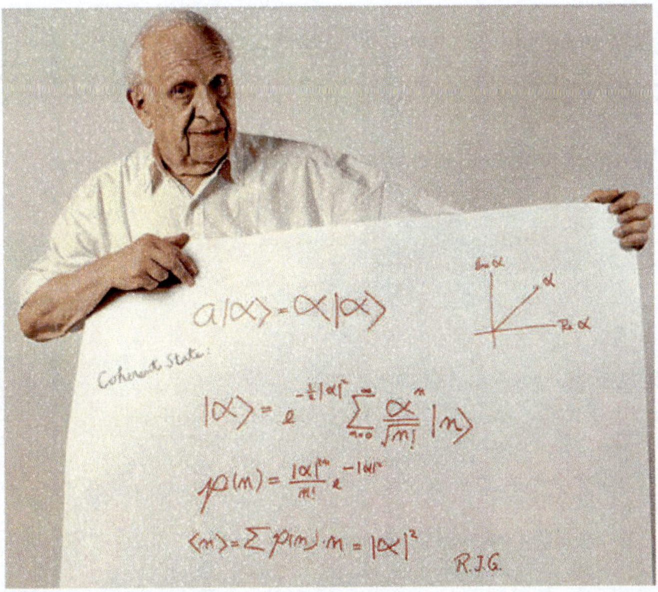

Fig. 3.6 Roy Glauber displaying a poster in his own handwriting with the states named after him. Courtesy of Volker Steger, Science Photo Library, 2006

importance because of its close resemblance to a classical EM wave. The states are also referred to as *quasi-classical states* or *Glauber states* [22] (see Fig. 3.6). Coherent states are defined by the following linear combination of number states $|m\rangle_k$ [21, 25, 26]

$$|\alpha\rangle_k = \frac{1}{e^{|\alpha_k|^2/2}} \sum_{m=0}^{\infty} \frac{\alpha_k^m}{\sqrt{m!}} |m\rangle_k = \frac{1}{e^{|\alpha_k|^2/2}} \sum_{m=0}^{\infty} \frac{(\alpha_k \mathbf{a}_k^\dagger)^m}{m!} |0\rangle_k, \tag{3.78}$$

where on the far right we have shown how the state is created from the zero-point state by use of the creation operator $\mathbf{a}_k^\dagger$.

In the following we shall drop the mode label k for convenience, keeping in mind that we deal with only a single k mode. Here $\langle\alpha|\alpha\rangle = 1$ and α has the meaning of a *complex field amplitude* whose real and imaginary parts both form continuous ranges. The states $|\alpha\rangle$ satisfy the *right* and *left* eigenvalue relations

$$\mathbf{a}_{p\mathbf{k}}\,|\alpha\rangle = \alpha\,|\alpha\rangle \tag{3.79}$$

$$\langle\alpha|\,\mathbf{a}_{p\mathbf{k}}^\dagger = \langle\alpha|\,\alpha^*. \tag{3.80}$$

The degeneracy parameter given by (3.69) is calculated as

$$n_{\text{coh}} = \langle\,\alpha|\mathbf{a}_{p\mathbf{k}}^\dagger\,\mathbf{a}_{p\mathbf{k}}|\alpha\,\rangle = \langle n\rangle = |\alpha|^2\,. \tag{3.81}$$

The probability of finding m photons in the mode is given by a *Poisson probability distribution*, obtained from (3.78) as[4]

$$\boxed{P(m, \langle n\rangle) = |\langle\, m\,|\,\alpha\,\rangle|^2 = \exp\left(-|\alpha|^2\right)\frac{|\alpha|^{2m}}{m!} = e^{-\langle n\rangle}\frac{\langle n\rangle^m}{m!}}. \tag{3.82}$$

It is normalized according to $\sum_m P(m, \langle n\rangle) = 1$. Hence we can simply write a coherent state as.

$$|\alpha\rangle = \sum_m \sqrt{P(m, \langle n\rangle)}\,|m\rangle. \tag{3.83}$$

Figure 3.7 illustrates distributions for three values of $\langle n\rangle$. We see from Fig. 3.7 that for a coherent state the occupation probability peaks approximately at $m = \langle n\rangle$, and its distribution variance is also $\langle n\rangle$.

---

[4] For large values of $\langle n\rangle$ the Poissonian probability (3.82) can also be written in the approximate Gaussian form

$$P(m, \langle n\rangle) \approx \frac{1}{\sqrt{2\pi\langle n\rangle}} \exp\left(-\frac{(m-\langle n\rangle)^2}{2\langle n\rangle}\right).$$

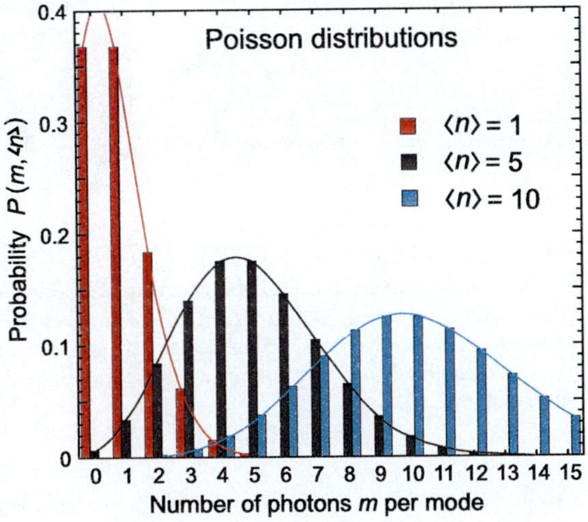

Fig. 3.7 Poisson probability distributions $P(m, \langle n \rangle)$ according to (3.82), representing the probability of the number of photons m in different *coherent* states $|\alpha\rangle$ containing $|\alpha|^2 = \langle n \rangle$ photons

> A *coherent state* consists of a Poisson probability distribution centered around $\langle n \rangle$ photons, and $\langle n \rangle$ is also the degeneracy parameter or average number of photons per mode.

The presence of adjacent number states that are coupled by the lowering and raising operators $\mathbf{a}$ and $\mathbf{a}^\dagger$ in the expression (3.61) for the electric field operator cause the mean field to be finite. To see this we write the complex number $\alpha = |\alpha|\, e^{i\theta}$ and write the field operator $\mathbf{E}$ as

$$\mathbf{E} = \mathbf{E}_0 \left[\mathbf{a}_{pk}\, e^{-i\chi} + \mathbf{a}_{pk}^{\dagger}\, e^{i\chi} \right]. \tag{3.84}$$

We obtain

$$S = \langle \mathbf{E} \rangle = \langle \alpha | \mathbf{E} | \alpha \rangle = 2E_0\, |\alpha|\, \cos(\chi - \theta) = \sqrt{\langle n \rangle}\, E_1\, \cos(\chi - \theta). \tag{3.85}$$

The variance of the field is obtained as

$$\left\langle (\Delta E)^2 \right\rangle = E_0^2 = \frac{E_1^2}{4} \tag{3.86}$$

and the field uncertainty is

$$\Delta E = E_0 = \frac{E_1}{2}, \tag{3.87}$$

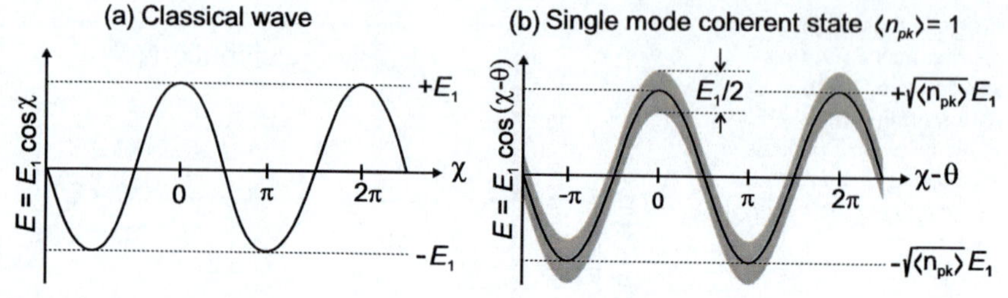

Fig. 3.8 Graphic representations of the electric field expectation value and the phase and amplitude uncertainties for two cases. **a** Real field of a classical EM wave given by (3.84) with $\mathbf{a}_{p\mathbf{k}} = \mathbf{a}^{\dagger}_{p\mathbf{k}} = 1$. **b** Expectation value given by (3.85) of the field of a single mode coherent state with $\langle n_{p\mathbf{k}} \rangle = 1$. The mean field amplitude is $E_1|\alpha| = E_1 \langle n_{p\mathbf{k}} \rangle^{1/2} = E_1$. The field uncertainty is given by that of the zero-point field, $E_1/2$

which is independent of $\langle n \rangle$ and according to (3.75) is just the field uncertainty of the zero-point field. The close correspondence between a classical field $E = 2E_0 \cos \chi$ and that of the corresponding coherent state for $\langle n_{p\mathbf{k}} \rangle = 1$ is shown in Fig. 3.8.

As expected, a classical EM wave has a well-defined phase χ and amplitude E_1. The signal S is well defined and has the value $S = E_1 \cos \chi$, and the field uncertainty ΔE is zero. The mean amplitude of a coherent state is given by $|\alpha| E_1 = \langle n \rangle^{1/2} E_1$ which for the case $\langle n \rangle = 1$ illustrated in Fig. 3.8b is just E_1. The field uncertainty is shown shaded.

As the mean photon number $\langle n \rangle$ of the coherent state increases, so does its amplitude $E_1 \langle n \rangle^{1/2}$, yet the field uncertainty given by (3.86) remains constant at the value $E_1/2$ of the zero-point field. Hence the coherent state becomes better defined in amplitude and phase and increasingly resembles a classical EM wave. It has the minimum uncertainty in amplitude and phase allowed by the uncertainty principle. The decreasing amplitude uncertainty with increasing number of photons $\langle n_{p\mathbf{k}} \rangle$ has been experimentally demonstrated by Breitenbach et al. [27].

> A single mode *coherent state* is a collective state that contains a mean number of photons $\langle n \rangle$. It is composed of number substates $|m\rangle$ whose occupation number is given by a Poisson probability distribution which peaks at $m = \langle n \rangle$. Its field amplitude is given by $\sqrt{\langle n \rangle}\, E_1$, and the uncertainty of the field is independent of $\langle n \rangle$ and is given by the uncertainty of the zero-point field $\Delta E_{ZP} = E_1/2$. With increasing $\langle n \rangle$, a coherent state increasingly resembles a classical EM wave.

3.4.4 Properties of Single Mode Chaotic States

Finally we discuss single mode chaotic states. In analogy to (3.83) they can be described by a probability distribution and single mode number states according to

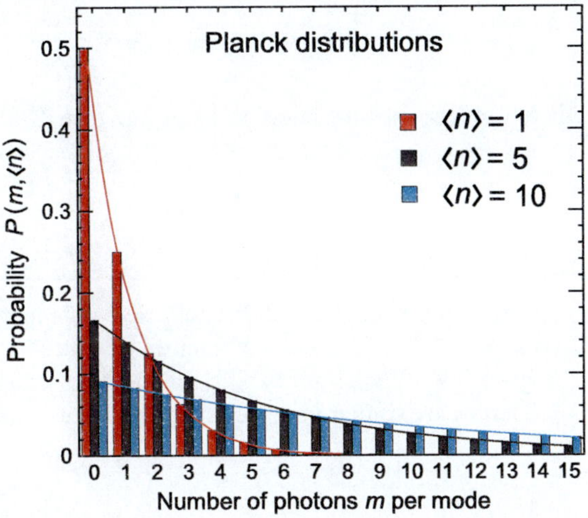

Fig. 3.9 Planck or Bose–Einstein distributions (3.89) expressing the probability of the number of photons m in different *chaotic* states $|\beta\rangle$ with different mean photon numbers $\langle n\rangle$

[21]

$$|\beta\rangle_k = \sum_{m=0}^{\infty} \sqrt{\frac{\langle n\rangle^m}{(1+\langle n\rangle)^{1+m}}}\,|m\rangle_k. \tag{3.88}$$

In the following we shall drop the mode label k for convenience. In this case the distribution $P(m, \langle n\rangle)$ is the *Planck probability function* given by

$$\boxed{P(m, \langle n\rangle) = \frac{\langle n\rangle^m}{(1+\langle n\rangle)^{1+m}}}. \tag{3.89}$$

It is also referred to as a *Bose–Einstein*, *thermal* or *geometric* distribution. Again $\sum_m P(m, \langle n\rangle) = 1$, and distributions for the same three values of $\langle n\rangle$ as in Fig. 3.7 are shown in Fig. 3.9.

We see that for a chaotic state, $m = 0$ always has the largest probability with a monotonic decrease with increasing m.

The degeneracy parameter given by (3.69) is calculated as

$$n_{\text{coh}} = \langle\, \beta|\mathbf{a}^{\dagger}_{p\mathbf{k}}\, \mathbf{a}_{p\mathbf{k}}|\beta\,\rangle = \langle n\rangle. \tag{3.90}$$

The signal and noise are given by averages of the corresponding expectation values for the number states. We have

$$S = \langle \mathbf{E}\rangle = \langle \beta|\mathbf{E}|\beta\rangle = 0. \tag{3.91}$$

Chaotic light cannot carry a coherent signal, and its phase is completely uncertain. The variance of the field defined by (3.71) is obtained as

$$\langle (\Delta E)^2 \rangle = \langle \beta | \mathbf{E}^2 | \beta \rangle = 2\, E_0^2 \left(\langle n \rangle + \frac{1}{2} \right) = \frac{E_1^2}{2} \left(\langle n \rangle + \frac{1}{2} \right), \tag{3.92}$$

where we have used the notation $E_1 = 2E_0$ (see (3.13)). The field uncertainty is,

$$\Delta E = \sqrt{2}\, E_0 \sqrt{\langle n \rangle + \frac{1}{2}} = \frac{E_1}{\sqrt{2}} \sqrt{\langle n \rangle + \frac{1}{2}}. \tag{3.93}$$

A single mode *chaotic state* is a collective state that contains a mean number of photons $\langle n \rangle$. It is composed of number substates $|m\rangle$ whose occupation number is given by a Planck probability distribution which peaks at $m = 0$. Its field properties are similar to those of number states with n replaced by $\langle n \rangle$.

3.5 Photon Modes, Density of States, and Coherence Volume

In this section we come back to the issue of the normalization volume V, which arises when classical and quantum properties of fields are linked as first discussed in Sect. 3.2.5 and encountered again in Sect. 3.3.1 in the quantization of the vector potential and field given by (3.55) and (3.61). Our discussion will lead to the fundamental concepts of the *modes*, the *density of states*, and *coherence volume* of the EM field.

3.5.1 Photon Modes

We have previously *box normalized* the vector potential, by use of a large cubic box of volume $L^3 = V$. This is a standard quantum mechnical procedure found in the literature [19, 21, 28, 29]. This normalization is somewhat awkward since the volume is left undefined and simply assumed to be large. It must therefore drop out when the expectation values of physical observables are calculated. Let us therefore take a closer look at the box normalization concept.

In deriving (3.50) we have implicitly assumed that the vector potential obeys periodic boundary conditions, i.e. that it has the same value on opposite faces of the cubic box, i.e. for the x direction

$$\mathbf{A}(0, y, z, t) = \mathbf{A}(L, y, z, t)\,. \tag{3.94}$$

For the $\mathbf{A}$ field given by (3.50) we have chosen the plane wave basis functions,

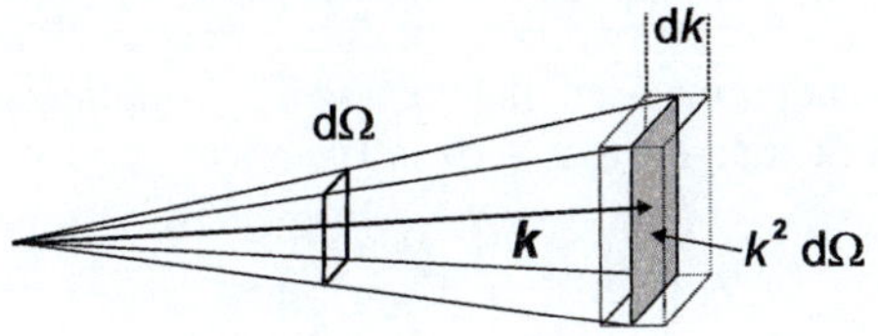

Fig. 3.10 Illustration how the three-dimensional differential volume element **dk** for the scattered radiation field, shown as a dotted volume, can be replaced by $dk\,k^2\,d\Omega$

$$\frac{1}{C}\boldsymbol{\epsilon}_p\,\mathrm{e}^{\mathrm{i}\mathbf{k}\cdot\mathbf{r}} \quad p = 1, 2\,, \tag{3.95}$$

with *box normalization* $C = \sqrt{L^3} = \sqrt{V}$. The box normalization results in a *discrete set* of wavevectors discussed below. In contrast, a normalization of the plane wave basis functions given by (3.95) over all space corresponds to a *continuous set* of wavevectors. The continuum normalization is accomplished by a Dirac δ-function normalization where $C = \sqrt{(2\pi)^3}$ in (3.95) [29].

For the field to satisfy (3.94) with $C = \sqrt{L^3} = \sqrt{V}$, the wave vector components must be multiples of $2\pi/L$

$$\mathbf{k} = \frac{2\pi}{L}\begin{pmatrix} n_x \\ n_y \\ n_z \end{pmatrix}. \tag{3.96}$$

One obtains a complete discrete set of transverse orthonormal basis functions and wave vectors when the numbers n_x, n_y, and n_z run through all whole numbers $n_x, n_y, n_z = 0, \pm1, \pm2, \pm3, \ldots$. In order to obtain the total number of states in a finite volume V we would simply sum all values of **k**, i.e. over all values of n_x, n_y, and n_z. For a box that fills all space $L \to \infty$ we need to replace the sum by an integral, $(\sum_k)/L^3 \to (\int \mathbf{dk})/(2\pi)^3$, but we can simply write for the differential number of states or "modes" in the interval **dk**,

$$\mathrm{dM} = \left(\frac{L}{2\pi}\right)^3 \mathbf{dk}. \tag{3.97}$$

Note that this expression is for a single polarization. As already discussed in conjunction with (3.55) there are actually two polarization states for each **k** state and one therefore sometimes finds the above expression multiplied by 2. We will keep this in mind for later.

We can now express the three-dimensional reciprocal space volume element **dk** in (3.97) by use of Fig. 3.10 in terms of the absolute value $k = |\mathbf{k}|$ and the solid angle $d\Omega = d\Omega_k$ around a given photon propagation direction **k**. The shown rectangular shape of the solid angle and the box for the volume are actually unimportant, and we could have used a cylindrical cone, for example. The important thing is that the solid angle element $d\Omega = d\Omega_k$ is centered around the wavevector direction **k**.

By expressing $\mathbf{dk} = dk\,k^2\,d\Omega$ and use of $k = \omega/c$ we can restate (3.97) as follows.

The number of radiation modes in the volume V, per polarization state and per solid angle $\mathrm{d}\Omega$, in the range k to $k + \mathrm{d}k$ is given by

$$\frac{\mathrm{d}M_k}{\mathrm{d}\Omega} = \frac{V}{8\pi^3} k^2 \, \mathrm{d}k. \tag{3.98}$$

In the angular frequency range ω_k to $\omega_k + \mathrm{d}\omega_k$ we obtain with $k^2 \, \mathrm{d}k = (\omega_k^2/c^3)\mathrm{d}\omega_k$

$$\frac{\mathrm{d}M}{\mathrm{d}\Omega \, \mathrm{d}\omega_k} = \frac{V}{8\pi^3} \frac{\omega_k^2}{c^3}. \tag{3.99}$$

The number of modes per polarization, per unit emission angle per unit energy per unit volume is therefore given by the general expression

$$\boxed{\frac{\mathrm{d}M}{\mathrm{d}\Omega \, \mathrm{d}\omega_k \, \mathrm{d}V} = \frac{\omega_k^2}{8\pi^3 c^3}}, \tag{3.100}$$

which can be used to convert the number of modes for different cases. For example, if we are interested in the number of modes emitted into a solid angle of 4π, we simply multiply the right side by this solid angle.

3.5.2 *Number of Modes per Unit Energy*

The number of modes per unit energy (dimension [1/energy]) is particularly important. It is often called the *density of states*. It may be defined in general for any kind of particle and by use of (3.98) may be written as

$$\rho(\mathcal{E}) = \frac{\mathrm{d}M_k}{\mathrm{d}\mathcal{E}} = \frac{V}{8\pi^3} k^2 \frac{\mathrm{d}k}{\mathrm{d}\mathcal{E}} \mathrm{d}\Omega. \tag{3.101}$$

If we include polarization for x-rays or spin for electrons, it needs to be multiplied by 2. The dispersion factor $\mathrm{d}k/\mathrm{d}\mathcal{E}$ is however different for x-rays and electrons. For x-rays there is a linear relationship between energy and wavevector $\mathcal{E} = \hbar\omega = \hbar c \, k$, whereas for electrons there is a quadratic relationship between the kinetic energy of the electron and its wavevector $\mathcal{E}_{\mathrm{kin}} = \hbar^2 k^2/(2m_{\mathrm{e}})$. For photons we can evaluate $\mathrm{d}k/\mathrm{d}\mathcal{E} = 1/\hbar c$ and state our final result as follows.

The *density of states* of the radiation field per unit energy and per polarization mode is given by (dropping the subscripts k for convenience),

$$\rho(\hbar\omega) = \frac{V\omega^2}{8\pi^3\hbar c^3}\,\mathrm{d}\Omega = \frac{V}{\hbar\omega\lambda^3}\,\mathrm{d}\Omega. \tag{3.102}$$

It has the dimension [1/energy]. From the last expression we see that within a volume $V = \lambda^3$ there is one photon of energy $\hbar\omega$ per unit solid angle $\mathrm{d}\Omega = 1$ and per polarization mode.

3.5.3 Number of Modes per Unit Volume

Above, we have obtained expressions for the number of modes per solid angle, (3.99), and per energy, (3.101). Combining (3.101) and (3.102) we can now obtain an expression for the number of modes per unit volume V. For a given polarization we obtain

$$\frac{\mathrm{d}M}{\mathrm{d}V} = \frac{\rho(\hbar\omega)\hbar\omega}{\mathrm{d}V} = \frac{4\pi}{\lambda^3}. \tag{3.103}$$

This means that for each photon of energy $\hbar\omega$ there are 4π modes per volume $V = \lambda^3$ per polarization state or 8π modes for both polarization states.

At this point it is interesting to make contact with Planck's radiation formula which gives the spectral energy density emitted by a body of temperature T. The "spectral" aspect can be defined in different ways by normalizing the energy of the radiation to the "spectral unit" of interest, e.g. the frequency range $\mathrm{d}\nu$, the angular frequency range $\mathrm{d}\omega$, the wavelength range $\mathrm{d}\lambda$ or the photon energy range $\mathrm{d}(\hbar\omega)$. In the latter formulation, the radiation energy density $\mathcal{E}_{\mathrm{rad}}$ per unit volume $\mathrm{d}V$ and per photon energy bandwidth $\mathrm{d}(\hbar\omega)$ for the two fundamental polarization states in the entire solid angle $\mathrm{d}\Omega = 4\pi$ is given by

$$\frac{\mathcal{E}_{\mathrm{rad}}}{\mathrm{d}V\,\mathrm{d}(\hbar\omega)} = \frac{\omega^3}{\pi^2 c^3}\,\frac{1}{\mathrm{e}^{\hbar\omega/(k_{\mathrm{B}}T)} - 1} = \frac{8\pi}{\lambda^3}\,\frac{1}{\mathrm{e}^{\hbar\omega/(k_{\mathrm{B}}T)} - 1}. \tag{3.104}$$

We see that the fundamental prefactor in Planck's formula is nothing but the number of modes in the volume λ^3. The modes get increasingly occupied as a function of temperature, as reflected by Planck's thermal excitation function.

3.5.4 Coherence Volume per Mode

The number of modes in the volume V can also be expressed in terms of the photon coherence volume V_{coh} for a single mode k, according to

$$M_\omega = \frac{V}{8\pi^3}\frac{\omega^2\,\mathrm{d}\omega}{c^3} = \frac{V}{\lambda^3}\frac{\mathrm{d}\omega}{\omega}\mathrm{d}\Omega = \frac{V}{\lambda^3}\frac{\Delta\lambda}{\lambda} = \frac{V}{V_{\mathrm{coh}}}. \tag{3.105}$$

We have therefore identified that each **k** mode of the radiation field has an associated coherence volume $V_{\mathrm{coh}} = V_k$.

For a given propagation direction **k** and each of the two elementary polarization states, the *coherence volume per mode* **k** is given by

$$V_{pk} = \lambda^3 \frac{\lambda}{\Delta\lambda} = \lambda^3 \frac{\hbar\omega}{\Delta(\hbar\omega)}. \tag{3.106}$$

We see that the volume is proportional to λ^3, as expected (also see Sect. 4.2.6 below). The resolving power or inverse bandwidth can be written either in a wavelength or photon energy notation $\lambda/\Delta\lambda = \hbar\omega/\Delta(\hbar\omega)$, where $\Delta\lambda$ or $\Delta(\hbar\omega)$ depends on the shape of the photon distribution function as discussed in Sect. 4.4.2.

The coherence volume may be written as the product of the lateral coherence area A_{coh} and the longitudinal coherence length according to

$$V_{\mathrm{coh}} = A_{\mathrm{coh}}\ell_{\mathrm{coh}}, \quad \text{where } A_{\mathrm{coh}} = \frac{\lambda^2}{\mathrm{d}\Omega}, \quad \text{and } \ell_{\mathrm{coh}} = \frac{\lambda^2}{\Delta\lambda} = \lambda\frac{\hbar\omega}{\Delta(\hbar\omega)}. \tag{3.107}$$

We shall come back to the importance of the coherence volume in Chap. 4 where we discuss the coherence concept in more detail and its relationship with the brightness of a source.

3.5.5 *Summary*

In the above section we have clarified the concept of *"photon modes"*. This concept is usually not discussed in books about x-rays since conventional x-ray sources do not have more than one x-ray photon in the same mode. We have also derived an expression for the photon density of states or more precisely the number of modes of the photon field per unit energy. This expression will be of importance in deriving the final state density $\rho(\mathcal{E}_f)$ which occurs in the Kramers-Heisenberg-Dirac theory for the transition rate. Finally, we have introduced the concept of the *coherence volume* of an x-ray beam. We will see below that the number of photons in the coherence volume or *degeneracy parameter* is directly related to the brightness of a source. The derived concepts are of fundamental importance in the treatment of radiation from x-ray lasers. We can summarize as follows.

- The *modes* of the quantized electromagnetic fields are defined by the wavevector directions $\mathbf{k}$ and two possible photon polarizations p. The so-defined modes provide a complete description since the wavevector magnitude also defines the photon energy according to $\hbar\omega_k = \hbar c|\mathbf{k}|$.
- The *photon density of states* is defined as the number of modes per unit energy.
- The *number of modes per unit volume* constitutes the prefactor in Planck's radiation formula.
- The *coherence volume* contains indistinguishable photons. Because photons are Bosons, the coherence volume may contain an unlimited number of photons. The number of photons in the coherence volume corresponds to the number of photons in the same mode and is called the *photon degeneracy parameter*.

3.6 Link of Classical and Quantum Properties of Radiation

Electromagnetic waves whose frequencies extend into the microwave region cause oscillations in an antenna or cavity that can be visualized on an oscilloscope. However, there are still no detectors fast enough to measure the rapidly oscillating fields of visible light with frequencies greater than 10^{14} Hz, or even x-rays with frequencies up to 10^{19} Hz, as a function of time. For x-rays one therefore uses photodetectors which are sensitive to the light intensity or the average of the *square of the electric field*, whose classical cycle-average is given by (3.17) or

$$\langle |\mathbf{E}(\mathbf{r}, t)|^2 \rangle = 2\,|\mathbf{E}_0|^2. \tag{3.108}$$

All other properties like the energy, momentum, flux, derived in Sect. 3.2.3 depend on this quantity. Below we shall take a look at these properties from a quantum point of view and compare them to the classical expressions. To do so, we need to calculate the expectation value of the squared field operator with appropriate wavefunctions. They consist of a generalization of the single mode number states introduced above.

3.6.1 Multi-mode Number States

Single-mode states correspond to a single photon energy $\hbar\omega = \hbar c|k|$. The quantum mechanical time–energy correlation or the classical Fourier transform relationship between time and energy then demands that they are *time independent*. In practice, however, we are often interested in the description of beams that contain photons with a range of photon energies and when transformed from the frequency to time domain lead to a time dependence. States which can be represented as products of the individual substates $|n_{p\mathbf{k}}\rangle$ are referred to as *multi-mode number states*. They are expressed as

$$|\Pi_{n_{p\mathbf{k}}}\rangle = \prod_{\mathbf{k}_i}\prod_{p_i} |n_{p_i\mathbf{k}_i}\rangle = |n_{p_1\mathbf{k}_1}\rangle|n_{p_2\mathbf{k}_1}\rangle|n_{p_1\mathbf{k}_2}\rangle|n_{p_2\mathbf{k}_2}\rangle... = |..., n_{p\mathbf{k}}, ...\rangle\,, \quad (3.109)$$

where p_1 and p_2 are the two polarization basis states.

The action of the creation and annihilation operators on such multi-mode states is restricted to the specific mode $p\mathbf{k}$ of the operator. It raises or lowers only the occupation number of one specific mode by unity, while leaving all other occupation numbers unchanged, according to

$$\begin{aligned}\mathbf{a}_{p\mathbf{k}}\,|\Pi_{n_{p\mathbf{k}}}\rangle &= \mathbf{a}_{p\mathbf{k}}\,|..., n_{p\mathbf{k}}, ...\rangle = \sqrt{n_{p\mathbf{k}}}\,|..., n_{p\mathbf{k}}-1, ...\rangle\\ \mathbf{a}^\dagger_{p\mathbf{k}}\,|\Pi_{n_{p\mathbf{k}}}\rangle &= \mathbf{a}^\dagger_{p\mathbf{k}}\,|..., n_{p\mathbf{k}}, ...\rangle = \sqrt{n_{p\mathbf{k}}+1}\,|..., n_{p\mathbf{k}}+1, ...\rangle\,. \quad (3.110)\end{aligned}$$

We have

$$\begin{aligned}\langle\Pi_{n_{p\mathbf{k}}}|\mathbf{a}_{p\mathbf{k}}\mathbf{a}^\dagger_{p\mathbf{k}}\,|\Pi_{n_{p\mathbf{k}}}\rangle &= n_{p\mathbf{k}}+1\\ \langle\Pi_{n_{p\mathbf{k}}}|\mathbf{a}^\dagger_{p\mathbf{k}}\mathbf{a}_{p\mathbf{k}}\,|\Pi_{n_{p\mathbf{k}}}\rangle &= n_{p\mathbf{k}}\,, \quad (3.111)\end{aligned}$$

and therefore the sum rule

$$\left\langle\sum_{p\mathbf{k}} N_{p\mathbf{k}}\right\rangle = \left\langle\Pi_{n_{p\mathbf{k}}}\left|\sum_{p\mathbf{k}}\mathbf{a}^\dagger_{p\mathbf{k}}\,\mathbf{a}_{p\mathbf{k}}\right|\Pi_{n_{p\mathbf{k}}}\right\rangle = \sum_{p\mathbf{k}} n_{p\mathbf{k}} = n_{\mathrm{ph}}, \quad (3.112)$$

where n_{ph} is the total number of photons. We shall now use these states to calculate the quantum mechanical expectation values for the key classical properties of photon beams such as the energy, momentum, intensity, and flux discussed in Sect. 3.2.3.

3.6.2 *Expectation Value of the Squared Electric Field*

Owing to the inability of measuring the EM field directly at optical and higher frequencies, the key quantity that is related to the measured intensity is the expectation value of the absolute value squared of the electric field $\langle|\mathbf{E}(\mathbf{r},t)|^2\rangle$. In calculating the quantum mechanical expectation value we use the expression of the electric field operator given by 3.61. The quantum mechanical expectation value of the square of the electric field is calculated with the *multi-mode* number states according to

$$\langle|\mathbf{E}(\mathbf{r},t)|^2\rangle = \langle\Pi_{n_{p\mathbf{k}}}|\,|\mathbf{E}(\mathbf{r},t)|^2|\,|\Pi_{n_{p\mathbf{k}}}\rangle\,. \quad (3.113)$$

We obtain

$$\langle \Pi_{n_{pk}} | \, |\mathbf{E}(\mathbf{r},t)|^2 \, | \Pi_{n_{pk}} \rangle = \frac{\hbar}{2\epsilon_0 V} \Bigg\langle \Pi_{n_{pk}} \Bigg| \sum_{uv} \sum_{\mathbf{k}_1 \mathbf{k}_2} \sqrt{\omega_{\mathbf{k}_1} \omega_{\mathbf{k}_2}}$$
$$\left[\boldsymbol{\epsilon}_u \, \mathbf{a}_{u\mathbf{k}_1} \, e^{\mathrm{i}(\mathbf{k}_1 \cdot \mathbf{r} - \omega_{\mathbf{k}_1} t)} + \boldsymbol{\epsilon}_u^* \, \mathbf{a}_{u\mathbf{k}_1}^\dagger \, e^{-\mathrm{i}(\mathbf{k}_1 \cdot \mathbf{r} - \omega_{\mathbf{k}_1} t)} \right]$$
$$\cdot \left[\boldsymbol{\epsilon}_v \, \mathbf{a}_{v\mathbf{k}_2} \, e^{\mathrm{i}(\mathbf{k}_2 \cdot \mathbf{r} - \omega_{\mathbf{k}_2} t)} + \boldsymbol{\epsilon}_v^* \, \mathbf{a}_{v\mathbf{k}_2}^\dagger \, e^{-\mathrm{i}(\mathbf{k}_2 \cdot \mathbf{r} - \omega_{\mathbf{k}_2} t)} \right] \Bigg| \Pi_{n_{pk}} \Bigg\rangle .$$

We have explicitly indicated the dot product between the two square brackets. This expression can be evaluated with the rules given by (3.110). Each of the terms in the sum acts on only one of the substates of $|\Pi_{n_{pk}}\rangle$ at a time, and only the diagonal matrix elements contribute to the sums. The two non-vanishing matrix elements each contain a product of a creation and annihilation operator and are given by (3.111). The phases associated with the non-vanishing terms fall out and the dot products of the unit polarization vectors are unity. We then obtain

$$\langle \Pi_{n_{pk}} | |\mathbf{E}(\mathbf{r},t)|^2 | \Pi_{n_{pk}} \rangle = \frac{\hbar}{2\epsilon_0 V} \sum_i (2n_{p_i \mathbf{k}_i} + 1) \, \omega_{\mathbf{k}_i} = \frac{1}{\epsilon_0 V} \sum_i \left(n_{p_i \mathbf{k}_i} + \frac{1}{2} \right) \hbar \omega_{\mathbf{k}_i} . \tag{3.114}$$

By subtracting the energy of the vacuum state $\frac{1}{2} \sum_i \hbar \omega_{\mathbf{k}_i}$ we obtain

$$\langle \Pi_{n_{pk}} | |\mathbf{E}(\mathbf{r},t)|^2 | \Pi_{n_{pk}} \rangle = \frac{1}{\epsilon_0 V} \sum_i n_{p_i \mathbf{k}_i} \, \hbar \omega_{\mathbf{k}_i} = \frac{n_{\mathrm{ph}} \, \hbar \omega}{\epsilon_0 V} . \tag{3.115}$$

To obtain the final expression in terms of the total number of photons n_{ph}, we have used (3.112) and assumed that all n_{pk} photons that may be in different modes i have the same energy $\hbar\omega$. We are now in a position to relate the classical amplitude of the electric field given by (3.17) and the quantum mechanical amplitude given by (3.115).

The expectation value of the electric field of an EM wave can be expressed by linking the classical and quantum mechanical expressions according to

$$\langle |\mathbf{E}(\mathbf{r},t)|^2 \rangle = 2|\mathbf{E}_0|^2 = \frac{n_{\mathrm{ph}} \, \hbar \omega}{\epsilon_0 V} , \tag{3.116}$$

where n_{ph} is the total number of photons of energy $\hbar\omega$ contained in a volume V.

We can readily write down the following relations between the classical and quantum mechanical expressions. The energy contained in the field, classically given by (3.22), is

$$\boxed{\text{Energy:} \quad \langle \mathcal{E} \rangle = 2\epsilon_0 V |\mathbf{E}_0|^2 = n_{\text{ph}}\, \hbar\omega.} \tag{3.117}$$

We note that we could have obtained the total energy directly as the expectation value of the Hamiltonian (3.54).[5] The classical power (3.23) becomes

$$\boxed{\text{Power:} \quad \mathcal{P}_0 = 2\epsilon_0\, c\, A\, |\mathbf{E}_0|^2 = \frac{n_{\text{ph}}\, \hbar\omega}{t}.} \tag{3.118}$$

For the *momentum* of an EM wave, classically given by (3.24), we obtain

$$\boxed{\text{Momentum:} \quad \langle \mathbf{P} \rangle = \frac{2\,\epsilon_0 V\, |\mathbf{E}_0|^2}{c} = \frac{n_{\text{ph}}\, \hbar\omega}{c} = n_{\text{ph}} \hbar\, |\mathbf{k}|.} \tag{3.119}$$

Similar to the photon energy, we could have obtained the photon momentum as the expectation value of the photon momentum operator.[6]

The *intensity*, classically given by (3.25) is given by

$$\boxed{\text{Intensity:} \quad I_0 = 2\epsilon_0\, c\, |\mathbf{E}_0|^2 = \frac{n_{\text{ph}}\, \hbar\omega\, c}{V}.} \tag{3.120}$$

The *photon flux*, classically given by (3.26) is

$$\boxed{\text{Flux:} \quad \Phi_0 = \frac{2\,\epsilon_0\, c\, |\mathbf{E}_0|^2}{\hbar\omega} = \frac{n_{\text{ph}}\, c}{V}.} \tag{3.121}$$

Finally, we obtain for the *fluence*, whose classical expression is (3.27)

$$\boxed{\text{Fluence:} \quad \mathcal{F} = \frac{2\,\epsilon_0\, V\, |\mathbf{E}_0|^2}{A} = \frac{n_{\text{ph}}\, \hbar\omega}{A}.} \tag{3.122}$$

---

[5] The total energy of the radiation field is

$$\mathcal{E} = \langle \mathcal{H}_{\text{rad}} \rangle = \langle \Pi_{n_{p\mathbf{k}}} | \mathcal{H}_{\text{rad}} | \Pi_{n_{p\mathbf{k}}} \rangle = \langle \Pi_{n_{p\mathbf{k}}} | \sum_{p\mathbf{k}} \hbar\omega_{\mathbf{k}}\, \mathbf{a}^{\dagger}_{p\mathbf{k}} \mathbf{a}_{p\mathbf{k}} | \Pi_{n_{p\mathbf{k}}} \rangle$$
$$= \sum_{p\mathbf{k}} \hbar\omega_{\mathbf{k}}\, n_{p\mathbf{k}} = n_{\text{ph}}\, \hbar\omega.$$

[6] The photon momentum operator, normalized to the vacuum momentum, is given by $\mathbf{P} = \sum_{p\mathbf{k}} \hbar\mathbf{k}\, \mathbf{a}^{\dagger}_{p\mathbf{k}}\, \mathbf{a}_{p\mathbf{k}}$ so that $\langle \mathbf{P} \rangle = n_{\text{ph}} \hbar |\mathbf{k}|$.

References

1. P.A.M. Dirac, Proc. Roy. Soc. A **114**, 243 (1927)
2. P.A.M. Dirac, Proc. Roy. Soc. Lond. A **114**, 710 (1927)
3. B.J. Hunt, Phys. Today **65**, 48 (2012)
4. J. Stöhr, H.C. Siegmann, *Magnetism: From Fundamentals to Nanoscale Dynamics* (Springer, Heidelberg, 2006)
5. A. Kirilyuk, A.V. Kimel, T. Rasing, Rev. Mod. Phys. **82**, 2731 (2010)
6. J.D. Jackson, *Classical Electrodynamics* (Wiley, New York, 1999)
7. D. Attwood, A. Sakdinawat, *X-Rays and Extreme Ultraviolet Radiation: Principles and Applications*, 2nd edn. (Cambridge University Press, Cambridge, 2017)
8. J. Als-Nielsen, D. McMorrow, *Elements of Modern X-Ray Physics*, 2nd edn. (Wiley, New York, 2011)
9. E. Hecht, *Optics*, 4th ed. (Addison-Wesley, Reading, 2001)
10. A. Hofmann, *The Physics of Synchrotron Radiation* (Cambridge University Press, Cambridge, 2004)
11. M. Faraday, Phil. Trans. Roy. Soc. Lond. **1**, 136 (1846)
12. L. Pasteur, Annales Chimie Physique **28**, 56 (1850)
13. J. Stöhr, *NEXAFS Spectroscopy* (Springer, Heidelberg, 1992)
14. G.N. Lewis, Nature **118**, 874 (1926)
15. A. Einstein, Ann. Phys. **17**, 132 (1905)
16. E. Fermi, Rev. Mod. Phys. **4**, 87 (1932)
17. M. Planck, Verh. Deut. Phys. Ges. **13**, 138 (1911)
18. P.W. Milonni, *The Quantum Vacuum: An Introduction to Quantum Electrodynamics* (Academic Press, San Diego, 1994)
19. L.I. Schiff, *Quantum Mechanics*, 3rd edn. (McGraw-Hill, New York, 1968)
20. F. Mandl, G. Shaw, *Quantum Field Theory* (Wiley, New York, 1993)
21. R. Loudon, *The Quantum Theory of Light*, 3rd edn. (Clarendon Press, Oxford, 2000)
22. G. Grynberg, A. Aspect, C. Fabre, *Introduction to Quantum Optics, From the Semi-Classical Approach to Quantized Light* (Cambridge University Press, Cambridge, 2010)
23. D.A. Steck, *Quantum and Atom Optics*. Available online at http://steck.us/teaching (2017)
24. R. Glauber, Optical coherence and photon statistics. in *Quantum Optics and Electronics*, ed. by A.B.C. de Witt, C. Cohen-Tannoudji (Gordon and Breach, New York, 1965)
25. R.J. Glauber, Phys. Rev. **130**, 2529 (1963)
26. M. Fox, *Quantum Optics: An Introduction* (Oxford University Press, Oxford, 2006)
27. G. Breitenbach, S. Schiller, J. Mlynek, Nature **387**, 471 (1997)
28. A. Messiah, *Quantum Mechanics* (Wiley, New York, 1958)
29. J.J. Sakurai, *Modern Quantum Mechanics, Revised* (Addison-Wesley, Reading, Mass., 1994)
30. A. Fresnel, Œuvres **1**, 731 (1822)
31. A. Einstein, O. Stern, Ann. Phys. **40**, 551 (1913)
32. J. Mehra, H. Rechenberg, Found. Phys. **29**, 91 (1999)
33. C.C. Gerry, P.L. Knight, *Introductory Quantum Optics* (Cambridge University Press, Cambridge, 2005)
34. M.O. Scully, M.S. Zubairy, *Quantum Optics* (Cambridge University Press, Cambridge, 1997)
35. P.A.M. Dirac, *Quantum Mechanics*, 4th edn. (Oxford University Press, Oxford, 1958)

Chapter 4
Brightness and Coherence

4.1 Introduction

In modern x-ray science, the concept of brightness or brilliance plays an important role as a measure of the quality of x-ray sources. We intuitively associate brightness with the intensity that falls into the eye of an observer. Today, the term "bright" or "brilliant" is even used in different connotations, like the description of exceptional intellectual abilities.[1] One goal of this chapter is to provide a rigorous discussion of the brightness of a light source, in general, and of the more specific concepts of *average brightness* used to describe synchrotron radiation (SR) and *peak brightness* utilized for the description of XFEL sources.

It is well known that the improvement in brightness of SR sources is associated with increasing coherence. The detailed link of brightness and coherence, however, is not obvious since we typically define "coherence" as the ability of waves to create an interference pattern, while we associate "brightness" with photon density. Hence the connection between brightness and coherence goes to the very heart of the wave-particle duality of light and touches on Dirac's famous and puzzling statement that each photon can only interfere with itself. The elucidation and connection between the seemingly different concepts of brightness and coherence is a key goal of this chapter.

In particular, we will show that the conventional concepts of brightness and coherence are equivalent descriptions of light in the *first order* of the fundamental theory of light and matter, quantum electrodynamics (QED). Before the advent of XFELs, this first order description of light was all that was needed in x-ray science. The reason is that even the most advanced synchrotron radiation sources produced less than a single photon within a source volume of order λ^3 as illustrated in Fig. 1.3. Hence

[1] When the concept of brightness was first introduced to support the proposal for third generation synchrotron sources in the USA, there was a funny exchange between a scientist in the audience and the speaker (both shall remain unnamed). The question "who needs brightness?" was answered "Well, it's a good thing to have. Some of us have it and some of us don't".

J. Stöhr, *The Nature of X-Rays and Their Interactions with Matter*, Springer Tracts in Modern Physics 288, https://doi.org/10.1007/978-3-031-20744-0_4

the question of correlations between photons, e.g. whether they can work together or interfere, did not have to be addressed. This is what is behind Dirac's famous statement that applied to the light sources available at his time.

Figure 1.3 also shows that a new way of thinking is required when considering science with XFELs since an incredible number of up to 10^{10} photons are born within the minimum coherence volume of order λ^3. This requires an improved description of the nature of light, called *quantum optics*, which was developed in the 1960s in the wake of the formulation of QED and the invention of the maser and laser. Quantum optics extends Dirac's first order formulation of the quantum properties of light, based on non-interacting individual photons, through an infinite series of photon-photon correlations underlying QED.

In this chapter we shall only discuss the conventional or first order description of light, in which the wave and quantum concepts yield the same results. This allows a "semi-classical" description of light where, based on the wave-particle duality, different experiments may be conveniently explained in either the wave or photon picture. The key to the first order quantum treatment is that interactions between photons are neglected. They emerge in a second order and higher order treatment of coherence that will be discussed in Chap. 5.

4.2 Brightness and Coherence

4.2.1 *Introduction to the Concept of Brightness*

The concept of *spectral brightness*, $\mathcal{B}$, was first used in radiometry and astronomy to describe the brightness of stars. Sources that emit isotropically are referred to as *Lambertian sources* after Johann Heinrich Lambert who in 1760 considered the angular emission properties of primary and secondary (i.e. scattering) sources. It can be shown that thermal or blackbody sources are Lambertian [1]. The total energy $\mathcal{E}=n_{\rm ph}\hbar\omega$ per spectral interval $\Delta(\hbar\omega)$ emitted by such sources is given by

$$\frac{\mathcal{E}}{\Delta(\hbar\omega)} = \mathcal{B}\,\mathrm{d}A\,\mathrm{d}t\,\mathrm{d}\Omega, \tag{4.1}$$

where the proportionality factor $\mathcal{B}$ is called *spectral brightness* or *spectral radiance*. It is a fundamental measure of how bright a star appears to an observer at a point on earth. Here $\mathrm{d}A$ is the unit source area, $\mathrm{d}t$ the unit time of observation, and $\mathrm{d}\Omega$ the solid angle element of emission or observation. Since stars are typically envisioned as spheres, they emit uniformly into the 4π solid angle.

If the radiating surface is flat, the emission of a Lambertian source is no longer isotropic but has a $\cos\theta$ distribution around the surface normal, known as Lambert's cosine emission law. The intensity is maximum in the forward direction and has a node for $\theta=\pm 90°$. For a *flat* Lambertian source, (4.1) changes to [2]

$$\frac{\mathcal{E}}{\Delta(\hbar\omega)} = \mathcal{B}\,\mathrm{d}A\,\mathrm{d}t\,\mathrm{d}\Omega\,\cos\theta, \tag{4.2}$$

where θ is the polar angle of the emission direction from the surface normal of the flat radiating surface. The so-called obliquity factor $\cos\theta$ accounts for the projection of the flat surface seen by an observer at angle θ from the surface normal.

Because of the narrow emission cone of synchrotron radiation sources around the forward direction, where $\cos\theta = 1$, the brightness defined in (4.1) is a true measure of the intrinsic quality of such sources. The brightness concept was introduced for the description of synchrotron radiation by K.-J. Kim in 1986 [3] to describe the improvement provided by third generation sources.[2] For synchrotron radiation sources we can write the brightness in terms of different quantities as

$$\begin{aligned}\mathcal{B} &= \underbrace{\frac{n_{\mathrm{ph}}\hbar\omega}{\mathrm{d}t\,\Delta(\hbar\omega)}}_{\text{spectral power } \mathcal{P}_\omega}\frac{1}{A_{\mathrm{s}}\,\mathrm{d}\Omega} = \underbrace{\frac{n_{\mathrm{ph}}\hbar\omega}{\mathrm{d}t\,\Delta(\hbar\omega)\,A_{\mathrm{s}}}}_{\text{spectral intensity } I_\omega}\frac{1}{\mathrm{d}\Omega} = \underbrace{\frac{n_{\mathrm{ph}}\hbar\omega}{\mathrm{d}t\,A_{\mathrm{s}}}}_{\text{intensity } I}\frac{1}{\mathrm{d}\Omega\,\Delta(\hbar\omega)}\\ &= \underbrace{\frac{n_{\mathrm{ph}}}{\mathrm{d}t\,A_{\mathrm{s}}}}_{\text{flux } \phi}\frac{1}{\Delta\Omega\,\Delta(\hbar\omega)/\hbar\omega} = \underbrace{\frac{n_{\mathrm{ph}}}{\mathrm{d}\Omega\,A_{\mathrm{s}}\,\mathrm{d}t\,\Delta(\hbar\omega)/\hbar\omega}}_{\text{photons in phase space volume}}.\end{aligned} \tag{4.3}$$

The brightness expression is based on the concept of *space-time separability* since it factors into lateral or spatial properties expressed through $A_{\mathrm{s}}\,\mathrm{d}\Omega$ and longitudinal or temporal properties expressed through $\mathrm{d}t\,\Delta(\hbar\omega)/\hbar\omega$. This separability allows the distinction of *lateral or spatial coherence* and *longitudinal or temporal coherence*. The last expression furthermore links brightness to the number of photons per phase space volume. The understanding of the link of brightness, coherence, and photon density is a key goal of the present chapter, with emphasis on an x-ray perspective of synchrotron and XFEL radiation.

4.2.2 *Formal Definition of Brightness*

In x-ray science, the most-used concept is the photon *flux* or number of photons per unit area per unit time and the related *intensity* which is the photon flux times the photon energy, as discussed in Sect. 3.2.3. *Spectral brightness* or brilliance is a more fundamental property that describes the quality of an x-ray source and the properties of the emitted radiation. We can picture the x-rays emitted from an electron bunch as illustrated in Fig. 4.1.

[2] This concept greatly helped in making the case for the construction of the Advanced Light Source (ALS) and the Advanced Photon Source (APS) in the USA. Relative to the four orders of magnitude increase of x-ray brightness, the modest flux increase offered by the competing proposal for the construction of a spallation neutron source led to the higher priority funding selection of the x-ray sources.

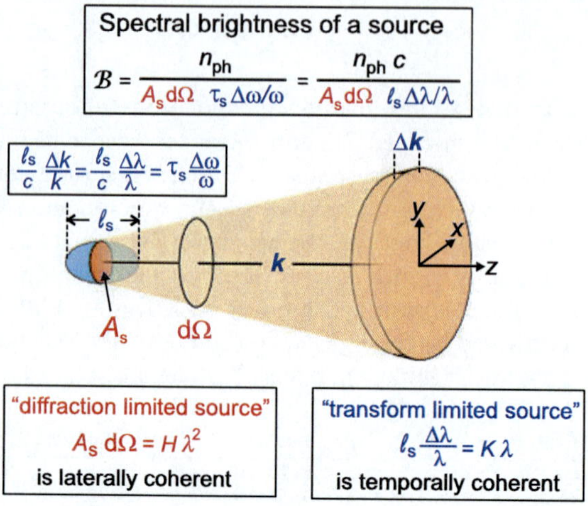

Fig. 4.1 Illustration of the spectral brightness concept to the lateral (red) and longitudinal (blue) properties of a source. The peak brightness is defined as the number of photons n_{ph} per source size A_s, per solid angle of emission $d\Omega$ (units [steradian]), per temporal pulse length τ_s, per relative bandwidth $\Delta\omega/\omega$. In practice, the area–solid angle product $A_s \, d\Omega$, and the orthogonal time-bandwidth product $\tau_s \, \Delta\omega/\omega$ are determined by both electron and photon distribution functions. Underneath we have identified the diffraction and transform limits for a coherent source. The factors H and K depend on the shape of the intensity distribution of the source and are given by (4.19) and (4.20), below

For simplicity we have drawn the source volume resembling an ellipsoid of lateral area A_s and length ℓ_s. In general, the source volume may be expressed as a product of an *effective* area and length that contain the generated energy, which leads to the shape dependent factors H and K in the stated diffraction and transform limits in the figure. The lateral source area A_s and the solid angle of emission $d\Omega$ define the transverse source properties.[3] The solid angle $d\Omega$, illustrated in Fig. 4.2, has units of steradian (sr), so that the maximum solid angle of emission from a point is 4π sr. A light source has a minimum coherence area of $\simeq \lambda^2$, and emission into the cone $d\Omega \simeq 1$ sr corresponds to coherent emission.

The source length ℓ_s defines the x-ray pulse length which is taken to be the length of the electron bunch. The uncertainty in the length of the emitted wavevector or wavelength $\lambda = 2\pi/k$ determines the relative bandwidth. The uncertainty in photon wavelength $\Delta\lambda$ and photon angular frequency $\Delta\omega = 2\pi\Delta\nu$ and energy $\Delta(\hbar\omega)$ is linked according to $\Delta\lambda = \Delta(c/\nu) = c\Delta\nu/\nu^2$. The *relative bandwidth* may be written equivalently as

$$\frac{\Delta k}{k} = \frac{\Delta\lambda}{\lambda} = \frac{\Delta\omega}{\omega} = \frac{\Delta(\hbar\omega)}{\hbar\omega} = \frac{\tau_0}{\tau_{coh}}. \tag{4.4}$$

[3] The solid angle of unit [sr] and the planar angle of unit [rad] are defined as ratios of two areas or lengths and are actually dimensionless. The names are introduced to distinguish between dimensionless quantities of different nature.

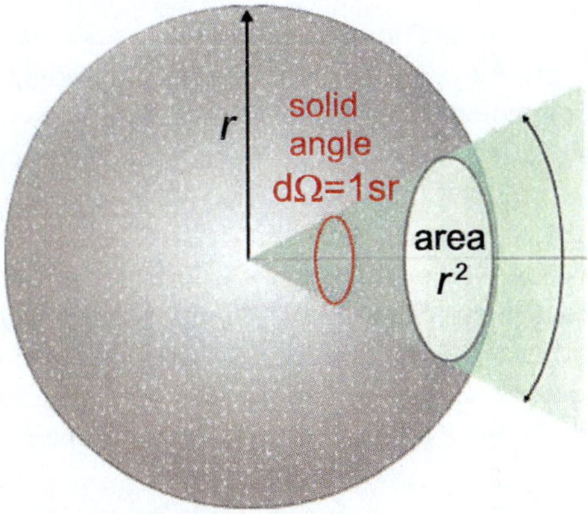

Fig. 4.2 Illustration of a cone of solid angle of $d\Omega = 1$ in units of steradian (sr), corresponding to the coherence cone of a point source. In practice, $d\Omega = 1$ can be considered the coherence cone of a source of minimum coherence area λ^2. Also shown is the associated *total* planar opening angle $\vartheta \simeq 65.5°$. In general, the solid angle $d\Omega$ is related to the total planar opening angle ϑ about the central axis according to $d\Omega = 4\pi \sin^2(\vartheta/4)$, which for small angles becomes $d\Omega \simeq \pi(\vartheta/2)^2$

where $\tau_0 = \lambda/c$, and the wavelength and wavevector are related according to $\lambda = 2\pi/k$.

The spectral brightness, identified by the boxed expression in Fig. 4.1, contains the fundamental uncertainty relations in space, identified in red, and time, shown in blue. Owing to the transverse nature of light it assumes separability of the lateral or spatial and longitudinal or temporal degrees of freedom. With these considerations we can define as follows.

The *spectral brightness* of a photon source is defined as

$$\mathcal{B} = \frac{n_{ph}\, c}{A_s\, d\Omega\, \ell_s\, \Delta\lambda/\lambda} = \frac{n_{ph}}{\underbrace{A_s\, d\Omega}_{\text{space-momentum uncertainty}} \quad \underbrace{\tau_s\, \Delta\omega/\omega}_{\text{time-energy uncertainty}}}. \tag{4.5}$$

Here n_{ph} is the number of photons. The spatial part is determined by the lateral source area A_s, and the dimensionless solid angle of emission $d\Omega$ (in sr). $d\Omega = 1$ (in sr) defines the coherence cone of a source of area λ^2, which has a total planar opening angle of $\vartheta \simeq 66°$ as shown in Fig. 4.2. The time τ_s is the temporal duration corresponding to a length $\ell_s = c\,\tau_s$, and $\Delta\omega/\omega = \Delta\lambda/\lambda$ is the dimensionless relative bandwidth of the emitted radiation.

By convention, the spectral x-ray brightness is stated by use of the small angle approximation $d\Omega = \pi\vartheta^2/4$ in units of [photons/s/mm^2/mrad^2/0.1% bandwidth], where the correspondence between $d\Omega$ with units [sr] and the *total* planar opening angle ϑ with units [rad] is illustrated in Fig. 4.2.

By use of (4.4) and $c/\ell_s = 1/dt$, the spectral brightness definition (4.5) agrees with the radiological one given by (4.3). We see that spectral brightness is defined abstractly as the number of photons in a six-dimensional phase space volume, comprised of a three-dimensional spatial volume $V = A_s\,\ell_s$ that contains the photons, the two-dimensional solid angle $\mathrm{d}\Omega$ into which photons are emitted and the one-dimensional bandwidth $\Delta\lambda/\lambda$ of the radiation.

Spectral brightness is conserved in phase space, which is a general consequence of Liouville's phase space theorem. For example, through focusing one can create a smaller image of the source size A_s, but this comes at the expense of the divergence angle $\mathrm{d}\Omega$, since their product is constant. Another example is the reduction of the emitted bandwidth $\Delta\omega$ by use of a monochromator at the expense of an increased pulselength τ_s.

Spectral brightness of a source is a conserved quantity in phase space and cannot be improved by optical elements. Imperfect optics will, however, lead to its degradation.

We note upfront that the so-defined concept of brightness is only a first order concept in the ultimate theory of light, QED, as will become apparent later.

4.2.3 *Average Brightness*

In (4.5) we have identified by underbrackets two orthogonal phase space contributions. Assuming x-ray propagation in the z direction, the first bracket defines the "lateral" source properties given by the lateral source area A_s, typically specified by the lengths along two perpendicular axes x and y, defined to lie in the ring plane and perpendicular to it. The solid angle of emission is then expressed as $\mathrm{d}\Omega = \pi\theta_x\theta_y$, where θ_x and θ_y are the *half* planar emission angles measured from the x and y axes in the two orthogonal x-z and y-z planes, respectively. The product $A_s\mathrm{d}\Omega$ defines the *emittance* of the source which is the key property of storage ring sources. From a SR perspective, the time-bandwidth part of the brightness expression indicated by the second underbracket in (4.5) is simply averaged and the so-called *average brightness* is defined as

$$\mathcal{B}_{\text{avg}} = \frac{F}{A_s\,\mathrm{d}\Omega}, \quad \text{where} \quad F = \frac{n_{\text{ph}}}{\Delta t\,\Delta\omega/\omega}. \tag{4.6}$$

Here F is the *spectral photon flux* with dimension [photons/time/bandwidth], to be distinguished from the *areal photon flux* Φ with the dimension of [photons/time/area].

The average brightness $\mathcal{B}_{\text{avg}}$ of a storage ring can be increased by a limited amount through reducing the time–energy uncertainty in (4.5). This is done by making each electron radiate longer wavetrains by sending them through an undulator. In practice, undulators reduce $\Delta\omega$ by a factor of about 100 relative to a bending magnet, with an increase of F by the same amount. Since most storage rings have similar cycle repetition rates $R \simeq 500\,\mathrm{MHz}$ and pulse lengths $\tau_p \simeq 100\,\mathrm{ps}$, one defines the emission time Δt in units of [s] as the time over which photons are emitted by the electron bunches per ring cycle, $\Delta t \simeq 5 \times 10^{-2}\,\mathrm{s}$, and by convention one assumes $\Delta\omega/\omega = 10^{-3}$ (0.1%).

As indicated by underbrackets in (4.5), the achievable lateral or average brightness is ultimately determined by uncertainty relations. In storage ring sources, the minimum time-bandwidth product $\tau_s\Delta\omega/\omega$ can only be reduced (without loss of flux) by narrowing the bandwidth $\Delta\omega$ through undulators. The real improvement arises through the reduction of the product $A_s\mathrm{d}\Omega$, i.e. through smaller beam emittances [4]. Owing to the fact that the area-angle product enters in the denominator, the brightness has an upper bound. Its limit is given by Heisenberg's space-momentum uncertainty principle, and when this limit is reached, the source becomes diffraction limited and all emitted photons are laterally coherent.

Shortly after Heisenberg's publication of the space-momentum uncertainty principle [5], Kennard [6] first proved the modern inequality which links the *standard deviation* of position, r, with that in momentum, p. For this reason, Gaussian *intensity* distributions of rms width σ^I naturally define the minimum uncertainty distributions in position, r and momentum, p, which can also be written in terms of the wavevector, $q = p/\hbar$, and angle $\theta = q/k = q\lambda/2\pi$ according to

$$\sigma_r^I\,\sigma_p^I \ge \frac{\hbar}{2}, \quad \text{or} \quad \sigma_r^I\,\sigma_q^I \ge \frac{1}{2}, \quad \text{or} \quad \sigma_r^I\,\sigma_\theta^I \ge \frac{\lambda}{4\pi}. \tag{4.7}$$

For coherent Gaussian *intensity* distributions with rms widths σ_r^I and σ_θ^I, the first underbracket in (4.5) is given by

$$\boxed{\text{Gaussian diffraction limit:}\;\; A_s\,\mathrm{d}\Omega = 2\pi(\sigma_r^I)^2\,2\pi(\sigma_\theta^I)^2 \ge \left(\frac{\lambda}{2}\right)^2}. \tag{4.8}$$

The minimum of the product, expressed by the equal sign, defines the so-called *diffraction limit* according to $A_{\mathrm{coh}}\,\mathrm{d}\Omega_{\mathrm{coh}} = \lambda^2/4$ (see Fig. 4.1). This expression means that a source area of $\lambda^2/4$ emits into a diffraction limited solid angle $\Omega_{\mathrm{coh}} = 1$. When the diffraction limit is reached, all n_{ph} photons are laterally coherent and the average brightness has its maximum

$$\mathcal{B}_{\mathrm{avg}}^{\mathrm{max}} = \frac{4\,F}{\lambda^2}. \tag{4.9}$$

In practice, only a fraction of the photons emitted by today's storage rings are laterally coherent due to the fact that the lateral size of electron bunches in storage rings cannot be compressed below the uncertainty area associated with a single electron illustrated in Fig. 2.23b. Such a source may be described to have a laterally *quasi-homogeneous* intensity distribution [7], where the intensity changes rather homogeneously with distance while the coherence changes much faster. If we denote the single-electron coherence area as A^{μ}_{coh} and the total lateral beam area as A_{s}, the average brightness (4.6) is reduced by the laterally coherent fraction

$$h_c = \frac{A^{\mu}_{\mathrm{coh}}}{A_{\mathrm{s}}} \tag{4.10}$$

and becomes

$$\mathcal{B}_{\mathrm{avg}} = \frac{A^{\mu}_{\mathrm{coh}}}{A_{\mathrm{s}}} \mathcal{B}^{\mathrm{max}}_{\mathrm{avg}} . \tag{4.11}$$

The photon energy dependence of the laterally coherent fraction is plotted in Fig. 2.24d.

4.2.4 *Peak Brightness*

The second underbracket in (4.5) defines the "longitudinal" or temporal source properties, i.e. those along the beam direction. It has become of prime interest with the advent of XFELs where all photons are crammed into an ultrashort pulse of $\tau_{\mathrm{s}} \simeq 1$–100 fs duration. One now uses the actual pulse duration and refers to $\mathcal{B}$ as the *peak brightness*. The reduction of the time-bandwidth product $\tau_{\mathrm{s}} \Delta\omega/\omega$ at XFELs over storage rings leads to a much larger value of the peak relative to the average brightness.

In analogy to the minimum area-angle relation (4.7), the corresponding minimum time-frequency Gaussian intensity distributions are given by

$$\sigma_t^I \, \sigma_{\mathcal{E}}^I \geq \frac{\hbar}{2}, \quad \text{or} \quad \sigma_t^I \, \sigma_\omega^I \geq \frac{1}{2}, \quad \text{or} \quad \sigma_t^I \, \frac{\sigma_\omega^I c}{\omega} \geq \frac{\lambda}{4\pi}. \tag{4.12}$$

The second underbracket in (4.5) can then be written as

$$\tau_{\mathrm{s}} \frac{\Delta\omega}{\omega} = \sqrt{2\pi}\sigma_t^I \, \frac{\sqrt{2\pi}\sigma_\omega^I}{\omega} = 2\pi\sigma_t^I \, \frac{\sigma_{\mathcal{E}}^I}{\hbar\omega} \geq \frac{\lambda}{2c}. \tag{4.13}$$

By relating the effective Gaussian pulse length $\tau_{\mathrm{s}} = \sqrt{2\pi}\sigma_t^I$ and effective energy width $\Delta\mathcal{E} = \sqrt{2\pi}\hbar\sigma_\omega^I$ we have

$$\boxed{\text{Gaussian transform limit:}\ \tau_{\mathrm{s}} \, \Delta\mathcal{E} \geq h/2}. \tag{4.14}$$

The minimum product, given by the equal sign, is now expressed as a *time-bandwidth product* that defines the so-called *transform limit* (see Fig. 4.1). When it is reached, all $n_{\rm ph}$ photons have become temporally coherent. The expression (4.13) states that a minimum pulse length of $\tau_{\rm s} = \lambda/(2c)$ emits a spectrum of unit bandwidth $\Delta\omega/\omega$. With increasing coherent pulse length the emitted bandwidth $\Delta\omega/\omega$ is proportionally reduced.

The peak brightness for a source with spatiotemporal Gaussian distributions is given by

$$\mathcal{B}_{\rm peak} = \frac{n_{\rm ph}}{A_{\rm s}\, {\rm d}\Omega\, \tau_{\rm s}\, \Delta\omega/\omega} \le \frac{8\, n_{\rm ph}^{\rm coh}\, c}{\lambda^3} = \mathcal{B}_{\rm peak}^{\rm max}. \tag{4.15}$$

The maximum value of the peak brightness is expressed by the equal sign. It corresponds to the minimum spatiotemporal uncertainty product where all photons are (first order) coherent $n_{\rm ph} = n_{\rm ph}^{\rm coh}$. The number of coherent photons in this limit is the *photon degeneracy parameter* to be discussed in more detail in Sect. 4.2.7 below.

4.2.5 Brightness Reduction Through Partial Coherence

In practice, conventional SASE and even self-seeded pulses are not temporally coherent or transform limited and are said to be *partially coherent*. In general, all sources that are not fully coherent fall under the category of "partial coherence". This difficult and broad concept covers all kinds of light behavior between the extremes, referred to as completely coherent and incoherent or chaotic. We will come back to the description of chaoticity below and at this point illustrate partial coherence through the properties of SASE XFEL pulses.

A macro-pulse of a SASE XFEL of total length $\tau_{\rm p} = \tau_{\rm s}$ consists of M coherent microspikes or modes of duration $\tau_{\rm coh}^{\mu}$ as previously illustrated in Fig. 2.31, where $M \sim \tau_{\rm p}/\tau_{\rm coh}^{\mu}$. In this case, the peak brightness is reduced by the temporally coherent faction

$$\text{Temporal coh. fraction:}\;\; k_c = \frac{\tau_{\rm coh}^{\mu}}{\tau_{\rm p}} \simeq \frac{1}{M}. \tag{4.16}$$

For example, a SASE pulse of $\tau_{\rm p} \simeq 100$ fs consists of $M \simeq 100$ spikes, so that the peak brightness is reduced from its transform-limited value by a factor of $\simeq 100$.

The behavior of partially coherent sources in time and space is illustrated in Fig. 4.3. In Fig. 4.3a we illustrate the concept of *partial temporal coherence*. We have assumed a total pulse length $\tau_{\rm p}$, defined by the statistical average over many pulses, indicated by a smooth envelope function. Each pulse consists of a spiky substructure of average width $\tau_{\rm coh}^{\mu}$ indicated in red. The pulse contains a temporally coherent fraction of photons $\tau_{\rm coh}^{\mu}/\tau_{\rm p}$. In the limit $\tau_{\rm coh}^{\mu} \ll \tau_{\rm p}$ the source is called temporally

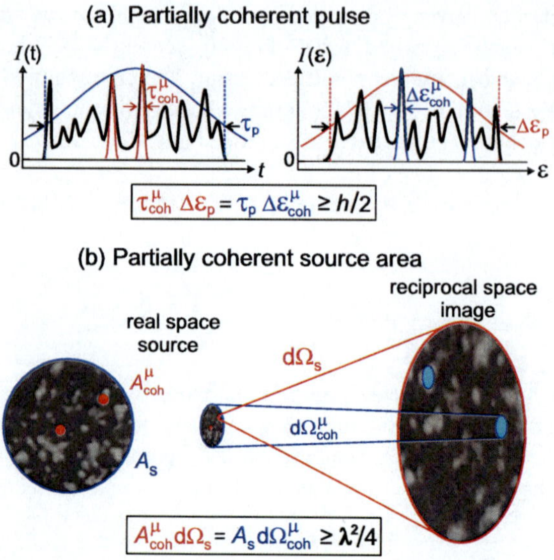

Fig. 4.3 **a** Illustration of the intensity distribution in a pulse emitted by a temporally partially coherent source. The macroscopic temporal intensity distribution in the pulse of length τ_p consists of microscopic coherent spikes of length $\tau^{\mu}_{\mathrm{coh}} < \tau_p$, and the pulse has a similar structure in the energy domain. The complementary uncertainty product relationship in the time and energy domain is $\tau^{\mu}_{\mathrm{coh}} \Delta\mathcal{E}_p = \tau_p \Delta\mathcal{E}^{\mu}_{\mathrm{coh}} \geq h/2$ underneath, assuming Gaussian lineshapes . **b** The equivalent concept of a spatially partially coherent source. The macroscopic intensity distribution across the source area A_s contains microscopic coherent regions of size $A^{\mu}_{\mathrm{coh}} < A_s$. The complementary uncertainty product relationship in the spatial and angular domain is shown underneath, assuming effective Gaussian areas

incoherent or chaotic, and in the limit $\tau^{\mu}_{\mathrm{coh}} = \tau_p$ it is coherent. Underneath we have indicated the time–energy uncertainty relation (4.14) for Gaussian lineshapes.

In Fig. 4.3b the corresponding concept of *partial spatial coherence* is illustrated. When averaged over a pulse length, the source has an area A_s that contains micro-coherence regions of size A^{μ}_{coh}, indicated by red circles. For simplicity we have assumed all regions to have circular shapes. The source then contains a spatially coherent fraction of photons, given in analogy to the temporal coherent fraction (4.16) by,

$$\text{Lateral coh. fraction:} \quad h_c = \frac{A^{\mu}_{\mathrm{coh}}}{A_s} \tag{4.17}$$

In the limit $A^{\mu}_{\mathrm{coh}} \ll A_s$ the source is called spatially incoherent or chaotic, and in the limit $A^{\mu}_{\mathrm{coh}} = A_s$ it is coherent. At the bottom of Fig. 4.3b, we have indicated the area–solid angle uncertainty product (4.8) by assuming effective Gaussian areas.

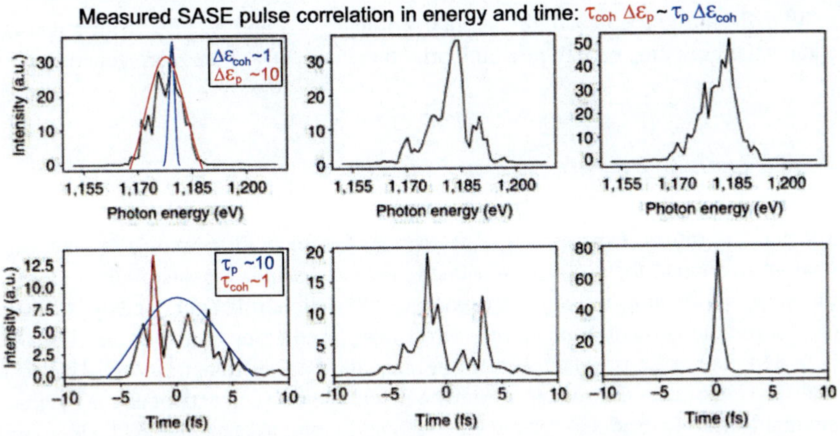

Fig. 4.4 Measured SASE pulse structures in energy (top row) and time (bottom row) for three exemplary shots [8]. The XFEL pulses of ∼7 fs FWHM and a nominal photon energy of 1180 eV had similar total energies of from left to right 36.5 μJ, 39.5 μJ, and 40 μJ. In the left column we have superimposed approximate red and blue curves on the original data to indicate the boxed relationship given in Fig. 4.3a

The peak brightness of a partially coherent source is therefore reduced according to

$$\mathcal{B}_{\text{peak}} = k_c\, h_c\, \mathcal{B}^{\text{max}}_{\text{peak}}. \tag{4.18}$$

The correlation of temporal and energy widths illustrated in Fig. 4.3a are born out by experimental results by Hartmann et al. [8] shown in Fig. 4.4. In the left column of the figure we have indicated the relationship between the temporal and energy widths of the SASE structure by superimposing approximate red and blue curves on the original data shown as black curves. The temporal structure of the SASE pulses was obtained with attosecond resolution via angular streaking of Ne 1s photoelectrons produced by the x-rays. The photoelectrons are modulated by the continuously rotating electric field of a circularly polarized infrared ($\lambda = 10.6\,\mu$m) laser, and depending on the amplitude and phase of the streaking field at the time of ionization, the photoelectrons are either accelerated or decelerated. The induced momentum change results in modulations of the photoelectron energy and angular distribution that is measured with an array of time-of-flight detectors.

It is important to realize that the corresponding patterns illustrated in Fig. 4.3 in conjugate variables of time–energy and space–reciprocal space are not linked by a simple Fourier transform. The reason is that they correspond to *intensity* patterns where the phase information that is contained in the fields has been lost.[4]

[4] The case shown in Fig. 4.3b is related to the well-known “phase problem” in x-ray crystallography where the reciprocal space intensity diffraction pattern cannot simply be Fourier transformed into a real space image as discussed in Sect. 8.8.

With the introduction of the coherent fractions τ_{coh}^{μ}/τ_p and A_{coh}^{μ}/A_s in the temporal and spatial domains, we can now link brightness and coherence more generally.

4.2.6 *Link of Brightness and Coherence*

The peak brightness expression (4.15) assumes Gaussian distributions in space and time. In particular, the prefactor 8 is due to the expressions for the minimum Gaussian space-momentum uncertainty product or diffraction limit expressed by (4.8) and the minimum Gaussian time–energy uncertainty product or transform limit (4.14). Gaussian distributions are convenient because the Fourier transform of a Gaussian is also a Gaussian and from a quantum mechanics point of view they define the minimum uncertainty product. They are also typically used to characterize the temporal and spatial distributions of synchrotron and XFEL radiation.

More generally, one may define shape-dependent dimensionless *coherence factors*, where H is determined by the functional form of the lateral (spatial) emission profile and K by the longitudinal (temporal) profile. The expressions leading to the diffraction and transform limits are then given by the following generalized forms, previously stated in Fig. 4.1.

The *diffraction limit* for a source of arbitrary intensity distribution function is given by the minimum value of the *area–solid angle* product

$$A_s \, d\Omega \geq H \lambda^2. \tag{4.19}$$

The *transform limit* is the minimum value of the so-called *time-bandwidth* product

$$\ell_s \frac{\Delta\lambda}{\lambda} \geq K \lambda \quad \text{or} \quad \tau_s \, \Delta\mathcal{E} \geq K \, h, \tag{4.20}$$

where we have used $\ell_s = \tau_s c$ and $\frac{\Delta\lambda}{\lambda} = \frac{\Delta\mathcal{E}}{\hbar\omega}$ (see (4.4)) and $h = 4.136\,\text{eV fs}$ is Planck's constant. The coherence factors H and K depend on the shape of the longitudinal and transverse intensity distribution functions in the source, given below by (4.25) and (4.24).

We can then write down the following generalized expression for the peak brightness.

If we define the *coherence volume* through shape-dependent dimensionless lateral (spatial) and longitudinal (temporal) coherence factors H and K as

$$V_{\text{coh}} = H\,K\lambda^3\,, \tag{4.21}$$

then the photons born in V_{coh} are emitted into a *unit solid angle* $\mathrm{d}\Omega = 1$ and have *unit bandwidth* $\Delta\omega/\omega = 1$. Defining the laterally and temporally coherent fractions of photons by factors

$$h_c = \frac{A^{\mu}_{\text{coh}}}{A_s}, \quad \text{and} \quad k_c = \frac{\tau^{\mu}_{\text{coh}}}{\tau_p}\,, \tag{4.22}$$

where the pulse length is also the source length, $\tau_p = \tau_s$, the master equation for peak brightness reads

$$\boxed{\mathcal{B}_{\text{peak}} = \frac{h_c}{H}\frac{k_c}{K}\frac{n_{\text{ph}}\,c}{\lambda^3}}. \tag{4.23}$$

The right side shows that the peak brightness of a source reflects the number of coherent photons $n_{\text{coh}} = h_c k_c n_{\text{ph}}$ of wavelength λ that are born in the coherence volume $V_{\text{coh}} = H\,K\lambda^3$ and emitted into a unit solid angle with unit bandwidth. Brightness therefore reflects the number density of coherent photons of a given wavelength λ or alternatively the *energy density* of the distribution of photons of energy $\hbar\omega$.

The parameters H and K depend on the shape of the spatial and temporal distribution functions and definition of their widths. In this book we will follow the definition of distribution widths used in statistical optics [9]. They are based on a powerful theorem involving the absolute values squared of a Fourier transform pair known as *Parseval's theorem* which assures conservation of energy associated with a Fourier transformation as discussed in Appendices A.4.1 and A.4.4 (also see Sect. 4.4.2 below).

For *coherent* sources, the factors H and K follow from Fourier relations between the *field* distributions. For incoherent sources, the factors are calculated from Fourier transforms between *intensities* according to the *spatial* van Cittert–Zernike [10, 11] theorem (see Sect. 4.6) and the related *temporal* Wiener–Khintchine [12] theorem, which play an important role in statistical optics [9, 13].

The relations for a coherent source are given by the following general expressions

Coherent source :

$$K = \frac{\left|\int_{-\infty}^{\infty} \langle |E(t)|^2\rangle \, \mathrm{d}t\right|^2}{\langle |E(0)|^2\rangle \left|\int_{-\infty}^{\infty} \sqrt{\langle |E(t)|^2\rangle} \, \mathrm{d}t\right|^2} \tag{4.24}$$

$$H = \frac{\left|\iint_{-\infty}^{\infty} \langle |E(\mathbf{r})|^2\rangle \, \mathbf{dr}\right|^2}{\langle |E(0)|^2\rangle \left|\iint_{-\infty}^{\infty} \sqrt{\langle |E(\mathbf{r})|^2\rangle} \, \mathbf{dr}\right|^2}. \tag{4.25}$$

The corresponding expressions for a partially coherent source are

Partially coherent source :

$$K = \frac{\int_{-\infty}^{\infty} \left|\langle |E(t)|^2\rangle\right|^2 \, \mathrm{d}t}{\langle |E(0)|^2\rangle \int_{-\infty}^{\infty} \langle |E(t)|^2\rangle \, \mathrm{d}t} \tag{4.26}$$

$$H = \frac{\iint_{-\infty}^{\infty} \left|\langle |E(\mathbf{r})|^2\rangle\right|^2 \, \mathbf{dr}}{\langle |E(0)|^2\rangle \iint_{-\infty}^{\infty} \langle |E(\mathbf{r})|^2\rangle \, \mathbf{dr}}. \tag{4.27}$$

The triangular brackets indicate a *statistic average*. If the fields are described classically with well-defined amplitudes, the triangular brackets become a cycle average. Using the above expressions, the spatial parameters H and h_c and temporal parameters K and k_c for Gaussian and flat-top source intensity distributions for both the coherent and chaotic cases are given in Table 4.1.

Table 4.1 Area–solid angle products H and time–energy products K for flat-top and Gaussian distributions that are either coherent or chaotic

	Coherent				Partially coherent			
Distribution	H	h_c	K	k_c	H	h_c	K	k_c
Flat-top	1	1	1	1	1	$\frac{A^{\mu}_{\mathrm{coh}}}{A_{\mathrm{s}}}$	1	$\frac{\tau^{\mu}_{\mathrm{coh}}}{\tau_{\mathrm{p}}}$
Gaussian	$\frac{1}{4}$	1	$\frac{1}{2}$	1	$\frac{1}{2}$	$\frac{A^{\mu}_{\mathrm{coh}}}{2\pi(\sigma_r^I)^2}$	$\frac{1}{\sqrt{2}}$	$\frac{\tau^{\mu}_{\mathrm{coh}}}{\sqrt{2\pi}\sigma_t^I}$

The *spatial* flat-top source distributions apply either for a box-like area, $A_{\mathrm{s}} = d^2$, or a circular area $A_{\mathrm{s}} = \pi R^2$, and the effective Gaussian area is $A_{\mathrm{s}} = 2\pi(\sigma_r^I)^2$ (see (4.8)). The micro coherence areas are denoted A^{μ}_{coh}. The *temporal* effective pulse length for a flat-top distribution is $\tau_{\mathrm{p}} = \ell_{\mathrm{p}}/c$ and for a Gaussian distribution $\tau_{\mathrm{p}} = \sqrt{2\pi}\sigma_t^I$ (see (4.13) with $\tau_{\mathrm{s}} = \tau_{\mathrm{p}}$). The microscopic coherence times are denoted $\tau^{\mu}_{\mathrm{coh}}$

4.2.7 Photon Degeneracy Parameter

The generalized peak brightness expression (4.23) for a coherent source, $h_c = k_c = 1$, is seen to be directly proportional to the coherent photon density, and it determines the *degeneracy parameter* according to

$$n_{\text{coh}} = \frac{\mathcal{B}_{\text{peak}}}{c} \underbrace{H\,K\,\lambda^3}_{V_{\text{coh}}}. \tag{4.28}$$

According to Table 4.1 the coherence volume differs for flat-top (box volume) and Gaussian distributions according to

$$\text{Flat-top: } V_{\text{coh}} = \lambda^3, \quad \text{Gaussian: } V_{\text{coh}} = \frac{\lambda^3}{8}. \tag{4.29}$$

The number of coherent photons per unit volume is linked to the brightness by (4.28). In this expression we have written the spatial and temporal uncertainties expressed by (4.19) and (4.20) in terms of the birth or coherence volume V_{coh} which corresponds to setting the solid angle of emission and spectral bandwidth to unity, i.e. $d\Omega = \Delta\omega/\omega = 1$. Note that only the space-angle and time-frequency *products* are preserved for a given distribution shape. One may therefore define the birth volumes differently which lead to a concomitant change of the solid emission angle and bandwidth, since their product is conserved. In Fig. 2.23, we have defined the coherent birth area as $\lambda^2/4$, while one may also chose the value λ^2/π [14] which corresponds to the maximum resonant atomic interaction cross section with emission of a spherical wave, derived by Breit and Wigner [15, 16].

> The maximum achievable brightness determines the maximum degeneracy parameter and reflects complete classical coherence. For a given wavelength, coherence can therefore be simply defined in terms of photon density or photon energy density instead of the conventional concept of wave interference.
>
> The equivalence of *coherence* and *photon energy density* is a consequence of the quantization of the EM field. In quantum mechanics, the classical concept of a phase relation between waves does not exist and is replaced by the concept of photon density (see Sect. 3.5).

4.2.8 Brightness of Storage Rings and XFELs

As an example, let us link the degeneracy parameter of a state-of-the art XFEL with its peak brightness. We assume that the XFEL is operated in a low charge mode of 10^8 electrons (0.016 nC) per bunch that produces a coherent pulse of order of 1 fs length containing about 10^{10} photons, as previously shown in Fig. 1.11. Since the photons are coherent, the degeneracy parameter is also 10^{10} (photons). These photons are born in a Gaussian coherence volume $V_{\text{coh}} = \lambda^3/8$ and are emitted into a unit solid angle, $\text{d}\Omega = 1$, with unit or 100% bandwidth (BW), $\Delta\omega/\omega = 1$. We are interested in the corresponding peak brightness in the conventional units [photons/s/mm^2/rad^2/0.1%BW] at a wavelength of $\lambda = 0.1$ nm (12.4 keV).

The Gaussian coherence volume is then $V_{\text{coh}} = \lambda^3/8 = 1.25 \times 10^{-22}$ mm^3, and the solid angle $\text{d}\Omega = 1$ sr corresponds to the product of two equal orthogonal planar *half* angles θ with units [rad], measured from the cone axis, that span the same cone, given in the small angle approximation (see footnote 3 and caption of Fig. 4.2 where $\theta = \vartheta/2$) by $\text{d}\Omega \simeq \pi\theta^2$. The unit conversion in the brightness expression is thus 1 [sr] = π [rad^2]. The 100% BW is simply reduced by 10^3 to obtain the proper brightness units. Inserting into the peak brightness expression (4.5) we obtain with $c = 3 \times 10^{11}$ mm/s,

$$\begin{aligned} \mathcal{B}_{\text{peak}} &= \frac{n_{\text{coh}}\, c}{V_{\text{coh}}\, \text{d}\Omega_{\text{coh}}\, \Delta\omega/\omega} \\ &= \frac{10^{10} * 3 \times 10^{11}}{1.25 \times 10^{-22} * \pi \times 10^6 * 10^3}\, \frac{\text{photons}}{\text{s}\,\text{mm}^2\,\text{mrad}^2\,0.1\%\text{bandwidth}} \\ &= 0.76 \times 10^{34}\, \frac{\text{photons}}{\text{s}\,\text{mm}^2\,\text{mrad}^2\,0.1\%\text{bandwidth}}. \end{aligned} \tag{4.30}$$

Using the correlation of brightness and degeneracy parameter expressed by (4.30) we have plotted in Fig. 4.5 the average and peak brightness numbers for synchrotron radiation sources and the peak brightness for an XFEL source. On the right (black scale) we show the corresponding degeneracy parameter. For comparison we have also indicated the values for sunlight ($n_{\text{coh}} \simeq 3 \times 10^{-4}$) and a common He:Ne laser (1 milliwatt power, $n_{\text{coh}} \simeq 3 \times 10^9$) at a wavelength of 633 nm [13].

Figure 4.5 illustrates that before the advent of XFELs, x-ray beams contained on average less than one photon per coherence volume, especially since beam line losses are not included in our numbers. It was therefore justified to neglect all interaction processes with matter where two or more coherent photons worked together. X-ray interactions occurred one-photon-at-a-time!

The emitted intensity is typically stated in different units for synchrotron radiation, where it is specified as [number of photons of energy $\hbar\omega$ per time per area], and XFELs, where the intensity is given in units adopted from the laser community of [mJ/cm^2/fs]. Let us compare how the photon numbers in modern synchrotron experiments convert into XFEL intensities. To do so, we assume a photon energy of 1000 eV and assume that a third generation synchrotron undulator source produces about 10^6 photons of 10 eV (1%) bandwidth in a 100 ps long pulse that may be focused without loss to a $10 \times 10\,\mu\text{m}^2$ spot. With the conversion 1 eV= 1.6×10^{-16} mJ this gives an intensity of 1.6×10^{-6} mJ/cm^2/fs. This number compares to a typical XFEL intensity of order 10^3 mJ/cm^2/fs [18] for a comparable photon energy and bandwidth. The XFEL intensity is thus larger by a factor of about 10^9, which roughly agrees with Fig. 4.5.

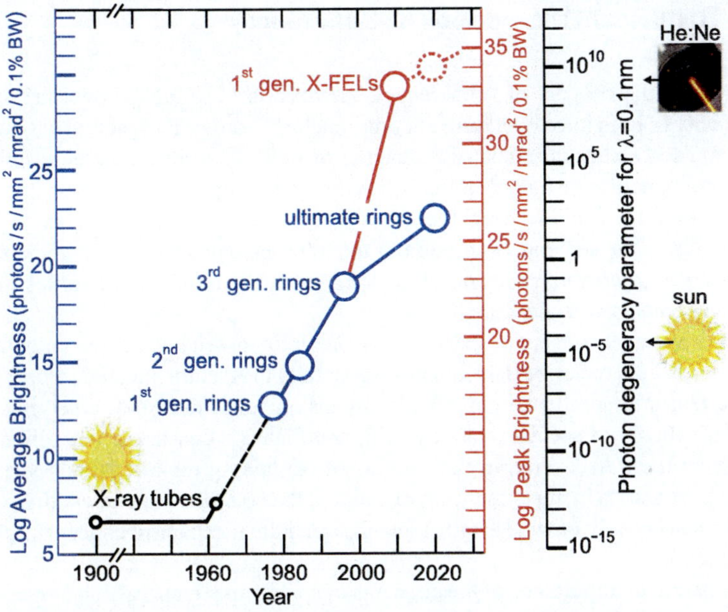

Fig. 4.5 Historical and projected future increase in average brightness of storage rings (blue) and peak brightness of XFELs (red) [17]. On the right we have also given in black the number of photons per coherence volume or degeneracy parameter, calculated by assuming a source of wavelength $\lambda = 0.1$ nm. Note that, in practice, the achievable degeneracy parameter is lowered by beam line losses. On the right we indicate the degeneracy parameter of sunlight and a He:Ne laser at a wavelength of 633 nm [13]

4.2.9 Summary

The *brightness* of a source is a fundamental measure of its quality. For synchrotron radiation sources, one typically states the *average brightness* while for XFEL sources the *peak brightness*.

A source is *diffraction limited* if the quantity $A_s d\Omega$ has the minimum value allowed by the uncertainty principle. A source is *transform limited* when the product $\tau_s \Delta\omega/\omega$ has the minimum value.

All photons emitted by a diffraction and transform-limited source are coherent. They are born within a minimum volume of order λ^3 and upon propagation remain contained in the *coherence volume* whose lateral area expands by Ω_{coh}. For a partially coherent or incoherent source, the number of photons in the coherence volume, called the *degeneracy parameter*, is smaller than the total number of emitted photons.

4.3 Historical Descriptions of Coherence

Historically, the concept of "coherence" has been used to break down the infinite complexity of light into some form of comprehensible order. This picture has evolved over time and so has our view of the nature of light. In practice, the description of the complex nature of light requires simplifying assumptions. The most important is *space-time separability* which has its origin in the observed transverse nature of light. It allows the separate discussion of the transverse or spatial properties of light in terms of *spatial coherence* and of the longitudinal or temporal properties of light in terms of *temporal coherence*.

The convenient concept of space-time separability is bridged by two fundamental constants that characterize light and more generally the electro magnetic world. They are the constant speed of light c with dimension [distance/time], emerging from Einstein's theory of special relativity [19], and Planck's constant h with dimension [energy × time], discovered in the process of explaining the black-body spectrum [20, 21]. As shown by the dimension of c and h, the common parameter "time" links the classical concept of wavelength λ and the quantum mechanical concept of photon energy hc/λ.[5]

The practical importance of the concepts of spatial and temporal coherence is that they provide the foundation for understanding the observed behavior of light through diffraction and spectroscopy experiments. We can state as follows.

> The concept of "coherence" provides a way to describe the observed phenomenon of light diffraction.
> In practice, we assume *space-time separability*, leading to the concepts of spatial and temporal coherence as the foundation for understanding the measured behavior of light through diffraction and spectroscopy experiments.
> The two fundamental constants c [distance/time] and h [energy×time] provide the link between the classical concept of "length", expressed by λ, and the quantum mechanical concept of "photon energy", given by hc/λ.

The concept of space-time separability greatly facilitates the development of a mathematical description of coherence. Below we will start by briefly outlining the geometrical concept of coherence dating back to Huygens, expressed through the magical Fresnel–Huygens principle. In its simplest form, coherence is related to the ability of waves to be in or out of phase, explaining the phenomenon of interference and diffraction.

In Sect. 4.4 we will then introduce the more powerful formulation of coherence through the Fourier transform method. It is the most common way to link time and

[5] In quantum mechanics, time is a *parameter* and is not expressed as the expectation value of an operator. This leads to a fundamental difference between the space-momentum uncertainty principle and the often used concept of a time-frequency uncertainty relation whose origin is somewhat problematic in quantum mechanics [15].

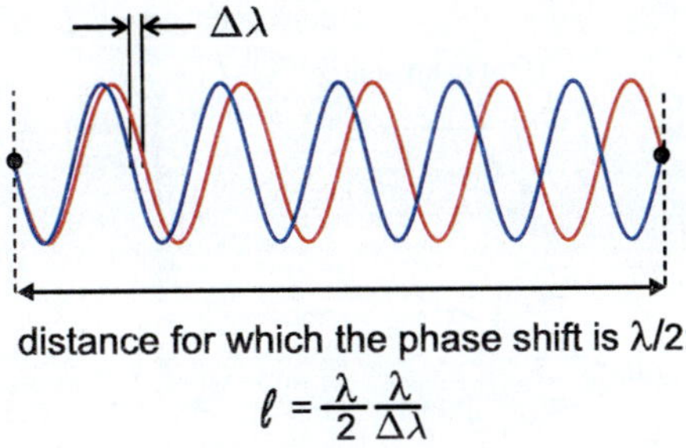

Fig. 4.6 Illustration of the length ℓ over which two waves that differ in wavelength by $\Delta\lambda$ acquire a relative phase shift of $\lambda/2$

energy in 1D and area and angle in 2D which we have implicitly used earlier to define the brightness of light.

The third step in our discourse is the topic of Sect. 4.5. It involves an advanced formulation of "coherence" by including fluctuations of the measured light intensity, described by statistical optics through the spatiotemporal correlations between fields. In quantum mechanics, the fields are replaced by photon probability amplitudes under the assumption that photons are independent and non-interacting. The formulations of "coherence" emerging from statistical optics and quantum mechanics are equivalent and provide the link of the seemingly different concepts of coherence and brightness discussed earlier.

Our final step requires a whole new chapter that will follow in Chap. 5. It discusses coherence within the complete framework of QED. It builds on the quantum mechanical formulation discussed in Sect. 4.5, which corresponds to the first order in QED, through an infinite series of increasing order. In QED, higher order coherence is manifested in correlations between an increasing number of photons. In the second order, corresponding to the correlation of two photons, the wave-particle duality already breaks down and we will specifically explore this all-important transition.

4.3.1 *Geometrical Coherence Concepts in Time and Space*

Classical coherence assumes a continuous light wave of wavelength λ with oscillation time $\tau_0 = \lambda/c$ or cycle frequency $\nu = 1/\tau_0$ that has no beginning and no end. Temporal coherence may be defined[6] by considering when two such waves of slightly different wavelengths launched in-phase at the same point become out of phase by $\lambda/2$ as illustrated in Fig. 4.6.

[6] The simple geometrical coherence condition $\Delta\lambda = \lambda/2$ used here [22] is different from the condition $\Delta\lambda = \lambda/(2\pi)$ used in the Fraunhofer approximation and that obtained from the Fourier transform for a coherent flat-top field distribution, $\Delta\lambda = \lambda$, but is the same as for a coherent Gaussian distribution (see Table 4.1).

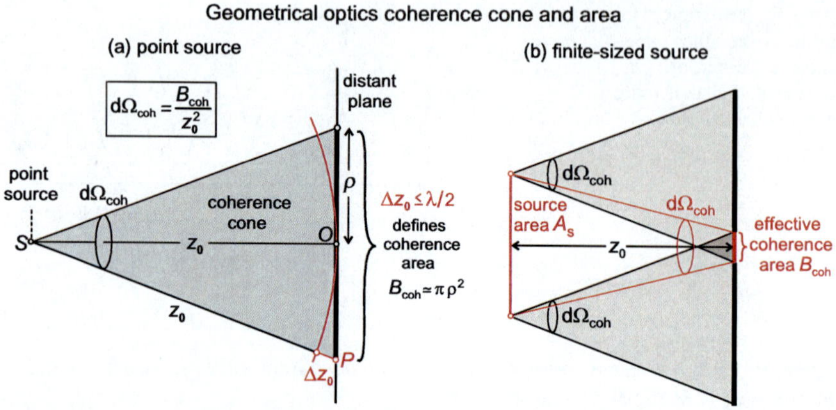

Fig. 4.7 Illustration of lateral coherence propagation in geometrical optics. **a** A *point source* is envisioned to emit a coherent spherical wave. In a distant plane, the spherical wave can be approximated by a coherent plane wave if the path difference $\Delta z_0 \leq \lambda$. Then the area B_{coh} constitutes the *coherence area* in a plane at distance z_0 and the corresponding solid angle $\mathrm{d}\Omega_{\text{coh}} = B_{\text{coh}}/z_0^2$ the *coherence cone*. **b** For a finite-sized source, the coherence cone and area indicated in red are reduced, since they are determined by the overlap of the point source coherence cones

The *longitudinal* coherence condition for a source then becomes

$$\ell_{\text{coh}} \leq \frac{\lambda}{2} \frac{\lambda}{\Delta\lambda}, \quad \text{or } \tau_{\text{coh}} \leq \frac{\tau_0}{2} \frac{\lambda}{\Delta\lambda}, \tag{4.31}$$

where ℓ_{coh} and $\tau_{\text{coh}} = \ell_{\text{coh}}/c$ are the longitudinal coherence length and time. When written in terms of the uncertainty product according to (4.20), we find $K = 1/2$, which is the value for a coherent Gaussian distribution in Table 4.1.

The geometrical optics concept of *lateral* coherence is illustrated in Fig. 4.7.

As shown in Fig. 4.7a for a *point source*, one considers emission of a spherical wave (red) in accordance with the Fresnel–Huygens principle that traverses a plane at a large distance z_0, oriented perpendicular to the optical axis $\overline{SO}$. The wavefront has to travel an extra distance Δz_0 to the point P relative to the point O in the distant plane. By geometry we obtain

$$\Delta z_0 = \sqrt{z_0^2 + \rho^2} - z_0. \tag{4.32}$$

For large z_0 we can simplify the square root by use of the so-called *paraxial approximation*[7]

$$\sqrt{z_0^2 + \rho^2} = z_0 + \frac{\rho^2}{2z_0}. \tag{4.33}$$

[7] Defining θ as the planar opening angle of the emitted radiation from the optical axis as shown in Fig. 4.7, we have $\tan\theta = \rho/z_0$ and the paraxial approximation is safely valid for $\theta \leq 25°$.

Using the same maximum wavelength difference of $\lambda/2$ as for the temporal case, the *phase shift* in the distant plane can be neglected if[8]

$$\Delta z_0 = \frac{\rho^2}{2z_0} \le \frac{\lambda}{2}. \tag{4.35}$$

Then all points within the area

$$B_{\mathrm{coh}} = \pi\rho^2 = \pi\lambda z_0 \tag{4.36}$$

are coherently illuminated. This corresponds to the geometrical coherence cone of a point source seen by an observer at large distance

$$\boxed{\text{Observer coherence cone}: \mathrm{d}\Omega_{\mathrm{coh}} = \frac{B_{\mathrm{coh}}}{z_0^2} = \frac{\pi\lambda}{z_0}}. \tag{4.37}$$

By inspection of Fig. 4.7b we can also reverse the *observer* coherence cone, defined by looking back at the source, into a coherence cone associated with a source of area A. Within the *source* coherence cone, the emitted fields add coherently within a 3D solid angle cone given by

$$\boxed{\text{Source coherence cone}: \mathrm{d}\Omega_{\mathrm{coh}} = \frac{\lambda^2}{A}}. \tag{4.38}$$

The so-defined source coherence cone is the maximum opening angle in the forward direction within which the fields emitted as spherical waves from points within the source area A can be treated as coherent by geometry alone, even if the source itself is chaotic!

The formulation of an uncertainty relation of the form (4.19) in geometrical optics leads to the problem of how to define the size of a point source. In principle, a "point source" has to have a finite area of order of λ^2 since radiation cannot propagate in space if it is not defined over a distance of order of the wavelength. As illustrated in Fig. 4.7b, the point source coherence cone concept is nevertheless useful in qualitatively explaining the reduction of the coherence area in the far plane when the source size is increased.

For a finite-size source, the coherence area in the far plane is reduced since it corresponds to the overlap of the cones from all points in the source plane. This is illustrated in Fig. 4.7b through the overlap of the cones from the most distant points

[8] In practice, one often uses a more stringent coherence requirement than (4.35) to define the coherence area, given by

$$\Delta z_0 = \frac{\rho^2}{2z_0} \le \frac{\lambda}{2\pi}. \tag{4.34}$$

This so-called Fraunhofer approximation is discussed in Sect. 8.4.3.

on the periphery of the source area. With increasing source size the source becomes increasingly "incoherent" in accord with Fig. 4.3b.

The formal limitations of geometrical optics can be overcome by Fourier optics, discussed next.

4.4 Fourier Optics Description of Coherent Sources

The Fourier transform, named after the French mathematician and physicist Jean-Baptiste Joseph Fourier (1768–1830), allows the transformation of coherent fields in the time domain to the frequency domain and similarly from the spatial domain to the momentum domain. The link between fields in conjugate variables allows the establishment of uncertainty relations which enter into the brightness definition (4.5) or its generalization expressed by (4.23) based on the generalized uncertainty relations (4.19) and (4.20).

The uncertainty relations derived by the classical method of Fourier transformations are special cases of the more general ones encountered in the more powerful statistical and quantum optics formalism discussed in the next section. It is not by accident that Fourier optics and quantum mechanics lead to similar uncertainty relations. In fact, Heisenberg was led to the discovery of quantum mechanics by replacing classical Fourier components by matrix elements.

In Fourier optics, the fluctuating fields in statistical optics and the fluctuating photon numbers in quantum optics are approximated by assuming classical fields of stable amplitudes and phases, and the measured intensity corresponds to an average over a full cycle of the field as discussed in Sect. 3.2.3. In its basic formulation, Fourier optics thus treats the properties of *coherent fields*. The more difficult concept of partially coherent fields will be treated in Sect. 4.5.

4.4.1 *Fourier Transformation of Temporal Coherence*

The Fourier transformation is defined in Appendix A.4. Here we briefly outline the 1D Fourier transform between coherent electric fields in time and frequency (or energy) domains under preservation of dimension (see Appendix A.4.2).

Starting with a coherent temporal field pulse

$$E(t) = E_0 \, f(t), \tag{4.39}$$

where $f(t)$ is a *dimensionless* temporal distribution function of unit peak value and E_0 is the peak value of the field with dimension ([V/m] in SI units, the corresponding field of the same dimension in the frequency domain is given by the Fourier transform

$$E(\omega) = E_0 \underbrace{\frac{1}{C} \int_{-\infty}^{\infty} f(t)\,\mathrm{e}^{-\mathrm{i}\omega t}\,\mathrm{d}t}_{F(\omega)}, \text{ where } C = \int_{-\infty}^{\infty} f(t)\,\mathrm{d}t. \tag{4.40}$$

The normalization constant C with dimension [time] assures that the fields in the time and frequency domains have the same dimension. The frequency distribution function $F(\omega)$ of angular frequency $\omega = 2\pi\nu$ (dimension [angle/time]) is also *dimensionless*, and its inverse Fourier transform reproduces $E(t)$ according to

$$E(t) = E_0 \frac{C}{2\pi} \int_{-\infty}^{\infty} F(\omega)\,\mathrm{e}^{\mathrm{i}\omega t}\,\mathrm{d}\omega. \tag{4.41}$$

Parseval's theorem then links the associated intensities (fields squared) according to

$$\int_{-\infty}^{\infty} |E(t)|^2\,\mathrm{d}t = \frac{C^2}{2\pi} \int_{-\infty}^{\infty} |E(\omega)|^2\,\mathrm{d}\omega. \tag{4.42}$$

It assures conservation of energy when transforming from one to the conjugate domain.

4.4.1.1 Effective Distribution Widths

Since Parseval's theorem assures energy conservation between the integrated intensity distributions, it is convenient to define their *effective widths* in the time Δ_t and energy ($\Delta_\mathcal{E} = \hbar\Delta_\omega$) domains according to

$$\Delta_t = \frac{\int_{-\infty}^{\infty} |E(t)|^2\,\mathrm{d}t}{|E(t{=}0)|^2}, \text{ and } \Delta_\mathcal{E} = \frac{\int_{-\infty}^{\infty} |E(\hbar\omega)|^2\,\mathrm{d}\hbar\omega}{|E(\hbar\omega{=}0)|^2}. \tag{4.43}$$

The time-bandwidth product $\Delta_t\Delta_\omega$, which defines the "transform limit", is given by

$$\boxed{\Delta_t\Delta_\omega = 2\pi K, \text{ or } \Delta_t\Delta_\mathcal{E} = h\,K}, \tag{4.44}$$

where h=4.136 eV fs, and

$$K = \frac{\left|\int_{-\infty}^{\infty} |E(t)|^2\,\mathrm{d}t\right|^2}{\left|E(t{=}0)\int_{-\infty}^{\infty} E(t)\,\mathrm{d}t\right|^2}. \tag{4.45}$$

In the following we consider three examples.

4.4.1.2 Flat-Top Temporal Pulse

A temporal flat-top electric field pulse of width τ_p and its transform in the frequency domain are given by

$$E(t) = E_0 \begin{cases} 1 & -\tau_p/2 < t < \tau_p/2 \\ 0 & \text{otherwise} \end{cases} \Longrightarrow E(\omega) = E_0 \frac{\sin(\omega\,\tau_p/2)}{\omega\,\tau_p/2}. \tag{4.46}$$

The corresponding intensity distributions are given by the absolute values squared,

$$|E(t)|^2 = E_0^2 \begin{cases} 1 & -\tau_p/2 < t < +\tau_p/2 \\ 0 & \text{otherwise} \end{cases} \Longrightarrow |E(\omega)|^2 = E_0^2 \left[\frac{\sin(\omega\,\tau_p/2)}{\omega\,\tau_p/2}\right]^2. \tag{4.47}$$

Energy conservation between the time and frequency domains is assured according to Parseval's theorem (4.42) according to

$$\int_{-\infty}^{\infty} |E(t)|^2\,\mathrm{d}t = \frac{C^2}{2\pi} \int_{-\infty}^{\infty} |E(\omega)|^2\,\mathrm{d}\omega, \tag{4.48}$$

where $E(t)$ and $E(\omega)$ have the same dimension and $C = \tau_p$.

Parseval's theorem also defines the *effective widths* of the intensity distributions according to

$$\Delta_t = \frac{\int_{-\infty}^{\infty} |E(t)|^2\,\mathrm{d}t}{|E_0|^2} = \tau_p \tag{4.49}$$

and

$$\Delta_\omega = \frac{\int_{-\infty}^{\infty} |E(\omega)|^2\,\mathrm{d}\omega}{|E_0|^2} = \frac{2\pi}{\tau_p}. \tag{4.50}$$

The time-bandwidth product (4.44) is obtained as

$$\Delta_t \Delta_\omega = 2\pi = 2\pi K, \quad \Rightarrow \quad K = 1. \tag{4.51}$$

This is the K value in Table 4.1.

4.4.1.3 Gaussian Temporal Pulse

The temporal field pulse and its transform are

$$E(t) = E_0 \exp\left[-\frac{t^2}{2{\sigma_t}^2}\right] \Longrightarrow E(\omega) = E_0 \exp\left[-\frac{\omega^2}{2\left[\frac{1}{\sigma_t}\right]^2}\right] \tag{4.52}$$

and the corresponding intensities are

$$|E(t)|^2 = E_0^2 \exp\left[-\frac{t^2}{2\left[\frac{\sigma_t}{\sqrt{2}}\right]^2}\right] \Longrightarrow |E(\omega)|^2 = E_0^2 \exp\left[-\frac{\omega^2}{2\left[\frac{1}{\sqrt{2}\sigma_t}\right]^2}\right], \tag{4.53}$$

where we have explicitly indicated the rms widths by square brackets. Energy conservation between the time and frequency domains is fulfilled by Parseval's theorem with $C = \sqrt{2\pi}\sigma_t$. By evaluation of (4.49) and (4.50) we obtain $\Delta_t = \sqrt{\pi}\sigma_t$ and $\Delta_\omega = \sqrt{\pi}/\sigma_t$, so that

$$\Delta_t \Delta_\omega = \pi = 2\pi K, \quad \Rightarrow \quad K = \frac{1}{2}. \tag{4.54}$$

The same K value is given in Table 4.1.

4.4.1.4 Exponential Temporal Decay

Our third example is the Fourier transform of exponential temporal decay, encountered when an excited state in an atom decays by emission of an electromagnetic wave whose energy distribution has a Lorentzian shape. The temporal field and its transform are

$$E(t) = E_0 \exp\left[-\frac{t}{2\tau}\right], \; t \geq 0 \Longrightarrow E(\omega) = E_0 \frac{1/(2\tau)}{-\mathrm{i}\omega + 1/(2\tau)}, \tag{4.55}$$

where τ is referred as the *natural lifetime* of the decaying state. The corresponding intensities are

$$|E(t)|^2 = E_0^2 \exp\left[-\frac{t}{\tau}\right], \; t \geq 0 \Longrightarrow |E(\omega)|^2 = E_0^2 \frac{1/(2\tau)^2}{\omega^2 + 1/(2\tau)^2}. \tag{4.56}$$

The dimensional constant in Parseval's theorem is obtained as $C = 2\tau$. By evaluation of (4.49) and (4.50) we obtain $\Delta_t = \tau$ and $\Delta_\omega = \pi/(2\tau)$, so that

$$\Delta_t \Delta_\omega = \frac{\pi}{2} = 2\pi K, \quad \Rightarrow \quad K = \frac{1}{4}. \tag{4.57}$$

This is the K value in Table 4.1.

4.4.2 Time-Bandwidth Product Definitions

The time-bandwidth products expressed by (4.51), (4.54), and (4.57) are based on *effective widths* which are defined through Parseval's theorem. They preserve energy between the integrated intensity distributions in the time and frequency domains. In the following we briefly mention different time-bandwidth definitions found in the literature. In the optical laser literature it is customary to define the time-bandwidth product differently by use of the *FWHM* of the *main peaks* in the distribution functions, ignoring side peaks if present. This definition leads to different values for K, as given in Table 4.2.

Let us illustrate the difference by considering the Fourier transformation between flat-top and sinc distributions given by (4.47). Owing to the reversibility of the Fourier transform the flat-top or sinc functions may be expressed in either the time or frequency domains as illustrated in Fig. 4.8. In (a) we use the label x to indicate either t or $\hbar\omega$ and the conjugate label q for $\hbar\omega$ or t is used in (b). The transformation from the time into the energy domain is expressed by the bottom scale and that from the energy into the time domain by the top scale.

For convenience we have expressed the frequency in terms of the photon energy in eV and start with a flat-top width of either 1 eV or 1 fs, as shown in Fig. 4.8a. The Fourier transform into the corresponding conjugate domain q is shown as the black curve in Fig. 4.8b, representing the absolute value of the conjugate *field* distribution $|f(q)|$. The red curves represent the field-squared or "intensity" distributions.

The figure illustrates different distribution widths, indicated in different colors. The widths of the sinc^2 intensity distribution in Fig. 4.8b illustrate the key difference between two often used time-bandwidth definitions, yielding different factors K. In the laser literature one typically uses the FWHM of the central intensity peak, given by the "FWHM main peaks" column in Table 4.2. It corresponds to the FWHM Γ_q of the red intensity distribution of the sinc^2 function, numerically given by its width at half its maximum value, Γ_q =3.66 eV or fs.

Table 4.2 Time-bandwidth products K_Γ and K_Δ for intensity distributions associated with coherent field distributions, where Planck's constant is $h = 4.136\,\text{eV fs}$

	FWHM main peaks $K_\Gamma = \Gamma_t \Gamma_\mathcal{E}/h$	Effective widths $K_\Delta = \Delta_t \Delta_\mathcal{E}/h$
Flat-top	0.886	1
Gaussian	0.441	1/2
Lorentzian	0.142	1/4

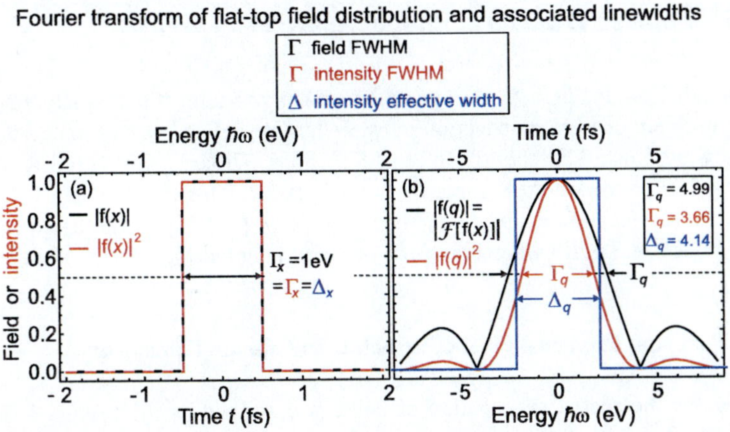

Fig. 4.8 **a** Flat-top field (black) and squared field or intensity (red) distributions in the x domain, where owing to the reversibility of the Fourier transform we may have $x = t$ or $\hbar\omega$, as indicated on the bottom and top abscissa. The flat-top distributions for the field and intensity are the same with unit width and unit areas. **b** Normalized Fourier transform $\mathcal{F}$ of the flat-top field distribution in terms of the conjugate variable q shown in black as the absolute value $|f(q)| = |\mathcal{F}[f(x)]|$ with unit peak value, and its square in red. The FWHM of the main peaks of the sinc and sinc^2 functions is obtained as the shown numerical values $\Gamma_q = 4.99\,\mathrm{eV}$ or fs (field) and $\Gamma_q = 3.66\,\mathrm{eV}$ or fs (intensity). In contrast, the effective width $\Delta_q = 4.14\,\mathrm{eV}$ or fs, indicated in blue, conserves power and includes contributions from the outer wings, as discussed in the text. Note that the areas under the red Gaussian and blue flat-top curves are the same

Our definition in (4.51) satisfies Parseval's theorem of energy conservation used in statistical optics [9], as expressed by the "effective width" column in Table 4.2. It is the width Δ_q shown in blue in Fig. 4.8b given by $\Delta_q = 4.14\,\mathrm{eV}$ (or fs). It is slightly larger than Γ_q because it accounts for the intensity in the weaker outer wings through Parseval's theorem. The outer wings are neglected when using the FWHM method which therefore does not rigorously conserve energy.

Unfortunately, there is yet another time-bandwidth expression found in the literature. It is the most famous one, used to describe the exponential temporal decay of excited electronic or nuclear states in atoms. It will be discussed in conjunction with x-ray emission in Sect. 12.2.3. In this case, the time-bandwidth product also follows from the intensity relation (4.56). However, the temporal width τ is taken as the $1/e$ decay time of the excited electronic state, while the Lorentzian energy width of the intensity of the emitted line is the so-called *natural linewidth* (FWHM) Γ with the time-bandwidth product given by (see Sect. 12.2.7)

$$\tau\,\Gamma = \hbar. \tag{4.58}$$

This relationship corresponds to yet another value of $K = 1/(2\pi) = 0.159$ than those in Table 4.2.

4.4.3 Fourier Transformation of Spatial Coherence

Historically, the mathematical Fourier formalism was almost immediately applied to the problem of light propagation in *space* by Fresnel and was extended through the work of Gustav Kirchhoff (1824–1887), John William Strutt (Lord Rayleigh) (1842–1919), and Arnold Sommerfeld (1868–1951). This development led to the important *Rayleigh-Sommerfeld diffraction formula* [23, 24], as discussed in more detail in Sect. 8.3.4. It will emerge from our treatment below.

4.4.3.1 From Huygens–Fresnel Principle to Fourier Transform

Classically, the spatial propagation of fields is described by the Huygens–Fresnel principle, stating that every point on a plane wave front may be thought of as the origin of a secondary spherical wave. To illustrate the origin of the Fourier description of wave propagation, we consider Fig. 4.9.

We assume a source with area A_s that is illuminated by an incoming plane wave and according to the Fresnel–Huygens all points within the source area may then be thought to emit spherical waves. We are interested in how the spherical waves emitted by all source points P add up at a point X in a distant plane, separated by a distance z_0 from the source plane. Neglecting polarization, the electric field of the spherical waves emitted at a point P is then given at the point X of coordinate $\boldsymbol{\rho}$ in the detector plane by

$$E(\boldsymbol{\rho}) = E(\mathbf{r})\,\frac{f}{r_{\mathrm{PX}}}\,\underbrace{\mathrm{e}^{\mathrm{i}[\alpha_{\mathrm{P}}+kr_{\mathrm{PX}}]}}_{\phi_{\mathrm{PX}}}\,. \tag{4.59}$$

The spherical wave field, emitted from a source point P of coordinate $\mathbf{r}$ with amplitude $E(\mathbf{r})$, decreases with the inverse separation r_{PX} from the source to the detection point. In order to conserve the dimension of the field (in SI units expressed as [V/m]), we have introduced a factor f of dimension [length]. The magic behind the Huygens–Fresnel principle is revealed by the fact that no atoms are needed to convert a plane wave into a spherical wave through a scattering process. It is assumed to magically happen at all points on the plane wave front with unit conversion efficiency $f = 1$. This science fiction like conversion actually occurs for true atomic scattering centers as discussed in Sect. 8.3.3. The quantity $E(\mathbf{r})f$ then characterizes the atomic source strength at points $\mathbf{r}$ and f is the atomic scattering length defined in Sect. 6.2.3.

In (4.59) we have specifically included a birth phase $\alpha_{\mathrm{P}} = \alpha(\mathbf{r})$ of the spherical wave at points P through the factor $\mathrm{e}^{\mathrm{i}\alpha_{\mathrm{P}}}$. The propagation results in another phase factor $\mathrm{e}^{\mathrm{i}kr_{\mathrm{PX}}}$, whose phase at the detector point X depends on the wavelength or frequency through the wavevector $k = 2\pi/\lambda = \omega/c$ and the traveled distance r_{PX}. The total phase factor may be decomposed as

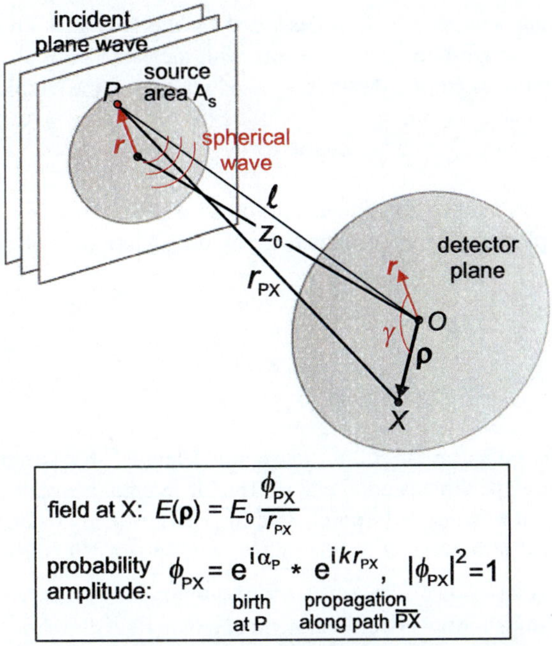

Fig. 4.9 Illustration of the propagation of the field $E(\mathbf{r})$ from a point P with coordinate $\mathbf{r}$ in a source to the field $E(\boldsymbol{\rho})$ at point X of coordinate $\boldsymbol{\rho}$ in a distant detector plane. The points are separated by a large distance r_{PX}, with z_0 denoting the separation between the planes. We picture the source of area A_s to be a secondary source illuminated by a plane wave that is converted with unit conversion efficiency at each source point P into a spherical wave according to the Fresnel–Huygens principle. Underneath we illustrate that the spherical wave emitted from different points with constant source strength $f E_0$ with $f = 1$ may be written in terms of a probability amplitude ϕ_{PX} for propagation P→X that contains birth and propagation phases

$$\phi_{PX} = \underbrace{e^{i\alpha_P}}_{\text{birth}} \underbrace{e^{ikr_{PX}}}_{\text{propagation}} . \tag{4.60}$$

It is also indicated at the bottom of Fig. 4.9. The dimensionless quantity ϕ_{PX}, with the normalization $|\phi_{PX}|^2 = 1$, is called the *single photon probability amplitude* in Feynman's formulation of quantum mechanics [25–28]. This concept was used by Feynman in his famous "Lectures on Physics" [29] to explain the double-slit experiments with electrons. The so-defined probability amplitude is valid, however, only under the assumption that possible photon-photon and electron-electron interactions are neglected [28]. This assumption means that the photon degeneracy parameter is less than unity. For this case, covered by conventional quantum mechanics or first order QED, the classical wave and quantum probability amplitude treatments give identical results [14, 30] (also see Sect. 4.8 below).

For a *coherent* source, which is described by the Huygens–Fresnel principle, spherical waves are emitted from all points with the same birth phase, so that the probability amplitudes from different source points P have the simple form

$$\phi_{\mathrm{PX}} = \mathrm{e}^{ikr_{\mathrm{PX}}}. \tag{4.61}$$

The fields and probability amplitudes add up at a given point X of coordinate $\boldsymbol{\rho}$ in the detector plane which is accounted for by integration over all source points $\mathbf{r}$ according to

$$E(\boldsymbol{\rho}) = C \iint_{-\infty}^{\infty} E(\mathbf{r}) \frac{\mathrm{e}^{ikr_{\mathrm{PX}}}}{r_{\mathrm{PX}}} \, \mathrm{d}\mathbf{r}. \tag{4.62}$$

We have introduced a constant C of dimension [1/length] that is required for both sides to have the same dimension of a field. The 2D integral is over the entire source plane, although in practice only points within A_s contribute. To determine C, we calculate the field at a point O on the optical axis in Fig. 4.9, separated from the source point P by a distance $\ell = \sqrt{z_0^2 + r^2}$. Assuming $E(\mathbf{r}) = E_0$ across the source and integrating in cylindrical coordinates we obtain

$$E(\rho = 0) = C\, E_0\, 2\pi \int_0^{\infty} \frac{\mathrm{e}^{ik\sqrt{z_0^2+r^2}}}{\sqrt{z_0^2 + r^2}} \, r\, \mathrm{d}r = \underbrace{\mathrm{i}\lambda\, C E_0}_{E_0} \mathrm{e}^{ikz_0}. \tag{4.63}$$

On the right we have indicated that the amplitude of the on-axis field must remain equal to the that of the incident plane wave so that we obtain[9]

$$C = \frac{1}{\mathrm{i}\lambda}. \tag{4.64}$$

This yields (4.62) in the form

$$\boxed{E(\boldsymbol{\rho}) = \frac{1}{\mathrm{i}\lambda} \iint_{-\infty}^{\infty} E(\mathbf{r}) \frac{\mathrm{e}^{ikr_{\mathrm{PX}}}}{r_{\mathrm{PX}}} \, \mathrm{d}\mathbf{r}}. \tag{4.65}$$

Our result is the *Rayleigh-Sommerfeld diffraction formula* discussed in Sect. 8.3.4.

[9] Note that the integration in (4.63) extends to infinity. If only a limited source area is considered, then the on-axis field at point O is not E_0 due to "edge effects" (see Sect. 8.3.3).

At large source-detector plane separation z_0, we can express the path length r_{PX} in Fig. 4.9 by assuming $r, \rho \ll z_0$, in the paraxial approximation, and obtain

$$r_{PX} = z_0 \left(1 + \frac{r^2}{2z_0^2} + \frac{\rho^2}{2z_0^2} - \frac{\mathbf{r} \cdot \boldsymbol{\rho}}{z_0^2}\right). \tag{4.66}$$

In (4.65), the distance r_{PX} enters in both the amplitude and the exponential phase. The $1/r_{PX}$ amplitude term only leads to a weak gradual change of the field strength in the distant detector plane, and since z_0 is much larger than all other distances, we can approximate the denominator in (4.65) by $r_{PX} = z_0$.

This approximation is inadequate for the *phase* term which enters exponentially and contains the key interference information. The integral in (4.65) significantly simplifies if we can assume $kr^2/(2z_0) = \pi r^2/(\lambda z_0) \ll 1$, corresponding to the assumption that at large z_0 the source appears small. For large z_0 we can also drop the term $\rho^2/(2z_0^2)$ in the exponential, and with these simplifications of the amplitude and phase terms we obtain

$$E(\boldsymbol{\rho}) = \frac{1}{\mathrm{i}\,\lambda\, z_0}\, \mathrm{e}^{ikz_0} \iint_{-\infty}^{\infty} E(\mathbf{r})\, \mathrm{e}^{-\mathrm{i}\frac{k}{z_0}\mathbf{r}\cdot\boldsymbol{\rho}}\, \mathrm{d}\mathbf{r}. \tag{4.67}$$

This expression, representing the *Fraunhofer approximation* of the Rayleigh-Sommerfeld diffraction formula (see Sect. 8.4.3), links the fields, $E(\mathbf{r})$, assumed to be coherently emitted at source points to the fields, $E(\boldsymbol{\rho})$, detected at a distant point through a *Fourier transform*. We can rewrite it in the more familiar *Fourier form* of conjugate variables r and $q = \rho k/z_0$

$$E(\mathbf{q}) = \frac{1}{\mathrm{i}\,\lambda\, z_0}\, \mathrm{e}^{ikz_0} \iint_{-\infty}^{\infty} E(\mathbf{r})\, \mathrm{e}^{-\mathrm{i}\mathbf{r}\cdot\mathbf{q}}\, \mathrm{d}\mathbf{r}. \tag{4.68}$$

4.4.3.2 Example of a Circular Flat-Top Source

For a *flat-top field* distribution within a circular area $A_s = \pi R^2$, the symmetrical field distribution with $r = \sqrt{x^2 + y^2}$ within the source area is expressed by

$$E(r) = \begin{cases} E_0 & r \le R \\ 0 & \text{otherwise.} \end{cases} \tag{4.69}$$

With $\mathbf{r} \cdot \boldsymbol{\rho} = r\,\rho \cos\gamma$ (see Fig. 4.9) and integrating (4.67) in cylindrical coordinates we obtain

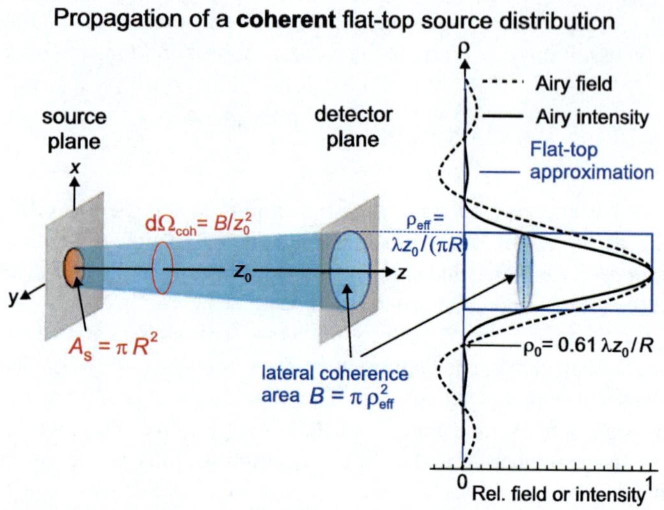

Fig. 4.10 Field and intensity distributions emitted by a uniform flat-top source of lateral area $A_s = \pi R^2$. The pattern in the detection plane is the Airy field distribution (4.70), shown as a function of ρ in the normalized form $2J_1(x)/x$, where $x = kR\rho/z_0$ (dashed-black) and its square, the normalized intensity distribution (4.71) (black). The Airy functions are zero for $\rho = \rho_0 = 0.61\lambda z_0/R$. Also shown in blue is a flat-top distribution of width $\rho_{\rm eff} = \lambda z_0/(\pi R)$, which contains the same power as the Airy intensity distribution. All distributions are normalized by their peak value and all have the same areas

$$E(\rho) = \frac{E_0}{\mathrm{i}\lambda z_0} \mathrm{e}^{\mathrm{i}kz_0} \int_0^R \int_0^{2\pi} \mathrm{e}^{-\mathrm{i}\frac{k}{z_0} r\rho\cos\gamma}\, r\,\mathrm{d}\gamma\,\mathrm{d}r$$
$$= E_0 \frac{\pi R^2}{\mathrm{i}\lambda z_0} \mathrm{e}^{\mathrm{i}kz_0} \frac{2J_1(kR\rho/z_0)}{kR\rho/z_0}. \tag{4.70}$$

Here the function $2J_1(x)/x$ is the well-known normalized Airy distribution function.[10] The "intensity" distribution in the detector plane, i.e. the diffraction pattern, is obtained as a function of the distance ρ or the momentum transfer q from the optical axis as

$$|E(\rho)|^2 = E_0^2 \left[\frac{\pi R^2}{\lambda z_0}\right]^2 \left[\frac{2J_1(kR\rho/z_0)}{kR\rho/z_0}\right]^2 \quad \text{or} \quad |E(\mathbf{q})|^2 = E_0^2 \left[\frac{\pi R^2}{\lambda z_0}\right]^2 \left[\frac{2J_1(qR)}{qR}\right]^2. \tag{4.71}$$

The functions $|2J_1(x)/x|$ and $[2J_1(x)/x]^2$ are plotted in Fig. 4.10.

Since there is no loss of energy in free space propagation, the integrated intensity distribution in the source plane (the power) must be the same as that in the detector plane. From the integral over the detector plane intensity distribution,

[10] It is named after the British astronomer and mathematician Sir George Biddell Airy (1801–1892).

$$2\pi \int_0^\infty \left[\frac{2J_1(kR\rho/z_0)}{kR\rho/z_0} \right]^2 \rho\, d\rho = \frac{\lambda^2 z_0^2}{\pi R^2} \tag{4.72}$$

we find that the "power" in the detector plane $\mathcal{P}_{\text{det}} = E_0^2 \pi R^2$ is indeed the same as that in the source plane $\mathcal{P}_{\text{s}} = E_0^2 \pi R^2$.

A convenient way to envision the power associated with an arbitrary peaked intensity distribution is to write it in form of an effective flat-top area times the peak intensity. For a peaked symmetric intensity distribution $I(\rho)$, the *effective area* is then defined as

$$A_{\text{eff}} = \frac{2\pi \int_0^\infty I(\rho)\, \rho\, \mathrm{d}\rho}{I(\rho=0)}, \tag{4.73}$$

so that the power is simply $A_{\text{eff}} I(\rho=0)$. For example, for a circular source of radius R with a flat-top intensity distribution we get the expected value $A_{\text{eff}} = \pi R^2$.

By use of this effective area concept, the power in the detector plane may be expressed in terms of an effective area B, shown in blue in Fig. 4.10, given by

$$B = \frac{\lambda^2 z_0^2}{A_{\text{s}}} = \pi \rho_{\text{eff}}^2. \tag{4.74}$$

The power in the detector plane is then simply the area under the effective flat-top distribution shown in blue in Fig. 4.10 since it is the same as the area under the black Airy intensity distribution. By geometry, the solid angle of emission is given by $\mathrm{d}\Omega_{\text{coh}} = B/z_0^2$, so that

$$A_{\text{s}}\, \mathrm{d}\Omega_{\text{coh}} = \lambda^2. \tag{4.75}$$

Comparison of this area×solid angle product with (4.19) yields $H = 1$, which is the same value as obtained from (4.25) and listed in Table 4.1.

4.4.3.3 Example of a Circular Gaussian Source

For a source whose emitted ***fields*** have a symmetrical lateral 2D Gaussian field distribution in real space $r^2 = x^2 + y^2$ and in reciprocal space $q = \sqrt{q_x^2 + q_y^2}$ we have

$$E(r) = E_0 \exp\left[-\frac{r^2}{2\sigma_r{}^2} \right] \Longrightarrow E(q) = E_0 \frac{\mathrm{e}^{\mathrm{i}kz_0}}{\mathrm{i}\lambda z_0} 2\pi\sigma_r^2 \exp\left[-\frac{q^2}{2\left[\frac{1}{\sigma_r}\right]^2} \right]. \tag{4.76}$$

The corresponding *intensity* distributions are obtained by squaring the fields

$$I(r)=I_0 \exp\left[-\frac{r^2}{2\left[\frac{\sigma_r}{\sqrt{2}}\right]^2}\right] \Longrightarrow I(q)=I_0 \frac{\sigma_r^4\, 4\pi^2}{\lambda^2 z_0^2} \exp\left[-\frac{q^2}{2\left[\frac{1}{\sigma_r\sqrt{2}}\right]^2}\right], \tag{4.77}$$

where we have explicitly indicated the rms widths by square brackets. They are related to the rms widths of the *field* distributions by

$$\sigma_r^I = \frac{\sigma_r}{\sqrt{2}} \quad \text{and} \quad \sigma_q^I = \frac{1}{\sigma_r\sqrt{2}} = \frac{1}{2\,\sigma_r^I}, \tag{4.78}$$

which gives $\sigma_r^I \sigma_q^I = 1/2$ previously expressed by (4.7).

The uncertainty product $A_s d\Omega_{coh}$ in (4.8) also follows from our present treatment. The effective area of the intensity distribution in (4.77) is given by $A_{eff} = 2\pi(\sigma_r^I)^2$, and the effective solid angle is $d\Omega_{eff} = 2\pi(\sigma_\theta^I)^2 = \lambda^2(\sigma_q^I)^2/(2\pi)$ (see 4.7). We obtain with (4.78)

$$A_{eff}\, d\Omega_{eff} = 2\pi(\sigma_r^I)^2\, 2\pi(\sigma_\theta^I)^2 = \lambda^2\,(\sigma_r^I)^2\,(\sigma_q^I)^2 = \frac{\lambda^2}{4}, \tag{4.79}$$

which is the same as (4.8). By comparison with (4.19) we find $H = 1/4$ which is the value listed in Table 4.1.

In Fig. 4.11 we illustrate the propagation of the intensity distribution from the source to the detector plane given by (4.77) in the effective area and solid angle form

$$I(r)=I_0 \exp\left[-\frac{\pi r^2}{A_{eff}}\right] \Longrightarrow I(\rho)=I_0 \frac{4\,A_{eff}^2}{\lambda^2 z_0^2} \exp\left[-\frac{\pi \rho^2}{\left(\frac{\lambda^2 z_0^2}{4\,A_{eff}}\right)}\right]. \tag{4.80}$$

The effective expanded area in the detector plane shown in blue is seen to be $B = \lambda^2 z_0^2/(4A_{eff})$.

4.5 Statistical and Quantum Description of Partially Coherent Sources

As discussed in Sect. 3.2.3, in classical electro magnetism and optics, self-interference of the fields E and E^*, represented by the product $E^*E = EE^*$, defines the intensity at a point in space or time. It is represented by the cycle average of the Poynting vector by $I = \epsilon_0 c\langle E^* E\rangle$.

In practice, the fields and associated intensities fluctuate at a single point as a function of time or at a single time as a function of distance, as illustrated in Fig. 4.3.

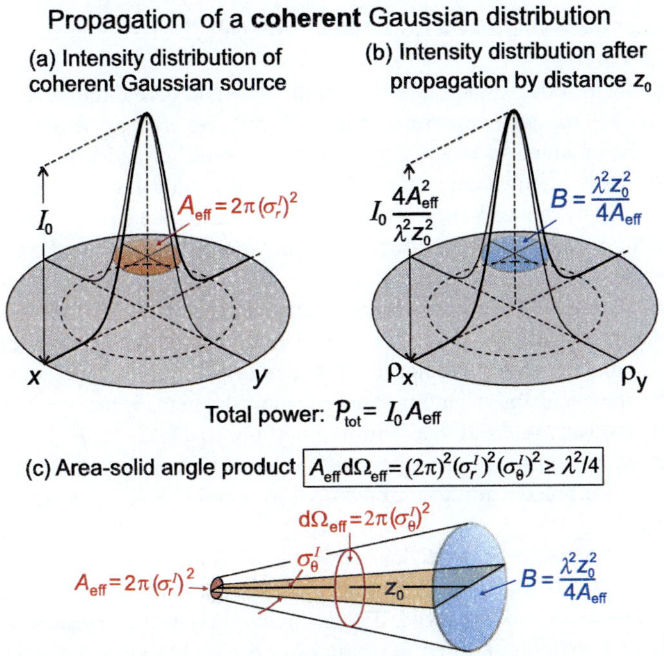

Fig. 4.11 **a** *Intensity* distribution of a two-dimensional Gaussian source distribution according to the left side of (4.80) whose power or area under the distribution is given by $I_0 A_{\rm eff}$ where $A_{\rm eff} = 2\pi(\sigma_r^I)^2$. **b** Expanded intensity distribution in real space ($\boldsymbol{\rho} = \mathbf{q}z_0/k$) after propagation by a distance z_0 from the source given by the right side of (4.80). The power in the detector plane is preserved according to $4I_0 A_{\rm eff}^2 B/(\lambda^2 z_0^2) = I_0 A_{\rm eff}$. **c** Illustration of the area–solid angle product given by (4.79)

In the quantum picture, these intensity fluctuations are the result of the photon-based graininess of light. Such fluctuating fields or photon numbers can be accounted for by statistical averages which provide the probabilities of observing different behavior. Such a probabilistic description underlies both statistic optics and quantum optics.

To account for a more realistic behavior of light, which deviates from classical waves with stable amplitudes and phases, one introduces the concept of correlation functions between fields at *different* times or points. The simplest case considers correlations between *fields* at two different spatial points at the same time or at two different times at the same point in space. In quantum mechanics, based on indivisible photons, one instead considers the probability amplitudes associated with different paths of a single photon. The probabilistic linking of fields at different points or of alternative paths of a single photon to the same detection point requires that the detection process of light is specified. Remarkably, this can be ignored in the classical formalism, which simply calculates an intensity $I = \epsilon_0 c\langle E^* E\rangle$ at a given point without specifying how it is measured. We shall discuss below how to include the detection process.[11]

We already introduced the link between the field and probability amplitude concepts in Fig. 4.9, formally expressed through (4.59). We will see below that statistical optics and quantum mechanics give the same result, provided that correlations between different photons are neglected. This statistical/quantum equivalence is of course reflected by the well-known wave-particle duality. We will also show that the measurement of two-point field correlations provides a powerful method to characterize the coherence properties of a source.

The fluctuating fields exhibit *average* statistical correlations which are denoted by triangular brackets as $\langle \mathbf{E}^*(\mathbf{x}_1, t_1)\mathbf{E}(\mathbf{x}_2, t_2)\rangle$. In the corresponding quantum formulation, the complex fields are replaced by operators $\langle \mathbf{E}^-(\mathbf{x}_1, t_1)\mathbf{E}^+(\mathbf{x}_2, t_2)\rangle$. The key issue in the statistical and quantum descriptions of the fields is the evaluation of the correlation brackets, referred to as formation of the *ensemble average* in statistical optics and the calculation of the expectation value of the joint operators in quantum mechanics. As discussed in the introduction, in practice, one assumes space-time separability.

The correlation between two fields is called a "second order" correlation in statistical optics [7, 9, 13], and since the product defines an "intensity", the correlation is also called by the somewhat awkward name "*mutual intensity*", meaning an "intensity" based on two fields at different positions or times. The mutual intensity differs from the conventionally defined single point intensity in that it is complex [9].

In contrast, the quantum theory of light is based on the elementary photon building blocks, and since the "intensity" $\langle \mathbf{E}^-(\mathbf{x}_1, t_1)\mathbf{E}^+(\mathbf{x}_2, t_2)\rangle$ is related to the energy of a single photon, it is called a "first order" correlation in quantum optics [33–35]. This definition has the advantage that the order of the correlation in quantum optics directly reflects the number of photons involved. We will see that this numbering of "order", introduced by Glauber [32] for the description of *orders of coherence*, naturally maps on the infinite series of perturbation orders in QED. Conventional quantum mechanics (QM) is then first order in QED and deals with the behavior of independent single photons, as expressed by Dirac's famous statement.[12]

> In this book we shall exclusively use the quantum optics definition of "order" of correlation functions and coherence. First order describes the correlation between two fields or the behavior of individual non-interacting photons.

Using this definition, the *first order* correlation function in *statistical optics* is defined as [13]

---

[11] The detection process was introduced semi-classically by Mandel et al. [31] in statistical optics and was included *ab initio* and played a key role in Glauber's formulation of quantum optics [32].

[12] At Dirac's time, the degeneracy parameter of all optical sources was significantly less than one, so that his statement correctly described the fact that interference between different photons could be neglected.

$$\Gamma^{(1,1)}(\mathbf{x}_1, t_1, \mathbf{x}_2, t_2) = \left\langle \mathbf{E}^*(\mathbf{x}_1, t_1)\mathbf{E}(\mathbf{x}_2, t_2) \right\rangle, \tag{4.81}$$

where the triangular brackets on the right refer to a statistical *ensemble average.*

The *quantum optics* definition for the spatio-temporal *first order* correlation function is similarly defined as [36, 37]

$$G^{(1)}(\mathbf{x}_1, t_1, \mathbf{x}_2, t_2) = \left\langle \mathbf{E}^-(\mathbf{x}_1, t_1)\mathbf{E}^+(\mathbf{x}_2, t_2) \right\rangle. \tag{4.82}$$

The triangular brackets on the right now denote the quantum mechanical *expectation value,* which depends on the ordering of the operators, since $\mathbf{E}^+$ and $\mathbf{E}^-$ do not commute. According to the quantum convention the operator on the right acts first. The placement of the photon destruction operator $\mathbf{E}^+(\mathbf{x}_2, t_2)$ on the right, referred to as *normal ordering*, expresses the fact that light is detected by destruction of photons through the photoelectric effect [36, 37]. The normal ordering also assures that no zero-point fluctuations of the EM field contribute to photo-excitations, which would require the creation operator $\mathbf{E}^-$ to act first.

In the following we shall use the same label for the first order correlation function in statistical optics as in quantum optics (i.e. $\Gamma^{(1,1)} = G^{(1)}$). This will allow us to seamlessly go between the quantum and statistical expressions by replacing $\mathbf{E}^- \leftrightarrow \mathbf{E}^*$ and $\mathbf{E}^+ \leftrightarrow \mathbf{E}$, but we have to keep in mind the different meaning of the triangular brackets, as will be discussed in more detail below.

4.5.1 Degree of First Order Coherence

In practice, it is often convenient to define the following normalized versions of the correlation function $G^{(1)}$, called the *complex degree of first order coherence.* In practice, one also assumes *space-time separability.* While such a separation of longitudinal=temporal and lateral=spatial degrees of freedom of the light is *not valid* in general [9], it holds for a large class of sources that are cross-spectrally pure [38].

The "same time", $t_1 = t_2$, degree of first order *spatial* coherence is a complex number defined as

$$g^{(1)}(\mathbf{x}_1, \mathbf{x}_2) = \frac{\langle \mathbf{E}^-(\mathbf{x}_1)\, \mathbf{E}^+(\mathbf{x}_2) \rangle}{\sqrt{\langle \mathbf{E}^-(\mathbf{x}_1)\, \mathbf{E}^+(\mathbf{x}_1) \rangle \langle \mathbf{E}^-(\mathbf{x}_2)\, \mathbf{E}^+(\mathbf{x}_2) \rangle}}, \tag{4.83}$$

and the "same point", $\mathbf{x}_1 = \mathbf{x}_2$, degree of first order *temporal coherence* is defined analogously as

$$g^{(1)}(t_1, t_2) = \frac{\langle \mathbf{E}^-(t_1)\, \mathbf{E}^+(t_2) \rangle}{\sqrt{\langle \mathbf{E}^-(t_1)\, \mathbf{E}^+(t_1) \rangle \langle \mathbf{E}^-(t_2)\, \mathbf{E}^+(t_2) \rangle}}, \tag{4.84}$$

where $0 \leq |g^{(1)}| \leq 1$. The limit $|g^{(1)}| = 1$ corresponds to classical "coherence" or complete first order coherence. It is obtained when the correlation brackets in the numerators can be factored. The other limit, $g^{(1)} = 0$ formally corresponds to "incoherence" or chaoticity with the intermediate case representing partial coherence.

4.5.2 Statistical Forms of Light Behavior

For the description of laterally partially coherent sources, Wolf et al. [13, 39–41] introduced the concept of *quasi-homogeneous* behavior, typically accompanied by assuming that the source is quasi-monochromatic. For such sources the areas of constant intensity are larger than those of constant coherence. If one further assumes that the coherence regions may be described by the same distribution function, one may write (4.83) in the form

$$g^{(1)}(\mathbf{r}_1, \mathbf{r}_2) = g^{(1)}(\mathbf{r}_1 - \mathbf{r}_2) = g^{(1)}(\mathbf{r}'). \tag{4.85}$$

For example, a blackbody source which has a wide frequency spectrum and is unpolarized is actually not truly chaotic in the sense that it contains small regions with a sinc function $g^{(1)}(r') = \sin(kr')/(kr')$ field-field correlation between two points separated by r' [1, 13, 40]. The best description of a quasi-monochromatic source which emits a characteristic line is through an Airy correlation function of the form $2J_1(kr')/(kr')$ [1, 40, 42]. A synchrotron or XFEL x-ray source, on the other hand, has an intrinsic microscopic Gaussian distribution function that causes the coherent "spikes" within the total pulse as shown in Figs. 2.22 and 2.31. On average, the macroscopic pulses also have a Gaussian distribution.

The quasi-homogeneous spatial concept has an analogue in the temporal concept of *quasi-stationary* behavior [9, 13]. It formally applies to a continuously emitted stream of light. When a measurement is made at a given spatial point with quasi-stationary light over a finite time interval $\tau = t_2 - t_1$ that is much longer than the coherence time τ_{coh}, the average intensity fluctuations are independent of the start time t_1 and only depend on the length of the measurement time $\tau = t_2 - t_1$. In this case, (4.84) becomes[13]

$$g^{(1)}(t_1, t_2) = g^{(1)}(t_2 - t_1) = g^{(1)}(\tau). \tag{4.86}$$

This concept can be extended to pulsed synchrotron and XFEL sources, which are in principle non-stationary, if the data collection time extends over a large number of pulses [43].

[13] The statistical condition of *ergodicity* imposes a further restriction. The ergodic theorem states that *time averages* over τ are indistinguishable from and interchangeable with *ensemble averages* over the total time duration of the beam. For light that is *quasi-stationary ergodic*, the correlation function can be evaluated as a time integral over a time interval longer than the fluctuation or coherence time.

When the light is temporally stationary and laterally homogeneous, the *degree of first order coherence* may be written as

$$\boxed{g^{(1)}(\mathbf{r}_1, t_1, \mathbf{r}_2, t_2) = \underbrace{\frac{\langle E^*(\mathbf{r}_1)E(\mathbf{r}_1 + \mathbf{r}')\rangle}{\langle |E(\mathbf{r}_1)|^2\rangle}}_{g^{(1)}(\mathbf{r}')} \underbrace{\frac{\langle E^*(t_1)E(t_1 + \tau)\rangle}{\langle |E(t_1)|^2\rangle}}_{g^{(1)}(\tau)}}. \tag{4.87}$$

This key expression exhibits space-time separability so that we can consider the first order correlation functions $g^{(1)}(\mathbf{r}')$ and $g^{(1)}(\tau)$, independently. The two functions furthermore depend only on a single variable, $\mathbf{r}'$ or τ, given by the difference of two variables. The reduction to a single variable follows from the assumption that the light is "temporally stationary" and "laterally homogeneous". This means that the average light intensity changes smoothly over the macroscopic source size, whereas the coherence is allowed to change over smaller distances. In essence, the formalism reduces the very complex phenomenon of *partial coherence* to the correlation between two fields either at the same point at different times or at the same time at different points.

In practice, one is interested how the spatial correlation function $g^{(1)}(\mathbf{r}')$ of a source propagates to a distant detector plane, and how the temporal correlation function $g^{(1)}(\tau)$ can be converted into the frequency or energy domain. For coherent sources this was accomplished through Fourier transformation of the temporal field distribution in Sect. 4.4.1 and spatial field distribution in Sect. 4.4.3.

For partially coherent light, the desired transformation can again be accomplished through a Fourier transformation according to two important theorems formulated in the 1930s. The first one, dealing with the propagation of spatial coherence, is the *van Cittert–Zernike (VCZ) theorem* [10, 11]. It describes through a Fourier transform of $g^{(1)}(\mathbf{r}')$ how the *spatial* correlation function of a *partially coherent* source propagates to a distant detection plane. In contrast to the *coherent* source case, where the *field* of the source is Fourier transformed, the VCZ theorem effectively transforms an intensity, since $g^{(1)}(\mathbf{r}')$ comprises a product of fields. The related *Wiener–Khintchine (WK) theorem* [12] similarly allows transformation of the *temporal* correlation functions $g^{(1)}(\tau)$ into the energy domain.

From an x-ray point of view, the VCZ theorem is particularly important since it underlies imaging and diffraction.[14] We shall derive it from a statistical optics point of view in Sect. 4.6. In Sect. 4.7 we will derive it quantum mechanically. This also serves as an introduction to the description of diffraction in terms of photon creation and destruction operators. We will not discuss the WK theorem here since its formulation of the Fourier transform of $g^{(1)}(\tau)$ in (4.87) is analogous to that of $g^{(1)}(\mathbf{r}')$ by the VCZ theorem [9, 13]. This is also reflected by the similarity of (4.26) for a partially *temporally* coherent source derived by means of the WK theorem, and (4.27) for a partially *laterally* coherent source derived with the VCZ theorem.

---

[14] Frits Zernike was awarded The Nobel Prize in Physics 1953 "for his demonstration of the phase contrast method, especially for his invention of the phase contrast microscope.".

4.5.3 *The Concept of Partial Lateral Coherence*

We start with a general discussion of lateral coherence. Upon closer inspection, even the concept of a laterally "incoherent" source is non-trivial. It strictly means that the fields originating from all points in a source area are uncorrelated, which is often called "chaotic" behavior. In textbooks, a thermal source is usually given as an example of an incoherent source, where no correlations are assumed to exist between the fields emitted at two different points (atoms). This picture is flawed because correlations between free fields (i.e. "radiation") must exist over the dimension of at least a wavelength [42, 44]. For the field to propagate away from the generating electric charge rather than remain attached to it (the very distinction between "acceleration" and "velocity" fields discussed in Sect. 2.6), it needs to be defined at least over one wavelength. In this sense, a completely incoherent source does not exit since it does not emit radiation!

It is interesting that we nevertheless regularly use the point source concept in optics in form of the Huygens–Fresnel principle. The utility of the point source concept comes from the fact that a point source is envisioned to radiate a perfect uniform-amplitude spherical wave with an outward traveling phase front and uniform intensity everywhere on the sphere. This is seen by representing a point source by a 3D δ-function. For a Cartesian source point at the origin we have $\delta^{(3)}(\mathbf{r}) = \delta(x, y, z)$ which in spherical coordinates has the form and spatial integral value (see Appendix A.3)

$$\delta^{(3)}(\mathbf{r}) = \frac{1}{4\pi r^2}\,\delta(r) \;\Rightarrow\; \int_0^{2\pi}\!\!\int_0^{\pi}\!\!\int_0^{\infty} \delta^{(3)}(\mathbf{r})\, r^2 \sin\theta \,\mathrm{d}r\,\mathrm{d}\theta\,\mathrm{d}\phi = 1. \tag{4.88}$$

If we now assume that a field of amplitude E_0 is launched at a point with strength $f E_0$, where f is the scattering length of the source, then the spherical wave $f E_0/r$ will have a constant field across the entire spherical surface according to (4.88). The concept of spherical waves around all possible source points (e.g. even points in an illuminated hole where there are no real atoms that can scatter) has been a great aid since it produces the experimentally observed diffraction pattern. We need to remember, however, that this magically simple concept actually works because it reflects an average over all possible directional single photon emission processes. Spherical waves simply approximate the sum over many random emission events.

What we actually mean by the concept of an incoherent source is the *limiting case* of a *partially coherent* source where the coherence area shrinks to the size of the wavelength. We may call this the chaotic limit of partial coherence. A chaotic source is thus the limit of a partially coherent source whose coherence regions become vanishingly small relative to the total source area. The most different light behavior is then described by coherence and chaoticity. In quantum mechanics this behavior is described by the so-called coherent states with a Poisson distribution shown in Fig. 3.7 and chaotic states with a Planck or Bose–Einstein distribution shown in Fig. 3.9.

Mathematically, the description of *lateral* chaoticity is accomplished by a trick. The field correlation around each source point is taken to be a 2D Dirac δ-function [9]. For example, the δ-function at a point $\delta^{(2)}(\mathbf{r}) = \delta(x, y)$ with $r = \sqrt{x^2 + y^2}$ and assuming cylindrical symmetry is given by

$$\delta^{(2)}(\mathbf{r}) = \frac{1}{2\pi r}\delta(r) \;\Rightarrow\; \int_0^{2\pi}\int_0^{\infty} \delta^{(2)}(\mathbf{r})\, r\, \mathrm{d}r\, \mathrm{d}\phi = 1. \tag{4.89}$$

In practical applications, the 2D δ-function may be approximated by different peaked distribution functions [23] whose microscopic effective coherence areas A^{μ}_{coh} become vanishingly small, $A^{\mu}_{\text{coh}} \to 0$. For example, by use of a cylindrically symmetric 2D Gaussian we can write

$$\delta^{(2)}(\mathbf{r}) = \lim_{A^{\mu}_{\text{coh}}\to 0} \frac{1}{A^{\mu}_{\text{coh}}} \exp\left[-\frac{\pi r^2}{A^{\mu}_{\text{coh}}}\right] \tag{4.90}$$

As the effective coherence area $A^{\mu}_{\text{coh}} = 2\pi\sigma_{\mu}^2$, where σ_{μ} is the Gaussian rms width, becomes infinitesimally small, the peak value goes to infinity under preservation of the unit integrated area according to (4.89). We will use this trick below.

From the above discussion it is clear that the concept of *partial coherence*, extending from δ-function like coherence regions to the entire source area emitting coherently, occupies the central role in our understanding of coherence. As shown in Fig. 4.3, we may envision a partially coherent source to consist of local coherence regions that are small relative to the total source area. The spatial location of these regions within the global source area fluctuates in time. The description of partial lateral coherence was largely developed by Wolf and collaborators in a series of papers in the 1970s [13, 39–41], building on a pioneering 1968 paper by Walther [1].

We can summarize as follows.

> An *incoherent* or *chaotic* source is defined as the limiting case of all partially coherent sources, whose coherence volume reduces to a value of order λ^3. In practice, this is mathematically expressed by a Dirac δ-function of the form (4.88).

4.6 The van Cittert–Zernike Theorem: Propagation of Field Correlations

The theorem is derived in statistical optics by considering the propagation of the *field correlations* between different points in a source area to a distant detection plane. Unlike the *coherent* source case, which can be treated by assuming the equivalence of all source points, as shown in Fig. 4.9, the general case of *partial coherence* needs to specifically consider the correlation of birth phases between different source points. This leads to the consideration of two source points illustrated in Fig. 4.12. One also considers two possible detection points, which allows the distinction of detection at the same point and two different points. Mathematically, this corresponds to linking the first order correlation function $G^{(1)}(\mathbf{r}_A, \mathbf{r}_B)$ between two source points to the correlation function $G^{(1)}(\boldsymbol{\rho}_1, \boldsymbol{\rho}_2)$ between one or two detection points. In statistic optics one refers to a two-field correlation function as the "*mutual intensity*", so that the VCZ theorem is said to express the propagation of the mutual intensity [7, 9].

For convenience we shall assume that the two planes are separated by a large distance and ignore the field polarization. This assumption facilitates the complete statistical optics derivation [7, 9] without loss of the underlying physics. It also corresponds to the geometry used in experimental studies of coherence propagation, which we will illustrate in Sect. 4.9 through an example. The coordinates that link the correlation function in the source plane, $G^{(1)}(\mathbf{r}_A, \mathbf{r}_B) = \langle \mathbf{E}^*(\mathbf{r}_A)\, \mathbf{E}(\mathbf{r}_B)\rangle$ and the detection plane, $G^{(1)}(\boldsymbol{\rho}_1, \boldsymbol{\rho}_2) = \langle \mathbf{E}^*(\boldsymbol{\rho}_1)\, \mathbf{E}(\boldsymbol{\rho}_2)\rangle$ are illustrated Fig. 4.12.

In principle, the mutual intensity defined by the product of conjugate fields at two points can be calculated without considering the detection process, similar to the classical single point intensity. In practice, it is measured by inserting two pinholes at positions 1 and 2, located symmetrically from the optical axis, and one then measures the intensity of the interference pattern created by the two secondary pinhole sources as a function of their separation $\rho' = |\boldsymbol{\rho}_2 - \boldsymbol{\rho}_1|$ (see Fig. 4.18). This procedure is based on the classical Huygens–Fresnel principle, where a field can change direction at any point due to secondary emission of a spherical wave.

One may also insert two mirrors at points 1 and 2 as shown in Fig. 4.12 and measure the interference of the reflected fields with a point detector. By choosing $r_{1D}=r_{2D}$ in Fig. 4.12, the additional propagation paths appear as additional field propagation phases and fall out according to

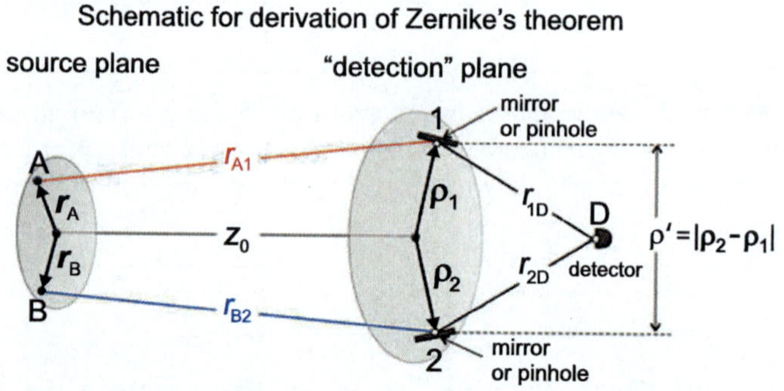

Fig. 4.12 Extension of Fig. 4.9 to the propagation from two source points A and B with coordinates $\mathbf{r}_A$ and $\mathbf{r}_B$ to two points 1 and 2 with coordinates $\boldsymbol{\rho}_1$ and $\boldsymbol{\rho}_2$ in a "detection" plane separated by z_0. The spatial correlation between fields in the detection plane, the "mutual intensity", is measured through their interference at a detector D, as shown. One may use two mirrors at points 1 and 2 and choosing the additional path lengths as $r_{1D}=r_{2D}$. In practice, the mirrors are typically replaced by pinholes and the intensity of the interference pattern created by the two secondary pinhole sources is measured as a function of their separation $\rho' = |\boldsymbol{\rho}_2 - \boldsymbol{\rho}_1|$

$$E_D^*(\boldsymbol{\rho}_1)\, E_D(\boldsymbol{\rho}_2) = E_1^*(\boldsymbol{\rho}_1)\mathrm{e}^{-\mathrm{i}kr_{1D}}\, E_2(\boldsymbol{\rho}_2)\mathrm{e}^{\mathrm{i}kr_{2D}} = E_1^*(\boldsymbol{\rho}_1)\, E_2(\boldsymbol{\rho}_2) \tag{4.91}$$

The two-point source and detection case shown in Fig. 4.12 may be constructed from the single point case in Fig. 4.9 expressed by (4.67). We obtain the following expression under the assumption of large z_0

$$\big\langle E^*(\boldsymbol{\rho}_1)E(\boldsymbol{\rho}_2)\big\rangle = \frac{1}{\lambda^2 z_0^2}\left\langle \iiiint_{-\infty}^{\infty} E^*(\mathbf{r}_\mathrm{A})E(\mathbf{r}_\mathrm{B})\,\mathrm{e}^{-\mathrm{i}\frac{2\pi}{\lambda z_0}(\mathbf{r}_\mathrm{B}\cdot\boldsymbol{\rho}_2-\mathbf{r}_\mathrm{A}\cdot\boldsymbol{\rho}_1)}\,\mathrm{d}\mathbf{r}_\mathrm{A}\,\mathrm{d}\mathbf{r}_\mathrm{B}\right\rangle. \tag{4.92}$$

This expression assumes the "same-time" detection of the fields at the two points $\boldsymbol{\rho}_1$ and $\boldsymbol{\rho}_2$, and in general the triangular brackets denote a statistical ensemble average. In practice one assumes that the temporal fluctuations in the fields are quasi-stationary, so that the triangular brackets represent an average over a time interval, and the average does not depend on the start time of the measurement. This allows one to exchange the spatial integration and the temporal average to obtain the far-field form of the VCZ theorem.[15]

The van Cittert–Zernike theorem expresses the propagation of coherence from a finite-sized *partially coherent* source to two points in a detection plane at large distance z_0.

$$\underbrace{\big\langle E^*(\boldsymbol{\rho}_1, z_0)E(\boldsymbol{\rho}_2, z_0)\big\rangle}_{G^{(1)}(\boldsymbol{\rho}_1,\boldsymbol{\rho}_2,z_0)}$$

$$= \frac{1}{\lambda^2 z_0^2}\iiiint_{-\infty}^{\infty} \underbrace{\big\langle E^*(\mathbf{r}_\mathrm{A},0)E(\mathbf{r}_\mathrm{B},0)\big\rangle}_{G^{(1)}(\mathbf{r}_\mathrm{A},\mathbf{r}_\mathrm{B},0)}\,\mathrm{e}^{-\mathrm{i}\frac{2\pi}{\lambda z_0}(\mathbf{r}_\mathrm{B}\cdot\boldsymbol{\rho}_2-\mathbf{r}_\mathrm{A}\cdot\boldsymbol{\rho}_1)}\,\mathrm{d}\mathbf{r}_\mathrm{A}\,\mathrm{d}\mathbf{r}_\mathrm{B}$$

$$= \frac{\langle |E_0|^2\rangle}{\lambda^2 z_0^2}\iiiint_{-\infty}^{\infty} g^{(1)}(\mathbf{r}_\mathrm{A},\mathbf{r}_\mathrm{B},0)\,\mathrm{e}^{-\mathrm{i}\frac{2\pi}{\lambda z_0}(\mathbf{r}_\mathrm{B}\cdot\boldsymbol{\rho}_2-\mathbf{r}_\mathrm{A}\cdot\boldsymbol{\rho}_1)}\,\mathrm{d}\mathbf{r}_\mathrm{A}\,\mathrm{d}\mathbf{r}_\mathrm{B}. \tag{4.94}$$

It is seen to essentially corresponds to a 2D Fourier transform of the intensity distribution in the source plane. Since the integrals over the source coordinates are proportional to the squared source area A_s^2, the factor $1/(\lambda^2 z_0^2)$ in front of the integrals becomes a dimensionless propagation factor $\propto A_s^2/(\lambda^2 z_0^2)$ describing the intensity propagation to the detector plane.

---

[15] Sometimes the term $\rho^2/(2z_0^2)$ in (4.66), which was neglected in the exponential phase factor of (4.67), is kept. This leads to an additional factor on the right side of (4.94) given by [9, 14]

$$\mathrm{e}^{\mathrm{i}\frac{\pi}{\lambda z_0}(\rho_2^2-\rho_1^2)}. \tag{4.93}$$

This shows that, in general, the "mutual intensity" is actually complex unlike the conventional real intensity.

For the general case of partial coherence, the integrations in (4.94) need to be performed numerically for an assumed source coherence function $G^{(1)}(\mathbf{r}_A, \mathbf{r}_B, 0)$. In order to demonstrate the VCZ theorem we consider a partially coherent source with *cylindrical symmetry* around the optical axis in the convenient form of a Schell model [7, 9, 13].

4.6.1 van Cittert–Zernike Theorem for a Schell Model Source

The Schell model approximation allows us to separate the source distribution into independent intensity and coherence functions. For the case of a planar circular source with a flat-top intensity profile we can write the correlation function in (4.94) as

$$G^{(1)}(\mathbf{r}_A, \mathbf{r}_B) = \underbrace{\langle |E_0|^2\rangle \begin{Bmatrix} 1 & 0 \le r \le R \\ 0 & \text{otherwise} \end{Bmatrix}}_{\text{intensity } \langle |E(\mathbf{r})|^2 \rangle} \underbrace{\exp\left[-\frac{\pi(\mathbf{r}_B - \mathbf{r}_A)^2}{A^{\mu}_{\text{coh}}}\right]}_{\text{coherence } g^{(1)}(\mathbf{r}_A, \mathbf{r}_B)}. \tag{4.95}$$

Here $\langle |E_0|^2\rangle$ is the time-averaged value of the intensity distribution which is assumed to be approximately constant across the source of area πR^2. We have taken the coherence function to have a Gaussian shape with an effective microscopic coherence area $A^{\mu}_{\text{coh}} = 2\pi(\sigma^I_{\mu})^2$. The assignment of the second underbracket in (4.95) is verified by calculating the degree of first order spatial coherence given by

$$g^{(1)}(\mathbf{r}_A, \mathbf{r}_B) = \frac{G^{(1)}(\mathbf{r}_A, \mathbf{r}_B)}{\sqrt{G^{(1)}(\mathbf{r}_A, \mathbf{r}_A)\, G^{(1)}(\mathbf{r}_B, \mathbf{r}_B)}} = \exp\left[-\frac{\pi(\mathbf{r}_B - \mathbf{r}_A)^2}{A^{\mu}_{\text{coh}}}\right]. \tag{4.96}$$

As expected, $g^{(1)}(\mathbf{r}_A, \mathbf{r}_B)$ is entirely determined by the coherence part of (4.95) and has the following chaotic and first order coherent limits

$$\text{Coherent:} \quad \lim_{A^{\mu}_{\text{coh}} \to \infty} g^{(1)}(\mathbf{r}_A, \mathbf{r}_B) = 1 \tag{4.97}$$

$$\text{Chaotic:} \quad \lim_{A^{\mu}_{\text{coh}} \to 0} g^{(1)}(\mathbf{r}_A, \mathbf{r}_B) = A^{\mu}_{\text{coh}}\, \delta(\mathbf{r}_B - \mathbf{r}_A). \tag{4.98}$$

For the coherent case, the entire source area is coherent, while in the chaotic case, the microscopic coherence areas have an average effective size A^{μ}_{coh}. The 2D δ-function $\delta(\mathbf{r}_B - \mathbf{r}_A)$ in (4.98) is defined through its dimensionless unit integrated area according to (4.89).

In the following we will illustrate the VCZ theorem for a *partially coherent* Schell model source in the coherent and chaotic limits for the case of a circular source with a *flat-top* and a *Gaussian* intensity distribution. As expected, we will recover the results of Sect. 4.4.3 for the coherent source cases.

4.6.2 Coherent Limit of a Circular Flat-Top Source

For the *coherent case* the entire source area emits coherently and the flat-top (constant) intensity profile of the source has the simple form (4.97)

$$g^{(1)}(\mathbf{r}_A, \mathbf{r}_B) = \begin{Bmatrix} 1, & 0 \le r_A, r_B \le R \\ 0, & \text{otherwise.} \end{Bmatrix} \tag{4.99}$$

For our case of a cylindrical intensity distribution about the optical axis, defined as the surface normal of the flat circular source, the integration in (4.94) is evaluated in cylindrical coordinates. We rewrite the vector dot products in terms of the angular projection angles $\mathbf{r}_B \cdot \boldsymbol{\rho}_2 = r_B \rho_2 \cos \gamma_2$ and $\mathbf{r}_A \cdot \boldsymbol{\rho}_1 = r_A \rho_1 \cos \gamma_1$, and the integrations over r_A, γ_1 and r_B, γ_2 can be performed separately as

$$G^{(1)}(\boldsymbol{\rho}_1, \boldsymbol{\rho}_2) = \frac{\langle |E_0|^2 \rangle}{\lambda^2 z_0^2} \int_0^{2\pi} \int_0^R e^{-i\frac{2\pi}{\lambda z_0} r_B \rho_2 \cos \gamma_2} \, d\gamma_2 \, r_B \, dr_B$$
$$\times \int_0^{2\pi} \int_0^R e^{i\frac{2\pi}{\lambda z_0} r_A \rho_1 \cos \gamma_1} \, d\gamma_1 \, r_A \, dr_A. \tag{4.100}$$

This yields,

$$G^{(1)}(\boldsymbol{\rho}_1, \boldsymbol{\rho}_2) = \left[\frac{\pi R^2}{\lambda z_0}\right]^2 \langle |E_0|^2 \rangle \frac{2J_1(kR\rho_1/z_0)}{kR\rho_1/z_0} \frac{2J_1(kR\rho_2/z_0)}{kR\rho_2/z_0}, \tag{4.101}$$

where the expression $2J_1(x)/x$ is the normalized Airy disk function already encountered in (4.70). The correlation function in the detector plane is seen to factor, so that $G^{(1)}(\boldsymbol{\rho}_1, \boldsymbol{\rho}_2) = \sqrt{G^{(1)}(\boldsymbol{\rho}_1, \boldsymbol{\rho}_1) G^{(1)}(\boldsymbol{\rho}_2, \boldsymbol{\rho}_2)}$. This has the important consequence that the Airy structure still contained in $G^{(1)}(\boldsymbol{\rho}_1, \boldsymbol{\rho}_2)$ is normalized out in the degree of first order coherence and inserting into (4.83) we obtain

$$g^{(1)}(\boldsymbol{\rho}_1, \boldsymbol{\rho}_2) = 1. \tag{4.102}$$

We now consider a two-field correlation measurement in the geometry shown in Fig. 4.13a. We arrange our detector to eliminate the path differences $r_{1D} = r_{2D}$ defined in Fig. 4.12 so that the "same time" condition assumed in (4.83) is fulfilled. Because of the cylindrical symmetry about the optical axis, we chose the points $\boldsymbol{\rho}_1, \boldsymbol{\rho}_2$ as shown in Fig. 4.13 (a) at the same distance from the optical axis, i.e. $\rho = |\boldsymbol{\rho}_1| = |\boldsymbol{\rho}_2|$. We can then distinguish the "1-point intensity" case $\boldsymbol{\rho}_1 = \boldsymbol{\rho}_2$ and "mutual intensity" case $\boldsymbol{\rho}_1 = -\boldsymbol{\rho}_2$, which from (4.101) are seen to be the same according to

$$I^{(1)}(\rho) = \epsilon_0 c \, G^{(1)}(\boldsymbol{\rho}_2, \boldsymbol{\rho}_2) = \epsilon_0 c \, G^{(1)}(-\boldsymbol{\rho}_2, \boldsymbol{\rho}_2). \tag{4.103}$$

Here we have kept the superscript "(1)" to indicate the "first order" coincident (mutual) intensity since we will later distinguish it from the second order coincident intensity denoted by a superscript "(2)" discussed in Chap. 5. We obtain (4.101) in the form

$$\boxed{I^{(1)}(\rho) = I_0 \left[\frac{\pi R^2}{\lambda z_0}\right]^2 \left[\frac{2J_1\left(kR\rho/z_0\right)}{kR\rho/z_0}\right]^2}. \tag{4.104}$$

Our result is that previously obtained for a coherent flat-top source, given by (4.71) and shown in Fig. 4.10. For convenience it is again shown in blue in Fig. 4.13b together with the black flat-top distribution which contains the same power.

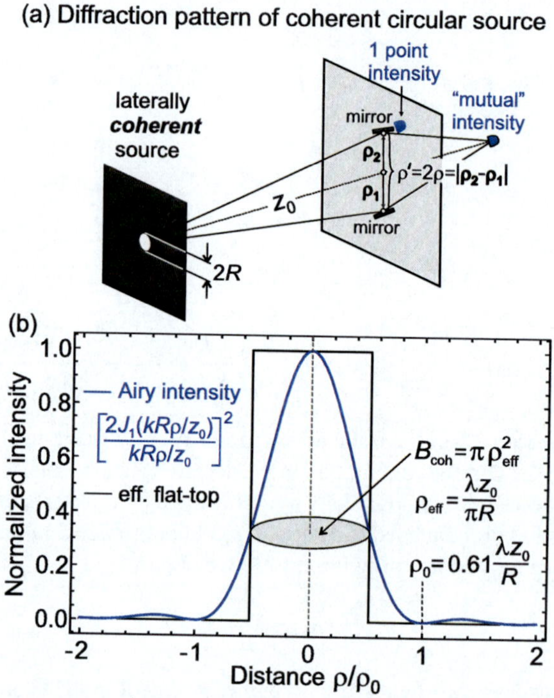

Fig. 4.13 **a** Assumed geometry of a *coherent* flat-top source of area πR^2. As indicated in blue, we distinguish the conventional intensity measured at a single point in the detection plane from the "mutual" intensity due to the correlation of the fields from two points that are symmetrically placed from the optical axis ($\boldsymbol{\rho}_1 = -\boldsymbol{\rho}_2$) and separated by $\rho' = |\boldsymbol{\rho}_2 - \boldsymbol{\rho}_1| = 2\rho$. The "mutual" intensity is measured by scanning two mirrors that allows the fields to interfere at a detector. **b** 1D projection of the normalized Airy intensity distribution (blue) according to (4.104). In black we show the effective flat-top distribution whose area $B_{\rm eff} = \pi\rho_{\rm eff}^2$ contains the same power as the Airy intensity distribution. The Rayleigh diffraction limit $\rho_0 = 0.61\lambda z_0/R$ is defined by the first zero of the Airy intensity

We find that for the coherent source case the two detector geometries shown in blue in Fig. 4.13a give the same result. Hence the conventionally measured intensity at single points in the detector plane gives the same pattern as the "mutual" intensity due to correlating the fields from two different points.

4.6.3 *Chaotic Limit of a Circular Flat-Top Source*

For the *chaotic* case we consider the same geometry as for the coherent case. In Fig. 4.14a, we distinguish two detection cases by different colors. The conventional intensity is measured at a single point in the detection plane by the green detector, while the correlation between two fields is measured as the mutual intensity by use of two mirrors by the red detector. In principle, the coordinates $\boldsymbol{\rho}_1$ and $\boldsymbol{\rho}_2$ are defined relative to the optical axis in the detection plane. However, at large distance from the source the intensity distribution is approximately constant around the optical axis and it does not matter whether the coordinate origin in the detection plane is slightly shifted.

We can express the first order correlation function of the source intensity distribution by (4.98) to obtain the VCZ formula (4.94) in the form

$$G^{(1)}(\boldsymbol{\rho}_1, \boldsymbol{\rho}_2) = \frac{\langle |E_0|^2 \rangle}{\lambda^2 z_0^2} \iiiint_{-\infty}^{\infty} A^{\mu}_{\mathrm{coh}}\, \delta(\mathbf{r}_B - \mathbf{r}_A)\, \mathrm{e}^{-\mathrm{i}\frac{2\pi}{\lambda z_0}(\mathbf{r}_B \cdot \boldsymbol{\rho}_2 - \mathbf{r}_A \cdot \boldsymbol{\rho}_1)}\, d\mathbf{r}_A\, d\mathbf{r}_B. \quad (4.105)$$

This expression can be simplified by use of the sifting property[16] of the 2D δ-function to simplify the integrals. By integrating in cylindrical coordinates as in (4.100) we obtain

$$G^{(1)}(\boldsymbol{\rho}_1, \boldsymbol{\rho}_2, z_0) = \langle |E_0|^2 \rangle \frac{A^{\mu}_{\mathrm{coh}} A_{\mathrm{eff}}}{\lambda^2 z_0^2} \frac{2J_1(kR\rho'/z_0)}{kR\rho'/z_0}, \quad (4.107)$$

where the effective source area is $A_{\mathrm{eff}} = \pi R^2$. We can write this expression similarly to (4.104) in the mutual intensity form with $\rho' = |\boldsymbol{\rho}_2 - \boldsymbol{\rho}_1|$, as

$$\boxed{\text{Coherent fraction}: \; I^{(1)}(\rho') = I_0 \frac{A^{\mu}_{\mathrm{coh}} A_{\mathrm{eff}}}{\lambda^2 z_0^2} \frac{2J_1(kR\rho'/z_0)}{kR\rho'/z_0}}. \quad (4.108)$$

[16] It is expressed by

$$\iint_{-\infty}^{\infty} f(\mathbf{r})\, \delta(\mathbf{r} - \mathbf{r}_0)\, d\mathbf{r} = f(\mathbf{r}_0). \quad (4.106)$$

This expresses the far-field version of the VCZ theorem for a circular *chaotic* source [10, 11]. The so-defined intensity is the *coherent fraction* of the total intensity in the distant plane. Its normalized form is plotted as a magenta curve in Fig. 4.14 as a function of $\rho' = |\boldsymbol{\rho}_2 - \boldsymbol{\rho}_1|$. For comparison we also show in blue the pattern for the coherent case plotted as a function of ρ/ρ_0, taken from Fig. 4.13.

In contrast to the coherent source (Fraunhofer) expression (4.101), the correlation function does not factor into functions of $\boldsymbol{\rho}_1$ and $\boldsymbol{\rho}_2$ but depends on the separation $\rho' = |\boldsymbol{\rho}_2 - \boldsymbol{\rho}_1|$ between two points in the detector plane. Inserting (4.107) into (4.83) with $\boldsymbol{\rho}_i = \mathbf{x}_i$, we obtain the degree of first order coherence

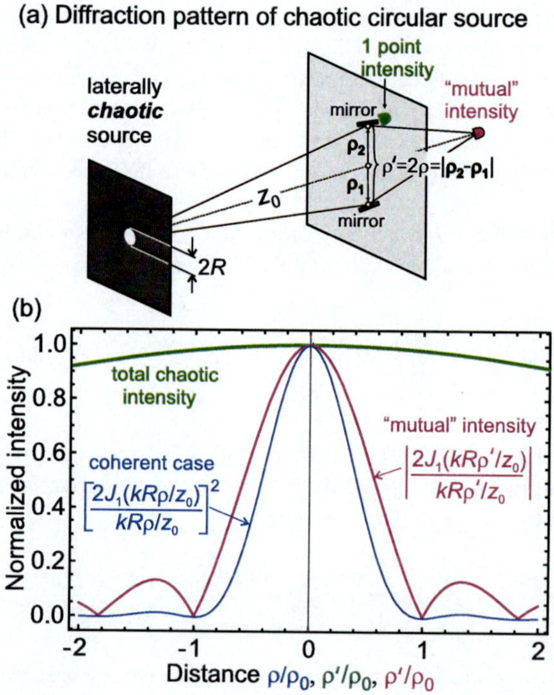

Fig. 4.14 **a** Detection of the intensity emitted by a chaotic flat-top source of area πR^2. We distinguish the conventional intensity measured at a single point in the detection plane (green) from the "mutual" intensity (magenta) due to correlation of the fields from two points, symmetrically placed from the optical axis ($\boldsymbol{\rho}_1 = -\boldsymbol{\rho}_2$) and separated by $\rho' = |\boldsymbol{\rho}_2 - \boldsymbol{\rho}_1| = 2\rho$. The magenta "mutual" intensity is measured by scanning two mirrors that allows the fields to interfere at the detector. **b** Total conventional normalized intensity (green), given by (4.113), plotted as a function of ρ'/ρ_0, where $\rho_0 = 0.61\lambda z_0/R$. Normalized mutual intensity (magenta), given by (4.109) as a function ρ'/ρ_0. It represents the coherent fraction of the green curve. The thin blue curve is the diffraction pattern of a coherent source as a function of ρ/ρ_0, taken from Fig. 4.13

$$g^{(1)}(\rho') = \frac{2J_1(kR\rho'/z_0)}{kR\rho'/z_0}. \tag{4.109}$$

It is seen that the absolute value of the degree of first order coherence is unity only on-axis $\rho' = 0$ but decreases away from it, reflecting the falloff the coherent fraction.

The green curve in Fig. 4.14b represents the *total intensity* distribution, measured as for the case of a coherent source by scanning a single detector (shown green in Fig. 4.14a) as a function of its separation ρ from the optical axis or eliminating one of the mirrors. The intensity distribution shown in green is given by

$$\boxed{\text{Total intensity}: \quad I^{(1)}_{\text{tot}}(\rho') = I_0 \frac{A^{\mu}_{\text{coh}} \pi R^2}{\lambda^2 z_0^2} \frac{1}{\left[1 + (\rho'/z_0)^2\right]^2}} \tag{4.110}$$

The last term arises from the fact that the detector plane is flat and causes the curvature of the green curve in Fig. 4.14b. By integrating over the entire flat detector plane in cylindrical coordinates we obtain the *total power* which has to be equal to the power emitted by the source area, $\mathcal{P}_s = I_0 \pi R^2$, according to

$$\mathcal{P}^{\text{tot}}_{\text{det}} = I_0 \frac{A^{\mu}_{\text{coh}} \pi R^2}{\lambda^2 z_0^2} \underbrace{\int_0^{2\pi}\int_0^{\infty} \frac{1}{\left[1 + (\rho'/z_0)^2\right]^2} \rho' \, \mathrm{d}\rho' \, \mathrm{d}\phi}_{\pi z_0^2} = \frac{\pi A^{\mu}_{\text{coh}}}{\lambda^2} \underbrace{I_0 \pi R^2}_{\mathcal{P}_s}. \tag{4.111}$$

Hence power conservation between the source and detector plane requires

$$A^{\mu}_{\text{coh}} = \frac{\lambda^2}{\pi}. \tag{4.112}$$

This area is hence the characteristic coherence area of a chaotic source. Its size is of order of the wavelength, as expected, since emitted wave fields need to be defined at least over this dimension. Interestingly, it corresponds to the maximum resonant atomic interaction cross section with emission of a spherical wave, derived by Breit and Wigner [15, 16] as discussed in Sect. 6.5.1. Inserting (4.112) into (4.110) we obtain

$$I^{(1)}_{\text{tot}}(\rho') = I_0 \frac{R^2}{z_0^2} \frac{1}{\left[1 + (\rho'/z_0)^2\right]^2}, \tag{4.113}$$

and according to (4.111) we have *total power* conservation $\mathcal{P}^{\text{tot}}_{\text{det}} = \mathcal{P}_s = I_0 \pi R^2$ as required.

We can now compare the total power with that contained in the coherent intensity fraction (4.108) obtained by

$$\mathcal{P}_{\rm det}^{\rm coh} = I_0 \frac{A_{\rm coh}^{\mu} \pi R^2}{\lambda^2 z_0^2} \underbrace{\int_0^{2\pi}\int_0^{\infty} \frac{2J_1(kR\rho'/z_0)}{kR\rho'/z_0} \rho' \, d\rho' \, d\phi}_{\frac{\lambda^2 z_0^2}{\pi R^2}} = I_0 A_{\rm coh}^{\mu} = I_0 \frac{\lambda^2}{\pi}. \quad (4.114)$$

We find that the coherent power fraction is given as expected by

$$\frac{\mathcal{P}_{\rm det}^{\rm coh}}{\mathcal{P}_{\rm det}^{\rm tot}} = \frac{A_{\rm coh}^{\mu}}{A_{\rm eff}} = \frac{\lambda^2/\pi}{\pi R^2}. \quad (4.115)$$

This is the value given in Table 4.1 with $A_s = A_{\rm eff} = \pi R^2$. With increasing coherence of the source, the intensity distribution (4.108) will go over into (4.104). The magenta curve in Fig. 4.14b will narrow, and its peak value will increase by a factor $A_{\rm eff}/A_{\rm coh}^{\mu}$.

4.6.4 *Coherent and Chaotic Limits of a Circular Gaussian Source*

For a Schell model circular source with Gaussian intensity distribution, the first order correlation function can be expressed similarly to (4.95) as

$$G^{(1)}(\mathbf{r}_A, \mathbf{r}_B) = \underbrace{\langle |E_0|^2 \rangle \exp\left[-\frac{\pi r^2}{2\pi(\sigma_r^I)^2}\right]}_{\text{intensity}\,\langle |E(\mathbf{r})|^2\rangle} \underbrace{\exp\left[-\frac{\pi(\mathbf{r}_B - \mathbf{r}_A)^2}{A_{\rm coh}^{\mu}}\right]}_{\text{coherence } g^{(1)}(\mathbf{r}_A,\mathbf{r}_B)}, \quad (4.116)$$

where $A_{\rm eff} = 2\pi(\sigma_r^I)^2$ is the effective source area and $A_{\rm coh}^{\mu}$ the effective size of the coherent areas within $A_{\rm eff}$. The incoherent intensity propagation is described by

$$I(r) = I_0 \exp\left[-\frac{\pi r^2}{A_{\rm eff}}\right] \Longrightarrow I(\rho') = I_0 \frac{A_{\rm eff} A_{\rm coh}^{\mu}}{\lambda^2 z_0^2} \exp\left[-\frac{\pi(\rho')^2}{\left(\frac{\lambda^2 z_0^2}{2A_{\rm eff}}\right)}\right], \quad (4.117)$$

where $\rho' = 2\rho$ (see Fig. 4.14a). The propagation is illustrated in Fig. 4.15. The incoherent source case may be directly compared to the coherence source case described by (4.80) and shown in Fig. 4.11.

If the intensity distribution in the detector plane is measured by scanning a single point detector as shown in green in Fig. 4.14a, we obtain the green nearly-constant 3D intensity distribution in Fig. 4.15b in analogy to the green curve in Fig. 4.14b. If the mutual intensity is measured as illustrated in magenta in Fig. 4.14a, we obtain the 3D magenta intensity distribution in Fig. 4.15b in analogy to the magenta curve in Fig. 4.14b.

The magenta intensity distribution in Fig. 4.15 corresponds to the *coherent fraction* within the total nearly constant green distribution. It has an effective area B in the

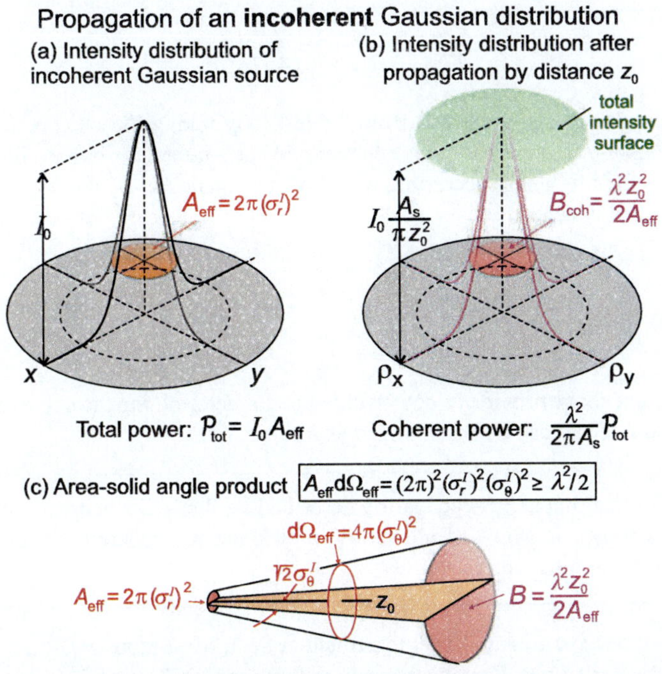

Fig. 4.15 **a** *Intensity* distribution in the plane of a two-dimensional *incoherent Gaussian* source with rms width σ_r^I and effective source area $A_{\rm eff} = 2\pi(\sigma_r^I)^2$. The total emitted power or area under the distribution is given by $\mathcal{P}_{\rm tot} = I_0 A_{\rm eff}$. **b** The total emitted intensity in the detector plane is shown in green and is flat similar to the green curve in Fig. 4.14. The *coherent* intensity fraction is shown in magenta, and the associated coherent power is $\mathcal{P}_{\rm coh} = \mathcal{P}_{\rm tot}\lambda^2/(2\pi A_{\rm eff})$. **c** Illustration of the area–solid angle product that enters into the source brightness expression. The detector plane area B contains the coherent fraction

detection plane which is seen to be a factor of 2 larger than the B value for a coherent source in Fig. 4.11.

We can now compare the *coherent* intensity distributions in the detection plane for a coherent source (4.80) and an incoherent source (4.117)

$$I(\rho) = I_0 \frac{4\,A_{\rm eff}^2}{\lambda^2 z_0^2} \exp\left[-\frac{\pi\rho^2}{\left(\frac{\lambda^2 z_0^2}{4\,A_{\rm eff}}\right)}\right], \quad \mathcal{P}_{\rm det}^{\rm coh} = 2\pi\int_0^\infty I(\rho)\,\rho\,{\rm d}\rho = I_0\,A_{\rm eff} = I_0\,2\pi(\sigma_r^I)^2$$

$$I(\rho') = I_0 \frac{A_{\rm eff}\,A_{\rm coh}^{\mu}}{\lambda^2 z_0^2} \exp\left[-\frac{\pi(\rho')^2}{\left(\frac{\lambda^2 z_0^2}{2A_{\rm eff}}\right)}\right], \quad \mathcal{P}_{\rm det}^{\rm coh} = 2\pi\int_0^\infty I(\rho')\,\rho'\,{\rm d}\rho' = I_0\,\frac{A_{\rm coh}^{\mu}}{2}. \qquad (4.118)$$

On the right side we have also given their integrated values, the coherent power. We find that the coherent power in the detection plane differs by

$$\frac{\mathcal{P}_{\mathrm{det}}^{\mathrm{coh}}}{\mathcal{P}_{\mathrm{det}}^{\mathrm{tot}}} = \frac{A_{\mathrm{coh}}^{\mu}}{2A_{\mathrm{eff}}} = \frac{A_{\mathrm{coh}}^{\mu}}{4\pi(\sigma_r^I)^2}. \tag{4.119}$$

The same value is also obtained from Table 4.1 by taking the ratio of the coherent fractions h_c/H for a partially coherent and for a coherent source. Since h_c/H determines the brightness according to (4.23), we again see the correspondence of brightness and coherence.

4.6.5 *Summary*

Our treatment above provides a powerful recipe for determining the coherence property of a source and can be summarized as follows.

> The mutual intensity propagation described by the VCZ theorem and measured as shown in Fig. 4.14 allows us to pick out the coherent fraction of a partially coherent source.

This will be experimentally demonstrated in Sect. 4.9. Before we do so, we want to derive the van Cittert–Zernike theorem in first order QED.

4.7 Quantum Derivation of the van Cittert–Zernike Theorem

In quantum mechanics, corresponding to first order QED, the VCZ coherence propagation problem is cast into the language of alternative paths of *single photons* from source to detector. We have seen in Sect. 4.4.3.1 that the propagation of classical wave fields is analogous to the quantum formulation of propagation in terms of Feynman's photon probability amplitudes. In Feynman's formulation of quantum mechanics, photons may be born at different source points with different phases $\alpha_{\mathrm{P}} = \alpha(\mathbf{r})$ leading to phase factors $\mathrm{e}^{\mathrm{i}\alpha_{\mathrm{P}}}$. For coherent sources, the phase factors are identical for all source points and may be set to unity. This formulation naturally gives the same result for the photon arrival probability distribution in a distant detector place as calculated classically as the intensity of the arriving wave fields, defined as $I \propto E\,E^* = E^*E$.

More generally, the concept of position-dependent birth phases of single photons is equivalent to the formulation in statistic optics where birth phases of spherical waves at different points are either taken to be the same (coherent case) or enter in the form of a Dirac δ-function as discussed in Sects. 4.5.3 and 4.6.1. It is therefore straightforward to derive the van Cittert–Zernike theorem by use of the equivalent phase-dependent formulations within statistical optics [7, 9] or Feynman's quantum formulation as done in [14]. Below, we shall take a different quantum approach that replaces the birth phases of classical fields or Feynman's probability amplitudes by an operator-based formulation, introduced by Dirac [45] and extended to higher order by Glauber [32, 36, 37].

In Dirac's quantum mechanics, the wave fields **E** and **E*** are replaced by quantum operators $\mathbf{E}^{+}$ and $\mathbf{E}^{-}$ as expressed by (3.61). Photons are created at source points by creation operators $\mathbf{a}^{\dagger}$ (associated with $\mathbf{E}^{-}$) and detected (destroyed) in the detector plane by annihilation operators **a** (associated with $\mathbf{E}^{+}$). The quantum theory of photon propagation, discussed below, replaces position-dependent *birth phases* by matrix elements of operators of the form $\mathbf{a}_{\mathrm{X}}^{\dagger}\mathbf{a}_{\mathrm{Y}}$ that can create different quantum states of light at point X and destroy them at point Y. After birth, the photon *propagation* from X to Y is described the same as in classical or statistical optics by a wavelength and distance-dependent geometrical phase factor.

This formalism incorporates the quantum nature of birth and destruction of photons through intra-atomic electronic processes. Photon *creation* corresponds to an electronic down transition in an excited source atom, with conversion of electronic to photon energy. Photon *destruction* corresponds to the transfer of the photon energy back to the electronic system through an up transition in a detector atom (photoelectric effect). Since photon propagation from a source distribution to a distant plane is equivalent to the calculation of the diffraction pattern of the source, our formalism corresponds to the calculation of diffraction in a photon picture.

> The quantum formulation of the propagation of *single photons* from a source area to a distant detector plane via all possible paths corresponds to the classical calculation of the diffraction pattern of the source by wave interference.

Let us take a look how "*quantum diffraction*" works. We rephrase the classical illustration of the problem previously shown in Fig. 4.12 by use of the coordinates illustrated in Fig. 4.16. Single photons may be created at different positions $\mathbf{r}_{\mathrm{A}}, \mathbf{r}_{\mathrm{B}}$ with associated emission directions $\mathbf{k}_{\mathrm{A}}, \mathbf{k}_{\mathrm{B}}$ to distant points 1 and 2. It is important that the birth only creates single photons, with the two arbitrary birth locations just being alternatives. The photons are not detected at points 1 and 2 but may alternatively pass through them on the way to a single photon detector D where they are destroyed through conversion into photoelectrons. The mirrors at points 1 and 2 allow the photons (like particles) to be redirected to the detector D. We again eliminate the influence of the additional paths lengths from points 1 and 2 to the detector by choosing $r_{1\mathrm{D}} = r_{2\mathrm{D}}$.

4.7.1 The Case of Two Source Points

Our quantum formulation starts with the first order correlation function (4.82) in the "same time" form. For a given 1-photon state of the radiation field, $|\Psi\rangle$, the triangular brackets are evaluated according to

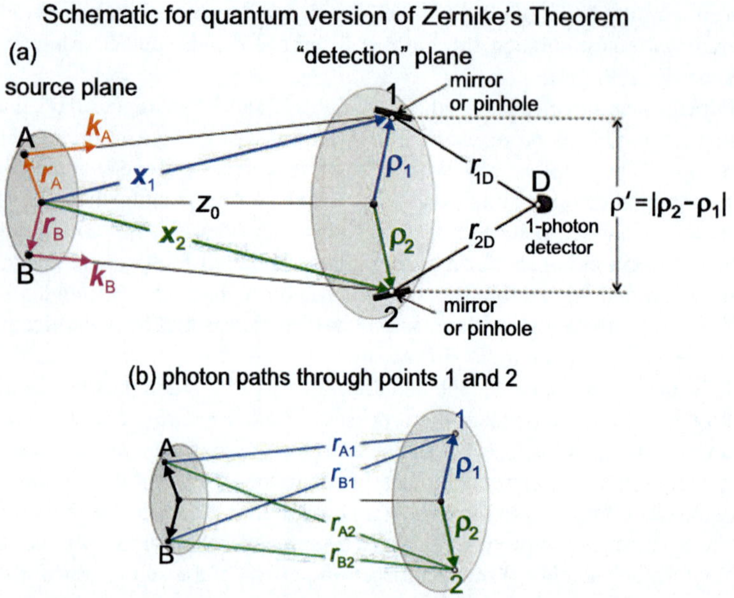

Fig. 4.16 Schematic used for the derivation of the quantum version of the VCZ theorem. **a** Coordinates linking two possible source points A and B with coordinates $\mathbf{r}_A$ and $\mathbf{r}_B$ to two points 1 and 2 with coordinates $\boldsymbol{\rho}_1$ and $\boldsymbol{\rho}_2$ in a "detection" plane separated by z_0. Two mirrors at 1 and 2 assure that a photon can reach the one-photon detector D via different paths, with the additional path lengths chosen as $r_{1D}=r_{2D}$ to eliminate added differential path length contributions. **b** Possible photon paths from points A and B through points 1 and 2 that may lead to the detector

$$
\begin{aligned}
G^{(1)}(\mathbf{x}_1, \mathbf{x}_2) &= \langle \Psi | \mathbf{E}^-(\mathbf{x}_1) \mathbf{E}^+(\mathbf{x}_2) | \Psi \rangle \\
&= \langle |E_0|^2 \underbrace{\langle \Psi | \mathbf{E}^-(\mathbf{x}_1) | 0 \rangle}_{\text{photon creation}} \times \underbrace{\langle 0 | \mathbf{E}^+(\mathbf{x}_2) | \Psi \rangle}_{\text{photon destruction}},
\end{aligned}
\tag{4.120}
$$

In the last line we have inserted a complete set of states, conveniently chosen as the zero-point states that cover all modes according to the closure relation $|0\rangle\langle 0| = 1$. This expression reveals that from a quantum point of view, photons are created out of the zero-point field and that the detection process corresponds to the reverse or conjugated process where the photon decays back into the zero-point field.

The order of the two operators in (4.120), referred to as "normal ordering" [13, 46], is intimately related to the photoelectric measurement of the photons [13, 47, 48], characterized by the destruction operator $\mathbf{E}^+$. More generally, in normal ordering of many operators, all creation operators are placed on the left and all destruction operators on the right. It assures that no zero-point fluctuations of the EM field come into play, i.e. that the detector will not respond when the field is in the vacuum state $|n\rangle = |0\rangle$.

For our case, we need to consider two alternative paths from birth points A or B to each of the points 1 and 2, as shown in blue and green in Fig. 4.16b. In the coordinates $\mathbf{x}_1$ and $\mathbf{x}_2$ of (4.120) this expression thus becomes

$$
\begin{aligned}
G^{(1)}(\mathbf{x}_1,\mathbf{x}_2) &= \langle \Psi \,|\mathbf{E}^-(\mathbf{x}_1)\,\mathbf{E}^+(\mathbf{x}_2)|\,\Psi\rangle \\
&= \langle |E_0|^2\rangle \Big\langle \Psi \Big| \underbrace{\frac{1}{2}\Big(\mathbf{a}^\dagger_{\mathbf{k}_A}\,\mathrm{e}^{-\mathrm{i}\mathbf{k}_A\cdot\mathbf{x}_1} + \mathbf{a}^\dagger_{\mathbf{k}_B}\,\mathrm{e}^{-\mathrm{i}\mathbf{k}_B\cdot\mathbf{x}_1}\Big)}_{\text{1-photon creation amplitude}} \Big| 0 \Big\rangle \\
&\times \Big\langle 0 \Big| \underbrace{\frac{1}{2}\Big(\mathbf{a}_{\mathbf{k}_A}\,\mathrm{e}^{\mathrm{i}\mathbf{k}_A\cdot\mathbf{x}_2} + \mathbf{a}_{\mathbf{k}_B}\,\mathrm{e}^{\mathrm{i}\mathbf{k}_B\cdot\mathbf{x}_2}\Big)}_{\text{1-photon destruction amplitude}} \Big| \Psi \Big\rangle.
\end{aligned}
\tag{4.121}
$$

The quantities identified by underbraces in (4.121) are *single photon* probability amplitudes in the creation and destruction operator formulation, explicitly given by,

$$
\begin{aligned}
\Phi^*(\mathbf{x}_1) &= \frac{1}{2}\Big(\mathbf{a}^\dagger_{\mathbf{k}_A}\,\mathrm{e}^{-\mathrm{i}\mathbf{k}_A\cdot\mathbf{x}_1} + \mathbf{a}^\dagger_{\mathbf{k}_B}\,\mathrm{e}^{-\mathrm{i}\mathbf{k}_B\cdot\mathbf{x}_1}\Big) \\
\Phi(\mathbf{x}_2) &= \frac{1}{2}\Big(\mathbf{a}_{\mathbf{k}_A}\,\mathrm{e}^{\mathrm{i}\mathbf{k}_A\cdot\mathbf{x}_2} + \mathbf{a}_{\mathbf{k}_B}\,\mathrm{e}^{\mathrm{i}\mathbf{k}_B\cdot\mathbf{x}_2}\Big)
\end{aligned}
\tag{4.122}
$$

In first order, the birth of two photons at the same point at the same time is forbidden. Such a scenario corresponds to a second order quantum treatment discussed in Sect. 5.5 of the next chapter. For our discrete two-source-point problem, single photon probability amplitudes for alternative births at points A and B are first summed and only then is the detection probability calculated as the absolute value squared of the conjugate amplitudes according to the rules of quantum mechanics. This yields

$$
\boxed{G^{(1)}(\mathbf{x}_1,\mathbf{x}_2) = \big\langle |E_0|^2\big\rangle \big\langle \Phi^*(\mathbf{x}_1)\Phi(\mathbf{x}_2)\big\rangle = \big\langle |E_0|^2\big\rangle \big\langle \mathbf{P}^{(1)}(\mathbf{x}_1,\mathbf{x}_2)\big\rangle}, \tag{4.123}
$$

where we have defined the unitless first order detection probability operator $\mathbf{P}^{(1)}(\mathbf{x}_1,\mathbf{x}_2)$ whose expectation value is the quantum mechanical detection probability. We write the operator as

$$
\boxed{\mathbf{P}^{(1)}(\mathbf{x}_1,\mathbf{x}_2) = \frac{1}{4}\Big\{\mathbf{X}+\mathbf{Y}\Big\}}. \tag{4.124}
$$

Here $\mathbf{X}$ contains only pairs of creation and destruction operators with the same wavevector mode, while $\mathbf{Y}$ contains pairs of creation and destruction operators in different modes according to

$$\mathbf{X} = \mathbf{a}^{\dagger}_{\mathbf{k}_A}\mathbf{a}_{\mathbf{k}_A}e^{i\mathbf{k}_A\cdot(\mathbf{x}_2-\mathbf{x}_1)} + \mathbf{a}^{\dagger}_{\mathbf{k}_B}\mathbf{a}_{\mathbf{k}_B}e^{i\mathbf{k}_B\cdot(\mathbf{x}_2-\mathbf{x}_1)} \tag{4.125}$$

$$\mathbf{Y} = \mathbf{a}^{\dagger}_{\mathbf{k}_A}\mathbf{a}_{\mathbf{k}_B}e^{i(\mathbf{k}_B\cdot\mathbf{x}_2-\mathbf{k}_A\cdot\mathbf{x}_1)} + \mathbf{a}^{\dagger}_{\mathbf{k}_B}\mathbf{a}_{\mathbf{k}_A}e^{i(\mathbf{k}_A\cdot\mathbf{x}_2-\mathbf{k}_B\cdot\mathbf{x}_1)}. \tag{4.126}$$

$\mathbf{P}^{(1)}$ may also be expressed in the coordinates $\boldsymbol{\rho}_i$ of the detection points defined in Fig. 4.16. For large source-detector separation z_0 and wavevectors of equal magnitude $k = |\mathbf{k}_A| = |\mathbf{k}_B| = 2\pi/\lambda$, the conversion of the coordinates follows from (4.66) and is given by the simple relation

$$\mathbf{k}_I \cdot \mathbf{x}_i = k\left(z_0 - \frac{\mathbf{r}_I \cdot \boldsymbol{\rho}_i}{z_0}\right), \tag{4.127}$$

where I=A,B and $i=1,2$ [14]. Our goal is to calculate the detection probability $\langle \mathbf{P}^{(1)}(\boldsymbol{\rho}_1, \boldsymbol{\rho}_2)\rangle$ in the "detection" plane through which a photon must pass to get to the detector. The two-point detection probability operator in (4.124) becomes

$$\mathbf{P}^{(1)}(\boldsymbol{\rho}_1, \boldsymbol{\rho}_2) = \frac{1}{4}\Big\{\mathbf{a}^{\dagger}_{\mathbf{k}_A}\mathbf{a}_{\mathbf{k}_A}e^{-i\frac{k}{z_0}\mathbf{r}_A\cdot(\boldsymbol{\rho}_2-\boldsymbol{\rho}_1)} + \mathbf{a}^{\dagger}_{\mathbf{k}_B}\mathbf{a}_{\mathbf{k}_B}e^{-i\frac{k}{z_0}\mathbf{r}_B\cdot(\boldsymbol{\rho}_2-\boldsymbol{\rho}_1)}$$
$$+\mathbf{a}^{\dagger}_{\mathbf{k}_A}\mathbf{a}_{\mathbf{k}_B}e^{i\frac{k}{z_0}(\mathbf{r}_A\cdot\boldsymbol{\rho}_1-\mathbf{r}_B\cdot\boldsymbol{\rho}_2)} + \mathbf{a}^{\dagger}_{\mathbf{k}_B}\mathbf{a}_{\mathbf{k}_A}e^{i\frac{k}{z_0}(\mathbf{r}_B\cdot\boldsymbol{\rho}_1-\mathbf{r}_A\cdot\boldsymbol{\rho}_2)}\Big\}. \tag{4.128}$$

In our quantum operator formulation, diffraction patterns are calculated as the expectation value of the detection probability operator as $\langle\Phi|\mathbf{P}^{(1)}(\boldsymbol{\rho}_1, \boldsymbol{\rho}_2)|\Phi\rangle$, where Φ represents different one-photon states of light that can be created in the source. We can state as follows.

> In the wave or probability amplitude formulation, the diffraction pattern in the detection plane depends on birth phases within the source area. In the quantum operator formulation, the diffraction pattern is determined by the expectation value of a probability operator, evaluated with different quantum states of light that can be created in the source of interest. This formalism naturally covers all cases of "partial coherence" in statistical optics, which are specifically represented by different quantum states of light. Different diffraction patterns are therefore direct signatures of different quantum states of light.

4.7.1.1 One-Photon Patterns of Coherent and Chaotic Two-Point Sources

As an illustrative example, we calculate the patterns emitted by a two-point source, where Φ is taken to be a coherent or chaotic single photon quantum state of light.

In practice, the two cases are encountered if two holes in a screen are illuminated from behind by a laterally incoherent but temporally quasi-coherent (quasi-monochromatic) source of degeneracy parameter less than 1 that is located either at a large distance (coherent illumination) or short distance (chaotic).

For convenience we assume $\mathbf{r}_A = -\mathbf{r}_B \parallel \boldsymbol{\rho}_1 \parallel \boldsymbol{\rho}_2$ in Fig. 4.16 and denote the two source point separation as $\ell = |\mathbf{r}_A - \mathbf{r}_B| = 2r$. We now compare the single point correlation case, defined by $\boldsymbol{\rho}_1 = \boldsymbol{\rho}_2$, where the detector D is moved into the detection plane or one of the mirrors is eliminated, to the two-point correlation case, defined by $\boldsymbol{\rho}_1 = -\boldsymbol{\rho}_2$.

The expectation value of the detection probability operator (4.128) may be calculated for different *two-mode one-photon quantum states* of light that are discussed in more detail in Chap. 16. Here we restrict ourselves to the simplest two-mode one-photon states given by

$$\boxed{|\phi_{N=1}\rangle_{\mathbf{k}_A,\mathbf{k}_B} = \frac{1}{\sqrt{2}}\Big[e^{i\varphi}|0\rangle_{\mathbf{k}_A}|1\rangle_{\mathbf{k}_B} + |1\rangle_{\mathbf{k}_A}|0\rangle_{\mathbf{k}_B}\Big]} \tag{4.129}$$

with the normalization,

$$\langle\phi_{N=1}|\phi_{N=1}\rangle_{\mathbf{k}_A,\mathbf{k}_B} = 1. \tag{4.130}$$

Hence the states contain a *single* photon, $N = 1$, in *two* modes $\mathbf{k}_A$ and $\mathbf{k}_B$, or half a photon per mode. In our general discussion of different quantum states of light in Chap. 16 we will see that the states (4.129) are substates of the more general *collective* quantum states of light, which contain an *average number* $\langle n\rangle = 1$ photons, rather than the *definitive* number $N = 1$ (see Sect. 5.6). The coherent and chaotic states of the form (4.129) are given by

$$|\phi_{\mathrm{coh1}}\rangle_{\mathbf{k}_A,\mathbf{k}_B} = \frac{1}{\sqrt{2}}\Big[|0\rangle_{\mathbf{k}_A}|1\rangle_{\mathbf{k}_B} + |1\rangle_{\mathbf{k}_A}|0\rangle_{\mathbf{k}_B}\Big] \tag{4.131}$$

$$|\phi_{\mathrm{chal}}\rangle_{\mathbf{k}_A,\mathbf{k}_B} = \frac{1}{\sqrt{2}}\Big[e^{i\varphi}|0\rangle_{\mathbf{k}_A}|1\rangle_{\mathbf{k}_B} + |1\rangle_{\mathbf{k}_A}|0\rangle_{\mathbf{k}_B}\Big]. \tag{4.132}$$

The matrix elements of the probability operator (4.128) with the states (4.131) and (4.132) are given in Table 4.3. They need to be averaged over many observations, which corresponds to a temporal average over many birth processes. Such an average is expressed by $\langle e^{-i\varphi}\rangle$. For *coherent* emission of photons we have $\langle e^{-i\varphi}\rangle = 1$ for all source points. For *chaotic* emission, this only happens when the photons are born at the same point, i.e. $\mathbf{r}_A = \mathbf{r}_B$, while there is a random phase relationship between different points leading to $\langle e^{-i\varphi}\rangle = 0$.

For the one-photon *coherent* state all matrix elements have the same value 1/2 and after some algebra we obtain

$$\left\langle \mathbf{P}^{(1)}(\boldsymbol{\rho}, \boldsymbol{\rho})\right\rangle_{\text{coh1}} = \left\langle \mathbf{P}^{(1)}(\boldsymbol{\rho}, -\boldsymbol{\rho})\right\rangle_{\text{coh1}} = \frac{1}{2}\cos^2\left[\frac{k\,\ell}{2z_0}\,\rho\right]. \tag{4.133}$$

The subscript denotes the "coherent one-photon state" (4.131). The pattern is seen to be the same for the two detection scenarios.

For the chaotic case, the two off-diagonal matrix elements in Table 4.3 average to zero, and now the two detection cases give the different answers,

$$\left\langle \mathbf{P}^{(1)}(\boldsymbol{\rho}, \boldsymbol{\rho})\right\rangle_{\text{cha1}} = \frac{1}{4} \tag{4.134}$$

and

$$\left\langle \mathbf{P}^{(1)}(\boldsymbol{\rho}, -\boldsymbol{\rho})\right\rangle_{\text{cha1}} = \frac{1}{4}\cos\left[\frac{k\,\ell}{z_0}\,\rho\right]. \tag{4.135}$$

The two cases are illustrated in terms of the detection geometries in blue and red colors in Fig. 4.17a, and the patterns for the two geometries are compared and plotted in corresponding colors in (b). The coherent pattern is the same for both detection schemes, while the chaotic patterns are different.

The numerical prefactors of 1/2 in (4.133) and 1/4 in (4.134) and (4.135) arise from the use of the specific one-photon quantum states (4.131) and (4.132). Their normalization means that they effectively contain only 1/2 photon per mode. For the coherent case, the number of emitted and detected photons per mode is the same, resulting in the factor of 1/2. For the chaotic case, the number of detected photons is 1/2 of the number emitted, yielding a factor of 1/4.

The somewhat awkward counting of fractional photon numbers does not occur when instead of the specific substates (4.129) with $N = 1$, the proper *collective* states are used which contain a distribution with an *average* number of $\langle n \rangle = 1$. This is discussed in Sect. 5.6 and illustrated in Fig. 5.11 for the coherent one- and two-photon collective states.

Table 4.3 Matrix elements (expectation values) of the four first order coherence operators $\mathbf{O}^{(1)}$, consisting of normally ordered one-photon creation and annihilation operators with the quantum states (4.131) and (4.132)

Operator $\mathbf{O}^{(1)}$	$\langle \phi_{\text{coh1}} \vert \mathbf{O}^{(1)} \vert \phi_{\text{coh1}} \rangle$	$\langle \phi_{\text{cha1}} \vert \mathbf{O}^{(1)} \vert \phi_{\text{cha1}} \rangle$
$\mathbf{a}^\dagger_{\mathbf{k}_A}\mathbf{a}_{\mathbf{k}_A}$	$\frac{1}{2}$	$\frac{1}{2}$
$\mathbf{a}^\dagger_{\mathbf{k}_B}\mathbf{a}_{\mathbf{k}_B}$	$\frac{1}{2}$	$\frac{1}{2}$
$\mathbf{a}^\dagger_{\mathbf{k}_A}\mathbf{a}_{\mathbf{k}_B}$	$\frac{1}{2}$	$\frac{1}{2}\mathrm{e}^{\mathrm{i}\varphi}$
$\mathbf{a}^\dagger_{\mathbf{k}_B}\mathbf{a}_{\mathbf{k}_A}$	$\frac{1}{2}$	$\frac{1}{2}\mathrm{e}^{-\mathrm{i}\varphi}$

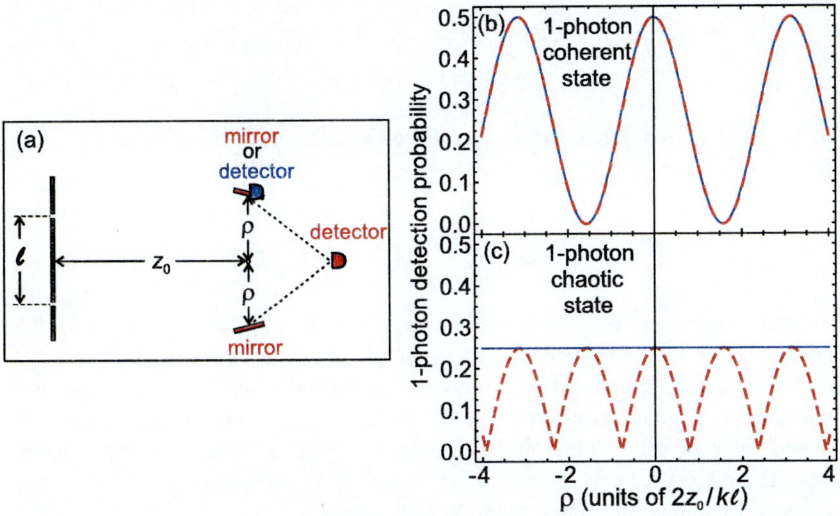

Fig. 4.17 **a** Geometries for recording the *first order* diffraction patterns of a source consisting of two points separated by $|\mathbf{r}_A - \mathbf{r}_B| = \ell$ that is either coherent or chaotic. **b** Patterns if the two source points emit single photons coherently, given by (4.133). The color of the two patterns corresponds to the two detection and scanning scenarios in (**a**). **c** Patterns if the two source points emit chaotically, given by (4.134) (blue) and the absolute value of (4.135) (dashed red). The color of the two patterns corresponds to the two detection and scanning scenarios in (**a**)

4.7.2 *Finite Area Sources*

For a *finite-sized source*, we need to integrate the probability amplitudes (4.122) over all points in the source area before we calculate the detection probability operator. Using the approximation for a large source-detector separation, (4.127) we obtain

$$\Phi^*(\boldsymbol{\rho}_1) = \frac{e^{-ikz_0}}{2A_{\text{eff}}} \left(\int_{\text{plane}} \mathbf{a}^\dagger_{\mathbf{k}_A} e^{i\frac{k}{z_0}\mathbf{r}_A\cdot\boldsymbol{\rho}_1}\, d\mathbf{r}_A + \int_{\text{plane}} \mathbf{a}^\dagger_{\mathbf{k}_B} e^{i\frac{k}{z_0}\mathbf{r}_B\cdot\boldsymbol{\rho}_1}\, d\mathbf{r}_B \right) \tag{4.136}$$

and

$$\Phi(\boldsymbol{\rho}_2) = \frac{e^{ikz_0}}{2A_{\text{eff}}} \left(\int_{\text{plane}} \mathbf{a}_{\mathbf{k}_A} e^{-i\frac{k}{z_0}\mathbf{r}_A\cdot\boldsymbol{\rho}_2}\, d\mathbf{r}_A + \int_{\text{plane}} \mathbf{a}_{\mathbf{k}_B} e^{-i\frac{k}{z_0}\mathbf{r}_B\cdot\boldsymbol{\rho}_2}\, d\mathbf{r}_B \right). \tag{4.137}$$

The unitless one-photon probability operator then consists of four terms of the general form,

$$\mathbf{P}^{(1)}(\boldsymbol{\rho}_1, \boldsymbol{\rho}_2) = \frac{1}{A_{\text{eff}}^2} \iint\limits_{\text{plane}} \mathbf{a}_{\mathbf{k}_X}^{\dagger} \mathbf{a}_{\mathbf{k}_Y} \mathrm{e}^{\mathrm{i}\frac{k}{z_0}(\mathbf{r}_X \cdot \boldsymbol{\rho}_1 - \mathbf{r}_Y \cdot \boldsymbol{\rho}_2)} \mathrm{d}\mathbf{r}_X \, \mathrm{d}\mathbf{r}_Y \tag{4.138}$$

and the total one-photon *detection probability* is calculated according to

$$\boxed{\left\langle \mathbf{P}^{(1)}(\boldsymbol{\rho}_1, \boldsymbol{\rho}_2) \right\rangle = \frac{1}{4\, A_{\text{eff}}^2} \sum_{X,Y} \iint\limits_{\text{plane}} \left\langle \phi_1 \left| \mathbf{a}_{\mathbf{k}_X}^{\dagger} \mathbf{a}_{\mathbf{k}_Y} \right| \phi_1 \right\rangle \mathrm{e}^{\mathrm{i}\frac{k}{z_0}(\mathbf{r}_X \cdot \boldsymbol{\rho}_1 - \mathbf{r}_Y \cdot \boldsymbol{\rho}_2)} \mathrm{d}\mathbf{r}_X \, \mathrm{d}\mathbf{r}_Y}. \tag{4.139}$$

The sum consists of four permutations with X = A or B and Y = A or B. The expectation value is calculated for a given one-photon quantum state of light $|\phi_1\rangle_{\mathbf{k}_A,\mathbf{k}_B}$ of the form (4.129) produced by the source (see Table 4.3) and is then averaged over many independent observations as $\langle \mathrm{e}^{-\mathrm{i}\varphi} \rangle$. The latter accounts for the fact that a single observation may still exhibit interference due to a well-defined difference in phases φ, but this difference may average to zero over many observations [49]. Our result (4.139) is the quantum version of the VCZ theorem (4.94).

4.7.2.1 Application to Circular Flat-Top Source

We again consider the case of a circular flat-top source with $A_{\text{eff}} = \pi R^2$. For the *coherent* case, the matrix elements of the first order operators in Table 4.3, corresponding to the four combinations of points X and Y in (4.139), have the same value of 1/2 and can be pulled out of the integral. The four integrals factor in the integration variables r_X and r_Y, and each integral has the Fourier form

$$\frac{1}{\pi R^2} \int_0^R \int_0^{2\pi} \mathrm{e}^{\pm \mathrm{i}\frac{k}{z_0} r \rho \cos\gamma} \, r \, \mathrm{d}\gamma \, \mathrm{d}r = \frac{2 J_1(kR\rho/z_0)}{kR\rho/z_0}. \tag{4.140}$$

We obtain for the *coherent* state (4.131)

$$\left\langle \mathbf{P}^{(1)}(\boldsymbol{\rho}_1, \boldsymbol{\rho}_2) \right\rangle_{\text{coh1}} = \frac{1}{2} \frac{2 J_1(kR\rho_1/z_0)}{kR\rho_1/z_0} \frac{2 J_1(kR\rho_2/z_0)}{kR\rho_2/z_0}. \tag{4.141}$$

For the *chaotic* state (4.132), the two-point matrix elements in Table 4.3 with $\langle \mathrm{e}^{-\mathrm{i}\varphi} \rangle = 0$ reveal that the integrals are only non-zero if the source points X and Y are the same. The integration then involves a correlation between points that can be expressed by a δ-function of the form $\delta(\mathbf{r}_X - \mathbf{r}_Y)$ and by normalizing to the number of four different permutations of X and Y we obtain

$$\left\langle \mathbf{P}^{(1)}(\boldsymbol{\rho}_1, \boldsymbol{\rho}_2) \right\rangle_{\text{cha1}} = \frac{1}{4} \frac{2 J_1(kR|\boldsymbol{\rho}_2 - \boldsymbol{\rho}_1|/z_0)}{kR|\boldsymbol{\rho}_2 - \boldsymbol{\rho}_1|/z_0} \tag{4.142}$$

Our results for the coherent case agree with (4.101) and for the chaotic case with (4.107) where $\rho' = |\boldsymbol{\rho}_2 - \boldsymbol{\rho}_1|$, except for the power-conserving propagation prefactors. Note that for the chaotic case, the pattern (4.142) represents only the coherent fraction of the total distribution.

4.8 Comparison of Classical and Quantum Diffraction

In our treatment of the van Cittert–Zernike intensity propagation, we encountered three different formulations of the powerful theorem. They expresses the propagation of the intensity or photons from a partially coherent source to a distant detection plane, yielding the diffraction pattern of the source. The first formulation is based on conventional classical Huygens–Fresnel wave fields, the second on Feynman's photon probability amplitudes, and the third, on Dirac's photon creation and destruction operators for different one-photon quantum states of light. In the following we briefly summarize the methods to see why they yield the same result.

4.8.1 *Link of Waves and One-Photon Probability Amplitudes*

The direct link between classical wave fields and quantum mechanical one-photon probability amplitudes is seen from (4.59), describing propagation from a point P with coordinate $\mathbf{r}$ in the source plane to a point X with coordinate $\boldsymbol{\rho}$ in the detection plane

$$E(\boldsymbol{\rho}) = \underbrace{E(\mathbf{r})\,\frac{f}{r_{\mathrm{PX}}}}_{\text{field amplitude}} \underbrace{\underbrace{\mathrm{e}^{\mathrm{i}\alpha_{\mathrm{P}}}}_{\text{birth phase}} \underbrace{\mathrm{e}^{\mathrm{i}k r_{\mathrm{PX}}}}_{\text{propagation phase}}}_{\phi_{\mathrm{PX}}} \,. \tag{4.143}$$

The factor f of dimension [length] assures preservation of the field amplitude (SI units [V/m]), compensating for the $1/r_{\mathrm{PX}}$ dependence of the emitted spherical wave. The quantity ϕ_{PX} is the dimensionless photon probability amplitude. This formulation naturally reveals the link between the classical wave field and probability amplitude formulations.

In both cases, the fields or probability amplitudes from different source points are first summed at a given detection point P. For a distant detection plane, this leads to a Fourier integral over a finite area source. The absolute values squared of the fields or probability amplitudes yield the classical *intensity* distribution $\propto |E(\boldsymbol{\rho})|^2$ or the photon *detection probability* $\propto |\phi_{\mathrm{PX}}|^2$ distribution. The shapes of the so expressed diffraction patterns are the same, revealing the direct correspondence of the classical wave and quantum mechanical probability amplitude formulations [14, 30].

4.8.2 Link of Probability Amplitude and Operator Formalisms

A seemingly different quantum description is based on replacements of the classical fields on Dirac's field operators (see Sect. 3.3.3). The birth phases contained in Feynman's photon probability amplitude as $e^{i\alpha(\mathbf{r})}$, are instead contained in the possible quantum states of light that are created by the phaseless creation operator $\mathbf{a}^\dagger$ in the source, and destroyed by the operator $\mathbf{a}$. We shall refer to the two quantum methods as *Feynman's* probability amplitude and *Glauber's* operator formulation, reflecting Glauber's important role in extending Dirac's 1-photon to a multi-photon description (see Chap. 5).

Let us compare the two formulations for alternative one-photon emission from two source points A and B and detection at a single point P at large distance. This case corresponds to the special VCZ case, with the coordinates defined in Fig. 4.16a, and illustrated as the blue detection scenario in Fig. 4.17. Feynman's total probability amplitude reads

$$\phi_{\rm tot} = \phi_{\rm AX} + \phi_{\rm BX} = e^{i\alpha}\, e^{ikr_{\rm AX}} + e^{i\beta}\, e^{ikr_{\rm BX}}, \tag{4.144}$$

where we have defined $\alpha_{\rm A} = \alpha$ and $\alpha_{\rm B} = \beta$, and the diffraction pattern is given by the 1-photon detection probability,

$$\boxed{\text{Feynman}: \; \langle \phi_{\rm tot}|\phi_{\rm tot}\rangle = 2 + \left\langle e^{i(\alpha-\beta)}\right\rangle e^{ik(r_{\rm AX}-r_{\rm BX})} + \left\langle e^{i(\beta-\alpha)}\right\rangle e^{ik(r_{\rm BX}-r_{\rm AX})}}\,. \tag{4.145}$$

The triangular brackets on the right side reflect an average over many independent observations.

We now compare this result to that obtained with the operator method of Sect. 4.7. We first express the geometry-dependent probability detection operator (4.128) for our single point detection case $\boldsymbol{\rho} = \boldsymbol{\rho}_1 = \boldsymbol{\rho}_2$ as

$$\mathbf{P}^{(1)}(\boldsymbol{\rho}) = \frac{1}{4}\left\{ \mathbf{a}^\dagger_{\mathbf{k}_{\rm A}} \mathbf{a}_{\mathbf{k}_{\rm A}} + \mathbf{a}^\dagger_{\mathbf{k}_{\rm B}} \mathbf{a}_{\mathbf{k}_{\rm B}} + \mathbf{a}^\dagger_{\mathbf{k}_{\rm A}} \mathbf{a}_{\mathbf{k}_{\rm B}} e^{i\frac{k}{z_0}(\mathbf{r}_{\rm A}-\mathbf{r}_{\rm B})\cdot\boldsymbol{\rho}} + \mathbf{a}^\dagger_{\mathbf{k}_{\rm B}} \mathbf{a}_{\mathbf{k}_{\rm A}} e^{i\frac{k}{z_0}(\mathbf{r}_{\rm B}-\mathbf{r}_{\rm A})\cdot\boldsymbol{\rho}} \right\}. \tag{4.146}$$

In order to link to the propagation notation used in (4.145), we rewrite this expression in the paraxial approximation with I=A,B as (see Fig. 5.6),

$$\frac{k}{z_0}\mathbf{r}_{\rm I}\cdot\boldsymbol{\rho} \simeq kz_0 - kr_{\rm IX} \;\Rightarrow\; \frac{k}{z_0}(\mathbf{r}_{\rm A}-\mathbf{r}_{\rm B})\cdot\boldsymbol{\rho} \simeq k(r_{\rm BX}-r_{\rm AX}). \tag{4.147}$$

We obtain

$$\mathbf{P}^{(1)}(\boldsymbol{\rho}) = \frac{1}{4}\left\{ \mathbf{a}^\dagger_{\mathbf{k}_{\rm A}} \mathbf{a}_{\mathbf{k}_{\rm A}} + \mathbf{a}^\dagger_{\mathbf{k}_{\rm B}} \mathbf{a}_{\mathbf{k}_{\rm B}} + \mathbf{a}^\dagger_{\mathbf{k}_{\rm A}} \mathbf{a}_{\mathbf{k}_{\rm B}} e^{ik(r_{\rm BX}-r_{\rm AX})} + \mathbf{a}^\dagger_{\mathbf{k}_{\rm B}} \mathbf{a}_{\mathbf{k}_{\rm A}} e^{ik(r_{\rm AX}-r_{\rm BX})} \right\}. \tag{4.148}$$

We now evaluate the expectation value of this operator with the two-mode one-photon state given by (4.129) or

$$|\phi\rangle = \frac{1}{\sqrt{2}}\Big[e^{i\varphi}\,|0\rangle_{\mathbf{k}_A}|1\rangle_{\mathbf{k}_B} + |1\rangle_{\mathbf{k}_A}|0\rangle_{\mathbf{k}_B}\Big]. \tag{4.149}$$

By utilizing the matrix elements in Table 4.3 and indicating their average over many observations, we have

$$\langle\phi|\mathbf{a}^{\dagger}_{\mathbf{k}_A}\mathbf{a}_{\mathbf{k}_A}|\phi\rangle = \langle\phi|\mathbf{a}^{\dagger}_{\mathbf{k}_B}\mathbf{a}_{\mathbf{k}_B}|\phi\rangle = \frac{1}{2} \tag{4.150}$$

$$\langle\phi|\mathbf{a}^{\dagger}_{\mathbf{k}_A}\mathbf{a}_{\mathbf{k}_B}|\phi\rangle = \frac{1}{2}\left\langle e^{i\varphi}\right\rangle, \quad \langle\phi|\mathbf{a}^{\dagger}_{\mathbf{k}_B}\mathbf{a}_{\mathbf{k}_A}|\phi\rangle = \frac{1}{2}\left\langle e^{-i\varphi}\right\rangle \tag{4.151}$$

The diffraction pattern in the operator formalism is then given by

$$\boxed{\text{Glauber:}\;\; \left\langle \mathbf{P}^{(1)}(\rho)\right\rangle = \frac{1}{8}\left\{2 + \left\langle e^{-i\varphi}\right\rangle e^{ik(r_{AX}-r_{BX})} + \left\langle e^{i\varphi}\right\rangle e^{ik(r_{BX}-r_{AX})}\right\}}. \tag{4.152}$$

For the coherent and chaotic source cases, which differ in the average $\langle e^{i\varphi}\rangle$ being either 1 or 0 (see Sect. 4.7.1.1), our result (4.152) is the same as that shown in blue in Fig. 4.17.

To facilitate their direct comparison, we have arranged the terms in (4.152) and (4.145) in the same order. For $\varphi = \beta - \alpha$, we find that the two detection patterns are the same, except for a normalization prefactor of 1/8. This arbitrary prefactor arises from the different normalizations used.[17] With these considerations, our comparison demonstrates the equivalence of the Feynman and Glauber formalisms, pointed out before in [14].

4.9 X-Ray Measurement of First Order Lateral Coherence

We now illustrate use of the VCZ theorem to determine the coherence function $G^{(1)}(\rho')$. In practice, it is easier to use pinholes than the mirrors shown in Fig. 4.14a, and one may similarly measure the interference intensity on the optical axis as a function of pinhole separation. By use of a position sensitive detector, the entire interference pattern created by the pinholes may be measured at the same time. The pattern is of course Young's double-slit diffraction pattern, which will be discussed in detail for different quantum states of light in Chap. 16.

[17] For example, the total probability amplitude (4.144) is not normalized in contrast to the two-mode one-photon quantum state (4.149).

For a coherent source, the double-slit pattern will exhibit detailed interference fringes that depend on the separation of the pinholes and are contained in an envelope function that is determined by the size of the pinholes (see Fig. 16.4). With decreasing coherence of the double-slit illumination, the fringes in the far-field pattern will gradually disappear until no fringes are left for incoherent illumination. The contrast or *visibility* of the fringe pattern may hence be used as a measure of coherence. Since the double-slit fringe pattern is independent of the incident intensity, which only improves the statistics, it is essentially self-normalizing and therefore its utilization to determine the coherence function of the source is quite convenient from an experimental point of view.

A measurement of the first order degree of lateral coherence for a soft x-ray undulator beam line is shown in Fig. 4.18 [50].

Of interest in the measurement was the lateral coherence of the emitted radiation in the horizontal orbit plane of the storage ring in which the electron beam was also wiggled by the undulator. The coherence was changed by an aperture of width a as shown in Fig. 4.18a. A monochromator was used to make the radiation quasi-temporally coherent by selecting a photon energy of 924 eV ($\lambda = 1.34$ nm) with a FWHM energy bandwidth of 1.5 eV ($\lambda/\Delta\lambda = 600$). Lithographically fabricated double pinholes of 500 nm diameter and different separations d (corresponding to ρ' in Fig. 4.14a) were used to map out the lateral coherence function. Scanning electron microscope images of the pinholes are also shown. The full double-slit fringe patterns were recorded by a CCD detector, as schematically indicated by the blue pattern at the top right of Fig. 4.14a, corresponding to an aperture width of $a = 13\,\mu$m and a pinhole separation $d = 10\,\mu$m.

The determination of $|g^{(1)}(d)|$ from the pinhole diffraction pattern is illustrated in Fig. 4.14b for an aperture setting of $a = 13\,\mu$m and a pinhole separation $d = 23\,\mu$m. The visibility of the fringes is defined by the indicated values of I_{max} and I_{min}, and it determines the degree of first order coherence [7, 9]

$$|g^{(1)}(d)| = \frac{I_{\mathrm{max}} - I_{\mathrm{min}}}{I_{\mathrm{max}} + I_{\mathrm{min}}}. \tag{4.153}$$

This yields $|g^{(1)}| = 0.51$ for the shown pattern. In Fig. 4.14c we show as black circles linked by a black curve the values of $|g^{(1)}|$ for the aperture setting $a = 13\,\mu$m, determined for five double pinholes of different separations d. The blue ($d = 10\,\mu$m) and red ($d = 23\,\mu$m) points correspond to the shown blue and red interference patterns in (**a**) and (**b**). The shape of the black $|g^{(1)}(\rho)|$ curve in Fig. 4.14c is a Gaussian of rms width $\sigma \simeq 20\,\mu$m, as indicated.[18] We note that for a completely coherent source,

[18] Following Mandel [47], one typically defines the 2D coherence area of a cylindrically symmetric source as [9]

$$A_{\mathrm{coh}} = 2\pi \int_0^\infty |g^{(1)}(\rho)|^2 \rho \,\mathrm{d}\rho. \tag{4.154}$$

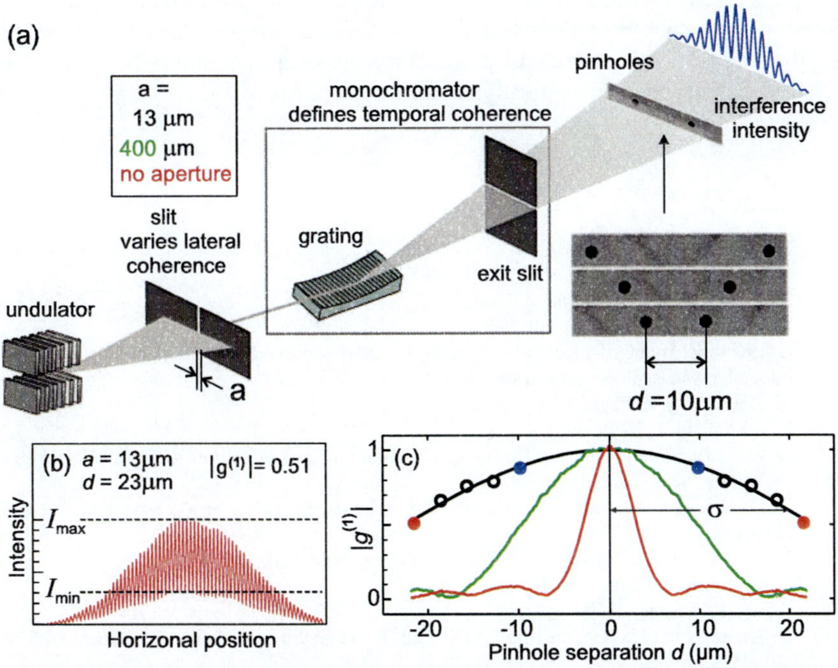

Fig. 4.18 **a** Schematic of the measurement of the horizontal degree of coherence of a soft x-ray undulator beamline as a function of horizontal aperturing by a slit of width a [50]. The degree of coherence $|g^{(1)}(d=\rho')|$ was measured through the change of the interference intensity pattern produced by two pinholes as a function their separation d as shown on the upper right. Also shown are scanning electron microscopy images of three double pinholes fabricated by ion beam milling and a line cut (blue curve) of the interference pattern for $a = 13\,\mu\text{m}$ and $d = 10\,\mu\text{m}$. **b** Line cut of the interference pattern for $a = 13\,\mu\text{m}$ and $d = 23\,\mu\text{m}$ and definition of $I_{\max}$ and $I_{\min}$ which determine $|g^{(1)}(d)|$ according to (4.153). **c** Degree of lateral coherence $|g^{(1)}(d)|$ defined through the visibility of the diffraction pattern for $a = 13\,\mu\text{m}$ and five pinhole separations d, shown as circles linked by a black Gaussian curve. The blue ($d = 10\,\mu\text{m}$) and red ($d = 23\,\mu\text{m}$) filled circles correspond to the shown blue and red interference patterns in **(a)** and **(b)**. The continuous green and red curves in **(c)** for aperture widths a specified in the same color on the top left of **(a)** were obtained by use of a non-redundant array as discussed in [50]

there would still be a slight curvature due to the flat detection plane, as discussed for the green curve in Fig. 4.14b.

The continuous green and red curves in Fig. 4.18c correspond to different settings of the aperture width a, given by the values on the top left in (a). They were obtained by use of a non-redundant array [50] that consisted of lithographically prepared slits of equal heights and different widths along the direction of diffraction. Analysis of the NRA diffraction pattern, after calibration against the simpler diffraction pattern of the two pinholes, allowed the derivation of the entire coherence function as a function of distance from the optical axis. For the case without aperturing, shown as a red curve, $|g^{(1)}(d)|$ assumes the shape of the magenta curve in Fig. 4.14b and the effective coherence width is greatly reduced to $2\sigma \simeq 5\mu$m.

References

1. A. Walther, J. Opt. Soc. Am. **58**, 1256 (1968)
2. E. Wolf, J. Opt. Soc. Am. **68**, 6 (1978)
3. K.J. Kim, Nucl. Instrum. Meth. Phys. Res. A **246**, 71 (1986)
4. K.J. Kim, Z. Huang, R. Lindberg, *Synchrotron Radiation and Free-Electron Lasers* (Cambridge University Press, Cambridge, 2017)
5. W. Heisenberg, Z. Phys. **43**, 172 (1927)
6. E.H. Kennard, Z. Phys. **44**, 326 (1927)
7. E. Wolf, *Introduction to the Theory of Coherence and Polarization of Light* (Cambridge University Press, Cambridge, 2007)
8. N. Hartmann, G. Hartmann, R. Heider, M.S. Wagner, M. Ilchen et al., Nat. Phot. **12**, 215 (2018)
9. J.W. Goodman, *Statistical Optics*, 2nd edn. (Wiley, New York, 2015)
10. P.H. van Cittert, Phys. Rev. B **1**, 201 (1934)
11. F. Zernike, Physica **5**, 785 (1938)
12. N. Wiener, Acta Mathematica **55**, 117 (1930); A. Khintchine, Mathematische Annalen **109**, 604 (1934)
13. L. Mandel, E. Wolf, *Optical Coherence and Quantum Optics* (Cambridge University Press, Cambridge, 1995)
14. J. Stöhr, Adv. Opt. Phot. **11**, 215 (2019)
15. J.J. Sakurai, *Modern Quantum Mechanics*, Revised. (Addison-Wesley, Reading, MA, 1994)
16. G. Breit, E. Wigner, Phys. Rev. **49**, 519 (1936)
17. J. Stöhr: Phys. Rev. Lett. **118**, 024801 (2017)
18. Z. Chen, D.J. Higley, M. Beye, M. Hantschmann, V. Mehta, O. Hellwig et al., Phys. Rev. Lett. **121**, 137403 (2018)
19. A. Einstein, Ann. Physik **17**, 891 (1905)
20. M. Planck, Verh. Deut. Phys. Ges. **2**, 202 (1900)
21. M. Planck, Verh. Deut. Phys. Ges. **2**, 237 (1900)
22. D. Attwood, A. Sakdinawat, *X-Rays and Extreme Ultraviolet Radiation: Principles and Applications*, 2nd edn. (Cambridge University Press, Cambridge, 2017)
23. J.W. Goodman, *Introduction to Fourier Optics*, 4th edn. (W. H. Freeman, New York, 2017)
24. T.M. Pritchett, A.D. Trubatch, Am. J. Phys. **72**, 1026 (2004)
25. R.P. Feynman, Rev. Mod. Phys. **20**, 367 (1948)
26. R.P. Feynman, Phys. Rev. **84**, 108 (1951)
27. R.P. Feynman, A.R. Hibbs, *Quantum Mechanics and Path Integrals* (McGraw-Hill Inc., New York, 1965)
28. R.P. Feynman, *QED: The Strange Theory of Light and Matter* (Princeton University Press, Princeton, NJ, 1985)
29. R.P. Feynman, R.B. Leighton, M. Sands, *The Feynman Lectures on Physics, Vol. III* (Addison-Wesley, Reading, 1965)
30. J. Liu, G. Zhang, Phys. Rev. A **82**, 013822 (2010)
31. L. Mandel, E.C.G. Sudarshan, E. Wolf, Proc. Phys. Soc. **84**, 435 (1964)
32. R. Glauber, Optical coherence and photon statistics, in *Quantum Optics and Electronics*, ed. by A.B.C. de Witt, C. Cohen-Tannoudji (Gordon and Breach, New York, 1965)
33. R. Loudon, *The Quantum Theory of Light*, 3rd edn. (Clarendon Press, Oxford, 2000)

34. M.O. Scully, M.S. Zubairy, *Quantum Optics* (Cambridge University Press, Cambridge, 1997)
35. C.C. Gerry, P.L. Knight, *Introductory Quantum Optics* (Cambridge University Press, Cambridge, 2005)
36. R.J. Glauber, Phys. Rev. **130**, 2529 (1963)
37. R.J. Glauber, Phys. Rev. **131**, 2766 (1963)
38. L. Mandel, J. Opt. Soc. Am. **51**, 1342 (1961)
39. E.W. Marchand, E. Wolf, J. Opt. Soc. Am. **64**, 1219 (1974)
40. W.H. Carter, E. Wolf, J. Opt. Soc. Am. **65**, 1067 (1975)
41. W.H. Carter, E. Wolf, J. Opt. Soc. Am. **67**, 785 (1977)
42. M.J. Beran, G.B. Parrent, *Theory of Partial Coherence* (Prentice-Hall, Englewood Cliffs, NJ, 1964)
43. I.A. Vartanyants, A. Singer, New J. Phys. **12**, 035004 (2010)
44. I.H. Deutsch, J.C. Garrison, Phys. Rev. A **43**, 2498 (1991)
45. P.A.M. Dirac, *Quantum Mechanics*, 4th edn. (Oxford University Press, Oxford, 1958)
46. K.E. Cahill, R.J. Glauber, Phys. Rev. **177**, 1857 (1969)
47. L. Mandel, Proc. Phys. Soc. (London) **72**, 1037 (1958)
48. L. Mandel, Proc. Phys. Soc. (London) **74**, 223 (1959)
49. L. Mandel, Rev. Mod. Phys. **71**, S274 (1999)
50. R. Rick, A. Scherz, W.F. Schlotter, D. Zhu, J. Lüning, J. Stöhr, Opt. Lett. **34**, 650 (2009)

Chapter 5
The Complete Description of Light: Higher Order Coherence

5.1 Introduction

As outlined in the Introduction of this book, the theoretical description of light was revolutionized in the 1960s through the development of *quantum optics*, pioneered by R. J. Glauber [1–3], and *statistical optics* largely developed by E. Wolf and L. Mandel [4, 5]. Mandel also made many pioneering experimental contributions to quantum optics. Both treatments abandon the deterministic nature of stable classical waves for the concept of probability amplitudes, following the earlier lead by Max Born[1]. The theoretical development occurred in the wake of the invention of the maser and laser in the late 1950s and the Hanbury Brown–Twiss experiment in 1956 [9–11].

One of the most important aspects of the quantum theory of light is Glauber's treatment of *coherence* through photon correlation functions of increasing order n, with $n = 1, 2 ...,$ which is an extension of the first order, $n = 1$, treatment in Sects. 4.5 and 4.6.[2] While statistical or quantum optics yield the same results within first order, this no longer holds in second and higher orders, where only the quantum description agrees with all observations [12–17]. The quantum treatment furthermore has the fundamental advantage that it is directly rooted in quantum electrodynamics. With this formalism, different states of light may be created through excitations out of "darkness" represented by the quantum mechanical zero-point field.

In stark contrast to the optical regime, the description of shorter wavelength "light" in the form of x-rays has continued to be based on first order coherence theory. In the lowest order, coherence can simply be achieved by insertion of apertures and monochromators into the beam path. X-ray spectroscopy experiments, which

[1] Born's contributions to the statistical interpretation of quantum mechanics are outlined in his 1954 Nobel lecture *The Statistical Interpretation of Quantum Mechanics*, available as a PDF file on the official website of the Nobel Prize. Also see [6–8].

[2] A wonderful account of this development is given in Glauber's Nobel lecture available on the website of the Nobel Prize.

J. Stöhr, *The Nature of X-Rays and Their Interactions with Matter*, Springer Tracts in Modern Physics 288, https://doi.org/10.1007/978-3-031-20744-0_5

measure the frequency dependence, have been treated quantum mechanically using Dirac's field quantization and Kramers-Heisenberg-Dirac perturbation theory [18–20]. Light-matter interactions then occur probabilistically as an average over many single photon processes, one-photon-at-a-time. In contrast, x-ray interference and diffraction are still described by wave concepts pioneered in the early nineteenth century by Young and Fresnel and extended to x-rays a hundred years later by von Laue and the Braggs [21]. In the past there simply has been no need for using higher order quantum descriptions of x-rays.

In this chapter we extend the conventional x-ray concepts of brightness and coherence, presented in the previous chapter, to higher order and discuss the propagation of coherence leading to higher order diffraction patterns. In the first order treatment of Chap. 4 it was possible to simply calculate the "intensity" as the absolute value squared of the electric field in a distant plane, without specifying how the intensity is measured. In this sense, the wave-based first order treatment of light is simply an extension of that of longer wavelength radio-frequency waves. Closer inspection, however, reveals that field oscillations which in the radio-frequency range can be detected by antennas and visualized by an oscilloscope are too fast to be observed directly in the optical and x-ray regions. What we refer to as "light" in this book requires a detection process based on the *photoelectric effect* which is a quantum phenomenon.

The detailed consideration of the photon detection process becomes increasingly important with increasing order in QED, where different photons may be detected at *different points* is space. Since "interference" of two signals is reflected by their product, such multi-point interference may be observed by *coincident detection* of photons, which in practice is accomplished by multiplication of two electronic signals. For this reason we will also discuss photon detection schemes and the response times of detectors relative to the fundamental oscillation time defined by the wavelength and the coherence time, which is a multiple of the single cycle time.

One of the remarkable things emerging from the higher order treatment is that the x-ray interference or diffraction pattern may change when higher order photon-photon correlations are included. The crucial step is from the first to second order coherence where conventional field-field interference, underlying first order coherence, is extended to photon-photon interference, described by second order coherence. At the end of this chapter we also discuss the extension of the coherence concept to orders $n > 2$. This reveals that the conventional diffraction limit may be overcome for sources of higher order coherence.

More generally, we show that quantum optics beautifully describes the historical description of light, classically pictured in terms of "light rays" which exhibit rectilinear propagation, to "spherical light waves" that can propagate simultaneously into different directions and diffract around corners, to "plane light waves" that again propagate on straight trajectories like rays.

5.2 The Existence of Photons and Their Interference

The quantum theory of light is based on the existence of photons which is often said to be proven by the photoelectric and Compton effects. However, both effects can be explained by assuming classical electromagnetic waves with quantum properties associated only with the atomic and electronic structure of matter in the interaction process [22]. In the following we discuss a beautiful conceptually simple quantum optics experiment proving the existence of indivisible single photons and then discuss pioneering experiments demonstrating the interference of the amplitudes of two photons, today referred to as "two-photon interference".

5.2.1 The Grangier–Roger–Aspect Experiment

The experiment that convincingly proved the existence of photons as elementary single units was performed in 1986 by Grangier, Roger, and Aspect (GRA) [23] and is illustrated in Fig. 5.1.

The GRA experiment utilized two photons emitted consecutively in an atomic cascade decay, with the first photon serving as a trigger to mark the arrival of the second photon, emitted about 4.7 ns later. The second photon entered the shown detection scheme where it could choose between two paths through a beam splitter leading to two single photon detectors, gated to the arrival of the first trigger photon. The second photon was found to only register in one of the two detectors with

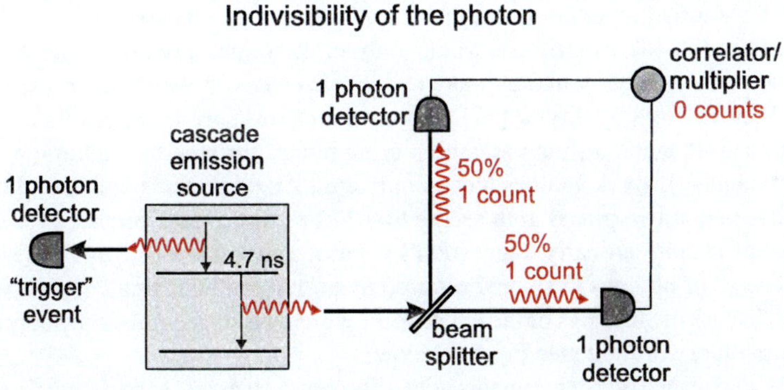

Fig. 5.1 Single photon detection experiment of Grangier, Roger, and Aspect [23], demonstrating the quantum concept of indivisible single photons. Two photons are produced in cascade emission with the first acting as the trigger for detection of the second one, emitted 4.7 ns later. The second photon is incident on a beam splitter, and the coincidence counts are observed at the two output ports of the beam splitter. Only single photon counts are observed in the two detectors with zero cross-correlation coincidence counts

zero cross-correlation coincident counts. This confirmed the quantum mechanical prediction of indivisible single photons.

In a wave picture, it takes the constructive interference of two fields, represented by the product EE^*, to account for photon creation and destruction (detection). For a single photon, only detector counts 0 and 1 are possible, corresponding to the classical picture of destructive or constructive interference between two fields. From a quantum point of view, a detector count of 1 corresponds to constructive interference of the *probability amplitude* of a single photon with itself.

We have chosen to begin with the GRA experiment because of its conceptual simplicity although it was not performed until 1986, about thirty years after the pioneering Hanbury Brown–Twiss (HBT), which caused a paradigm shift in our thinking of the nature of light. We shall discuss it next.

5.2.2 The Hanbury Brown–Twiss Experiment

Together with the development of the maser and laser it was the Hanbury Brown–Twiss (HBT) experiment conducted with thermal light sources in the mid-1950s that played an important role in the development of statistical optics as outlined in a recent book by Wolf [24] and especially quantum optics, as recalled by Glauber in his Nobel lecture [25]. In the first two years after the publication of the original HBT paper in 1956, the results caused quite a controversy as reviewed by Silva and Freire [26].

While the experiment was explained by HBT through classical wave interference, it appeared to contradict the quantum notion of indivisible photons. The problem with the photon interpretation was that it was believed that the low intensity thermal sources used in the experiment would only emit single photons within a narrow window of time. This thinking was clearly reflected in a well-known experiment carried out in 1909 by Taylor [27] who used an extremely weak gas flame source to demonstrate that a diffraction pattern accumulates through the addition of single photon counts. If weak sources indeed only emit single photons at a time, how can it be detected at two places at the same time? The controversy persisted for several years [26] despite an early suggestion by Purcell that the effect might be due to "clustering" of photons [28], now referred to as "photon bunching". Below we will briefly discuss the findings of the experiment which has its natural explanation in the quantum theory of light (see Fig. 5.8 below).

HBT performed two experiments with different sources and experimental arrangements shown in Fig. 5.2a, b. The experiment in (a) used the star Sirius [9, 29] as a source and that in (b) utilized a laboratory mercury arc lamp [11, 30]. In both cases, the thermal sources were quasi-monochromatic and the coincident arrival of two photons was detected at two spatial points as a function of their separation. One of the problems at the time was the rather long response time of the detectors which was much longer than the coherence time of the photons determined by their spectral bandwidth, best pictured as the temporal width of the photon wavepackets. Since the

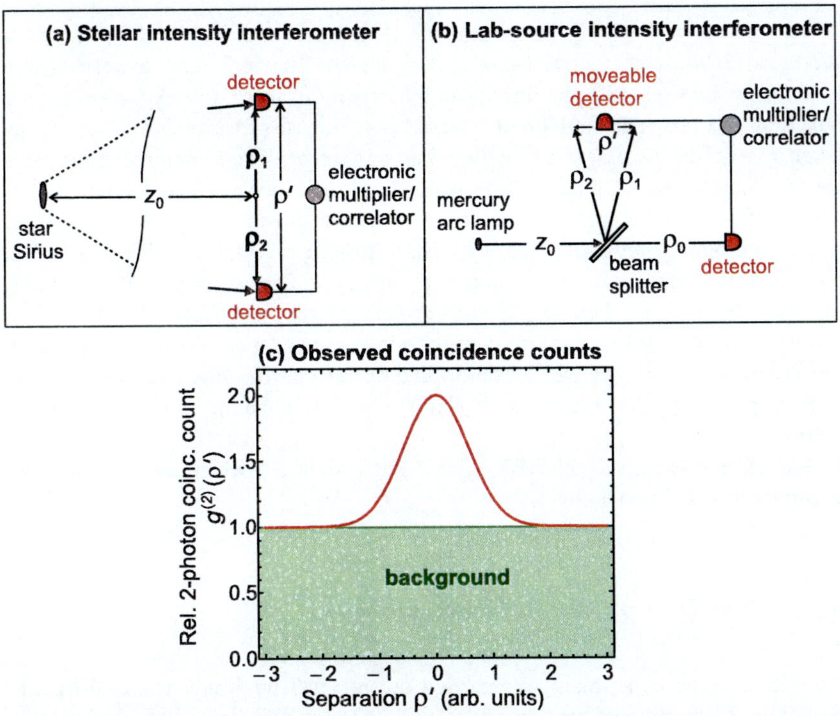

Fig. 5.2 **a** Illustration of the Hanbury Brown–Twiss intensity interferometer, used to measure the correlation of light emitted by the star Sirius ($\lambda \simeq 500$ nm) [9, 29]. The coincident counts in two one-photon detectors were measured as a function of the detector separation $\rho' = |\boldsymbol{\rho}_2 - \boldsymbol{\rho}_1|$ which was varied up to about 10 m. **b** Experimental arrangement for a laboratory-based experiment with a mercury arc lamp ($\lambda = 436$ nm). In this case the spatial or lateral correlation was measured by scanning one of the detectors perpendicular to its incident beam, as shown [11, 30]. The source to detector distances were about 2.5 m and ρ' was changed up to about 10 mm. **c** Schematic dependence of the normalized coincidence counts in both experiments as a function of the path length $\rho' = |\boldsymbol{\rho}_2 - \boldsymbol{\rho}_1|$. The coincident counts represents the degree of second order coherence $g^{(2)}(\rho')$ according to (5.7) below. The shown red curve, representing preferential "photon bunching" at small detector distances, is taken to be a Gaussian but, in general, depends on the macroscopic shape of the source

detector response time significantly exceeded the coherence time, the identification of the true coincidence signal required excellent signal-to-noise data as reviewed by Wolf [24].

For their experiment with light emitted by Sirius, HBT used the schematic arrangement shown in Fig. 5.2a. The measurement allowed the determination of the angular size of the star. The experiment with a mercury arc lamp used the scheme in Fig. 5.2b, where the two single photon detectors, joined by a coincidence circuit, are placed at the exit ports of a beam splitter. For convenience, only one of the detectors was scanned perpendicular to its incident beam to determine the relative *spatial correlation* of the two photons.

The results of the HBT experiment are sketched in Fig. 5.2c, and the two contributions to the coincidence signal are indicated. The red curve represents true correlations between two photons at small detector distances, referred to as "photon bunching". Its shape depends on the macroscopic intensity distribution of the source, taken here to be a circular source with a Gaussian intensity distribution. For a chaotic circular flat-top source, the red curve would have the shape of an Airy function [31] (see Fig. 5.8 below). For chaotic sources, the random emission of photons also leads to a background, shown in green, that arises from accidental coincident arrival of two photons that were born at different points and different times in a completely uncorrelated way, but their path lengths to the detectors just happen to compensate for their different birth times. This contribution does not depend on the detector separation and is constant. We will see below that the red curve in Fig. 5.2c represents the degree of second order coherence $g^{(2)}$ defined in (5.4) in the form (5.7), as discussed below.

We will come back to the HBT experiment in Sect. 5.5 where we explain it as a two-photon interference effect.

5.2.3 The Hong–Ou–Mandel Experiment

Another seminal experiment was carried out in 1987 by Hong, Ou, and Mandel (HOM) [32]. It demonstrated an effect that we today refer to as *two-photon interference*. While the "bunching peak" in the HBT experiment may be interpreted as constructive two-photon interference, or more precisely the *constructive* interference of the amplitudes of two photons, the HOM experiment was more specific.

It used a source that simultaneously emitted pairs of two photons, also referred to as *entangled biphotons* [33], by use of a process called *spontaneous parametric down conversion* (SPDC) [34–36]. In the non-linear SPDC process, widely used for quantum optics experiments [16, 17, 37, 38], a crystal without inversion symmetry is used to simultaneously generate two lower energy photons through fission of an incident photon.[3] Instead of a peak due to constructive two-photon interference, it revealed a "dip" due to *destructive* interference when the probability of coincident arrival of two photons in two detectors was measured.

An important aspect of the HMO experiment was the demonstration of achieving a time resolution in the relative delay of two photons that was comparable to their coherence time of about 100 fs. Through precision movement of a beam splitter, the experiment changed the relative length of the paths of the two photons to two detectors that recorded the coincident arrival probability as a function of the relative delay of the two photons. The detection window was a few nanoseconds, but the incident flux was kept low enough to ensure that only single photon pairs arrived at the two detectors. The result is most easily understood by describing the two photons as wavepackets

[3] In the x-ray regime, diamond single crystals are typically used as the non-linear medium [39–42].

whose widths define their coherence time. The joint detection probability is then simply given by wavepacket overlap.

The situation encountered in the experiment was independently shown in 1987 by Fearn and Loudon to arise theoretically when the transmission of the two photons through a lossless beam splitter was treated quantum mechanically [43]. We shall not discuss the quantum theory of a beam splitter which is presented in references [14, 37, 44, 45] and in our discussion below only utilize the fact that there is no phase shift for the transmitted beam, but the reflected beam is phase shifted by $\pi/2$.

5.2.3.1 A Complete Version of the HOM Experiment

The HOM experiment has been prominently reviewed in the literature [16, 17, 37, 38]. For educational purposes we here present a more complete version of the experiment that links it to the HBT experiment. It was carried out in 2003 by Di Giuseppe et al. [46] and used a different experimental scheme introduced by Shih and Sergienko [47], schematically shown in Fig. 5.3.

Pairs of two orthogonally polarized, collinearly propagating photons of 700 nm wavelength were produced by an SPDC process. A polarization dependent delay stage allowed their wavepackets to be shifted relative to each other with femtosecond resolution. The photons then entered a beam splitter with equal transmission for the two polarizations, allowing them to either propagate together to the same output port or to different ports. Two 45° polarizers behind the beam splitter served to establish polarization indistinguishability when both photons propagated to the same detector. They also added to the $\pi/2$ beam splitter phase shift between reflected and transmitted paths so that the total phase shift between the wavepackets arriving at the two one-photon coincidence detectors was π. The two-photon amplitudes could therefore interfere constructively at a single two-photon detector (blue and light blue) or destructively at two one-photon detectors (gray).

The coherence time of the biphotons was not increased by use of energy filters, but accidental coincidences were avoided by using a low-power pump beam so that the time between biphoton arrivals exceeded the coincidence detection window of $\simeq$1 ns. Since photons need to be conserved in the transmission through a lossless beam splitter, there are three detection probabilities, labeled $P(1, 1)$, $P(2, 0)$, and $P(0, 2)$ in Fig. 5.3. The results of the normalized detection probabilities as a function of delay time τ are shown in Fig. 5.3b. The three arrival probabilities were found to be complementary, as required and occurred in 50% ($P(1, 1)$), 25% ($P(2, 0)$), and 25% ($P(0, 2)$) of all cases. The data follow the theoretical expectation (dashed lines) with deviations resulting from imperfect alignment and asymmetric polarization losses in the system.

When the travel (arrival) times of the two photons are different by more than their coherence time, the photons are *distinguishable* and the probabilities are constant according to $P(1, 1) = P(2, 0) + P(0, 2)$. At reduced delay times, the increasing overlap of the two-photon wavepackets changes the probability $P(1, 1) \rightarrow 0$ and concomitantly $P(2, 0) + P(0, 2) \rightarrow 2$. This may be interpreted as *constructive*

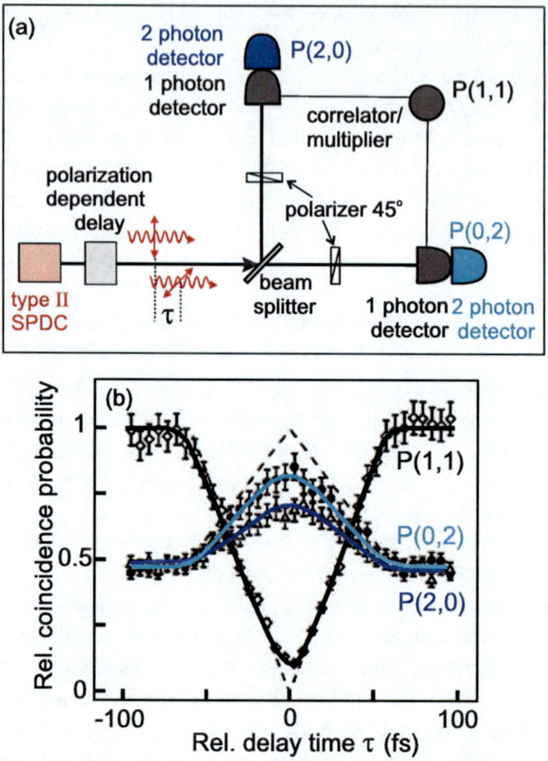

Fig. 5.3 **a** Experimental arrangement used by Di Giuseppe et al. [46]. Two collinear orthogonally polarized photons of 700 nm wavelength are emitted by a SPDC source, and a beam splitter is used to direct them to two-photon number resolving detectors positioned at the two output ports. Two 45° polarizers behind the beam splitter served to establish polarization indistinguishability when both photons propagated to the same detector, leading to the detection probabilities $P(2, 0)$ (blue) or $P(0, 2)$ (light blue). The detection probability $P(1, 1)$ describes the case where the two photons are detected in coincidence by two different detectors (gray). **b** Experimental results for the three detection probabilities as a function of delay time τ between the two photons. The dashed lines are the theoretically expected results

interference of the two-photon probability amplitudes at one port and *destructive* interference at the other. The destructive interference yielding $P(1, 1) \rightarrow 0$ is a consequence of the opposite phase of the two one-photon wavepackets at the two exit ports.

Comparison of the entangled biphoton results in Fig. 5.3b at $\tau = 0$ with the single photon results in Fig. 5.1 reveals a striking similarity. The single photon counts are simply replaced by two-photon counts. In Dirac's formulation, one might say that similar to individual photons, the entangled biphotons created in SPDC form a collective new quanton state that may only interfere with itself. Then *constructive* biphoton self-interference corresponds to two counts and *destructive* biphoton self-interference to 0 counts.

5.3 Light and Detection Timescales

Before the advent of XFELs, it is apparent from Fig. 4.5 that experiments were naturally carried out by detecting one-photon-at-a-time. The present availability of up to 10^{10} photons in femtosecond pulses requires a brief review of photon and detector timescales. Such a discussion is important in light of future experiments that will undoubtedly explore the diversity of x-ray states that can be created from a quantum optics point of view. For example, in the optical region different quantum states of light with different coherence properties have been created and studied, like coherent states, squeezed states or entangled states (see Sect. 16.2). Such studies will necessarily follow at x-ray energies in the future.

In the past, the exploration of quantum optical phenomena has been largely restricted to the infrared and visible range, where light created by lasers may be used to create different quantum states in a secondary process. As shown for the two-photon case above, higher order quantum measurements typically involve "simultaneous" detection of two or more photons, even at different points in space, within a coincidence window defined by the response time of detectors. For this reason it is important to understand the relative timescales associated with photons such as the fundamental *oscillation time* defined by the wavelength and the *coherence time*, which is a multiple of the oscillation time, in comparison with the time constants associated with detectors. An important part of quantum optics experiments is optical elements such as beam splitters, energy filters, polarizers, and fast detectors. While such optical elements exist in the visible region, their development in the x-ray region remains a challenge. However, it is envisioned that the development of suitable x-ray optics is just a question of time.

In the following we will review the fundamental timescales associated with "light" and detectors.

5.3.1 "Light" Timescales

The most fundamental time is set by the *oscillation time* of the EM field, $\tau_0 = \lambda/c$, during which the photon wave moves forward by one wavelength λ, as illustrated in red in Fig. 5.4. It is determined by the wavelength itself and is the inverse cycle frequency $\tau_0 = 1/\nu$. For x-rays of 1 keV (1.2 nm) we have $\tau_0 \simeq 0.4 \times 10^{-18}$ s or 0.4 as and for optical photons of about 1 eV the time is 0.4×10^{-15} s or 0.4 fs. A self-propagating field that can separate from a charge, referred to in classical electromagnetism as an acceleration field [48], needs to be defined over at least one wavelength or one oscillation time. Also, from a wave point of view one may envision that electronic transitions are induced during a time $\sim \tau_0/(2\pi)$ over which the electric field amplitude is the largest and the vector potential changes most rapidly. In the x-ray regime this time is < 0.1 as and can be taken as "instantaneous".

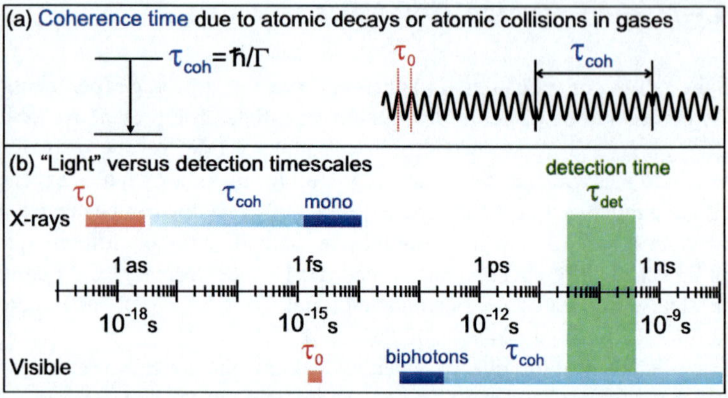

Fig. 5.4 **a** Coherence time τ_{coh}, which is a multiple of the cycle time $\tau_0 = \lambda/c$, associated with electronic decays in atoms, determined by the lifetime of the upper electronic state (left) or in thermal sources by the average length of wavetrains emitted between collisions, indicated by vertical lines. **b** Typical ranges of cycle times (red) and coherence times (blue) encountered for optical and x-ray sources. In green we show the range of detection times, mostly determined by electronic jitter as discussed in the text

The second timescale is set by the (longitudinal) *coherence time* τ_{coh}, illustrated in Fig. 5.4a. When light is created by atomic decays, the coherence time reflects the time over which the decay probability has decreased to $1/e$ its original value. It is equivalent to the *natural lifetime* of the excited state and related to the energy FWHM of the emitted *Lorentzian* line Γ according to $\tau_{coh} = \hbar/\Gamma$ (see Sect. 12.2.7 and Fig. 12.5). In conventional thermal optical sources, light is emitted independently by excited atoms. The emitted wavetrain of a particular frequency may be interrupted by an atomic collision. If the collision is *elastic*, the emission will resume with the same frequency but a different phase until another collision occurs, as schematically shown by vertical lines in Fig. 5.4a. The coherence time is then given by the average time intervals between collisions. The existence of collision intervals adds frequency Fourier components that cause a finite energy bandwidth of *Gaussian* shape. Typical coherence times of conventional optical sources are in the $10^{-12} - 10^{-9}$ s range [12] as indicated in light blue in Fig. 5.4b.

For SR sources, the emitted x-ray pulse of approximately 100 ps length consists of incoherent contributions from individual electrons which radiate over a time of $\sim 10^{-18}$ s for a bending magnet source and $\sim 10^{-16}$ s for an undulator. For SASE XFELs the pulse consists of coherent spikes with a typical coherence time of $\tau_{coh} \sim 1$ fs.

The range of coherence times is shown in blue in Fig. 5.4b. For optical sources the shown range extends from conventional sources to optical lasers with coherence times as long as ~ 1 ms. The dark blue region indicates the range for biphotons created by SPDC. For x-ray sources, the light blue range covers the inherent source coherence times up to the timescales reached by insertion of a monochromator shown in dark blue.

5.3.2 Detector Timescales

The third timescale of interest is associated with the detection process. The most relevant of the various timescales associated with the detection process itself and the electronics depends on the experiment. Detector timescales also need to be considered in conjunction with other detector characteristics such as the detection efficiency, dark count background, linearity of response, and the ability to provide photon energy and photon number resolution.

Photon detectors involve the conversion of the energy of a photon into an electrical signal. The timing characteristics of a detector are determined not only by the photon/electron conversion process but also by the associated electronics. Today, the two most-used detectors in quantum optics are semiconducting avalanche photodiodes (APDs) and superconducting single-photon detectors (SSPDs) [49–52].

APDs are typically made of Si, Ge, or InGaAs and offer low dark count rates, high detection efficiencies, and high count rates. Most APDs can operate at room temperature and due to their low expense and common availability have increasingly replaced photomultiplier tubes and microchannel plates [49].

In SSPDs, the detector element (typically a NbN nanowire) has to be kept well below the superconducting critical temperature, adding cost and experimental complexity at the benefit of the fastest response times (see below). Since the energy gap across which electrons are exited is about two to three orders of magnitude lower in a superconductor than in a semiconductor, for the same incident photon energy the avalanche electron charge is considerable larger in SSPDs than APDs.

The most important detector timescales that determine the total detection time τ_{det} indicated in green in Fig. 5.4b may be summarized as follows.

- The *dead or recovery time*, τ_{dead}, is the interval after a detection event during which the detector is unable to reliably register a new incoming pulse. The main contribution to τ_{dead} may be either the detector element itself or the counting electronics. The dead time limits the maximum count rate of the detector, which is of order 10 MHz for APDs and 1 GHz for SSPDs [49].
- The *jitter time*, τ_{jit}, corresponds to the average interval between the absorption of a photon and the generation of an electrical output pulse. The timing jitter of the detector sets the minimum time interval for gating or time-stamping and typically sets the minimum clock time of a photon counting experiment. Typical values for τ_{jit} are about 400 ps for APDs and only about 30 ps for SSPDs [49] as shown in green in Fig. 5.4b.
- Position sensitive detectors are typically charge coupled devices (CCDs) that accumulate the locally deposited charge in individual pixels. The *frame readout time*, τ_{frame}, is the time required to completely read out all pixels. For x-ray imaging applications, for example, τ_{frame} is a few milliseconds [53–55].

Typically, APDs and SSPDs only give a binary response of 0 and 1 so that a multi-photon pulse triggers the same output signal as a single photon. This limitation is overcome by *photon number resolving detectors* which can distinguish between the arrival of single and more photons within a time window of about a nanosecond [50,

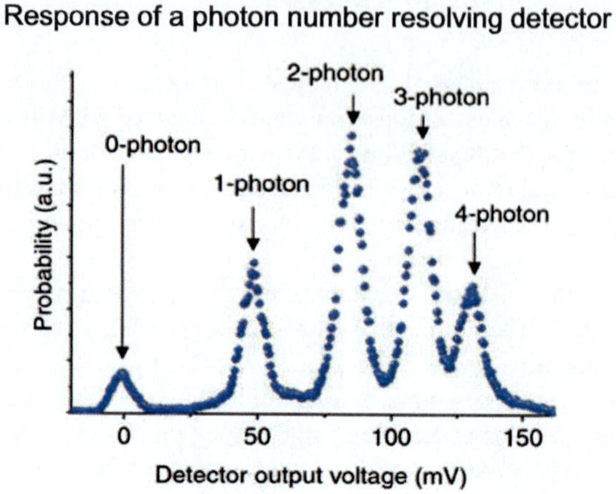

Fig. 5.5 Response of a photon number resolving detector consisting of an electric field modulated silicon avalanche photodiode. Shown are output voltage pulses for the case of a detected photon flux of 4.8 photons per pulse [57]

56, 57]. An example of the photon number resolving capabilities of an electric field modulated APD is illustrated in Fig. 5.5. Photon number resolving detectors have been less used for quantum optics experiments relative to conventional APDs, but we have encountered an application of such a detector in the experiment shown in Fig. 5.3.

According to Fig. 5.4b, the shortest detection time $\tau_{\rm det}$, mostly determined by $\tau_{\rm jit}$, falls into the range of optical coherence times, but it is the orders of magnitude longer than x-ray coherence times. For SR and XFEL sources one may however utilize their pulsed nature to correlate signals observed at different positions within the duration of each pulse, with a temporal average obtained by summing over many pulses as pioneered by Vartanyants and collaborators [58–60].

5.4 The Description of Second Order Coherence

5.4.1 Introduction

While the concept of *first order* coherence, based on the correlation of fields, is expressed as $G^{(1)}(\mathbf{x}_1, t_1, \mathbf{x}_2, t_2) = \langle E^-(\mathbf{x}_1, t_1)E^+(\mathbf{x}_2, t_2)\rangle$ according to (4.82), the concept of *second order* coherence is based on the correlation of four field operators expressed by

$$G^{(2)}(\mathbf{x}_1, t_1, \mathbf{x}_2, t_2) = \left\langle \mathbf{E}^-(\mathbf{x}_1, t_1)\mathbf{E}^-(\mathbf{x}_2, t_2)\mathbf{E}^+(\mathbf{x}_2, t_2)\mathbf{E}^+(\mathbf{x}_1, t_1)\right\rangle . \tag{5.1}$$

Since according to the semi-classical relation the energy of a single photon is proportional to $|E|^2$, the second order case involves the correlation between two photons. While second order coherence can still be formally defined in statistical optics as a four field or two intensity correlation, the quantum treatment is more general. Glauber extended the treatment of coherence in terms of a power series description [2, 3, 61, 62], where *n*th order coherence represents correlation between *n* photons, the hallmark of lasers. This extension is discussed in Sect. 5.7 below.

5.4.2 Second Order Spatial Coherence Functions

In the optical region, experiments are often performed in the time-frequency domain. In the x-ray regime temporal correlation experiments are much more difficult because of the short x-ray coherence times relative to the much longer response times of today's detectors, illustrated in Fig. 5.4.

Measurement of the *lateral* x-rays coherence properties is aided by the expansion of the coherence areas. With today's position sensitive detectors containing pixels of about 10 μm size and use of source-detector distances around 10 m spatial correlation experiments are straightforward [63, 64] (see Fig. 5.10 below). We are therefore mainly interested in the spatial behavior of light, which is described by the "same-time" *second order lateral coherence* function,

$$G^{(2)}(\mathbf{x}_1, \mathbf{x}_2) = \big\langle \mathbf{E}^-(\mathbf{x}_1)\mathbf{E}^-(\mathbf{x}_2)\mathbf{E}^+(\mathbf{x}_2)\mathbf{E}^+(\mathbf{x}_1)\big\rangle . \tag{5.2}$$

Below we shall also rewrite $G^{(2)}(\mathbf{x}_1, \mathbf{x}_2)$ in terms of the expectation value of the dimensionless probability detection operator $\mathbf{P}^{(2)}(\mathbf{x}_1, \mathbf{x}_2)$ as

$$\boxed{G^{(2)}(\mathbf{x}_1, \mathbf{x}_2) = \langle |E_0|^2\rangle^2 \big\langle \mathbf{P}^{(2)}(\mathbf{x}_1, \mathbf{x}_2)\big\rangle ,} \tag{5.3}$$

where $\mathbf{P}^{(2)}$ is defined in analogy to the first order operator $\mathbf{P}^{(1)}$ in (4.123).

The dimensionless *degree of second order spatial coherence* is defined in analogy to the first order expression $g^{(1)}(\mathbf{x}_1, \mathbf{x}_2)$, given by (4.83). However, in contrast to the complex degree of first order coherence, the degree of second order is a *real* quantity defined as

$$\begin{aligned} g^{(2)}(\mathbf{x}_1, \mathbf{x}_2) &= \frac{\big\langle \mathbf{E}^-(\mathbf{x}_1)\mathbf{E}^-(\mathbf{x}_2)\mathbf{E}^+(\mathbf{x}_2)\mathbf{E}^+(\mathbf{x}_1)\big\rangle}{\langle \mathbf{E}^-(\mathbf{x}_1)\mathbf{E}^+(\mathbf{x}_1)\rangle \langle \mathbf{E}^-(\mathbf{x}_2)\mathbf{E}^+(\mathbf{x}_2,)\rangle} . \\ &= \frac{G^{(2)}(\mathbf{x}_1, \mathbf{x}_2)}{G^{(1)}(\mathbf{x}_1, \mathbf{x}_1)\, G^{(1)}(\mathbf{x}_2, \mathbf{x}_2)} = \frac{\big\langle \mathbf{P}^{(2)}(\mathbf{x}_1, \mathbf{x}_2)\big\rangle}{\big\langle \mathbf{P}^{(1)}(\mathbf{x}_1, \mathbf{x}_1)\big\rangle \big\langle \mathbf{P}^{(1)}(\mathbf{x}_2, \mathbf{x}_2)\big\rangle} . \end{aligned} \tag{5.4}$$

The field operators are again written in normal order. The quantum mechanically allowed range $0 \le g^{(2)}(\mathbf{x}_1, \mathbf{x}_2) < \infty$ is larger than the classically allowed range of $1 \le g^{(2)}(\mathbf{x}_1, \mathbf{x}_2) < \infty$ [14]. The quantum optics description can thus account for cases $g^{(2)} < 1$ that do not exist in statistical optics. This arises from the fact that the photon aspect of light is not properly accounted for in statistical optics. The observation of $g^{(2)} < 1$ associated with photon antibunching, for example, has been one of the triumphs of the quantum approach [65, 66].

Complete lateral second order coherence is defined by the *simultaneous* identities

$$g^{(2)}(\mathbf{x}_1, \mathbf{x}_2) = 1 \text{ and } g^{(1)}(\mathbf{x}_1, \mathbf{x}_2) = 1. \tag{5.5}$$

There are cases, for example the Hanbury Brown–Twiss experiment at large detector separation, where $g^{(2)} = 1$ due to accidental correlations (coincidences), with the lack of true second order coherence being reflected by $g^{(1)} < 1$.

5.4.3 Link of First and Second Order Coherence

Before we discuss the propagation of the degree of second order coherence, let us briefly review some well-known facts about the relationship between $g^{(2)}$ and the degree of first order coherence $g^{(1)}$. Interestingly, a general link can only be established for chaotic and second order coherent light. For first order coherent light there is no general link to its second order properties.

5.4.3.1 $g^{(2)}$ for Chaotic Light

For *chaotic* light, the second order coherence properties can be expressed in terms of the first order properties through the so-called Reed theorem [67] or complex Gaussian moment theorem [4, 68]. Under the assumption that the light is laterally homogeneous, the theorem states

$$G^{(2)}(\mathbf{x}_1, \mathbf{x}_2) = G^{(1)}(\mathbf{x}_1, \mathbf{x}_1) G^{(1)}(\mathbf{x}_2, \mathbf{x}_2) + |G^{(1)}(\mathbf{x}_1, \mathbf{x}_2)|^2. \tag{5.6}$$

The degree of second order coherence is expressed by the normalized version of (5.6), often referred to as the Siegert relation [69]

$$g^{(2)}(\mathbf{x}_1, \mathbf{x}_2) = 1 + |g^{(1)}(\mathbf{x}_1, \mathbf{x}_2)|^2. \tag{5.7}$$

The first term on the right side of (5.6) and (5.7) corresponds to accidental coincidences in the two detectors at $\mathbf{x}_1$ and $\mathbf{x}_2$, with the second term reflecting true coincidences.

5.4.3.2 $g^{(2)}$ for First Order Coherent Light

If the light is *first order coherent*, there is no unique link between first and second order coherence. This is readily revealed by comparing the first and second order coherence properties of different single mode quantum states [12]. The number, chaotic, and coherent states are all first order coherent, but only the coherent states are second order coherent. Hence by knowledge of the first order coherence alone, one cannot say anything about the degree of second order coherence!

5.4.3.3 $g^{(2)}$ for Second Order Coherent Light

Since *second order coherent* light is also first order coherent, we have the simple relations

$$G^{(2)}(\mathbf{x}_1, \mathbf{x}_2) = G^{(1)}(\mathbf{x}_1, \mathbf{x}_1) G^{(1)}(\mathbf{x}_2, \mathbf{x}_2) \tag{5.8}$$

and

$$g^{(2)}(\mathbf{x}_1, \mathbf{x}_2) = g^{(1)}(\mathbf{x}_1, \mathbf{x}_2) = 1. \tag{5.9}$$

These relations follow from the *factorization properties* of second order coherent light into first order coherent light and then from the factorization of first order coherent light.

5.5 The Propagation of Second Order Coherence

The treatment of coherence propagation in second order QED proceeds similar to the first order case discussed in Sect. 4.7. We start by considering possible propagation scenarios of two photons from two points in a source plane to two points in a distant detector plane.

5.5.1 *Propagation of Two Photons from Two Source to Two Detection Points*

We follow Fig. 5.6 and consider photon paths from two points A and B in a source plane to two points 1 and 2 in a detector plane at distance z_0. We assume that the photons have the same energy or absolute values of their wavevectors k.

The single photon probability amplitudes linking birth and detection points $\phi_{\mathrm{X}j}$ (X = A, B and $j = 1, 2$) are indicated in the figure. They consist of creation operators for source points A and B and propagation terms of the form $\mathrm{e}^{-\mathrm{i}\mathbf{k}_X \cdot \mathbf{x}_j}$. We consider the case of large distance z_0 where we may approximate the separations between source and detector points by use of the paraxial approximation (4.66) as

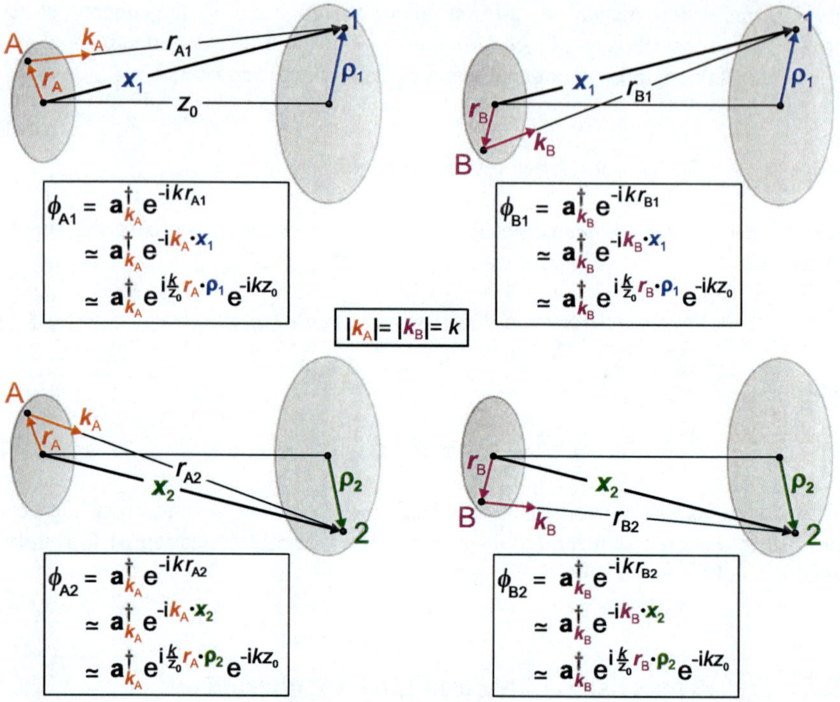

Fig. 5.6 Illustration of coordinates for the calculation of second order coherence propagation. The listed four *one-photon* probability amplitudes $\phi_{\rm Xj}$ link two points X=A, B in a source plane and two points $j = 1, 2$ in a detector plane. For large distance z_0 between the planes, the amplitudes can be expressed in the shown alternative forms. The amplitudes contain photon creation operators $a^\dagger_{\mathbf{k}_{\rm X}}$ at point X. In all cases, the light is assumed to have the same wavelength $\lambda = 2\pi/k$, where $k = |\mathbf{k}_{\rm X}|$. The one-photon amplitudes are used to construct two-photon amplitudes as discussed in the text

$$r_{\rm Xj} = z_0 \left(1 + \frac{r_{\rm X}^2}{2z_0^2} + \frac{\rho_j^2}{2z_0^2} - \frac{\mathbf{r}_{\rm X} \cdot \boldsymbol{\rho}_j}{z_0^2} \right) \simeq z_0 - \frac{\mathbf{r}_{\rm X} \cdot \boldsymbol{\rho}_j}{z_0}. \tag{5.10}$$

On the far right we have furthermore made the Fraunhofer approximation $|\mathbf{r}_{\rm A} - \mathbf{r}_{\rm B}|, |\boldsymbol{\rho}_1 - \boldsymbol{\rho}_2| \ll z_0$, which holds at large z_0. This leads to the boxed expressions for the single photon probability amplitudes in Fig. 5.6 which will allow us to seamlessly link expressions in different coordinates.

We ignore polarization and following Fig. 5.6 express the second order coherence function (5.2) by the same time wavevector-dependent creation and destruction operators $\mathbf{E}^+(\mathbf{x}_j, t)$ and $\mathbf{E}^-(\mathbf{x}_j, t)$ at two detector coordinates $\mathbf{x}_1$ (point 1) and $\mathbf{x}_2$ (point 2) in terms of contributions from two source points A and B as

$$
\begin{aligned}
G^{(2)}(\mathbf{x}_1,\mathbf{x}_2) &= \langle \Psi \,|\mathbf{E}^-(\mathbf{x}_1)\,\mathbf{E}^-(\mathbf{x}_2)\,\mathbf{E}^+(\mathbf{x}_2)\,\mathbf{E}^+(\mathbf{x}_1)|\,\Psi\rangle \\
&= \langle |E_0|^2\rangle^2 \left\langle \Psi \left| \underbrace{\frac{1}{2}\left(\mathbf{a}^\dagger_{\mathbf{k}_A}\mathrm{e}^{-i\mathbf{k}_A\cdot\mathbf{x}_1} + \mathbf{a}^\dagger_{\mathbf{k}_B}\mathrm{e}^{-i\mathbf{k}_B\cdot\mathbf{x}_1}\right)\left(\mathbf{a}^\dagger_{\mathbf{k}_A}\mathrm{e}^{-i\mathbf{k}_A\cdot\mathbf{x}_2} + \mathbf{a}^\dagger_{\mathbf{k}_B}\mathrm{e}^{-i\mathbf{k}_B\cdot\mathbf{x}_2}\right)}_{\text{2 photon creation amplitude } \Phi^*(\mathbf{x}_1,\mathbf{x}_2)} \right| 0 \right\rangle \\
&\quad \times \left\langle 0 \left| \underbrace{\frac{1}{2}\left(\mathbf{a}_{\mathbf{k}_A}\mathrm{e}^{i\mathbf{k}_A\cdot\mathbf{x}_2} + \mathbf{a}_{\mathbf{k}_B}\mathrm{e}^{i\mathbf{k}_B\cdot\mathbf{x}_2}\right)\left(\mathbf{a}_{\mathbf{k}_A}\mathrm{e}^{i\mathbf{k}_A\cdot\mathbf{x}_1} + \mathbf{a}_{\mathbf{k}_B}\mathrm{e}^{i\mathbf{k}_B\cdot\mathbf{x}_1}\right)}_{\text{2 photon destruction amplitude } \Phi(\mathbf{x}_1,\mathbf{x}_2)} \right| \Psi \right\rangle .
\end{aligned}
\tag{5.11}
$$

Multiplying out, we obtain the two-photon probability amplitude

$$
\begin{aligned}
\Phi(\mathbf{x}_1,\mathbf{x}_2) = \frac{1}{4}\Big\{ &\mathbf{a}_{\mathbf{k}_A}\mathbf{a}_{\mathbf{k}_A}\mathrm{e}^{i\mathbf{k}_A\cdot\mathbf{x}_2}\,\mathrm{e}^{i\mathbf{k}_A\cdot\mathbf{x}_1} + \mathbf{a}_{\mathbf{k}_A}\mathbf{a}_{\mathbf{k}_B}\,\mathrm{e}^{i\mathbf{k}_A\cdot\mathbf{x}_2}\mathrm{e}^{i\mathbf{k}_B\cdot\mathbf{x}_1} \\
&+ \mathbf{a}_{\mathbf{k}_B}\mathbf{a}_{\mathbf{k}_A}\mathrm{e}^{i\mathbf{k}_B\cdot\mathbf{x}_2}\,\mathrm{e}^{i\mathbf{k}_A\cdot\mathbf{x}_1} + \mathbf{a}_{\mathbf{k}_B}\mathbf{a}_{\mathbf{k}_B}\mathrm{e}^{i\mathbf{k}_B\cdot\mathbf{x}_2}\,\mathrm{e}^{i\mathbf{k}_B\cdot\mathbf{x}_1}\Big\}.
\end{aligned}
\tag{5.12}
$$

The second order correlation function $G^{(2)}(\mathbf{x}_1,\mathbf{x}_2)$ given by (5.11) may be rewritten in terms of the unitless *second order* detection probability operator $\mathbf{P}^{(2)}(\mathbf{x}_1,\mathbf{x}_2)$ defined in (5.3) as

$$
G^{(2)}(\mathbf{x}_1,\mathbf{x}_2) = \langle |E_0|^2\rangle^2 \left\langle \Phi^*(\mathbf{x}_1,\mathbf{x}_2)\,\Phi(\mathbf{x}_1,\mathbf{x}_2)\right\rangle = \langle |E_0|^2\rangle^2 \left\langle \mathbf{P}^{(2)}(\mathbf{x}_1,\mathbf{x}_2)\right\rangle . \tag{5.13}
$$

Using the fact that pairs of creation and pairs of destruction operators commute, i.e. $\mathbf{a}^\dagger_{\mathbf{k}_A}\mathbf{a}^\dagger_{\mathbf{k}_B} = \mathbf{a}^\dagger_{\mathbf{k}_B}\mathbf{a}^\dagger_{\mathbf{k}_A}$ and $\mathbf{a}_{\mathbf{k}_A}\mathbf{a}_{\mathbf{k}_B} = \mathbf{a}_{\mathbf{k}_B}\mathbf{a}_{\mathbf{k}_A}$, we obtain the following result for the *second order* detection *probability operator*

$$
\mathbf{P}^{(2)}(\mathbf{x}_1,\mathbf{x}_2) = \frac{1}{16}\Big\{\mathbf{A}+\mathbf{B}+\mathbf{C}+\mathbf{D}\Big\}. \tag{5.14}
$$

It may be compared to the first order one $\mathbf{P}^{(1)}(\mathbf{x}_1,\mathbf{x}_2)$ given by (4.124). The four operators $\mathbf{A}$, $\mathbf{B}$, $\mathbf{C}$, $\mathbf{D}$ are explicitly given by

$$
\begin{aligned}
\mathbf{A} &= \mathbf{a}^\dagger_{\mathbf{k}_A}\mathbf{a}^\dagger_{\mathbf{k}_B}\mathbf{a}_{\mathbf{k}_A}\mathbf{a}_{\mathbf{k}_B} + \mathbf{a}^\dagger_{\mathbf{k}_A}\mathbf{a}^\dagger_{\mathbf{k}_B}\mathbf{a}_{\mathbf{k}_A}\mathbf{a}_{\mathbf{k}_B} \\
&\quad + \mathbf{a}^\dagger_{\mathbf{k}_A}\mathbf{a}^\dagger_{\mathbf{k}_B}\mathbf{a}_{\mathbf{k}_A}\mathbf{a}_{\mathbf{k}_B}\mathrm{e}^{-i(\mathbf{k}_A-\mathbf{k}_B)\cdot(\mathbf{x}_1-\mathbf{x}_2)} \\
&\quad + \mathbf{a}^\dagger_{\mathbf{k}_A}\mathbf{a}^\dagger_{\mathbf{k}_B}\mathbf{a}_{\mathbf{k}_A}\mathbf{a}_{\mathbf{k}_B}\mathrm{e}^{i(\mathbf{k}_A-\mathbf{k}_B)\cdot(\mathbf{x}_1-\mathbf{x}_2)} \\
&= \mathbf{a}^\dagger_{\mathbf{k}_A}\mathbf{a}^\dagger_{\mathbf{k}_B}\mathbf{a}_{\mathbf{k}_A}\mathbf{a}_{\mathbf{k}_B}\,\{2+2\cos[(\mathbf{k}_A-\mathbf{k}_B)\cdot(\mathbf{x}_1-\mathbf{x}_2)]\} \\
&= 4\,\mathbf{a}^\dagger_{\mathbf{k}_A}\mathbf{a}^\dagger_{\mathbf{k}_B}\mathbf{a}_{\mathbf{k}_A}\mathbf{a}_{\mathbf{k}_B}\cos^2\left[\frac{1}{2}(\mathbf{k}_A-\mathbf{k}_B)\cdot(\mathbf{x}_1-\mathbf{x}_2)\right]
\end{aligned}
\tag{5.15}
$$

$$\begin{aligned}\mathbf{B} = {} & \mathbf{a}^\dagger_{\mathbf{k}_A}\mathbf{a}^\dagger_{\mathbf{k}_A}\mathbf{a}_{\mathbf{k}_A}\mathbf{a}_{\mathbf{k}_A} + \mathbf{a}^\dagger_{\mathbf{k}_B}\mathbf{a}^\dagger_{\mathbf{k}_B}\mathbf{a}_{\mathbf{k}_B}\mathbf{a}_{\mathbf{k}_B} \\ & + \mathbf{a}^\dagger_{\mathbf{k}_A}\mathbf{a}^\dagger_{\mathbf{k}_A}\mathbf{a}_{\mathbf{k}_B}\mathbf{a}_{\mathbf{k}_B}\mathrm{e}^{-\mathrm{i}(\mathbf{k}_A-\mathbf{k}_B)\cdot(\mathbf{x}_1+\mathbf{x}_2)} \\ & + \mathbf{a}^\dagger_{\mathbf{k}_B}\mathbf{a}^\dagger_{\mathbf{k}_B}\mathbf{a}_{\mathbf{k}_A}\mathbf{a}_{\mathbf{k}_A}\mathrm{e}^{\mathrm{i}(\mathbf{k}_A-\mathbf{k}_B)\cdot(\mathbf{x}_1+\mathbf{x}_2)}\end{aligned} \tag{5.16}$$

$$\begin{aligned}\mathbf{C} = {} & \left[\mathbf{a}^\dagger_{\mathbf{k}_A}\mathbf{a}^\dagger_{\mathbf{k}_A}\mathbf{a}_{\mathbf{k}_A}\mathbf{a}_{\mathbf{k}_B} + \mathbf{a}^\dagger_{\mathbf{k}_A}\mathbf{a}^\dagger_{\mathbf{k}_B}\mathbf{a}_{\mathbf{k}_B}\mathbf{a}_{\mathbf{k}_B}\right]\mathrm{e}^{-\mathrm{i}(\mathbf{k}_A-\mathbf{k}_B)\cdot\mathbf{x}_1} \\ & + \left[\mathbf{a}^\dagger_{\mathbf{k}_A}\mathbf{a}^\dagger_{\mathbf{k}_B}\mathbf{a}_{\mathbf{k}_A}\mathbf{a}_{\mathbf{k}_A} + \mathbf{a}^\dagger_{\mathbf{k}_B}\mathbf{a}^\dagger_{\mathbf{k}_B}\mathbf{a}_{\mathbf{k}_A}\mathbf{a}_{\mathbf{k}_B}\right]\mathrm{e}^{\mathrm{i}(\mathbf{k}_A-\mathbf{k}_B)\cdot\mathbf{x}_1}\end{aligned} \tag{5.17}$$

$$\begin{aligned}\mathbf{D} = {} & \left[\mathbf{a}^\dagger_{\mathbf{k}_A}\mathbf{a}^\dagger_{\mathbf{k}_A}\mathbf{a}_{\mathbf{k}_A}\mathbf{a}_{\mathbf{k}_B} + a^\dagger_{\mathbf{k}_A}\mathbf{a}^\dagger_{\mathbf{k}_B}\mathbf{a}_{\mathbf{k}_B}\mathbf{a}_{\mathbf{k}_B}\right]\mathrm{e}^{-\mathrm{i}(\mathbf{k}_A-\mathbf{k}_B)\cdot\mathbf{x}_2} \\ & + \left[\mathbf{a}^\dagger_{\mathbf{k}_A}\mathbf{a}^\dagger_{\mathbf{k}_B}\mathbf{a}_{\mathbf{k}_A}\mathbf{a}_{\mathbf{k}_A} + \mathbf{a}^\dagger_{\mathbf{k}_B}\mathbf{a}^\dagger_{\mathbf{k}_B}\mathbf{a}_{\mathbf{k}_A}\mathbf{a}_{\mathbf{k}_B}\right]\mathrm{e}^{\mathrm{i}(\mathbf{k}_A-\mathbf{k}_B)\cdot\mathbf{x}_2}\,.\end{aligned} \tag{5.18}$$

These expressions consist of normally ordered products of two creation and two destruction operators which obey the quantum mechanical commutation relations (3.64). Since the order of two adjacent creation or destruction operators may be switched, we have listed the operators with wavevector $\mathbf{k}_A$ first.

5.5.2 *Fundamental Two-Photon Patterns of Two-Point Sources*

We proceed similar to the one-photon case discussed in Sect. 4.7. We evaluate the expectation value of the operators in (5.14) with different two-photon quantum states. The general two-mode quantum states of light that contain either the specific number $N = 2$ of photons or the average number of $\langle n\rangle = 2$ photons will be discussed in Sect. 16.4. Here we restrict ourselves to the simplest two-mode two-photon states of the general form[4]

$$\boxed{|\phi_{N=2}\rangle_{\mathbf{k}_A,\mathbf{k}_B} = a\,\mathrm{e}^{\mathrm{i}\varphi}|1\rangle_{\mathbf{k}_A}|1\rangle_{\mathbf{k}_B} + b\left(\mathrm{e}^{\mathrm{i}\phi}|0\rangle_{\mathbf{k}_A}|2\rangle_{\mathbf{k}_B} + |2\rangle_{\mathbf{k}_A}|0\rangle_{\mathbf{k}_B}\right),} \tag{5.19}$$

where the coefficients a and b are real and fulfill the normalization $a^2 + 2b^2 = 1$.

As illustrative examples we calculate the detection probability (diffraction patterns) of two-point sources that emit four different types of quantum states of light. We consider the two-mode two-photon quantum states (5.20)–(5.23) which are limiting cases of (5.19).

[4] If we introduce three different phasefactors, we can always eliminate one because a common phasefactor is unimportant.

$$|\phi_{coh2}\rangle_{\mathbf{k_A},\mathbf{k_B}} = \frac{1}{2}\left\{\sqrt{2}|1\rangle_{\mathbf{k_A}}|1\rangle_{\mathbf{k_B}} + |0\rangle_{\mathbf{k_A}}|2\rangle_{\mathbf{k_B}} + |2\rangle_{\mathbf{k_A}}|0\rangle_{\mathbf{k_B}}\right\} \tag{5.20}$$

$$|\phi_{num2}\rangle_{\mathbf{k_A},\mathbf{k_B}} = |1\rangle_{\mathbf{k_A}}|1\rangle_{\mathbf{k_B}} \tag{5.21}$$

$$|\phi_{ent2}\rangle_{\mathbf{k_A},\mathbf{k_B}} = \frac{1}{\sqrt{2}}\left\{|0\rangle_{\mathbf{k_A}}|2\rangle_{\mathbf{k_B}} + |2\rangle_{\mathbf{k_A}}|0\rangle_{\mathbf{k_B}}\right\} \tag{5.22}$$

$$|\phi_{cha2}\rangle_{\mathbf{k_A},\mathbf{k_B}} = \frac{1}{\sqrt{3}}\left\{e^{i\varphi}|1\rangle_{\mathbf{k_A}}|1\rangle_{\mathbf{k_B}} + e^{i\phi}|0\rangle_{\mathbf{k_A}}|2\rangle_{\mathbf{k_B}} + |2\rangle_{\mathbf{k_A}}|0\rangle_{\mathbf{k_B}}\right\} \tag{5.23}$$

The labels of the states describe their fundamental nature, denoted as "coherent", "entangled", "number", and "chaotic" states. The origin of the labeling will become more apparent when we discuss their origin and experimental creation in Sect. 16.4. Table 5.1 lists the matrix elements $\langle\phi_x|\mathbf{O}|\phi_x\rangle = \langle\mathbf{O}\rangle_x$ of the various second order operators (5.15)–(5.18). After the matrix elements were evaluated, the phase factors were averaged over many observations which resulted in the zero matrix elements in the table. If the matrix elements are written by inclusion of the phase factors, they directly correspond to the birth phases in Feynman's formulation of the two-photon diffraction pattern [38, 70]. We can therefore extend the equivalence of Feynman's probability amplitude and Glauber's quantum operator formulation shown for the first order one-photon case Sect. 4.8.2 to the second order two-photon case. In fact, the equivalence holds to all orders, and we can state as follows.

> Feynman's probability amplitude formulation of multi-photon diffraction [38, 70] is equivalent to Glauber's operator formulation discussed here. The birth phases of Feynman's probability amplitudes directly correspond to operator matrix elements for different quantum states of light.

Table 5.1 Matrix elements $\langle\phi_x|\mathbf{O}^{(2)}|\phi_x\rangle = \langle\mathbf{O}^{(2)}\rangle_x$ of second order coherence operators $\mathbf{O}^{(2)}$ with the states (5.20)–(5.23)

Terms§	Operators $\mathbf{O}^{(2)}$	$\langle\mathbf{O}^{(2)}\rangle_{coh2}$	$\langle\mathbf{O}^{(2)}\rangle_{num2}$	$\langle\mathbf{O}^{(2)}\rangle_{ent2}$	$\langle\mathbf{O}^{(2)}\rangle_{cha2}$
A	$\mathbf{a}^\dagger_{\mathbf{k_A}}\mathbf{a}^\dagger_{\mathbf{k_B}}\mathbf{a}_{\mathbf{k_A}}\mathbf{a}_{\mathbf{k_B}}$	$\frac{1}{2}$	1	0	$\frac{1}{3}$
B	$\mathbf{a}^\dagger_{\mathbf{k_A}}\mathbf{a}^\dagger_{\mathbf{k_A}}\mathbf{a}_{\mathbf{k_A}}\mathbf{a}_{\mathbf{k_A}}$	$\frac{1}{2}$	0	1	$\frac{2}{3}$
	$\mathbf{a}^\dagger_{\mathbf{k_B}}\mathbf{a}^\dagger_{\mathbf{k_B}}\mathbf{a}_{\mathbf{k_B}}\mathbf{a}_{\mathbf{k_B}}$	$\frac{1}{2}$	0	1	$\frac{2}{3}$
	$\mathbf{a}^\dagger_{\mathbf{k_A}}\mathbf{a}^\dagger_{\mathbf{k_A}}\mathbf{a}_{\mathbf{k_B}}\mathbf{a}_{\mathbf{k_B}}$	$\frac{1}{2}$	0	1	0
	$\mathbf{a}^\dagger_{\mathbf{k_B}}\mathbf{a}^\dagger_{\mathbf{k_B}}\mathbf{a}_{\mathbf{k_A}}\mathbf{a}_{\mathbf{k_A}}$	$\frac{1}{2}$	0	1	0
C, D	$\mathbf{a}^\dagger_{\mathbf{k_A}}\mathbf{a}^\dagger_{\mathbf{k_A}}\mathbf{a}_{\mathbf{k_A}}\mathbf{a}_{\mathbf{k_B}}$	$\frac{1}{2}$	0	0	0
	$\mathbf{a}^\dagger_{\mathbf{k_A}}\mathbf{a}^\dagger_{\mathbf{k_B}}\mathbf{a}_{\mathbf{k_B}}\mathbf{a}_{\mathbf{k_B}}$	$\frac{1}{2}$	0	0	0
	$\mathbf{a}^\dagger_{\mathbf{k_A}}\mathbf{a}^\dagger_{\mathbf{k_B}}\mathbf{a}_{\mathbf{k_A}}\mathbf{a}_{\mathbf{k_A}}$	$\frac{1}{2}$	0	0	0
	$\mathbf{a}^\dagger_{\mathbf{k_B}}\mathbf{a}^\dagger_{\mathbf{k_B}}\mathbf{a}_{\mathbf{k_A}}\mathbf{a}_{\mathbf{k_B}}$	$\frac{1}{2}$	0	0	0

§ Note that $\mathbf{a}^\dagger_{\mathbf{k_A}}\mathbf{a}^\dagger_{\mathbf{k_B}} = \mathbf{a}^\dagger_{\mathbf{k_B}}\mathbf{a}^\dagger_{\mathbf{k_A}}$ and $\mathbf{a}_{\mathbf{k_A}}\mathbf{a}_{\mathbf{k_B}} = \mathbf{a}_{\mathbf{k_B}}\mathbf{a}_{\mathbf{k_A}}$
The latter are normally ordered products of two creation and two destruction operators and are listed in the same sequence as in (5.15)–(5.18).

5.5.2.1 Two-Photon Coherent State

The *coherent* two-mode two-photon state is produced by illuminating the two secondary source points quasi-monochromatically by a distant source so that the two points fall into the coherence cone of the illuminating source or simply by a laser. The first and second order diffraction patterns of such a source have been studied by Shimizu et al. [71] using either a monochromatized halogen lamp or a Ti:sapphire laser to illuminate the points. The so-produced two-photon quantum state is given by the following form of (5.19)

$$|\phi_{\mathrm{coh2}}\rangle_{\mathbf{k}_A,\mathbf{k}_B} = \frac{1}{2}\left\{\sqrt{2}|1\rangle_{\mathbf{k}_A}|1\rangle_{\mathbf{k}_B} + |0\rangle_{\mathbf{k}_A}|2\rangle_{\mathbf{k}_B} + |2\rangle_{\mathbf{k}_A}|0\rangle_{\mathbf{k}_B}\right\}. \tag{5.24}$$

In this case all phase factors are absent and all matrix elements have the same values of $1/2$. The expectation value of (5.14) is evaluated as[5]

$$\begin{aligned}
&\left\langle \mathbf{P}^{(2)}(\mathbf{x}_1,\mathbf{x}_2)\right\rangle_{\mathrm{coh2}} \\
&= \frac{1}{8}\Bigg\{ \underbrace{\cos^2\left[\frac{1}{2}(\mathbf{k}_A-\mathbf{k}_B)\cdot(\mathbf{x}_1-\mathbf{x}_2)\right]}_{\langle\mathbf{A}\rangle} + \underbrace{\cos^2\left[\frac{1}{2}(\mathbf{k}_A-\mathbf{k}_B)\cdot(\mathbf{x}_1+\mathbf{x}_2)\right]}_{\langle\mathbf{B}\rangle} \\
&\quad \underbrace{+2\cos\left[\frac{1}{2}(\mathbf{k}_A-\mathbf{k}_B)\cdot(\mathbf{x}_1-\mathbf{x}_2)\right]\cos\left[\frac{1}{2}(\mathbf{k}_A-\mathbf{k}_B)\cdot(\mathbf{x}_1+\mathbf{x}_2)\right]}_{\langle\mathbf{C+D}\rangle}\Bigg\}.
\end{aligned} \tag{5.27}$$

We see that the term $\langle\mathbf{C}+\mathbf{D}\rangle$ is just the interference terms of the amplitudes $\sqrt{\langle\mathbf{A}\rangle}$ and $\sqrt{\langle\mathbf{B}\rangle}$ since $\langle\mathbf{C}+\mathbf{D}\rangle = 2\sqrt{\langle\mathbf{A}\rangle\langle\mathbf{B}\rangle}$. In the coordinates $\mathbf{r}_X$ and $\boldsymbol{\rho}_j$ we have

$$\begin{aligned}
&\left\langle \mathbf{P}^{(2)}(\boldsymbol{\rho}_1,\boldsymbol{\rho}_2)\right\rangle_{\mathrm{coh2}} \\
&= \frac{1}{8}\Bigg\{ \cos^2\left[\frac{k}{2z_0}(\mathbf{r}_A-\mathbf{r}_B)\cdot(\boldsymbol{\rho}_1-\boldsymbol{\rho}_2)\right] + \cos^2\left[\frac{k}{2z_0}(\mathbf{r}_A-\mathbf{r}_B)\cdot(\boldsymbol{\rho}_1+\boldsymbol{\rho}_2)\right] \\
&\quad +2\cos\left[\frac{k}{2z_0}(\mathbf{r}_A-\mathbf{r}_B)\cdot(\boldsymbol{\rho}_1-\boldsymbol{\rho}_2)\right]\cos\left[\frac{k}{2z_0}(\mathbf{r}_A-\mathbf{r}_B)\cdot(\boldsymbol{\rho}_1+\boldsymbol{\rho}_2)\right]\Bigg\}.
\end{aligned} \tag{5.28}$$

---

[5] We use the identities,

$$1+\cos[(\mathbf{k}_A-\mathbf{k}_B)\cdot\mathbf{x}] = 2\cos^2\left[\frac{1}{2}(\mathbf{k}_A-\mathbf{k}_B)\cdot\mathbf{x}\right] \tag{5.25}$$

and

$$\cos[\mathbf{k}\cdot\mathbf{x}_1] + \cos[\mathbf{k}\cdot\mathbf{x}_2] = 2\cos\left[\frac{1}{2}\mathbf{k}\cdot(\mathbf{x}_1-\mathbf{x}_2)\right]\cos\left[\frac{1}{2}\mathbf{k}\cdot(\mathbf{x}_1+\mathbf{x}_2)\right]. \tag{5.26}$$

.

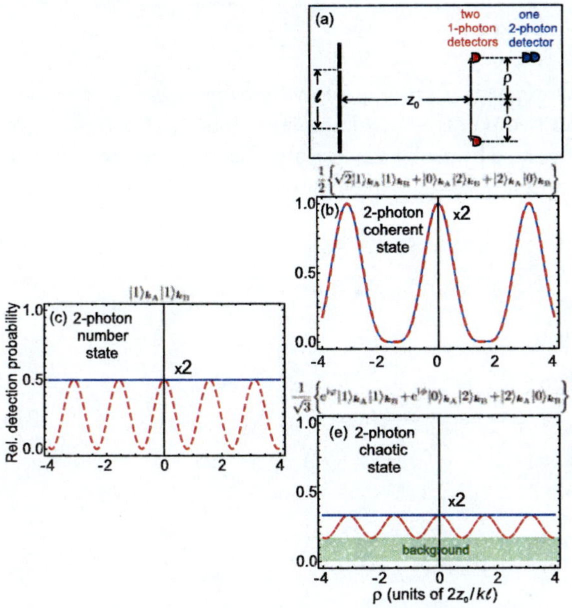

Fig. 5.7 **a** Two geometries for recording the *second order* diffraction patterns of a two-point source that emits the four different types of two-mode two-photon quantum states in (5.20)–(5.23). The detectors record the arrival of two photons in coincidence ("same time") either at the same point (blue) or different points (red). They are scanned symmetrically around the optical axis and parallel to the axis containing the two source points, separated by $\ell = |\mathbf{r}_A - \mathbf{r}_B|$. The color of the patterns in the lower panels **(b)**–**(e)** correspond to the two detection and scanning scenarios in **(a)**, and in all cases we have multiplied the detection probability by a factor of 2. **b** Patterns (5.29) if the two source points emit coherently. **c** Patterns (5.32) if the two source points emit only single photons corresponding to the indicated number state. **d** Patterns (5.35) if the two source points only emit photon pairs corresponding to the indicated entangled state. **e** Patterns (5.38) if the two source points emit chaotically

For the geometry shown in Fig. 5.7 (a) we have $\mathbf{r}_A \parallel \mathbf{r}_B \parallel \boldsymbol{\rho}_1 \parallel \boldsymbol{\rho}_2$ and denoting the two source point separation as $\ell = |\mathbf{r}_A - \mathbf{r}_B|$, we obtain

$$\left\langle \mathbf{P}^{(2)}(\boldsymbol{\rho}_1, \boldsymbol{\rho}_2)\right\rangle_{\text{coh2}} = \frac{1}{2}\cos^2\left[\frac{k\,\ell}{2z_0}\,\rho_1\right]\cos^2\left[\frac{k\,\ell}{2z_0}\,\rho_2\right]. \tag{5.29}$$

We obtain the same result for both detection scenarios $\boldsymbol{\rho} = \boldsymbol{\rho}_1 = \boldsymbol{\rho}_2$ and $\boldsymbol{\rho} = \boldsymbol{\rho}_1 = -\boldsymbol{\rho}_2$. The result multiplied by 2 is plotted in Fig. 5.7b.

5.5.2.2 Two-Photon Number State

The two-photon number state represents the situation where only *single photons* are simultaneously emitted from two points which in practice consist of two quantum dots. This case can be implemented by use of a laser pulse that triggers two quantum dots or two trapped atoms or ions to simultaneously emit single photons [72–75]. The state is given by

$$|\phi_{\rm num2}\rangle_{\mathbf{k}_A,\mathbf{k}_B} = |1\rangle_{\mathbf{k}_A}|1\rangle_{\mathbf{k}_B}. \tag{5.30}$$

For the geometry in Fig. 5.7a we obtain

$$\left\langle \mathbf{P}^{(2)}(\boldsymbol{\rho}_1, \boldsymbol{\rho}_2)\right\rangle_{\rm num2} = \frac{1}{4}\cos^2\left[\frac{k}{2z_0}(\mathbf{r}_A - \mathbf{r}_B)\cdot(\boldsymbol{\rho}_1 - \boldsymbol{\rho}_2)\right] \tag{5.31}$$

or

$$\left\langle \mathbf{P}^{(2)}(\boldsymbol{\rho}, \boldsymbol{\rho})\right\rangle_{\rm num2} = \frac{1}{4}, \quad \left\langle \mathbf{P}^{(2)}(\boldsymbol{\rho}, -\boldsymbol{\rho})\right\rangle_{\rm num2} = \frac{1}{4}\cos^2\left[\frac{k\,\ell}{z_0}\,\rho\right]. \tag{5.32}$$

The probability multiplied by 2 is plotted in Fig. 5.7c.

5.5.2.3 Two-Photon Entangled State

The entangled two-photon state is produced by illuminating a suitable thin crystal, placed before the two pinholes that through spontaneous parametric down conversion produces two spatially entangled photons, each with half the incident photon energy [33]. The "entangled biphoton" diffraction case has been extensively studied in the literature [16, 38, 70]. The state is given by

$$|\phi_{\rm ent2}\rangle_{\mathbf{k}_A,\mathbf{k}_B} = \frac{1}{\sqrt{2}}\left\{|0\rangle_{\mathbf{k}_A}|2\rangle_{\mathbf{k}_B} + |2\rangle_{\mathbf{k}_A}|0\rangle_{\mathbf{k}_B}\right\}. \tag{5.33}$$

Its structure suggests through the appearance of the zero-point quantum state that-photon entanglement may be mediated by the quantum vacuum.
We obtain

$$\left\langle \mathbf{P}^{(2)}(\boldsymbol{\rho}_1, \boldsymbol{\rho}_2)\right\rangle_{\rm ent2} = \frac{1}{4}\cos^2\left[\frac{k}{2z_0}(\mathbf{r}_A - \mathbf{r}_B)\cdot(\boldsymbol{\rho}_1 + \boldsymbol{\rho}_2)\right]. \tag{5.34}$$

The entangled state gives the opposite result of the number state

$$\left\langle \mathbf{P}^{(2)}(\boldsymbol{\rho}, \boldsymbol{\rho})\right\rangle_{\rm ent2} = \frac{1}{4}\cos^2\left[\frac{k\,\ell}{z_0}\,\rho\right], \quad \left\langle \mathbf{P}^{(2)}(\boldsymbol{\rho}, -\boldsymbol{\rho})\right\rangle_{\rm ent2} = \frac{1}{4}. \tag{5.35}$$

The pattern multiplied by 2 is plotted in Fig. 5.7d.

5.5.2.4 Two-Photon Chaotic State

The two-photon chaotic state is created by illuminating the slits with a quasi-monochromatic thermal source [76] that is close to the pinholes so that they are outside the coherence cone of the source. In practice, it is convenient to use higher intensity "pseudo-thermal" light generated by shining a laser on a rotating ground glass plate [77]. Such light exhibits both phase and intensity fluctuations with *Bose–Einstein* counting statistics [78]. It has been employed for double-slit diffraction by several groups [79–81]. The *chaotic* two-mode two-photon state is given by the following form of (5.19)

$$|\phi_{\text{cha2}}\rangle_{\mathbf{k}_A,\mathbf{k}_B} = \frac{1}{\sqrt{3}}\left\{ e^{i\varphi}|1\rangle_{\mathbf{k}_A}|1\rangle_{\mathbf{k}_B} + e^{i\phi}|0\rangle_{\mathbf{k}_A}|2\rangle_{\mathbf{k}_B} + |2\rangle_{\mathbf{k}_A}|0\rangle_{\mathbf{k}_B}\right\}. \quad (5.36)$$

The detection probability is

$$\left\langle \mathbf{P}^{(2)}(\boldsymbol{\rho}_1, \boldsymbol{\rho}_2)\right\rangle_{\text{cha2}} = \frac{1}{12}\left\{1 + \cos^2\left[\frac{k}{2z_0}(\mathbf{r}_A - \mathbf{r}_B)\cdot(\boldsymbol{\rho}_1 - \boldsymbol{\rho}_2)\right]\right\}. \quad (5.37)$$

For the geometry in Fig. 5.7 (a) we obtain

$$\left\langle \mathbf{P}^{(2)}(\boldsymbol{\rho}, \boldsymbol{\rho})\right\rangle_{\text{cha2}} = \frac{1}{6}, \quad \left\langle \mathbf{P}^{(2)}(\boldsymbol{\rho}, -\boldsymbol{\rho})\right\rangle_{\text{cha2}} = \frac{1}{12}\left\{1 + \cos^2\left[\frac{k\ell}{z_0}\rho\right]\right\}. \quad (5.38)$$

The result multiplied by 2 is plotted in Fig. 5.7 (e). In this case the interference modulation shown by the red curve is reduced and superimposed on a background of equal size indicated by green shading.

5.5.2.5 Summary

The $N = 2$ number $|\phi_{\text{num2}}\rangle$ and entangled $|\phi_{\text{ent2}}\rangle$ states form basis states from which the coherent and chaotic states may be constructed as linear combinations. Their patterns are complete opposites, corresponding to the exchange of the red and blue detection schemes in Fig. 5.7a. The *number state* is difficult to produce in practice and requires special source preparation [72–75]. In contrast, the *entangled state* $|\phi_{\text{ent2}}\rangle$ may be produced by parametric down conversion in a crystal illuminated by a laser. Its properties have been extensively studied in experiments involving beam splitters and double-slit diffraction [16, 17, 37, 38]. It incorporates the essence of quantum behavior since it is maximally entangled and plays a prominent role in quantum information science [82–85]. The complementary behavior of the two states holds a central position in quantum optics since their diffraction patterns cannot be explained by the wave formalism.

The two-photon *coherent state* $|\phi_{\text{coh2}}\rangle$ corresponds to an *in-phase* addition of the number and entangled states. From Table 5.1 and (5.27) we see that in this case all photon amplitudes linking the source and detector plane maximally interfere giving rise to non-zero interference terms C and D. It is the only state that is completely symmetrical with respect to the detection scenarios.

The pattern of the *chaotic state* $|\phi_{\text{cha2}}\rangle$ is determined by different relative contributions of the number and entangled states giving rise to a reduced interference modulation superimposed on a background. The modulation and background size quantitatively scales with the square of the coefficients expressing the number and entangled state contributions.

5.5.3 The New Paradigm of Quantum Diffraction

Our above discussion allows us to introduce the following important new paradigm for understanding diffraction.

> Within the fundamental theory of light, QED, diffraction images are direct signatures of different quantum states of light. The broad and difficult concept of "partial coherence" in wave optics can then be succinctly defined through the degrees of coherence of specific quantum states in different orders of QED.

We shall see in Chap. 16 that this paradigm is beautifully verified when results of modern versions of Young's double-slit experiment are performed by illumination with differently modified laser light and photon-based detection.

5.5.4 Finite Source Areas

For a finite-sized source, we need to sum (integrate) over the contributions from all source points and also consider their correlations. The two point sums in round brackets in (5.11) are now replaced by integrals over all source points, so that we have

$$\mathbf{P}^{(2)}(\boldsymbol{\rho}_1, \boldsymbol{\rho}_2) = \frac{1}{16} \sum_{\mathrm{U,V,X,Y}} \underbrace{\frac{1}{A_{\text{eff}}^2} \left(\int\limits_{\text{plane}} \mathbf{a}^\dagger_{\mathbf{k}_\mathrm{U}} e^{\mathrm{i}\frac{k}{z_0}\mathbf{r}_\mathrm{U}\cdot\boldsymbol{\rho}_1} d\mathbf{r}_\mathrm{U} \right) \left(\int\limits_{\text{plane}} \mathbf{a}^\dagger_{\mathbf{k}_\mathrm{V}} e^{\mathrm{i}\frac{k}{z_0}\mathbf{r}_\mathrm{V}\cdot\boldsymbol{\rho}_2} d\mathbf{r}_\mathrm{V} \right)}_{2\,\text{photon creation}\ \Phi^*(\boldsymbol{\rho}_1,\boldsymbol{\rho}_2)}$$

$$\times \underbrace{\frac{1}{A_{\text{eff}}^2} \left(\int\limits_{\text{plane}} \mathbf{a}_{\mathbf{k}_\mathrm{X}} e^{-\mathrm{i}\frac{k}{z_0}\mathbf{r}_\mathrm{X}\cdot\boldsymbol{\rho}_1} d\mathbf{r}_\mathrm{X} \right) \left(\int\limits_{\text{plane}} \mathbf{a}_{\mathbf{k}_\mathrm{Y}} e^{-\mathrm{i}\frac{k}{z_0}\mathbf{r}_\mathrm{Y}\cdot\boldsymbol{\rho}_2} d\mathbf{r}_\mathrm{Y} \right)}_{2\,\text{photon destruction}\ \Phi(\boldsymbol{\rho}_1,\boldsymbol{\rho}_2)}, \tag{5.39}$$

where the sum consists of sixteen permutations with U,V,X,Y= A or B. The integrated 2-photon *probability amplitudes* have the form,

$$\Phi(\rho_1,\rho_2) = \frac{1}{A_{\text{eff}}^2} \iint\limits_{\text{plane}} \mathbf{a}_{\mathbf{k}_X} \mathbf{a}_{\mathbf{k}_Y}\, \mathrm{e}^{-\mathrm{i}\frac{k}{z_0}(\mathbf{r}_X\cdot\rho_1+\mathbf{r}_Y\cdot\rho_2)}\, \mathrm{d}\mathbf{r}_X\, \mathrm{d}\mathbf{r}_Y. \tag{5.40}$$

They resemble the one-photon *probability operators* (4.138) which in the same notation read

$$\mathbf{P}^{(1)}(\rho_1,\rho_2) = \frac{1}{A_{\text{eff}}^2} \iint\limits_{\text{plane}} \mathbf{a}_{\mathbf{k}_X}^{\dagger} \mathbf{a}_{\mathbf{k}_Y}\, \mathrm{e}^{\mathrm{i}\frac{k}{z_0}(\mathbf{r}_X\cdot\rho_1-\mathbf{r}_Y\cdot\rho_2)} \mathrm{d}\mathbf{r}_X\, \mathrm{d}\mathbf{r}_Y. \tag{5.41}$$

The key difference lies in the Hermitian conjugate of one of the operators and the associated change of sign in the exponential impulse response function. We can state as follows.

For a finite-size source, the expectation value of the second order probability operator (5.39) is evaluated according to

$$\left\langle \mathbf{P}^{(2)}(\rho_1,\rho_2)\right\rangle. = \frac{1}{16\,A_{\text{eff}}^4} \sum_{\text{U,V,X,Y}} \iiiint\limits_{\text{plane}} \langle \mathbf{a}_{\mathbf{k}_U}^{\dagger} \mathbf{a}_{\mathbf{k}_V}^{\dagger} \mathbf{a}_{\mathbf{k}_X} \mathbf{a}_{\mathbf{k}_Y}\rangle\, \mathrm{e}^{\mathrm{i}\frac{k}{z_0}(\mathbf{r}_U-\mathbf{r}_X)\cdot\rho_1}\, \mathrm{e}^{\mathrm{i}\frac{k}{z_0}(\mathbf{r}_V-\mathbf{r}_Y)\cdot\rho_2}\, \mathrm{d}\mathbf{r}_U\, \mathrm{d}\mathbf{r}_V\, \mathrm{d}\mathbf{r}_X\, \mathrm{d}\mathbf{r}_Y. \tag{5.42}$$

For a given quantum state, only matrix elements corresponding to certain index permutations U,V,X,Y=A,B will have non-zero values (see Table 5.1) and they will determine the integration.

5.5.5 *Second Order Pattern of a Circular Flat-Top Source*

As an example we consider a circular flat-top source emitting either one of the four quantum states in Table 5.1.

For the two-photon *coherent* state $|\phi_{\text{coh2}}\rangle$ given by (5.20) all matrix elements have the same value $\langle \mathbf{a}_{\mathbf{k}_U}^{\dagger} \mathbf{a}_{\mathbf{k}_V}^{\dagger} \mathbf{a}_{\mathbf{k}_X} \mathbf{a}_{\mathbf{k}_Y}\rangle = 1/2$, independent of all sixteen permutations X,Y,U,V=A,B. They can hence be pulled out of the integral which then factors in the $\mathbf{r}_I$ (I=U,V,X,Y) coordinates. By use of the integral (4.140) we obtain,

$$\left\langle \mathbf{P}^{(2)}(\boldsymbol{\rho}_1, \boldsymbol{\rho}_2)\right\rangle_{\text{coh2}} = \frac{1}{2}\left[\frac{2J_1(kR\rho_1/z_0)}{kR\rho_1/z_0}\right]^2\left[\frac{2J_1(kR\rho_2/z_0)}{kR\rho_2/z_0}\right]^2. \tag{5.43}$$

For the two-photon *number* state $|\phi_{\text{num2}}\rangle$ only the four matrix elements of term **A** in Table 5.1 are non-zero and have the same value $\langle \mathbf{a}^\dagger_{\mathbf{k}_U}\mathbf{a}^\dagger_{\mathbf{k}_V}\mathbf{a}_{\mathbf{k}_X}\mathbf{a}_{\mathbf{k}_Y}\rangle = 1$ for the four permutations U,V,X,Y=A,B within term **A**. The integrals in (5.42) then involve a correction between points that can be expressed by $\delta(\mathbf{r}_U, \mathbf{r}_X)$ and $\delta(\mathbf{r}_v, \mathbf{r}_y)$, and by normalizing to the number of permutations we obtain

$$\left\langle \mathbf{P}^{(2)}(\boldsymbol{\rho}_1, \boldsymbol{\rho}_2)\right\rangle_{\text{num2}} = \frac{1}{4\,A_{\text{eff}}^2}\iint\limits_{A_{\text{eff}}} e^{i\frac{k}{z_0}(\mathbf{r}_A-\mathbf{r}_B)\cdot(\boldsymbol{\rho}_1-\boldsymbol{\rho}_2)}\, d\mathbf{r}_A d\mathbf{r}_B \tag{5.44}$$

where the factor 1/4 reflects that **A** is only one of the four terms **A** – **D**. This is evaluated as

$$\left\langle \mathbf{P}^{(2)}(\boldsymbol{\rho}_1, \boldsymbol{\rho}_2)\right\rangle_{\text{num2}} = \frac{1}{4}\left[\frac{2J_1(kR|\boldsymbol{\rho}_1-\boldsymbol{\rho}_2|/z_0)}{kR|\boldsymbol{\rho}_1-\boldsymbol{\rho}_2|/z_0}\right]^2. \tag{5.45}$$

For the two-photon *entangled* state $|\phi_{\text{ent2}}\rangle$ only the matrix elements of term **B** are non-zero and have the value 1. In analogy to the number state evaluation we obtain

$$\left\langle \mathbf{P}^{(2)}(\boldsymbol{\rho}_1, \boldsymbol{\rho}_2)\right\rangle_{\text{ent2}} = \frac{1}{4\,A_{\text{eff}}^2}\iint\limits_{A_{\text{eff}}} e^{i\frac{k}{z_0}(\mathbf{r}_A-\mathbf{r}_B)\cdot(\boldsymbol{\rho}_1+\boldsymbol{\rho}_2)}\, d\mathbf{r}_A d\mathbf{r}_B \tag{5.46}$$

which evaluates to

$$\left\langle \mathbf{P}^{(2)}(\boldsymbol{\rho}_1, \boldsymbol{\rho}_2)\right\rangle_{\text{ent2}} = \frac{1}{4}\left[\frac{2J_1(kR|\boldsymbol{\rho}_1+\boldsymbol{\rho}_2|/z_0)}{kR|\boldsymbol{\rho}_1+\boldsymbol{\rho}_2|/z_0}\right]^2. \tag{5.47}$$

For the two-photon *chaotic* state $|\phi_{\text{cha2}}\rangle$ only matrix elements of the terms **A** and **B** in Table 5.1 are non-zero. The term **A** matrix elements are 1/3 of those of the number state. The term **B** only contributes a constant due to the fact that only the first two of its matrix element are non-zero. The constant contribution of **B** is of the same size as that of term **A** so that we obtain

$$\left\langle \mathbf{P}^{(2)}(\boldsymbol{\rho}_1, \boldsymbol{\rho}_2)\right\rangle_{\text{cha2}} = \frac{1}{12}\left\{1+\left[\frac{2J_1(kR|\boldsymbol{\rho}_1-\boldsymbol{\rho}_2|/z_0)}{kR|\boldsymbol{\rho}_1-\boldsymbol{\rho}_2|/z_0}\right]^2\right\}. \tag{5.48}$$

Aside from the prefactor,[6] this is the famous Hanbury Brown–Twiss result [9–11]. The HBT result schematically shown in Fig. 5.2 assumed a Gaussian shaped "bunching peak" which for a circular flat-top source is simply replaced by an Airy function.

We have plotted the geometry and the patterns for the four cases in Fig. 5.8a–e. In (f) we have used a log scale to emphasize the change in the outer Airy rings between the two-photon *coherent* state pattern in (b) shown in black, with the pink pattern representing the shape of the "incoherent" patterns in (c)–(e). This more clearly reveals the change in both the width of the main peak and the relative intensity of the outer rings.

5.5.6 Power Conservation in Two-Photon Diffraction

The patterns in Fig. 5.8 nicely reveal the shapes of the two-photon patterns. However, they do not account for energy or photon number conservation between the emitted and detected photons. Energy or power conservation was an important part of the classical Fourier optics treatments in Sect. 4.4.3 and Zernike's more general formalism in Sect. 4.6. In particular, the integrated intensity of the coherent one-photon pattern (4.104) and of the total chaotic pattern (4.110) reflect power-conservation between the source and detector planes. In the following we shall discuss how one may similarly derive power-conserving intensity expressions for the second order or two-photon case.

The classical energy (3.22) or intensity (3.25) is linked to the photon picture by the semi-classical relations (3.117) and (3.120). For the one-photon case, energy conservation can readily be expressed in terms of the quantum mechanical first order coherence function $G^{(1)}$ since it has the dimension of field-squared, $|E|^2$. The first order intensity is then given by the simple relation (4.103) or

$$I^{(1)} = \epsilon_0 c\, G^{(1)} = \epsilon_0 c\, \langle |E_0|^2 \rangle \langle \mathbf{P}^{(1)} \rangle. \tag{5.49}$$

In contrast, the second order correlation functions $G^{(2)} = \langle |E_0|^2 \rangle^2 \langle \mathbf{P}^{(2)} \rangle$ have the dimension of $|E|^4$ and for a complete description of the second order case we need to discuss how energy or power conservation is handled. In the following we shall consider power conservation for the *coherent source* case which is facilitated by the limited spatial extent of the diffracted pattern. The lack of an incoherent constant background that extends to infinity facilitates integration of the intensity pattern, required for consideration of power conservation.

---

[6] We note that when (5.48) is normalized to obtain the degree of second order coherence $g^{(2)}(\rho_1, \rho_2)$, the prefactor 1/12 does not normalize out to unity since with the matrix elements $\langle \Phi_{\text{cha2}} | \mathbf{a}_{\mathbf{k}}^{\dagger} \mathbf{a}_{\mathbf{k}} | \Phi_{\text{cha2}} \rangle = 1$ and $\langle \Phi_{\text{cha2}} | \mathbf{a}_{\mathbf{k}}^{\dagger} \mathbf{a}_{\mathbf{k}'} | \Phi_{\text{cha2}} \rangle = 0$ one obtains $\langle \phi_{\text{cha2}} | \mathbf{P}^{(1)}(\boldsymbol{\rho}_j, \boldsymbol{\rho}_j) | \phi_{\text{cha2}} \rangle^2 = 1/2$. The proper HBT result $g^{(2)}(0,0) = 2$ is however obtained for the *collective chaotic state* discussed in Chap. 16.

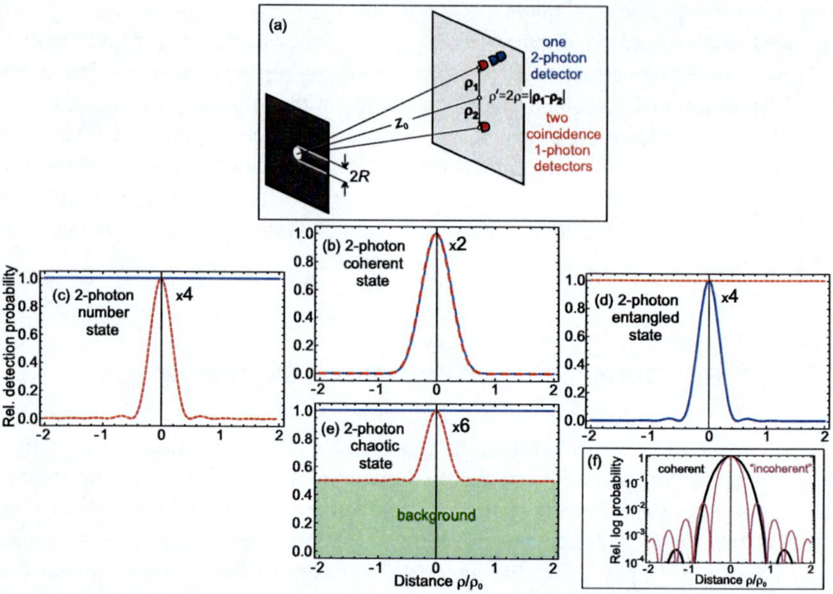

Fig. 5.8 **a** Two geometries for recording the *second order* diffraction patterns of a circular flat-top source that emits the four different types of two-mode two-photon quantum states (5.20)–(5.23). The detectors record the arrival of two photons in coincidence ("same time") either at the same point (blue) or different points (red). They are scanned symmetrically around the optical axis. The color of the patterns in the lower panels **(b)**–**(e)** corresponds to the two detection and scanning scenarios in **(a)**, and the scaling factors of the detection probabilities are indicated. **b** Patterns (5.43) for coherent emission, where $\rho_0 = 0.61\lambda z_0/R$ is the position of the first node (see Fig. 4.13). **c** Patterns (5.45) if all source points emit only single number state photons. **d** Patterns (5.47) if the two source points only emit entangled photon pairs. **e** Patterns (5.48) if the two source points emit chaotically. **f** Log scale comparison of the two-photon coherent state pattern in **(b)** shown in black, with the interference patterns in the "incoherent" cases **(c)**–**(e)** in magenta

In practice, the two-photon intensity $I^{(2)}$ is a measure of the number of photon pairs and one needs to consider how it is related to $G^{(2)}$. Denoting the two-photon peak intensity of the cylindrically symmetric source as $I_s^{(2)}(0)$, we can define the effective source area in analogy to the first order case given by (4.73) as

$$A_{\text{eff}}^{(2)} = \frac{2\pi \int_0^\infty I_s^{(2)}(r)\, r\, \mathrm{d}r}{I_s^{(2)}(0)}. \tag{5.50}$$

The emitted two-photon power is then $\mathcal{P}_s^{(2)} = I_s^{(2)}(0) A_{\text{eff}}^{(2)}$.

In the *detector plane*, we similarly can write $\mathcal{P}_d^{(2)} = I_d^{(2)}(0) B_{\text{eff}}^{(2)}$, where $I_d^{(2)}(0)$ is the peak intensity and $B_{\text{eff}}^{(2)}$ is the effective area given by

$$B_{\text{eff}}^{(2)} = \frac{2\pi \int_0^\infty I_{\text{d}}^{(2)}(\rho)\,\rho\,\text{d}\rho}{I_{\text{d}}^{(2)}(0)} = \frac{2\pi \int_0^\infty G_{\text{d}}^{(2)}(\rho,\rho)\,\rho\,\text{d}\rho}{G_{\text{d}}^{(2)}(0,0)} = \frac{2\pi \int_0^\infty \left\langle \mathbf{P}_{\text{d}}^{(2)}(\rho,\rho)\right\rangle \rho\,\text{d}\rho}{\left\langle \mathbf{P}_{\text{d}}^{(2)}(0,0)\right\rangle}. \tag{5.51}$$

Power conservation between the source and detector plane requires

$$\underbrace{\mathcal{P}_{\text{s}}^{(2)} = I_{\text{s}}^{(2)}(0)\,A_{\text{eff}}^{(2)}}_{\text{source plane}} = \underbrace{\mathcal{P}_{\text{d}}^{(2)} = I_{\text{d}}^{(2)}(0)\,B_{\text{eff}}^{(2)}}_{\text{detector plane}}. \tag{5.52}$$

Our power-conserving two-photon intensity in the detector plane for a cylindrically symmetric source is given by

$$\boxed{I_{\text{d}}^{(2)}(\rho) = \underbrace{I_{\text{s}}^{(2)}(0)\,A_{\text{eff}}^{(2)}}_{\mathcal{P}_{\text{s}}^{(2)}}\ \frac{\left\langle \mathbf{P}_{\text{d}}^{(2)}(\rho,\rho)\right\rangle}{B_{\text{eff}}^{(2)}\left\langle \mathbf{P}_{\text{d}}^{(2)}(0,0)\right\rangle}.} \tag{5.53}$$

When both sides are integrated over the detector plane, we directly obtain power conservation by use of (5.51). The same relationship holds by replacing the real space distance ρ by the momentum transfer $q = \rho\, k/z_0$.

5.5.6.1 Power Conservation for a Coherent Flat-Top Source

Let us compare the 1-photon and 2-photon diffraction patterns of a circular flat-top source of the same area $A_{\text{eff}}^{(2)} = A_{\text{eff}}^{(1)} = \pi R^2$. The coherent 1-photon pattern is given by (4.104) as

$$\boxed{I^{(1)}(\rho) = I_0^{(1)}\left[\frac{\pi R^2}{\lambda z_0}\right]^2 \left[\frac{2J_1\,(kR\rho/z_0)}{kR\rho/z_0}\right]^2} \tag{5.54}$$

where the superscript (1) specifically denotes the 1-photon nature of the intensities. The effective area of the pattern in the detector plane is given by (4.74) or,

$$B_{\text{eff}}^{(1)} = \frac{\lambda^2 z_0^2}{\pi R^2} \tag{5.55}$$

The coherent 2-photon pattern for detection at $\boldsymbol{\rho}_1 = \boldsymbol{\rho}_2 = \boldsymbol{\rho}$, is given by (5.43) as

$$\left\langle \mathbf{P}^{(2)}(\boldsymbol{\rho},\boldsymbol{\rho})\right\rangle_{\text{coh2}} = \frac{1}{2}\left[\frac{2J_1(kR\rho/z_0)}{kR\rho/z_0}\right]^4 \tag{5.56}$$

Its effective area in the detector plane, $B_{\text{eff}}^{(2)}$, is obtained from (5.51) as

$$B_{\text{eff}}^{(2)} = 2\pi \int_0^{\infty} \left[\frac{2J_1(kR\rho/z_0)}{kR\rho/z_0}\right]^4 \rho \, d\rho = \left(1 - \frac{16}{3\pi^2}\right) \frac{\lambda^2 z_0^2}{\pi R^2} = \frac{B_{\text{eff}}^{(1)}}{2.176} \tag{5.57}$$

The power-conserving second order intensity expression (5.53) becomes,

$$\boxed{I^{(2)}(\rho) = I_0^{(2)} \underbrace{\frac{1}{1-16/(3\pi^2)}}_{2.176} \left[\frac{\pi R^2}{\lambda z_0}\right]^2 \left[\frac{2J_1(kR\rho/z_0)}{kR\rho/z_0}\right]^4} \tag{5.58}$$

The superscript (2) denotes the 2-photon nature of the intensities. In order to compare the patterns we have plotted (5.54) and (5.58) with $\pi R^2/(\lambda z_0) = 1$ in Fig. 5.9 in the normalized forms,

$$\mathcal{P}^{(n)}(\rho) = \frac{I^{(n)}(\rho)}{I_0^{(n)}} \tag{5.59}$$

This expresses the detection probability of $n = 1$ or $n = 2$ photons as a function of ρ relative to the corresponding number of photons emitted from the flat-top source.

The two diffraction patterns are plotted in Fig. 5.9 (a) on a logarithmic scale to emphasize the change in the outer Airy rings. The patterns have the same integrated area, reflecting power conservation between source and detection planes.

In Fig. 5.9 (b) we replot the central part of the patterns on a linear scale to emphasize the corresponding change in peak value and effective distribution areas required by power conservation. The effective detector plane areas are linked by (5.57). The power conserving factor $1/(1-16/(3\pi^2))$ slightly differs from 2 because the central peak of the 2-photon Airy distribution also grows at the expense of the outer Airy rings of the 1-photon case. For the 1-photon pattern the outer rings contain 16.2% of the total power, while they contain only 0.2% for the 2-photon case.

5.5.7 *X-Ray Measurements of Second Order Lateral Coherence*

Measurements of the second order coherence functions are particularly interesting for x-ray sources since they are typically "incoherent".[7] HBT-like measurements for synchrotron radiation were first proposed by Shuryak [87] as early as 1975 but were

[7] In the first order, any source can be made laterally coherent by a suitable aperture. A higher order coherent "x-ray" source can presently be created only in the XUV range through external laser seeding of the FEL process [86].

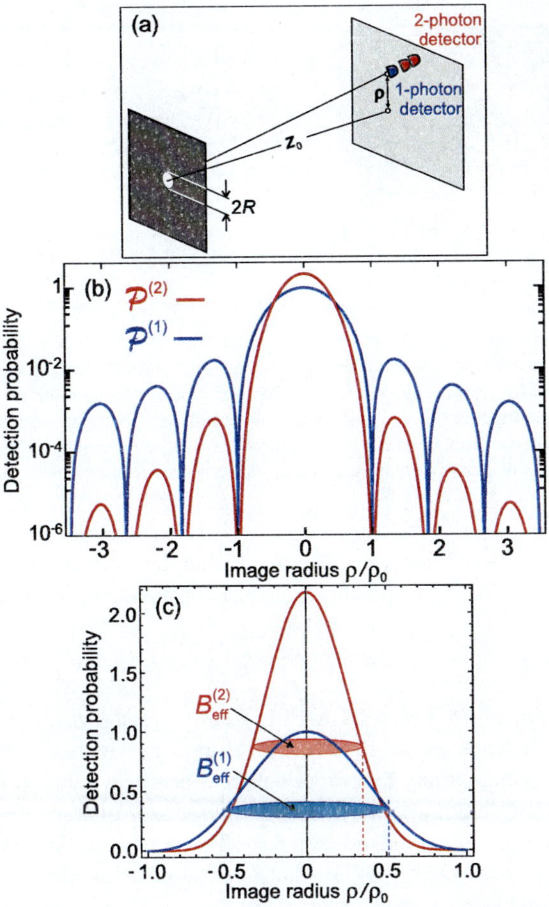

Fig. 5.9 **a** Detection geometry. **b** 1D projection of the first order diffraction probability (Airy, blue) and the second order pattern (Airy squared, red), calculated according to (5.59) where $\rho_0 = 0.61\lambda z_0/R$ is the first zero of the Airy pattern. Both 2D patterns contain the same total power. The fractional power in the central Airy disk extending to $\rho \leq \rho_0$ increases from 83.8% (blue curve) to 99.8% (red curve). **c** The central intensity plotted on a linear scale. The two-photon pattern (red) has an effective area $B^{(2)}_{\rm eff} = B^{(1)}_{\rm eff}/2.176$ and a peak intensity that is larger by 2.176

only implemented around 2000 for both synchrotron and XFEL sources [59, 60, 63, 64, 88].

By assuming space-time separability, second order correlation function measurements can in principle be performed in both the time and spatial domains. However, as illustrated in Fig. 5.4 the short coherence time of x-ray pulses relative to the shortest detector response time does not allow measurements in the time domain. Therefore one measures in the spatial domain as originally done by HBT under the condition

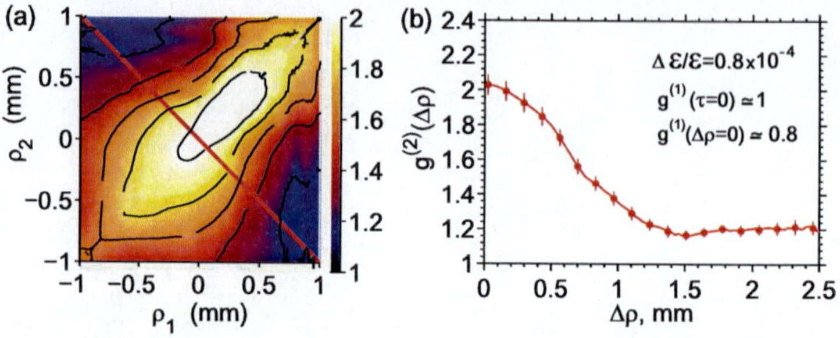

Fig. 5.10 Measurement of the degree of second order spatial coherence at the FLASH X-FEL facility [63]. The beam was monochromatized to a narrow bandwidth of $\Delta\mathcal{E}/\mathcal{E} = 0.8 \times 10^{-4}$ so that it was essentially temporally coherent. **a** Contour plot of the normalized measured spatial intensity correlation function $g^{(2)}(\rho_1, \rho_2)$, where the orthogonal coordinates ρ_1 and ρ_2 are relative to the beam center in the distant plane of the CCD detector. The image contains about 10^4 single-shot intensity profiles. **b** Intensity correlation function $g^{(2)}(\Delta\rho)$ taken along the red line in (**a**)

of monochromaticity or temporal coherence. Using the notation of Fig. 5.2 where ρ' denotes the separation of two points in the detector plane, we have the relation

$$g^{(2)}(\rho') = 1 + |g^{(1)}(\rho')|^2. \tag{5.60}$$

Results obtained at the FLASH X-FEL facility are shown in Fig. 5.10 [63], in the notation $\rho' = \Delta\rho = |\boldsymbol{\rho}_2 - \boldsymbol{\rho}_1|$. In the measurement a monochromator was used to create a nearly temporally first order coherent beam as required for the validity of (5.60). The correlation measurements at pulsed X-FEL facilities are naturally gated by the duration of the individual pulses. The normalized intensity in Fig. 5.10 consisted of an average over about 10^4 pulses of about 50 fs duration.

In the notation of Fig. 5.10 we have

$$g^{(2)}(\Delta\rho) = 1 + |g^{(1)}(\Delta\rho)|^2 = 1 + \exp\left[-\frac{\pi(\Delta\rho)^2}{B_{\mathrm{eff}}}\right]. \tag{5.61}$$

From a Gaussian fit of the red curve in Fig. 5.10 (b) one may derive the effective coherence area $B_{\mathrm{eff}} = 2.7\,\mathrm{mm}^2$ in the detector plane. Furthermore, renormalizing the red curve to corresponds to unity at large $\Delta\rho$, as required by (5.61), the peak value $g^{(2)}(\Delta\rho{=}0)$ is reduced to about 1.64 so that the degree of first order lateral coherence is obtained as $|g^{(1)}(\Delta\rho{=}0)| \simeq \sqrt{0.64} = 0.8$.

Finally, we mention that the measurement of second order correlation functions between photons has been extended to elementary particles [89], electrons [90, 91] and macromolecules [92]. The preferential bunching behavior of Bosons versus the antibunching behavior of Fermions has also been experimentally demonstrated [93].

5.6 Detection of First and Second Order Coherence Patterns

We have seen in Chap. 4 that in the classical wave field description one does not have to specify how the first order diffraction intensity is measured. In the first order quantum treatment, we only considered the two-mode one-photon state (4.131) of the form

$$|\phi_{\rm coh1}\rangle_{\mathbf{k}_A,\mathbf{k}_B} = \frac{1}{\sqrt{2}}\Big[|0\rangle_{\mathbf{k}_A}|1\rangle_{\mathbf{k}_B} + |1\rangle_{\mathbf{k}_A}|0\rangle_{\mathbf{k}_B}\Big]. \tag{5.62}$$

This quantum state gives the same pattern for a single detection point, expressed by (4.141) with $\boldsymbol{\rho}_1 = \boldsymbol{\rho}_2$, as the classical result given by (4.71). The detection of the pattern of this state, however, requires use of a single photon detector, i.e. a photon number resolving detector as illustrated in Fig. 5.5. It therefore does not explain why the detection process can be ignored in the classical description.

Similarly, our derivation of the second order pattern in Sect. 5.5 only considered the two-mode two-photon state (5.20) given by

$$|\phi_{\rm coh2}\rangle_{\mathbf{k}_A,\mathbf{k}_B} = \frac{1}{2}\Big\{\sqrt{2}|1\rangle_{\mathbf{k}_A}|1\rangle_{\mathbf{k}_B} + |0\rangle_{\mathbf{k}_A}|2\rangle_{\mathbf{k}_B} + |2\rangle_{\mathbf{k}_A}|0\rangle_{\mathbf{k}_B}\Big\}. \tag{5.63}$$

The pattern of this state, given by (5.43), is shown in Fig. 5.8b, the black curve in Fig. 5.8f or the red curve in Fig. 5.9. The pattern is picked out by a two-photon number resolving detector or two one-photon detectors linked by a coincidence circuit. Both detection schemes are based on same time detection of two photons, either at the same point or two points. The question arises whether such detection schemes are always required or whether a charge integrating detector that counts all incident photons may be conveniently used in some cases.

In order to clarify and resolve these issues, we shall in the following broaden our discussion of *coherent* quantum states beyond the simplest states containing only one or two photons. This more general quantum derivation of the first and second order detection probabilities will provide the key to their measurement by use of different types of detectors. A more general discussion of various quantum states of light, including chaotic and entangled states, will be given in Sect. 16.4.

5.6.1 The Coherent Quantum States of Light

The states (5.62) and (5.63), which contain specific numbers of photons, are substates of the general two-mode *collective coherent state*, also called a *Glauber state*. It is constructed from two single mode coherent states of the form (3.78) and may be written as a linear combination of substates which contain definite number of photons, according to

$$|\Phi_{\text{coh}}\rangle_{\mathbf{k}_A,\mathbf{k}_B} = |\alpha\rangle_{\mathbf{k}_A}|\alpha\rangle_{\mathbf{k}_B} = \sum_{N=0}^{\infty} \underbrace{\frac{2^{N/2}\,\alpha^N}{\sqrt{N!}\,e^{|\alpha|^2}}}_{c_N^{\alpha}} |\phi_{\text{cohN}}\rangle_{\mathbf{k}_A,\mathbf{k}_B}. \tag{5.64}$$

Here α is a complex number, and we have assumed that the two modes each contain the same average number $\langle n\rangle$ photons, given by $|\alpha_{\mathbf{k}_A}|^2 = |\alpha_{\mathbf{k}_B}|^2 = |\alpha|^2 = \langle n\rangle$. The coefficients c_N^{α} fulfill the normalization

$$\sum_{N=0}^{\infty} |c_N^{\alpha}|^2 = 1. \tag{5.65}$$

The squared coefficients represent probabilities for the contributions of the two-mode coherent substates $|\phi_{\text{cohN}}\rangle_{\mathbf{k}_A,\mathbf{k}_B}$ in (5.64), given by

$$\begin{aligned}
|\phi_{\text{cohN}}\rangle_{\mathbf{k}_A,\mathbf{k}_B} &= \frac{1}{2^{N/2}} \sum_{n=0}^{N} \sqrt{\frac{N!}{n!\,(N-n)!}}\; |n\rangle_{\mathbf{k}_A} |N-n\rangle_{\mathbf{k}_B} \\
&= \underbrace{|0\rangle_{\mathbf{k}_A}|0\rangle_{\mathbf{k}_B}}_{|\phi_{\text{coh0}}\rangle_{\mathbf{k}_A,\mathbf{k}_B}} + \underbrace{\frac{1}{\sqrt{2}}\Big[|0\rangle_{\mathbf{k}_A}|1\rangle_{\mathbf{k}_B} + |1\rangle_{\mathbf{k}_A}|0\rangle_{\mathbf{k}_B}\Big]}_{|\phi_{\text{coh1}}\rangle_{\mathbf{k}_A,\mathbf{k}_B}} \\
&\quad + \underbrace{\frac{1}{2}\Big[\sqrt{2}|1\rangle_{\mathbf{k}_A}|1\rangle_{\mathbf{k}_B} + |0\rangle_{\mathbf{k}_A}|2\rangle_{\mathbf{k}_B} + |2\rangle_{\mathbf{k}_A}|0\rangle_{\mathbf{k}_B}\Big]}_{|\phi_{\text{coh2}}\rangle_{\mathbf{k}_A,\mathbf{k}_B}} + \ldots .
\end{aligned} \tag{5.66}$$

The substates have a *binomial form* and contain a definite number of photons $N = 0, 1, 2 \ldots \infty$. We have specifically identified by underbrackets the previously introduced $N = 1$ and $N = 2$ substates (5.62) and (5.63). The $N = 0$ substate does not contain any photons.

The collective two-mode coherent states $|\Phi_{\text{coh}}\rangle_{\mathbf{k}_A,\mathbf{k}_B}$ in (5.64) contain an average number of $|\alpha|^2 = \langle n\rangle$ photons per mode. The average number is determined by a Poisson probability distribution of specific numbers of photons, contained in number bins as illustrated in Fig. 5.11 for the cases $\langle n\rangle = 1$ and $\langle n\rangle = 2$. The two-mode probability distributions are similar to the one-mode ones shown in Fig. 3.7.

> The *collective* coherent or Glauber state (5.66) accounts for the inherent quantum mechanical uncertainty through a Poisson distribution in the number of photons as illustrated in Fig. 5.11.

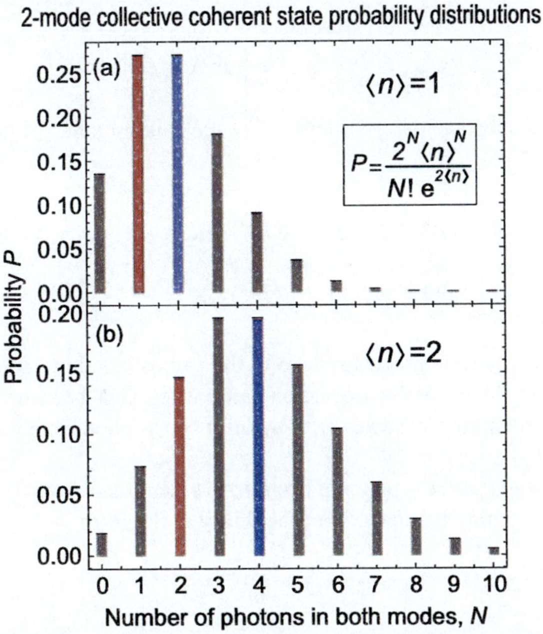

Fig. 5.11 Substructure of the collective coherent state (5.64) containing an average number of photons per mode $|\alpha|^2 = \langle n\rangle = 1$ and $|\alpha|^2 = \langle n\rangle = 2$. The number substates N occur with a probability $P = |c_N^\alpha|^2$ given by a Poisson distribution. The red bars identify the substates with $N = \langle n\rangle$. The distributions peak approximately at $N = 2\langle n\rangle$, indicated in blue, corresponding to the total number of photons in both modes

The substates $|\phi_{\text{cohN}}\rangle_{\mathbf{k}_A,\mathbf{k}_B}$ defined by (5.66) containing a total of $N = 0, 1, 2...\infty$ photons in both modes occur with the indicated Poisson probability distribution $P = |c_N^\alpha|^2$. The red bars identify the substates with $N = \langle n\rangle$. The blue bars identify the total number of photons in both modes, $N = 2\langle n\rangle$, reflecting the approximate peak values of the distributions (also see Fig. 16.3 below).

Remarkably, in both first and second order, the collective states $|\Phi_{\text{coh}}\rangle$ and their substates $|\phi_{\text{cohN}}\rangle$ give diffraction patterns of the same shape, independent of $\langle n\rangle$ and N, respectively. Thus the simplest $N = 1$ and $N = 2$ substates (5.62) and (5.63) already yield the representative first and second order patterns.

This arises from the fact that in all cases all four first order coherence operators $\mathbf{O}^{(1)}$ defined in Table 4.3 and all sixteen second order operators $\mathbf{O}^{(2)}$ in Table 5.1 have the same matrix elements for a given coherent quantum state, and only the size of the matrix element changes for different states. The matrix elements for the collective coherent states $|\Phi_{\text{coh}}\rangle$ depend on the average number of photons they contain according to

$$\langle \Phi_{\text{coh}} | \mathbf{O}^{(1)} | \Phi_{\text{coh}} \rangle = \langle \mathbf{O}^{(1)} \rangle_{\text{coh}} = \langle n \rangle$$
$$\langle \Phi_{\text{coh}} | \mathbf{O}^{(2)} | \Phi_{\text{coh}} \rangle = \langle \mathbf{O}^{(2)} \rangle_{\text{coh}} = \langle n \rangle^2 \tag{5.67}$$

and those of the substates $|\phi_{\text{cohN}}\rangle$ depend on the total number of photons N they contain as

$$\langle \phi_{\text{cohN}} | \mathbf{O}^{(1)} | \phi_{\text{cohN}} \rangle = \langle \mathbf{O}^{(1)} \rangle_{\text{cohN}} = \frac{N}{2}$$
$$\langle \phi_{\text{cohN}} | \mathbf{O}^{(2)} | \phi_{\text{cohN}} \rangle = \langle \mathbf{O}^{(2)} \rangle_{\text{cohN}} = \frac{N(N-1)}{4}. \tag{5.68}$$

For $N = 1, 2$ the matrix elements reduce to the values listed in Tables 4.3 and 5.1. This shows that the shape of the diffraction patterns remains the same, while the total detection probability simply scales with the number of photons, either $\langle n \rangle$ or N, as expected.

The evolution of coherent patterns from first ($m = 1$) to second ($m = 2$) order is best revealed by writing the detection probability in the form

$$\left\langle \phi_{\text{x}} | \mathbf{P}^{(m)}(\boldsymbol{\rho}_1, \boldsymbol{\rho}_2) | \phi_{\text{x}} \right\rangle = \langle \mathbf{O}^{(m)} \rangle_{\text{x}} \, \mathcal{G}^{(m)}(\boldsymbol{\rho}_1, \boldsymbol{\rho}_2). \tag{5.69}$$

It factors into the position- and order-dependent *shape* of the pattern $\mathcal{G}^{(m)}(\boldsymbol{\rho}_1, \boldsymbol{\rho}_2)$ and an overall order-dependent scaling factor $\langle \mathbf{O}^{(m)} \rangle_{\text{x}}$, given for $m = 1, 2$ by (5.67) and (5.68).

For the example of a circular flat-top source, the patterns given by (4.141) and (5.43) have scaling factors (matrix elements) $\langle \mathbf{O}^{(1)} \rangle_{\text{coh1}} = \langle \mathbf{O}^{(2)} \rangle_{\text{coh2}} = 1/2$ and for single point detection the shape evolves according to

$$\mathcal{G}^{(1)}_{\text{coh1}}(\boldsymbol{\rho}, \boldsymbol{\rho}) = \left[\frac{2J_1(kR\rho/z_0)}{kR\rho/z_0} \right]^2 \Longrightarrow \mathcal{G}^{(2)}_{\text{coh2}}(\boldsymbol{\rho}, \boldsymbol{\rho}) = \left[\frac{2J_1(kR\rho/z_0)}{kR\rho/z_0} \right]^4. \tag{5.70}$$

The second order pattern is simply the squared first order pattern. The *coherent* first and second order patterns are also the same for single point detection, (ρ, ρ), and two-point detection, $(\rho, -\rho)$.

5.6.2 *Implication for the First Order Pattern*

The classical treatment of a coherent source gives a diffraction intensity pattern, calculated according to $I \propto E^*E = EE^*$. Remarkably, the detection process remains unspecified. The classical coherence picture corresponds to the quantum assumption of coherent one-photon states in the form of the coherent substate of the form (5.62), with $N = 1$ photon, or the collective coherent state of the form (5.64) with an average number $\alpha^2 = \langle n \rangle = 1$ of photons. The one-photon coherent quantum states are produced by photon sources with a photon degeneracy parameter less than 1.

If the source produces single photon coherent states with $N = 1$, they arrive as individual photons, so that a photon number resolving detector would measure the same $N = 1$ arrival events as a detector that cannot distinguish between different numbers of arriving photons, e.g. a charge integrating detector. If the source emits collective coherent states with $\langle n \rangle = 1$, a photon number resolving detector would pick out the $N = 1$ substate, represented by the red bar in Fig. 5.11a, while a charge integrating detector would naturally average over the entire Poisson distribution in that figure. The so recorded quantum patterns are the same. This is the deeper explanation why the detection process for a coherent classical source does not need to be specified.

5.6.3 *Implications for the Second Order Pattern*

The second order coherent pattern involves the detection of two photons corresponding to two-photon substates of the form (5.63), with $N = 2$ photons, or the collective coherent state of the form (5.64) with an average number $\alpha^2 = \langle n \rangle = 2$ photons. When the degeneracy parameter exceeds unity, different substates with $N \geq 2$ are typically created. It is then necessary to pick out the pattern of the specific $N = 2$ substate by *coincidence detection* of two photons.

The same coincidence method is required when the emitted states are collective coherent states with $\langle n \rangle \geq 2$, like those emitted by an optical laser. One then needs to detect the simultaneous arrival of two photons at the same point or of two photons arriving at two points at the same time. Such a measurement picks out the $N = 2$ substate defined by the arrival of photon pairs.

There is an important special case where two-photon coincidence detection is *not* required. If a source emits only the specific collective coherent state with an average number of $\langle n \rangle = 2$ photons per mode, the number of photons arriving at the detector ranges from $N = 1$ up to about $N = 10$ with the probabilities for different N shown in Fig. 5.11b. This special collective state may be imaged in two alternative ways. The two-photon coincidence detection scheme will pick out the $N = 2$ substate defined by the arrival of photon pairs. On the other hand, a CCD detector will sum over all arriving numbers of photons and thus record the entire collective $\langle n \rangle = 2$ pattern. The so recorded quantum patterns are the same.

5.6.3.1 Evolution of the Degree of Second Order Coherence

For the chaotic case, the degree of second order coherence defined in (5.4) contains characteristic interference structure which was first observed by HBT. As mentioned in conjunction with Fig. 5.2(c) the red curve in that figure quantitatively represents $g^{(2)}(\rho_1, \rho_2)$, which for a chaotic source is given by (5.7).

In contrast, for a coherent source $g^{(2)}$ is constant since the numerator in (5.4) factors into the denominator which contains the same structure. Hence any structure is normalized out. However, $g^{(2)}$ may still assume values that differ from unity, which is a necessary condition for second order coherence according to (5.9). A particularly interesting case is the evolution of the value of $g^{(2)}$ of the state $|\phi_{\text{cohN}}\rangle$ with increasing number N of photons.

For the coherent case, the degrees of first and second order coherence, i.e. $m = 1, 2$, may be expressed as

$$g_{\text{x}}^{(m)} = \frac{\left\langle \mathbf{O}^{(m)} \right\rangle_{\text{x}}}{\left\langle \mathbf{O}^{(1)} \right\rangle_{\text{x}}^{m}} \tag{5.71}$$

where x labels the quantum state of interest. This allows us to calculate $g_{\text{x}}^{(m)}$ for $m = 1, 2$ by use of the matrix elements given by (5.67) and (5.68). As expected, we obtain for the collective coherent state (5.64)

$$g_{\text{coh}}^{(1)} = g_{\text{coh}}^{(2)} = 1, \tag{5.72}$$

while for the N-photon substate (5.66) we have

$$g_{\text{cohN}}^{(1)} = 1, \text{ and } g_{\text{cohN}}^{(2)} = 1 - \frac{1}{N}. \tag{5.73}$$

While the degree of first order coherence is unity for both states Φ_{coh} and ϕ_{cohN}, this is not the case for the degrees of second order coherence. Therefore only the collective coherent states (5.64) are second order coherent, independent of $\langle n \rangle$, while $g_{\text{cohN}}^{(2)}$ approaches second order coherence only for large N. This is plotted in Fig. 5.12.

The figure extends our conclusion reached in Sect. 4.2.6 about the meaning of coherence. We now see that in both first and second orders, coherence increases with photon number density.

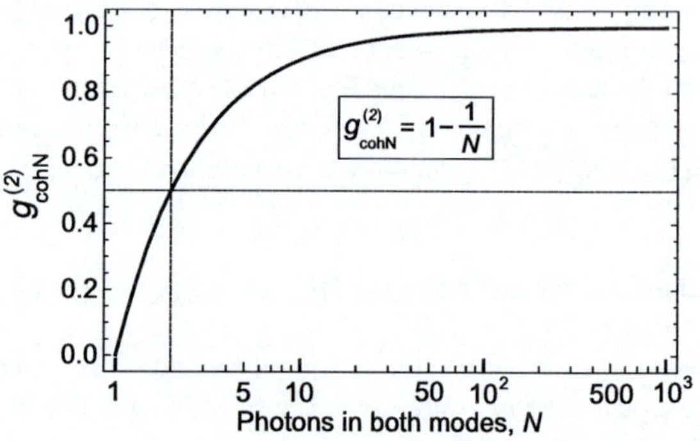

Fig. 5.12 Evolution of the degree of second order coherence with occupation number N according to (5.73) for the substates $|\phi_{\text{cohN}}\rangle$ defined by (5.66). Second order coherence is only reached in the limit of large N

5.6.3.2 Observation of Change from First to Second Order Diffraction

In Fig. 5.13 we show a 2D rendition of the experimentally observed evolution of the 2D diffraction pattern from first order to second order. It supports the calculated patterns previously shown in Fig. 5.9.

While the experimental realization [94, 95] will be discussed in Sect. 15.9.1.1 after we have properly accounted for the non-linear interactions of photons with atoms, we here just present the result. In the experiment a thin film is resonantly excited by incident x-rays from a XFEL tuned to the atomic resonance. At low intensity, the excited atoms decay back to the ground state by spontaneously emitting photons. This results in the conventional first order Airy pattern previously shown as the black curve in Fig. 4.10, which is shown again on a log scale as a black curve in Fig. 5.13a.

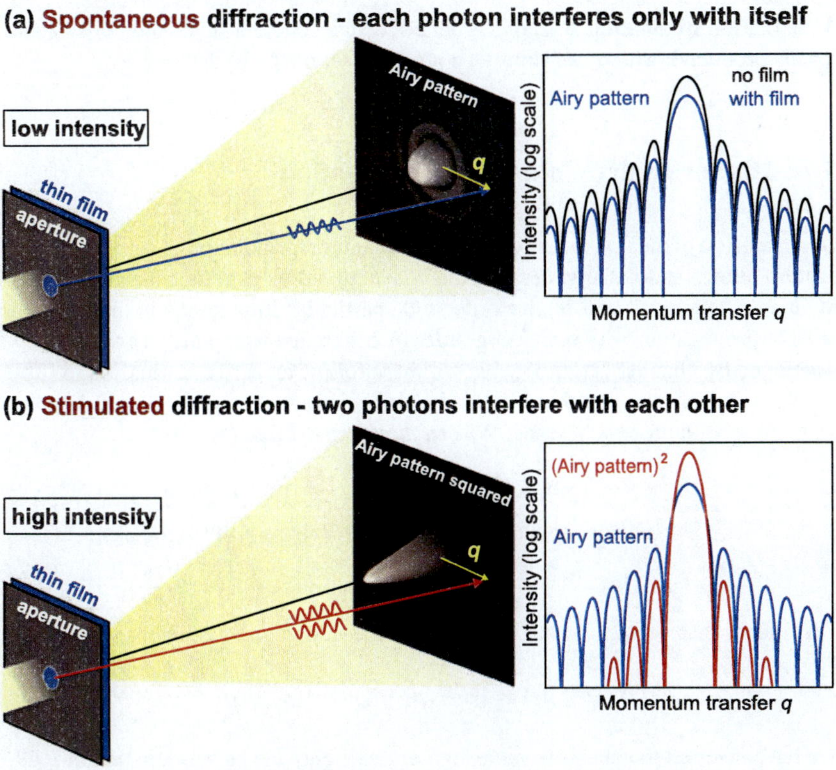

Fig. 5.13 Comparison of the diffraction patterns of a thin film in a circular aperture that is coherently illuminated at either low or high incident intensity. **a** The black intensity pattern is that of the aperture without the film and the blue pattern the conventional (spontaneous) pattern in the presence of the film. **b** The red pattern is the stimulated pattern of the film. In all cases, the incident photon energy is the same, tuned to a core-to-valence resonance in the uniform film. Both patterns can be recorded with the charge integrating CCD detector, without two-photon coincidence detection

In the presence of the film the diffracted intensity is simply dimmed due to absorption of a fraction of the incident photons, as shown by the blue curve.

If now the intensity is increased to a point where the *spontaneous* decays of the excited atoms in the film, which are driven by the virtual photons of the zero-point field (see Sect. 3.3.2), are replaced by *stimulated* decays, driven by real photons in the incident beam, the diffracted radiation is increasingly emitted into the forward direction. This corresponds to the simple notion that in stimulated decays the driving photons clone themselves and the emission direction is no longer random but in the direction of the incident photons [96, 97]. With increasing number of incident photons, the spontaneous first order coherent response is transformed into a stimulated second order coherent one.

The first order blue pattern in Fig. 5.13b changes into the second order red pattern, in accord with Fig. 5.9. The experiment utilized a charge integrating CCD detector *without* coincidence detection. This is in agreement with our discussion in Sect. 5.6.3. At saturation stimulation, a second order collective coherent state with $\langle n \rangle = 2$ photons is produced, which can simply be recorded with a CCD detector.

5.7 Higher Order Coherence Propagation

Our extension of the first to the second order quantum formulation of coherence propagation allows us to also generalize the conventional brightness concept discussed in Chap. 4 to second and higher orders. Of particular importance is the evolution of light propagation with increasing order of coherence which utilizes the concepts developed by Glauber [2, 3, 61]. We will here follow an earlier discussion in [38].

We start from the fact that a second order coherent source has a unit degree of first and second order coherence. We can therefore express (5.4) as

$$g^{(2)}(\boldsymbol{\rho}_1, \boldsymbol{\rho}_2) = \frac{G^{(2)}(\boldsymbol{\rho}_1, \boldsymbol{\rho}_2)}{G^{(1)}(\boldsymbol{\rho}_1, \boldsymbol{\rho}_1)\, G^{(1)}(\boldsymbol{\rho}_2, \boldsymbol{\rho}_2)} = \frac{\langle \mathbf{P}^{(2)}(\boldsymbol{\rho}_1, \boldsymbol{\rho}_2)\rangle}{\langle \mathbf{P}^{(1)}(\boldsymbol{\rho}_1, \boldsymbol{\rho}_1)\rangle \langle \mathbf{P}^{(1)}(\boldsymbol{\rho}_2, \boldsymbol{\rho}_2)\rangle} = 1. \tag{5.74}$$

This means that we can alternatively write for a second order coherent source

$$G^{(2)}(\boldsymbol{\rho}_1, \boldsymbol{\rho}_2) = \left| G^{(1)}(\boldsymbol{\rho}_1, \boldsymbol{\rho}_1)\, G^{(1)}(\boldsymbol{\rho}_2, \boldsymbol{\rho}_2) \right|. \tag{5.75}$$

We have inserted the absolute value sign on the right side because, in general, $G^{(1)}$ may be complex. Following Glauber [1], we define *n*th-order coherence through the equivalence of the correlation function $G^{(n)}(\boldsymbol{\rho}_1, \boldsymbol{\rho}_2, ...\boldsymbol{\rho}_n)$ and a product of n first order correlation functions according to

$$G^{(n)}(\rho_1, \rho_2 ... \rho_n) = \left| \prod_n G^{(1)}(\rho_n, \rho_n) \right|$$

or

$$\left\langle \mathbf{P}^{(n)}(\rho_1, \rho_2 ... \rho_n) \right\rangle = \left| \prod_n \left\langle \mathbf{P}^{(1)}(\rho_n, \rho_n) \right\rangle \right|. \tag{5.76}$$

Similar to the detection symmetry of the second order coherent case by either using a single two-photon detector of two symmetrically placed one-photon detectors, the pattern of an nth order coherent source will be the same if one uses a single n photon detector that measures $G^{(n)}(\rho, \rho, ...\rho)$ or n symmetrically placed one-photon detectors. It is therefore straightforward to generalize our expressions for the diffraction pattern, which reflects how the emitted radiation pattern expands from the source.

Similar to (5.74), the degree of nth-order coherence is given by

$$g^{(n)}(\rho_1, \rho_2, ...\rho_n) = \frac{\left\langle \mathbf{P}^{(n)}(\rho_1, \rho_2, ...\rho_n) \right\rangle}{\left| \prod_n \left\langle \mathbf{P}^{(1)}(\rho_n, \rho_n) \right\rangle \right|} \tag{5.77}$$

which is unity for an nth-order coherent source. In this case the interference structure contained in both the numerators and denominators is normalized out. In order to reveal the interference structure it is then convenient to instead plot the normalized quantity $\left\langle \mathbf{P}^{(n)}(\rho, \rho, ...\rho) \right\rangle / \left\langle \mathbf{P}^{(n)}(0, 0, ...0) \right\rangle$ which we shall discuss now.

5.7.1 Higher Order Patterns of a Coherent Source

The first and second order matrix elements given by (5.67) and (5.68) may be generalized to all orders m and are given by

$$\left\langle \Phi_{\text{coh}} \left| \mathbf{O}^{(m)} \right| \Phi_{\text{coh}} \right\rangle = \langle n \rangle^m \tag{5.78}$$

and for the substates we have

$$\left\langle \phi_{\text{cohN}} \left| \mathbf{O}^{(m)} \right| \phi_{\text{cohN}} \right\rangle = \frac{1}{2^m} \frac{N!}{(N-m)!} \tag{5.79}$$

which are valid for all orders of m and are linked by the sum rule,[8]

[8] We also have

$$\frac{1}{2^m} \sum_{N=m}^{\infty} \frac{N!}{(N-m)!} |c_N^{\alpha}|^2 = \frac{1}{2^m} \left[\sum_{N=0}^{\infty} N |c_N^{\alpha}|^2 \right]^m \tag{5.80}$$

.

$$\sum_{N=m}^{\infty} \frac{1}{2^m} \frac{N!}{(N-m)!} |c_N^{\alpha}|^2 = \langle n \rangle^m \tag{5.81}$$

In analogy to the second order case described by (5.53), the power-conserving intensity distribution for an nth-order coherent source is given by the following relationships.

The interference pattern of an nth-order coherent flat source that is symmetrical around the optical axis is given by

$$I^{(n)}(\rho) = \underbrace{I_s^{(n)}(0)\, A_{\text{eff}}}_{\mathcal{P}_s^{(n)}} \frac{\langle \mathbf{P}^{(n)}(\rho, \rho, ...\rho)\rangle}{B_{\text{eff}}^{(n)} \langle \mathbf{P}^{(n)}(0, 0, ...0)\rangle} = I_s^{(n)}(0) \frac{A_{\text{eff}}}{B_{\text{eff}}^{(n)}} \mathcal{G}^{(n)}(\rho) \tag{5.82}$$

where the effective area of the detected pattern is given in analogy to the twos-photon expression (5.51) by

$$B_{\text{eff}}^{(n)} = \frac{2\pi \int_0^{\infty} \langle \mathbf{P}^{(n)}(\rho, \rho, ...\rho)\rangle\, \rho\, \mathrm{d}\rho}{\langle \mathbf{P}^{(n)}(0, 0, ...0)\rangle} = 2\pi \int_0^{\infty} \mathcal{G}^{(n)}(\rho)\, \rho\, \mathrm{d}\rho \tag{5.83}$$

As an example we discuss the case of a circular flat-top source of effective source area $A_{\text{eff}} = \pi R^2$. By generalization of the $n = 1, 2$ cases we have the normalized n photon detection probability

$$\frac{\langle \mathbf{P}^{(n)}(\boldsymbol{\rho}, \boldsymbol{\rho}, ...\boldsymbol{\rho})\rangle}{\langle \mathbf{P}^{(n)}(0, 0, ...0)\rangle} = \left[\frac{2J_1(kR\rho/z_0)}{kR\rho/z_0}\right]^{2n}. \tag{5.84}$$

The nth-order coherence area in the detector plane becomes

$$B_{\text{eff}}^{(n)} = 2\pi \int_0^{\infty} \left[\frac{2J_1(kR\rho/z_0)}{kR\rho/z_0}\right]^{2n} \rho\, d\rho \simeq \frac{\lambda^2 z_0^2}{n\,\pi R^2} = \frac{B_{\text{eff}}^{(1)}}{n}. \tag{5.85}$$

The approximate equal sign accounts for the fact that for small n there is a small deviation from the $1/n$ dependence as expressed by (5.57) for $n = 2$. We note that for a source with a symmetrical 2D *Gaussian* distribution the relationship $B_{\text{eff}}^{(n)} = B_{\text{eff}}^{(1)}/n$ holds exactly [38].

The power-conserving intensity is obtained from (5.53) in terms of the real space distance ρ or momentum transfer q as

$$I^{(n)}(\rho) \simeq n I_0 \frac{A_{\text{eff}}}{B_{\text{eff}}^{(1)}} \left[\frac{2J_1(kR\rho/z_0)}{kR\rho/z_0}\right]^{2n} \text{ or } I^{(n)}(q) \simeq n I_0 \frac{A_{\text{eff}}}{B_{\text{eff}}^{(1)}} \left[\frac{2J_1(qR)}{qR}\right]^{2n}. \tag{5.86}$$

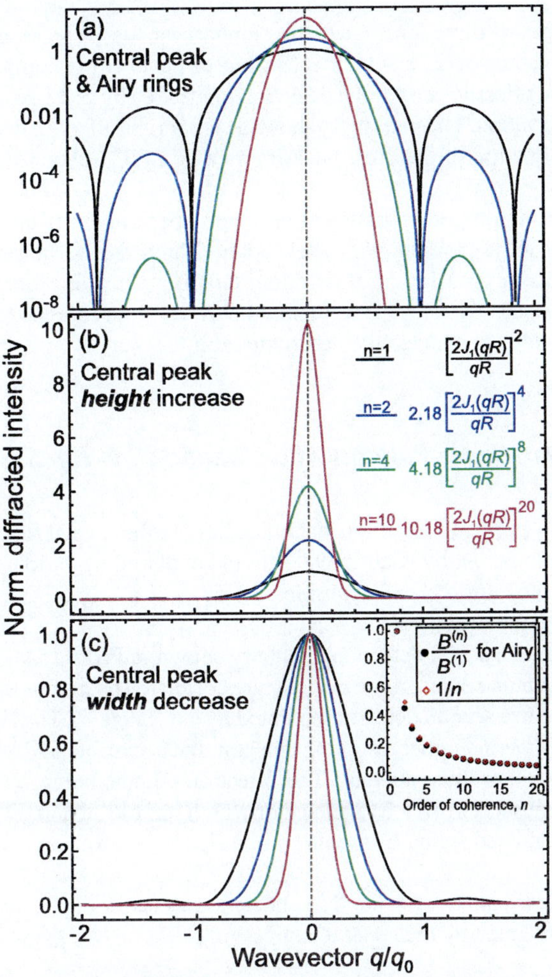

Fig. 5.14 **a** 1D projection of the first order Airy diffraction pattern (black) with increasing order of coherence n (colored), given by (5.86) in the normalized form $I_0 A_{\mathrm{eff}}/B^{(1)}_{\mathrm{eff}} = 1$. The patterns are plotted as a function of q/q_0 where $q_0 = 1.22\pi/R$ is the first node of the Airy pattern. **b** Increase of the central intensity on a linear scale with order n and **c** concomitant decrease of the effective width (area) of the diffraction patterns. The inset in (**c**) shows a comparison of a $1/n$ dependence (red) to the exact reduction of the effective Airy area $B^{(n)}_{\mathrm{eff}}/B^{(1)}_{\mathrm{eff}}$ (black)

The evolution of the pattern with n is shown in Fig. 5.14, plotted as function of momentum transfer q/q_0 where $q_0 = 0.61(2\pi/R)$ is the first zero of the Airy pattern. It extends the $n = 2$ case shown in Fig. 5.9.

The panels (a)–(c) in Fig. 5.14 emphasize different aspects. Panel (a), plotted on a log intensity scale, best reveals the increasing confinement of the overall pattern around the forward direction. Panel (b) emphasizes the increase in central peak intensity by a factor of $\simeq n$, which according to panel (c) is compensated by the reduction of the effective area of the detection plane area by $\simeq 1/n$ as a consequence of power conservation. The inset in (c) shows a comparison of a $1/n$ dependence (red) to the exact reduction of the effective Airy area $B_{\rm eff}^{(n)}/B_{\rm eff}^{(1)}$ obtained by integration in (5.85).

This may be interpreted as growth of the central peak intensity due to the forward interference of all n photons with each other, which due to power conservation reduces the beam expansion by diffraction. Glauber's higher order description of light therefore opens the door for resolution below the conventional diffraction limit [38], which was long considered an absolute limit.

5.7.2 Higher Order Patterns of a Chaotic Flat-Top Source

Before going on we need to mention that higher order effects exist for chaotic sources as well, first pointed out by Richter [98, 99] and explored by various authors [100–104]. In the following we shall consider the case of a chaotic flat-top source as discussed by Pearce et al. [104].

We consider the special detection geometry shown in Fig. 5.15a.

The single photon counts in one detector are correlated by a coincidence measurement with those in a second detector separated by a distance ρ'. The latter detector is assumed to be photon number resolving detector that is capable of distinguishing the arrival of one, two, or three photons. The detection scheme in Fig. 5.15 may also be applied to a position sensitive CCD detector which has the advantage that the entire pattern is accumulated frame-by-frame by correlation of pixel counts in each frame [104].

The coincident patterns for the three detection cases in Fig. 5.15a are shown in (b), and they are given by the general expression [104]

$$\boxed{g_{\rm cha}^{(n)}(\boldsymbol{\rho}_1, \boldsymbol{\rho}_2, ...\boldsymbol{\rho}_n) = (n-1)!\left\{1 + (n-1)\underbrace{\left[\frac{2J_1(kR|\boldsymbol{\rho}_1 - \boldsymbol{\rho}_2|/z_0)}{kR|\boldsymbol{\rho}_1 - \boldsymbol{\rho}_2|/z_0}\right]^2}_{g^{(1)}(|\boldsymbol{\rho}_1 - \boldsymbol{\rho}_2|)}\right\}.} \tag{5.87}$$

As indicated by the underbracket, the patterns can be written in terms of the degree of first order coherence $g^{(1)}(|\boldsymbol{\rho}_1 - \boldsymbol{\rho}_2|)$ previously derived in (4.109)) which is determined by the van Cittert–Zernike theorem as discussed in Sect. 4.6.3. In particular, the central value of the patterns scales as

$$g_{\rm cha}^{(n)}(\boldsymbol{\rho}, \boldsymbol{\rho}, ...\boldsymbol{\rho}) = n! \tag{5.88}$$

The two-photon coincidence case, shown as a red curve, is just the Hanbury Brown–Twiss result for the degree of second order coherence. With increasing n the

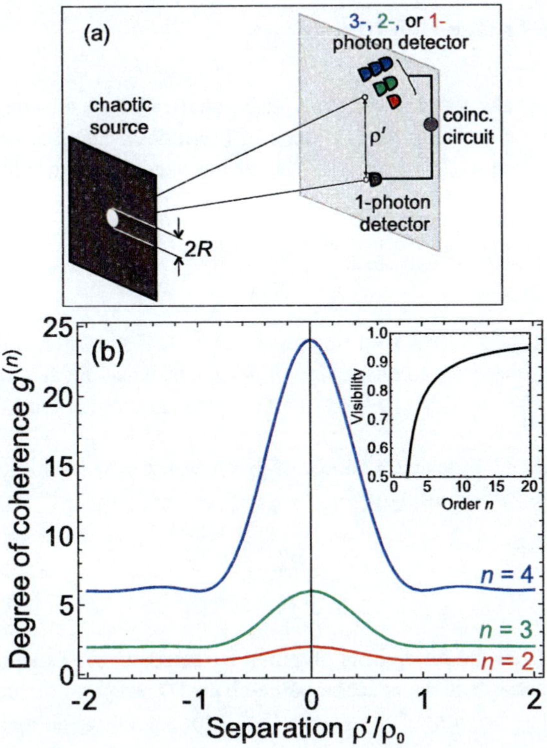

Fig. 5.15 **a** Assumed detection geometry for measurement of the second (red), third (green), and fourth (blue) order degree of coherence of a circular chaotic source. In all cases the coincident intensity between a single one-photon detector and a single n photon detector ($n = 1, 2, 3$) is measured as a function of separation ρ'. One may also use a CCD detector by use of correlation of counts in pixels separated by ρ' [104]. **b** The degree of second order coherence given by (5.87) for $n = 2, 3, 4$. The inset shows the increase in visibility of the central interference peak with n, given by (5.89)

central interference bump has the *same shape* but increases in size relative to the background. The visibility of the central peak is given by

$$V^{(n)} = \frac{g^{(n)}_{\max} - g^{(n)}_{\min}}{g^{(n)}_{\max} + g^{(n)}_{\min}} = \frac{n! - (n-1)!}{n!} = 1 - \frac{1}{n}. \tag{5.89}$$

It increases from 50% to 100% with increasing order n as shown in the inset of Fig. 5.15 (b). In contrast to the coherent case, the shape and width of the central structure ("bunching" peak) remain unchanged from the two-photon HBT result.

5.8 Higher Order Brightness

The increasing peaking of the coherent state pattern with increasing order n shown in Fig. 5.14 can be cast into the language of brightness. The *first order* brightness formulation in Sect. 4.2.6 resulted in the generalized peak brightness expression (4.23) or

$$\mathcal{B}^{(1)}_{\mathrm{peak}} = \frac{h_c}{H}\frac{k_c}{K}\frac{n_{\mathrm{ph}}\,c}{\lambda^3} \tag{5.90}$$

where for clarity we have added the superscript "(1)" for "first order". It reflects the number of coherent photons $n_{\mathrm{coh}} = h_c\,k_c\,n_{\mathrm{ph}}$ of wavelength λ that are born in the coherence volume $V_{\mathrm{coh}} = H\,K\lambda^3$ and emitted into a unit solid angle with unit bandwidth.

With increasing coherence, a source of given area A_{eff} will emit into a coherence cone that is reduced according to

$$\mathrm{d}\Omega^{(n)}_{\mathrm{eff}} = \frac{B^{(n)}_{\mathrm{eff}}}{z_0^2} = \frac{\lambda^2}{n\,A_{\mathrm{eff}}}, \tag{5.91}$$

where the result on the far right is obtained by inserting the expression (5.85) for $B^{(n)}_{\mathrm{eff}}$. The 2D spatial factor n is accompanied by a 1D temporal factor $\sqrt{n}$, so that in nth order the peak brightness expression (5.90) for a coherent source ($h_c = k_c = 1$) takes the form

$$\mathcal{B}^{(n)}_{\mathrm{peak}} \leq \sqrt{n^3}\,\frac{n^{\mathrm{coh}}_{\mathrm{ph}}\,c}{H\,K\,\lambda^3}. \tag{5.92}$$

According to (5.91) the spatial part of the brightness expression is determined by the product

$$\boxed{A_{\mathrm{eff}}\,\mathrm{d}\Omega^{(n)}_{\mathrm{eff}} = \frac{\lambda^2}{n}.} \tag{5.93}$$

This relation may be envisioned in two ways, as illustrated in Fig. 5.16.

In Fig. 5.16a we consider a *point-like atomic source* whose size is taken as $A^{\mu} = \lambda^2$. The emphasis in this case is on the change of the solid emission angle with the order of coherence n. This corresponds to writing (5.93) as

$$\underbrace{\lambda^2}_{A^{\mu}}\;\underbrace{\frac{1}{n}}_{\mathrm{d}\Omega^{(n)}} = \frac{\lambda^2}{n}, \quad n \geq 1. \tag{5.94}$$

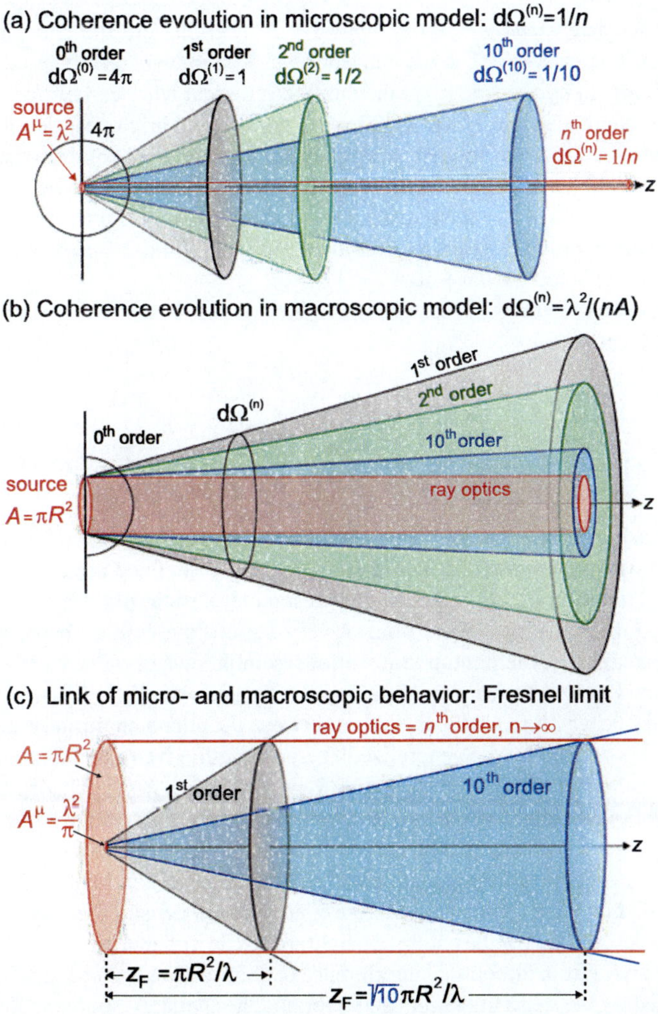

Fig. 5.16 Illustration of the generalized concept of coherence. At a given wavelength λ or photon energy $\hbar\omega$, spatial coherence is simply the energy density of the emitted photons, i.e. the number of photons contained in the coherence cone $\mathrm{d}\Omega^{(n)}$ which decreases with order of coherence n according to (5.93). **a** Microscopic model where the coherence area in a source $A^{\mu} = \lambda^2$. For a chaotic source of 0th order of coherence, the photons are emitted randomly into a solid angle of 4π. For each order of coherence $n \geq 1$ the emission cone is reduced by a factor of 2, so that for a nth order coherent source the solid angle of emission is reduced to $1/n$ as expressed by (5.94). **b** For an extended microscopic source of area A, the total emission cone is defined by the overlap of the microscopic coherence cones with $\mathrm{d}\Omega^{(n)}$ given by (5.95). **c** The transition from microscopic to macroscopic behavior, corresponds to the Fourier limit. It is defined by the intersection of the ray optics beam radius with the first order quantum coherence cone at the Fresnel length $z_{\mathrm{F}} = z_0 = \pi R^2/\lambda$

We may associate a *chaotic source* with $n = 0$. It emits into the entire $\mathrm{d}\Omega = 4\pi$ solid angle. The case $n = 1$ is the conventional (first order) coherence case, which is usually derived by requiring that the emitted spherical wave can be approximated within the coherence cone by a plane wave. In our original brightness definition (4.5) a source of area λ^2 corresponded to a dimensionless coherent solid emission angle of $\mathrm{d}\Omega = 1$ as shown in Fig. 4.2. The coherence cone is then reduced by a factor of 2 to $\mathrm{d}\Omega^{(1)} = 1$. The case $n = 2$ expresses the solid angle of emission of a second order coherent source which is again reduced by a factor of 2 to $\mathrm{d}\Omega^{(2)} = 1/2$. For an nth order coherent source we have $\mathrm{d}\Omega^{(n)} = 1/n$.

Figure 5.16b illustrates the case of a *macroscopic source* of area $A = \pi R^2$, where (5.93) is written as

$$A \underbrace{\frac{\lambda^2}{n\,A}}_{\mathrm{d}\Omega^{(n)}} = \frac{\lambda^2}{n}, \quad n \geq 1. \tag{5.95}$$

With increasing n, the total coherence cone of the source, determined by the overlap of the microscopic atomic cones, is then increasingly confined around the forward direction. The photon paths increasingly resemble particle-like trajectories along the forward direction and in the limit of very large n the emitted beam no longer expands. Amazingly, the photon trajectories resemble rays of light, which may also be envisioned as a wave front propagating like a plane wave.

In Fig. 5.16c we illustrate the transition between the microscopic and macroscopic behavior. It is defined by the generalized Fresnel length, given as a function of order n for a circular flat-top source of radius R by $z_{\mathrm{F}} = \sqrt{n}\pi R^2/\lambda$. This relation also defines the Fourier limit, where as a function of source size R, the ray-like behavior of light changes into wave-like behavior, discussed in more detail below.

Figure 5.16 supports a simple picture of brightness corresponding to the density of the emitted photons. Brightness, coherence, and photon density therefore constitute alternative descriptions of the nature of a light source. In a quantum picture, increasing coherence arises from directional interference of photons. We will see that this picture is supported by the experimental observation that a secondary source, consisting of a thin film in an aperture, may be changed from first to second order coherent as shown in Fig. 5.13 and discussed in more detail in Sect. 15.9.1.

> In space-time, brightness, coherence, and photon density constitute alternative descriptions of the properties of a light source.

5.9 From Rays to Waves to Photons to Rays—Going Full Circle

At the end of this chapter we summarize the presented QED description of the nature of light and show how it maps onto the historical evolution from "ray optics" to "wave optics" to "quantum optics". Remarkably, the description of light goes full circle, with a ray-like description re-emerging in highest order QED.

The historical evolution of the description of light has largely been based on the spatial changes of light beams introduced by objects placed into their paths. The macroscopic "*light ray*" or geometrical optics description utilizes the concepts of classical mechanics and assumes light to travel on straight trajectories like a particle beam. The interaction patterns of light beams with objects are similar to those of a macroscopic beam of particles that propagates along straight lines and may be attenuated or blocked by objects in its path. The addition of the classical laws of refraction and reflection at interfaces of matter with different densities allows the beam to assume new rectilinear propagation directions.

The "*light wave*" description further allows light to propagate into geometrically forbidden regions. This puzzling behavior was prominently discussed by Young and Fresnel around 1800 and is famously evidenced by the Arago-Fresnel-Poisson bright spot behind an opaque beam stop, discussed in Sect. 8.6.2. The rectilinear propagation laws are modified by the Huygens–Fresnel (HF) principle, where the plane perpendicular to the propagation direction is decomposed into "points". The HF principle leaves the size of the points unspecified, but according to our discussion in Sect. 4.5.3 it may be associated with the size of a wavelength. The wavelength λ and cycle time τ_0 are linked through one of the fundamental constants of physics, the speed of light $c = \lambda/\tau_0$. An incident plane wave can be decomposed into "wavelets" emerging from every point of the wavefront. The wavelets expand like spherical waves and their interference accounts for diffraction. If the detector plane is far away, the field interference pattern is the Fourier transform of the shape of the source. The detected intensity is the absolute value squared of the so calculated diffracted field.

The first *order QED description* is that in conventional quantum mechanics. It considers light to be composed of individual *non-interacting* photons, as famously articulated by Dirac. The concept of the photon leads to a second fundamental physical constant, Planck's constant h, which links photon energy and wavelength according to $\mathcal{E} = hc/\lambda$. Wave and particle aspects are furthermore linked through the minimum photon birth volume $V_{\text{coh}} \sim \lambda^3$, which defines the quantum mechanical coherence volume. The birth volume is separable into orthogonal space and time coordinates with separate uncertainty principles between conjugate variables as expressed in the brightness expression (4.5). Only the product of the birth area A_s and solid angle of wave expansion $d\Omega$ is well defined (see (4.8)), and so is the product of birth time τ_s and bandwidth $\Delta\omega/\omega$ (see (4.13)). The equivalence of the wave and first order quantum description of diffraction is the essence of the puzzling wave-particle duality.

The second *order QED description* allows an effective correlation between photons, not allowed in first order. The simplest second order case consists of two photons, whose correlation is imprinted at birth and can be described through different two-photon quantum states whose characteristic patterns are illustrated in Fig. 5.7. Our discussion of the second order case leads to the important new paradigm stated in Sect. 5.5.3 that within QED, diffraction images are direct signatures of different quantum states of light. While the wave theory basically distinguishes coherent, partially coherent, and chaotic light, the quantum formulation is much more specific. The fuzzy term "partially coherent" can now be specified in terms of the coherence properties of different quantum states. Such states may not only be defined in terms of specific number of photons but come in the form of collective quantum states, defined through their probabilistic distribution of numbers of photons in substates.

The *higher order QED description* of light is particularly interesting for the case of the collective coherent state introduced by Glauber. The evolution of the pattern with increasing order of coherence, n, shown in Figs. 5.14 and 5.16, reveals an increasing collimation of the emitted beam. When the diffraction limit is defined in terms of Heisenberg's uncertainty principle as the product of the source area A and solid angle of emission $d\Omega$, the product is shown to decrease as $1/n$. The behavior in first order arises from the linearity of quantum mechanics, yielding an uncertainty principle that, in principle, is only valid in first order and hence can be overcome [38]. The disappearance of diffraction effects with increasing order of coherence n may be interpreted as increasing localization and particle-like behavior of light within the collective state. At large n, the image of the source in a distant detector plane becomes that of the source itself, similar to the ray optics result. The presented results reveal that increasing degrees of coherence of a source can be pictured simply as increasing density of the emitted photons.

The historical evolution of understanding of the spatial nature of light is expressed by the single equation (5.93) which summarizes the development of lateral coherence. This equation, linking the source area A, the solid angle of emission of order of coherence n, $d\Omega^{(n)}$, and the wavelength λ, reveals the following fascinating cycle in our understanding of light:

$$\begin{aligned}
\text{Geometrical lightray concept:} \quad & A\, d\Omega^{(n)} = 0 \\
\text{Wave-photon ambiguity:} \quad & A\, d\Omega^{(1)} = \lambda^2 \\
\text{General QED expression:} \quad & A\, d\Omega^{(n)} = \frac{\lambda^2}{n} \\
\text{QED-lightray link:} \quad & \lim_{n\to\infty} A\, d\Omega^{(n)} = 0
\end{aligned}$$

References

1. R. Glauber, Optical coherence and photon statistics. in *Quantum Optics and Electronics*, ed. by A.B.C. de Witt, C. Cohen-Tannoudji (Gordon and Breach, New York, 1965)
2. R.J. Glauber, Phys. Rev. **130**, 2529 (1963)
3. R.J. Glauber, Phys. Rev. **131**, 2766 (1963)
4. L. Mandel, E. Wolf, Rev. Mod. Phys. **37**, 231 (1965)
5. L. Mandel, E. Wolf, *Optical Coherence and Quantum Optics* (Cambridge University Press, Cambridge, 1995)
6. M. Born, Z. Phys. **37**, 863 (1926)
7. M. Born, Z. Phys. **38**, 803 (1926)
8. M. Born, *My Life: Recollections of a Nobel Laureate* (Taylor and Francis, New York, 1978)
9. R.H. Brown, R.Q. Twiss, Nature **178**, 1046 (1956)
10. R.H. Brown, R.Q. Twiss, Proc. Roy. Soc. Lond. A **242**, 300 (1957)
11. R.H. Brown, R.Q. Twiss, Proc. Roy. Soc. Lond. A **243**, 291 (1958)
12. R. Loudon, *The Quantum Theory of Light*, 3rd edn. (Clarendon Press, Oxford, 2000)
13. M.O. Scully, M.S. Zubairy, *Quantum Optics* (Cambridge University Press, Cambridge, 1997)
14. C.C. Gerry, P.L. Knight, *Introductory Quantum Optics* (Cambridge University Press, Cambridge, 2005)
15. G. Grynberg, A. Aspect, C. Fabre, *Introduction to Quantum Optics, From the Semi-classical Approach to Quantized Light* (Cambridge University Press, Cambridge, 2010)
16. Y. Shih, *An Introduction to Quantum Optics: Photon and Biphoton Physics* (CRC, Boca Raton, Florida, 2011)
17. Z.Y.J. Ou, *Quantum Optics for Experimentalists* (World Scientific, Singapore, 2017)
18. H.A. Kramers, W. Heisenberg, Z. Phys. **31**, 681 (1925)
19. P.A.M. Dirac, Proc. Roy. Soc. A **114**, 243 (1927)
20. P.A.M. Dirac, Proc. Roy. Soc. Lond. A **114**, 710 (1927)
21. J. Als-Nielsen, D. McMorrow, *Elements of Modern X-Ray Physics*, 2nd edn. (Wiley, New York, 2011)
22. G. Greenstein, A.G. Zajonc, *The Quantum Challenge: Modern Research on the Foundations of Quantum Mechanics*, 2nd edn. (Jones and Bartlett, Sudbury, MA, 2006)
23. P. Grangier, G. Roger, A. Aspect, Europhys. Lett. **1**, 173 (1986)
24. E. Wolf, *Introduction to the Theory of Coherence and Polarization of Light* (Cambridge University Press, Cambridge, 2007)
25. R.J. Glauber, Rev. Mod. Phys. **78**, 1267 (2006)
26. I. Silva, J.O. Freire, Hist. Stud. Nat. Sci. **43**, 453 (2013)
27. G.I. Taylor, Proc. Camb. Phil. Soc. **15**, 114 (1909)
28. E.M. Purcell, Nature **178**, 1449 (1956)
29. R.H. Brown, R.Q. Twiss, Proc. Roy. Soc. Lond. A **248**, 222 (1958)
30. R.H. Brown, R.Q. Twiss, Nature **177**, 27 (1956)
31. A.B. Haner, N.R. Isenor, Am. J. Phys. **38**, 748 (1970)
32. C.K. Hong, Z.Y. Ou, L. Mandel, Phys. Rev. Lett. **59**, 2044 (1987)
33. D.N. Klyshko, *Photons and Nonlinear Optics* (Gordon and Breach, New York, 1988)
34. D.N. Klyshko, LJETP **6**, 490 (1967)
35. S.E. Harris, M.K. Oshman, R.L. Byer, Phys. Rev. Lett. **18**, 732 (1967)
36. D. Magde, H. Mahr, Phys. Rev. Lett. **18**, 905 (1967)
37. Z.Y.J. Ou, *Multi-photon Quantum Interference* (Springer, Heidelberg- New York, 2007)
38. J. Stöhr, Adv. Opt. Photonics **11**, 215 (2019)
39. Y. Yoda, T. Suzuki, X.W. Zhang, K. Hirano, S. Kikuta, J. Synchrotron Rad. **5**, 980 (1998)
40. B. Adams, P. Fernandez, W.K. Lee, G. Materlik, D.M. Mills, D.V. Novikov, J. Synchrotron Rad. **7**, 81 (2000)
41. K. Tamasaku, K. Sawada, E. Nishibori, T. Ishikawa, Nat. Phys. **7**, 705 (2011)
42. S. Shwartz, R.N. Coffee, J.M. Feldkamp, Y. Feng, J.B. Hastings, G.Y. Yin, S.E. Harris, Phys. Rev. Lett. **109**(013), 602 (2012)

43. H. Fearn, R. Loudon, Opt. Commun. **64**, 485 (1987)
44. R.A. Campos, B.E.A. Saleh, M.C. Teich, Phys. Rev. A **40**, 1371 (1989)
45. C.H. Holbrow, E. Galvez, M.E. Parks, Am. J. Phys. **70**, 260 (2002)
46. G. Di Giuseppe, et al., Phys. Rev. A **68**, 063 817 (2003)
47. Y.H. Shih, A.V. Sergienko, Phys. Lett. A **186**, 29 (1994)
48. J.D. Jackson, *Classical Electrodynamics* (Wiley, New York, 1999)
49. R. Hadfield, Nat. Photon. **3**, 696 (2009)
50. M.D. Eisaman, J. Fan, A. Migdall, S.V. Polyakov, Rev. Sci. Instrum. **82**, 071 101 (2011)
51. C.J. Chunnilall, I.P. Degiovanni, S. Kück, I. Müller, A.G. Sinclair, Opt. Eng. **53**, 081 910 (2014)
52. I.E. Zadeh, et al., APL Photonics **2**, 111 301 (2017)
53. S.M. Gruner, E.F. Eikenberry, M.W. Tate, X-ray detectors, in *Crystallography of Biological Molecules*, ed. by M.G. Rossmann, E. Arnold (Kluwer Acad. Publ., Dordrecht, 2001), p. 143
54. B. Henrich, A. Bergamaschi, C. Broennimann, R. Dinapoli, E. Eikenberry, I. Johnson, M. Kobas, P. Kraft, A. Mozzanica, B. Schmitt, Nucl. Instrum. Meth. A **607**, 247 (2009)
55. C. Milne et al., Appl. Sci. **7**, 720 (2017)
56. B. Cabrera, R. Clarke, P. Colling, A. Miller, S. Nam, R. Romani, Appl. Phys. Lett. **73**, 735 (1998)
57. O. Thomas, Z.L. Yuan, A. Shields, Nat. Commun. **3**, 644 (2012)
58. I.A. Vartanyants, A. Singer, New J. Phys. **12**, 035 004 (2010)
59. I.A. Vartanyants, A. Singer, Coherence properties of third-generation synchrotron sources and free-electron lasers. in *Synchrotron Light Sources and Free-Electron Lasers. Accelerator Physics, Instrumentation and Science Applications*, ed. by E. Jaeschke, S. Khan, J. Schneider, J. Hastings (Springer, Amsterdam, 2016), p. 821
60. O.Y. Gorobtsov, et al., Phys. Rev. A **95**, 023 843 (2017)
61. U.M. Titulaer, R.J. Glauber, Phys. Rev. **140**, B676 (1965)
62. U.M. Titulaer, R.J. Glauber, Phys. Rev. **145**, 1041 (1966)
63. A. Singer, et al., Phys. Rev. Lett. **111**, 034 802 (2013)
64. A. Singer, et al., Phys. Rev. Lett **113**, 064 801 (2014)
65. H.J. Kimble, M. Dagenais, L. Mandel, Phys. Rev. Lett. **39**, 691 (1977)
66. H. Paul, Rev. Mod. Phys. **54**, 1061 (1982)
67. I.S. Reed, IRE Trans. Inf. Theory **8**, 194 (1962)
68. J.W. Goodman, *Statistical Optics*, 2nd edn. (Wiley, New York, 2015)
69. A.J.F. Siegert, MIT Rad. Lab. Rep. 465 (1943)
70. J. Liu, G. Zhang, Phys. Rev. A **82**, 013 822 (2010)
71. R. Shimizu, K. Edamatsu, T. Itoh, Phys. Rev. A **74**, 013 801 (2006)
72. C. Santori, D. Fattal, J. Vučković, G.S. Solomon, Y. Yamamoto, Nature **419**, 594 (2002)
73. R.B. Patel, A.J. Bennett, I. Farrer, C.A. Nicoll, D.A. Ritchie, A.J. Shields, Nat. Photonics **4**, 632 (2010)
74. E.B. Flagg, A. Muller, S.V. Polyakov, A. Ling, A. Migdall, G.S. Solomon, Phys. Rev. Lett. **104**, 137 401 (2010)
75. A. Neuzner, M. Körber, O. Morin, S. Ritter, G. Rempe, Nat. Photonics **10**, 303 (2016)
76. Y.H. Zhai, X.H. Chen, D. Zhang, L.A. Wu: Phys. Rev. A **72**, 043 805 (2005)
77. W. Martienssen, E. Spiller, Am. J. Phys. **32**, 919 (1964)
78. F.T. Arecchi, Phys. Rev. Lett. **15**, 912 (1965)
79. G. Scarcelli, A. Valencia, Y. Shih, Phys. Rev. A **70**, 051 802(R) (2004)
80. J. Xiong, D.Z. Cao, F. Huang, H.G. Li, X.J. Sun, K. Wang, Phys. Rev. Lett. **94**, 173 601 (2005)
81. Y. Zhou, F.L. Li, B. Bai, H. Chen, J. Liu, Z. Xu, H. Zheng, Phys. Rev. A **95**, 053 809 (2017)
82. D. Bouwmeester, A.K. Ekert, A. Zeilinger (Eds.), *The Physics of Quantum Information: Quantum Cryptography, Quantum Teleportation, Quantum Computation* (Springer, Heidelberg, 2000)
83. J.M. Raimond, M. Brune, S. Haroche, Rev. Mod. Phys. **73**, 565 (2001)
84. R. Horodecki, P. Horodecki, M. Horodecki, K. Horodecki, Rev. Mod. Phys. **81**, 865 (2009)
85. J.W. Pan, Z.B. Chen, C.Y. Lu, H. Weinfurter, A. Zeilinger, M. Żukowski, Rev. Mod. Phys. **84**, 777 (2012)

86. O.Y. Gorobtsov et al., Nat. Commun. **9**, 4498 (2018)
87. E.V. Shuryak, Sov. Phys. JETP **40**, 30 (1975)
88. R.Z. Tai et al., Rev. Sci. Instrum. **71**, 1256 (2000)
89. G. Baym, Acta Phys. Polon. B **29**, 1839 (1998)
90. M. Henny et al., Science **284**, 296 (1999)
91. W.D. Oliver et al., Science **284**, 299 (1999)
92. P. Geyer, U. Sezer, J. Rodewald, L. Mairhofer, N. Dörre, P. Haslinger, S. Eibenberger, C. Brand, M. Arndt, Phys. Scr. **91**, 063 007 (2016)
93. T. Jeltes et al., Nature **445**, 402 (2007)
94. B. Wu, T. Wang, C.E. Graves, D. Zhu, W.F. Schlotter, J.J. Turner, O. Hellwig, Z. Chen, H.A. Dürr, A. Scherz, J. Stöhr, Phys. Rev. Lett. **117**, 027 401 (2016)
95. Z. Chen, D.J. Higley, M. Beye, M. Hantschmann, V. Mehta, O. Hellwig et al., Phys. Rev. Lett. **121**, 137 403 (2018)
96. J. Stöhr, A. Scherz, Phys. Rev. Lett. **115**, 107 402 (2015)
97. J. Stöhr, Phys. Rev. Lett. **118**, 024 801 (2017)
98. T. Richter, Phys. Rev. A **42**, 1817 (1990)
99. T. Richter, Quantum Opt. **3**, 115 (1991)
100. J. Liu, Y. Shih, Phys. Rev. A **79**, 023 819 (2009)
101. T. Iskhakov, A. Allevi, D. Kalashnikov, V. Sala, M. Takeuchi, M. Bondani, M. Chekhova, Eur. Phys. J. Special Topics **199**, 127 (2011)
102. D.G. Monticone et al., Phys. Rev. Lett. **113**, 143 602 (2014)
103. C. Thiel, T. Bastin, J. Martin, E. Solano, J. von Zanthier, G.S. Agarwal, Phys. Rev. Lett. **99**, 133 603 (2007)
104. M.E. Pearce, T. Mehringer, J. von Zanthier, P. Kok, Phys. Rev. A **92**, 043 831 (2015)

Part II
Semi-classical Theory of X-Ray Interactions with Matter

Chapter 6
Semi-classical Response of Atoms to Electromagnetic Fields

6.1 Introduction

The complete treatment of the interactions of x-rays with charges and spins requires quantum theory, yet many phenomena of interest can be accounted for by use of semi-classical concepts which can also be cast into simple pictures. In this chapter we follow the historical development and describe the interactions of x-rays with matter by using a classical field or wave-based description of light and the simplest quantum description of matter in terms of individual atoms and their electronic shells. This semi-classical description will lend itself to an extension in the following chapter, based on linking the fundamental *atomic* building blocks through interatomic bonds that to a good approximation only changes the *valence* electronic structure. In contrast to optical photons, the interaction of x-rays with matter may be cast into the language of atomic-like interaction parameters, known as atomic scattering factors.

We start with the simplest interaction process, where the oscillating electromagnetic **E**- and **H**-fields wiggle the electrons and spin, respectively. The reradiated fields by the oscillating electrons and spins are what we call *Thomson scattering*. We describe this process in terms of the incident and radiated fields and in the process introduce the concept of the x-ray *scattering length* and *scattering cross section*, which is extended from individual electrons and spins to their collective response in atoms.

We then discuss processes involving the intra-atomic shell structure leading to *x-ray absorption* and emission and the closely related process of *resonant scattering*. These processes involve excitation of core electrons and are today typically pictured in a quantum picture of light where a photon transfers its energy to a bound electron, i.e. the photoelectric effect. We will here describe this interaction as a field-driven excitation. This picture will later allow us to describe non-linear x-ray interactions as strong-field effects, which corresponds to cooperative multi-photon processes in the quantum theory of light.

J. Stöhr, *The Nature of X-Rays and Their Interactions with Matter*, Springer Tracts in Modern Physics 288, https://doi.org/10.1007/978-3-031-20744-0_6

The conventional wave-based interaction of light with matter is founded on the important *Born approximation* which we will illustrate. It is often used to describe only non-resonant Thomson scattering, but we will extend it to include resonant effects. The Born approximation then includes dispersion corrections, historically referred to as "anomalous" x-ray scattering, which facilitates the description of resonant interactions. We also review the extension of the fundamental *optical theorem* to the x-ray range.

We continue with the discussion of the famous *Kramers-Kronig relations* that provide a quantitative link between the real and imaginary part of the electronic response functions. These relations have proven invaluable in conjunction with experimental data, but one needs to realize that the classical treatment by itself does not provide accurate values for the response functions. This is overcome only by quantum theory discussed in later chapters.

In the last section we present the Henke-Gullikson formulation of the x-ray response, which today underlies comprehensive web-based data. It covers the atomic response from approximately 100 eV to 10 keV and self-consistently fulfills the Kramers-Kronig relations and important photon energy dependent sum rules. We will illustrate these sum rules for the electromagnetic response of Fe atoms from the infrared to the hard x-ray range.

6.2 X-Ray Thomson Scattering by Electrons and Spins

We start by considering the simplest and most fundamental form of x-ray scattering by the charge and spin of a single *free electron* without loss of energy and conservation of momentum [1, 3]. This process is called *Thomson scattering*. The scattering processes proceed in a simple intuitive picture where the electric field wiggles the charge and the magnetic field wiggles the spin, and the so-created electric and magnetic dipoles reradiate the electromagnetic wave. Since scattering processes involving higher multi-pole radiation are much weaker than the lowest order dipole processes, our formalism gives a good description of the general scattering processes.

6.2.1 Response of an Electron and Its Spin to the Incident Field

We assume an incident polarized EM wave derived from Maxwell's wave equation given by (3.13) and describe the electric and magnetic parts of the field as

$$\mathbf{E}(\mathbf{r}, t) = \boldsymbol{\epsilon}\, E_0\, \mathrm{e}^{-\mathrm{i}(\omega t - \mathbf{k}\cdot\mathbf{r})} , \tag{6.1}$$

$$\mathbf{B}(\mathbf{r}, t) = \frac{1}{c}(\mathbf{k}_0 \times \boldsymbol{\epsilon}) E_0\, \mathrm{e}^{-\mathrm{i}(\omega t - \mathbf{k}\cdot\mathbf{r})} . \tag{6.2}$$

This notation of the fields corresponds to omission of the complex conjugate term that induces stimulated decays. By convention, the polarization vector $\boldsymbol{\epsilon}$ is the unit electric field vector $\boldsymbol{\epsilon} = \mathbf{E}/E$ and the incident wave direction is specified by the unit wavevector $\mathbf{k}_0 = \mathbf{k}/k$. For illustration purposes we shall in the following assume a linearly polarized EM wave (where $\boldsymbol{\epsilon}$ is real).

We shall describe the incident and scattered fields by unprimed ($\mathbf{E}, \mathbf{B}$) and primed ($\mathbf{E}', \mathbf{B}'$) fields, respectively, and use a simple intuitive picture for the scattering process. The incident EM wave first induces an accelerated motion of the charge and spin that results in an oscillating electric or magnetic dipole. The dipole then reradiates the E, M and the polarization and amplitude of the radiated wave define the scattered intensity.

The incident EM wave exhibits a force $\mathbf{F} = m_e\mathbf{a}$ on the charge leading to an acceleration $\mathbf{a} = -e\mathbf{E}/m_e$ in the direction of the electric field $\mathbf{E}$. Since the x-ray electric field oscillates rapidly in time according to $\mathbf{E}(t) = \mathbf{E}_0 e^{-i\omega t}$, where $\omega \sim 10^{18}$ s^{-1} in the x-ray region, the charge motion is that of an electric dipole. Since the acceleration $\mathbf{a}$ is related to the second time derivative $\ddot{\mathbf{p}} = -e\mathbf{a}$ of the electric dipole moment $\mathbf{p}(t) = -e\,\mathbf{x}(t)$, we obtain

$$\mathbf{p}(t) = -\frac{e^2}{m_e\omega^2}\,\mathbf{E}_0\,e^{-i\omega t}. \tag{6.3}$$

For spin scattering we consider how the magnetic field acts on the spin. When a spin is initially oriented at an angle from a fixed magnetic field $\mathbf{H} = \mathbf{B}/\mu_0$, the spin precesses around the field according to the torque equation $\mathbf{T} = \hbar\,\mathrm{d}\mathbf{s}(t)/\mathrm{d}t$

$$\frac{\mathrm{d}\mathbf{s}(t)}{\mathrm{d}t} = -\frac{e}{m_e}\,\mathbf{s}(t) \times \mathbf{B} \tag{6.4}$$

as illustrated in Fig. 6.1. In (6.4), the spin $|\mathbf{s}| = 1/2$ is a number without units.

For a linearly polarized EM wave the magnetic field $\mathbf{B}(t) = \mathbf{B}_0 e^{-i\omega t}$ oscillates along an axis in space and changes sign periodically. According to (6.4) a sign change in $\mathbf{B}(t)$ leads to a sign change of the torque $\mathbf{T} = \hbar\,\mathrm{d}\mathbf{s}(t)/\mathrm{d}t$ direction and thus the vector $\mathbf{s}$ rapidly precesses back-and-forth on a cone around the axis of $\mathbf{B}(t)$, as shown in Fig. 6.1. The component $\mathbf{s}_0$ is approximately constant, while the small perpendicular component $\mathbf{s}_d$ oscillates with the frequency of the EM wave and represents a magnetic dipole. With $\mathbf{m} = -2\mu_B\mathbf{s}_d$ and $\mu_B = e\hbar\mu_0/2m_e$ we obtain for the oscillating magnetic dipole moment

$$\mathbf{m}(t) = \mathrm{i}\,\frac{e^2\hbar\mu_0}{\omega m_e^2}\,\mathbf{s} \times \mathbf{B}_0\,e^{-i\omega t}, \tag{6.5}$$

where $\mathbf{s} \sim \mathbf{s}_0$ is the initial spin direction, whereas the electric dipole oscillates in the direction of $\mathbf{E}$, and the magnetic dipole oscillates in a direction that is perpendicular to $\mathbf{B}$ and $\mathbf{s}$. This leads to different polarization effects in the scattering as discussed below.

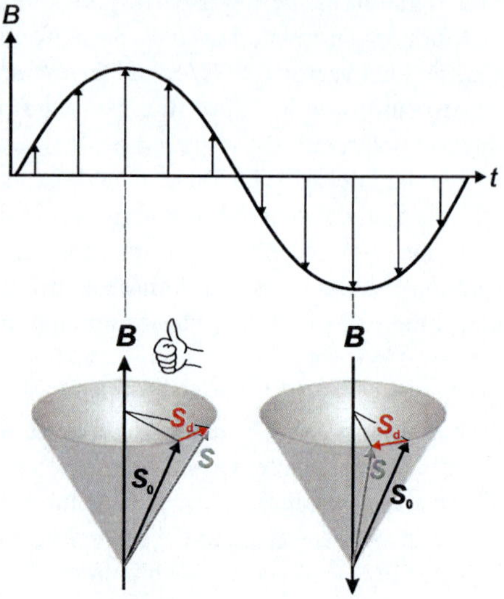

Fig. 6.1 Precessional motion of a spin in the magnetic field $\mathbf{B}(t)$ of a linearly polarized EM wave, according to the right hand rule. When the oscillating field changes direction, so does the precession direction of s. If we write $\mathbf{s} = \mathbf{s}_0 + \mathbf{s}_d$, the perpendicular component $\mathbf{s}_d$ changes sign periodically and therefore represents an oscillating magnetic dipole

The induction of electric and magnetic dipole moments is the first step in the x-ray scattering process. The second step is simply reradiation of EM waves by the oscillating electric and magnetic dipoles.

6.2.2 The Scattered Dipole Field

In general, the radiated dipole fields look somewhat different in the region near the dipole and far from it. Since we are interested in the scattered radiation, we need the fields in the spatial region far from the dipole. This is usually called the "wave region" in textbooks since the radiated dipolar fields have a rather simple spherical wave form $e^{ik'r}/r$. The relevant dipole fields have been derived by Jackson [1], and we shall use his results for convenience.[1]

[1] In some textbooks [2–4] the right sides of (6.7) and (6.8) are written in terms of the second time derivative of the dipole moment $\ddot{\mathbf{p}}(t^*)$ at the retarded time $t^* = t - r/c$, e.g. for (6.7)

$$\mathbf{E}'(t) = \frac{1}{4\pi\epsilon_0 c^2}\,\frac{1}{r}\,\left[\mathbf{k}'_0 \times (\mathbf{k}'_0 \times \ddot{\mathbf{p}}(t^*))\right]_{t^*}. \tag{6.6}$$

The retarded time accounts for the fact that it takes a finite time r/c for the EM wave to travel the distance r from the oscillating dipole to the observer. Since we consider a point charge with a very small oscillation amplitude, we can assume that the retardation is constant over the source volume. We then have $\ddot{\mathbf{p}}(t^*) = (e^2/m_e)\mathbf{E}(t^*)$ and with $k' = \omega/c$ and $\mathbf{E}(t^*) = \mathbf{E}_0 e^{-i\omega(t-r/c)} = \mathbf{E}(t)e^{ik'r}$ we obtain $\ddot{\mathbf{p}}(t^*) = -\omega^2\mathbf{p}(t)e^{ik'r}$. This accounts for the different forms of (6.7) and (6.6).

The fields radiated by an electric dipole $\mathbf{p}(t)$ are

$$\mathbf{E}'(t) = \frac{\omega^2}{4\pi\epsilon_0 c^2}\frac{e^{ik'r}}{r}[\mathbf{k}_0' \times \mathbf{p}(t)] \times \mathbf{k}_0' \tag{6.7}$$

$$\mathbf{B}'(t) = \frac{1}{c}\mathbf{k}_0' \times \mathbf{E}'(t) = \frac{\omega^2}{4\pi\epsilon_0 c^3}\frac{e^{ik'r}}{r}[\mathbf{k}_0' \times \mathbf{p}(t)]. \tag{6.8}$$

The fields radiated by a magnetic dipole $\mathbf{m}(t)$, defined as that arising from a current loop

$$\mathbf{m} = -\frac{e\,\mu_0}{2}(\mathbf{r} \times \mathbf{v}) \tag{6.9}$$

have a similar form [1].[2]

The fields radiated by a magnetic dipole $\mathbf{m}(t)$ are

$$\mathbf{B}'(t) = \frac{\omega^2}{4\pi c^2}\frac{e^{ik'r}}{r}[\mathbf{k}_0' \times \mathbf{m}(t)] \times \mathbf{k}_0' \tag{6.11}$$

$$\mathbf{E}'(t) = c\,\mathbf{B}'(t) \times \mathbf{k}_0' = -\frac{\omega^2}{4\pi c}\frac{e^{ik'r}}{r}[\mathbf{k}_0' \times \mathbf{m}(t)]\,. \tag{6.12}$$

The two scattering processes are illustrated in Fig. 6.2, and they clearly show the dipolar nature of the electric and magnetic scattering processes.

It is now simple to derive the scattered field for charge scattering by combining (6.3) and (6.7), and we obtain

$$\mathbf{E}'(t) = -\frac{1}{4\pi\epsilon_0}\frac{e^2}{m_e c^2}\frac{e^{ik'r}}{r}[\mathbf{k}_0' \times \mathbf{E}(t)] \times \mathbf{k}_0'. \tag{6.13}$$

Similarly we can combine (6.5) and (6.12) to obtain the spin scattering amplitude, remembering that $c^2 = 1/\epsilon_0\mu_0$,

$$\mathbf{E}'(t) = i\frac{1}{4\pi\epsilon_0}\frac{e^2}{m_e c^2}\frac{\hbar\omega}{m_e c^2}\frac{e^{ik'r}}{r}[\mathbf{s} \times (\mathbf{k}_0 \times \mathbf{E}(t))] \times \mathbf{k}_0'. \tag{6.14}$$

[2] Again we could have written (6.12) in the retarded form

$$\mathbf{E}'(t) = \frac{1}{4\pi c}\frac{1}{r}\left[\mathbf{k}_0' \times \ddot{\mathbf{m}}(t^*)\right]_{t^*}. \tag{6.10}$$

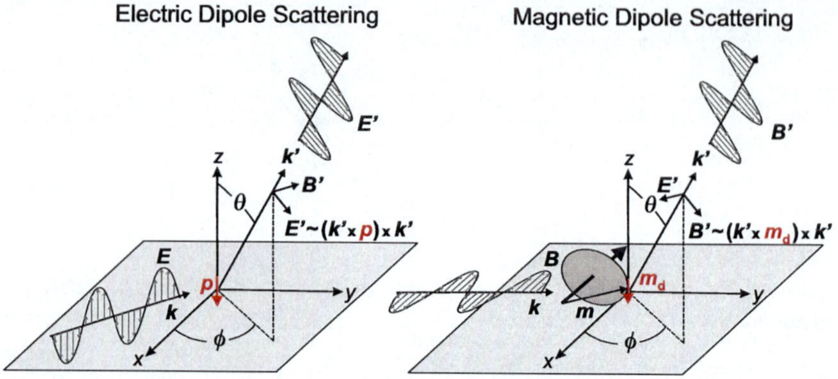

Fig. 6.2 Mechanisms of x-ray dipole scattering from a charge and a spin. Note that charge scattering in the forward direction does not rotate the polarization direction, while spin scattering rotates it by 90°

There are important differences in the polarization, phase, and amplitude of charge and spin scattering. We note the following important points:

- Charge scattering causes a phase shift of π (minus sign in (6.13)) between the incident **E** and scattered **E**′ fields.
- Spin scattering introduces a phase shift of $\pi/2$ (factor i in (6.14)) between the incident **B** and scattered **B**′ fields.
- The spin scattering amplitude is reduced by a factor of $\hbar\omega/m_e c^2$ relative to the charge scattering amplitude, where $m_e c^2 = 511$ keV.
- Charge scattering in the forward direction does not rotate the polarization, while spin scattering rotates it by 90° (see Fig. 6.2).

6.2.3 *Thomson Scattering Length and Cross Sections*

We now touch base with conventional scattering theory by defining the scattering length of a single electron. We have seen that the scattered wave has a spherical wave form $e^{ik'r}/r$. In order to describe the amplitude as well as the polarization of the scattered wave we assume that we can determine it by a suitable analyzer. We define the measured polarization direction by $\boldsymbol{\epsilon}'$, which by geometry is perpendicular to the scattered wave vector $\mathbf{k}_0'$. This allows us to define the scattering length.

The scattering length

$$f(\epsilon, \epsilon') = -\frac{r\,e^{-ik'r}}{E}\,\mathbf{E}' \cdot \epsilon' \tag{6.15}$$

contains all the important information on the scattering process, namely the scattered amplitude and polarization. It is in general a complex quantity with the dimension [length].

In general, (6.15) should be written using the complex conjugate vector ϵ'^* to also account for the use of imaginary polarization vectors as used for circularly polarized x-rays. By use of (6.13) we obtain the electric dipole scattering factor for a single electron

$$f_e(\epsilon, \epsilon') = \frac{1}{4\pi\epsilon_0}\frac{e^2}{m_e c^2}\,\epsilon \cdot \epsilon' = r_0\,\epsilon \cdot \epsilon' . \tag{6.16}$$

Here we have introduced a new quantity, r_0, which is called the *classical electron radius* or the *Thomson scattering length*. We shall see below that the area r_0^2, apart from a geometrical factor, gives the x-ray scattering cross section of a single electron. It is therefore a very important reference number for describing the scattering amplitude of all multi-electron systems.

The *Thomson scattering length* is also called the *classical electron radius* and is given by

$$r_0 = \frac{1}{4\pi\epsilon_0}\frac{e^2}{m_e c^2} = 2.82 \times 10^{-6}\text{nm} . \tag{6.17}$$

For spin dependent dipole scattering we obtain from (6.14)

$$f_s(\epsilon, \epsilon') = -\mathrm{i}\,r_0\,\frac{\hbar\omega}{m_e c^2}\,\mathbf{s} \cdot (\mathbf{k}_0 \times \epsilon) \times (\mathbf{k}_0' \times \epsilon') . \tag{6.18}$$

The so-defined charge and spin dependent scattering lengths are closely related to the concept of a *cross section* which has the dimension of [area], typically expressed in units of barn (b) ($1\,\text{b} = 10^{-28}\,\text{m}^2$).[3] The scattering length defined by (6.15) is related to the differential scattering cross section according to

$$\boxed{\frac{\mathrm{d}\sigma_{\text{scat}}}{\mathrm{d}\Omega} = |f(\epsilon, \epsilon')|^2} . \tag{6.19}$$

[3] The unit of "barn" was created during the Manhattan Project to describe the cross-sectional area of a typical nucleus.

The differential Thomson scattering cross section for a single electron is obtained with (6.16) as

$$\left[\frac{d\sigma_T}{d\Omega}\right]_e = r_0^2\, |\epsilon \cdot \epsilon'|^2 = r_0^2\, \sin^2\theta, \tag{6.20}$$

where θ is defined in Fig. 6.2. The total scattering cross section is obtained by angular integration which yields the *Thomson cross section* for scattering by a single electron

$$[\sigma_T]_e = r_0^2 \int_0^{2\pi}\int_0^{\pi} \sin^2\theta\ \sin\theta\ d\theta\ d\phi = \frac{8\pi}{3}\, r_0^2\,. \tag{6.21}$$

The factor $8\pi/3$ arises from the dipolar nature of the emitted radiation pattern which has a node along the incident electric field vector. Hence only two of three directions in a Cartesian coordinate system contribute and the isotropic value of 4π is reduced by 2/3.

The Thomson cross section

$$[\sigma_T]_e = \sigma_e = \frac{8\pi}{3}\, r_0^2 = 0.665 \times 10^{-28}\,\text{m}^2 = 0.665\,\text{barn} \tag{6.22}$$

is the total scattering cross section from the charge of a *single electron*. One can conveniently remember that the x-ray scattering cross section for a single electron is about 1 barn.

For the differential scattering cross section of a single spin we obtain from (6.18)

$$\left[\frac{d\sigma_T}{d\Omega}\right]_s = r_0^2 \left(\frac{\hbar\omega}{m_e c^2}\right)^2\ |\mathbf{s}\cdot(\mathbf{k}_0 \times \epsilon) \times (\mathbf{k}_0' \times \epsilon')|^2 \tag{6.23}$$

The differential cross sections (6.20) and (6.23) explicitly contain the polarization dependence of the scattered intensity, also seen in Fig. 6.2. The charge scattering intensity is the largest when ϵ and ϵ' are parallel, which is the case anywhere within the gray plane in Fig. 6.2. As one goes out of the plane the intensity has the typical $\cos^2\psi = \sin^2\theta$ (where $\psi = 90° - \theta$) dependence often discussed in textbooks.

By angular integration of (6.23) we obtain the total spin scattering cross section. Introducing unit vectors associated with the **B** fields $\epsilon_B = \mathbf{B}/B$ and $\epsilon_B' = \mathbf{B}'/B'$ we can rewrite the polarization dependent term $\mathbf{s}\cdot(\mathbf{k}_0' \times \epsilon') \times (\mathbf{k}_0 \times \epsilon) = \epsilon_B' \cdot (\epsilon_B \times \mathbf{s}) = -\,|\mathbf{s}|\sin\theta$ and obtain with $|\mathbf{s}| = 1/2$,

$$[\sigma_T]_s = \frac{8\pi}{3}\left(\frac{\hbar\omega}{2m_e c^2}\right)^2 r_0^2 \tag{6.24}$$

(6.21) and (6.24) clearly show that for typical x-ray energies charge scattering is stronger by orders of magnitude than spin scattering. For example, assuming a photon energy of 10 keV, we have $(\hbar\omega/2m_ec^2)^2 \simeq 10^{-4}$ and the weak spin scattering signal will only be observable in very special circumstances. An example is an antiferromagnetic lattice where the spin structure has a different periodicity than the charge structure and weak magnetic scattering peaks exist at locations in reciprocal space where charge scattering peaks are absent [5–7].

When the photon energy $\hbar\omega$ is equal to twice the rest energy of the electron $2m_ec^2 = 1.022$ MeV, we see that magnetic and charge scattering have the same cross section. The corresponding x-ray wavelength $\lambda/2\pi = \lambda_C = \hbar c/(m_ec^2) = 3.86 \times 10^{-3}$ Å is called the Compton scattering length [4], and the ratio of the Thomson and *Compton scattering length* is the *fine structure constant* $\alpha_f = r_0/\lambda_C \simeq 1/137$.

6.2.4 Thomson Scattering By An Atom

The above discussion of elastic scattering by a single electron may readily be extended to that by a multi-electron atom. We limit our discussion to electric dipole scattering illustrated on the left in Fig. 6.2. In such scattering processes all electrons remain confined to their electronic shells and are just wiggled by the incident EM wave. Our treatment of atomic scattering will be extended in Sect. 6.3.2 to include resonant scattering where core electrons are promoted to empty states and their de-excitation results in the emission of scattered x-rays.

The extension of scattering from a single electron and an atom is facilitated by the fact that the atomic size is small enough that the wavefront of any incident wave can be assumed to be flat, so that we can assume a plane wave with a wavevector $\mathbf{k}$. We observe the scattered wave in a direction $\mathbf{k}'$ where $|\mathbf{k}| = |\mathbf{k}'|$ because the scattering process is assumed to be elastic. Since at x-ray energies the wavelength is comparable to the size of an atom, we need to take into account path length differences of the incident x-rays as they pass through the finite atomic volume to the detector, as illustrated in Fig. 6.3. The small path lengths differences may then affect the scattered amplitude through interference effects.

The incident plane wave of wavevector $\mathbf{k}$ is scattered from different points in the atomic volume (shown shaded in gray) into a direction $\mathbf{k}'$, corresponding to a momentum transfer $\mathbf{Q} = \mathbf{k}' - \mathbf{k}$. The amplitudes of the scattered wave from the center and an arbitrary position $\mathbf{r}$ in the atomic volume exhibit a relative phase shift $(\mathbf{k}_0' - \mathbf{k}_0) \cdot \mathbf{r}$, indicated in Fig. 6.3. The $\mathbf{Q}$-dependent total atomic scattering factor is obtained by integration over the atomic electron density $\rho(\mathbf{r})$ and is called the *atomic form factor*, which in the past has also been called the "atomic structure factor" or the "atomic scattering factor". It is defined as a *positive dimensionless number* that reflects the number of electrons per atom according to

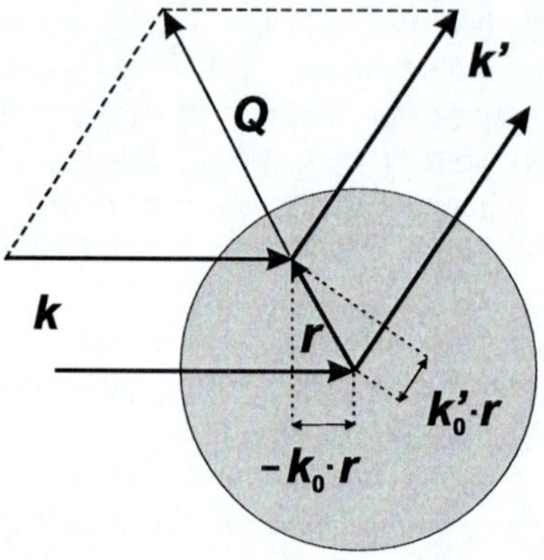

Fig. 6.3 Scattering from an atom. The incident plane wave of wavevector $\mathbf{k}$ is scattered from different points in the atomic volume (shown shaded in gray) into a direction $\mathbf{k}'$ corresponding to a momentum transfer $\mathbf{Q} = \mathbf{k}' - \mathbf{k}$. The wave scattered from the center and an arbitrary position $\mathbf{r}$ experience a path difference $\mathbf{k}_0' \cdot \mathbf{r} - \mathbf{k}_0 \cdot \mathbf{r}$, where the subscripts "0" denote unit vectors, leading to a $\mathbf{Q}$-dependent phase shift of the scattered field

$$\text{The atomic form factor:}\; F^0(\mathbf{Q}) = -\frac{1}{e}\int \rho(\mathbf{r})\, \mathrm{e}^{\mathrm{i}\mathbf{Q}\cdot\mathbf{r}} \mathrm{d}\mathbf{r} \tag{6.25}$$

The atomic form factor accounts for the *collective* $\mathbf{Q}$-dependent response of all Z electrons in an atom without taking into consideration intershell electronic transitions, as indicated by the superscript "0". It is $-1/e$ times the Fourier transform of the (negative) atomic electron density $\rho(\mathbf{r})$ or the Fourier transform of the (positive) number density of electrons.

The Thomson scattering length of an atom is then given by

$$\text{Thomson atomic scattering length:}\; f^0(\mathbf{Q}) = r_0\, (\boldsymbol{\epsilon} \cdot \boldsymbol{\epsilon}')\, F^0(\mathbf{Q}). \tag{6.26}$$

Again the superscript "0" indicates that we have not taken resonant excitations into account. As expressed by (6.30), the same equation without the superscript also accounts for resonance effects. For forward scattering, $\mathbf{Q} = 0$, the polarization factor is unity, and for an atom in the middle of the periodic table with $Z \simeq 30$, the scattering length is about $f^0 = Zr_0 \simeq 10^{-4}$ nm. By use of (6.19) the differential scattering cross section is given by

$$\left(\frac{\mathrm{d}\sigma}{\mathrm{d}\Omega}\right)_{\text{atom}} = r_0^2\, |\boldsymbol{\epsilon} \cdot \boldsymbol{\epsilon}'|^2\, |F^0(\mathbf{Q})|^2. \tag{6.27}$$

We can summarize as follows.

> The *atomic form factor* $F^0(\mathbf{Q})$ is a positive number that reflects the number of electrons per atom.

The Thomson *atomic scattering length* $f^0(\mathbf{Q})$ with dimension [length] is the product of the atomic form factor with the Thomson scattering length r_0 per electron and a polarization term.
As for the case of a single electron, Thomson scattering by an atom introduces a 180° phase shift, but in the forward direction the polarization is not rotated.

At high x-ray energy, when the wavelength λ becomes small relative to the atomic size, the waves scattered from different parts of the atom interfere destructively and cause the form factor to vanish, $F^0(\mathbf{Q}) = 0$. In the opposite limit, when the wavelength λ becomes large relative to the atomic size, there is no path length difference and different volume elements scatter in phase. Thus in the soft x-ray region where λ is of the order of 1 nm, the form factor becomes equal to the number of electrons, $F^0(\mathbf{Q}) = Z$. This also happens for any photon energy in the forward scattering direction where $\mathbf{Q} = 0$ by geometry.

In the soft x-ray region the *Thomson cross section per atom* is given by Z^2 times the Thomson cross section per electron given by (6.22), or

$$[\sigma_\mathrm{T}]_\mathrm{atom} = \frac{8\pi}{3} r_0^2 Z^2 = \sigma_e Z^2 \tag{6.28}$$

where $\sigma_e = 0.665$ barn.

6.3 X-Ray Resonant Scattering and Absorption

So far, we have not considered the different binding energies of the atomic electrons, i.e. the shell structure of the atom. It is clear that some kind of "resonance effects" will arise when the x-ray energy is close or equal to the characteristic energies of the shells. Owing to their strong dependence on wavelength, shell-dependent effects were originally referred to as "dispersion" phenomena. Since the shell structure itself is a quantum phenomenon, the description of these effects, which give rise to pronounced absorption and scattering resonances, is best treated by quantum theory as introduced in Chap. 9.

Here we give a semi-classical treatment that goes back to a famous paper written in early 1925 by Kramers and Heisenberg [8]. The semi-classical treatment of resonance or "dispersion" effects utilized Kramers' idea to replace the atom by oscillators vibrating with the frequencies of the Bohr model of the atom. The importance of the

Kramers-Heisenberg dispersion formula is reflected by the fact that it is equivalent to the quantum mechanical formulation of resonant scattering derived in 1927 by Dirac [9] (see Chap. 13).[4]

6.3.1 Dispersion Corrections to the Atomic Scattering Factor

In the x-ray region, scattering near absorption edges has historically been called "anomalous" because it strongly varies with energy. At hard x-ray energies, near edge resonances are less pronounced since they are energetically smeared out due to shorter lifetimes of the created inner shell holes. In contrast, resonance effects are of great importance in soft x-ray studies since they yield greatly enhanced cross sections, are used for elemental and bonding specificity, and through their polarization dependence allow the determination of preferential orientations of the charge and spin of the valence electrons [11, 12].

In the x-ray region, resonance or dispersion effects are treated by the introduction of a complex scattering factor which is added to the **Q**-dependent atomic form factor [4, 13]. The real part of the additional factor is denoted $F'(\hbar\omega)$ and imaginary part $F''(\hbar\omega)$, and one neglects their **Q**-dependence due to the small extent of the core shell involved in the resonant excitation relative to the wavelength. In the literature, the signs of $F'(\hbar\omega)$ and $F''(\hbar\omega)$ are not uniquely defined. We use the following sign convention which follows from our description of the phase of the propagating wave.

The total *atomic scattering factor* is a dimensionless number corresponding to the number of electrons. It contains three contributions given by

$$F(\mathbf{Q},\omega) = F^0(\mathbf{Q}) + F'(\omega) - \mathrm{i}\, F''(\omega)\,, \tag{6.29}$$

where $F^0(\mathbf{Q}) + F'(\omega)$ and $F''(\omega)$ are *positive* and $F'(\omega)$ is *negative* (see Fig. 6.4). The corresponding total atomic scattering *length* is given in accordance with (6.26) by

$$f(\mathbf{Q},\omega) = (\boldsymbol{\epsilon}\cdot\boldsymbol{\epsilon}')\, r_0\, F(\mathbf{Q},\omega) = f^0(\mathbf{Q}) + f'(\omega) - \mathrm{i}\, f''(\omega)\,. \tag{6.30}$$

In the high energy limit $\omega \to \infty$ and forward scattering $\mathbf{Q} \to 0$ the real and imaginary parts of the scattering factor satisfy important sum rules as discussed in Sect. 6.8 and illustrated in Fig. 6.11. In this context, it is important to comment on our sign convention for the real part.

---

[4] Heisenberg later stated that at the time of submission in January 1925 they "almost had matrix mechanics ... without knowing it". Heisenberg submitted his matrix formulation of quantum mechanics [10] in July 1925, and later said "I always regretted that Kramers never got the Nobel Prize".

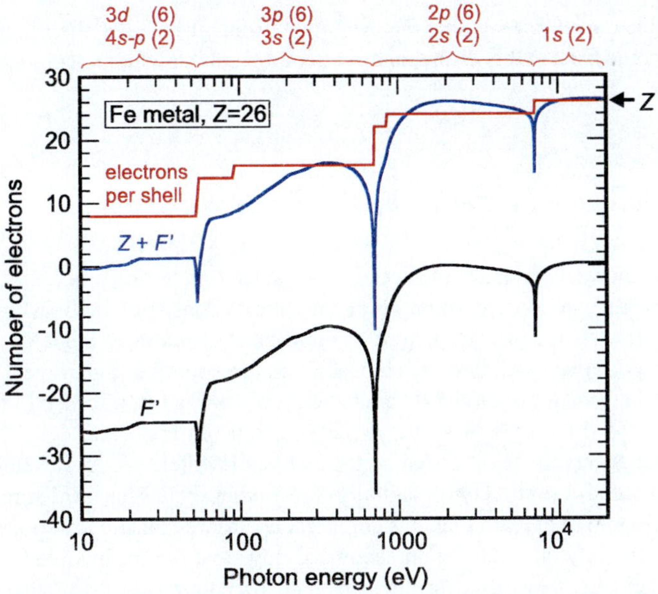

Fig. 6.4 Illustration of how the real part of the scattering factor for Fe with $Z = 26$ is composed of a constant atomic form factor $F^0(\mathbf{Q}) = Z = 26$ and a negative energy dependent dispersion correction $F'(\omega)$. The red curve assumes that with increasing photon energy the scattering increases proportionally to the number of electrons per atom that can be excited with unit probability by the incident photons

The energy dependence of the negative dispersion correction $F'(\omega)$ and the positive total real part $F^0(\mathbf{Q}) + F'(\omega)$ is illustrated in Fig. 6.4 for the case of Fe with $Z = 26$, assuming forward scattering so that $F^0(\mathbf{Q}) = Z$.

With increasing photon energy excitations from deeper core shells become possible, which leads to the well-known absorption edges. We shall see below that absorption and scattering are closely related through the Kramers-Kronig relations, and the jumps in absorption are also reflected by changes in the real part $Z + F'(\omega)$ around the edges, which resemble the derivative of the edge jump. As shown in Fig. 6.4, the total real part increases with photon energy until it becomes constant, equal to Z, for energies that exceed all core electron binding energies. This is in fact one of the energy dependent sum rules.

The figure shows how this comes about as the sum of a negative energy dependent factor $F'(\omega)$ that subtracts from to the constant factor $F^0(\mathbf{Q}) = Z$ at the highest energy.[5] Also shown in the figure as a red curve is a simple model which assumes that scattering increases proportionally to the electrons that can be excited by the incident photons with *unit* excitation probability. The discrepancy between the red

[5] We shall see below that the sum $Z + F'(\omega)$ is the web-tabulated Henke-Gullikson factor f_1.

and the blue curves reveals that the excitation probability is not shell independent. We shall come back to this discrepancy in Sect. 6.8, in particular in conjunction with Fig. 6.11.

6.3.2 Resonant Scattering by Atomic Core Electrons

In this section we discuss the absorption and resonant scattering processes in a semi-classical picture. It consists of picturing an atom to consist of shells with different energies. In a semi-classical picture, a multi-electron atom is modeled as a collection of harmonic oscillators. Each oscillator corresponds to a particular quantum mechanical resonant excitation–de-excitation process of an atomic electron. We start by considering the simple two-level model illustrated in Fig. 6.5a.

The excitation process is driven by the incident EM field which elevates an electron from a filled core shell into a partially empty outer shell. This is the simple model used by Einstein to explain Planck's empirical law for the blackbody spectrum [14]. However, we only consider "spontaneous" decays that do not involve the incident radiation field, i.e. we assume the intensity to be low enough that "stimulated" decays do not occur. In this model, the *x-ray absorption* process consists of a field-driven excitation of an electron (filled circle) from a core level to an empty state (empty circle). A *resonant scattering* process also includes the subsequent "spontaneous"

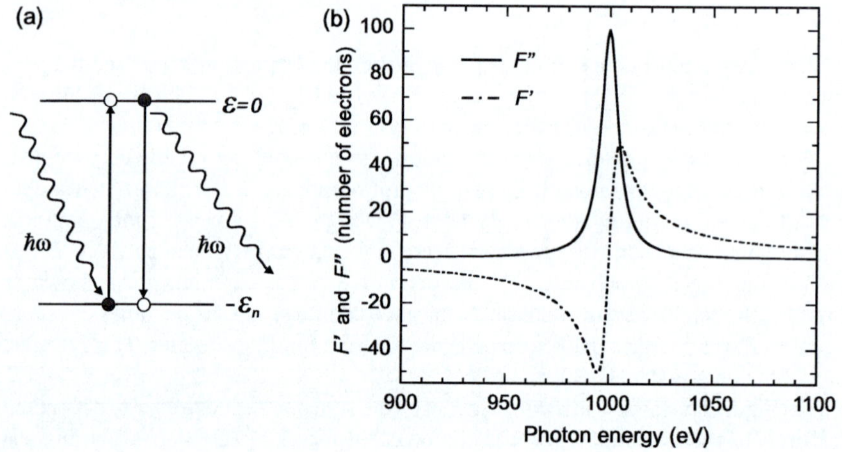

Fig. 6.5 **a** Resonant process in a simple one-electron picture. It consists of absorption of a photon through excitation of a core electron (filled circle) to an empty excited state (open circle), followed by a de-excitation process with photon emission. The process involves an energy difference $\mathcal{E}_n$, yielding a single resonance. **b** Energy dependent resonance factors $F'(\hbar\omega)$ and $F''(\hbar\omega)$ corresponding to an excitation of a core electron of binding energy $-\mathcal{E}_n = 1000\,\text{eV}$ and resonance width $\Gamma_n = \hbar\Delta_n = 10\,\text{eV}$, according to (6.42) and (6.43)

decay of the excited electron back into the created core hole. With a certain probability, the decay process will result in emission of a photon with the same energy $\hbar\omega = \mathcal{E}_n$ as that of the incident photon (elastic process). The photon emitted during decay will have random directions and therefore involve a momentum transfer.

The bound or core electron is modeled as a forced harmonic oscillator. The equation of motion of its coordinates $\mathbf{x}$ is then given by

$$\frac{\mathrm{d}^2\mathbf{x}}{\mathrm{d}t^2} + \Delta_n \frac{\mathrm{d}\mathbf{x}}{\mathrm{d}t} + \omega_n^2 \mathbf{x} = -\frac{e\mathbf{E}_0}{m_\mathrm{e}} \mathrm{e}^{-\mathrm{i}\omega t}, \tag{6.31}$$

where ω_n is the oscillator resonance frequency and Δ_n a characteristic damping constant. We have neglected the much smaller Lorentz term $-(e/m_\mathrm{e})(\mathbf{v} \times \mathbf{B})$ which is of the order v/c. Since the incident EM field $\mathbf{E} = \mathbf{E}_0\mathrm{e}^{-\mathrm{i}\omega t}$ impresses its frequency onto the charge, the displacement $\mathbf{x}$ and its time derivatives will all have the same $\mathrm{e}^{-\mathrm{i}\omega t}$ time dependence and the time derivative simply gives $-\mathrm{i}\omega$. This yields

$$\mathbf{x} = \frac{1}{\omega^2 - \omega_n^2 + \mathrm{i}\omega\Delta_n} \frac{e\mathbf{E}_0}{m_\mathrm{e}} \tag{6.32}$$

and with $\mathbf{p} = -e\,\mathbf{x}$ and (6.7) and (6.15) we obtain for the frequency dependent scattering length of a bound electron n with resonance frequency ω_n

$$f_n(\omega) = r_0\, F_n(\omega)\, \boldsymbol{\epsilon} \cdot \boldsymbol{\epsilon}'\,. \tag{6.33}$$

The dimensionless factor $F_n(\omega)$ contains the important frequency dependence and is given by

$$F_n(\omega) = \frac{\omega^2}{\omega^2 - \omega_n^2 + \mathrm{i}\omega\Delta_n}\,. \tag{6.34}$$

The angle integrated cross section is calculated as $\sigma = \int |f_n(\omega)|^2 \mathrm{d}\Omega$, and we obtain the following general expression for the scattering cross section

$$\boxed{\sigma_{\mathrm{scat}} = \frac{8\pi}{3} r_0^2 \frac{\omega^4}{(\omega^2 - \omega_n^2)^2 + (\omega\Delta_n)^2}}\,. \tag{6.35}$$

6.3.3 *Distinction of Thomson, Rayleigh, and Resonant Scattering*

Depending on the photon energy $\hbar\omega$ relative to the atomic resonance energies $\hbar\omega_n$ and associated decay linewidths $\hbar\Delta$ in (6.35), one distinguishes the following three *single-electron* cross sections.

For ω, $\Delta_n \ll \omega_n$ we obtain the *Rayleigh scattering* cross section

$$\sigma_{\text{scat}}^{\text{R}} = \frac{8\pi}{3} r_0^2 \frac{\omega^4}{\omega_n^4}. \tag{6.36}$$

For $\omega \simeq \omega_n$ we have *resonant scattering* and by use of $\omega^2 - \omega_n^2 = (\omega + \omega_n)(\omega - \omega_n) \simeq 2\omega(\omega - \omega_n)$ the cross section can be written as

$$\sigma_{\text{scat}}^{\text{res}} = \frac{2\pi}{3} r_0^2 \frac{\omega_n^2}{(\omega - \omega_n)^2 + (\Delta_n/2)^2}, \tag{6.37}$$

and for $\omega \gg \omega_n \gg \Delta_n$ we obtain the familiar *Thomson scattering* cross section of a single electron

$$\sigma_{\text{scat}}^{\text{T}} = \frac{8\pi}{3} r_0^2. \tag{6.38}$$

The famous Rayleigh cross section with a ω^4 frequency dependence explains the blue color of the sky when looking up and the red color of the setting sun [13]. At resonance $\omega = \omega_n$, the resonant scattering cross section is enhanced by a factor ω_n^2/Δ_n^2 over the non-resonant Thomson cross section which in the soft x-ray region is a factor of about $10^4 - 10^6$.

Thomson scattering refers to the frequency *independent* scattering when the photon frequency is larger than typical resonance frequencies of bound electrons. *Rayleigh scattering* refers to the limit when the photon frequency is much less than the characteristic electron frequencies, and it is described by the famous ω^4 dependence of the scattering cross section.

In the following we shall take a look at the relationship between x-ray resonant scattering and absorption.

6.4 The Classical Link of Resonant Scattering and Absorption

The frequency dependent resonance factor $F_n(\omega)$ in (6.34) produces a single resonance at each frequency ω_n, corresponding to core electron excitation energies $\mathcal{E}_n = \hbar\omega_n$. For convenience let us rewrite $F_n(\omega)$ in terms of energy variables, the photon energy $\hbar\omega$, the resonance width $\Gamma_n = \hbar\Delta_n$ and the resonance position $\mathcal{E}_n = \hbar\omega_n$. The resonance factor in (6.34) is then

$$F_n(\hbar\omega) = \frac{(\hbar\omega)^2}{(\hbar\omega)^2 - \mathcal{E}_n^2 + \mathrm{i}\hbar\omega\Gamma_n} . \tag{6.39}$$

Since $\Gamma_n \ll \mathcal{E}_n$, we can rewrite (6.39) as

$$F_n(\hbar\omega) \simeq 1 + F_n'(\hbar\omega) - \mathrm{i}F_n''(\hbar\omega). \tag{6.40}$$

The real part is given by

$$F_n'(\hbar\omega) = \frac{\mathcal{E}_n^2\,[(\hbar\omega)^2 - \mathcal{E}_n^2]}{\left[(\hbar\omega)^2 - \mathcal{E}_n^2\right]^2 + (\hbar\omega\Gamma_n)^2} , \tag{6.41}$$

which for energies near the resonance, $\hbar\omega \sim \mathcal{E}_n$, can be approximated by use of $(\hbar\omega)^2 - \mathcal{E}_n^2 = (\hbar\omega + \mathcal{E}_n)(\hbar\omega - \mathcal{E}_n) \simeq 2\hbar\omega(\hbar\omega - \mathcal{E}_n)$ to give [15]

$$F_n'(\hbar\omega) = \frac{\mathcal{E}_n}{2}\,\frac{\hbar\omega - \mathcal{E}_n}{(\hbar\omega - \mathcal{E}_n)^2 + (\Gamma_n/2)^2} . \tag{6.42}$$

The imaginary part is obtained as

$$F_n''(\hbar\omega) = \frac{\mathcal{E}_n}{\Gamma_n}\,\frac{(\Gamma_n/2)^2}{(\hbar\omega - \mathcal{E}_n)^2 + (\Gamma_n/2)^2} . \tag{6.43}$$

The latter represents a Lorentzian lineshape characterized by a position $\mathcal{E}_n$ and full width at half maximum (FWHM) Γ_n, height $\mathcal{E}_n/\Gamma_n$, and area $\pi\mathcal{E}_n/2$. The functions $F_n'(\hbar\omega)$ and $F_n''(\hbar\omega)$ are plotted in Fig. 6.5b for $\mathcal{E}_n = 1000$ eV and $\Gamma_n = 10$ eV. Because of its lineshape, $F_n'(\hbar\omega)$ is often called the *dispersion term*. Our formalism reveals an important relationship between the real and imaginary parts which according to (6.42) and (6.43) can be expressed as

$$\boxed{F_n'(\hbar\omega) = \frac{2(\hbar\omega - \mathcal{E}_n)}{\Gamma_n}\,F_n''(\hbar.\omega) .} \tag{6.44}$$

This link, valid only for the resonance region where $\hbar\omega \sim \mathcal{E}_n$, is a special case of the more general Kramer-Kronig formalism, discussed in more detail below.

We have already mentioned that the imaginary term i $F''_n(\hbar\omega)$ in (6.40) represents absorption of energy from the EM wave. It is therefore related to the x-ray absorption cross section $\sigma_{\rm abs}$ for the core electron n. Since we have not yet discussed the theory of x-ray absorption, we introduce the absorption cross section by means of the simple concept of *oscillator strength*[6]. The idea behind this concept is that, on average, the oscillator strength or excitation probability is $g_n = 1$ for each atomic electron n so that the total oscillator strength of an atom is simply the total number of electrons, Z. As shown in Fig. 6.11, this assertion, known as the Thomas-Reiche-Kuhn sum rule, is confirmed in the high energy limit when the photon energy exceeds the binding energy of all core shells.

> *The oscillator strength* g_n of an electronic transition n is a dimensionless number that defines its strength.
> The *Thomas-Reiche-Kuhn* sum rule states that the total oscillator strength for an atom, summed over all shells, is equal to the number of electrons of the atom, $\sum_n g_n = Z$.

Beautiful examples of large oscillator strengths are shown in Fig. 6.6, which shows the x-ray absorption spectrum of the optical recording material $Fe_{70.5}Tb_{24.5}Co_{5.0}$. In the figure the pronounced resonances at the L-edge of Fe and the M-edge of Tb correspond to $2p \rightarrow 3d$ transitions and $3d \rightarrow 4f$ transitions, respectively.

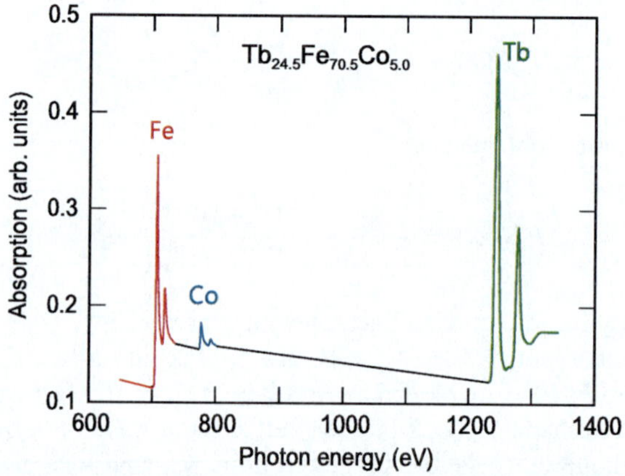

Fig. 6.6 X-ray absorption spectrum of $Fe_{70.5}Tb_{24.5}Co_{5.0}$ showing resonances at the $L_{3,2}$-edges of Fe and Co ($2p_{3/2}, 2p_{1/2} \rightarrow 3d$ transitions) and the $M_{5,4}$-edges ($3d_{5/2}, 3d_{3/2} \rightarrow 4f$ transitions). The large resonance intensities are due to the degeneracy of the core states: 4 ($p_{3/2}$), 2 ($p_{1/2}$), 6 ($d_{5/2}$), and 4 ($d_{3/2}$) and the large number of empty d and f final states

[6] The oscillator strength is usually denoted f, but we shall use the symbol g here to avoid a notation conflict with the scattering factor f.

The oscillator strength is related to the *x-ray absorption cross section* $\sigma_{abs}(\hbar\omega)$ according to

$$\sigma_{abs}(\hbar\omega) = C \frac{dg_n}{d\hbar\omega}, \tag{6.45}$$

where $C = \pi\, h\, c\, r_0 = 1.098 \times 10^8$ barn eV. Since the resonance factor $F_n''(\hbar\omega)$ in (6.43) also represents a resonance lineshape due to damping or absorption, it is quantitatively related to the x-ray absorption cross section. Furthermore, the resonant scattering cross section given by (6.37) can be written in terms of $F_n'(\hbar\omega)$ and $F_n''(\hbar\omega)$, and we have the following relationships.

Classically, the *resonant x-ray absorption cross section* is given by

$$\sigma_{abs} = 2\,\lambda\, r_0 \underbrace{\frac{\mathcal{E}_n}{\Gamma_n} \frac{(\Gamma_n/2)^2}{(\hbar\omega - \mathcal{E}_n)^2 + (\Gamma_n/2)^2}}_{F_n''} = 2\,\lambda\, f_n'' \tag{6.46}$$

and the *resonant scattering cross section* (6.37) can be written as

$$\sigma_{scat}^{res} = \frac{8\pi}{3} r_0^2 \underbrace{\frac{\mathcal{E}_n^2}{\Gamma_n^2} \frac{(\Gamma_n/2)^2}{(\hbar\omega - \mathcal{E}_n)^2 + (\Gamma_n/2)^2}}_{|F_n'|^2 + |F_n''|^2} = \frac{8\pi}{3} \left[(f_n')^2 + (f_n'')^2 \right]. \tag{6.47}$$

For the scattering length expressions on the right we have used (6.33) and the angular integration factor $8\pi/3$ is that given by (6.21). To include the non-resonant Thomson contribution (6.28), we need to substitute $(f_n')^2 \rightarrow (f_n' + r_0 Z)^2$.

6.4.1 Relative Size of Thomson, Resonant Scattering, and Absorption

In our semi-classical treatment, the absorption and scattering cross sections (6.46) and (6.47) have peak values that may be compared with each other and the non-resonant Thomson cross section (6.28) according to

$$\sigma_{abs} = 2\,\lambda\, r_0 \frac{\mathcal{E}_n}{\Gamma_n} \quad \text{and} \quad \sigma_{scat}^{res} = \underbrace{\frac{8\pi}{3} r_0^2\, Z^2}_{\sigma_T} \frac{\mathcal{E}_n^2}{Z^2 \Gamma_n^2} = \frac{8\pi}{3} \frac{\mathcal{E}_n}{\Gamma_n} \frac{r_0}{2\lambda} \sigma_{abs}^{max}. \tag{6.48}$$

Inserting typical soft x-ray numbers $\lambda = 1.2$ nm, corresponding to $\mathcal{E}_n = 1000$ eV, and a resonance width $\Gamma_n \simeq 1$ eV, and use of $r_0 = 2.8 \times 10^{-6}$ nm, one would predict values $\sigma_{\rm abs}^{\rm max} \simeq 7 \times 10^7$ barn and $\sigma_{\rm scat}^{\rm max} \simeq 7 \times 10^5$ barn. These values are too large relative to the measured values by a factor of about 10 for absorption and a factor of 100 for scattering.

The reason for the discrepancy is the underlying assumption that the oscillator strength for the excitation of each electron is 1. We shall see below in conjunction with Fig. 6.11 that this is incorrect. As the photon energy is increased, the oscillator strength jumps at excitation thresholds (edges) proportional to the number of electrons that become available for excitation, but the oscillator strength of the outer shells is reduced and the average oscillator strength only catches up to the number of 1 per electrons at higher energies. In the high energy limit where all atomic electrons can be excited, one indeed obtains an average value of 1 per electrons, as expressed by the Thomas-Reiche-Kuhn sum rule value of Z. We shall see later that the discrepancy between experimental and classical cross sections is overcome by quantum mechanical methods. We can summarize as follows.

> The *resonant* x-ray absorption and x-ray scattering cross sections predicted classically from the assumption that the oscillator strength per electron is unity are too large by factors of about 10 and 100, respectively, relative to the experimentally measured maximum cross sections of order $\sigma_{\rm abs}^{\rm max} \le 10^7$ barn and $\sigma_{\rm scat}^{\rm max} < 10^4$ barn.

6.5 The First Born Approximation

We have seen above that x-ray scattering, even resonant scattering, is much weaker than x-ray absorption. This fact underlies the *first Born approximation* which is of great importance in modern scattering theory [4, 16]. We shall not formally derive how it emerges from the general Lippmann-Schwinger scattering theory but rather refer the reader to quantum mechanics texts such as Sakurai [16]. Here we will discuss the Born approximation for the case of x-ray scattering by an atom.

We will extend the treatment of single atom scattering to that of thin atomic sheets and solids in Sect. 7.5.2. When applied to multiple scattering centers, the first Born approximation naturally leads to only considering *single scattering* events since the low probability of a scattering event renders repeated scattering very unlikely. This is also referred to as a *kinematic* treatment of scattering. Multiple scattering can be treated by extension of the first order Born approximation to higher order, and this is referred to as the *dynamical* scattering theory. We shall treat such a case in Sect. 7.8 where we discuss the propagation of the field through a thick sample.

Our emphasis here is on a simple physical model of the atomic scattering process illustrated in Fig. 6.7.

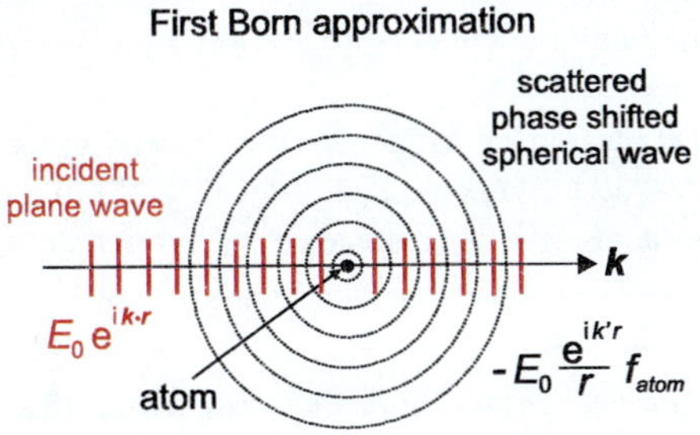

Fig. 6.7 Illustration of scattering in the first Born approximation. A plane wave of wavelength λ (red) is scattered by a single atom, resulting in a nearly unchanged incident wave plus a weak spherical wave (dotted black). For Thomson scattering, the atomic scattering length $f_{\text{atom}} = f^0(\mathbf{Q})$ is related according to (6.26) to the *real* atomic form factor and the spherical wave has a phase shift of 180°, as shown. For resonant scattering, the phase shift of the spherical wave is determined by the phase of the total complex scattering factor $f_{\text{atom}} = f^0(\mathbf{Q}) + f'(\omega) - \mathrm{i}\, f''(\omega)$. At resonance $f_{\text{atom}} \simeq -\mathrm{i}\, f''(\omega)$ and the spherical wave has a phase shift of 90°

With reference to Fig. 6.7, the first Born approximation can be stated as follows.

The *first Born approximation* gives the total elastically ($|\mathbf{k}| = |\mathbf{k}'|$) scattered field of an incident plane wave by an atomic scattering center, under the assumption of weak scattering and that the observation point is at a distance much larger than the size of the scattering object. For an incident field $E_{\text{in}} = E_0 \exp(\mathrm{i}\mathbf{k}\cdot\mathbf{r})$ propagating in the direction $\mathbf{k}$, after the scattering event, the total field $E_{\text{tot}}(\mathbf{r})$ at a distant observation point $\mathbf{r}$ in the direction $\mathbf{k}'$, involving momentum transfer $\mathbf{Q} = \mathbf{k}' - \mathbf{k}$, is given by

$$E_{\text{tot}}(\mathbf{r}) = \underbrace{E_0\, \mathrm{e}^{\mathrm{i}\mathbf{k}\cdot\mathbf{r}}}_{\text{plane wave: } E_{\text{in}}} \underbrace{-\, E_0 \frac{\mathrm{e}^{\mathrm{i}k'r}}{r} f(\mathbf{Q}, \omega)}_{\text{spherical wave: } E_{\text{scat}}} . \tag{6.49}$$

As indicated by underbrackets, the total field consists of a "strong" or zeroth order part, which is the original plane wave propagating in the direction $\mathbf{k}$ but evaluated at the position $\mathbf{r}$, and a "weak" or first order scattered part that consists of an outgoing (plus sign of exponential) spherical wave from the atomic center. The phase shift of the outgoing spherical wave depends on the atomic scattering length given by (6.30).

The scattering length $f(\mathbf{Q})$ is in general given by (6.30) as $f(\mathbf{Q}, \omega) = f^0(\mathbf{Q}) + f'(\omega) - \mathrm{i}\, f''(\omega)$. For *hard* x-rays, one usually includes only the first (Thomson) term $f^0(\mathbf{Q}) \propto r_0 Z$ in the first Born approximation without the dispersion terms. The scattered Born field is then determined by the *real* atomic scattering length

$f^0(\mathbf{Q}) \propto r_0 Z$, and the outgoing spherical wave has a phase shift of π (minus sign) as shown in Fig. 6.7.

For *soft* x-rays it is customary to include the dispersion terms in the first Born approximation, and $f(\mathbf{Q}, \omega) = f_a = r_0 Z + f'(\omega) - \mathrm{i}\, f''(\omega)$ is the momentum independent atomic scattering length. At resonance, $f_a \simeq -\mathrm{i}\, f''(\omega)$ (see Fig. 6.5) is then *imaginary* and the scattered wave has a phase shift of $\pi/2$ (factor i).

For single atom *non-resonant Thomson scattering*, the scattering factor $f(\mathbf{Q}, \omega) \sim r_0 Z$ is real and there is a 180° phase shift of the scattered relative to the incident wave.
For single atom *resonant scattering*, the scattering factor $f(\mathbf{Q}, \omega) \sim -\mathrm{i}\, f''(\omega)$ is imaginary and there is a 90° phase shift of the scattered relative to the incident wave.

The scattered intensity calculated from the Born approximation is readily linked to the atomic scattering cross section. To see this, we square the expression for the scattered field represented by the second term in (6.49)

$$|E_{\text{scat}}|^2 = \frac{|E_0|^2}{r^2}\, |f(\mathbf{Q})|^2. \tag{6.50}$$

Solving for the squared scattering length, we obtain the differential atomic scattering cross section

$$\frac{\mathrm{d}\sigma_{\text{scat}}}{\mathrm{d}\Omega} = |f(\mathbf{Q})|^2 = \frac{|E_{\text{scat}}|^2 r^2}{|E_0|^2}. \tag{6.51}$$

6.5.1 *The Breit–Wigner Atomic Cross Section*

In 1936 Breit and Wigner considered the cross section for particle scattering by a spherical potential [16, 17]. When this formalism is applied to photon scattering by an atom, the scattering cross section can be written in terms of outgoing partial waves of angular momentum ℓ as

$$\sigma^{\ell}_{\text{scat}} = \frac{\lambda^2}{\pi}\,(2\ell + 1)\, \frac{(\Gamma/2)^2}{(\hbar\omega - \mathcal{E}_0)^2 + (\Gamma/2)^2}. \tag{6.52}$$

The Lorentzian form assumes that the scattering is maximized when the incident photon energy $\hbar\omega$ becomes equal to an atomic resonance energy $\mathcal{E}_0$. The cross section decreases away from resonance with a characteristic energy width (FWHM) Γ that reflects the damping of the excitation. The case $\ell = 0$ corresponds to *s-wave scattering* or emission of a spherical outgoing wave. The same assumption underlies the first Born approximation and the Huygens–Fresnel principle. The maximum scattering cross section, which occurs at resonance, is then given by the remarkably simple result

$$\text{Maximum } s\text{-}wave \text{ scattering cross section}: \sigma_{\text{scat}} = \frac{\lambda^2}{\pi}. \tag{6.53}$$

The maximum s-wave atomic scattering cross section is independent of the internal structure of the atom and only depends on the wavelength!

The maximum lowest order ($\ell = 0$) *Breit–Wigner* cross section assumes the existence of a *resonant* atomic response. For this case we have previously derived both the atomic scattering and absorption cross sections, which correspond to $f' = 0$ in Fig. 6.5. Ignoring polarization effects (i.e. replacing $8\pi/3$ by 4π), the resonant elastic scattering cross section (6.47) is then given by

$$\sigma_{\text{scat}} = 4\pi \left[(f')^2 + (f'')^2 \right] = 4\pi \left(f''\right)^2 \tag{6.54}$$

and is linked to the absorption cross section (6.46) according to

$$\sigma_{\text{scat}} = \sigma_{\text{abs}} \frac{2\pi}{\lambda} f''. \tag{6.55}$$

The resonant Breit–Wigner result (6.53) corresponds to $f'' = \lambda/2\pi$, so that the absorption and scattering cross sections become the same

$$\sigma_{\text{abs}} = \sigma_{\text{scat}} = \frac{\lambda^2}{\pi}. \tag{6.56}$$

This is most easily visualized in a photon picture. An incident photon that is lost ("absorbed") in the forward direction due to the atomic interaction must occur somewhere else. It is hence "scattered" out of the incident direction. Loss in the forward direction, described by σ_{abs}, means gain in another direction, described by σ_{scat}, which simply corresponds to photon conservation.

The link between absorption and resonant scattering expressed by the Breit–Wigner result for a single atom is more generally expressed by the *optical theorem* [18], which will be discussed in Sect. 7.6.

6.6 The Kramers-Kronig Relations

Our formalism for the description of absorption and scattering, summarized by (6.46) and (6.47), allows a simple formulation of the processes in terms of the scattering lengths f_n' and f_n'', with absorption linked to the imaginary part f_n'' and resonant scattering with $(f_n')^2 + (f_n'')^2$. Furthermore we have seen in Sect. 6.4 that absorption and resonant scattering are related, which follows from the fundamental link between f_n' and f_n'' given by (6.44). This correlation was first derived by Ralph de Laer Kronig [19] and Hendrik Anthony Kramers [20] and is expressed by the *Kramers-Kronig relations.*[7]

In physics, the Kramers-Kronig relations were first used to show that a signal cannot travel faster than the velocity of light c in any medium for which the dispersion relation is satisfied. More generally, it is a consequence of *causality*, meaning that for a time-independent interaction, an output function (e.g. a scattered wave) is a linear functional of an "input" function (e.g. the primary wave) and cannot begin before the input function is applied.

The Kramers-Kronig relations hold not only for resonant processes. In fact, inspection of Fig. 6.6 reveals that each absorption spectrum also exhibits a step-like jump in the cross section which underlies any resonant enhancement. Such an absorption step always exists due to the fact that a new excitation channel opens up and, of course, is the historical origin of the term "absorption edge". The single oscillator model for a two-level system discussed above cannot account for the existence of an "edge". We therefore need a more general formalism.

We start by establishing a more general connection between the imaginary part of the resonance factor F_n'' and the x-ray absorption cross section σ_{abs} by use of Fig. 6.8a.

The general x-ray absorption or scattering process consists of excitation of a core electron (filled circle) with binding energy $-\mathcal{E}_n$ to an empty state (open circle) that lies in a continuum of empty states (shaded). In practice, the empty states are continuum states above an energy defined as $\mathcal{E} = 0$, which corresponds to the vacuum level for a free atom or molecule or for a solid is taken to be the Fermi level (which lies below the vacuum level by the work function).

As illustrated in Fig. 6.8b a step in an absorption cross section or spectrum may be described by a continuous series of resonances, starting at a threshold energy $\mathcal{E}_n$, and the decay above the edge can be accounted for by slowly decreasing weight factors, i.e. oscillator strengths g_n^i of overlapping resonances. We can then express $F_n''(\hbar\omega)$ as a continuous sum of oscillators with a threshold value $\mathcal{E}_n$. Instead of the shown bar diagram in Fig. 6.8 we can use Lorentzians for the individual oscillators, as given by (6.46) and at a given photon energy $\hbar\omega$ above a threshold energy $\mathcal{E}_n$ we can write

$$F_n''(\hbar\omega) = \int_{\mathcal{E}_n}^{\infty} \frac{\mathrm{d}g_n}{\mathrm{d}\mathcal{E}} \frac{\mathcal{E}}{\Gamma_n} \frac{(\Gamma_n/2)^2}{(\hbar\omega - \mathcal{E})^2 + (\Gamma_n/2)^2} \, \mathrm{d}\mathcal{E} \,. \tag{6.57}$$

By considering that the integral is only significant when the integration variable $\mathcal{E}$ approaches the photon energy $\hbar\omega$ and with $\mathrm{d}g_n/\mathrm{d}\mathcal{E} = \sigma_n^{abs}(\mathcal{E})/C$ according to (6.45), this integrates to

[7] In mathematics these relations are known under the names Sokhotski–Plemelj theorem and Hilbert transform.

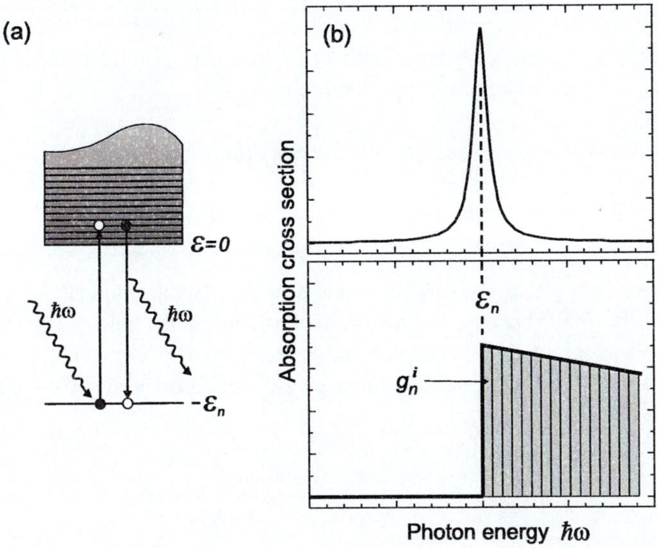

Fig. 6.8 **a** Schematic of a resonant process involving excitation of a core electron (filled circle) to an empty state (open circle) that lies in a continuum of empty states (shaded), followed by the reverse de-excitation process. **b** Comparison of x-ray absorption spectra for two different cases. The top case corresponds to a single core electron oscillator with resonance energy $\hbar\omega = \mathcal{E}_n$ giving rise to a single resonance peak in absorption (compare Fig. 6.5). The case shown at the bottom corresponds to excitations to empty continuum states as indicated in (**a**). This leads to a step function for $\hbar\omega \geq \mathcal{E}_n$ which can be approximated by a infinite series of oscillators g_n^i with $0 \leq i < \infty$ and slowly decreasing weight as shown in gray

$$F_n''(\hbar\omega) = \frac{\sigma_{\rm abs}(\hbar\omega)}{2\,\lambda\, r_0}\left[\frac{1}{2} + \frac{1}{\pi}\,\arctan\left(\frac{\hbar\omega - \mathcal{E}_n}{\Gamma_n/2}\right)\right], \tag{6.58}$$

which is a step function that jumps from 0 to a height $H = \sigma_{\rm abs}/(2\lambda r_0)$ at the resonance position $\mathcal{E}_n$, i.e.

$$\begin{aligned} F_n''(\hbar\omega) &= 0 && \text{if } \hbar\omega < \mathcal{E}_n \\ F_n''(\hbar\omega) &= \frac{\sigma_{\rm abs}(\hbar\omega)}{2\,\lambda\, r_0} && \text{if } \hbar\omega > \mathcal{E}_n. \end{aligned} \tag{6.59}$$

We can therefore generalize the link between $F_n''(\hbar\omega)$ and the resonant x-ray absorption cross section $\sigma_{\rm abs}(\hbar\omega)$ given by (6.46) as follows.

The energy dependent x-ray absorption cross section $\sigma_{abs}(\hbar\omega)$ and the scattering factor $F''_n(\hbar\omega)$ are linked by the general relation

$$\sigma_{abs}(\hbar\omega) = 2\lambda\, r_0\, F''_n(\hbar\omega), \tag{6.60}$$

where resonance effects are included in the form of both σ_{abs} and F''_n.

We shall see below that tabulations or plots of $F''_n(\hbar\omega)$ are typically given in the step-like form (6.59) because the resonance structure at the absorption edge varies from sample to sample, while the step-like structure is always present.

We now turn to $F'_n(\hbar\omega)$ which for a single oscillator is given by the general expression (6.41) or

$$F'_n(\hbar\omega) = \frac{\mathcal{E}_n^2\,[(\hbar\omega)^2 - \mathcal{E}_n^2\,]}{\left[(\hbar\omega)^2 - \mathcal{E}_n^2\,\right]^2 + (\hbar\omega\Gamma_n)^2}\,. \tag{6.61}$$

We can generalize to the case of a series of oscillators by the concepts used for $F''_n(\hbar\omega)$ above. In analogy to (6.57) we obtain

$$F'_n(\hbar\omega) = \int_{\mathcal{E}_n}^{\infty} \frac{\mathrm{d}g_n}{\mathrm{d}\mathcal{E}}\, \frac{\mathcal{E}^2[(\hbar\omega)^2 - \mathcal{E}^2]}{[(\hbar\omega)^2 - \mathcal{E}^2]^2 + (\hbar\omega\Gamma_n)^2}\, \mathrm{d}\mathcal{E}\,. \tag{6.62}$$

We now substitute $\mathrm{d}g_n/\mathrm{d}\mathcal{E} = \sigma_{abs}(\mathcal{E})/C$ according to (6.45) and use the fact that the integral in (6.62) is essentially independent of the damping term $(\hbar\omega\Gamma_n)^2$. This is not obvious but has been shown through a detailed analysis by Parratt and Hempstead [21]. We obtain one of the Kramers-Kronig relations.

The Kramers-Kronig relation allows the determination of $F'_n(\hbar\omega)$ from the measured absorption cross section $\sigma_n^{abs}(\mathcal{E})$ according to

$$F'_n(\hbar\omega) = \frac{1}{C}\, P\!\!\int_0^{\infty} \frac{\mathcal{E}^2\,\sigma_n^{abs}(\mathcal{E})}{(\hbar\omega)^2 - \mathcal{E}^2}\, \mathrm{d}\mathcal{E}\,, \tag{6.63}$$

where $C = \pi\, h\, c\, r_0 = 1.098 \times 10^8$ barn eV.

In (6.63) we have replaced the lower integration limit $\mathcal{E}_n$ by zero because the cross section $\sigma_n^{abs}(\mathcal{E})$ and hence the integral is zero for $\mathcal{E} < \mathcal{E}_n$. In practice, the integration is tricky because of the singularity for $\mathcal{E} = \hbar\omega$. This is indicated by the "$P\!\int$" symbol, meaning "principal value" integral. Today this task is readily accomplished with modern mathematical software packages. In textbooks two Kramers-Kronig relations are typically found that relate the two resonance factors. The first one is obtained from (6.63) by use of (6.60)

$$F'_n(\hbar\omega) = \frac{2}{\pi} P\!\!\int_0^\infty \frac{\mathcal{E}\, F''_n(\mathcal{E})}{(\hbar\omega)^2 - \mathcal{E}^2}\, \mathrm{d}\mathcal{E} \,, \tag{6.64}$$

and the second one allows the opposite transformation

$$F''_n(\hbar\omega) = \frac{2\,\hbar\omega}{\pi} P\!\!\int_0^\infty \frac{F'_n(\mathcal{E})}{(\hbar\omega)^2 - \mathcal{E}^2}\, \mathrm{d}\mathcal{E} \,. \tag{6.65}$$

Again, the "$P\!\int$" symbol means principal value integral.

6.7 The Henke-Gullikson Formalism

The semi-classical formalism developed above is largely phenomenological and based on crude approximations for the shell-dependent oscillator strengths. As discussed in conjunction with Fig. 6.4 and at the end of Sect. 6.4 we know that the shell-dependent oscillator strengths do not simply scale with the number of the electrons in the shell that become available for excitation at an absorption edge. Also, the predicted resonant enhancement at the excitation thresholds over the step-like non-resonant cross sections is much too large.

For this reason, the semi-classical theory needed to be combined with experimental data to account for cross sections over an extended energy range. This procedure relied on the use the Kramers-Kronig relations (6.64) and (6.65) to link the real and imaginary parts of the electronic response and to establish a quantitative link between scattering and absorption cross sections. We shall see later in Chaps. 9–13 that this deficit is overcome by quantum mechanical methods, but we continue here by presenting the Henke-Gullikson formalism which conveniently allows the calculation of the non-resonant electronic response over a large energy range through web-tabulated scattering factors that are closely related to the ones introduced above.

At x-ray energies, we are interested in resonant scattering by *atoms* and we have to consider contributions from electrons in different subshells that are characterized by different binding energies. Although we have not yet discussed *spin-orbit coupling*, which is done in Sect. 9.2.2, it is included in the nomenclature in Table 6.1 which summarizes the designation of the most prominent absorption edges typically used in conjunction with x-ray scattering and absorption.

It is clear that the description of the atomic response over a large photon energy range needs to include the different edges where a new excitation channel opens up. As shown in Fig. 6.4, the experimental steps at the different edges, however, do not accurately follow the simple notion that they should be proportional to the number of electrons that become available for excitation. In order to have a better description one therefore uses experimental data where the resonance region itself

Table 6.1 Labeling of important core shells and absorption edges

Core shells n	Absorption edges	Subgroups of main edges	Spin-orbit subshells $n\ell_j$, $j = \ell \pm 1/2$	Number of electrons per subshell
1	K	K	1s	2
2	L	L_1	2s	2
		L_2	$2p_{1/2}$	2
		L_3	$2p_{3/2}$	4
3	M	M_1	3s	2
		M_2	$3p_{1/2}$	2
		M_3	$3p_{3/2}$	4
		M_4	$3d_{3/2}$	4
		M_5	$3d_{5/2}$	6

is excluded and simply mimicked by a step. The scattering factors are then fitted to the non-resonant energy dependent data and consistency is established by use of the Kramers-Kronig relations and sum rules that apply for the integrated response up to the highest photon energies where all core shells can be excited. These sum rules will be discussed in Sect. 6.8.

Here we will present the convenient formalism developed and tabulated by Henke, Gullikson, and co-workers [15, 22] for the soft x-ray region. At the heart of the soft x-ray formulation is the expression (6.29) or

$$F(\mathbf{Q}, \omega) = F^0(\mathbf{Q}) + F'(\omega) - \mathrm{i}\, F''(\omega)\,, \tag{6.66}$$

which is simplified by use of $F^0(\mathbf{Q} \sim 0, \hbar\omega) = Z$ for soft x-rays and by grouping the real and imaginary parts we obtain the Henke-Gullikson formulation.

The Henke-Gullikson factors f_1 and f_2 are defined in the limit $\mathbf{Q} \to 0$ as

$$F(\mathbf{Q}, \omega) = f_1(\hbar\omega) \; - \; \mathrm{i}\, f_2(\hbar\omega)\,, \tag{6.67}$$

$$f_1(\hbar\omega) = Z + F'(\hbar\omega)\,, \tag{6.68}$$

$$f_2(\hbar\omega) = F''(\hbar\omega). \tag{6.69}$$

The factors f_1 and f_2 are in units of number of electrons. According to (6.30), the polarization dependent resonant scattering length is given by

$$f(\omega) = (\boldsymbol{\epsilon} \cdot \boldsymbol{\epsilon}') \Big[\underbrace{r_0\, Z + f'(\omega)}_{r_0 f_1(\omega)} - \mathrm{i} \underbrace{f''(\omega)}_{r_0 f_2(\omega)} \Big]. \tag{6.70}$$

The minus sign for the imaginary part in (6.67) is consistent with a mathematical representation of forward wave propagation expressed by $\exp[-\mathrm{i}(\omega t - \mathbf{k}\cdot\mathbf{r})]$. With the Henke-Gullikson factors the total x-ray scattering cross section (6.47) becomes

$$\boxed{\sigma_{\mathrm{scat}} = \frac{8\pi}{3} r_0^2 \left[f_1^2 + f_2^2\right] = \frac{8\pi}{3}\left[(r_0 Z + f')^2 + (f'')^2\right],} \tag{6.71}$$

which has a Thomson and resonant scattering contribution. The x-ray absorption cross section is

$$\boxed{\sigma_{\mathrm{abs}} = 2\lambda r_0\, f_2 = 2\lambda\, f''.} \tag{6.72}$$

As an example we have plotted in Fig. 6.9 the $f_1(\hbar\omega)$ and $f_2(\hbar\omega)$ factors for Fe metal in the 100–10,000eV range, taken from the Center for x-ray Optics website. The data are only approximate in the vicinity of the L ($\sim$ 710 eV) and K ($\sim$ 7110 eV) edges, but they reliably describe the non-resonant values and the overall energy dependence. The energy dependence of $f_2(\hbar\omega)$ closely resembles that of the x-ray absorption cross section with jumps at the L-edge and K-edge positions. The energy dependence of $f_1(\hbar\omega)$ shows very strong negative resonances at the L-edge and K-edge positions. Starting at the highest energy we find $f_1 = Z$=26, as expected. With decreasing energy $f_1(\hbar\omega)$ drops by about two electrons at the K-edge position (2 K-shell electrons) and by another eight electrons at the L-edge position (8 L-shell electrons), as expected from a simple model.

On the right side of the same figure we have plotted the x-ray absorption cross sections and scattering cross sections according to (6.72) and (6.71), respectively. The important information contained in this plot is the significant larger size of the x-ray absorption cross section and its strong energy dependence. In accordance with the behavior of $f_2(\hbar\omega)$ the absorption cross section decreases greatly with increasing photon energy. Experimentally the x-ray absorption cross section is found to be approximately proportional to Z^4 and to falloff with $(\hbar\omega)^{-3}$ [4]. The scattering cross section is nearly energy independent, except near the absorption edges. We can summarize as follows.

> The non-resonant absorption cross section is approximately proportional to Z^4 and decreases with photon energy as $(\hbar\omega)^{-3}$.
> The non-resonant (Thomson) scattering cross section is nearly energy independent.

The Henke-Gullikson factors neglect the **Q**-dependence in the scattering factors $F(\mathbf{Q}, \hbar\omega) = F(\hbar\omega)$. The polarization dependence of the scattering cross section given by $|\boldsymbol{\epsilon}\cdot\boldsymbol{\epsilon}'|^2$ gives a factor $8\pi/3$ when integrated over all emission directions, which is contained in the Thomson cross section $\sigma_e = (8\pi/3)r_0^2$ in (6.71).[8] The

[8] The factor of $8\pi/3$ arises from the fact that the scattered intensity has a node along the incident polarization $\boldsymbol{\epsilon}$. Since only two of the three orthogonal Cartesian directions contribute to the scattered intensity, the isotropic angular integration factor 4π is therefore modified by a factor $2/3$.

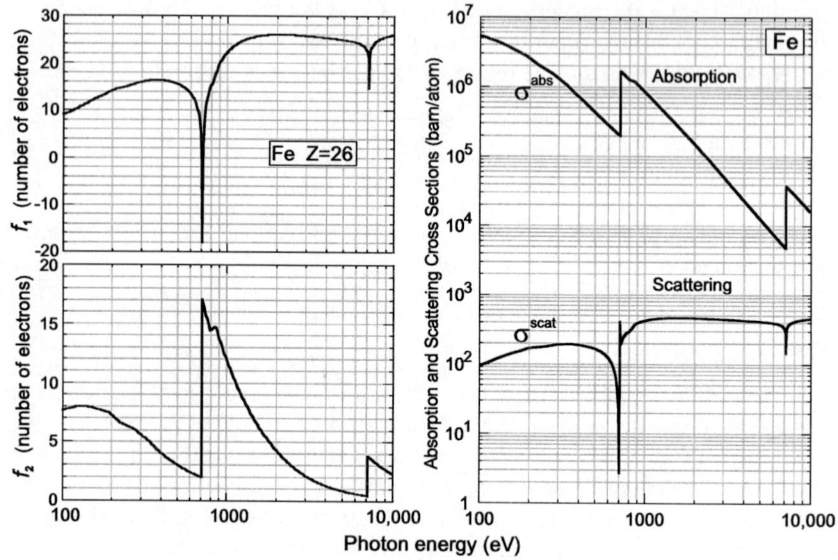

Fig. 6.9 Left: Tabulated energy dependent resonance factors f_1 and f_2 for Fe metal [22]. The structures around 700 eV are due to the L-edge, those around 7100 eV arise from the K-edge. Right: Energy dependent cross sections for Fe. The x-ray absorption (top) and scattering (bottom) cross sections were calculated from the scattering factors according to (6.72) and (6.71), respectively

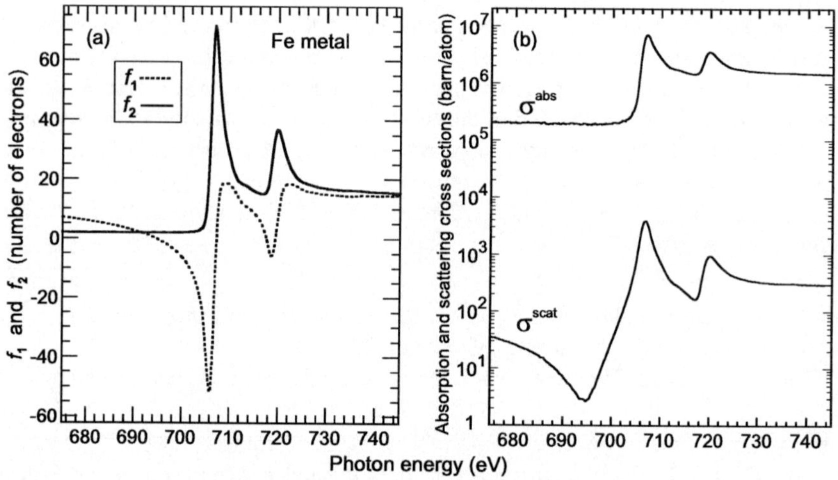

Fig. 6.10 **a** Measured energy dependent resonance factors f_1 and f_2 for the $L_{3,2}$-edges of Fe metal ($2p_{3/2}, 2p_{1/2} \rightarrow 3d$ transitions) [23]. **b** Energy dependent cross sections for x-ray absorption (top) and scattering (bottom) calculated from the scattering factors according to (6.72) and (6.71), respectively. Note that we have plotted $f_1(\hbar\omega) = Z + F'(\hbar\omega)$ where $Z = 26$, while $F'(\hbar\omega)$ is plotted in Fig. 6.4 of [23]

neglect of the **Q**-dependence leads to scattering cross sections that are good approximations in the soft x-ray region, but they are increasingly too large at higher energies. This is because the form factors are the largest in the forward direction and falloff with increasing scattering angles [4].

The non-resonant Henke-Gullikson scattering factor may be replaced in the resonant region by substituting the measured response for a given molecule or solid, and as an example we show in Fig. 6.10 the factors in the vicinity of the L_3 and L_2 edges of Fe metal as measured by Kortright and Kim [23]. As in Fig. 6.9 we show both the scattering factors and cross sections.

6.8 Atomic Shell Sum Rules

The energy dependence of the scattering factors obeys two important sum rules. The first one is the well-known Thomas-Reiche-Kuhn sum rule for the energy integrated oscillator strength which can be written by means of (6.45) as follows.

The *Thomas-Reiche-Kuhn sum rule* states that the total energy integrated oscillator strength of an atom obeys the relation

$$\sum_{n=1}^{Z} g_n = \frac{1}{C}\int_0^\infty \sigma_{\text{abs}}(\hbar\omega)\,\mathrm{d}\hbar\omega = \frac{2}{\pi}\int_0^\infty \frac{f_2(\hbar\omega)}{\hbar\omega}\,\mathrm{d}\hbar\omega = Z\,, \tag{6.73}$$

where

$$C = \pi\, h\, c\, r_0 = 109.89\,\text{Mb eV}. \tag{6.74}$$

According to the sum rule, the integrated area under the energy dependent cross section, normalized by the constant C is equal to the number of electrons in the atom. To check the sum rule, we have plotted in Fig. 6.11a the experimentally determined cross section $\sigma_{\text{abs}}(\hbar\omega)$ for Fe metal over a very large range.

The sum rule integrated oscillator strength according to (6.73) is shown as a black line in Fig. 6.11b. At the highest energy, where all electronic shell contribute, the sum rule integration gives $Z = 25$ close to the actual $Z = 26$ value for Fe. The largest sum rule contribution originates from the spectral region 50 to 2000 eV which contains about 70% of the total oscillator strength. As indicated in red in Fig. 6.11b, it contains the absorption edges of the 16 electrons in the $3p$, $2s$ ($M_{3,2,1}$-edges) and $2p$, $2s$ ($L_{3,2,1}$-edges) core shells corresponding to 62% of the total number of electrons.

When comparing the contributions from the individual electronic shells that can be excited with increasing energy, shown as a red curve in Fig. 6.11b, the observed increase in atomic oscillator strength with photon energy (black curve) is too low at

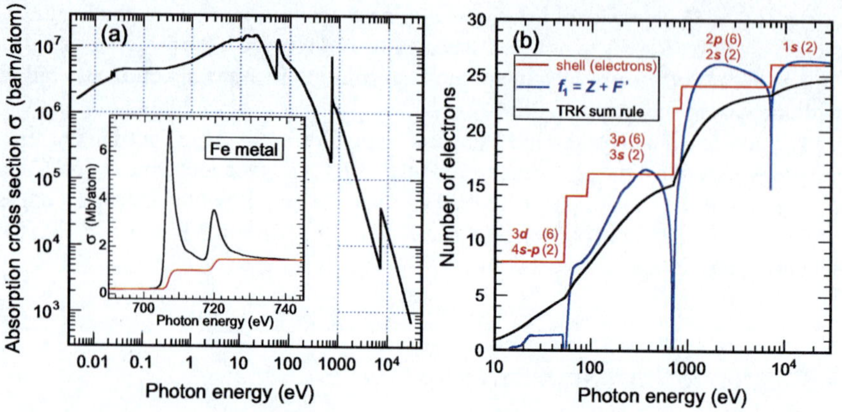

Fig. 6.11 **a** Photon absorption cross section for Fe metal over an extended range, synthesized from many sources and including measured data at the Fe L-edge [23]. The inset shows the detailed cross section around the Fe L-edges, where we have shown in red the step-like contributions (excitations to continuum states) which increase proportional to the number of core electrons in the $2p_{3/2}$ and $2p_{1/2}$ subshells, respectively. **b** Illustration of sum rules. In red we have plotted the number of electrons that can be excited at the respective absorption edge positions, indicated by the steps. We have assumed that each excited electron contributes an oscillator strength of unity. The black curve is the energy integral of the cross section in (**a**) according to the Thomas-Reiche-Kuhn sum rule (6.73), and the blue curve is a plot of $f_1(\hbar\omega)$ which obeys the sum rule (6.75)

low energies and catches up only later to that expected from an oscillator strength of 1 per electron (red curve), irrespective of which atomic shell it occupies. In particular, the outer shells contribute less than 1 per electron to the oscillator strength while the inner shells contribute more than 1 per electron. This is in good accord with shell-by-shell photoelectric cross-section calculations [24]. It has the remarkable consequence that intense hard x-ray FEL pulses remove electrons sequentially from the inner to the outer shells and can therefore produce exotic "hollow atoms" [25, 26].

The second sum rule is associated with the real part $f_1(\hbar\omega)$ of $F(\mathbf{Q}, \hbar\omega)$. In the limit $\hbar\omega \gg \mathcal{E}_n$ and for forward scattering, $\mathbf{Q} = 0$ we have the following sum rule for the atomic form factor.

The atomic form factor sum rule for $\hbar\omega \gg \mathcal{E}_n$ and $\mathbf{Q} \rightarrow 0$ states

$$F(\mathbf{Q}\rightarrow 0, \hbar\omega\rightarrow\infty) = f_1(\infty) = Z\,. \tag{6.75}$$

This sum rule is plotted as a blue curve in Fig. 6.11b, and the high energy limit accurately gives $Z = 26$.

References

1. J.D. Jackson, *Classical Electrodynamics* (Wiley, New York, 1999)
2. L.D. Landau, E.M. Lifshitz, *The Classical Theory of Fields* (Butterworth-Heinemann, Oxford, 2002)
3. R.P. Feynman, R.B. Leighton, M. Sands, *The Feynman Lectures on Physics*, Vol. II (Addison-Wesley, Reading, 1964)
4. J. Als-Nielsen, D. McMorrow, *Elements of Modern X-Ray Physics*, 2nd edn. (Wiley, New York, 2011)
5. F. de Bergevin, M. Brunel, Phys. Lett. A **39**, 141 (1972)
6. F. de Bergevin, M. Brunel, Acta Cryst. A **37**, 314 (1981)
7. S.W. Lovesey, S.P. Collins, *X-Ray Scattering and Absorption by Magnetic Materials* (Clarendon Press, Oxford, 1996)
8. H.A. Kramers, W. Heisenberg, Z. Phys. **31**, 681 (1925)
9. P.A.M. Dirac, Proc. Roy. Soc. London A **114**, 710 (1927)
10. W. Heisenberg, Z. Phys. **33**, 879 (1925)
11. J. Stöhr, *NEXAFS Spectroscopy* (Springer, Heidelberg, 1992)
12. J. Stöhr, H.C. Siegmann, *Magnetism: From Fundamentals to Nanoscale Dynamics* (Springer, Heidelberg, 2006)
13. D. Attwood, A. Sakdinawat, *X-Rays and Extreme Ultraviolet Radiation: Principles and Applications*, 2nd edn. (Cambridge University Press, Cambridge, 2017)
14. A. Einstein, Verh. Deut. Phys. Ges. **18**, 318 (1916)
15. B.L. Henke, E.M. Gullikson, J.C. Davis, At. Data Nucl. Data Tables **54**, 181 (1993)
16. J.J. Sakurai, *Modern Quantum Mechanics*, Revised. (Addison-Wesley, Reading, Mass., 1994)
17. G. Breit, E. Wigner, Phys. Rev. **49**, 519 (1936)
18. R.G. Newton, American. J. Phys. **44**, 639 (1976)
19. R. de L. Kronig: J. Opt. Soc. Am. **12**, 547 (1926)
20. H.A. Kramers, Atti Cong. Intern. Fisici **2**, 545 (1927)
21. L.G. Parratt, C.F. Hempstead, Phys. Rev. **94**, 1593 (1954)
22. https://henke.lbl.gov/optical_constants/
23. J.B. Kortright, S.K. Kim, Phys. Rev. B **62**, 12–216 (2000)
24. J.J. Yeh, I. Lindau, At. Data Nucl. Data Tables **32**, 1 (1985)
25. L. Young et al., Nature **466**, 56 (2010)
26. G. Doumy et al., Phys. Rev. Lett. **106**, 083 002 (2011)

Chapter 7
Semi-classical Response of Solids to Electromagnetic Fields

7.1 Introduction

In the previous chapter we discussed the classical description of x-rays with the key building blocks of the electromagnetic (EM) world, electrons, spins, and atoms. Here we extend this treatment to materials. We start by discussing the static response of materials to electric and magnetic fields and then discuss how the static concepts can be extended to include frequency dependent fields. Historically, the response of materials to electromagnetic fields has been cast into the language of the complex refractive index, which is related to the electric permittivity and magnetic permeability. Here we extend this low-frequency formalism to the entire frequency range from microwaves to x-rays. In the process we discuss the important transition from the microwave to the optical regime whose link has been historically impeded by the so-called THz gap, the frequency region extending from hundreds of GHz to tens of THz. This is the very range where spin motion can no longer follow the frequency of the magnetic field of the EM waves and the refractive index becomes independent of the permeability.

We then discuss the extension of the optical parameters ("constants") into the x-ray region and establish their link with the x-ray scattering factors introduced in the previous chapter. In order to do so we consider the transmission of x-rays through a thin film and describe the processes of absorption and scattering from complementary dielectric response and scattering points of view. We distinguish incoherent and coherent scattering by a solid, using the geometrical optics concepts of coherence of Sect. 4.3.1. As a pedagogical example we review the Huygens–Fresnel principle for the case of scattering by a thin film in a circular aperture.

Finally, we discuss the polarization dependence of the x-ray response of materials in the optical constant and scattering factor formalisms, emphasizing the interaction with magnetic materials like the Faraday effect and magnetic circular dichroism.

J. Stöhr, *The Nature of X-Rays and Their Interactions with Matter*, Springer Tracts in Modern Physics 288, https://doi.org/10.1007/978-3-031-20744-0_7

7.2 Static Magnetic and Electric Fields Inside Materials

The conventional link of the magnetic field $\mathbf{H}$ and the magnetic induction $\mathbf{B}$ given by $\mathbf{B} = \mu_0\mathbf{H}$ applies only in vacuum and to a good approximation in air. In order to describe the fields existing in materials, a third field vector is introduced, the *magnetization* $\mathbf{M}$. In this book we express all quantities in SI units [1] (see Appendix A.1). The magnitude of $\mathbf{M}$ is then defined as the volume density of magnetic moments, and we have the following relationship.

The three magnetic field vectors are related according to

$$\mathbf{B} = \mu_0\mathbf{H} + \mathbf{M}. \tag{7.1}$$

$\mathbf{M}$ is measured in the same units $[\mathrm{V\,s\,m^{-2}}] \equiv [\mathrm{T}]$ as $\mathbf{B}$, $\mathbf{H}$ has units $[\mathrm{A\,m^{-1}}]$ and $\mu_0 = 4\pi \times 10^{-7}\,\mathrm{V\,s\,A^{-1}m^{-1}}$ is the permeability of the vacuum.

This relationship is general. In the absence of an externally applied magnetic field, the field $\mathbf{H} = \mathbf{H}_\mathrm{d}$ inside a magnetic sample is called the demagnetizing field, indicating that it points in the opposite direction of $\mathbf{M}$. When a uniformly magnetized magnetic body is placed into an additional external magnetic field $\mathbf{H}_\mathrm{ext}$, the field $\mathbf{H}$ is given by $\mathbf{H} = \mathbf{H}_\mathrm{d} + \mathbf{H}_\mathrm{ext}$.

According to (7.1), $\mathbf{M}$ determines the maximum $\mathbf{B}$-field that can be generated by a fully magnetized body, i.e. outside a "permanent" magnet. In the molecular field approximation, $\mathbf{M}$ is envisioned as the origin of the Weiss molecular field $\mathbf{H}_\mathrm{W}$.[1] The magnetizations for the three important ferromagnetic metals Fe, Co, and Ni are listed in Table 7.1.

All three magnetic field vectors must have the same symmetry because they occur as a sum in the same equation (7.1). That is $\mathbf{B}$, $\mathbf{H}$, and $\mathbf{M}$ each define a sense of rotation and are *axial vectors*. They change sign upon time reversal. The symmetry properties of the magnetic field vectors are based on experimental observations.

In isotropic *paramagnetic* and *diamagnetic* substances we have the simple linear relationship.

Table 7.1 Saturation magnetizations for the ferromagnetic elemental metals Fe, Co, and Ni at 4.2 K in [Tesla] [2]

Metal	M [T]
Fe (bcc)	2.199
Co (hcp)	1.834
Ni (fcc)	0.665

[1] The Weiss molecular field $\mu_0\mathbf{H}_\mathrm{W}$ inside a magnetic material is of order 10^3 T and should not be confused with the demagnetizing field $\mu_0\mathbf{H}_\mathrm{d}$ which is only of order 1 T.

$$\mathbf{B} = \mu\,\mathbf{H} = \mu_r\,\mu_0\,\mathbf{H}\,, \tag{7.2}$$

where μ is the *absolute magnetic permeability* with the same units as μ_0. The *relative magnetic permeability* μ_r is dimensionless and >1 for paramagnetic but <1 for diamagnetic media and in such systems typically differs from unity by a factor of about 10^{-5}.

In general, μ and μ_r of a medium are complex and second rank tensors. We shall only consider isotropic media where they are scalars. In order to describe the interaction of the rapidly changing **H**-field of the EM wave with the magnetization of a material we have to consider the frequency dependence of **M**. This requires treatment of the permeability as a complex quantity as discussed in more detail in Sect. 7.3 below, where we shall indicate the complex nature by a tilde and we have $\tilde{\mu} = \tilde{\mu}_r\,\mu_0$. At low frequencies, $\tilde{\mu}_r = \mu_r$ is real to a good approximation and in ferromagnetic materials μ_r is typically much larger than 1. Mu-metal (approximately 75% nickel, 15% iron, plus copper and molybdenum) can have relative permeabilities of 80,000–100,000 which makes it very effective at screening static or low-frequency magnetic fields.

The relation between the magnetization **M** and the magnetic field **H** defines the often quoted *magnetic susceptibility* χ_m. Combining (7.1) and (7.2) we obtain

$$\mathbf{M} = (\mu_r - 1)\mu_0\,\mathbf{H} = \chi_m\,\mu_0\,\mathbf{H}. \tag{7.3}$$

The magnetic susceptibility $\chi_m = \mu_r - 1$ is dimensionless. For paramagnetic and ferromagnetic media we have $\chi_m > 0$ and for diamagnetic media $\chi_m < 0$.

Analogous equations exist for the electric field vector **E** and the dielectric field or *dielectric displacement* **D**. In vacuum and in many materials of interest we have $\mathbf{D} = \epsilon_0\mathbf{E}$, where $\epsilon_0 = 8.86 \times 10^{-12}\,\mathrm{A\,s\,V^{-1}\,m^{-1}}$ is the dielectric constant or electric permittivity of the vacuum. To generally describe the electric fields in matter, the *dielectric polarization* **P** is introduced which is the density of electric dipoles.

The three electric vectors are related by the equation

$$\mathbf{D} = \epsilon_0\mathbf{E} + \mathbf{P}, \tag{7.4}$$

D and **P** have units $[\mathrm{A\,s\,m^{-2}}]$, and $\epsilon_0 = 8.86 \times 10^{-12}\,\mathrm{A\,s\,V^{-1}\,m^{-1}}$ is the dielectric constant or permittivity of the vacuum.

A material with a quasi-permanent electric charge or electric polarization **P** is called an *electret*, a name coined by Oliver Heaviside in 1885 from "*electr*icity" and "magn*et*". In air, electrets attract stray charges from the surroundings which soon neutralize the surface charges. Some materials also display *ferroelectricity*, i.e. they exhibit external electric field hysteresis.

Similar to the relationship (7.2) for the magnetic vectors we have

$$\mathbf{D} = \epsilon\, \mathbf{E} = \epsilon_r\, \epsilon_0\, \mathbf{E}. \tag{7.5}$$

Here ϵ is the *absolute electric permittivity* and has the same units as the *electric permittivity of free space* ϵ_0, while the *relative electric permittivity* ϵ_r is dimensionless. Like the relative permeability μ_r, in general, the relative permittivity ϵ_r is complex and a tensor. We shall only consider isotropic substances, but in order to account for its frequency dependence we shall in Sect. 7.3 below consider the complex nature of ϵ_r, indicated as $\tilde{\epsilon}_r$, also called the *complex dielectric function*,

The *electric susceptibility* χ_e is defined in analogy to (7.3) between the dielectric polarization **P** and the electric field **E** according to

$$\mathbf{P} = (\epsilon_r - 1)\epsilon_0\, \mathbf{E} = \chi_e\, \epsilon_0\, \mathbf{E}. \tag{7.6}$$

The electric susceptibility $\chi_e = \epsilon_r - 1$ is dimensionless.

In contrast to the magnetic field vectors, the electric field vectors do not indicate a sense of rotation, they only define a direction and do not change sign when the time is inverted. The electric field vectors are *polar vectors*.

7.3 Frequency Response of Materials: Microwaves to X-Rays

We consider a material exhibiting both a static magnetic polarization **M** given by (7.3) and dielectric polarization **P** given by (7.6) and then consider its response with increasing frequency.

7.3.1 Permittivity: Electric Field Response

According to Maxwell's equations, time-varying magnetic fields are always accompanied by electric fields. Classically, a material's ability to prevent field penetration by internal screening is described for the *electric* fields by the complex permittivity and for the magnetic fields by the complex permeability. The complex frequency dependent dimensionless *relative permittivity*, $\tilde{\epsilon}_r$, is usually written as[2]

$$\tilde{\epsilon}_r(\omega) = \epsilon_1(\omega) + \mathrm{i}\,\epsilon_2(\omega)\,. \tag{7.7}$$

The real part ϵ_1 is related to the stored energy within the medium, and the imaginary part ϵ_2 is related to the dissipation of energy within the medium. For metals, key parameters that determine the frequency dependence of $\tilde{\epsilon}_r(\omega)$ are the relaxation time,

[2] We have used the phase convention that the time evolution of the electromagnetic wave is given by $\exp(-\mathrm{i}\omega t)$.

τ_E, of the conduction electrons to the E-field excitation and the plasma frequency ω_P. In metals, electromagnetic waves with frequencies below ω_P are strongly absorbed because the electrons are able to respond fast enough to screen the incident electric field. They are then re-emitted, and the reflectivity is near unity. For waves with frequencies above ω_P the electrons are no longer able to follow and screen the electric field, and the waves can penetrate into the metal. In most metals, the plasma frequency is in the ultraviolet (several eV), making them shiny (reflective) in the visible range.

At frequencies $\omega \ll \omega_P$, one can describe $\tilde{\epsilon}_r(\omega)$ by the following expressions for the real and imaginary parts[3]

$$\epsilon_1(\omega) = 1 - \frac{\omega_P^2\,\tau_E^2}{1+\omega^2\,\tau_E^2} \text{ and } \epsilon_2 = \frac{\omega_P^2\,\tau_E}{\omega(1+\omega^2\,\tau_E^2)}, \tag{7.9}$$

where τ_E (units [s]) is the conduction electron relaxation time in response to the E-field and ω_P is the plasma frequency (units [rad/s]). When evaluating (7.9), one needs to be careful because all frequencies ω are *angular* frequencies (units [rad/s]) which are related to the *cycle* frequencies ν (units of [Hz]$\equiv$[s^{-1}]) by $\omega = 2\pi\nu$. Typically, one plots the permittivity as a function of the cycle frequency so that one needs to write $\omega_P = 2\pi\nu_P$ and $\omega = 2\pi\nu$ in (7.9) with the frequencies ν_P and ν in units of [Hz]. The real and imaginary parts obey the Kramers-Kronig relation (see Sect. 6.6), and we have plotted the frequency dependent quantities ϵ_1 and ϵ_2 for Fe metal in Fig. 7.1. In the figure we have indicated the characteristic E-field induced relaxation rate of the electrons where $-\epsilon_1 = \epsilon_2$. For Fe with $\tau_E = 34$ fs [3] we have $1/2\pi\tau_E = 4.6$ THz. The gray shading indicates the region where the response is dominated by the imaginary part $|\epsilon_2|$.

7.3.2 *Permeability: Magnetic Field Response*

Similar to the E-field response inside a material, which is expressed by the complex relative permittivity, the **H**-field response is characterized by the dimensionless complex *relative permeability*. It is written as

$$\tilde{\mu}_r(\omega) = \mu_1(\omega) + \mathrm{i}\,\mu_2(\omega), \tag{7.10}$$

---

[3] At frequencies below about 1 THz, ϵ_1 and ϵ_2 are also related to the components of the complex Drude conductivity given by

$$\tilde{\sigma}(\omega) = \sigma_1(\omega) + \mathrm{i}\,\sigma_2(\omega) = \sigma_0\left(\frac{1}{1+\omega^2\,\tau^2} + \mathrm{i}\,\frac{\omega\,\tau}{1+\omega^2\,\tau^2}\right), \tag{7.8}$$

where σ_0 is the DC electrical conductivity and τ the Drude relaxation time, usually assumed to be $\tau = \tau_E$.

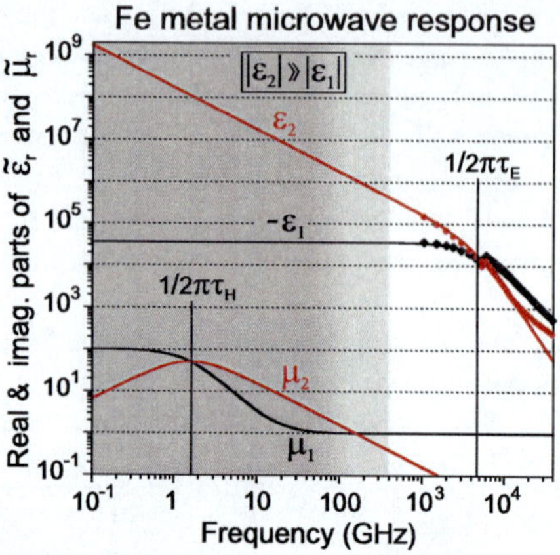

Fig. 7.1 Frequency dependence of the real and imaginary parts of the dimensionless relative permittivity, ϵ_1 and ϵ_2, and relative permeability, μ_1 and μ_2, for ferromagnetic Fe metal. The red circles and black diamonds are derived from experimental data by Ordal et al. [3, 4]. The permittivity curves were calculated according to (7.9) with the parameters $\omega_P/2\pi = 885\,\text{THz}$ and $\tau_E = 34\,\text{fs}$ [3]. The permeability curves were calculated according to (7.11) with the parameters $\mu_r^0 = 100$ and $\tau_H = 100\,\text{ps}$. The gray shading indicates the region where the response is dominated by the imaginary part $|\epsilon_2|$

and the ratio of the imaginary to the real part $\mu_2(\omega)/\mu_1(\omega)$ of the complex permeability is called the *loss tangent* and provides a measure of how much power is lost in a material versus how much is stored. When losses occur, **B** lags behind **H**.

The general frequency dependence of the permeability is quite complicated. Losses can be due to different mechanisms and are dependent on the composition, the microstructure and macroscopic shape of the magnetic material. Damping phenomena may arise from eddy currents, magnetic domain wall motions, and damping of the precession of the magnetization [5–7]. While the resonant behavior of $\tilde{\mu}_r$ or the related AC susceptibility $\tilde{\chi} = \tilde{\mu}_r - 1$ can be probed in the GHz regime under application of an external magnetic field by ferromagnetic resonance, the general behavior of $\tilde{\mu}_r$ for electromagnetic fields up to the optical regime is seldom discussed because of its complexity.

From experiment one knows two limits. Gradual alignment of the random magnetic domains of a sample by application of an **H**-field yields the real DC permeability μ_r^0, determined from the initial slope of the magnetization curve. For Fe, for example, one finds $\mu_r^0 \simeq 100$. We take this real DC permeability μ_r^0 as our constant low-frequency value for $\tilde{\mu}_r$, with a negligibly small imaginary part. The other limit is the well-studied optical regime starting at photon energies around 1 eV or frequencies $\geq 250\,\text{THz}$. Here $\tilde{\mu}_r$ is also known to be real and has the simple real value 1, even for magnetic materials. That is why the **H**-field response is typically not even considered in electromagnetic response theories at optical and higher frequencies.

By considering the intrinsic low-frequency response of a ferromagnetic material to **H**-fields one may write down a general response equation for the real and imaginary parts of $\tilde{\mu}_r$. The key parameter is the **H**-field response time of the magnetic moments in the sample, assumed to be aligned in local domains that point into different directions. The **H**-field applies torque on the atomic moments and the characteristic response time τ_H is that associated with the magnetic anisotropy energy which favors the original magnetic alignment in the domains. Typical relaxation times are $\tau_H = 100\,\mathrm{ps}$ corresponding to a change of $\tilde{\mu}_r$ in the range between about 1 GHz and 1 THz. One may quite generally write $\tilde{\mu}_r(\omega)$ in terms of the following Kramers-Kronig-related real and imaginary parts

$$\mu_1(\omega) = 1 + \frac{\mu_r^0 - 1}{1 + \omega^2\,\tau_H^2} \quad \text{and} \quad \mu_2(\omega) = \frac{\omega\,\tau_H\,(\mu_r^0 - 1)}{1 + \omega^2\,\tau_H^2}, \tag{7.11}$$

where μ_r^0 is the real DC permeability and τ_H the spin relaxation time in response to the **H**-field. The expected frequency dependence of the quantities ϵ_1 and ϵ_2 for $\mu_r^0 = 100$ and $\tau_H = 100\,\mathrm{ps}$, characteristic of Fe metal, is plotted in the lower half of Fig. 7.1, where we have also indicated by a vertical line at $\mu_1 = \mu_2$ the characteristic **H**-field induced relaxation rate of the spins $1/2\pi\tau_H = 1.6\,\mathrm{GHz}$.

7.3.3 *From Permittivity and Permeability to Optical Parameters*

Above the microwave regime, at frequencies exceeding about 1 THz, the propagation of electromagnetic (EM) fields into an isotropic magnetic medium is typically described by the *complex refractive index*, $\tilde{n}$, composed of a real part, the *real refractive index*, n, and an imaginary part, the *extinction coefficient*, κ, according to [8, 9]

$$\boxed{\tilde{n}^2(\omega) = [n(\omega) + \mathrm{i}\,\kappa(\omega)]^2 = \frac{\tilde{\epsilon}(\omega)\,\tilde{\mu}(\omega)}{\epsilon_0\mu_0} = \tilde{\epsilon}_r(\omega)\tilde{\mu}_r(\omega).} \tag{7.12}$$

The quantities n and κ are usually referred to as the "optical constants". Owing to their strong energy dependence, we will here refer to them instead as the *optical parameters*. They describe the photon energy dependent refraction and absorption of the fields, respectively. The master equation (7.12) links $\tilde{n}$ to the complex *electric permittivity* $\tilde{\epsilon}$ and *magnetic permeability* $\tilde{\mu}$. From (7.12), we obtain the general expressions

$$n^2 - \kappa^2 = \epsilon_1\mu_1 - \epsilon_2\mu_2, \quad 2\,n\,\kappa = \epsilon_1\mu_2 + \epsilon_2\mu_1. \tag{7.13}$$

The real and imaginary parts of the refractive index are then given by

$$n = \sqrt{\frac{\sqrt{(\epsilon_1\mu_1 - \epsilon_2\mu_2)^2 + (\epsilon_1\mu_2 + \epsilon_2\mu_1)^2} + \epsilon_1\mu_1 - \epsilon_2\mu_2}{2}} \quad (7.14)$$

and

$$\kappa = \sqrt{\frac{\sqrt{(\epsilon_1\mu_1 - \epsilon_2\mu_2)^2 + (\epsilon_1\mu_2 + \epsilon_2\mu_1)^2} - \epsilon_1\mu_1 + \epsilon_2\mu_2}{2}}. \quad (7.15)$$

For *non-magnetic materials* we have $\mu_1 = 1$ and $\mu_2 = 0$ so that

$$n^2 - \kappa^2 = \epsilon_1, \quad 2n\kappa = \epsilon_2 \quad (7.16)$$

and

$$\boxed{\text{Non-magnetic: } n = \sqrt{\frac{\sqrt{\epsilon_1^2 + \epsilon_2^2} + \epsilon_1}{2}}, \quad \kappa = \sqrt{\frac{\sqrt{\epsilon_1^2 + \epsilon_2^2} - \epsilon_1}{2}}.} \quad (7.17)$$

The real part n of the complex refractive index, and the imaginary part κ, which characterizes extinction or absorption are plotted for Fe metal in Fig. 7.2. The limited available experimental data [3, 4] are shown as blue and red circles, while the solid blue and red curves are calculated by use of (7.14) and (7.15) with the permittivities and permeabilities plotted in Fig. 7.1.

The solid black curves in Fig. 7.2 are calculated assuming *non-magnetic* Fe according to (7.17) using the permittivity values in Fig. 7.1. They merge into the blue and red magnetic curves above about 10^3 GHz =1 THz, revealing that the magnetic permeability ceases to be important. Since the permeability reflects the response of the **H**-field component of the EM wave, we see that in the THz region, the influence of the **H**-field ceases to be important and the EM field response can be phrased in terms of that of the **E**-field alone.

> At frequencies *above* about 1 THz, the complex refractive index $\tilde{n}$ and the optical parameters n and κ are determined by the electrical permittivity alone. The material response to an EM wave can then be phrased in terms of the **E**-field alone and the effect of the **H**-field can be neglected.

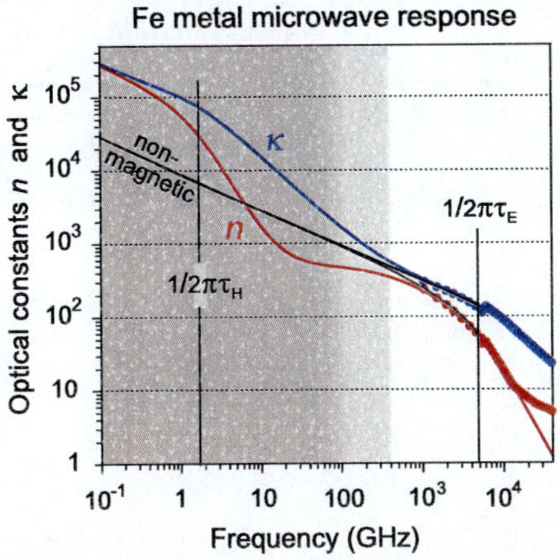

Fig. 7.2 Frequency dependence of the optical parameters n and κ. The blue and red circles are from [3, 4]. The solid red and blue curves are calculated by use of (7.14) and (7.15) with the permittivities and permeabilities in Fig. 7.1. The solid black curves are calculated assuming *non-magnetic* Fe according to (7.17) with the permittivity values plotted in Fig. 7.1. Note that they merge into the magnetic curves above about 10^3 GHz. The gray shaded area is that shown in Fig. 7.1

7.3.4 Penetration Depth of EM Waves: From Microwaves to X-Rays

We briefly want to follow up on the important point that above the microwave region (above 1 THz $\simeq$ 4 meV) the interaction of EM waves with matter no longer depends on the **H** component of the EM wave and can be described through the frequency dependent response of the electrons to the **E**-field. The permeability then becomes unimportant, and the optical parameter description describes the dielectric response to the **E**-field. One also changes notation, renaming $\kappa = \beta$ which describes the important absorptive response that determines the penetration depth into materials.

Above the microwave regime, the characteristic penetration depth or absorption length ℓ_x at a given wavelength λ is defined as the distance over which the intensity decreases by a factor $1/e$ or about 35% (see Sect. 7.4.2 below) as

$$\ell_x = \frac{\lambda}{4\pi\beta}. \tag{7.18}$$

It is half of the so-called *skin depth* typically used in the microwave region.

In order to get an overall perspective of the change of ℓ_x, we have plotted it for Fe metal in Fig. 7.3 over the entire spectral range from microwaves to x-rays, spanning about ten orders of magnitude.

Here the transition from the microwave region to higher frequencies is of particular interest. In the microwave region, the absorption length in Fe is dominated by the

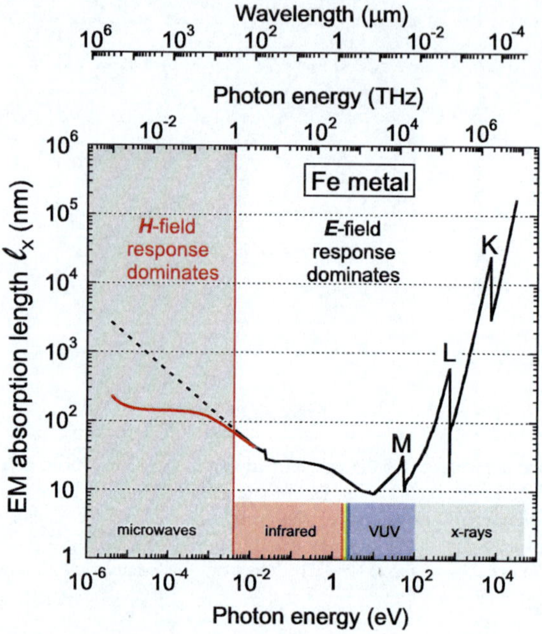

Fig. 7.3 Characteristic $1/e$ absorption length ℓ_x of electromagnetic radiation for Fe metal from microwaves to x-rays. The approximate name of the different spectral regions is indicated at the bottom. Below about 1 THz (top scale) the response of ferromagnetic Fe is dominated by the interaction of the EM **H**-field with the magnetization, shown as a red curve. The dashed black curve neglects the **H**-field response and considers only the **E** interaction with the electrons which completely dominates in the optical and x-ray regions (solid black curve)

interaction of the EM **H**-field with the magnetization, as shown by the red curve. The **H**-field contributes through the DC magnetic permeability (also see Fig. 7.1), resulting in enhanced absorption and reduced penetration length relative to the dashed black line which ignores magnetic effects.

The important phenomenon may be understood in a simple physical picture as follows. A frequency of 1 THz corresponds to an EM cycle time of 1 ps. At sub-ps cycle times, the **H**-field oscillations become so fast that the spins can no longer follow because the **H**-field induced torque on the spins requires transfer of angular momentum to the lattice [1]. This however becomes bottlenecked in the range below about 1 ps [10, 11]. At shorter EM cycle times only the electrons can respond and their response follows the **E**-field.

In the following we will consider the behavior above the microwave range, above the red demarkation line in Fig. 7.3 and consider both the refractive and absorptive response determined by $n(\omega)$ and $\kappa(\omega) = \beta(\omega)$.

7.3.5 Optical Parameters: From the THz to the X-Ray Regime

Above the microwave region, the dielectric sample response is then phrased in terms of the dimensionless complex *refractive index* $\tilde{n}(\omega)$, composed by its dispersive real part n and imaginary absorptive part $\kappa = \beta$ (sometimes denoted $n = n'$ and $\kappa = n''$). In the optical region the response is described by (7.12), or

$$\boxed{\text{Optics: } \tilde{n}(\omega) = n(\omega) + \mathrm{i}\,\beta(\omega)} \tag{7.19}$$

where n and β ($n, \beta > 0$) are the dimensionless energy dependent optical parameters. This formulation can be extended to the x-ray regime, and as an example we show in Fig. 7.4 the characteristic change of n and β from the THz to the x-ray regime for Fe metal. The figure shows that at photon energies above about 10 eV, $n(\omega)$ approaches unity and in fact becomes slightly less than 1 for Fe metal. This has the important consequence that below a certain "critical" grazing incidence angle x-rays are totally reflected. In the ultraviolet and x-ray regions one conveniently uses the following

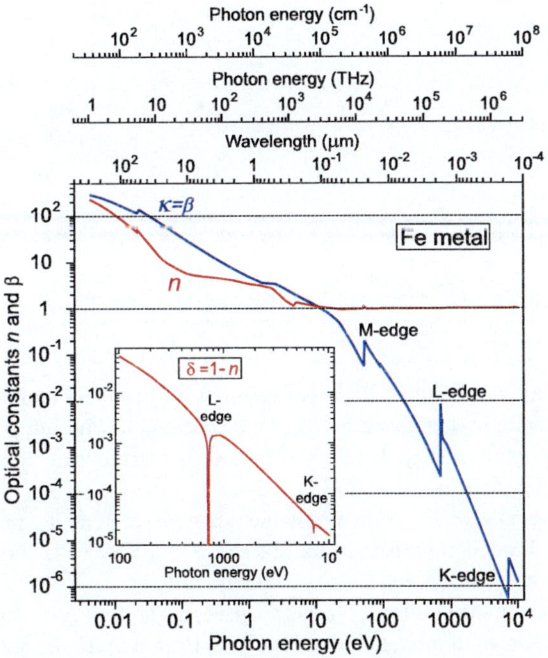

Fig. 7.4 Optical parameters n and β for Fe metal derived from experimental data [1, 12, 13]. Due to the large energy range we show various energy scales and also the wavelength scale in μm. Near the L-edge we have included the pronounced resonant contributions as discussed below. Near the L-edge we have included the resonant contributions (see Fig. 7.12). In the inset we show the energy dependent refractive index $\delta = 1 - n$, defined in (7.20)

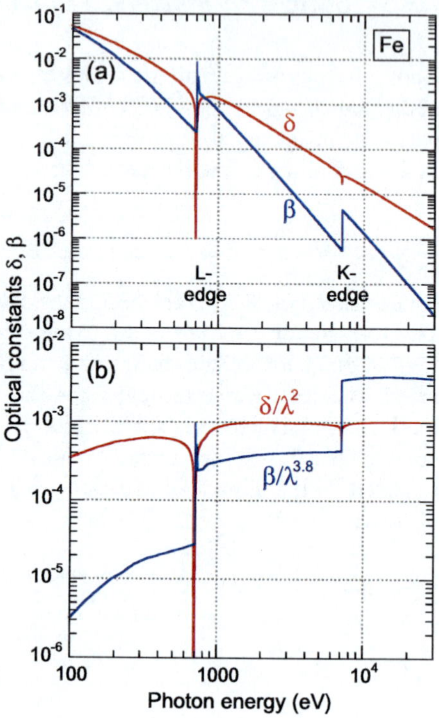

Fig. 7.5 **a** Comparison of the magnitudes of $\delta = 1 - n$ and β for Fe metal, derived from the data in Fig. 7.4 above 100 eV. Note that δ becomes negative right before the L-edge position (not shown). **b** Scaling of δ and β with wavelength λ, which at higher photon energies increasingly becomes $\beta \propto \lambda^4$ and $\delta \propto \lambda^2$

terminology [9, 14]

$$\boxed{\text{X-rays: } \tilde{n}(\omega) = 1 - \delta(\omega) + \mathrm{i}\,\beta(\omega),} \tag{7.20}$$

where the real part is redefined as $n = 1 - \delta$. The choice of the positive sign of the imaginary part in (7.19) and (7.20) is consistent with our phase choice for propagating electromagnetic waves, given by (3.11). According to Fig. 7.4, $\beta \geq 0$ decreases strongly with photon energy, leading to greater penetration of x-ray than optical radiation into materials as shown in Fig. 7.3.

The replacement $n = 1 - \delta$ is reasonable when $|\delta| \ll 1$, so it can be regarded as a correction. A direct comparison of the sizes of β and δ for Fe is shown Fig. 7.5a in the energy region > 100 eV where n is close to unity.

Outside the resonance regions, $\delta(\omega)$ is positive, but at the L_3 absorption resonance it is negative as shown in more detail in Fig. 7.12. Both β and δ decrease with photon energy, and above about 1000 eV the absorptive index β typically becomes smaller than the dispersive index δ. In the hard x-ray region they scale as $\beta \propto \lambda^4$ and $\delta \propto \lambda^2$, as illustrated in Fig. 7.5b.

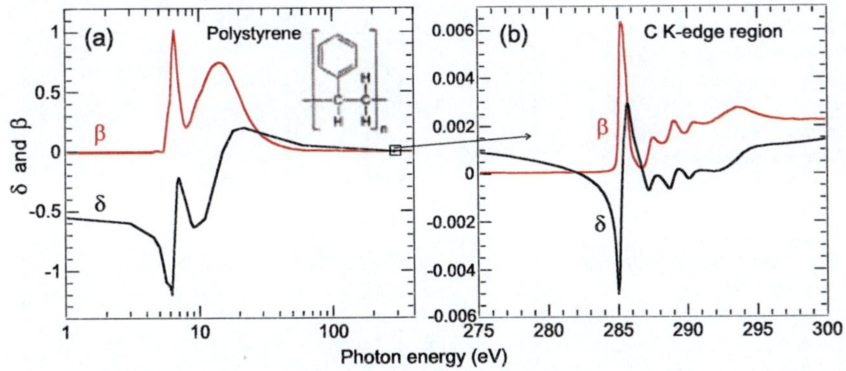

Fig. 7.6 **a** Energy dependence of the optical parameters β and δ for polystyrene over a wide photon energy range [17]. **b** Enlarged region near the C K-edge (see Footnote 4)

Another example of the energy dependence of δ and β from the optical to the x-ray range is shown for polystyrene (PS) in Fig. 7.6a, and in (b) we show the details around the C K-edge region.[4]

In PS, β is about zero in the optical region since no optical transitions are possible across the large band gap of about 4.2 eV. Transitions from the highest occupied (HOMO) to the lowest unoccupied (LUMO) molecular orbitals become visible in the 4–10 eV region. Both δ and β become small and smooth above about 100 eV, until the carbon K-shell excitation channel opens up around 285 eV. The rich fine structure in δ and β near the C K-shell absorption threshold is shown enlarged in Fig. 7.6b. It results from transition from the K-shell to empty molecular orbitals [18].

In Fig. 7.7 we show the values of β and δ for a hydrocarbon of composition C_8H_8, which is the same as PS. The plot is arranged similar to that for Fe in Fig. 7.5 and also shows the scaling of the β and δ with wavelength λ. It shows that at a photon energy of 10 keV δ is about a factor of 10^3 larger than β. This has important consequences for x-ray protein crystallography, which utilizes x-rays around 10 keV. The absorptive (β) response, which results in the creation of photo- and Auger electrons that cause deleterious radiation damage, is greatly reduced relative to the dispersive (δ) response which enters on an equal footing as non-resonant Thomson ($\propto r_0 Z$) scattering. It also makes it advantageous to utilize phase contrast imaging ($\Delta\delta$) [19] rather than absorption contrast ($\Delta\beta$) imaging, typically used in the soft x-ray region [20]. We can summarize as follows.

The effective medium response of matter to x-rays, described by the semi-empirical and phenomenological optical parameters, simplifies in the x-ray

[4] The data shown in Fig. 7.6b were measured with somewhat lower photon energy resolution than those published in [15] (also see Suppl. Mat. in [16]).

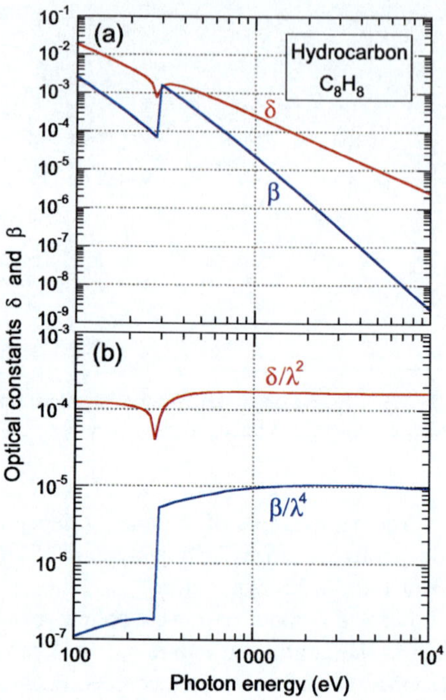

Fig. 7.7 **a** Comparison of the magnitudes of $\delta = 1 - n$ and β for a hydrocarbon above 100 eV. We have ignored the detailed resonant contributions near the C K-edge which depends on C-C bond hybridization. **b** Scaling of δ and β with wavelength λ, which in the x-ray region is well approximated by $\beta \propto \lambda^4$ and $\delta \propto \lambda^2$

relative to the optical regime. With increasing photon energy, the absorptive response, characterized by β, decreases and at energies around 1 keV become weaker than the dispersive response, expressed by δ. In the hard x-ray region, the response scales with wavelength as $\beta \propto \lambda^4$ and $\delta \propto \lambda^2$.

7.4 Dielectric Response Formulation of X-Ray Absorption

Here we derive three equivalent formulations for the exponential x-ray absorption law based on the phenomenological effective medium response, introduced above as an extension of the low-frequency response.

7.4.1 *Optical Parameter Formulation*

X-ray absorption corresponds to the attenuation of electric fields transmitted through a sample whose electronic response is expressed by the complex refractive index written in the form of (7.20). We assume that the incident field propagates in vacuum in the z direction and traverses a sample in the form of a film in the x-y plane of thickness d along z. For simplicity we shall ignore the time dependence of the field and consider only the change of the field as a function of distance z in the sample. The incident field is then given by

$$\mathbf{E}_{\text{in}}(z) = \mathbf{E}_0 \left[\mathrm{e}^{\mathrm{i}kz} + \mathrm{e}^{-\mathrm{i}kz} \right] = 2\,\mathbf{E}_0 \cos(kz). \tag{7.21}$$

The first term in the square brackets causes electronic excitations in the sample, while the second term causes stimulated de-excitations from the excited state. Since we assume that all atoms in the sample are originally in their ground state, the second term is not active and can be omitted. The starting point for our description is therefore an incident field of the form

$$\mathbf{E}_{\text{in}} = \mathbf{E}_0\, \mathrm{e}^{\mathrm{i}kz}. \tag{7.22}$$

The attenuation of the field upon transmission through a sample is illustrated in Fig. 7.8.

We define the entry point of the electric field in the sample as $z = 0$ and assume a non-magnetic sample so that the refractive index is determined by the response of the charge in the sample, only. After having traversed a distance z in the sample, the field is given by

$$\mathbf{E} = \mathbf{E}_0\, \mathrm{e}^{\mathrm{i}\tilde{n}kz} = \underbrace{\mathbf{E}_0\, \mathrm{e}^{\mathrm{i}kz}}_{\mathbf{E}_{\text{in}}} \; \underbrace{\mathrm{e}^{-\mathrm{i}\delta kz}}_{\text{phase shift}} \; \underbrace{\mathrm{e}^{-\beta kz}}_{\text{reduction}} \tag{7.23}$$

so that for a sample thickness d the *field* emerging from the sample, i.e. the transmitted field, is given by

$$\boxed{\mathbf{E}_{\text{trans}} = \mathbf{E}_{\text{in}} \; \underbrace{\mathrm{e}^{-\mathrm{i}\delta kd}}_{\text{dispersion}} \cdot \underbrace{\mathrm{e}^{-\beta kd}}_{\text{absorption}} .} \tag{7.24}$$

where we have indicated the dispersive contribution $\mathrm{e}^{-\mathrm{i}\delta kd} = \mathrm{e}^{-\mathrm{i}2\pi\delta d/\lambda}$, leading to a phase shift $\Delta\phi = -2\pi\delta d/\lambda$ in the transmitted relative to the incident field, and

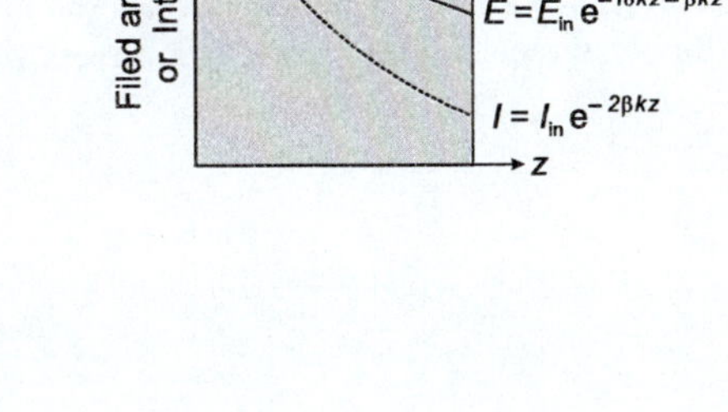

Fig. 7.8 Decay of the amplitude E and intensity $I \propto |E|^2$ of an incident wave $E_{\text{in}} = E_0\, \mathrm{e}^{\mathrm{i}kz}$ during transmission through a homogeneous sample with complex index of refraction $\tilde{n} = 1 - \delta + \mathrm{i}\beta$

the real absorptive contribution which reduces the incident amplitude E_0 by a factor $e^{-\beta kd} = e^{-2\pi\beta d/\lambda}$.

The dispersive phase shift in the transmitted field arises from the electronic fields created inside the sample in response to the incident field. The incident field causes a charge oscillation in the material, and the oscillating dipoles in turn generate a (scattered) field that is phase shifted and contributes to the total transmitted field. The real exponential causes a decrease of the incident field amplitude E_{in} in the material determined by the *absorptive* coefficient β.

By forming the absolute value squared of the transmitted field (7.24) we obtain the law for the transmitted intensity, defined by (3.120) as $I = 2\epsilon_0 c\,|\mathbf{E}|^2$.

$$\boxed{I_{\mathrm{trans}} = I_0\,e^{-2\beta\,k\,z}} \tag{7.25}$$

7.4.2 Absorption Coefficient Formulation

The absorption law given by (7.25) may also be derived by defining a characteristic length ℓ_x in a material over which the transmitted intensity decreases by a factor $1/e$ or about 35%, corresponding to $2\beta kz = 1$ in (7.25). One then calls the quantity

$$\mu_x = \frac{1}{\ell_x} = 2k\beta = \frac{4\pi\beta}{\lambda} \tag{7.26}$$

with the dimension [1/length] the *linear absorption coefficient*. In this formulation the intensity is attenuated by the amount $\mu_x \mathrm{d}z$ as it traverses the thin sheet of thickness $\mathrm{d}z$ at a depth z from the surface so that

$$-\frac{\mathrm{d}I(z)}{\mathrm{d}z} = \mu_x\,I(z). \tag{7.27}$$

The differential equation has the solution $I(z) = A\,e^{-\mu_x z}$ and by choosing $A = I_0$ as the incident intensity at the point $z = 0$ we obtain

$$\boxed{I_{\mathrm{trans}} = I_0\,e^{-\mu_x z}} \tag{7.28}$$

By use of (7.26) this is seen to be an equivalent formulation of (7.25).

The energy dependence of the absorption coefficient μ has previously been shown for water in Fig. 1.16 and for Co metal and CoO in Fig. 1.17. In the insulator CoO, for example, absorption is absent in the optical range due to the existence of a 2.5 eV band gap. A similar response is observed in Fig. 7.9 for the absorption coefficient μ of polystyrene, whose optical parameters were previously shown in Fig. 7.6.

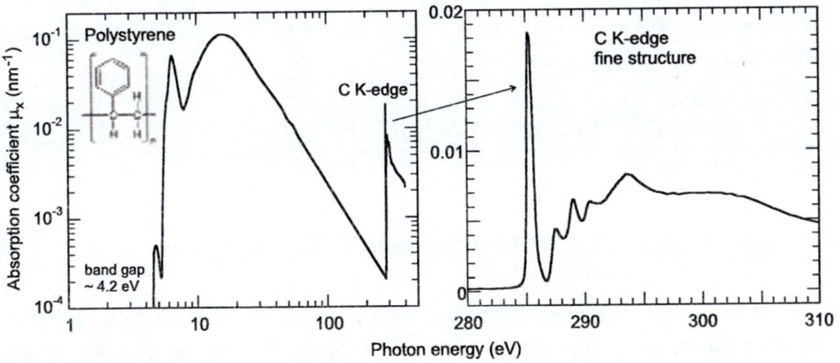

Fig. 7.9 Energy dependence of the absorption coefficient μ_x for polystyrene from the optical to the x-ray region and near the C K-edge [15, 17]. Also see Fig. 7.6 for the corresponding optical parameters

7.4.3 *Beer-Lambert Formulation*

We finally discuss another equivalent formulation of the exponential absorption law expressed by (7.25) and (7.28). It is referred to as the *Beer-Lambert law* in honor of the two scientists who first suggested it and reads[5]

$$\boxed{I_{\text{trans}} = I_0 \, e^{-\sigma_{\text{abs}} \rho_{\text{a}} z}\,.} \tag{7.29}$$

It is expressed in terms of the atomic absorption cross section σ_{abs} and the atomic number density ρ_{a}.

We previously linked the cross section σ_{abs} to the imaginary part of the atomic scattering factor by (6.72) as

$$\boxed{\sigma_{\text{abs}} = 2\lambda r_0 f_2 = 2\lambda f'' = \frac{4\pi}{\lambda \rho_{\text{a}}}\beta\,.} \tag{7.30}$$

We will formally establish the relation between f_2 (and f'') and β in the next sections below which will prove the equivalence of (7.25) and (7.29). The atomic number density ρ_{a} is calculated from the tabulated atomic mass density, ρ_{m} [mass/volume], the atomic mass number, N_{M} [mass/mol], and Avogadro's number $N_{\text{A}} = 6.02214 \times 10^{23}$ [atoms/mol] according to

$$\rho_{\text{a}} = \frac{\rho_{\text{m}} N_{\text{A}}}{N_{\text{M}}}\,. \tag{7.31}$$

---

[5] Historically, Lambert found that the absorbance is proportional to the thickness d of the sample, and Beer inferred that the absorbance is proportional to the concentration of the sample (ρ_{a}).

In summary, the absorption law for a sample of thickness d can therefore be written in the equivalent forms

$$\boxed{I_{\text{trans}} = I_0\,\mathrm{e}^{-2\beta\,k\,d} = I_0\,\mathrm{e}^{-\mu_x\,d} = I_0\,\mathrm{e}^{-\sigma_{\text{abs}}\,\rho_a\,d} = I_0\,\mathrm{e}^{-2\lambda\,r_0\,f_2\,\rho_a\,d}\,.} \tag{7.32}$$

7.5 From Dielectric to Atom-Based X-Ray Response

In the following sections we will link the description of x-ray interactions with matter, framed above in the framework of dielectric response through the phenomenological optical parameters, to an atom-based formalism based on the x-ray scattering factors introduced in the previous Chap. 6. The link between the two formulations is suggested by the formal similarity of the expressions for the total atomic scattering *factor* expressed by the $\mathbf{Q}$ independent form of (6.29)

$$F(\omega) = Z + F'(\omega) - \mathrm{i}\,F''(\omega) \tag{7.33}$$

and the expression for the refractive index (7.20), or

$$\tilde{n}(\omega) = 1 - \delta(\omega) + \mathrm{i}\,\beta(\omega). \tag{7.34}$$

One might therefore expect that the energy dependent real terms $Z + F'(\omega)$ and $1 - \delta(\omega)$ and the imaginary terms $F''(\omega)$ and $\beta(\omega)$ can be related.

In order to do so, we first need to extend the formulation of the atomic response in terms of the scattering factors, formulated for single atoms in the last chapter, to the collective response of many atoms. We will accomplish this through an atom-by-atom building block description of the x-ray response, starting from single atoms to a thin sheet and then to a thick film or a solid. We first briefly revisit the atom-based formalism of scattering introduced in Chap. 6.

7.5.1 *Brief Review of X-Ray Scattering Factors*

The *complex atomic scattering factor* given in units of [number of electrons per atom] is defined by (6.29) or

$$F(\mathbf{Q}, \omega) = F^0(\mathbf{Q}) + F'(\omega) - \mathrm{i}\,F''(\omega). \tag{7.35}$$

Here the energy independent *atomic form factor* $F^0(\mathbf{Q})$ describes the angle or $\mathbf{Q}$-dependence of the x-ray interaction with the entire atomic charge density. The two energy dependent "anomalous" scattering factors $F'(\omega)$ and $F''(\omega)$ describe the strong energy dependent deviation of scattering near absorption edges.[6]

The related *complex atomic scattering length* of dimension [length] is obtained by multiplying the atomic scattering factor with the Thomson scattering length $r_0 = e^2/4\pi\epsilon_0 m_e c^2 = 2.818 \times 10^{-15}$ m/electron, and accounting for the polarization dependence of scattering by a factor $\boldsymbol{\epsilon} \cdot \boldsymbol{\epsilon}'$, so that

$$\begin{aligned} f(\mathbf{Q}, \omega) &= f^0(\mathbf{Q}) + f'(\omega) - \mathrm{i}\, f''(\omega) \\ &= (\boldsymbol{\epsilon} \cdot \boldsymbol{\epsilon}')\, r_0 \left[F^0(\mathbf{Q}) + F'(\omega) - \mathrm{i}\, F''(\omega)\right]. \end{aligned} \tag{7.36}$$

In the soft x-ray range or for scattering in the forward direction, where the momentum transfer $\mathbf{Q}$ is small, we have $f^0(\mathbf{Q} \to 0) = Z$, where Z is the number of electrons per atom. This allows one to rewrite (7.35) in terms of the web-tabulated *Henke-Gullikson factors* [13, 21] $f_1(\omega)$ and $f_2(\omega)$ according to (6.67) or[7]

$$F(\omega) = \underbrace{Z + F'(\omega)}_{f_1(\omega)} - \mathrm{i}\, \underbrace{F''(\omega)}_{f_2(\omega)}. \tag{7.37}$$

The corresponding atomic scattering lengths are then

$$f_a(\omega) = (\boldsymbol{\epsilon}' \cdot \boldsymbol{\epsilon}) \left[\underbrace{r_0\, Z + f'(\omega)}_{r_0 f_1(\omega)} - \mathrm{i}\, \underbrace{f''(\omega)}_{r_0 f_2(\omega)} \right]. \tag{7.38}$$

7.5.2 From Single Atoms to Atomic Sheets

We start our building block approach of atomic response by considering a thin atomic sheet, as illustrated in Fig. 7.10. We express the transmission of the field through the sheet through atomic scattering instead of the dielectric response in Sect. 7.4.

As shown in (a) we assume an incident EM wave of wavelength λ that is larger than the thickness Δ of the atomic sheet so that the interaction may be described in the dipole approximation. We assume that the incident wave has a planar wavefront across a homogeneously illuminated sheet area, defined by a circular aperture of area $A = \pi R^2$, that contains a number of atoms N_a defined by the atomic number density $\rho_a = N_a/(A\Delta)$. The sheet is thus coherently illuminated.

We now consider *elastic* scattering of the atoms in the sheet by use of the first Born approximation discussed in Sect. 6.5, and are interested in the scattered field at a point O on the optical axis, at a large distance z_0 from the sheet. The scattered Born

[6] The term "anomalous" has been maintained today in the important "multiple-wavelength *anomalous* diffraction" (MAD) technique in x-ray crystallography as discussed in Sect. 8.11.

[7] Unfortunately, the historical notation is somewhat confusing because the scattering *factors* f_1 and f_2 and the scattering *lengths* f' and f'' in (7.36) are both denoted by a small f.

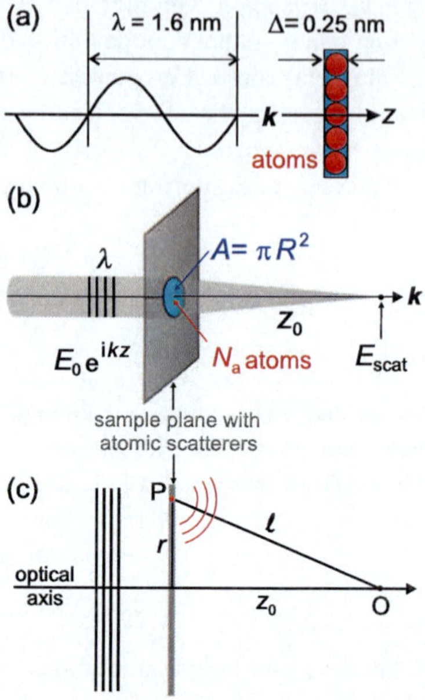

Fig. 7.10 **a** X-rays of wavelength λ traverse a thin atomic sheet of thickness Δ. **b** The x-rays are described by a plane wave $E_{\text{in}} = E_0 \exp(\mathrm{i}kz)$ incident at normal incidence on an area $A = \pi R^2$ of the thin sheet containing N_a atoms and we are interested in the field E_{scat} at the point O on the optical axis at a distance z_0. **c** Definition of key distances used for the derivation of the field E_{scat}

field consists of contributions from the N_a atoms in the illuminated sheet area A and we have to integrate over the contributions from all points P in the sample plane, separated from our observation point O by distances $\ell = \sqrt{z_0^2 + r^2}$ as illustrated in Fig. 7.10c. Since the fields are forward scattered with $\mathbf{Q} \simeq 0$, we can denote the atomic scattering length as $f(\mathbf{Q}) = f_\text{a}$ and obtain the scattered field at point O by integrating in cylindrical coordinates as

$$E_{\text{scat}}^{\text{O}} = -E_0 \frac{N_\text{a}}{\pi R^2} f_\text{a}\, 2\pi \int_0^R \frac{\mathrm{e}^{\mathrm{i}k\sqrt{z_0^2+r^2}}}{\sqrt{z_0^2 + r^2}}\, r\, \mathrm{d}r\,. \tag{7.39}$$

We obtain

$$E_{\text{scat}}^{\text{O}} = \underbrace{-\mathrm{i}E_0 \frac{N_\text{a} f_\text{a} \lambda}{\pi R^2}\, \mathrm{e}^{\mathrm{i}kz_0}}_{\text{sheet scattering}} + \underbrace{\mathrm{i}E_0 \frac{N_\text{a} f_\text{a} \lambda}{\pi R^2}\, \mathrm{e}^{\mathrm{i}k\sqrt{R^2+z_0^2}}}_{\text{edge diffraction}}\,. \tag{7.40}$$

As indicated by underbrackets, the on-axis scattered field arises from the contributions of all atoms in the sheet, plus a contribution from the atoms at the perimeter of the aperture at distance R. The perimeter atoms cause an edge diffraction effect when the fields are calculated off-axis, leading to the well-known Airy rings. The edge diffraction contribution disappears with increasing R due to the increasingly smaller fraction of perimeter atoms.[8]

The forward scattered *field* by the sheet (without edge diffraction) is seen to be proportional to the number of atoms so that the scattered *intensity* scales with N_a^2, revealing the coherent superposition of the fields due to the cylindrical symmetry about the optical axis. The total on-axis Born field, given by (6.49), is obtained by adding the incident field to the scattered field, and we obtain the following total field at point O

$$E_{\text{trans}}^{\text{O}} = \underbrace{E_0\, \mathrm{e}^{ikz_0}}_{E_{\text{trans}}^{\text{hole}}} \underbrace{- \mathrm{i}\, \frac{N_a f_a \lambda}{A} E_0\, \mathrm{e}^{ikz_0}}_{E_{\text{scat}}^{\text{sheet}}} \,. \tag{7.42}$$

The new label "trans" indicates that the field at O can also be taken to represent the *transmitted field* in the forward direction. Like the scattered field from a single atom, it contains an unscattered contribution from the incident field transmitted through the "hole" (no film) plus a scattered contribution from the atoms in the film.

7.5.3 *Atomic Scattering Factors: Born Approximation Versus Huygens–Fresnel Principle*

The magical Huygens–Fresnel (HF) principle allows the calculation of the diffraction pattern of an aperture whose open area is vacuum so that there are no true scattering centers within the aperture hole. Historically, this caused little concern because light propagation was envisioned to require a medium, the infamous "aether", so that an aperture could be imagined as being filled by the aether medium. In modern physics, the idea that a point scatterer converts a plane wave into spherical waves is the essence of the first Born approximation discussed in Sect. 6.5. In the x-ray region, where scattering is due to atoms, one may thus compare the Born scattering of real atoms with the HF scattering of virtual atoms (or points) of the same number density. In the following we will compare the two cases following Fig. 7.11.

Born scattering is described by (7.42) where the atomic scattering factors f_a express the conversion efficiency of a plane into a spherical wave. In HF scattering, the real atoms are replaced by "points" that convert a plane wave into a spherical wave with *unit efficiency*, corresponding to a virtual scattering length f_v. The size

[8] This is formally seen by extending the integration in (7.39) to $R \to \infty$ so that only the sheet scattering remains according to

$$E_{\text{scat}}^{\text{O}} = -E_0 \frac{N_a f_a}{\pi R^2}\, 2\pi \int_0^{\infty} \frac{\mathrm{e}^{\mathrm{i}k\sqrt{z_0^2+r^2}}}{\sqrt{z_0^2+r^2}}\, r\, \mathrm{d}r = \underbrace{-\mathrm{i}E_0 \frac{N_a f_a}{\pi R^2}\, \lambda\, \mathrm{e}^{ikz_0}}_{\text{sheet scattering}} \,. \tag{7.41}$$

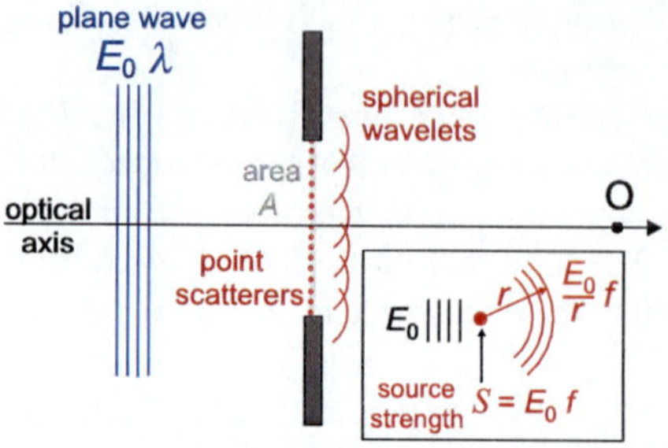

Fig. 7.11 Illustration of the Huygens–Fresnel principle of conversion of a plane wave of amplitude E_0 and wavelength λ into spherical wavelets (restricted to the forward half of the total 4π solid angle) that originate in a plane of area A shown in gray. The red points represent either atoms in a thin sheet or virtual scattering centers of the same density in an empty plane of size A. We are interested in the field transmitted on the optical axis which is again a plane wave of amplitude E^0_{scat}, as indicated. In the inset we picture the conversion of a plane wave of amplitude E_0 into a spherical wave $E = E_0 f/r$, arising from a source of strength $S = E_0 f$ where f is the scattering length. In the text we distinguish f_a for an atom and f_v for a virtual point scatterer

of the "points" is left unspecified in the HF principle, but from an x-ray point of view one may envision the points as hypothetical spheres of atomic size. We simply replace the real N_a atomic scattering centers in the thin sheet by virtual atoms of the same density in a circular hole of the same size A. The HF transmitted field at an on-axis point O is then given by the transmitted Born field (7.42), written in terms of the virtual scattering length f_v as

$$E^{\text{O}}_{\text{trans}} = \underbrace{E_0\,\mathrm{e}^{ikz_0}}_{E^{\text{hole}}_{\text{trans}}} \underbrace{-\,\mathrm{i}\,\frac{N_a\, f_v \lambda}{A} E_0\,\mathrm{e}^{ikz_0}}_{E^{\text{hole}}_{\text{scat}}} = \underbrace{E_0\,\mathrm{e}^{ikz_0}}_{E^{\text{hole}}_{\text{trans}}} - \underbrace{\mathrm{i}\,\frac{N_a\, f_v \lambda}{A}}_{=1} E_0\,\mathrm{e}^{ikz_0}\,. \tag{7.43}$$

where the second term now reflects the scattering of the virtual atoms with scattering length f_v. On the far right side we have invoked that in HF scattering, there is complete conversion efficiency of a plane into a spherical wave so that the superposition of the spherical waves at an on-axis point must have a value equal to the incident field. Hence we find

$$\boxed{\text{Huygens–Fresnel scattering length: } f_v = \frac{A}{\mathrm{i}\lambda N_a}\,.} \tag{7.44}$$

As expected, the virtual sources of atomic size have a purely imaginary scattering length. In Fig. 7.11 the scattering length is denoted generically as f, and we have $f = f_a$ for real atoms and $f = f_v$ for virtual atoms. As illustrated in the inset of Fig. 7.11, the scattering length characterizes the efficiency of converting the incident plane wave into scattered spherical waves with a source strength $S = E_0 f$. Since

E_0 has the dimension [V/m] and f has the dimension [length], the source strength S has the dimension [V].[9]

We have derived a remarkable result. We can describe the conversion of an incident plane wave into spherical waves by either real atoms in a thin sheet with scattering lengths f_a or by virtual atoms of the same density in an open aperture with scattering length f_v.

The on-axis "*transmitted field*" through a thin sheet of area A given by (7.42) can be pictured as arising from scattering by the real atoms in the sheet with scattering length f_a and Huygens–Fresnel virtual atoms of the same density and scattering lengths f_v, according to

$$E_{\text{trans}}^{\text{tot}} = \underbrace{\mathrm{i}\,\frac{N_a \lambda\, f_v}{A}\, E_0\, \mathrm{e}^{\mathrm{i}kz_0}}_{\text{virtual atoms: } E_{\text{scat}}^{\text{v}}} \underbrace{-\,\mathrm{i}\,\frac{N_a \lambda\, f_a}{A}\, E_0\, \mathrm{e}^{\mathrm{i}kz_0}}_{\text{real atoms: } E_{\text{scat}}^{\text{atom}}} . \tag{7.45}$$

The conversion efficiency of the incident plane wave into spherical waves by the real and virtual atoms is determined by the respective scattering lengths f_a and f_v. The virtual and real scattering contributions have opposite sign, revealing that opposite to the real atoms, the virtual atoms do not induce a 180° scattering phase shift.

Let us put in some numbers and compare the real atomic (Thomson) scattering length for Fe atoms $f_a = Zr_0 \simeq 7 \times 10^{-5}$ nm to that of virtual scattering centers with the same density in a plane. With the atomic density $\rho_a = 84.9$ atoms/nm^3 for Fe metal, a single layer has a thickness given by the atomic diameter $d = 0.23$ nm and contains $N_a/A = 19.3$ atoms/nm^2. For a wavelength of 1.75 nm (Fe L_3-edge) we obtain from (7.44) $|f_v| \simeq 3 \times 10^{-2}$ nm. The Born atomic scattering factor is smaller than the virtual HF scattering factor by $|f_a/f_v| \simeq 2 \times 10^{-3}$. This number is in good accord with expectations based on the Fe L_3-edge fluorescence yield of 6×10^{-3} [22] which expresses the atomic x-ray emission probability in random directions, classically corresponding to the emission of a spherical wave.

7.5.4 Scattering Phase Shifts

The forward scattered wave by the atomic sheet exhibits a different phase than that forward scattered by a single atom discussed in Sect. 6.5. For a single atom, the scat-

---

[9] As discussed in Sect. 3.2.3, a synchrotron radiation source generates an electromagnetic wave that is strongly forward directed and can be described as a plane wave with a field amplitude $E_0 \sim 100$ V/m. If this field scatters off an atom in the middle of the periodic table ($Z \sim 30$) with an atomic Thomson scattering length of $f = Zr_0 \simeq 10^{-4}$ nm, the associated source strength is $S \simeq 10^{-11}$ V.

tered field for non-resonant Thomson scattering is described by the atomic scattering factor $f_a = r_0 Z + f' - \mathrm{i} f'' \simeq r_0 Z$, and the forward scattered field is given by the Born approximation (6.49) with $r = z_0$ as

$$E_{\text{trans}}^{\text{atom}} = \underbrace{E_0\, \mathrm{e}^{ikz_0}}_{\text{plane wave: } E_{\text{in}}} \underbrace{- \frac{r_0 Z}{z_0} E_0\, \mathrm{e}^{ikz_0}}_{\text{forw. scatt. wave: } E_{\text{scat}}} . \tag{7.46}$$

Comparison with the forward scattered *Thomson* field of an atomic sheet given by (7.42) or

$$E_{\text{trans}}^{\text{sheet}} = \underbrace{E_0\, \mathrm{e}^{ikz_0}}_{\text{plane wave: } E_{\text{in}}} \underbrace{- \mathrm{i} r_0 Z \frac{N_a \lambda}{A} E_0\, \mathrm{e}^{ikz_0}}_{\text{Thomson: } E_{\text{scat}}^{\text{sheet}}} \tag{7.47}$$

reveals a relative phase shift difference of 90° (factor i) for the two cases. This is a geometric effect arising from the integration over the atomic plane.

For *resonant* scattering, the atomic scattering factor reduces to the resonant form $f_a = r_0 Z + f' - \mathrm{i} f'' \simeq -\mathrm{i} f''$ (see Fig. 6.5) so that (7.42) becomes

$$E_{\text{trans}}^{\text{sheet}} = E_0\, \mathrm{e}^{ikz_0} \underbrace{- f'' \frac{N_a \lambda}{A} E_0\, \mathrm{e}^{ikz_0}}_{\text{Resonant: } E_{\text{scat}}^{\text{sheet}}} . \tag{7.48}$$

Now the forward scattered wave is 180° out of phase with the incident field.

Our above results apply for scattering by atomic charges and are independent of the incident polarization. We can summarize the scattering phase shifts as follows.

> For forward scattering by atomic charges, the phase shifts upon scattering are different for non-resonant and resonant scattering and also for scattering from a single atom and an atomic thin sheet. The following statements hold independent of the incident polarization.
> For non-resonant *Thomson scattering*, the scattered wave is phase shifted relative to the incident wave by 180° for a single atom but by 90° for an atomic sheet.
> For *resonant scattering*, the scattered wave is phase shifted relative to the incident wave by 90° for a single atom but 180° for an atomic sheet.

For the case of a thin atomic sheet, which is of primary interest in the present chapter, the collective forward scattering phase shift of 180° in resonance scattering will result in destructive interference of the weak forward scattered wave with the strong incident wave. This leads to a reduction of the intensity in the forward direction, which we typically associate with a loss due to "absorption". We will use this important forward scattering loss in Sect. 7.8 to show that one can indeed derive the exponential absorption law for a thick film by the consecutive addition of thin sheet losses in the forward direction.

7.5.5 Link of X-Ray Scattering Factors and Optical Parameters

Having derived the forward scattered fields by an atomic sheet, we are now ready to establish the fundamental relationship between the atomic scattering factors and the optical parameters. In contrast to the optical range, where the small photon energy causes an interaction mostly with the weakly bound valence electrons, the interaction in the x-ray regime is primarily with the atomic core electrons. They typically constitute a larger fraction of the total number of atomic electrons Z and most importantly have larger interaction cross sections as shown in Fig. 1.27. The latter point is beautifully demonstrated by the fact that high-intensity and short x-ray pulses can strip atoms from the inside-out, leaving behind hollow atoms with only the valence electrons remaining [23, 24].

One may therefore formulate the interaction of x-rays with matter in terms of the response of the total of Z electrons per atom and their atomic number density ρ_a. The atom-based treatment is a very good approximation over most of the x-ray region with the exception of the narrow region around absorption edges, where large resonances are observed that depend on the detailed valence electronic structure. The x-ray scattering factors found in the literature cover only the non-resonant spectral ranges as discussed in Sect. 6.7. They may then be supplemented near absorption edges by inclusion of core to valence resonances, either empirically from measured spectra or through quantum mechanical calculations introduced in Chap. 9.

We may compare two formulations of the field transmitted through a thin sheet. The first is based on an atomic scattering factor-based approach. It expresses the transmitted Born field as (7.42) or

$$E_{\text{trans}}^{\text{Born}} = E_0\, \mathrm{e}^{ikz_0} \left\{ 1 - \mathrm{i}\, \frac{N_a \lambda}{A}\, f_a \right\} . \tag{7.49}$$

For forward scattering the atomic scattering factor given by (7.38) simplifies since $\boldsymbol{\epsilon} \cdot \boldsymbol{\epsilon}' = 1$ and may be expressed as $f_a = r_0[f_1 - \mathrm{i} f_2]$. The atomic area density in the thin sheet can be written as $N_a/A = \rho_a \Delta$, where ρ_a is the atomic volume density and Δ the sheet thickness. Then (7.49) becomes

$$\boxed{E_{\text{trans}}^{\text{Born}} = E_0\, \mathrm{e}^{ikz} \left\{ 1 - [\mathrm{i} f_1 + f_2]\, r_0\, \rho_a \lambda\, \Delta \right\} .} \tag{7.50}$$

This expression can now be compared by one given in terms of the refractive index or optical parameters, given by (7.24). By replacing the arbitrary thickness d of the film by $\Delta \ll d$, the exponential reduces to a linear response expression in δ

and β, and with $k = 2\pi/\lambda$ we obtain

$$E_{\text{trans}}^{\text{index}} = E_0\, e^{ikz} \left\{ 1 - [i\delta + \beta] \frac{2\pi\,\Delta}{\lambda} \right\}. \tag{7.51}$$

Comparison of the real and imaginary parts in (7.50) and (7.51) gives us the desired fundamental link between optical parameters and the atomic x-ray scattering factors.

The optical parameters and scattering factors and lengths are related according to

$$\delta(\omega) = \frac{\lambda^2 \rho_a}{2\pi} r_0\, f_1(\omega) = \frac{\lambda^2 \rho_a}{2\pi} \left[r_0 Z + f'(\omega) \right], \tag{7.52}$$

$$\beta(\omega) = \frac{\lambda^2 \rho_a}{2\pi} r_0\, f_2(\omega) = \frac{\lambda^2 \rho_a}{2\pi}\, f''(\omega), \tag{7.53}$$

where δ and $\beta \geq 0$ are real and dimensionless numbers, the scattering factors f_1 and $f_2 \geq 0$ are in units of number of electrons, and the scattering lengths f' and f'' have the dimension [length].

To illustrate the relative size of the scattering factors, scattering lengths, and optical parameters, we show in Fig. 7.12 the values for Fe metal in the L-edge region.

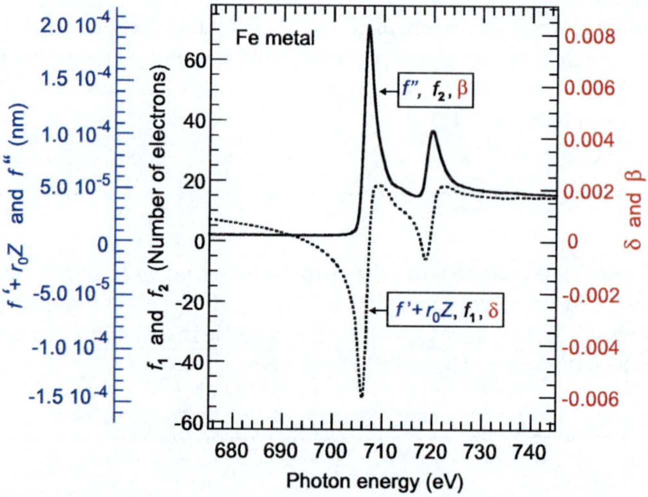

Fig. 7.12 Energy dependent scattering lengths $f' + r_0 Z$ and f'' (blue), scattering factors f_1 and f_2 (black), and optical parameters δ and β (red) near the $L_{3,2}$-edges of Fe metal, related through (7.52) and (7.53). Data are those of Fig. 6.10 from [25]

For a sample composed of different atoms, we can write the following relationship between the complex refractive index $\tilde{n}(\omega)$ in (7.20) and the scattering factors

$$\tilde{n}(\omega) = 1 - \delta(\omega) + \mathrm{i}\,\beta(\omega) = 1 - \frac{r_0\lambda^2}{2\pi}\sum_j \rho_j\left(Z + F'_j(\omega) - \mathrm{i}F''_j(\omega)\right), \quad (7.54)$$

where r_0 is the classical electron radius, λ is the wavelength, and ρ_j is the number density of atomic species j (atoms/length^3), so that the right hand side of (7.54) is dimensionless, as required.

7.6 The Optical Theorem

The optical theorem expresses energy conservation between the incident beam, said to propagate into the "forward direction", and that lost in and emerging from the sample in all directions. It was first stated independently in 1871 by Wolfgang Sellmeier and William Strutt (later Lord Rayleigh). The interesting history of the optical theorem has been reviewed by Newton [26]. As the name implies, the optical theorem was originally established in the *visible region*. Over the years the improved understanding and development of quantum theory have led to its clarification and repeated "reinvention".

Formally, the theorem may be expressed through the imaginary part $f'(\omega)$ of the atomic scattering length given by (6.30) as $f(\mathbf{Q},\omega) = f^0(\mathbf{Q}) + f'(\omega) - \mathrm{i}\,f''(\omega)$. In its most general form, the *optical theorem* reads [26]

$$\boxed{\text{Optical theorem:}\ \ \mathrm{Im}\, f(\mathbf{Q}{=}0) = f'' = \frac{k}{4\pi}\,\sigma_{\mathrm{tot}}\,.} \quad (7.55)$$

where $k = 2\pi/\lambda$. The left side of (7.55) describes the radiation loss (imaginary part) experienced in the forward direction, $\mathbf{k}' = \mathbf{k}$, corresponding to no momentum transfer $\mathbf{Q} = 0$ (see Fig. 6.3). The total cross section σ_{tot} on the right side describes all interactions with the sample that lead to a loss of energy from that re-emerging in the forward direction.

Classically, the imaginary part of a response function describes how a physical system dissipates energy when it is out of phase with the driving force. For example, energy may be lost from the photon system to the electronic system (or lattice) so that f'' is associated with "absorption" of radiation. Then (7.55) with $\sigma_{\mathrm{tot}} = \sigma_{\mathrm{abs}}$ is just our previous semi-classical *x-ray* expression (6.46) $2\lambda f'' = \sigma_{\mathrm{abs}}$.

In the *optical* region, the response of matter strongly varies with its valence electronic structure, and "absorption" may be defined in different ways [27]. For gases and insulators the energy of visible light is typically less than the ionization potential or band gap. Dissipative processes where energy is irreversibly transferred to the electronic system are then absent. In these cases "absorption" may be defined as loss of photons in the forward direction due to out-of-beam scattering into 4π. Hence "absorption" and scattering cross sections are the same, as for the Breit–Wigner cross section (6.56). The "absorption" loss in the forward direction is then compensated by the gain of the energy scattered out of the beam direction, and we have $\sigma_{tot} = \sigma_{abs} = \sigma_{scat}$ in (7.55).

The situation is different when electrons in the sample can be excited by visible light as in metals. Then true dissipative energy losses exist, and absorption needs to be distinguished from scattering. In the optical range some of the incident energy is irreversibly lost by transfer to the electronic or phonon system. In the x-ray regime, the incident radiation always has enough energy to create core holes that may be filled through Auger decay. Hence irreversible energy transfer from the photon to the electron system always exists. Then σ_{tot} in (7.55) takes the interesting general form $\sigma_{tot} = \sigma_{abs} + \sigma_{scat}$.

In the following section we will discuss this interesting case and show that, in practice, the optical theorem in the x-ray range is described by

$$\mathrm{Im}\, f(\mathbf{Q}=0) = \frac{k}{4\pi}\,\sigma_{tot} = \frac{1}{2\lambda}\,(\sigma_{abs} + \sigma_{scat}) \simeq \frac{\sigma_{abs}}{2\lambda}\,. \tag{7.56}$$

The right side then expresses the fact that in the x-ray region absorption completely dominates so that the scattering cross section σ_{scat} in (7.56) can be neglected and (6.46) is obtained.

7.7 Coherent Versus Incoherent X-Ray Scattering from Solids

In our previous discussion of x-ray scattering we considered the special case of forward scattering along the optical axis. In the classical description of x-ray scattering, the fields scattered into the forward direction are in phase corresponding to their coherent superposition. More generally, the field scattered into an arbitrary direction by a solid sample depends on the relative phases between the fields scattered by the individual atoms, and one can distinguish two general cases.

Incoherently scattered fields have random relative phases and their intensity (field squared) scales linearly with the number of scattering centers.
Coherently scattered fields have the same phase and their intensity scales quadratically with the number of scattering centers.

7.7.1 *Incoherent Versus Coherent Field Superposition*

Let us consider how the scattered fields from N_a atoms in the incident beam add up to the scattered intensity. In our discussion we follow Fig. 7.13a. We are interested in the difference in the intensity scattered into the forward direction **k** relative to the intensity scattered into a solid angle $d\Omega$ around a direction $\mathbf{k}' \neq \mathbf{k}$. We may refer to the two cases as "in-beam" and "out-of-the beam" scattering. We consider the scattered intensity from a thin sheet sample containing N_a atoms in the illuminated area A, measured by a detector at a distance r that is very large compared to the illuminated sample area.

To facilitate our discussion, we start with the two atom case illustrated in Fig. 7.13b. The two atoms are separated by a distance ℓ whose spherical wave fields add at the point of a distant detector D. The path length differences Δ_ℓ from the atoms to the detector lead to a phase shift factor $\exp(ik\Delta_\ell)$, so that the total scattered field at the detector in the Born approximation at the detector position **r** is the sum of the phase-shifted fields from the atoms.

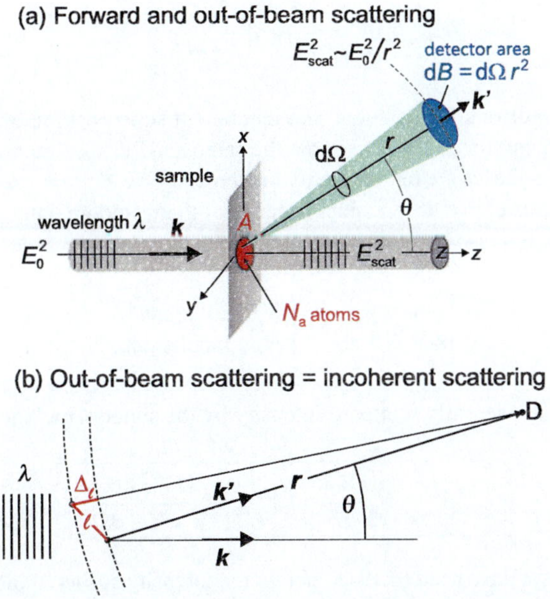

Fig. 7.13 **a** Geometry for "in-beam" forward scattering into direction **k** from an illuminated area A of an atomic sheet versus "out-of-beam" scattering into a solid angle $d\Omega$ around direction $\mathbf{k}'$. **b** Simplified case of scattering of an incident plane wave by two atoms separated by a distance ℓ into the direction $\mathbf{k}'$ of a distant point detector located at a distance $r \gg \ell$. At the detector position, the spherical wave fields scattered by the atoms have a phase difference Δ_ℓ due to the different path lengths

For soft x-rays, we can assume that the atomic scattering length f_a is *independent* of the scattering direction. This corresponds to two assumptions. First, that the atomic form factor (6.25) is not **Q**-dependent, i.e. that the wavelength is much longer that the atomic dimension. Second, that the polarization of the incident light is ignored, i.e. that the fields are treated as scalars and the angular integration simply gives a factor 4π.[10]

We can readily extend the two atom case to N_a atoms by using the label ℓ for the phase shifts of the N_a interfering fields and obtain

$$E_{\text{scat}}(\mathbf{r}) = -\frac{E_0}{r} f_a \, e^{ikr} \sum_{\ell=1}^{N_a} e^{ik\Delta_\ell} . \tag{7.57}$$

Here we have used the fact that for elastic scattering $k' = k$ and $r \gg \Delta_l$. The latter condition allowed us to approximate $r + \Delta_l \simeq r$ in the denominator, but we needed to keep the full expression for the phases since they enter exponentially. In linear response, the scattered intensity is given by

$$|E_{\text{scat}}(\mathbf{r})|^2 = \frac{E_0^2}{r^2} |f_a|^2 \left| \sum_{\ell=1}^{N_a} e^{ik\Delta_\ell} \right|^2 . \tag{7.58}$$

The phase term differs for coherent and incoherent superposition of the fields. For coherent superposition all the phases are the same and the absolute value of the sum yields N_a^2. If the phases are random, only the terms $e^{ik\Delta_n} e^{-ik\Delta_n} = 1$ remain while the interference terms $e^{ik\Delta_n} e^{-ik\Delta_m}$ cancel because of their random nature. We therefore obtain

$$\left| \sum_{\ell=1}^{N_a} e^{ik\Delta_\ell} \right|^2 = \begin{cases} N_a^2 & \text{coherent} \\ N_a & \text{incoherent} . \end{cases} \tag{7.59}$$

Indicating the incoherently scattered intensity by the superscript "inc", we obtain

$$\boxed{|E_{\text{scat}}^{\text{inc}}|^2 = N_a \frac{|E_0|^2}{r^2} |f_a|^2 = N_a \frac{|E_0|^2}{r^2} \frac{d\sigma_{\text{scat}}}{d\Omega} .} \tag{7.60}$$

In the last step we have used (6.19) to obtain the intensity scattered into a solid angle $d\Omega$ around $\mathbf{k}'$. For the calculation of the incoherently scattered intensity we can thus ignore the orientation of the film and just calculate the *intensity* scattered by a single atom and then multiply by the number of atoms N_a in the incident beam.

[10] Inclusion of polarization effects would instead give a factor $\Omega = 8\pi/3$ since the scattered intensity has a node along the polarization vector.

By considering the definitions of the different beam parameters, we can write (7.60) in terms of the following general relationship between the intensities and number of photons per solid angle $\mathrm{d}\Omega$

$$\frac{|E_{\mathrm{scat}}^{\mathrm{inc}}|^2}{|E_0|^2} = \frac{I_{\mathrm{scat}}}{I_0} = \frac{n_{\mathrm{scat}}}{n_0}\frac{A}{\mathrm{d}\Omega\, r^2}, \tag{7.61}$$

where n_0 is the number of incident and n_{scat} the number of scattered photons. We can then state as follows.

For an incident beam containing n_0 photons within the beam cross-sectional area A that illuminates N_{a} atoms in a sample, the number of photons that are incoherently scattered into $\mathrm{d}\Omega$ is given by

$$\frac{n_{\mathrm{scat}}}{\mathrm{d}\Omega} = n_0 \frac{N_{\mathrm{a}}}{A}\frac{\mathrm{d}\sigma_{\mathrm{scat}}}{\mathrm{d}\Omega}. \tag{7.62}$$

7.7.2 X-Ray Forward Scattering and Absorption

The optical theorem links what we refer to as x-ray absorption and "forward scattering". Forward scattering means that the fields are considered only within the coherence cone of the radiation emerging from the sample (see Fig. 4.7). One may then distinguish the radiation that remains within the incident beam cross-sectional area from that scattered out of the beam. Below we discuss the "in-beam" forward scattered field and intensity and in the next section consider the "out-of-beam" scattered fields.

By use of (7.52) and (7.53) we may introduce the convenient short notation

$$C = \left[r_0 Z + f' - \mathrm{i} f''\right] \rho_{\mathrm{a}} \lambda d = [\delta - \mathrm{i}\beta]\, kd\,, \tag{7.63}$$

where C is complex and scales linearly with the sample thickness d. At the exit plane of the sample the field and intensity transmitted into the forward direction are then given by the exponential attenuation law (7.24) which holds to *all orders* in C or thickness d. At a distance z behind the sample the forward scattered fields are in phase and confined to the coherence cone and we have

$$E_{\mathrm{trans}} = E_0\, \mathrm{e}^{\mathrm{i}kz}\, \mathrm{e}^{-\mathrm{i}C}, \quad I_{\mathrm{trans}} = I_0\, \mathrm{e}^{-2\lambda f'' \rho_{\mathrm{a}} d} = I_0\, \mathrm{e}^{-2\beta kd}\,. \tag{7.64}$$

The exponential in the transmitted field can be expanded to different orders in f'' or β. The corresponding intensity will be correct only to the *same* order as the field. In *first order*, we obtain in the scattering length notation[11]

$$\text{First order}: E_{\text{trans}} = E_0\, \mathrm{e}^{\mathrm{i}kz}\, [1 - \mathrm{i}\, C]\,, \quad I_{\text{trans}} = I_0 \Big[1 - \underbrace{2\lambda f''\, \rho_{\text{a}}\Delta}_{\sigma_{\text{abs}}}\Big]. \tag{7.65}$$

This expression shows that within first order or linear response, the transmitted intensity contains a loss term (minus sign) that can be attributed to absorption, as indicated by the absorption cross section σ_{abs} previously given by (6.72).

In a scattering picture, the loss term in (7.65) may be interpreted as *destructive* interference of the incident field with the *spontaneously* forward scattered field created by radiative decay of a core hole. The process scales linearly with the number of atoms, and for a thin film the scattered field is 180° phase shifted as discussed in Sect. 7.5.4.

We are now interested in calculating the transmitted intensity to *second order* in the optical parameters. We then need to start with the transmitted field expressed to second order

$$\text{Second order}: \; E_{\text{trans}} = E_0\, \mathrm{e}^{\mathrm{i}kz}\, \mathrm{e}^{-\mathrm{i}C} \simeq E_0\, \mathrm{e}^{\mathrm{i}kz} \left[1 - \mathrm{i}C + \frac{(\mathrm{i}C)^2}{2}\right]. \tag{7.66}$$

The transmitted intensity in the scattering length notation is obtained as[12]

$$\text{Second order}: I_{\text{trans}} = I_0\, \mathrm{e}^{-2\lambda f'' \rho_{\text{a}}\Delta} \simeq I_0 \Big\{ 1 - \underbrace{2\lambda f''\, \rho_{\text{a}}\Delta}_{\text{1st ord. abs. loss}} + \underbrace{2\lambda^2 (f'')^2 \rho_{\text{a}}^2 \Delta^2}_{\text{coh. forw. scat. gain = 2nd ord. abs.}} \Big\}. \tag{7.67}$$

The intensity now contains an additional second order contribution that scales with the number of atoms squared N_{a}^2 since $\rho_{\text{a}}\Delta = N_{\text{a}}/A$. It is simply the second order term in the conventional exponential absorption law, and its positive sign corresponds to a gain that is absent in first order. We can summarize as follows.

[11] We have ignored diffraction effects from the edge of the illuminated area A, but they can be neglected if we use a large beam diameter.

[12] Note that the expression for the second order transmitted intensity (7.67) cannot be obtained if we used only the *first* order field expression (7.65). If we did so and naively kept the second order term, the intensity would have the incorrect second order term $\lambda^2 \rho_{\text{a}}^2 \Delta^2 \left[(r_0 Z + f')^2 + (f'')^2\right]$.

In *first order* or linear response, the transmission loss of the intensity scales *linearly* with the number of atoms N_a in the beam. It may be interpreted either as absorption loss, or equivalently, as a loss caused by *destructive* interference of the incident field with the 180° phase shifted field that is *spontaneously* forward scattered by the atomic sheet.

In *second order*, there is a gain in transmission that scales with N_a^2. It may be interpreted either as the second order correction term within the exponential absorption law, or equivalently, as a gain caused by *constructive* interference of the incident field with an in-phase forward scattered field.

7.7.3 The Total Transmitted Intensity

Our above discussion reveals the fascinating correspondence between the first and second order terms in the exponential absorption law and *resonant scattering processes* in the forward direction. For a more complete description of the radiation emerging from the sample, we need to also consider the fraction scattered out of the forward direction of the beam. For an atom this is described by the atomic scattering cross section σ_{scat} given by (6.71) or

$$\sigma_{\text{scat}} = \frac{8\pi}{3}\left[(r_0 Z + f')^2 + (f'')^2\right]. \tag{7.68}$$

For a sample with randomly distributed atoms, the fields scattered in different directions will superimpose with random phases and the scattered intensity will scale *linearly* with the number of atoms N_a. In contrast to the "in beam" coherent forward scattered intensity, the out-of-beam scattered intensity is *incoherent* and given by

$$I_{\text{scat}}^{\text{inc}} = I_0 \frac{N_a}{A} \underbrace{\frac{8\pi}{3}\left[(r_0 Z + f')^2 + (f'')^2\right]}_{\sigma_{\text{scat}}}. \tag{7.69}$$

The factor $8\pi/3$ differs from the factor 4π for isotropic emission and accounts for the fact that the field is not a scalar but a transverse vector field with no elastic scattering into the direction of **E**. In principle, part of the incoherently scattered intensity is also emitted as a gain into the forward direction of the beam, but we ignore it here. The total intensity emerging from the sample may then be written up to second order in the scattering factors as follows.

The total intensity transmitted by a sample of atomic areal density $N_a/A = \rho_a\Delta$, where Δ is much smaller than an x-ray absorption length, is given up to *second order* in the atomic scattering lengths by

$$I_{\text{trans}} = I_0 \Bigg\{ 1 - \underbrace{2\lambda f'' \rho_a \Delta}_{\text{lin. abs. loss}} - \underbrace{\frac{8\pi}{3}\left[(r_0 Z + f')^2 + (f'')^2\right]\rho_a \Delta}_{\text{out-of-beam loss Thomson+resonant scat.}} + \underbrace{2\lambda^2 (f'')^2 \rho_a^2 \Delta^2}_{\text{coh. forw. scat. gain =2nd order abs. term}} \Bigg\}. \tag{7.70}$$

The out-of-beam scattering term has a minus sign, meaning that it corresponds to a "loss" from the in-beam transmitted intensity. We note that in the optical region, where electronic transitions may be absent in atoms or band gap materials, the out-of-beam scattered intensity is sometimes treated like an "absorption" loss [27].

Our treatment of absorption and scattering to the same second order is of fundamental importance since it reveals the different dependence on the number of atoms in the beam in the second order ($\propto N_a^2$) and first order ($\propto N_a$) terms. Equation (7.70) will serve as the starting point for the formulation of diffraction imaging in the next chapter and the introduction of stimulated forward scattering in Chaps. 13 and 15.

7.7.4 Relative Size of the Absorbed and Scattered Intensities

The cross sections σ_{abs} and σ_{scat} are the conventional absorption and scattering cross sections for a single atom $N_a/A = 1$, previously given by (6.72) and (6.71) and shown for Fe atoms in Fig. 6.9. We can now update this figure by inclusions of the pronounced resonant contributions at the L-edges for Fe metal from Fig. 7.12, and the result is shown in Fig. 7.14.

The figure clearly reveals that absorption dominates at 100 eV by about five orders of magnitude over x-ray scattering, while the difference decreases to less than two orders of magnitude at 10 keV. The response reflects the optical theorem in the form (7.56). Near absorption thresholds there is strong resonance behavior in both absorption and scattering. The details of non-resonant versus resonant response, indicated in red and blue, are best treated quantum mechanically as done in Sect. 9.5. There we also show in Fig. 9.5 an annotated version of Fig. 7.14 that distinguishes the different contributions.

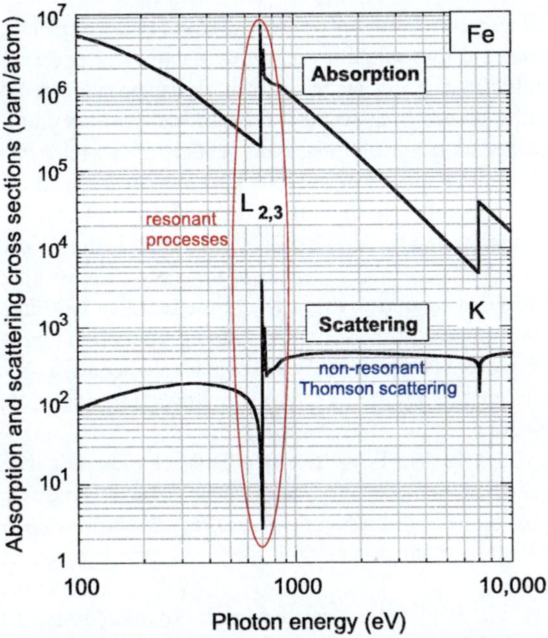

Fig. 7.14 Angle integrated absorption and scattering cross sections per atom for Fe metal including the detailed resonant contributions from Fig. 7.12 which are absent in Fig. 6.9. The absorption cross section σ_{abs} is defined in (7.65) and the scattering cross section $\sigma_{\text{scat}}^{\text{inc}}$ in (7.69). Resonant processes near the $L_{2,3}$ edge are highlighted by a red oval, and we have also indicated in blue the nearly energy independent behavior of Thomson scattering

In the following section we will take the next building block step and add the atomic sheets to a film of arbitrary thickness.

7.8 The Response of a Thick Sample: Dynamical Scattering Theory

The transmitted intensity given by (7.70) is valid only for a thin film sample of at most one absorption length thickness. The loss due to true absorption is described through the second and last terms which represent the second order expansion of the exponential absorption law given by (7.32). We will now show that the full exponential absorption law for a sample of arbitrary thickness may be derived by a series of contributions from atomic sheets that are stacked together. The theory is called the *Darwin-Prins theory* after Darwin [28, 29] who in 1914 developed self-consistent equations that describe wave propagation between periodic layers and Prins [30] who in 1930 extended the theory by including absorption through a complex refractive index.

In scattering theory one distinguishes so-called *kinematical* from *dynamical* treatments. The kinematical approximation assumes that the scattering is so weak that the probability of multiple scattering events in a sample can be neglected. With increasing probability of multiple scattering events the dynamical scattering limit is approached. If we were concerned with conventional x-ray scattering by a thick sample, kinematic scattering theory would typically be sufficient because of the weak atomic scattering cross sections, corresponding to the third term in (7.70) and illustrated in Fig. 7.14. In contrast, the transmission through a thin atomic sheet is dominated by the second term in (7.70). It reflects the first order absorption response or in a scattering picture a coherent forward scattering event where the incident field interferes with the spontaneously 180° phase shifted forward scattered field. This type of absorption or coherent scattering event is not weak. When considered sequentially through a stack of sheets, it thus requires a multiple scattering treatment that constitutes a particularly simple and beautiful example of *dynamical scattering*.

The essence of the Darwin-Prins treatment is that by keeping the sheet thickness small one only needs to consider the first order or linear response of the field given by (7.65). Multiple scattering in a sample of arbitrary thickness $d = N\Delta$ is then built up through sheet-by-sheet single coherent forward scattering events, up to the total number of sheets N.

Since we have not previously discussed the x-ray reflectivity of solids, we start with a brief review of *Snell's reflectivity law* from a classical point of view, based on how the propagation of a plane wave is changed in a dielectric continuous medium. The other sections are devoted to a discussion of x-ray reflection and transmission from an atomic scattering/absorption point of view for a sample of finite thickness.

7.8.1 Snell's Law and Total X-Ray Reflection

Snell's law is named after the Dutch astronomer and mathematician Willebrord Snellius (1580–1626), although it is now known that the law was first discovered by the Persian mathematician and physicist Ibn Sahl and documented in an optical treatise in 984 [31]. It is typically derived from purely classical considerations, namely how a plane EM wave responds to a change of the complex refractive index $\tilde{n}$. From a modern point of view, this treatment is based on two key assumptions: (i) that the x-rays can be described as a classical plane EM wave and (ii) that the sample can be treated as a continuous (homogeneous) medium, with all sample properties lumped into a single parameter, $\tilde{n}$. In the x-ray range this problem is greatly simplified by the fact that according to (7.20), the real part of the refractive index $n(\omega) = 1 - \delta(\omega) \simeq 1$, where δ is strongly energy dependent as plotted for Fe in Fig. 7.5 and for polystyrene in Fig. 7.7.

Snell's law is illustrated in Fig. 7.15 for non-resonant x-rays traversing from vacuum into a material.

We follow the convention usually used for x-rays [9, 13, 14] with the beam incident at an angle θ measured from the surface *plane*. With the geometric definitions in

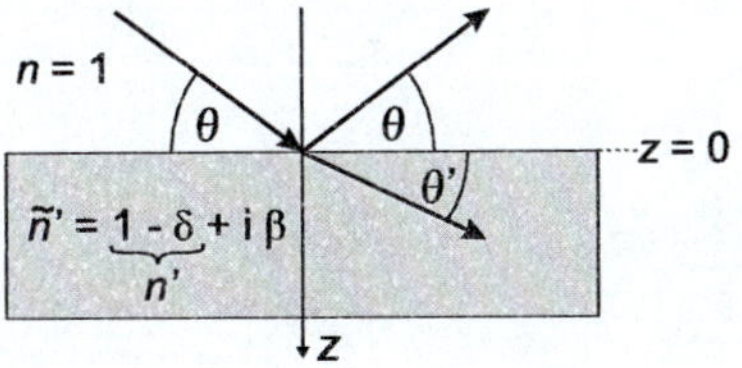

Fig. 7.15 Illustration of Snell's law for x-rays incident at an angle θ from vacuum with $n = 1$ onto a dielectric with complex refractive index $\tilde{n}' = 1 - \delta + i\beta$ and real index $n' = 1 - \delta$. It is customary in the x-ray region to define the incidence angle θ as measured from the surface plane of the sample. As shown in Fig. 7.5, we have $\delta > 0$ except near resonances. When the incidence angle θ is less than a critical angle $\theta_c = \sqrt{2\delta}$, the x-rays are totally reflected

Fig. 7.15, Snell's law reads

$$\frac{\cos\theta}{\cos\theta'} = \frac{n'}{n} = 1 - \delta(\omega)\,. \tag{7.71}$$

For non-resonant x-rays we have $\delta > 0$ and the incident beam is refracted away from the surface normal as shown. Therefore at small grazing incidence angles that are below a critical angle, $\theta \leq \theta_c$, x-rays are totally reflected. The critical angle θ_c in units of [rad] is determined by the real part of the refractive index according to [9, 14]

$$\theta_c = \sqrt{2\delta}\,. \tag{7.72}$$

7.8.2 *Darwin-Prins Dynamical Scattering Theory*

In our discussion of the transmitted and reflected intensity from a solid, we want to go beyond the simple continuous medium description of the sample and consider an atom-based response. We will, however, maintain that the sample consists of homogeneously distributed atoms. We will classically describe the x-rays as plane waves, keeping in mind that this is equivalent to assuming that the x-rays are completely coherent with infinite lateral and longitudinal coherence lengths. In practice this is only fulfilled if the incident field is laterally and longitudinally coherent over the illuminated sample volume.

We consider a soft x-ray beam of wavelength $\lambda \simeq 2$ nm, traversing a ultrathin sheet of thickness $\Delta \simeq 0.2$ nm, corresponding to the thickness of a single atomic layer, as illustrated in Fig. 7.15b. The soft x-ray condition $\lambda \gg \Delta$ simplifies our treatment since Bragg diffraction between atomic planes does not exist and Compton scattering can be neglected. We describe the atomic response by scattering factors $f(\mathbf{Q}) = f_a$ that are angle independent. This is a good approximation below about 1000 eV [32]. We restrict our discussion to the case $\theta \gtrsim 10°$, so that we do not have to worry about total x-ray reflection and can conveniently use the same propagation direction and angle

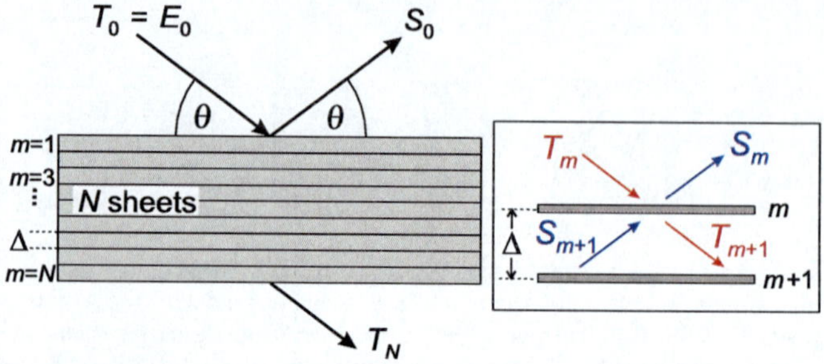

Fig. 7.16 Geometry and definition of the incident field E_0 and the reflected (denoted S) and transmitted (denoted T) fields for a sample composed of N identical thin sheets of thickness $\Delta \ll \lambda$. On the right we show the definition of the fields inside the solid for two arbitrary layers m and $m+1$ with a center-to-center separation Δ

θ outside and inside the sheet. This also applies when we stack sheets to form a thin film of thickness $d = N\Delta$, where N is the number of sheets, as shown in Fig. 7.16. We choose the surface normal as our z direction with $z=0$ chosen right before the top layer.

We start by considering the field scattered by a single sheet in the Born approximation. For our definition of θ shown in Fig. 7.16, the geometrical phase change between two successive sheets is given by

$$\phi = \frac{2\pi\,\Delta}{\lambda}\,\sin\theta = k\Delta\sin\theta\,. \tag{7.73}$$

The *transmitted* field at the position Δ behind the layer is given by the incident field plus the forward scattered field, and according to (7.50) we have

$$\begin{aligned} E_1^{\text{trans}}(z=\Delta) &= \left(1 - \mathrm{i}\,\frac{N_{\mathrm{a}}\lambda f_{\mathrm{a}}}{A}\right) E_0(z=0)\,\mathrm{e}^{\mathrm{i}\phi} \\ &= (1 - \mathrm{i}\,\rho_{\mathrm{a}}\lambda f_{\mathrm{a}}\Delta)\,E_0(z=0)\,\mathrm{e}^{\mathrm{i}\phi}\,. \end{aligned} \tag{7.74}$$

The exponential factor accounts for the geometrical path length difference or phase shift upon propagation by the distance Δ. Similarly, the *reflected* field is given by the back-scattered field by the layer, and since the layer scatters equally in the forward and backward direction, we have at the position $z \le 0$ before the layer

$$E_0^{\text{refl}}(z=0) = -\mathrm{i}\,\rho_{\mathrm{a}}\lambda f_{\mathrm{a}}\Delta\,E_0(z=0)\,. \tag{7.75}$$

We have distinguished transmitted and reflected fields, and for clarity have used subscripts 0 and 1 to denote at which position the fields are considered. Keeping in mind the correlation between subscripts and positions $1 \Leftrightarrow z=0$ and $2 \Leftrightarrow z=\Delta$, we can now generalize our above discussion to more than a single sheet.

7.8.2.1 The Darwin-Prins Difference Equations

We assume that the incident field is a plane wave. This is equivalent to assuming that the field has an infinite lateral and longitudinal coherence length. In order to keep things simple, we shall for brevity use subscripts only and omit the position dependence. Also, we shall use the convenient short form definition (7.63) or

$$C = \rho_a \lambda \Delta\, f_a = \rho_a \lambda \Delta\, r_0 (f_1 - \mathrm{i} f_2) = (\delta - \mathrm{i}\beta)\, k\Delta\,. \tag{7.76}$$

In our new notation (7.74) and (7.75) become

$$1\text{ sheet}: E_1^{\text{trans}} = (1 - \mathrm{i}\, C)\, E_0\, \mathrm{e}^{\mathrm{i}\phi}, \quad \text{and} \quad E_0^{\text{refl}} = -\mathrm{i}\, C\, E_0\,. \tag{7.77}$$

We now add a second sheet. The field transmitted through the first sheet is then reflected by the second sheet. If we denote the reflected intensity at the position in-between the sheets as E_1^{refl}, the total reflected field by two layers is obtained by considering the transmission of E_1^{refl} back through the first layer and adding it to the field reflected by the first layer, i.e.

$$2\text{ sheets}: E_0^{\text{refl}} = -\mathrm{i}\, C\, E_0 + (1 - \mathrm{i}\, C)\, \mathrm{e}^{\mathrm{i}\phi}\, E_1^{\text{refl}}\,. \tag{7.78}$$

Instead of being transmitted back through the first layer, the field E_1^{refl} can also be reflected by it. It then contributes to the transmitted field E_1^{trans} at the position between the two sheets, and we have

$$2\text{ sheets}: E_1^{\text{trans}} = (1 - \mathrm{i}\, C)\, \mathrm{e}^{\mathrm{i}\phi}\, E_0 - \mathrm{i}\, C\, \mathrm{e}^{2\mathrm{i}\phi}\, E_1^{\text{refl}}\,. \tag{7.79}$$

The last exponential factor accounts for the geometrical path length difference between the two components to E_1^{trans}. Realizing that in our labeling the incident field E_0 can also be denoted E_0^{trans}, we can generalize (7.78) and (7.79) to an arbitrary pair of layers m and $m + 1$ in a solid. We now change our notation to that often used in the literature [13, 14]

$$\begin{aligned} T_m &= E_m^{\text{trans}}\,; \quad T_{m+1} = E_{m+1}^{\text{trans}} \\ S_m &= E_m^{\text{ref}}\,; \quad S_{m+1} = E_{m+1}^{\text{ref}}\,, \end{aligned} \tag{7.80}$$

which is illustrated in Fig. 7.16. We obtain the following *Darwin-Prins* difference equations.

The transmitted and reflected fields associated with two arbitrary layers m and $m+1$ as illustrated on the right side of Fig. 7.15b are linked by the *Darwin-Prins* difference equations according to

$$S_m = -\mathrm{i}\,C\,T_m + (1-\mathrm{i}\,C)\,\mathrm{e}^{\mathrm{i}\phi}\,S_{m+1} \tag{7.81}$$

$$T_{m+1} = (1-\mathrm{i}\,C)\,\mathrm{e}^{\mathrm{i}\phi}\,T_m - \mathrm{i}\,C\,\mathrm{e}^{2\mathrm{i}\phi}\,S_{m+1}\,. \tag{7.82}$$

These self-consistent equations are valid for soft as well as hard x-rays. For hard x-rays, their solution for a semi-infinite perfect single crystal is the basis for the dynamical theory of x-ray diffraction. The theory determines the important *Darwin width* of the monochromatic intensity obtained by Bragg diffraction from a perfect crystal. For soft x-rays, where Bragg diffraction does not occur, the above equations can be directly used to calculate the reflectivity of a solid of arbitrary thickness, and we will do so below.

We can significantly simplify these equations under the following assumptions:

- Normal incidence $\theta = 90°$, so that $\phi = k\Delta$.
- The atoms are uniformly distributed within the sample, and the incident wavelength is larger than the distances between atoms. This eliminates Bragg diffraction.
- The scattering process maintains a well-defined phase relationship between the incident and scattered field. This is fulfilled only for *elastic* scattering where scattering by each sheet leads to a phase shift of 180° (minus sign) relative to the incident field as derived in Sect. 7.5.4.

7.8.3 The Transmitted Field and Intensity

The *Darwin-Prins* difference equations readily yield the transmitted field. To see how this works, we calculate the transmitted field for an increasing number of stacked sheets. In our notation, the incident field before sheet 1 is

$$T_0 = E_0\,. \tag{7.83}$$

and the field after sheet 1 in the absence of a second sheet is

$$T_1 = (1-\mathrm{i}\,C)\,\mathrm{e}^{\mathrm{i}k\Delta}\,E_0\,. \tag{7.84}$$

If we add a second sheet, the transmitted field after sheet 1 is changed only slightly by the reflected intensity from sheet 2 that is again reflected by sheet 1 to propagate into the direction of the transmitted field. We have according to (7.82)

$$T_1 = (1-\mathrm{i}\,C)\,\mathrm{e}^{\mathrm{i}k\Delta}\,E_0 - \mathrm{i}\,C\,\mathrm{e}^{2\mathrm{i}k\Delta}\,S_1\,. \tag{7.85}$$

But from (7.82) we can see that the reflected contribution S_1 is small. It is given by

$$S_1 = \frac{S_0 + \mathrm{i}\, C\, T_0}{(1 - \mathrm{i}\, C)\, \mathrm{e}^{\mathrm{i}k\Delta}} \simeq (S_0 + \mathrm{i}\, C\, T_0)\, \mathrm{e}^{-\mathrm{i}k\Delta} \simeq 0\,, \tag{7.86}$$

where we have used $S_0 = -\mathrm{i}\, C\, E_0$. Hence we can neglect the reflected contributions from now on. With

$$T_1 \simeq (1 - \mathrm{i}\, C)\, \mathrm{e}^{\mathrm{i}k\Delta}\, E_0\,, \tag{7.87}$$

we get the field after 2 sheets as

$$T_2 \simeq (1 - \mathrm{i}\, C)\, \mathrm{e}^{\mathrm{i}k\Delta}\, T_1 \simeq (1 - \mathrm{i}\, C)^2\, \mathrm{e}^{2\mathrm{i}k\Delta}\, E_0\,. \tag{7.88}$$

As we progress through additional layers, we get another multiplicative transmission factor $(1 - \mathrm{i}\, C) \exp(\mathrm{i}k\Delta)$ for each layer, and after N sheets we have

$$T_N \simeq (1 - \mathrm{i}\, C)^N\, \mathrm{e}^{N\mathrm{i}k\Delta}\, E_0\,. \tag{7.89}$$

We see that each sheet contributes a multiplicative factor $(1 - \mathrm{i}\, C)\, \mathrm{e}^{\mathrm{i}k\Delta}$, where C is given by (7.76). Since the absolute value of C is of order 10^{-2} (see Fig. 7.12 for Fe), we can make the substitution $(1 - \mathrm{i}\, C)^N = (\mathrm{e}^{-\mathrm{i}C})^N$ in (7.89). In fact, a plot of the absolute values of the two expressions as a function of N shows that they are indistinguishable from each other. Hence for the total film thickness $d = N\Delta$ we obtain with (7.76) the transmitted field in terms of the optical parameters as

$$\boxed{T_N = E_{\text{trans}} \simeq E_0\, \mathrm{e}^{\mathrm{i}kd}\, \mathrm{e}^{-\mathrm{i}(\delta - \mathrm{i}\beta)kd} = E_0\, \mathrm{e}^{\mathrm{i}kd}\, \mathrm{e}^{-\mathrm{i}\lambda r_0 (f_1 - \mathrm{i} f_2)\rho_a d}}\,, \tag{7.90}$$

which is identical to our previous result (7.23). For a film of finite thickness $d = N\Delta$, the *linear* transmission response due to forward scattering from each sheet of thickness Δ evolves into an *exponential* response due to propagation phase shifts between the N sheets. The transmitted intensity is obtained as

$$I_{\text{trans}} \simeq I_0\, \mathrm{e}^{-2\beta kd} = I_0\, \mathrm{e}^{-2\lambda r_0 f_2 \rho_a d}\,, \tag{7.91}$$

which is the same as our previous derived *absorption* law (7.32). We see that the link of absorption and coherent forward scattering discussed for a thin sheet in Sect. 7.7.2 can be extended to a thick sample, if we consider dynamical coherent atomic forward scattering, with the scattering of each layer treated by the first order Born approximation.

7.8.4 The Reflected Field in the Soft X-Ray Region

Similarly, the total reflected field S_0 before the sample is calculated from the Darwin-Prins equations by adding the (small) contributions from the N elementary sheets, with the total sample thickness given by $d = N\Delta$. We find for the dependence of the *reflected* field as a function of the number of sheets

$$\begin{aligned} N=1: &\quad S_0 = -\mathrm{i}\,C E_0 \\ N=2: &\quad S_0 = -\mathrm{i}\,C E_0 + (1-\mathrm{i}\,C)\,\mathrm{e}^{\mathrm{i}k\Delta}\,S_1 \\ N=3: &\quad S_0 = -\mathrm{i}\,C E_0\left[1+(1-\mathrm{i}\,C)^2\,\mathrm{e}^{2\mathrm{i}k\Delta}\right] + (1-\mathrm{i}\,C)^2\,\mathrm{e}^{2\mathrm{i}k\Delta}\,S_2\,. \end{aligned} \tag{7.92}$$

For an arbitrary number of N sheets we have

$$S_0 = -\mathrm{i}\,C E_0 \left\{ \sum_{\ell=0}^{N} (1-\mathrm{i}\,C)^{2\ell}\,\mathrm{e}^{2\ell\,\mathrm{i}k\Delta} \right\} + (1-\mathrm{i}\,C)^{N-1}\,\mathrm{e}^{(N-1)\mathrm{i}k\Delta}\,S_{N-1}\,. \tag{7.93}$$

The reflected field S_{N-1} arises from the last sheet N. According to (7.82), it is given by

$$S_{N-1} = -\mathrm{i}\,C\,T_{N-1} + (1-\mathrm{i}\,C)\,\mathrm{e}^{\mathrm{i}k\Delta}\,S_N = -\mathrm{i}\,C\,T_{N-1}\,, \tag{7.94}$$

where we have made use of the fact that there is no reflected field S_N from the vacuum behind sheet N. By use of (7.89) we have

$$T_{N-1} \simeq (1-\mathrm{i}\,C)^{N-1}\,\mathrm{e}^{(N-1)\mathrm{i}k\Delta}\,E_0\,. \tag{7.95}$$

This allows us to write the reflected field for N sheets as

$$\begin{aligned} E_{\mathrm{refl}} = S_0 &= -\mathrm{i}\,C E_0 \left\{ \sum_{\ell=0}^{N} (1-\mathrm{i}\,C)^{2\ell}\,\mathrm{e}^{2\ell\,\mathrm{i}k\Delta} \right\} - \mathrm{i}\,C E_0\,(1-\mathrm{i}\,C)^{2(N-1)}\,\mathrm{e}^{2(N-1)\mathrm{i}k\Delta} \\ &\simeq E_0 \frac{(\beta+\mathrm{i}\delta)k\Delta}{\mathrm{e}^{2\,\mathrm{i}k\Delta}-1} \left\{1-\mathrm{e}^{2\mathrm{i}kN\Delta}[1-\mathrm{i}(\delta-\mathrm{i}\beta)k\Delta]^{2N}\right\}, \end{aligned} \tag{7.96}$$

where in the last step we have expressed C by (7.76), evaluated the sum as a power series in the optical parameters, and kept only the first order terms. For $\Delta \ll \lambda/2\pi$ or $k\Delta \ll 1$ we obtain our final result.[13]

[13] This agrees with the result in [13] for an N layer film, given in terms of the reflectivity of a semi-infinite film, given by (7.98).

The *reflected field* of a film of total thickness $d = N\Delta$ at normal incidence is given by

$$E_{\text{refl}} \simeq E_0 \frac{\delta - \mathrm{i}\beta}{2} \left\{1 - \mathrm{e}^{2ikd(1-\delta+\mathrm{i}\beta)}\right\} . \tag{7.97}$$

For a semi-infinite solid ($d \to \infty$) the reflected intensity $\propto |E_{\text{refl}}|^2$ oscillates because of the exponential term but assumes the limiting value

$$I_{\text{refl}} \simeq I_0 \frac{\delta^2 - \beta^2}{4} . \tag{7.98}$$

Because of the small value of the optical parameters, which according to Fig. 7.12 are $\delta, \beta < 10^{-2}$ even in the resonance regions, the reflected intensity is very small. It is therefore neglected in the exponential absorption law.

7.9 Polarization Dependent Absorption: Dichroism

We start this section with an historical overview of the discovery of dichroic or polarization dependent effects which occurred by use of visible light. We then discuss a simple classical description of the experimentally observed effects based on an effective medium response through phenomenologically modified optical parameters. This formalism, which may also be extended into the x-ray region, however, fails to link the effects to the atomic, electronic, and spin dependent structure of matter. This requires quantum theory as will be discussed in Chap. 11 and Sect. 13.6.

7.9.1 History of Polarization Dependent Effects

Today the term *dichroism* refers to the change of the polarization of light upon transmission or reflection by matter. The first documented observation of a dichroic effect goes back to 1669 when the Dutch scientist Erasmus Bartholinus (1625–1698) observed the phenomenon of double refraction in calcspar (calcite). In his 1690 book *Traité de la lumiére*, Huygens reported that the two rays arising from refraction in calcite may be extinguished when passing through a second calcite crystal. The next milestone came another 100 years later, when in 1808 the Frenchman Étienne Louis Malus (1775–1812) used calcite to show that light may become polarized upon reflection by a glass window pane. Malus also first used the term “polarization” but did not attempt to interpret the phenomenon. In 1811 Arago [33] discovered the phenomenon of *optical activity*, consisting of a rotation of the polarization direction upon transmission through crystalline quartz, and around the same time Biot extended the measurements to vapors and liquids [34]. In 1815 the Scottish physicist David

Brewster (1781–1868) discovered the relationship between the index of refraction and the angle of total reflection resulting in complete linearly polarized light.

The phenomenon of light polarization was put on a firm scientific footing by Fresnel in the period 1818–1822. He suggested that light is a transverse wave and introduced and linked the concepts of linearly and circularly polarized light, discussed in Sect. 3.2.6. Fresnel also attributed optical activity to different indices of refraction of the left and right components. Fresnel's explanation was supported by an important experiment carried out by Louis Pasteur in 1850 [35]. He observed that the organic compound tartrate, when synthesized in a laboratory, was optically inactive, unlike the tartrate extracted from grapes. He then succeeded in separating the left- and right-handed crystals from each other and showed that each was optically active with opposite rotations of the polarization. Despite Fresnel's general explanation of optical activity and Pasteur's recognition that chiral molecules give rise to optical activity, it was not until 1895 that Cotton [36] performed what might be called the first *circular dichroism* experiment. He discovered that natural light becomes partially circularly polarized when traversing a chiral medium. The importance of the concepts of handedness or chirality is reflected by the fact that all 20 biologically active amino acids are left handed, while amino acids made in the laboratory are 50% left and 50% right handed. Owing to the fact that many drugs need to be synthesized in a single chiral state, the topic of optical activity has remained pertinent in the fields of synthetic and structural chemistry, biochemistry, and in pharmaceutical science [37].

Originally, the observation of light polarization effects was restricted to non-magnetic samples. In 1846 Faraday discovered a direct connection between magnetism and light polarization, the magneto-optical or *Faraday effect* [38]. It refers to the change of light polarization in *transmission* through a magnetized material. The same effect in *reflection* was discovered in 1877 by the Scottish physicist John Kerr (1824–1907) [39] and is called the magneto-optical Kerr effect in his honor. As discussed earlier, these magneto-optical effects do not arise from the interaction of the EM **H**-field with the sample but indirectly from interaction of the **E**-field which couples to the charge and then indirectly via the spin-orbit interaction to the spins or magnetization. Today, the magneto-optical Kerr effect (MOKE), which because of its reflection geometry is widely applicable, remains an important tool in magnetism research [40, 41]. In addition, the availability of short and intense optical laser pulses has created the new field of ultrafast magnetization dynamics [42, 43].

7.9.2 Chiral Versus Magnetic Orientation

The key difference of polarization effects observed for chiral and magnetic materials results from the basic symmetry concepts of inversion symmetry (parity) and time reversal symmetry. This is beautifully demonstrated by the conceptually simple light transmission experiments through a chiral and magnetic sample shown in Fig. 7.17.

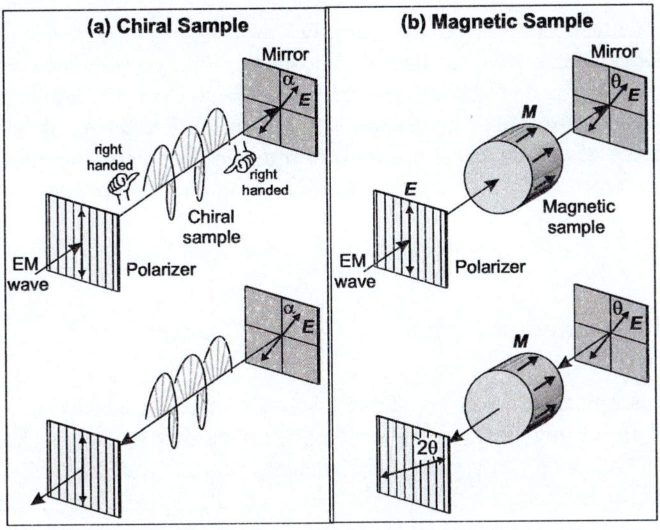

Fig. 7.17 Comparison of transmission of a linearly polarized electromagnetic wave through a right-handed chiral in (**a**) and a magnetic sample in (**b**). In both cases the linearly polarized wave experiences a rotation of its polarization upon transmission through the sample by rotation angles α and θ, respectively. The beams are then reflected by a mirror that preserves the polarization, and sent back through the respective samples. The right-handed chiral sample always rotates the polarization clockwise when looking into the propagation direction (thumb direction) so that the polarization vector retraces its original path after reflection and arrives back unchanged. The magnetic sample also rotates the polarization clockwise on its first pass through the sample with magnetization direction **M** pointing into the propagation direction. After reflection the propagation direction is opposite to **M** and the polarization vector rotates anticlockwise. The light arrives back with the polarization vector rotated by 2θ

In Fig. 7.17a, linearly polarized light is transmitted through a right-handed chiral sample. The handedness is illustrated by the right hand rule, with the thumb pointing into a given direction and the fingers indicating the rotation direction of the helix (see Sect. 3.2.6). The chiral sample causes a rotation of the polarization of the transmitted light by an angle α. The wave is reflected by a mirror which preserves the linear polarization direction. Close inspection reveals that the reflected wave also encounters a right-handed helix, and therefore, the polarization is rotated back into the original direction. This example shows that the chirality of a sample is preserved upon its rotation in space. Hence optical activity may also be observed in liquids composed of randomly rotated molecules of the same chirality.

As shown in Fig. 7.17b, a magnetic sample looks different to the EM wave on its first and second pass. The magnetization **M** is a vector, and it matters whether this vector points into or opposite to the propagation direction of the EM wave. If the polarization vector rotates clockwise by θ on the first path, it will rotate anticlockwise by θ on the second path. The two **E**-vector rotations by the sample add up, and the final rotation angle will be 2θ.

The different behavior of the chiral and magnetic samples in Fig. 7.17 is a direct consequence of the opposite symmetry properties of the two samples. Chirality is associated with a spatial twist of the charge best described by a helix. A chiral sample *lacks a center of inversion*. In contrast, a magnetic direction is defined through the direction of current flow through a wire loop. The existence of a magnetic field direction hence corresponds to *breaking the time reversal symmetry* of current flow. The existence of chirality is therefore linked to the symmetry concept of "space reversal" symmetry or parity and a magnetic direction to the symmetry concept of "time reversal" symmetry.

7.9.3 Fundamental Forms of X-Ray Dichroism

On a microscopic level the origin of dichroic behavior of a material originates from the spatial anisotropy of the charge or the spin. In cases where the dichroic effect depends on the charge, only, one speaks of charge or *natural dichroism*.

> *X-ray* **"natural"** *dichroism* is caused by the spatial anisotropy of the charge in the absence of spin alignment.
>
> – X-ray natural linear dichroism—**XNLD**—arises from a spatially anisotropic charge distribution.
> – X-ray natural circular dichroism—**XNCD**—is associated with a spatial handedness of the charge distribution.

The dichroic effect originating from *spin* orientation is referred to as *magnetic dichroism*. In general one distinguishes the existence of a preferential spin *direction* as in a ferromagnet or ferrimagnet and the existence of an alignment axis, where the same number of spins point in opposite directions as in an antiferromagnet. There are two important types of magnetic dichroism.

> *X-ray* **"magnetic"** *dichroism* is due to spin alignment and the coupling of the spin to the charge via the spin-orbit interaction.
>
> – X-ray magnetic linear dichroism—**XMLD**—arises from an anisotropic distortion of the charge induced by a preferred spin *direction or axis*. It exists for all matter with an intrinsic spin orientation or an **H**-field induced spin orientation.
> – X-ray magnetic circular dichroism—**XMCD**—arises from the existence of a preferred spin *direction*. It exists in ferromagnets and ferrimagnets and paramagnets subjected to a large **H**-field.

In the following we shall briefly discuss the first x-ray demonstrations of the different effects illustrated in Fig. 7.18, which complements our earlier Fig. 1.19.

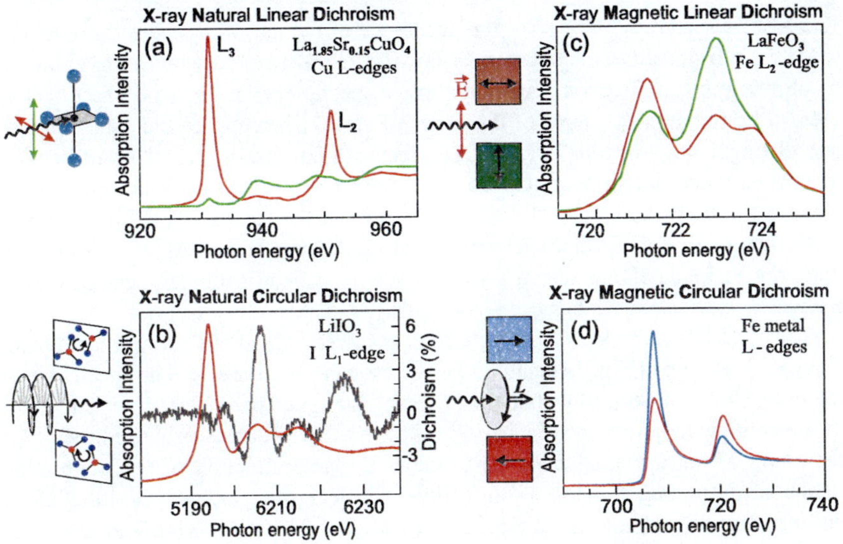

Fig. 7.18 Four important types of dichroism. **a** X-ray natural linear dichroism spectra of $La_{1.85}Sr_{0.15}CuO_4$ near the Cu L-edge [44]. The resonances are due to transitions to the highest energy unfilled $d_{x^2-y^2}$ orbital. **b** X-ray absorption spectrum (red) of single crystal $LiIO_3$ and the difference spectrum (gray), the x-ray natural circular dichroism spectrum, obtained from absorption spectra with left and right circularly polarized x-rays, incident along a special crystalline axis [45]. **c** Magnetic linear dichroism spectrum of an epitaxial thin film of antiferromagnetic $LaFeO_3$ with the **E**-vector aligned parallel and perpendicular to the antiferromagnetic axis [46]. The splitting of the L_2 resonance is due to multiplet effects. **d** X-ray magnetic circular dichroism spectrum around the L_3 and L_2 edges of Fe metal. The photon angular momentum was aligned parallel or antiparallel to the magnetization direction of the sample [47]

The XNLD effect was first demonstrated in 1981 [48, 49]. Strong XNLD effects are found in the so-called near edge x-ray absorption fine structure (NEXAFS) spectra low-Z molecules, macromolecules, polymers, or nematic liquid crystals [18, 50]. Such systems contain directional covalent bonds leading to polarization dependent x-ray absorption spectra like that illustrated in Fig. 1.19 for benzene molecules weakly bonded to a Ag(110) surface. In Fig. 7.18a we show the XNLD effect for the Cu L-edge of $La_{1.85}Sr_{0.15}CuO_4$ [44]. The single crystal sample has a layered structure. The Cu atom shown in black is surrounded by four in-plane O atoms and two out-of-plane O atoms. If we define the x, y plane of our coordinate system to lie in the plane of the layers (shown in gray), the in-plane $d_{x^2-y^2}$ orbital is unfilled, and when the **E** lies in the $(x - y)$ plane, a large peak-like transition is observed to this orbital. This resonance is absent when **E** is oriented perpendicular to the plane, as shown, since there are no empty states in the perpendicular direction. One may envision the polarization vector as a search light for empty valence orbitals, the so-called search light effect [18].

An XNCD effect was first observed in 1998 [45, 51]. It is illustrated in Fig. 7.18b for the iodine L_1-edge x-ray absorption spectrum (red) of single crystal $LiIO_3$ [52].

The XNCD spectrum, defined as the difference of two absorption spectra obtained with left and right circularly polarized x-rays, is shown as a gray curve. It was obtained by aligning the single crystal sample along a special crystalline axis. The effect is seen to be remarkably large, of the order of several percent. In the x-ray region, one distinguishes two types of XNCD effects which arise in higher order beyond the dipole approximation, as will be discussed in Sect. 11.4. The one observed by Goulon and coworkers [45, 51, 52] is only present in oriented crystalline samples and is due to an electric dipole/electric quadrupole interference effect. The other is the x-ray analogue of optical activity and is due to an electric dipole/magnetic dipole interference effect, and it has been studied in [53–57].

The XMLD effect was first predicted and observed in 1986 [58, 59]. A more recent example is shown in Fig. 7.18c at the Fe L_2 resonance in $LaFeO_3$. The spectra show the difference in absorption when the **E**-vector is aligned parallel and perpendicular to the antiferromagnetic axis in $LaFeO_3$ [46]. XMLD is extensively used today for the study of antiferromagnets, in particular, the determination of the orientation of the antiferromagnetic axis in thin films and near surfaces, and the imaging of antiferromagnetic domains. The principal difference between XNLD and XMLD is that the latter exists only in the presence of magnetic alignment, and hence it vanishes at temperatures above the Neél or Curie temperature, or for paramagnets in the absence of an external magnetic field.

The XMCD effect was first reported in 1987 [60]. It is illustrated in Fig. 7.18d for the Fe L-edge in Fe metal. The effect is maximum when the x-ray angular momentum is parallel and antiparallel to the magnetic moment of the sample. The effect is seen to be very large at the resonance positions and is opposite at the L_3 and L_2 edges. This directly reflects the opposite sign of the spin component at the two edges, $j = l + s$ at the L_3-edge and $j = l - s$ at the L_2 edge. The observation of XMCD requires the presence of a preferred magnetic direction. It is therefore zero for antiferromagnets.

There are other more complicated types of "magnetic" dichroism where charge and spin effects are both present [52, 61]. For example, x-ray *magnetochiral dichroism* arises from axial spin alignment and a chiral charge distribution [62]. X-ray *non-reciprocal linear dichroism* arises from charge chirality that is induced by an axial spin alignment, and the effect is parity odd and time odd [63]. We shall not discuss these cases here.

7.10 Natural Dichroism and Orientational Order

For the phenomenological description of the response of charge distributions in solids to *linearly* polarized light, i.e. XNLD, one considers the orientation of the **E** polarization vector relative to different orientations of the sample, assumed to be non-magnetic. In crystalline materials, an anisotropy of the valence charge distribution exists when the lattice has lower than cubic symmetry. A beautiful example of this case is the layered compound $La_{1.85}Sr_{0.15}CuO_4$ whose strongly anisotropic absorption spectrum is shown in Fig. 7.18a. Other examples of strongly anisotropic systems are molecules bonded to surfaces, since they typically have a preferred bond orientation [18].

In the following we want to develop a phenomenological description of the general case of *orientational order*. It means that, on average, a sample contains preferential orientations of interatomic bonds or molecular groups. This may exist even if the sample would appear "amorphous" from an x-ray crystallography point of view. The situation is encountered for many "soft matter" systems including liquids, gels, liquid crystals, polymers, and biological materials. Their exploration was pioneered by Pierre-Gilles de Gennes (1932–2007), who was awarded the 1991 Nobel Prize in physics for his work. Orientational order is conveniently treated by a formalism developed in the optical region [64, 65] that describes charge distributions in terms of their projections onto three axes in a Cartesian coordinate system, so-called *orientation factors*.

> Systems without crystallographic order or translational atomic order may nevertheless possess *orientational order*. It is defined through the existence of a spatial symmetry axis $\hat{\mathbf{u}}$, called "director", along which there is a preferential alignment of internuclear axes or bonds. There may also be additional orientational order in the plane perpendicular to the director, which in its simplest form is called *biaxial* order.

While the spaghetti-like chains in bulk polymers typically arrange themselves in random directions, the randomness disappears through confinement in thin films or the presence of a surface or interface. This leads to preferential orientation of the molecular groups forming the chain backbone and attached functional groups, e.g. phenyl rings. The orientation may also be enhanced in a specific direction by rubbing a polymer. Remarkably, this low-tech technique has been extensively used to create a template for the alignment of liquid crystals, utilized to modulate light transmission in high-tech flat panel displays [50, 66] (also see Fig. 11.6). In the following we shall first discuss how orientational order in rubbed polymers may be detected by XNLD and quantified by a phenomenological framework [50, 67, 68].

Figure 7.19a and b illustrates an XNLD absorption measurement of a rubbed polyimide film, produced for the alignment of liquid crystals in flat panel displays by Japan Synthetic Rubber [68]. The geometry of the experiment is defined in (a) and the NEXAFS spectra for three orientations of the **E** polarization vector relative to the coordinate system are shown in (b), indicated by corresponding colors. The spectra were recorded by Auger electron yield detection which is preferentially sensitive to the surface of the rubbed polymer film [18, 50].

The observed peaks labeled π_1, π_2, and π_3 correspond to specific electronic transitions on the identified C atoms to empty molecular π^* orbitals [68] (also see Fig. 1.18). From the angular dependence of the peak intensities one can derive the preferential orientation of the corresponding π orbitals. For example, peak 1 originates from $1s \rightarrow \pi^*$ electronic transitions on the central four C atoms in the phenyl rings and has maximum intensity when the **E**-vector is parallel to the phenyl π system, i.e. when

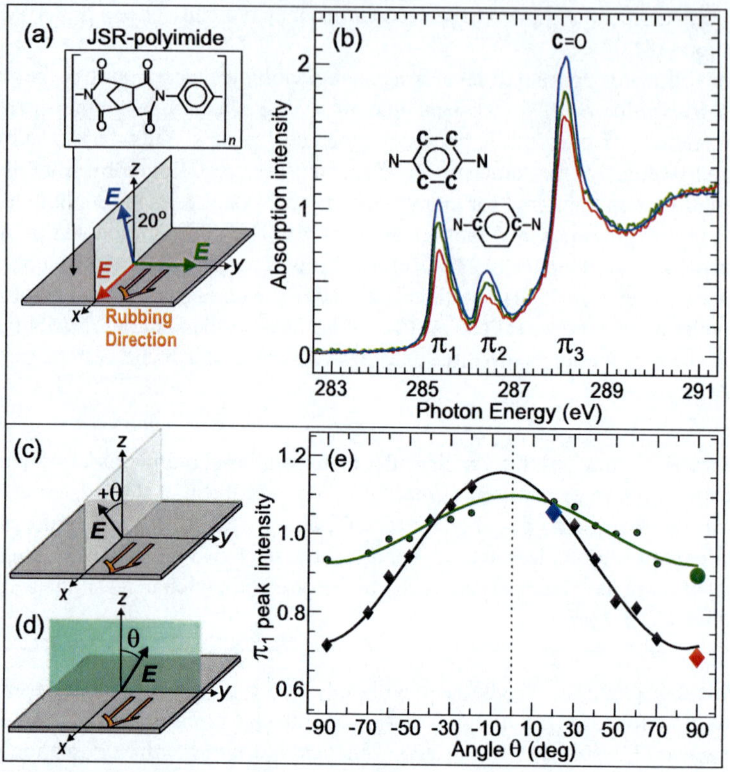

Fig. 7.19 Molecular orientation of functional groups with a polyimide film of the structure shown in (**a**) rubbed in the x-direction. **b** X-ray absorption spectra for the three the **E** polarization vector orientations in (**a**) [68]. The three identified resonances correspond to K-edge transition $1s \rightarrow \pi*$ to empty π^* orbitals associated with the identified C atoms in the polymer. **c** and **d** define the **E**-vector orientation angles θ in two orthogonal planes used for the measurement of the π_1 peak intensity plotted in (**e**). We have also identified the three **E** orientations for the spectra in (**b**) by bigger symbols with the corresponding colors

E is perpendicular to the plane of the phenyl ring [18]. The figure demonstrates that all π orbitals are preferentially oriented along the surface normal (blue), and there are more π orbitals oriented along y (green) than along the rubbing direction x (red).

A more detailed dependence of the π_1 resonance intensity as a function of the **E** orientation is shown in Figs. 7.19c–d as a function of tilt angle θ of **E** from the surface normal z in the $(x - z)$ (light gray) and $(y - z)$ (light green) planes. The curves in (e) shown in green and black colors are theoretical fits to a model we shall discuss now.

7.10.1 Orientation Factors and Saupe Matrix

Orientational order may be treated by concepts originally developed in the late 1950s by Maier and Saupe for the description of the interactions and ordering within liquid crystals (LCs) [69]. The formalism is based on the introduction of a *tensor order parameter*, also called ordering matrix or *Saupe matrix*, which has been extensively used for the description of the orientational order of nematic LCs [64, 65, 70], polymer surfaces [50, 67, 68], and even amorphous carbon surfaces [71]. We will see in Sect. 11.3.2 that this phenomenological description is closely related to the quantum formulation of XNLD in terms of a quadrupolar distribution.

The tensor order parameter description is based on the following assumptions for the "soft matter" system.

- The system consists of individual molecular units which are anisotropic (e.g. chains and rings).
- The centers of the molecular units exhibit no long range structural order.
- On average, the molecular units have a preferred spatial orientation along a symmetry direction $\hat{\mathbf{u}}$, called the *director*.
- The alignment along $\hat{\mathbf{u}}$ and $-\hat{\mathbf{u}}$ is indistinguishable.
- There may also be orientational order in the plane perpendicular to the director.

In general, the average distribution of the molecular units in the LC is described by a molecular distribution function $O(\vartheta, \varphi)$ which depends on the polar and azimuthal angles ϑ and φ in a molecular Cartesian coordinate system (x', y', z') defined below. It is often sufficient to characterize the molecular distribution by the projections of $O(\vartheta, \varphi)$ onto the axes (x', y', z'). The normalized projections are called *orientation factors* [65]. This formalism is suited for the description of polarization dependent optical and XNLD experiments like those shown in Fig. 7.19. XNLD measurements cannot determine the complete distribution function $O(\vartheta, \varphi)$ but only its projection.

In practice, the molecular frame is typically rotated relative to a Cartesian frame (x, y, z) fixed to the macroscopic shape of the sample. The two reference frames are in general linked by three Euler rotation angles describing a matrix rotation. For our rubbing case, the sample and molecular frames are linked by only a single rotation angle γ, as shown in Fig. 7.20a. The sample frame is chosen with z along the surface normal and x along the rubbing direction.

For an *unrubbed film* the sample frame (x, y, z) and molecular frame (x', y', z') are the same. The orientation factors in the surface plane are the same, $f_x = f_y$, but typically differ from the one along the surface normal, f_z as illustrated in Fig. 7.20b. The rubbing process leads to a rotation of the molecular orientation function about the perpendicular y axis, and the sample and molecular frames are now different as shown in Fig. 7.20c. This manifests itself in Fig. 7.19b by an inequivalence of the measured XNLD intensities along x, y and approximately along z, shown as red, green and blue curves. Also, the angular change about the z direction in the $(x - y)$ plane shown as the black curve in Fig. 7.19e is asymmetrical, while it is symmetrical about z in the $(y - z)$ plane as shown by the green curve in Fig. 7.19e.

For the rubbed polymer, XNLD measurements determine the orientation factors $f_{x'}$, $f_{y'}$, $f_{z'}$ onto the molecular frame axes (x', y', z') as shown in Fig. 7.20c. They are defined as the projections of the bond distribution function $O(\vartheta, \varphi)$ according to

$$f_{x'} = \int_0^{2\pi}\int_0^{\pi} \sin^2\vartheta \, \cos^2\varphi \; O(\vartheta, \varphi) \; \sin\vartheta \, d\vartheta \, d\varphi \tag{7.99}$$

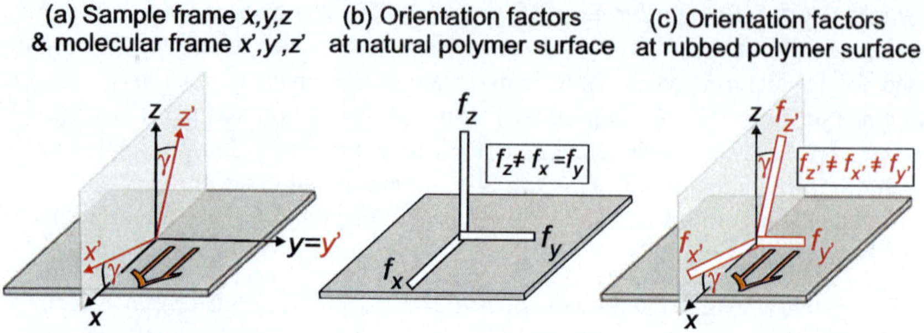

Fig. 7.20 **a** Sample reference frame (x, y, z), defined by the macroscopic sample shape and rubbing direction, versus molecular frame (x', y', z') defined by the diagonal Saupe matrix (7.102). **b** Typical orientation factors $f_z \neq f_x = f_y$ for *unrubbed* polymer surfaces for which the sample and molecular frames are the same. **c** Orientation factors $f_{z'} \neq f_{x'} \neq f_{y'}$ in the molecular frame for a typical rubbed polymer surface, where the Saupe matrix is given by (7.105)

$$f_{y'} = \int_0^{2\pi}\int_0^{\pi} \sin^2\vartheta \, \sin^2\varphi \, O(\vartheta, \varphi) \, \sin\vartheta \, d\vartheta \, d\varphi \tag{7.100}$$

$$f_{z'} = \int_0^{2\pi}\int_0^{\pi} \cos^2\vartheta \, O(\vartheta, \varphi) \, \sin\vartheta \, d\vartheta \, d\varphi \, . \tag{7.101}$$

One can envision these orientation factors as the number of bonds along the molecular frame x', y', and z' axes, respectively, and for a normalized distribution $\int O(\vartheta, \varphi) \, d\Omega = 1$ we have $f_{x'} + f_{y'} + f_{z'} = 1$.

In the molecular frame one can define a diagonal ordering or Saupe matrix according to

$$\mathbf{Q}(x', y', z') = \begin{pmatrix} \frac{1}{2}(3f_{x'} - 1) & 0 & 0 \\ 0 & \frac{1}{2}(3f_{y'} - 1) & 0 \\ 0 & 0 & \frac{1}{2}(3f_{z'} - 1) \end{pmatrix} = \begin{pmatrix} -\frac{1}{2}(S + \eta) & 0 & 0 \\ 0 & -\frac{1}{2}(S - \eta) & 0 \\ 0 & 0 & S \end{pmatrix} . \tag{7.102}$$

Here we have introduced the *uniaxial order parameter* S and the *biaxiality* η [65] according to

$$S = \frac{1}{2}(3f_{z'} - 1) \tag{7.103}$$

and

$$\eta = \frac{3}{2}(f_{y'} - f_{x'}) \tag{7.104}$$

For $\eta = 0$ the distribution is *uniaxial* and cylindrically symmetric about the z' axis. For finite η the system has different projections along all three Cartesian axes, and we have a *biaxial* system. For reference, typical nematic liquid crystals are described by a uniaxial distribution $\eta = 0$ and order parameters in the range $0.4 < S < 0.7$ [70].

In general, the Saupe matrix is not diagonal and one may define the molecular frame (x', y', z') as that where the Saupe matrix is diagonal. Since the molecular frame is rotated from the sample frame (x, y, z), the Saupe matrix will no longer be diagonal in the latter. For our case it is given by

$$\mathbf{Q}(x, y, z) = \begin{pmatrix} -\frac{1}{2}(S+\eta)\cos^2\gamma + S\sin^2\gamma & 0 & \frac{1}{2}(S+\eta)\sin\gamma\cos\gamma + S\sin\gamma\cos\gamma \\ 0 & -\frac{1}{2}(S-\eta) & 0 \\ \frac{1}{2}(S+\eta)\sin\gamma\cos\gamma + S\sin\gamma\cos\gamma & 0 & -\frac{1}{2}(S+\eta)\sin^2\gamma + S\cos^2\gamma \end{pmatrix}, \tag{7.105}$$

where γ is defined in Fig. 7.20a. In general, the Saupe matrix is real, symmetric, and traceless, and the trace is invariant under a unitary transformation which transforms one coordinate system into another. These properties can be verified by inspection of (7.102) and (7.105).

7.10.2 Determination of Orientation Factors

The angular plot of the measured intensities in Fig. 7.19e reveals that the scans in the perpendicular planes (black and green) do not have the same value for $\theta = 0$ (z direction). This is due to the fact that data were recorded on a wiggler beam line which produces slightly elliptically polarized radiation. For convenience we used the description of the degree of linear polarization given by (3.35) which in our case was $P = 0.82$ [68] (see blue curve in Fig. 2.26). If the sample is rotated so that $\mathbf{E}^{\|}$ lies in the $(x - z)$ sample plane which contains the rubbing direction, the XNLD intensity is given by [50, 68]

$$\begin{aligned} I_{xz}(\theta) &= I_0\left[A^{\|} + B^{\|}\sin^2\theta + C^{\|}\sin 2\theta\right] \\ A^{\|} &= P\,(f_{x'}\sin^2\gamma + f_{z'}\cos^2\gamma) + (1-P)\,f_{y'} \\ B^{\|} &= P\left\{(f_{x'} - f_{z'})\left(\cos^2\gamma - \sin^2\gamma\right)\right\} \\ C^{\|} &= -\frac{\tan 2\gamma}{2}\,B^{\|}\,. \end{aligned} \tag{7.106}$$

When $\mathbf{E}^{\parallel}$ is rotated in the $(y - z)$ plane perpendicular to the rubbing direction, the intensity distribution is

$$\begin{aligned} I_{yz}(\theta) &= I_0\left[A^{\perp} + B^{\perp}\sin^2\theta\right] \\ A^{\perp} &= P\left(f_{x'}\sin^2\gamma + f_{z'}\cos^2\gamma\right) + (1-P)\left(f_{x'}\cos^2\gamma + f_{z'}\sin^2\gamma\right) \\ B^{\perp} &= P\left\{f_{y'} - f_{x'}\sin^2\gamma - f_{z'}\cos^2\gamma\right\}. \end{aligned} \tag{7.107}$$

In practice, one fits the data to derive the parameters $A^{\parallel}$, $B^{\parallel}$, $C^{\parallel}$, $A^{\perp}$, $B^{\perp}$ which also yields γ through the last expression in (7.106) and then calculates the orientation factors according to

$$f_{x'} = \frac{A^{\perp} + B^{\parallel}\left(1 + \frac{\sin^2\gamma}{P\cos 2\gamma}\right)}{I_{\text{tot}}} \tag{7.108}$$

$$f_{y'} = \frac{A^{\parallel} + B^{\perp}}{I_{\text{tot}}} \tag{7.109}$$

$$f_{z'} = \frac{A^{\perp} + B^{\parallel}\left(1 - \frac{\cos^2\gamma}{P\cos 2\gamma}\right)}{I_{\text{tot}}} \tag{7.110}$$

$$I_{\text{tot}} = \frac{3}{2}(A^{\parallel} + A^{\perp}) + \frac{3P - 1}{2P}(B^{\parallel} + B^{\perp}). \tag{7.111}$$

The last expression follows from the normalization condition $f_{x'} + f_{y'} + f_{z'} = 1$. The polarization factor can also be directly obtained as

$$P = \frac{B^{\perp} - B^{\parallel}}{A^{\parallel} - A^{\perp} + B^{\perp} - B^{\parallel}} \tag{7.112}$$

The orientation factors for the rubbed PI surface deduced from experiment are summerized in Fig. 7.21a. The molecular π bonds become asymmetrical in the rubbed surface plane and also exhibit an out-of-plane tilt angle γ, as indicated in the figure by the orientation of the yellow π orbital of a phenyl ring. When a liquid crystal is deposited on the surface, the π system of the LC locks to the π system of the polymer surface, which has been termed a "maximum overlap model" of the π charge densities [50]. The preferential bond orientation in the polymer surface acts as a template for LC alignment as shown in Fig. 7.21b, and the LC director exhibits both an in-plane orientation along the rubbing direction and an out-of-plane so-called *pretilt* angle γ, which plays an important role when the LC orientation is rotated by an applied voltage in a LC display [50, 66, 71] (also see Fig. 1.18).

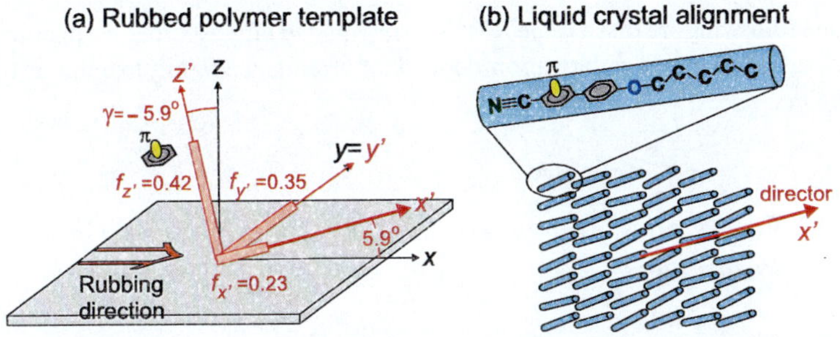

Fig. 7.21 **a** Orientation factors deduced from the experimental data in Fig. 7.19 which reveal the asymmetry of π bonds near the rubbed PI surface, as illustrated schematically by the preferred direction of phenyl π orbitals (yellow). **b** Corresponding alignment of liquid crystal rod-shaped molecules when placed on the rubbed surface. The orientation of the LC molecules results from locking of the LC π system (yellow) to that of the polymer surface

7.11 Magnetic Dichroism and Faraday Rotation

Similar to the phenomenological treatment of *charge orientation* in XNLD, one may describe *spin orientation* which leads to the XMCD effect. A preferred spin direction is created by use of an external magnetic field. It is required even for ferromagnets to produce a macroscopic magnetization direction by aligning the randomly spontaneously magnetized (through the exchange interaction) microscopic domains. In contrast to the charge case where the "director" only defines an *axis*, spins are truly directional vectors, reflecting a broken time reversal symmetry. For the description of XMCD one may also extend a semi-classical formalism which was developed to describe the optical Faraday and Kerr effects [41].

The interaction of x-rays and matter happens on the atomic scale through the coupling of the photon and electronic degrees of freedom. Since the magnetic field of the photon plays no role as discussed in Sect. 7.3.3, the interaction is determined by the coupling of the **E**-vector to the atomic charge. The coupling to the atomic spin is only indirect through the spin-orbit interaction, which aligns the atomic spin and orbital angular momenta to an atomic magnetic moment. In a simple picture, one would intuitively expect that there is a difference in the coupling or absorption for circularly polarized light if its angular momentum points in the same or opposite direction of the atomic magnetic moments. This is indeed the case and is the origin of the XMCD effect.

This effect is closely related to the magneto-optical Faraday and Kerr effects which are measured with linearly polarized light, since linearly polarized light corresponds to a coherent superposition of equal "right" and "left" circular components. When viewed in a linear polarization basis, the preferential weakening of one of the circular components then leads to a polarization rotation.

In the following we discuss the XMCD and Faraday effects through a phenomenological introduction of polarization dependent atomic scattering lengths or optical parameters.

7.11.1 Polarization Dependent Scattering Length and Optical Parameters

Within the semi-classical framework the link of the experimentally observed effects to the fundamental atomic structure and the related electronic and spin structure remains hidden and is lumped into phenomenological parameters. A true understanding of all dichroism effects requires a quantum theoretical formalism, which will be developed in Chap. 11.

We can, however, treat polarization dependent effects in our semi-empirical formalism by simply adding a polarization label p, that takes the values $p = 0$ for linear polarization, and $p = \pm 1$ for circular polarization, to both the x-ray scattering length and optical parameters that were linked in Sect. 7.5.5, so that

$$f^p = r_0 Z + f'^p - \mathrm{i} f''^p = \frac{2\pi}{\lambda^2 \rho_a} \left(\delta^p - \mathrm{i}\beta^p\right) . \tag{7.113}$$

In particular, our previous expressions (7.52) and (7.53) then become

$$\begin{aligned} \delta^p &= \frac{\lambda^2 \rho_a}{2\pi} \left(r_0 Z + f'^p\right) \\ \beta^p &= \frac{\lambda^2 \rho_a}{2\pi} f''^p . \end{aligned} \tag{7.114}$$

This formalism is particularly useful for the description of the interaction of circularly polarized light with ferromagnetic materials.

> The semi-classical description of the x-ray response of matter may be phrased in terms of phenomenological energy, $\hbar\omega$, and polarization, p, dependent scattering lengths or optical parameters. Their values are based on experimental measurements of either the absorptive parts f''^p or β^p or refractive parts f'^p or $\delta^p(\hbar\omega)$. Knowledge of one part allows the determination of the other by use of the Kramers-Kronig formalism.

7.11.2 Phenomenological Model

For the quantitative derivation of magneto-optical effects one starts by considering x-rays that are purely circularly polarized since light of plus or minus angular momenta and a magnetic sample of plus or minus magnetization directions are both described by the same angular momentum basis states. We can conveniently utilize the description of light polarization in Sect. 3.2.6 and assume light propagation into the z direction in a Cartesian coordinate system (x, y, z). The unit polarization vectors are then defined according to (3.30) or

$$\boldsymbol{\epsilon}_x = \frac{1}{\sqrt{2}}(\boldsymbol{\epsilon}_- - \boldsymbol{\epsilon}_+)\,, \quad \boldsymbol{\epsilon}_y = \frac{\mathrm{i}}{\sqrt{2}}(\boldsymbol{\epsilon}_- + \boldsymbol{\epsilon}_+)\,,$$
$$\boldsymbol{\epsilon}_+ = -\frac{1}{\sqrt{2}}\left(\boldsymbol{\epsilon}_x + \mathrm{i}\boldsymbol{\epsilon}_y\right)\,, \quad \boldsymbol{\epsilon}_- = \frac{1}{\sqrt{2}}\left(\boldsymbol{\epsilon}_x - \mathrm{i}\boldsymbol{\epsilon}_y\right)\,. \tag{7.115}$$

They are illustrated in Fig. 7.22 and may be labeled by their angular momentum projection onto the z propagation direction (also see Fig. 3.4). Our definition directly maps onto the quantum description presented in Chap. 11, where the unit polarization vectors are expressed in terms of first order spherical tensors (see Table A.2).

To emphasize the XMCD effect we assume that there is cylindrical symmetry about the incident beam direction in both charge and spin distributions, so that there are no other dichroic effects arising from different **E** orientations in the $x-y$ plane.[14] With this assumption we can distinguish the three cases of the incident light polarization $\boldsymbol{\epsilon}_p$ through the labels $p = 0, \pm$, where $p = 0$ corresponds to linearly or naturally polarized light with the **E**-vector somewhere in the $x-y$ plane.

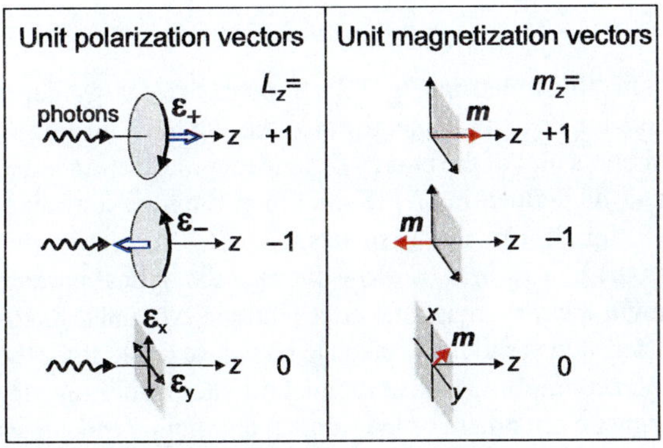

Fig. 7.22 Definitions of unit polarization and magnetization vectors

[14] If there was a preferential spin or charge orientation direction $\hat{\mathbf{o}}$ in the $x-y$ plane, there would also be a linear dichroism effects for alignments of the polarization vector $\mathbf{E} \parallel \hat{\mathbf{o}}$ versus $\mathbf{E} \perp \hat{\mathbf{o}}$.

In Fig. 7.22 we also show three basic orientations of the unit magnetization vector $\hat{\mathbf{m}}$ in the same coordinate system. Similar to the photon angular momentum projections L_z we also define magnetization projections $m_z = \hat{\mathbf{m}} \cdot \mathbf{e}_z$ onto the x-ray propagation direction z. We then distinguish the three cases $m_z = 0, \pm 1$. The value $m_z = 0$ reflects a random magnetic orientation in the $x-y$ plane or a non-magnetic sample.

Of interest to us is how the incident polarized field is transmitted through a sample for different relative orientations of the x-ray polarization, described by unit vectors $\boldsymbol{\epsilon}_p$, and atomic magnetic moments, described by unit vectors $\hat{\mathbf{m}}$. In the following we shall phrase our phenomenological model in terms of phase or sign definitions that will directly map onto the quantum treatment in Chap. 11.

For our assumed case of cylindrical magnetic and charge symmetry around the x-ray propagation direction z, we only need to distinguish three polarization notations. The case $\boldsymbol{\epsilon}_0$ denotes linear polarization or natural polarization with $\mathbf{E}$ somewhere in the $x-y$ plane, and $\boldsymbol{\epsilon}_+$ and $\boldsymbol{\epsilon}_-$, respectively, denote circular polarization with angular momentum (photon spin) in the $+z$ and $-z$ directions. We can then define polarization dependent optical parameters for the three polarizationcases by use of these polarization labels

$$\begin{aligned} \tilde{n} &= 1 - \delta^0 + \mathrm{i}\,\beta^0 \\ \tilde{n}^{\pm} &= 1 - \delta^{\pm} + \mathrm{i}\,\beta^{\pm}, \quad \delta^{\pm} = \delta^0 \pm \Delta\delta\,, \quad \beta^{\pm} = \beta^0 \pm \Delta\beta\,. \end{aligned} \tag{7.116}$$

We also have

$$\Delta\beta = \frac{\beta^+ - \beta^-}{2}, \quad \Delta\delta = \frac{\delta^+ - \delta^-}{2} \tag{7.117}$$

$$\beta^0 = \frac{\beta^+ + \beta^-}{2}, \quad \delta^0 = \frac{\delta^+ + \delta^-}{2}\,. \tag{7.118}$$

Our definitions are illustrated in Fig. 7.23 through the energy dependent dichroic optical parameters for Fe and Co metals (also see Fig. 7.12 for Fe).

Figure 7.24 shows a plot of the energy dependent polarization independent optical parameters δ^0 and β^0 defined in (7.118) and their dichroic differences $\Delta\delta$ and $\Delta\beta$ defined in (7.117) for Fe, Co, and Ni metals.

Table 7.2 lists the L_3 resonant (peak) values for the optical parameters β^0 and δ^0 and their maximum dichroic magnetic contributions $\Delta\beta$ and $\Delta\delta$, corresponding to Fig. 7.24. The listed values should be taken to be approximate since they are sensitive to the experimental conditions, such as monochromator resolution, degree of circular polarization, magnetic alignment of the sample, or whether transmission or electron yield detection is used [72].

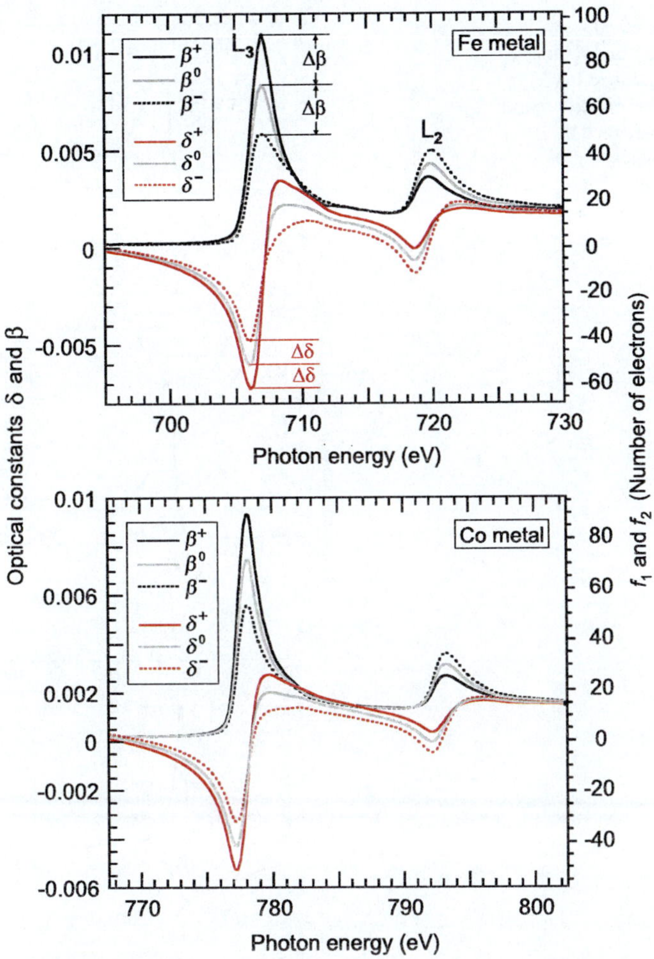

Fig. 7.23 Polarization and magnetization direction dependent optical parameters δ and β as a function of x-ray energy for Fe and Co metals. On the left we define the labels "+", "–", and "0" for the corresponding x-ray polarizations and magnetic directions. The main figure shows the energy dependent optical parameters β and δ for the three polarization labels for Fe and Co metal (left ordinate) and the related scattering factors f_1 and f_2 (right ordinate)

7.11.3 Transmission of Circularly Polarized X-Rays: XMCD

For magnetic systems with the magnetization direction aligned parallel to the x-ray propagation direction it is convenient to work with circular basis states because they are eigenstates that are not changed upon transmission through the magnetically orientated sample. Only the field amplitude is changed.

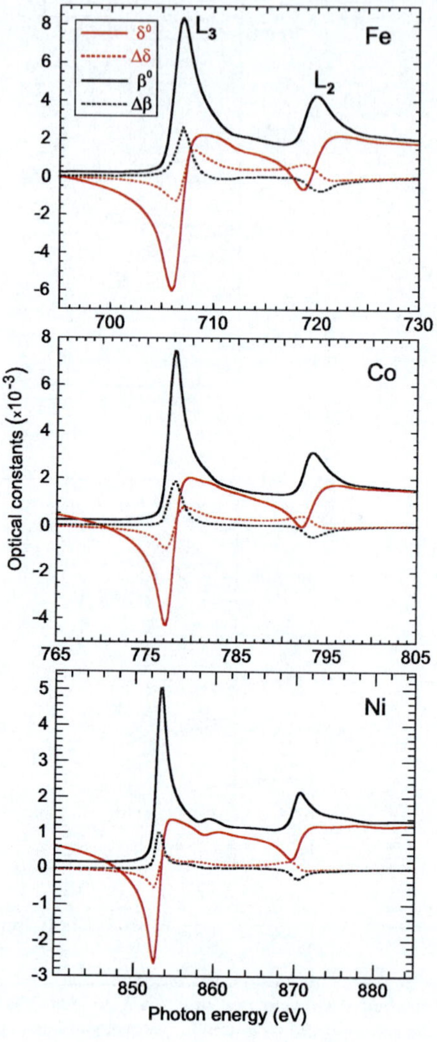

Fig. 7.24 Optical parameters δ^0 and β^0 for Fe, Co, and Ni defined in (7.118) and their dichroic differences $\Delta\delta$ and $\Delta\beta$ defined in (7.117) as a function of photon energy

In a **circular polarization basis**, the incident polarization is preserved upon transmission through a magnetic sample aligned along the x-ray propagation direction and only the amplitude of the field is reduced.

For a given polarization direction $p = \pm$ for incident circularly polarized light and magnetization direction $\hat{m}_z$, the transmitted fields are obtained by rewriting (7.23) by use of (7.117) and (7.118) as

Table 7.2 Optical parameters and their magnetic dichroism contributions at the L_3 resonance positions of Fe, Co, and Ni metals, as shown in Fig. 7.24

	ρ_a [atoms/nm^3]	$\mathcal{E}_0$ [eV]	λ_0 [nm]	β^0	δ^0	$\Delta\beta$	$\Delta\delta$
Fe	84.9	707	1.75	0.0084	−0.0060	0.0026	−0.0013
Co	90.9	778	1.59	0.0075	−0.0042	0.0019	−0.0010
Ni	91.4	853	1.45	0.0050	−0.0028	0.0010	−0.0005

Listed are the atomic number densities ρ_a, the resonance energies $\mathcal{E}_0$ and wavelengths λ_0 , the maximum values β^0 (on resonance) and δ^0 (below resonance) and the maximum dichroic contributions $\Delta\beta$ (on resonance) and $\Delta\delta$ (below resonance), assuming propagation along the magnetization direction

$$\boxed{\mathbf{E}^p_{\text{trans}} = E_0\, \mathrm{e}^{\mathrm{i}kd} \mathrm{e}^{-(\mathrm{i}\overline{\delta}+\overline{\beta})kd}\, \boldsymbol{\epsilon}_p\, \mathrm{e}^{-p\,\hat{m}_z(\mathrm{i}\Delta\delta+\Delta\beta)\,kd}\,.} \tag{7.119}$$

Here, $\overline{\delta} = \delta^0$ and $\overline{\beta} = \beta^0$ reflect polarization averages according to (7.118), and we assumed that for both polarizations the incident field amplitude has the same value E_0. For linear polarization $p = 0$ or for $\hat{m}_z = 0$ there is no magnetic effect. For circular polarization the *transmitted* intensity only depends on the whether p and $\hat{m}_z$ have the same or opposite signs. The transmitted *intensity* is obtained from the fields as

$$\boxed{I^{p,\hat{m}_z}_{\text{trans}} = I^p_0\, \mathrm{e}^{-2(\overline{\beta}+p\,\hat{m}_z\Delta\beta)kd}\,.} \tag{7.120}$$

The polarization and magnetization labels p and $\hat{m}_z$ with possible values ± 1 enter equivalently. In our sign convention the transmitted intensity is smaller for $p = \hat{m}_z$ and larger for $p = -\hat{m}_z$. The opposite is the case for the *absorbed* intensity since

$$I^{p,\hat{m}_z}_{\text{abs}} = I^p_0 - I^{p,\hat{m}_z}_{\text{trans}}\,. \tag{7.121}$$

When normalized to the same incident intensity ($I^+_0 = I^-_0 = I_0$), we define the XMCD difference intensity as

$$\Delta I_{\text{XMCD}} = I^{p=\hat{m}_z}_{\text{trans}} - I^{p=-\hat{m}_z}_{\text{trans}} = I^{p=-\hat{m}_z}_{\text{abs}} - I^{p=\hat{m}_z}_{\text{abs}}\,. \tag{7.122}$$

For our sign convention we obtain

$$\boxed{\Delta I_{\text{XMCD}} = -2I_0\, \mathrm{e}^{-2\overline{\beta}kd} \sinh[2\Delta\beta\, kd]\,.} \tag{7.123}$$

The sign of the difference signal is therefore determined by the sign of $\Delta\beta$. Since $\Delta\beta > 0$ at the L−3-resonance of Fe, Co, and Ni, the dichroic contrast ΔI_{XMCD} is negative, in agreement with the usual sign convention [1].

7.11.4 *Transmission of Linearly Polarized X-Rays: XMLD*

In order to see what happens to *linearly* polarized light upon transmission through a sample with magnetization direction $\hat{\mathbf{m}}$ parallel to the x-ray propagation direction, we assume that the x-ray **E**-vector lies along an arbitrarily chosen x axis. The trick is that we can deduce the unknown transmission behavior of linearly polarized light in the sample from the known transmission of circularly polarized light discussed above.

To do so we use the relationships between the unit polarization vectors $\boldsymbol{\epsilon}_\pm$ for circular and $\boldsymbol{\epsilon}_x, \boldsymbol{\epsilon}_y$ for linear light given by (7.115). For incident x-rays linearly polarized along x we can write the transmitted field in terms of two equal circular components given by (7.119) according to

$$\mathbf{E}_{\text{trans}} = \frac{1}{\sqrt{2}}\left(\mathbf{E}^{-}_{\text{trans}} - \mathbf{E}^{+}_{\text{trans}}\right)$$
$$= \frac{1}{\sqrt{2}} E_0\, \mathrm{e}^{\mathrm{i}kd}\, \mathrm{e}^{-(\mathrm{i}\delta^0+\beta^0)kd}\left\{\boldsymbol{\epsilon}_-\, \mathrm{e}^{\hat{m}_z(\mathrm{i}\Delta\delta+\Delta\beta)kd} - \boldsymbol{\epsilon}_+\, \mathrm{e}^{-\hat{m}_z(\mathrm{i}\Delta\delta+\Delta\beta)kd}\right\}. \tag{7.124}$$

Note that the optical parameters are still defined in terms of the circular polarization response given by (7.117) and (7.118). The transmitted field has the form of an elliptically polarized wave of the general form (3.32). It can be rewritten in the linear polarized form (3.31) by use of (7.115) with the result

$$\boxed{\mathbf{E}_{\text{trans}} = E_0\, \mathrm{e}^{\mathrm{i}kd}\mathrm{e}^{-(\mathrm{i}\delta^0+\beta^0)kd}\left\{\boldsymbol{\epsilon}_x \cos[(\Delta\delta - \mathrm{i}\Delta\beta)kd] + \boldsymbol{\epsilon}_y \hat{m}_z \sin[(\Delta\delta - \mathrm{i}\Delta\beta)kd]\right\}.} \tag{7.125}$$

The incident field, polarized along $\boldsymbol{\epsilon}_x$, is converted into an elliptically polarized transmitted field with complex x and y components. The first term due to the charge response preserves the polarization along x, while the second magnetic term, which depends on $\hat{m}_z$, rotates it by 90°. The latter gives rise to the famous Faraday rotation discussed in more detail below.

In contrast to (7.123), the transmitted intensity for incident *linearly* polarized x-rays is found to be

$$I_{\text{trans}} = I_0\, \mathrm{e}^{-2\beta^0 kd}. \tag{7.126}$$

It is independent of $\Delta\beta$, and there is no magnetic dichroism effect.

If instead one measures the XAS spectrum for a sample of magnetization $\hat{m}_z$ by adding the intensities obtained with $\pm$ helicities, one obtains from (7.120)

$$I_{\text{trans}} = \frac{I^{p=\hat{m}_z}_{\text{trans}} + I^{p=-\hat{m}_z}_{\text{trans}}}{2} = I_0\, \mathrm{e}^{-2\beta^0 kd} \cosh(2\Delta\beta kd) \simeq I_0\, \mathrm{e}^{-2\beta^0 kd}. \tag{7.127}$$

On the far right we have expanded $\cosh(2\Delta\beta kd) = 1 + 2(\Delta\beta)^2k^2d^2$ and only kept the first order term in $\Delta\beta$. This result is the same as (7.126).

7.11.5 *Faraday Rotation*

The details of the x-ray Faraday rotation are best demonstrated by considering the transmitted field given by (7.125) for a thin film whose thickness is about one x-ray absorption length or $d \simeq 20\,\text{nm}$. One may then utilize the fact that according to Fig. 7.24 the resonant dichroic parameters $\Delta\delta$, $\Delta\beta$, which account for the magnetic response, are smaller than the non-dichroic parameters δ^0, β^0 representing the charge response. This allows us to expand the cos and sin functions in (7.125) to first order in $\Delta\delta$, $\Delta\beta$ to obtain,

$$\boxed{\mathbf{E}_{\text{trans}} \simeq E_0\,\mathrm{e}^{ikd}\mathrm{e}^{-(\mathrm{i}\delta^0+\beta^0)kd}\left\{\boldsymbol{\epsilon}_x + \boldsymbol{\epsilon}_y\,\hat{m}_z(\Delta\delta - \mathrm{i}\Delta\beta)kd\right\}} \quad (7.128)$$

For a magnetic sample of orientation $\hat{m}_z = \pm 1$ along the propagation direction z, the transmitted light has a real component along x and a complex component along y. Its effect on the polarization of the transmitted field may be nicely illustrated by considering the changes of the real, $\Delta\delta$, and imaginary, $\Delta\beta$, dichroic contributions near a resonance. We consider the case of the L_3 resonance of Co metal shown enlarged in Fig. 7.25a.

Right before the resonance we have $\Delta\delta < 0$, $\Delta\beta \simeq 0$, indicated by ① in Fig. 7.25a. Then (7.128) becomes

$$\boxed{\mathbf{E}_{\text{trans}} \simeq E_0\,\mathrm{e}^{ikd}\mathrm{e}^{-(\mathrm{i}\delta^0+\beta^0)kd}\left\{\boldsymbol{\epsilon}_x + \boldsymbol{\epsilon}_y\,\hat{m}_z\Delta\delta\,kd\right\}\,.} \quad (7.129)$$

The transmitted field has real x and y components so that the light remains *linearly polarized*, but the polarization vector has rotated by an angle φ, called the *Faraday rotation angle*, given by

$$\text{Faraday rotation}: \ \tan\varphi = \frac{E_y}{E_x} = \hat{m}_z kd\Delta\delta\,. \quad (7.130)$$

On the bottom left in Fig. 7.25 we show the case ① of $\Delta\beta \simeq 0$, $\Delta\delta < 0$ under the assumption $\hat{m}_z = -1$. The negative values of both $\Delta\delta$ and $\hat{m}_z$ cause φ to be positive as shown in the figure.

The other limiting case of (7.128) occurs at the resonance position where $\Delta\delta = 0$, $\Delta\beta > 0$, as indicated by ② in Fig. 7.25a. The transmitted field becomes

$$\boxed{\mathbf{E}_{\text{trans}} \simeq E_0\,\mathrm{e}^{ikd}\mathrm{e}^{-(\mathrm{i}\delta^0+\beta^0)kd}\left\{\boldsymbol{\epsilon}_x - \mathrm{i}\boldsymbol{\epsilon}_y\,\hat{m}_z\Delta\beta\,kd\right\}\,.} \quad (7.131)$$

In this case the y component is imaginary and the transmitted light has become *elliptically polarized.* The **E**-vector endpoint rotates on an ellipse with the rotation sense

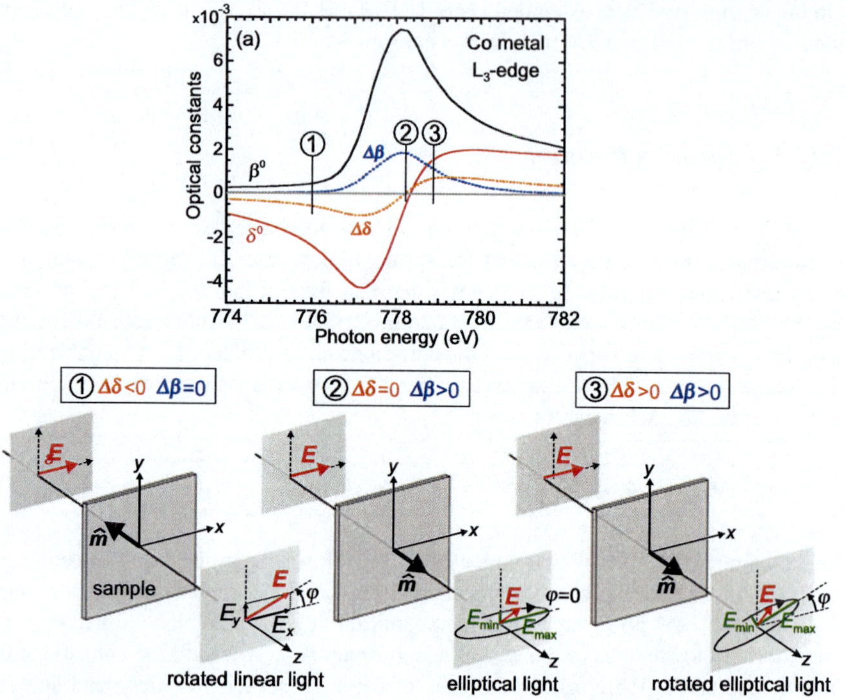

Fig. 7.25 **a** Optical parameters δ^0 and β^0 and their magnetic dependences $\Delta\delta$ and $\Delta\beta$ for the L_3 resonance of Co metal. Below we show the transmitted field polarization for three photon energies indicated in (**a**). At energy ① we have $\Delta\delta < 0$, $\Delta\beta \simeq 0$. We show the polarization for $\hat{m}_z = -1$ given by (7.129). At energy ② we have $\Delta\delta = 0$, $\Delta\beta > 0$, and we show the case $\hat{m}_z = +1$ according to (7.131). For energy ③ we have $\Delta\delta$, $\Delta\beta > 0$ and we show the case for $\hat{m}_z = +1$ described by (7.128)

defined by the sign of the imaginary component. The relative size of the principal axes of the ellipse define the *ellipticity* ϵ according to

$$\text{Ellipticity}: \ \tan\epsilon = \frac{|\mathbf{E}_y|}{|\mathbf{E}_x|} = \frac{E_{\text{min}}}{E_{\text{max}}} = \Delta\beta\, kd\,. \tag{7.132}$$

The elliptical light may be described as having either a dominant linear or circular component. Owing to the imaginary y component, the Faraday rotation angle averages to zero over a full **E**-vector rotation (one wave cycle). In the lower middle of Fig. 7.25 we show the case ②, where $\Delta\delta = 0$, $\Delta\beta > 0$ and assume $\hat{m}_z = +1$. The rotation sense of **E** is indicated by a black arrow, so that the transmitted light has a net angular momentum $L_z = -\hbar$.

Finally, we show the case indicated by ③ in (a) with $\Delta\beta$, $\Delta\delta > 0$ on the lower right of Fig. 7.25 under the assumption $\hat{m}_z = +1$. This case is expressed by (7.128), and the transmitted light is elliptically polarized. The major axis of the ellipse is rotated by a Faraday angle φ which is positive, as for case ① due to the signs of $\Delta\delta$

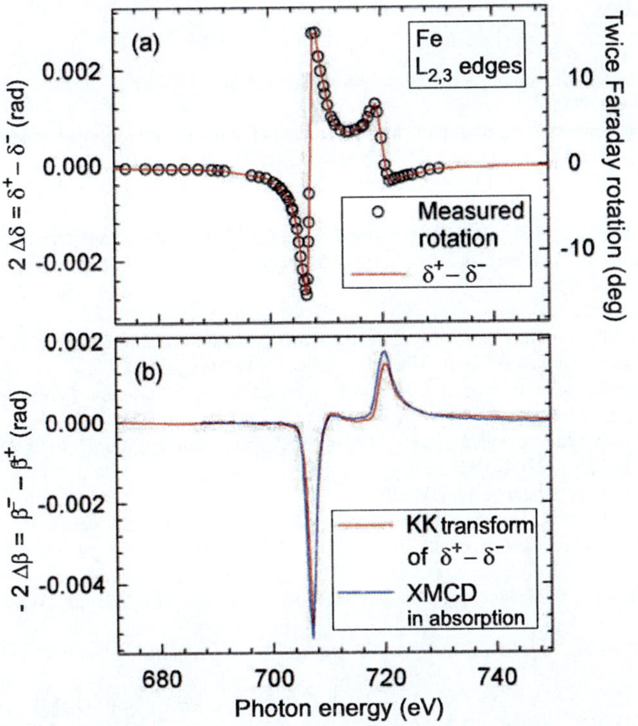

Fig. 7.26 Measurement of the photon energy dependence of $\delta^+ - \delta^-$ and $\beta^- - \beta^+$ by Kortright and Kim [25] for Fe metal. **a** Faraday rotation measured through a 32 nm Fe film at 30° grazing incidence. On the left ordinate $2\Delta\delta = \delta^+ - \delta^-$ is given in radians and on the right the corresponding rotation angle in degrees. Note that it corresponds to twice the Faraday rotation angle. Solid circles are measured data-points, and the line is a smooth fit of the data by (7.130). **b** Experimental XMCD spectrum (blue curve) in comparison with the Kramers-Kronig transformed data in (**a**) (red curve)

and $\hat{m}_z$. The rotation sense of **E** indicated by a black arrow results in a net angular momentum $L_z = -\hbar$.

The strongly energy dependent behavior of the dichroic optical parameters in conjunction with possible values $\hat{m}_z = \pm 1$ results in a rich polarization dependence of the transmitted field illustrated in Fig. 7.25. The behavior has been studied experimentally for Fe, Co, and Ni metals by several groups [25, 73–75]. As an example, we show in Fig. 7.26 the measurement of $2\Delta\delta = \delta^+ - \delta^-$ and $-2\Delta\beta = \beta^- - \beta^+$ for Fe metal by Kortright and Kim [25], which are seen to support the theory.

References

1. J. Stöhr, H.C. Siegmann, *Magnetism: From Fundamentals to Nanoscale Dynamics* (Springer, Heidelberg, 2006)
2. P. Bruno, Physical origins and theoretical models of magnetic anisotropy, in *Magnetismus von Festkörpern und Grenzflächen* (KFA Jülich, Jülich, 1993), p. 24.1
3. M.A. Ordal, R.J. Bell, J.R.W. Alexander, L.A. Newquist, M.R. Querry, Appl. Opt. **27**, 1203 (1988)
4. M.A. Ordal, R.J. Bell, J.R.W. Alexander, L.L. Long, M.R. Querry, Appl. Opt. **24**, 4493 (1985)
5. F. Jorgensen, *The Complete Handbook of Magnetic Recording* (TAB Books, Mc-Graw Hill, New York, 1995)
6. R.M. Bozorth, *Ferromagnetism* (Wiley-IEEE Press, New York, 1993)
7. B.D. Cullity, *Introduction to Magnetic Materials* (Addison-Wesley, Reading, 1972)
8. J.D. Jackson, *Classical Electrodynamics* (Wiley, New York, 1999)
9. D. Attwood, A. Sakdinawat, *X-Rays and Extreme Ultraviolet Radiation: Principles and Applications*, 2nd edn. (Cambridge University Press, Cambridge, 2017)
10. I. Tudosa, C. Stamm, A.B. Kashuba, F. King, H.C. Siegmann, J. Stöhr, G. Ju, B. Lu, D. Weller, Nature **428**, 831 (2004)
11. C. Dornes et al., Nature **565**, 209 (2019)
12. D.W. Lynch, W.R. Hunter, An introduction to the data for several metals, in *Handbook of Optical Constants, Vol. II*, ed. by E.D. Palik (Academic Press, 1997), pp. 341–419
13. B.L. Henke, E.M. Gullikson, J.C. Davis, At. Data Nucl. Data Tables **54**, 181 (1993)
14. J. Als-Nielsen, D. McMorrow, *Elements of Modern X-Ray Physics*, 2nd edn. (Wiley, New York, 2011)
15. J.M. Virgili, Y. Tao, J.B. Kortright, N.P. Balsara, R.A. Segalman, Macromolecules **40**, 2092 (2007)
16. H. Yan, C. Wang, A.R. McCarn, H. Ade, Phys. Rev. Lett. **110**, 177401 (2013)
17. T. Inagaki, E.T. Arakawa, R.N. Hamm, M.W. Williams, Phys. Rev. B **15**, 3243 (1977)
18. J. Stöhr, *NEXAFS Spectroscopy* (Springer, Heidelberg, 1992)
19. S.C. Mayo, A.W. Stevenson, S.W. Wilkins, Materials **937** (2012)
20. C. Jacobsen, *X-Ray Microscopy* (Cambridge University Press, Cambridge, 2020)
21. https://henke.lbl.gov/optical_constants/
22. M.O. Krause, J. Phys. Chem. Ref. Data **8**, 307 (1979)
23. L. Young et al., Nature **466**, 56 (2010)
24. G. Doumy et al., Phys. Rev. Lett. **106**, 083002 (2011)
25. J.B. Kortright, S.K. Kim, Phys. Rev. B **62**, 12216 (2000)
26. R.G. Newton, American. J. Phys. **44**, 639 (1976)
27. R. Loudon, *The Quantum Theory of Light*, 3rd edn. (Clarendon Press, Oxford, 2000)
28. C.G. Darwin, Phil. Mag. **27**, 315 (1914)
29. C.G. Darwin, Phil. Mag. **27**, 675 (1914)
30. J.A. Prins, Z. Physik **63**, 477 (1930)
31. R. Rashed, *A Pioneer in Anaclastics: Ibn Sahl on burning mirrors and lenses*, Isis **81**, 464 (1990)
32. B.L. Henke et al., J. X-Ray Sci. Technol. **2**, 17 (1990)
33. D.F.M. Arago, Mem. Inst. Fr. (Paris) **12**, 93 (1811)
34. J.B. Biot, Mem. Inst. Fr. (Paris) **13**, 218 (1812)
35. L. Pasteur, Annales de chimie et de physique **28**, 56 (1850)
36. A. Cotton, C. R. Acad. Sci. (Paris) **120**, 989 (1895)
37. L.D. Barron, *Molecular Light Scattering and Optical Activity*, 2nd edn. (Cambridge University Press, Cambridge, 2004)
38. M. Faraday, Phil. Trans. Roy. Soc. London **1**, 136 (1846)
39. M. Faraday, Philos. Mag. **3**, 321 (1877)
40. A. Hubert, R. Schaefer, *Magnetic Domains* (Springer, Berlin, 1998)

41. Z.Q. Qiu, S.D. Bader, Rev. Sci. Instrum. **71**, 1243 (2000)
42. B. Hillebrands, K. Ounadjela (eds.), Spin Dynamics in Confined Magnetic Structures I. (Springer, Berlin, 2002)
43. B. Hillebrands, K. Ounadjela (eds.), Spin Dynamics in Confined Magnetic Structures II. (Springer, Berlin, 2003)
44. C.T. Chen, L.H. Tjeng, J. Kwo, H.L. Kao, P. Rudolf, F. Sette, R.M. Fleming, Phys. Rev. Lett. **68**, 2543 (1992)
45. J. Goulon, C. Goulon-Ginet, A. Rogalev, V. Gotte, C. Malgrange, C. Brouder, J. Chem. Phys. **108**, 6394 (1998)
46. J. Lüning, F. Nolting, A. Scholl, H. Ohldag, J. Seo, J. Fompeyrine, J.P. Locquet, J. Stöhr, Phys. Rev. B **67**, 214433 (2003)
47. C.T. Chen, Y.U. Idzerda, H.J. Lin, N.V. Smith, G. Meigs, E. Chaban, G.H. Ho, E. Pellegrin, F. Sette, Phys. Rev. Lett. **75**, 152 (1995)
48. J. Stöhr, K. Baberschke, R. Jaeger, R. Treichler, S. Brennan, Phys. Rev. Lett. **47**, 381 (1981)
49. J. Stöhr, R. Jaeger, Phys. Rev. B **26**, 4111 (1982)
50. J. Stöhr, M.G. Samant, J. Elec. Spectrosc. Relat. Phenom. **98**, 189 (1999)
51. L. Alagna, T. Prosperi, S. Turchini, J. Goulon, A. Rogalev, C. Goulon-Ginet, C.R. Natoli, R.D. Peacock, B. Stewart, Phys. Rev. Lett. **80**, 4799 (1998)
52. J. Goulon, A. Rogalev, F. Wilhelm, N. Jaouen, C. Goulon-Ginet, C. Brouder, J. Phys.: Condens. Matter **15**, S633 (2003)
53. S. Turchini, N. Zema, S. Zennaro, L. Alagna, B. Stewart, R.D. Peacock, T. Prosperi, J. Am. Chem. Soc. **126**, 4532 (2004)
54. S. Villaume, P. Norman, Chirality **21**, E13 (2009)
55. S. Bernadotte, A.J. Atkins, C.R. Jacob, J. Chem. Phys. **137**, 204106 (2012)
56. Y. Izumi, M. Tanabe, A. Imazu, A. Mimoto, M. Tanaka, A. Agui, T. Muro, K. Nakagawa, J. Am. Chem. Soc. **138**, 074305 (2013)
57. O. Takahashi, M. Kimoto, L.G.M. Pettersson, Chem. Phys. **450–451**, 109 (2015)
58. B.T. Thole, G. van der Laan, G.A. Sawatzky, Phys. Rev. Lett. **55**, 2086 (1985)
59. G. van der Laan, B.T. Thole, G.A. Sawatzky, J.B. Goedkoop, J.C. Fuggle, J.M. Esteva, R. Karnatak, J.P. Remeika, H.A. Dabkowska, Phys. Rev. B **34**, 6529 (1986)
60. G. Schütz, W. Wagner, W. Wilhelm, P. Kienle, R. Zeller, R. Frahm, G. Materlik, Phys. Rev. Lett. **58**, 737 (1987)
61. A. Rogalev, F. Wilhelm, N. Jaouen, J. Goulon, J. P. Kapler, in *Magnetism: A Synchrotron Radiation Approach*, ed. by E. Beaurepaire, H. Bulou, F. Scheurer, J.-P. Kappler. Lecture Notes in Physics, vol. 697 (Springer, Berlin Heidelberg New York, 2006), p. 71
62. J. Goulon, A. Rogalev, F. Wilhelm, C. Goulon-Ginet, P. Carra, D. Cabaret, C. Brouder, Phys. Rev. Lett. **88**, 237401 (2002)
63. J. Goulon, A. Rogalev, C. Goulon-Ginet, G. Benayoun, L. Paolasini, C. Brouder, C. Malgrange, P.A. Metcalf, Phys. Rev. Lett. **85**, 4385 (2000)
64. J. Michl, E.W. Thulstrup, *Spectroscopy with Polarized Light* (VCH Publishers Inc., New York, 1995)
65. E.W. Thulstrup, J. Michl, *Elementary Polarization Spectroscroscopy* (VCH Publishers Inc., New York, 1989)
66. K. Takatoh, M. Hasegawa, M. Koden, N. Itoh, R. Hasegawa, M. Sakamoto, *Alignment Technologies and Applications of Liquid Crystal Devices* (Taylor and Francis, London, 2005)
67. K. Weiss, C. Wöll, E. Böhm, B. Fiebranz, G. Forstmann, B. Peng, V. Scheumann, D. Johannsmann, Macromolecules **31**, 1930 (1998)
68. J. Stöhr, M.G. Samant, A. Cossy-Favre, J. Díaz, Y. Momoi, S. Odahara, T. Nagata, Macromolecules **31**, 1942 (1998)
69. W. Maier, A. Saupe, Z. Naturfor. A **13**, 564 (1958)
70. P.G. De Gennes, J. Prost, *The Physics of Liquid Crystals* (Clarendon Press, Oxford, 1993)
71. J. Stöhr, M.G. Samant, J. Lüning, A.C. Callegari, P. Chaudhari, J.A. Doyle, J.A. Lacey, S.A. Lien, S. Purushothaman, J.L. Speidell, Science **292**, 2299 (2001)
72. R. Nakajima, J. Stöhr, Y. Idzerda, Phys. Rev. B **59**, 6421 (1999)

73. H.C. Mertins, F. Schäfers, X.L. Cann, A. Gaupp, W. Gudat, Phys. Rev. B **61**, R874 (2000)
74. J. Kuneš, P.M. Oppeneer, H.C. Mertins, F. Schäfers, A. Gaupp, W. Gudat, P. Novaák, Phys. Rev. B **64**, 174417 (2001)
75. H.C. Mertins, S. Valencia, A. Gaupp, W. Gudat, P.M. Oppeneer, C.M. Schneider, Appl. Phys. A **80**, 1011 (2005)

Chapter 8
Classical Diffraction and Diffractive Imaging

8.1 Introduction and Chapter Overview

In contrast to conventional light, x-rays have allowed us to see the invisible. Invisibility may be caused by the lack of penetration of visible light, preventing us to see below the outermost skin of matter. It also comes in the form of objects that are smaller than the wavelength of visible light, due to the diffraction limit. x-rays are best known because they can overcome both these limitations, prominently exploited over the last 100+ years in medical x-ray imaging and in x-ray crystallography. Today we still distinguish between the two different types of x-ray imaging which strongly overlap on the nanometer length scale [1].

Conventional x-ray imaging provides a real space picture of a sample with a spatial resolution limited by image forming lenses. Although the term "x-ray microscopy" is still used, today's spatial resolution is well below the micrometer size. The strength of real space imaging is that the recorded image directly reflects the micro- or nanostructure of the sample. It is typically the preferred technique when the structures of interest are larger than about 100 nm, which is about 10 times larger than the state-of-the-art 10 nm resolution [2].

The present chapter starts with a short review of the development and capabilities of real space x-ray microscopy in Sect. 8.2. We discuss modern implementations of x-ray microscopes and emphasize the utilization of tuning the x-ray energy for atomic and chemical specificity and use of the polarization for the study of charge and spin-based orientational order on the nanoscale.

Section 8.3 introduces *diffractive imaging* or *lensless imaging* where the limitation of x-ray lenses is overcome with the spatial resolution only limited by the wavelength itself, i.e. the diffraction limit. Now the recorded "images" are the reciprocal space diffraction patterns. Diffractive x-ray imaging is typically employed in the range below about 100 nm down to the atomic size of about 0.1 nm. The history of its theoretical formulation, based on the classical wave theory and the Huygens–Fresnel principle, is outlined with the key result being the Rayleigh-Sommerfeld diffraction formula.

J. Stöhr, *The Nature of X-Rays and Their Interactions with Matter*, Springer Tracts in Modern Physics 288, https://doi.org/10.1007/978-3-031-20744-0_8

In practice, approximations need to be made in applications of the Rayleigh-Sommerfeld formula, and in Sect. 8.4 we discuss consecutive simplifications that lead to the distinction of near-field Fresnel and far-field Fraunhofer diffraction. The Fraunhofer diffraction formula is particularly important since it links the *field* (not intensity) distributions in the source and detector planes through a Fourier transform.

In Sect. 8.5 we touch upon one of the most profound scientific debates that continues to this day, the wave-particle ambiguity. The wave formulation of diffraction is still nearly exclusively taught in conjunction with the magical Huygens–Fresnel principle. This formulation, which allows light to go around corners, expressed by Sommerfeld as the deviation from rectilinear propagation, has been remarkably successful in explaining diffraction phenomena over a wavelength range from meters to sub-nanometers, covering more than ten orders of magnitude. The wave-particle ambiguity is briefly discussed by outlining the description of diffraction through Feynman's probability amplitudes and de Broglie's pilot wave theory.

In Sect. 8.6 we discuss consequences of the Fraunhofer diffraction formula, such as the Rayleigh diffraction limit, the Arago-Fresnel-Poisson bright spot, and Babinet's intriguing theorem.

In Sect. 8.7 we formulate diffraction by matter in the Fraunhofer approximation. The atomic response is phrased in terms of polarization dependent optical parameters, introduced in Sect. 7.11.1, whose dichroic dependence is taken up to second order. Emphasis is placed on resonant effects at absorption edges that allow the detection of orientational order in nanoscale charge and magnetic domains and their distinction through polarization. The derived form of the diffraction pattern up to second order in the scattering factors or optical parameters will be utilized in last Sect. 8.11 where its close relation is shown to the modern *Multi-wavelength Anomalous Diffraction* or MAD technique.

The remaining part of the chapter is devoted to how the reciprocal space diffraction pattern can be inverted into a real space image. We start by elucidating the origin of the famous "*phase problem*" in Sect. 8.8 through a simple example that clarifies the all-important effect of the phase in reconstruction of the real space image.

Section 8.9 is devoted to the discussion of the most elegant phasing technique for 2D structures, *Fourier transform holography* (FTH). We show how the holographic intensity image can be directly inverted to obtain a real image of the object. We present an analytical solution of the phase problem for a double-slit object in the presence of a laterally displaced reference source and apply the technique for reconstructing the image of nanoscale magnetic domains.

Section 8.10 gives an overview of non-holographic solutions to the phase problem. We review the historical development toward overcoming the phase problem in x-ray crystallography, which has certainly resulted in one of the great scientific accomplishment of the twentieth century, the elucidation of the atomic structure of matter. We only outline the physical principles of modern phasing techniques without discussing computer algorithms. The interested reader is referred to the recent comprehensive book by Jacobsen [1].

Crystallography relies on the presence of periodicity of the lattice in crystals or the periodicity of artificial stacks of macromolecules (see Fig. 1.14), resulting in diffraction patterns dominated by sharp Bragg peaks [3–5]. We discuss the evolution of phasing from so-called direct methods, pioneered by Hauptman and Karle in the 1950s [6] to modern techniques. They depend on the increased scattering by heavy atoms that are either native or substituted, referred to as *multiple isomorphous replacement* or MIR. Of particular importance is the *Multi-wavelength Anomalous Diffraction* or MAD technique which exploits the change of the scattering phase at the heavy atom absorption edges.

The section also discusses diffractive imaging of samples without long range periodic order. The Bragg peaks are then replaced by a so-called *speckle pattern*, and new phasing techniques are used which rely on computer programs with iterative algorithms that include specific boundary conditions also called "constraints" to guide the solution. We point out that in practice, all diffraction patterns are not sampled continuously but discretely because of the finite pixel size of detectors and discuss the requirement of sampling the diffraction patterns finely enough without loss of information. The section is concluded with a brief review of *ptychography*, which relies on sampling the diffraction pattern of an object through overlapping regions. The overlap provides a consistency check or constraint in the iterative phase retrieval.

Section 8.11 discusses in more detail resonant phasing techniques related to MAD because they go the heart of x-ray matter interactions. They naturally evolved when tunable synchrotron radiation became available and gained additional momentum with the increase of coherent intensity at modern sources. We present the MAD formalism and extend it to the case of non-crystalline samples, emphasizing the beautiful interplay near resonances of the absorptive imaginary and phase-related real part of the atomic scattering factors or the related optical parameters.

8.2 Real Space X-Ray Imaging

In the following section we briefly review *real space* x-ray microscopy techniques, with emphasis on the use of atomic resonances that provide elemental, chemical, and charge and spin sensitivity. The combination is often called *x-ray spectro-microscopy* which has been extensively covered in other books and review articles [1, 5, 7–9]. By now this technique is routinely used as evidenced by the nanoscale images presented in the Introduction of our book in Sect. 1.4.3. We will not cover the details of x-ray lenses and microscopes, limiting our discussions to the three main versions of spectro-microscopy used today.

The rich nanoscale landscape of interest in x-ray imaging is illustrated in Fig. 8.1. It extends from the optical diffraction limit to the atomic length scale.

The great advantage of real space x-ray imaging is that one can directly produce pictures of invisible objects that are smaller than the wavelength of visible light or the diffraction limit, indicated by the rainbow in Fig. 8.1. Polarization dependent x-ray microscopy also allows us to record images of spin-based structures that our eyes cannot distinguish even if they are larger than the diffraction limit. In some cases the nanostructures lack long range order like the "grains" in materials shown in the

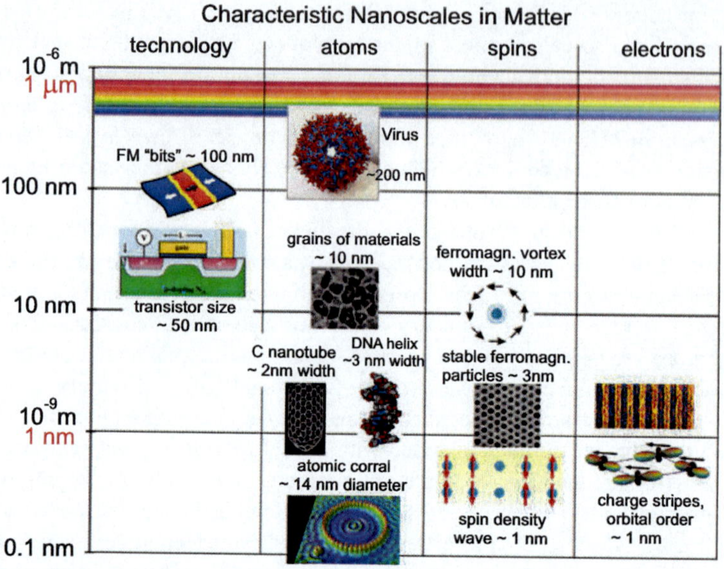

Fig. 8.1 Overview of structures of interest in the nanoworld, ranging from the optical diffraction limit (top) to the atomic size (bottom)

center of the figure. Also of interest are images of transient structures in response to a stimulation as shown in Fig. 1.22, or emerge as domains during fleeting moments in equilibrium, as in water [10]. Today, the observation of nanoscale dynamics down to the femtosecond timescale (see Fig. 1.12) is becoming increasingly important [8].

8.2.1 X-Ray Microscopes

The spatial resolution of real space x-ray microscopy has significantly improved over the years as illustrated in Fig. 8.2, taken from a review by Kirz and Jacobsen [11], which outlines the historical development (also see [1]).

As indicated in the figure, the development of photon-in/photon-out soft x-ray microscopy has been driven by the improvement in Fresnel zone plates (FZPs). In the hard x-ray range, the manufacturing of FZPs becomes more challenging since larger thicknesses are required due to increased transmission through the opaque regions. Their use has therefore been augmented by use of Kirkpatrick-Baez mirror pairs, multilayer diffractive optics, or compound refractive lenses. In the latter case one utilizes the fact that at x-ray energies the index of refraction is slightly less than unity and the refractive index δ is larger than the absorptive index β, as shown in Figs. 7.5 and 7.7. The optical glass lens in air is therefore replaced by an x-ray "air lens" incorporated into a low-Z material with low absorption [12]. More recently, multilayer lenses have been fabricated to create 2D focal spots below 10 nm [2].

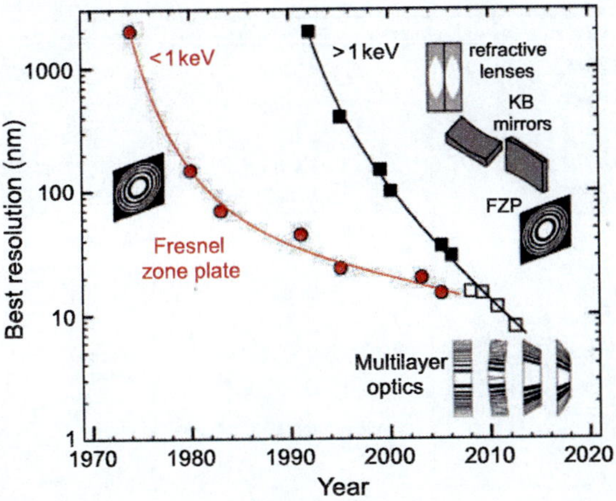

Fig. 8.2 Improvement of real space x-ray imaging methods with time [11]. For soft x-rays (<1 keV) the improvement in spatial resolution is due to the use of Fresnel zone plates (FZP) (red circles and curve). In the hard x-ray range, the improvements due to FZPs, Kirkpatrick-Baez (KB) mirrors, and compound refractive optics are indicated by black squares, while recent advances through multilayer optics of various shapes are indicated by open black squares

> Conventional x-ray microscopy relies on reflective, refractive, or diffractive elements ("lenses") to form an enlarged image of a object. In practice, the spatial resolution is limited by the lenses and is typically larger than the fundamental diffraction limit determined by the wavelength.

In the soft x-ray region, transmission imaging which requires thin film samples may be replaced by utilizing electron lenses to image photoelectrons emitted by solid samples in *photoemission electron microscopy* (PEEM). Also, for soft x-ray imaging one typically combines microscopy (spatial resolution) with spectroscopy (energy resolution) in so-called *spectro-microscopy*. Such studies allow the investigation of the atomic, chemical, or spin dependent structure of regions down to nanometer dimensions. Three common experimental schemes for soft x-ray spectro-microscopy studies are illustrated in Fig. 8.3. The techniques have been reviewed in more detail in [1, 13, 14].

In *scanning x-ray microscopy* (STXM), illustrated in Fig. 8.3a, a monochromatic x-ray beam is focused to the smallest possible spot size and the x-ray intensity transmitted through the scanned sample is monitored as a function of the focused beam position on the sample. One may also measure the fluorescent x-ray or electron intensity from the sample, but measurement of the transmitted intensity is most common. In this approach the energy resolution is given by the monochromator in the beam line (not shown) and the spatial resolution is determined by the size of the x-ray

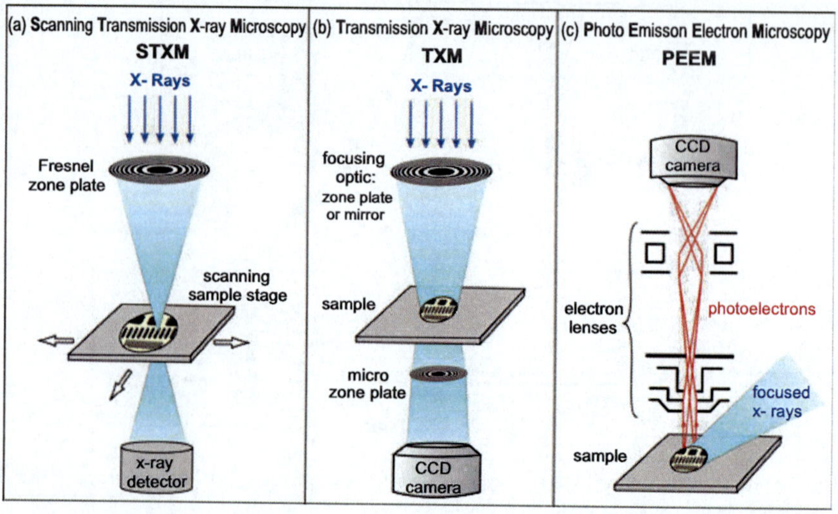

Fig. 8.3 Schematic of three x-ray microscopy methods as discussed in the text. **a** In scanning transmission x-ray microscopy, STXM, a monochromatic x-ray beam is focused to a small x-ray spot and the sample is scanned relative to the x-ray focal spot. **b** In transmission x-ray microscopy, TXM, the incident beam is focused by a condenser zone plate or a mirror onto the sample and a microzone plate generates a magnified image of the illuminated sample area. **c** In photoemission electron microscopy,PEEM, the x-rays are typically focused by a shaped mirror and a magnified image of the illuminated area is created by imaging the emitted photoelectrons with an electron microscope column

spot. In practice, zone plate focusing is used since it produces the smallest spot sizes, in the range of 20–30 nm. In this case the x-rays only need to be laterally coherent over the size of the zone plate (<10 mm). Because the focal length of the zone-plate lens changes with photon energy, for spectroscopic studies the sample position also needs to be scannable along the beam direction. STXM is "bulk" sensitive, in the sense that the transmitted intensity is determined by the entire thickness of the sample. By changing the transmission angle through the sample, the complete 3D structure of small samples may be obtained by *x-ray tomography*.

Transmission x-ray microscopy (TXM), shown in Fig. 8.3b, is a full field imaging technique. The incident beam is focused onto the sample by a condenser zone plate or a mirror to match the field of view of a microzone plate that generates a magnified image of the illuminated area onto a CCD camera. A condenser zone plate is typically used in conjunction with a *bending magnet* beam line whose lateral coherence cone is smaller than the entire zone plate. The total focal area on the sample then consists of an incoherent superposition of different coherently illuminated zone-plate regions.[1] The zone plate also monochromatizes the incident polycromatic beam with moderate resolving power ($E/\Delta E \approx 200$). For *undulator* beam lines, a monochro-

[1] This is similar to the case in Fig. 8.5b

matic incident beam is used in conjunction with a focusing mirror. In this case, use of a condenser zone plate is unfavorable since the lateral undulator coherence cone typically covers the hole zone plate. This leads to coherent illumination of the sample area and causes diffractive speckle structure that is superimposed on the image created by the microzone plate. As in STXM, the spatial resolution is determined by the microzone plate and a resolution of 15 nm has been obtained [15].

Photoemission electron microscopy (PEEM), illustrated in Fig. 8.3c, is based on x-rays-in/electrons-out. The sample is illuminated by a monochromatic x-ray beam that is typically focused by a mirror to match the field of view of a photoelectron microscope column. The electron imaging column may be entirely electrostatic or contain magnetic focusing elements. The energy resolution is determined by an x-ray monochromator in the beam line, and the spatial resolution is determined by the electron optics. Modern PEEM microscopes incorporate an electron energy filter to reduce chromatic aberrations and increase the spatial resolution to a value comparable to that in STXM and TXM [14].

Owing to the fierce competition of x-ray microscopy with electron microscopy and other imaging techniques down to atomic resolution, it is important to find application areas where x-rays give unique information. One of these areas is modern magnetism research, which is based on atomically engineered multi-component nanostructures and their ultrafast response [8]. Such structures and real technological devices can be picked apart by x-rays, and their functioning can be observed down to the femtosecond timescale.

8.2.2 *Polarization Dependent Microscopy*

We finish this section by showing in Fig. 8.4 the first element specific real space microscopy images with contrast originating from polarization dependent resonant excitations (see Sect. 1.4.3). In (a) we show the PEEM image of ferromagnetic bits written on a CoPtCr recording disk, recorded by Co L_3 resonant excitation and use of the x-ray magnetic circular dichroism (XMCD) effect [16]. The image in (b) reveals preferential radial orientational order in a cross section of a Kevlar (poly(p-phenylene terephthalamide)) fiber, recorded by STXM in transmission through a thin slice and use of x-ray natural linear dichroism (XNLD) effect at the C π^* resonance [17] (also see [18].) In (c) we show the antiferromagnetc domain structure of a $LaFeO_3$ film on $SrTiO_3(001)$ recorded by PEEM utilizing the x-ray magnetic linear dichroism (XMLD) effect at the Fe L_3 resonance [19].

Today, real space x-ray microscopy is often combined with the intrinsic time structure of synchrotron sources to record the temporal evolution of magnetic nanostructures on the 50–100 ps timescale, previously illustrated in Fig. 1.22 and discussed in Sect. 11.7 (also see [20–24]).

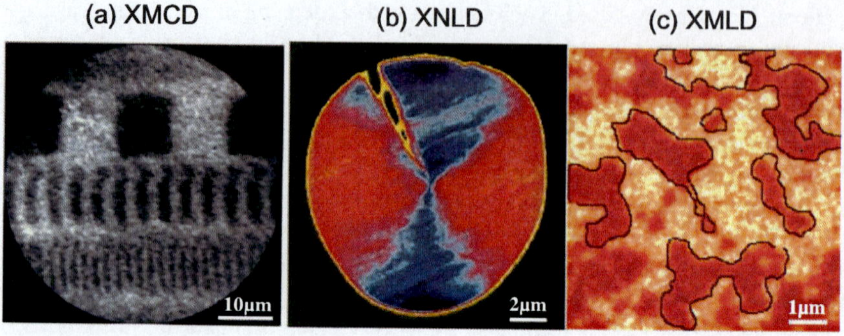

Fig. 8.4 First x-ray images recorded with resonant polarization dependent contrast. **a** Magnetic bits in a CoPtCr recording disk recorded with PEEM using resonant Co L_3-edge excitation and x-ray magnetic circular dichroism contrast [16]. **b** Cross-sectional image of a Kevlar fiber due to its radially symmetric internal structure, revealed by STXM images recorded with x-ray natural linear dichroism contrast at the C K-edge π^* resonance [17]. **c** Images of antiferromagnetic domains in a $LaFeO_3$ film by PEEM by use of x-ray magnetic linear dichroism contrast at the Fe L_3-edge [19]

8.3 Diffractive Imaging

8.3.1 Introduction to Diffractive Imaging

Coherent diffractive imaging (CDI) is based on the elimination of image forming "lenses" to achieve a spatial resolution which is only limited by the wavelength, i.e. the diffraction limit. To appreciate the essence of CDI in the x-ray region, we follow Fig. 8.5 which links the conventional (first order) coherence properties of an x-ray source, discussed in Chap. 4 to the diffraction of x-rays by different samples.

We first consider a single crystal inserted into a pseudo-monochromatic x-ray beam with $\Delta\lambda/\lambda \ll 1$ as shown in Fig. 8.5a. The effective coherence volume at the sample is then defined by the longitudinal coherence length $\ell_{coh} = (\lambda/\Delta\lambda)\lambda \gg \lambda$ and the effective coherence areas B_{eff} (see Fig. 4.7b), shown as red circles in the figure. Because all unit cells are identical, the unit cell interference patterns are identical and add up into intense diffraction spots. The remarkable thing about x-ray crystallography is that it works even if the x-ray source is globally laterally *incoherent*. All that is required is that the x-ray beam contains small coherent regions of the size of the crystallographic unit cell. As discussed in Sect. 4.5.3 this is the case even for "incoherent" sources, since the birth volume of size $\simeq \lambda^3$ expands on propagation and one therefore just has to assure that the sample is placed at a sufficient distance from the source (also see Fig. 4.3).

Such diffraction patterns from crystals have been recorded ever since the early work by van Laue and the Braggs around 1912, mentioned in Sect. 1.3.1. What we refer to as *Bragg diffraction* today is the enhancement of the intensity scattered into certain directions caused by a periodic arrangement of atoms in the unit cell,

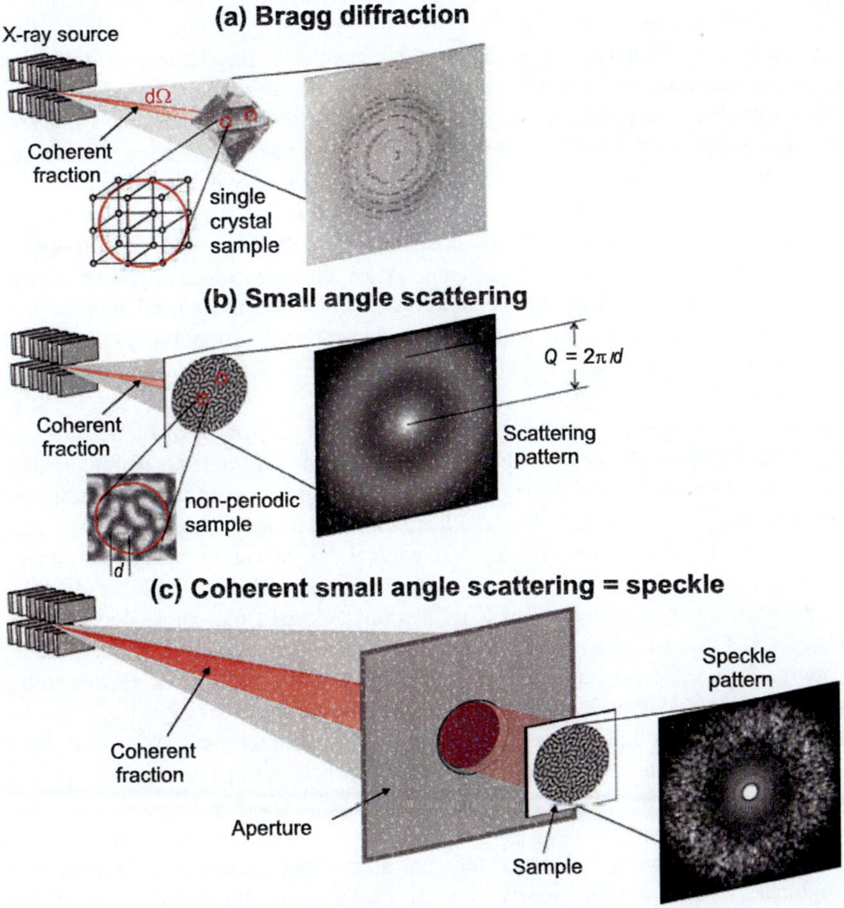

Fig. 8.5 Concepts of lateral source coherence and x-ray diffraction for different types of samples. The undulator source is assumed to be temporally coherent. **a** Single crystal Bragg diffraction. The incident beam has a wavelength λ comparable to atomic dimensions and is laterally coherent only within tiny solid angle cones $d\Omega$ shown in red. The position sensitive detector is positioned to accept large diffraction angles. **b** Sample with non-periodic nanoscale domains. The x-ray wavelength is assumed to be longer than the atomic spacing (no Bragg diffraction), and the beam is coherent only over small regions indicated by red circles which are of order of the characteristic domain size. The small angle scattering pattern at a distant detector is the incoherent superposition of the coherent patterns from different sample regions identified by red circles. **c** Same beam conditions as in **(b)** but the sample has been moved away from the source increasing the size of the coherence area to the entire illuminated area. The entire sample is now coherently illuminated, and the "speckle pattern" is a coherent superposition from all regions of the sample

which is furthermore enhanced because of the periodicity of unit cells in a single crystal. Bragg diffraction is nothing but the constructive interference of fields that are Thomson scattered by atoms ordered in a periodic arrangement. The real space crystal structure may then be solved by inversion of the reciprocal space image as discussed in Sects. 8.8 and 8.10. Bragg diffraction has revolutionized our knowledge of atomic structure, as illustrated for protein crystallography in Fig. 1.14.

In practice, the diffraction pattern emerges from many single photon events as shown in Fig. 8.10 and one may ask why only the coherence volume is important and the photon birth phases do not enter. After all, they are an important part of Feynman's single photon probability amplitude formulation introduced in Sect. 4.4.3. The reason is that in a quantum description of conventional diffraction, single photons only interfere with themselves, so that the relative birth phases of different photons do not matter.

We next consider a sample with non-periodic atomic arrangement, as shown in Fig. 8.5b. The interference patterns from small areas in the sample are no longer identical, and therefore intense diffraction spots will no longer exist. We have indicated the coherence areas as red circles, which are assumed to be larger than the characteristic nanostructures shown in the inset. In general, the internal sample structure may correspond to regions of different electron density as in a polymer [25] or domains of different magnetic orientations [8]. The conventional small angle x-ray scattering (SAXS) pattern [5] corresponds to the incoherent superposition of the coherent patterns from different sample regions of the size of the red circles. Owing to the different structural or orientational units in the different sample regions the scattering pattern is smeared out. For a sample with an average characteristic size d associated with the nanoregions, the scattering pattern consists of a ring-like intensity pattern as shown in Fig. 8.5b, located at a momentum transfer $Q = 2\pi/d$.

By moving the sample away from the source, as shown in Fig. 8.5c, one increases the effective coherence volume at the cost of reduced intensity. At modern high brightness undulator x-ray sources one may coherently illuminate a sample that has a lateral size of the order of $10\,\mu$m and still have sufficient intensity. Then the scattered waves from all regions within a $10\,\mu$m area, selected by an aperture, can interfere. The scattered intensity distribution, the "speckle pattern", recorded by an imaging detector contains the complete information on the real space structure within the coherently illuminated area. In contrast, the SAXS pattern in Fig. 8.5b is the averaged speckle pattern and contains only statistical information about the sample structure.

Figure 8.5 reveals the essence and natural evolution of what today encompasses the term *"coherent diffractive imaging"*. It may be summarized as follows.

> *Coherent diffractive imaging* extends x-ray crystallography, which can reveal the atomic structure of periodic unit cells in matter, to the determination of non-periodic structures, ranging from atomic to microscopic dimensions. This evolution became possible in practice only through the increase of x-ray brightness, which has allowed coherent illumination of large sample regions without detrimental losses of intensity.
>
> The spatial resolution of CDI is limited in principle only by the wavelength, i.e. the diffraction limit. The central challenge of CDI is the inversion of the reciprocal into a real space image, which requires advanced solutions of the "phase problem" discussed in Sect. 8.8.

We start our treatment of diffraction with a quick review of its history.

8.3.2 Historical Development of Diffraction Theories

The description of wave propagation was greatly influenced by the Dutch physicist Christiaan Huygens (1629–1695) and the French physicist Augustin-Jean Fresnel (1788–1827), and their work today is honored by what is called the *Huygens–Fresnel principle*. It is founded on two basic assumptions, Huygens' 1678 assertion that light is a wave[2] and that every point of disturbance can be considered to be a source of a spherical wave, and Fresnel's 1816 assertion that these spherical waves may interfere. Figure 8.6 shows a page from Huygens' notebook which illustrates his early thinking about curvature, a key concept associated with deviation from rectilinear propagation.

The important phenomenon of diffraction was first accurately recorded through observations in 1665 by the Italian scientist Francesco Maria Grimaldi who also coined the word "diffraction" (from Latin diffractus, broken into pieces). Diffraction was put on a mathematical foundation by Gustav Kirchhoff (1824–1887) in 1883 [26] and extended by John William Strutt (Lord Rayleigh) (1842–1919) [27] and Arnold Sommerfeld (1868–1951) [28].[3] Sommerfeld, in particular, derived the first truly satisfactory solution of a diffraction problem in 1896 [28] by treating the case of a plane wave incident on an infinitesimally thin, perfectly conducting plane. While the Fresnel–Kirchhoff integral equation uses the input field directly to calculate the diffracted field, it can be derived from the more general Rayleigh-Sommerfeld theory

---

[2] This led to the well-known controversy with Newton. Both Huygens and Newton exchanged ideas in their early thinking about curvature (called "crookedness" by Newton) which required new mathematical developments, and it is documented that Newton highly respected Huygens' thinking. However, they parted ways in their thinking about the wave or particle nature of light which led to a suppression of Huygens' ideas for a considerable time because of Newton's influence.

[3] Lord Rayleigh won the 1904 Nobel Prize, a prize that eluded Sommerfeld as the most nominated person ever and the Ph.D. supervisor of many future Nobel Prize winners.

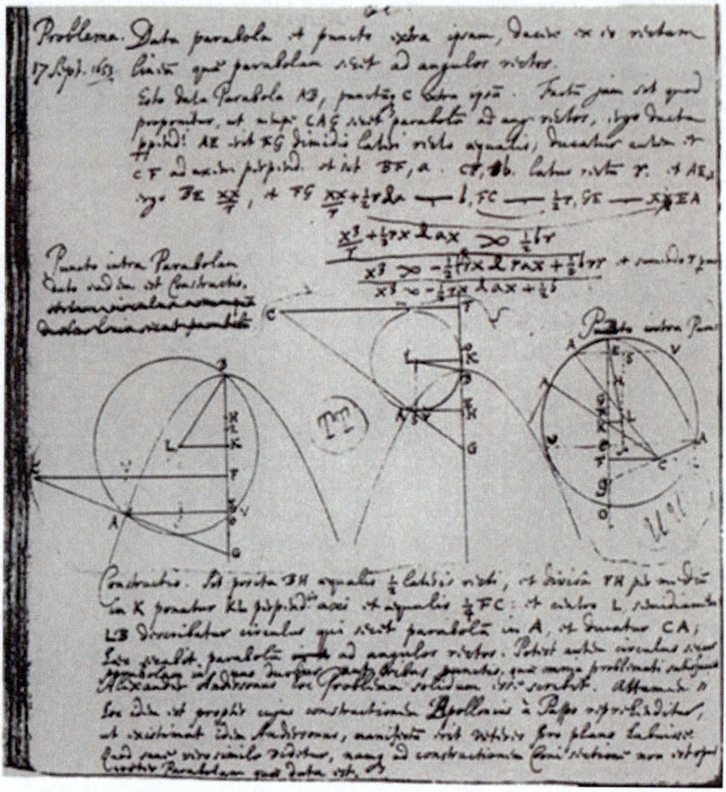

Fig. 8.6 A page from Christiaan Huygens' notebook, written in Latin and dated September 17, 1653, showing his early thinking about the concept of curvature. This later led to his ideas of light as secondary spherical wavelets which originate from all points in the light path

which expresses the diffracted field in terms of the Fourier transform of the input field [29, 30].

Today, diffraction is typically calculated by means of the Rayleigh-Sommerfeld diffraction formula discussed below, which is a scalar theory that neglects the vector nature of the EM fields. The justification for the scalar theory comes from the fact that the EM wave equation, derived from Maxwell's equation, applies equally for the electric and magnetic fields, and the wave behavior can therefore be described through a single scalar wave equation. The scalar treatment neglects the field modifications of the orthogonal **E**- and **H**-fields at the boundary and therefore implicitly assumes that the boundary regions are small relative to the total size of the disturbing object. It is interesting that from a more intuitive point of view, first expressed in 1802 by Thomas Young, diffraction by an object placed into the path of a wave can be viewed as the interference of part of the wave that is transmitted unperturbed through the object and a modified part from the boundary of the object. This idea is confirmed by our previous discussion in Sect. 7.5.2 where we showed that the field transmitted through a well-defined circular area in a thin film can be decomposed in a "film" and "edge" contribution, with the latter arising from the circular rim of the illuminated area.

The Rayleigh-Sommerfeld theory has been proven to give good results under two conditions:

- The wavelength must be much smaller than the lateral size of the object causing the wave disturbance (e.g. an aperture or sample domain).
- The fields diffracted by the disturbing object are calculated at a position that is many wavelengths away. Depending on the observation distance one distinguishes the "near-field" Fresnel and "far-field" Fraunhofer regimes, as discussed below.

With the discovery of x-rays, early diffraction experiments with well-ordered crystals [31] could also be explained by the wave nature of x-rays, as clearly illustrated by the Ewald sphere construction [32]. The Braggs insisted however that both wave and particle properties of x-rays were required [33]. Today, x-ray diffraction is still taught in textbooks by use of the wave concept [5].

8.3.3 *The Huygens–Fresnel Principle*

We start our discussion of diffraction by stating the amazing Huygens–Fresnel (HF) principle. It is based on combining the concepts of spherical waves centered on source points put forward by Huygens in *Traité de la Lumiere* which was completed in 1678 and published in Leyden in 1690 and the concept of their interference added by Fresnel [34] around 1800 (see Sect. 8.6.2 below). The principle can be stated as follows.[4]

> The *Huygens–Fresnel principle* states that each point on a planar wave front can be envisioned as a conversion center of the plane wave into a spherical wave. The spherical waves, however, only propagate into the forward 2π solid angle and are referred to as spherical wavelets. The total intensity that propagates forward is then determined by the interference of all wavelets and is calculated as the absolute value squared of the summed wavelet amplitudes.

As discussed earlier, over the centuries, many famous scientists worked on exploring this concept mathematically and experimentally in more detail.

[4] The principle is in accord with modern quantum electrodynamics, where any disturbance created in a sufficiently small region of *isotropic space* or in an isotropic medium propagates from that region, on average, in all radial directions.

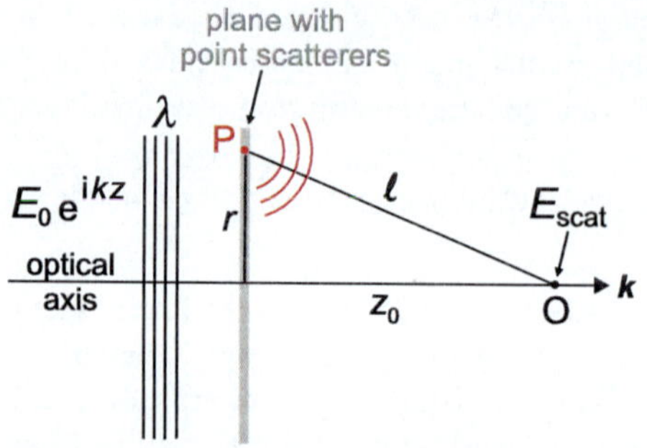

Fig. 8.7 Scattering by a plane wave $E_0 \exp(ikz)$ of wavelength $\lambda = 2\pi/k$ propagating along the optical axis $\mathbf{k} \parallel \mathbf{z}$ by a point scatterer positioned at a point P in a plane perpendicular to $\mathbf{k}$ at a distance r from the optical axis. We are interested in calculating the scattered field E_{scat} on the optical axis at a point O at distance z_0 from the source plane due to the interference of spherical waves from all points P in the plane

In the following we shall set the stage for the entire chapter on diffraction by a relatively simple[5] calculation by use of the HF principle, following Fig. 8.7 (also see Fig. 7.10c).

We consider a uniform plane wave field of amplitude E_0 that is incident on a plane (shown in gray). The essence of the HF principle is the complete conversion of the incident plane wave into spherical wavelets at all points $P(r)$ which then propagate into the forward 2π solid angle. All points P in the plane therefore constitute new secondary source points. The wavelet fields are added up at a point O on the optical axis, and the absolute value squared of the summed fields gives the desired intensity at point O. The cylindrical symmetry about the optical axis allows us to conveniently perform the integration over the plane in cylindrical coordinates. We shall see below that the integration is simple for on-axis points O but non-trivial for off-axis observation points, so that approximations have to be made.

The fictitious source points in the Huygens–Fresnel theory bear a strong resemblance with the case of real atomic source points in a thin sheet treated in Sect. 7.5.3 by the first Born approximation. The real atomic scattering centers in the Born treatment are replaced by virtual source points in the HF theory, and the small part of the incident field scattered as a spherical wave by real atoms is replaced by a complete conversion of the incident field into spherical waves. This illustrates the vexing aspect of the HF theory which appears non-physical.

As shown in Fig. 8.7, a spherical wave from a general point P in the plane has to travel a distance $\ell = \sqrt{z_0^2 + r^2}$ to the on-axis point O. The total transmitted on-axis field E_O at a distance z_0 behind a plane is obtained by integration over all sample points P. We start with a plane that extends to $\pm\infty$ from the optical axis. Since the plane integration introduces a factor with dimension [area], we write the integral by use of a to-be-determined normalization factor C, and by integration in cylindrical

[5] At least today if one uses modern computer programs like "Mathematica" to solve the integrals.

coordinates, we obtain the scattered field at O as

$$E_{\mathrm{scat}}(\mathrm{O}) = \underbrace{C E_0\, 2\pi \int_0^\infty \frac{\mathrm{e}^{\mathrm{i}k\sqrt{z_0^2+r^2}}}{\sqrt{z_0^2+r^2}}\, r\, \mathrm{d}r}_{2\pi \mathrm{i}\, \mathrm{e}^{\mathrm{i}kz_0}/k} = \underbrace{\mathrm{i}\,\lambda\, C\, E_0}_{E_0}\, \mathrm{e}^{\mathrm{i}kz_0} \tag{8.1}$$

On the right we have indicated by underbrackets that the on-axis transmitted field has to remain the incident field since there is no real sample. Hence we find, $C = 1/(\mathrm{i}\lambda)$. The properly normalized form of (8.1) with $r^2 = x^2 + y^2$ is therefore given by

$$\boxed{E_{\mathrm{scat}}(\mathrm{O}) = -\frac{\mathrm{i}}{\lambda} \int_{\mathrm{sample}} E_0 \frac{\mathrm{e}^{\mathrm{i}k\sqrt{z_0^2+x^2+y^2}}}{\sqrt{z_0^2+x^2+y^2}}\, \mathrm{d}x\, \mathrm{d}y.} \tag{8.2}$$

This on-axis form is a special case of the Rayleigh-Sommerfeld formula for the diffracted field from a sample, given by (8.8) below. As previously discussed in Sect. 7.5.4, the integration produces a factor $-\mathrm{i}$ corresponding to a 90° phase shift.

We can readily obtain the following solutions for the on-axis field if we assume cylindrically symmetric source areas about the optical axis.

[1] Infinite plane:

$$E_{\mathrm{trans}}(\mathrm{O}) = \frac{E_0}{\mathrm{i}\lambda}\, 2\pi \int_0^\infty \frac{\mathrm{e}^{\mathrm{i}k\sqrt{z_0^2+r^2}}}{\sqrt{z_0^2+r^2}}\, r\, \mathrm{d}r = \underbrace{E_0 \mathrm{e}^{\mathrm{i}kz_0}}_{\text{open area}} \tag{8.3}$$

[2] Circular central hole of radius R in opaque screen:

$$E_{\mathrm{trans}}(\mathrm{O}) = \frac{E_0}{\mathrm{i}\lambda} 2\pi \int_0^R \frac{\mathrm{e}^{\mathrm{i}k\sqrt{z_0^2+r^2}}}{\sqrt{z_0^2+r^2}}\, r\, \mathrm{d}r = \underbrace{E_0 \mathrm{e}^{\mathrm{i}kz_0}}_{\text{open area}} - \underbrace{E_0\, \mathrm{e}^{\mathrm{i}k\sqrt{R^2+z_0^2}}}_{\text{edge at } R} \tag{8.4}$$

[3] Circular central opaque beam stop of radius R:

$$E_{\mathrm{trans}}(\mathrm{O}) = \frac{E_0}{\mathrm{i}\lambda} 2\pi \int_R^\infty \frac{\mathrm{e}^{\mathrm{i}k\sqrt{z_0^2+r^2}}}{\sqrt{z_0^2+r^2}}\, r\, \mathrm{d}r = \underbrace{E_0\, \mathrm{e}^{\mathrm{i}k\sqrt{R^2+z_0^2}}}_{\text{edge at } R} \tag{8.5}$$

4 Circular open zone of width δ in opaque screen:

$$E_{\text{trans}}(\text{O}) = \frac{E_0}{\mathrm{i}\lambda} 2\pi \int_R^{R+\delta} \frac{\mathrm{e}^{\mathrm{i}k\sqrt{z_0^2+r^2}}}{\sqrt{z_0^2+r^2}}\, r\,\mathrm{d}r = \underbrace{E_0\,\mathrm{e}^{\mathrm{i}k\sqrt{R^2+z_0^2}}}_{\text{edge at } R} - \underbrace{E_0\,\mathrm{e}^{\mathrm{i}k\sqrt{(R+\delta)^2+z_0^2}}}_{\text{edge at } R+\delta} \tag{8.6}$$

Some of the cases are linear superpositions of two others. For example, the complementary cases 2 and 3 add up to 1. As indicated by underbrackets, for a finite area $A = \pi R^2$ the boundary region gives rise to an edge diffraction field. The edge contribution exactly cancels for an infinite plane without boundaries.

8.3.4 *The Rayleigh-Sommerfeld Diffraction Formula*

The above discussion provides a natural introduction to the general Rayleigh-Sommerfeld diffraction formula. We consider a coherent EM wave of wavelength λ and field strength E_0 that is incident on a "sample" of lateral dimension $2R > \lambda$. The "sample" may be either an open aperture or a thin film. The axes of our Cartesian coordinate system (x, y, z) are chosen as shown in Fig. 8.8, with x-ray propagation along z. The source field $E_S(x, y, 0)$ that gives rise to the diffraction pattern in the detector plane is given by a generalized version of (7.24)

$$\boxed{\text{Source field: } E_S(x, y, 0) = E_0(x, y)\,\mathrm{e}^{-[\mathrm{i}\delta(x,y)+\beta(x,y)]k\,d(x,y)}.} \tag{8.7}$$

All fields are treated as scalars as discussed earlier. The source field defines the amplitude of the scattered spherical waves at $z = 0$ according to the Huygens–Fresnel principle. For an open aperture, the "sample" thickness is $d(x, y) = 0$ and $E_S(x, y, 0) = E_0$ is constant in the x–y plane of the aperture. For a thin film, the source field $E_S(x, y, 0)$ is the attenuated incident field in the exit plane of the sample according to (8.7), where we have explicitly indicated that the optical constants δ and β and the sample thickness d are dependent on position (x, y).

According to the Huygens–Fresnel principle, we consider each point $\mathbf{r} = (x, y, 0)$ in the exit plane of the sample as the source of a spherical EM wave, and with reference to Fig. 8.8 we start our mathematical treatment of diffraction with the Rayleigh-Sommerfeld formula, describing the interference pattern of photons that are elastically ($|\mathbf{k}| = |\mathbf{k}'|$) scattered by a "sample" [30].[6]

The *Rayleigh-Sommerfeld formula* expresses the diffracted field E_{diff} at a detector point $P'(\mathbf{r}') = P'(\mathbf{z}_0 + \boldsymbol{\rho})$ in terms of the source fields $E_S(\mathbf{r})$ from all points

[6] We have omitted (set to unity) a so-called *obliquity factor* which varies in different theories of x-ray diffraction.

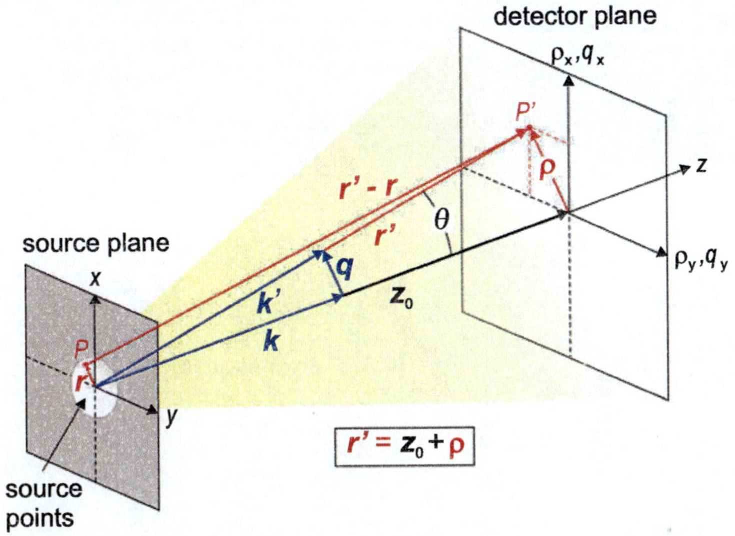

Fig. 8.8 Illustration for deriving the Fresnel and Fraunhofer diffraction formulas from the Rayleigh-Sommerfeld equation. We consider the diffracted fields at points $P'(\mathbf{r}') = P'(\mathbf{z}_0 + \boldsymbol{\rho})$ in the detector plane that arise from the fields originating from points $P(x, y, 0)$ in the source plane

P at $\mathbf{r} = (x, y, 0)$ according to [30],

$$E_{\text{diff}}(\mathbf{r}') = -\frac{\mathrm{i}}{\lambda} \int_{\text{sample}} E_{\text{S}}(\mathbf{r}) \frac{\mathrm{e}^{\mathrm{i}k|\mathbf{r}'-\mathbf{r}|}}{|\mathbf{r}'-\mathbf{r}|}\, \mathrm{d}\mathbf{r} \tag{8.8}$$

where the integration is over all source points $\mathbf{r}$ illuminated by the incident beam.

The Rayleigh-Sommerfeld formula is the coherent source limit of Zernike's propagation theorem for partial coherence (see Sect. 4.6), which expresses the relationship between the field correlation function $\langle E^*(\mathbf{r}_1)E(\mathbf{r}_2)\rangle$ between two source points $\mathbf{r}_1$ and $\mathbf{r}_2$ in the source plane and the correlation function $\langle E^*(\boldsymbol{\rho}_1)E(\boldsymbol{\rho}_2)\rangle$ between two points $\boldsymbol{\rho}_1$ and $\boldsymbol{\rho}_2$ in the detector plane. The two cases are compared in Fig. 8.9.

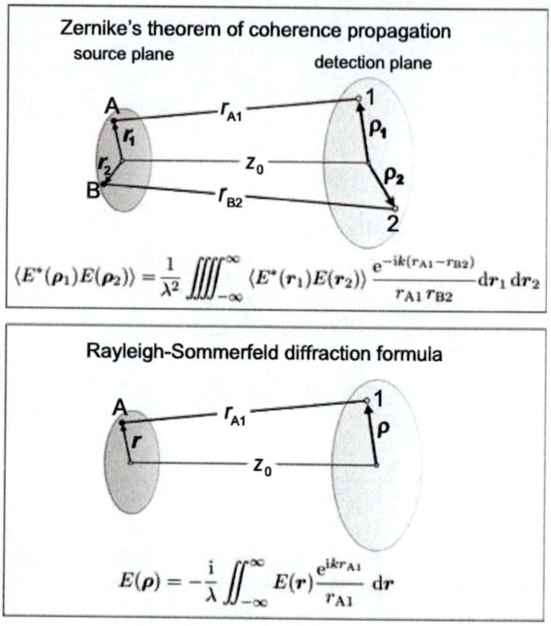

Fig. 8.9 Comparison of the Rayleigh-Sommerfeld formula (8.8) for a coherent source and Zernike's propagation formula (4.94) for a partially coherent source

8.4 Approximate Solutions of the Rayleigh-Sommerfeld Formula

The Rayleigh-Sommerfeld formula (8.8) which according to Fig. 8.8 gives the total diffracted field after integration over the source coordinates $\mathbf{r}$ at a position $\mathbf{r}' = \mathbf{z}_0 + \boldsymbol{\rho}$ has no general analytical solution. The difficulty arises when off-axis positions $\boldsymbol{\rho} \neq 0$ are considered. Hence in practice one has to make approximations. By geometry we can express the coordinates in (8.8) as

$$|\mathbf{r}' - \mathbf{r}| = z_0\sqrt{1 + \frac{x^2 + y^2}{z_0^2} + \frac{x'^2 + y'^2}{z_0^2} - 2\frac{x'x + y'y}{z_0^2}}, \tag{8.9}$$

and it is this expression which may be simplified by increasingly stricter approximations.

8.4.1 *Paraxial Approximation*

The most important simplification of this expression is the *paraxial approximation* already discussed in Sect. 4.3.1, which yields

$$|\mathbf{r}' - \mathbf{r}| \simeq z_0 \left[1 + \frac{x^2 + y^2}{2z_0^2} + \frac{x'^2 + y'^2}{2z_0^2} - \frac{x'x + y'y}{z_0^2} \right] \tag{8.10}$$

In (8.8) the distance $|\mathbf{r}' - \mathbf{r}|$ enters in the amplitude through a factor $1/|\mathbf{r}' - \mathbf{r}|$ and exponentially through the phase term $e^{ik|\mathbf{r}'-\mathbf{r}|}$. The *amplitude* term only leads to a weak gradual change of the field strength in the distant detector plane, and since z_0 is much larger than all other distances appearing in (8.10), the denominator in (8.8) may be simplified as $|\mathbf{r}' - \mathbf{r}| = z_0$ to a very good approximation. This approximation is inadequate for the *phase* term, however, which contains the key interference information. The simplification of the phase term is typically accomplished by two further approximations, referred to as the Fresnel and Fraunhofer approximations. The two approximations distinguish what is referred to as "near-field" *Fresnel diffraction* and "far-field" *Fraunhofer diffraction.*[7]

8.4.2 Fresnel Diffraction

From a phase perspective, the most important interference information is contained in the last term $(x'x + y'y)/z_0^2$ in (8.10). Its mixed coordinates clearly indicate interference, and it needs to be retained. The third term $(x'^2 + y'^2)/(2z_0^2)$ in (8.10) does not depend on the sample or integration coordinates and can be pulled in front of the integral in (8.8). Also it will be of no consequence when the intensity distribution in the detector plane is calculated since it will disappear upon forming the absolute value squared of the field.

The second term in (8.9) corresponds to a phase shift $\exp[ik(x^2 + y^2)/(2z_0)]$ that depends on the sample integration coordinates. It is retained in Fresnel diffraction so that (8.8) becomes

Fresnel diffraction:

$$E_{\text{diff}}(x', y', z_0) = \frac{e^{ikz_0}}{i\lambda\, z_0}\, e^{i\frac{k}{2z_0}(x'^2+y'^2)} \times \int_{\text{source}} E_S(x, y, 0)\, e^{i\frac{k}{2z_0}(x^2+y^2)}\, e^{-i\frac{k}{z_0}(x'x+y'y)} dx\, dy \tag{8.11}$$

The *Fresnel approximation* is valid in the *near-field* region.[8]

---

[7] The name is in honor of Joseph von Fraunhofer (1787–1826) although he was not actually involved in the development of the theory.

[8] The integral in the Fresnel expression can be evaluated for a constant incident field $E_S(x, y, 0) = E_0$ similar to (8.3) as follows,

8.4.3 *Fraunhofer Diffraction*

The integral in (8.11) significantly simplifies by omission of the first phase shift term $\exp[\mathrm{i}k(x^2+y^2)/(2z_0)]$ which originates from the finite size of the source area x^2+y^2. The neglect of this term corresponds to the *Fraunhofer approximation* which with $k = 2\pi/\lambda$ may be written as

$$\text{Fraunhofer condition for source:} \quad \frac{\pi(x^2+y^2)}{z_0\lambda} \ll 1. \tag{8.13}$$

For large z_0 we can also approximate $\mathrm{e}^{\mathrm{i}\frac{k}{2z_0}(x'^2+y'^2)} \simeq 1$, so that (8.11) simplifies to

$$E_{\text{diff}}(x', y', z_0) = \frac{\mathrm{e}^{\mathrm{i}kz_0}}{\mathrm{i}\lambda\, z_0} \int\limits_{\text{source}} E_S(x, y, 0)\, \mathrm{e}^{-\mathrm{i}\frac{k}{z_0}(x'x+y'y)}\, \mathrm{d}x\, \mathrm{d}y. \tag{8.14}$$

8.4.4 *Fourier Theorem*

Following Fig. 8.8, the field E_{diff} in the detector plane may be expressed in terms of the real coordinates (x', y', z_0), or alternatively in terms of the momentum transfer $\mathbf{q} = \mathbf{k} - \mathbf{k}'$, and with $q^2 = q_x^2 + q_y^2$ we have

$$\frac{x'}{z_0} = \frac{q_x}{k}, \quad \text{and} \quad \frac{y'}{z_0} = \frac{q_y}{k}. \tag{8.15}$$

We can therefore write the field (8.14) in the detector plane as a function of momentum transfer as

---

$$E_{\text{diff}}(x', y', z_0) = \frac{\mathrm{e}^{\mathrm{i}kz_0}}{\mathrm{i}\lambda\, z_0}\, \mathrm{e}^{\mathrm{i}\frac{k}{2z_0}(x'^2+y'^2)} E_0 \underbrace{\int_{-\infty}^{\infty}\int_{-\infty}^{\infty} \mathrm{e}^{\mathrm{i}\frac{k}{2z_0}(x^2+y^2)}\, \mathrm{e}^{-\mathrm{i}\frac{k}{z_0}(x'x+y'y)}\mathrm{d}x\, \mathrm{d}y}_{\mathrm{i}\lambda z_0\, \mathrm{e}^{-\mathrm{i}\frac{k}{2z_0}(x'^2+y'^2)}}$$

$$= E_0\, \mathrm{e}^{\mathrm{i}kz_0}, \tag{8.12}$$

which is just the incident plane wave. Note that the integral is the Fourier transform of the function $\exp[\mathrm{i}k(x^2+y^2)/(2z_0)]$ which is neglected in the Fraunhofer approximation by assuming that the exponent is negligibly small so that $\exp[\mathrm{i}0] = 1$.

Fraunhofer diffraction:

$$E_{\text{diff}}(q_x, q_y) = \underbrace{\frac{e^{ikz_0}}{i\lambda z_0}}_{C} \int_{\text{source}} E_S(x, y, 0)\, e^{-i(q_x x + q_y y)}\, dx\, dy. \tag{8.16}$$

This is equivalent to the result (4.68) previously obtained in Sect. 4.4.3 by use of Feynman's probability amplitude formulation. The prefactor C preserves the field dimension and accounts for propagation from the source to the distant detector plane, i.e. the falloff of the field amplitude with distance z_0 from the sample.

The integral in (8.16) is nothing but the Fourier transform of the field distribution $E_S(x, y, 0)$ in the exit plane of the sample. We have derived the following important result.

The field distribution at a distant detector is the Fourier transform of the electric field distribution in the exit plane of a sample. We symbolically write this as

$$E_{\text{diff}}(q_x, q_y) = C\,\mathcal{F}\{E_S(x, y, 0)\}\,. \tag{8.17}$$

The reverse transformation also holds and is given by

$$E_S(x, y, 0) = C^{-1}\,\mathcal{F}^{-1}\big\{E_{\text{diff}}(q_x, q_y)\big\}\,. \tag{8.18}$$

Here C is defined in (8.16).

The intensity of the Fraunhofer diffraction pattern is given by

$$|E_{\text{diff}}(q_x, q_y)|^2 = |C|^2\ |\mathcal{F}\{E_S(x, y, 0)\}|^2\,. \tag{8.19}$$

While the Fourier transform of the *fields* is reversible, this is not the case for the intensities. The loss of the phase of the field when the intensity is calculated leads to the "phase problem" discussed below in Sects. 8.8 and 8.10.

8.5 Diffraction and Wave-Particle Duality

In the following two sections we will briefly address the wave-particle duality of light which may appear puzzling when trying to formulate diffraction in a photon picture.

8.5.1 Feynman's Probability Amplitudes

During his Ph.D. thesis, Louis de Broglie first developed the idea that even particles are accompanied by so-called phase waves [36], later referred to as *pilot waves* (see below). De Broglie's conjecture of a wave-particle duality for electrons [37] was proven in 1927 by the Davisson and Germer experiment of electron diffraction by crystals [38].

Around the same time, Dirac developed his quantum theory and formally related the concepts of wave fields and photons through quantizing the EM field [39–42]. His quantum theory which laid the foundation for quantum electrodynamics, fully developed about twenty years later by Feynman, Schwinger, and Tomonaga, was the basis for his famous and bold statement [43] *"...each photon interferes only with itself. Interference between different photons never occurs"*. Dirac's statement was supported by an earlier experiment in 1909 by Sir Geoffrey Ingram Taylor [44] who used a very weak gas flame as a lightsource to form a diffraction pattern of a needle. At the lowest intensity, on average, only one-photon-at-a-time hit the needle. The interference pattern was found to be the same as a function of intensity. The theoretical discussion and experimental verification of one-photon-at-a-time diffraction has a long history [45].

The equivalence of light and electron diffraction was famously discussed in Feynman's "Lectures on Physics" [46]. Rather than expressing interference as the superposition of classical EM waves, he rephrased the problem as the superposition of quantum mechanical *probability amplitudes*, as introduced in Sect. 4.4.3.1. Feynman did not actually carry out a diffraction calculation, but it is apparent from Sect. 4.4.3 that his probability amplitudes have the same form as classical fields. His treatment for electrons, however, hinged on the assumption that the probability amplitudes for photons and electrons are the same. This point was only clarified later [47], and it holds only in first order QED.

Since Feynman's lectures were given in the timeframe 1961–63, he must not have been aware that the double-slit experiment was actually performed with electrons in 1961 by Jönssen [48]. Since then Jönssen's experiment has been repeated and improved by several groups [49, 50], and in Fig. 8.10 we show the result of Bach et al. [50].

The observed pattern in Fig. 8.10 is granular since each incident electron produces only one-electron count in one of the pixels in the detector plane. Note that the corresponding photon diffraction pattern exhibits a similar granularity [52]. The pattern becomes increasingly visible as the cumulative one-electron counts in different pixels probabilistically "fill in" the diffraction pattern, eventually evolving into the yellow intensity curve in the limit of recording a large number of counts. The yellow curve is given by the intensity distribution

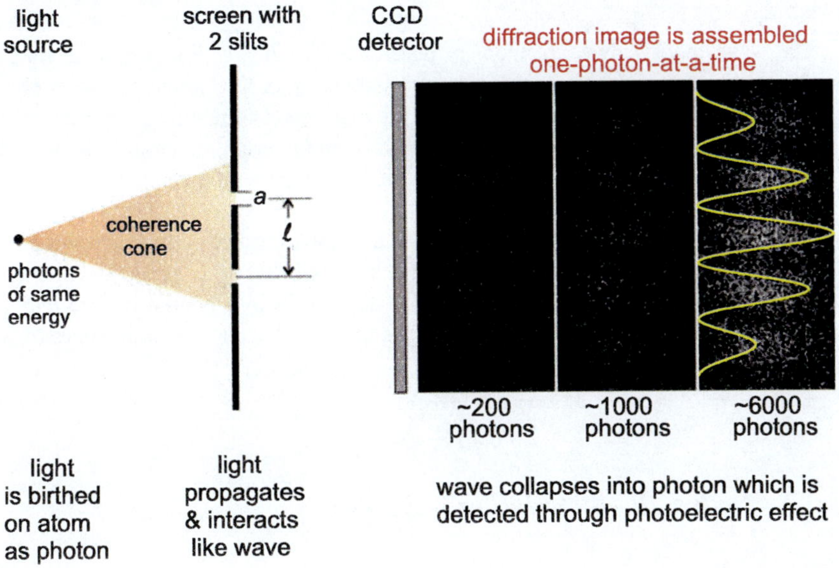

Fig. 8.10 Assembly of the double-slit diffraction pattern, one-photon-at-a-time. The granular patterns on the right were actually recorded with single electrons [50], but the photon case is the same [51]. The yellow curve is the intensity distribution in the limit of recording a large number of counts

$$I(q) = 4\, I_0 \underbrace{\frac{a^2}{\lambda z_0} \operatorname{sinc}^2\left[\frac{qa}{2}\right]}_{\text{single slit pattern}} \underbrace{\cos^2\left[\frac{q\ell}{2}\right]}_{\text{double slit modulation}} . \tag{8.20}$$

Here a is the width of the two slits whose centers are separated by ℓ, and q is the momentum transfer, measured from the center in the diffraction plane. It is related to the distance z_0 between the source and detector planes and the lateral distance ρ from the center of the detector plane by $q = k\rho/z_0$.

These experiments have provided clear evidence that the diffraction pattern builds up probabilistically through single "particle" detector clicks. This fact remains hidden in the classical theory which only gives an intensity distribution without making a statement about its measurement. We note that single particle interference is observed even for macromolecules [53]. We can state as follows.[9]

[9] Feynman's formulation of diffraction follows the general principle underlying quantum mechanical processes. The system evolves from the initial state (source) to the final state (detector) through all possible intermedicate states (slits). All probability amplitudes that connect the initial and final states via intermediate states are first summed. Only then does one take the absolute value squared of the summed amplitudes to obtain the detection probability.

The wave theory gives an intensity distribution or diffraction pattern in the detector plane without specifying how it is measured. The granular nature of the pattern can only be accounted for by the inclusion of quantum processes. In principle, granularity may be introduced through either the quantum substructure of light (photons) or matter (atoms), but the full quantum treatment includes both.
The wave formulation may be understood as a lowest order description within QED, as previously discussed in Chap. 5. This will be more directly demonstrated in Chap. 16, where we treat x-ray diffraction in a photon picture and show that diffraction patterns are the signatures of different quantum states of light.

8.5.2 *De Broglie–Bohm Pilot Wave Theory*

Another less utilized approach for the description of diffraction is based on alternative formulations of quantum mechanics by Erwin Madelung (1881–1972) [54], Louis de Broglie [55], and David Bohm (1917–1992) [56, 57], referred to as the *pilot wave* or *quantum trajectory* method [58–60]. A key ingredient of the theory is Bohmian mechanics [56, 57] which through a deterministic or "causal" interpretation of quantum mechanics can provide information about particle trajectories. The particles perform quantum mechanical fluctuations around the trajectories that give rise to Heisenberg's position-momentum uncertainty. The fluctuations may be viewed as arising from the interaction of the particles with the stochastic zero-point field and also lead to a finite coherence length.

An example of calculated quantum trajectories is shown in Fig. 8.11 for the case of cold neutron ($\lambda = 1.93$ nm) diffraction from a single 92 μm slit [59]. The calculation reveals well-defined trajectories that produce the diffraction pattern.

In particular, for any given initial condition at the sample, there exists only one Bohmian trajectory taking the particle to the detector. This semi-classical behavior has been associated with differences between the relative intensities of the Airy rings in the so-calculated and experimentally observed diffraction patterns [61, 62].

8.6 Consequences of Fraunhofer Diffraction

A particular important application of the Fraunhofer diffraction formula is the case of a homogeneous *circular source*. The existence of diffraction has important implications for the ability to clearly distinguish objects by use of light with a given wavelength. Ultimately the diffraction phenomenon will blur objects that are comparable in size to the wavelength and even a perfect lens cannot get around the diffraction

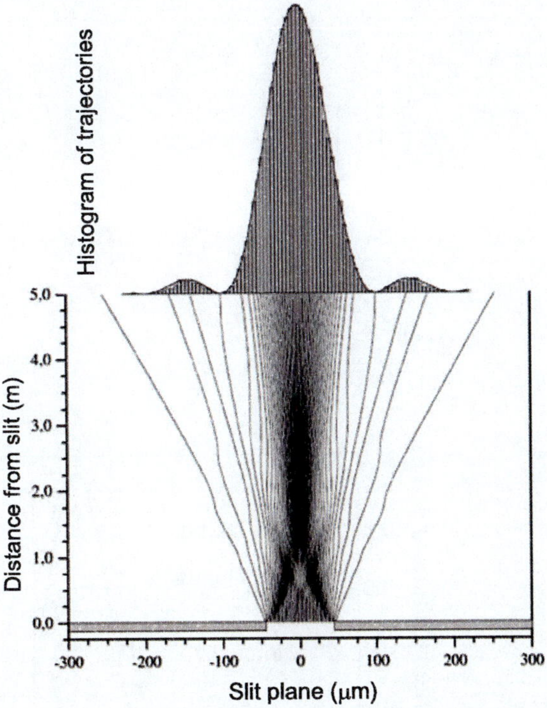

Fig. 8.11 Bohm quantum trajectories for the case of cold neutron ($\lambda = 1.93\,\text{nm}$) diffraction from a single $92\,\mu\text{m}$ slit [59]

blurring as pointed out by Ernst Abbe in 1873 [63] and Lord Rayleigh in 1879 [64]. In the following we shall take a closer look at the origin of the diffraction limit, introduced in conjunction with brightness in Sect. 4.2.2. We discuss the diffraction limit through the example of a circular light source or secondary light source formed by a circular aperture, previously treated in Sect. 4.4.3.

8.6.1 The Diffraction Limit

The central dilemma of wave optics or first order diffraction theory is best illustrated by the finite width of the central Airy disk in a distant detector plane produced by a uniformly illuminated circular aperture, as illustrated and defined in Fig. 8.12a. In astronomy, for example, the circular aperture size corresponds to the telescope aperture which limits the resolution due to diffraction.

The diffraction pattern of a uniformly illuminated circular aperture of area πR^2 in a detector plane at a large distance z_0 is the Airy intensity pattern shown as a black curve in Fig. 8.12b. It is a special case of (8.19) and is given by (4.71), which may be rewritten as

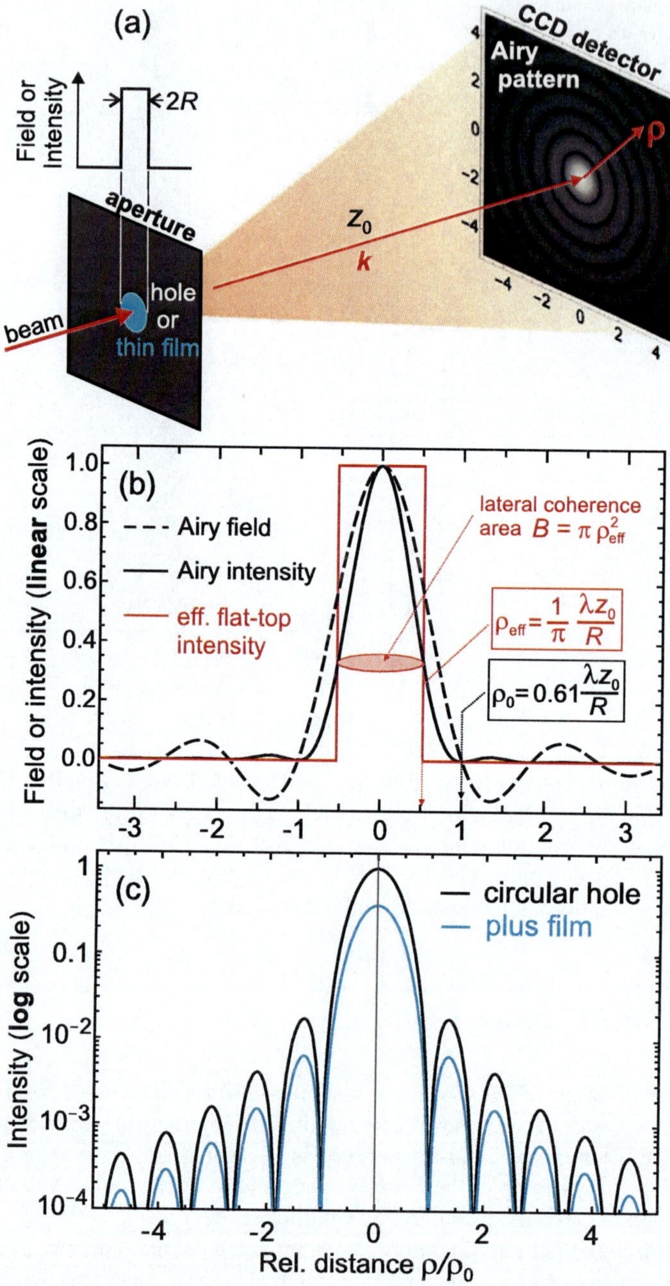

Fig. 8.12 **a** Assumed geometry of a flat-top secondary source of area πR^2 that is *coherently* illuminated. The inset above the aperture shows the 1D projection of the flat-top field and intensity distribution. **b** Central part of the normalized field (dashed black) and intensity (black) Airy patterns for a hole and effective flat-top intensity distribution (red), plotted on a linear scale. The areas under all curves are the same. The Rayleigh diffraction limit $\rho_0 = 0.61\lambda z_0/R$ is defined by the first zero of the Airy intensity. **c** Airy intensity distribution plotted on a logarithmic scale for a hole and an inserted film that transmits $1/e$ of the intensity

$$I(\rho, z_0) = I_0 \left[\frac{\pi R^2}{\lambda z_0}\right]^2 \left[\frac{2J_1\left(2\pi R\rho/(\lambda z_0)\right)}{2\pi R\rho/(\lambda z_0)}\right]^2 = I_0 \left[\frac{\pi R^2}{\lambda z_0}\right]^2 \left[\frac{2J_1\left(1.22\pi\rho/\rho_0\right)}{1.22\pi\rho/\rho_0}\right]^2. \tag{8.21}$$

As illustrated in Fig. 8.12b, the first zero of the diffracted intensity is located at a position $\rho_0 = 0.61(\lambda z_0/R)$ or momentum transfer $q_0 = 0.61(2\pi/R)$, which is the so-called *Rayleigh diffraction* limit. The dashed black curve is the corresponding field distribution function given by (4.70). Also shown as a red curve is an effective flat-top distribution of radius $\rho_{\text{eff}} = (1/\pi)\lambda z_0/R$. The integrated area under all three curves of normalized unit central intensity is the same[10] and given by the red area in Fig. 8.12b given by

$$B_{\text{eff}} = \pi\rho_{\text{eff}}^2 = \frac{\lambda^2 z_0^2}{\pi R^2}. \tag{8.23}$$

The effective area B_{eff} is also the effective coherence area and defines the solid angle coherence cone in geometrical optics according to $\mathrm{d}\Omega = B_{\text{eff}}/z_0^2 = \pi\rho_{\text{eff}}^2/z_0^2$.

We also illustrate in Fig. 8.12c in light blue that the intensity of the pattern in the detector plane is reduced if a uniform film is inserted into the aperture. The reduction is given by the absorption in the film, described by a factor $\exp(-\mu d)$, where μ is the linear absorption coefficient (see (7.26)) of the material and d its thickness. In the figure we have chosen $\mu d = 2\beta k d = 1$. In the presence of a thin film the transmitted field distribution is given by

$$E(\rho) = -\mathrm{i}E_0\, \mathrm{e}^{\mathrm{i}kz_0}\, \mathrm{e}^{-(\mathrm{i}\delta+\beta)kd}\, \frac{\pi R^2}{\lambda z_0}\, \frac{2J_1(kR\rho/z_0)}{kR\rho/z_0} \tag{8.24}$$

and the corresponding diffraction pattern in the detector plane is

$$\boxed{I(\rho) = I_0 \underbrace{\mathrm{e}^{-2\beta kd}}_{\text{attenuation}} \underbrace{\frac{\pi^2 R^4}{\lambda^2 z_0^2}}_{\text{propagation}} \underbrace{\left[\frac{2J_1(kR\rho/z_0)}{kR\rho/z_0}\right]^2}_{\text{Fourier transform of aperture}}.} \tag{8.25}$$

As expected, a homogeneous film in an aperture does not change the diffraction pattern but merely reduces its intensity through absorption. The dispersive properties of the film do not influence the diffraction pattern at all.

---

[10] We have

$$\int_0^{2\pi}\!\!\int_0^{\infty} \frac{2J_1(kR\rho/z_0)}{kR\rho/z_0}\,\rho\,\mathrm{d}\rho\,\mathrm{d}\phi = \int_0^{2\pi}\!\!\int_0^{\infty} \left[\frac{2J_1(kR\rho/z_0)}{kR\rho/z_0}\right]^2 \rho\,\mathrm{d}\rho\,\mathrm{d}\phi = \frac{\lambda^2 z_0^2}{\pi R^2}. \tag{8.22}$$

8.6.1.1 Rayleigh Criterion

To illustrate Rayleigh's criterion, let us a assume that we want to record an image of a star using an aberration-corrected telescope with a circular light collecting lens of area πR^2. Although the distant star is a laterally incoherent (or chaotic) source, the light will be coherent over the finite aperture of the telescope πR^2. In practice, a lens with refractive index $n = 1$ is placed right behind the telescope aperture with a focal length equal to the aperture-detector distance z_0. What is measured at the focus of the lens is simply the Airy pattern of the circular aperture of the telescope [65]. The central part of the Airy disk has a radius $\rho_0 = 0.61\lambda z_0/R$ (see Fig. 8.12), and if we define the *numerical aperture* of the lens as $\mathcal{N} = R/z_0$, we can write the Rayleigh criterion as[11]

$$\rho_0 = 0.61\frac{\lambda}{\mathcal{N}}. \tag{8.26}$$

In principle one can reduce ρ_0 by making the wavelength smaller or increasing the numerical aperture $\mathcal{N}$ of the lens as is done by using large diameter telescopes in astronomy. The resolution of the image, however, is fundamentally limited by the wavelength.

In the following sections we illustrate two well-known related phenomena, observed when a circular aperture (hole) in an opaque screen is replaced by inverting the open and opaque areas, i.e. a circular opaque beam stop surrounded by nothing. This leads to the famous *Arago-Fresnel-Poisson bright spot* in the center behind the beam stop and the fascinating *Babinet's principle*.

8.6.2 The Arago-Fresnel-Poisson Bright Spot

A famous case of light diffraction is the replacement of a circular hole in an opaque screen, as shown in Fig. 8.12, by the opposite, namely a circular beam stop surrounded by nothing, as illustrated in Fig. 8.13a, b. In both cases the incident beam is assumed to have a lateral cross section much larger than the hole or beam stop.

The beam stop case has an interesting history. In 1818, Fresnel submitted a paper to a competition of the French Academy of Sciences on the wave theory of light. As a member of the evaluation committee, Siméon Poisson studied Fresnel's theory and thought that he had found a flaw. He argued that as a consequence of Fresnel's theory there should be a bright spot on the optical axis in the shadow of a circular opaque beam stop. Dominique Arago performed an experiment and succeeded in observing the predicted spot.[12] As a consequence, Fresnel won the competition.

---

[11] One can also express ρ_0 in terms of the *f-number* $f = z_0/(2R)$ typically used to describe the lens properties in photography.

[12] Arago later realized that the spot had already been observed by J.–N. Delisle in 1715 and G.F. Maraldi in 1723, a century earlier.

To see the origin of the bright spot, we assume that the incident light is temporally and laterally coherent. This is fulfilled if the incident light is quasi-monochromatic and the beam stop lies well inside the lateral coherent cone of a source. We consider a circular hole and equivalent beam stop of size $A = \pi R^2$. The on-axis field behind the hole in an opaque screen can be calculated without approximation and is given by (8.4), or

$$E_{\text{trans}}^{\text{hole}}(\mathrm{O}) = \underbrace{E_0 \mathrm{e}^{ikz_0}}_{\text{open area}} - \underbrace{E_0\, \mathrm{e}^{ik\sqrt{R^2+z_0^2}}}_{\text{edge at } R}. \tag{8.27}$$

The field for the complementary case of a circular opaque beam stop is given by (8.5), or

$$E_{\text{trans}}^{\text{stop}}(\mathrm{O}) = \underbrace{E_0\, \mathrm{e}^{ik\sqrt{R^2+z_0^2}}}_{\text{edge at } R}. \tag{8.28}$$

This result directly verifies the existence of a field on the optical axis in the shadow of the beam stop! It arises solely from edge diffraction and gives rise to the Arago-Fresnel-Poisson bright spot.

We also find the on-axis sum rule for the fields

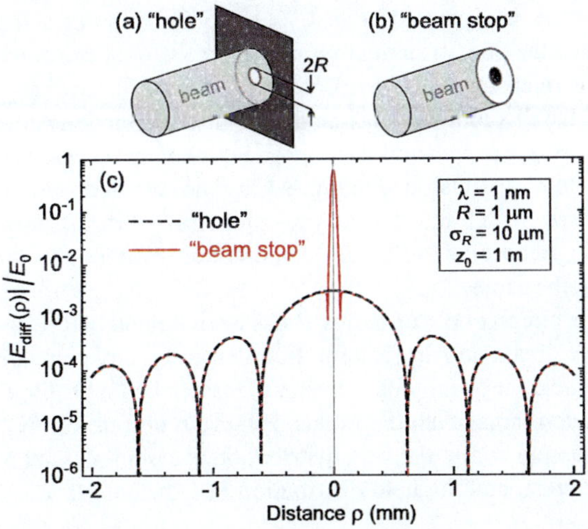

Fig. 8.13 **a** Two opposite cases, where either a circular *hole* in a screen is coherently illuminated or in (**b**) a circular *beam stop* surrounded by nothing. **c** The diffraction pattern shown as a dashed black line was calculated according to (8.30) and is the usual Airy pattern. The pattern shown as a red line was calculated by use of (8.31). We used the same listed parameters for both calculations. The central red spike in the center arises from the finite size of the incident beam with a broad Gaussian intensity distribution, as discussed in Sect. 8.6.3 in conjuction with Babinet's principle

$$\boxed{E_{\mathrm{trans}}^{\mathrm{hole}} + E_{\mathrm{trans}}^{\mathrm{stop}} = E_0 \mathrm{e}^{ikz_0}.} \tag{8.29}$$

When the fields with and without the beam stop at the central point O are added, they simply give the incident field. This is the expected result when neither the hole nor the beam stop are present. The edge contributions to the field from the hole and the beam stop, which give rise to diffraction, have exactly the same magnitude but opposite phase, so that they cancel.

8.6.3 Babinet's Principle

Babinet's principle [66], named after the French physicist Jacques Babinet (1794–1872), can be stated as follows.

> If a sample consists of well-defined regions (domains) with different light transmission, then the diffraction pattern will be the same if the response of the regions is inverted.

An example is a magnetic sample which contains microregions with different spin orientations. If we use circularly polarized light, then the two types of regions will transmit differently as discussed in Sect. 7.11, leading to a diffraction pattern. If we now switch either the photon angular momentum (polarization) or the magnetization directions in the microregions, the diffraction pattern will not change.

To illustrate the theorem we consider the far-field diffraction patterns of a circular hole in a large opaque screen and the opposite case of an opaque circular beam stop surrounded by nothing, illustrated in Fig. 8.13a. The shown cases constitute a simple example where we switch the "black" (opaque) and "white" (transparent) response of our "sample". According to Babinet's theorem the diffraction patterns for the two cases should be the same.

The case of a circular hole of radius R has been treated before, and in the Fraunhofer regime we obtain an Airy pattern. In order to directly compare the two cases, we assume an incident beam with a lateral Gaussian beam profile $\mathrm{e}^{-r^2/(2\sigma_R^2)}$ whose rms width is much larger than the radius of our circular object, $\sigma_R \gg R$. Because of the large Gaussian width, the field distribution is essentially flat over our circular area πR^2 of interest, and the field distribution in a distant plane is given by (4.70) or,

$$\begin{aligned} E_{\mathrm{diff}}^{\mathrm{hole}}(\rho) &= -\frac{\mathrm{i}}{\lambda z_0} E_0\, \mathrm{e}^{ikz_0\left(1+\frac{\rho^2}{2z_0^2}\right)} \int_0^R \int_0^{2\pi} \exp\left[-\frac{r^2}{2\sigma_R^2}\right] \mathrm{e}^{-ik\frac{r\rho\cos\gamma}{z_0}}\, r\,\mathrm{d}\gamma\,\mathrm{d}r \\ &\simeq -\mathrm{i}\frac{\pi R^2}{\lambda z_0} E_0\, \mathrm{e}^{ikz_0\left(1+\frac{\rho^2}{2z_0^2}\right)} \underbrace{\frac{2J_1(kR\rho/z_0)}{kR\rho/z_0}}_{\text{Airy function}}, \end{aligned} \tag{8.30}$$

where we could have eliminated the Gaussian function since it is essentially constant over the hole.

For the beam stop case, the diffracted fields are calculated the same way, but the integration limits in (8.30) are changed according to

$$E_{\text{diff}}^{\text{stop}}(\rho)=-\frac{\mathrm{i}}{\lambda z_0}E_0\,\mathrm{e}^{ikz_0\left(1+\frac{\rho^2}{2z_0^2}\right)}\int\limits_R^{\infty}\int\limits_0^{2\pi}\exp\left[-\frac{r^2}{2\sigma_R^2}\right]\mathrm{e}^{-\mathrm{i}k\frac{r\rho\cos\gamma}{z_0}}\,r\,\mathrm{d}\gamma\,\mathrm{d}r. \tag{8.31}$$

The diffracted fields for the two cases, obtained by numerical integration are shown in Fig. 8.13c for the listed assumed parameters.

While for the hole case, the incident Gaussian beam size of rms width $\sigma_R = 10R$ has no effect since only its central nearly constant part illuminates the hole, the beam size enters for the beam stop case since it defines the outer contour of the incident beam that passes outside the beam stop. The broad incident Gaussian beam shape will create its own reciprocal space diffraction pattern, which is the narrow Gaussian red spike in the center of Fig. 8.13c. This artifact will disappear when the width of the beam σ_R is increased. We see that the diffraction patterns, shown as the black dashed Airy pattern for the hole and the red one for the beam stop, are indeed the same.

8.7 Formulation of Polarization Dependent Diffractive Imaging

In line with the emphasis of the present book on the nature of x-rays, in particular the topic of coherence, the following sections focus on coherent diffractive imaging with x-rays, also referred to a lensless x-ray imaging. This type of imaging renders a reciprocal space diffraction pattern that in most cases needs to be inverted into real space. There are cases where this is not necessary. For example, if one is interested in the structure evolution with temperature or time on a particular length scale, one may simply measure the response at a certain momentum transfer that corresponds to a certain length scale in the sample.

In the following we consider the case of a thin film sample that is inhomogeneous and consists of "domains" in the $x-y$ plane of a sample that is illuminated along z. In general, "domains" arise from various kinds of sample inhomogeneities such as atomic composition, crystallographic structure, electronic structure, bond orientation, or magnetic orientation and may even be defined by artificially constructed regions like electronic or magnetic microdevices. They imprint themselves on the transmitted fields through changes in thickness $d(x, y)$ or through polarization or energy dependent optical constants $\delta^p(x, y)$ and $\beta^p(x, y)$. The optical constants and the sample thickness enter completely equivalently in the product $(\beta + \delta)kd$ that determines the transmitted field.

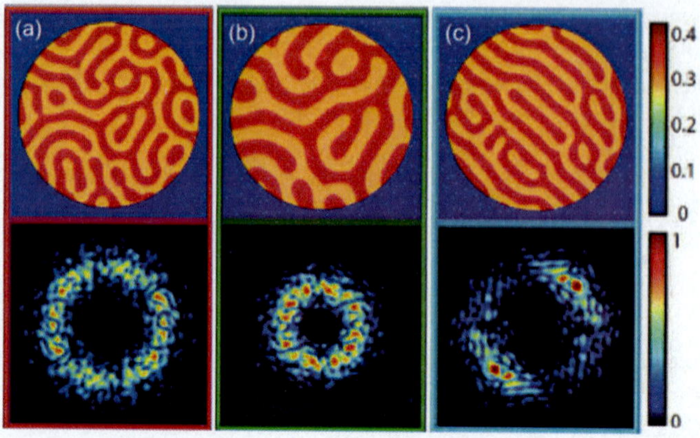

Fig. 8.14 Comparison of samples with different domain structures (top row) and the corresponding speckle diffraction patterns (bottom row). The field amplitudes emerging from the domains in the exit plane of the sample have changed on transmission, and the diffraction pattern is given by the Fourier transform of the field pattern. In the bottom row we have ignored the Airy diffraction rings that would arise from the shown circular shape of samples. From B. Wu, Ph.D. dissertation, Stanford University, 2012

The general relationship between the size and spatial orientation of domains and the corresponding reciprocal space diffraction patterns is illustrated in Fig. 8.14.

The shown domain and corresponding diffraction patterns are representative of different types of domains, like "charge domains" in polymer films with phase separated regions or "magnetic domains" in a ferromagnetic film with different magnetization directions. The diffraction patterns arise from different transmissions of the incident x-rays through the two types of domains. The different field amplitudes emerging from the regions in the exit plane of the sample then interfere to the shown intensity patterns. In the bottom row of Fig. 8.14 we have omitted for simplicity the Airy diffraction rings that would arise from the circular shapes of the illuminated regions (top row). Comparison of panels (a) and (b) demonstrates the inverse relationship between the sizes of the randomly oriented domains and the ring-like speckle diffraction patterns. Panel (c) illustrates the effect of preferential domain orientations as stripes and the preferential orientation of the speckle pattern perpendicular to the stripes.

In addition, "domains" may evolve reversibly in time like technological devices who are designed to reliably undergo an astronomical number of repetitive cycles. In such cases one can typically add images taken at a given time during the triggered temporal cycle. In other cases, a system may evolve spontaneously and irreversibly over a transient timescale down to femtoseconds. One then needs to capture a single snapshot image which is only possible with XFELs.

The key for the interpretation of the polarization dependent pattern is the lack of interference of the two orthogonal basis states in either a linear basis, ϵ_x, ϵ_y, or circular basis, ϵ_+, ϵ_-. For "domains" that differ due to their *charge properties* alone, the incident polarization is preserved upon transmission, if we ignore higher order effects beyond the dipole approximation, such as the XNCD effect discussed in Sect. 3.2.6. For *magnetic* "domains", the transmitted polarization is more complicated as discussed in Sect. 7.11. We can state as follows.

> The most versatile diffractive imaging approach relies on *polarization dependent resonant scattering*, which besides elemental and chemical specificity also allows the separation of *charge* and *spin*-based structures.
> In all cases, one first considers the amplitude and polarization dependent fields emerging locally from the exit plane of the sample. Their interference then determines the measured intensity diffraction pattern according to the Rayleigh-Sommerfeld diffraction formula (8.8), usually approximated in the far-field Fraunhofer form (8.16).
> Fields from different domains only interfere if they have the *same* polarization.

In the following we distinguish and separately discuss diffractive imaging by charge and magnetic domains.

8.7.1 Charge Domains with Orientational Order

We start by considering a *non*-magnetic film containing regions or domains that transmit x-rays differently. In diffractive imaging, the intensity distribution measured in the detector plane arises from the interference of the differently transmitted fields emerging from the domains. The exit plane of the film acts as the source of fields whose amplitude and phase are defined by transmission through the domains. We first consider a film that consists of regions that have the same chemical composition but differ in their *orientational order* introduced in Sect. 7.10.

We consider a sample with two types of charge regions as schematically illustrated in Fig. 8.15 and assume that the incident beam is monochromatic and laterally homogeneous and coherent over the sample region of interest. It is defined by a circular aperture of area $A = \pi R^2$ that is much larger than the domains. The effect of the aperture may be eliminated by increasing its size, so that its diffraction pattern is compressed into the centrally transmitted intensity.

We assume that the x-rays are incident along the surface normal z and that there are two kinds of domains distinguished through their chemical bond projections along the in-plane x and y axes of our coordinate system. Hence for *linear polarization* there is a difference in transmission through the domains due to the x-ray natural linear dichroism or XNLD effect. An example is a polymer film with orientational order as discussed in Sect. 7.10. Domains with different charge directors $\hat{\mathbf{u}}_x$ and $\hat{\mathbf{u}}_y$ would lead to different projections in the (x, y)-plane onto the $\mathbf{E}$-vector and hence different field transmissions. It can be shown that the optimum domain contrast is achieved if the $\mathbf{E}$-vector is aligned in the "forbidden geometry", perpendicular to the projection of one of the directors.[13] If the directors have equal projections along z, there is no diffraction contrast for circular polarized or "unpolarized" x-rays, because the random **E**-vector orientation in the x, y plane would lead to equivalent transmission through the domains.

For constant film thickness, the domain pattern $\hat{u}(x, y)$ of the sample is thus imprinted on the x-ray transmission pattern through the polarization dependent optical constants β and δ. In a phenomenological model, we distinguish the macroscopic transmission of a sample measured without spatial resolution in an x-ray absorption experiment (e.g. the XAS detector in Fig. 8.15) from the microscopic transmission. The microscopic difference could of course be observed by real space transmission x-ray microscopy, but we are here interested in the domain diffraction pattern in reciprocal space.

The transmitted intensity seen by the XAS detector in Fig. 8.15 would be an average of the polarization dependent parameters of the individual domains in the $x-y$ plane of the sample. If there is a different number N_x of domains with director projections along x, than N_y with projections along y, the *macroscopic* transmission would be an average described by

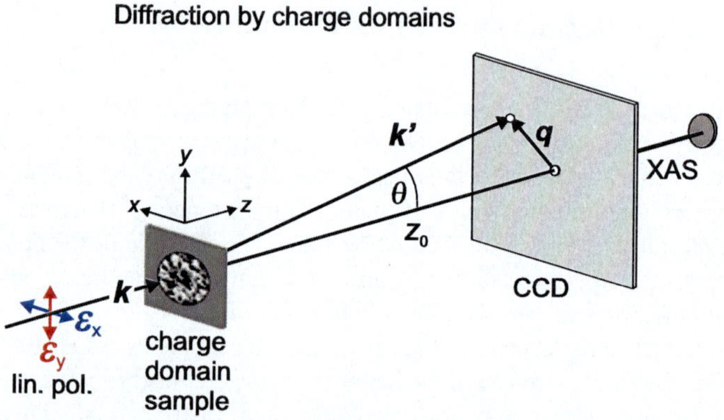

Fig. 8.15 Geometry for coherent diffraction imaging of charge domains. The x-rays are propagating into the z direction and are linearly polarized either along x or y. They traverse a sample of thickness d consisting of two types of charge domains, indicated in black and white. We have assumed that the illuminated area is defined by a circular aperture. The diffraction pattern is recorded by a position sensitive detector (CCD) at large distance z_0 from the exit plane of the sample. The scattered radiation is characterized by a momentum transfer **q**. A central hole in the CCD transmits the undiffracted beam ($q = 0$) to a separate detector which measures the average transmitted intensity in a conventional x-ray absorption (XAS) geometry

[13] The XNLD *diffraction* contrast has the same origin as that encountered in the *real space microscopy* image in Fig. 8.4b which arises from orthogonal directors in the horizontal and vertical directions in the cross-sectioned fiber. It is also related to the XMLD contrast from *antiferromagnetic* domains with different directors, as shown in Fig. 1.21b.

$$\overline{\beta} = \frac{N_x\beta_x + N_y\beta_y}{N_x + N_y}, \quad \overline{\delta} = \frac{N_x\delta_x + N_y\delta_y}{N_x + N_y}. \tag{8.32}$$

For $N_x \neq N_y$ there will be a finite *macroscopic* difference in *transmission* through the entire sample. It would be detectable by a change in total transmission as a function of **E**-vector orientation. For a given **E**-vector orientation there is a *microscopic* difference in domain transmission. It is responsible for *diffraction*. The *microscopic* difference between two domains is described by

$$\Delta\beta = \frac{\beta_x - \beta_y}{2}, \quad \Delta\delta = \frac{\delta_x - \delta_y}{2}. \tag{8.33}$$

For a film of constant thickness d, the transmitted field in the exit plane ($z_0 = 0$ in Fig. 8.15) has the position-dependent form of (7.90)

$$E_{\text{trans}}(x, y) = E_0\, \mathrm{e}^{ikd}\, \mathrm{e}^{-[\mathrm{i}\overline{\delta}+\overline{\beta}]kd}\, \mathrm{e}^{-\hat{u}(x,y)[\mathrm{i}\Delta\delta+\Delta\beta]kd}. \tag{8.34}$$

Here we have expressed the domain pattern in the binary form $\hat{u}(x, y) = \pm 1$, with the actual contrast at a given photon energy depending on the values of $\Delta\delta(\hbar\omega)$ and $\Delta\beta(\hbar\omega)$. A microscopic contrast will only exist when $\hbar\omega$ is tuned close to a molecular resonance (see Fig. 7.19). We can rewrite the last term as[14]

$$\mathrm{e}^{-\hat{u}(x,y)(\mathrm{i}\Delta\delta+\Delta\beta)\,kd} = \cosh[(\mathrm{i}\Delta\delta+\Delta\beta)kd] - \underbrace{\hat{u}(x, y)\sinh[(\mathrm{i}\Delta\delta+\Delta\beta)kd]}_{\text{domains}}, \tag{8.36}$$

where we have used that for $\hat{u} = \pm 1$ we have $\cosh(\hat{u}x) = \cosh x$ and $\sinh(\hat{u}x) = \hat{u}\sinh x$. The total diffracted field is seen to consist of two parts. The symmetric cosh term does not depend on the domain pattern, while the asymmetric term sinh reflects the domain contribution.

The diffracted field in the far field is calculated by use of the Fraunhofer diffraction formula (8.16) that involves a Fourier transform so that the diffracted field pattern in reciprocal space is given by

$$E_{\text{diff}}(q_x, q_y) = \frac{\mathrm{e}^{ikz_0}}{\mathrm{i}\lambda z_0} \iint_{-\infty}^{\infty} E_{\text{trans}}(x, y)\, \mathrm{e}^{-\mathrm{i}(q_x x + q_y y)}\, \mathrm{d}x\, \mathrm{d}y, \tag{8.37}$$

---

[14] We could have linearized (8.34) by approximating

$$\mathrm{e}^{-\hat{u}(x,y)[\mathrm{i}\Delta\delta+\Delta\beta]kd} \simeq 1 - \hat{u}(x, y)[\mathrm{i}\Delta\delta + \Delta\beta]kd, \tag{8.35}$$

which would lead to the same final result (8.43) below. In general, however, care has to be exercised when using linear approximations for the fields which may lead to incorrect interference terms when the absolute value squared or intensity is calculated (see Sect. 7.7.2). This is not the case here because $|\mathrm{i}\Delta\delta + \Delta\beta|^2$ gives a result that is correct to second order in $\Delta\delta$ and $\Delta\beta$, due to the complex form.

where the integrals are zero outside the illuminated sample area (aperture). The diffracted field becomes

$$E_{\mathrm{diff}}(\mathbf{q}) = E_0 \frac{\mathrm{e}^{\mathrm{i}kz_0}}{\mathrm{i}\lambda z_0} \, \mathrm{e}^{\mathrm{i}kd} \, \mathrm{e}^{-(\mathrm{i}\bar{\delta}+\bar{\beta})kd} \, D(\mathbf{q}), \tag{8.38}$$

where

$$D(\mathbf{q}) = \cosh[(\mathrm{i}\Delta\delta+\Delta\beta)kd] \underbrace{\iint_{-\infty}^{\infty} \mathrm{e}^{-\mathrm{i}(q_x x+q_y y)} \, \mathrm{d}x \, \mathrm{d}y}_{D_A(\mathbf{q})}$$
$$- \sinh[(\mathrm{i}\Delta\delta+\Delta\beta)kd] \underbrace{\iint_{-\infty}^{\infty} \hat{u}(x,y) \, \mathrm{e}^{-\mathrm{i}(q_x x+q_y y)} \, \mathrm{d}x \, \mathrm{d}y}_{D_{\hat{u}}(\mathbf{q})}. \tag{8.39}$$

Here $D_A(\mathbf{q})$ is the Fourier transform of the aperture in Fig. 8.15 and $D_{\hat{u}}(\mathbf{q})$ the Fourier transform of the domain pattern. They both have the dimension [area] because of the integration. It is important to realize that the aperture contribution is a *non-resonant* interference pattern, while the domain pattern strongly changes by tuning through a molecular resonance of the film. The diffraction pattern is given by the absolute value squared of the diffracted field or $|E_{\mathrm{diff}}(\mathbf{q})|^2$, and we see that there is interference between the two terms in (8.39).

In calculating the diffraction pattern we assume $|\Delta\delta| < |\bar{\delta}|$ and $|\Delta\beta| < |\bar{\beta}|$, i.e. that the dichroic resonance effect is smaller than the polarization independent resonance effect. We can then expand the diffraction pattern $|E_{\mathrm{diff}}(\mathbf{q})|^2$ to second order in $\Delta\delta$ and $\Delta\beta$. In addition we write the complex patterns as $D_A = |D_A| \, \mathrm{e}^{\mathrm{i}\phi_A}$ and $D_{\hat{u}} = |D_{\hat{u}}| \, \mathrm{e}^{\mathrm{i}\phi_u}$ and define the $\mathbf{q}$-dependent phase difference as $\phi = \phi_A - \phi_u$. This yields the following result.[15]

---

[15] In second order of $\Delta\delta$ and $\Delta\beta$ the cosh and sinh expressions in (8.36) are evaluated as

$$|\cosh[(\mathrm{i}\Delta\delta+\Delta\beta)kd]|^2 \simeq 1 + [-(\Delta\delta)^2 + (\Delta\beta)^2]k^2d^2 \simeq 1 \tag{8.40}$$

$$|\sinh[(\mathrm{i}\Delta\delta+\Delta\beta)kd]|^2 \simeq [(\Delta\delta)^2 + (\Delta\beta)^2]k^2d^2, \tag{8.41}$$

and the cross term is

$$- D_A D_{\hat{u}}^*[-\mathrm{i}\Delta\delta + \Delta\beta]kd - D_A^* D_{\hat{u}}[\mathrm{i}\Delta\delta + \Delta\beta]kd = -2|D_A D_{\hat{u}}|(\Delta\delta \sin\phi + \Delta\beta \cos\phi)kd. \tag{8.42}$$

For incident linearly polarized light, the diffracted intensity from *charge domains* within an illuminated sample area, distinguished by energy or polarization dependent differences in optical parameters $\Delta\delta$ and $\Delta\beta$ up to second order, is given by

$$I_{\text{diff}}(\mathbf{q}) = \frac{I_0}{\lambda^2 z_0^2}\,\mathrm{e}^{-2\bar{\beta}kd}\Big\{|D_A(\mathbf{q})|^2 + |D_{\hat{u}}(\mathbf{q})|^2\left[(\Delta\delta)^2 + (\Delta\beta)^2\right]k^2d^2 - 2\,|D_A(\mathbf{q})||D_{\hat{u}}(\mathbf{q})|\,(\Delta\delta\,\sin\phi + \Delta\beta\,\cos\phi)kd\Big\} \tag{8.43}$$

$\phi = \phi_A - \phi_u$ is the $\mathbf{q}$-dependent phase shift between the complex amplitudes $D_A = |D_A|\,\mathrm{e}^{i\phi_A}$ and $D_{\hat{u}} = |D_{\hat{u}}|\,\mathrm{e}^{i\phi_u}$.

In the forward direction, $\mathbf{q} = 0$, (8.43) needs to reduce to the transmitted intensity measured in XAS spectroscopy. The term $|D_A(\mathbf{q} = 0)|^2$ in (8.43) is just the on-axis transmitted intensity. For a circular aperture of area $A = \pi R^2$, for example, $|D_A(\mathbf{q})|^2$ is the Airy pattern

$$|D_A(\mathbf{q})|^2 = \left|\iint_{-\infty}^{\infty} \mathrm{e}^{-\mathrm{i}(q_x x + q_y y)}\,\mathrm{d}x\,\mathrm{d}y\right|^2 = A^2\left[\frac{2J_1(qR)}{qR}\right]^2 \tag{8.44}$$

so that the first term in (8.43) at $\mathbf{q} = 0$ becomes

$$I_{\text{diff}}(\mathbf{q} = 0) = I_0\,\frac{A^2}{\lambda^2 z_0^2}\,\mathrm{e}^{-2\bar{\beta}kd}. \tag{8.45}$$

For an equal number of domains, the second term in (8.43) can be neglected and the last term averages to zero. The propagation factor from the sample to the detector $A^2/(\lambda^2 z_0^2)$ is unity in the forward direction.[16] Then (8.45) reduces to the Beer-Lambert law for the transmitted intensity (7.32).

8.7.2 Charge Domains of Different Chemical Composition

In practice, charge domains may arise from areas of different thickness, atomic composition, or bond orientation, which all lead to differential attenuations of the incident field. For example, domains with different atomic compositions or different molecular groups, studied by Ade and Hitchcock [25], are distinguished by their

[16] The coherently forward scattered intensity is confined to the coherence cone $\mathrm{d}\Omega = \lambda^2/A$ or to the area $B = \lambda^2 z_0^2/A$ at distance z_0. In the absence of diffraction, the transmitted beam retains the size of the incident beam so that $B = A$ and $A^2 = \lambda^2 z_0^2$.

resonance energies (see Fig. 1.18). When the beam is tuned to an atom- or bond-specific resonance that exists in one but not the other domain, the relative transmission contrast could be very large.

For the description of this case we again follow Fig. 8.15, but now assume that there is no orientational order in our sample so that the transmission is independent of polarization. The diffractive imaging contrast is strongly photon energy dependent and will mostly disappear when the incident energy is tuned away from any absorption resonances in the sample and be maximum when the atoms in one domain are resonant and those in the other are not.

The fields at the XAS detector in Fig. 8.15 would be the average of those transmitted through the compositionally different domains. If there are different numbers N_1 and N_2 of the two types of domains, the *macroscopic* transmission is described by the average of the optical constants

$$\overline{\beta} = \frac{N_1\beta_1 + N_2\beta_2}{N_1 + N_2}, \quad \overline{\delta} = \frac{N_1\delta_1 + N_2\delta_2}{N_1 + N_2}. \tag{8.46}$$

The *microscopic* transmission change depends on the differences of the optical parameters in the domains

$$\Delta\beta = \frac{\beta_1 - \beta_2}{2}, \quad \Delta\delta = \frac{\delta_1 - \delta_2}{2}. \tag{8.47}$$

In this case the phenomenological description of is just the same as for the orientational order case, with the polarization dependence of the optical parameters replaced by the photon energy dependence.

8.7.3 Diffraction by Ferromagnetic Domains

Next we turn our attention to diffraction by ferromagnetic domains. We again assume that the sample is coherently illuminated along z through a circular aperture of area $A = \pi R^2$ and that the sample itself has a uniform composition and thickness d. The case is illustrated in Fig. 8.16.

The inhomogeneity of the sample is assumed to be entirely due to ferromagnetic domains of opposite magnetic orientations $\hat{m}_z(x, y) = \pm 1$ which are much smaller than the aperture. In this case we can take advantage of our description of the transmission of polarized light through a magnetically oriented film (i.e. a single domain) in Sect. 7.11. We will see that the description of magnetic domain diffraction resembles the case of diffraction from charge domains with orientational order described in Sect. 8.7.1. For the magnetic case, however, there is a diffraction effect for both linear and circular polarized x-rays. The effect of the aperture may again be eliminated by increasing its size so that its diffraction pattern is compressed into the centrally transmitted intensity. In the present case, however, it serves an interesting

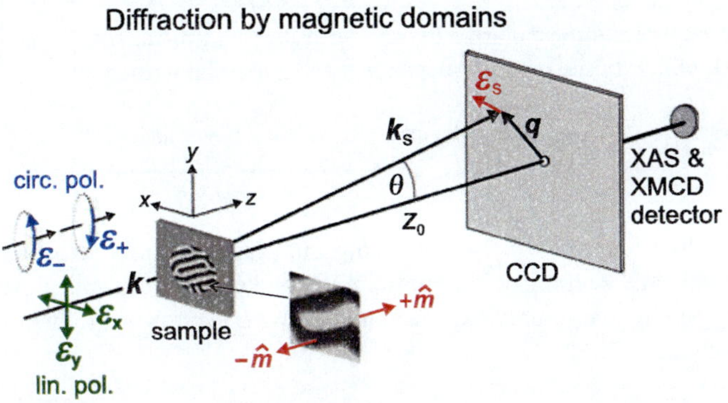

Fig. 8.16 Geometry for coherent diffraction imaging of magnetic domains. The incident x-rays are propagating into the z direction and have a polarization ϵ_p, where $p = \pm$ for circular and $p = 0$ for linear polarization. They traverse a sample of thickness d whose unit magnetization direction $\hat{\mathbf{m}}$ in the two domains is in opposite directions, $+\hat{m}_z$ and $-\hat{m}_z$ as indicated in red. The scattered radiation is characterized by a momentum transfer $\mathbf{q}$ and polarization ϵ_s, and its diffracted intensity is recorded with a position sensitive detector (CCD) located a distance z_0 from the exit plane of the sample. A central hole in the CCD allows the measurement of the transmitted XAS and XMCD intensity ($\mathbf{q} = 0$) by a separate detector. For XMCD measurements the entire sample may be uniformly magnetized along $+\hat{m}_z$ or $-\hat{m}_z$

purpose since it represent, charge scattering in contrast to magnetic scattering from the domains. We can therefore address the question whether charge and magnetic scattering can be separated.

8.7.3.1 Incident Circular Polarization

Following Sect. 7.11.2 we describe the film response phenomenologically by optical parameters in terms of the circular polarization response $\pm$ and labeling the number of domains $N_\pm$ by their polarization response, the average response of the sample is given in analogy to (8.46) by

$$\overline{\beta} = \frac{N_+\beta_+ + N_-\beta_-}{N_+ + N_-}, \quad \overline{\delta} = \frac{N_+\delta_+ + N_-\delta_-}{N_+ + N_-}. \tag{8.48}$$

For $N_+ \neq N_-$ there will be a macroscopic XMCD transmission effect, detectable by change of the incident circular polarization. In addition, for a given polarization there is a *microscopic* difference in domain transmission which leads to *diffraction*, which we describe as

$$\Delta\beta = \frac{\beta_+ - \beta_-}{2}, \quad \Delta\delta = \frac{\delta_+ - \delta_-}{2}. \tag{8.49}$$

For incident *circularly* polarized x-rays with $p = \pm$, the transmitted field is given by (7.119), which in the position-dependent form may be written as

$$\mathbf{E}^{p}_{\text{trans}}(x, y) = E_0\, \mathrm{e}^{\mathrm{i}kd}\, \mathrm{e}^{-(\mathrm{i}\bar{\delta}+\bar{\beta})kd} \underbrace{\epsilon_p\, \mathrm{e}^{-p\,\hat{m}_z[\mathrm{i}\Delta\delta(x,y)+\Delta\beta(x,y)]kd}}_{\text{domains}}. \tag{8.50}$$

We now follow Sect. 8.7.3.2 by calculating the diffracted field in the Fraunhofer approximation and keeping only terms up to second order in the optical parameters. In analogy to the charge-only expressions (8.39) the aperture and domain patterns are then given by

$$D_A(\mathbf{q}) = \iint\limits_{-\infty}^{\infty} \mathrm{e}^{-\mathrm{i}(q_x x + q_y y)}\, \mathrm{d}x\, \mathrm{d}y$$

$$D_{\hat{m}_z}(\mathbf{q}) = \iint\limits_{-\infty}^{\infty} \hat{m}_z(x, y)\, \mathrm{e}^{-\mathrm{i}(q_x x + q_y y)}\, \mathrm{d}x\, \mathrm{d}y. \tag{8.51}$$

The integral $D_A(\mathbf{q})$ is the Fourier transform of the illuminated shape function (e.g. a circular aperture), while $D_{\hat{m}_z}(\mathbf{q})$ is the Fourier transform of the magnetization pattern.

By expressing $D_A = |D_A|\,\mathrm{e}^{\mathrm{i}\phi_1}$ and $D_{\hat{m}_z} = |D_{\hat{m}_z}|\,\mathrm{e}^{\mathrm{i}\phi_2}$ and defining $\phi = \phi_1 - \phi_2$, and the field up to second order in $\Delta\delta$ and $\Delta\beta$, we obtain the following expression for incident *circular polarization* $p = \pm$ in analogy to (8.43).

For incident *circular polarization*, $p = \pm$ and unit magnetization directions $\hat{m}_z = \pm$, the diffracted intensity from *magnetic domains*, distinguished by energy and polarization dependent differences in optical parameters $\Delta\delta$ and $\Delta\beta$ within an illuminated sample area A, is given by

$$I^{p}_{\text{diff}}(\mathbf{q}) = \frac{I^{p}_{0}}{\lambda^2 z_0^2}\, \mathrm{e}^{-2\bar{\beta}kd} \Big\{ |D_A(\mathbf{q})|^2 + |D_{\hat{m}_z}(\mathbf{q})|^2 \left[(\Delta\delta)^2 + (\Delta\beta)^2\right] k^2 d^2 - 2p\hat{m}_z |D_A(\mathbf{q})||D_{\hat{m}_z}(\mathbf{q})| (\Delta\delta \sin\phi + \Delta\beta \cos\phi) kd \Big\}. \tag{8.52}$$

This magnetic domain expression which includes the aperture is valid only for circular polarization.

In the forward direction, $\mathbf{q}=0$, we have $|D_A(\mathbf{q}=0)| = A$ as previously discussed in conjunction with (8.45).

If the sample consists of an equal number of oppositely oriented magnetic domains or random domains, the second term can be neglected and the last term averages to zero, so that

$$I^p_{\text{diff}}(\mathbf{q}{=}0) = I^p_0 \, \frac{A^2}{\lambda^2 z_0^2} \, \mathrm{e}^{-2\overline{\beta}\, kd}. \tag{8.53}$$

With $A^2 = \lambda^2 z_0^2$ this reduces to the transmitted XAS intensity given by (7.127).

If the sample consists of a *single magnetic domain* of optical parameter $\overline{\beta} \pm \Delta\beta$, the second term in (8.52) can be neglected and in the last term the relative phase is $\phi = 0$. Hence we have $\sin\phi = 0$ and $\cos\phi = \pm 1 = \hat{m}_z$ and also $|D_{\hat{m}_z}(\mathbf{q}{=}0)| = A$. With $A^2 = \lambda^2 z_0^2$ we obtain

$$I^p_{\text{diff}}(\mathbf{q}{=}0) = I^p_0 \, \mathrm{e}^{-2\overline{\beta} kd} \left\{1 - 2\, p\, \hat{m}_z \Delta\beta\, kd\right\}. \tag{8.54}$$

The terms in curly brackets are the first order corrections due to $\Delta\beta$ in the general expression (7.120).

8.7.3.2 Incident Linear Polarization

For incident *linear* light, we utilize the results of Sect. 7.11.4 and start with expression (7.125) for the transmitted field in a linear polarization basis in the form,

$$\mathbf{E}_{\text{trans}} = E_0 \, \mathrm{e}^{ikd} \mathrm{e}^{-(\mathrm{i}\overline{\delta}+\overline{\beta})kd} \left\{\boldsymbol{\epsilon}_x \cos[(\Delta\delta - \mathrm{i}\Delta\beta)kd] + \boldsymbol{\epsilon}_y \hat{m}_z \sin[(\Delta\delta - \mathrm{i}\Delta\beta)kd]\right\}. \tag{8.55}$$

Here we have replaced the single-domain linear polarization response expressed by (7.118) by the domain average in (8.56) according to

$$\begin{aligned} \beta^0 &= \frac{\beta_+ + \beta_-}{2} \Rightarrow \overline{\beta} = \frac{N_+\beta_+ + N_-\beta_-}{N_+ + N_-} \\ \delta^0 &= \frac{\delta_+ + \delta_-}{2} \Rightarrow \overline{\delta} = \frac{N_+\delta_+ + N_-\delta_-}{N_+ + N_-}. \end{aligned} \tag{8.56}$$

The individual dichroic domain responses $\Delta\delta$ and $\Delta\beta$ are still given by (7.117) or (8.49).

We approximate (8.55), which is valid to all orders in $\Delta\delta$ and $\Delta\beta$, to second order and write it in the position-dependent form

$$\begin{aligned} \mathbf{E}_{\text{trans}}(x, y) &= E_0 \, \mathrm{e}^{ikd} \mathrm{e}^{-(\mathrm{i}\overline{\delta}+\overline{\beta})kd} \\ &\quad \times \left\{\boldsymbol{\epsilon}_x \left[1 - \frac{1}{2}(\Delta\delta - \mathrm{i}\Delta\beta)^2 k^2 d^2\right] + \boldsymbol{\epsilon}_y \hat{m}_z (\Delta\delta - \mathrm{i}\Delta\beta) kd\right\} \\ &\simeq E_0 \, \mathrm{e}^{ikd} \, \mathrm{e}^{-(\mathrm{i}\overline{\delta}+\overline{\beta})kd} \left\{\boldsymbol{\epsilon}_x + \boldsymbol{\epsilon}_y \hat{m}_z (\Delta\delta - \mathrm{i}\Delta\beta) kd\right\}. \end{aligned} \tag{8.57}$$

Note that the second order term in the expansion of $\sin[(\Delta\delta - \mathrm{i}\Delta\beta)kd]$ is zero, so that the expression is valid to second order in the optical parameters. The diffracted field given by (8.37) is obtained as

$$\mathbf{E}_{\mathrm{diff}}(\mathbf{q}) = E_0 \frac{\mathrm{e}^{\mathrm{i}kz_0}}{\mathrm{i}\lambda z_0}\, \mathrm{e}^{\mathrm{i}kd}\, \mathrm{e}^{-(\mathrm{i}\overline{\delta}+\overline{\beta})kd} \left\{ \boldsymbol{\epsilon}_x D_A(\mathbf{q}) + \boldsymbol{\epsilon}_y (\Delta\delta - \mathrm{i}\Delta\beta) kd\, D_{\hat{m}_z}(\mathbf{q}) \right\}. \tag{8.58}$$

The diffracted intensity is given as follows.

For incident *linear polarization*, $p = 0$, the diffracted intensity from *magnetic domains* is given by

$$I^0_{\mathrm{diff}}(\mathbf{q}) = \frac{I_0}{\lambda^2 z_0^2}\, \mathrm{e}^{-2\overline{\beta}\, kd} \Big\{ \underbrace{|D_A(\mathbf{q})|^2}_{\text{charge}} + \underbrace{|D_{\hat{m}_z}(\mathbf{q})|^2 \left[(\Delta\delta)^2 + (\Delta\beta)^2 \right] k^2 d^2}_{\text{magnetic domains}} \Big\}. \tag{8.59}$$

There is a key difference of this pattern relative to that for circular polarization (8.52) in that there is no interference between the first (charge) term and the second (domain) term due to orthogonal polarizations.

In the limit $\mathbf{q}=0$, (8.59) reduces to the XAS intensity. In this case we have $D_A(\mathbf{q}=0)=1$ and either $D_{\hat{m}_z}(\mathbf{q}=0)=\hat{m}_z$ for a single-domain sample or $D_{\hat{m}_z}(\mathbf{q}=0)=0$ for a sample with equal number of oppositely oriented magnetic domains. In both cases we obtain from (8.59) within the central beam defined by $A^2 = \lambda^2 z_0^2$

$$I^0_{\mathrm{diff}}(\mathbf{q}=0) = I^0_0\, \mathrm{e}^{-2\overline{\beta}\, kd}, \tag{8.60}$$

which is the Beer-Lambert transmission law.

8.7.4 *Separation of Charge and Magnetic Contrast*

The different diffraction patterns of a sample with magnetic domains, illuminated with linear and circular incident polarization, are illustrated in Fig. 8.17 by a simulation.

In (a) we show the assumed sample domain pattern that is illuminated through a circular aperture. The aperture produces an Airy diffraction pattern arising from the charge contrast at its edge, which is present in the diffraction patterns shown in (b). For incident *linearly* polarized light the pattern given by (8.59) reveals separate Airy and magnetic patterns, while for *circularly* polarized light, the magnetic pattern and aperture pattern are intertwined according to (8.52) and cannot be separated. Note that the magnetic domains are not perfect stripe domains so that weaker magnetic speckles are also present besides the dominant red ones.

If the same domain pattern shown in Fig. 8.17a was due to charge domains, e.g. in polymer blends [25], one would observe the same diffraction pattern as the one

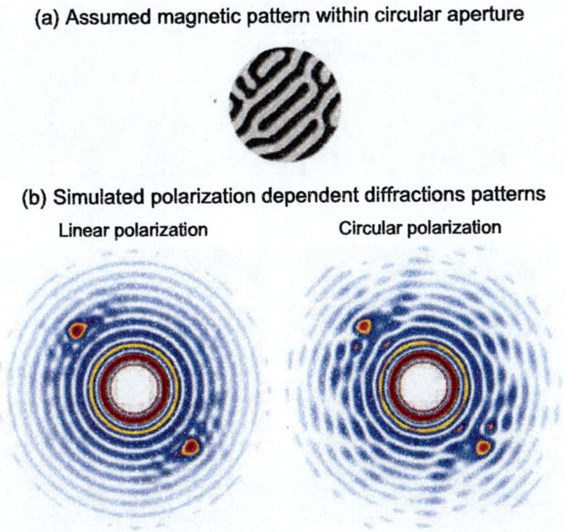

Fig. 8.17 **a** Assumed illuminated sample area containing domains, with opposite magnetization directions along the incident beam. **b** Simulated diffraction patterns for incident linearly and circularly polarized light. The central intensity has been omitted within the white circular region. For linear polarization, charge and magnetic diffraction patterns superimpose separately. For circular polarization, there is interference between charge and magnetic diffraction, visible as modulation of the circular Airy rings. From B. Wu, Ph.D. dissertation, Stanford University, 2012

recorded with circular light in Fig. 8.17b. For the charge case, interference between the aperture and charge domain pattern is always present.

A practical demonstration of the separation of the charge and magnetic patterns recorded for a sample similar to that shown in Fig. 8.17 is given in Fig. 8.18 [67].

According to (8.55) the incident linearly polarized x-rays, with polarization $\boldsymbol{\epsilon}_x$, scattered by the charge aperture remain polarized along x (blue), while the scattered fields by the domains rotate the polarization by $\pm 90°$ along y (red). There is no interference between the charge and magnetically scattered fields due to their orthogonality. The magnetic speckle pattern (red frame) in Fig. 8.18 shows that the worm domains have a preferential orientation so that the most intensity is observed perpendicular to the stripe-like domains.

We may summarize as follows.

> Comparison of (8.52) and (8.59) reveals that in the absence of a beam defining aperture, the diffraction pattern of ferromagnetic domains is the same for circularly and linearly polarized x-rays. In the presence of an aperture, it is advantageous to use linearly polarized light for magnetic diffraction, since the charge and magnetic patterns do not interfere and can be separated.

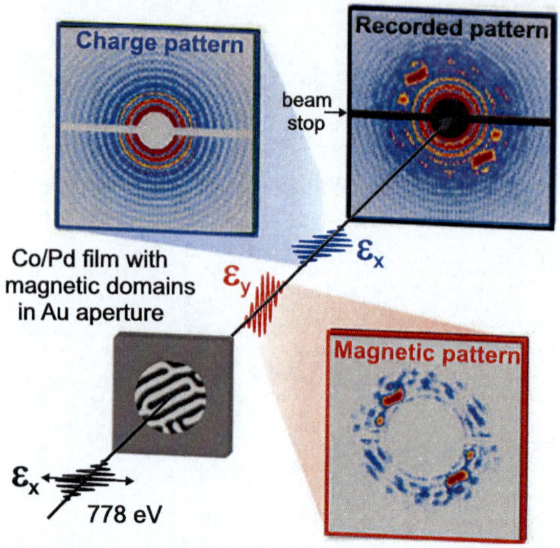

Fig. 8.18 Diffraction of incident *linearly* polarized x-rays by a ferromagnetic Co/Pd multilayer film, positioned behind a circular aperture [67]. The sample contains ferromagnetic worm domains with opposite magnetization directions along the x-ray propagation direction. The incident photon energy is resonantly tuned to the Co L_3-edge (778 eV) yielding the shown diffraction pattern (black frame), recorded with a CCD, described by (8.59). The "charge" Airy pattern (blue frame) of the circular aperture may be separated from the "magnetic" speckle pattern (red frame)

8.8 Illustration of the Phase Problem in X-Ray Diffractive Imaging

Finding solutions of the phase problem has undergone many advances over the last 100 years. Starting with structure determinations of macroscopic single crystals, the field progressed to samples consisting of stacked microscopic crystals (e.g. protein crystallography) resulting in structure determinations that clearly constitute one of the great scientific achievements during the twentieth century. More recently, with the advent of brighter more coherent sources, diffractive imaging has been extended to non-crystalline samples with some sort of order on the *nanoscale* (see Fig. 8.5). It has been suggested [68] that diffractive imaging of non-crystalline samples (like single proteins or viruses) can even be extended to the *atomic* scale, but this challenge still remains.

The topic of diffractive x-ray imaging is too broad and complex to be adequately covered in this book. It also involves advanced computational methods and tricks because of its general mathematical complexity. For this reason we will limit ourselves to a pedagogical discussion of the underlying physics of "phasing" and how an analytical solution can be obtained for the simplest case of diffraction from two atoms, slits or holes. For a more extensive treatment, the interested reader is referred

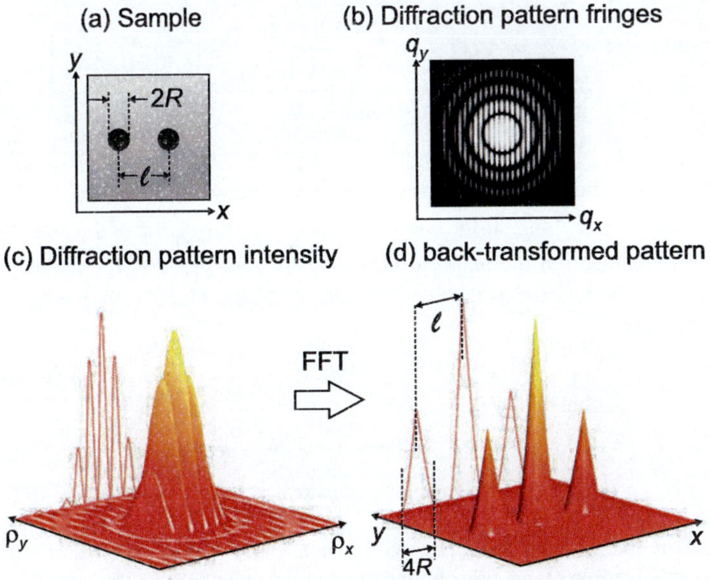

Fig. 8.19 **a** Assumed sample of two circular holes of radius R, separated by a distance ℓ along x. **b** 2D diffraction pattern at large distance z_0 as a function of momentum transfer q_x and q_y. **c** Diffraction pattern given by (8.61) ($\rho_i = q_i z_0/k$) for $\ell = 6R$ with the intensity shown vertically. **d** Fourier transform of the pattern in (**c**) back into the source plane, consisting of three conical peaks separated by ℓ with a base width $4R$. The central conical peak arises from the autocorrelation of a single hole with itself, the two side peaks from the cross-correlation of the two holes. The three-peak structure is a simple example of a Patterson map. From W. F. Schlotter, Ph.D. dissertation, Stanford University, 2007

to books and reviews elsewhere [1, 5, 69–71]. After we have developed a basic understanding of the problem, we will briefly come back to the historical developments in Sect. 8.10.

The origin of the phase problem is that the diffraction pattern is an *intensity* pattern where at a given point in the detector plane the interference information contained in the phases of the complex scattered fields *disappears* when the absolute value squared of the fields, the intensity, is calculated. While the *complex field* distribution in the detection plane that cannot be measured can be inverted through a Fourier transform to yield the field distribution at the exit plane of the sample, the detected *real intensity* pattern cannot be Fourier transformed back to obtain the intensity pattern or real space structure of the sample.

The problem is illustrated in Fig. 8.19 for the simple case of the diffraction pattern of two identical holes of radius R, whose centers are separated by a distance ℓ as shown in (a).

The *double hole* 2D diffraction pattern shown in Fig. 8.19b is given

$$\text{Double hole: } I(\rho) = 4\, I_0 \underbrace{\left[\frac{\pi R^2}{\lambda z_0}\right]^2 \left[\frac{2J_1(kR\rho/z_0)}{kR\rho/z_0}\right]^2}_{\text{single hole pattern}} \underbrace{\cos^2\left[\frac{k\ell\rho_x}{2z_0}\right]}_{\text{double hole modulation}}. \tag{8.61}$$

Here R is the radius of the holes, z_0 the distance between the source and detector planes, and $\rho = \sqrt{\rho_x^2 + \rho_y^2}$ is the lateral distance from the center of the detector plane which is related to the momentum transfer q by $\rho = qz_0/k$. The pattern exhibits similar fringe modulation as the *two-slit* pattern (8.20), which with $q = k\rho/z_0$ reads for comparison

$$\text{Double slit: } I(\rho) = 4\, I_0 \underbrace{\frac{a^2}{\lambda z_0}\text{sinc}^2\left[\frac{ka\rho}{2z_0}\right]}_{\text{single slit pattern}} \underbrace{\cos^2\left[\frac{k\ell\rho}{2z_0}\right]}_{\text{double slit modulation}}. \tag{8.62}$$

When the two-hole intensity pattern shown in Fig. 8.19b, c is Fourier transformed back into the sample plane, one obtains three cone-like peaks as shown in Fig. 8.19d. The middle peak arises from the "autocorrelation" of the *single hole* pattern.[17] The two identical side peaks arise from the "cross-correlation" between the *two holes*. The base width of the cone-like peaks is the convolution widths of the two holes.[18]

The backtransformed diffraction pattern in Fig. 8.19d differs significantly from the two-hole image, although it contains information on the distances R and ℓ associated with the holes and also shows that the holes are aligned along the x direction. It is a simple example of a so-called *Patterson map* [72, 73] that serves as a starting point for solving crystal structures to this day. In our case, the sample consists of two pinholes instead of the usual atomic scattering centers.

> In general, the Fourier transform of the measured intensity diffraction pattern of an object does *not* reproduce the desired real space structure of the object but corresponds to a so-called Patterson map. The reconstruction of the real space object requires a special experimental arrangement as in holography, or additional information that is obtained by "phasing techniques".

In the following sections we discuss different phasing techniques. We emphasize the physical formulation of the solution, if possible, in analytical form. An example is holography, which will be discussed first. In most cases, however, phasing techniques utilize the maximum information available and through so-called constraints guide iterative computational techniques to find a unique solution.

---

[17] The Greek derived word "auto" means self, so that autocorrelation means correlation with itself.

[18] By application of the convolution theorem, the shape of the conical peaks is given by the convolution of two flat-top circular intensity functions and it deviates from a true cone by flaring out at the bottom.

8.9 Fourier Transform Holography: FTH

Arguably the most elegant phasing technique for 2D structures is Fourier transform holography. In contrast to other solutions to the phase problem, which depend on iterative computational methods of phase retrieval, the holographic intensity image can be directly inverted by a Fourier transform to obtain a real image of the object of interest.

Holography was invented in 1948 by Dennis Gabor [74], who later received the Nobel Prize in Physics for it. The name is derived from the Greek word holos, meaning whole, and grafe meaning writing, to indicate that the method can completely determine the real space structure of the object. Opposite to common perception, where a hologram is envisioned to contain the entire 3D structure of an object, a hologram is typically recorded with a directional beam and contains very little 3D information. The most important application of holography is in 2D imaging.

Gabor's original concept was *in-line* holography, where the object is placed into a coherent incident beam, resembling a plane wave, of lateral extent larger than the specimen of interest. It continues outside the specimen to the detection plane, where it interferes with the fraction of the incident wave that is propagating through and modulated by the object of interest. The resolution is limited by the spatial resolution of the detector since the interference pattern exists at large angles where it possesses a high spatial frequency that needs to be resolved by pixels of a detector that are smaller than the fringe spacing.

The modern success of holography is to a large extent associated with the invention of *off-axis* holography, often called *Fourier transform holography*, in the 1960s by Leith and Upatnieks [75, 76], Stroke [77, 78], and Winthrop and Worthington [79], building on earlier work by Lohmann [80] in 1956. In this case, a spherical reference wave emitted by a small pinhole-like source that is laterally offset from the object interferes with the object waves. The essence of Fourier transform holography is phase encoding the conventional diffraction pattern of a sample of unknown structure, called the "object", with a "reference" beam. The reference beam introduces fringes with spatial frequencies inversely proportional to the distance between points on the object and the point of the reference source. The fringe information is used to reconstruct the real space object by Fourier inversion of the hologram.

In the x-ray region, a point-like reference source may be created by the small focal area of a zone plate, as first explored with a laboratory x-ray source by Reuter and Mahr [81] and later successfully implemented with an undulator source by McNulty et al. [82, 83], who could reconstruct the object with a resolution of 60 nm. Another way of creating the reference source is by drilling a tiny hole of about 100 nm into a thin film by use of a focused ion beam. This method was used by Eisebitt et al. [84, 85] in 2004, benefitting from the advances in nanoscale fabrication (see Sect. 8.9.3 below).

8.9.1 Illustration of FTH

For simplicity of the analytical formalism and its graphical illustration, we consider the famous Young's double-slit case, whose diffraction pattern in the plane perpendicular to the slit directions is given by (8.62). We shall step through the 1D holography formalism below, noting that it can be readily applied to the 2D case of two circular holes, with the replacement of the 1D sinc by the 2D Airy function [see (8.61)].

We represent a single slit of width w by

$$\text{slit}_w(x) = \begin{cases} 1 & \text{for } |x| < w/2 \\ 0 & \text{for } |x| > w/2. \end{cases} \tag{8.63}$$

We arrange two slits symmetrically around $x = 0$ and assume that they emit fields of real magnitude E_0 with flat-top distributions. The fields emerging from the slits of width $w = a$ are given by

$$\begin{aligned} E_1(x - \ell/2) &= E_0 \,\text{slit}_a(x - \ell/2) \\ E_2(x + \ell/2) &= E_0 \,\text{slit}_a(x + \ell/2). \end{aligned} \tag{8.64}$$

The fields E_1 and E_2 create the *intensity* diffraction pattern of our *object.*

We now add a *reference* field that is created by a third narrower slit of width $b \ll a$ that is offset from the center $x = 0$ by a distance ℓ_{ref} and given by

$$E_3(x + \ell_{\text{ref}}) = E_{\text{ref}} \,\text{slit}_b(x + \ell_{\text{ref}}). \tag{8.65}$$

The total emitted field becomes

$$E_{\text{tot}}(x) = E_0 \,\text{slit}_a(x - \ell/2) + E_0 \,\text{slit}_a(x + \ell/2). + E_{\text{ref}} \,\text{slit}_b(x + \ell_{\text{ref}}). \tag{8.66}$$

The diffracted field in a plane at large distance z_0 is given by the Fraunhofer diffraction formula (8.16) and consists of the Fourier transform of the total source field denoted $\mathcal{F}\{E_{\text{tot}}(x)\}$. By utilization of the linearity of the Fourier transform expressed by the Fourier *addition theorem,*

$$\boxed{\text{Addition theorem:}\; \mathcal{F}\{g + f\} = \mathcal{F}\{g\} + \mathcal{F}\{f\}} \tag{8.67}$$

the holographic diffracted field is obtained as

$$E(q) = E_0 \,\mathcal{F}\{\text{slit}_a(x - \ell/2)\} + E_0 \,\mathcal{F}\{\text{slit}_a(x + \ell/2)\} + E_{\text{ref}} \,\mathcal{F}\{\text{slit}_b(x + \ell_{\text{ref}})\}. \tag{8.68}$$

We now use the Fourier *shift theorem* in the form

$$\boxed{\text{Shift theorem}: \ \mathcal{F}\{g(x)\} = G(q_x) \ \Rightarrow \ \mathcal{F}\{g(x \pm \ell/2)\} = G(q_x)\,\mathrm{e}^{\pm \mathrm{i} q_x \ell/2}} \tag{8.69}$$

to obtain

$$E(q) = E_0\, \mathcal{F}\{\mathrm{slit}_a(x)\}\left[\mathrm{e}^{-\mathrm{i}q\ell/2} + \mathrm{e}^{+\mathrm{i}q\ell/2}\right] + E_{\mathrm{ref}}\, \mathcal{F}\{\mathrm{slit}_b(x)\}\,\mathrm{e}^{-\mathrm{i}q\ell_{\mathrm{ref}}}. \tag{8.70}$$

Including the propagation factor from the source to the detector plane [see (8.16)], the Fourier transform is

$$\mathcal{F}\{\mathrm{slit}_w[x]\} = \frac{w}{\sqrt{\lambda z_0}}\, \mathrm{sinc}\left[\frac{qw}{2}\right]. \tag{8.71}$$

The holographic intensity pattern is given by

$$|E(q)|^2 = \underbrace{4E_0^2\, \frac{a^2}{\lambda z_0}\, \mathrm{sinc}^2\left[\frac{qa}{2}\right]\cos^2\left[\frac{q\ell}{2}\right]}_{\text{object intensity pattern}: o^2(q)} + \underbrace{E_{\mathrm{ref}}^2\, \frac{b^2}{\lambda z_0}\, \mathrm{sinc}^2\left[\frac{qb}{2}\right]}_{\text{reference intensity pattern}: r^2(q)}$$

$$+ \boxed{\underbrace{2E_0\, \frac{a}{\sqrt{\lambda z_0}}\, \mathrm{sinc}\left[\frac{qa}{2}\right]\cos\left[\frac{q\ell}{2}\right]}_{\text{object field pattern}: o(q)}}\ \underbrace{2E_{\mathrm{ref}}\, \frac{b}{\sqrt{\lambda z_0}}\, \mathrm{sinc}\left[\frac{qb}{2}\right]\cos[q\ell_{\mathrm{ref}}]}_{\text{reference field pattern}: r(q)}. \tag{8.72}$$

The first two intensity patterns identified by underbraces are also referred to as the *autocorrelations* of the fields, while the product of the two-field patterns in the second line is called the *cross-correlation.*

The boxed part of the total expression holds the key to holographic recovery of the object in real space. Since it is the diffracted field of the object, its reverse Fourier transform into real space is the desired image of the object. All we have to do is figure out how to separate it from the rest.

First, the Fourier addition theorem assures that the two *autocorrelations* functions and the *cross-correlation* function can be backtransformed separately. By choice of the distance $\ell_{\mathrm{ref}} > \ell$, this also provides a means to spatially separate the real space location of the backtransform of the all-important cross-correlation function from that of the two backtransformed autocorrelations.

Before we illustrate this we need to point out another important aspect of the backtransform of the entire hologram $\mathcal{F}^{-1}\{|E(q)|^2\}$. It follows from the Fourier transform *convolution theorem* which for our real functions in (8.72) states as follows.

Defining the convolution between two functions by the symbol $\star$, we can write the inverse Fourier transform of the product of two real functions, $h(q)g(q)$, in terms of the Fourier convolution theorem

$$\mathcal{F}^{-1}[h(q)\, g(q)] = \mathcal{F}^{-1}[h(q)] \star \mathcal{F}^{-1}[g(q)] = H(x) \star G(x), \tag{8.73}$$

where the convolution is given by

$$H(x) \star G(x) = \int_{-\infty}^{\infty} H\left(x'\right)\, G\left(x - x'\right)\, \mathrm{d}x'. \tag{8.74}$$

This tells us that the Fourier backtransforms of both the autocorrelation and cross-correlation intensities in (8.72) correspond to a *convolution of the real space structures*. There are three convolutions involved. Convolution of the two-slit object with itself will result in three conical peaks, just like the convolution of the two-circular holes with themselves, i.e. the Patterson map, in Fig. 8.19c. The single reference slit convolution with itself will be a single conical peak. Finally, the convolution of the reference slit with the two-slit object will result in two peaks which for $b \ll a$ will reproduce the two-slit structure with a resolution limited by the finite size b of the reference slit. In the limit that the reference slit width approaches a Dirac δ-function, the object will be exactly reproduced.

8.9.1.1 Backtransform of the Hologram into Real Space

Because of the difficulty of evaluating the inverse transform of (8.72) analytically, we will illustrate it pictorially. Another reason for this approach is that in practice, holographic reconstruction always involves a pixelated image, typically recorded with a CCD detector.

For our illustration, we choose our object as two slits of width $a = 0.5$ and separation $\ell = 2$, symmetrically arranged around $x = 0$ by $\pm\ell/2$. A narrower reference slit of width $b = 0.05$ is placed off-center at $\ell_{\mathrm{ref}} = 6$, as illustrated in Fig. 8.20a. The slits act as sources with flat-top intensity distributions, as shown.

The total holographic intensity pattern $|E(q)|^2$, given by (8.72), for $E_0 = E_{\mathrm{ref}} = 1$ and $\lambda z_0 = 1$ is shown in Fig. 8.20b over a large range of $-80 \leq q \leq 80$, and its central part is shown in the inset enlarged over a range $-10 \leq q \leq 10$.

To facilitate Fourier inversion, we put the holographic intensity distribution shown in Fig. 8.20b on a 1D grid of a large number of 32,000 pixels over the range $-80 \leq q \leq 80$. Because the diffraction pattern is in reciprocal space, the large q range assures optimum *resolution*, so that it is limited only by the size of the reference slit as discussed above. On the other hand, the pixel size determines the *field of view* of the backtransformed image, which is therefore very large for our small pixels. The

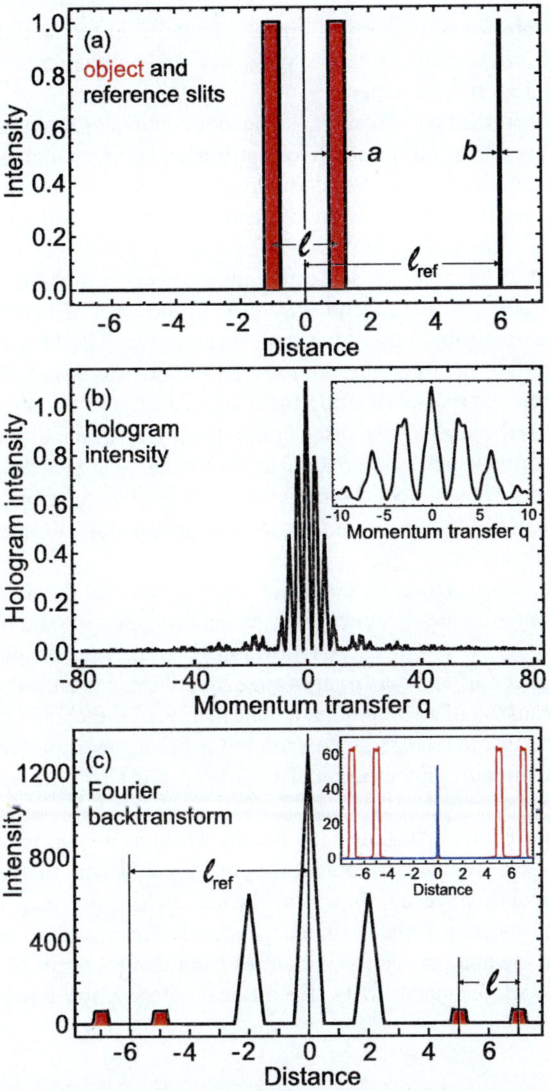

Fig. 8.20 **a** Assumed two-slit object with separation $\ell = 2$ and width $a = 0.5$ plus a shifted reference slit at $\ell_{\text{ref}} = 6$ and width $b = 0.05$, acting as flat-top intensity sources. **b** Interference pattern or intensity hologram as a function of q given by (8.72), assuming $E_0 = E_{\text{ref}} = 1$ and $\lambda z_0 = 1$. The inset shows the enlarged central structure. **c** Fourier backtransform of the entire intensity hologram shown in black. It consists of a superposition of the backtransformed two autocorrelation intensities $o^2(q)$ and $r^2(q)$ plus the backtransformed cross-correlation intensity $o(q)r(q)$ in (8.72). The total backtransform is dominated by the three triangular peak Patterson map with an added small central spike due to the backtransformed reference intensity $r^2(q)$, shown in blue in the inset. The reconstructed two-slit object consists of mirror images that are displaced from the center by ℓ_{ref}, highlighted by a red fill and shown enlarged in red in the inset

central part of total backtransformed image (about 1% of the total field of view) is shown in black in Fig. 8.20c. Our pedagogical simple slit example can readily be programmed to reveal these aspects.[19]

The backtransformed pattern in Fig. 8.20c looks confusing at first sight. We know, however, that it must have three additive contributions corresponding to the separate backtransforms of the two autocorrelation intensities $o^2(q)$ and $r^2(q)$ of the object and reference, plus the backtransform of the cross-correlation intensity $o(q)r(q)$ in (8.72).

We can readily identify the separate contributions by individually backtransforming the parts $o^2(q)$, $r^2(q)$, and $o(q)r(q)$. In the inset of Fig. 8.20c, we show the two smallest contributions to the total black curve. The blue central spike is the backtransform of the reference autocorrelation intensity $r^2(q)$. The red curve is the backtransformed cross-correlation intensity $o(q)r(q)$ of the object and the reference. The intensity scale of the inset shows that it is quite small relative to that of the total black curve which is dominated by three triangular peaks. They arise from the backtransform of the object autocorrelation intensity $o^2(q)$ which is nothing but the Patterson map of the two-slit object, in complete analogy to that shown for the 2D two-hole case in Fig. 8.19d.

The desired reconstruction of the *object* alone is reflected by the small displaced red-shaded rectangular areas, shown enlarged as a red curve in the inset. It consists of two mirror images, symmetrically positioned around the image center by the reference slit offset ℓ_{ref}. The twin images arise from the fact that the cross-correlation term actually consists of two conjugate contributions. The choice of a reasonably large ℓ_{ref} assures that the desired reconstructed twin images do not overlap with the autocorrelation structure in the center. The offset ℓ_{ref} is why "off-axis" holography works so well.

For our simple slit case,"in-line" holography works as well. If we place the reference slit right in-between the two object slits, i.e. $\ell_{\mathrm{ref}} = 0$, with the same parameters as before we obtain the result shown in Fig. 8.21. The twin images of the object now overlap and appear together with the autocorrelation images in the center. Their separation from the autocorrelation structures is however a result of our simple 1D slit case. In general, one needs to use the off-axis holography approach.

8.9.2 *Improved Reference Beams*

The extension from a single to multiple reference beams is illustrated in Fig. 8.22 [86, 87]. The figure shows how the offset distances and positions of the reference

[19] Details of detector-limited sampling are discussed in W. F. Schlotter, Ph.D. dissertation, Chap. 2, Stanford University, 2007.

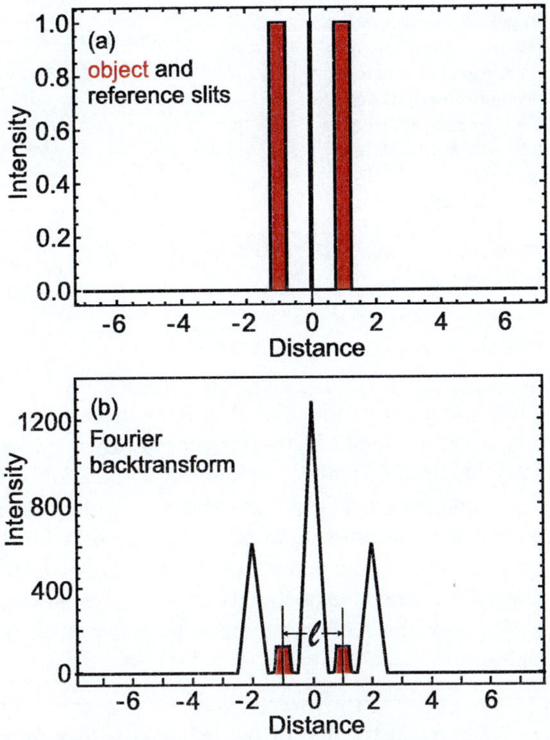

Fig. 8.21 **a** Assumed two-slit arrangement as in Fig. 8.20a but with $\ell_{\mathrm{ref}} = 0$. **b** Backtransform of the holographic diffraction pattern. In contrast to the pattern in Fig. 8.20c, the reconstructed two-slit object mirror images, identified by red shading, now overlap and occur in the center between the dominant three triangular Patterson map peaks

holes relative to the object, in form of an Ampelmännchen,[20] need to be chosen to avoid overlap of the twin images for each reference hole with each other and the central autocorrelation images. The different object images can be added for improved signal-to-noise, which has been utilized for single-shot imaging of nanostructures with femtosecond XFEL pulses [88].

Another extension of off-axis non-iterative FTH is the replacement of the reference pinhole by a slit, whose fabrication by focused ion beam (FIB) milling is less sensitive to exposure time. One then utilizes emergence of the reference wave from the sharp edge of the slit or an L-shaped slit. The slit width determines the resolution in the direction perpendicular to the slit, and the sharpness of the terminal edge determines the resolution along the slit. The technique was suggested by Guizar- Sicairos and Fienup [89, 90] and termed holography with extended reference by autocorrelation linear differential operation (HERALDO).

The differential holographic encoding technique was implemented for soft-ray imaging by Zhu and Guizar-Sicairos et al. [91, 92]. They compared the results obtained with the smallest reference pinhole of 40 nm that could be produced in

[20] The shown Ampelmännchen figure was used on pedestrian crosswalks in East Germany before unification, as a green light signal to indicate "walk". It was introduced to us by Andreas Scherz, a native Berliner.

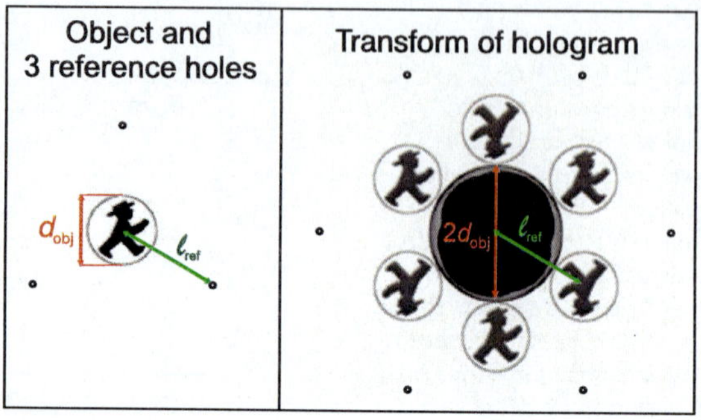

Fig. 8.22 Demonstration of the use of multi-reference holes and choice of their positions to avoid overlap of the twin images for each reference hole with each other and the central autocorrelation images. From W. F. Schlotter, Ph.D. dissertation, Stanford University, 2007

a fabrication series with that for a 20 nm wide L-shaped slit, whose fabrication was less sensitive to the FIB milling process. By use of an L-shaped slit they achieved a resolution of 16 nm by synthesizing images in the Fourier domain from a single diffraction pattern. Direct comparison of the HERALDO results with images of the same sample obtained by diffractive imaging with iterative phase retrieval or real space microscopy with state-of-the-art zone-plate STXM and TXM microscopes showed a comparable or slightly improved resolution.

We finish this section by stating three key attributes of Fourier transform holography.

> The reconstruction of the recorded hologram by Fourier inversion is fast, simple, and robust.
> The *spatial resolution* of the reconstructed image is determined by the reference hole size, if the maximum momentum transfer is kept large.
> The *field of view* of the reconstructed real space image is determined by the spatial frequency sampling of the hologram in reciprocal space, given by the pixel size of the CCD detector.

8.9.3 Application of FTH: Magnetic Domains

The FTH formalism for magnetic domain diffraction is an extension of that discussed in Sect. 8.7.3 with addition of the reference hole contribution. We consider the circular polarization case treated in Sect. 8.7.3.1. The diffraction pattern (8.52) then contains additional self-interference and cross-interference terms due to the reference hole.

The backtransformed diffraction pattern will consist of autocorrelation and cross-correlation images of the reference hole, the central aperture, and the domain pattern inside. The Fourier transforms of the fields emerging from the three structures are separated through the addition theorem. The offset ℓ_{ref} of the reference hole, assumed to be along x, is accounted for by the shift theorem, so that the total diffracted field is

$$E_{\mathrm{diff}}(\mathbf{q}) = E_0 \frac{\mathrm{e}^{ikz_0}}{\mathrm{i}\lambda z_0} \mathrm{e}^{ikd} \, \mathrm{e}^{-(\mathrm{i}\bar{\delta}+\bar{\beta})kd} \, D(\mathbf{q}), \tag{8.75}$$

where $D(\mathbf{q})$ is the total polarization dependent diffracted field given by the Fourier transform of the transmitted field in the exit plane of the sample. By linearizing the exponential transmission (see (8.35)) $D(\mathbf{q})$ depends on the relative alignment of the circular polarization helicity $p = \pm$ and the unit magnetization vector orientation $\hat{m}_z = \pm 1$ along the x-ray propagation direction (see Sect. 7.11.3), and we have

$$D(\mathbf{q}) = \underbrace{\mathrm{e}^{-\mathrm{i}q_x \ell_{\mathrm{ref}}} \int_{\mathrm{reference}} \mathrm{e}^{-\mathrm{i}(q_x x + q_y y)} \, \mathrm{d}x \, \mathrm{d}y}_{D_R(\mathbf{q})} + \underbrace{\int_{\mathrm{aperture}} \mathrm{e}^{-\mathrm{i}(q_x x + q_y y)} \, \mathrm{d}x \, \mathrm{d}y}_{D_A(\mathbf{q})}$$
$$- p\hat{m}_z(\mathrm{i}\Delta\delta + \Delta\beta)kd] \underbrace{\int_{\mathrm{aperture}} \hat{m}_z(x, y) \, \mathrm{e}^{-\mathrm{i}(q_x x + q_y y)} \, \mathrm{d}x \, \mathrm{d}y}_{D_{\hat{m}_z}(\mathbf{q})}. \tag{8.76}$$

The FTH intensity pattern is then given by $|E_{\mathrm{diff}}(\mathbf{q})|^2$ which has auto- and cross-correlation interference contributions. Our earlier pattern without the reference hole contribution (8.52) is then modified according to

$$I^p_{\mathrm{diff}}(\mathbf{q}) = \frac{I^p_0}{\lambda^2 z_0^2} \mathrm{e}^{-2\bar{\beta}kd}$$
$$\times \Big\{ |D_R(\mathbf{q})|^2 + |D_A(\mathbf{q})|^2 + |D_{\hat{m}_z}(\mathbf{q})|^2 \left[(\Delta\delta)^2 + (\Delta\beta)^2 \right] k^2 d^2$$
$$-2p\,\hat{m}_z |D_A(\mathbf{q})||D_{\hat{m}_z}(\mathbf{q})|(\Delta\delta \sin\phi + \Delta\beta \cos\phi)kd$$
$$\underbrace{+2|D_R(\mathbf{q})| \cos[q_x \ell_{\mathrm{ref}}] \Big[|D_A(\mathbf{q})| - 2p\,\hat{m}_z |D_{\hat{m}_z}(\mathbf{q})|(\Delta\delta \sin\phi + \Delta\beta \cos\phi)kd \Big]}_{\text{cross-correlation for image reconstruction}} \Big\}. \tag{8.77}$$

When this holographic diffraction pattern is transformed back into real space, the autocorrelation images due to the second line and the cross-correlation image arising from the third line will overlap in the center. The last underbraced term will create twin images, offset by ℓ_{ref} from the center, which reflect the desired real space structure of the central aperture and the domain pattern inside it.

The formalism was utilized for imaging the nanoscale magnetic domain structure of a magnetic multilayer by Eisebitt et al. [85]. As illustrated in Fig. 8.23, the incident circularly polarized x-rays from an undulator source were monochromatized by a spherical grating monochromator that determined the longitudinal coherence length. At the used photon energy 778 eV (wavelength 1.59 nm), corresponding to the Co L_3 resonance, the longitudinal coherence length was 1.6 µm. The lateral coherence length was improved by spatial filtering of the beam from the monochromator by a

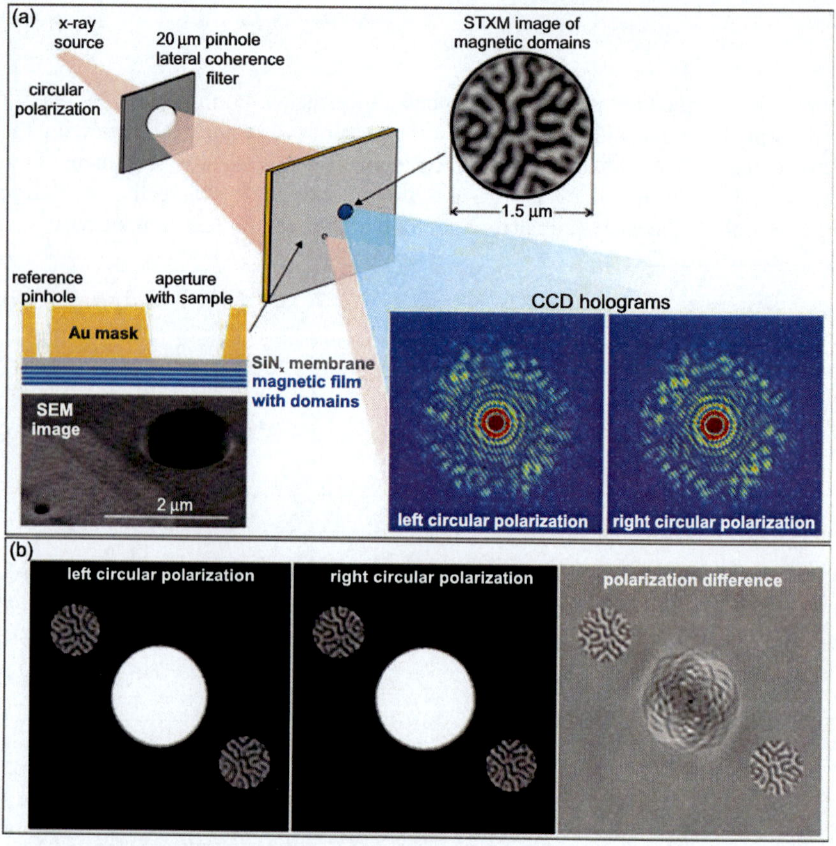

Fig. 8.23 **a** Experimental schematic. Monochromatized circular polarized x-rays are incident on a mask–sample structure after spatial coherence filtering. The object and reference beams are defined by the mask, and the resulting polarization dependent holograms, recorded with a CCD detector for $p = \pm$ at the Co L_3 resonance, are shown in the lower right. The lower left inset shows the layered sample structure and an electron microscopy image. The upper right inset shows a STXM image of the magnetic worm domain structure within the sample aperture. **b** Fourier transforms of the holograms recorded with opposite circular polarizations and their difference. From [85] and W. F. Schlotter, Ph.D. dissertation, Stanford University, 2007

circular aperture of 20 μm diameter. The coherent central Airy disk of the transmitted beam (see Fig. 4.18) had a transverse coherence length of 9.1 μm at the sample position, where it illuminated a circular Au transmission mask of 1.5 μm diameter directly in front of the multilayer sample.

The key element was the integrated reference, aperture plus sample design illustrated in Fig. 8.23a. It was fabricated by use of a 100-nm-thick Si_3N_4 membrane on a Si support frame. The front side of the membrane contained a 600-nm-thick Au film, the other side a magnetic multilayer film of 50 Co(0.4 nm)/Pt(0.7 nm) bilayers on a 20-nm-thick Pt base layer. The multilayer which contained magnetic worm domains with magnetization directions perpendicular to the plane was capped with a 2 nm Pt layer to prevent corrosion. A focused ion beam was then used to cut the circular aperture out of the gold film, down to the Si_3N_4 membrane. At a distance of 3 μm from the center of the aperture, a circular pinhole was drilled through the entire mask–sample structure by a focused ion beam. The high-aspect ratio pinhole had a conical shape along the beam direction with a minimum diameter of 100 nm at its exit. The diffraction pattern hologram is expressed by (8.77). It was recorded with an in-vacuum charge-coupled device (CCD) camera, with the photon energy tuned to the peak of the Co L_3 resonance, where $\Delta\delta \simeq 0$ and $\Delta\beta \simeq 2 \times 10^{-3}$ (see Fig. 7.25a).

Since no beam stop was used, the central part of the holograms shown on the lower right of Fig. 8.23a was dominated by the Airy pattern from the circular aperture which had approximately 10^5 times the intensity of the outer part of the pattern. The Airy pattern is modulated by a high-frequency speckle pattern arising from the interference of the centrally transmitted field with that from the offset reference pinhole. Because of the limited dynamic range of the CCD camera, 50 image frames were acquired, each with a 10 s exposure time.

The Fourier transform of the hologram is shown in Fig. 8.23b for opposite circular polarizations, with the two conjugate images, corresponding to the underbraced term in (8.77) separated from each other and offset from the central intensity arising from the other terms in (8.77). On circular polarization reversal of the incident soft x-rays, the magnetic domain contrast inverts, in accordance with (8.77). It would also invert if for a given x-ray polarization the magnetic orientation of all domains was inverted. The polarization *difference* can be used to enhance the magnetic contrast and suppress any non-magnetic contributions as shown. Note that Babinet's principle (see Sect. 8.6.3) does not apply to the magnetic domain case because the diffraction contrast is based on the relative *directions* of angular momentum vectors of x-rays and spins.

The 10–90% contrast change in the 1D image profiles (not shown) had a width of only 50 nm, smaller than the minimum 100 nm diameter of the reference pinhole. It is likely that the conical shape of the pinhole, tapered because of the 8° convergence of the focused ion beam, acted as a capillary waveguide, and reduced the effective focal spot size that determines the spatial resolution.

8.10 Non-holographic Solutions to the Phase Problem

Below we give a brief history of x-ray crystallography which elucidates how it was possible to overcome the phase problem in x-ray diffraction after the early work by von Laue, the Braggs, and others.

8.10.1 Brief History of X-Ray Crystallography

During the first 20 years of x-ray diffraction, relatively few crystal structures were solved. Solutions involved a "trial and error" approach based on Bragg's insight that diffraction peaks arise from reflections from periodic lattice planes in crystals and Paul Ewald's link of the "sphere of reflection" with the reciprocal lattice [93]. Experimental diffraction patterns of simple crystals with cubic lattices such as the alkali halides and diamond were compared with those calculated for envisioned structures, without the need for phase information. This trial and error method, however, rapidly stopped working for more complicated crystal structures.

Remarkably, some of the modern solutions of phasing diffraction patterns were already envisioned in the late 1920s. As early as 1925, Mark and Szilard [94] recognized the importance of phase changes near absorption edges, which many years later led to the *Single-wavelength Anomalous Diffraction* (SAD) and *Multi-wavelength Anomalous Diffraction* (MAD) techniques. Another idea was the utilization of phase changes associated with the isomorphic substitution of lighter by heavier atoms first demonstrated in 1927 by Cork [95],[21] which foreshadowed the development of *Multiple Isomorphous Replacement* (MIR). These ideas, however, were not fully developed at the time, and we will come back to them later.

One of the key advances in x-ray crystallography came in 1934/35 when A. L. Patterson developed a method, now called the *Patterson map*, that serves as a starting point for solving crystal structures to this day [72, 73]. The Patterson map contains peaks at positions of all possible interatomic vectors between the atoms of a crystal. The inverse Fourier transform of the two pinhole intensity pattern, shown by the three peak structure in Fig. 8.19c, is a simple example of a Patterson map with the pinholes representing atomic sources. The map contains information on possible distances as well as directions of the diffracting holes. As is readily revealed by this simple example, additional information is required to truly solve the structure. One important aspect of the Patterson map is that it is dominated by the vectors between heavy atoms in a crystal. The combination of the Patterson map with the MIR technique proved especially valuable and was utilized by Max Perutz and coworkers in completing the structure determination of hemoglobin in 1954 [96].

Arguably the most important theoretical development of crystallography started in the 1950s, leading to so-called *direct methods*. The work was pioneered by Herbert A. Hauptman and Jerome Karle between 1950 and 1956 and was awarded the 1985 Nobel Prize in Chemistry [6]. The method is termed "direct" because it allows the structure determination directly from the data by comprehensive computer algorithms whose implementation became possible with the advent of increasingly powerful computers. An important conceptual contribution was made in 1952 by David Sayre [97], who considered the case where a crystal does not contain heavy atoms but only

[21] Cork investigated the structure of crystals of a family called alums consisting of the symbolic chemical formula $A{\cdot}B{\cdot}(SO_4)_2{\cdot}12H_2O$, where "A" is a monovalent cation like K, Rb, Cs or Th and "B" a trivalent metallic ion like Al, Cr, or Fe. Cork also pointed out that S may be replaced by Se or Te, which later became an important use of MIR in protein crystallography.

light atoms like C, N, O, and H. He then showed that one may utilize the peaked electron density at atomic positions to exploit the similarity between the electron density and its square.

Any valid inversion of the diffraction pattern by direct methods has to give a density map on the *smallest* length scale of positive sign. Therefore any negative values do not reflect the real sample structure. On the *largest* length scale, the sample has a finite size and an inversion of the diffraction pattern should show no structure outside the sample boundary. These are the two most fundamental "constraints" on any structure solution. Today, the so-obtained complicated real space macromolecular structures may be visualized in different forms, ranging from early "ball and stick" models to the representation of larger structural motives by helix-shaped ribbons, as reviewed by Kaushik and Rath [98].

8.10.2 MIR, SAD, and MAD Image Reconstruction

The MIR and MAD techniques, whose deep roots were mentioned above, combine chemistry-based (atomic modification) and physics-based (x-ray tuning) solutions to the phase problem. The history of MIR has been reviewed by Rossmann [69] and that of MAD by Ramaseshan [70] and Hendrickson [71].

From an x-ray point of view, the suggestion of MAD by Mark and Szilard [94] in 1925 brought together the fields of x-ray diffraction and x-ray spectroscopy. At the time, the change of x-ray response near resonances was known and referred to as "dispersion", leading to the Kramers-Heisenberg dispersion formula [99] in 1925 and the Kramers Kronig relations soon thereafter [100, 101]. The marriage of the two fields is reflected by the work of Hönl [102], who in 1933 used atomic wave functions to obtain oscillator strengths of x-ray absorption resonances that were then used to calculate dispersive atomic scattering factor corrections.

The use of resonances for x-ray diffraction was further developed in the early 1950s by Bijvoet [103, 104] and Okaya et al. [105]. The two-wavelength MAD method was demonstrated in 1971 by Hoppe and Jakubowski [106] who determined the phase of Fe scatterers by use of two-photon energies below and above the Fe K-edge at 7,112 eV, corresponding to the characteristic Co K_α (6,930 eV) and Ni K_α (7,478 eV) emission lines. A similar approach was used in 1977 by Phillips et al. [107] who conveniently used tunable synchrotron radiation below and above the Fe K-edge.

On the theory side, Karle [108] and Hendrickson [109–111] played an important role in the formulation of SAD and MAD. Because of its modern importance, we will discuss the theoretical formulation of MAD in more detail in Sect. 8.11. SAD experiments are a subset of MAD experiments. The small anomalous SAD signal from intrinsic S atoms was shown to be sufficient to phase the structure of the small protein crambin in 1981 by Hendrickson and Teeter [112]. Another milestone is the use of MAD to determine the structure by use of the Se K-edge in selenomethionyl proteins by Hendrickson et al. [110, 113].

The heavy atom resonant scattering used in MAD may be envisioned as a holographic reference for the scattering by all atoms in the crystal. In practice, it utilizes the change of the phase of the heavy atom reference wave by recording diffraction patterns at different incident energies around the heavy atom absorption edge. The practical implementation of MAD was greatly aided by the availability of synchrotron radiation sources which not only provided higher intensity but readily allowed tuning of the x-ray energy to atom specific absorption edges. This has led to the dominant use of MAD for the determination of *de novo* crystal structures of biological macromolecules [71] (see Fig. 1.14).

8.10.3 Sampling of the Diffraction Pattern

An important issue in all phasing techniques is that the diffraction pattern is recorded as a mosaic by use of a CCD detector with finite pixel size. This means that in practice, the inversion of diffraction patterns into a real space image involves discrete rather than continuous sampling by use of a discrete Fourier transform [114]. The pixel size has to be small enough to avoid signal overlap or *aliasing*. The latter is avoided by fulfilling the so-called *Nyquist* criterion. Nyquist's paper of 1928 [115] considered when the information contained in a continuous temporal (telegraph) signal can be recovered when the signal is measured with a finite frequency bandwidth. In other words, when instead of the infinite number of sinusoidal Fourier components only a finite number are known. The work was extended by Kotelnikov in 1933 [116] and Shannon in 1949 [117], resulting in a theorem which today goes by various combinations of author names. The theorem appears to have been proven already by Whittaker in 1915 [118]. Nyquist stated his theorem by considering the sampling of a continuous temporal function of time, $f(t)$, as follows. If the maximum cycle frequency of $f(t)$ is B (which corresponds to the bandwidth in Hertz), the function $f(t)$ is completely determined if it is sampled at temporal points that are $1/(2B)$ or less apart.

In diffraction, the Nyquist criterion defines the maximum pixel size that can be used to record the continuous theoretical diffraction pattern without loss of information. It states that the maximum pixel size is determined by the inverse of twice the size of the diffracting object.

Let us illustrate the powerful criterion by considering our earlier example of the holographic diffraction pattern of a two-slit object and a pinhole reference, discussed in Sect. 8.9.1. For convenience we show the geometry again in Fig. 8.24a.

In our case, the highest frequency in the reciprocal space hologram is determined by the fringes produced by the object-reference interference term. It is the part of the total hologram pattern (8.72) given in normalized form by

$$f(q) = 4\,a\,b\,\mathrm{sinc}\left[\frac{qa}{2}\right]\mathrm{sinc}\left[\frac{qb}{2}\right]\cos\left[\frac{q\ell}{2}\right]\cos[q\ell_{\mathrm{ref}}]. \tag{8.78}$$

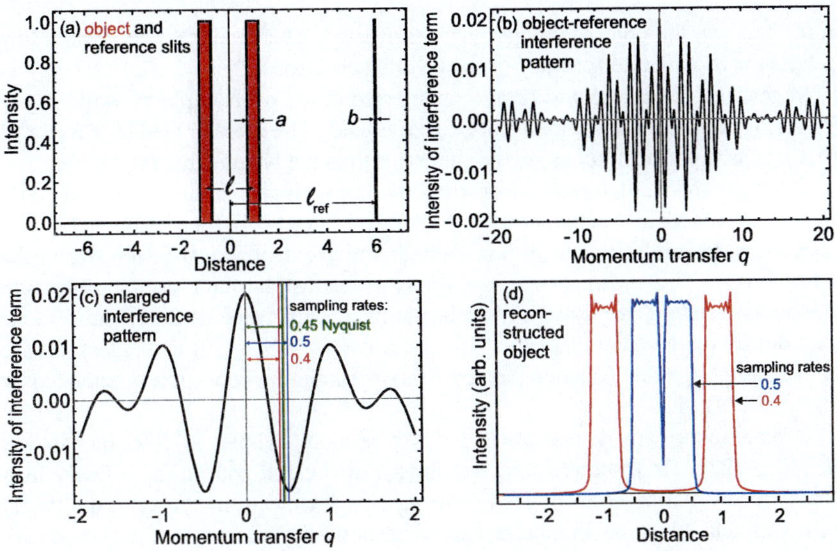

Fig. 8.24 **a** Object and reference geometry taken from Fig. 8.20. We have assumed $a = 0.5$, $\ell = 2$, $b = 0.01$, and $\ell_{\text{ref}} = 6$. **b, c** Interference pattern (8.78) on different q scales, with the maximum Nyquist sampling frequency $[\Delta q]_{\text{Ny}} = 0.45$ given by (8.79) indicated in green in (**c**). **d** Backtransform of (8.78) for two sampling intervals, showing the reconstructed object, with only one of the twin images shown. The two-slit object is only reproduced when the sampling interval satisfies the Nyquist criterium $\Delta q \leq 0.45$, as shown for the red reconstruction, where $\Delta q = 0.4$. For a larger sampling interval $\Delta q = 0.5$, the reconstructed blue pattern does not exhibit the proper separation between the slits ℓ_{ref}

The highest frequency in q is determined by the modulation due to the two cos terms. It arises from the largest real space distance $\ell_{\text{ref}} + \ell/2$ in Fig. 8.24a,[22] which in our case is $B = \ell_{\text{ref}} + \ell/2 = 7$. According to the Nyquist criterion, the reciprocal space pattern must be sampled at Δq intervals that satisfy

$$\Delta q \leq [\Delta q]_{\text{Ny}} = \frac{2\pi}{2B} = \frac{\pi}{7} = 0.45. \tag{8.79}$$

In practice, the pixels of our detector, located at a distance z_0 from the sample, must then have a size $\Delta \rho \leq [\Delta q]_{\text{Ny}} z_0/k$.

In Fig. 8.24d we show backtransforms of the intensity patterns in (b) given by (8.78) for two sampling intervals, yielding the blue and red reconstructed images of the object (only one of the twin images is shown). If the pattern is sampled at intervals $\Delta q = 0.5 > [\Delta q]_{\text{Ny}}$, as indicated in blue in (c) and (d), the reconstructed pattern does not exhibit the proper separation between the slits ℓ_{ref} due to aliasing. If the sampling interval is reduced slightly to $\Delta q = 0.4 < [\Delta q]_{\text{Ny}}$ as indicated in red

[22] This is directly revealed by the Fourier transform of the two cos terms, which exhibits peaks at distances $\ell_{\text{ref}} - \ell/2$, ℓ_{ref} and $\ell_{\text{ref}} + \ell/2$, as expected.

in (c) and (d), the object is properly reconstructed. The small change in sampling reveals the remarkable accuracy of the Nyquist criterion.

In general, the case $\Delta q < [\Delta q]_{\mathrm{Ny}}$ is referred to as "oversampling" while $\Delta q > [\Delta q]_{\mathrm{Ny}}$ is called "undersampling". A signal is said to be oversampled by a factor of N if it is sampled at N times smaller intervals than the Nyquist interval.

Direct methods attempt to reconstruct the object image from the intensity of a single wavelength diffraction pattern. As shown in Fig. 8.21b for the simple two-slit case, the autocorrelation function or Patterson map extends to twice the size of the actual image. This means that if one wants to reconstruct the object starting with the autocorrelation, the pattern must be sampled with twice as many intervals as required by the Nyquist criterion. This led Bates in 1982 [119] to suggest the use of oversampling the diffraction pattern. The oversampling concept was extended by Sayre in 1991 [120].

Oversampling was demonstrated for non-periodic samples in 1999 by Miao et al. [121, 122]. They showed that sampling a diffraction pattern more finely than the Nyquist criterion corresponds to surrounding the electron density of the object with a no-density region. Finer sampling corresponds to an increase of the no-density region, and when its size exceeds the electron density region, sufficient information is recorded to retrieve the phase. This work emphasized the importance of the isolation of the object from the non-diffracting surrounding region when only a single pattern is recorded. In practice, oversampling is used today together with iterative phase retrieval algorithms [1, 123–126].

8.10.4 Ptychography

The concept of *ptychography* was introduced by Hegerl and Hoppe in 1970 [127]. It exploits a *convolution* in reciprocal space between the Fourier transform of an entire object, $\mathcal{F}\{O(\mathbf{r})\}$, with that of a finite part of the object, $\mathcal{F}\{P(\mathbf{r}-\mathbf{R})\}$, typically defined by a movable aperture. The name was coined by Hegerl and Hoppe by substituting the German word for convolution "falten" by the Greek word "$\pi\tau\upsilon\xi$=ptycho" (similar to Gabor's substitution of "whole" by the Latin transliteration "holo" of the Greek word "$o\lambda o$"). The name thus describes a solution to the phase problem that is essentially based on the Fourier convolution theorem.

Remarkably, ptychography was nearly forgotten for almost four decades, mostly because of the limitation of coherent intensity emitted by electron and x-ray sources. It re-emerged through the work of Rodenburg and Faulkner in 2004 [128, 129] in conjunction with the use of iterative phasing algorithms. Owing to the increased coherent intensity of modern x-ray sources it was soon adopted by the synchrotron radiation community [130].

The history of ptychography within the broader efforts to overcome the phase problem has been reviewed by Rodenburg [131]. In his definition, ptychography is based on the use of at least two diffraction patterns, recorded with a relative lateral shift of object and illumination area. In practice, all that is required is a source of coherent radiation, an aperture, and a method of moving the aperture a known distance across the sample, which today is achieved by use of piezoelectric devices.

We cannot adequately cover ptychography here and refer the interested reader to the recent book by Jacobsen [1]. We restrict ourselves to some important points. In its simplest form, an extended object is coherently illuminated and a moveable aperture defines a restricted spatial function $P(\mathbf{r}-\mathbf{R})$ that describes the transmitted field. By shifting the aperture to two or more positions, $\mathbf{R}$, different regions of the object $O(\mathbf{r})$ are sampled. Importantly, the aperture is moved so that there is *overlap* between the sampled regions. This way the diffraction pattern of the entire object can be obtained with the overlapping regions acting as a link or constraint that is incorporated into a suitable iterative phase retrieval algorithm. The overlap constraint helps with convergence problems in the phase retrieval, especially when the object is complex. For a single aperture position, Faulkner and Rodenberg's ptychography algorithm [128, 129] reduces to the conventional phase retrieval one of Fienup [123].

From a physics point of view, ptychography exploits a convolution in *reciprocal space* between the Fourier transform of the object $\mathcal{F}\{O(\mathbf{r})\}$ and that of a finite-area function $\mathcal{F}\{P(\mathbf{r}-\mathbf{R})\}$, whose position is scanned across the bigger object. The diffracted intensity distribution has the form

$$I(\mathbf{q}) \propto |\mathcal{F}\{P(\mathbf{r}-\mathbf{R})\, O(\mathbf{r})\}|^2 = |\mathcal{F}\{P(\mathbf{r}-\mathbf{R})\} \star \mathcal{F}\{O(\mathbf{r})\}|^2 \,. \tag{8.80}$$

On the right, we recognize the convolution theorem (8.73) which gave the technique its name. Ptychography derives the phase knowledge by recording two or more diffraction patterns which are related by the convolution theorem. For a more detailed discussion of x-ray ptychography and its increasing number of applications, the reader is referred to a recent reviews by Pfeiffer [130]. The comprehensive book of Jacobsen [1] also discusses phase retrieval algorithms.

In the following section we come back to the MAD technique with a formal treatment, owing to its great importance in protein structure determinations and its use in reconstructing the speckle patterns of non-periodic nanoscale objects.

8.11 Multiple-Wavelength Anomalous Diffraction—MAD

The technique of *multiple-wavelength anomalous diffraction* (MAD) has revolutionized atomic scale structure determinations of macromolecules (see Fig. 1.14). The term "anomalous" refers to the fact that historically diffraction experiments were typically performed with fixed energy laboratory x-ray sources that induced only non-resonant Thomson scattering by the sample. When the x-ray energy was close to the absorption edge of certain atoms, the changed scattering was termed "anomalous" (see Sect. 7.5.1). Today, "anomalous" scattering is used synonymously with "resonant" scattering. Thomson scattering is the x-ray response of the entire atomic electron cloud, which is largely independent of photon energy, while the

MAD method utilizes intra-atomic transitions that give rise to strong atom-specific resonance behavior (see Fig. 7.14).

8.11.1 MAD of Macromolecular Crystals

The difference in scattering contributions from non-resonant Thomson and resonant MAD scattering is schematically illustrated in Fig. 8.25a, taken from the work of Hendrickson [110]. In order to enhance the signal from the molecular units they are stacked together to form a periodic arrangement or crystal (see Fig. 8.5). Each molecule in a crystal might contain a few thousand light atoms (such as C, N, and O) and few additional heavy atoms which are either native to the molecule (like Fe in hemoglobin) or are artificially introduced by isomorphous replacement (like Se for S). The light atoms are not shown individually in Fig. 8.25a but are contained inside the shown macromolecular envelopes. Heavy atoms are shown as circles. When the photon energy is *not* tuned to any of the absorption edges of the atoms, all atoms create Thomson scattering which is enhanced by the crystal periodicity in certain directions, resulting in Bragg peaks, as shown in Fig. 8.5a.

When the photon energy is tuned to the heavy atom absorption edge, enhanced resonance scattering occurs, as shown for the atomic scattering factors f'' and f' of Se in Fig. 8.25b. This enhancement is symbolized by replacement of the open circles for weaker Thomson scattering in the left panel of Fig. 8.25a by solid circles in the middle panel. On the right the sum of all contributions is shown.

The phasing of the diffraction pattern is facilitated by measurement of the total diffraction data at a few wavelengths (usually 3–5) around the absorption edge of the heavy atoms. From the set of measurements it is then possible to extract the structure of the heavy atoms in the molecular unit cell. The determination of the locations of these heavy atoms then serves as a reference for building up a model for the molecular units.

8.11.2 Formulation of MAD in Protein Cystallography

In crystals, the measured diffraction intensity is proportional to the squared amplitude of the *molecular structure factor*

$$|F^{\text{mol}}(\mathbf{q})|^2 = \left| \sum_j^{\text{atoms}} f_j(\mathbf{q}) \exp(\mathrm{i}\, \mathbf{q} \cdot \mathbf{r}_j) \right|^2 , \tag{8.81}$$

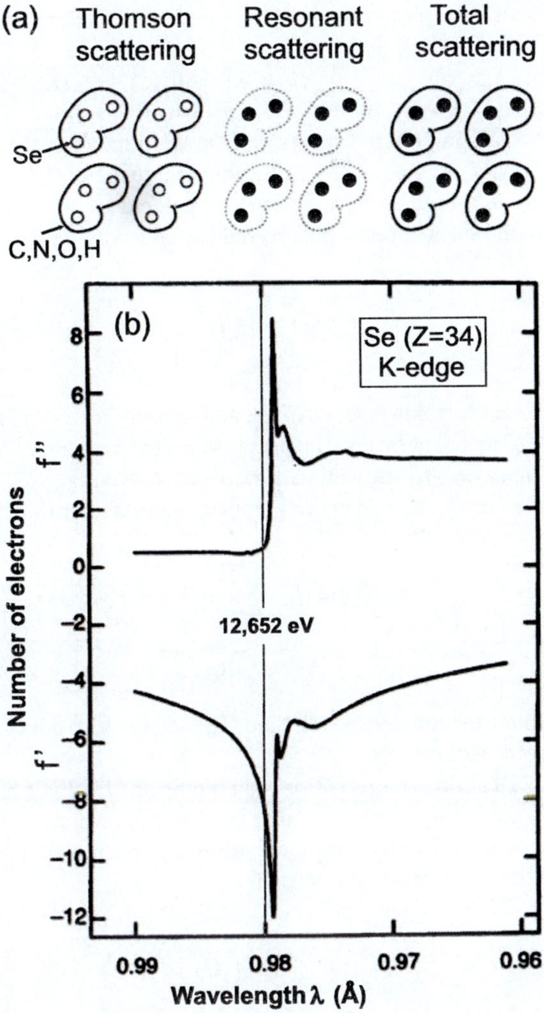

Fig. 8.25 **a** Schematic of scattering contributions from a macromolecular crystal, whose stacked molecular units consist of many light atoms such as C, N, O, and H, contained inside the macromolecular envelopes, plus heavy atoms like Se shown as circles. Conventional non-resonant Thomson scattering is created by all atoms in a molecule. Resonant scattering occurs when the photon energy is tuned to the absorption edge of the heavy atoms (see panel (**b**) below). The enhanced resonance scattering of the heavy atoms is symbolized by solid circles relative to the weaker Thomson scattering indicated by open circles. On the right the sum of all contributions is shown. **b** Illustration of the resonance enhancement of the atomic scattering factors f'' and f' at the K-edge of Se in a crystal of thioredoxin selenomethionyl thioredoxin from *Escherichia coli* [110]

where $\mathbf{q}$ is the momentum transfer. Here we have denoted the *molecular* structure factor as $F^{\mathrm{mol}}(\mathbf{q})$ which consists of the summed contributions from the individual atoms, denoted $f_j(\mathbf{q})$ and given in units of number of electrons per atom.

In MAD, one utilizes the strong difference of the scattering contributions of heavy atoms, e.g. Fe or Se, which serve as resonant scatterers, and those from the lighter atoms, typically C, N, and O. For the *heavy resonant scatterers*, denoted with a subscript "H" for "heavy", the momentum, $\mathbf{q}$, and wavelength, λ, dependent atomic scattering factors can be decomposed into three parts as discussed in Sect. 7.5.1. We write the atomic scattering factors in units of number of electrons given by (7.35), with the replacement $F(\mathbf{Q}, \omega) \to f(\mathbf{q}, \lambda)$ in our present notation as[23]

$$f_{\mathrm{H}}(\mathbf{q}, \lambda) = f_{\mathrm{H}}^{0}(\mathbf{q}) + f_{\mathrm{H}}'(\lambda) - \mathrm{i}\, f_{\mathrm{H}}''(\lambda). \tag{8.82}$$

Leaving out the superscript "mol" in the molecular structure factor in (8.81) for brevity, the *heavy-atom* contribution, denoted $F_{\mathrm{H}}(\mathbf{q})$, consists of a sum over those atoms whose scattering vectors $\mathbf{q}$ coincide with the reciprocal sublattice vectors $\mathbf{G}$ in the stacked molecular crystal expressed by

$$F_{\mathrm{H}}(\mathbf{G}) = \left[f_{\mathrm{H}}^{0} + f_{\mathrm{H}}'(\lambda) - \mathrm{i}\, f_{\mathrm{H}}''(\lambda)\right] \sum_{j}^{\text{atoms H}} \mathrm{e}^{\mathrm{i}\,\mathbf{G}\cdot\mathbf{r}_j}. \tag{8.83}$$

Here the $\mathbf{q}$-dependence of the atomic form factors $f_{\mathrm{H}}^{0}(\mathbf{q})$ has been absorbed into the $\mathbf{G}$-dependent factor. The heavy atoms have a non-resonant, wavelength-independent *Thomson*, f_{H}^{0}, as well as a *resonant* scattering, $f_{\mathrm{H}}'(\lambda) - \mathrm{i}\, f_{\mathrm{H}}''(\lambda)$, contribution. We then rewrite the heavy (H) atom structure factor (8.83) in the form

$$F_{\mathrm{H}}(\mathbf{G}) = \underbrace{f_{\mathrm{H}}^{0} \sum_{j}^{\text{H atoms}} \mathrm{e}^{\mathrm{i}\,\mathbf{G}\cdot\mathbf{r}_j}}_{F_{\mathrm{HT}}(\mathbf{G})} + \underbrace{F_{\mathrm{HT}}(\mathbf{G}) \left\{ \frac{f_{\mathrm{H}}'(\lambda) - \mathrm{i}\, f_{\mathrm{H}}''(\lambda)}{f_{\mathrm{H}}^{0}} \right\}}_{F_{\mathrm{HR}}(\mathbf{G})}, \tag{8.84}$$

where the subscripts "HT" and "HR" distinguish heavy atom Thomson and resonance contributions.

Denoting the *low-Z* or light atoms as "L", their part of the molecular structure factor consists only of *Thomson* scattering, which occurs when the scattering vector $\mathbf{q}$ coincides with a reciprocal sublattice vector $\mathbf{G}$ associated with any of those atoms. With our labeling it is written as

$$F_{\mathrm{LT}}(\mathbf{G}) = f_{\mathrm{L}}^{0} \sum_{n}^{\text{L atoms}} \mathrm{e}^{\mathrm{i}\,\mathbf{G}\cdot\mathbf{r}_n}. \tag{8.85}$$

We can now group the light and heavy atom contributions by distinguishing Thomson and resonant scattering. The total *Thomson* contribution from the light and heavy atoms is given by $F_{\mathrm{HT}}(\mathbf{G}) + F_{\mathrm{LT}}(\mathbf{G})$ or

$$\boxed{\text{Thomson:}\quad F_{\mathrm{T}}(\mathbf{G}) = |F_{\mathrm{T}}|\, \mathrm{e}^{\mathrm{i}\phi_{\mathrm{T}}} = |F_{\mathrm{HT}} + F_{\mathrm{LT}}|\, \mathrm{e}^{\mathrm{i}\phi_{\mathrm{T}}}.} \tag{8.86}$$

In conventional crystallography, only this Thomson scattering contribution, which through interference creates Bragg peaks, is measured. Even the Thomson contri-

---

[23] The present notation differs from that in Sect. 6.2.4, where we used a capital letter, $F(\mathbf{q})$ for the atomic scattering or form *factor* in units of [number of electrons], while a small letter f was used for the scattering *length* $f(\mathbf{q}) = r_0 F(\mathbf{q})$.

bution which scales with the atomic number Z is dominated by the heavy atoms. For a heavy atom like Se ($Z = 34$), the diffracted intensity $|F_{\rm HT}|^2$ is enhanced over that of C ($Z = 6$) $|F_{\rm LT}|^2$ by about a factor of 30. The reconstruction of the Thomson diffraction pattern typically starts with a Patterson map which emphasizes the Se positions and then proceeds by computationally intensive iterative phase retrieval, as in the direct method developed by Hauptman and Karle [6].

The MAD phasing formalism specifically hones in on the positions of the heavy atoms through their *resonant* contributions. By changing the wavelength one can now record several diffraction patterns whose difference is utilized in the phasing process. The resonant molecular structure factor of the heavy atoms is obtained from (8.84) as

$$\text{Resonant:}\quad F_{\rm R}(\mathbf{G}) = |F_{\rm HR}|\mathrm{e}^{\mathrm{i}\phi_{\rm R}} = |F_{\rm HT}| \left\{ \frac{f'_{\rm H}(\lambda) - \mathrm{i}\, f''_{\rm H}(\lambda)}{f^0_{\rm H}} \right\} \mathrm{e}^{\mathrm{i}\phi_{\rm R}}. \tag{8.87}$$

The *total* molecular structure factor is given by the sum of (8.86) and (8.87) as

$$F_{\rm tot}(\mathbf{G}) = |F_{\rm T}|\,\mathrm{e}^{\mathrm{i}\phi_{\rm T}} + |F_{\rm HT}| \left\{ \frac{f'_{\rm H}(\lambda) - \mathrm{i}\, f''_{\rm H}(\lambda)}{f^0_{\rm H}} \right\} \mathrm{e}^{\mathrm{i}\phi_{\rm R}}. \tag{8.88}$$

The diffracted intensity is proportional to $|F_{\rm tot}(\mathbf{G})|^2$. Leaving out the explicit $\mathbf{G}$-dependence of the structure factors for brevity, the detected intensity can be written in the following parameterized form [5, 71, 111].

The diffracted MAD intensity consists of a total *Thomson* (denoted T) molecular structure factor contribution from the light (L) and heavy (H) atoms $F_{\rm T}(\mathbf{G})$, given by (8.86). The Thomson contribution is phase shifted by $\phi_{\rm T} - \phi_{\rm R}$ relative to the *resonant* contribution from the heavy atoms $F_{\rm HR}(\mathbf{G})$. The latter is linked by (8.87) to $|F_{\rm HT}|$ which appears in the MAD intensity expression used for structure determinations. The total MAD pattern is given by

$$\begin{aligned} I_{\rm diff}(\mathbf{G}) \propto |F_{\rm tot}(\mathbf{G})|^2 = |F_{\rm T}|^2 &+ a(\lambda)\,|F_{\rm HT}|^2 \\ &+ b(\lambda)\,|F_{\rm T}|\,|F_{\rm HT}|\,\cos[\phi_{\rm T} - \phi_{\rm R}] \\ &+ c(\lambda)\,|F_{\rm T}|\,|F_{\rm HT}|\,\sin[\phi_{\rm T} - \phi_{\rm R}]. \end{aligned} \tag{8.89}$$

The wavelength-dependent parameters

$$a(\lambda) = \frac{[f'_{\rm H}(\lambda)]^2 + [f''_{\rm H}(\lambda)]^2}{(f^0_{\rm H})^2}, \quad b(\lambda) = \frac{2 f'_{\rm H}(\lambda)}{f^0_{\rm H}}, \quad c(\lambda) = -\frac{2 f''_{\rm H}(\lambda)}{f^0_{\rm H}} \tag{8.90}$$

are entirely determined by the response of the *heavy* atoms. They can be determined in a separate x-ray absorption measurement of the sample which yields

$f''_H(\lambda)$, supplemented by a Kramers-Kronig determination of $f'_H(\lambda)$ (Sect. 6.6). The wavelength-independent Thomson response, f^0_H, is available in tabulated form [132–134]. One is left with three unknowns, $|F_T|$, $|F_{HT}|$ and $\phi_T - \phi_R$, which are determined from diffraction patterns recorded at three or more wavelengths near the heavy atom resonance (see Fig. 8.25).

The molecular structure factor (8.88) contains important symmetry information that allows one to distinguish whether a system satisfies *Friedel's rule*, $|F(\mathbf{G})|^2 = |F(-\mathbf{G})|^2$ or not. This arises from the equivalence of the transformations $\mathbf{G} \rightarrow -\mathbf{G}$ and $[\phi_T - \phi_R] \rightarrow -[\phi_T - \phi_R]$, so that the symmetry is directly reflected by the even cos and odd sin terms in (8.89) [5].

8.11.3 MAD Imaging of Non-crystalline Samples

The extension of MAD to *non-periodic nanostructures* was treated and demonstrated by Scherz et al. [135]. The key difference between the diffraction pattern of crystals and non-periodic objects is that for crystals the pattern consists of well-defined bright spots, while for non-periodic objects the patterns is much weaker and smeared out, typically consisting of intensity speckles, as illustrated in Fig. 8.5.

The non-periodic charge or magnetic patterns are given by the formalism developed in Sect. 8.7, which is closely related to the MAD formalism discussed above. In this section we show the close relationship between the MAD diffraction pattern for a macromolecular crystal, (8.89), and that of a film containing nanoscale charge domains given by (8.43).

In order to see the correspondence we convert the optical parameter form in (8.43) to the scattering factor description in (8.89). The conversion utilizes the link established in Sect. 7.5.5. For dimensional clarification, the *scattering factors* are in units of [number of electrons per atom], which differ from the *scattering lengths* of dimension [length] simply by the single-electron Thomson scattering length r_0. In the following we will use the notation f' and f'' for the resonant scattering factors.

We can then convert the *resonant* optical parameters $\Delta\delta$ and $\Delta\beta$, defined in (8.33), which are multiplied by kd in (8.43) according to

$$\Delta\delta\, kd = f' \underbrace{r_0 \lambda \rho_a d}_{C} \quad \text{and} \quad \Delta\beta\, kd = f'' C. \tag{8.91}$$

Here ρ_a is the number density of atoms and the dimensionless number $C = r_0\lambda\rho_a d$ reflects the number of atoms in the volume $r_0\lambda d$. In addition to the differential domain transmission described by f' and f'' we account for the averaged absorption of all domains by[24]

[24] The corresponding parameter $\overline{\delta}\, kd = \overline{f'} C$ does not enter in the transmitted intensity.

$$\overline{\beta}\, kd = \overline{f''}\, C. \tag{8.92}$$

By explicitly stating the phase shift in (8.43) as $\phi = \phi_A - \phi_u$, the charge diffraction pattern (8.43) takes the form

$$I^{\text{film}}_{\text{diff}}(\mathbf{q}) = \frac{I_0}{\lambda^2 z_0^2} e^{-2C\overline{f''}} \Big\{ |D_A(\mathbf{q})|^2 + C^2\big[(f')^2 + (f'')^2\big]\,|D_{\hat{u}}(\mathbf{q})|^2 \\ -2f'\,C\;|D_A(\mathbf{q})|\,|D_{\hat{u}}(\mathbf{q})|\;\sin[\phi_A - \phi_u] \\ -2f''\,C\;|D_A(\mathbf{q})|\,|D_{\hat{u}}(\mathbf{q})|\;\cos[\phi_A - \phi_u] \Big\}. \tag{8.93}$$

The factor in front of the curly braces is just an intensity scaling factor, with $1/(\lambda^2 z_0^2)$ accounting for propagation and conservation of dimension since $|D_A|$ and $|D_{\hat{u}}|$ have the dimension [area], while the average attenuation factor $e^{-2C\overline{f''}}$ is independent of the sample microstructure and simply scales the overall intensity.

For direct comparison we write the MAD expression (8.89) by insertion of the parameters given by (8.90). Since Thomson scattering is constant, we then multiply the right side of this expression by f^0, which simply scales the intensity. By also dropping the subscript "H" of the scattering factors we obtain

$$I^{\text{mol}}_{\text{diff}}(\mathbf{q}) \propto \Big\{ \left(f^0\right)^2 |F_{\text{T}}(\mathbf{q})|^2 + \big[(f')^2 + (f'')^2\big]\;|F_{\text{HT}}(\mathbf{q})|^2 \\ +2f' f^0\;|F_{\text{T}}(\mathbf{q})|\;|F_{\text{HT}}(\mathbf{q})|\;\cos[\phi_{\text{T}} - \phi_{\text{R}}] \\ -2f'' f^0\;|F_{\text{T}}(\mathbf{q})|\;|F_{\text{HT}}(\mathbf{q})|\;\sin[\phi_{\text{T}} - \phi_{\text{R}}] \Big\}. \tag{8.94}$$

We can now compare the key structure dependent parts, identified by curly braces in (8.93) and (8.94).

The obvious formal resemblance becomes more quantitative when remembering that in (8.93), $|D_A(\mathbf{q})|$ reflects the contribution from the atomless aperture hole, while the corresponding term $f^0|F_{\text{T}}(\mathbf{q})|$ in (8.94) reflects non-resonant Thomson scattering. The two expressions thus differ in the nature of the *non-resonant* terms. Their relationship emerges by comparing the aperture contribution in Fig. 8.15 to the non-resonant Thomson contribution in 8.25a. Adding a non-resonant Thomson scattering background for the film, expressed by $f^0 r_0 \lambda \rho_a d = f^0 C$ in analogy to the resonant contribution (8.91), would result in the replacement $D_A(\mathbf{q}) \to f^0 C D_A(\mathbf{q})$ in (8.93). We would then have

$$I^{\text{film}}_{\text{diff}}(\mathbf{q}) \propto C^2 \Big\{ (f^0)^2\,|D_A(\mathbf{q})|^2 + \big[(f')^2 + (f'')^2\big]\,|D_{\hat{u}}(\mathbf{q})|^2 \\ -2f' f^0\;|D_A(\mathbf{q})|\,|D_{\hat{u}}(\mathbf{q})|\;\sin[\phi_A - \phi_u] \\ -2f'' f^0\;|D_A(\mathbf{q})|\,|D_{\hat{u}}(\mathbf{q})|\;\cos[\phi_A - \phi_u] \Big\}. \tag{8.95}$$

Since C^2 is a scaling factor, we factored it out of the curly braces to better reveal the close similarity with (8.94).

There is, however, a remaining curious difference between the corresponding sin and cos phase terms and signs in (8.94) and (8.95). This is explained by our discussion of phase shifts upon resonant scattering in Sect. 7.5.4. While the resonantly scattered waves directly emerge from individual atoms in a molecule, they originate from the exit plane of the film after the incident wave has traversed it. As pointed out in Sect. 7.5.4, there is a $\pi/2$ phase shift between the atomic and film cases. This is also revealed by the factor $-\mathrm{i}$ in the Rayleigh-Sommerfeld diffraction formula (8.8), corresponding to a 90° phase shift. The difference in the sin and cos phase factors and signs then follows from the trigonometric identities $\mp\sin[\phi] = \cos[\phi \pm \pi/2]$ and $\pm\cos[\phi] = \sin[\phi \pm \pi/2]$.

Hence real space image reconstruction by MAD of molecular crystals with atomic resolution and of non-periodic domain structures with nanometer resolution proceeds similarly. In the following we will present an example of applying MAD for diffractive imaging of a non-periodic nanostructured sample.

8.11.4 Implementation of MAD for Non-periodic Samples

We now apply the non-periodic MAD formalism by considering the experimental geometry and nanostructured sample shown in Fig. 8.26, studied by Scherz *et al.* [135]. We are interested in imaging polystyrene (PS) spheres of 90 and 300 nm in size deposited on a Si_3N_4 membrane inside a field-of-view defining circular aperture of 1.2 μm diameter. For imaging of the PS spheres with an isotropic charge (bond) distribution, the x-ray polarization is unimportant. Of interest here is the reconstruction of the real sample image, shown by an SEM image on the top right, from the shown diffraction pattern recorded with a CCD detector. This is accomplished by use of the strong energy dependence of the optical parameters shown on the lower left. We note that the photon energy scale in Fig. 8.26 has been corrected to agree with that shown in Fig. 7.6 and the peak values of the optical parameters are smaller since they were obtained by use of the same polystyrene spheres and with the same energy resolution used in the imaging experiment.

The intensity of the interference pattern for the experimental configuration shown in Fig. 8.26 is given by (8.43) in the optical parameter formulation. It can be separated into non-resonant and resonant components as discussed in Sect. 8.11.3. We can therefore rewrite (8.43) in the notation used by Scherz et al. [135] by denoting the non-resonant (N) aperture pattern $D_A = D_{\mathrm{N}}$ and the resonant (R) pattern by $D_{\hat{u}} = D_{\mathrm{R}}$. Similarly, we denote the $\mathbf{q}$-dependent phase shift between the complex amplitudes $D_{\mathrm{N}} = |D_{\mathrm{N}}|\,\mathrm{e}^{\mathrm{i}\phi_{\mathrm{N}}}$ and $D_{\mathrm{R}} = |D_{\mathrm{R}}|\,\mathrm{e}^{\mathrm{i}\phi_{\mathrm{R}}}$ as $\Delta\phi = \phi_{\mathrm{N}} - \phi_{\mathrm{R}}$. We further abbreviate by defining new energy dependent parameters $\tilde{\delta} = \delta k d$ and $\tilde{\beta} = \beta k d$, where δ and β are shown in Fig. 8.26 to obtain (8.43) in the form,

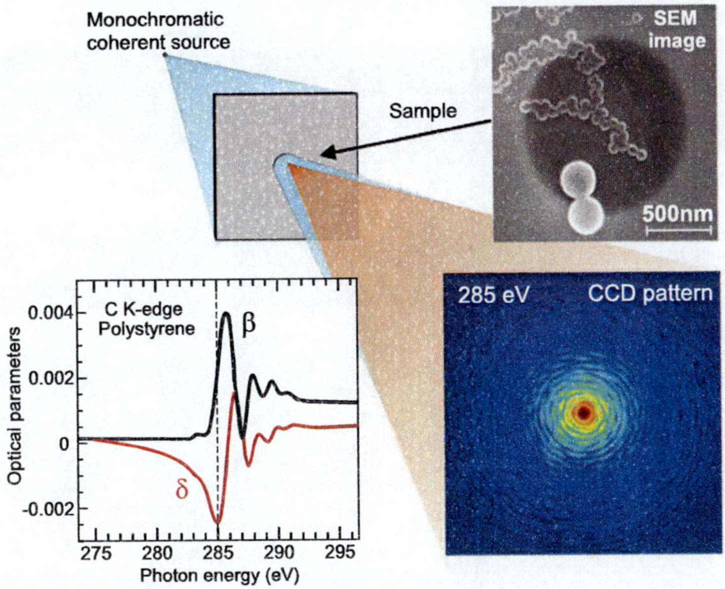

Fig. 8.26 Schematic experimental setup for MAD diffractive imaging. The sample consisted of polystyrene spheres, of 90 and 300 nm in size dispersed from aqueous solution on a Si_3N_4 membrane, with the field of view defined by a circular 1.2 μm diameter aperture created by deposition of a 800 nm thick Au film on the back of the membrane. An SEM image of the sample is shown on the top right. It was illuminated with a x-ray beam from an undulator source that was made spatially coherent with an aperture (not shown) and a monochromator allowed tuning the photon energy around the C K-edge of polystyrene. MAD phasing exploits the energy dependent interference of the resonant exit wave (orange) with the non-resonant exit wave (light blue). The center portions of the interference patterns recorded with a CCD detector at a photon energy of 285 eV are shown on a logarithmic scale. The strong energy dependence of the optical parameters of polystyrene is shown on the lower left [135]

$$I_{\mathrm{diff}}(\mathbf{q}) = \frac{I_0}{\lambda^2 z_0^2}\, \mathrm{e}^{-2\overline{\beta}kd} \Big\{ |D_{\mathrm{N}}(\mathbf{q})|^2 + |D_{\mathrm{R}}(\mathbf{q})|^2 \left(\tilde{\delta}^2 + \tilde{\beta}^2\right) \\ -2\,|D_{\mathrm{N}}(\mathbf{q})||D_{\mathrm{R}}(\mathbf{q})| \left(\tilde{\delta}\sin[\Delta\phi] + \tilde{\beta}\cos[\Delta\phi]\right) \Big\}. \quad (8.96)$$

Here $\overline{\beta}$ is an energy independent absorption parameter that accounts for the transmission loss due to the sample support (e.g. a Si_3N_4 membrane).

To illustrate the distinct change of the diffraction pattern when the photon energy is changed from below to the resonance energy, we show in Fig. 8.27, simulated and measured diffraction patterns for the sample structure in Fig. 8.26. Below the resonance the patterns are the usual Airy pattern of a circular aperture which near the PS C K-shell resonance becomes modulated by diffraction from the PS spheres and, importantly, the symmetry is changed from even to odd centrosymmetric. From the energy scale in Fig. 8.26 it is seen that the photon energy of 285 eV coincides with

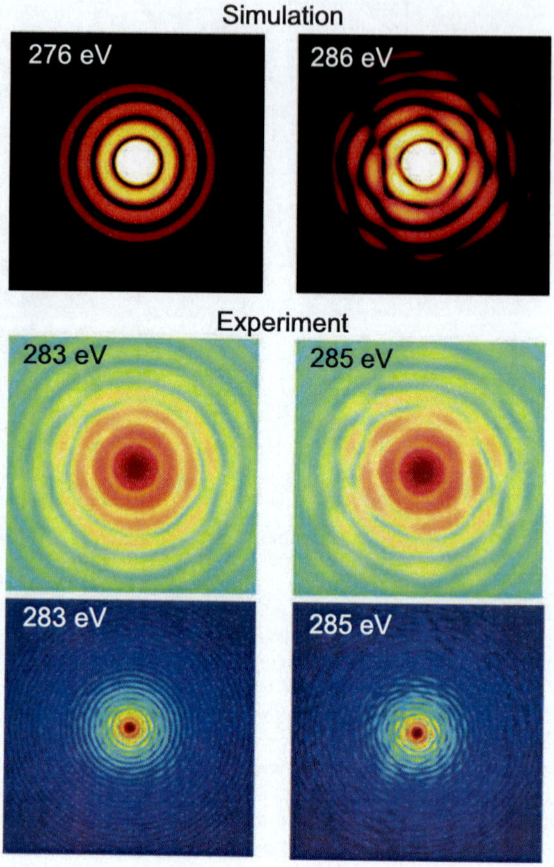

Fig. 8.27 On top we show the simulated central part of the diffraction pattern of the object in Fig. 8.26 at the two listed energies, calculated with (8.96). Underneath we show measured diffraction patterns at two different energies, for the central (green) and extended (blue) **q** range. Note the change from the mostly circular aperture pattern recorded below the resonance to a speckled pattern near the resonance. From D. Zhu, Ph.D. dissertation, Stanford University, 2010, and [135]

the resonance in δ and the pattern is therefore largely due to a phase contrast (see Fig. 8.30 below).

From knowledge of the optical parameters, we can derive analytical expressions for the three unknown parameters $|D_{\mathrm{N}}|$, $|D_{\mathrm{R}}|$ and $\Delta\phi$ under the assumption that the sample is very thin. If at a selected energy both $\tilde{\delta} = \delta k d \ll 1$ and $\tilde{\beta} = \beta k d \ll 1$, we can neglect the quadratic term in the optical parameters and obtain the simpler expression

$$I_{\mathrm{diff}}(\mathbf{q}, \hbar\omega) \propto |D_{\mathrm{N}}(\mathbf{q})|^2 - 2|D_{\mathrm{N}}(\mathbf{q})|\,|D_{\mathrm{R}}(\mathbf{q})| \left\{\tilde{\delta}(\hbar\omega)\sin[\Delta\phi] + \tilde{\beta}(\hbar\omega)\cos[\Delta\phi]\right\}. \tag{8.97}$$

Here we have explicitly identified the **q** and $\hbar\omega$ dependent terms. An analytical solution can then be obtained from normalized intensity measurements at three different photon energies with associated optical parameters denoted $\tilde{\delta}_i = \tilde{\delta}(\hbar\omega_i)$ and

$\tilde{\beta}_i = \tilde{\beta}(\hbar\omega_i)$

$$I_1 = |D_N|^2 - 2|D_N|\,|D_R| \left\{\tilde{\delta}_1 \sin[\Delta\phi] + \tilde{\beta}_1 \cos[\Delta\phi]\right\}$$
$$I_2 = |D_N|^2 - 2|D_N|\,|D_R| \left\{\tilde{\delta}_2 \sin[\Delta\phi] + \tilde{\beta}_2 \cos[\Delta\phi]\right\}$$
$$I_3 = |D_N|^2 - 2|D_N|\,|D_R| \left\{\tilde{\delta}_3 \sin[\Delta\phi] + \tilde{\beta}_3 \cos[\Delta\phi]\right\}. \tag{8.98}$$

This directly yields the analytical solutions for the three unknown parameters $|D_R|$, $|D_N|$, and $\Delta\phi$

$$|D_R| = \frac{|D_N|^2 - I_1}{2|D_N|\left(\tilde{\delta}_1 \sin[\Delta\phi] + \tilde{\beta}_1 \cos[\Delta\phi]\right)} \tag{8.99}$$

$$|D_N| = \left[\frac{(I_1\tilde{\delta}_2 - I_2\tilde{\delta}_1)\sin[\Delta\phi] + (I_1\tilde{\beta}_2 - I_2\tilde{\beta}_1)\cos[\Delta\phi]}{(\tilde{\delta}_2 - \tilde{\delta}_1)\sin[\Delta\phi] + (\tilde{\beta}_2 - \tilde{\beta}_1)\cos[\Delta\phi]}\right]^{1/2} \tag{8.100}$$

$$\Delta\phi = \arctan\left\{\frac{(I_1 - I_3)(\tilde{\beta}_2 - \tilde{\beta}_1) - (I_1 - I_2)(\tilde{\beta}_3 - \tilde{\beta}_1)}{(I_1 - I_2)(\tilde{\delta}_3 - \tilde{\delta}_1) - (I_1 - I_3)(\tilde{\delta}_2 - \tilde{\delta}_1)}\right\}. \tag{8.101}$$

Depending on the sample, one may take advantage of computationally separating even and odd centrosymmetric components of the recorded images, expressed by the cos and sin terms in (8.97). The symmetry change with photon energy for our sample is evident from the patterns in Fig. 8.27. This allows simplification of the reconstruction problem by use of patterns at only two energies near the resonance, as utilized by Scherz *et al.* [135].

In practice, the analytical solution to the reconstruction expressed by (8.99)–(8.101) or by the symmetry-based two-energy method is only approximate since the thin-sample approximation is typically not fulfilled. For example, at an energy of 285 eV ($\lambda = 4.35$ nm), the optical constants for PS have the values indicated by the vertical line in Fig. 8.26, or $\beta \simeq \delta \simeq 0.002$. For the thick PS spheres of $d = 300$ nm thickness this corresponds to $\tilde{\beta} \simeq \tilde{\delta} = 0.87$ which does not justify neglect of the quadratic terms. One therefore uses the linear analytical solution as an input to iterative image reconstruction. Figure 8.28 demonstrates the rapid improvement of the reconstructed image with the number of iterations which ultimately reproduces the object with a resolution of about 25 nm, limited by the statistics of the diffraction patterns at the highest q values [135].

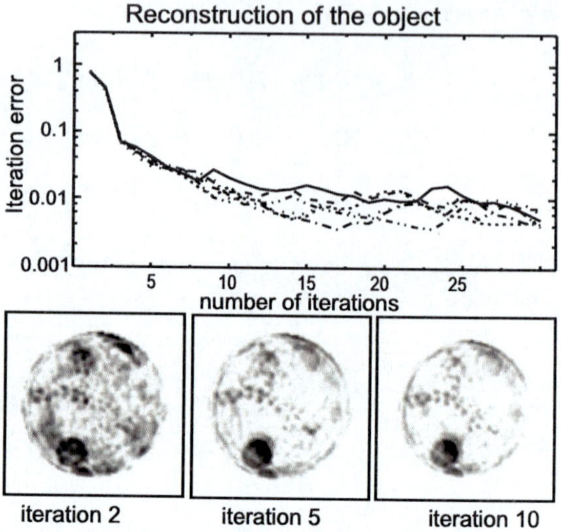

Fig. 8.28 Convergence of the object image reconstruction, characterized by decreasing iteration error. The upper plot shows how five independent runs of the iterative reconstruction algorithm followed a very reproducible trajectory to convergence. Shown below are the reconstructed images of the sample in Fig. 8.26 at iterations 2, 5, and 10. From D. Zhu, Ph.D. dissertation, Stanford University, 2010, and [135]

8.11.4.1 Reference-Guided Phase Retrieval

We briefly mention another phasing method, which from a FTH point of view overcomes the limited resolution due to the size of the reference source, and from a MAD point of view speeds up the phase retrieval. The technique, referred to as *reference-guided phase retrieval* (RPR), implemented by Zhu et al. [136], adds a reference hole covered with a film of resonant atoms to the geometry shown in Fig. 8.26. By tuning the photon energy across the resonance of the reference film, both the intensity and phase of the reference beam can then be changed. In particular, the transmission of the reference film which exponentially changes with β and film thickness can be changed by many orders of magnitude before and at the peak of the absorption resonance as illustrated in Fig. 8.29 for a 200-nm-thick Co film. The reference beam can then be either utilized or eliminated [136].

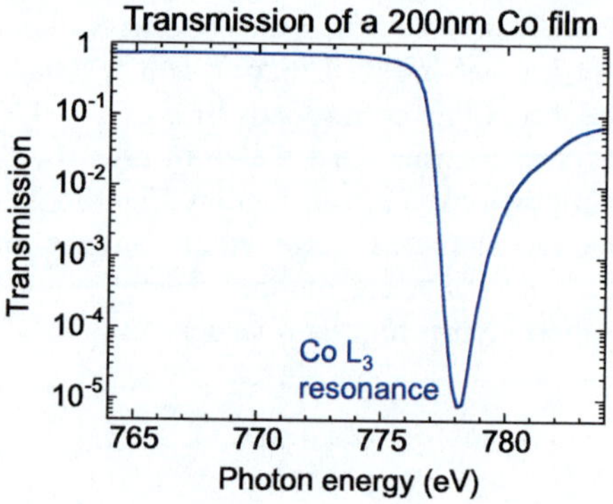

Fig. 8.29 Illustration of the transmission of a 200 nm Co film near the L_3 resonance, which changes by about five orders of magnitude within a few eV

When the diffraction pattern of the object and reference is recorded below the resonance, the reference film transmission is high and one obtains the conventional holographic diffraction pattern. At resonance, the reference contribution is essentially absent since the film has negligible transmission, and one obtains the diffraction pattern of the object alone. The two patterns are readily scaled by the known change in wavelength. The backtransform of the hologram recorded below the resonance already contains low-resolution twin images of the object. One can use these images in a reference-guided phase retrieval of the object pattern recorded on resonance. The reconstruction process is greatly speeded up [136].

8.11.5 Phase Contrast Imaging: Combining FTH and MAD

We previously showed in Fig. 8.23 the application of FTH to magnetic domain imaging. In that case the incident energy was tuned to the peak of the Co L_3 resonance, where the transmitted field is strongly attenuated by absorption. More generally, the diffracted intensity, given by (8.77), depends on both the dichroic differences $\Delta\beta$ and $\Delta\delta$, as previously shown in Fig. 7.25a for the transmitted intensity.

In some cases it is advantageous, e.g. for XFEL imaging, to reduce absorption as much as possible since the energy transferred to the sample in the absorption process heats the electronic system, which may lead to damage (see Sect. 15.2 below). This is accomplished by maximizing the contrast that arises from $\Delta\delta$ and minimizing that of $\Delta\beta$.

This so-called *phase imaging* was demonstrated by Scherz et al. [137]. The experimental arrangement was similar to that in Fig. 8.23 and the sample was a magnetic multilayer consisting of 125 Co(0.25 nm)/Pd(0.9 nm) bilayers which similarly contained magnetic worm domains with magnetization directions perpendicular to the plane. The striking results obtained by tuning the incident energy through the Co L_3 absorption resonance are shown in Fig. 8.30.

At a given photon energy, the shown real space images are obtained by first subtracting holograms recorded with opposite circular helicities. The phase contrast is particularly sensitive to properly centering the two holograms at $q = 0$, since this location differentiates between odd and even symmetry behavior of components of the total holographic pattern. This symmetry is revealed in (8.77) by the odd symmetry $\sin[\phi]$ term associated with $\Delta\delta$ and the even symmetry $\cos[\phi]$ term associated with $\Delta\beta$. Once the holograms for the two polarizations are properly centered, the difference is backtransformed into real space. The *real part* of the backtransform gives the *absorption contrast* shown on top in Fig. 8.30 and the *imaginary part* the *phase contrast* shown at the bottom. The reconstructed images are plotted on the same contrast scale.

The observed behavior closely follows the energy dependence of $\Delta\beta$ and $\Delta\delta$, shown in the center of Fig. 8.30. Before the absorption peak, at position Ⓐ, $\Delta\beta \simeq 0$ and $\Delta\delta$ is negative. There is almost no absorption but finite phase contrast. At position Ⓑ, $\Delta\beta$ and $\Delta\delta$ have about the same magnitude but opposite signs. This causes the

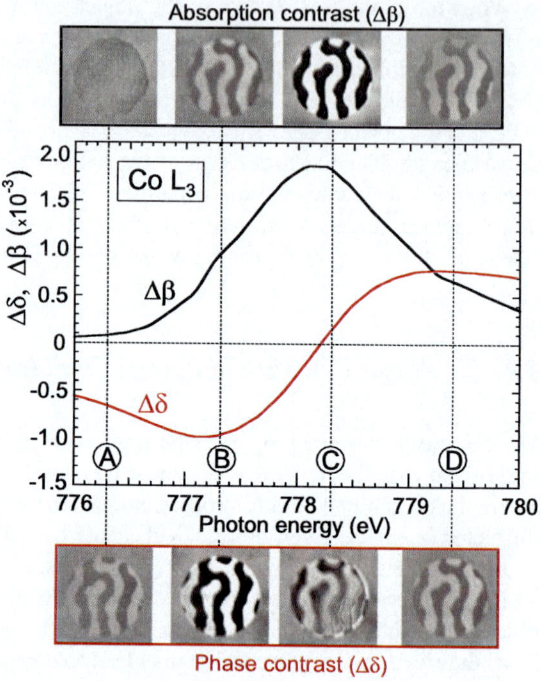

Fig. 8.30 Results of an experiment similar to that in Fig. 8.23, performed by tuning the incident energy through the Co L_3 absorption resonance. Shown are the real (top row) and imaginary (bottom row) parts of the backtransformed difference between holograms recorded with opposite circular polarization as a function of photon energy across the Co L_3 resonance [137]. The contrast follows the energy dependence of the dichroic differences $\Delta\delta$ and $\Delta\beta$

image contrast to be inverted. At position Ⓒ, $\Delta\beta$ has its maximum while $\Delta\delta$ is close to zero, and the absorption contrast dominates. At position Ⓓ, $\Delta\beta$ and $\Delta\delta$ have comparable magnitudes and the same sign. The contrast is therefore the same and not inverted, unlike case Ⓑ.

As mentioned above, the great advantage of phase imaging is mitigation of radiation damage since for comparable image quality the radiation dose may be substantially reduced (about a factor of 10). As such, the technique should be of great value for radiation sensitive samples, such as biological specimens or polymers.

References

1. C. Jacobsen, *X-Ray Microscopy* (Cambridge University Press, Cambridge, 2020)
2. S. Bajt et al., Light Sci. Appl. **7**, 17162 (2018)
3. B.E. Warren, *X-Ray Diffraction* (Dover Publications, New York, 1999)
4. C. Hummond, *The Basics of Crystallography and Diffraction*, 4th edn. (Oxford University Press, Oxford, 2015)
5. J. Als-Nielsen, D. McMorrow, *Elements of Modern X-Ray Physics*, 2nd edn. (Wiley, New York, 2011)
6. H. Hauptman, J. Karle, *Solution of the Phase Problem. I. The Centrosymmetric Crystal*, A.C.A. Monograph No. 3, The Letter Shop, Wilmington (1953)

7. D. Attwood, A. Sakdinawat, *X-Rays and Extreme Ultraviolet Radiation: Principles and Applications*, 2nd edn. (Cambridge University Press, Cambridge, 2017)
8. J. Stöhr, H.C. Siegmann, *Magnetism: From Fundamentals to Nanoscale Dynamics* (Springer, Heidelberg, 2006)
9. A.P. Hitchcock, J. Elec. Spectrosc. Relat. Phenom. **200**, 49 (2015)
10. A. Nilsson, F. Perakis, "X-ray studies of water", in *Synchrotron Light Sources and Free-Electron Lasers: Accelerator Physics, Instrumentation and Science Applications*, ed. by E.J. Jaeschke, S. Khan, J.R. Schneider, J.B. Hastings (Springer, New York, 2020)
11. J. Kirz, C. Jacobsen, J. Phys.: Conf. Ser. **186**, 012001 (2009)
12. B. Lengeler, C.G. Schroer, M. Kuhlmann, B. Benner, T.F. Günzler, O. Kurapova, F. Zontone, A. Snigirev, I. Snigireva, J. Phys. D: Appl. Phys. **38**, A218 (2005)
13. B. Kaulich, P. Thibault, A. Gianoncelli, M. Kiskinova, J. Phys.: Condens. Matter **23**, 083002 (2011)
14. J. Feng, A. Scholl, "Photoemission electron microscopy", in *Springer Handbook of Microscopy*, ed. by P.W. Hawkes, J.C.H. Spence (Springer, New York, 2019)
15. W. Chao, B.D. Harteneck, J.A. Liddle, E.H. Anderson, D.T. Attwood, Nature **435**, 1210 (2005)
16. J. Stöhr, Y. Wu, B.D. Hermsmeier, M.G. Samant, G.R. Harp, S. Koranda, D. Dunham, B.P. Tonner, Science **259**, 658 (1993)
17. H. Ade, H. Hsiao, Science **262**, 1427 (1993)
18. A.P. Smith, H. Ade, Appl. Phys. Lett. **69**, 3833 (1996)
19. F. Nolting, A. Scholl, J. Stöhr, J.W. Seo, J. Fompeyrine, H. Siegwart, J.P. Locquet, S. Anders, J. Lüning, E.E. Fullerton, M.F. Toney, M.R. Scheinfein, H.A. Padmore, Nature **405**, 767 (2000)
20. S.B. Choe, Y. Acremann, A. Scholl, A. Bauer, A. Doran, J. Stöhr, H.A. Padmore, Science **304**, 420 (2004)
21. Y. Acremann, J.P. Strachan, V. Chembrolu, S.D. Andrews, T. Tyliszczak, J.A. Katine, M.J. Carey, B.M. Clemens, H. Siegmann, J. Stöhr, Phys. Rev. Lett. **96**, 217202 (2006)
22. P. Fischer, H. Ohldag, Rep. Prog. Phys. **78**, 094501 (2015)
23. H. Stoll et al., Front. Phys. **3**, 26 (2015)
24. S. Woo, K. Litzius, B. Krüger et al., Nat. Mater. **15**, 501 (2016)
25. H. Ade, A.P. Hitchcock, Polymer **49**, 643 (2008)
26. G. Kirchhoff, Ann. Physik **254**, 663 (1883)
27. L. Rayleigh, Philos. Mag. **43**, 259 (1897)
28. A. Sommerfeld, Math. Ann. **47**, 317 (1896)
29. T.M. Pritchett, A.D. Trubatch, Am. J. Phys. **72**, 1026 (2004)
30. J.W. Goodman, *Introduction to Fourier Optics*, 4th edn. (W. H. Freeman, New York, 2017)
31. W. Friedrich, P. Knipping, Experimental X-ray diffraction patterns were first reported on June 8, 1912 , Sitz. Ber. Bayer. Akad. Wiss. 311–322 (1912). In 1913 the results were published by W. Friedrich, P. Knipping and M. Laue, Ann. Phys. **346**, 971 (1913), with Laue writing the theoretical and the two others the experimental part of the paper
32. P.P. Ewald, Z. Phys. **14**, 465 (1913)
33. W.H. Bragg, W.L. Bragg, Proc. R. Soc. London A **88**, 428 (1913)
34. A. Fresnel, Mémoires de l'Académie Royale des Sciences de l'Institut de France p. 339 (1818)
35. F. Zernike, Physica **5**, 785 (1938)
36. L. de Broglie, Nature **112**, 540 (1923)
37. L. de Broglie, Ann. Phys. (Paris) **3**, 22 (1925)
38. C. Davisson, L.H. Germer, Phys. Rev. **303**, 705 (1927)
39. P.A.M. Dirac, Proc. Roy. Soc. A **114**, 243 (1927)
40. P.A.M. Dirac, Proc. Roy. Soc. London A **114**, 710 (1927)
41. P.A.M. Dirac, Proc. Roy. Soc. A **117**, 610 (1928)
42. P.A.M. Dirac, Proc. Roy. Soc. A **118**, 351 (1928)
43. P. A. M. Dirac, *Quantum Mechanics*, 4th edn. (Oxford University Press, 1958), p. 9
44. G.I. Taylor, Proc. Camb. Phil. Soc. **15**, 114 (1909)
45. C. Santori, D. Fattal, Y. Yamamoto, *Single Photon Devices and Applications* (Wiley, Weinheim, 2010)

46. R.P. Feynman, R.B. Leighton, M. Sands, *The Feynman Lectures on Physics*, vol. III (Addison-Wesley, Reading, 1965)
47. R.P. Feynman, *QED: The Strange Theory of Light and Matter* (Princeton University Press, Princeton NJ, 1985)
48. C. Jönsson, Zeit. Phys. **161**, 454 (1961); English translation: Am. J. Phys. **42**, 4 (1974)
49. A. Tonomura, J. Endo, T. Matsuda, T. Kawasaki, H. Ezawa, Am. J. Phys. **57**, 117 (1989)
50. R. Bach, D. Pope, S.H. Liou, H. Batelaan, New J. Phys. **15**, 033018 (2013)
51. T.L. Dimitrova, A. Weis, Am. J. Phys. **76**, 137 (2008)
52. E. Hecht, *Optics*, 4th edn. (Addison-Wesley, Reading, 2001)
53. S. Eibenberger, S. Gerlich, M. Arndt, M. Mayor, J. Tüxen, Phys. Chem. Chem. Phys. **15**, 14696 (2013)
54. E. Madelung, Z. Phys. **40**, 332 (1926)
55. L. de Broglie, Comp. Rend. Acad. Sci. **183**, 447 (1926)
56. D. Bohm, Phys. Rev. **85**, 166 (1952)
57. D. Bohm, Phys. Rev. **85**, 180 (1952)
58. D.M. Greenberger, M.A. Horne, A. Zeilinger, Phys. Today **48**, 22 (1993)
59. A.S. Sanz, F. Borondo, S. Miret-Artés, J. Phys.: Condens. Matter **14**, 6109 (2002)
60. A.S. Sanz, S. Miret-Artés, *A Trajectory Description of Quantum Processes I* (Springer, Heidelberg, 2012)
61. K. Jung, J. Phys.: Conf. Ser. **442**, 012060 (2013)
62. P. Chen, H. Kleinert, Elect. J. Theor. Phys. **13**, 1 (2016)
63. E. Abbe, Archiv Mikrosk. Anatomie **9**, 413 (1873)
64. L. Rayleigh, Philos. Mag. **8**, 261 (1879)
65. E. Wolf, J. Opt. Soc. Am. **68**, 6 (1978)
66. A. Babinet, Compt. Rend. Acad. Sci. **4**, 638 (1837)
67. B. Wu, T. Wang, C.E. Graves, D. Zhu, W.F. Schlotter, J.J. Turner, O. Hellwig, Z. Chen, H.A. Dürr, A. Scherz, J. Stöhr, Phys. Rev. Lett. **117**, 027401 (2016)
68. R. Neutze, R. Wouts, D. van der Spoel, E. Weckert, J. Hajdu, Nature **406**, 752 (2000)
69. M.G. Rossmann, *The Molecular Replacement Method* (Gordon and Breach, New York, 1972)
70. S. Ramaseshan, Curr. Sci. **65**, 644 (1993)
71. W.A. Hendrickson, Q. Rev, Biophysics **47**, 49 (2014)
72. A.L. Patterson, Phys. Rev. **46**, 372 (1934)
73. A.L. Patterson, Z. Kristallographie **90**, 517 (1935)
74. D. Gabor, Nature **161**, 777 (1948)
75. E. Leith, J. Upatnieks, JOSA **52**, 1123 (1962)
76. E. Leith, J. Upatnieks, JOSA **53**, 1377 (1963)
77. G.W. Stroke, D.G. Falconer, Phys. Lett. **13**, 306 (1964)
78. G.W. Stroke, Appl. Phys. Lett. **6**, 201 (1965)
79. J.T. Winthrop, C.R. Worthington, Phys. Lett. **15**, 124 (1965)
80. A. Lohmann, Opt. Acta **3**, 97 (1956)
81. B. Reuter, H. Mahr, J. Phys. E **9**, 746 (1976)
82. I. McNulty, J. Kirz, C. Jacobsen, E.D. Anderson, M.R. Howells, D.P. Kern, Science **256**, 1009 (1992)
83. I. McNulty, Nucl. Instr. and Meth. A **347**, 170 (1994)
84. S. Eisebitt, M. Lörgen, W. Eberhardt, J. Lüning, S. Andrews, J. Stöhr, Appl. Phys. Lett. **84**, 3373 (2004)
85. S. Eisebitt, W.F.S.J. Lüning, M. Lörgen, O. Hellwig, W. Eberhardt, J. Stöhr, Nature **432**, 885 (2004)
86. W.F. Schlotter et al., Appl. Phys. Lett. **89**, 163112 (2006)
87. W.F. Schlotter, J. Lüning, R. Rick, K. Chen, A. Scherz, S. Eisebitt, C.M. Günther, W. Eberhardt, O. Hellwig, J. Stöhr, Opt. Lett. **32**, 3110 (2007)
88. T. Wang et al., Phys. Rev. Lett. **108**, 267403 (2012)
89. M. Guizar-Sicairos, J.R. Fienup, Opt. Express **15**, 17592 (2007)
90. M. Guizar-Sicairos, J.R. Fienup, Opt. Lett. **33**, 2668 (2008)

91. M. GuizarspsSicairos, D. Zhu, J.R. Fienup, B. Wu, A. Scherz, J. Stöhr, Opt. Lett. **35**, 928 (2010)
92. D. Zhu, M. Guizar-Sicairos, B. Wu, A. Scherz et al., Phys. Rev. Lett. **105**, 043901 (2010)
93. P.P. Ewald, Annal. Phys. **369**, 253 (1921)
94. H. Mark, L. Szilard, Z. Phys. **33**, 688 (1925)
95. J.M. Cork, Lond. Edinb. Dubl. Philos. Mag. J. Sci. **4**, 688 (1927)
96. D.W. Green, V.M. Ingram, M.F. Perutz: Proc. Roy. Soc. (London) **A 225**, 287 (1954)
97. D. Sayre, Acta Crystallogr. **5**, 60 (1952)
98. S. Kaushik, S.L. Rath, Protein structure visualization, in *Encyclopedia of Bioinformatics and Computational Biology*, ed. by S. Ranganathan, M. Gribskov, K. Nakai, C. Schönbach (Academic Press, Oxford, 2019), p. 520
99. H.A. Kramers, W. Heisenberg, Z. Phys. **31**, 681 (1925)
100. R. de L. Kronig: J. Opt. Soc. Am. **12**, 547 (1926)
101. H.A. Kramers, Atti Cong. Intern. Fisici **2**, 545 (1927)
102. H. Hönl, Z. Phys. **84**, 1 (1933)
103. J.M. Bijvoet, A.F. Peerdeman, A.J. van Bommel, Nature **168**, 271 (1951)
104. J.M. Bijvoet, Nature **173**, 888 (1954)
105. Y. Okaya, R. Pepinsky, Phys. Rev. **103**, 1645 (1956)
106. W. Hoppe, V. Jakubowski, The determination of phases of erythrocruorin using the two-wavelength method with iron as anomalous scatterer, in ed. by S. Ramaseshan, S. Abrahams, *Anomalous Scattering* (Munksgåd, Copenhagen, 1971)
107. J.C. Phillips, A. Wlodawer, J.M. Goodfellow, K.D. Watenpaugh, L.C. Sieker, L.H. Jensen, K.O. Hodgson, Acta Cryst. **A33**, 445 (1977)
108. J. Karle, Int. J. Quant. Chem. Symp. **7**, 357 (1980)
109. W.A. Hendrickson, Trans. Am. Cryst. Assoc. **21** (1985)
110. W.A. Hendrickson, J.R. Horton, D.M. LeMaster, EMBO J. **9**, 665 (1991)
111. W.A. Hendrickson, Science **254**, 51 (1991)
112. W.A. Hendrickson, M. Teeter, Nature **290**, 107 (1981)
113. W. Yang, W. Hendrickson, R.C.Y. Satow, Science **249**, 1398 (1990)
114. J.W. Goodman, *Fourier Transforms Using Mathematica* (SPIE Press, Bellingham, WA, 2020)
115. H. Nyquist, Trans. AIEE. **47**, 617 (1928)
116. V.A. Kotel'nikov, *On the transmission capacity of ether and wire in electro-communications*, Material for the First All-Union Conference on Questions of Communication, Izd. Red. Upr. Svyazi RKKA (in Russian) (1933)
117. C.E. Shannon, Proc. Inst. Radio Engrs., N.Y. **37**, 10 (1949)
118. E.T. Whittaker, On the Functions Which are Represented by the Expansions of the Interpolation Theory. Proc. Royal Soc. Edinburgh. **35**, 181 (1915)
119. R.H.T. Bates, Optik (Stuttgart) **61**, 247 (1982)
120. D. Sayre, in ed. by H. Schenk, *Direct Methods of Solving Crystal Structure*, NATO ASI Series B (Physics), vol. 274 (Plenum Press, New York, 1991), p. 353
121. J.W. Miao, P. Charalambous, J. Kirz, D. Sayre, Nature **400**, 342 (1999)
122. J.W. Miao, J. Kirz, D. Sayre, Acta Crystal. D **56**, 1312 (2000)
123. J. Fienup, Appl. Opt. **21**, 2758 (1982)
124. D. Sayre, H.N. Chapman, Acta Crystallogr. A **51**, 237 (1995)
125. S. Marchesini, H. He, H.N. Chapman, S.P. Hau-Riege, A. Noy, M.R. Howells, U. Weierstall, J.C.H. Spence, Phys. Rev. B **68**, 140101 (2003)
126. S. Marchesini, Rev. Sci. Instr. **78**, 011301 (2007)
127. R. Hegerl, W. Hoppe, Berichte der Bunsengesellschaft für physikalische Chemie **74**, 1148 (1970)
128. J.M. Rodenburg, H.M.L. Faulkner, Appl. Phys. Lett. **85**, 4795 (2004)
129. H.M.L. Faulkner, J.M. Rodenburg, Phys. Rev. Lett. **93**, 023903 (2004)
130. F. Pfeiffer, Nature Photon. **12**, 9 (2018)
131. J.M. Rodenburg, Adv. Imag. Electron. Phys. **150**, 87 (2008)
132. B.L. Henke, E.M. Gullikson, J.C. Davis, At. Data Nucl. Data Tables **54**, 181 (1993)

133. C.T. Chantler, J. Phys. Chem. Ref. Data **29**, 597 (2000)
134. https://henke.lbl.gov/optical_constants/
135. A. Scherz, D. Zhu, R. Rick, W.F. Schlotter, S. Roy, J. Lüning, J. Stöhr, Phys. Rev. Lett. **101**, 076101 (2008)
136. D. Zhu, B. Wu, R. Rick, J. Stöhr, A. Scherz, Opt. Lett. **34**, 2604 (2009)
137. A. Scherz et al., Phys. Rev. B **76**, 214410 (2007)

Part III
Quantum Theory of Weak Interactions

Chapter 9
Quantum Formulation of X-Ray Interactions with Matter

9.1 Introduction and Overview

The previous three chapters, forming Part II of this book, described the interaction of x-rays with matter in terms of classical EM waves. The classical wave theory quantitatively yields the Thomson scattering cross section, and scattering and absorption are linked through the Kramers-Kronig formalism. It needs to be recognized, however, that the classical approach does not allow the *ab initio* calculation of resonant scattering and absorption. In particular, the Kramers-Kronig transformation between absorption and scattering relies on the availability of experimental data. This shortcoming is overcome by the use of quantum theory, which allows the calculation of x-ray scattering factors or optical constants from first principles. It also provides a deeper appreciation of the true photon nature of light within the complete framework of QED.

This chapter introduces the quantum theoretical formulation of photon-matter interactions, constituting the first of five chapters in Part III of our book. The quantum approach has its roots in a foundational paper by Kramers and Heisenberg in 1925 [1], written just before the formulation of quantum theory. Two years later Dirac quantized the EM field and used first order perturbation theory to derive the quantum mechanical expression for x-ray absorption and emission [2], later called by Fermi "Golden rule number 2" [3]. Shortly thereafter Dirac rederived the Kramers-Heisenberg formula in the second order quantum formulation [4]. We shall here use this modern quantum version of the theory and, giving credit to its architects, refer to it as the *Kramers-Heisenberg-Dirac* (KHD) theory.

The basic premise of the KHD theory is that the photon-matter interaction may be described as a weak perturbation of the equilibrium state of matter. This allows one to view photons as a weak probe that can provide information on the unperturbed ground state of matter. The validity of this "weak field" perturbation approach is not obvious since the energy of a single x-ray photon is comparable to intra-atomic binding energies, and after absorption of a photon the atom is left in a highly excited state. The key assumption underlying the KHD theory is that the interaction process

J. Stöhr, *The Nature of X-Rays and Their Interactions with Matter*, Springer Tracts in Modern Physics 288, https://doi.org/10.1007/978-3-031-20744-0_9

may be assumed to be quasi-instantaneous or time independent, so that the influence of the temporal evolution of the excited electronic system can be neglected. The KHD theory has been remarkably successful in describing all x-ray experiments during the first one hundred years of x-ray science, including those at the brightest synchrotron radiation sources.

We note upfront that the KHD theory will eventually break down with increasing photon density which leads to two fundamental effects. First, when there are enough photons to excite the majority of atoms in a sample, photons at the back end of a pulse may experience a change of the electronic system. Secondly, at high intensity there are necessarily more than a single photon in a mode. These photons are indistinguishable and can act together to cause non-linear effects. This case will be treated in the last three chapters forming Part IV of our book.

The present chapter provides an introduction of the foundation and formulation of the KHD theory. We first derive the complete photon-matter interaction Hamiltonian and by estimating the relative size of different contributions reduce it to the two leading terms which are linear and quadratic, respectively, in the vector potential (or electric field). We next outline the assumptions underlying the quantum mechanical calculation of transition rates that are expressed by the key result of the chapter, the KHD formula. We then give a brief overview of how the first order part of the KHD formula is used to calculate transition rates for x-ray absorption, x-ray emission, and Thomson scattering. We finish the chapter by discussing the application of the second order part of the KHD formula for the description of spontaneous and stimulated resonant scattering and two-photon absorption and photoemission.

9.2 The Photon-Matter Interaction Hamiltonian

In order to describe the interaction of x-rays with atomic matter by means of quantum mechanics we need to consider the relative size of the energies associated with the radiation itself and the atomic building blocks of matter, so that we can formulate an appropriate Hamiltonian that allows us to treat the interaction through a perturbation approach. Furthermore, we have already discussed in a semi-classical picture that processes such as absorption, emission, and scattering have different interaction strengths or cross sections. Quantum mechanically, they therefore need to be treated in different orders of perturbation. We start by considering the typical relative size of energies that will allow us to establish a hierarchy in perturbation order.

9.2.1 The Pauli Equation Including the EM Field

In deriving the Hamiltonian that describes the interaction of the EM fields with the electron charge and spin we need to start with a quantum mechanical description that goes beyond the basic Schrödinger equation which does not include spin. Our starting point will be the *Pauli equation* for an atom (labeled "at") which incorporates the spin into the non-relativistic time-dependent Schrödinger equation [5] in terms of a sum of the electron (e) and spin (s) contributions, according to

$$i\hbar\frac{\partial\psi(\mathbf{r},t)}{\partial t} = \underbrace{\left[\frac{\mathbf{p}^2}{2m_\mathrm{e}} + V_\mathrm{at}(\mathbf{r}) + \frac{e\hbar}{m_\mathrm{e}}\mathbf{s}\cdot\mathbf{B}^*\right]}_{\mathcal{H}_\mathrm{at}=\mathcal{H}^\mathrm{e}_\mathrm{at}+\mathcal{H}^\mathrm{s}_\mathrm{at}}\psi(\mathbf{r},t). \tag{9.1}$$

The solution of the Pauli equation for the ground state of a magnetic sample is very complicated and only possible by making approximations for the atomic potential energy $V_\mathrm{at}(\mathbf{r})$, into which we have lumped many terms [5].

The atomic Pauli equation (9.1) can be generalized to include terms due to an external EM field, similar to the general Dirac equation. This involves adding the photon Hamiltonian $\mathcal{H}_\mathrm{rad}$ given by (3.52) or

$$\mathcal{H}_\mathrm{rad} = \sum_{p\mathbf{k}} \hbar\omega_k\left(a^\dagger_{p\mathbf{k}}\,a_{p\mathbf{k}} + \frac{1}{2}\right). \tag{9.2}$$

As discussed by Strange [5], another modification is necessary, which consists of the replacement of the usual momentum term $\mathbf{p}$ by $\mathbf{p}+e\mathbf{A}$. This produces the correct motion of a charged particle with momentum $\mathbf{p}$ in the presence of a time-dependent vector potential $\mathbf{A}=\mathbf{A}(\mathbf{r},t)$ where $\mathbf{p}$ and $\mathbf{A}$ commute. The total Hamiltonian for the atom in an EM field becomes

$$\mathcal{H} = \frac{1}{2m_\mathrm{e}}\mathbf{p}^2 + \frac{e}{m_\mathrm{e}}\mathbf{p}\cdot\mathbf{A} + \frac{e^2}{2m_\mathrm{e}}\mathbf{A}^2 + V_\mathrm{at}(\mathbf{r}) + \frac{e\hbar}{m_\mathrm{e}}\mathbf{s}\cdot\mathbf{B}^* + \mathcal{H}_\mathrm{rad}. \tag{9.3}$$

We now take a closer look at the spin dependent term $e\hbar(\mathbf{s}\cdot\mathbf{B}^*)/m_\mathrm{e}$.

9.2.2 *Evaluation of the Spin Dependent Part of the Pauli Equation*

In our treatment of the spin dependent term we use of the general expression for the electric field (3.37)

$$\mathbf{E}(\mathbf{r},t) = -\nabla\Phi(\mathbf{r},t) - \frac{\partial\mathbf{A}(\mathbf{r},t)}{\partial t} \tag{9.4}$$

and write the term $\mathbf{B}^*$ in the presence of both an atomic electrostatic potential and a radiation field as [6]

$$\mathbf{B}^* = \mathbf{B} + \frac{1}{2m_e c^2}\left[\mathbf{p} \times (\nabla\Phi + \dot{\mathbf{A}})\right]. \tag{9.5}$$

With the same substitution $\mathbf{p} \rightarrow \mathbf{p} + e\mathbf{A}$ as above we obtain

$$\begin{aligned}\mathbf{B}^* &= \mathbf{B} + \frac{1}{2m_e c^2}\left[(\mathbf{p} + e\mathbf{A}) \times (\nabla\Phi + \dot{\mathbf{A}})\right] \\ &\simeq \nabla \times \mathbf{A} + \frac{e}{2m_e c^2}(\mathbf{A} \times \dot{\mathbf{A}}) + \underbrace{\frac{1}{2m_e c^2}(\mathbf{p} \times \nabla\Phi)}_{\text{spin-orbit}}.\end{aligned} \tag{9.6}$$

In the last step we have used $\mathbf{B} = \nabla \times \mathbf{A}$ according to (3.40), and in the evaluation of the cross product have omitted terms linear in $\mathbf{A}$ (and $\dot{\mathbf{A}}$) and kept only those quadratic in $\mathbf{A}$ or independent of $\mathbf{A}$. The reason for omitting the linear terms in $\mathbf{A}$ is that they represent smaller scattering contributions than the term $\mathbf{A} \times \dot{\mathbf{A}}$. As indicated, the last term gives rise to the spin-orbit interaction between the electrons in the atom given by

$$\mathcal{H}_{\text{so}} = \frac{e\hbar}{2m_e^2 c^2}\,\mathbf{s} \cdot (\mathbf{p} \times \nabla\Phi) = \xi_{nl}(r)\,\mathbf{s} \cdot \mathbf{l}. \tag{9.7}$$

On the right we give the usual form of the spin-orbit interaction, where the expectation value $\xi_{nl} = \langle \xi_{nl}(r) \rangle$ is the *spin-orbit coupling constant*. In the context of this book, we are particularly interested in the values of ξ_{nl} for the $2p$ core and $3d$ valence shells of the later transition metal atoms, which are listed in Table 9.1 [7, 8]. In particular the $2p$ core shell is spin-orbit split into $2p_{1/2}$ (L_2) and $2p_{3/2}$ (L_3) states separated by $3\xi_{2p}/2$. The $3d$ valence shell spin-orbit interaction is of great importance in magnetic transition metal compounds, since it determines the orientation of the easy magnetization axis relative to the lattice (the magneto-crystalline anisotropy) [6].

Table 9.1 Spin-orbit coupling constants for the $2p$ core and $3d$ valence shells of the transitions metal ions [8]

Ion config.	ξ_{2p} (eV)	Ion config.	ξ_{3d} (meV)
Cr^{2+} p^5d^5	5.67	Cr^{2+} d^4	30
Mn^{2+} p^5d^6	6.85	Mn^{2+} d^5	40
Fe^{2+} p^5d^7	8.20	Fe^{2+} d^6	52
Co^{2+} p^5d^8	9.75	Co^{2+} d^7	66
Ni^{2+} p^5d^9	11.51	Ni^{2+} d^8	83
Cu^{2+} p^5d^9	13.23	Cu^{2+} d^9	102

Note that for $s = 1/2$, the spin-orbit splitting between the levels $j_+ = s + l$ and $j_- = s - l$ in a shell of angular momentum l is given by the Landé interval rule $\Delta\mathcal{E} = j_+\xi_{nl} = (2l + 1)\xi_{nl}/2$. The Cu values are from [9]

9.2.3 The Complete Interaction Hamiltonian

We can now rewrite (9.3) in terms of atomic contributions, photon contributions, and atom-photon interactions as

$$\mathcal{H} = \underbrace{\frac{\mathbf{p}^2}{2m_e} + V_{at}(\mathbf{r}) + \mathcal{H}_{so}}_{\text{atom}} + \underbrace{\mathcal{H}_{rad}}_{\text{photons}}$$
$$\underbrace{+\frac{e}{m_e}\mathbf{p}\cdot\mathbf{A} + \frac{e^2}{2m_e}\mathbf{A}^2 + \frac{e\hbar}{m_e}\mathbf{s}\cdot(\nabla\times\mathbf{A}) + \frac{e^2\hbar}{2m_e^2c^2}\mathbf{s}\cdot(\mathbf{A}\times\dot{\mathbf{A}})}_{\text{atom-photon interaction}}\,. \qquad (9.8)$$

By underbrackets we have identified the pure atomic, pure photon, and mixed photon-atomic part, with the latter constituting the complete electromagnetic interaction Hamiltonian. It contains electron and photon terms. The electron contributions arise from the charge[1] $q = -e$, mass m_e, momentum $\mathbf{p}$ or spin $\mathbf{s}$, while the photon terms are completely described by the time-dependent vector potential $\mathbf{A}$, given in operator form by (3.55). Let us put a box around our final result.

The complete electromagnetic interaction Hamiltonian is given by

$$\mathcal{H}_{int} = \underbrace{\frac{e^2}{2m_e}A^2}_{\mathcal{H}_1} + \underbrace{\frac{e}{m_e}\mathbf{p}\cdot\mathbf{A}}_{\mathcal{H}_2} + \underbrace{\frac{e\hbar}{m_e}\mathbf{s}\cdot(\nabla\times\mathbf{A})}_{\mathcal{H}_3} + \underbrace{\frac{e^2\hbar}{2m_e^2c^2}\mathbf{s}\cdot(\mathbf{A}\times\dot{\mathbf{A}})}_{\mathcal{H}_4}\,. \qquad (9.9)$$

The terms $\mathcal{H}_1$ and $\mathcal{H}_4$ are *quadratic* in $\mathbf{A}$, while $\mathcal{H}_2$ and $\mathcal{H}_3$ are *linear* in $\mathbf{A}$. According to (3.56) the vector potential $\mathbf{A}(\mathbf{r}, t) = \mathbf{A}^{ab} + \mathbf{A}^{em}$ can cause absorption or photon destruction and emission or photon creation. If the Hamiltonian depends linearly on $\mathbf{A}$, the interaction can either destroy *or* create photons, but not both. Hence it can be used to describe the independent processes of x-ray absorption and emission. If the Hamiltonian is proportional to $\mathbf{A}^2$, it can both destroy and create photons, which corresponds to scattering.

The question arises whether we need to consider all four interaction Hamiltonians, or whether some can be neglected. We expect that the spin dependent interactions are weaker as we have seen in comparison with Thomson charge and spin scattering in Sect. 6.2.3. Also, we have seen in Sects. 7.3.3 and 7.3.4 that the direct coupling of the $\mathbf{B}$ field with the spins can be neglected at optical and higher frequencies. Let us therefore take a look at the relative size of the interactions by estimating the expectation values of the Hamiltonians.

[1] Our result agrees with that given in the literature [5, 10, 11], but some authors define the electron charge as $q = e$, while we define it as $q = -e$.

9.2.4 Relative Size of the Interactions

In order to estimate the relative size of the energies corresponding to the four interaction Hamiltonians, we phrase all expressions in terms of the amplitude of the incident electric field $|E|$. We simply write $|A| = |E|/\omega$ and $|B| = |E|/c$. Further we assume $|s| = 1$ which corresponds to a typical atomic magnetic moment of $|m| = 2\mu_B$ and express the expectation value of the momentum operator according to $\langle p \rangle = m_e\omega\langle r \rangle$, where $\langle r \rangle$ is the expectation value of the core orbital involved in the electronic transition. We then obtain

$$\begin{aligned}|\mathcal{H}_1| &= \frac{e^2}{2m_e\omega^2}\,E^2\\ |\mathcal{H}_2| &= e\,\langle r \rangle\,|E|\\ |\mathcal{H}_3| &= \frac{e\hbar}{m_e c}\,|E|\\ |\mathcal{H}_4| &= \frac{e^2\hbar}{2m_e^2c^2\omega}\,E^2. \end{aligned} \tag{9.10}$$

The Hamiltonians $\mathcal{H}_1$ and $\mathcal{H}_4$ are quadratic in the field and one readily obtains

$$|\mathcal{H}_4| = \frac{e^2\hbar}{2m_e^2c^2\omega}E^2 = \underbrace{\frac{e^2}{2m_e\omega^2}\,E^2}_{|\mathcal{H}_1|}\,\frac{\hbar\omega}{m_e c^2}. \tag{9.11}$$

The two interactions correspond to Thomson scattering by charges, $\mathcal{H}_1$, and by spins, $\mathcal{H}_4$. Spin scattering contains an additional factor $\hbar\omega/(m_e c^2)$, which due to the large value of $m_e c^2 \simeq 511\,\text{keV}$ is much weaker than charge scattering. Spin dependent x-ray Thomson scattering has been measured in a beautiful tour de force experiment for antiferromagnetic crystals by de Bergevin and Brunel [12, 13]. We shall neglect $\mathcal{H}_4$ in the following.

Next we compare the Hamiltonians $\mathcal{H}_2$ and $\mathcal{H}_3$ which are linear in the field and responsible for driving excitations of atomic core electrons. We obtain

$$|\mathcal{H}_3| = \frac{e\,\hbar}{m_e c}\,E = \underbrace{e\,\langle r \rangle\,E}_{|\mathcal{H}_2|}\,\frac{\hbar\omega}{m_e c^2}\,\frac{\lambda}{2\pi\,\langle r \rangle}. \tag{9.12}$$

Again, we find the reduction factor $\hbar\omega/(m_e c^2)$ for the spin dependent Hamiltonian $\mathcal{H}_3$, but there is also another dimensionless factor $\lambda/(2\pi\,\langle r \rangle)$ that depends on the ratio of the wavelength relative to the effective size of the atomic core shell $\langle r \rangle$. We can estimate this value by means of the electronic binding energy of the core shell. We use a hydrogenic model where the H 1s binding or Rydberg energy of $\mathcal{E}_B = 13.6\,\text{eV}$

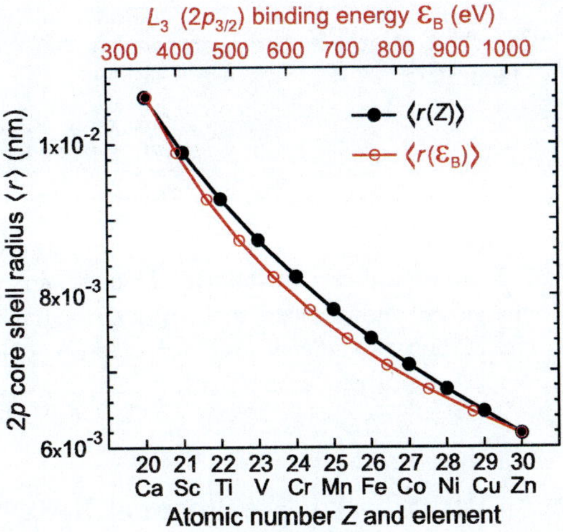

Fig. 9.1 Core radius $\langle r\rangle$ of the $2p_{3/2}$ core shell in the $3d$ transition metals, calculated in a hydrogenic model according to (9.14). We have plotted $\langle r\rangle$ both as a function of Z (black curve, lower scale) and $2p_{3/2}$ (L_3) binding energy $\mathcal{E}_B$ (red curve, upper scale)

and the Bohr radius a_0 are linked according to[2]

$$\mathcal{E}_B = \frac{m_e e^4}{2\hbar^2(4\pi\epsilon_0)^2}, \quad a_0^2 = \frac{\hbar^4(4\pi\epsilon_0)^2}{m_e^2 e^4}. \tag{9.13}$$

This leads to the generalized relation

$$\langle r\rangle^2 = \frac{\hbar^2}{2m_e\mathcal{E}_B}. \tag{9.14}$$

We have plotted $\langle r\rangle$ for the $2p_{3/2}$ core shell of the important $3d$ transition metals in Fig. 9.1. The values are found to be within a factor of 2 of those obtained with atomic Hartree–Fock calculations plotted in Fig. 8–10 by Cowan [7]. They are also approximately equal to the radial $2p_{3/2} \rightarrow 3d$ radial matrix element $\mathcal{R}$ plotted in Fig. 12.11.

We can now estimate the value of the desired factor in (9.12) by using the values $\hbar\omega = \mathcal{E}_B$ and similarly for λ and obtain for the $3d$ transition metals,

$$\frac{\hbar\omega}{m_e c^2}\frac{\lambda}{2\pi\langle r\rangle} \simeq 5\times 10^{-2}. \tag{9.15}$$

Hence we will in the following also neglect the Hamiltonian $\mathcal{H}_3$. We can state as follows.

---

[2] The expression of the Rydberg energy in terms of the electron's rest mass and charge and Planck's constant was derived by Bohr in 1913 [14], constituting a seminal step in understanding the atomic structure through spectroscopy.

The interaction between x-rays and matter can be described entirely by the Hamiltonian

$$\mathcal{H}_{\text{int}} = \underbrace{\frac{e^2}{2m_e} A^2}_{\substack{\text{Thomson} \\ \text{scattering}}} + \underbrace{\frac{e}{m_e} \mathbf{p} \cdot \mathbf{A}}_{\substack{\text{photoelectric} \\ \text{transitions}}} . \tag{9.16}$$

It does not explicitly include the electron spin, but the coupling to the electronic charge still allows access to the spin through the spin-orbit coupling. This gives rise to the x-ray magnetic dichroism effects discussed in Sect. 7.9.

9.3 Perturbation Treatment of X-Ray Scattering and Absorption

If the interaction of the EM field with an atomic sample, described by (9.9), was as strong as the electronic interactions within the sample itself, we would have to consider the terms in the interaction Hamiltonian on an equal footing with the atomic and solid state ones. Fortunately this is not the case, and in the following section we shall discuss the concept of perturbation theory, whose essence is the fundamental assumption that the interactions of the EM fields with charges and spins are "weak".

9.3.1 *On the Use of Time-Dependent Perturbation Theory*

When the total Hamiltonian describing a system depends on time, the concept of discrete energy levels and stationary wave functions must be modified. The essence of approximate methods for the description of electronic excitations is to assume that the system is mostly time independent and that the time evolution can be approximately treated by assuming that it is either (i) a small change due to the weakness of the interaction, (ii) slowly varying with time (adiabatic approximation), or (iii) changing over a very short time interval (sudden approximation).

Method (i) is similar to the concepts used in time independent perturbation theory where the total Hamiltonian is written in two parts, $\mathcal{H} = \mathcal{H}_0 + \mathcal{H}'$ and the "smaller" part $\mathcal{H}'$ is treated as a perturbation of $\mathcal{H}_0$. Time-dependent perturbation theory makes the same Ansatz. Here the time dependence is entirely contained in the "weaker" part $\mathcal{H}'$ which acts as a time-dependent perturbation of the stationary Hamiltonian $\mathcal{H}_0$. Let us explore whether the interaction Hamiltonian $\mathcal{H}_{\text{int}}$ given by (9.16) fulfills the requirement that it is "weak" relative to the electronic ground state Hamiltonian.

From consideration of the semi-classical interaction of an EM wave with charges and spins, discussed in the last chapter, it is intuitively clear why elastic Thomson

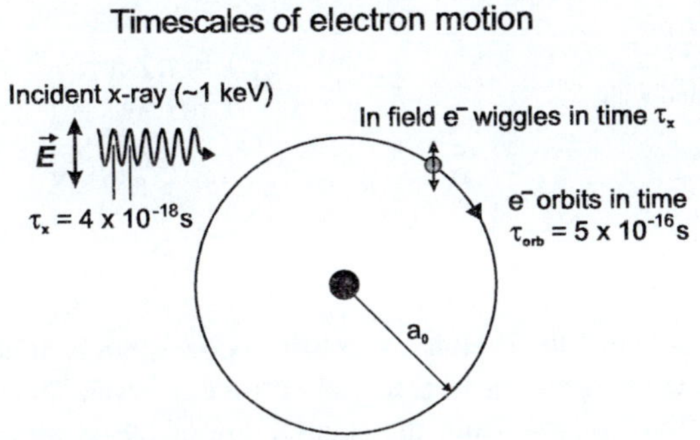

Fig. 9.2 Simple picture of an x-ray scattering process by an electron in a hydrogen atom. In the scattering process the electron is wiggled at the frequency of the incident electric field. At $h\nu = 1000$ eV we have $\tau_x = 1/\nu = 4 \times 10^{-18}$ s, which is much shorter than the orbit time of the electron $\tau_{\mathrm{orb}} = 5 \times 10^{-16}$ s. Therefore the scattering process takes place over a small fraction of the electron's orbit. Even for the most intense x-ray sources today the electric field of the EM wave is still relatively weak, with an amplitude of only $|\mathbf{E}| \sim 250$ V/m. This is many orders of magnitude less that the atomic electric field $|\mathbf{E}| \sim 10^{11}$ V/m, and the electron is therefore barely distorted from its orbit

scattering by an atom can be viewed as a small perturbation. Let us consider a scattering process by an EM wave off an electron in an atom, as illustrated in Fig. 9.2.

For simplicity we use a hydrogen atom which has an electron bound by about $E_{\mathrm{B}} = 10$ eV, orbiting with the Bohr radius of $a_0 = 0.5 \times 10^{-10}$ m and an orbit time of about $\tau_{\mathrm{orb}} = 5 \times 10^{-16}$ s, estimated from the relation $\tau_{\mathrm{orb}}\, E_{\mathrm{B}} = h = 4.136 \times 10^{-15}$ eV s, connecting energy and time. At the site of the electron the electric field of the nucleus is $|\mathbf{E}| = e/(4\pi\varepsilon_0 a_0^2) \sim 5 \times 10^{11}$ V/m. In Sect. 3.2.4 we have discussed the typical size of fields found in today's state-of-the-art x-ray beams. They were readily calculated from the emitted intensity at third generation synchrotron sources, given by the Poynting vector. At an x-ray energy of 1000 eV, corresponding to a wave period of $\tau_x = 4 \times 10^{-18}$ s we obtained field strengths of $|\mathbf{E}| \sim 250$ V/m, which are about nine orders of magnitude weaker than atomic fields. This leads to the picture shown in Fig. 9.2, where the EM wave wiggles the electron with a frequency that is large relative to the orbit frequency and an amplitude $A_{\mathrm{osc}} = e|\mathbf{E}|/(m_e\omega^2)$ that is very small relative to the classical orbit radius. Hence the scattering process can be treated by perturbation theory, both, because it is weak and because it is fast.

It is more difficult to give a simple intuitive picture why an x-ray absorption event can also be viewed as a small perturbation, since excitation or ejection of a photoelectron from an atom involves a rather large energy transfer. The reason for our intuitive difficulty is that x-ray absorption is a quantum mechanical process and our simple classical thinking breaks down. If we compare x-ray scattering and x-ray absorption from a quantum mechanical point of view, they are rather similar. In one case a photon comes in, scatters and goes out, in the other, a photon comes in, scatters, and an electron goes out. From this picture one would therefore expect that both cases can be described in the same perturbative fashion.

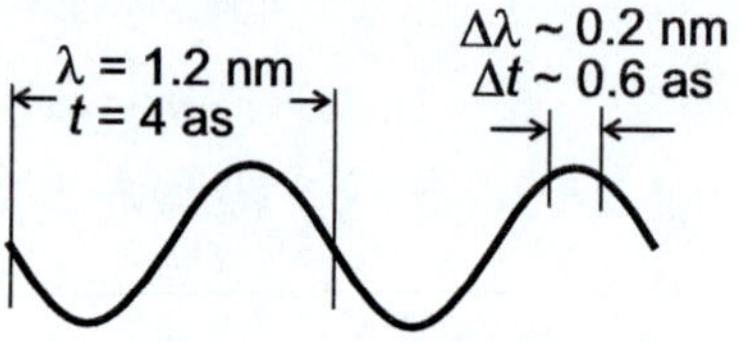

Fig. 9.3 Illustration of spatial and temporal dimensions associated with a 1 keV photon

Most important is that the excitation probability per atom is quite small, and the x-ray absorption event occurs over a very short time interval. The small excitation probability per atom follows from the small value of the x-ray absorption cross section, which has the approximate area of a core shell and is therefore much smaller than the total area of an atom. The interaction time can be estimated from the time that it takes the photon to cross the diameter of the inner shell, or about 10^{-18} s or 1 as. A similar time estimate is obtained from the oscillation time of the EM field, considering that the field that drives the excitation is large only for a fraction of its cycle period as illustrated in Fig. 9.3. For a photon energy of 1 keV (λ =1.2 nm), the field is large over a time window of order $\Delta t = \lambda/(2\pi c) \simeq 0.6$ as.

The interaction process is so fast that it is over before the other electrons in the atom can adjust to the change.

> The perturbation treatment of the atomic response to x-rays is largely based on the fact the interaction process with a single photon is so fast (<1 as) that it is not influenced by all atomic processes that occur on longer timescales.

Consideration of the relative speed of processes is a common concept used to make approximations in physics. For example, it is the basis for the *Born–Oppenheimer approximation* [15] that states that electronic motion (of order 10^{-18}–10^{-15} s) is so much faster than nuclear motion (of order 10^{-12} s) that the electronic changes can be treated by assuming constant nuclear positions. In photoemission spectroscopy one uses the so-called *sudden approximation* which allows a description of the excitation process that is one-electron in nature [16]. The excitation of the "active" electron is viewed as "sudden" relative to the relaxation times of the other "passive" electrons. Another manifestation of the "speed" approximation is encountered in the "probe-before-destroy" concept, proven so valuable in single-shot protein imaging experiments with very intense X-FEL pulses [17, 18]. If the pulse length is kept short enough (of order 10 fs), the image can be recorded before the atoms have time to move, which proceeds with the speed of sound ($\simeq$ 1 nm/ps).

9.4 Kramers-Heisenberg-Dirac Perturbation Theory

The x-ray absorption and x-ray scattering cross sections are both calculated by consideration of the time-dependent perturbation of the sample by the EM field, described by the Hamiltonian (9.16). The time-dependent EM field induces transitions between an initial state $|i\rangle$ and final state $|f\rangle$ where both states contain an electronic and a photon part. In time-dependent perturbation theory the time evolution of the system from a time t_0 in the ground state $|i\rangle$ to a time t in a final state $|f\rangle$ can be envisioned to proceed in multiple steps.

In the first step, the electronic and photon parts of the system are decoupled. The electronic ground state $|\mathbf{a}\rangle$ evolves in an exactly predictable manner under the influence of the time *independent* Hamiltonian $\mathcal{H}_0$ according to[3]

$$|a(t_1)\rangle = e^{-\mathrm{i}(t_1-t_0)\mathcal{H}_0/\hbar}\,|a(t_0)\rangle. \tag{9.19}$$

At a time $t_2 = \tau$, the interaction (second step) takes place, and under the perturbation $\mathcal{H}_{\text{int}}$ the system is taken from an electronic state $|a(\tau)\rangle$ with energy E_a to an electronic state $|b(\tau)\rangle$ with energy E_b. We can also write this transition in a more general form $\langle f|\mathcal{H}_{\text{int}}|i\rangle$, where the states $|i\rangle = |a\rangle|n_1\rangle$ and $|f\rangle = |b\rangle|n_2\rangle$ are composite electronic and photon states with photon numbers n_1 and n_2, respectively. In the third step the electronic state $|b\rangle$ evolves entirely predictable again, similar to (9.19),

$$|b(t_3)\rangle = e^{-\mathrm{i}(t_3-\tau)\mathcal{H}_0/\hbar}\,|b(\tau)\rangle. \tag{9.20}$$

In the second order, the transition is made in two steps through an intermediate state m by a process $\langle f|\mathcal{H}_{\text{int}}|m\rangle\langle m|\mathcal{H}_{\text{int}}|i\rangle$. In the third order there are two intermediate states with an evolution characterized by $\langle f|\mathcal{H}_{\text{int}}|l\rangle\langle l|\mathcal{H}_{\text{int}}|j\rangle\langle j|\mathcal{H}_{\text{int}}|i\rangle$ and so on. Each step represents a higher order of perturbation. The formalism is developed in most standard quantum mechanics books, and we will not repeat it here but simply state the result up to second order, which we will refer to as the Kramers-Heisenberg-Dirac formula.

[3] This comes about as follows. Assume a time-dependent state $|\alpha\rangle$ that satisfies the time-dependent Schrödinger equation of a time independent Hamiltonian $\mathcal{H}$

$$i\hbar\frac{\mathrm{d}}{\mathrm{d}t}|\alpha\rangle = \mathcal{H}|\alpha\rangle\,. \tag{9.17}$$

This integrates to

$$|\alpha(t)\rangle = |\alpha(0)\rangle\, e^{-\mathrm{i}\mathcal{H}t/\hbar} = |\alpha(0)\rangle\, e^{-\mathrm{i}E_\alpha t/\hbar}. \tag{9.18}$$

9.4.1 The Kramers-Heisenberg-Dirac Formula

The Kramers-Heisenberg-Dirac formula constitutes the starting point for the quantitative calculation of electronic x-ray processes under the assumption of the Born–Oppenheimer approximation and can be summarized as follows.

The Kramers-Heisenberg-Dirac formula gives the *transition probability per unit time*, $\mathcal{W}_{if}$, from a state i to a state f up to second order according to

$$\mathcal{W}_{if} = \frac{2\pi}{\hbar}\left|\langle f|\mathcal{H}_{\text{int}}|i\rangle + \sum_m \frac{\langle f|\mathcal{H}_{\text{int}}|m\rangle\langle m|\mathcal{H}_{\text{int}}|i\rangle}{\mathcal{E}_i - \mathcal{E}_m}\right|^2 \rho(\mathcal{E}_f)\,\delta(\mathcal{E}_f - \mathcal{E}_i). \tag{9.21}$$

The dimension of $\mathcal{W}_{if}$ is [1/time]. The dimension of $\hbar$ is [energy×time], that of the absolute value squared expression is [energy^2], and $\rho(\mathcal{E}_f)$ is the density per unit energy of dimension [1/energy]. The tricky Dirac δ-function conserves energy and is formally defined only through its unitless integrated property (see Appendix A.3). It does not enter in the overall dimension of (9.21).

$\mathcal{H}_{\text{int}}$ is the interaction Hamiltonian, and the wave functions $|i\rangle$ and $|f\rangle$ are products of electronic and photon states and the energies $\mathcal{E}_j$ are sums of electronic and photon energies. The sum is over all possible intermediate states of energy $\mathcal{E}_{\text{m}}$. If the final states $|f\rangle$ for the *first order* term and the *second order* term are the same, the two terms can interfere.

The first term in the sum is sometimes referred to as "Fermi's golden rule" in reference to Fermi's review of the quantum theory of radiation [3]. The second term gives the transition probability from $|i\rangle$ to $|f\rangle$ via a range of *intermediate states* $|m\rangle$, which are also products of electronic and photon states. The sum is over all possible intermediate states of energy $\mathcal{E}_{\text{m}}$ through which the system can pass in a virtual sense that does not require energy conservation until the final state is reached. Since the first and second order terms are summed before the absolute value is calculated, there can be interference. In practice, this occurs when the interaction Hamiltonian in the first and second terms is different (otherwise the first term dominates), and the final states $|f\rangle$, composed of the product of the electronic final state and the final photon state, are the same. This is encountered in a purely elastic scattering process that starts from the electronic and vibrational ground state, and both the first order elastic *Thomson* scattering process and the second order elastic *resonant* scattering process return the system back to the initial state [19].

The final two terms in (9.21) represent the density of final states per unit energy (dimension [1/energy]), and Dirac's δ-function assures energy conservation. Its value and dimension are defined only through its integral over energy (see Appendix A.3), so that the transition rate $\mathcal{W}_{if}$ represents the transition probability per unit time.

The KHD formula assumes an excitation process that occurs within an ultra-short time window under neglect of the detailed temporal evolution. The quasi-instantaneous events may be envisioned to occur when the incident electric field that drives the interaction process is large, as shown in Fig. 9.3. As discussed earlier, the perturbation approach may be justified by the fact that even the largest interaction process, x-ray absorption, typically requires an excitation attempt by more than 100 photons before an excitation occurs, and that the excitation of the core electron is so fast that the other electrons have difficulty to respond.

The dominant Thomson and photoelectric terms in the interaction Hamiltonian (9.9) given by $\mathcal{H}_1$ and $\mathcal{H}_2$ may be written in the alternative vector potential and field forms introduced in Sect. 3.3.3

$$\mathcal{H}_{\text{int}} = \underbrace{\frac{e^2}{2m_e} A^2}_{\text{Thomson}} + \underbrace{\frac{e}{m_e} \mathbf{p} \cdot \mathbf{A}}_{\text{Photoelectric}}$$
$$= \frac{e^2}{2m_e\omega^2} E^2 - \mathrm{i}\frac{e}{m_e\omega} \mathbf{p} \cdot \mathbf{E} = \frac{e^2}{2m_e\omega^2} E^2 - e\,\mathbf{r} \cdot \mathbf{E}. \tag{9.22}$$

In rewriting the expression we have replaced the vector potential given by 3.55 by the electric field operator given by (3.61), according to $\mathbf{A} = -\mathrm{i}\mathbf{E}/\omega$, and the *momentum operator* by the *length operator* according to $\mathbf{p} = m_e \, \mathrm{d}\mathbf{r}/\mathrm{d}t = m_e \, \omega \, \mathbf{r}$. This is an example of the common concept of operator equivalents in quantum mechanics and corresponds to replacing the Coulomb gauge by the Göppert-Mayer gauge [20, 21].

The replacement of the vector potential **A** by the *electric* field operator **E** is possible because the interaction of the EM **H** field with the spins in a sample is extremely weak, of order $\hbar\omega/(m_e c^2)$. This allowed us to neglect the smaller terms $\mathcal{H}_3$ and $\mathcal{H}_4$ in the Hamiltonian (9.9). The negligible contribution of the *magnetic* field **H** in the x-ray region may be understood in a physical picture by the requirement of angular momentum conservation. At frequencies beyond the THz range, any excitation of the spin system requires that the associated change in angular momentum can be transferred to another reservoir, which in solids is the lattice. With increasing photon energy, the required angular momentum transfer is increasingly bottlenecked. This situation has been discussed in Sect. 7.3.3 in terms of the refractive index, which at optical and higher frequencies is only determined by the electric permittivity $\tilde{\epsilon}$ and no longer by the magnetic permeability $\tilde{\mu}$.

In practice, the KHD formula is evaluated by assuming an interaction Hamiltonian described by one of the terms in (9.22), where the operators $\mathbf{A} = \mathbf{A}^+ + \mathbf{A}^-$ or $\mathbf{E} = \mathbf{E}^+ + \mathbf{E}^-$ are written in terms of photon destruction (superscript +) and creation (superscript −) operators according to (3.56) or (3.61). The photon part of the product states $|i\rangle$ and $|f\rangle$ is then evaluated to leave only matrix elements between electronic states. We will show how this is done in the following chapters. The present chapter is intended as an overview of how different x-ray/matter interactions are described by the first and second order terms in the KHD formula.

9.5 Overview of First Order Processes

The basic quantum mechanical processes described by the first order term in (9.21) are x-ray absorption, x-ray emission, and x-ray Thomson scattering, illustrated in Fig. 9.4. In the following we shall outline the underlying concepts and the special absorption and emission cases, where the excitation process is resonant. In this case the atom is not ionized by excitation of a core electron into continuum states, but the core electron is promoted into a bound state, where it remains trapped over its lifetime.

9.5.1 *X-Ray Absorption*

The strongest x-ray interaction with matter is *x-ray absorption*, schematically shown in Fig. 9.4a. The process is described by the interaction Hamiltonian $\propto \mathbf{p} \cdot \mathbf{A}^{+} \propto \mathbf{r} \cdot \mathbf{E}^{+}$ in (9.22) which is linear in the vector potential or field. In a quantum picture, the incident electric field or vector potential acts on an atomic core electron, with energy transfer from an incident photon, which gets destroyed, to an atomic core electron, which gets excited. In the figure we have assumed that the incident photon energy is larger than the atomic ionization potential, so that the electron is liberated as a

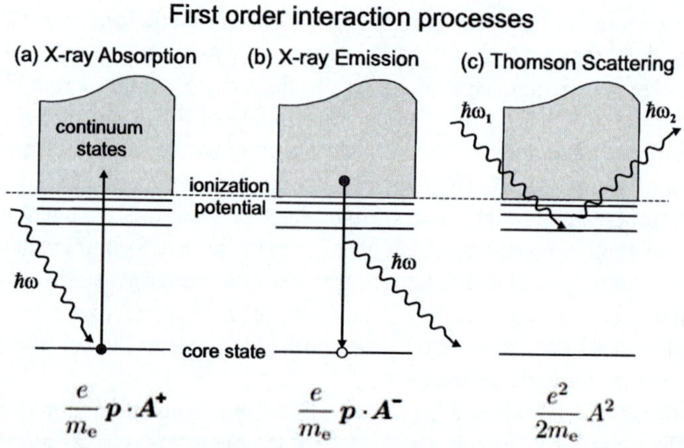

Fig. 9.4 Illustration of first order quantum mechanical processes in the interaction of a photon with the electronic charge of an atom. The atom is depicted as having discrete energy levels consisting of filled electronic shells and empty Rydberg states close to the ionization potential (IP) or vacuum level and a continuum of empty states above the IP. Electrons are shown as filled circles, holes as open circles. X-ray absorption and emission are first order processes where an x-ray is absorbed (destroyed) or emitted (created), described by the action of the destruction operator $\mathbf{A}^{+}$ and creation operator $\mathbf{A}^{-}$ (see (3.58)) in the photoelectric Hamiltonian $\propto \mathbf{p} \cdot \mathbf{A}$ in (9.22). Non-resonant Thomson scattering is also a first order process but involves the interaction Hamiltonian $\propto A^2$ in (9.22)

photoelectron. When at a given photon energy, the kinetic energy of the emitted photoelectron is analyzed by an electron spectrometer, the measured *photoemission* spectrum will exhibit peaks reflecting electrons excited from different core shells. The intensities of the peaks directly reflect the ionization cross sections of the core shells. *X-ray absorption spectroscopy* (XAS) measures a photon energy dependent cross section that is the sum of all photoemission cross sections of shells with energy less than the photon energy.

It is important to realize that although the total *absorption* cross section in the x-ray range decreases with photon energy as shown in Fig. 6.11a, the shell-specific *photoemission* cross sections behave the opposite way [22]. As illustrated for Ne in Fig. 1.27, at x-ray energies that exceed the binding energies of the innermost $1s$ electrons, the interaction cross section is the largest for the innermost core shell and smallest for the valence shell.

For x-rays, there is a fundamental distinction between absorption and scattering, which becomes somewhat blurred for atoms in the optical region. Since atoms often have ionization potentials (IPs) of order of 10 eV, optical photons ($\simeq$ 1 eV) cannot induce electronic excitations. "Absorption" of incident photons is then sometimes associated with scattering of photons out of the incident beam, resulting in an "absorption" loss in the forward direction [23]. While this loss mechanism also exists for x-rays, it is clearly distinguished conceptually and quantitatively. The x-ray absorption process always involves an *irreversible* conversion of some or all of the energy of incident photons into electrons, which are ejected either through the direct photoemission process or the indirect Auger process.

Atomic ionization cross sections change smoothly with photon energy and are tabulated as atom-specific quantities [24]. The tabulation assumes step-like changes at absorption thresholds and ignores the existence of resonant-like structures near the IPs. Structures in the cross sections above the IPs are also ignored. They may arise from multi-electron excitations, and in molecular and condensed matter samples oscillations are present due to backscattering of photoelectrons from the nearest neighbors, referred to as "Extended X-ray Absorption Fine Structure" or EXAFS [25–27].

9.5.2 Resonant X-Ray Absorption

In practice, the spectral regions over which the *absorption cross section* varies smoothly are interrupted by step-like increases at absorption "edges", which often exhibit strong resonances as shown in Figs. 1.18 or 6.6. The fine structure at core shell absorption thresholds, referred to as "X-ray Absorption Near Edge Structure" (XANES) or "near edge x-ray absorption fine structure" (NEXAFS) [16], corresponds to excitation of core electrons to empty valence states. It is therefore strongly dependent on the bonding environment of the selected atoms as discussed in Sect. 1.4.2.

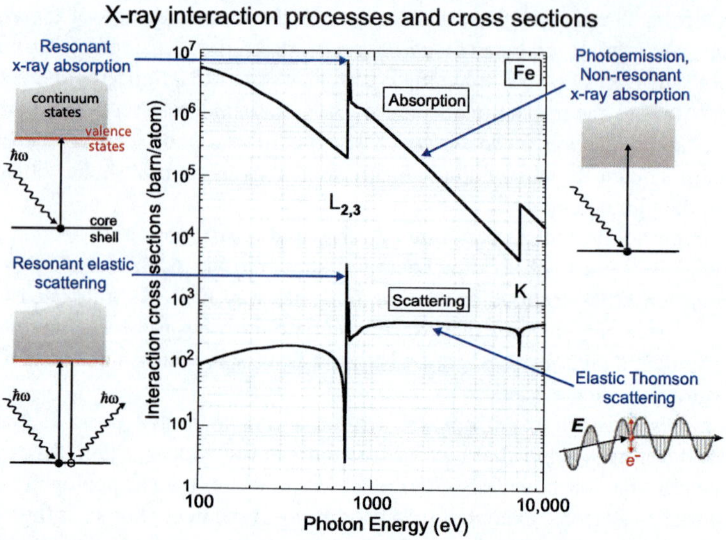

Fig. 9.5 Illustration of the fundamental absorption and scattering cross sections per atom ($1\,\mathrm{b}=10^{-28}\,\mathrm{m}^2$) as a function of photon energy, obtained from measurements of Fe metal between resonances [24] and around the L-edge resonances [6, 28]. The K-edge does not exhibit a resonance [29] and is approximated by a step. The associated photon interaction processes are also illustrated

Non-resonant and resonant x-ray absorption processes are identified in the upper half of Fig. 9.5 by schematics that have been added to our earlier Fig. 7.14. The shown cross sections σ of dimension [area] are given in units of barn (b) where $1\,\mathrm{b}=10^{-28}\,\mathrm{m}^2$. They are obtained from the KHD transition probability per unit time $\mathcal{W}_{if}$ by normalization to the incident photon flux Φ_0, given by (3.121), according to,

$$\sigma = \frac{\mathcal{W}_{if}}{\Phi_0}. \tag{9.23}$$

9.5.3 X-Ray Emission

The quantum mechanical x-ray emission process is illustrated in Fig. 9.4b. As discussed in Chap. 1, x-ray emission spectroscopy (XES) has a long history dating back to Moseley's 1913 discovery of the characteristic change of the emitted $K\alpha$ and $K\beta$ photon energies for different elements [30], illustrated in Fig. 1.8. Like x-ray absorption, the XES process is also a first order process, calculated by evaluation of the first term in the KHD formula (9.21), i.e. Fermi's golden rule, with the interaction Hamiltonian $\propto \mathbf{p}\cdot\mathbf{A}^- \propto \mathbf{r}\cdot\mathbf{E}^-$ in (9.22).

The first order description is possible because one does not specify how the core hole state, which is the initial state of the process, has been produced. Hence it describes different cases where core holes have been produced by the interaction of photons or charged particles with a sample. The quantum process describes the *spontaneous* creation of a photon out of the zero-point vacuum of virtual photons by use of the energy contained in the exited electronic state.

In the x-ray region, the filling of the core hole always proceeds by two competing *spontaneous* processes, resulting in emission of either a photon or Auger electron. In the soft x-ray region, the Auger process dominates while at hard x-ray energies x-ray emission is more likely [31]. The *first order* transition rate calculated by use of the KHD formula reflects only the dipolar radiative decay probability resulting in a photon. The intrinsically linked Auger decay channel is usually included only in an *ad hoc* fashion by defining a total decay rate that consists of the sum of the calculated dipolar radiative rate plus an Auger rate. The latter is obtained from more complicated electronic structure calculations [32] or from compiled semi-empirical values [33, 34].

If Auger decay is included in the *ab initio* formulation of x-ray emission, the treatment requires a second order formalism as discussed in Sect. 12.2.5. When the excitation step is included, one speaks of *resonant inelastic x-ray scattering* (RIXS). This process is also described by a second order formalism as discussed in Chap. 13.

9.5.4 X-Ray Thomson Scattering

The process shown in Fig. 9.4c represents *non-resonant* or *Thomson scattering*. More specifically, we distinguish elastic and inelastic Thomson scattering. In inelastic Thomson scattering, the scattered photon has a reduced energy, with the energy difference used for atomic core or valence excitations [35].

In quantum theory, Thomson scattering is also a first order process described by the first term of the KHD formula (9.21) but with the A^2 (E^2) interaction Hamiltonian in (9.22). The Thomson *scattering cross section* is considerably weaker than the x-ray absorption cross section and exhibits only a weak photon energy dependence as shown in the lower half of Fig. 9.5. Since the Hamiltonian is proportional to $\mathbf{A}^2$, the scattering process may be pictured as the destruction of a photon by interaction with atomic electrons and recreation of a photon. Thomson scattering involves no specific electronic shells and scales with the total number of electrons per atom, Z.

At hard x-ray energies, where the momentum of the photon needs to be taken into account, the scattering direction is angle dependent, as expressed by the atomic form factor discussed in Sect. 6.2.4. In the soft x-ray region, Thomson scattering becomes independent of the scattering direction and photons are scattered into random directions. In quantum theory, the random scattering direction, described classically by an outgoing spherical wave, is a consequence of the photon recreation process out of the random modes of the zero-point quantum vacuum.

The most important consequence of Thomson scattering is Bragg diffraction, which is just the coherent superposition of elastic Thomson scattering from atoms in a crystal.

The strong variation of the scattering cross sections near absorption thresholds in the lower half of Fig. 9.5 was historically called "anomalous scattering" [36]. Today, it is typically called *resonant scattering* since it involves a resonant absorption process followed by an x-ray emission process. The full description of resonant scattering requires a second order formulation as will be discussed next.

9.6 Overview of Second Order Processes

Second order x-ray processes are treated quantum mechanically by the second term in the Kramers-Heisenberg-Dirac formula (9.21). It involves the product of two matrix elements of the interaction operator $\mathbf{p} \cdot \mathbf{A}$, where $\mathbf{A} = \mathbf{A}^+ + \mathbf{A}^-$. Since the second order formula involves three states, the initial, intermediate, and final states, the KHD formula applies to a variety of electronic processes, which may proceed via real or virtual intermediate states $|m\rangle$. As the name implies, second order processes involve two photons. One distinguishes the response of the atomic system to incident non-interacting single photons and the joint action of two incident photons. Figure 9.6 illustrates three examples of second order interaction processes.

9.6.1 *Spontaneous X-Ray Resonant Scattering*

In spontaneous x-ray resonant scattering, illustrated in Fig. 9.6a, the double matrix element $\langle f|\mathcal{H}_{\text{int}}|m\rangle\langle m|\mathcal{H}_{\text{int}}|i\rangle$ in (9.21) is evaluated with the photoelectric operator in (9.22), in the form $\langle f|\mathbf{p} \cdot \mathbf{A}^-|m\rangle\langle m|\mathbf{p} \cdot \mathbf{A}^+|i\rangle$. In this case one only considers the actions of individual non-interacting incident photons, so that the initial state $|i\rangle$ contains a single photon $|n\rangle = |1\rangle$ and a core electron in state $|a\rangle$, i.e. $|i\rangle = |1, a\rangle$.

In the first step, the core electron gets excited to an unfilled electronic state $|c\rangle$ and the photon gets destroyed (destruction operator $\mathbf{A}^+$). The created intermediate state $|m\rangle$ then consists of the product of an excited electronic state $|c\rangle$ and a photon state $|n\rangle = |0\rangle$ that contains no real photons but only the quantum vacuum.

In the second step, the electronic energy stored in the intermediate state $|m\rangle = |0, c\rangle$ can be used to *spontaneously* create a photon (creation operator $\mathbf{A}^-$) out of the quantum vacuum, corresponding to the decay of the state $|m\rangle$ into the final state $|f\rangle$. The final state then contains a single photon in an electronic state $|b\rangle$, i.e. $|f\rangle = |1, b\rangle$. If $|b\rangle = |a\rangle$, we have *spontaneous elastic* resonant scattering and the case $|b\rangle \neq |a\rangle$ describes *spontaneous inelastic* resonant scattering.

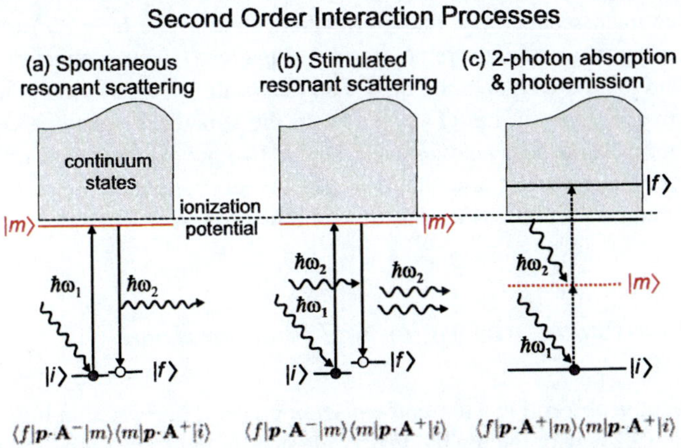

Fig. 9.6 Illustration of three kinds of second order interaction processes. In all cases, the system evolves from an initial state $|i\rangle$ to a final state $|f\rangle$ through an intermediate state $|m\rangle$, where all states are products of photon number states and electronic states. **a** In *spontaneous* x-ray resonant scattering an incident photon excites a core electron to a partially empty valence state. The created intermediate state $|m\rangle$ contains no photon, but a photon is spontaneously created out of the zero-point quantum vacuum during the decay of $|m\rangle$ to the final state $|f\rangle$. **b** In *stimulated* x-ray resonant scattering the intermediate state contains at least one other incident photon. This photon, which may have a different energy from the absorbed photon, drives or "stimulates" the emission of a second photon during the decay of $|m\rangle$ to the final state $|f\rangle$. The emitted two photons constitute a cloned pair. **c** In two-photon absorption or photoemission, one incident photon excites the atom to a virtual intermediate state $|m\rangle$ and a second incident photon provides the energy to excite the atom further to a bound or continuum state $|f\rangle$. Underneath we give the corresponding double matrix elements in the KHD formula

9.6.2 Stimulated X-Ray Resonant Scattering

In stimulated x-ray resonant scattering, illustrated schematically in Fig. 9.6b, the double matrix element in (9.21) has the same form as for spontaneous resonant scattering, $\langle f|\mathbf{p}\cdot\mathbf{A}^-|m\rangle\langle m|\mathbf{p}\cdot\mathbf{A}^+|i\rangle$. The key difference is that one considers the combined action of two incident photons. This case was first treated by Maria Göppert-Mayer in 1931 [21], who derived the expressions for two-photon absorption and photoemission and stimulated Rayleigh/Raman scattering, long before such processes could be observed. Such events are typically very rare because they require a high density of suitable photons in the same interaction volume of the sample. In the x-ray regime, it took the advent of XFELs to observe stimulated decays [37, 38].

As in spontaneous resonant scattering, a core electron gets excited in a first step to an unfilled electronic state $|c\rangle$ and one of the incident photons in the number state $|n\rangle$ gets destroyed (destruction operator $\mathbf{A}^+$). The created intermediate state $|m\rangle$ then consists of the product of an excited electronic state $|c\rangle$ and a photon state $|n-1\rangle$ that contains one less photon.

In the coupled second step, the intermediate state $|m\rangle = |n-1, c\rangle$ may decay through the action of one of the remaining $n-1$ photons (creation operator $\mathbf{A}^-$). The decay is said to be driven or *stimulated*. The final state $|f\rangle$ then contains n photons in an electronic state $|b\rangle$, i.e. $|f\rangle = |n, b\rangle$. In the stimulated decay processes, the photon that drives the decay clones itself, so that two photons emerge with the same polarization, energy and phase. If $|b\rangle = |a\rangle$, we have *stimulated elastic* resonant scattering and the case $|b\rangle \neq |a\rangle$ describes *stimulated inelastic* resonant scattering.

9.6.3 Two-Photon Absorption and Photoemission

The two-photon absorption and photoemission process, illustrated in Fig. 9.6c, corresponds to (9.21) with the double matrix element of the form $\langle f|\mathbf{p}\cdot\mathbf{A}^+|m\rangle\langle m|\mathbf{p}\cdot\mathbf{A}^+|i\rangle$. The essence of two-photon absorption is that two photons have to act together, in what Göppert-Mayer called an "Elementarakt", an elementary non-separable act. The detection of two-photon absorption in the x-ray regime became possible only by use of XFELs [39].

References

1. H.A. Kramers, W. Heisenberg, Z. Phys. **31**, 681 (1925)
2. P.A.M. Dirac, Proc. Roy. Soc. A **114**, 243 (1927)
3. E. Fermi, Rev. Mod. Phys. **4**, 87 (1932)
4. P.A.M. Dirac, Proc. Roy. Soc. London A **114**, 710 (1927)
5. P. Strange, *Relativistic Quantum Mechanics—With Applications in Condensed Matter and Atomic Physics* (Cambridge University Press, Cambridge, 1998)
6. J. Stöhr, H.C. Siegmann, *Magnetism: From Fundamentals to Nanoscale Dynamics* (Springer, Heidelberg, 2006)
7. R.D. Cowan, *The Theory of Atomic Structure and Spectra* (University of California Press, Berkeley, 1981)
8. G. van der Laan, B. T. Thole, Phys. Rev. B **43**, 13401 (1991)
9. S. Stepanow, A. Mugarza, G. Ceballos, P. Moras, J.C. Cezar, C. Carbone, P. Gambardella, Phys. Rev. B **82**, 014405 (2010)
10. M. Blume, J. Appl. Phys. **57**, 3615 (1985)
11. M. Altarelli, Resonant x-ray scattering: a theoretical introduction, in *Magnetism: A Synchrotron Radiation Approach. Springer Lecture Notes in Physics*, vol. 697, ed. by E. Beaurepaire, H. Bulou, F. Scheurer, J.P. Kappler (Springer, Berlin, 2006), p. 201
12. F. de Bergevin, M. Brunel, Phys. Lett. A **39**, 141 (1972)
13. F. de Bergevin, M. Brunel, Acta Cryst. A **37**, 314 (1981)
14. N. Bohr, Phil. Mag. **26**, 1 (1913)
15. M. Born, J.R. Oppenheimer, Ann. Physik **389**, 457 (1927)
16. J. Stöhr, *NEXAFS Spectroscopy* (Springer, Heidelberg, 1992)
17. R. Neutze, R. Wouts, D. van der Spoel, E. Weckert, J. Hajdu, Nature **406**, 752 (2000)
18. C. Bostedt, S. Boutet, D.M. Fritz, Z. Huang, H.J. Lee, H.T. Lemke, A. Robert, W.F. Schlotter, J.J. Turner, G.J. Williams, Rev. Mod. Phys. **88**, 015007 (2016)
19. A. Pietzsch, Y.P. Sun, F. Hennies, Z. Rinkevicius, H.O. Karlsson, T. Schmitt et al., Phys. Rev. Lett. **106**, 153004 (2011)

20. G. Grynberg, A. Aspect, C. Fabre, *Introduction to Quantum Optics, From the Semi-Classical Approach to Quantized Light* (Cambridge University Press, Cambridge, 2010)
21. M. Göppert-Mayer, Ann. Phys. (Leipzig) **401**, 273 (1931)
22. J.J. Yeh, I. Lindau, At. Data Nucl. Data Tables **32**, 1 (1985)
23. R. Loudon, *The Quantum Theory of Light*, 3rd edn. (Clarendon Press, Oxford, 2000)
24. B.L. Henke, E.M. Gullikson, J.C. Davis, At. Data Nucl. Data Tables **54**, 181 (1993)
25. D.C. Koningsberger, E.R. Prins, *X-Ray Absorption: Principles, Applications, Techniques of EXAFS, SEXAFS and XANES* (Wiley, New York, 1988)
26. J.J. Rehr, R.C. Albers, Rev. Mod. Phys. **72**, 621 (2000)
27. G. Bunker, *Introduction to XAFS: A Practical Guide to X-Ray Absorption Fine Structure Spectroscopy* (Cambridge University Press, Cambridge, 2010)
28. C.T. Chen, Y.U. Idzerda, H.J. Lin, N.V. Smith, G. Meigs, E. Chaban, G.H. Ho, E. Pellegrin, F. Sette, Phys. Rev. Lett. **75**, 152 (1995)
29. S. Pizzini, A. Fontaine, E. Dartyge, C. Giorgetti, F. Baudelet, J.P. Kappler, P. Boher, F. Giron, Phys. Rev. B **50**, 3779 (1994)
30. H.G.J. Moseley, Phil. Mag. **26**, 1024 (1913)
31. A.C. Thompson et al., *X-Ray Data Booklet*, 3rd edn. (LBNL, Berkeley, 2009). Available at: http://xdb.lbl.gov
32. W. Bambynek, B. Crasemann, R.W. Fink, H.U. Freund, H. Mark, C.D. Swift, R.E. Price, P.V. Rao, Rev. Mod. Phys. **44**, 716 (1972)
33. M.O. Krause, J. Phys. Chem. Ref. Data **8**, 307 (1979)
34. J.H. Hubbell, P.N. Trehan, N. Singh, B. Chand, D. Mehta, M.L. Garg, R.R. Garg, S. Singh, S. Puri, J. Phys. Ref. Data **23**, 339 (1994)
35. W. Schülke, *Electron Dynamics by Inelastic X-Ray Scattering* (Oxford University Press, Oxford, 2007)
36. J. Als-Nielsen, D. McMorrow, *Elements of Modern X-Ray Physics*, 2nd edn. (Wiley, New York, 2011)
37. N. Rohringer et al., Nature **481**, 488 (2012)
38. B. Wu, T. Wang, C.E. Graves, D. Zhu, W.F. Schlotter, J.J. Turner, O. Hellwig, Z. Chen, H.A. Dürr, A. Scherz, J. Stöhr, Phys. Rev. Lett. **117**, 027401 (2016)
39. J. Szlachetko et al.: Sci. Rep. **6**, 33292 (2016)

Chapter 10
Quantum Theory of X-Ray Absorption Spectroscopy

10.1 Overview

As the first application of the Kramers-Heisenberg-Dirac theory we discuss the quantitative calculation of the x-ray absorption rate and cross section. X-ray absorption spectroscopy (XAS) is so fundamental and important that we devote the next two chapters to its various aspects. We will emphasize its dependence on the properties of light, such as the photon energy and polarization, and its dependence on the charge and spin of the sample. For most of our discussion the quantum theory will be phrased in terms an intuitive independent electron model, where electrons are excited from core states to partially empty valence states or into the continuum. Multi-electron correlation effects, giving rise to multiplet splittings in the observed spectra, will only be covered conceptually and through selected examples. Our emphasis will be on the pronounced resonant fine structure near absorption thresholds, known as NEXAFS or XANES, and we will not cover the extended x-ray absorption fine structure (EXAFS) above absorption edges but refer the reader to specific texts on the topic [1–3].

10.2 Quantum Formulation of X-Ray Absorption Spectroscopy (XAS)

The *first order* term in the KHD formula (9.21) with the interaction Hamiltonian written as $\mathcal{H}_{\text{int}} = -e\,\mathbf{r}\cdot\mathbf{E}$ in (9.22) describes x-ray absorption and emission, schematically shown in Fig. 9.4 a, b. The transition rate is given by

$$\mathcal{W}_{if} = \frac{2\pi}{\hbar}\,|\langle f|e\,\mathbf{r}\cdot\mathbf{E}|i\rangle|^2\,\delta(\mathcal{E}_f - \mathcal{E}_i)\,\rho(\mathcal{E}_f). \tag{10.1}$$

J. Stöhr, *The Nature of X-Rays and Their Interactions with Matter*, Springer Tracts in Modern Physics 288, https://doi.org/10.1007/978-3-031-20744-0_10

The field operator is given by $\mathbf{E} = \mathbf{E}^+ + \mathbf{E}^-$ according to (3.61), with $\mathbf{E}^+ \propto \mathbf{a}_{p\mathbf{k}}$ destroying photons in XAS and $\mathbf{E}^- \propto \mathbf{a}^\dagger_{p\mathbf{k}}$ creating photons in XES.

For the description of x-ray absorption one first writes the states $|i\rangle$ and $|f\rangle$ in (10.1) as products of electronic and photon states. Before the interaction we have $|i\rangle = |a\rangle|n_{p\mathbf{k}}\rangle$ and after the interaction $|f\rangle = |b\rangle|n_{p\mathbf{k}} - 1\rangle$, where $|a\rangle$ and $|b\rangle$ are electronic states. The matrix elements in the KHD formula can be factored into a photon and electronic part, and the photon part is evaluated as

$$\left|\langle n_{p\mathbf{k}} - 1|\mathbf{E}^+|n_{p\mathbf{k}}\rangle\right|^2 = \frac{n_{p\mathbf{k}}\hbar\omega}{2\epsilon_0 V_{p\mathbf{k}}} \tag{10.2}$$

The transition rate can then be expressed in terms of the electronic transition operator and wavefunctions as

$$\mathcal{W}^{\text{abs}}_{a\to b} = \underbrace{\frac{n_{p\mathbf{k}}c}{V_{p\mathbf{k}}}}_{\Phi_{p\mathbf{k}}} \underbrace{4\pi^2\hbar\omega_{p\mathbf{k}}\,\alpha_{\text{f}}\left|\langle b|\,\mathbf{r}\cdot\boldsymbol{\epsilon}_p\, e^{\text{i}\mathbf{k}\cdot\mathbf{r}}|a\rangle\right|^2 \rho(\mathcal{E}_b)\,\delta[\hbar\omega_{p\mathbf{k}} - (\mathcal{E}_b - \mathcal{E}_a)]}_{\sigma_{\text{XAS}}}. \tag{10.3}$$

Here $\Phi_{p\mathbf{k}}$ is the photon flux per mode with dimension [number/(area$\times$time)], σ_{XAS} the absorption cross section of dimension [area], $\boldsymbol{\epsilon}_p$ the unit photon polarization vector, $\mathcal{E}_b - \mathcal{E}_a$ the energy difference between the electronic states, $\rho(\mathcal{E}_b)$ the density of final states per unity energy of dimension [1/energy], and

$$\boxed{\alpha_{\text{f}} = \frac{e^2}{4\pi\epsilon_0\hbar c} \simeq \frac{1}{137}} \tag{10.4}$$

is the dimensionless *fine structure constant* in SI units which characterizes the strength of the photon-electron interaction.

The expression (10.3) still contains the photon wavevector $\mathbf{k}$ whose magnitude $k = 2\pi/\lambda$ characterizes the spatial extent λ of a wavecycle. The neglect of the $\mathbf{k}$-dependence underlies the famous *dipole approximation*

$$e^{\text{i}\mathbf{k}\cdot\mathbf{r}} = 1 + \text{i}\mathbf{k}\cdot\mathbf{r} + \ldots \simeq 1. \tag{10.5}$$

It turns out that this approximation, which assumes that the atomic size reflected by $\langle\mathbf{r}\rangle$ is small relative to the wavelength $\lambda = 2\pi/|\mathbf{k}|$, is remarkably good for the description of almost all aspects of XAS. We will, however, see in Sect. 11.4 that the dipole approximation breaks down for the description of XNCD, since the $\mathbf{k}$ dependence is required for x-rays to couple to the *spatial* "twist" in the atomic charge density for chiral systems.

In the dipole approximation we can write the absorption rate in the following form.

The polarization (p) dependent x-ray absorption cross section in the dipole approximation is given by

$$\sigma_{\mathrm{XAS}}^{p} = 4\pi^2 \hbar\omega\, \alpha_{\mathrm{f}} \left|\langle b|\, \mathbf{r}\cdot\boldsymbol{\epsilon}_p |a\rangle\right|^2 \rho(\mathcal{E}_b)\, \delta[\hbar\omega - (\mathcal{E}_b - \mathcal{E}_a)]. \qquad (10.6)$$

It depends on the fundamental dimensionless *fine structure constant* α_{f}, which characterizes the strength of the photon-electron interaction and the photon energy $\hbar\omega$, the transition probability given by the squared matrix element between electronic states $|a\rangle$ and $|b\rangle$, and the electronic density of final states per unit energy $\rho(\mathcal{E}_b)$ (dimension [1/energy]). The δ-function assures energy conservation. Spin degeneracy may be included in the states $|a\rangle$ and $|b\rangle$ or the density of final states.

10.2.1 Photon Flux, Intensity, and Absorption Cross Section

The incident photon flux Φ can be written in different forms. In practice, an x-ray beam incident on a sample of thickness d consists of a total number of photons n_0 within an illuminated area A (see Fig. 3.2), where the photons may be in different modes. The photon flux through the sample volume $V = Ad$ is then

$$\Phi_0 = c\,\frac{n_0}{V}. \qquad (10.7)$$

When the volume is reduced to the single-mode coherence volume given by (3.62) which contains only photons in the same mode $p\mathbf{k}$, we have with $V_{p\mathbf{k}}$ given by (3.106),

$$\Phi_{p\mathbf{k}} = c\,\frac{n_{p\mathbf{k}}}{V_{p\mathbf{k}}} = \frac{n_{p\mathbf{k}}c}{\lambda^3}\,\frac{\Delta_{p\mathbf{k}}}{\hbar\omega}, \qquad (10.8)$$

where $\Delta_{p\mathbf{k}}$ is the energy bandwidth. The flux can be pictured as the number of photons that flows with the speed of light through a specified volume according to

$$\boxed{\Phi = \frac{n_{p\mathbf{k}}c}{V_{p\mathbf{k}}} = \frac{n_0 c}{V} = \frac{I_0}{\hbar\omega} = \frac{2\,\epsilon_0\, c\, E_0^2}{\hbar\omega}}. \qquad (10.9)$$

We have also expressed the flux in terms of the incident intensity I_0 with dimension [energy/area/time] and field amplitude E_0 [V/distance].

10.3 Non-resonant Absorption: Excitation into Continuum States

The cross-section (10.6) describes non-resonant as well as resonant absorption processes, which are distinguished by the description of the final electronic states b reached by the transition. In non-resonant processes the electron is excited into continuum states, corresponding to photoemission. The cross-section expression for continuum excitation has been derived in [4–7], and the specific atomic shell-dependent cross sections have been calculated and tabulated by Yeh and Lindau [8]. In the following we shall outline the cross-section derivation for this case and then consider the more interesting resonant cross sections, owing to their atomic, chemical, and spin dependent specificity.

10.3.1 Wavefunctions

Wavefunctions for atomic spin orbitals may be written in the form [7, 9–11]

$$\Phi_{n,l,m,m_s} = R_{nl}(r)\, Y_{l,m}\, m_s, \tag{10.10}$$

where $R_{nl}(r)$ has the dimension $[1/\sqrt{\text{length}^3}]$ and satisfies the orthogonality condition and normalization

$$\langle R_{nl}(r)|R_{n'l}(r)\rangle = \int_0^\infty R_{nl}(r)R_{n'l}(r)\, r^2 \mathrm{d}r = \delta_{nn'}. \tag{10.11}$$

We note that the radial functions, obtained by solving the Schrödinger equation, are sometimes defined as $P(r) = r R(r)$ [9, 11]. The simplest analytical functions are modified functions for hydrogen where Z is taken to have the effective value $Z^* < Z$ to mimic screening effects [9]. For the description of K ($1s \rightarrow 2p$) and L ($2p \rightarrow 3d$ and $2p \rightarrow 4s$) shell excitations, the most important hydrogenic wavefunctions are[1]

$$\begin{aligned}
R_{1s}(r) &= 2\left(\frac{Z^*}{a_0}\right)^{3/2} \mathrm{e}^{-Z^* r/a_0} \\
R_{2p}(r) &= \frac{r}{2\sqrt{6}}\left(\frac{Z^*}{a_0}\right)^{5/2} \mathrm{e}^{-Z^* r/2a_0} \\
R_{3d}(r) &= \frac{4\,r^2}{81\sqrt{30}}\left(\frac{Z^*}{a_0}\right)^{7/2} \mathrm{e}^{-Z^* r/3a_0}.
\end{aligned} \tag{10.12}$$

---

[1] These functions are those given by Cowan [9] in Table 3–1, converted to our notation. The normalization is readily evaluated by means of the integral $\int_0^\infty r^m\, \mathrm{e}^{-ar}\, \mathrm{d}r = m!/a^{m+1}$.

where

$$a_0 = 4\pi\epsilon_0 \frac{\hbar^2}{m_e e^2} \tag{10.13}$$

is the Bohr radius $a_0 = 0.529 \times 10^{-10}$ m.

The key in the cross-section derivation is the description of the wavefunction of the photoelectron moving away from the atom. It is not limited to a finite volume and is influenced by the Coulomb interaction with the left-behind ion. The photoelectron wavefunction depends on the kinetic energy of the outgoing photoelectron, given in terms of the photoelectron wavevector **q** (with units [1/length]) by

$$\mathcal{E}_{\rm kin} = \frac{\hbar^2}{2m_e} q^2. \tag{10.14}$$

In general, a free electron state can be described by a function that is constructed from either a plane wave or spherical wave basis.[2] In analogy to the bound state wavefunction (10.10) we can write the continuum free electron wavefunction as

$$\Phi_{q,l,m,m_s} = R_{q,l}(r)\, Y_{l,m}\, m_s'. \tag{10.16}$$

The radial part can be expressed in terms of spherical Bessel functions $j_l(qr)$ [12] (also see (10.15))

$$R_{q,l}(r) = \mathrm{i}^l \sqrt{\frac{2m_e q}{\pi\hbar^2}}\, j_l(qr). \tag{10.17}$$

The angular momentum l of the photoelectron wave is determined by the dipole selection rule that connects the core function of angular momentum ℓ_c with the Bessel function of angular momentum $l = \ell_c \pm 1$. The product $q\,r$ is dimensionless and so is the Bessel function. The absolute value squared of the radial function, $|R_{q,l}(r)|^2$, has dimension [1/(energy×length^3)].

The spherical Bessel functions $j_l(qr)$ that characterize the final states in K and L edge dipole excitations correspond to the lowest angular momenta $l=0$, $l=1$ and $l=2$, respectively, and are given by[3]

[2] A plane wave can be expressed as an infinite sum of spherical waves by the well-known expansion,

$$\mathrm{e}^{\mathrm{i}\mathbf{q}\cdot\mathbf{r}} = 4\pi \sum_{l=0}^{\infty} \sum_{m=-l}^{l} \mathrm{i}^l\, j_l(qr)\, Y_{l,m}(\theta_q, \phi_q)\, Y^*_{l,m}(\theta_r, \phi_r), \tag{10.15}$$

which depends on the spherical angles of the electron wave vector **q** and the position vector **r**.

[3] The spherical Bessel functions obey the closure relation

$$\int_0^\infty j_L(ar)\, j_L(br)\, r^2 \mathrm{d}r = \frac{1}{2a^2}\,\delta(a-b) \tag{10.18}$$

$$j_0(qr) = \frac{\sin qr}{qr}$$
$$j_1(qr) = \frac{\sin qr}{q^2r^2} - \frac{\cos qr}{qr}$$
$$j_2(qr) = \left(\frac{3}{q^3r^3} - \frac{1}{qr}\right)\sin qr - \frac{3}{q^2r^2}\cos qr. \qquad (10.21)$$

10.3.2 Continuum Cross Section

The x-ray absorption cross section is determined by dipole transitions from a core spin orbital to continuum states. The atomic core states have the form (10.10), $\Phi_{\text{core}} = \Phi_{n,\ell_c,m_c,m_s}$, of angular momentum ℓ_c, whose radial function $R_{n\ell_c}(r)$ is given by (10.12). The continuum state Φ_{q,l,m,m'_s} is given by (10.16) and its radial part $R_{q,l}(r)$ by (10.17). The dipole operator does not act on spin so that $m'_s = m_s$ and we can drop the spin label and multiply the transition rate by 2 to account for the two spin states. The orbital momenta are linked by the dipole selection rule $l = \ell_c \pm 1$. The polarization ($p = 0, +, -$) dependent x-ray absorption cross-section (10.6) becomes

$$\sigma^p_{\text{XAS}} = 8\pi^2\hbar\omega\,\alpha_{\text{f}} \sum_{m_c=-\ell_c}^{\ell_c} \left|\langle \Phi_{q,l,m}|\,\mathbf{r}\cdot\boldsymbol{\epsilon}_p|\Phi_{n,\ell_c,m_c}\rangle\right|^2 \rho(\mathcal{E}_{\text{kin}})\,\delta[\hbar\omega - (\mathcal{E}_{\text{kin}} + \mathcal{E}_{\text{c}})]. \qquad (10.22)$$

The energy balance reflected by the δ-function now contains the electron kinetic energy $\mathcal{E}_{\text{kin}}$ and core level binding energy $\mathcal{E}_{\text{c}}$ (taken positive), both referenced to the vacuum level. Their sum is equal to the photon energy $\hbar\omega$.

The density of final states per unit energy $\rho(\mathcal{E}_{\text{kin}})$ in (10.22) is taken care of by the normalization of the final state wavefunction. It yields

$$\int_{-\infty}^{\infty} \rho(\mathcal{E}_{\text{kin}})\,\delta(\mathcal{E}_{\text{kin}} - \mathcal{E})\,\mathrm{d}\mathcal{E} = 2\int_0^{\infty} \langle \Phi_{q,l,m}|\Phi_{q,l,m}\rangle\,\delta(\mathcal{E}_{\text{kin}} - \mathcal{E})\,\mathrm{d}\mathcal{E} = 2 \qquad (10.23)$$

---

orthogonality relation,

$$\int_0^{\infty} j_L(ar)\,j_{L'}(ar)\,\mathrm{d}r = \frac{\pi}{2a(2L+1)}\,\delta(L, L') \qquad (10.19)$$

and satisfy the asymptotic condition for $r \to \infty$

$$j_l(qr) \longrightarrow \frac{1}{qr}\cos\left[qr - \frac{\pi(l+1)}{2}\right]. \qquad (10.20)$$

Table 10.1 Angular momentum dipole matrix elements $\langle L, M|\, C_p^{(1)}\, |l, m\rangle$

$\langle l+1, m\|\, C_0^{(1)}\, \|l, m\rangle =$	$\sqrt{\frac{(l+1)^2-m^2}{(2l+3)(2l+1)}}$
$\langle l-1, m\|\, C_0^{(1)}\, \|l, m\rangle =$	$\sqrt{\frac{l^2-m^2}{(2l-1)(2l+1)}}$
$\langle l+1, m+1\|\, C_1^{(1)}\, \|l, m\rangle =$	$\sqrt{\frac{(l+m+2)(l+m+1)}{2(2l+3)(2l+1)}}$
$\langle l-1, m+1\|\, C_1^{(1)}\, \|l, m\rangle =$	$-\sqrt{\frac{(l-m)(l-m-1)}{2(2l-1)(2l+1)}}$
$\langle l+1, m-1\|\, C_{-1}^{(1)}\, \|l, m\rangle =$	$\sqrt{\frac{(l-m+2)(l-m+1)}{2(2l+3)(2l+1)}}$
$\langle l-1, m-1\|\, C_{-1}^{(1)}\, \|l, m\rangle =$	$-\sqrt{\frac{(l+m)(l+m-1)}{2(2l-1)(2l+1)}}$

Non-listed matrix elements are zero. The matrix elements are real, so that $\langle L, M|C_p^{(1)}|l, m\rangle^* = \langle L, M|C_p^{(1)}|l, m\rangle = (-1)^p \langle l, m|C_{-p}^{(1)}|L, M\rangle$.
The matrix elements $\langle L, M|\, C_p^{(1)}|l, m\rangle$ are identical to the coefficients $c^{k=1}(LM; lm)$ tabulated by Condon and Shortley [13] and Slater [14] and include the reduced matrix element $\langle l+1||C^{(1)}||l\rangle = -\langle l||C^{(1)}||l+1\rangle = \sqrt{l+1}$ [11] listed in Appendix A.8.

which properly reflects the two possible spin states per electron.

The polarization dependent matrix element is evaluated by writing the dipole operator as

$$\mathbf{r}\cdot\boldsymbol{\epsilon}_p = r\, C_p^{(1)}. \tag{10.24}$$

Here $C_p^{(l)}$ with $p = 0, +, -$ are the first order $l = 1$ (dipolar) subset of Racah's spherical tensors $C_m^{(l)}$ [11], which are normalized versions of the spherical harmonics $Y_{l,m} = |l, m\rangle$ and are given up to $l = 4$ in Table A.2 in Appendix A.5. This notation facilitates evaluation of the polarization dependent transition matrix elements, $\langle L, M|\, C_p^{(1)}|l, m\rangle$, which are listed in Table 10.1. The radial part of the matrix element $\mathcal{R} = \langle b|r|a\rangle$ may be factored.

When averaging over polarizations, the total matrix element is obtained as

$$\begin{aligned}
&\frac{1}{3}\sum_{m_c,p} |\langle\, \Phi_{q,l,m}|\, \mathbf{r}\cdot\boldsymbol{\epsilon}_p|\Phi_{n,\ell_c,m_c}|^2 \\
&= \frac{m_e q}{3\pi\hbar^2}\sum_{m_c,p} |\langle j_l(qr)|\, r\,|R_{n\ell_c}(r)\rangle|^2\, |\langle l, m|\, C_p^{(1)}|\ell_c, m_c\rangle|^2 \\
&= \frac{m_e q}{3\pi\hbar^2}\left\{\ell_c\, |\langle j_{\ell_c-1}(qr)|\, r\,|R_{n\ell_c}(r)\rangle|^2 + (\ell_c+1)|\langle j_{\ell_c+1}(qr)|\, r\,|R_{n\ell_c}(r)\rangle|^2\right\}.
\end{aligned} \tag{10.25}$$

The angular momentum prefactors ℓ_c and ℓ_c+1 are the reduced angular matrix elements of the allowed $\ell_c \to \ell_c-1$ and $\ell_c \to \ell_c+1$ dipole transitions. Denoting the radial matrix elements as

$$\mathcal{R}^2_{\ell_c\pm1} = |\langle j_{\ell_c\pm1}(qr)|\, r\, |R_{n\ell_c}(r)\rangle|^2 \tag{10.26}$$

and inserting into (10.22) we obtain[4]

$$\boxed{\sigma_{\mathrm{XAS}} = \frac{8\pi}{3}\hbar\omega\,\alpha_{\mathrm{f}}\frac{2m_{\mathrm{e}}q}{\hbar^2}\left\{\ell_c\,\mathcal{R}^2_{\ell_c-1} + (\ell_c+1)\mathcal{R}^2_{\ell_c+1}\right\}}. \tag{10.28}$$

The dimensions are better seen by rewriting this expression in the form

$$\sigma_{\mathrm{XAS}} = \frac{8\pi}{3}\hbar\omega\,\alpha_{\mathrm{f}}\frac{q^3}{\mathcal{E}_{\mathrm{kin}}}\left\{\ell_c\,\mathcal{R}^2_{\ell_c-1} + (\ell_c+1)\mathcal{R}^2_{\ell_c+1}\right\}. \tag{10.29}$$

We see that in order for σ_{XAS} to have the proper dimension [length^2], the squared matrix element $\mathcal{R}^2$ has the dimension [length^5] (see footnote 4).

10.3.3 Simple Model Calculation

The non-resonant XAS cross section from a specific core shell into continuum states given by (10.29) is also the photoemission spectroscopy (PES) cross section. It needs to be evaluated by use of good atomic wavefunctions. An approximate analytical solution, however, can be obtained by modified hydrogenic wavefunctions like those given by (10.12), where Z is taken to have an effective value $Z^* < Z$ to account for screening effects. To illustrate the results obtained in such an approximation, we consider the K-shell cross section of Ne and L-shell cross section of Cu.

For the K-shell, we have $\ell_c = 0$ and only the channel $\ell_c \to \ell_c + 1$ is allowed. By use of hydrogenic core and plane wave final state wavefunctions we obtain

$$\sigma_{\mathrm{XAS}} = \sigma_{\mathrm{PES}} = \frac{8\pi\alpha_{\mathrm{f}}\hbar\omega}{3}\,\frac{1}{\mathcal{E}_{\mathrm{kin}}}\,\underbrace{\frac{256\,(Z^*)^5 a_0^7\, q^5}{\left[(a_0 q)^2 + (Z^*)^2\right]^6}}_{\widetilde{\mathcal{R}}^2}. \tag{10.30}$$

Here a_0 is the Bohr radius given by (10.13) and q the electron momentum, which is related by (10.14) to the kinetic energy of the photoelectron relative to the vacuum level. The first term has the dimension energy, and the middle term [1/energy] and the redefined radial matrix element $\widetilde{\mathcal{R}}^2$, distinguished by a tilde superscript, has the

---

4 Our expression in SI units is the same as that derived by Cooper [4, 5] in atomic units given by

$$\sigma_{\mathrm{XAS}} = \frac{8\pi}{3}\,\hbar\omega\,\alpha_{\mathrm{f}}\,a_0^2\left\{\ell_c\,R^2_{\ell_c-1} + (\ell_c+1)R^2_{\ell_c+1}\right\}. \tag{10.27}$$

Here a_0 is the Bohr radius, and the squared radial matrix elements denoted R^2 (instead of $\mathcal{R}$ in (10.28)) have the dimension [1/energy].

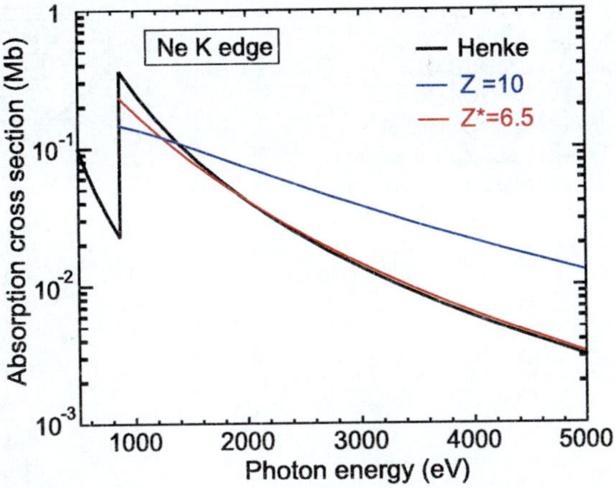

Fig. 10.1 K-edge excitation cross section for Ne, calculated with (10.30). By use of hydrogenic wavefunctions, the actual number of electrons $Z = 10$ in Ne needs to be reduced to $Z^* = 6.5$ to account for screening effects in order to better approximate the energy dependence

usual dimension [length^2]. The K-shell cross section for Ne calculated with (10.30) for $Z^* = Z = 10$ (blue curve) and $Z^* = 6.5$ (red curve) is shown in Fig. 10.1 in comparison with the one obtained from the Henke-Gullikson tables (black curve).

For the L-shell, $2p$ ($\ell_c = 1$) core elections may be dipole excited into either εs ($l = 0$) or εd ($l = 2$) continuum states. The cross section is obtained as

$$\sigma_{\mathrm{XAS}} = \frac{8\pi\alpha_f\hbar\omega}{3}\,\frac{1}{\mathcal{E}_{\mathrm{kin}}}\,\underbrace{\frac{12\,(Z^*)^7 a_0^5 q^3}{\left[4(a_0 q)^2+(Z^*)^2\right]^8}\left\{\underbrace{2048\left[(Z^*)^2-4(a_0 q)^2\right]^2}_{2p\to\varepsilon s}+\underbrace{512^2 a_0^4 q^4}_{2p\to\varepsilon d}\right\}}_{\tilde{\mathcal{R}}^2}. \tag{10.31}$$

The cross Sect. (10.31) is plotted in Fig. 10.2, decomposed into the two dipole allowed channels. We show the results obtained for the true $Z = Z^* = 29$ value (blue) and also for $Z^* = 15$ (red) which most closely approximates the empirical cross section shown in black, obtained from the Henke-Gullikson compilation [15, 16].

Although, the ambiguity of the Z^* value limits the reliability of such calculations, they still provide insight into the relative strength of the two dipole channels, which shows the dominance of the $2p \to \varepsilon d$ channel in agreement with more sophisticated calculations [17].

10.3.4 The Core Level Photoemission Spectrum and Its Linewidth

The core level photoemission spectrum consists of a measurement of the number of emitted electrons as a function of their kinetic energy. The measured peak is

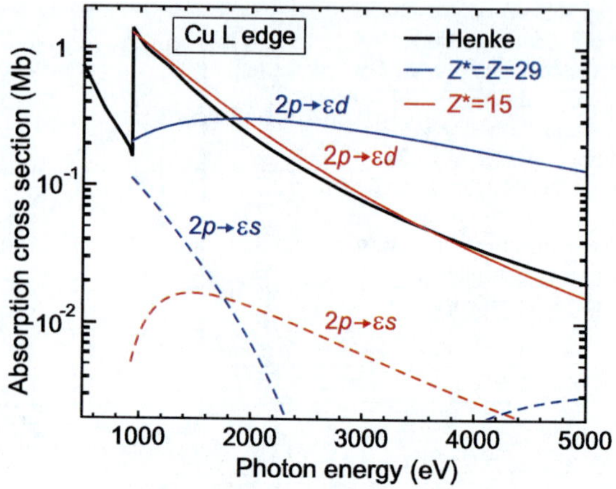

Fig. 10.2 Cross section for L-edge excitation in Cu metal, calculated with (10.31). The two allowed dipole excitation channels $2p \rightarrow \varepsilon s$ and $2p \rightarrow \varepsilon d$ are shown separated for the two indicated choices of effective Z^* values

centered at $\mathcal{E}_{\mathrm{kin}}^0$, determined by the difference of incident photon energy $\hbar\omega$ and the core electron ionization potential (IP), i.e. $\mathcal{E}_{\mathrm{kin}}^0 = \hbar\omega - IP$. The distribution of the number of electrons around $\mathcal{E}_{\mathrm{kin}}^0$ defines the intrinsic PES linewidth (FWHM) $\Delta\mathcal{E}_{\mathrm{kin}}$. If the instrumental (Gaussian) resolution is negligibly small, the intrinsic Lorentzian PES linewidth has the value $\Delta\mathcal{E}_{\mathrm{kin}} = \Gamma$, where Γ is related to the lifetime of the core hole τ_Γ according to the uncertainty relation $\Gamma = \hbar/\tau_\Gamma$. We will see that this fundamental atom-specific relation also plays a key role in resonant x-ray absorption (see Sect. 10.4.1) and in x-ray emission (see Sect. 12.2.7).

> The intrinsic Lorentzian core level photoemission linewidth has the (FWHM) value Γ, where Γ, listed in Table 10.2 for selected core shells, is related to the lifetime τ_Γ of the core hole by the uncertainty relation $\Gamma = \hbar/\tau_\Gamma$.

The core level PES spectrum is formally defined as the number of photoemitted electrons N_{PES}, per solid angle, per time, and per kinetic energy according to

$$\frac{\mathrm{d}^3 N_{\mathrm{PES}}}{\mathrm{d}\Omega\, \mathrm{d}t\, \mathrm{d}\mathcal{E}_{\mathrm{kin}}} = \Phi_0\, \sigma_{\mathrm{XAS}} \underbrace{\frac{2}{\pi\Gamma} \frac{(\Gamma/2)^2}{(\mathcal{E}_{\mathrm{kin}} - \mathcal{E}_{\mathrm{kin}}^0)^2 + (\Gamma/2)^2}}_{\rho(\mathcal{E}_{\mathrm{kin}})} . \tag{10.32}$$

Here Φ_0 is the incident photon flux, σ_{abs} the continuum absorption (= photoemission) cross section, $\mathcal{E}_{\mathrm{kin}}^0 = \hbar\omega - IP$ the central kinetic energy of the photoemission peak, and Γ (dimension [energy]) the FWHM of a Lorentzian distribution function determined by the core hole lifetime. The distribution function $\rho(\mathcal{E}_{\mathrm{kin}})$ is normalized to give unity when integrated over kinetic energy

$$\int_{-\infty}^{\infty} \rho(\mathcal{E}_{\text{kin}})\, \mathrm{d}\mathcal{E}_{\text{kin}} = 1. \tag{10.33}$$

In the following Section we will consider the excitation of core electrons into a well-defined resonant state, consisting of an atomic Rydberg orbital or an unfilled molecular orbital. In this case, we have a resonant x-ray absorption process where the excited electron remains trapped and is not photo-emitted. We will see that, remarkably, both the core level PES spectrum and the resonant XAS spectrum have the same minimum widths given by the total decay width Γ associated with the core hole.

10.4 Resonant X-Ray Absorption

The resonant x-ray absorption case is illustrated in Fig. 10.3 for the case of transitions between two well-defined electronic states. For this case, the absorption rate (10.6) expresses the transition rate between two sharp non-degenerate levels, with energy conservation expressed by the Dirac δ-function, defined through its sifting property

$$\int_{-\infty}^{\infty} \rho(\mathcal{E}_b)\, \delta(\hbar\omega - (\mathcal{E}_b - \mathcal{E}_a))\, \mathrm{d}\mathcal{E}_b = 1. \tag{10.34}$$

The final state in XAS is an excited state with a core hole and our formulation so far has neglected that the core hole has a finite lifetime. In quantum mechanics, the lifetime τ of the core hole state gives rise to an uncertainty of the excitation energy, given by the uncertainty relation $\Gamma = \hbar/\tau$. This causes a finite energy width in the measurement of any absorption line.

10.4.1 Natural Linewidth of XAS Resonances

In the x-ray regime, the core hole may be spontaneously filled by two types of decay processes during which the energy stored in the excited electronic state is converted either into emitted x-rays or Auger electrons. Since the final electronic configurations reached by x-ray and Auger decays differ, the two emission channels do not interfere and the spontaneous decay rate for x-ray emission $\Gamma^{\mathrm{X}}/\hbar$ and for Auger emission $\Gamma^{\mathrm{A}}/\hbar$ simply add to the total decay rate $\Gamma/\hbar$, as illustrated in Fig. 10.3. We therefore obtain for the total decay energy width Γ

$$\Gamma = \Gamma^{\mathrm{X}} + \Gamma^{\mathrm{A}}. \tag{10.35}$$

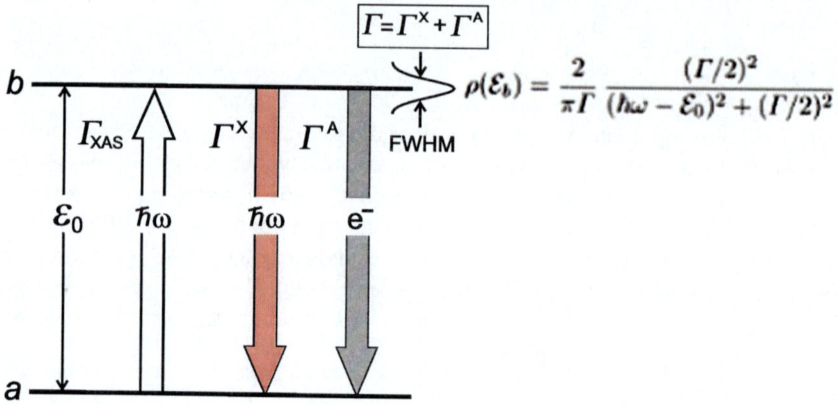

Fig. 10.3 Schematic of excitation and de-excitation processes in a two-level system described by an electronic energy separation of $\mathcal{E}_0 = \mathcal{E}_b - \mathcal{E}_a$. The probability of x-ray absorption depends on the so-called detuning energy $|\hbar\omega - \mathcal{E}_0|$. The absorption rate $\Gamma_{\mathrm{XAS}}/\hbar$ reflects the total "up" rate. Once the electron is in the excited state $|b\rangle$ it can decay either by x-ray emission with a rate $\Gamma^{\mathrm{X}}/\hbar$ or Auger decay with a rate $\Gamma^{\mathrm{A}}/\hbar$. The total decay energy width is given by $\Gamma = \Gamma^{\mathrm{X}} + \Gamma^{\mathrm{A}}$ and, as indicated, corresponds to the FWHM of a Lorenzian

Through this equation, we have introduced the so-called *natural linewidth* Γ in an *ad hoc* fashion, as typically done in the literature. We shall come back to the meaning of (10.35) later and again in conjunction with the linewidth in x-ray emission in Sect. 12.2.5. Since excited electronic states decay exponentially in time, the corresponding lineshapes in the energy domain are Lorentzians, and by convention Γ represents the Lorentzian FWHM. The widths reflect the total decay rates into a given core hole and constitute a polarization average.

In Table 10.2 we have listed semi-empirical values of Γ for the K-shell of low Z atoms and the L_3 shell of the 3*d* transition metals.

The listed Γ values are taken from the tabulation of Krause [19]. Other tabulations [20, 21] give values that may differ by about 20%. This uncertainty reflects possible errors introduced by unfolding the intrinsic Lorentzian from instrumental Gaussian lineshape contributions, comparison of linewidths measured by different experimental techniques (x-ray emission, photoemission, Auger emission, and x-ray absorption), as well as differences in atomic linewidths due to chemical bonding or oxidation states [22]. Also listed are the corresponding core shell binding energies [18], which correspond to the absorption thresholds of the K and L_3 shells. They approximately reflect the core shell resonance energies $\mathcal{E}_0 \simeq \mathcal{E}_B$.

We have plotted the dependence of the natural energy widths Γ as a function of the binding energies listed in Table 10.2 in Fig. 10.4 for the 1*s* (K-edge) and $2p_{3/2}$ (L_3-edge) core shells [18]. For the limited energy range of the two atomic series there is an approximately linear relation. More generally, for the K-shell of all elements in the periodic table, the scaling of Γ with binding energy $\mathcal{E}_0$ is approximately $\Gamma \propto \mathcal{E}_0^2 Z^4$ [23].

Table 10.2 Core shell resonance or binding energies $\mathcal{E}_0 \simeq \mathcal{E}_B$ [18] and natural Lorentzian decay linewidths (FWHM) Γ, for the K-shell of low Z atoms and L_3 shell of the 3d transition metals [19, 20]

K-shell				L_3-shell			
Z	Element	$\mathcal{E}_0$ [eV]	Γ [meV]	Z	Element	$\mathcal{E}_0$ [eV]	Γ [meV]
6	C	284	100	20	Ca	346	170
7	N	410	130	21	Sc	399	190
8	O	543	160	22	Ti	454	220
9	F	697	200	23	V	512	240
10	Ne	870	240	24	Cr	574	270
11	Na	1071	300	25	Mn	639	320
12	Mg	1303	360	26	Fe	707	360
13	Al	1559	420	27	Co	778	430
14	Si	1839	480	28	Ni	853	480
15	P	2149	530	29	Cu	933	560
16	S	2472	590	30	Zn	1022	650
17	Cl	2833	640				
18	Ar	3206	680				

Figure 10.4 shows a plot of Γ versus $\mathcal{E}_0$

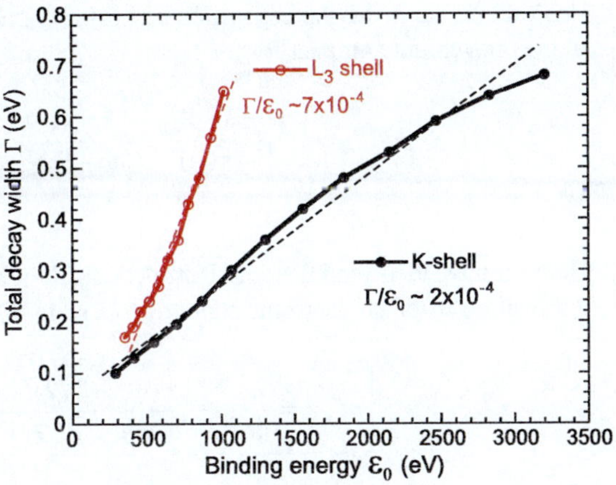

Fig. 10.4 Plot of the tabulated values for Γ versus resonance (binding) energies $\mathcal{E}_0$ for the K- and L_3-shells in Table 10.2

10.4.2 *The Natural Shape of XAS Resonances*

We may utilize the integration property (10.34) of the singular δ-function and replace it by a finite width Lorentzian lineshape function with the same integrated property. In principle, each empty spin orbital will give a unit integrated contribution. For a single final state we can then express the final state energy density as

$$\rho(\mathcal{E}_b) = \frac{2}{\pi\Gamma} \frac{(\Gamma/2)^2}{(\hbar\omega - (\mathcal{E}_b - \mathcal{E}_a))^2 + (\Gamma/2)^2}. \tag{10.36}$$

The normalized Lorentzian function has the dimension [1/energy], a FWHM energy width Γ, given by (10.35), and a resonance peak height of $2/\pi\Gamma$. It has unit area when integrated over energy according to

$$\int_{-\infty}^{\infty} \rho(\mathcal{E}_b)\, \delta(\hbar\omega - [\mathcal{E}_b - \mathcal{E}_a])\, \mathrm{d}(\hbar\omega) = 1. \tag{10.37}$$

The so-defined energy density $\rho(\mathcal{E}_b)$ is per spin and accounts for each spin dependent transition to substates within the total manifold of final states denoted by $|b\rangle$ in the absorption cross Sect. (10.6). The case where transitions to many different final states occur is therefore implicity included in the matrix element. If the matrix elements to all final states are the same, one may also simply multiply the expression for $\rho(\mathcal{E}_b)$ by the number of empty states, as will be discussed in conjunction with the calculation of the matrix elements.

By denoting the resonance energy as $\mathcal{E}_0 = \mathcal{E}_b - \mathcal{E}_a$ and use of the dipole approximation, we can write the absorption rate (10.3) or absorption cross Sect. (10.6) in the convenient compact form

$$\mathcal{W}^{\mathrm{abs}}_{a\to b} = \Phi_{\mathrm{pk}} \underbrace{\frac{\lambda^2}{\pi} \frac{\Gamma^p_{ab}}{\Gamma} \frac{(\Gamma/2)^2}{(\hbar\omega - \mathcal{E}_0)^2 + (\Gamma/2)^2}}_{\sigma^p_{\mathrm{XAS}}}. \tag{10.38}$$

Here we have identified the polarization dependent XAS cross section given for a transition from an electronic state $|a\rangle$ to a state $|b\rangle$ by

$$\boxed{\sigma^p_{\mathrm{XAS}} = \frac{\lambda^2}{\pi} \frac{\Gamma^p_{ab}}{\Gamma} \frac{(\Gamma/2)^2}{(\hbar\omega - \mathcal{E}_0)^2 + (\Gamma/2)^2}}. \tag{10.39}$$

The polarization dependent *dipole transition energy width* Γ^p_{ab} (dimension [energy]) is given by

$$\boxed{\Gamma^p_{ab} = \underbrace{\frac{8\pi^2\hbar\omega}{\lambda^2}}_{\substack{\text{photon}\\ \text{part}}} \underbrace{\alpha_{\mathrm{f}}}_{\text{coupling}} \underbrace{|\langle b|\mathbf{r}\cdot\boldsymbol{\epsilon}_p|a\rangle|^2}_{\substack{\text{atomic part}\\ \text{includes spin}}}}. \tag{10.40}$$

The form of the dipolar matrix element nicely exhibits the photon and electronic contributions coupled by the fine structure constant.

According to (10.38) the XAS rate is determined by the ratio of two transition rates, $\Gamma^p_{ab}/\hbar$ and $\Gamma/\hbar$. The rate $\Gamma^p_{ab}/\hbar$ is the *dipole transition rate* that follows from first order perturbation theory, while according to (10.35), the rate $\Gamma/\hbar$ is the total decay rate. We can identify Γ^{X} in (10.35) as the dipolar rate Γ^p_{ab}. This leads to a remarkable result.

> Although only the photon channel is detected in an x-ray absorption experiment, the minimum measured XAS linewidth has a "ghost" contribution from the Auger channel. The origin of this astounding fact will be discussed in more detail in Sect. 12.2.5.

Our formulation of x-ray absorption in terms of the two rates $\Gamma^p_{ab}/\hbar$ and $\Gamma/\hbar$ is similar to that used for the description of Mössbauer resonance scattering by Hannon and Trammell [24]. This reveals the similarity between resonant electronic x-ray and nuclear γ-ray processes. In both cases there is a non-radiative linewidth contribution, with the Auger channel for x-rays replaced by the conversion electron channel for γ-rays, also emphasized by Röhlsberger [25].

Our formulation also directly reduces to the optical Weisskopf-Wigner result [26] for the total decay linewidth Γ, which we will show now.

10.4.3 *Natural Linewidth of Optical Versus X-Ray Transitions*

In the *optical region*, dissipative losses due to ionization followed by Auger-like decays may be absent. In this case "absorption" corresponds to out-of-beam scattering.[5] One can then derive a quantum mechanical expression for Γ from (10.40). By inserting the expression for α_{f}, specifying the incident polarization along x, and accounting for the spin dependence of the matrix element by a factor of 2, one obtains the Weisskopf-Wigner result for *spontaneous emission* [26, 28]

$$\Gamma = \Gamma_{ab} = \frac{1}{4\pi\epsilon_0}\frac{4\omega^3}{3c^3}\underbrace{|\langle b|e\,x|a\rangle|^2}_{\text{without spin}} = \frac{16\pi^2}{3}\,\alpha_{\mathrm{f}}\,\hbar\omega\,\frac{|\langle b|x|a\rangle|^2}{\lambda^2}. \tag{10.41}$$

It is apparent that for $\Gamma_{ab} = \Gamma$ the absorption rate and cross section given by (10.38) are maximized and σ^p_{XAS} becomes a special case of the *resonant atomic* Breit–Wigner cross section (6.52) [12, 29] discussed in Sect. 6.5.1.

---

[5] This is how Dirac pictured optical "absorption", as discussed in Chapt. VII Sect. 53 in his book [27].

10.4.4 The Dipole Matrix Element

The polarization dependent transition matrix elements in (10.40) may be written in terms of *Racah's spherical tensor operators* with the angular wavefunctions written in the form of normalized spherical harmonics according to [11]

$$C_m^{(\ell)} = \sqrt{\frac{4\pi}{2\ell+1}}\, Y_{\ell,m}(\theta,\phi)\,, \qquad \left(C_m^{(\ell)}\right)^* = (-1)^m C_{-m}^{(\ell)}. \tag{10.42}$$

Because of their importance in modern angular momentum theory, we have listed these operators for $\ell = 0, 1, 2, 3, 4$ in the Appendix in Table A.2. The orthogonality of the spherical harmonics is expressed by

$$\langle Y_{l,m}|Y_{l',m'}\rangle = \langle\, l, m\, | l', m'\,\rangle = \delta_{l,l'}\, \delta_{m,m'}, \tag{10.43}$$

and the spherical tensors leads to the normalization sum rules

$$\sum_{m=-\ell}^{\ell} |Y_{\ell,m}|^2 = \frac{2\ell+1}{4\pi}\,, \qquad \sum_{m=-\ell}^{\ell} |C_m^{(\ell)}|^2 = 1. \tag{10.44}$$

The dipole matrix elements may then be factored into a radial and angular part given by $\ell = 1$ and $m = p = 0, \pm 1$ assuming the role of the polarization label, according to

$$\boxed{\langle b|\mathbf{r}\cdot\boldsymbol{\epsilon}_p|a\rangle = \mathcal{R}\,\langle b|\hat{\mathbf{r}}\cdot\boldsymbol{\epsilon}_p|a\rangle = \mathcal{R}\,\langle b|C_p^{(1)}|a\rangle}\,. \tag{10.45}$$

Here $\hat{\mathbf{r}}$ is a unit vector, and $\mathcal{R} = \langle b|r|a\rangle$ is the radial transition matrix element linking two electronic shells with radial wavefunctions $R_{n,c}$ and $R_{n',L}$ defined according to (10.10) and (10.11), given by,

$$\boxed{\mathcal{R} = \langle b|r|a\rangle = \int_0^\infty R_{n',L}^*\, r\, R_{n,c}\, r^2\, \mathrm{d}r}\,. \tag{10.46}$$

$\mathcal{R}$ has the dimension [length], and we shall assume it to be constant over the resonance width. The radial matrix element is typically calculated by means of atomic Hartree Fock self-consistent field approaches [9] where the final state configuration contains the core hole. Solid-state effects can be ignored for the core states and may be taken into account for the valence states by renormalizing them within their appropriate Wigner-Seitz spheres [30].

We will discuss the empirically derived radial matrix elements for the L-shell $2p \rightarrow 3d$ excitation of the $3d$ transition metals in Sect. 10.5.3 (see Fig. 10.15) and for both the K-shell $1p \rightarrow 2p$ and L-shell in Sect. 12.4.1 where we combine x-ray absorption and emission data (see Fig. 12.11).

The spherical tensor notation in (10.45) facilitates evaluation of the polarization dependent transition matrix elements when the angular parts of the wavefunctions are written in spherical harmonics $Y_{\ell,m} = |\ell, m\rangle$ [31]. The non-zero angular transition matrix elements in the dipole approximation, $\langle L, M|\, C_p^{(1)}|l, m\rangle$, between the spherical harmonics states denoted $Y_{l,m} = |l, m\rangle$ are listed in Table 10.1.

10.4.5 One-Electron/Hole Model

In the simplest intuitive picture of resonant x-ray absorption, a photon transfers its energy to a core electron and the electron is excited into an unoccupied electronic state. In this so-called *one-electron* model one simply follows what happens to the photoelectron. The matrix elements (10.45) then connect filled electronic core states of the form $|a\rangle = |n, l, m, s_z\rangle$ with empty valence states $|b\rangle = |n', L, M, s_z'\rangle$. The model ignores all other "passive" electrons in the atom during the excitation process. The only coupling of angular momenta of electrons that is included in this scheme is the spin-orbit coupling between the spin and orbital angular momenta of the single electron. Couplings of angular momenta between different electrons are neglected.

The essence of this one-electron picture is illustrated in Fig. 10.5 a for the L-edge $2p \rightarrow 3d$ x-ray absorption spectrum of a $3d$ transition metal atom.

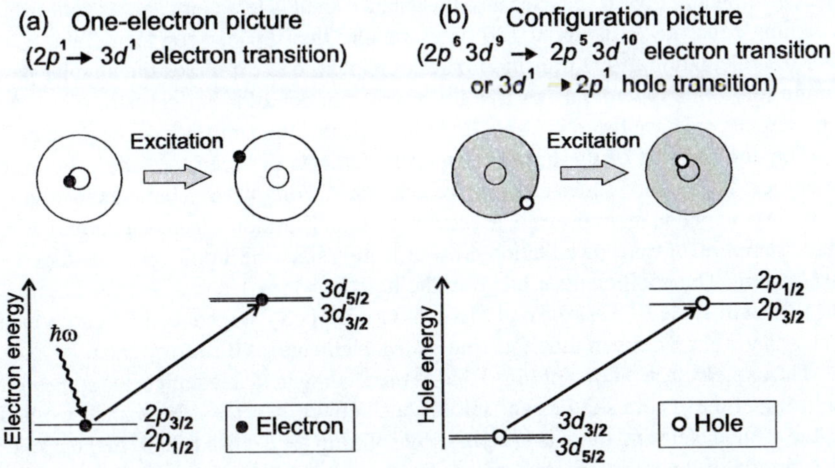

Fig. 10.5 Description of resonant L-edge x-ray absorption in two pictures, a one-electron picture and **b** d^9 configuration picture. Each energy level is labeled by its quantum numbers $n, l, j = l \pm s$. In **a** the quantum numbers label one-electron states, in **b** they label hole states for the special case of a $3d^9$ electronic ground state, containing nine electrons or equivalently one hole. The two pictures are equivalent for the description of the electronic transitions, but the sign of the angular part of the dipole matrix elements may be different, as discussed in the text

Table 10.3 Spin-orbit wave functions for p states

One-elec. label	$\|j, m_j\rangle$ basis		$\|\ell, m, s, m_s\rangle$ basis
ℓ_j	j	m_j	$Y_{\ell,m}\uparrow$ or $\downarrow$
$p_{\frac{1}{2}}$	$\frac{1}{2}$	$+\frac{1}{2}$	$\frac{1}{\sqrt{3}}\left(-Y_{1,0}\uparrow +\sqrt{2}\,Y_{1,+1}\downarrow\right)$
		$-\frac{1}{2}$	$\frac{1}{\sqrt{3}}\left(-\sqrt{2}\,Y_{1,-1}\uparrow +Y_{1,0}\downarrow\right)$
$p_{\frac{3}{2}}$	$\frac{3}{2}$	$+\frac{3}{2}$	$Y_{1,+1}\uparrow$
		$+\frac{1}{2}$	$\frac{1}{\sqrt{3}}\left(\sqrt{2}\,Y_{1,0}\uparrow +Y_{1,+1}\downarrow\right)$
		$-\frac{1}{2}$	$\frac{1}{\sqrt{3}}\left(Y_{1,-1}\uparrow +\sqrt{2}\,Y_{1,0}\downarrow\right)$
		$-\frac{3}{2}$	$Y_{1,-1}\downarrow$

The electron spin wavefunctions are $|s, 1/2\rangle = \uparrow$ (spin-up) and $|s, -1/2\rangle = \downarrow$ (spin-down)

In this model the "initial state" is envisioned as an electron in the $2p$ core state. The active electron has an angular momentum $l = 1$ and spin $s = 1/2$, so that the spin-orbit coupling given by (9.7) produces two energy states with $j = l \pm s$. The states $2p_{3/2}$ with $j = 3/2$ and $2p_{1/2}$ with $j = 1/2$ are separated by $\Delta\mathcal{E} = 3\xi_{2p}/2$, with ξ_{2p} given in Table 9.1. The core spin-orbit splitting gives rise to the prominent L_3 ($2p_{3/2} \rightarrow 3d$) and L_2 ($2p_{1/2} \rightarrow 3d$) edges in the experimental spectra in Fig. 7.23.

In the "final state" the active electron is located in the $3d$ shell with angular momentum $l = 2$ and spin $s = 1/2$. The spin-orbit coupling again produces two energy states with $j = l \pm s$, i.e. the states $3d_{5/2}$ with $j = 5/2$ and $3d_{3/2}$ with $j = 3/2$, as shown in Fig. 10.5a. Because the valence shell is less compact, its spin-orbit coupling constant ξ_{3d} is about 100 times smaller than ξ_{2p} as seen from Table 9.1. For the calculation of the transition matrix element, one expresses the angular part of the spin-orbit spit core states as a linear combination of spherical harmonics. As an example, we have listed in Table 10.3 the m_j spin-orbit substates of the p states.[6]

The angular part of the transition matrix elements between the spin-orbit core states and the valence states can then be calculated from matrix elements involving core and valence states expressed in terms of spherical harmonic basis functions under the assumption of spin conservation in the transition since the dipole operator does not act on spin. The coefficients a and b in the functions $|jm_j\rangle = a|l, m\uparrow\rangle + b|l, m+1\downarrow\rangle$ listed in Table 10.3 are the real Clebsch–Gordon or $3j$ Wigner coefficients which enter as weight factors in the transition matrix elements, as discussed later.

The one-electron picture of Fig. 10.5a is misleading in that it depicts the spin-orbit splitting of the p core shell as an "initial state" effect. In reality the p shell is filled in the ground state and there is no spin-orbit splitting for a closed shell. In the proper description of the x-ray absorption process, based on a *configuration picture* shown in Fig. 10.5b, an atom is excited from a ground or initial state configuration to an excited or final state configuration. In discussing transitions between configurations one typically omits all closed subshells since they are spherically symmetrical and their net angular momentum is zero [9]. Listing only the active shells, resonant L-

[6] In Appendix A.7 we give the spin-orbit functions of all s, p, and d states in Table A.5.

edge absorption is described by an initial state electron configuration $2p^6\,3d^n$ and a final state configuration $2p^5\,3d^{n+1}$. The problem with the one-electron description now emerges. Instead of properly including the couplings of all angular momenta within the open $3d^n$ valence shell in the ground state and the couplings of electrons in the open $2p$ and $3d$ shells in the final state configuration $2p^5\,3d^{n+1}$, the one-electron picture ignores all these couplings.

Instead one considers the simplest case where the initial state contains a filled p^6 shell and nine electrons in the d shell. Now one pictures the d^9 electron configuration in a hole picture, which is simply a d^1 hole configuration. In the configuration model, the initial state is then the d^1 hole configuration. The final state p^5d^{10} has a filled d-shell, and a p^5 electron is equivalent to a p^1 hole configuration. Hence for a d^9 ground state, L-edge spectra in the configuration picture are described by a transition between the hole configurations $d^1 \rightarrow p^1$. The one-electron model has turned into an inverted one-hole model as shown in Fig. 10.5b. Now the spin-orbit splitting of the p shell and the presence of a core hole are properly described as final state effects.

For the description of valence states that contain more than single electrons or holes one may simply account for the number of electrons or holes by a filling factor. For example, a d^2 valence electron configuration with eight holes will have a four times larger resonant cross section than a d^8 electron configuration with two holes. This is in remarkable agreement with experimental L edge spectra of the elementary $3d$ transition metals as shown in Fig. 10.16, below. The power of the one-electron/hole model lies in the description of solids with weak electron correlations. In such cases, one may simply describe the valence properties in an independent electron band-like model as discussed in Sects. 10.5.3 and 11.5.1.1.

> The *one-electron model* or its inverted one-hole scheme only considers the intrinsic coupling of the spin and angular momentum of independent electrons and ignores multi-electron coupling effects. Its wide spread use is based on its extension to the description of x-ray absorption in weakly correlated solids, with the valence electrons treated in an independent electron model by band or density functional theory.

The different signs of the charge of electrons and holes lead to some differences that are easily accounted for. For example, the energy order of the j states is inverted in the two schemes in Fig. 10.5 because electrons and holes have opposite spin.[7] Also the polarization dependent angular matrix elements may have opposite signs, which however, disappears when the transition probability is calculated as the absolute value squared.

[7] This is expressed by Hund's third rule stating that the electronic ground state has the minimum possible j value for a less than half filled shell and the maximum possible j value for a more than half full shell.

10.4.6 Polarization Dependence of the Angular Transition Matrix Element

The angular transition matrix elements may similarly be given in a one-electron picture or the upside-down hole picture. The only difference is that certain matrix elements change sign because of the inverted motion of electrons and holes in Fig. 10.5. This is readily seen from the form of the matrix elements in Table 10.1 which depend on the direction of the transition according $\langle L, M|C_p^{(1)}|l, m\rangle = (-1)^p \langle l, m|C_{-p}^{(1)}|L, M\rangle$ for $p=+1, 0, -1$. For convenience we give the values for the non-zero angular matrix elements between *s*, *p*, and *d* states in Fig. 10.6 (also see Fig. A.8).

The tensor matrix elements listed in Fig. 10.6 obey sum rules which follow from the orthogonality of the spherical harmonics and the quantum mechanical rule that only *indistinguishable* paths between initial and final states can interfere. Since all transitions in the figure involve individual lower and upper states, the listed amplitudes do not interfere. The individual and summed transition probabilities are therefore just the square of the listed amplitudes. The listed transition matrix elements do not include the *spin* degeneracy which is added by consideration of strict *spin conservation* in dipole transitions. For a non-magnetic sample or a magnetic sample

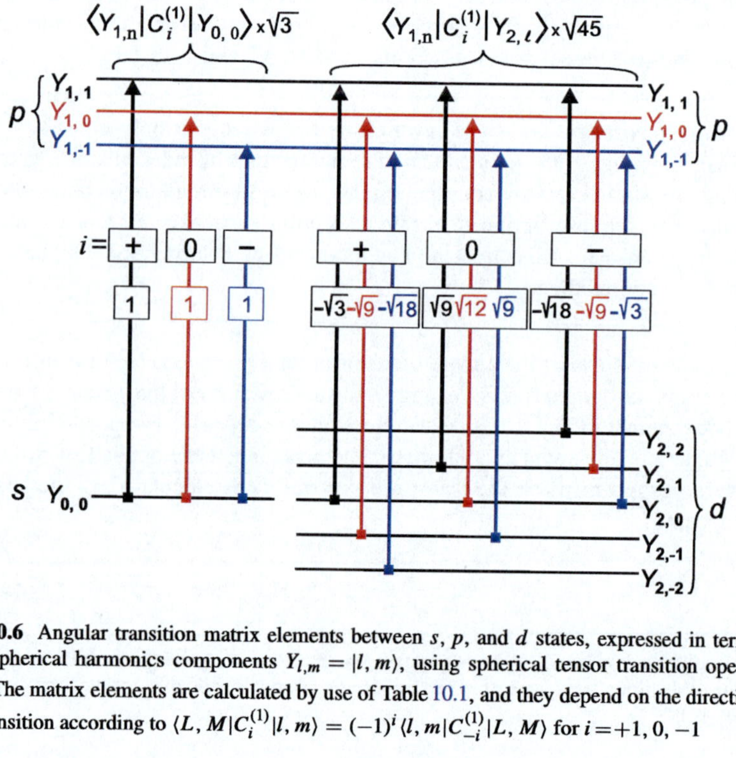

Fig. 10.6 Angular transition matrix elements between *s*, *p*, and *d* states, expressed in terms of their spherical harmonics components $Y_{l,m} = |l, m\rangle$, using spherical tensor transition operators $C_i^{(1)}$. The matrix elements are calculated by use of Table 10.1, and they depend on the direction of the transition according to $\langle L, M|C_i^{(1)}|l, m\rangle = (-1)^i \langle l, m|C_{-i}^{(1)}|L, M\rangle$ for $i=+1, 0, -1$

with random domains one may therefore calculate the core to valence transition probability based on *orbital degeneracy*, only, and then account for the spin degeneracy by multiplying the final result by 2.

10.4.7 Sum Rules for the Angular Transition Matrix Element

The transition probabilities increase with the total number of ways that transitions may occur. In practice, this is determined by the degeneracies and occupancy of the $2c+1$ core and $2L+1$ valence orbitals plus the three possible polarization orientations. In a Cartesian coordinate system, the polarization dependent matrix elements given in Fig. 10.6 correspond to the values $p=\pm1$ for circularly polarized photons with positive and negative helicities propagating into the z direction ($\mathbf{k} \parallel z$), and $p=0$ describes linearly polarized photons with polarization vector $\mathbf{E} \parallel z$ as illustrated in Fig. 10.7.

We first state a general theorem for the absolute value squared of various types of sums of dipole matrix elements

$$\boxed{\left|\sum \langle L, M|C_q^{(1)}|c, m\rangle\right|^2 = \sum \left|\langle L, M|C_q^{(1)}|c, m\rangle\right|^2}. \tag{10.47}$$

There are no cross terms, and we can treat all individual angular momentum matrix elements as orthogonal. The polarization dependent transition probabilities

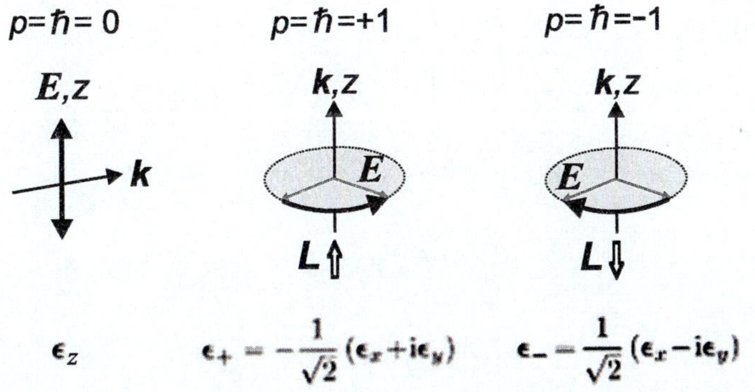

Fig. 10.7 Three orthogonal polarized photon states corresponding to the polarization labels $p = 0, +1, -1$ which also reflect the photon angular momenta $0 + \hbar, -\hbar$ according to Fig. 3.4a. In a Cartesian (x, y, z) coordinate system the three cases correspond to linearly polarized photons with their E-vector along z and circularly polarized photons for $\mathbf{k}$ along z and the shown angular momentum directions $\mathbf{L}$. Underneath we have identified the associated unit polarization vectors, defined in (3.30)

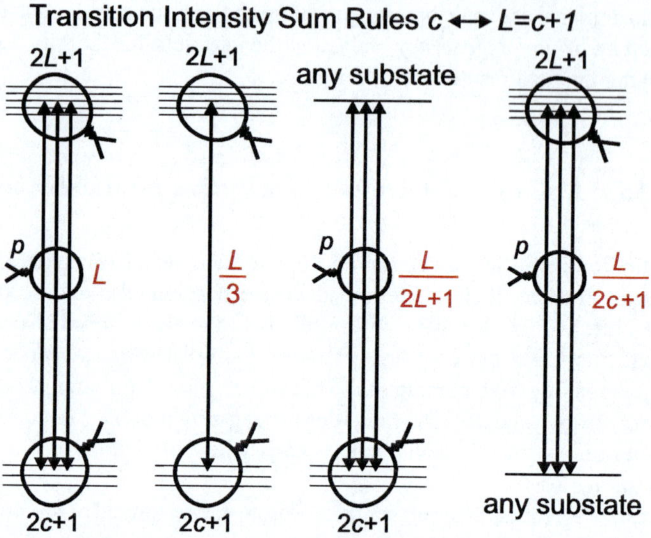

Fig. 10.8 Sum rules for the square of the angular transition matrix element for transitions between core states with angular momentum c, m and a total number of $2c+1$ core orbitals and valence states with angular momentum L, M, and $2L+1$ orbitals, assuming $L=c+1$. The x-ray polarization is represented by a label $p=0,\pm1$. All sums over polarization and manifolds are indicated graphically by a loop. When a sum is performed over at least two of the three labels M, m, p, the transition matrix element is independent of polarization and yields the sum rule value given in red

$|\langle L, M|C_p^{(1)}|c, m\rangle|^2 = |\langle c, m|C_{-p}^{(1)}|L, M\rangle|^2$ then fulfill the following sum rules (without spin)

$$\sum_{m,M,p} |\langle L, M|C_p^{(1)}|c, m\rangle|^2 = L \tag{10.48}$$

$$\sum_{m,M} |\langle L, M|C_p^{(1)}|c, m\rangle|^2 = \frac{L}{3} \tag{10.49}$$

$$\sum_{m,p} |\langle L, M|C_p^{(1)}|c, m\rangle|^2 = \frac{L}{2L+1} \tag{10.50}$$

$$\sum_{M,p} |\langle L, M|C_p^{(1)}|c, m\rangle|^2 = \frac{L}{2c+1}. \tag{10.51}$$

By inclusion of spin we need to consider that the dipole operator does not act on spin, so that only spin-conserving transitions are allowed. To include spin, one therefore needs to first calculate the matrix elements according to the above orbital degeneracy sum rules and then multiply by a factor of 2. The sum rules are graphically illustrated in Fig. 10.8, where a loop indicates a sum over m, M, or p.

Table 10.4 Mathematical description of s, p and d orbitals in terms of spherical harmonics $Y_{l,m_l} = |l, m\rangle$

$s =$	$\frac{1}{\sqrt{4\pi}} =$	$Y_{0,0}$
$p_x =$	$\sqrt{\frac{3}{4\pi}}\frac{x}{r} =$	$\frac{1}{\sqrt{2}}(Y_{1,-1} - Y_{1,+1})$
$p_y =$	$\sqrt{\frac{3}{4\pi}}\frac{y}{r} =$	$\frac{i}{\sqrt{2}}(Y_{1,-1} + Y_{1,+1})$
$p_z =$	$\sqrt{\frac{3}{4\pi}}\frac{z}{r} =$	$Y_{1,0}$
$d_{xy} =$	$\sqrt{\frac{15}{4\pi}}\frac{xy}{r^2} =$	$\frac{i}{\sqrt{2}}(Y_{2,-2} - Y_{2,+2})$
$d_{xz} =$	$\sqrt{\frac{15}{4\pi}}\frac{xz}{r^2} =$	$\frac{1}{\sqrt{2}}(Y_{2,-1} - Y_{2,+1})$
$d_{yz} =$	$\sqrt{\frac{15}{4\pi}}\frac{yz}{r^2} =$	$\frac{i}{\sqrt{2}}(Y_{2,-1} + Y_{2,+1})$
$d_{x^2-y^2} =$	$\sqrt{\frac{15}{16\pi}}\frac{(x^2-y^2)}{r^2} =$	$\frac{1}{\sqrt{2}}(Y_{2,-2} + Y_{2,+2})$
$d_{3z^2-r^2} =$	$\sqrt{\frac{5}{16\pi}}\frac{(3z^2-r^2)}{r^2} =$	$Y_{2,0}$

The orbitals o_i are real and normalized according to $\int |o_i|^2\, d\Omega = \int_0^{2\pi}\int_0^{\pi} |o_i|^2 \sin\theta\, d\theta\, d\phi = \langle o_i|o_i\rangle = 1$

By use of (10.40), (10.45), and (10.49) we obtain by including spin

$$\Gamma^p_{c\to L} = \frac{8\pi^2\hbar\omega\,\alpha_f}{\lambda^2}\mathcal{R}^2_{c\to L}\, 2\sum_{m,M} |\langle L, M|C^{(1)}_p|c, m\rangle|^2 = \frac{16\pi^2\hbar\omega\,\alpha_f}{3\lambda^2}\mathcal{R}^2_{c\to L}\, L. \quad (10.52)$$

This expression assumes that all $2(2c+1)$ core spin-orbital substates are filled and all $2(2L+1)$ valence substates are empty. As shown by the last expression, there is no polarization dependence in this case (sum rule (10.49)). This is a consequence of the fact that closed shells are spherically symmetric as expressed by the sum rules (10.44).

Sum rules also exist when the complex spherical harmonics are combined into the real molecular orbitals listed in Table 10.4, which is typically done to describe bonding of atoms in molecules and solids. Transition matrix elements between the $p_{3/2}$ and $p_{1/2}$ core states and the d_i orbitals are shown in Fig. A.8.

The symmetry properties of different sums of orbitals listed in Tables 10.3 and 10.4 are illustrated in Fig. 10.9, where we have plotted charge densities (absolute values squared of the wavefunctions) for selected cases. The top rows show that the charge densities of sums of the p orbitals and their spin-orbit split $p_{1/2}$ and $p_{3/2}$ submanifolds are all spherically symmetric. The spherical symmetry of the spin-orbit split submanifolds allows us to separately consider sum rules for the transitions from the $2p_{1/2}$ (L_2) and $2p_{3/2}$ (L_3) core states.

Similar symmetry rules apply for the d orbitals, shown in the lower rows of Fig. 10.9, and also for their spin-orbit split $d_{3/2}$ and $d_{5/2}$ submanifolds (not shown). The spherical symmetry of the summed d orbitals is preserved even when they are split into the well-known e_g and t_{2g} substates by a cubic ligand field, as shown in the figure. As a consequence we may state the following rules for the polarization dependence of core to valence transitions.

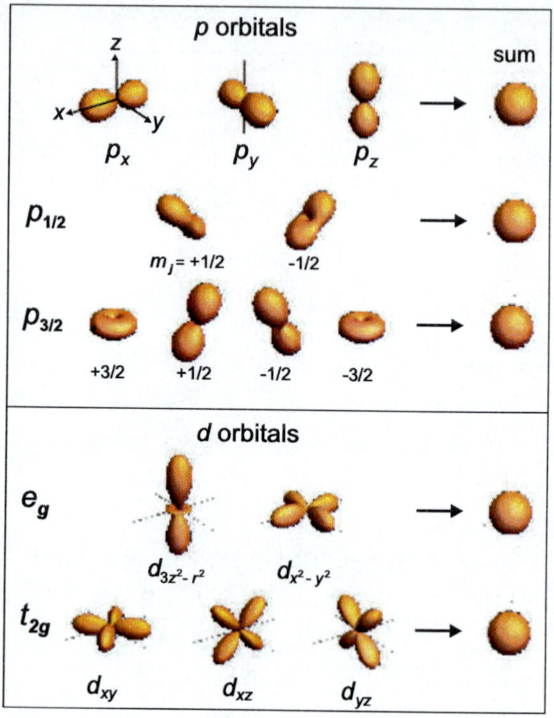

Fig. 10.9 Illustration of the *charge densities* of various wavefunctions in Tables 10.3 and 10.4 and their sums. On top we show that the charge densities of sums of the p and their spin-orbit split $p_{1/2}$ and $p_{3/2}$ submanifolds are all spherically symmetric. In the lower panel we show the five d_i orbital densities whose sum is again spherically symmetric. The spherical symmetry is preserved even when they are split into the e_g and t_{2g} irreducible representations by a cubic ligand field

> There is *no polarization* dependence of core to valence transitions whenever the charge densities of the core states and valence states are spherically symmetric. This applies for all free atoms and molecules and for solids with cubic symmetry and random bond orientations due to polycrystallinity or lack of orientational order in soft matter. In the presence of magnetic order, the polarization dependence also vanishes for samples with random orientations of magnetic domains.

More generally, the polarization average is defined as

$$\Gamma_{c\to L} = \frac{1}{3}\sum_p \Gamma^p_{c\to L} = \frac{1}{3}\left\{\Gamma^0_{c\to L} + \Gamma^{+1}_{c\to L} + \Gamma^{-1}_{c\to L}\right\}, \qquad (10.53)$$

and if the transition matrix element is polarization independent, we have $\Gamma^0_{c\to L} = \Gamma^{+1}_{c\to L} = \Gamma^{-1}_{c\to L}$ and therefore

$$\Gamma_{c\to L} = \Gamma^0_{c\to L} = \frac{1}{2}\left\{\Gamma^{+1}_{c\to L} + \Gamma^{-1}_{c\to L}\right\}. \qquad (10.54)$$

In cases where an orientational average of the sample exists, (10.52) may be generalized to the case where the valence states are partially filled. In this case we may define an effective number of empty or "hole" states given by the fraction $N_h/(2(2L+1))$. The valence charge then remains effectively spherically symmetric, and we can rewrite (10.52) and the associated cross section in the following form that we will consider in the next section.

For a sample where an orientational average can be made and only a fraction of the valence states $N_h/(2(2L+1))$ is unfilled, the resonant XAS cross section and the transition matrix element are given by

$$\sigma_{\mathrm{XAS}} = \frac{\lambda^2}{\pi} \frac{\Gamma_{c\to L}}{\Gamma} \frac{(\Gamma/2)^2}{(\hbar\omega - \mathcal{E}_0)^2 + (\Gamma/2)^2}, \tag{10.55}$$

where including spin we have

$$\Gamma_{c\to L} = \frac{8\pi^2\hbar\omega\,\alpha_f}{\lambda^2} \mathcal{R}^2_{c\to L} \frac{L\,N_h}{3(2L+1)}. \tag{10.56}$$

In the following we shall compare experimental data with this quantum theoretical expression for the x-ray absorption cross section.

10.5 Resonant XAS in Experiment and Theory

In this section we present selected experimental K-shell absorption spectra for low-Z atoms and L-shell spectra for the 3d transition metals. Our choice of these cases is due to their practical importance, their large cross sections, and sharp absorption lines [31–34].

10.5.1 K-Shell Resonance in the Low-Z Atom Ne

The K-shell absorption spectrum of Ne atoms consists of a series of Rydberg resonances that smoothly merge into a continuum around the ionization potential (IP) as illustrated in Fig. 10.10, which shows the experimentally determined cross section [35].

We first consider the linewidths of the Rydberg resonances. The experimental data shown as circular points are plotted as the absorption cross section on a logarithmic scale to better reveal the fitted Voigt lineshapes (see Appendix A.2.3). The cross section of each XAS Rydberg resonance is described by (10.55) with specific dipole transition matrix elements $\Gamma_{1s\to np}$ and FWHM linewidths $\Gamma = \Gamma(np)$. The natural

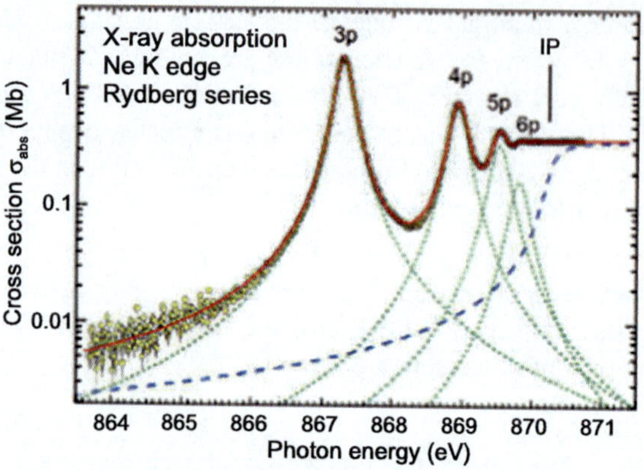

Fig. 10.10 High resolution x-ray absorption spectrum of the K edge of Ne, revealing the series of Rydberg resonances that merge into the ionization potential (IP) of 870.2 eV [35]. The cross section is plotted on a logarithmic scale to emphasize the Voigt profiles of near Lorentzian shapes of the Rydberg resonances, shown as dotted green lines and the continuum step shown as a dashed blue line

linewidths $\Gamma(np)$ were determined by Müller et al. [35] from Voigt lineshape fits with an experimental Gaussian resolution function 32 meV as $\Gamma(3p) = 248(2)$ meV , $\Gamma(4p) = 260(3)$ meV and $\Gamma(5p) = 297(5)$ meV, respectively. The $3p$ value is in accord with the semi-empirical value of 240 meV in Table 10.2 and compares with the Ne $1s$ core level photoemission linewidth of 270 meV (see footnote 1 in [36]) and the Kα x-ray emission linewidth value of 270 meV [37] (see Fig. 12.6).

Next we consider the value of the dipole matrix element $\Gamma_{1s\rightarrow 3p}$ for the first Rydberg resonance in Fig. 10.10. The final state has $3p$ character, and all six $3p$ ($L = 1$) spin orbitals are empty so that $N_\mathrm{h} = 2(2L + 1) = 6$. The peak value of the cross section (10.55) is obtained as

$$\sigma_{\mathrm{XAS}}^{\mathrm{res}} = \frac{\lambda^2}{\pi}\frac{\Gamma_{1s\rightarrow 3p}}{\Gamma} = \frac{16\pi\,\hbar\omega\,\alpha_\mathrm{f}}{3\Gamma}\,\mathcal{R}_{1s\rightarrow 3p}^2. \tag{10.57}$$

Using the resonant wavelength of $\hbar\omega = 867.3$ eV ($\lambda = 1.43$ nm) and the experimental values $\Gamma = 248$ meV and the measured peak value $\sigma_{\mathrm{XAS}}^{\mathrm{res}} = 2\,\mathrm{Mb} = 2 \times 10^{-4}\,\mathrm{nm}^2$ we can deduce the value of the radial matrix element to be

$$\mathcal{R}_{1s\rightarrow 3p}^2 = \frac{3\Gamma}{16\pi\,\hbar\omega\,\alpha_\mathrm{f}}\,\sigma_{\mathrm{XAS}}^{\mathrm{res}} = 4.7 \times 10^{-7}\,\mathrm{nm}^2 \tag{10.58}$$

or $\mathcal{R}_{1s\rightarrow 3p} = 6.8 \times 10^{-4}$ nm. We will see in Sect. 12.4 (see Fig. 12.11) that this $1s \rightarrow 3p$ radial matrix element is smaller than the $1s \rightarrow 2p$ one by a factor of about 6. This is expected due to the smaller radial overlap of the $1s$ core state with the more extended $3p$ state relative to the more compact $2p$ state.

The dipolar transition matrix element is evaluated to be $\Gamma_{1s\rightarrow 3p} = 7.6 \times 10^{-2}$ meV. The ratio of $\Gamma_{1s\rightarrow 3p}/\Gamma \simeq 3 \times 10^{-4}$ reveals the well-known fact that the dominant linewidth contribution originates from the Auger decay channel.

10.5.2 K-Shell Resonances in the N_2 and O_2 Molecules

Two other fundamental examples of resonant x-ray absorption spectra are the strong K-shell XAS resonances in the N_2 and O_2 molecules. We shall later in Sect. 13.10.1 also discuss their high-resolution resonant x-ray emission spectra. They are described by a resonant scattering process where the XAS final state is an intermediate state in the scattering process.

10.5.2.1 N_2 Molecule

We start with the well-known XAS spectrum of N_2, whose fine structure was first observed by electron energy loss spectroscopy [38] in the late 1970s [39, 40]. It was reproduced with x-rays only about 10 years later [41].

The electronic orbital structure of N_2, its spin dependent electron filling, and the origin of the vibrational fine structure in the K-shell $2\sigma_u \rightarrow 1\pi_g^*$ resonance are illustrated in Fig. 10.11 together with experimental data recorded by Prince et al. [42].

The N_2 XAS spectrum is determined by transitions illustrated with increasing detail in Figs. 10.11a–c between the electronic ground state, whose lowest vibrational state $\nu = 0$ is the only one occupied at room temperature and the shown vibrational states in the electronic final state [43]. The experimental XAS spectrum [42] is shown in Figs. 10.11d. A detailed lineshape fit yielded a core hole lifetime width of $\Gamma = 115 \pm 4$ meV [21, 42], which compares to the semi-empirical value of 130 meV in Table 10.2.

The origin of the relative intensities of the vibrational XAS resonances, separated by $\simeq 230$ meV, is illustrated in (c). There we show the vibrational wavefunctions of the electronic ground and final state in the internuclear distance dependent Morse potentials [45], taken from [43, 44]. The gray areas in (c) represent the integrals of the products of the vibrational wavefunctions of the $\nu = 0$ ground state with those of the excited states $\nu = 0, 1, 2....$ The square of the gray areas is the so-called *Franck–Condon* overlap factors [46] which determine the relative intensities of the vibrational XAS resonances.

10.5.2.2 O_2 Molecule

For comparison, we show in Fig. 10.12 the case of O_2, again illustrating the spin dependent filling of the molecular orbitals, and the origin of the same K-shell $1\sigma_u \rightarrow$

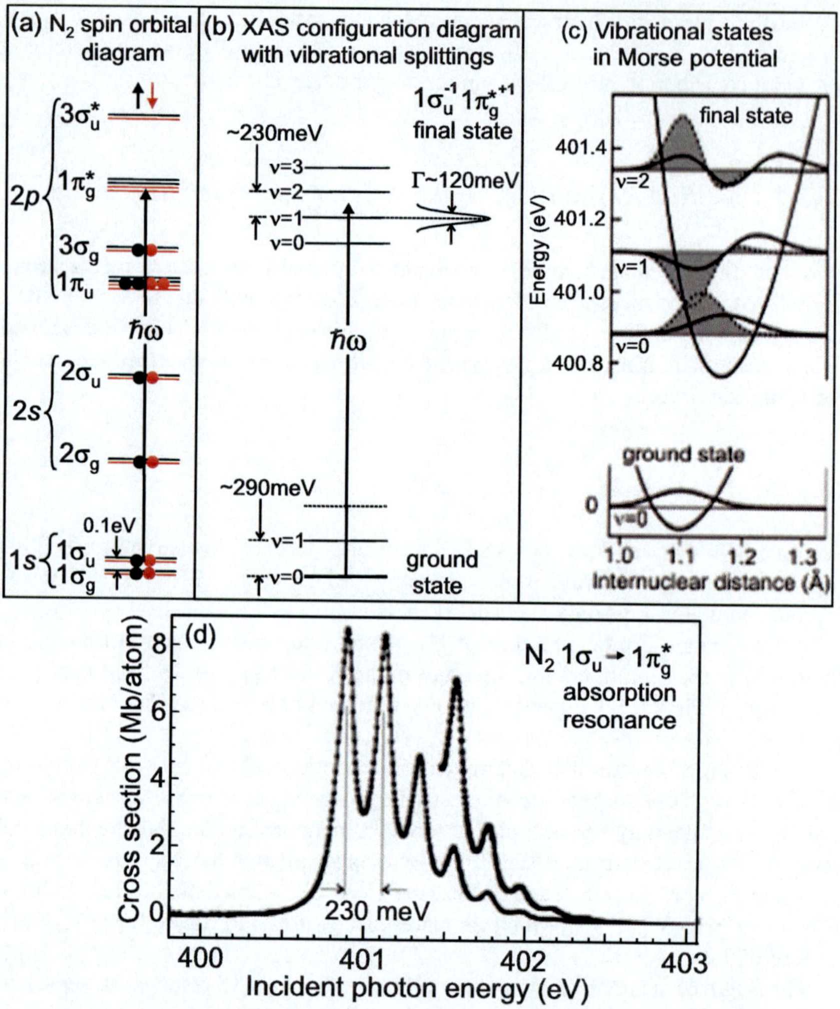

Fig. 10.11 **a** Molecular orbital diagram of the N_2 molecule, with spin dependent filling in the ground state indicated by black (spin-up) and red (spin-down) circles. The dipole and parity allowed $1\sigma_u \rightarrow 1\pi_g^*$ spin-conserving transition is also shown. **b** XAS transition diagram in a configuration model. With increasing photon energy transitions may be observed to all vibrational states in the final state with a hole in the $1\sigma_u$ and an added electron in the $1\pi_g^*$ orbital. The natural Lorentzian linewidth (FWHM) of the final states is $\Gamma \simeq 120\,\text{meV}$ [21, 42]. **c** Vibrational wavefunctions of the ground state and the first three vibrational final states, with shaded areas indicating products of ground and final state functions [43, 44]. **d** High-resolution XAS spectrum of the $1\sigma_u \rightarrow 1\pi_g^*$ resonance in N_2 [42]. The intensity scale has been converted to the cross section per atom. The vibrational splitting of the resonances is $\simeq 230\,\text{meV}$, and their relative intensities are given by the square of the gray areas in (**c**)

$1\pi_g^*$ resonance as in N_2, using experimental data of [21]. For a precise calibration of the energy of the $1\sigma_u \rightarrow 1\pi_g^*$ and higher energy Rydberg transitions we refer the reader to the interesting article by Leutenegger et al. [47].

The electronic orbital structure of O_2, its spin dependent electron filling, and the origin of the K-shell resonance are illustrated in Figs. 10.12a, b. The $1s$ core state is split into the $1\sigma_u$ and $1\sigma_g$ molecular components with a splitting of 1.11 eV [48]. The molecular orbital structure of N_2 and O_2 is very similar, the main difference being the filling of the $1\pi_g^*$ orbital which is empty in N_2 and in O_2 contains two electrons of the same spin according to Hund's rule. The molecule is hence paramagnetic. The dipole operator does not act on spin so that the two orbitals involved in a transition must have the same spin. We have arbitrarily assumed excitations of "spin-down" electrons in Figs. 10.12a. Since the dipole operator has odd parity, parity conservation demands that the lowest energy allowed transition is also parity odd [32]. The electronic transition of interest indicated by a red arrow connects the $1\sigma_u$ spin-down K-shell orbital and the empty $1\pi_g^*$ orbital of the same spin.

The experimental spectrum is shown in Fig. 10.12c [21], plotted in terms of the x-ray absorption cross section per atom. The cross section was obtained by conversion of the oscillator strength of the inner shell electron energy loss spectra (ISEELS) [40] and (A. Hitchcock private communication). The spectrum reveals partially unresolved vibrational finestructure[8] which is more clearly revealed by the shown fitted Voigt profiles consisting of Gaussian 55 ± 10 meV and Lorentzian 149 ± 10 meV FWHM components. The implied natural linewidth of the $1s$ core hole state of $\Gamma_{1s} = 149 \pm 10$ meV agrees rather well with the semi-empirical value of 160 meV in Table 10.2.

10.5.3 L-Shell Resonances in 3d Transition Metal Atoms

Other important resonances involve the so-called white-lines[9] in the L absorption spectra of the $3d$ transition metals. They were already discussed in Chapt. 7 and shown in Figs. 7.12 and 7.24, and their study has played an important role in modern magnetism studies of thin films and nanostructures [31].

We note that with the advent of low energy FELs and up-converted laser sources that cover a limited spectral range up to about 100 eV (see Fig. 1.4) with femtosecond time resolution, it has become popular to study the dynamical processes of $3d$ transition metals by M-shell ($3p \rightarrow 3d$) absorption spectroscopy in the 30–80 eV range [50, 51], instead of L-shell ($2p \rightarrow 3d$) spectroscopy in the 300–1000 eV, discussed here.

The advantage of L-edge studies is the about eight times larger $2p$ spin-orbit splitting which leads to a reduction of multi-electron correlation effects. The L-shell resonances also exhibit larger resonances of smaller widths, since M-shell

[8] A higher resolution O_2 XAS spectrum has been recorded by Strocov et al. [49].

[9] In early studies the intense resonances appeared as white lines on photographic films.

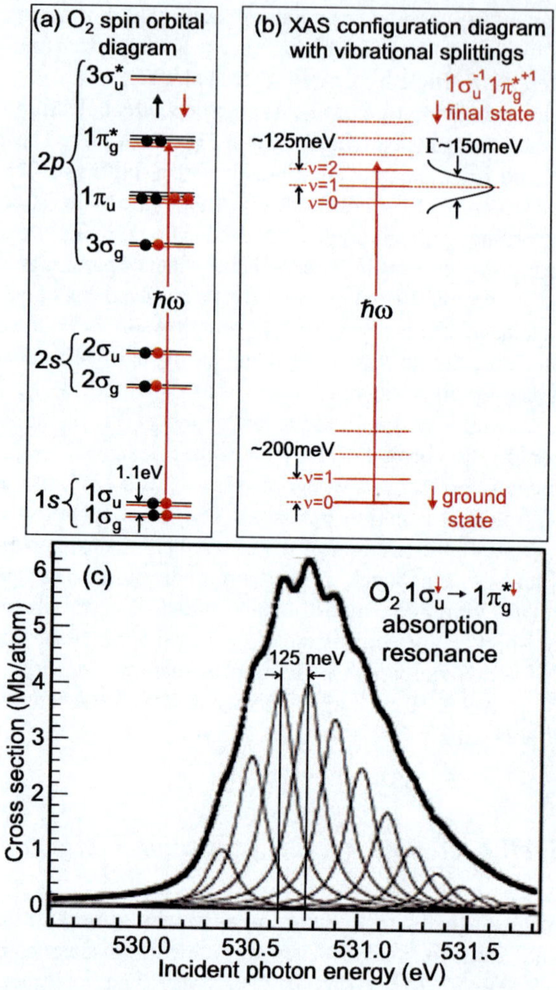

Fig. 10.12 **a** Molecular orbital diagram of the O_2 molecule, with spin dependent orbital filling in the ground state indicated by black (spin-up) and red (spin-down) circles. Indicated in red is the $1\sigma_u \rightarrow 1\pi_g^*$ transition of interest, where an electron from the K-shell $1\sigma_u$ orbital (arbitrarily assumed to be spin-down) is dipole exited to the lowest unfilled $1\pi_g^*$ orbital of the same spin. **b** XAS transition diagram in a configuration model. At room temperature only the lowest vibrational ground state is filled and with increasing photon energy transitions may be observed to all vibrational states in the final state with a hole in the spin down $1\sigma_u$ and an added electron in the $1\pi_g^*$ orbital. **c** High-resolution x-ray absorption spectrum of the $1\sigma_u \rightarrow 1\pi_g^*$ resonance in the O_2 molecule [21], with the intensity scale converted to the cross section per atom. The vibrational fine structure with splitting $\sim$ 125 meV has been fitted with the shown Voigt profiles containing instrumental Gaussian and Lorentzian lifetime contributions [21]. The O_2 K-shell natural linewidth is obtained from the width of the vibrational lineshapes to be $\Gamma = 150$ meV

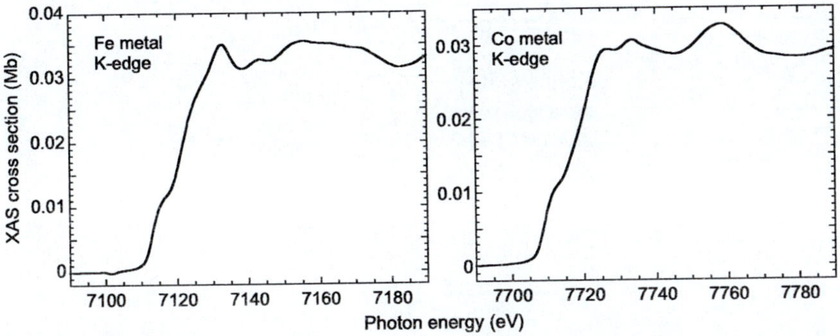

Fig. 10.13 K-shell XAS spectra of Fe and Co metal (data from Alain Fontaine private communication). We have subtracted a pre-edge background and put the data on an absolute cross section scale that only reflects the K-shell contribution by comparison with tabulated cross sections [15]

resonances are lifetime broadened due to the presence of super-Coster-Kronig decays, which occur between shells of the same principal quantum number [52, 53]. For example, the natural M_3 linewidth for Co is $\Gamma = 2.85\,\mathrm{eV}$ compared to the L_3 width of $\Gamma = 0.43\,\mathrm{eV}$. The difference between the M and L spectra of the $3d$ transition metals is most beautifully demonstrated by the spectra of the elementary atoms and cations in [54–56].

L-shell XAS spectra of the transition metals also have an advantage over K-shell spectra ($1s \rightarrow 4p$ transitions) in that their strong resonant structure is absent at the K-shell thresholds as illustrated for Fe and Co and in Fig. 10.13. Furthermore, the absolute K-shell cross section, also indicated in that figure, is about two orders of magnitude weaker than for the L-edges. This is more directly apparent from Fig. 7.14 which shows the Fe XAS cross section over an extended energy scale.

For L-edges, electrons from the $2p_{1/2}$ (L_2) and $2p_{3/2}$ (L_3) core states are excited to empty $3d$ valence states. In a simple density of states model the valence shell can be described in terms of the number of filled "electron" orbitals, n_e, and empty "hole" orbitals, n_h, so that for each spin we have $n_e + n_h = 2L + 1$, as shown in Fig. 10.14.

The model assumes that at a given energy relative to the Fermi level, the density of states contains equal contributions from all angular momentum substates $-L \leq M \leq L$ due to hybridization (see Fig. 11.10). This is a reasonable approximation for metallic bonding even if the lattice deviates somewhat from cubic symmetry as in hexagonal close packed (hcp) Co metal. This mixing of different angular momentum substates M eliminates any polarization dependence. The spin dependence is also averaged out if the sample contains microscopic magnetic domains of random orientation.

We consider the total resonant cross section for $2p \rightarrow 3d$ excitations, i.e. the added L_2 and L_3 contributions. The angular parts of the transition matrix elements follow from the sum rules (10.48)–(10.51). The spin-inclusive expression is simply

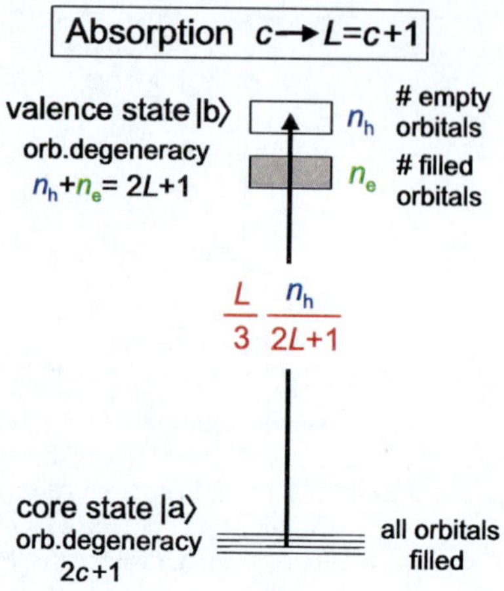

Fig. 10.14 Probability of dipole transitions from a core manifold of angular momentum c to a valence manifold of angular momentum $L = c + 1$. The core manifold consists of $2c + 1$ $(-c \leq m_c \leq c)$ orbitals and the valence manifold of $2L + 1$ orbitals between which spin-conserving transitions can occur. For the valence orbitals we have assumed that a "hole" fraction $n_h/(2L + 1)$ of orbitals is empty and a fraction $n_e/(2L + 1)$ is filled with electrons. The transition probability is independent of polarization and given in red

a factor of 2 larger since there are two identical charge scenarios for each spin, and spin-flip transitions are dipole forbidden. In anticipation of the discussion of magnetic effects in Sect. 11.5 we can include spin in the total number of filled and empty states by replacing the number of orbitals by the number of *spin orbitals* according to $N_h = 2n_h$ and $N_e = 2n_e$, so that $N_e + N_h = 2(2L + 1)$ reflects the total substates of a shell of angular momentum L. The total polarization averaged dipole transition width is then obtained from (10.56) by considering the partial occupancy $N_h/(2(2L + 1))$ of the valence shell as

$$\Gamma_{c\to L} = \frac{8\pi^2\hbar\omega\,\alpha_f}{\lambda^2}\,\mathcal{R}^2_{c\to L}\,\frac{LN_h}{3(2L+1)}, \tag{10.59}$$

and the resonant XAS cross section (10.55) becomes

$$\sigma_{XAS} = \frac{8\pi\,\hbar\omega\,\alpha_f}{\Gamma}\,\frac{(\Gamma/2)^2}{(\hbar\omega-\mathcal{E}_0)^2+(\Gamma/2)^2}\,\mathcal{R}^2_{c\to L}\,\frac{LN_h}{3(2L+1)}. \tag{10.60}$$

We are also interested in the energy integrated value of the resonant cross section which has the units [Mb eV]. This number can be determined from experimental data. We approximate the energy integration of the resonant cross section by integrating over the Lorentzian lineshape at a constant central energy to obtain

$$\int_0^\infty \sigma_{\mathrm{XAS}}\, \mathrm{d}(\hbar\omega) = \underbrace{4\pi^2 \hbar\omega\, \alpha_{\mathrm{f}}\, \mathcal{R}^2_{c\to L} \frac{L}{3(2L+1)}}_{C_{c\to L}} N_{\mathrm{h}}. \tag{10.61}$$

This is the same expression given in [31]. It is important to realize that N_{h} reflects the number of empty states in the *electronic ground state*, i.e. in the absence of a core hole. This is a consequence of the fact that in the proper configuration model discussed in conjunction with Fig. 10.5 the core hole state is a final state effect. This is also expressed by the so-called initial and final states rules in x-ray absorption and emission spectroscopies [57, 58].

For the $2p \to 3d$ L-shell resonances of the transition metal atoms the radial transition matrix element is then given by

$$\mathcal{R}^2_{2p\to 3d} = \frac{15}{8\pi^2 \hbar\omega\, \alpha_{\mathrm{f}}}\, C_{2p\to 3d}. \tag{10.62}$$

We can use this expression and the conversion $1\,\mathrm{nm}^2 = 10^4\,\mathrm{Mb}$ to determine the values of $\mathcal{R}_{2p\to 3d}$ for the $3d$ transition metal atoms by use of the constant theoretical value $C_{2p\to 3d} = 16\,\mathrm{Mb\,eV}$ given by Thole and van der Laan for the series Ca ($Z = 20$) through Cu ($Z = 29$) [59]. The result is plotted in Fig. 10.15 as black circles.

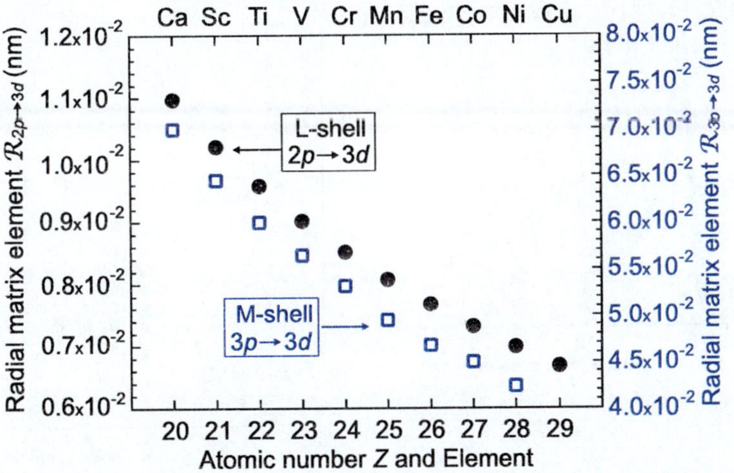

Fig. 10.15 Radial transition matrix elements $\mathcal{R}$ in [nm] for the $2p \to 3d$ (L-shell, black circles) and $3p \to 3d$ (M-shell, blue squares) excitations in Ca (Z = 20) to Cu (Z = 29). The L-shell matrix elements were calculated according to (10.62) with $\hbar\omega$ taken as the $2p_{3/2}$ binding energies with the theoretical value $C_{2p\to 3d} = 16\,\mathrm{Mb\,eV}$ of Thole and van der Laan [59]. The M shell matrix elements were obtained from Hartree–Fock calculations for single charge cations by Hansen et al. [55]

We also show in Fig. 10.15 as blue squares the Z dependence of the related $3p \rightarrow 3d$ (M-edge) radial matrix element. It was obtained from Hartree–Fock calculations by Hansen et al. [55].[10] The $3p \rightarrow 3d$ radial matrix element shows the same trend with Z but is larger by a factor of about 6 than the $2p \rightarrow 3d$ one, owing to the greater overlap of core and valence states, as expected.

10.5.4 L-Shell XAS Intensities and Valence Shell Occupation

According to (10.61) the integrated intensity of the L-shell XAS intensities scales with the number of empty $3d$ states or holes. With increasing Z from Ca (Z=20) to Zn ($Z = 30$) the total number of $3d$ spin orbitals, $2(2L + 1) = 10$, is increasingly filled so that the white line intensity should gradually descrease. This is indeed observed as illustrated in Fig. 10.16a where we show the measured XAS spectra of the $3d$ metals [60], normalized for comparison to the same value above the resonances.

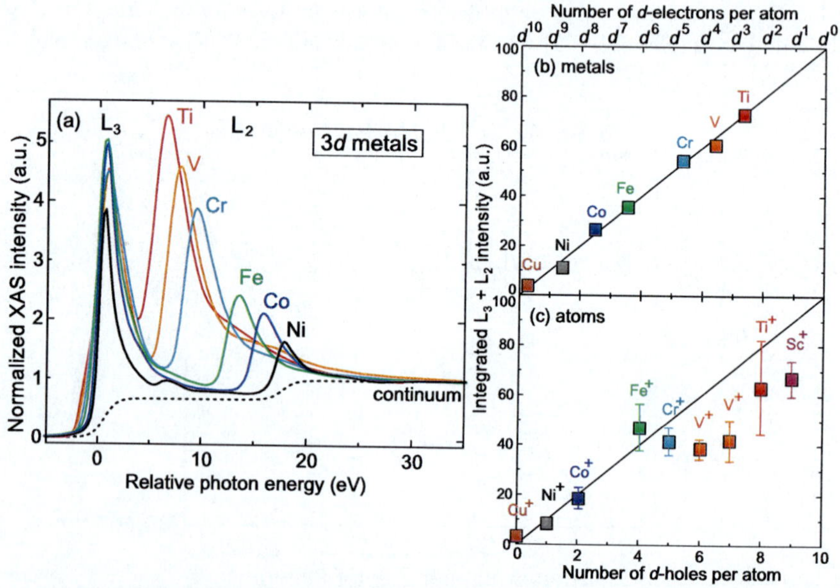

Fig. 10.16 **a** XAS absorption spectra of the $3d$ transition metals Ti–Ni [60], normalized to the same continuum jump above the edges and an energy scale relative to the inflection point of the absorption onset. The spectra reveal the change of the L_3–L_2 spin-orbit splitting and the change in the total resonance "white-line" L_3+L_2 intensity. **b** Total resonance intensities versus number of $3d$ holes in metals and **c** in individual trapped cations [56]. Also see Fig. 10.5 in [61]

[10] We have converted the radial matrix element given in [55] in atomic units and multiplied by the reduced angular matrix element $\sqrt{L} = \sqrt{2}$ to correspond to our convention (10.46).

In Fig. 10.16b we also show the change of the integrated L_3 plus L_2 "white line" intensities with the number of valence holes [31] for the elementary metals and in (c) for cations in the form of singly ionized atoms [56]. The same linear dependence was also reported in [61] using electron energy loss spectroscopy.

It is apparent from Fig. 10.16a that the intensity ratio between the L_3 and L_2 resonances changes significantly between Ni and Ti. In a one-electron model the L_3/L_2 intensity ratio is 2/1, reflecting the relative degeneracy $2j+1=4$ for $j=3/2$ (L_3) and $2j+1=2$ for $j=1/2$ (L_2). The deviation was theoretically shown to be due to multi-electron correlation effects [59, 62] that shuffle intensities between the two resonances. This demonstrates the inadequacy of the one-electron theory. Remarkably, the sum of the L_3 and L_2 resonance intensities appears to be proportional to the number of 3*d* holes, in both one-electron and multi-electron theories.

For XAS spectra of solids, the quantitative determination of the integrated white line intensities is complicated by an underlying background, which is typically modeled as the dashed step-like curve in Fig. 10.16a. This complicates the determination of the factors $C_{2p\to 3d}$, defined in (10.61). The value $C_{2p\to 3d}\simeq 11\,\text{Mb}\,\text{eV}$ determined from XAS and XMCD data for Fe, Co, and Ni metals, for example, where the underlying step intensity was subtracted [31, 63], is smaller than the value $C_{2p\to 3d}=16\,\text{Mb}\,\text{eV}$ obtained from atomic calculations by Thole and van der Laan [59]. This may indicate that some of the resonant atomic oscillator strength is transferred into the continuum "background".

10.5.5 Resonant Lineshapes in Atoms and Solids

Theoretically, the absorption cross section is given by (10.60) and the resonances therefore should have the natural FWHM Lorentzian linewidth Γ listed in Table 10.2 and plotted in Fig. 10.4. In atoms the resonant widths beautifully exhibit the natural Lorentzian shape as seen for Ne in Fig. 10.10. In contrast the lineshapes and width in solids typically deviate significantly from Lorentzian forms. In the early days this could be attributed to limited Gaussian instrumental energy resolution, but modern spectra recorded with instrumental resolution well below the natural lifetime width still exhibit broader lines. This is revealed, for example, by the spectra in Fig. 10.16. In this case, experimentally observed widths are broadened by the spread of the 3*d* valence states caused by the band structure. As a consequence, the natural $2p\to 3d$ Lorenzian lineshape is transformed into a broader peak of reduced intensity, with preservation of the integrated resonance cross section.

The broadening of resonances in solids is illustrated for the L_3 resonances for the elemental metals Fe, Co, and Ni in Fig. 10.17.

The experimental XAS cross sections shown as black curves were measured with linear polarization and polycrystalline samples with random magnetic domains to eliminate any polarization dependence. We have subtracted a small pre-edge background of about 0.2 Mb due to outer shell contributions, which underlies the resonant structure as shown for Fe in Fig. 7.14. The measured lineshapes are best accounted for by a fit with Voigt lineshapes shown in red which consist of Lorentzian and Gaussian components indicated on the right. The Voigt function is a convolution of the theoretical Lorentzian profile of natural linewidth Γ, and a Gaussian, whose FWHM Γ_G and peak experimental cross section, σ_0, are determined by fit of the Voigt profile (see Appendix A.2.3)

$$\sigma_{XAS}^{exp}(\hbar\omega) = \sigma_0 \frac{\mathrm{Re}\left\{\exp\left[-\frac{a^2(\hbar\omega-\mathcal{E}_0-\mathrm{i}\Gamma/2)^2}{2\Gamma_G^2}\right]\mathrm{erfc}\left[\frac{a(\mathrm{i}(\hbar\omega-\mathcal{E}_0)+\Gamma/2)}{\sqrt{2}\Gamma_G}\right]\right\}}{\exp\left[\frac{a^2\Gamma^2}{8\Gamma_G^2}\right]\mathrm{erfc}\left[\frac{a\Gamma}{2\sqrt{2}\Gamma_G}\right]}, \qquad (10.63)$$

where $a = 2\sqrt{\ln 4} = 2.355$ links the Gaussian FWHM, Γ_G, and rms width, σ_G, by $\Gamma_G = a\sigma_G$.

The red experimental Voigt profiles may then be converted into the shown theoretical blue Lorentzian profiles under the assumption that the areas under the curves are preserved. The peak values of the experimental and converted theoretical cross sections are seen to differ significantly, and so do the widths of the associated Voigt and Lorentzian lineshapes. Both the red Voigt and blue Lorentzian resonances in Fig.

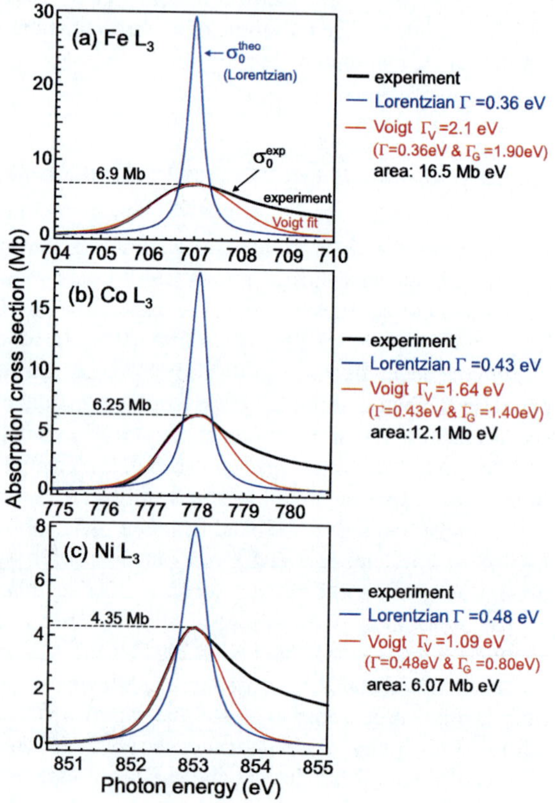

Fig. 10.17 Experimental L_3 absorption cross sections for Fe, Co, and Ni metals (black curves) as a function of photon energy. In all cases a small background of about 0.2 Mb resulting from absorption contributions from outer shells has been subtracted. Superimposed in red are Voigt lineshape fits consisting of the listed Lorentzian components of FWHM, Γ, and Gaussian components of FWHM, Γ_G, that are listed on the right. Shown in blue are Lorentzians of natural linewidth Γ whose integrated cross sections are equal to those of the red Voigt lineshapes

Table 10.5 Parameters for the L_3 resonances of Fe, Co, and Ni metals. Listed are the atomic number densities ρ_a, the resonance energies $\mathcal{E}_0$ and wavelengths λ_0, the number of 3*d* valence shell holes N_h and electrons N_e, and the polarization averaged peak XAS cross section σ_0^{exp}

	ρ_a $\frac{\text{atoms}}{\text{nm}^3}$	$\mathcal{E}_0$ eV	λ_0 nm	N_h	N_e	σ_0^{exp} Mb	Γ_x^{exp} meV	σ_0^{theo} Mb	Γ_x^{theo} meV	Γ meV
Fe	84.9	707	1.75	3.5	6.5	6.9	0.26	29.2	1.08	360
Co	90.9	778	1.59	2.5	7.5	6.25	0.34	18.0	0.96	430
Ni	91.4	853	1.45	1.5	8.5	4.35	0.32	8.02	0.575	480

The listed dipole transition widths Γ_x^{exp} are obtained from the experimental cross sections by use of (10.55). The listed peak cross sections σ_0^{theo} correspond to the blue curves in Fig. 10.17 with the assumption that all oscillator strength is contained in a Lorentzian of FWHM Γ listed in the last column. The values Γ_x^{theo} correspond to these cross sections. The theoretical values may be taken as those of free individual atoms

10.17 for a given element have the same integrated L_3 resonance values, which are 16.5 Mb eV (Fe), 12.1 Mb eV (Co), and 6.1 Mb eV (Ni). One may associate the blue theoretical cross sections with those of free individual atoms.

The key parameters are summarized in Table 10.5 which lists the atomic number densities ρ_a, the resonance energies and wavelengths, the number of 3*d* valence shell holes N_h and electrons N_e, and the polarization averaged peak XAS cross section σ_0^{exp}, corresponding to the peak values of the black or red curves in Fig. 10.17) [64]. By use of (10.55) they yield the listed *dipolar* matrix elements Γ_x^{exp}. Similarly, the peak Lorentzian cross sections σ_0^{theo} of the blue curves correspond to the theoretical values Γ_x^{theo}. The last column lists the total natural (dipole plus Auger) decay widths Γ from Table 10.2.

10.5.6 Dipole Matrix Element, Oscillator Strength, and Sum Rules

The energy integrated resonant absorption cross section is related to the quantum mechanical *oscillator strength* *g* of a resonance introduced in Sect. 6.4. For a dipole transition $a \rightarrow b$ of energy $\mathcal{E}_b - \mathcal{E}_a = \mathcal{E}_0$, it is defined including polarization as [7]

$$g_{a\rightarrow b}^{p} = \frac{2m_e\,\mathcal{E}_0}{\hbar^2}\,|\langle b|\mathbf{r}\cdot\boldsymbol{\epsilon}_p|a\rangle|^2 = -g_{b\rightarrow a}^{p}. \tag{10.64}$$

The energy integrated *resonance cross section* is related to the dimensionless oscillator strength by

$$\int_0^\infty \sigma_{\mathrm{XAS}}^{p}(\hbar\omega)\,\mathrm{d}(\hbar\omega) = \underbrace{\frac{2\pi^2\hbar^2\alpha_f}{m_e}}_{C}\,g_{a\rightarrow b}^{p} = \frac{\lambda^2}{2}\,\Gamma_{ab}^{p}, \tag{10.65}$$

where $C = \pi \, h \, c \, r_0 = 109.89\,\text{Mb eV}$. The last expression linking the oscillator strength g^p to the transition width Γ^p_{ab} only holds for a *resonance*, while the link of the integrated cross section to the oscillator strength is more general. It also holds for the non-resonant case as discussed by Egerton [65]. By integrating the absorption cross section up to a certain energy above the absorption edge one obtains a g value that represents the energy dependent fraction of the total oscillator strength of the shell.

As an example, we consider the oscillator strength contained in the $3p$ K-shell Rydberg resonance of Ne (see Fig. 10.10). By use of (10.65) and the value $\Gamma_{1s\rightarrow 3p} = 7.6 \times 10^{-2}$ meV at $\lambda = 1.43$ nm, we find an oscillator strength of

$$g_{1s\rightarrow 3p} = \frac{\lambda^2}{2C} \Gamma_{1s\rightarrow 3p} = 7.1 \times 10^{-3}. \tag{10.66}$$

Thus of the maximum oscillator strength of 2 for the two electrons in the K-shell, only a fraction of about 0.35% is contained in the lowest K-shell Rydberg resonance.

Another example is the "white line" L_3 resonance in Fe metal at $\lambda = 1.75$ nm with a value $\Gamma_{2p_{3/2}\rightarrow 3d} = 1.08$ meV (see Table 10.5), which corresponds to an oscillator strength of

$$g_{2p_{3/2}\rightarrow 3d} = \frac{\lambda^2}{2C} \Gamma_{2p_{3/2}\rightarrow 3d} = 0.15 \tag{10.67}$$

which is about 3.8% of the maximum oscillator strength 4 for the four electrons in the $2p_{3/2}$ shell. Our value may be compared to the experimental value $g_{2p\rightarrow 3d} = 0.14$ per d valence hole obtained for single Mn^{5+} ions ($3d^2$ or 8 d valence holes) [66]. To do so, we account for the 3.5 d valence holes in our case (see Fig. 10.16) and for the missing L_2 contribution by a factor of 3/2. This gives $g_{2p\rightarrow 3d} = 0.064$ per $3d$ valence hole for Fe metal which is smaller than the value for free Mn ions. This again suggests that in the metal some of the resonant oscillator strength is transferred to the non-resonant continuum.

10.5.6.1 Shell-Specific Versus Thomas-Reiche-Kuhn Sum Rule

The integrated L-shell white line cross section given by (10.61) or

$$\int_0^\infty \sigma_{\text{XAS}} \, \mathrm{d}\hbar\omega = C_{2p\rightarrow 3d} \, N_\text{h} \tag{10.68}$$

links the area under the resonant $2p$-shell (6 electrons) absorption cross section to the number of empty states or holes N_h in the $3d$ shell. The energy integrated *resonant* L-shell cross section is a fraction of the total integrated resonant plus continuum cross section which enters into the Thomas-Reiche-Kuhn (TRK) sum rule (6.73) and links it to the total number of atomic electrons Z according to

$$\int_0^\infty \sigma_{\mathrm{XAS}}^{\mathrm{tot}} \, \mathrm{d}\hbar\omega = C\,Z. \tag{10.69}$$

The TRK sum rule was previously discussed in Sect. 6.8 in conjunction with Fig. 6.11. If the electrons in all shells had unit oscillator strengths, the summed oscillator strengths would naturally be the number of electrons. According to (10.64) we would then have the oscillator strength per electron state $|a\rangle$ to all final states $|b\rangle$

$$\sum_b g_{a\to b} = \sum_b \frac{2m_\mathrm{e}\,\mathcal{E}_0}{\hbar^2}\,|\langle b|r|a\rangle|^2 = 1, \tag{10.70}$$

and the resonant dipole matrix element given by (10.40) per electron would be

$$\sum_b \Gamma_{ab} = \frac{8\pi^2\alpha_\mathrm{f}\,\mathcal{E}_0}{\lambda^2}\,\frac{\hbar^2}{2m_\mathrm{e}\,\mathcal{E}_0} = \alpha_\mathrm{f}\,\frac{(\hbar\omega)^2}{m_\mathrm{e}c^2}. \tag{10.71}$$

This beautifully simple result expresses the resonant dipole matrix element per electron in terms of the dimensionless fine structure constant, representing the photon-electron coupling, and the ratio of the square of the x-ray energy and the rest mass energy of the electron $m_\mathrm{e}c^2 = 511\,\mathrm{keV}$.

Figure 6.11b shows that, in practice, the black TRK sum rule curve lies below the red curve at all energies and catches up only later. The gradual increase with photon energy reveals that most of the oscillator strength is not contained in discrete resonances at the excitation thresholds but rather extends into the continuum, in good accord with our earlier discussion.

10.6 Multi-electron Formalism: Multiplet Structure

In general, the one-electron and configuration pictures are not equivalent because the latter includes all possible couplings between angular momenta in the ground and excited states. This leads to splittings of absorption resonances (and photoemission peaks) due to *exchange and correlation effects*, so-called *multiplet structure*, as extensively discussed by van der Laan and Thole [67, 68] and de Groot and Kotani [33, 69]. The description of the valence electron states in different models, such as density functional band theory, weak and strong ligand field theory, and molecular orbital theory, has been reviewed by Stöhr and Siegmann [31].

Multiplet splittings are typically encountered in the L absorption spectra of free transition metal atoms and ions [54–56, 70] and when transition metal ions are embedded in inorganic oxides and coordination compounds [33, 34, 59, 62, 67–

69]. In the configuration picture, the XAS spectrum consists of transitions from the electronic ground configuration consisting of a filled $2p^6$ core shell and a partially empty $3d^N$ shell with $10 - N$ holes to an excited configuration where one electron has been removed from the core shell and added to the $3d$ shell according to

$$\text{Electron excitation picture:} \quad 2p^6\, 3d^N \rightarrow 2p^5\, 3d^{N+1}. \tag{10.72}$$

If the $3d$ shell is more than half full, it simplifies things to use the concept of holes instead of electrons and taking care of signs arising from the opposite charge. For example, Hund's third rule states that the spin-orbit interaction changes sign and the LF splitting is upside down. With these rules we can use instead of (10.72) the following *hole* excitation scheme

$$\text{Hole excitation picture:} \quad 2p^0\, 3d^{10-N} \rightarrow 2p^1\, 3d^{9-N}. \tag{10.73}$$

In the *configuration picture* one distinguishes two limiting coupling schemes for the spin and angular momenta of the individual electrons which may be in the same shell or different shells. The Russell-Saunders (or LS) coupling scheme describes the case where the Coulomb and exchange interactions between the electrons are larger than the spin-orbit interaction. The total spin and orbital momenta, denoted by capital letters, are then defined as the *vector sum* of those of the individual electrons, indicated by small letters, according to $\mathbf{S} = \sum_i \mathbf{s}_i$ and $\mathbf{L} = \sum_i \mathbf{l}_i$. The possible multiplet configurations are denoted by *term symbols*

$$^{2S+1}L_J. \tag{10.74}$$

The values $L = 0, 1, 2, 3, 4, \ldots$ are denoted in spectroscopic notation as $S, P, D, F, G, \ldots$, and the simplest two electron spin states are the well-known $2S + 1 = 1$ $(S = 0)$ singlet and $2S + 1 = 3$ $(S = 1)$ triplet states. The smaller spin-orbit interaction couples the spin and orbital momenta according to

$$\mathbf{J} = \mathbf{L} + \mathbf{S}, \tag{10.75}$$

where the vector addition results in possible J values $L - S \leq J \leq L + S$.

When the spin-orbit coupling is stronger than the Coulomb and exchange interactions, one uses the jj-coupling scheme, where the individual electron spins and angular momenta are coupled first according to $\mathbf{j}_i = \mathbf{l}_i + \mathbf{s}_i$ and the total angular momentum is then given by

$$\mathbf{J} = \sum_i \mathbf{j}_i. \tag{10.76}$$

In the one-electron model there is no summation over i and there is no distinction between LS and jj coupling.

For multi-electron systems, one typically encounters a situation in between the two limiting cases. In so-called *intermediate coupling*, one expresses electronic states as a linear combination of those in either the LS or jj coupling scheme. The two schemes can be transformed into each other through vector coupling coefficients given by the Wigner $3j$, $6j$, and $9j$ symbols [9, 11, 33, 68], today implemented in different programming languages.[11]

10.6.1 Evolution of One-Electron to Multiplet Theory

To demonstrate the evolution of the one-electron to the multi-electron or multiplet theory, we discuss the L-edge spectra of transition metal ions with d-shell electron occupations $3d^9$ and $3d^8$ which are encountered for the Ni^+ and Co^+ ions. These fundamental cases illustrate both the essence of the multiplet theory and how it merges into the limiting case of the one-electron theory.

10.6.1.1 Ground State d^9

We start with the case in Fig. 10.5 and assume that the ground state has the configuration $1s^2 2s^2 2p^6 3s^2 3p^6 3d^9 4s^0$ corresponding to Ni^+ ions. All closed shells contribute zero angular momenta $J = L = S = 0$ so that the angular momenta in the ground configuration are determined by the nine electrons or the single hole in the $3d$ shell. For the single hole, the labels $l, s,$ and j are also the term labels L, S, J and the lowest ${}^{2S+1}L_J$ terms are ${}^2D_{5/2}$ and ${}^2D_{3/2}$, as shown in Fig. 10.18a. Similarly, the possible states in the excited state configuration $2p^5 3d^{10}$ are ${}^2P_{3/2}$ and ${}^2P_{1/2}$.

The spin-orbit splitting in the ground and final states is obtained by use of the spin-orbit coupling constants in Table 9.1. The splitting is given by $\Delta_{so} = 3\xi_{nl}/2$, and the $3d^9$ values given in Table 9.1 yield those given in Fig. 10.18a. Since the ground state s–o splitting, $\Delta_{so} \simeq 0.15\,\mathrm{eV}$ is larger than the thermal energy at room temperature ($k_B T = 0.026\,\mathrm{eV}$), only the ${}^2D_{5/2}$ will typically be occupied. This causes the absence of the ${}^2D_{5/2} \rightarrow {}^2P_{1/2}$ transition which is forbidden by the dipole selection rule $\Delta J = 0, \pm 1$. In general the dipole selection rules are as follows.

[11] The description of angular momentum coupling between electrons or holes goes back to a formalism developed by the mathematicians Clebsch and Gordan in the late 1800s for the coupling of *two* angular momenta. The Clebsch–Gordon formalism is today handled by Wigner's equivalent tensor formalism ($3j$ symbols) developed in the 1930s (original unpublished, [71]) [9, 11]. The formalism was extended by Wigner and Racah [72, 73] to the coupling of three and four angular momenta, expressed by the $6j$ and $9j$ symbols, respectively [9, 11].

Transitions between electronic states $|S, L, J, M_J\rangle$ and $|S', L', J', M'_J\rangle$ are governed by the dipole selection rules

$$\begin{aligned}\Delta S &= S' - S = 0\\ \Delta L &= L' - L = 0, \pm 1\\ \Delta J &= J' - J = 0, \pm 1\\ \Delta M_J &= M'_J - M_J = p = 0, \pm 1.\end{aligned} \tag{10.77}$$

where p denotes the x-ray polarization (angular momentum).

The theory is beautifully verified by the absorption spectrum for trapped Ni^+ ions shown in black in Fig. 10.18b recorded by Hirsch et al. [56]. In contrast, the Ni metal spectrum shown in red exhibits a pronounced L_2 resonance. As discussed in Sect. 10.5.3, the formation of valence bands in Ni metal involves an energy of $\simeq$3 eV which is considerably larger than the 3d spin-orbit splitting of $\simeq$0.15 eV. Hence the ground state in Ni metal contains both $^2D_{5/2}$ and $^2D_{3/2}$ states.

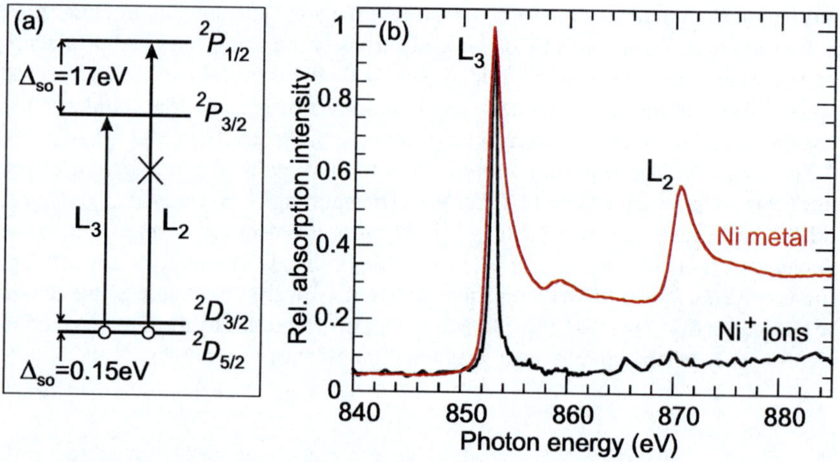

Fig. 10.18 **a** L-edge transitions between initial and final states for a Ni^+ $2p^6\,3d^9\,4s^0$ ground state configuration. Both initial and final states are spin-orbit split and at room temperature only the lowest $^2D_{5/2}$ state is occupied, so that the $^2D_{5/2} \rightarrow {}^2P_{1/2}$ (L_2 edge) is dipole forbidden. **b** Absorption spectrum (black) for the case in (**a**) recorded with trapped Ni^+ ions [56] (and K. Hirsch, private communication) shifted by +3.5 eV to match the peak position of the Ni metal spectrum shown in red [74]. In the latter the L_2 edge is observed due to band mixing of the $^2D_{5/2}$ and $^2D_{3/2}$ initial states

10.6.1.2 Ground State d^8

As expected, for the one-hole case, the multiplet theory gives the same result as the one-hole theory. The difference between the two treatment emerges when considering the case of two valence holes, i.e. the case of the ground electron state $2p^6\,3d^8$. The final state then has the $2p^5\,3d^9$ electron or $2p\,3d$ hole configuration. The hole picture is thus considerably simpler, and for the $3d^2 \rightarrow 2p\,3d$ hole transition we need to take into account the coupling or correlation effects between two holes in both the ground and excited states.

For our discussion we follow Fig. 10.19 and first consider the ground state with two $3d$ holes. In LS coupling we have five possible states with term values $\{^3F,\ ^3P,\ ^1G,\ ^1D,\ ^1S\}$ [33]. According to Hund's rules, the ground state is ^3F.

> Hund's rules [75] state as follows. For a given set of possible multiplets, the ground state is characterized by, (1) the maximum value of the total spin S allowed by the exclusion principle, and (2) the maximum value of the orbital angular momentum L consistent with the value of S. (3) The value of the total angular momentum J for the lowest state is equal to $|L - S|$ when the shell is less than half full and $L + S$ when the shell is more than half full.
> When the shell is half full, the application of the first rule gives $L = 0$, so that $J = S$. Also, for a set of possible multiplets, the term with the highest in S, and among those the highest in L, has the lowest energy. Generally, states of higher multiplicity have lower energies.

Similarly, the possible LS combinations for the final states (one d hole and one p hole) are $\{^3F,\ ^3D,\ ^3P,\ ^1F,\ ^1D,\ ^1P\}$. The LS coupling terms for the ground and excited states resulting from the Coulomb and exchange interactions are illustrated in Fig. 10.19, which also gives their energetic separations calculated by Hartree–Fock theory [33, 59, 62, 67].

If the s–o coupling is weak relative to the Coulomb and exchange interaction, it leads to states $^{2S+1}L_J$ with $L - S \leq J \leq L + S$ and the allowed J states are split according to

$$\mathcal{E}_{\rm so}(J) = \frac{\xi_{nl}}{2}\,[J(J+1) - L(L+1) - S(S+1)]\,. \tag{10.78}$$

This leads to the splitting of the ground states illustrated in Fig. 10.19. At room temperature only the lowest state $^3F_4$ is populated, reducing the number of allowed transitions.

The effect of the s–o interaction on the excited states is more complicated since the $2p$ s–o interaction is larger ($\xi_{2p} \simeq 10\,$eV in Table 9.1) than the exchange and Coulomb interaction ($\simeq 5\,$eV) and mixes different LS terms, so that L and S cease to be good quantum numbers. This is illustrated in Fig. 10.19, where the final states that

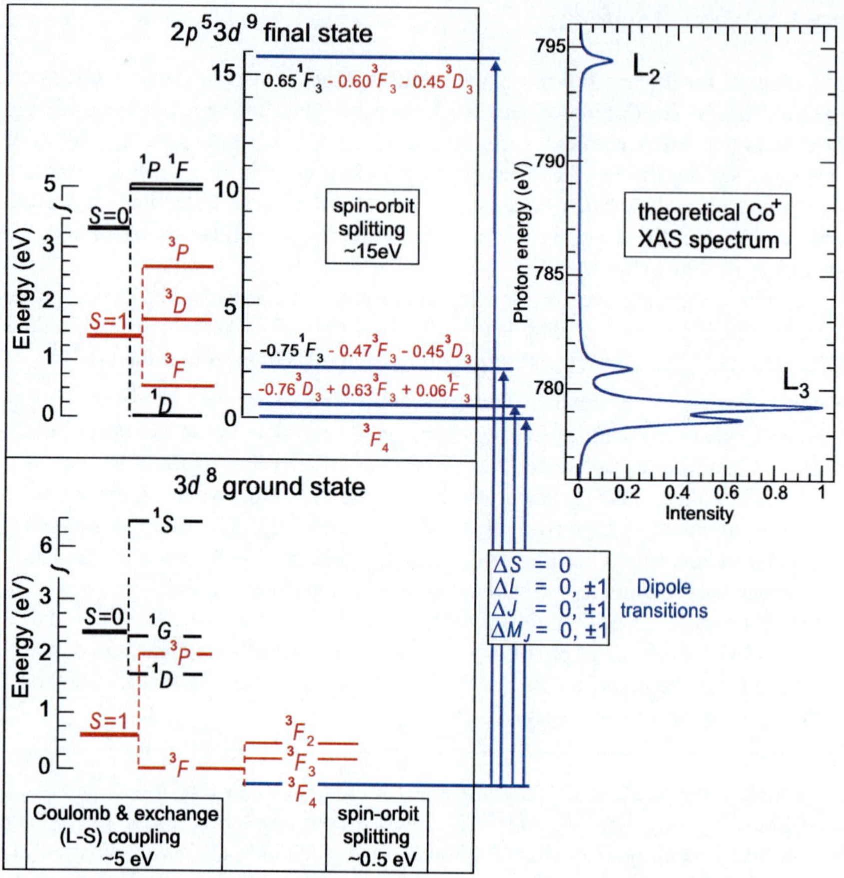

Fig. 10.19 Multiplet terms and splitting for a transition metal ion with a $2p^6\,3d^8$ electronic ground state and a $2p^5\,d^9$ final state. The level splitting evolves from left to right through the Coulomb and exchange interaction, yielding the terms ^{2S+1}L, which in the presence of s–o interaction results in terms $^{2S+1}L_J$. In the ground state J remains a good quantum number because the spin-orbit interaction in the $3d$ shell is small, while the large s–o interaction in the $2p$ core shell mixes $^{2S+1}L_J$ terms [33]. The mixed final states that can be reached by dipole transitions from the $^3F_4$ ground state are listed (F. de Groot, private communication). The resulting transitions and the calculated absorption spectrum are shown on the right

can be reached by dipole transitions from the $^3F_4$ ground state are listed. They consist of linear combinations of different LS-terms of the same J.[12] The transitions allowed by the dipole selection rule are indicated in blue, and the calculated absorption

[12] The simplest s–o coupling of two angular momenta is the one-electron case discussed in the previous section, resulting in the splitting of the angular momentum $l = 1$ through the spin angular momentum $s = \pm 1/2$ into $j = l + s$, ($p_{3/2}$) and $j = l - s$, ($p_{1/2}$) states. The resulting m_j spin-orbit substates can then be written as linear combinations of the angular (spherical harmonics) and

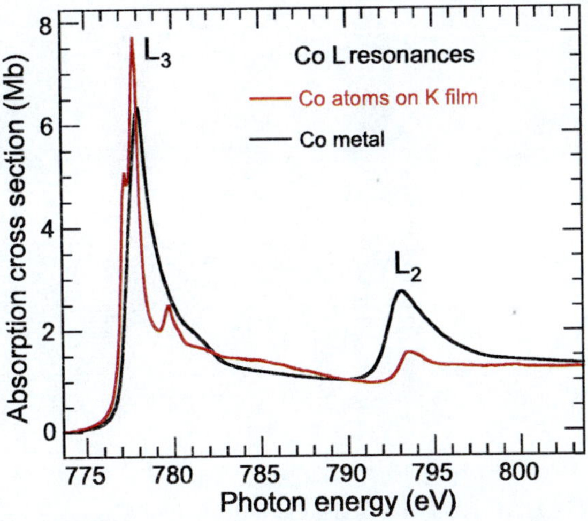

Fig. 10.20 Evolution of Co L-shell resonances from Co atoms deposited on a K film in a d^8 configuration to those in Co metal with an approximate $d^{7.5}$ configuration where the multiplet fine structure is no longer observed due to band formation [77] (and P. Gambardella private comm.)

spectrum is shown on the right. The theory beautifully explains the spectra observed for individual trapped Co^+ ions [56] and Co atoms isolated as impurities on alkali films [76, 77]. This is illustrated in Fig. 10.20, where we show the difference in the L-edge resonances observed for Co atoms on a K film (red) and those in bulk Co metal (black) [77].

10.6.1.3 Evolution from LS to jj Coupling for d^8

The complete transition from LS coupling to jj coupling for the d^8 ground state has been calculated by van der Laan and Thole [62] as shown in Fig. 10.21. As indicated by arrows on the right, the LS coupling spectrum gradually evolves into the jj coupling one with increasing s–o coupling and decreasing Coulomb and exchange coupling.[13]

Here the evolution of the spectrum is shown between two extreme cases. If the Coulomb and exchange interactions are zero, the spectrum (top trace) shows one-electron like behavior with two peaks separated by the $2p$ s–o splitting. In the other extreme of zero $2p$ core s–o splitting but strong Coulomb and exchange splitting, the spectrum (bottom trace) is that calculated in pure L–S coupling. It consists of two peaks, corresponding to the dipole allowed $^3F \rightarrow {}^3F$ and $^3F \rightarrow {}^3D$ transitions. In the intermediate region, where both the $2p$ core spin-orbit and $2p$–$3d$ Coulomb and exchange splitting are present, a more complicated L-edge spectrum is found [59, 62].

---

the spin states, with the expansion coefficients given by the Clebsch–Gordon or Wigner $3j$ coupling coefficients given in Table 10.3.

[13] For a given electronic configuration, the states for each J block can be transformed between LS and jj coupling through $9j$ recoupling coefficients (see (10.5) in [78]).

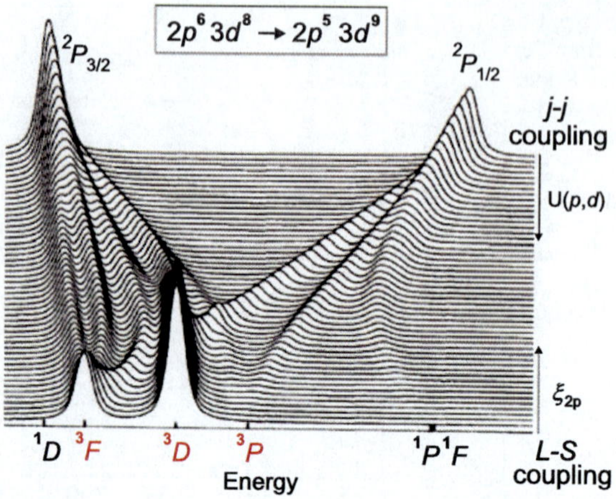

Fig. 10.21 Calculated L-edge spectra for $2p^6\,3d^8$ ground state configuration with a ${}^3F_4$ ground state and a $2p^5\,3d^9$ electronic final state configuration [59, 62] as shown in Fig. 10.19. The horizontal scale corresponds to excitation energy, while the $2p$–$3d$ Coulomb and exchange interaction $U(p,d)$ and the $2p$ spin-orbit coupling ξ_{2p} are varied vertically. The top trace corresponds to pure $2p$ spin-orbit coupling ($U(p,d)=0$) and the lowest trace to pure L–S coupling ($\xi_{2p}=0$)

The above discussion has been limited to spectra of atoms not subjected to any anisotropic electric or magnetic interactions. In the absence of an exchange or external magnetic field, the degeneracy of the ${}^3F_4$ ground state in the M_J substates $-J \leq M_J \leq J$ is not lifted and the XAS spectrum is independent of polarization since the ${}^3F_4$ state is spherically symmetric.

In the presence of a strong magnetic field, an unequal thermal population of the M_J substates in the the ${}^3F_4$ groundstate will cause a polarization dependence due to the x-ray magnetic circular (XMCD) and linear (XMLD) dichroism effects. In the presence of (electrostatic) ligand fields of lower than cubic symmetry, the valence *charge* density of the ground state will become anisotropic, leading to a polarization dependence due to the x-ray natural linear dichroism (XNLD) effect. These effects will be discussed in the following chapter.

References

1. D.C. Koningsberger, E.R. Prins, *X-Ray Absorption: Principles, Applications, Techniques of EXAFS, SEXAFS and XANES* (Wiley, New York, 1988)
2. J.J. Rehr, R.C. Albers, Rev. Mod. Phys. **72**, 621 (2000)
3. G. Bunker, *Introduction to XAFS: A Practical Guide to X-ray Absorption Fine Structure Spectroscopy* (Cambridge University Press, Cambridge, 2010)
4. J.W. Cooper, Phys. Rev. **128**, 681 (1962)
5. S.T. Manson, J.W. Cooper, Phys. Rev. **165**, 126 (1968)
6. U. Fano, J.W. Cooper, Rev. Mod. Phys. **40**, 441 (1968)

7. H.A. Bethe, E.E. Salpeter, *Quantum Mechanics of One- and Two-Electron Atoms* (Plenum, New York, 1977)
8. J.J. Yeh, I. Lindau, At. Data Nucl. Data Tables **32**, 1 (1985)
9. R.D. Cowan, *The Theory of Atomic Structure and Spectra* (University of California Press, Berkeley, 1981)
10. S. Blundell, *Magnetism in Condensed Matter* (Oxford Univ. Press, New York, 2001)
11. B.W. Shore, D.H. Menzel, *Principles of Atomic Spectra* (Wiley, New York, 1968)
12. J.J. Sakurai, *Modern Quantum Mechanics*, Revised. (Addison-Wesley, Reading, Mass., 1994)
13. E.U. Condon, G.H. Shortley, *The Theory of Atomic Spectra* (Cambridge University Press, Cambridge, 1963)
14. J.C. Slater, *Quantum Theory of Atomic Structure*, vol. II (McGraw-Hill, New York, 1960)
15. B.L. Henke, E.M. Gullikson, J.C. Davis, At. Data Nucl. Data Tables **54**, 181 (1993)
16. https://henke.lbl.gov/optical_constants/
17. H. Ebert, J. Stöhr, S.S.P. Parkin, M. Samant, A. Nilsson, Phys. Rev. B **53**, 16–067 (1996)
18. A.C. Thompson et al.: *X-Ray Data Booklet, Third Edition* (Available at: http://xdb.lbl.gov, LBNL, Berkeley, 2009)
19. M.O. Krause, J. Phys. Chem. Ref. Data **8**, 307 (1979)
20. J.L. Campbell, T. Papp, At. Data and Nucl. Data Tables **77**(1), 1 (2001)
21. K. Prince, M. Vondráček, J. Karvonen, M. Coreno, R. Camilloni, L. Avaldi, M. de Simone, J. Elec. Spectrosc. Relat. Phenom. **101–103**, 141 (1999)
22. P.A. Brühwiler, O. Karis, N. Mårtensson, Rev. Mod. Phys. **74**, 703 (2002)
23. M.O. Krause, J.H. Oliver, J. Phys. Chem. Ref. Data **8**, 329 (1979)
24. J.P. Hannon, G.T. Trammell: Hyperfine Int. **123/124**, 127 (1999)
25. R. Röhlsberger: Fortschr. Phys. p. 1 (2012)
26. V. Weisskopf, E. Wigner, Z. Physik **63**, 54 (1930)
27. P.A.M. Dirac, *Quantum Mechanics*, 4th edn. (Oxford University Press, Oxford, 1958)
28. M.O. Scully, M.S. Zubairy, *Quantum Optics* (Cambridge University Press, Cambridge, 1997)
29. G. Breit, E. Wigner, Phys. Rev. **49**, 519 (1936)
30. D.H. Pearson, C.C. Ahn, B. Fultz, Phys. Rev. B **47**, 8471 (1993)
31. J. Stöhr, H.C. Siegmann, *Magnetism: From Fundamentals to Nanoscale Dynamics* (Springer, Heidelberg, 2006)
32. J. Stöhr, *NEXAFS Spectroscopy* (Springer, Heidelberg, 1992)
33. F. de Groot, A. Kotani, *Core Level Spectroscopy of Solids* (CRC Press, Boca Raton, USA, 2008)
34. S.P. Cramer, *X-Ray Spectroscopy with Synchrotron Radiation: Fundamentals and Applications* (Springer, Berlin, 2020)
35. A. Müller et al., Astrophys. Journal **836**, 166 (2017)
36. S. Svensson, E.B.N. Mårtensson, P.Å. Malmquist, U. Gelius, K. Siegbahn, Physica Scripta **14**, 141 (1976)
37. H. Ågren, J. Nordgren, L. Selanger, C. Nordling, K. Siegbahn, J. Elec. Spectrosc. Relat. Phenom. **14**, 27 (1978)
38. R.F. Egerton, *Electron Energy-Loss Spectroscopy in the Electron Microscope*, 3rd edn. (Springer, Heidelberg, 2011)
39. G.C. King, F.H. Read, M. Tronc, Chem. Phys. Lett. **52**, 50 (1977)
40. A.P. Hitchcock, C.E. Brion, J. Elec. Spectrosc. Relat. Phenom. **18**, 1 (1980)
41. C.T. Chen, Y. Ma, F. Sette, Phys. Rev. B **40**, 6737 (1989)
42. K. Prince et al., J. Synchrotron Rad. **5**, 565 (1998)
43. L. Kjellsson, V. Ekholm, M. Agåker, C. Såthe, A. Pietzsch, H.O. Karlsson, N. Jaouen, A. Nicolaou, M. Guarise, C. Hague, J. Lüning, S. Chiuzbăian, J.E. Rubensson, Phys. Rev. A **103**, 022–812 (2021)
44. L. Kjellsson, *X-ray spectroscopy on diatomic and cationic molecules*, Ph.D. thesis, Department of Physics and Astronomy, Uppsala University (2021)
45. P.M. Morse, Phys. Rev. **34**, 57 (1929)
46. E.U. Condon, Phys. Rev. **32**, 858 (1928)

47. M.A. Leutenegger et al., Phys. Rev. Lett. **125**, 243–001 (2020)
48. M. Larsson, P. Baltzer, S. Svensson, B. Wannberg, N. Mårtensson, A.N. de Brito, N. Correia, M.P. Keanej, M. Carlsson-Göthe, L. Karlsson, J. Phys. B **23**, 1175 (1990)
49. V.N. Strocov, T. Schmitt, U. Flechsig, T. Schmidt, A. Imhof, Q. Chen, J. Raabe, R. Betemps, D. Zimoch, J. Krempasky, X. Wang, M. Grioni, A. Piazzalunga, L. Patthey, J. Synchrotron Rad. **17**, 631 (2010)
50. F. Siegrist et al., Nature **571**, 240 (2019)
51. M. Hofherr et al., Sci. Adv. **6**, 8717 (2020)
52. E.J. McGuire, Phys. Rev. A **5**, 1043 (1972)
53. W. Bambynek, B. Crasemann, R.W. Fink, H.U. Freund, H. Mark, C.D. Swift, R.E. Price, P.V. Rao, Rev. Mod. Phys. **44**, 716 (1972)
54. M. Martins, K. Godehusen, T. Richter, P. Wernet, P. Zimmermann, J. Phys. B: At. Mol. Opt. Phys. **39**, R79 (2006)
55. J.E. Hansen, H. Kjeldsen, F. Folkmann, M. Martins, J.B. West, J. Phys. B: At. Mol. Opt. Phys. **40**, 293 (2007)
56. K. Hirsch, V. Zamudio-Bayer, F. Ameseder, A. Langenberg, J. Rittmann, M. Vogel, T. Möoller, B. v. Issendorff, J.T. Lau: Phys. Rev. A **85**, 062–501 (2012)
57. A. Nilsson, N.Mårtensson: Physica B **208/209**, 19 (1995)
58. A. Nilsson, J. Stöhr, T. Wiell, M. Aldén, P. Bennich, N. Wassdahl, M. Samant, S.S.P. Parkin, N. Mårtensson, J. Nordgren, B. Johansson, H.L. Skriver, Phys. Rev. B **54**, 2917 (1996)
59. B.T. Thole, G. van der Laan, Phys. Rev. B **38**, 3158 (1988)
60. Andreas Scherz: Ph.D. dissertation Freie Universiät (Berlin, 2004). Available on website http://www-ssrl.slac.stanford.edu/stohr
61. J. Graetz, C.C. Ahn, H. Ouyang, P. Rez, B. Fultz, Phys. Rev. B **69**, 235–103 (2004)
62. G. van der Laan, B.T. Thole, Phys. Rev. Lett. **60**, 1977 (1988)
63. J. Stöhr, R. Nakajima, IBM J. Res. Develop. **42**, 73 (1998)
64. J. Stöhr, A. Scherz, Phys. Rev. Lett. **115**, 107–402 (2015)
65. R.F. Egerton, Ultramicroscopy **50**, 13 (1993)
66. C. Blancard, D. Cubaynes, S. Guilbaud, J.M. Bizau, Phys. Rev. A **96**, 013–410 (2017)
67. G. van der Laan, B.T. Thole, Phys. Rev. B **43**, 13–401 (1991)
68. G. van der Laan: Hitchhiker's guide to multiplet calculations, in *Magnetism: A Synchrotron Radiation Approach*, ed. by E. Beaurepaire, H. Bulou, F. Scheurer, J.P. Kappler (Springer, Berlin, 2006), p. 143
69. F. de Groot, Coord. Chem. Rev. **249**, 31 (2005)
70. K. Hirsch, J.T. Lau, P.Klar, A. Langenberg, J. Probst, J. Rittmann, M. Vogel, V. Zamudio-Bayer, T. Möller, B. v. Issendorff: J. Phys. B: At. Mol. Opt. Phys **42**, 154–029 (2009)
71. E. P. Wigner, *On the matrices which reduce the Kronecker products of representation of S.R. groups*, ed. by L.C. Biedenharn, H. van Dam. Quantum Theory of Angular Momentum. (Acad. Press, 1965), pp. 87–133
72. G. Racah, Phys. Rev. **62**, 438 (1942)
73. G. Racah, Phys. Rev. **63**, 367 (1943)
74. R. Nakajima, J. Stöhr, Y. Idzerda, Phys. Rev. B **59**, 6421 (1999)
75. F. Hund, Z. Phys. **51**, 759 (1928)
76. P. Gambardella, S.S. Dhesi, S. Gardonio, C. Grazioli, P. Ohresser, C. Carbone, Phys. Rev. Lett. **88**, 047–202 (2002)
77. H. Brune, P. Gambardella, Surf. Sci. **603**, 1812 (2009)
78. K.T. Moore, G. van der Laan, Rev. Mod. Phys. **81**, 235 (2009)

Chapter 11
Quantum Theory of X-Ray Dichroism

11.1 Overview

This chapter is devoted to the quantum formulation of the phenomenon of x-ray dichroism, i.e. the polarization dependence of resonant core to valence transitions. In quantum theory, the four types of polarization dependent absorption effects, XNLD, XMLD, XNCD, and XMCD, whose historical development has been outlined in Sect. 7.9 (see Fig. 7.18), arise from non-spherically symmetric charge or spin densities in the atomic volume. Owing to the tunability of photon energy and polarization at modern x-ray sources, dichroic studies have evolved into important and in many cases unique applications of x-rays. Here we treat this topic with emphasis on fundamental concepts that are illustrated by experimental results chosen to complement those in previous books [1–3].

The breath of dichroism is illustrated by applications ranging from surface and polymer science to spin dependent effects associated with *localized* spins on atoms as well as *delocalized* itinerant spins. XMCD studies of the behavior of itinerant spins are particularly challenging since they involve very small transient (pulsed) signals in nanostructured samples, giving rise to phenomena such as spin exchange scattering underlying giant magnetoresistance (GMR) or the creation of spin polarization through spin-orbit torques in the spin Hall effect (SHE).

11.2 Introduction to the Quantum Theory of Dichroism

Quantum mechanically, the dichroic response arises from asymmetries in the polarization dependent transition matrix elements. The "magnetic" XMLD and XMCD effects are due to the spin-orbit interaction that couples the spin and charge degrees of freedom, while the "natural" XNLD and XNCD effects arise from anisotropies or spatial twists of the atomic charge density induced by bonding.

J. Stöhr, *The Nature of X-Rays and Their Interactions with Matter*, Springer Tracts in Modern Physics 288, https://doi.org/10.1007/978-3-031-20744-0_11

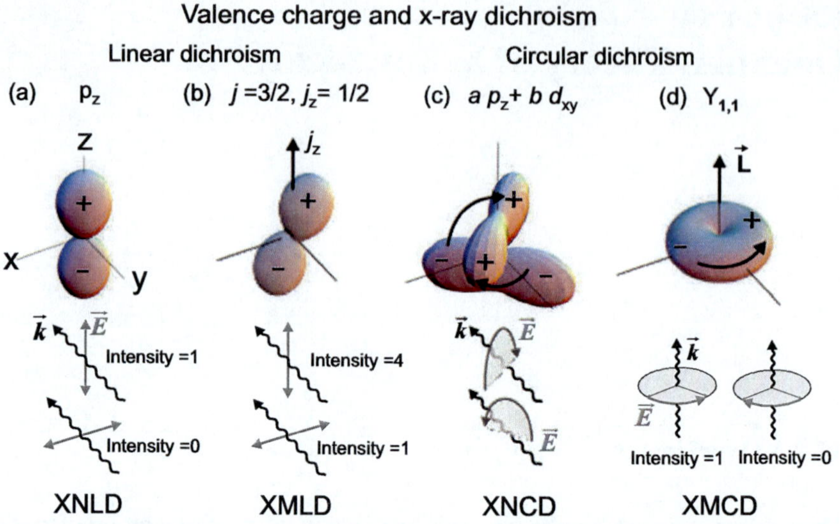

Fig. 11.1 Four important types of dichroism, illustrated for transitions from spherically symmetric *s* core states (not shown) to *p* valence states. The illustration shows specific valence states that give rise to the four prominent types of dichroism indicated in the Figure. The anisotropy of the charge arises either from bonding alone (natural dichroism) or is induced by the existence of spin alignment (magnetic dichroism) in the presence of an external magnetic field or an internal exchange field which is sensed by the photon **E**-field through the spin-orbit coupling of the atomic charge

In a molecular orbital picture, the fundamental origin of the four effects may be envisioned as illustrated in Fig. 11.1 for the simplest case of K-shell transitions from a spherically symmetric *s* core shell to a *p* valence shell.

XNLD measures the spatial anisotropy of the *p* valence shell, illustrated for the p_z orbital in Fig. 11.1a. XMLD, shown in Fig. 11.1b, also measures the spatial anisotropy of the *p* shell but in the presence of spin alignment and the spin-orbit interaction. This is illustrated for a simple $j = 3/2$, $j_z = m_j = 1/2$ state, formed by vector coupling of $l = 1$ and $s = 1/2$, with spin alignment along the z axis (see Table 10.3). Note that for XMLD, the maximum orbital amplitude does not necessarily lie parallel (or perpendicular) to the spin direction.

The origins of the circular dichroism effects are shown in Fig. 11.1c and d, using the distinction of angular momentum and helicity of circularly polarized light in Fig. 3.4. As shown in (c), XNCD may arise from the *spatial handedness* or chirality induced into the *p* valence shell by mixing with *d* orbitals when the ligand field lacks inversion symmetry. Then a p_i orbital can mix with a d_i orbital, as illustrated for the $p_z + d_{xy}$ orbital. The resulting orbital has opposite chirality along the x and y directions and no chirality along z. The transition intensity differs if the *spatial* rotation direction of the **E**-vector for circularly polarized light is the same or opposite to the spatial twist of the chiral valence orbital. The effect does not exist in the dipole approximation which neglects any **k**-dependent *spatial* rotation of the field as it traverses the atomic volume.

In contrast, the XMCD effect arises from the *temporal* rotation of the EM field which gives circular light its angular momentum, as shown in Fig. 11.1d. While the XNCD effect is due to the lack of *spatial inversion symmetry*, XMCD is due to the lack of *time reversal symmetry*. This is illustrated through the $Y_{1,1}$ orbital of a p valence shell which possesses an angular momentum $\hbar$ in the direction shown in Fig. 11.1d. In this case the transition intensity to the $Y_{1,1}$ orbital depends on the relative alignment of angular momentum of the circularly polarized light (rotation in time at the same point), and the angular momentum of the valence charge. The effect is opposite for transitions to the $Y_{1,-1}$ orbital.

The semi-classical and quantum descriptions of light interactions with anisotropic materials give the same answers. The quantum description is more fundamental and, in principle, does not require the use of semi-empirical data obtained by experiment. For example, the semi-classical theory describes charge asymmetries through a dielectric tensor whose components are determined semi-empirically. For x-rays this semi-empirical dielectric tensor is replaced by a quadrupolar tensor that reflects the anisotropy of the valence charge within the atomic volume. The expectation value of the quadrupole operator calculated by use of quantum mechanical wave-functions then directly expresses various dichroic charge effects. This formulation may be extended to include the anisotropy of spin densities through the inclusion of spin-orbit coupling.

In the following sections we illustrate the quantum mechanical calculation of dichroic effects, with emphasis on the XNLD and XMCD effects due to their practical importance in studies of chemical bonding and magnetism.

More generally, we need to denote the polarization dependent dipole operators not just by the polarization label p but an additional Cartesian coordinate label $\alpha = x, y, z$ determined by the sample. For linearly polarized light α specifies the $\mathbf{E}$ direction relative to the charge distribution in the sample and for circularly polarized light indicates the propagation direction $\mathbf{k}$ relative to the magnetization direction of the sample. The completely defined polarization operators are then given by

$$\boxed{P_\alpha^p = \epsilon_p^\alpha \cdot \mathbf{r}}, \tag{11.1}$$

where $p = 0, \pm 1$ (abbreviated $p = 0, +, -$) defines the photon angular momentum and $\alpha = x, y, z$ defines a coordinate direction. The so-defined operators are listed in Table 11.1.

The sums over p in the previously stated sum rules (10.48)–(10.51) and (10.53) specifically assumed $\alpha = z$. More generally, the polarization sums or averages can equivalently be expressed by use of any three orthogonal operators in Table 11.1. For a given direction α there are three possible angular momenta $p = 0, \pm 1$, and for a given p there are three directions $\alpha = x, y, z$.

Table 11.1 Polarization dependent dipole operators P_α^p, expressed in terms of Racah tensors $C_m^{(l)}$ where α indicates the direction of **k** or **E** and $p = 0, +1, -1$ characterizes the photon angular momentum

Linear Polarization			
$\mathbf{E} \parallel x:$	$P_x^0 =$	x	$= r\,\frac{1}{\sqrt{2}}\left[C_{-1}^{(1)} - C_{+1}^{(1)}\right]$
$\mathbf{E} \parallel y:$	$P_y^0 =$	y	$= r\,\frac{\mathrm{i}}{\sqrt{2}}\left[C_{-1}^{(1)} + C_{+1}^{(1)}\right]$
$\mathbf{E} \parallel z:$	$P_z^0 =$	z	$= r\,C_0^{(1)}$
Circular Polarization			
$\mathbf{k} \parallel x:$	$P_x^+ =$	$-\frac{1}{\sqrt{2}}(y + \mathrm{i}z)$	$= -r\,\frac{\mathrm{i}}{\sqrt{2}}C_0^{(1)} - r\,\frac{\mathrm{i}}{2}\left[C_{-1}^{(1)} + C_{+1}^{(1)}\right]$
	$P_x^- =$	$\frac{1}{\sqrt{2}}(y - \mathrm{i}z)$	$= -r\,\frac{\mathrm{i}}{\sqrt{2}}C_0^{(1)} + r\,\frac{\mathrm{i}}{2}\left[C_{-1}^{(1)} + C_{+1}^{(1)}\right]$
$\mathbf{k} \parallel y:$	$P_y^+ =$	$-\frac{1}{\sqrt{2}}(z + \mathrm{i}x)$	$= -r\,\frac{1}{\sqrt{2}}C_0^{(1)} - r\,\frac{\mathrm{i}}{2}\left[C_{-1}^{(1)} - C_{+1}^{(1)}\right]$
	$P_y^- =$	$\frac{1}{\sqrt{2}}(z - \mathrm{i}x)$	$= r\,\frac{1}{\sqrt{2}}C_0^{(1)} - r\,\frac{\mathrm{i}}{2}\left[C_{-1}^{(1)} - C_{+1}^{(1)}\right]$
$\mathbf{k} \parallel z:$	$P_z^+ =$	$-\frac{1}{\sqrt{2}}(x + \mathrm{i}y)$	$= r\,C_{+1}^{(1)}$
	$P_z^- =$	$\frac{1}{\sqrt{2}}(x - \mathrm{i}y)$	$= r\,C_{-1}^{(1)}$

11.3 X-Ray Natural Linear Dichroism—XNLD

We start our discussion of the polarization dependent x-ray absorption intensity with the case of XNLD and assume a non-magnetic system to emphasize the pure charge character of the effect without complications due to spin. In Sect. 7.9.3 we have already mentioned the "*search light effect*" as an easy way to visualize the polarization dependence. This is the most common way to describe the XNLD effect, and we shall therefore start by deriving its physical origin.

The polarization dependent x-ray absorption rate and cross section given by (10.38) are determined by the dipole matrix elements (10.45) involving the polarization operators in Table 11.1. For linear polarization with **E** along $\alpha = x, y$ or z we can write the cross section of a transition from the core shell with angular momenta c, m to an unfilled valence shell with angular momenta L, M. Expressing the matrix elements (10.45) in terms of the linear ($p = 0$) polarization operators P_α^0 ($\alpha = x, y, x$) in Table 11.1, the cross section 10.39 for *linear polarization* takes the form

$$\sigma_{\text{abs}}^{\alpha,0} = \underbrace{\frac{8\pi\hbar\omega\alpha_{\text{f}}}{\Gamma}\frac{(\Gamma/2)^2}{(\hbar\omega - \mathcal{E}_0)^2 + (\Gamma/2)^2}\mathcal{R}^2}_{\mathcal{A}} \underbrace{\sum_{m,M}\left|\langle L, M|\frac{P_\alpha^0}{r}|c, m\rangle\right|^2}_{|\langle b|\epsilon_0^\alpha\cdot\mathbf{r}/r|a\rangle|^2}, \quad (11.2)$$

where the underbrackets identify a polarization *independent* term $\mathcal{A}$ and the polarization dependent dipole matrix element in the form (11.1), where we have used (10.47) to factor the sum and squared matrix element. By use of Table 11.1, this expression may be written as

$$\sigma_{\text{abs}}^{\alpha,0} = \mathcal{A}\sum_{m,M}\left|\left\langle L, M\right|\sum_{p=0,\pm 1} e_\alpha^p C_p^{(1)}\left|c, m\right\rangle\right|^2, \quad (11.3)$$

where the coefficients e_α^p are those in Table 11.1, i.e. $e_x^\pm = \mp 1/\sqrt{2}$, $e_x^0 = 0$, $e_y^\pm = i/\sqrt{2}$, $e_y^0 = 0$, and $e_z^\pm = 0$, $e_z^0 = 1$ and fulfill the sum rule $\sum_p |e_\alpha^p|^2 = 1$.

To picture the polarization dependent transition intensities, we can express the transition probabilities from spherically symmetric core states to the real p and d molecular orbitals, whose wavefunctions are listed in Table 10.4 in terms of their linear combinations of complex spherical harmonics. Their charge densities, corresponding to the absolute values squared of the wavefunctions, are illustrated in Fig. 10.9. Also shown in that figure is how the anisotropic charge densities of the individual p and d orbitals become spherically symmetric upon their combinations. For the d orbitals the t_{2g} and e_g irreducible representations that are energetically separated in cubic symmetry furthermore consist of balanced in-plane $(x-y)$ and out-of-plane (z) charge densities whose average is spherically symmetric.

11.3.1 The Search Light Effect

According to (11.2) the angular dependence of the absorption cross section is determined by the absolute value squared of the core to valence transition matrix elements. For the most important K-edge and L-edge excitations, the transitions of interest are the $1s \rightarrow 2p_i$ and $2p \rightarrow nd_i$ $(n = 3, 4, 5)$ transitions. In both cases the *core* shells are spherically symmetric as illustrated in Fig. 10.9.

The polarization dependence then arises entirely from the anisotropic charge density of the individual p_i and d_i *valence* orbitals. The angular dependence of the transition probabilities, which may be derived from the matrix elements in Fig. 10.6 (also see Fig. A.8), can be pictured along different **E**-vector directions relative to the orientation of the p and d orbital charge densities as shown in Fig. 11.2.

For all $1s \rightarrow 2p_i$, $(i = x, y, z)$ transitions we have a transition intensity 1/3 for $\mathbf{E} \parallel x, y, z$, respectively, and zero otherwise. The polarization averaged intensity per p orbital is $L/(2L+1) = 1/3$, and the total $1s \rightarrow 2p$ intensity is $L = 1$ per spin in agreement with (10.48). For the L-edge $(2p \rightarrow 3d)$ we have the polarization

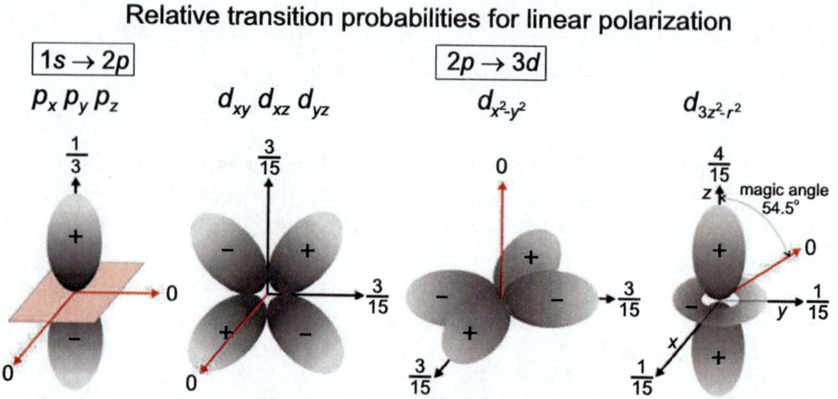

Fig. 11.2 Polarization dependent transition probabilities per spin $|\langle b|\boldsymbol{\epsilon}_0^\alpha \cdot \mathbf{r}/r|a\rangle|^2$ ($\alpha = x$, y or z) defined in (11.2) for the case of p and d orbitals. The listed relative intensities along the three coordinate axes are those for the **E**-vector aligned along the axes. The intensities are proportional to the orbital densities along the **E** direction, expressed by the search light effect discussed in the text. For $1s \to 2p$ K-shell excitation the transition probabilities are zero if **E** lies in the shown nodal *plane*, indicated in red. For the $2p \to d$ L-shell transitions, the excitation probabilities vanish if **E** lies along a nodal direction indicated in red

averaged intensity per d orbital $L/(2L+1) = 2/5$ and the total $2p \to 3d$ intensity is $L = 2$ per spin in agreement with (10.48). The $d_{3z^2-r^2}$ orbital shown on the right has nodal directions at the "magic angle", 54.5° from the z axis. In all cases, the summed polarization dependent transition probabilities in Fig. 11.2 follow the sum rules (10.48)–(10.51).

For the K-edge, we readily recognize the foundation of the *search light effect* [1] which explains the polarization dependence in the following simple picture. The photoelectron is ejected from the spherically symmetric core state along the direction of the **E**-vector. The **E**-vector acts as a "search light" that senses the hole density of the valence orbital. The transition probability scales directly with the orbital density along **E**. For K-shell excitations the transition probability (intensity) vanishes when **E** lies in the *nodal plane*, as indicated in red on the left of Fig. 11.2. For L-edges the transition intensity is zero if the **E**-vector lies along a *nodal axis*, also indicated in red in the figure.

In Fig. 11.3 we illustrate a strong search light polarization effect for two examples. In (a) we show XNLD spectra for oriented benzene (C_6H_6) monolayers on a Ag(110) surface [4], and in (b) for Cu-phthalocyanine (Cu-Pc) monolayers adsorbed on Ag(100) [5, 6]. In both cases the molecules are bonded with their planes parallel to the surface. For benzene the empty π^* carbon orbitals are perpendicular and the σ^* orbitals parallel to the ring plane, while in Cu-Pc the unfilled $d_{x^2-y^2}$ Cu orbital (see Fig. 11.4 below) is parallel to the molecular plane. According to the search light effect, transition intensities are maximized when the **E** polarization vector points along the unfilled valence orbitals. The small residual L_3 intensity in the red spec-

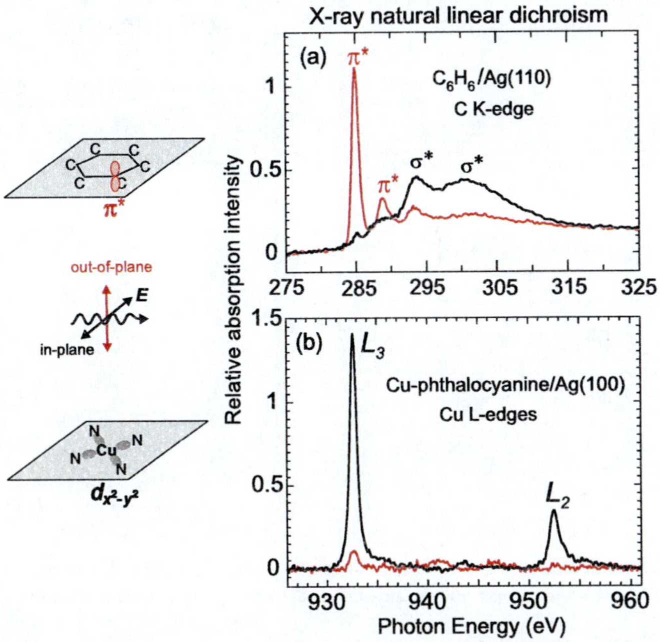

Fig. 11.3 Examples of large contrast XNLD spectra and illustration of the search light effect. **a** K-shell spectra of benzene (C_6H_6) chemisorbed in a lying-down geometry on Ag(110) [4] (also see Fig. 1.19), showing the linear dichroism associated with $1s \rightarrow 2p$ transitions for empty π^* and σ^* orbitals of the C atoms. **b** L-shell spectra of Cu-phthalocyanine also bonded lying down on Ag(110) [5, 6], revealing the strong dichroism of $2p \rightarrow 3d$ transitions to the half filled $d_{x^2-y^2}$ orbital on Cu atoms (see Fig. 11.4 below)

trum in Fig. 11.3b is due to the fact that the **E**-vector was actually at a finite angle of 20° from the normal of the Cu-Pc plane. The XNLD effect is similar to that observed in the layered transition metal compound $La_{1.85}Sr_{0.15}CuO_4$ [7] previously shown in Fig. 7.18.

11.3.2 XNLD and the Quadrupolar Valence Charge Density

The absorption cross section for linear ($p = 0$) polarization (11.3) expresses the dipole matrix elements in the one-electron form, in terms of transitions from core states, $|c, m\rangle$, to valence states $|L, M\rangle$. The anisotropy arises from that of the valence states. One can replace the transition matrix element by an expression for the anisotropic charge density of the valence shell, alone. It is expressed in terms of its deviation from a constant spherically symmetric charge distribution, which corresponds to the quadrupole moment of the charge density. The derivation of this result utilizes

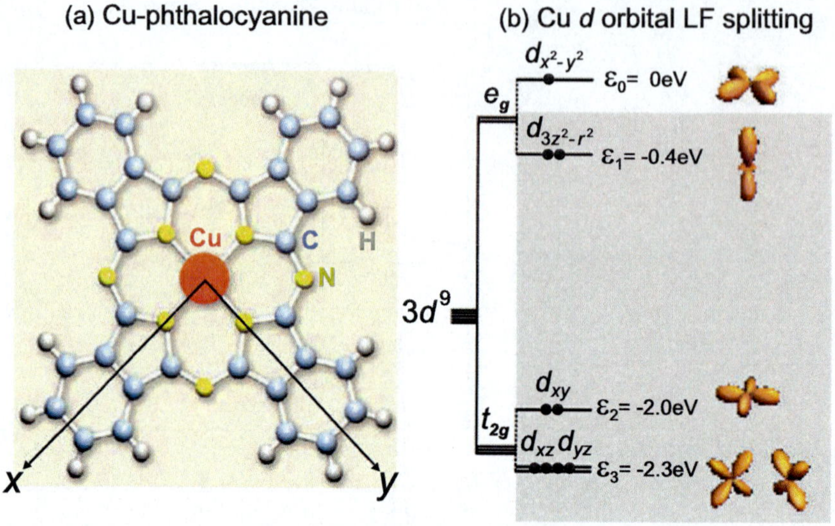

Fig. 11.4 **a** Structural model and coordinate axes of Cu-phthalocyanine. **b** Ligand field splitting of its 3*d* orbitals [5], where the *z* coordinate axis is taken along the normal of the molecular $x-y$ plane in **(a)**

angular momentum sum rules which underlie the derivation of the famous XMCD sum rules [8–13]. The key step in the derivation is to rewrite the squared matrix element so that the final state that contains the core hole can be eliminated by use of quantum mechanical *closure relations* according to

$$\sum_m |\langle L, M|C_p^{(1)}|c, m\rangle|^2 = \sum_m \langle L, M|C_p^{(1)}|c, m\rangle\langle L, M|C_p^{(1)}|c, m\rangle^*$$
$$= (-1)^p \langle L, M|C_p^{(1)} \underbrace{\sum_m |c, m\rangle\langle c, m|}_{\text{projection}} C_{-p}^{(1)}|L, M\rangle. \quad (11.4)$$

The projection summation does not yield unity, but it is related to the unit closure relation by an L-dependent expression, which allows its elimination as discussed in Appendix A.8.3. This leaves matrix elements of a product of first order tensors $C_p^{(1)} C_{-p}^{(1)}$ that connect only valence states. The first order tensor product can be expressed in terms of second order tensors $C_L^{(2)}$ by the relations given in Table A.3, which in turn define the components of the quadrupole tensor of the charge. It has a particularly simple principal axis form, given by,[1]

[1] The diagonal quadrupole tensor corresponds to the *tensor order parameter* or *Saupe matrix* [14–16] discussed in Sect. 7.10.1, utilized in NEXAFS studies of the orientation of nematic liquid crystals and polymer surfaces [17].

$$Q_{xx} = 1 - 3\sin^2\theta\cos^2\phi = C_0^{(2)} - \sqrt{\frac{3}{2}}\left(C_2^{(2)} + C_{-2}^{(2)}\right), \tag{11.5}$$

$$Q_{yy} = 1 - 3\sin^2\theta\sin^2\phi = C_0^{(2)} + \sqrt{\frac{3}{2}}\left(C_2^{(2)} + C_{-2}^{(2)}\right), \tag{11.6}$$

$$Q_{zz} = 1 - 3\cos^2\theta = -2\,C_0^{(2)}, \tag{11.7}$$

with a vanishing trace, $Q_{xx} + Q_{yy} + Q_{zz} = 0$.

After some lengthy algebra one obtains the following alternative expression for the polarization dependent XAS cross section [10, 12, 13]

$$\boxed{\sigma_{\mathrm{abs}}^{\alpha,0} = \mathcal{A}\,\frac{L}{6L+3}\left(1 - \frac{2L+3}{2L}\,\langle\,\phi_L|Q_{\alpha\alpha}|\phi_L\,\rangle\right)}. \tag{11.8}$$

For the K- and L-edges, the valence states $|\phi_L\,\rangle$ with angular momentum L are the p or d orbitals in Table 10.4 or their linear combinations. This leads to the pictorial description shown in Fig. 11.5. The vanishing trace of $Q_{\alpha\alpha}$ is seen to be fulfilled by adding the three values along the orthogonal axes in the figure. An average over α then yields the cross section and matrix elements given by (10.55) and (10.56), and the angle and energy integrated x-ray absorption cross section are determined by the radial transition matrix element $\mathcal{R}_{c\rightarrow L}$ and the number of unoccupied valence states N_h.

The quadrupolar formulation shows that the core hole state is eliminated by the closure relation and the polarization dependence is entirely determined by the charge distribution of the valence orbitals.

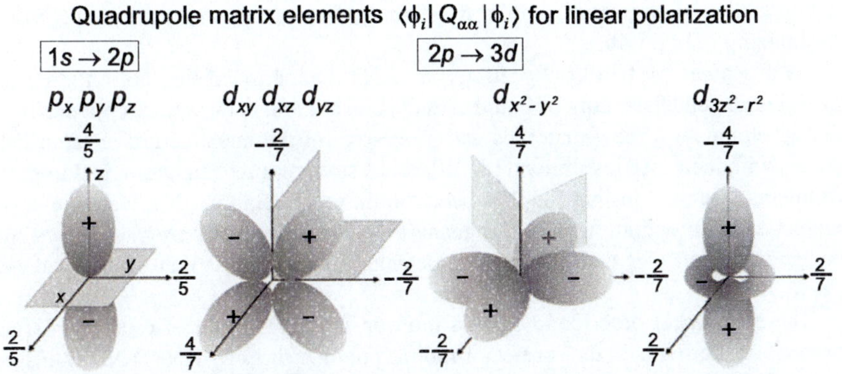

Fig. 11.5 Quadrupole matrix elements $\langle p_z|Q_{\alpha\alpha}|p_z\rangle$, $\langle d_{yz}|Q_{\alpha\alpha}|d_{yz}\rangle$, $\langle d_{x^2-y^2}|Q_{\alpha\alpha}|d_{x^2-y^2}\rangle$, and $\langle d_{3z^2-r^2}|Q_{\alpha\alpha}|d_{3z^2-r^2}\rangle$ for $\alpha \parallel x$, y or z. Also shown are nodal planes of the charge densities. The quadrupole matrix elements for other p and d orbitals follow from symmetry arguments

> The angle and energy integrated x-ray absorption cross section do not depend on the core hole and are directly related to the valence shell properties in the electronic *ground state*.

In particular, we obtain for the K-edge ($L = 1$)

$$\sigma_{\text{abs}}^{\alpha,0} = \mathcal{A}\,\frac{1}{9}\left(1 - \frac{5}{2}\,\langle\phi_{2p}|Q_{\alpha\alpha}|\phi_{2p}\rangle\right), \tag{11.9}$$

where ϕ_{2p} is one of the p_i orbitals in Table 10.4. The L-edge ($L = 2$) intensity is given by

$$\sigma_{\text{abs}}^{\alpha,0} = \mathcal{A}\,\frac{2}{15}\left(1 - \frac{7}{4}\,\langle\phi_{3d}|Q_{\alpha\alpha}|\phi_{3d}\rangle\right), \tag{11.10}$$

where ϕ_{3d} is one of the d_i orbitals in Table 10.4. The quadrupole moments for the representative p_i and d_i orbitals are given in Fig. 11.5. Inserting the values for the matrix elements listed in Fig. 11.5 into (11.9) and (11.10) gives the transition intensities in Fig. 11.2.

11.3.3 Application of XNLD

We have already illustrated several applications of XNLD, such as the determination of the orientation of molecules bonded to surfaces in Fig. 1.19a and the orientation of planes in layered single crystals in Fig. 7.18a. In Sect. 7.10 we discussed the determination of orientational order in polymer films and showed the use of XNLD for imaging in Fig. 8.4b.

At this point we will briefly discuss an XNLD experiment that not only led to the scientific understanding of liquid crystal (LC) alignment on surfaces but enabled the development of new materials and processes for the manufacture of large flat panels with increased resolution [18, 19]. It is a story of direct technological impact following scientific insight that has been told in more detail in [20]. Here we only emphasize a key point, namely that many complex phenomena are caused by tiny statistical imbalances that can only be detected by techniques that offer exquisite sensitivities.

The experiment described below is built on the observation of relatively large orientational order near the surface of rubbed polymer films by XNLD NEXAFS as shown in Fig. 7.19. As mentioned in that context, the low technology rubbing process was originally used to anchor liquid crystals sandwiched between two polymer films rubbed in perpendicular directions, as illustrated in Fig. 11.6a. The understanding of the rubbing-induced preferential bond orientation [17] led to the notion that such orientation may also be induced in other covalently bonded materials such as hydrogenated "diamond-like" carbon films, by replacing the rubbing process by directional ion beam irradiation. Rather than mechanically pulling polymer chains into the rubbing direction, an ion beam could simply be imagined to preferentially break bonds oriented perpendicular to its impact direction over those oriented along its direction.

That this is indeed the case was demonstrated by the XNLD experiment illustrated in Fig. 11.6b–c. Angle dependent NEXAFS spectra for an ion beam irradiated diamond-like carbon film revealed a tiny asymmetry in the orientation of the C=C π orbitals (oriented perpendicular to the C=C internuclear axis). As illustrated in Fig. 11.6b–c, the searchlight effect reveals that π orbitals are aligned slightly more along the directions of **E** indicated in red. There is a preferential in-plane and out-of-plane orientation of the π bonds revealed by their projections onto the **E**-vector. This bond asymmetry is induced through bond breaking by the ion beam as schematically illustrated in Fig. 11.6d by considering phenyl ring like structural units. The rings with their plane perpendicular (gray) are more destroyed than the rings with their plane parallel (red) to the ion beam direction. The relatively small statistical asymmetry of bonding is sufficient to act as a template for the alignment of LCs placed on the surface [19].

11.4 X-Ray Natural Circular Dichroism—XNCD

As mentioned before, XNCD is a higher order effect that requires inclusion of the **k** dependence in the transition matrix element and therefore goes beyond the dipole approximation. Keeping the lowest order **k** term in the transition matrix element in (10.3) we obtain

$$\begin{aligned} M &= \langle b|\,(\mathbf{r}\cdot\boldsymbol{\epsilon}_p)\,(1+\mathrm{i}\mathbf{k}\cdot\mathbf{r})\,|a\rangle \\ &= \underbrace{-\frac{1}{e}\,\langle b|\,\boldsymbol{\mathcal{D}}\cdot\boldsymbol{\epsilon}_p\,|a\rangle}_{\text{electric dipole}} \\ &\quad \underbrace{-\frac{1}{e\,c\,\mu_0}\,\langle b|\boldsymbol{\mathcal{M}}\cdot(\mathbf{k}_0\times\boldsymbol{\epsilon}_p)|a\rangle}_{\text{magnetic dipole}} \\ &\quad \underbrace{+\mathrm{i}\frac{\omega}{6ec}\big\langle b\,\big|\mathbf{k}_0\cdot\boldsymbol{\mathcal{Q}}\cdot\boldsymbol{\epsilon}_p\,\big|\,a\big\rangle}_{\text{electric quadrupole}}, \end{aligned} \tag{11.11}$$

where $\boldsymbol{\epsilon}_p$ is the unit x-ray polarization vector and $\mathbf{k}_0$ the unit x-ray wavevector. Here

$$\boldsymbol{\mathcal{D}} = -e\,\mathbf{r} \tag{11.12}$$

is the electric dipole operator,

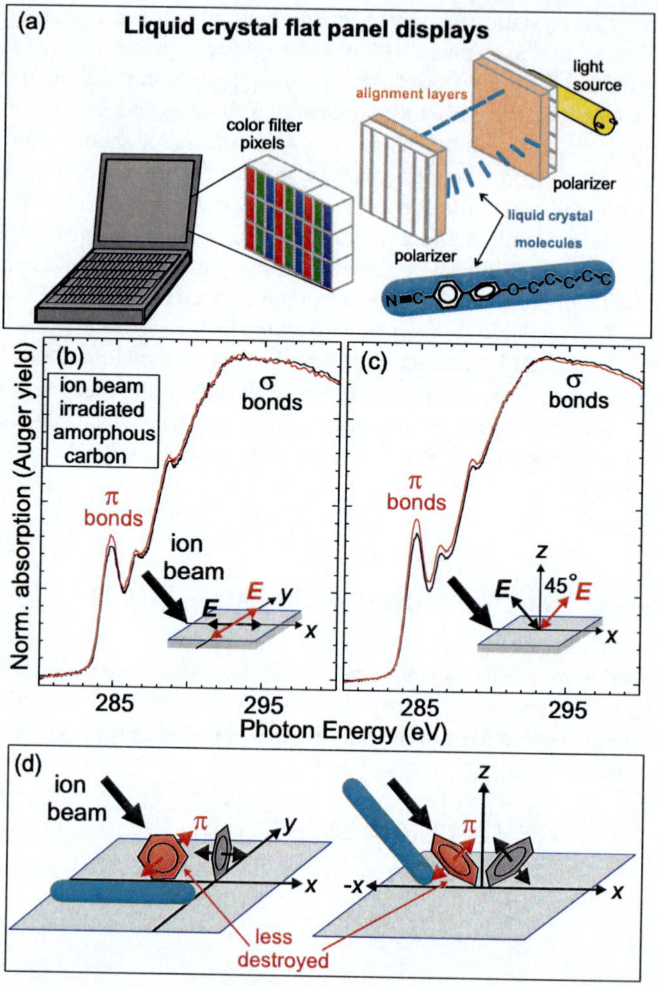

Fig. 11.6 **a** Illustration of some key components of liquid crystal (LC) flat panel displays discussed in the text. **b–c** Angle dependent NEXAFS spectra for a diamond-like carbon film, after irradiation with a directional ion beam, revealing preferential bond alignment [18]. In particular, the π orbitals of C=C double bonds are seen to be aligned slightly more along the direction of **E** indicated in red. **d** Asymmetry induced through bond breaking by the ion beam, illustrated through the preferential orientation of planes (σ bonds) and their normals (π bonds) in structural units represented by phenyl rings. The rings with their plane perpendicular (gray) are more destroyed than the rings with their plane parallel (red) to the ion beam direction

$$\mathcal{M} = -\frac{e\mu_0}{2m_e}(\mathbf{L} + 2\mathbf{S}) \tag{11.13}$$

the magnetic dipole operator, and

$$\mathcal{Q} = -e\,(3\,\mathbf{r}_\alpha \mathbf{r}_\beta - r^2) \tag{11.14}$$

the electric quadrupole operator of the charge, which differs from $Q_{\alpha\beta}$ in (11.5)–(11.7) by e. To see that the terms in (11.11) indeed have the proper dimension of [length], we rewrite it as follows.

The complete x-ray absorption transition matrix element up to first order in **k** is given by

$$M = \left\langle b \left| \mathbf{r} \cdot \boldsymbol{\epsilon}_p + \frac{1}{2m_e c}(\mathbf{L} + 2\mathbf{S}) \cdot (\mathbf{k}_0 \times \boldsymbol{\epsilon}_p) + \frac{i\omega}{2c}(\mathbf{k}_0 \cdot \mathbf{r})(\mathbf{r} \cdot \boldsymbol{\epsilon}_p) \right| a \right\rangle. \tag{11.15}$$

The three terms have the required dimension of [length]. This is seen from the SI dimensions of the different quantities which according to Appendix A.1 are $\mathbf{r}$ [length], $\mathbf{L}, \mathbf{S}$ [energy×time], $m_e c$ [energy×time/length], and $\omega/c = k$ [1/length]. The angular dependence of the three terms is seen to differ significantly.

The total matrix element is valid for all photon energies and is routinely used in the optical range to account for *optical activity* [16]. The individual processes associated with the three transition matrix elements are illustrated in Fig. 11.7.

In an electric dipole transition, illustrated on the left of Fig. 11.7 for a $p \rightarrow d$ core to valence transition the electron moves from the core shell with principal quantum number n to a valence shell with principal quantum number n' and there is no selection rule linking n and n'.

Magnetic dipole transitions are only allowed within a shell with the same principal quantum number n and are illustrated in the middle of Fig. 11.7 for an *optical* $p_y \rightarrow p_z$ transition within the $2p$ valence orbitals of C, N, or O atoms in molecules. Now the electronic transition corresponds to an electron motion that mimics a rotation, as indicated by the red arrow, similar to electron motion that creates an orbital magnetic moment. In an atomic picture the magnetic dipole operator does not depend on r and the radial integral vanishes for $n \neq n'$. Hence in the x-ray region, magnetic dipole transitions alone have a very small probability that only arises through core-valence orbital mixing in higher order, e.g. of the $1s$ core and $2p$ valence orbitals.

Finally, electric quadrupole transitions are not directional and electron excitations can occur into different directions.[2] We show the case for a $s \rightarrow d_{yz}$ excitation because d orbitals have a natural quadrupolar shape. The importance of quadrupole

[2] Quadrupole transitions are related to light that carries multiple units of $\hbar$ angular momentum, so-called orbital angular momentum beams with helical wavefronts. In quantum optics, such beams

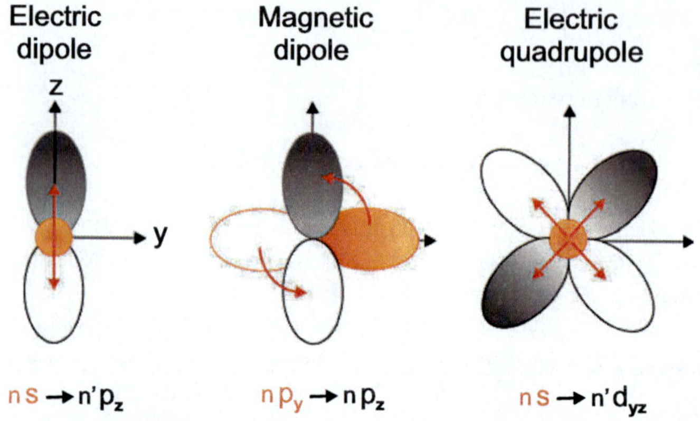

Fig. 11.7 Schematic illustration of electric dipole transitions, magnetic dipole transitions, and electric quadrupole transitions. In all cases the excitation is from the orange colored orbital to the black and white orbital. The electron excitation path is indicated by the red arrows and reflects the nature of the transition operators, i.e. an electric dipole, a magnetic dipole, and an electric quadrupole. The shading of orbitals reflects the nodal pattern, i.e. the plus and minus signs of the angular part of the wavefunctions

transitions rapidly increases from the visible to the soft x-ray region. It further increases from the soft x-ray to the hard x-ray region, but the increase is somewhat diminished by the shrinking radius of the core shells.

11.4.1 The Two Types of XNCD

When the absorption cross section defined by the underbracket in (10.3) is calculated by forming the absolute value squared of the total matrix element (11.15), both the pure magnetic dipole and quadrupole terms will be small. Their *cross-terms* with the large electric dipole matrix element will then dominate the non-dipolar cross section. In the calculation of XNCD, the cross-terms involving the three operators $\mathbf{r} \cdot \boldsymbol{\epsilon}_p$, $\mathbf{k}_0 \times \boldsymbol{\epsilon}_p$, and $\mathbf{k}_0 \cdot \mathbf{r}$ are expressed in spherical tensors [22] (see Appendix A.5). While the electric dipole operator has odd parity, the quadrupole and magnetic dipole operators are of even parity. The two *odd* parity cross-terms with the electric dipole operator make x-ray absorption sensitive to charge chirality or handedness which is also of odd parity. This gives rise to two types of XNCD effects that may be observed.

The first XNCD effect originates from the interference of the electric dipole/ magnetic dipole terms, and it has been studied in [23–27]. It exists even for randomly oriented samples like chiral molecules. The development of molecular chirality is

are distinguished from circularly polarized beams, which are said to have only a "spin angular momentum" [21].

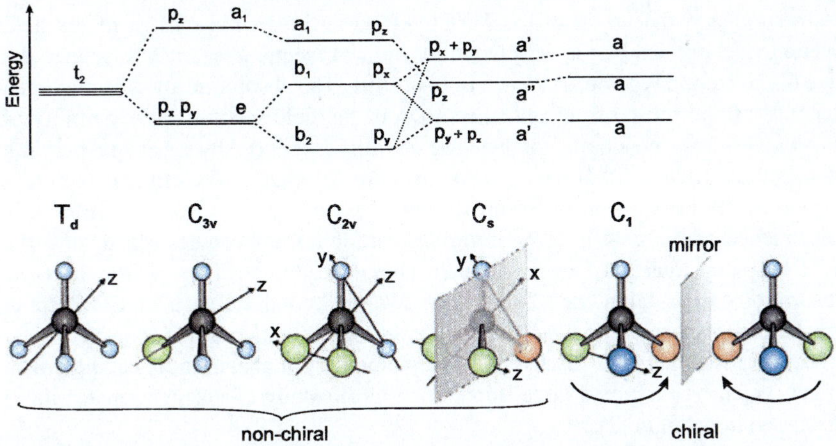

Fig. 11.8 Evolution of a molecule of the form AB_4 and tetrahedral symmetry to lower symmetry by substitution of atoms. As the symmetry is lowered the ligand field Hamiltonian becomes increasingly complicated. Assuming that the central atom (e.g. C) possesses *p* valence electrons we also show the evolution of the ligand field induced splitting of the *p*-orbitals. When the symmetry is lower than C_{2v}, the p_x, p_y, and p_z orbitals are no longer good eigenfunctions and can mix. When the symmetry is lowered to C_1, the last intramolecular mirror plane is removed. Then the molecule and its mirror image are no longer superimposable, as shown, and the molecule has become chiral

illustrated Fig. 11.8 by the development out of the tetrahedral structure associated with the chemical formula AB_4, a well-known example being methane CH_4. Successive substitution of the B atoms leads to a sequence of structures with increasingly lower symmetry, until finally, the last mirror plane still present in C_s symmetry has disappeared and we obtain the shown chiral structure of C_1 symmetry. The molecule and its mirror image are no longer superimposable, and the molecule has become chiral. An example is bromochlorofluoromethane, CHFClBr, where three of the hydrogen ligands in CH_4 have been replaced by different halogens.

Intuitively we associate the concept of *chirality* with the concept of *helicity* because it is easy to visualize the difference between a right-handed and a left-handed helix. For example, proteins contain almost entirely "left-handed" amino acids, while nucleic acids, starch, glycogen, etc., contain sugars that are all "right handed". However, a molecule does not have to exhibit a complete helical turn to be chiral and may simply lack a reflection plane as shown in Fig. 11.8 or lack an inversion center.

The second type of XNCD arises from the interference of the electric dipole/electric quadrupole channels, and it has been thoroughly investigated in [22, 28–33]. It is determined by the symmetry of the quadrupole tensor which in its diagonal form has a vanishing trace. This leads to a vanishing angle integrated response, and this type of XNCD effect requires oriented samples, as illustrated for single crystals of $LiIO_3$ in Fig. 7.18.

In summary, the origin of the XNCD effect is the *spatial* change of the field over a cycle, defined by the wavelength or the wavevector $k = 2\pi/\lambda$ relative to the size of the atomic volume, as shown in Fig. 11.9a. The absorption cross section then depends on the sense of the spatial rotation of the field relative to the chirality of the charge across the atomic volume. The chirality is induced by the spatial bonding to neighbors (ligand field) which lacks *inversion symmetry*. The effect is therefore related to the basic symmetry property called *parity*. If the spatial variation or **k** dependence of the field is taken into consideration, one becomes sensitive to the relative spatial "twists" of the field and the charge, leading to differences in the x-ray absorption cross section, the XNCD effect. We shall not discuss the XNCD effect in greater depth since, in practice, the calculations are rather challenging.

In the following will discuss the complementary magnetic circular dichroism effect which has assumed large importance in the study of magnetic materials as discussed in detail in [2].

11.5 X-Ray Magnetic Circular Dichroism—XMCD

The key difference between the *charge* related XNCD and *spin* dependent XMCD effect is illustrated in Fig. 11.9. In both cases one utilizes the rotation of the **E**-vector for circularly polarized x-rays, but the effects are differentiated by its motion in space and time as previously illustrated in Fig. 3.4.

The key to the description of the XMCD effect is that it exists even in the dipole approximation. This arises from the fact that the atom can be assumed to be small relative to a field or wave cycle, and the point-like atom still senses the *temporal* rotation of the **E**-vector as the wave passes by. Hence the interaction is sensitive to the *time reversal* properties of the atom, i.e. the direction of the magnetic moments, relative to the time reversal properties of the x-rays, i.e. the direction of the photon angular momentum. XMCD arises indirectly via two mechanisms, the electric field couples to the charge, and the charge couples to the spin via the *spin-orbit coupling*. Since it exists in lowest order where the **k** dependence can be neglected, XMCD can be considerably larger than XNCD.

11.5.1 Key Concepts of Magnetism and Magnetic Alignment

In order to derive the quantum mechanical origin of XMCD, we need to first briefly review basic concepts used to describe the complex phenomenon of "magnetism" [2]. It fundamentally arises from correlation effects between electrons, but owing to their complexity we have developed simpler concepts to describe the properties of magnetic materials based on the notion of atomic magnetic moments. In quantum theory, the moments arise from the intrinsic spin of electrons and orbital moments associated with electron "orbits" in atoms. The essence of magnetism is then expressed

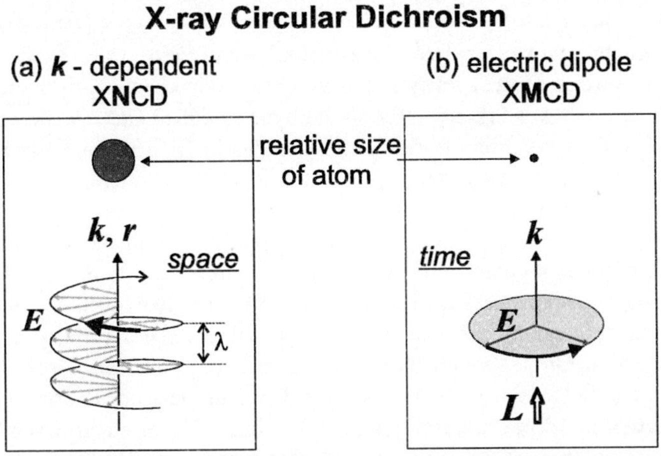

Fig. 11.9 Illustration of the size of the absorbing atom relative to the x-ray wavelength for circular polarized light, where the rotation of the **E** is shown in space and time. Note that the rotation sense of **E** is opposite in space and time. **a** First order **k**-dependent transitions giving rise to XNCD. Here the **k**-dependent twist of the **E** over the atomic volume senses any chirality of charge induced by bonding to neighbors with no inversion symmetry around the absorbing atom. **b** In the dipole approximation the rotation of **E** in time at a given point defines the photon angular momentum **L**. The XMCD effect arises from the relative orientation of **L** and the magnetic moment (angular momentum) of the absorbing atom

through the size and the coupling of these spin and orbital moments. One distinguishes "momenta" expressed in units of $\hbar$ from "magnetic moments" of the same magnitude but opposite sign, given in units of the Bohr magneton μ_B (see Appendix A.1). In brief, one uses the following concepts.

One starts from the premise that the electronic charges and spins form *separate and independent* thermodynamic reservoirs. Within the *charge* reservoir, order and spatial anisotropy may be created by interatomic bonding, typically visualized in terms of molecular orbitals. Within the *spin* reservoir, atomic spins align parallel for ferromagnets and antiparallel for antiferromagnets when the so-called *exchange energy* overcomes the thermal disorder energy. The exchange coupling which keeps spins parallel may be thought of as an extremely strong molecular alignment field of order 10^3-10^4 Tesla that only acts on the spin moments and separates the two spin states by an energy of order of 1 eV. Owing to the large exchange field, magnetism may exist well above room temperature.

In *paramagnets*, magnetic alignment may be created by application of an external field which acts on both spin and orbital moments. In practice, external fields are of order 1–10 T, and a 1 Tesla field only splits the spin states by an energy of about 1 meV. For spin alignment to occur one therefore needs to cool the sample to low temperatures of a few Kelvin to overcome thermal disorder (10 K$\sim$1 meV). In both para- and ferromagnets, a preferred alignment of the spin system relative to the macroscopic shape of the sample may arise from the dipole-dipole interaction of the atomic magnetic moments, referred to as the magnetic *shape anisotropy*.

The separate charge and spin systems can talk to each other only through the *spin-orbit interaction*. It was first observed through the so-called *fine structure* in atomic spectra and the observation of additional splittings in applied magnetic fields, the Zeeman effect. Both were originally interpreted by the Bohr and Sommerfeld schools in terms of orbital substructure, since the concept of the electron spin emerged only later [2]. Today, we know that the fine structure arises from the s–o coupling of the spin and orbital angular momenta, described by the Hamiltonian (9.7). The s–o interaction is envisioned as the vector coupling of a point-like intrinsic electron spin to an angular momentum associated with a spatially extended orbit of an electron around an atom. Orbital moments which point in the direction perpendicular to the orbit plane only exist if there is an imbalance in electron motion in opposite rotation senses, i.e. if time reversal symmetry is broken. Closed electron shells always have zero orbital moments due to their spherical symmetry. There is strict angular momentum conservation between the spin and orbital degrees of freedom.

In molecules and solids, electron orbits and the associated orbital moments may have a preferred direction arising from bonding (charge) anisotropies, so that the orbital moment is "locked" into a preferential direction. This gives rise to the so-called *magnetocrystalline* anisotropy (MCA). It often competes with the shape anisotropy mentioned earlier. The MCA direction may be atomically engineered, and this is utilized in magnetic data storage to orient the magnetic up and down "bits".

In order to understand how x-rays couple to magnetic spin and orbital moments, we also need to briefly introduce the essence of the fundamental magnetic exchange and spin-orbit interactions.

11.5.1.1 The Magnetic Exchange Interaction: Spin Moments

In non-magnetic solids whose valence electrons form an energy band, the band dispersion across the Brilloin zone is characterized by spaghetti-like states whose energy changes as a function of the electron wavevector **k** (also called crystal momentum). When the **k**-dependent states are summed (integrated) over the Brillouin zone, one obtains the electronic density of states (DOS) as illustrated in Fig. 11.10a for the d valence band of transition metals [2].

The filled electron states of total number N_e/atom are separated by the Fermi energy (E_F) from the empty hole states of total number N_h/atom. The sum of the number of electrons and holes per atom of a valence shell of angular momentum L (=2 for the d-shell) must be equal to the total degeneracy of the shell, including spin, given by $N_e + N_h = 2(2L + 1)$. This may be schematically represented by the *Stoner model* [34, 35] that specifically separates spin-up and spin-down electrons, as illustrated Fig. 11.10b. Here "spin-up" and "spin-down" reflect the states of opposite direction relative to a direction in space, the quantization axis. In the Stoner model one typically does not worry about the exact shape of the DOS and we will simply represent it by a box, as previously done in Fig. 10.14.

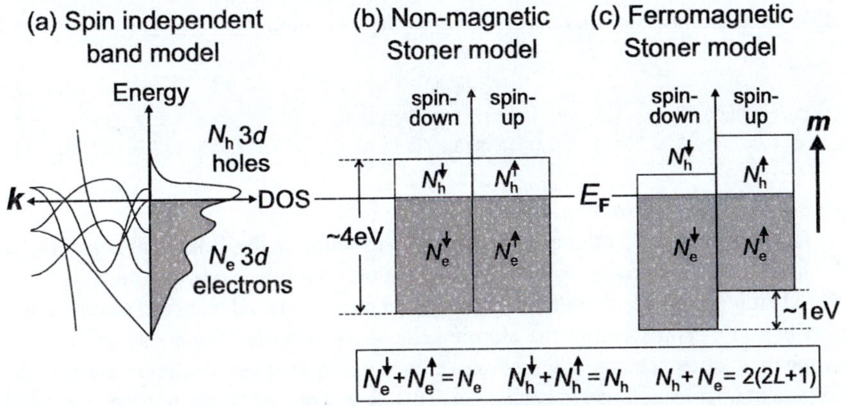

Fig. 11.10 Models for the 3*d* valence band in transition metals. **a** Spin independent band model, where the **k** integrated band structure at each energy gives the density of states (DOS). Each state has a degeneracy of 2 due to spin. The filled electron states of total number N_e/atom are separated by the Fermi energy (E_F) from the empty hole states of total number N_h/atom. The sum $N_e + N_h$ is equal to the total number of spin orbitals of angular momentum $L = 2$ given by $2(2L + 1) = 10$. The band width is approximately 4 eV. **b** Schematic DOS of a *non-magnetic* 3*d* metal in the Stoner model, which separates the DOS into independent equivalent spin-up and spin-down bands. **c** Same for a ferromagnetic 3*d* metal. Now the center of gravity of the spin-up and spin-down bands are shifted relative to each other by the exchange interaction, which is about 1 eV. By convention, the atomic magnetic moments **m** have the opposite directions from the atomic spins ($\mathbf{m} = -2\mathbf{s}$), and hence point into the directions of the spin-up holes

In a non-magnetic material, the separation of up and down spins is artificial since there is no direction relative to which there is a spin imbalance. The total number of filled electron states per atom $N_e = N_e^{\uparrow} + N_e^{\downarrow}$ is then given by an equal number of up and down spins and similarly for the number of empty or hole states $N_h = N_h^{\uparrow} + N_h^{\downarrow}$, as shown in Fig. 11.10b. Hence there is no net difference in electron spins (or hole spins) and no magnetic moment per atom.

In a ferromagnet, the spin-up and spin-down bands are defined relative to a magnetization direction **M** along which the atomic magnetic moments **m** are aligned. The magnetization is the sum of all atomic moments that are kept parallel within a region by the magnetic exchange interaction. A macroscopic magnetic material is typically composed of small "spontaneously magnetized" regions (domains) where the atomic moments are aligned. In order to create a macroscopically magnetized sample, the domains with random orientations need to be aligned by an external field. Within an ordered domain, the exchange interaction causes a relative shift of order 1 eV of the spin-up and spin-down bands, as shown in Fig. 11.10c. In the process, the spin dependent occupation of electron and hole states changes to preserve the total number of electrons per atom.

The beauty of the Stoner model is that one can easily understand and theoretically define some key properties of a magnetic material, arising from the different occupations of the spin-up and-down bands. The total number of empty states or holes,

N_h, and the number of filled electron states, N_e, are related according to

$$N_h = N_h^{\uparrow} + N_h^{\downarrow} \tag{11.16}$$

$$N_e = 2(2L + 1) - N_h. \tag{11.17}$$

These relations are summarized at the bottom of Fig. 11.10.

The spin magnetic moment per atom m_s in units of the Bohr magneton μ_B is defined as the integrated spin density over the atomic volume. In the context of XMCD it is important that the so-defined spin moment is independent of the spatial distribution of spins within the atomic sphere. Its value is exclusively determined by the magnetic exchange interaction which is minimized for parallel or antiparallel alignment of spins relative to each other. This definition implicitly assumes, as stated previously, that the spin and orbital degrees of freedom are decoupled, so that the value and direction of the spin moment are independent of the spin-orbit interaction.

According to the Stoner model, the spin moment is simply the difference in occupation between the two spin states, and it can be expressed in terms of the difference in electrons or holes which enter with opposite signs. In the notation of Fig. 11.10 it is given by

$$m_s = -2\frac{\mu_B}{\hbar}\langle \mathbf{s}_z \rangle = -2\mu_B \left(N_e^{\uparrow} - N_e^{\downarrow}\right) = 2\mu_B \left(N_h^{\uparrow} - N_h^{\downarrow}\right). \tag{11.18}$$

The difference in occupation arises when the expectation value of $\langle \mathbf{s}_z \rangle$ (units $[\hbar]$) is calculated with realistic band states. The spin moment hence reflects the imbalance between the number of electrons or holes in the two spin states. While the theoretical definition of the spin moment is unambiguous, care has to be exercised in the comparison of theoretical values and those determined by different experimental techniques.

11.5.1.2 The Spin-Orbit Interaction: Orbital Moments

As mentioned earlier, without the s–o interaction the charge and spin systems are independent. In analogy to the spin moment defined by (11.18), the orbital moment is given by

$$m_o = -\frac{\mu_B}{\hbar}\langle \mathbf{L}_z \rangle. \tag{11.19}$$

The triangular brackets in $\langle \mathbf{L}_z \rangle$ again denote the quantum mechanical expectation value (units of $\hbar$). For a full or empty shell we have $\langle \mathbf{L}_z \rangle = 0$ since there is a zero balance of states with $\pm L_z = \pm M$. For a single d electron with $L = 2$, the maximum orbital moment is given by the value of the largest orbital quantum number $\langle L_z \rangle = \pm M = \pm 2$, so that $|m_o| = 2\,\mu_B$ as illustrated in in Fig. 11.11a. This value is comparable to the spin moments in Co or Fe $m_s \approx 2\,\mu_B$.

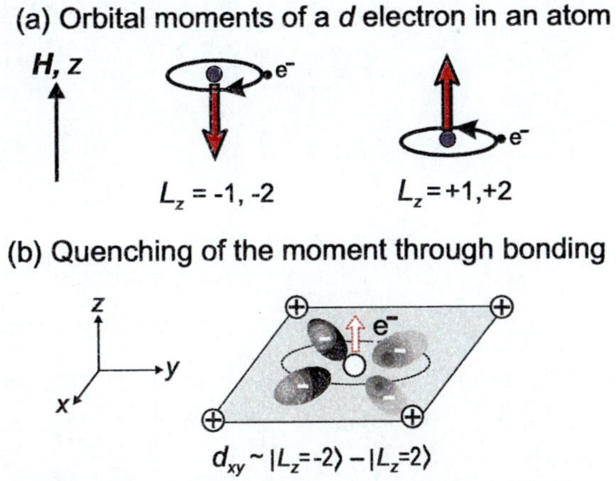

Fig. 11.11 Illustration of orbital moments in atoms and their quenching through the formation of molecular orbitals. **a** Atomic case. **b** Case of the d_{xy} orbital formed in a molecule or solid through bonding. Its preferentially ordered charge density breaks up the electron orbit and quenches the orbital moment

When an atom is embedded into a solid, the understanding of the existence of orbital moments is more complicated. The importance of orbital contributions to the magnetic properties of transition metal ions in insulators was thoroughly studied by means of electron paramagnetic resonance (EPR) in the 1960s as reviewed by Abragam and Bleaney [36]. Orbital contributions also played an important role in unraveling the localized versus delocalized behavior of transition metal impurities in metallic hosts, which in the 1960s and 1970s formed one of the key topics in condensed matter physics, associated with buzz words like Anderson localization and the Kondo effect as reviewed by Hirst [37].

If we picture the bonding between atoms through the formation of molecular orbitals shown in Fig. 10.9, one finds that all individual orbitals have zero orbital moments. This is readily verified by calculating the expectation value $\langle o_i|\mathbf{L}_z|o_i\rangle$ for any of the orbitals in Table 10.4. One may picture the destruction of the atomic orbital moments through bonding (i.e. the formation of molecular orbitals) as illustrated in Fig. 11.11b for the specific case of the d_{xy} orbital and an external **H**-field along z. An orbiting electron on the central atom will experience a Coulomb attraction near the corners of the bonding square where the positive neighbor ions are located, and the orbit is broken up. In textbooks, this is referred as "quenching" of orbital moments.

In solids, the ligand field (LF) splitting of the different orbitals is typically of order of a few eV. If we now turn on the intra-atomic s–o interaction in the valence shell, which for transition metal atoms is of order $\xi_{3d} \simeq 50\,\text{meV}$, one creates small orbital moments due to the mixing of LF orbitals. This is illustrated in Fig. 11.12, where we show the effect of mixing the LF orbitals $d_{xy}^{\uparrow}$ and $d_{x^2-y^2}^{\uparrow}$ by the s–o interaction.

For simplicity, we only show the s–o mixing of orbitals of the same spin ($\uparrow = +s_z$) and ignore the mixing of states of up and down spin, discussed later in Sect. 11.6. The perturbed ligand field orbitals are given by

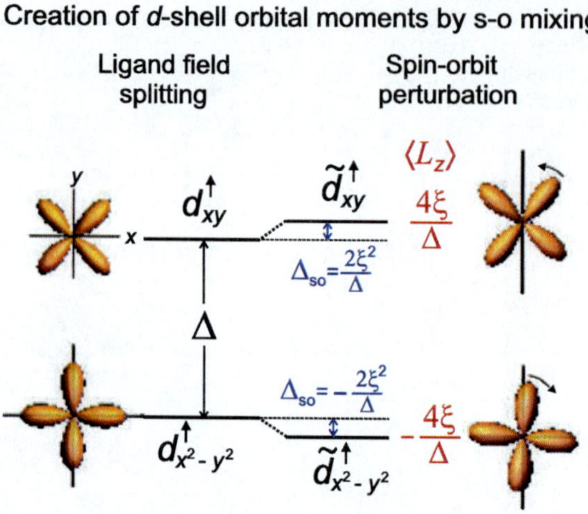

Fig. 11.12 Creation of orbital momenta $\langle \mathbf{L}_z \rangle$ by the spin-orbit (s–o) interaction. We assume spin-up alignment by an external field $\mathbf{H} \parallel z$. The pure spin-up ligand field orbitals $d^{\uparrow}_{xy}$ and $d^{\uparrow}_{x^2-y^2}$ have no orbital momenta, but they are induced in the $\tilde{d}^{\uparrow}_{xy}$ and $\tilde{d}^{\uparrow}_{x^2-y^2}$ orbitals, given by (11.20), through the s–o interaction. The orbitals possess angular momenta $\langle \mathbf{L}_z \rangle$ given by (11.21), as shown. On the right we indicate the spin-orbit-induced orbital rotation in the $x-y$ plane around $\mathbf{H} \parallel z$ by assuming $\xi/\Delta = 0.2$

$$|\tilde{d}^{\uparrow}_{xy}\rangle = |d^{\uparrow}_{xy}\rangle + \frac{\langle d^{\uparrow}_{x^2-y^2}|\xi \mathbf{L}_z \mathbf{s}_z|d^{\uparrow}_{xy}\rangle}{\mathcal{E}_{d_{xy}} - \mathcal{E}_{d_{x^2-y^2}}}\,|d^{\uparrow}_{x^2-y^2}\rangle = |d^{\uparrow}_{xy}\rangle - \mathrm{i}\frac{\xi}{\Delta}\,|d^{\uparrow}_{x^2-y^2}\rangle$$

$$|\tilde{d}^{\uparrow}_{x^2-y^2}\rangle = |d^{\uparrow}_{x^2-y^2}\rangle + \frac{\langle d^{\uparrow}_{xy}|\xi \mathbf{L}_z \mathbf{s}_z|d^{\uparrow}_{x^2-y^2}\rangle}{\mathcal{E}_{d_{x^2-y^2}} - \mathcal{E}_{d_{xy}}}\,|d^{\uparrow}_{xy}\rangle = |d^{\uparrow}_{x^2-y^2}\rangle - \mathrm{i}\frac{\xi}{\Delta}\,|d^{\uparrow}_{xy}\rangle \quad (11.20)$$

where we have used $\mathcal{E}_{d_{xy}} - \mathcal{E}_{d_{x^2-y^2}} = \Delta$, and the spin and angular momentum operator rules in Tables A.6 and A.7, namely $\langle\uparrow|\mathbf{s}_z|\uparrow\rangle = 1/2$, and $\mathbf{L}_z|d_{xy}\rangle = -2\mathrm{i}|d_{x^2-y^2}\rangle$, and $\mathbf{L}_z|d_{x^2-y^2}\rangle = 2\mathrm{i}|d_{xy}\rangle$ to obtain the expressions on the right. The s–o interaction shifts the upper orbital up in energy by $2\xi^2/\Delta$ and the lower orbital down by $-2\xi^2/\Delta$, and the perturbed orbitals have angular momenta $\langle \mathbf{L}_z \rangle$ given by

$$\left\langle \tilde{d}_{xy}\,|\mathbf{L}_z|\,\tilde{d}_{xy}\right\rangle = \frac{4\xi}{\Delta}, \quad \text{and} \quad \left\langle \tilde{d}_{x^2-y^2}\,|\mathbf{L}_z|\,\tilde{d}_{x^2-y^2}\right\rangle = -\frac{4\xi}{\Delta}. \quad (11.21)$$

These are the values listed in Fig. 11.12. In the 3*d* magnetic transition metals Fe, Co and Ni, for example, one finds orbital moments that are only about 5–10% of the size of spin moments [2, 13, 38, 39]. On the right of Fig. 11.12 we also illustrate how one may picture the creation of an orbital moment along *z* as arising from the s–o induced charge rotation in the $x-y$ plane.

In the following we shall show how spin and orbital magnetic moments can be determined by XMCD through measured resonance intensities. The key to the link of the magnetic properties of matter and XMCD intensities are the famous XMCD sum rules which we will discuss next. We start with the sum rule for the determination of the orbital magnetic moment, first derived by Thole et al. [8]. It is more robust and more easily derived than the sum rule associated with the spin magnetic moment [9], discussed later.

11.5.2 XMCD Sum Rule for the Orbital Moment

The XMCD effect only exists indirectly through the coupling of the **E**-vector to the charge which in turn is coupled to the spin through the spin-orbit interaction. We are interested in the determination of the spin and orbital magnetic moments, either of a ferromagnetic sample whose domains are aligned by an external magnetic field (typically 2 T), or a paramagnetic sample where a net spin orientation is induced by application of a large magnetic field (of order of 10 T). In the latter case one furthermore has to assure that the Zeeman splitting by the applied magnetic field is larger than the thermal energy so that such experiments are carried out at low temperature (of order of 1 K) [40].

For our discussion we consider a valence shell of angular momentum L (e.g. $L = 2$ for the $3d$ bands) and consider the band model illustrated in Fig. 11.10c. For convenience, we fix the sample magnetization direction **m** so that the XMCD measurements are performed by switching the incident circular polarization. We consider **k**-summed band states, projected on their spin s_z and orbital M quantum numbers, of the form

$$|\phi_{LM}(\mathcal{E})\rangle = |\phi^{\uparrow}_{LM}(\mathcal{E})\rangle + |\phi^{\downarrow}_{LM}(\mathcal{E})\rangle = c^{\uparrow}_{M}(\mathcal{E})\,|LM\uparrow\rangle + e^{\downarrow}_{M}(\mathcal{E})\,|LM\downarrow\rangle. \quad (11.22)$$

The coefficients $c^{\uparrow}_{M}(\mathcal{E})$ and $e^{\downarrow}_{M}(\mathcal{E})$ deviate from unity through mixing of states by the ligand field and s–o interactions. We have expressed the angular band states in a spherical harmonics basis $|L, M\rangle = |Y_{L,M}\rangle$, but one may also use a molecular orbital (MO) basis since the MOs are also orthogonal. When done so, the band model in the limit of no **k** dispersion (flat bands) and no exchange interaction reduces to a LF model.

The orbital momentum of the valence states at a given energy $\mathcal{E}$ is given by

$$\langle \mathbf{L}_z(\mathcal{E})\rangle = \sum_M \langle \phi_{LM}(\mathcal{E})|\mathbf{L}_z|\phi_{LM}(\mathcal{E})\rangle = \hbar \sum_M \left(|c^{\uparrow}_{M}(\mathcal{E})|^2 + |e^{\downarrow}_{M}(\mathcal{E})|^2\right) M. \quad (11.23)$$

In the absence of the s–o interaction the sum over M vanishes. When integrated over filled electron states or empty hole states, we obtain the total orbital momentum of the valence shell according to

$$\langle \mathbf{L}_z \rangle = - \int_{-\infty}^{E_F} \langle \mathbf{L}_z(\mathcal{E}) \rangle \, d\mathcal{E} = \int_{E_F}^{\infty} \langle \mathbf{L}_z(\mathcal{E}) \rangle \, d\mathcal{E}. \tag{11.24}$$

Owing to the opposite sign of electron and hole states, $\langle \mathbf{L}_z \rangle$ vanishes when the integration is performed over the entire band, which corresponds to the shell being completely empty or filled. The orbital moment therefore depends on the band filling.

It is this quantity, $\langle \mathbf{L}_z \rangle$, which we want to determine as the difference of resonant x-ray absorption cross sections measured with opposite circular polarizations. In XMCD, the valence shell L is picked out by the dominant $c \rightarrow L{=}c{+}1$ dipole transition channel (see Fig. 10.2). Sensitivity to the orbital momentum of the valence shell arises from its coupling to the photon angular momentum of the circularly polarized x-rays. XMCD measures the hole part of the valence shell angular momentum since only transitions to empty states are allowed.

The orbital moment sum rule is determined by the total contributions of all electrons in a given core shell of angular momentum c. The core s–o splitting separates the $c - s_z$ and $c + s_z$ (e.g. $p_{1/2}$ and $p_{3/2}$) states which differ only in the spin but not orbital contributions. While sensitivity to the spin magnetic moment therefore requires the separation of the intensities of the transition from the two s–o core states (see below), this is not required for the determination of the orbital moment of the valence shell. This fact makes the orbital moment determination of the valence shell more robust, and it can be measured even if the core shell s–o splitting is small so that the XAS resonances overlap or if there is no core s–o splitting as for the K-shell.

We consider *circularly polarized* x-rays incident parallel to the z axis of our coordinate system, described by dipole operators $P_z^{\pm} = rC_{\pm}^{(1)}$ in Table 11.1, with the sample magnetization also along z as in Fig. 11.10c. For our assumed geometry with $\mathbf{k} \parallel z$, the polarization dependent XAS cross section (10.39)

$$\sigma_{\mathrm{XAS}}^{p} = \frac{\lambda^2}{\pi} \frac{\Gamma_{ab}^{p}}{\Gamma} \frac{(\Gamma/2)^2}{(\hbar\omega - \mathcal{E}_0)^2 + (\Gamma/2)^2} \tag{11.25}$$

for circular polarization denoted by $p = \pm$ is specifically given by

$$\sigma_{\mathrm{XAS}}^{z,\pm} = \underbrace{\frac{8\pi\hbar\omega\alpha_{\mathrm{f}}}{\Gamma} \frac{(\Gamma/2)^2}{(\hbar\omega - \mathcal{E}_0)^2 + (\Gamma/2)^2} \mathcal{R}^2}_{\mathcal{A}} \sum_{M,m,s_z} \left| \langle \phi_{L,M}(\mathcal{E}) | C_{\pm}^{(1)} | c, m, s_z \rangle \right|^2 . \tag{11.26}$$

The underbracket identifies the polarization *independent* term $\mathcal{A}$ introduced in (11.2).

The key to the orbital sum rule is the following expression for the difference in matrix elements in (11.26), which is proven by use of the matrix elements in Table 10.1 for the dominant $c \rightarrow c + 1 = L$ dipole transition (e.g. $2p \rightarrow 3d$)

$$\sum_{M,m,s_z} |\langle \phi_{LM}(\mathcal{E})|C_1^{(1)}|c,m,s_z\rangle|^2 - |\langle \phi_{LM}(\mathcal{E})|C_{-1}^{(1)}|c,m,s_z\rangle|^2$$
$$= \frac{1}{2L+1} \sum_M \left(|c_M^{\uparrow}(\mathcal{E})|^2 + |e_M^{\downarrow}(\mathcal{E})|^2 \right) M = \frac{1}{2L+1} \frac{\langle \mathbf{L}_z(\mathcal{E})\rangle}{\hbar}. \quad (11.27)$$

This allows us to write (11.26) in terms of the cross-section difference (dropping the label z)

$$\sigma_{\text{XAS}}^{+} - \sigma_{\text{XAS}}^{-} = \mathcal{A} \frac{1}{2L+1} \frac{\langle \mathbf{L}_z(\mathcal{E})\rangle}{\hbar}. \quad (11.28)$$

We can eliminate the prefactor $\mathcal{A}$ by normalization to the polarization averaged cross section given by (10.55) or[3]

$$\sigma_{\text{XAS}} = \frac{\sigma_{\text{XAS}}^{+} + \sigma_{\text{XAS}}^{-} + \sigma_{\text{XAS}}^{0}}{3} = \mathcal{A} \frac{L\, N_{\text{h}}}{3(2L+1)}, \quad (11.30)$$

where the polarization average follows from (10.53). This gives the following sum rule expression by integration over energy, first derived by Thole et al. [8].[4]

The orbital magnetic moment of a valence shell of angular momentum L containing N_{h} empty states is determined by the energy integrated resonant absorption cross section for a $c \to c+1=L$ core to valence transition according to

$$\frac{\langle \mathbf{L}_z\rangle}{\hbar} = -\frac{m_{\text{o}}}{\mu_{\text{B}}} = L\, N_{\text{h}} \underbrace{\frac{\int (\sigma_{\text{XAS}}^{+} - \sigma_{\text{XAS}}^{-})\, \mathrm{d}\hbar\omega}{\int (\sigma_{\text{XAS}}^{+} + \sigma_{\text{XAS}}^{-} + \sigma_{\text{XAS}}^{0})\, \mathrm{d}\hbar\omega}}_{\text{from experiment}}. \quad (11.32)$$

Here we have used (11.19) to link $\langle \mathbf{L}_z\rangle$ and the orbital magnetic moment m_{o}. Application of the sum rule requires knowledge of the unfilled number of valence states, N_{h}.

---

[3] For the $3d$ transition metals, one may rewrite (11.30) by use of

$$\sigma_{\text{XAS}}^{0} = \frac{\sigma_{\text{XAS}}^{+} + \sigma_{\text{XAS}}^{-}}{2} \quad (11.29)$$

so that $\sigma_{\text{XAS}}^{\text{res}} = \sigma_{\text{abs}}^{0}$ which is more easily determined experimentally.

[4] For the transition $c \to c+1 = L$, the angular momentum prefactor in [8] reduces to

$$\frac{2L(L+1)}{c(c+1) - L(L+1) - 2} = -L \quad (11.31)$$

which together with their inverted sign $\sigma_{\text{XAS}}^{-} - \sigma_{\text{XAS}}^{+}$ reduces to our expression.

We will see below that the XMCD cross-section difference $\sigma^+_{\mathrm{XAS}} - \sigma^-_{\mathrm{XAS}}$ in the numerator eliminates any non-magnetic background. In contrast, the experimental determination of the resonant cross-section sum $\sigma^+_{\mathrm{XAS}} + \sigma^-_{\mathrm{XAS}} + \sigma^0_{\mathrm{XAS}}$ in the denominator is often complicated by the existence of a non-resonant background that needs to be subtracted. It is typically modeled as the dashed step-like curve in Fig. 10.16a, but this procedure may introduce sizable errors.

It may be advantageous to use a different concept based on the transferability of the radial transition matrix element and another sum rule link of the polarization averaged "white line" intensity with the number of holes suggested in [41] and expressed by (11.30). For the L absorption edge the integral over energy in the denominator of (11.32) may then be replaced by use of (10.61) in the form

$$\int_0^\infty \sigma^+_{\mathrm{XAS}} + \sigma^-_{\mathrm{XAS}} + \sigma^0_{\mathrm{XAS}} \,\mathrm{d}\hbar\omega = 3 \int_0^\infty \sigma^{\mathrm{res}}_{\mathrm{abs}} \,\mathrm{d}(\hbar\omega)$$
$$= \underbrace{4\pi^2 \hbar\omega\, \alpha_{\mathrm{f}}\, \mathcal{R}^2_{2p\to 3d}\, \frac{2}{15}}_{C_{2p\to 3d}} \, 3N_{\mathrm{h}} \tag{11.33}$$

which yields the simpler expression,

$$\frac{\langle \mathbf{L}_z \rangle}{\hbar} = -\frac{m_{\mathrm{o}}}{\mu_{\mathrm{B}}} = \frac{2}{3\,C_{2p\to 3d}} \underbrace{\int (\sigma^+_{\mathrm{XAS}} - \sigma^-_{\mathrm{XAS}})\,\mathrm{d}\hbar\omega,}_{\text{from experiment}} \tag{11.34}$$

where $C_{2p\to 3d} = 16\,\mathrm{Mb\,eV}$ is the theoretical value given by Thole and van der Laan for the series Ca ($Z = 20$) through Cu ($Z = 29$) [42]. The value $C_{2p\to 3d} \simeq 11\,\mathrm{Mb\,eV}$ determined from XAS and XMCD data for Fe, Co, and Ni metals, for example, where the underlying step intensity was subtracted [2, 41], is smaller than the value 16 Mb eV obtained from atomic calculations by Thole and van der Laan [42].

11.5.3 Experimental Studies of Orbital Magnetism

The capability of XMCD to directly determine orbital magnetic moments is unique. In the past their presence was only indirectly revealed by material dependent deviations of the *gyromagnetic factor* or *g-factor* from the pure spin value of $g = 2$ measured by magnetic resonance techniques [43, 44] or direction dependent differences in the saturation magnetization [45].

From the sum rule (11.32) we see that the orbital moment vanishes if the energy integrated difference in cross section $\int (\sigma^+_{\mathrm{XAS}} - \sigma^-_{\mathrm{XAS}})\,\mathrm{d}\hbar\omega = 0$. For the L absorption edges this means that the XMCD difference intensity vanishes when integrated over both the L_3 and L_2 resonances. In other words, the different intensities for the L_3 and

L_2 resonances need to have the same areas but opposite signs. A beautiful example of this case is given in Fig. 11.13a, where the Cu L-edge XMCD difference spectrum is shown for Cu that has become magnetic in an alloy with Co due to hybridization of the 3*d* valence states [46]. The L_3 and L_2 XMCD effects are seen to be opposite with approximately the same area under the curves. In this case the orbital moment vanishes because the 3*d* valence shell of Cu is nearly full.

Figure 11.13b reveals how the orbital moment of Co changes due to its bonding environment. The spectra were recorded by Gambardella *et al.* [47] using electron yield detection with the sample at $T = 10\,\mathrm{K}$ and magnetically saturated in an external magnetic field of $B = 7\,\mathrm{T}$. In order to emphasize the changes of the L_3 and L_2 XMCD intensities, the spectra have been normalized to the same L_2 area. The relative increase of the L_3 area is directly proportional to the increase in orbital moment which follows the intuitive model shown in Fig. 11.11. The LF induced quenching of the moment in bulk Co is increasingly removed by the decreasing number of bonds with neighbor

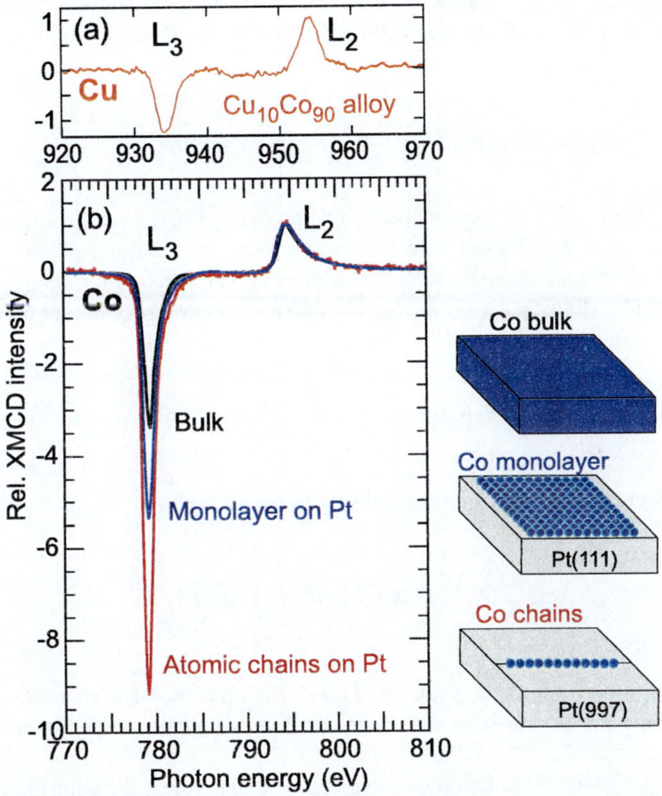

Fig. 11.13 **a** XMCD difference spectra of Cu atoms in a $Co_{10}Co_{90}$ alloy [46], **b** XMCD signal for Co atoms in bulk films, monolayers on Pt(111), and chains on Pt(997) [47]

Table 11.2 Orbital magnetic moments $\langle L_z \rangle$ measured by XMCD for Co atoms on Pt surfaces [47–49]

System	$\langle L_z \rangle$ [μ_B]
Co metal (hcp)	0.14
CoPt (111) ML	0.29
CoPt (997) ML	0.31
CoPt (997) chain	0.68 ± 0.05
CoPt (111) adatom	1.1 ± 0.1

atoms. The increase in orbital moments is quantified in Table 11.2 for Co on Pt surfaces [47, 48].

For the Co on Pt systems, the orbital moments were also linked with the magnetocrystalline anisotropy, which was directly determined from XMCD magnetization loops [47–49]. Finally we mention that enhanced orbital moments are also found in XMCD studies of gas phase atoms, dimers, and clusters [50, 51].

11.5.4 XMCD Sum Rule for the Spin Moment

For the derivation of the spin sum rule, first derived by Carra et al. [9] for the general case of an atom in a ligand field that is subjected to a magnetic field, we again describe the valence shell by the band states (11.35). The spin magnetic moment defined by (11.18) as $m_s = -2\,\mu_B \langle \mathbf{s}_z \rangle / \hbar$ is then theoretically defined by integrating the expression

$$\langle \mathbf{s}_z(\mathcal{E}) \rangle = \sum_M \langle \phi_{LM}(\mathcal{E}) | \mathbf{s}_z | \phi_{LM}(\mathcal{E}) \rangle = \frac{\hbar}{2} \sum_M \left(|c_M^{\uparrow}(\mathcal{E})|^2 - |c_M^{\downarrow}(\mathcal{E})|^2 \right) \tag{11.35}$$

over filled electron states or empty hole states according to

$$\langle \mathbf{s}_z \rangle = -\int_{-\infty}^{E_F} \langle \mathbf{s}_z(\mathcal{E}) \rangle \, \mathrm{d}\mathcal{E} = \int_{E_F}^{\infty} \langle \mathbf{s}_z(\mathcal{E}) \rangle \, \mathrm{d}\mathcal{E}. \tag{11.36}$$

As for the orbital moment, given by (11.24), the spin moment vanishes when the integration is performed over the entire band and therefore depends on the band filling.

In contrast to the orbital moment case, we need to separate the contributions from the two s–o split core states and calculate their cross sections separately. We again consider *circularly polarized* x-rays incident parallel to the z axis of our coordinate system, described by dipole operators $P_z^{\pm} = r C_{\pm}^{(1)}$ in Table 11.1, with the sample magnetization also along z as in Fig. 11.10c. The cross section for the two core s–o substates $j_+ = c + s_z$ and $j_- = c - s_z$ is then given by rewriting (11.26) as

$$[\sigma^{\pm}_{\text{abs}}]_{j\pm} = \mathcal{A} \sum_{M,m,s_z} \left| \langle \phi_{L,M}(\mathcal{E}) | C^{(1)}_{\pm} | j_\pm m_j \rangle \right|^2 , \tag{11.37}$$

where the states $|j_\pm m_j\rangle$ with $-m_j \leq j_\pm \leq m_j$ are written in terms of a linear combination of spherical harmonics as given in Table 10.3 for the $p_{j\pm}$ ($c = 1$) and $d_{j\pm}$ ($c = 2$) states. The matrix elements in (11.37) can be evaluated by use of Table 10.1 for the dominant $c \rightarrow c + 1 = L$ transitions (also see Figs. 10.6 and A.8).

From an experimental point of view it is difficult (and not done in practice) to keep track of the directions of the sample magnetization and the x-ray angular momenta which determine the sign of the XMCD difference effect. Furthermore it is a nuisance that the spin angular momentum $\mathbf{s}_z$ and the spin moment m_s have opposite signs (see (11.18)). It is therefore customary to define the experimental XMCD difference intensity to be negative at the lower energy and more intense j_+ resonance. This corresponds to the following theoretical sum rule link of the spin angular momentum $\mathbf{s}_z$ and the XMCD cross-section difference.[5]

The spin angular momentum $\langle \mathbf{s}_z \rangle$ of a valence shell of orbital angular momentum L containing N_{h} empty states may be obtained from the energy integrated resonant absorption cross sections of transitions from the separate s–o core states $j_+ = c + s_z$ and $j_- = c - s_z$ to the valence shell $L = c + 1$ according to

$$\frac{2\langle \mathbf{s}_z \rangle}{\hbar} + \frac{2(2L+3)}{L} \frac{\langle \mathbf{T}_z \rangle}{\hbar} = 3N_{\text{h}} \underbrace{\frac{\int_{j_+} (\sigma^{+}_{\text{XAS}} - \sigma^{-}_{\text{XAS}})\, \mathrm{d}\hbar\omega - \frac{L}{L-1} \int_{j_-} (\sigma^{+}_{\text{XAS}} - \sigma^{-}_{\text{XAS}})\, \mathrm{d}\hbar\omega}{\int_{j_+ + j_-} (\sigma^{+}_{\text{XAS}} + \sigma^{-}_{\text{XAS}} + \sigma^{0}_{\text{XAS}})\, \mathrm{d}\hbar\omega}}_{\text{from experiment}} . \tag{11.38}$$

Here $\langle \mathbf{T}_z \rangle$ is the expectation value of the intra-atomic magnetic dipole operator along the external **H**-field direction z, given by

$$\mathbf{T}_\alpha = [\mathbf{s} - 3\hat{\mathbf{r}}(\hat{\mathbf{r}} \cdot \mathbf{s}]_\alpha = \sum_\beta \mathbf{Q}_{\alpha\beta} \mathbf{s}_\beta , \tag{11.39}$$

[5] With $L = c + 1$ the angular momentum factors given in [9] reduce to

$$\frac{L(L+1) - 2 - c(c+1)}{3c} = \frac{2}{3}, \text{ and}$$

$$\frac{L(L+1)[L(L+1) + 2c(c+1) + 4] - 3(c-1)^2(c+2)^2}{6c\,L(L+1)} = \frac{6+4L}{3L}$$

where $\mathbf{Q}_{\alpha\beta}$ are the components of the quadrupole tensor whose diagonal elements are given by (11.5)–(11.7). Application of the sum rule requires knowledge of the unfilled number of valence states, N_{h}.

We see that the determination of the spin angular momentum or magnetic moment linked by $m_s = -2\langle \mathbf{s}_z \rangle \mu_{\mathrm{B}}/\hbar$ is complicated by the presence of the additional $\langle \mathbf{T}_z \rangle$ term in (11.38).

In the following section we develop a physical picture of the meaning of our mathematical orbital and spin moment sum rule expressions by considering a specific example, which also serves to test the theoretical predictions through experimental results.

11.6 Test of the Sum Rules: Cu-Phthalocyanine

For the exploration of the sum rule from a theoretical and experimental perspective, we discuss XMCD results for the paramagnetic molecule Cu-phthalocyanine (Cu-Pc) whose dichroic properties have been studied by Stepanow et al. [5, 6]. We have already discussed the strongly anisotropic XNLD spectra of Cu-Pc adsorbed as a monolayer on Ag(100) in conjunction with Fig. 11.3. The XNLD spectra of the same sample are determined by transitions from the s–o split $2p$ core states to the half empty $3d_{x^2-y^2}$ orbital in the molecule adsorbed with its plane parallel to the Ag substrate surface. Below we consider the quantitative correlation predicted by the sum rules between the XMCD effect and the orbital and spin magnetic moments.

11.6.1 Electronic Structure of Cu-Pc

Metal-Pcs, named after the Greek words for rock oil (naphtha) and blue (cyan) and nicknamed "magnetic blue", are flat molecules with a central metal atom surrounded by a network of N, C, and H atoms as illustrated in Fig. 11.4a. Cu-Pc is a model spin 1/2 system with an $2p^6\,3d^9$ ground state whose ligand field splitting of the Cu $3d$ orbitals is shown in Fig. 11.4b.

The single unpaired hole in the $3d_{x^2-y^2}$ orbital gives rise to paramagnetism, and spin alignment may be created at low temperature by application of a strong external magnetic field. The L-edge x-ray absorption transition diagram is illustrated in Fig. 11.14a, and the XMCD spectra recorded at normal incidence to the molecular plane, i.e. along z, in a field $\mathbf{H} \parallel z$ of value $B = \mu_0 H = 5\,$Tesla and a sample temperature of $T = 6\,$K [5] are shown in Fig. 11.14b. The difference spectrum is shown in (c).

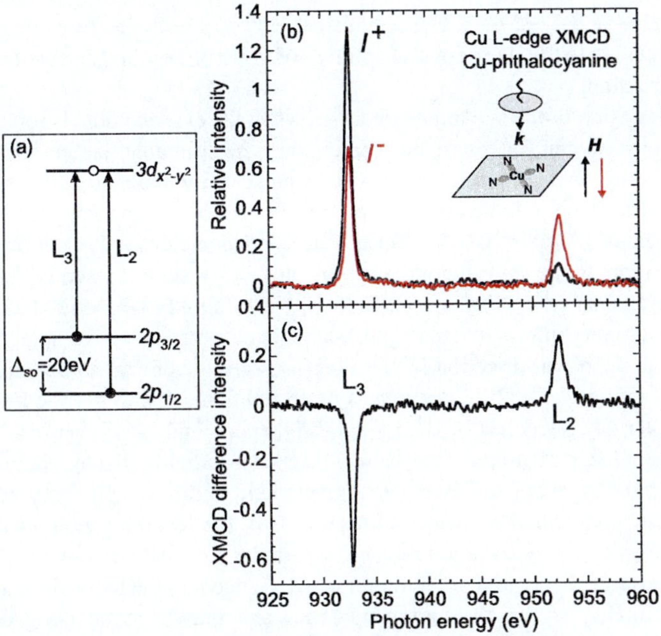

Fig. 11.14 **a** One-electron or hole model of $2p_{3/2}$ and $2p_{1/2}$ core to $3d_{x^2-y^2}$ valence transitions in Cu-phthalocyanine. **b** XMCD spectra of Cu-phthalocyanine, recorded for $\mathbf{k} \parallel \mathbf{H} \parallel z$, as indicated, at a field strength $B = \mu_0 H = 5$ Tesla and a sample temperature of $T = 6$ K [5]

Below we shall discuss the origin of the spin and orbital XMCD effects and their angular dependence. For the Cu-Pc case this is facilitated by the simple $2p^6\,3d^9$ electron or $2p^0\,3d^1$ hole ground state, which can conveniently be treated in a one-hole model shown in Fig. 11.14a. In this case the s–o effects, arising in the core and valence shell, can be separately accounted for and owing to the small s–o and Zeeman interactions in the $3d$ valence shell both interactions can be treated in perturbation. Our calculation below which treats the s–o coupling to first order in the coupling constant ξ will be seen to agree very well with the full calculation presented by Stepanow et al. [5]. Our treatment has the advantage of yielding analytical expressions that offer greater physical transparency.

The key to our calculation is to include the additional effects of the s–o coupling and applied external field on the partially empty $3d_{x^2-y^2}$ valence orbital which determines the electronic transitions as illustrated in Fig. 11.14a. Let us consider the relative size of the two interactions, which determines their perturbation treatment. If we ignore the s–o interaction, the $3d_{x^2-y^2}$ ground state has pure spin-1/2 character and in a field $B = \mu_0 H = 10$ Tesla exhibits a Zeeman splitting of $\Delta\mathcal{E} = 2\mu_B\, H \simeq 1$ meV. This Zeeman splitting may be compared to the s–o splitting of the unfilled $3d_{x^2-y^2}$ orbital, which for the LF structure shown in Fig. 11.4b may be estimated following our previous discussion in conjunction with Fig. 11.12. We obtain a value of order

$\Delta\mathcal{E} \simeq \xi^2/(\mathcal{E}_0 - \mathcal{E}_3) \sim 5\,\text{meV}$. Hence both interactions are smaller by a factor of more than 100 relative to the size of the LF splitting of order of 2 eV and can well be treated as a perturbation.

The tricky part of the s–o treatment arises when the external field is rotated. Then one needs to account for the rotated direction of the spin quantization axis relative to the LF coordinate system (x, y, z). This is most easily understood by assuming a *ferromagnetic* sample. In this case, spins are held parallel by the strong inter-atomic exchange field of order 10^3-10^4 Tesla. The exchange field only acts on the spin system, leading to the splitting of the spin-up and-down states shown in Fig. 11.10c. Application of an external magnetic field $\mathbf{H}_{\text{ext}}$ may simply be pictured as aligning different domains into a macroscopic magnetization or collective spin direction. The spin quantization direction is therefore naturally given by the direction of the externally applied field $\mathbf{H}_{\text{ext}}$. The effects of orbital fields due to the s–o interaction, which are largely quenched by the LF splitting, appear only in higher order.

In *paramagnets*, of interest here, there is no exchange field and magnetic alignment is created only by an applied external magnetic field of order 1–10 Tesla, which acts on both spin and orbital moments. However, as in the ferromagnetic $3d$ transition metals, the spin moment also dominates in paramagnetic $3d$ compounds. One may therefore assume that to first order the spins are aligned parallel to the direction of the applied field $\mathbf{H}_{\text{ext}}$, so that the spin quantization axis may be taken along $\mathbf{H}_{\text{ext}}$. Spin rotations may be treated by use of Pauli's quantum mechanical 2D spinor formalism [2, 52] or classical Euler rotations of 3D spin vectors. We shall briefly show the equivalence of the two treatments by considering the s–o interaction.

11.6.2 Treatment of the Spin-Orbit Interaction

We first consider the case where the spin quantization axis is the same as the z axis of the (x, y, z) frame of the LF orbitals, defined in Fig. 11.4. In this frame the s–o interaction Hamiltonian (9.7) has the vector form

$$H_{\text{so}} = \xi\,(\mathbf{l}_x\mathbf{s}_x + \mathbf{l}_y\mathbf{s}_y + \mathbf{l}_z\mathbf{s}_z) = \xi\left(\mathbf{l}_z\,\mathbf{s}_z + \frac{1}{2}\mathbf{l}_+\,\mathbf{s}_- + \frac{1}{2}\mathbf{l}_-\,\mathbf{s}_+\right), \tag{11.40}$$

where the operators are related by [53]

$$\mathbf{l}_x = \frac{1}{2}(\mathbf{l}_+ + \mathbf{l}_-),\ \mathbf{l}_y = -\frac{\mathrm{i}}{2}(\mathbf{l}_+ - \mathbf{l}_-),\ \mathbf{s}_x = \frac{1}{2}(\mathbf{s}_+ + \mathbf{s}_-),\ \mathbf{s}_y = -\frac{\mathrm{i}}{2}(\mathbf{s}_+ - \mathbf{s}_-) \tag{11.41}$$

or

$$\mathbf{l}_+ = \mathbf{l}_x + \mathrm{i}\mathbf{l}_y,\quad \mathbf{l}_- = \mathbf{l}_x - \mathrm{i}\mathbf{l}_y,\quad \mathbf{s}_+ = \mathbf{s}_x + \mathrm{i}\mathbf{s}_y,\quad \mathbf{s}_- = \mathbf{s}_x - \mathrm{i}\mathbf{s}_y. \tag{11.42}$$

The matrix elements in the Cartesian (x, y, z)-basis are evaluated by use of Tables A.6 and A.7 in the Appendix or by the spherical basis expressions[6]

$$\begin{aligned}\mathbf{l}_z\,|\ell, m\rangle &= \hbar m\,|\ell, m\rangle\\ \mathbf{l}_\pm\,|\ell, m\rangle &= \hbar\sqrt{\ell(\ell+1)-m(m\pm 1)}\,|\ell, m\pm 1\rangle \end{aligned}\tag{11.44}$$

$$\begin{aligned}\mathbf{s}_z\,|s, m_s\rangle &= \hbar m_s\,|s, m_s\rangle\\ \mathbf{s}_\pm\,|s, m_s\rangle &= \hbar\sqrt{s(s+1)-m_s(m_s\pm 1)}\,|s, m_s\pm 1\rangle. \end{aligned}\tag{11.45}$$

The s–o interaction couples the charge and spin degrees of freedom of the sample, and the form given by (11.40) is correct only if the z axis in the frame of the LF orbitals can also be taken as the spin quantization axis. By use of the operator rules (11.44) and (11.45), we can derive the wavefunctions of the perturbed unfilled state $\tilde{d}_{x^2-y^2}$, proceeding similar to the calculation of the s–o perturbed states given by (11.20) in Fig. 11.4b. We obtain for $\mathbf{H} \parallel \mathbf{s} \parallel z$ with the notation $\uparrow = +s_z$ and $\downarrow = -s_z$

$$\begin{aligned}|\tilde{d}^{\,\mathbf{s}\parallel+z}_{x^2-y^2}\rangle = |\tilde{d}^{\uparrow}_{x^2-y^2}\rangle &= |d^{\uparrow}_{x^2-y^2}\rangle + \frac{\mathrm{i}\xi}{\mathcal{E}_0-\mathcal{E}_2}\,|d^{\uparrow}_{xy}\rangle + \frac{\xi}{2(\mathcal{E}_0-\mathcal{E}_3)}\Big[|d^{\downarrow}_{xz}\rangle - \mathrm{i}|d^{\downarrow}_{yz}\rangle\Big]\\ |\tilde{d}^{\,\mathbf{s}\parallel-z}_{x^2-y^2}\rangle = |\tilde{d}^{\downarrow}_{x^2-y^2}\rangle &= |d^{\downarrow}_{x^2-y^2}\rangle - \frac{\mathrm{i}\xi}{\mathcal{E}_0-\mathcal{E}_2}\,|d^{\downarrow}_{xy}\rangle - \frac{\xi}{2(\mathcal{E}_0-\mathcal{E}_3)}\Big[|d^{\uparrow}_{xz}\rangle + \mathrm{i}|d^{\uparrow}_{yz}\rangle\Big].\end{aligned}\tag{11.46}$$

By use of these functions and the $\mathbf{l}_\alpha$ $(\alpha = x, y, x)$ operator rules in Table A.7,[7] we obtain the following expectation values for the orbital momenta along z

$$\begin{aligned}\langle\tilde{d}^{\uparrow}_{x^2-y^2}|\mathbf{l}_z|\tilde{d}^{\uparrow}_{x^2-y^2}\rangle &= +\frac{4\xi}{\mathcal{E}_0-\mathcal{E}_2}\\ \langle\tilde{d}^{\downarrow}_{x^2-y^2}|\mathbf{l}_z|\tilde{d}^{\downarrow}_{x^2-y^2}\rangle &= -\frac{4\xi}{\mathcal{E}_0-\mathcal{E}_2}.\end{aligned}\tag{11.47}$$

For the states (11.46), corresponding to $\mathbf{s} \parallel \pm z$, we obtain the orbital momentum expectation values $\langle\mathbf{l}_x\rangle = \langle\mathbf{l}_y\rangle = 0$.

[6] For our $s = 1/2$ case we have

$$\begin{aligned}\left\langle \pm\frac{1}{2}\right|\mathbf{s}_z\left|\pm\frac{1}{2}\right\rangle &= \pm\frac{1}{2}\\ \left\langle +\frac{1}{2}\right|\mathbf{s}_+\left|-\frac{1}{2}\right\rangle &= 1\\ \left\langle -\frac{1}{2}\right|\mathbf{s}_-\left|+\frac{1}{2}\right\rangle &= 1.\end{aligned}\tag{11.43}$$

[7] In particular, $\mathbf{l}_z|d_{x^2-y^2}\rangle = 2\mathrm{i}|d_{xy}\rangle$ and $\mathbf{l}_z|d_{xy}\rangle = -2\mathrm{i}|d_{x^2-y^2}\rangle$.

11.6.2.1 Rotation of the Spin States in the Spinor Formalism

The LF states for spin alignment $\mathbf{s} \parallel \pm x$ and $\mathbf{s} \parallel \pm y$ are obtained by rotating the spin parts of the wavefunctions (11.46). To do so we associate the spin states along z with the spinor components according to

$$\uparrow = +s_z = \begin{pmatrix}1\\0\end{pmatrix} \text{ and } \downarrow = -s_z = \begin{pmatrix}0\\1\end{pmatrix}. \tag{11.48}$$

In the Pauli spinor formalism the rotated spin states for $\mathbf{s} \parallel \pm x$ and $\mathbf{s} \parallel \pm y$ can then be expressed as a linear combination of the states for $\mathbf{s} \parallel \pm z$ as[8]

$$s_{\pm x} = \frac{1}{\sqrt{2}}\Big(\uparrow \pm \downarrow\Big), \text{ and } s_{\pm y} = \frac{1}{\sqrt{2}}\Big(\uparrow \pm \mathrm{i} \downarrow\Big). \tag{11.49}$$

Upon rotation of $\mathbf{s} \parallel \pm x$, the states (11.46) become

$$|\tilde{d}^{\,\mathbf{s}\parallel+x}_{x^2-y^2}\rangle = \frac{1}{\sqrt{2}}\Big\{|\tilde{d}^{\,\uparrow}_{x^2-y^2}\rangle + |\tilde{d}^{\,\downarrow}_{x^2-y^2}\rangle\Big\} \tag{11.50}$$

$$|\tilde{d}^{\,\mathbf{s}\parallel-x}_{x^2-y^2}\rangle = \frac{1}{\sqrt{2}}\Big\{|\tilde{d}^{\,\uparrow}_{x^2-y^2}\rangle - |\tilde{d}^{\,\downarrow}_{x^2-y^2}\rangle\Big\} \tag{11.51}$$

and the rotates states for $\mathbf{s} \parallel \pm y$ are given by

$$|\tilde{d}^{\,\mathbf{s}\parallel+y}_{x^2-y^2}\rangle = \frac{1}{\sqrt{2}}\Big\{|\tilde{d}^{\,\uparrow}_{x^2-y^2}\rangle + \mathrm{i}\,|\tilde{d}^{\,\downarrow}_{x^2-y^2}\rangle\Big\} \tag{11.52}$$

$$|\tilde{d}^{\,\mathbf{s}\parallel-y}_{x^2-y^2}\rangle = \frac{1}{\sqrt{2}}\Big\{|\tilde{d}^{\,\uparrow}_{x^2-y^2}\rangle - \mathrm{i}\,|\tilde{d}^{\,\downarrow}_{x^2-y^2}\rangle\Big\}. \tag{11.53}$$

We obtain the following orbital momentum expectation values $\langle \mathbf{l}_x \rangle$ and $\langle \mathbf{l}_y \rangle$

$$\langle \tilde{d}^{\,\mathbf{s}\parallel\pm x}_{x^2-y^2}|\mathbf{l}_x|\tilde{d}^{\,\mathbf{s}\parallel\pm x}_{x^2-y^2}\rangle = \langle \tilde{d}^{\,\mathbf{s}\parallel\pm y}_{x^2-y^2}|\mathbf{l}_y|\tilde{d}^{\,\mathbf{s}\parallel\pm y}_{x^2-y^2}\rangle = \pm\frac{\xi}{\mathcal{E}_0-\mathcal{E}_3}. \tag{11.54}$$

As expected from Fig. 11.4a, we find the same expectation values along x and y because of the molecular C_{4v} symmetry of Cu-Pc. This result compares to our earlier one for $\langle \mathbf{l}_z \rangle$ given by (11.47) which in the same notation reads

$$\langle \tilde{d}^{\,\mathbf{s}\parallel\pm z}_{x^2-y^2}|\mathbf{l}_z|\tilde{d}^{\,\mathbf{s}\parallel\pm z}_{x^2-y^2}\rangle = \pm\frac{4\xi}{\mathcal{E}_0-\mathcal{E}_2}. \tag{11.55}$$

---

[8] See Sect. 8.4.2 in [2].

Comparison of the results reveals an anisotropic orbital angular momentum, with an approximately four times larger value in the direction normal to the molecular plan than in the molecular plane.

11.6.2.2 Vector Rotation of the S–O Hamiltonian

The above results for the anisotropy of $\langle \mathbf{l}_\alpha \rangle$ with $\alpha = x, y, z$ are a special case of an arbitrary orientation of the spin quantization axis relative to the LF frame. In the following we shall derive the same results by treating the rotation of the spin frame in a vector model through classical Euler angles. We then need to distinguish the Cartesian LF or charge frame x, y, z from the rotated spin frame which we shall denote as $\tilde{x}, \tilde{y}, \tilde{z}$, as illustrated in Fig. 11.15. The two frames are transformed into each other through Euler angles $\tilde{\theta}$ and $\tilde{\phi}$. Owing to the symmetry in the $x - y$ molecular plane, we may take $\tilde{y}$ to lie in the x-y–plane, as shown, so that the $(\tilde{x}, \tilde{y}, \tilde{z})$ frame is obtained from the (x, y, z) frame by a $\tilde{\phi}$ rotation about z and then a $\tilde{\theta}$ rotation about $\tilde{y}$.

Separation of the LF and spin frames allows us to write the s–o Hamiltonian for any $\mathbf{H}_{\text{ext}}$ direction in terms of the Euler rotation angles $\tilde{\theta}$ and $\tilde{\phi}$, defined in Fig. 11.15. For example, by use of the relations given at the bottom of Fig. 11.15 we obtain the following expressions for $\mathbf{H}_{\text{ext}}$ aligned along the directions x, y or z of the LF frame

$$\mathbf{H}_{\text{ext}}, \mathbf{s} \parallel z = \tilde{z}: \quad H_{\text{so}} = \xi \left(\mathbf{l}_x \mathbf{s}_{\tilde{x}} + \mathbf{l}_y \mathbf{s}_{\tilde{y}} + \mathbf{l}_z \mathbf{s}_{\tilde{z}} \right) \tag{11.56}$$

$$\mathbf{H}_{\text{ext}}, \mathbf{s} \parallel y = \tilde{z}: \quad H_{\text{so}} = \xi \left(-\mathbf{l}_x \mathbf{s}_{\tilde{y}} + \mathbf{l}_y \mathbf{s}_{\tilde{z}} - \mathbf{l}_z \mathbf{s}_{\tilde{x}} \right) \tag{11.57}$$

$$\mathbf{H}_{\text{ext}}, \mathbf{s} \parallel x = \tilde{z}: \quad H_{\text{so}} = \xi \left(\mathbf{l}_x \mathbf{s}_{\tilde{z}} + \mathbf{l}_y \mathbf{s}_{\tilde{y}} - \mathbf{l}_z \mathbf{s}_{\tilde{x}} \right). \tag{11.58}$$

We can now directly calculate the wavefunctions associated with these three forms of the s–o Hamiltonian. For (11.56) we have $(\tilde{x}, \tilde{y}, \tilde{z}) = (x, y, z)$ and the wavefunctions are given by (11.46) leading to the orbital momentum expectation values $\langle \mathbf{l}_z \rangle$ given by (11.55).

The expectation values $\langle \mathbf{l}_y \rangle$ and $\langle \mathbf{l}_x \rangle$ are calculated with the eigenfunctions of the Hamiltonians (11.57) and (11.58), respectively. For example, the eigenstate of (11.58) for $\mathbf{s} \parallel +x$, i.e. $\tilde{z} = x$, is obtained with the notations $\uparrow = +s_{\tilde{z}}$ and $\downarrow = -s_{\tilde{z}}$ as

$$|\tilde{d}^{\,\mathrm{s}\parallel +x}_{x^2-y^2}\rangle = |d^{\uparrow}_{x^2-y^2}\rangle - \frac{\mathrm{i}\,\xi}{\mathcal{E}_0 - \mathcal{E}_2} |d^{\downarrow}_{xy}\rangle + \frac{\xi}{2(\mathcal{E}_0 - \mathcal{E}_3)} \left[|d^{\downarrow}_{xz}\rangle - \mathrm{i} |d^{\uparrow}_{yz}\rangle \right]. \tag{11.59}$$

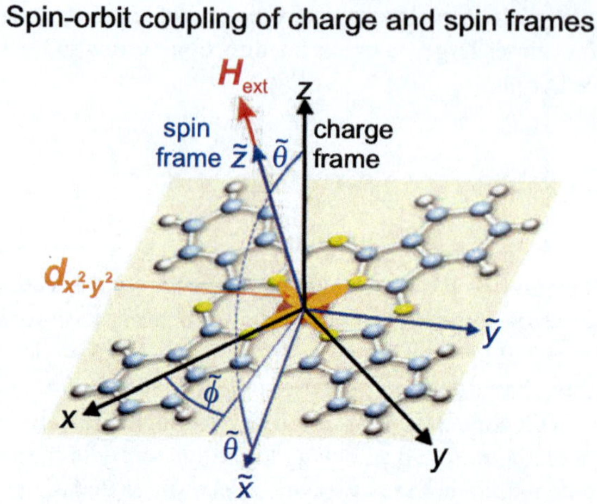

$$s_x = s_{\tilde{x}} \cos\tilde{\phi} \cos\tilde{\theta} - s_{\tilde{y}} \sin\tilde{\phi} + s_{\tilde{z}} \cos\tilde{\phi} \sin\tilde{\theta}$$
$$s_y = s_{\tilde{x}} \sin\tilde{\phi} \cos\tilde{\theta} + s_{\tilde{y}} \cos\tilde{\phi} + s_{\tilde{z}} \sin\tilde{\phi} \sin\tilde{\theta}$$
$$s_z = -s_{\tilde{x}} \sin\tilde{\theta} + s_{\tilde{z}} \cos\tilde{\theta}$$

Fig. 11.15 Relationship of the ligand field coordinate system of the sample (x, y, z) (black) used to describe the molecular orbitals and the rotated spin frame $(\tilde{x}, \tilde{y}, \tilde{z})$ (blue), whose quantization direction $\tilde{z}$ is assumed to lie along the externally applied magnetic field $\mathbf{H}_{\text{ext}}$. For Cu-Pc, the directions of the $\tilde{x}$ and $\tilde{y}$ axes in the plane perpendicular to H_{ext} can be arbitrarily chosen for symmetry reasons, and we take $\tilde{y}$ to lie in the x-y–plane, as shown. The spherical angles $\tilde{\theta}$ and $\tilde{\phi}$ then uniquely specify the direction of $\tilde{z}$ in the (x, y, z) frame. The $(\tilde{x}, \tilde{y}, \tilde{z})$ frame is obtained from the (x, y, z) system by a $\tilde{\phi}$ rotation about z and then a $\tilde{\theta}$ rotation about $\tilde{y}$. The relations between the spin coordinates in the two frames are given at the bottom

Similarly, we have for $\mathbf{s} \parallel -x$, i.e. $\tilde{z} = -x$,

$$|\tilde{d}^{\,\mathbf{s}\|-x}_{x^2-y^2}\rangle = |d^{\downarrow}_{x^2-y^2}\rangle + \frac{\mathrm{i}\,\xi}{\mathcal{E}_0 - \mathcal{E}_2}|d^{\uparrow}_{xy}\rangle - \frac{\xi}{2(\mathcal{E}_0 - \mathcal{E}_3)}\left[|d^{\uparrow}_{xz}\rangle + \mathrm{i}|d^{\downarrow}_{yz}\rangle\right]. \qquad (11.60)$$

By use of these wavefunctions and those for the cases $\mathbf{s} \parallel \pm y$, which we skip for brevity, we obtain the same orbital momentum expectation values given by (11.54).[9]

[9] It may be surprising that the same angular momentum expectation values, e.g. $\langle \mathbf{l}_x \rangle$, are calculated with the states (11.50) and (11.51), where $\uparrow = +s_z$ and $\downarrow = -s_z$ refer to the spin quantization axis z, and the seemingly different corresponding states (11.59) and (11.60), where $\uparrow = +s_{\tilde{z}}$ and $\downarrow = -s_{\tilde{z}}$ with the quantization axis defined by $\tilde{z} = x$. This arises from the fact that in the calculation of $\langle \mathbf{l}_x \rangle$ (and similarly for $\langle \mathbf{l}_y \rangle$) the spin parts of the states always satisfy the orthogonality relations $\langle \uparrow | \uparrow \rangle = \langle \downarrow | \downarrow \rangle = 1$ and $\langle \uparrow | \downarrow \rangle = \langle \downarrow | \uparrow \rangle = 0$.

11.6.3 Comparison of Orbital Momenta in Theory and Experiment

The states $|\tilde{d}^{\,s\|\alpha}_{x^2-y^2}\rangle$ given by (11.46) for $\alpha = \pm z$ and (11.59)–(11.60) for $\alpha = \pm x$ consist of the zeroth order states $|d_{x^2-y^2}\rangle$ plus first order s–o perturbation contributions from other LF states. With increasing field strengths $H_{\rm ext} \parallel \tilde{z}$ the spin degeneracy of the spin-up and-down states $\pm s_{\tilde{z}}$ will be lifted and with decreasing temperature the sample magnetization M will increase according to the Brillouin function

$$M = M_{\rm sat} \tanh\left[\frac{\mu_{\rm B}(2\langle s_{\tilde{z}}\rangle + \langle l_{\tilde{z}}\rangle)H_{\rm ext}}{k_{\rm B}T}\right]. \tag{11.61}$$

This expression includes the contributions from both the spin and orbital momenta in the applied field direction $\tilde{z}$. The dominant spin contribution is isotropic in first order, $2\langle s\rangle = 1\hbar$, while the orbital one differs according to (11.54) and (11.55). With the s–o parameter $\xi = 102\,$meV (see Table 9.1) and the LF splittings $\mathcal{E}_0 - \mathcal{E}_2 = 2.0\,$eV and $\mathcal{E}_0 - \mathcal{E}_3 = 2.3\,$eV given in Fig. 11.4 we obtain the numerical values

$$\langle \mathbf{l}_z\rangle = \frac{4\xi}{\mathcal{E}_0 - \mathcal{E}_2} = 0.204, \text{ and } \langle \mathbf{l}_x\rangle = \frac{\xi}{\mathcal{E}_0 - \mathcal{E}_3} = 0.044 \tag{11.62}$$

In Fig. 11.16a we give our theoretical results for $\langle \mathbf{l}_z\rangle$ and $\langle \mathbf{l}_x\rangle$ resulting from the s–o mixing of the different d_i LF orbitals, assuming spin alignment by an applied field. The signs of the listed orbital momenta correspond to the spin-up $|\tilde{d}^{\uparrow}_{x^2-y^2}\rangle$ orbital.

In Fig. 11.16b we show as blue squares the detailed angular dependence of the orbital momenta measured by Stepanow et al. [5], recorded at $B = \mu_0 H_{\rm ext} = 5\,$Tesla and a sample temperature of $T = 6\,$K. The solid blue curve corresponds to the theoretical expression

$$\langle \mathbf{l}\rangle = k\,\frac{M}{M_{\rm sat}}\left[\langle \mathbf{l}_z\rangle \cos^2\tilde{\theta} + \langle \mathbf{l}_x\rangle \sin^2\tilde{\theta}\right], \tag{11.63}$$

where $\tilde{\theta}$ is defined in Fig. 11.15. The curve was calculated with $M/M_{\rm sat} = 0.58$, reflecting the experimental parameters.[10] We also accounted for electron delocalization that is not included in a simple LF model by use of a covalency reduction factor $k = 0.9$ [5]. We find excellent agreement between experiment and theory, thus illustrating the power of the angle-dependent XMCD orbital momentum sum rule.

[10] This corresponds to a ground state splitting of $\Delta\mathcal{E} = \mu_{\rm B}(2\langle s\rangle + \langle l_z\rangle)H_{\rm ext} \simeq 0.7\,$meV.

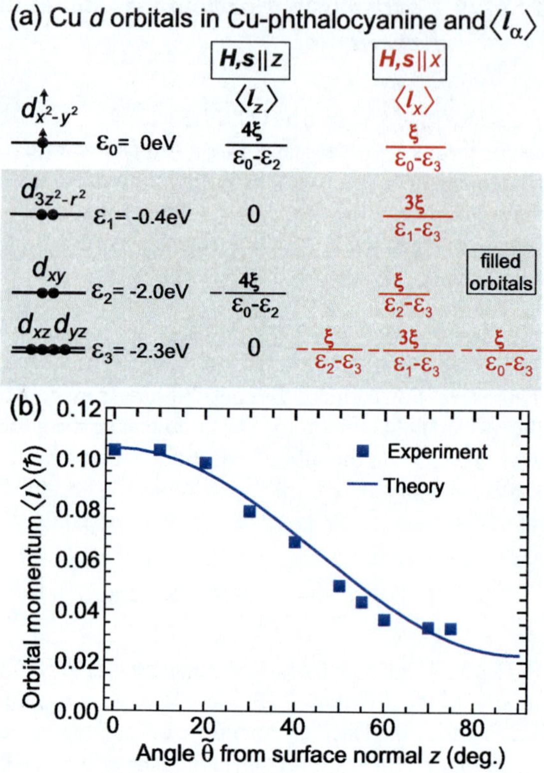

Fig. 11.16 **a** Orbital momenta $\langle \mathbf{l}_\alpha \rangle$ in units of $\hbar$ of the LF states in Cu-Pc in the presence of the s–o interaction (11.40) for alignment of the **H**-field either along the surface normal z, $\langle \mathbf{l}_z \rangle$, or parallel to the molecular plane, $\alpha = x$ or y, $\langle \mathbf{l}_x \rangle = \langle \mathbf{l}_y \rangle$. The listed values for $\langle \mathbf{l}_\alpha \rangle$ are for saturation spin alignment by the applied **H**-field assuming that the spin-up state $|\tilde{d}^{\uparrow}_{x^2-y^2}\rangle$ is occupied. **b** Measured angle dependence of $\langle \mathbf{l} \rangle$ at $B = \mu_0 H = 5\,$Tesla at $T = 6\,$K (squares) [5], and theory (blue line) according to (11.63)

11.6.4 Spin Momenta in Theory and Experiment

In analogy to the expressions for the angle dependent s–o Hamiltonian given by (11.56)–(11.58), the magnetic dipole operator is given by the following expressions for $\mathbf{k}, \mathbf{H}_{\text{ext}}, \mathbf{s} \parallel x$, y or z,

$$\mathbf{k}, \mathbf{H}_{\text{ext}}, \mathbf{s} \parallel z: \quad \mathbf{T}_z = Q_{zx}\mathbf{s}_{\tilde{x}} + Q_{zy}\mathbf{s}_{\tilde{y}} + Q_{zz}\mathbf{s}_{\tilde{z}} \tag{11.64}$$

$$\mathbf{k}, \mathbf{H}_{\text{ext}}, \mathbf{s} \parallel y: \quad \mathbf{T}_y = -Q_{yx}\mathbf{s}_{\tilde{y}} + Q_{yy}\mathbf{s}_{\tilde{z}} - Q_{yz}\mathbf{s}_{\tilde{x}} \tag{11.65}$$

$$\mathbf{k}, \mathbf{H}_{\text{ext}}, \mathbf{s} \parallel x: \quad \mathbf{T}_x = Q_{xx}\mathbf{s}_{\tilde{z}} + Q_{xy}\mathbf{s}_{\tilde{y}} - Q_{xz}\mathbf{s}_{\tilde{x}}. \tag{11.66}$$

Here Q_{ij} are the elements of the quadrupole tensor introduced in Sect. 11.3.2 and the spin quantization axis is denoted as $\tilde{z}$. The matrix elements are calculated by expressing the operators Q_{ij} by spherical tensors as given in Table 11.3.

By use of Table 11.3, we can rewrite (11.64)–(11.66) in the convenient operator forms

Table 11.3 Quadrupole operator elements expressed in terms of Racah's spherical tensors

$Q_{xx} = C_0^{(2)} - \sqrt{\frac{3}{2}}\left(C_2^{(2)} + C_{-2}^{(2)}\right)$	$Q_{xy} = Q_{yx} = \mathrm{i}\sqrt{\frac{3}{2}}\left(C_2^{(2)} - C_{-2}^{(2)}\right)$
$Q_{yy} = C_0^{(2)} + \sqrt{\frac{3}{2}}\left(C_2^{(2)} + C_{-2}^{(2)}\right)$	$Q_{xz} = Q_{zx} = \sqrt{\frac{3}{2}}\left(C_1^{(2)} - C_{-1}^{(2)}\right)$
$Q_{zz} = -2\,C_0^{(2)}$	$Q_{yz} = Q_{zy} = -\mathrm{i}\sqrt{\frac{3}{2}}\left(C_1^{(2)} + C_{-1}^{(2)}\right)$

$\boxed{\tilde{z} \parallel z:}$

$$\mathbf{T}_z = \underbrace{-2\,C_0^{(2)}}_{Q_{zz}}\,\mathbf{s}_{\tilde{z}} - \sqrt{\frac{3}{2}}\,C_{-1}^{(2)}\,\mathbf{s}_+ + \sqrt{\frac{3}{2}}\,C_1^{(2)}\,\mathbf{s}_-$$

$\boxed{\tilde{z} \parallel y:}$

$$\mathbf{T}_y = \Big[\underbrace{C_0^{(2)} + \sqrt{\frac{3}{2}}\left(C_2^{(2)} + C_{-2}^{(2)}\right)}_{Q_{yy}}\Big]\mathbf{s}_{\tilde{z}}$$

$$+\sqrt{\frac{3}{8}}\Big[-C_2^{(2)} + C_{-2}^{(2)} + \mathrm{i}\,C_1^{(2)} + \mathrm{i}\,C_{-1}^{(2)}\Big]\mathbf{s}_+ + \sqrt{\frac{3}{8}}\Big[C_2^{(2)} - C_{-2}^{(2)} + \mathrm{i}\,C_1^{(2)} + \mathrm{i}\,C_{-1}^{(2)}\Big]\mathbf{s}_-$$

$\boxed{\tilde{z} \parallel x:}$

$$\mathbf{T}_x = \Big[\underbrace{C_0^{(2)} - \sqrt{\frac{3}{2}}\left(C_2^{(2)} + C_{-2}^{(2)}\right)}_{Q_{xx}}\Big]\mathbf{s}_{\tilde{z}}$$

$$+\sqrt{\frac{3}{8}}\Big[C_2^{(2)} - C_{-2}^{(2)} - C_1^{(2)} + C_{-1}^{(2)}\Big]\mathbf{s}_+ + \sqrt{\frac{3}{8}}\Big[-C_2^{(2)} + C_{-2}^{(2)} - C_1^{(2)} + C_{-1}^{(2)}\Big]\mathbf{s}_-, \tag{11.67}$$

where the operators $\mathbf{s}_\pm$ refer to the spin frame $(\tilde{x}, \tilde{y}, \tilde{z})$. The expectation values of $\langle \mathbf{T}_\alpha \rangle$ are calculated with the wavefunctions for the corresponding field directions, which for $\alpha = z$ are given by (11.46) and for $\alpha = x$ by (11.59)–(11.60). In (11.67) we have indicated by underbrackets that $\langle \mathbf{T}_\alpha \rangle$ exists even in the absence of the s–o interaction. Then the atomic spin density can simply rotate into the $\mathbf{H}_{\mathrm{ext}}$ field direction, and the spin-flip contributions represented by the operators $\mathbf{s}_\pm$ are absent. The angular dependence of $\langle \mathbf{T}_\alpha \rangle$ then simply reflects the anisotropy of the charge distribution in the atomic sphere, expressed by the diagonal elements of the quadrupole tensor, $\mathbf{Q}_{\alpha\alpha}$.

In the middle column of Fig. 11.17a we show the calculated values of $\langle \mathbf{T}_z \rangle$ and $\langle \mathbf{T}_\parallel \rangle = \langle \mathbf{T}_x \rangle = \langle \mathbf{T}_y \rangle$ for the d_i orbitals in Cu-Pc in the *absence* of the s–o interaction but full spin alignment along $\mathbf{H}_{\mathrm{ext}}$. Of particular interest is the $d_{x^2-y^2}$ orbital probed in XMCD, for which we also give the s–o induced change of the dipole moment calculated with the spin-up states (11.46) $|\tilde{d}^{\uparrow}_{x^2-y^2}\rangle_{\mathbf{s}\parallel z}$ and the states (11.59)–(11.60)

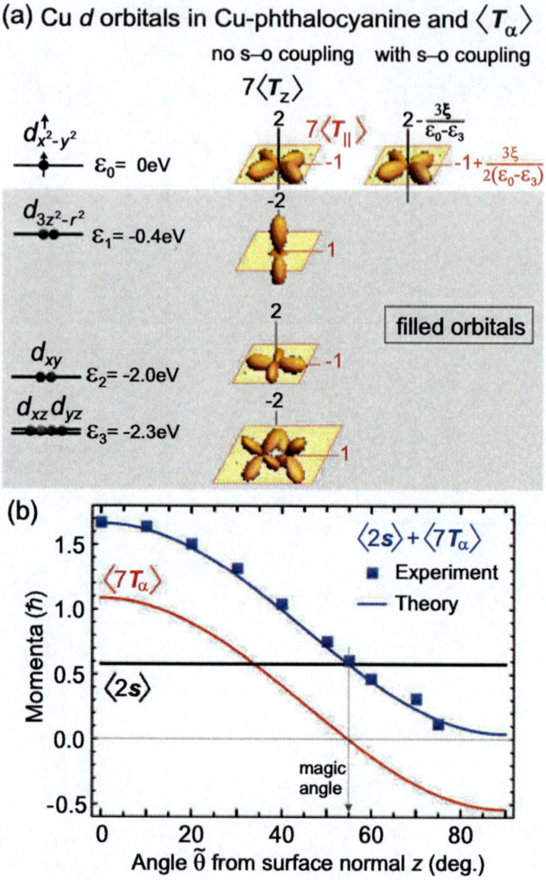

Fig. 11.17 **a** Orbital energies, calculated saturation values of $\langle 7\mathbf{T}_\alpha \rangle$ for the d_i orbitals of Cu-Pc without s–o coupling, assuming spin alignment along **H** and a spin-up ground state. For the orbital $|d^{\uparrow}_{x^2-y^2}\rangle$ probed by XMCD we also give on the right the $\langle 7\mathbf{T}_\alpha \rangle$ values in the presence of s–o coupling, given by $\langle 7\mathbf{T}_z \rangle = 2 - 3\xi/(\mathcal{E}_0 - \mathcal{E}_3) = 1.87$ and $\langle 7\mathbf{T}_\| \rangle = -1 + 3\xi/2(\mathcal{E}_0 - \mathcal{E}_3) = -0.93$. **b** Comparison of the measured (blue squares) [5] and calculated (blue curve) angular dependence of the sum rule values according to (11.68) with $M/M_{\rm sat} = 0.58$. The decomposition into pure spin (black) and dipolar (red) contributions is also shown. Note that $\langle 7\mathbf{T} \rangle = 0$ near the magic angle of $\tilde{\theta} = 54.7°$

$|\tilde{d}^{\uparrow}_{x^2-y^2}\rangle_{\mathbf{s}\|x}$, respectively. On the right of Fig. 11.17a we show the additional s–o contribution for the $|\tilde{d}^{\uparrow}_{x^2-y^2}\rangle$ state. It is seen to scale with the ratio of the s–o coupling constant ξ and the LF splitting $\mathcal{E}_0 - \mathcal{E}_i$, and we obtain $\langle 7\mathbf{T}_z \rangle = 2 - 3\xi/(\mathcal{E}_0 - \mathcal{E}_3) = 1.87$ and $\langle 7\mathbf{T}_\| \rangle = -1 + 3\xi/2(\mathcal{E}_0 - \mathcal{E}_3) = -0.93$ at magnetic saturation.

The angular dependence of the XMCD sum rule signal is expressed by

$$\langle 2\mathbf{s} + 7\mathbf{T} \rangle = \frac{M}{M_{\rm sat}} \left[\langle 2\mathbf{s} + 7\mathbf{T}_z \rangle \cos^2\tilde{\theta} + \langle 2\mathbf{s} + 7\mathbf{T}_x \rangle \sin^2\tilde{\theta} \right], \tag{11.68}$$

where $\tilde{\theta}$ is defined in Fig. 11.15. The spin momentum $\langle 2\mathbf{s}\rangle = 1$ becomes anisotropic only in second order of ξ. In our first order treatment the total moment $\langle 2\mathbf{s} + \mathbf{L}_\alpha\rangle$ is anisotropic only due to the anisotropy of the orbital moment. This gives rise to the anisotropy of the g-factor measured in electron paramagnetic resonance [54]. The calculated angular dependence of the XMCD sum rule signal (11.68) for Cu-Pc is shown as a blue curve in Fig. 11.17, where it is compared to the measured values shown as blue squares, assuming the experimental value $M/M_{\text{sat}} = 0.58$. In the figure we also show the theoretical decomposition into contributions of $\langle 2\mathbf{s}\rangle$ (black line) and $\langle 7\mathbf{T}_\alpha\rangle$ (red curve).

For the case of Cu-Pc we find excellent agreement between the theoretical XMCD sum rules and experiment. By use of the $\mathbf{T}_\alpha$ sum rule of Stöhr and König [10], one may eliminate the large angular anisotropy due to the $7\mathbf{T}_\alpha$ term by summing over three measurements with the circularly polarized light incident along the three orthogonal directions of $\mathbf{H}_{\text{ext}}$ along x, y, z or a single "magic angle", $\tilde{\theta} = 54.7°$, measurement. This is nicely revealed in Fig. 11.17, where the $7\mathbf{T}_\alpha$ term vanishes at $\tilde{\theta} \simeq 55°$.

In closing we note that Cu-Pc is a special case that can be described by one-electron theory. The s–o coupling in the core and valence shell can then be treated separately, with the small s–o coupling in the valence shell treated by perturbation theory. The XMCD sum rules are valid also for the more complicated cases where the s–o interaction in the core and valence shell can no longer be treated separately, since in the final state configuration the coupling of the core hole and multi-hole valence states results in multiplet effects. These more complicated cases have been discussed in more detail in [55–59].

11.7 Application of XMCD to the Study of Transient Spin Effects

Since its discovery by Schütz and collaborators [60] in 1987, XMCD spectroscopy has had significant impact in modern magnetism research, which around the same time was put on a new trajectory through the discovery of the giant magnetoresistance (GMR) effect independently discovered around 1988 by Albert Fert and collaborators [61] and Peter Grünberg and collaborators [62], who were awarded the Nobel Prize in Physics in 2007. Some of the impact of XMCD has been reviewed previously in [2, 63–66].

In contrast to bulk magnetic crystals, whose spin structure was largely solved by neutron scattering, modern applications of magnetism are based on artificially engineered thin films and nanostructures, utilized for writing and reading magnetic data bits in computers and servers. The power of XMCD for the investigation of such materials is founded on the following five x-ray capabilities.

- The large *x-ray intensity* available at synchrotron and XFEL sources facilitates short measurement times.
- The *photon energy tunability* allows the determination of atom and bond-specific contributions.

- *Polarization control* coupled with large resonant cross sections allows separation of charge and spin effects with high sensitivity.
- The *short wavelength* opens the door for imaging on the nanometer length scale, down to the atomic size.
- The *pulsed nature* of synchrotron and FEL sources provides temporal resolution, down to the intrinsic femtosecond timescale of changes in the electronic and spin structure.

Most applications of XMCD have been associated with picking apart the *static* magnetic properties of materials through spectroscopy and spatially resolved imaging of atom specific spin and orbital moments. The atomic moments associated with the valence electrons are either intrinsically created by the exchange interaction in magnetic systems, induced in non-magnetic atoms through hybridization with magnetic neighbors, or induced in paramagnets by application of external magnetic fields. The so-created static magnetic moments typically vary in size from 0.01 μ_B to a few μ_B [2, 63, 64].

In order to highlight other remarkable capabilities of XMCD we will present below its use to study the *transient* behavior of magnetic structures which underlies the function of magnetic devices. While *atomic moments* may be pictured as relatively large static spin imbalances localized within the atomic volume, the field of spin transport [67–69] is concerned with dynamic phenomena caused by application of a voltage. Spin currents may be created by preferential spin dependent *exchange scattering* when electrons flow through a ferromagnet or by *spin-orbit* scattering when a current flows through a heavy metal with large s–o coupling such as Pt. In the following we will present two state-of-the-art applications of XMCD that utilize all five x-ray capabilities outlined above.

11.7.1 Spin Accumulation in Cu upon Injection from Co

Our first example is related to current flow through the nanostructure shown previously in Fig. 1.22a. As shown in that figure, a spin polarized current created in one ferromagnetic (FM) layer, after transmission through a non-magnetic Cu decoupling layer, may exert a torque on the static atomic moments in another FM layer and change its magnetization *direction*. In the following we will show that one may actually directly measure the sign and magnitude of the spin current that flows into the Cu spacer layer. Such an experiment requires “x-rays at their best” by combining spectroscopic capabilities with time-resolved nanoscale imaging as reviewed in [2, 65, 66].

While spin dependent current or resistivity effects are typically studied by transport measurements, XMCD can provide more detailed information on the spin dependent scattering mechanisms. For example, the GMR effect arises from the change in resistivity between parallel and antiparallel orientations of the static magnetization directions in two FM layers in Fig. 1.22a. The thin non-magnetic Cu layer serves to decouple the magnetic orientations in the two FM layers so that they can be rotated relative to each other. The size of the GMR effect depends on the magnitude of the spin current in the Cu layer which, however, is not measured.

Early XMCD measurements revealed the existence of magnetic moments in Cu interface atoms adjacent to magnetic layers such as Co and Fe [46, 70, 71]. This effect previously illustrated in Fig. 11.13 for Cu atoms in a $Co_{10}Co_{90}$ alloy [46] arises from the hybridization of the $3d$ bands and is a static "proximity" effect. It exists without current flow.

In the presence of current flow from a FM layer into Cu, an additional much smaller *transient* effect occurs. When a charge current flows through the FM, the traveling electrons may scatter on the magnetic atoms, and according to Fermi's golden rule, the scattering cross section depends on the density of empty atomic states that the electrons can scatter into. But as illustrated in Fig. 11.10 there is an imbalance of localized empty spin-up and-down states. So naturally one kind of spin is scattered more. This argument hinges on the important fact that the spin is conserved in the scattering process, known as Mott's two current model [2], developed in the 1930s [72–74]. The charge current which is the sum of the two spins may be broken up into the two spin channels. When the current in the ferromagnet is injected into a non-magnet, there is preferential accumulation of one of the spins at the interface. The spin imbalance slowly disappears with distance, called the spin diffusion length, in the non-magnetic layer. By keeping it sufficiently thin, the imbalance or spin accumulation may be observed, as shown for Co/Cu interface in a pillar in Fig. 11.18 [75].

The measurement combined XMCD transmission microscopy (see Sect. 8.2) with the synchronized detection of the effect of current pulses as schematically illustrated in Fig. 11.18a. While the number of photons in the individual 100 ps x-ray pulses separated by 2.1 ns varies slightly, the average number of photons emitted during the 780 ns storage ring repetition time is constant and is therefore best used to determine the tiny XMCD signal difference between current on/off periods. The sample structure and current flow direction is shown in Fig. 11.18b. Of interest is the transient XMCD signal in the Cu layer due to the injection of a spin current from the adjacent Co layer.

Figure 11.18c shows the result for the Cu XMCD contrast, spatially resolved with a resolution of 35 nm, measured with opposite circular x-ray polarization directions and current density of 10^7 A/cm^2. It is interesting that such a large current density would result in melting of a typical Cu wire, but in the pillar configuration the large surface area allows sufficient heat dissipation. From the measurement, the transiently induced moment per Cu atom was derived to be $3 \times 10^{-5}\,\mu_B$. The measurement also revealed a linear dependence of the transient moment with current and that the sign of the spin current in Cu is the same as that in Co on the other side of the interface. The measurement reveals a remarkable XMCD sensitivity equivalent to the magnetic moment of only about 50 Fe atoms.

A related XMCD experiment also detected spin accumulation in Cu without flow of a charge current by use of "spin pumping", consisting of the creation of precessing magnetic moments by microwaves in an adjacent $Ni_{80}Fe_{20}$ permalloy layer [76]. On

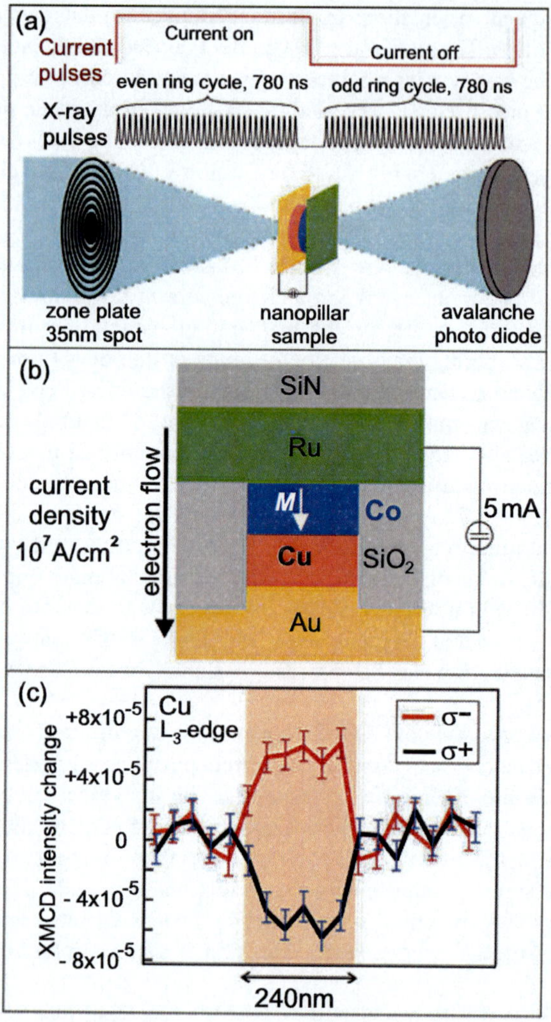

Fig. 11.18 **a** Timing sequence and synchronization between current pulses sent through the nanopillar sample (shown enlarged in (**b**)) and individual x-ray pulses (100 ps wide, separated by 2.1 ns) emitted by different electron bunches within the total 780 ns cycle time of the storage ring. The monochromatic x-rays of 200 meV bandwidth, tuned to the Cu L_3 resonance (933 eV), were focused by a zoneplate based transmission x-ray microscope to a 35 nm spot which was scanned across the sample, with the time and polarization dependent transmitted intensity measured by an avalanche photodiode [75]. **b** Schematic of the nanoscale pillar, with Ru and Au contacts, containing a Co based structure whose magnetization **M** was aligned along the pillar and a $\simeq$ 30-nm-thick Cu layer into which the spin polarized electrons generated in the Co layer were injected. **c** Normalized transient XMCD signal (current on/off), arising from spin accumulation in Cu, corresponding to a minute imbalance of spins (moment) on each Cu atom

the theory side, the transient spin effects in Cu are so small that they presently cannot be reliably calculated by first-principles theory. To avoid these problems, recent theoretical work instead investigated the Co/Pd bilayer system which yields larger effects due to the higher spin polarizability of Pd [77].

11.7.2 Spin-Orbit Induced Spin Currents in Pt, Injected into Co

Our second example of high sensitivity XMCD studies of spin current-induced effects deals with the spin Hall effect, predicted by Russian physicists Mikhail Dyakonov and Vladimir Perel in 1971 [78].[11] These scientists also introduced the notion of a pure spin current that is decoupled from a charge current.

For the Co/Cu system discussed above, the spin current is created through scattering of electrons on *magnetic* atoms. In contrast, a spin current may also be created through the spin-orbit interaction of flowing electrons with heavy atoms. This so-called *spin Hall effect* does not require the atoms to be magnetic, and in contrast to the conventional Hall effect, no magnetic field is required. Since the s–o interaction increases with atomic size, one uses materials consisting of heavy atoms such as Pt [80].

The spin dependent deflection of electrons flowing through a metallic Pt wire of rectangular cross section is illustrated in Fig. 11.19a. It leads to spin accumulation at the wire surface with directions indicated by red arrows **s**. The spin accumulation at the top Pt surface has been measured by Stamm et al. [81] by its effect on a ferromagnetic Co layer added on top, shown in blue.

The magnetization direction of Co was aligned in the direction of electron flow, **x**, by an external magnetic field $\mathbf{H}_{\text{ext}}$, as shown in the figure. The spin accumulation at the Pt surface leads to spin diffusion into Co, and owing to the perpendicular orientation of the injected spins **s** and the Co moments **m**, a torque is exerted on the direction of the Co moments. They rotate slightly toward the surface normal **z** from their original **x** direction, as indicated. The slightly increased component of **m** along z is detected by XMCD by use of the shown normal incidence "forbidden geometry" which is almost exclusively sensitive to the m_z component.

The change of the spin torque-induced component of the Co moments m_z with current through Pt is shown in Fig. 11.19b. By use of 50 kHz modulation of the current through Pt, it was possible to measure the tiny XMCD effect of order of 10^{-5} with a remarkable signal to noise ratio. Comparison of the current densities and dichroism effects with those in Fig. 11.18c reveals an even higher sensitivity. The measured XMCD signals were converted into the transient spin and orbital Co moments plotted as a function of current density in Fig. 11.19c, assuming that for metallic Co the spin density $\langle \mathbf{T}_z \rangle$ can be neglected. Both transiently induced Co moments were found to scale linearly with current density.

[11] The term "spin Hall effect" was introduced by J. E. Hirsch in 1999 who repredicted the effect [79].

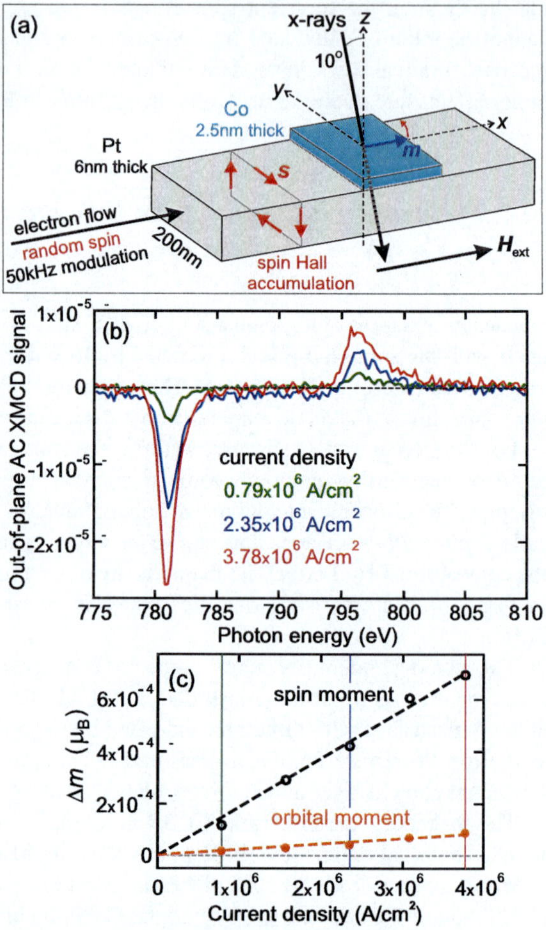

Fig. 11.19 **a** Experimental configuration used by Stamm et al. [81] to measure the spin Hall effect with XMCD. **b** Co XMCD difference signal along z, measured by 50 kHz modulation of the current through Pt with current density. **c** Extracted changes of the spin and orbital Co moment as a function of current density

11.8 X-Ray Magnetic Linear Dichroism—XMLD

11.8.1 Introduction

The XMLD effect was first predicted and observed in 1985/86 by Thole and van der Laan et al. [82, 83]. In contrast to *natural* linear dichroism or XNLD, which arises from an anisotropy of the valence charge, *magnetic* linear dichroism or XMCD only exists in the presence of spin alignment in a sample. It is defined as the difference in cross section or intensity of two spectra measured with the **E**-vector oriented parallel and perpendicular to the spin alignment axis. Since for linearly polarized light the **E** oscillates in time along an *axis*, the XMLD effect only depends on the relative orientations of the **E** axis and the magnetic axis of the sample. This leads to the dependence of XMLD on the square of the magnetization. In a simple picture the effect arises from a uniaxial distortion of the charge density relative to the spin axis as schematically shown in Fig. 11.20.

> The *XMLD effect* arises from a distortion of the atomic charge by the spin-orbit interaction relative to a spin alignment axis of the sample. The XMLD signal is defined as the difference of the XAS spectra recorded for parallel and perpendicular alignment of **E** relative to the spin axis.

Spin alignment may be created in paramagnets by application of an external magnetic field or is naturally present in magnetic materials due to the intrinsic exchange field. In practice, magnetic fields produced in the laboratory are much weaker (less than 10 Tesla) than the intrinsic exchange field in ferromagnets or collinear antiferromagnets (of order 10^3-10^4 Tesla). In practice, XMLD therefore has its most important applications in the study of antiferromagnets, which owing to their internal spin compensation do not exhibit an XMCD effect.

The real power of XMLD emerges mostly in the presence of multiplet structure where pronounced changes of the relative intensity of different peaks may be observed, as first shown theoretically by van der Laan and Thole [84]. As discussed in Sect. 10.6, the essence of multiplet structure is the coupling of spin and orbital angular momenta which in the final state involves those of both core and valence electrons. Since XMLD results from the coupling of the spin and orbital degrees of freedom in the sample, the effect is stronger in the presence of multiplet coupling than in one-electron theory. In the latter, the s–o couplings in the core and valence shells enter separately, and in transition metal ions the valence shell s–o coupling is weak relative to the larger ligand field effects. This typically results in relatively weak XMLD effects when multiplet structure is absent.

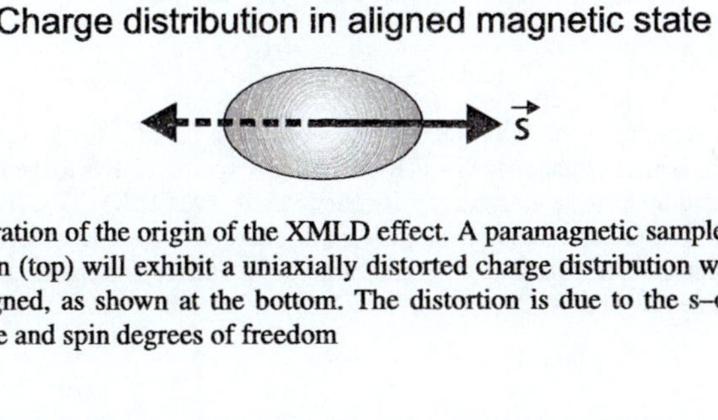

Fig. 11.20 Illustration of the origin of the XMLD effect. A paramagnetic sample with a spherical charge distribution (top) will exhibit a uniaxially distorted charge distribution when the spins are preferentially aligned, as shown at the bottom. The distortion is due to the s–o interaction that couples the charge and spin degrees of freedom

Materials may exhibit both a charge anisotropy due to the ligand field as well as a spin alignment direction, so that XNLD and XMLD effects may be simultaneously present. In theory one may distinguish two types of XNLD measurements [85] referred to as type I and type II XMLD which are illustrated in Fig. 11.21.

Type I XMLD is measured with fixed x-ray polarization **E** as the difference between rotation of the spin axis parallel and perpendicular to **E**. It eliminates any XNLD effect since the **E**-vector is fixed. It may be used for paramagnetic and ferromagnetic samples where the spin direction can be rotated by a strong field. In *antiferromagnets* one can, in principle, also "flop" the magnetization axis, but this typically requires fields larger than available. Type I also contains useful information on the magnetocrystalline anisotropy [86–88].

Type II XMLD measurements are performed when it is difficult or impossible to rotate the spin alignment direction. One then rotates the polarization vector and measures the difference between its parallel and perpendicular alignment with the spin axis. The interpretation of the results may be complicated by the existence of competing XMLD and XNLD effects.

11.8.2 Theoretical Formulation in One-Electron Theory

The XMLD absorption cross-section difference is obtained from two measurements with **E**-vector parallel and perpendicular to the spin or magnetic alignment axis

$$\Delta\sigma_{\mathrm{XMLD}} = \sigma^{\parallel}_{\mathrm{XAS}}(\hbar\omega) - \sigma^{\perp}_{\mathrm{XAS}}(\hbar\omega). \tag{11.69}$$

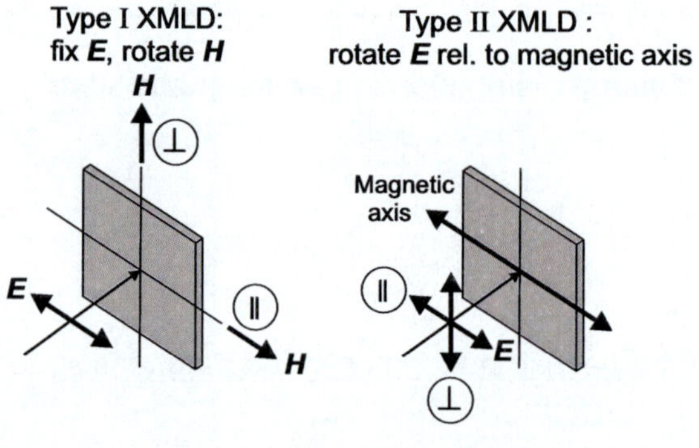

Fig. 11.21 The two methods to record XMLD spectra. Type I XMLD is a pure magnetic effect, measured with a fixed orientation of the linearly polarized **E**-vector relative to the sample whose spin orientation is rotated by a sufficiently strong magnetic field. Type II XMLD is typically measured when it is difficult to rotate the magnetic alignment direction in a sample as for antiferromagnets. One then rotates the **E**-vector instead. The measured cross sections are labeled ∥ and ⊥ with the XMLD signal defined by (11.69)

Assuming magnetic alignment along the z-axis, the energy dependent XMLD difference intensity is given by the difference of the intensities measured for $\mathbf{E} \parallel z$ and $\mathbf{E} \perp z$. The cross sections are written in analogy to those for XNLD (11.3) and XMCD (11.37) by use of the polarization dependent dipole operators in Table 11.1. For $\mathbf{E}$ along z and x, we obtain

$$\boxed{\Delta\sigma_{\mathrm{XMLD}} = \mathcal{A} \sum_{j,m_j} \left| \langle \Phi_d | C_0^{(1)} | p_j, m_j \rangle \right|^2 - \frac{1}{2} \left| \langle \Phi_d | C_{-1}^{(1)} - C_{+1}^{(1)} | p_j, m_j \rangle \right|^2}, \quad (11.70)$$

where $\mathcal{A}$ is defined in (11.2) and Φ_d is the wavefunction of the empty LF states to which transitions can occur.

11.8.3 XNLD Versus XMLD in Cu-Phthalocyanine

We will see later that the largest XMLD effects arise in the presence of multiplet effects since they lead to a large s–o mixing of the core and valence states as discussed in Sect. 10.6. Multiplet effect, however, is typically handled by computer programs at the expense of simple physical pictures that underlie one-electron theory. For this reason we shall first discuss a case where one-electron theory applies. We again consider the pedagogical case of Cu-Phthalocyanine (Cu-Pc), whose XMCD sum rule analysis was presented in Sect. 11.6. Here our goal is to put into perspective the large *natural* linear dichroism effect in Cu-Pc caused by the LF anisotropy, discussed in Sect. 11.3.1, in terms of the much weaker *magnetic* linear dichroism effect which is solely due to the s–o interaction. Our comparison disentangles the two effects and reveals the fundamental role of the s–o coupling in XMLD.

The L-edge spectrum of Cu-Pc shown in Fig. 11.3b reveals a pronounced polarization dependence associated with transitions to the partially empty orbital $\tilde{d}_{x^2-y^2}$. The small residual intensity in the red spectrum in Fig. 11.3b is mostly due to the fact that $\mathbf{E}$ was aligned at a finite angle of 20° from the normal z [5]. However, there is also a smaller s–o contribution due to admixtures of the pure LF orbitals d_{xy}, d_{xz}, d_{yz} to the $d_{x^2-y^2}$ orbital in the presence of an external magnetic field.

While it is experimentally challenging to disentangle the two effects for Cu-Pc due to the difficulty of aligning the E-vector exactly along z, this can be done theoretically. In this "forbidden" geometry, the slight s–o rotation of the pure LF $d_{x^2-y^2}$ to the $\tilde{d}_{x^2-y^2}$ orbital will lead to a type I XMLD effect, revealed by the emergence of a small intensity. To illustrate this effect, we align $\mathbf{E} \parallel z$ and differentiate between $\mathbf{H}_{\mathrm{ext}} \parallel z$ and $\mathbf{H}_{\mathrm{ext}} \parallel x$. Depending on the polarity of the $\mathbf{H}_{\mathrm{ext}}$ field, the ground state of the valence shell will either be the spin-up or spin-down state, and we consider the spin-up state in (11.46) for $\mathbf{H}_{\mathrm{ext}} \parallel +z$ given by

$$|\Phi^{\uparrow}_{\mathbf{H}_{\mathrm{ext}}\parallel +z}\rangle = |d^{\uparrow}_{x^2-y^2}\rangle + \frac{\mathrm{i}\,\xi}{\mathcal{E}_0 - \mathcal{E}_2} |d^{\uparrow}_{xy}\rangle + \frac{\xi}{2(\mathcal{E}_0 - \mathcal{E}_3)} \left[|d^{\downarrow}_{xz}\rangle - \mathrm{i} |d^{\downarrow}_{yz}\rangle \right] \quad (11.71)$$

Table 11.4 Squared angular transition matrix elements per spin between the $2p$ core states and the five d valence orbitals (also see Fig. 11.2)

Polarization	Core state	d_{xy}	d_{xz}	d_{yz}	$d_{x^2-y^2}$	$d_{3z^2-r^2}$
$\mathbf{E} \parallel z$	$2p$	0	$\frac{3}{15}$	$\frac{3}{15}$	0	$\frac{4}{15}$
$\mathbf{E} \parallel y$	$2p$	$\frac{3}{15}$	0	$\frac{3}{15}$	$\frac{3}{15}$	$\frac{1}{15}$
$\mathbf{E} \parallel x$	$2p$	$\frac{3}{15}$	$\frac{3}{15}$	0	$\frac{3}{15}$	$\frac{1}{15}$

and the spin-up state (11.59) for $\mathbf{H}_{\text{ext}} \parallel +x$

$$|\Phi^{\uparrow}_{\mathbf{H}_{\text{ext}}\parallel+x}\rangle = |d^{\uparrow}_{x^2-y^2}\rangle - \frac{\mathrm{i}\,\xi}{\mathcal{E}_0-\mathcal{E}_2}|d^{\downarrow}_{xy}\rangle + \frac{\xi}{2(\mathcal{E}_0-\mathcal{E}_3)}\left[|d^{\downarrow}_{xz}\rangle - \mathrm{i}|d^{\uparrow}_{yz}\rangle\right]. \tag{11.72}$$

The spin-down states will yield the same results, as will be replacing the directions x and y. We now use these valence states to evaluate the matrix elements in (11.70). This is facilitated by use of Table 11.4 which lists the matrix elements previously shown in Fig. 11.2.

We obtain the following result for $\boxed{\mathbf{H}_{\text{ext}} \parallel +z:}$

$$\mathbf{E} \parallel z: \quad \sum_{j,m_j}\left|\langle\Phi^{\uparrow}_{\mathbf{H}_{\text{ext}}\parallel+z}|C^{(1)}_0|p_j,m_j\rangle\right|^2 = \underbrace{0}_{\text{XNLD}} + \underbrace{\frac{2}{5}\left[\frac{\xi}{2(\mathcal{E}_0-\mathcal{E}_3)}\right]^2}_{\text{XMLD}} \tag{11.73}$$

and

$$\mathbf{E} \parallel x: \quad \frac{1}{2}\sum_{j,m_j}\left|\langle\Phi^{\uparrow}_{\mathbf{H}_{\text{ext}}\parallel+z}|C^{(1)}_{-1}-C^{(1)}_{+1}|p_j,m_j\rangle\right|^2$$

$$= \underbrace{\frac{1}{5}}_{\text{XNLD}} + \underbrace{\frac{1}{5}\left[\frac{\xi}{(\mathcal{E}_0-\mathcal{E}_2)}\right]^2 + \frac{1}{5}\left[\frac{\xi}{2(\mathcal{E}_0-\mathcal{E}_3)}\right]^2}_{\text{XMLD}}. \tag{11.74}$$

While the XNLD effect identified by the first underbracket is large, we find only a very small type II XMLD effect, since $\varepsilon = 102\,\text{meV}$ (see Table 9.1) and $\mathcal{E}_0-\mathcal{E}_2 = 2.0\,\text{eV}$ is similar to $\mathcal{E}_0-\mathcal{E}_3 = 2.3\,\text{eV}$. Furthermore, it turns out that for the rotated magnetic field $\mathbf{H}_{\text{ext}} \parallel +x$ we obtain the same result as for $\mathbf{H}_{\text{ext}} \parallel +z$, given for $\mathbf{E} \parallel z$ (11.73) and $\mathbf{E} \parallel x$ by (11.74). Hence there is no type I but only a small type II XMLD.

11.8.4 XMLD in Ferromagnetic Transition Metals

Stronger XMLD effects may be observed when the magnetic alignment field increases as in *ferromagnetic transition metals*, and even in non-magnetic transition metals where magnetism is induced by hybridization with adjacent ferromagnets, as observed by Schwickert et al. [89]. The origin of XMLD in ferromagnetic transition metals has been explained by use of band theory by Kuneš and Oppeneer [90]. Below we reduce their rigorous band model, which in essence is an independent electron model, to an atomic-like model illustrated in Fig. 11.22a which can be solved analytically in a one-electron formalism and contains the essence of the physics.

We assume that the spin-down valence states are filled and describe the empty spin-up states by an equal weighting of all d_i orbitals, reflecting an effective averaging of the **k**-dependent states over the Brillouin zone. Since the s–o interaction in the $3d$ valence shell is small ($\sim$ 50 meV) compared to the exchange interaction (1–2 eV), we neglect it.

In a ferromagnet, the strong exchange field of order of 10^3 Tesla which only acts on the spin leads to a small exchange splitting of the two individual $2p_{3/2}$ and $2p_{1/2}$ core states. The resulting individual m_j substates are no longer isotropic as illustrated

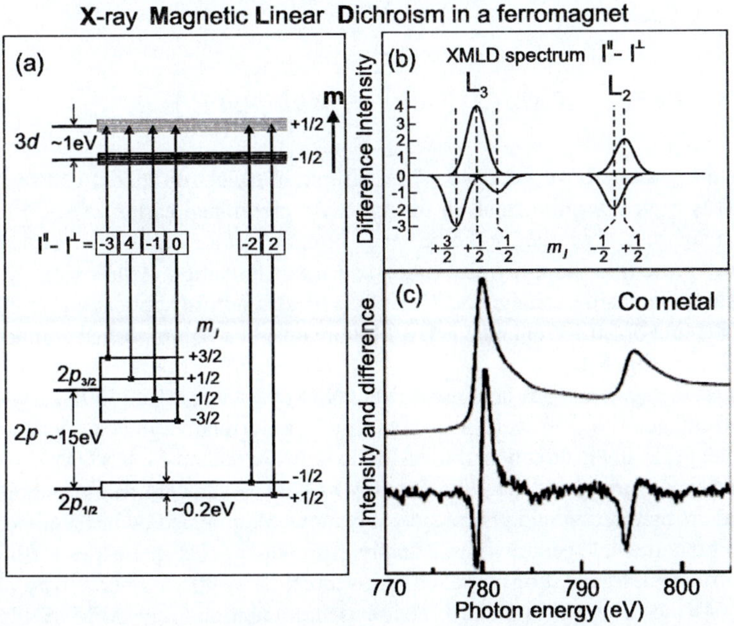

Fig. 11.22 **a** L-edge transitions from s–o split $2p_{3/2}$ and $2p_{1/2}$ core states ($\sim$15 eV) to $3d$ valence states split by a strong exchange field ($\sim$1 eV), with the magnetization direction **m** chosen so that the spin-down $3d$ states are filled and the spin-up states partially empty. The s–o splitting of the valence states has been neglected due to its small size ($\sim$50 meV). The $2p_{3/2}$ and $2p_{1/2}$ core states are further split into m_j substates ($\sim$0.2 eV) by a strong exchange field ($\sim 2\times10^3$ Tesla). Note the opposite order of m_j states for $p_{3/2}$, $l+s$ and $p_{1/2}$, $l-s$ because of the opposite sign of s. The transition intensities are calculated for polarization $\mathbf{E} \parallel z \parallel \mathbf{m}$ and $\mathbf{E} \parallel x \perp \mathbf{m}$ with the linear polarization operators in Table 11.1. The shown difference intensities multiplied (see Fig. A.8) by a factor of 90 correspond to the difference of the matrix elements in (11.70). **b** Schematic difference intensities according to **(a)**. **c** Original data of XMLD for Co metal by Schwickert et al. [89]

in Fig. 10.9. Owing to their separation of order 0.2 eV [90–92] and the associated difference in the transition matrix elements, the splitting changes the *lineshape* of the core to valence transitions as schematically shown in Fig. 11.22b. The transition intensity differences can be worked out by use of Fig. A.8. In practice, this effect is seen in the XMCD difference spectrum as a pronounced differential resonance lineshape at both the L_3 and L_2 edges. The same lineshape at the two edges is a consequence of the fact that for the $p_{3/2}$, $(l+s)$ and $p_{1/2}$, $(l-s)$ levels the signs of both s and m_j are inverted. The cross section (11.70) of the largest peak in the XMLD (difference) spectrum is $(4/90)\mathcal{A}$ which compares to the value $(2/9)\mathcal{A}$ for the XMCD difference obtained from (11.26). In the ferromagnetic metals the XMLD effect is hence considerably smaller than the XMCD effect.

In addition to x-ray absorption, XMLD spectra can also be measured by analyzing the polarization of the x-rays transmitted through the sample. Such magneto-optical polarization spectroscopy can be performed in several different geometries [93] and is complementary to x-ray absorption spectroscopy.

11.8.5 Enhanced XMLD Through Multiplet Effects

In our discussion above we have used a simple one-electron model and seen that XMLD is typically quite small in this case. As mentioned earlier, XMLD effects may be significantly larger in the presence of multiplet structure. Such structure is typically present in the x-ray absorption spectra of transition metal oxides [3, 94], which are often antiferromagnetic. One of the most beautiful applications of XMLD is the imaging of antiferromagnetic (AFM) domains, first accomplished around 2000 [95–99] (see Fig. 8.4c).

Following spectroscopy studies of cubic NiO by Alders et al. [100], early AFM domain images [98, 99] were interpreted by linking relative intensity changes of multiplet peaks to the direction of the AFM axis. More recently, it has been shown by detailed experimental and theoretical studies on NiO by van der Laan, Arenholz, and co-workers that the situation is considerably more complicated by the superposition of two fundamental spectra with distinctly different spectral intensities [101–104]. This led to a reinterpretation of the AFM axis direction by 90°, as reviewed by van der Laan [64] and shown in Fig. 1.21d. The revised orientation of the AFM axis in NiO is in agreement with that found by neutron scattering [105] and the *perpendicular* exchange coupling between AFM NiO and ferromagnetic Co theoretically predicted by Koon [106] and Schulthess and Butler [107].

Another family of antiferromagnets are Fe-based oxides. The linear polarization dependence of XAS spectra of Fe_2O_3 (hematite) was first investigated by Kuiper et al. [108] and revised more recently in both experiment and theory by Arenholz et al. [109]. As for the NiO case it was found that the XMLD spectra consist of a superposition of two fundamental multiplet spectra.

Similar multiplet structure as Fe_2O_3 is found in the much studied rare earth (R) orthoferrites, $RFeO_3$. As an example we show in Fig. 11.23 the strong XMLD effect

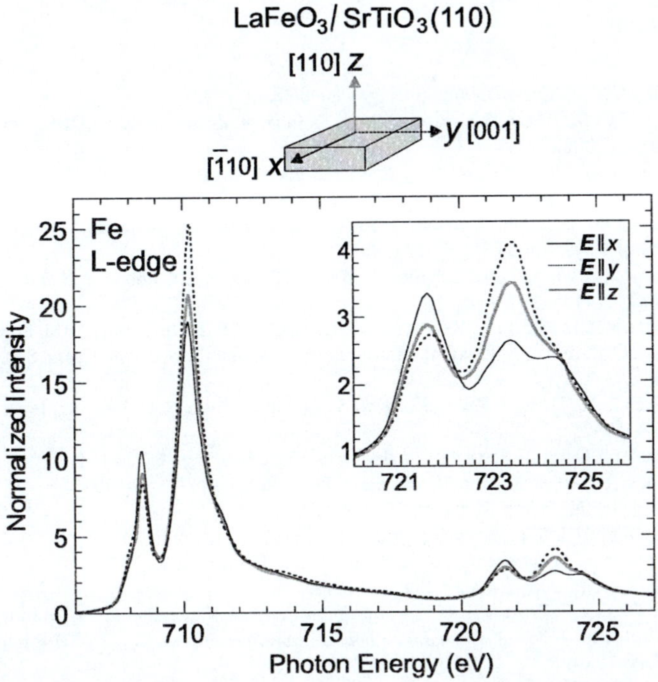

Fig. 11.23 Polarization dependent Fe $L_{2,3}$ XMLD spectra of $LaFeO_3/SrTiO_3$ (110) [110]. For the three shown spectra the electric field vector of the linearly polarized x-rays was oriented along the three indicated crystallographic axes

of the multiplet peaks in $LaFeO_3$ grown on cubic $SrTiO_3$ (110), taken from the work of Lüning et al. [110]. Owing to the local cubic coordination of the Fe by O atoms, the measured polarization dependence is of magnetic origin, also shown by its disappearance above the reduced Néel when doped with Sr. The $LaFeO_3/SrTiO_3$ (110) system was also used to demonstrate AFM domain imaging by Nolting and Scholl and collaborators [96, 97, 111]. Later studies revealed the sensitivity to the AFM axis orientation to strain [112] and the influence of different growth substrates (e.g. MgO) on domain size and exchange bias when coupled to ferromagnetic Co [113].

In summary, it appears that the enhanced XMLD effects in oxides arising from multiplet effects come at the price of spectral complexity, which typically requires a detailed comparison of experiment with theory.

References

1. J. Stöhr, *NEXAFS Spectroscopy* (Springer, Heidelberg, 1992)
2. J. Stöhr, H.C. Siegmann, *Magnetism: From Fundamentals to Nanoscale Dynamics* (Springer, Heidelberg, 2006)
3. F. de Groot, A. Kotani, *Core Level Spectroscopy of Solids* (CRC Press, Boca Raton, USA, 2008)
4. M. Weinelt, N. Wassdahl, T. Wiell, O. Karis, J. Hasselström, P. Bennich, A. Nilsson, J. Stöhr, M. Samant, Phys. Rev. B **58**, 7351 (1998)
5. S. Stepanow, A. Mugarza, G. Ceballos, P. Moras, J.C. Cezar, C. Carbone, P. Gambardella, Phys. Rev. B **82**, 014405 (2010)
6. S. Stepanow, A.L. Rizzini, C. Krull, J. Kavich, J.C. Cezar, F. Yakhou-Harris, P.M. Sheverdyaeva, P. Moras, C. Carbone, G. Ceballos, A. Mugarza, P. Gambardella, J. Am. Chem. Soc. **136**, 5451 (2014)
7. C.T. Chen, L.H. Tjeng, J. Kwo, H.L. Kao, P. Rudolf, F. Sette, R.M. Fleming, Phys. Rev. Lett. **68**, 2543 (1992)
8. B.T. Thole, P. Carra, F. Sette, G. van der Laan, Phys. Rev. Lett. **68**, 1943 (1992)
9. P. Carra, B.T. Thole, M. Altarelli, X. Wang, Phys. Rev. Lett. **70**, 694 (1993)
10. J. Stöhr, H. König, Phys. Rev. Lett. **75**, 3748 (1995)
11. A. Ankudinov, J.J. Rehr, Phys. Rev. B **51**, 1282 (1995)
12. J. Stöhr, J. Elec. Spectrosc. Relat. Phenom. **75**, 253 (1995)
13. J. Stöhr, J. Magn. Magn. Mater. **200**, 470 (1999)
14. P.G. De Gennes, J. Prost, *The Physics of Liquid Crystals* (Clarendon Press, Oxford, 1993)
15. E.W. Thulstrup, J. Michl, *Elementary Polarization Spectroscroscopy* (VCH Publishers Inc., New York, 1989)
16. J. Michl, E.W. Thulstrup, *Spectroscopy with Polarized Light* (VCH Publishers Inc., New York, 1995)
17. J. Stöhr, M.G. Samant, J. Elec. Spectrosc. Relat. Phenom. **98**, 189 (1999)
18. J. Stöhr, M.G. Samant, J. Lüning, A.C. Callegari, P. Chaudhari, J.A. Doyle, J.A. Lacey, S.A. Lien, S. Purushothaman, J.L. Speidell, Science **292**, 2299 (2001)
19. P. Chaudhari et al., Nature **411**, 56 (2001)
20. J. Stöhr, M.G. Samant, J. Lüning, Synchr. Rad. News **14**, 32 (2001)
21. A.M. Yao, M.J. Padgett, Adv. Opt. and Photon. **3**, 161 (2011)
22. C.R. Natoli, C. Brouder, P. Sainctavit, J. Goulon, C. Goulon-Ginet, A. Rogalev, Eur. Phys. J. B **4**, 1 (1998)
23. S. Turchini, N. Zema, S. Zennaro, L. Alagna, B. Stewart, R.D. Peacock, T. Prosperi, J. Am. Chem. Soc. **126**, 4532 (2004)
24. S. Villaume, P. Norman, Chirality **21**, E13 (2009)
25. S. Bernadotte, A.J. Atkins, C.R. Jacob, J. Chem. Phys. **137**, 204106 (2012)
26. Y. Izumi, M. Tanabe, A. Imazu, A. Mimoto, M. Tanaka, A. Agui, T. Muro, K. Nakagawa, J. Am. Chem. Soc. **138**, 074305 (2013)
27. O. Takahashi, M. Kimoto, L.G.M. Pettersson, Chem. Phys. **450–451**, 109 (2015)
28. C. Brouder, J. Phys., Condens. Matter **2**, 701 (1990)
29. J. Goulon, C. Goulon-Ginet, A. Rogalev, V. Gotte, C. Malgrange, C. Brouder, J. Chem. Phys. **108**, 6394 (1998)
30. L. Alagna, T. Prosperi, S. Turchini, J. Goulon, A. Rogalev, C. Goulon-Ginet, C.R. Natoli, R.D. Peacock, B. Stewart, Phys. Rev. Lett. **80**, 4799 (1998)
31. J. Goulon, A. Rogalev, C. Goulon-Ginet, G. Benayoun, L. Paolasini, C. Brouder, C. Malgrange, P.A. Metcalf, Phys. Rev. Lett. **85**, 4385 (2000)
32. J. Goulon, A. Rogalev, F. Wilhelm, C. Goulon-Ginet, P. Carra, D. Cabaret, C. Brouder, Phys. Rev. Lett. **88**, 237401 (2002)
33. J. Goulon, A. Rogalev, F. Wilhelm, N. Jaouen, C. Goulon-Ginet, C. Brouder, J. Phys.: Condens. Matter **15**, S633 (2003)

34. E.C. Stoner, Proc. Roy. Soc. (London) A **165**, 372 (1938)
35. E.C. Stoner, E.P. Wohlfarth, Phil. Trans. Roy. Soc. (London) A **240**, 599 (1948)
36. A. Abragam, B. Bleaney, *Electron Paramagnetic Resonance of Transition Ions* (Clarendon Press, Oxford, 1970)
37. L.L. Hirst, Adv. Phys. **27**, 231 (1978)
38. O. Eriksson, B. Johansson, R.C. Albers, A.M. Boring, M.S.S. Brooks, Phys. Rev. B **42**, 2707 (1990)
39. J. Trygg, B. Johansson, O. Eriksson, J.M. Wills, Phys. Rev. Lett. **75**, 2871 (1995)
40. S.P. Cramer, *X-Ray Spectroscopy with Synchrotron Radiation: Fundamentals and Applications* (Springer, Berlin, 2020)
41. J. Stöhr, R. Nakajima, IBM, J. Res. Develop. **42**, 73 (1998)
42. B.T. Thole, G. van der Laan, Phys. Rev. B **38**, 3158 (1988)
43. C. Kittel, Phys. Rev. **76**, 743 (1949)
44. J. Pelzl, R. Meckenstock, D. Spoddig, F. Schreiber, J. Pflaum, Z. Frait, J. Phys.: Condens. Matter **15**, S451 (2003)
45. G. Aubert, J. Appl. Phys. **39**, 504 (1968); F. Ducastelle, F. Cyrot-Lackmann, J. Physique (Paris) **C1**, 537 (1971); J. P. Rebouillat, IEEE Magn. **8**, 630 (1972); P. Escudier, Ann. Physique (Paris) **9**, 125 (1975); R. Pauthenet, J. Appl. Phys. **53**, 8187 (1982)
46. M.G. Samant, J. Stöhr, S.S.P. Parkin, G.A. Held, B.D. Hermsmeier, F. Herman, M. van Schilfgaarde, L.C. Duda, D.C. Mancini, N. Wassdahl, R. Nakajima, Phys. Rev. Lett. **72**, 1112 (1994)
47. P. Gambardella, A. Dallmeyer, K. Maiti, M.C. Malagoli, W. Eberhardt, K. Kern, C. Carbone, Nature **416**, 301 (2002)
48. P. Gambardella, S. Rusponi, M. Veronese, S.S. Dhesi, C. Grazioli, A. Dallmeyer, I. Cabria, R. Zeller, P.H. Dederichs, K. Kern, C. Carbone, H. Brune, Science **300**, 1130 (2003)
49. P. Gambardella, S. Rusponi, T. Cren, N. Weiss, H. Brune, C. R. Physique **6**, 75 (2005)
50. A. Langenberg, K. Hirsch, A. Ławicki, V. Zamudio-Bayer, M. Niemeyer, P. Chmiela, B. Langbehn, A. Terasaki, B.V. Issendorff, J.T. Lau, Phys. Rev. B **90**, 184420 (2014)
51. V. Zamudio-Bayer, K. Hirsch, A. Langenberg, A. Ławicki, A. Terasaki, B.V. Issendorff, J.T. Lau, J. Chem. Phys. **143**, 244318 (2015)
52. W. Pauli, Z. Phys. **43**, 601 (1927)
53. E. Merzbacher, *Quantum Mechanics* (Wiley, New York, 1970)
54. S. Lee, M. Yudkowsky, W.P. Halperin, M.Y. Ogawa, B.M. Hoffman, Phys. Rev. B **35**, 5003 (1987)
55. M. Komelj, C. Ederer, J.W. Davenport, M. Fänle, Phys. Rev. B **66**, 140407 (2002)
56. C. Ederer, M. Komelj, J.W. Davenport, M. Fänle, J. Elec. Spectrosc. Relat. Phenom. **130**, 97 (2003)
57. C. Ederer, M. Komelj, M. Fänle, Phys. Rev. B **68**, 052402 (2003)
58. J. Minár, S. Bornemann, O. Šipr, S. Polesya, H. Ebert, Appl. Phys. A **82** (2006)
59. O. Šipr, J. Minár, H. Ebert, Eur. Phys. Lett. **87** (2009)
60. G. Schütz, W. Wagner, W. Wilhelm, P. Kienle, R. Zeller, R. Frahm, G. Materlik, Phys. Rev. Lett. **58**, 737 (1987)
61. M.N. Baibich, J.M. Broto, A. Fert, F.N.V. Dau, F. Petroff, P. Eitenne, G. Creuzet, A. Friederich, J. Chazelas, Phys. Rev. Lett. **61**, 2472 (1988)
62. G. Binasch, P. Grünberg, F. Saurenbach, W. Zinn, Phys. Rev. B **39**, 4828 (1989)
63. E. Beaurepaire, H. Bulou, F. Scheurer, J.P. Kappler (eds.), *Magnetism: A Synchrotron Radiation Approach*. Springer Lecture Notes in Physics, vol. 697 (Springer, 2006)
64. G. van der Laan, J. Phys.: Conf. Ser. **430**, 012127 (2013)
65. P. Fischer, H. Ohldag, Rep. Prog. Phys. **78**, 094501 (2015)
66. H. Stoll et al., Front. Phys. **3**, 26 (2015)
67. S.D. Sarma, J. Fabian, X. Hu, I. Žutić, Superlattice Microstr. **27**, 289 (2000)
68. Y. Tsymbal, I. Žutić, *Handbook of Spin Transport and Magnetism* (CRC Press, 2012)
69. S. Maekawa, S. Valenzuela, E. Saitoh, T. Kimura, *Spin Current*, 2nd edn. (Oxford University Press, 2017)

70. S. Pizzini, A. Fontaine, C. Giorgetti, E. Dartyge, J.F. Bobo, M. Piecuch, F. Baudelet, Phys. Rev. Lett. **74**, 1470 (1995)
71. G.A. Held, M.G. Samant, J. Stöhr, S.S.P. Parkin, B.D. Hermsmeier, M. van Schilfgaarde, R. Nakajima, Z. Phys. B **100**, 335 (1996)
72. N.F. Mott, Proc. Phys. Soc. (London) **47**, 571 (1935)
73. N.F. Mott, Proc. Phys. Soc. (London), Ser. A **153**, 699 (1936)
74. N.F. Mott, Proc. Phys. Soc. (London), Ser. A **156**, 368 (1936)
75. R. Kukreja, S. Bonetti, Z. Chen, D. Backes, Y. Acremann, J. Katine, A.D. Kent, H.A. Dürr, H. Ohldag, J. Stöhr, Phys. Rev. Lett. **115**, 096601 (2015)
76. J. Ding, W. Zhang, M.B. Jungfleisch, J.E. Pearson, H. Ohldag, V. Novosad, A. Hoffmann, Phys. Rev. Res. **2**, 013262 (2020)
77. A. Marmodoro, S. Wimmer, O. Šipr, M. Ogura, H. Ebert, Phys. Rev. Res. **2**, 032067 (2020)
78. M.I. Dyakonov, V.I. Perel, Sov. Phys. JETP Lett. **13**, 467 (1971)
79. J.E. Hirsch, Phys. Rev. Lett. **83**, 1834 (1999)
80. A. Manchon, J. Železný, I. Miron, T. Jungwirth, J. Sinova, A. Thiaville, K. Garello, P. Gambardella, Rev. Mod. Phys. 91, 035004 (2019)
81. C. Stamm, C. Murer, Y. Acremann, M. Baumgartner, R. Gort, S. Däster, A. Kleibert, K. Garello, J. Feng, M. Gabureac, Z. Chen, J. Stöhr, P. Gambardella, Phys. Rev. B **100**, 024426 (2019)
82. B.T. Thole, G. van der Laan, G.A. Sawatzky, Phys. Rev. Lett. **55**, 2086 (1985)
83. G. van der Laan, B.T. Thole, G.A. Sawatzky, J.B. Goedkoop, J.C. Fuggle, J.M. Esteva, R. Karnatak, J.P. Remeika, H.A. Dabkowska, Phys. Rev. B **34**, 6529 (1986)
84. G. van der Laan, B.T. Thole, Phys. Rev. B **43**, 13401 (1991)
85. G.Y. Guo, H. Ebert, W.M. Temmerman, P.J. Durham, Phys. Rev. B **50**, 3861 (1994)
86. G. van der Laan, J. Phys.: Condens. Matter **10**, 3239 (1998)
87. G. van der Laan, Phys. Rev. B **57**, 5250 (1998)
88. G. van der Laan, Phys. Rev. Lett. **82**, 640 (1999)
89. M.M. Schwickert, G.Y. Guo, M.A. Tomaz, W.L. O'Brien, G.R. Harp, Phys. Rev. B **58**, R4289 (1998)
90. J. Kuneš, P.M. Oppeneer, Phys. Rev. B **67**, 024431 (2003)
91. H. Ebert, J. Phys.: Condens. Matter **1**, 9111 (1989)
92. L. Baumgarten, C.M. Schneider, H. Peteren, F. Schäfers, J. Kirschner, Phys. Rev. Lett. **65**, 492 (1990)
93. H.C. Mertins, S. Valencia, A. Gaupp, W. Gudat, P.M. Oppeneer, C.M. Schneider, Appl. Phys. A **80**, 1011 (2005)
94. F. de Groot, Coord. Chem. Rev. **249**, 31 (2005)
95. J. Stöhr, A. Scholl, T.J. Regan, S. Anders, J. Lüning, M.R. Scheinfein, H.A. Padmore, R.L. White, Phys. Rev. Lett. **83**, 1862 (1999)
96. A. Scholl, J. Stöhr, J. Lüning, J.W. Seo, J. Fompeyrine, H. Siegwart, J.P. Locquet, F. Nolting, S. Anders, E.E. Fullerton, M.R. Scheinfein, H.A. Padmore, Science **287**, 1014 (2000)
97. F. Nolting, A. Scholl, J. Stöhr, J.W. Seo, J. Fompeyrine, H. Siegwart, J.P. Locquet, S. Anders, J. Lüning, E.E. Fullerton, M.F. Toney, M.R. Scheinfein, H.A. Padmore, Nature **405**, 767 (2000)
98. H. Ohldag, A. Scholl, F. Nolting, S. Anders, F.U. Hillebrecht, J. Stöhr, Phys. Rev. Lett. **86**, 2878 (2001)
99. H. Ohldag, T.J. Regan, J. Stöhr, A. Scholl, F. Nolting, J. Lüning, C. Stamm, S. Anders, R.L. White, Phys. Rev. Lett. **87**, 247201 (2001)
100. D. Alders, L.H. Tjeng, F.C. Voogt, T. Hibma, G.A. Sawatzky, C.T. Chen, J. Vogel, M. Sacchi, S. Iacobucci, Phys. Rev. B **57**, 11623 (1998)
101. E. Arenholz, G. van der Laan, R.V. Chopdekar, Y. Suzuki, Phys. Rev. Lett. **98**, 197201 (2007)
102. E. Arenholz, G. van der Laan, F. Nolting, Appl. Phys. Lett. **93**, 162506 (2008)
103. H. Ohldag, G. van der Laan, E. Arenholz, Phys. Rev. B **79**, 052403 (2009)
104. G. van der Laan, N.D. Telling, A. Potenza, S.S. Dhesi, E. Arenholz, Phys. Rev. B **83**, 064409 (2011)
105. J.A. Borchers, R.W. Erwin, S.D. Berry, D.M. Lind, J.F. Ankner, E. Lochner, K.A. Shaw, D. Hilton, Phys. Rev. B **51**, 8276 (1995)

106. N.C. Koon, Phys. Rev. Lett. **78**, 4865 (1997)
107. T.C. Schulthess, W.H. Butler, Phys. Rev. Lett. **81**, 4516 (1998)
108. P. Kuiper, B.G. Searle, P. Rudolf, L.H. Tjeng, C.T. Chen, Phys. Rev. Lett. **70**, 1549 (1993)
109. E. Arenholz, G. van der Laan, R.V. Chopdekar, Y. Suzuki, Phys. Rev. B **74**, 094407 (2006)
110. J. Lüning, F. Nolting, A. Scholl, H. Ohldag, J. Seo, J. Fompeyrine, J.P. Locquet, J. Stöhr, Phys. Rev. B **67**, 214433 (2003)
111. A. Scholl, F. Nolting, J.W. Seo, H. Ohldag, J. Stöhr, S. Raoux, J.P. Locquet, J. Fompeyrine, Appl. Phys. Lett. **85**, 4085 (2004)
112. S. Czekaj, F. Nolting, L.J. Heyderman, P.R. Willmott, G. van der Laan, Phys. Rev. B **73**, 020401(R) (2006)
113. J.W. Seo, E.E. Fullerton, F. Nolting, A. Scholl, J. Fompeyrine, J.P. Locquet, J. Phys.: Condens. Matter **20**, 264014 (2008)

Chapter 12
Quantum Theory of X-Ray Emission and Thomson Scattering

12.1 Introduction and Overview

Following our discussion of XAS and its polarization dependence in the last two chapters we here continue with the discussion of two other important first order processes, namely *x-ray emission spectroscopy* (XES) and *Thomson scattering*, outlined in Sect. 9.5 and illustrated schematically in Fig. 9.4b, c. X-ray emission is the inverse of the x-ray absorption process, and both can be described within KHD theory by transitions between two states. The interaction Hamiltonian for the two processes is the same, given by $\mathcal{H}_{int} = -e\,\mathbf{r}\cdot\mathbf{E}$ in (9.22). In XES, one starts with an initial state that is an excited state with a core hole in an inner shell and considers its decay to another excited electronic state, where the core hole is now in an outer electronic shell, typically the valence shell.

Our treatment of XES covers the quantum mechanical derivation of the emission rate as well as the tricky problem of the emitted linewidth. We will show that, in general, the XES linewidth contains additive contributions from the initial and final core hole states. Particularly remarkable is that the XES linewidth is not solely determined by the dipolar transition width that describes *radiative* decays, but that it always has a "ghost contribution" from the *Auger* decay channel. We finish our discussion of XES with examples of fundamental XES experiments of atomic and molecular gases and solids.

We then discuss the quantitative separation of the radiative XES and non-radiative Auger channels, expressed by the so-called fluorescence yield. It is a measure of the total XES probability averaged over polarization, emission directions, and energies and is defined as the dimensionless ratio of the total x-ray emission probability relative to the total x-ray plus Auger emission probability. In the process, we discuss how our quantum mechanical XES results compare to tabulated semi-empirical values.

The last part of the present chapter is devoted to the quantum derivation of atomic *Thomson scattering*. Like XAS and XES, it is also described as a first order process

J. Stöhr, *The Nature of X-Rays and Their Interactions with Matter*, Springer Tracts in Modern Physics 288, https://doi.org/10.1007/978-3-031-20744-0_12

within KHD theory, but the interaction Hamiltonian is given by $e^2 E^2/(2m_e\omega^2)$ in (9.22). The quantum formalism is shown to give the same results as the semi-classical formalism for *elastic* Thomson scattering, previously discussed in Sect. 6.2.4.

While elastic Thomson scattering, which underlies Bragg diffraction, has historically had larger impact in x-ray science than *inelastic* Thomson scattering, we shall also discuss the latter. It is based on detecting the change in energy, direction, and polarization of the scattered photons. The lost energy reflects fundamental excitations in a sample, ranging from phonons and magnons to more energetic excitations of valence and even core electrons. As a particularly interesting application we discuss inelastic Thomson scattering in the form of so-called *x-ray Raman scattering* (XRS) also called *non-resonant inelastic x-ray scattering* (NIXS), which are shown to give the same results as XAS spectroscopy in the limit of small momentum transfer.

12.2 Quantum Formulation of X-Ray Emission Spectroscopy (XES)

12.2.1 XES History and Terminology

We start by defining what we mean by "x-ray emission spectroscopy", or XES for short. This is required because modern spectroscopy experiments are typically performed differently than early x-ray emission experiments. Following early studies by Barkla, Moseley, and Manne Siegbahn in the 1910–1920 period, high-resolution x-ray emission experiments were pioneered by the group of Kai Siegbahn at Uppsala University, Sweden, in the 1970s. In these studies, core holes were conveniently produced by electron impact ionization. With the advent of synchrotron radiation, x-ray emission studies increasingly involved better defined core excitations by x-rays. It became then important to distinguish different ways of preparing the core hole state since they required different quantum treatments.

In this book we define XES as the study of the emission process that follows after core electron *ionization*. The excitation process may involve electron or photon ionization, and the excitation process does not enter into the XES description. This case is distinguished from resonant excitations of core electrons into unfilled localized valence states, referred to as *resonant inelastic x-ray scattering* or RIXS. In RIXS, the excitation step is specifically included since the excited core electron influences the decay, as discussed in Chap 13. The key theoretical difference is that XES may be described by first order KHD theory, while RIXS requires a second order description.

We define the term "x-ray emission spectroscopy", or XES, to describe the case where the excitation process does not enter into the description of the atomic core hole decay process. The decay process starts from an *initial* state that contains a core hole and may have photons present in a polarization and wavevector mode $p\mathbf{k}$. The *final* XES state consists of one more photon and an electronic hole state in an outer atomic shell.

As in the formulation of XAS in Sect. 10.2, the quantum mechanical description of XES may be accomplished through the *first order* term in the KHD formula (9.21) and utilizes the interaction Hamiltonian written as $\mathcal{H}_{int} = -e\,\mathbf{r}\cdot\mathbf{E}$ in (9.22). The KHD transition rate is given by (10.1) or

$$\mathcal{W}_{if} = \frac{2\pi}{\hbar}\,|\langle f|e\,\mathbf{r}\cdot\mathbf{E}|i\rangle|^2\,\rho(\mathcal{E}_f)\,\delta(\mathcal{E}_f - \mathcal{E}_i). \tag{12.1}$$

In contrast to x-ray absorption, the general field operator given by $\mathbf{E} = \mathbf{E}^+ + \mathbf{E}^-$ is now given by the photon creation part $\mathbf{E}^- \propto \mathbf{a}^\dagger_{p\mathbf{k}}$.

The *initial* XES state $|i\rangle$ in (12.1) is a product state of the upper electronic state $|b\rangle$ and a photon number state $|n_{p\mathbf{k}}\rangle$ containing $n_{p\mathbf{k}}$ photons in the mode $p\mathbf{k}$. The decay may occur without the presence of any photons. The initial photon state is then the zero-point quantum vacuum $|n_{p\mathbf{k}}\!=\!0\rangle$. The *final* XES state is the product of the lower electronic state $|a\rangle$ and the photon state $|n_{p\mathbf{k}} + 1\rangle$ since a photon has been generated in the decay.

12.2.2 The Photon Part of the Transition Matrix Element

The photon part of the transition matrix element in (12.1) is evaluated in analogy to the XAS case (10.2) as

$$\left|\langle n_{p\mathbf{k}} + 1|\mathbf{E}^-|n_{p\mathbf{k}}\rangle\right|^2 = \frac{(n_{p\mathbf{k}}+1)\hbar\omega}{2\epsilon_0 V_{p\mathbf{k}}}, \tag{12.2}$$

where $V_{p\mathbf{k}}$ is given by (3.106). The factor $n_{p\mathbf{k}} + 1$ is the sum of the spontaneous and stimulated contributions to emission. Before we further evaluate (12.1) into separate photon and electronic state components, let us take a closer look at the fundamental and well-known term $1 + n_{p\mathbf{k}}$.

12.2.2.1 Spontaneous and Stimulated Decays: Zero-Point Field

The KHD theory conveniently utilizes the concept of labeling photons by their $p\mathbf{k}$ modes reflecting the quantization or division of the EM field into "coherent" fractions. The factor $1 + n_{p\mathbf{k}}$ means that emission of a photon may be induced two ways, either *spontaneously* by the zero-point (ZP) field, with a relative probability 1, or by stimulation of the decay by $n_{p\mathbf{k}}$ real photons, with a probability $n_{p\mathbf{k}}$. The distinction between stimulated and spontaneous emission was introduced by Einstein in 1917 [1, 2] in his derivation of Planck's blackbody radiation law [3, 4].

There is an important difference between the nature of the photons that are created in spontaneous and stimulated emission. In *spontaneous emission*, the photon may be emitted into any direction, owing to the random nature of the ZP field from which it is created. In contrast, *stimulated* emission is a cloning process that results in the emission of two indistinguishable photons which have been referred to as "cloned biphotons" [5, 6].

In *spontaneous* emission, photons are emitted into random directions and with random polarizations. The quantum mechanical average corresponds to the classical emission of a spherical wave.
In *stimulated* emission, the created photons represent clones of the incident ones and hence are emitted into the same mode, corresponding to the classical emission of a plane wave.

The factor $1+n_{p\mathbf{k}}$ reveals that the stimulated decay probability induced by a single real photon, $n_{p\mathbf{k}} = 1$, is the same as the spontaneous one induced by virtual photons of the ZP field which contains one half photon per mode $p\mathbf{k}$ or one photons per $\mathbf{k}$-mode. Hence the ZP field contains one photon per $\mathbf{k}$-mode.

We can express the quantized ZP field in analogy to (3.62) as

$$|E_{\mathrm{ZP}}^{k}|^2 = \frac{\hbar\omega}{2\epsilon_0 V_{p\mathbf{k}}}. \tag{12.3}$$

The incident field containing n_k photons in the same polarization mode is similarly given by

$$|E_0^k|^2 = n_k \frac{\hbar\omega}{2\epsilon_0 V_{p\mathbf{k}}}. \tag{12.4}$$

Therefore n_k is simply a measure of the relative field strength of the incident to zero-point field in the mode k according to

$$n_k = \frac{|E_0^k|^2}{|E_{\mathrm{ZP}}^k|^2}. \tag{12.5}$$

While it appears that it should be easy to overcome the ZP field and change spontaneous to stimulated emission by use of modern synchrotron sources, this is not so. The reason is that it takes an XFEL to increase the photon degeneracy parameter, defined in Sect. 4.2.7 as the number of photons in the same mode, i.e. $n_{p\mathbf{k}}$, to values larger than 1, as illustrated in Fig. 4.5.

12.2.3 XES Decay Time and Linewidth

Next we consider the factor $\delta(\mathcal{E}_f - \mathcal{E}_i)\,\rho(\mathcal{E}_f)$ in (12.1). The δ-function assures energy conservation in the emission process between all photon and electronic degrees of freedom. The density of states factor $\rho(\mathcal{E}_f)$ in (12.1) reflects the total density per unit energy of the *combined* photon and electronic systems after emission. We can, however, take advantage of the fact that energy is also conserved between the electronic and photon systems, so that we can separately consider them as we have done for the transition matrix element in Sect. 12.2.2.

For our discussion we adopt the notations used for the XAS case in Fig. 10.3. The electronic system is initially in the excited core hole state $|b\rangle$. It contains a certain stored energy $\mathcal{E}_b$. Since the core hole state can decay, its energy has an intrinsic uncertainty width. We first assume for simplicity that the electronic final state $|a\rangle$ is the electronic *ground state*. Since it does not decay, it has an infinitely long lifetime, and there is no uncertainty in its energy $\mathcal{E}_a$.

We consider the decay process illustrated by the electronic diagram shown in Fig. 12.1a. In the figure we have indicated an uncertainty Γ_b (dimension [energy]) in the initial state energy $\mathcal{E}_b$. In a semi-classical model, the energy uncertainty corresponds to a decay rate $\Gamma_b/\hbar$ (dimension [1/time]) of state $|b\rangle$. At this point, Γ_b is a parameter and we do consider its deeper meaning.

We keep in mind, however, that we later need to clarify two issues associated with our assumed model. First, we need to consider that the final state $|a\rangle$ is not the ground state but may be another excited electronic ground state of finite lifetime. Second, we need to address that core hole decays actually involve competing radiative and Auger decay channels, as previously discussed in Sect. 10.4.1 and illustrated in Fig. 10.3.

The wavefunction or probability amplitude of our two-level system at $t \geq 0$ is given by

$$\Phi(t) = \sqrt{\frac{\Gamma_b}{\hbar}}\; \mathrm{e}^{\mathrm{i}\,\omega_{ba} t}\, \mathrm{e}^{-\Gamma_b t/(2\hbar)}. \tag{12.6}$$

The system oscillates with the resonance frequency ω_{ba}, where $\mathcal{E}_{ba} = \hbar\omega_{ba}$ is the energy separation of the excited electronic state $|b\rangle$ from the ground state $|a\rangle$. At $t > 0$ the amplitude of the wavefunction decreases exponentially with time and integration of the probability $|\Phi(t)|^2$ over time shows that the sum of upper and lower state *populations* is conserved, as required,

$$\int_0^\infty |\Phi(t)|^2\, \mathrm{d}t = \frac{\Gamma_b}{\hbar} \int_0^\infty \mathrm{e}^{-\Gamma_b t/\hbar}\, \mathrm{d}t = 1. \tag{12.7}$$

The temporal wavefunction $\Phi(t)$ is given in the frequency domain by the Fourier transform

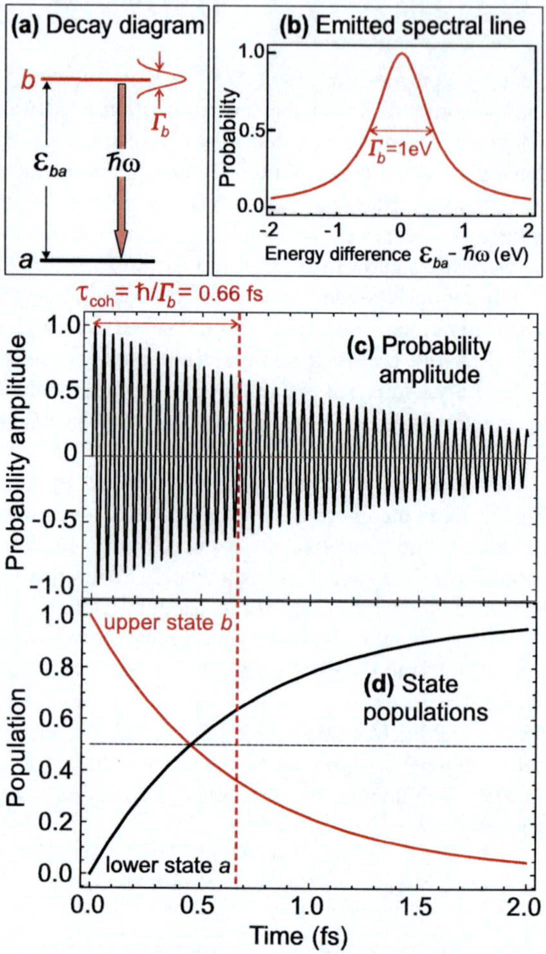

Fig. 12.1 **a** Decay diagram of an upper electronic state $|b\rangle$ with an energy uncertainty of Γ_b to a lower state $|a\rangle$. **b** Lorentzian shape of the emitted radiation of FWHM Γ_b. **c** Decrease of the emitted probability amplitude (12.6) in the form $\sin(\omega_{ba}t)\exp[-\Gamma_b t/(2\hbar)]$ for $\Gamma_b = 1\,\text{eV}$ and $\hbar\omega_{ba} = 100\,\text{eV}$ with time t in fs. The coherence time $\hbar/\Gamma_b$ over which the upper population decays to $1/e$ is shown in red. **d** Change of the upper and lower state populations with time

$$\Psi(\omega) = \sqrt{\frac{\Gamma_b}{\hbar}} \int_0^\infty e^{i\,\omega_{ba}t}\, e^{-\Gamma_b t/(2\hbar)}\, e^{-i\,\omega t}\, dt = \frac{\sqrt{\Gamma_b \hbar}}{i(\mathcal{E}_{ba} - \hbar\omega) + \Gamma_b/2}. \tag{12.8}$$

The associated probability distribution has a Lorentzian shape of FWHM Γ_b given by

$$|\Psi(\omega)|^2 = \frac{2\,h}{\pi\Gamma_b} \frac{(\Gamma_b/2)^2}{(\mathcal{E}_{ba} - \hbar\omega)^2 + (\Gamma_b/2)^2}. \tag{12.9}$$

Now we invoke energy conservation between the electronic and photon systems. Since the electronic system ends up in the zero-energy ground state, the expression (12.9) reflects the probability of the emitted photon energy distribution. It is illustrated in normalized form in Fig. 12.1b. The final state quantity $\rho(\mathcal{E}_f)\delta(\mathcal{E}_f - \mathcal{E}_i)$ in (12.1) can then be expressed in the form

$$\boxed{\rho(\mathcal{E}_f)\delta(\mathcal{E}_f - \mathcal{E}_i) = \frac{|\Psi(\omega)|^2}{h} = \frac{2}{\pi\Gamma_b}\frac{(\Gamma_b/2)^2}{(\mathcal{E}_{ba} - \hbar\omega)^2 + (\Gamma_b/2)^2}}. \tag{12.10}$$

It has the dimension [1/energy] and fulfills the normalization

$$\int_{-\infty}^{\infty} \rho(\mathcal{E}_f)\,\delta(\mathcal{E}_f - \mathcal{E}_i)\,\mathrm{d}(\hbar\omega) = 1 \tag{12.11}$$

corresponding to one emitted photon. The decay process may be pictured as illustrated in Fig. 12.1c, d with the probability amplitude oscillating between the excited and ground state during the emission process with concomitant change of the excited and ground state populations.

In the following two sections we extend our simple model to the cases that the final state is another excited electronic state and that there is also an Auger decay channel.

12.2.4 Decays to Excited Final States

We now consider that the lower electronic state is another excited state of *finite lifetime* that may decay further, which is the situation encountered in practice. This problem was first addressed from a quantum theoretical point of view in 1930 by Weisskopf and Wigner [7, 8].

We may treat the case of a finite lifetime of the lower state $|a\rangle$ by introducing an associated decay width Γ_a as illustrated in Fig. 12.2. It illustrates that we now have energy uncertainties expressed by Lorentzian density distributions in both upper and lower states, leading to an increased energy uncertainty (width) of the emitted photons.

From Fig. 10.4 we know that the decay width Γ of electronic states increases with their energy $\mathcal{E}_0$ above the ground state with an approximate correlation $\Gamma/\mathcal{E}_0 \approx 10^{-3}$. For a core hole energy of $\mathcal{E}_b \sim 1\,\mathrm{keV}$ in Fig. 12.2 we thus have $\Gamma_b \sim 1\,\mathrm{eV}$ and for a typical final state energy $\mathcal{E}_a \sim 1-10\,\mathrm{eV}$ we expect $\Gamma_a \sim 1-10\,\mathrm{meV}$. Hence in XES, the core hole decay width will typically dominate. In other words, core holes will decay much faster (femtoseconds) than valence holes (picoseconds). Let us nevertheless consider the general case following Weisskopf and Wigner [7, 8].

By inclusion of energy conservation, indicated by the length of the red arrow in Fig. 12.2, we need to integrate over all energies $\mathcal{E}_x$ and obtain the following expression instead of (12.10)

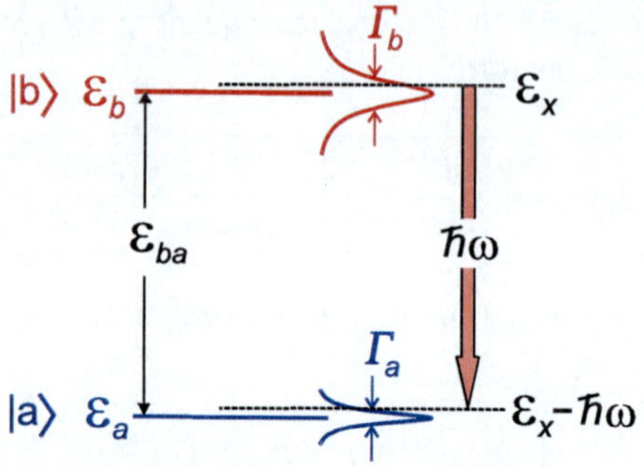

Fig. 12.2 Atomic level diagram for decay from an upper electronic state $|b\rangle$ with an associated Lorentzian FWHM Γ_b to a lower state $|a\rangle$ of Lorentzian FWHM Γ_a. We have indicated the energy uncertainties by Lorentzian density distributions in both upper and lower states. The red transition arrow conserves energy according to the δ-function in (12.1)

$$\frac{|\Psi(\omega)|^2}{h} = \frac{2}{\pi\Gamma_a}\frac{2}{\pi\Gamma_b}\int_{-\infty}^{\infty}\frac{(\Gamma_b/2)^2}{(\mathcal{E}_x-\mathcal{E}_b)^2+(\Gamma_b/2)^2}\frac{(\Gamma_a/2)^2}{(\mathcal{E}_x-\hbar\omega-\mathcal{E}_a)^2+(\Gamma_a/2)^2}\,\mathrm{d}\mathcal{E}_x. \tag{12.12}$$

This is seen to be a convolution of two Lorentzians which is another Lorentzian and with the notation $\mathcal{E}_{ba}=\mathcal{E}_b-\mathcal{E}_a$ we obtain instead of (12.10)

$$\boxed{\rho(\mathcal{E}_f)\delta(\mathcal{E}_f-\mathcal{E}_i) = \frac{2}{\pi(\Gamma_a+\Gamma_b)}\frac{(\Gamma_a+\Gamma_b)^2/4}{(\mathcal{E}_{ba}-\hbar\omega)^2+(\Gamma_a+\Gamma_b)^2/4}}. \tag{12.13}$$

Our new linewidth expression contains the sum of the upper and lower state decay widths. It reduces to our previous expression (12.10) in the limit $\Gamma_a=0$ where $|a\rangle$ is the stable ground state. In retrospect, our previous result for a ground state of infinite lifetime is seen to correspond to a convolution of the form (12.12) where the final state Lorentzian has the form of a δ-function.

12.2.5 Auger Contribution to the XES Linewidth

We now address a final issue in the derivation of the XES linewidth, namely that core hole decays involve both radiative and Auger channels, as previously introduced in an *ad hoc* fashion in Fig. 10.3. The decay problem is illustrated in detail in Fig. 12.3 and requires consideration of both radiative x-ray and non-radiative Auger decay channels of the initial core hole state $|b\rangle$ and final state $|a\rangle$. While radiative transitions are driven by the dipole operator, $e\,\mathbf{r}\cdot\mathbf{E}$, Auger transitions involve the Coulomb operator, e^2/r.

The proper quantum treatment of the XES linewidth requires taking account of both photon and electron emission channels by a *second order* treatment. Our first order description of XES in the KHD formalism worked because it tacitly used the

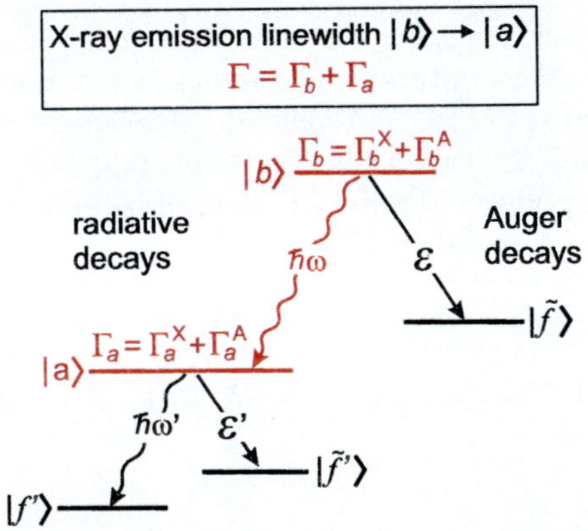

Fig. 12.3 Illustration of the radiative and Auger decay channels involved in the derivation of the x-ray emission linewidth of a dipole transition from an initial core hole state $|b\rangle$ to a final electronic state $|a\rangle$ (red) that is not the ground state. A given excited electronic state may decay either by emission of a photon (wiggly lines) of energy $\hbar\omega$ or Auger electron (straight arrows) of energy denoted $\mathcal{E}$. The measured x-ray emission width for the transition $b \to a$ shown in red is $\Gamma = \Gamma_b + \Gamma_a$ and contains additive dipolar and Auger contributions for both the initial and final states of the transition as indicated and given by (12.14)

second order linewidth result which is $\Gamma = \Gamma^{\mathrm{X}} + \Gamma^{\mathrm{A}}$. In the literature, the proper linewidth derivation is typically circumvented by adopting this result in an *ad hoc* fashion.

We also will not go through the detailed second order linewidth derivation and just state the key results.[1] It turns out that one may associate both an energy $\mathcal{E}_n$ and intrinsic energy width Γ_n with any given electronic state $|n\rangle$ where each Γ_n is again the sum of a radiative and Auger contribution, $\Gamma_n = \Gamma_n^{\mathrm{X}} + \Gamma_n^{\mathrm{A}}$. In Fig. 12.3, the total decay width from $|b\rangle$ to $|a\rangle$ is then given by the following remarkable result.

The XES linewidth of an electronic transition $|b\rangle \to |a\rangle$ is given by a Lorentzian of FWHM

$$\Gamma = \Gamma_a + \Gamma_b = \underbrace{\Gamma_a^{\mathrm{X}} + \Gamma_b^{\mathrm{X}}}_{\text{dipole width}} + \underbrace{\Gamma_a^{\mathrm{A}} + \Gamma_b^{\mathrm{A}}}_{\text{Auger width}}. \tag{12.14}$$

Although only the x-ray decay channel is measured, Auger decays contribute an inseparable "ghost" linewidth contribution. Similarly, the Auger linewidth contains a "ghost" radiative contribution.

The fascinating ghost contributions to the measured linewidth in the alternative decay channels arise from the quantum mechanical link established in the birth of

[1] I would like to thank Faris Gel'mukhanov for sending me his full quantum derivation of the x-ray emission linewidth.

the decay process. Such a quantum link through *birth* also occurs in spontaneous parametric down conversion, where the two born photons are entangled. Each of the so-created photons also carries a ghost contribution from the other photon, exploited today in so-called ghost imaging [9, 10]. The proof of the existence of the entangled ghost phenomenon was honored by the 2022 Nobel Prize in Physics awarded to Alain Aspect, John F. Clauser, and Anton Zeilinger.

12.2.6 Putting It All Together: The X-Ray Emission Rate and Width

We are now in a position to complete the evaluation of the KHD emission rate (12.1). The final state density is given by (12.10), with Γ given by our final result (12.14). In the derivation of the Lorentzian form we implicitly assumed energy conservation, so that the δ-function in (12.1) has already been incorporated. The transition matrix element is factored into a photon part, given by (12.2), and the electronic part may be written in the form (10.40) of the absorption case. Putting it all together we obtain as follows.

In the dipole approximation, the spontaneous plus stimulated XES rate is given by

$$\mathcal{W}^{\text{XES}}_{b\to a} = \underbrace{\frac{(1+n_{p\mathbf{k}})c}{V_{p\mathbf{k}}}}_{\text{Flux}, \Phi_{p\mathbf{k}}} \underbrace{\frac{\lambda^2}{\pi}\frac{\Gamma^p_{ba}}{\Gamma}\frac{(\Gamma/2)^2}{(\hbar\omega - \mathcal{E}_{ba})^2 + (\Gamma/2)^2}}_{\sigma^p_{\text{XES}}}, \tag{12.15}$$

where $\Gamma = \Gamma_a + \Gamma_b$ is given by (12.14). The generalized XES cross section including stimulated emission may be defined as

$$\boxed{\sigma^p_{\text{XES}} = \frac{\lambda^2}{\pi}\frac{\Gamma^p_{ba}}{\Gamma}\frac{(\Gamma/2)^2}{(\hbar\omega - \mathcal{E}_{ba})^2 + (\Gamma/2)^2}\left(1 + n_{p\mathbf{k}}\right)}, \tag{12.16}$$

where the radiative *dipolar width* is given by

$$\boxed{\Gamma^p_{ba} = \underbrace{\frac{8\pi^2\hbar\omega}{\lambda^2}}_{\substack{\text{photon}\\ \text{part}}} \underbrace{\alpha_{\text{f}}}_{\text{coupling}} \underbrace{|\langle a|\mathbf{r}\cdot\boldsymbol{\epsilon}_p|b\rangle|^2}_{\substack{\text{atomic part}\\ \text{includes spin}}}}. \tag{12.17}$$

The form of the dipolar matrix element reveals the photon and electronic contributions coupled by the fine structure constant, just like the corresponding XAS expression (10.40).

There is great similarity between the absorption rate $\mathcal{W}^{\text{abs}}_{a\to b}$ given by (10.38) and the emission rate $\mathcal{W}^{\text{em}}_{b\to a}$ (12.15). However, for XAS, the lower state $|a\rangle$ is the ground state of infinite lifetime and we have $\Gamma = \Gamma_b = \Gamma^{\text{X}}_b + \Gamma^{\text{A}}_b$. In XES, the total linewidth Γ is instead given by the full expression (12.14) or $\Gamma = \Gamma_a + \Gamma_b$. As discussed in Sect. 12.2.4, we typically have $\Gamma_b \gg \Gamma_a$, so we can summarize as follows.

The XES linewidth is typically dominated by that of the initial *core hole* state, Γ_b, which is larger than that of the final *valence hole* state, Γ_a. This is a consequence of the approximate scaling of the decay linewidth of a level with its binding energy (see Fig. 10.4).

The fundamental XES linewidth Γ that determines the decay rate and emission cross section in (12.16) is only observed when the emission spectrometer has an instrumental Gaussian resolution function of width much less than Γ. If this is not the case, the theoretical Lorentzian width needs to be convoluted with the Gaussian instrumental width, resulting in a broadened Voigt lineshape [11] (see Appendix A.2.3).

12.2.7 Atomic Decay Time: Core Hole-Clock

As discussed above, we typically have $\Gamma \simeq \Gamma_b \gg \Gamma_a$ in XES since the decay energy of the initial core hole state $|b\rangle$ is larger than that of the final state $|a\rangle$. The core hole width then represents an atom-specific lifetime τ_Γ according to the relation,

$$\tau_\Gamma = \frac{\hbar}{\Gamma} \tag{12.18}$$

The decay time τ_Γ is called the *natural* lifetime and constitutes an atom-specific reference or *"core hole clock"* time.[2] As illustrated in Fig. 12.1d, it represents the time interval during which the probability (or population) of the excited core hole state decays to $1/e$ of its initial value and as shown in Fig. 12.1c may be associated with the *coherence time* of the emitted radiation.

By use of (12.18) we can convert the atomic Γ values in Table 10.2 (also see Fig. 10.4) into characteristic atomic decay times which are shown in Fig. 12.4 for $1s$ (K-shell) and $2p_{3/2}$ (L_3 shell) core holes as a function of atomic number Z.

For the shown cases the dominant mechanism is Auger decay so that in (12.14) we have $\Gamma \simeq \Gamma^{\text{A}}_b \gg \Gamma^{\text{X}}_b$. In the figure we have also shown in red on the right scale the low probability that the decay occurs by radiative x-ray emission. This probability is called the *fluorescent yield* as discussed in Sect. 12.4.

---

[2] The name emerged from studies of electronic relation processes of core holes in adsorbates on surfaces [12, 13].

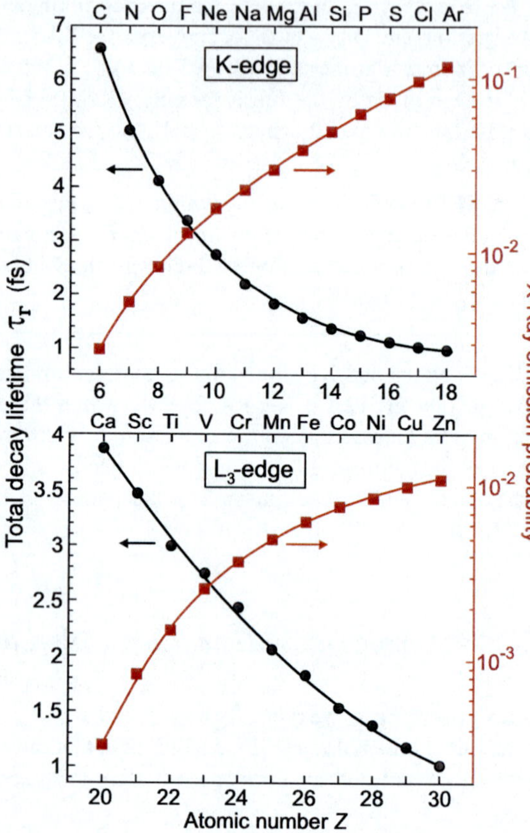

Fig. 12.4 K- and L_3-shell core hole clock life times (black, left scale) calculated by use of (12.18) from the Γ values in Table 10.2. The probability that the decay occurs by radiative x-ray emission is shown in red (right scale). The red curve is also the fluorescent yield as discussed in Sect. 12.4 and shown in Fig. 12.10

A beautiful experimental verification of (12.18) is shown in Fig. 12.5, measured by R. Röhlsberger using Mössbauer spectroscopy [14].

In the shown example, an excited nuclear state decays to a long-lived lower state by emission of a γ-ray photon or emission of a conversion electron. In complete analogy to an electronic state decay, the total lifetime of the excited state is determined by the sum of the radiative (photon emission) and non-radiative (electron emission) channels. Figure 12.5a shows the ^{57}Fe Mössbauer spectrum measured in the energy domain by the conventional transmission method. The time domain measurement, shown in (b) for the same sample, was carried out by exciting the 14.4125 keV Mössbauer transition by a narrowband synchrotron radiation pulse of about 100 ps, and then measuring the time dependence of the decaying intensity.

The decay linewidth in the energy domain is readily obtained from the conventional single-line ^{57}Fe Mössbauer spectrum of stainless steel shown in Fig. 12.5a. The spectrum is almost a perfect Lorentzian of FWHM $\Gamma = 18.9$ neV. The time domain measurement, shown in (b) for the same sample, yields a $1/e$ decay time of

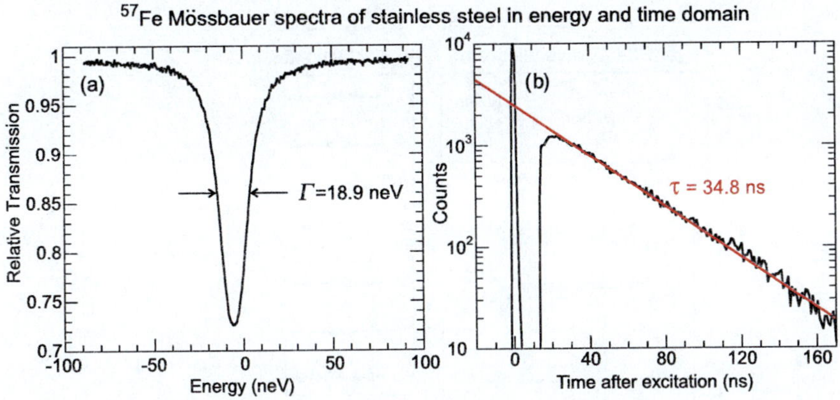

Fig. 12.5 **a** Mössbauer spectrum for stainless steel recorded in transmission for a 9.5-μm-thick stainless steel foil in the energy domain, yielding the FWHM value Γ. **b** Measurement of the time dependent excited state decay after excitation with narrowband synchrotron radiation pulses, with an exponential fit (red), yielding the $1/e$ decay time constant τ. The two quantities are related according to (12.18) with $\tau = \tau_\Gamma$. In both cases the measured linewidth is about four times the natural linewidth of 4.7 neV. Courtesy of R. Röhlsberger, DESY

$\tau = \tau_\Gamma = 34.8$ ns. By use of the value $\hbar = 658.2$ neV ns, the measured temporal and energetic widths can be directly converted into each other by (12.18).[3]

12.3 Fundamental X-Ray Emission Experiments

12.3.1 *K-shell Emission in Ne*

Emission spectra of *atoms* are particularly simple since they are not influenced by vibrational fine structure which is present in molecules. Furthermore, the decay width of the 1s core state can be directly determined from the measured spectrum since the decay typically ends in an electronic state with a long lifetime, so that to a very good approximation we have $\Gamma_f^X = \Gamma_f^A = 0$ in (12.14). Finally, the decay of the 1s core hole proceeds predominantly (~98%) through the Auger process so that the emission linewidth is $\Gamma \simeq \Gamma_i^A$, providing a beautiful demonstration that it is almost entirely due to the "ghost" linewidth contribution of the Auger channel.

As a pedagogical case we consider the K$\alpha_{1,2}$ x-ray emission spectrum of Ne atoms, as illustrated in Fig. 12.6. The Ne emission spectrum is also of interest because it has been utilized to demonstrate its amplification through stimulated emission [15] as previously shown in Fig. 1.29.

[3] Note that the natural lifetime τ_Γ over which the excited state population decays to $1/e$ of its original value differs from the often quoted (shorter) *half-life* $\tau_{1/2} = \ln 2\, \tau$ of radioactive decays.

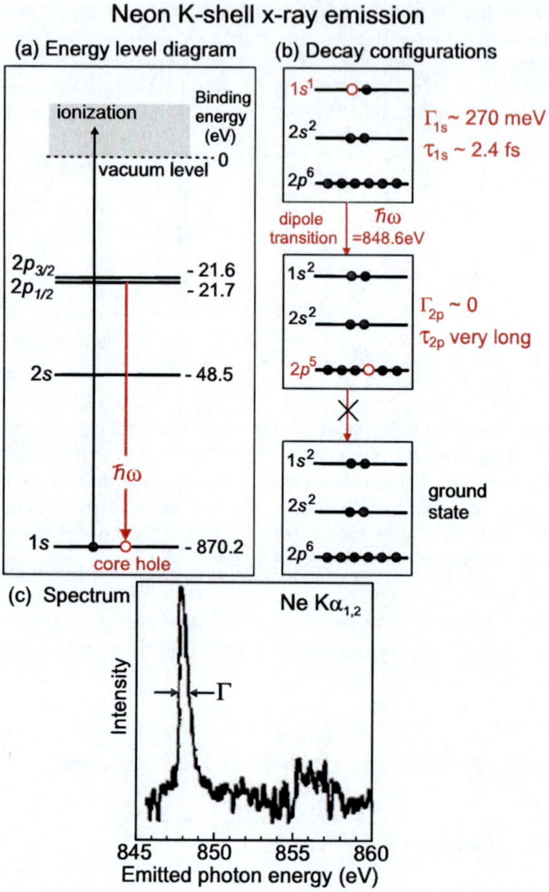

Fig. 12.6 Illustration of K-shell emission in Ne atoms. **a** Ne energy level diagram and levels involved in K$\alpha_{1,2}$ emission. **b** Diagram of the electron/hole configurations involved in the decay of the $1s$ core hole. Electrons are indicated by filled black circles and the hole by an open red circle. The upper configuration is the initial state in the x-ray emission process. It decays exponentially via a dipole transition to the middle configuration which is metastable with a long lifetime as discussed in the text. **c** K$\alpha_{1,2}$ emission spectrum recorded after electron impact ionization with an experimental resolution of 150 meV by Ågren et al. [16]. The $2p_{3/2}$ and $2p_{1/2}$ spin-orbit components separated by 100 meV are not resolved. The emitted $K\alpha_{1,2}$ doublet of $\hbar\omega \simeq 848.6$ eV has two unresolved components, each of linewidth $\Gamma_{1s} = 0.27 \pm 0.01$ eV

Figure 12.6a illustrates the origin of the Ne K-shell x-ray emission line ($K\alpha_{1,2}$) of $\hbar\omega \simeq 848.6$ eV. It corresponds to the decay of a Ne $1s$ core hole configuration to the $2p$ hole configuration, with spin-orbit components $2p_{1/2}$ and $2p_{3/2}$ separated by 100 meV resolved by high-resolution photoemisson (see Fig. 12.4 in [17]). The core decay in a configuration picture is shown in Fig. 12.6b. It illustrates that the final configuration containing a $2p$ hole is long lived [18]. This is apparent from the extremely small $2p_{1/2}$ and $2p_{3/2}$ photoemission linewidths $\Gamma_{2p} \ll 5$ meV shown in Fig. 12.2 of [17]. Hence we have $\Gamma_{2p} \simeq 0$, and the x-ray emission width is that of the initial state core hole $\Gamma = \Gamma_{1s}$.

The $K\alpha_{1,2}$ emission spectrum was measured after electron impact ionization by Ågren et al. [16], as shown in Fig. 12.6c. In this 1978 experiment the instrumental (Gaussian) resolution was a remarkable 150 meV. A theoretical fit of the emission spectrum that took into account the unresolved infinitely sharp $2p_{1/2}$ and $2p_{3/2}$ spin-orbit components with a 100 meV separation yielded a value of $\Gamma_{1s} = 270 \pm 10$ meV for the natural K-shell linewidth. This value compares to the values 248–297 meV derived from the width of the Rydberg XAS resonances in Fig. 10.10.

The related case of K-shell (binding energy 3206 eV) excitations in Kr atoms has also been studied. Again good agreement is found for the natural K-shell lifetime width Γ determined by Rydberg absorption studies [19] ($\Gamma = 680 \pm 30$ meV) and the $1s$ x-ray photoemission linewidth ($\Gamma = 675 \pm 20$ meV) [20].

12.3.2 *K-Shell Emission in N_2*

As a second example we discuss the K-shell emission in the N_2 molecule, illustrated in Fig. 12.7. Like the Ne spectrum in Fig. 12.6 it was measured after electron beam ionization by the pioneering Uppsala group [21].

The molecular orbitals and their relative energies determined by photoemission spectroscopy [25] are shown in Fig. 12.7a. In particular, we also show the small $1\sigma_u - 1\sigma_g$ core splitting of 97 ± 3 meV [26]. The N_2 level structure is that previously shown in Fig. 10.11a. Here we consider the scenario where the molecule is *ionized* through excitation of a $1s$ electron in either the $1\sigma_g$ or $1\sigma_u$ state into the continuum of empty states above the ionization potential (IP) indicated in blue. The created core hole may be of either spin or parity.

Figure 12.7b illustrates the initial and final state configurations of the XES process, arbitrarily assuming spin-down transitions. The two initial states with opposite parity $1s$ holes may decay via spin-conserving and parity allowed dipole transitions to the shown three final state configurations with holes in the $2\sigma_u$, $1\pi_u$, and $3\sigma_g$ orbitals. The latter have a long lifetime so that the minimum XES linewidth is determined by that of the $1s$ core hole width Γ_{1s}, alone. The width has the value $\Gamma_{1s} \simeq 120$ meV, determined from vibrationally resolved XAS [22] and PES spectra [23].

In Fig. 12.7c we show the pioneering N_2 K-shell spectrum recorded in 1973 [21], which used electron impact ionization to create the initial XES core hole state. It clearly shows the signature of the outermost $2\sigma_u$, $1\pi_u$, and $3\sigma_g$ occupied molecular

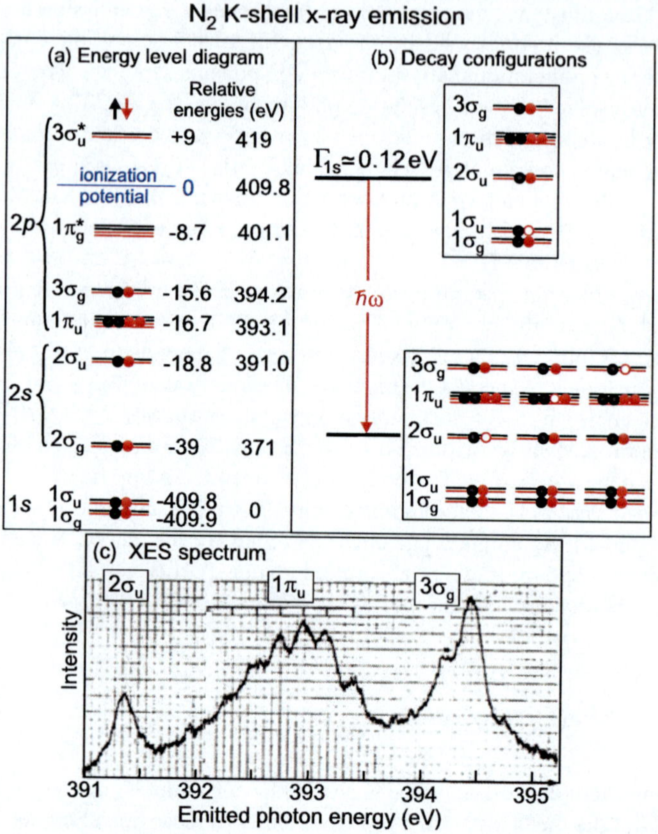

Fig. 12.7 K-shell x-ray emission in N_2 molecules. **a** N_2 energy level diagram and level filling in the ground state. The two compensated spin states are shown in different colors. **b** Diagram of the electron/hole configurations involved in Kα emission of the 1s core hole, assuming that the active hole is spin-down. Electrons are indicated by filled circles and holes by open circles. The upper configurations with opposite parity 1s holes may decay via dipole and parity allowed transitions to the shown three final state configurations with holes in the $2\sigma_u$, $1\pi_u$, and $3\sigma_g$ orbitals, respectively. The final hole configurations have a long lifetime so that the minimum XES linewidth is determined by that of the 1s core hole of $\Gamma_{1s} \simeq 120\,\text{meV}$ [22, 23]. **c** K-shell emission spectrum, recorded after electron impact ionization with an instrumental resolution of $\simeq 100\,\text{meV}$ by Werme et al. [21], revealing vibrational fine structure

orbitals originating from parity and dipole allowed transitions $2\sigma_u \to 1\sigma_g$, $1\pi_u \to 1\sigma_g$, and $3\sigma_g \to 1\sigma_u$. Owing to its high instrumental resolution of about 100 meV and the small intrinsic 1s core hole width of $\Gamma_{1s} = 120\,\text{meV}$, the XES spectrum also reveals the first observation of vibrational fine structure. This has been studied in more detail by use of resonant photon excitation (RIXS), in [27–30]. As discussed in Sect. 13.10.1, the RIXS case needs to be described by second order perturbation theory in contrast to the first order XES case.

12.3.3 L-Shell XES in 3d Metals

Denoting the total number of occupied spin orbitals by $N_e = 2n_e$, the polarization independent *spontaneous* emission width $\Gamma_{\rm XES}$, including spin, is obtained as

$$\Gamma_{\rm XES} = \frac{8\pi^2\hbar\omega\,\alpha_f}{\lambda^2}\mathcal{R}^2\frac{LN_e}{3(2L+1)(2c+1)}, \tag{12.19}$$

where $L = 2$ and $c = 1$. If the decay is stimulated, the incident polarization is preserved since the incident photons $n_{p\mathbf{k}}$ clone themselves. The above XES expression corresponds to the polarization averaged absorption width (10.59), which for direct comparison reads

$$\Gamma_{\rm XAS} = \frac{8\pi^2\hbar\omega\,\alpha_f}{\lambda^2}\mathcal{R}^2\frac{LN_h}{3(2L+1)}. \tag{12.20}$$

The ratio of the two transition rates is given by

$$\frac{\Gamma_{\rm XAS}}{\Gamma_{\rm XES}} = (2c+1)\frac{N_h}{N_e}, \tag{12.21}$$

where $N_h + N_e = 2(2L+1)$.

The relation (12.21) is consistent with a simple argument. For the $2p \rightarrow 1s$ K-shell case we have $L = 1$ and $c = 0$, and for a half filled valence shell, $N_h = N_e$, absorption and emission are predicted to have the same probability. In the absorption process a core *electron* is excited to the *half empty* valence shell. The emission process corresponds to replacing the core electron by a core *hole* that is excited to the *half filled* valence shell. For the same radial transition matrix element, the two processes must have the same probability as reflected by (12.21).

For the transition metals, the $3d$ shell gets increasingly filled and the XAS resonance intensities due to $2p \rightarrow 3d$ transitions becomes increasingly weaker from Ti to Ni as shown in Fig. 10.16 reflecting the relative change in the number of valence holes N_h and electrons N_e. The XES spectra of the elementary transition metals, recorded after photon excitation at an energy 30 eV above the L_2 resonance energy, are shown in Fig. 12.8 [31].

12.3.4 L_3-Shell XES in Cu Metal

As an example we consider in Fig. 12.9 the L_3-shell XAS and XES spectra of Cu metal.

In Fig. 12.9a we illustrate in an active electron picture the electronic states involved in the emission process, which complements the absorption case previously shown in Fig. 10.14. We account for the degeneracy and occupation of the upper valence state by assuming that of the total number of $2L + 1$ orbitals only the fraction $n_e/(2L + 1)$

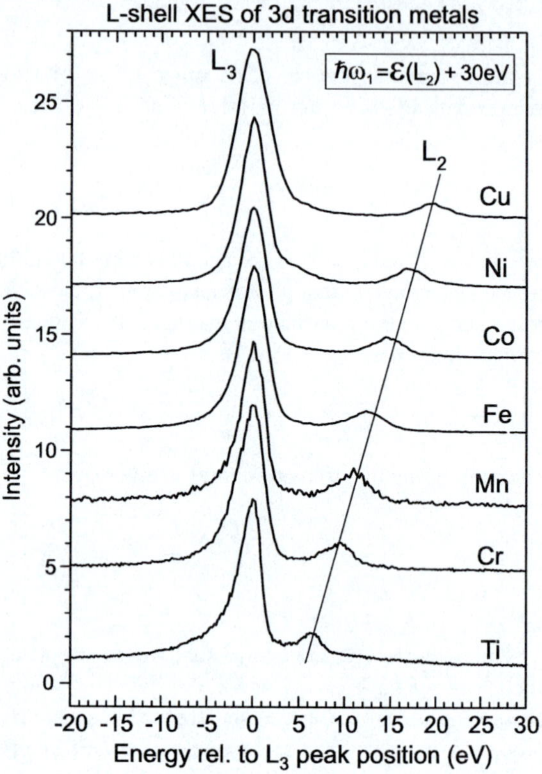

Fig. 12.8 L-shell XES spectra of the elementary 3*d* transition metals [31] (and A. Moewes private commun.)

is filled with electrons and can participate in the decay. The lower state of angular momentum *c* has a hole in the manifold of $2c + 1$ core orbitals which can be filled. The two possible spin scenarios are accounted for by multiplying the probability in Fig. 12.9a by a factor of 2.

In the following we are interested in the XES spectrum of Cu metal which has a nearly filled 3*d* shell, described by $N_h = 0.5$ and $N_e = 9.5$ [34]. This is reflected by the L-edge XAS spectrum shown in Fig. 12.9b [35] when compared to those of other transition metals in Fig. 10.16. The Cu L-edge XES spectrum is expected to be a signature of the filled 3*d* density of states (DOS).

In practice, one finds that the DOS is better resolved by x-ray photoemission (XPS) than XES. This is shown by comparison of the high-resolution (0.32 eV) XPS spectrum recorded at $\hbar\omega = 1486.6$ eV (monochromatized Al Kα radiation) [32] in Fig. 12.9c with the XES spectra in Fig. 12.9d recorded with a spectrometer resolution of 0.8 eV. Although the instrumental resolution in the two measurements was different, there are intrinsic reasons why the photoemission spectra more closely resemble the DOS.

In conventional XES, spectra of transition metals contain strong shake-up structure which causes a significant difference of the XES spectra and the DOS. This is

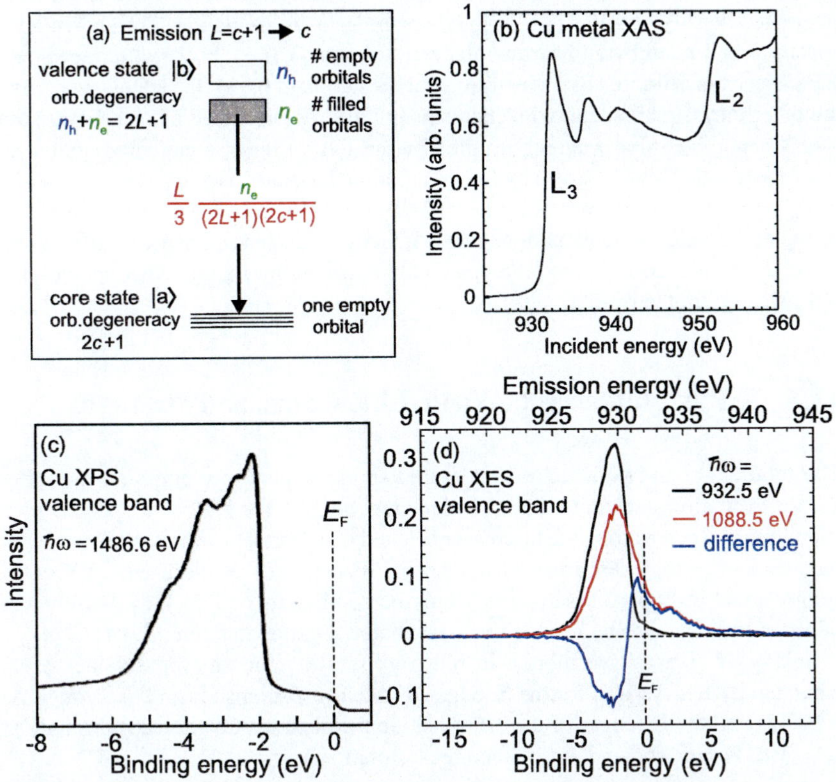

Fig. 12.9 **a** Angular part of the dipole matrix elements for transitions from valence orbitals of angular momentum $L = c + 1$ to a single core hole of angular momentum c. We have assumed that a fraction $n_e/(2L+1)$ of valence orbitals is filled with electrons. The emission probability is given in red. **b** L-shell x-ray absorption spectrum of fcc Cu metal [35]. **c** High-resolution (0.32 eV) XPS spectrum recorded at $\hbar\omega = 1486.6$ eV of the 3d valence band of Cu metal [32]. **d** Cu valence band recorded by XES with a spectrometer resolution of 0.8 eV after photon excitation with a bandwidth of 0.5 eV to empty states just above the Fermi level ($\hbar\omega = 932.5$ eV, thick black line) and excitations to states in the continuum ($\hbar\omega = 1088.5$ eV) [33] and M. Magnuson private communication. The spectra were normalized to each other by demanding that their integrated intensity is the same. The difference spectrum reveals the multi-electron shake-up structure present for high energy excitation

illustrated by the black and red XES spectra in Fig. 12.9d, where the core hole was produced with incident x-rays of bandwidth of 0.5 eV that were either tuned just above the L_3 threshold ($\hbar\omega = 932.5$ eV, black curve) or had an energy $\hbar\omega = 1088.5$ eV well above threshold [33]. The shown spectra were normalized to each other so that their integrated intensity is the same. The difference spectrum, shown in blue, reveals the multi-electron shake-up structure present for high energy excitation (also see [36]).

The resonantly excited XES spectrum (black) in Fig. 12.9d reveals less details in comparison with the XPS spectrum in Fig. 12.9c. Both spectra reflect the momentum integrated density of states [33, 37] and are expected to be similar. The shown XPS spectrum is limited by an instrumental resolution of 0.32 eV [32], while the shown XES spectrum is limited by an instrumental resolution of 0.8 eV [33]. However, there are other intrinsic broadening mechanisms in XES. As discussed in Sect. 12.2.6, the XES linewidth is fundamentally limited by the sum of the core hole and final state linewidths $\Gamma = \Gamma_a + \Gamma_b$ given by (12.14). For the L_3-edge in Cu the core hole width is $\Gamma_a = 0.56\,\text{eV}$ according to Table 10.2. The final state consists of a hole in the occupied $3d$ valence band, similar to XPS, and is expected to have a smaller width $\Gamma_b < \Gamma_a$. While the intrinsic XPS linewidth is limited by Γ_b, the XES linewidth is intrinsically limited by Γ_a.

12.4 X-Ray Fluorescence Yield, Linewidths, and Strengths

The total angle and polarization integrated spontaneous x-ray emission rate after a core excitation constitutes the *fluorescence yield* (FY), Y_f. For a specific core excitation in a given atom, the fluorescence yield is assumed to be an atomic property, independent of the chemical environment. A recent study of the L-edge FY of Ni suggests that this approximation may only be good to about 30% [38]. Differences of order 20% can also be found in values tabulated in the literature [39–41].

In Fig. 12.10 we have plotted in red (right scale) semi-empirical values of Y_f tabulated by Krause [40] for the K-edges of the low-Z atoms C ($Z=6$) through Ar ($Z=18$) and the L_3-edges of the $3d$ transition metal atoms Ca ($Z=20$) through Ar ($Z=30$). For the shown cases Auger decays dominate so that $\Gamma = \Gamma^\text{A} + \Gamma^\text{X} \simeq \Gamma^\text{A}$.

12.4.1 Radial Dipole Matrix Element

We can use the complementary resonant absorption case in Fig. 10.14 and corresponding emission case in Fig. 12.9a to obtain an empirical value for the radial dipole matrix element $\mathcal{R}$, defined in (10.46). By assuming resonant excitation and adding the active electron to the empty valence shell we obtain from (12.19)

$$\Gamma^\text{X} = \frac{8\pi^2\hbar\omega\,\alpha_\text{f}}{\lambda^2}\,\mathcal{R}^2\,\frac{L(N_\text{e}+1)}{3(2L+1)(2c+1)}. \tag{12.22}$$

Neglecting the energy difference between filled and empty valence states, which is small relative to the absorption and emission energies, we take $\hbar\omega$ and λ to correspond to the core hole binding energy. The fluorescence yield is then given by

$$Y_\text{f} = \frac{\Gamma^\text{X}}{\Gamma} = \frac{8\pi^2\hbar\omega\,\alpha_\text{f}}{\lambda^2\,\Gamma}\,\mathcal{R}^2\,\frac{L(N_\text{e}+1)}{3(2L+1)(2c+1)}. \tag{12.23}$$

The radial matrix element is obtained as

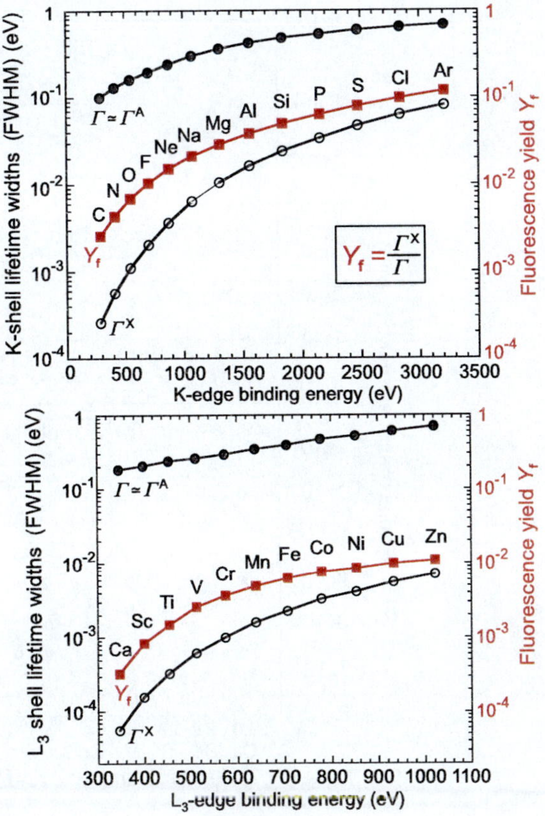

Fig. 12.10 Tabulated fluorescence yields Y_f shown in red (right scale) [40] for the K-shells of low Z atoms and L_3-shells for the $3d$ transition metals. Also plotted (left scale) are the total core hole decay widths Γ from Table 10.2 (solid black circles) and the dipolar transition widths Γ^X (open circles) derived from the relation $\Gamma^X = Y_f \Gamma$

$$\mathcal{R}^2 = \frac{\lambda^2 \Gamma}{8\pi^2 \hbar\omega\, \alpha_f} \frac{3(2L+1)(2c+1)}{L(N_e+1)} Y_f. \tag{12.24}$$

Using the polarization averaged fluorescence yields plotted in Fig. 12.10, we have plotted the radial dipolar matrix elements for the K-shell excitation in the low-Z atoms in Fig. 12.11a. The Z-dependence of $\mathcal{R}$ scales approximately as a_0/Z where $a_0 = 0.0529$ nm is the Bohr radius. This is expected since $\mathcal{R}$, given by (10.46), is determined by the spatial overlap with the core orbital.

The radial matrix element $2p \rightarrow 3d$ for the L-shell excitation in the transition metal atoms is plotted in Fig. 12.11b. In the figure, the blue squares are calculated according to (12.24) as for the K-edge in (a). Also shown as black circles are the theoretical values from Fig. 10.15. The red diamonds correspond to use of (12.24) with $N_e + 1$ replaced by $N_e + 0.5$, which is the so-called *transition state* model introduced by Slater [42], as discussed in [43]. The three methods agree very well for the heavy $3d$ metals Mn–Cu, but there are deviations for the elements Ca-Cr, as shown.

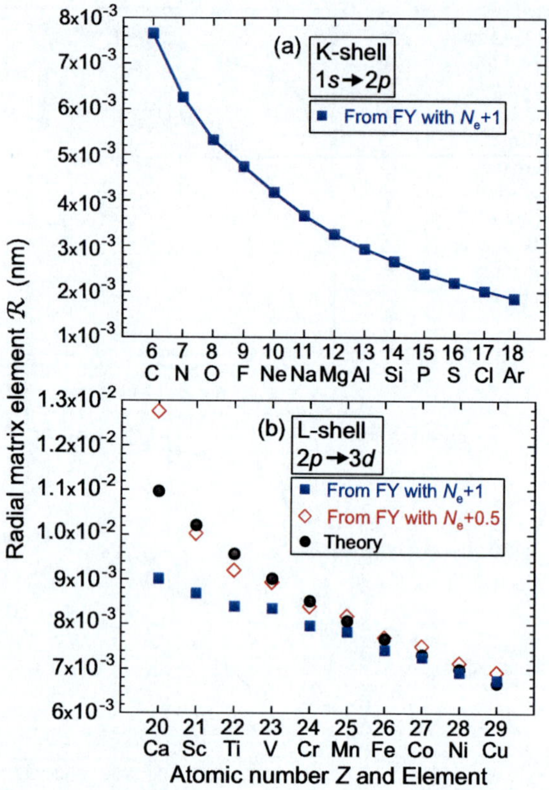

Fig. 12.11 **a** Radial transition matrix elements $\mathcal{R}$ in nm (1 nm^2 = 10^4 Mb) for the K-shell $1s \rightarrow 2p$ excitation in the low-Z atoms C through Ar, calculated according to (12.24) with $c = 0$, $L = 1$ with $4 \leq N_e \leq 16$ and the values for Γ and Y_f in Fig. 12.10. **b** Same for the L-shell $2p \rightarrow 3d$ excitation in Ca through Zn. The black circles are taken from Fig. 10.15. For the blue squares we have used (12.24) with $c = 1$, $L = 2$ and $0 \leq N_e \leq 10$. The red diamonds were calculated by substituting $N_e + 1$ by $N_e + 0.5$ in (12.24), corresponding to Slater's transition state model [42]

12.5 Quantum Theory of Thomson Scattering

The simplest form of x-ray scattering by atomic matter is Thomson scattering. It was introduced within the classical electrodynamics framework in Sect. 6.2, and its quantum treatment was briefly introduced in Sect. 9.5. The process may be elastic or inelastic and is illustrated in Fig. 12.12 for photon scattering by an atom.

The various Thomson scattering processes associated with different energy losses of the incident photons are illustrated in Fig. 12.13. *Elastic* Thomson scattering is one of the most important x-ray interaction processes with matter since it underlies Bragg diffraction. *Inelastic* Thomson scattering involves losses in photon energy due to excitations which are illustrated in the figure for solid samples. At the lowest energy ranging up to a few hundred meV, losses are due to excitations of phonons and

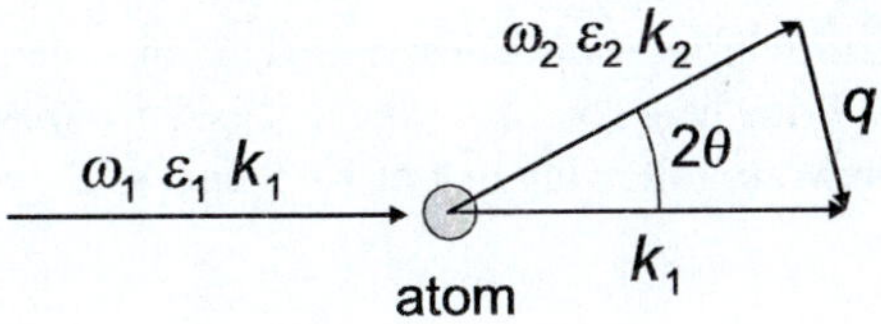

Fig. 12.12 Scattering geometry. An incident photon of energy $\hbar\omega_1$, polarization ϵ_1 and wavevector $\mathbf{k}_1$ is scattered by an angle 2θ, resulting in an outgoing photon of energy $\hbar\omega_2$, polarization ϵ_2, and wavevector $\mathbf{k}_2$. The scattering process transfers an energy $\hbar\omega = \hbar\omega_1 - \hbar\omega_2$ and a momentum $\mathbf{q} = \mathbf{k}_1 - \mathbf{k}_2$ to the sample

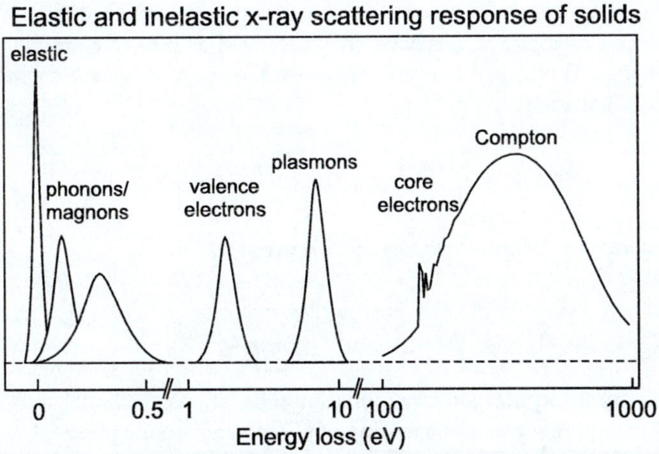

Fig. 12.13 Schematic overview of the elastic and inelastic scattering response of solids, showing typical excitations due to energy losses of the incident photons

magnons. Above about 1 eV valence electrons may be excited from filled to empty states and near about 10 eV collective excitations of plasmons may be observed. At higher energies core electrons may be excited to unfilled states, and these element-specific excitations typically overlap with a Compton scattering background.

In this book we will not discuss Compton scattering in detail,[4] but only show how it becomes increasingly important at large incident photon energies > 30 keV and scattering angles. For more details the reader is referred to [45–47].

In the following we consider in detail how the Thomson scattering of photons is treated in KHD theory. The classical and quantum approaches will be seen to give identical results. This is expected since within the KHD theory, Thomson scattering is a first order interaction process and we have seen in Chap. 5 that the general QED

[4] Compton scattering, discovered in 1923 [44] today, has arguably the most important application in radiobiology and radiation therapy.

framework reduces to the classical wave description in first order. Our quantum treatment of Thomson scattering here completes our discussion of *first order interactions* within the KHD framework, following that of XAS and XES.

12.5.1 Quantum Theoretical Formulation of Thomson Scattering

The quantum process of Thomson scattering is described by the energy diagram in Fig. 9.4c. It consists of a *non-resonant* elastic or inelastic scattering process on the total atomic charge and in space may be pictured as illustrated in Fig. 12.12. In quantum theory, Thomson scattering is described by the *first order* term in the KHD formula (9.21) given by

$$\mathcal{W}_{if} = \frac{2\pi}{\hbar} \, |\langle f\,|\mathcal{H}_{\rm int}|\,i\rangle|^2 \, \rho(\mathcal{E}_f)\,\delta(\mathcal{E}_f - \mathcal{E}_i) \tag{12.25}$$

with the interaction Hamiltonian (see (9.22)) given by

$$\mathcal{H}_{\rm int} = \frac{e^2}{2m_{\rm e}} A^2 = \frac{e^2}{2m_{\rm e}\omega^2} E^2. \tag{12.26}$$

As for the x-ray absorption and x-ray emission calculations, the initial and final states $|i\rangle$, and $|f\rangle$ are product states consisting of photon and atomic parts.

12.5.1.1 Evaluation of the Transition Matrix Element

The transition matrix element in (12.25) is evaluated with photon number states, and we denote the incident state as $|n_{p_1\mathbf{k}_1}\rangle$ and scattered state as $|m_{p_2\mathbf{k}_2}\rangle$. They differ in occupation $n \neq m$ and may differ in wave vector $\mathbf{k}_1 \neq \mathbf{k}_2$ and polarization $p_1 \neq p_2$. As we have done for XAS and XES, we denote the initial electronic state as $|a\rangle$ and final state as $|b\rangle$.

The complete initial product state consists of a photon and electronic part of the form $|i\rangle = |a, n_{p_1\mathbf{k}_1}, 0\rangle$ and has a total energy $\mathcal{E}_i = \mathcal{E}_a + n_{p_1\mathbf{k}_1}\,\hbar\omega_{\mathbf{k}_1}$. The final state contains one less photon in state $|n_{p_1\mathbf{k}_1}\rangle$ and a single photon $m_{p_2\mathbf{k}_2} + 1 = 1$ in state $|m_{p_2\mathbf{k}_2}\rangle$. It is written $|f\rangle = |b, n_{p_1\mathbf{k}_1} - 1, 1\rangle$ and has an energy $\mathcal{E}_f = \mathcal{E}_b + (n_{p_1\mathbf{k}_1} - 1)\hbar\omega_{\mathbf{k}_1} + \hbar\omega_{\mathbf{k}_2}$.

The matrix element in (12.25) then takes the form

$$\langle f\,|\mathcal{H}_{\rm int}|\,i\rangle = \left\langle b,\ n_{p_1\mathbf{k}_1}-1, 1 \left| \frac{e^2}{2m_{\rm e}} A^2 \right| a,\ n_{p_1\mathbf{k}_1}, 0 \right\rangle. \tag{12.27}$$

The vector potential **A** (or field **E**) operators are expressed in quantized form according to (3.55) (or (3.60)), and we have (note dot product)

$$\frac{e^2}{2m_e}A^2 = \frac{\hbar e^2}{4m_e\epsilon_0}\sum_{uv}\sum_{\mathbf{k}_1\mathbf{k}_2}\frac{1}{\sqrt{\omega_{\mathbf{k}_1}\omega_{\mathbf{k}_2}V_{\mathbf{k}_1}V_{\mathbf{k}_2}}}$$
$$\left[\boldsymbol{\epsilon}_u\, a_{u\mathbf{k}_1}\, e^{\mathrm{i}(\mathbf{k}_1\cdot\mathbf{r}-\omega_{\mathbf{k}_1}t)} + \boldsymbol{\epsilon}_u^*\, a_{u\mathbf{k}_1}^\dagger\, e^{-\mathrm{i}(\mathbf{k}_1\cdot\mathbf{r}-\omega_{\mathbf{k}_1}t)}\right]$$
$$\cdot\left[\boldsymbol{\epsilon}_v\, a_{v\mathbf{k}_2}\, e^{\mathrm{i}(\mathbf{k}_2\cdot\mathbf{r}-\omega_{\mathbf{k}_2}t)} + \boldsymbol{\epsilon}_v^*\, a_{v\mathbf{k}_2}^\dagger\, e^{-\mathrm{i}(\mathbf{k}_2\cdot\mathbf{r}-\omega_{\mathbf{k}_2}t)}\right]. \tag{12.28}$$

By inspection of possible matrix elements, one finds that only cross terms, consisting of pairs of creation and annihilation operators $a_{u\mathbf{k}_1}a_{v\mathbf{k}_2}^\dagger$ and their complex conjugates, can couple the initial and final photon number states in (12.27). The effective Thomson scattering Hamiltonian therefore consists of such cross terms and is given by

$$\mathcal{H}_\mathrm{T} = \frac{\hbar e^2}{4m_e\epsilon_0}\sum_{uv}\sum_{\mathbf{k}_1\mathbf{k}_2}\frac{1}{\sqrt{\omega_{\mathbf{k}_1}\omega_{\mathbf{k}_2}V_{\mathbf{k}_1}V_{\mathbf{k}_2}}}$$
$$\left[\boldsymbol{\epsilon}_u\cdot\boldsymbol{\epsilon}_v^*\, a_{u\mathbf{k}_1}a_{v\mathbf{k}_2}^\dagger + \boldsymbol{\epsilon}_u^*\cdot\boldsymbol{\epsilon}_v\, a_{u\mathbf{k}_1}^\dagger a_{v\mathbf{k}_2}\right] e^{\mathrm{i}(\mathbf{k}_1-\mathbf{k}_2)\mathbf{r}}\, e^{-\mathrm{i}(\omega_{\mathbf{k}_1}-\omega_{\mathbf{k}_2})t}.$$

The only non-vanishing matrix element, which occurs twice, is given by

$$\langle n_{p_1\mathbf{k}_1}-1,\ 1|\, a_{p_2\mathbf{k}_2}^\dagger a_{p_1\mathbf{k}_1}\, |n_{p_1\mathbf{k}_1},\ 0\rangle = \sqrt{n_{p_1\mathbf{k}_1}}. \tag{12.29}$$

The transition matrix element in (12.27) is obtained as

$$\langle b,\ n_{p_1\mathbf{k}_1}-1, 1\,|\mathcal{H}_\mathrm{T}|\, a,\ n_{p_1\mathbf{k}_1}, 0\rangle$$
$$= \frac{\sqrt{n_{p_1\mathbf{k}_1}}}{\sqrt{\omega_{\mathbf{k}_1}\omega_{\mathbf{k}_2}V_{\mathbf{k}_1}V_{\mathbf{k}_2}}}\frac{\hbar e^2}{2m_e\epsilon_0}\,[\boldsymbol{\epsilon}_{p_1}\cdot\boldsymbol{\epsilon}_{p_2}^*]\,\langle b|\, e^{\mathrm{i}(\mathbf{k}_1-\mathbf{k}_2)\mathbf{r}}\,|a\rangle. \tag{12.30}$$

12.5.1.2 The Density of Final States

Next we evaluate the product of the density of states and δ-function in (12.25). It is obtained by use of the photon density expression (3.102) or $\rho(\hbar\omega_{\mathbf{k}_2}) = V_{\mathbf{k}_2}\omega_{\mathbf{k}_2}^2\mathrm{d}\Omega_{\mathbf{k}_2}/(8\pi^3c^3\hbar)$ and by summing over all electronic final states $|b\rangle$ subject to the energy conserving δ-function $\delta(\mathcal{E}_f-\mathcal{E}_i) = \delta(\mathcal{E}_b-\mathcal{E}_a-\hbar\omega_{\mathbf{k}_1}+\hbar\omega_{\mathbf{k}_2}) = \delta(\mathcal{E}_{ba}-\hbar\omega_{\mathbf{k}_1}+\hbar\omega_{\mathbf{k}_2})$. We obtain

$$\rho(\mathcal{E}_f)\,\delta(\mathcal{E}_f-\mathcal{E}_i) = \frac{V_{\mathbf{k}_2}\omega_{\mathbf{k}_2}^2\mathrm{d}\Omega_{\mathbf{k}_2}}{8\pi^3c^3\hbar}\,\delta(\mathcal{E}_{ba}-\hbar\omega_{\mathbf{k}_1}+\hbar\omega_{\mathbf{k}_2}). \tag{12.31}$$

12.5.1.3 The Thomson Cross Section

We have now all ingredients to write down the Thomson scattering rate (12.25). It may be converted into the double differential Thomson scattering cross section according to

$$\frac{\mathrm{d}^2\sigma_\mathrm{T}}{\mathrm{d}\Omega_{\mathbf{k}_2}\,\mathrm{d}\hbar\omega_{\mathbf{k}_2}} = \frac{1}{\Phi_1}\frac{\mathrm{d}^2\mathcal{W}_{if}}{\mathrm{d}\Omega\,\mathrm{d}\hbar\omega_2} = \frac{V_{\mathbf{k}_1}}{n_{p_1\mathbf{k}_1}c}\frac{\mathrm{d}^2\mathcal{W}_{if}}{\mathrm{d}\Omega\,\mathrm{d}\hbar\omega_2}. \tag{12.32}$$

The differential Thomson scattering cross section is simply the scattering rate $\mathcal{W}_{if}$ normalized by the incident flux and accounting for the fraction of photons emitted into the solid angle $\mathrm{d}\Omega_2$ of a detector and its bandwidth $\mathrm{d}(\hbar\omega_2)$. Inserting the squared matrix element (12.30) and the final state density (12.31) into (12.32), we see that the two photon normalization volumes $V_{\mathbf{k}_1}$ and $V_{\mathbf{k}_2}$ fall out, and we obtain

$$\begin{aligned}\frac{\mathrm{d}^2\sigma_\mathrm{T}}{\mathrm{d}\Omega_{\mathbf{k}_2}\,\mathrm{d}\hbar\omega_{\mathbf{k}_2}} &= \frac{e^4}{16\pi^2\epsilon_0^2 m_\mathrm{e}^2 c^4}\frac{\omega_{\mathbf{k}_2}}{\omega_{\mathbf{k}_1}}|\boldsymbol{\epsilon}_1\cdot\boldsymbol{\epsilon}_2^*|^2\sum_b\left|\langle b|\,e^{\mathrm{i}(\mathbf{k}_1-\mathbf{k}_2)\cdot\mathbf{r}}\,|a\rangle\right|^2\delta(\mathcal{E}_{ba}-\hbar\omega_{\mathbf{k}_1}+\hbar\omega_{\mathbf{k}_2})\\ &= \underbrace{r_0^2\frac{\omega_{\mathbf{k}_2}}{\omega_{\mathbf{k}_1}}|\boldsymbol{\epsilon}_1\cdot\boldsymbol{\epsilon}_2^*|^2}_{\text{single electron scattering}}\underbrace{\sum_b\left|\langle b|\,e^{\mathrm{i}(\mathbf{k}_1-\mathbf{k}_2)\cdot\mathbf{r}}\,|a\rangle\right|^2\delta(\mathcal{E}_{ba}-\hbar\omega_{\mathbf{k}_1}+\hbar\omega_{\mathbf{k}_2})}_{\text{dynamic structure factor } S(\mathbf{k}_1-\mathbf{k}_2,\hbar\omega_{\mathbf{k}_1}-\hbar\omega_{\mathbf{k}_2})}.\end{aligned} \tag{12.33}$$

Here $r_0 = 2.82 \times 10^{-6}$ nm is the classical electron radius or Thomson scattering length, previously defined in (6.17), and we have identified two contributions by underbrackets. The first corresponds to *elastic* Thomson scattering off a single electron with dimension [area], given classically by (6.19). The second term is called the *dynamic structure factor* with the sum expressing the collective contributions of all electron states $|b\rangle$. It has the dimension [1/energy] through the δ-function, which conserves the dimensions of the left and right sides of the equation. It is seen to depend on the change in photon energy, $\hbar\omega_{\mathbf{k}_1} - \hbar\omega_{\mathbf{k}_2}$ and momentum, $\mathbf{q} = \mathbf{k}_1 - \mathbf{k}_2$, illustrated in Fig. 12.12. The inelastic formalism remains valid for no change in energy, i.e. the elastic case. We can summarize as follows.

The double differential **Thomson scattering cross section** which describes both elastic ($\omega_2 = \omega_1$) and inelastic ($\omega_2 \neq \omega_1$) processes with photon momentum transfer $\mathbf{q} = \mathbf{k}_1 - \mathbf{k}_2$ illustrated in Fig. 12.12 into a finite solid angle of detection $\mathrm{d}\Omega_2$ with a detector bandpass $\mathrm{d}\omega_2$ is given by

$$\begin{aligned}\frac{\mathrm{d}^2\sigma_\mathrm{T}}{\mathrm{d}\Omega_2\,\mathrm{d}\hbar\omega_2} &= r_0^2\frac{\omega_2}{\omega_1}|\boldsymbol{\epsilon}_1\cdot\boldsymbol{\epsilon}_2^*|^2\,S(\mathbf{q},\omega_1-\omega_2)\\ &= \alpha_\mathrm{f}^2\frac{\hbar^2c^2}{(m_\mathrm{e}c^2)^2}\frac{\omega_2}{\omega_1}|\boldsymbol{\epsilon}_1\cdot\boldsymbol{\epsilon}_2^*|^2\,S(\mathbf{q},\hbar\omega_1-\hbar\omega_2),\end{aligned} \tag{12.34}$$

where $r_0^2 = 0.079$ barn and $\alpha_f = e^2/4\pi\epsilon_0\,\hbar c \simeq 1/137$ is the dimensionless fine structure constant which is a fundamental measure of the photon-charge coupling (see (10.40)). The quantity

$$S(\mathbf{q}, \hbar\omega_1 - \hbar\omega_2) = \sum_b \left|\langle b|\, e^{i\mathbf{q}\cdot\mathbf{r}}\,|a\rangle\right|^2 \,\delta(\mathcal{E}_{ba} - \hbar\omega_1 + \hbar\omega_2) \tag{12.35}$$

is the *dynamic structure factor* of dimension [1/energy] since the δ-function is only defined by its dimensionless integral over energy. The difference $\hbar\omega_1 - \hbar\omega_2$ corresponds to an inelastic loss with momentum change $\mathbf{q} = \mathbf{k}_1 - \mathbf{k}_2$ shown in Fig. 12.12.

12.5.2 Elastic Thomson Scattering: Atomic Form Factor

For elastic scattering we have $\omega_1 = \omega_2$ and $a = b$, so that the differential non-resonant scattering cross section becomes

$$\frac{d\sigma_T}{d\Omega_2} = r_0^2\,|\boldsymbol{\epsilon}_1\cdot\boldsymbol{\epsilon}_2^*|^2 \left|\sum_{n=1}^{Z}\langle n|e^{i\mathbf{q}\cdot\mathbf{r}_n}|n\rangle\right|^2 = r_0^2\,|\boldsymbol{\epsilon}_1\cdot\boldsymbol{\epsilon}_2^*|^2\,|F^0(\mathbf{q}, Z)|^2. \tag{12.36}$$

Here n are the atomic electrons of total number Z, and $\mathbf{r}_n$ is the distance vector from the nucleus to the n-th electron. The quantity $F^0(\mathbf{q}, Z)$ is the dimensionless *atomic form factor* given by (6.25) or

$$F^0(\mathbf{q}, Z) = -\frac{1}{e}\int \rho(\mathbf{r})\, e^{i\mathbf{q}\cdot\mathbf{r}}\, d\mathbf{r}. \tag{12.37}$$

At soft x-ray energies where the wavelength λ is larger than the atomic diameter the form factor becomes equal to the number of electrons $F^0(\mathbf{q}) = F^0 = Z$, and by integration over emission angles (see (6.21)) we obtain for the polarization and emission angle integrated atomic Thomson cross section

$$\boxed{\sigma_T = \frac{8\pi}{3}\, r_0^2\, Z^2 = \sigma_e\, Z^2\,,} \tag{12.38}$$

where $\sigma_e = 0.665$ barn is the single-electron Thomson cross section. This agrees with the classical result given by (6.28).

12.5.3 Inelastic Thomson Scattering: Dynamical Structure Factor

The *dynamic structure factor*

$$S(\mathbf{q}, \Delta\hbar\omega) = \sum_b \left| \langle b | \, e^{\mathrm{i}\mathbf{q}\cdot\mathbf{r}} | \, a \rangle \right|^2 \, \delta(\mathcal{E}_{ba} - \hbar\omega_1 + \hbar\omega_2) \tag{12.39}$$

contains all atomic matter dependent information. It describes the excitation spectrum of the sample as a function of transferred momentum $\mathbf{q} = \mathbf{k}_1 - \mathbf{k}_2$ and energy change $\hbar\omega_1 - \hbar\omega_2$. The term "dynamic" reflects the fact that it may also be written in terms of a time dependent two-particle density correlation function. It then expresses the fluctuation-dissipation theorem that connects the dissipative response of a system to a perturbation with the equilibrium fluctuations of the system [46].

One would expect that the scattered intensity peaks at the elastic scattering condition $\hbar\omega_1 = \hbar\omega_2$ and $\mathbf{q} = 0$ gradually become weaker with increasing energy and momentum. This is seen by expanding the operator which couples different electronic states according to

$$e^{\mathrm{i}\mathbf{q}\cdot\mathbf{r}} = 1 + \mathrm{i}\mathbf{q}\cdot\mathbf{r} + \frac{1}{2}(\mathrm{i}\mathbf{q}\cdot\mathbf{r})^2 + \ldots. \tag{12.40}$$

The first *monopole* excitation term describes elastic forward scattering $\mathbf{q} = 0$ where electrons oscillate with the x-ray frequency and re-emit radiation of the same frequency in the same direction. The second term is the *electric dipole* term often called the *Raman scattering* term and corresponds to dipole excitations between states a and b, involving an energy transfer $\hbar\omega_1 - \hbar\omega_2$.[5] The third term corresponds to *magnetic dipole* and *electric quadrupole* excitations as discussed in Sect. 11.4, and in accordance with (11.11) can be written as

$$\frac{1}{2}(\mathrm{i}\mathbf{q}\cdot\mathbf{r})^2 = \underbrace{\frac{\mathrm{i}}{2}(\mathbf{r}\times\mathbf{p})\cdot(\mathbf{q}\times\boldsymbol{\epsilon}_p)}_{\text{magnetic dipole}} + \underbrace{\frac{\mathrm{i}}{2}(\mathbf{q}\cdot\mathbf{r})(\mathbf{p}\cdot\boldsymbol{\epsilon}_p) + (\mathbf{q}\cdot\mathbf{p})(\mathbf{r}\cdot\boldsymbol{\epsilon}_p)}_{\text{electric quadrupole}}. \tag{12.41}$$

Another way to derive the different excitations associated with $S(\mathbf{q}, \hbar\omega_1 - \hbar\omega_2)$ [48] is to expand the scattering operator in terms of spherical harmonics or spherical tensors according to

[5] Note that for absorption and emission, the dipole approximation corresponds to approximating the exponential by 1 as in (10.5), while in Thomson scattering it corresponds to the second term on the right in (12.40).

$$
\begin{aligned}
e^{\mathrm{i}\mathbf{q}\cdot\mathbf{r}} &= 4\pi \sum_{l=0}^{\infty}\sum_{m=-l}^{l} \mathrm{i}^l\, j_l(qr)\, Y^*_{l,m}(\mathbf{e_q})\, Y_{l,m}(\mathbf{e_r}) \\
&= \sum_{l=0}^{\infty}\sum_{m=-l}^{l} \mathrm{i}^l\,(2l+1)\, j_l(qr)\, C_m^{(l)*}(\theta_q,\phi_q)\, C_m^{(l)}(\theta_r,\phi_r),
\end{aligned} \tag{12.42}
$$

where $j_l(qr)$ are Bessel functions given for $l = 0, 1, 2$ by (10.21), and the unit vectors $\mathbf{e_q}$ and $\mathbf{e_r}$ define the coordinate frames of the position vector $\mathbf{r}$ and momentum transfer vector $\mathbf{q}$. Note that the scattering operator depends on both the *direction* (through the angle dependent functions $Y_{l,m}$) and the *magnitude* (through $j_l(qr)$). For a dipole transition, for example, the **q**-vector assumes the role of the **E**-vector in the determination of the angular dependence of the transitions and maximum inelastic excitations occur along **q**.

12.5.4 Core Shell Excitations: X-Ray Raman Scattering (XRS)

Inelastic Thomson scattering also occurs when the photon energy loss is due to core electron excitations. The process is similar to the loss in energy of high energy electrons in an electron microscope referred to as *electron energy loss spectroscopy* [49] or when used for core excitations as *inner shell electron energy loss spectroscopy* (ISEELS) [43, 50]. In the limit of small momentum transfer, where the dipole approximation holds, inelastic loss spectra recorded with electrons or photons are equivalent to XAS spectra.

The long history of inelastic x-ray spectroscopy has been reviewed by Bergmann, Glatzel, and Cramer [51]. Following the modern theoretical description by Mizuno and Ohmura [52] and its clear experimental demonstration by Suzuki [53] in 1967, it is often referred to by the name *x-ray Raman scattering* (XRS). Another name is *non-resonant inelastic x-ray scattering* (NIXS) [46]. In using the acronym NIXS we have to keep in mind that the “N” for “non-resonant” derives from the historical view that Thomson scattering is a first order “non-resonant” process that in its elastic form involves the entire atomic charge. When inelastic excitations of specific core shells are picked out by their characteristic loss energies, inelastic Thomson scattering can nevertheless provide information about resonant excitations involving specific atomic subshells.

XRS or NIXS is typically performed with incident hard x-rays around 10 keV and use of a crystal monochromator to select an incident energy bandpass, and the scattered radiation is analyzed by a crystal spectrometer. Experiments may be performed by scanning either the incident, $\hbar\omega_1$, or scattered, $\hbar\omega_2$, photon energy. The measured XRS energy loss extends from the characteristic XAS threshold of interest to larger values as illustrated for the C K-edge of graphite in Fig. 12.14 [51]. The spectrum was recorded as specified in the caption.

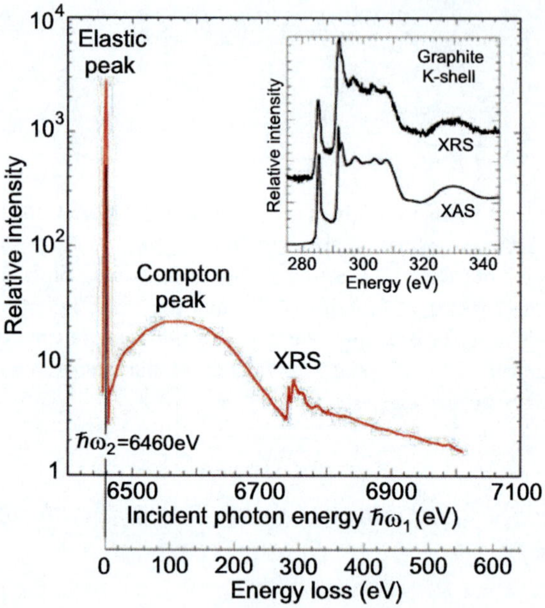

Fig. 12.14 Experimental inelastic photon energy loss spectrum of the carbon K-shell in graphite [51], recorded with an instrumental resolution of 2.5 eV and a range of momentum transfers $4\text{Å}^{-1} \leq q \leq 6\text{Å}^{-1}$. The loss spectrum reveals the elastic peak, the Compton scattering background peak, and the XRS loss. In the measurement the scattered photon energy was detected at a fixed energy $\hbar\omega_2 = 6460\,\text{eV}$ and the incident energy $\hbar\omega_1$ was scanned as indicated. The XRS spectrum in the inset was recorded with an instrumental resolution of 1 eV, while the XAS spectrum was recorded by electron yield detection with a resolution of 0.15 eV

In practice, losses associated with core electron binding energies (x-ray absorption edges) larger than about 1 keV become too weak. For this reason, most applications have involved losses associated with soft x-ray core excitations such as the C K-edge around 285 eV and O K-edge around 535 eV as reviewed in [46, 54–56].

For the study of low-Z elements like C and O, the dipole approximation is easily satisfied in practice. The $1s$ core shell radius $\langle r_{1s}\rangle$ is to a good approximation given by the radial dipole matrix element $\mathcal{R}_{1s}$ plotted in Fig. 12.11a which scales with Z approximately as a_0/Z, where $a_0 = 0.0529\,\text{nm}$ is the Bohr radius. For the O K-shell, for example, we have $\langle r_{1s}\rangle \simeq \mathcal{R}_{1s} \simeq 5\times10^{-2}\,\text{Å}$. The momentum transfer is given by $q^2 = [\omega_1^2 + \omega_2^2 - \omega_1\omega_2\cos(2\theta)]/c^2$, where θ is the angle between incident and scattered photons (see Fig. 12.17). At a relatively large scattering angle of $2\theta = 90°$, for example, and an incident energy of 10 keV we have a momentum transfer of about $q \simeq 7\text{Å}^{-1}$. This yields $\langle r_{1s}\rangle q = 0.35$, and the dipole approximation still holds. In this case, XRS gives results similar to XAS spectroscopy as shown in the inset of Fig. 12.14 and also demonstrated in [54, 55].

12.5.4.1 Optimum Range of Momentum Transfer in XRS

At finite but small wavevector transfer $q < 5\text{Å}^{-1}$, the XRS cross section is dominated by the dipole term $\propto \mathbf{q} \cdot \mathbf{r}$ in the expansion (12.40) or (12.42) of the matrix element of the dynamic structure factor. In order to illustrate the relative sizes of the monopole ($\ell = 0$), dipole ($\ell = 1$), and quadrupole ($\ell = 2$) terms in (12.42), we have calculated the expectation values

$$\langle R_{1s} | j_\ell(qr) | R_{2p} \rangle = \int_0^\infty R_{1s}\, j_\ell(qr)\, R_{2p}\, r^2\, \mathrm{d}r \tag{12.43}$$

corresponding to $1s \to 2p$ excitations in oxygen atoms ($Z = 8$). We used the hydrogenic wavefunctions given by (10.12) and the Bessel functions $j_\ell(qr)$ for $\ell = 0, 1, 2$ given by (10.21). The results for the radial probabilities $R^2_{n\ell}\, r^2$ and the corresponding multi-pole terms (12.43) are shown in Fig. 12.15, neglecting interference terms between the channels.

The multi-pole terms shown in Fig. 12.15 only account for the magnitudes, q, but not the directions, $\mathbf{q}$, of the scattered photons. The monopole term only contributes to the elastically forward scattered intensity around $\mathbf{q} = 0$, since inelastic scattering involves excited states that are orthogonal to the ground state, and can only be reached through the higher order multi-pole terms. We see that for relatively small momentum transfer $q \sim 5\,\text{Å}^{-1}$, the *dynamic structure factor* may be approximated by the dipolar $\ell = 1$ form

$$S(\mathbf{q}, \Delta\hbar\omega) = q^2 \sum_b |\langle b|\, \mathbf{e_q} \cdot \mathbf{r}\, |a\rangle|^2\, \delta(\mathcal{E}_{ba} - \hbar\omega_1 + \hbar\omega_2), \tag{12.44}$$

where $\mathbf{e_q}$ is a unit vector in the direction of $\mathbf{q}$.

There is another important reason for keeping the momentum transfer relatively small, i.e. $q < 5\,\text{Å}^{-1}$. It facilitates removal of the Compton scattering background that appears as a broad background in the inelastically scattered intensity, as seen

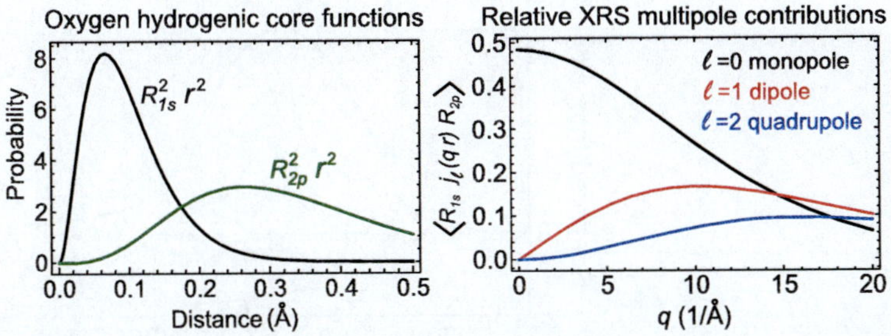

Fig. 12.15 Radial probabilities $R^2_{n\ell}\, r^2$ for oxygen atoms calculated with hydrogenic wavefunctions (10.12) and q-dependence of multi-pole terms (12.43) calculated with the Bessel functions $j_\ell(qr)$ for $\ell = 0, 1, 2$ given by (10.21). Note that the monopole term does not contribute to *inelastic* scattering

in Fig. 12.14. The change of the Compton background with momentum transfer is shown in more detail in Fig. 12.16 in conjunction with K-shell XRS spectrum of a diamond powder [57].

With increasing **q** the Compton peak shifts to larger loss energies [56]. For very large momentum transfer **q**, XRS evolves into Compton scattering [45, 56] and the XRS onset for a core shell excitation may be viewed as the threshold of its Compton profile as illustrated in Fig. 12.16 [57].

12.5.4.2 XRS Versus XAS Cross Sections

In the following we are interested in a quantitative comparison of the relative size of the XRS and XAS cross sections. We will see that the XRS cross section is only of order 10^{-10} of the corresponding soft x-ray absorption cross section, so that XRS studies greatly benefit from state-of-the-art synchrotron facilities or XFELs. Nevertheless, in some cases the cross-section disadvantage is acceptable because of the following unique capabilities of XRS.

> For the study of absorption fine structure in the soft x-ray region up to about 1 keV, x-ray Raman scattering or XRS may be used instead of conventional XAS. While the XAS cross section is significantly larger, XRS offers increased penetration lengths owing to the use of hard x-rays. This avoids the vacuum environment required for soft x-ray XAS and facilitates transmission studies of samples such as gases, liquids, and bulk materials and their investigations in extreme conditions such as at high pressure and temperature.

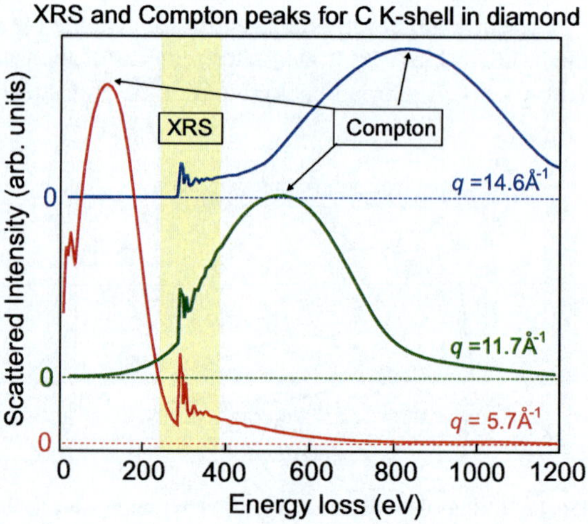

Fig. 12.16 q-dependence of the Compton background underlying XRS, illustrated by study of the C K-edge of diamond powder [57]

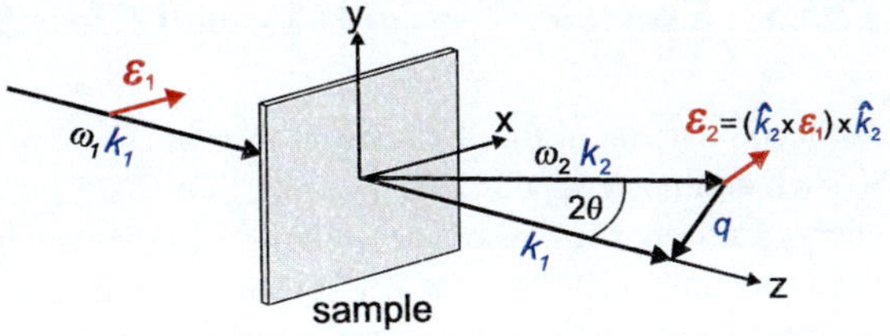

Fig. 12.17 Scattering geometry for *non-resonant* Raman scattering, indicating the wavevector and polarization dependence in the inelastic scattering process

For our quantitative description of XRS we envision the scattering process as schematically shown in Fig. 12.17. For incident linear polarization the polarization vectors are real and for charge scattering the scattered polarization direction is $\boldsymbol{\epsilon}_2 = (\hat{\mathbf{k}}_2 \times \boldsymbol{\epsilon}_1) \times \hat{\mathbf{k}}_2$, and with the notation of Fig. 12.17 we have $|\boldsymbol{\epsilon}_1 \cdot \boldsymbol{\epsilon}_2|^2 = \cos^2(2\theta)$ when $\mathbf{q}$ lies in the x–z plane and $|\boldsymbol{\epsilon}_1 \cdot \boldsymbol{\epsilon}_2|^2 = 1$ when $\mathbf{q}$ lies in the y–z plane.

When the dynamic structure factor is written in the dipolar form (12.44), the state $|a\rangle$ in the matrix element is the electronic ground state, and there is still a sum over all atomic shells $|b\rangle$. If we energy select a specific excitation $|a\rangle \rightarrow |b\rangle$ where $|b\rangle$ is a specific core hole state, we obtain an element-specific XRS cross section which describes the core hole production in a specific atomic shell. By selecting a specific final core hole state $|b\rangle$, we need to also consider its lifetime or decay width Γ_b. We may then replace the δ-function by a Lorentzian which integrates over energy to the same unit value (assuming a single final state) as the δ-function. The Lorentzian is given by (10.36) and has the dimension [1/energy]. The double differential cross section then becomes

$$\frac{\mathrm{d}^2\sigma_{\mathrm{XRS}}}{\mathrm{d}\Omega_2\, \mathrm{d}\hbar\omega_2} = r_0^2\, q^2\, \frac{\omega_2}{\omega_1}\, \frac{2}{\pi \Gamma_b}\, |\boldsymbol{\epsilon}_1 \cdot \boldsymbol{\epsilon}_2^*|^2 |\langle\, b |\, \mathbf{e}_{\mathbf{q}} \cdot \mathbf{r}\, | a \rangle|^2\, \frac{(\Gamma_b/2)^2}{(\mathcal{E}_{ba} - \hbar\omega_1 + \hbar\omega_2)^2 + (\Gamma_b/2)^2}. \tag{12.45}$$

This may be directly compared to the XAS cross section written in a similar form by use of (10.38) and (10.40) as

$$\sigma_{\mathrm{XAS}} = \frac{8\pi\, \hbar\omega\, \alpha_{\mathrm{f}}}{\Gamma_b}\, |\langle b|\boldsymbol{\epsilon}_1 \cdot \mathbf{r}|a\rangle|^2\, \frac{(\Gamma_b/2)^2}{(\mathcal{E}_{ba} - \hbar\omega)^2 + (\Gamma_b/2)^2}. \tag{12.46}$$

We see that apart from the unit polarization vector the squared marix element is the same in both cases. The difference in cross section is therefore simply given by the prefactors.

12.5.5 *Example: Typical O K-shell Cross Sections*

For a comparison of the XRS and XAS cross section we consider the case of the O K-shell and neglect polarization effects. For XRS we assume monochromatic high energy incident photons of energy $\hbar\omega_1 \simeq 10\,\text{keV}$ and bandwidths less than the O K-shell core hole width of $\Gamma_b \simeq 0.15\,\text{eV}$, and we measure the scattered photon intensity at an energy around $\hbar\omega_2 \simeq 9.5\,\text{keV}$ corresponding to the O K-edge loss. We assume a momentum transfer of $q \sim 5\,\text{Å}^{-1}$ where the Compton peak lies below the K-shell loss spectrum. With $r_0^2 = 0.079\,\text{barn} = 0.079 \times 10^{-10}\,\text{nm}^2 = 7.9 \times 10^{-10}\,\text{Å}^2$ we obtain a prefactor value in (12.45) of

$$r_0^2\, q^2 \frac{\omega_2}{\omega_1} \frac{2}{\pi \Gamma_b} \simeq 8.4 \times 10^{-8}. \tag{12.47}$$

For the XAS case we assume incident photons around $\hbar\omega = 535\,\text{eV}$, and with a core hole width of $\Gamma_b \simeq 0.15\,\text{eV}$ we obtain the prefactor in (12.46) as

$$\frac{8\pi\, \hbar\omega\, \alpha_f}{\Gamma_b} \simeq 6.5 \times 10^2. \tag{12.48}$$

The relative size of the differential XRS and XAS cross sections is, therefore,

$$\boxed{\frac{\frac{\mathrm{d}^2\sigma_{\text{XRS}}}{\mathrm{d}\Omega_2\,\mathrm{d}\omega_2}}{\sigma_{\text{XAS}}} \simeq 1.3 \times 10^{-10}}. \tag{12.49}$$

References

1. A. Einstein, Phys. Z. **18**, 121 (1917)
2. A. Einstein, Verh. Deut. Phys. Ges. **18**, 318 (1916)
3. M. Planck, Verh. Deut. Phys. Ges. **2**, 202 (1900)
4. M. Planck, Verh. Deut. Phys. Ges. **2**, 237 (1900)
5. J. Stöhr, Phys. Rev. Lett. **118**, 024–801 (2017)
6. J. Stöhr, Adv. Optics Photonics **11**, 215 (2019)
7. V. Weisskopf, E. Wigner, Z. Physik **63**, 54 (1930)
8. V. Weisskopf, E. Wigner, Z. Physik **65**, 18 (1930)
9. D.V. Strekalov, A.V. Sergienko, D.N. Klyshko, Y.H. Shih, Phys. Rev. Lett. **74**, 3600 (1995)
10. Y. Shih, *An Introduction to Quantum Optics: Photon and Biphoton Physics* (CRC Press, Boca Raton, Florida, 2011)
11. W. Voigt: Sitzungsbericht der Bayerischen Akad. Wissenschaften p. 25 (1912)
12. O. Björneholm, A. Nilsson, A. Sandell, B. Hernnäs, N. Mårtensson, Phys. Rev. Lett. **68**, 1892 (1992)
13. W. Wurth, P. Feulner, D. Menzel, Phys. Scripta **T41**, 213 (1992)
14. R. Röhlsberger, DESY, private communication. For a review of state-of-the-art Mössbauer spectroscopy see R. Röhlsberger, Fortschr. Phys. **61**, 360 (2013)

15. N. Rohringer et al., Nature **481**, 488 (2012)
16. H. Ågren, J. Nordgren, L. Selanger, C. Nordling, K. Siegbahn, J. Elec. Spectrosc. Relat. Phenom. **14**, 27 (1978)
17. P. Baltzer, B. Wannberg, L. Karlsson, M.C. Göthe, M. Larsson, Phys. Rev. A **45**, 4374 (1992)
18. A. Müller et al., Astrophys. J. **836**, 166 (2017)
19. T. Watanabe, Phys. Rev. **139**(A), 1747 (1965)
20. R. Püttner et al., Phys. Rev. Lett. **114**, 001–093 (2015)
21. L.O. Werme, B. Grennberg, J. Nordgren, C. Nordling, K. Siegbahn, Nature **242**, 453 (1973)
22. K. Prince, M. Vondráček, J. Karvonen, M. Coreno, R. Camilloni, L. Avaldi, M. de Simone, J. Elec. Spectrosc. Relat. Phenom. **101–103**, 141 (1999)
23. S.K. Semenov et al., J. Phys. B **39**, 375 (2006)
24. K. Prince et al., J. Synchrotron Rad. **5**, 565 (1998)
25. S. Svensson, M. Carlsson-Göthe, L. Karlsson, A. Nilsson, N. Mårtensson, U. Gelius, Physica Scripta **44**, 184 (1991)
26. U. Hergenhahn, O. Kugeler, A. Rüdel, E.E. Rennie, A.M. Bradshaw, J. Phys. Chem. A **105**, 5704 (2001)
27. M. Neeb, J.E. Rubensson, M. Biermann, W. Eberhardt, J. Elec. Spectrosc. Relat. Phenom. **67**, 261 (1994)
28. P. Glans, P. Skytt, K. Gunnelin, J.H. Guo, J. Nordgren, J. Elec. Spectrosc. Relat. Phenom. **82**, 193 (1996)
29. J.E. Rubensson et al., Phys. Rev. Lett. **114**, 001–133 (2015)
30. L. Kjellsson, V. Ekholm, M. Agåker, C. Såthe, A. Pietzsch, H.O. Karlsson, N. Jaouen, A. Nicolaou, M. Guarise, C. Hague, J. Lüning, S. Chiuzbăian, J.E. Rubensson, Phys. Rev. A **103**, 022–812 (2021)
31. E.Z. Kurmaev, A.L. Ankudinov, J.J. Rehr, L.D. Finkelstein, P.F. Karimov, A. Moewes, J. Elec. Spectrosc. Relat. Phenom. **148**, 1 (2005)
32. A.C. Miller, G.W. Simmons, Surf. Sci. Spectra **2**, 55 (1993)
33. M. Magnuson, N. Wassdahl, J. Nordgren, Phys. Rev. B **56**, 12–238 (1997)
34. J. Stöhr, H.C. Siegmann, *Magnetism: From Fundamentals to Nanoscale Dynamics* (Springer, Heidelberg, 2006)
35. H. Ebert, J. Stöhr, S.S.P. Parkin, M. Samant, A. Nilsson, Phys. Rev. B **53**, 16–067 (1996)
36. O. Karis, M. Magnuson, T. Wiell, M. Weinelt, N. Wassdahl, A. Nilsson, N. Mårtensson, E. Holmström, A.M.N. Niklasson, O. Eriksson, Phys. Rev. B **62**, R16-239 (2000)
37. A. Nilsson, J. Stöhr, T. Wiell, M. Aldén, P. Bennich, N. Wassdahl, M. Samant, S.S.P. Parkin, N. Mårtensson, J. Nordgren, B. Johansson, H.L. Skriver, Phys. Rev. B **54**, 2917 (1996)
38. M. Guerra, J.M. Sampaio, F. Parente, P. Indelicato, P. Hönicke, M. Müller, B. Beckhoff, J.P. Marques, J.P. Santos, Phys. Rev. A **97**, 042–501 (2018)
39. W. Bambynek, B. Crasemann, R.W. Fink, H.U. Freund, H. Mark, C.D. Swift, R.E. Price, P.V. Rao, Rev. Mod. Phys. **44**, 716 (1972)
40. M.O. Krause, J. Phys. Chem. Ref. Data **8**, 307 (1979)
41. J.H. Hubbell, P.N. Trehan, N. Singh, B. Chand, D. Mehta, M.L. Garg, R.R. Garg, S. Singh, S. Puri, J. Phys. Ref. Data **23**, 339 (1994)
42. J.C. Slater, K.H. Johnson, Phys. Rev. B **5**, 844 (1972)
43. J. Stöhr, *NEXAFS Spectroscopy* (Springer, Heidelberg, 1992)
44. A.H. Compton, Phys. Rev. **22**, 409 (1923)
45. M. Cooper, P. Mijnarends, N. Shiotani, N. Sakai, A. Bansil, *X-Ray Compton Scattering* (Oxford University Press, Oxford, 2004)
46. W. Schülke, *Electron Dynamics by Inelastic X-Ray Scattering* (Oxford University Press, Oxford, 2007)
47. J. Als-Nielsen, D. McMorrow, *Elements of Modern X-Ray Physics*, 2nd edn. (Wiley, New York, 2011)
48. M.W. Haverkort, A. Tanaka, L.H. Tjeng, G.A. Sawatzky, Phys. Rev. Lett. **99**, 257–401 (2007)
49. R.F. Egerton, *Electron Energy-Loss Spectroscopy in the Electron Microscope*, 3rd edn. (Springer, Heidelberg, 2011)

50. A.P. Hitchcock, C.E. Brion, J. Elec. Spectrosc. Relat. Phenom. **18**, 1 (1980)
51. U. Bergmann, P. Glatzel, S.P. Cramer, Microchem. J. **71**, 221 (2002)
52. Y. Mizuno, Y. Ohmura, J. Phys. Soc. Jpn. **22**, 445 (1967)
53. T. Suzuki, J. Phys. Soc. Jpn. **22**, 1139 (1967)
54. M.L. Gordon, D. Tulumello, G. Cooper, A.P. Hitchcock, P. Glatzel, O.C. Mullins, S.P. Cramer, U. Bergmann, J. Phys. Chem. A **107**, 8512 (2003)
55. A. Nilsson, D. Nordlund, I. Waluyo, N. Huang, H. Ogasawara, S. Kaya, U. Bergmann, L.-Å. Näslund, H. Öström, P. Wernet, K.J. Andersson, T. Schiros, L.G.M. Pettersson, J. Elec. Spectrosc. Relat. Phenom. **177**, 99 (2010)
56. C.J. Sahle, A. Mirone, J. Niskanen, J. Inkinen, M. Krisch, S. Huotari, J. Synchrotron Rad. **22**, 400 (2015)
57. S. Huotari, T. Pylkkänen, J.A. Soininen, J.J. Kas, K. Hämäläinen, G. Monaco, J. Synchrotron Rad. **19**, 106 (2012)

Chapter 13
Quantum Theory of X-Ray Resonant Scattering

13.1 Introduction and Overview

The goal of this chapter is to extend the *first order* KHD formalism describing x-ray absorption, emission, and Thomson scattering to the description of *second order* phenomena, most prominently encountered in *resonant* x-ray scattering. The general KHD formula (9.21) contains both Thomson (first order) and resonant scattering (second order) contributions, and they may even interfere because the respective scattering lengths need to be added first before the total scattered intensity is calculated as the absolute value squared.[1]

In this chapter we specifically discuss the second order term in the KHD formula (9.21) which entails the quantum theory of both energy dependent elastic (REXS) and inelastic (RIXS) processes. The theory will be illustrated through experimental examples. For simplicity and clarity we will omit the Thomson scattering term discussed in Sect. 12.5, which may be envisioned to contribute a polarization dependent but energy independent "background".

There are many advantages of using resonant scattering rather than conventional Thomson scattering. The most important ones are

- *Elastic* resonant scattering (REXS) is typically used for coherent *diffractive imaging*, while *inelastic* scattering (RIXS) is used for *spectroscopic studies* of the electronic and spin structure of the valence electrons, including its energy/momentum dispersion.
- RIXS differs from XES spectroscopy through the link of the excitation and emission processes and reduces to XES when the excitation process is non-specific, as for broadband incident x-rays.

[1] Interference between first and second order scattering amplitudes exists when the final states of the Thomson and resonant terms are the same. An example is a purely elastic scattering event where the final state is not only the electronic but also the vibrational ground state in a molecule. For this case, the polarization dependent interplay between the Thomson and resonant term can be seen from Fig. 1(c) of Phys. Rev. Lett. **106**, 153004 (2011).

J. Stöhr, *The Nature of X-Rays and Their Interactions with Matter*, Springer Tracts in Modern Physics 288, https://doi.org/10.1007/978-3-031-20744-0_13

- The intrinsic spectral resolution or measured linewidth of resonant scattering is *not* limited by the core hole widths of the intermediate states through which the electronic system passes from the initial to the final state. This leads to a significant advantage of RIXS over conventional XES.
- Through its dependence on the *incident* photon energy, resonant scattering is element and chemical state specific. Like Thomson scattering, the resonantly scattered intensity depends on polarization, momentum transfer, and energy loss.
- Resonant scattering offers enhanced cross sections over Thomson scattering and through its polarization dependence can separate charge and spin dependent processes. Spin dependent resonant scattering is much larger than spin dependent Thomson scattering.
- The resonant process links electronic excitation and decay processes that both obey dipole and parity selection rules, resulting in the selection of specific final states that cannot be reached in a single dipole transition from the electronic ground state. In other cases, the final states may be a subset of those that are reached in conventional XES spectroscopy.

With the advent of brighter x-ray sources, RIXS has attracted considerable attention in recent years because of its capability of providing unique information. RIXS is complementary to valence band photoemission and inelastic neutron scattering, yielding information about the low energy electronic and spin excitations in solids as previously illustrated in Fig. 12.13.

We mention upfront that resonant x-ray scattering which specifically measures the scattered photons has complementary analogues in resonant photoemission and Auger spectroscopies as reviewed in [1, 2]. We will not discuss the latter techniques which are based on the measurement of electrons, keeping with the "x-ray spirit" of this book.

13.2 Formulation of Resonant Scattering: REXS and RIXS

Resonant scattering, including both elastic, REXS, and inelastic, RIXS, processes, is described by the second order term in (9.21) which in the dipole approximation reads

$$\mathcal{W}^{\text{res}}_{i\to f} = \frac{2\pi e^2}{\hbar}\left|\sum_m \frac{\langle f|\mathbf{r}\cdot\mathbf{E}^-|m\rangle\langle m|\mathbf{r}\cdot\mathbf{E}^+|i\rangle}{\mathcal{E}_i-\mathcal{E}_m}\right|^2 \rho(\mathcal{E}_f)\,\delta(\mathcal{E}_f-\mathcal{E}_i). \quad (13.1)$$

Again, the states $|i\rangle$, $|m\rangle$, $|f\rangle$ are products of photon number states and electronic states. We shall label the latter $|a\rangle$, $|c\rangle$, $|b\rangle$, respectively.

The KHD formula shows that second order processes consist of two consecutive and *linked* first order XAS and XES processes. This "link" between the probability amplitudes of the two consecutive excitation and decay processes distinguishes it from first order XES. The second order process consists of a sum over all energetic paths through intermediate states m that link the initial and final states before the expectation value is formed as the absolute value squared. If different intermediate core hole states are involved, paths can interfere and such interference effects most clearly distinguish second order from two consecutive first order processes.

13.2.1 Evaluation of the Double Matrix Element

The resonant scattering rate in (13.1) is calculated by assuming that the incident flux may in general contain photons in two different modes $p_1\mathbf{k}_1$ and $p_2\mathbf{k}_2$ with corresponding energies $\hbar\omega_1$ and $\hbar\omega_2$. The case where all photons are in the same mode is a special case of the general formalism. The photons $n_{p_1\mathbf{k}_1}$ excite the electronic system from the ground state $|a\rangle$ of energy $\mathcal{E}_a$ to intermediate electronic states $|c\rangle$ of energy $\mathcal{E}_c$. The photons $n_{p_2\mathbf{k}_2}$ may then de-excite the system from states $|c\rangle$ to the final state $|b\rangle$ of energy $\mathcal{E}_b$. The process is elastic if $\hbar\omega_1 = \hbar\omega_2$ so that $|b\rangle = |a\rangle$ and inelastic if $\hbar\omega_2 < \hbar\omega_1$ so that the system ends up in an excited state with $\mathcal{E}_b > \mathcal{E}_a$.

The photon parts of the matrix elements in (13.1) are again evaluated by the action of the destruction and creation operators on *number states* as done in (10.2) and (12.2). According to (3.61), the absorption operator $\mathbf{E}^+ \propto a_{p_1\mathbf{k}_1}$ destroys photons in the state $p_1\mathbf{k}_1$ and the emission operator $\mathbf{E}^- \propto a^\dagger_{p_2\mathbf{k}_2}$ creates photons in the state $p_2\mathbf{k}_2$. The matrix elements in (13.1) are evaluated by considering the states involved. We have an initial state

$$|i\rangle = |a, n_1, n_2\rangle \tag{13.2}$$

with an energy $\mathcal{E}_i = \mathcal{E}_a + n_1\hbar\omega_1 + n_2\hbar\omega_2$. The final state is

$$|f\rangle = |b, n_1 - 1, n_2 + 1\rangle, \tag{13.3}$$

with energy $\mathcal{E}_f = \mathcal{E}_b + (n_1 - 1)\hbar\omega_1 + (n_2 + 1)\hbar\omega_2$. We can now find the intermediate states which give non-vanishing photon matrix elements in (13.1). There are only two kinds. First,

$$|m\rangle = |c, n_1 - 1, n_2\rangle \tag{13.4}$$

with energy $\mathcal{E}_m = \mathcal{E}_c + (n_1 - 1)\hbar\omega_1 + n_2\hbar\omega_2$. In this case the energy denominator in (13.1) is given by $\mathcal{E}_i - \mathcal{E}_m = \mathcal{E}_a - \mathcal{E}_c + \hbar\omega_1$. The second allowed intermediate state is virtual and given by

$$|m\rangle = |c, n_1, n_2 + 1\rangle \tag{13.5}$$

with energy $\mathcal{E}_m = \mathcal{E}_c + n_1\hbar\omega_1 + (n_2+1)\hbar\omega_2$. This virtual state contains two more photons than the other intermediate state. Now the energy denominator in (13.1) is given by $\mathcal{E}_i - \mathcal{E}_m = \mathcal{E}_a - \mathcal{E}_c - \hbar\omega_2$.

By explicit insertion of the states and energies for the two pathways and by making the dipole approximation, which eliminates the **k** dependence of the transition operator, the resonant scattering rate per solid emission angle and per emitted energy band width is obtained as [3, 4],

$$\frac{W^{\text{res}}_{a\to b}}{\mathrm{d}\Omega_2\,\mathrm{d}\hbar\omega_2} = \Phi_1 \sum_b^{\omega_2\le\omega_1} \frac{\omega_1\omega_2^3\,\hbar^2\alpha_f^2}{c^2}\left(n_{p_2\mathbf{k}_2}+1\right)$$
$$\times\left|\sum_c\left(\frac{\langle b|\mathbf{r}\cdot\boldsymbol{\epsilon}^*_{p_2}|c\rangle\langle c|\mathbf{r}\cdot\boldsymbol{\epsilon}_{p_1}|a\rangle}{\hbar\omega_1-(\mathcal{E}_c-\mathcal{E}_a)+\mathrm{i}\,\Gamma_c/2} - \frac{\langle b|\mathbf{r}\cdot\boldsymbol{\epsilon}_{p_1}|c\rangle\langle c|\mathbf{r}\cdot\boldsymbol{\epsilon}^*_{p_2}|a\rangle}{\hbar\omega_2+(\mathcal{E}_c-\mathcal{E}_a)}\right)\right|^2$$
$$\times\,\delta[(\mathcal{E}_b-\mathcal{E}_a)-(\hbar\omega_1-\hbar\omega_2)]\,. \tag{13.6}$$

Here Φ_1 is the incident photon flux (see (10.3)). This expression is valid for elastic as well as inelastic scattering. The important factor $n_{p_2\mathbf{k}_2}+1$ represents stimulated x-ray emission by the incident photons in state $|n_{p_2\mathbf{k}_2}\rangle$ with energy $\hbar\omega_2$. When $\mathcal{E}_a = \mathcal{E}_b$, the scattering is elastic, and when $\mathcal{E}_b \neq \mathcal{E}_a$, the scattering process is inelastic. The finite lifetime of the final state is neglected since energy conservation is simply assured by the Dirac δ-function between the initial state $|a\rangle$ and final states $|b\rangle$, independent of the intermediate core hole states $|c\rangle$. Note that the δ-function is only defined through its integration over energy which yields a unitless number, so that it formally has the dimension [1/energy]. The sum over intermediate electronic states $|c\rangle$ expresses paths through which the quantum mechanical system evolves from $|a\rangle$ to $|b\rangle$. We have inserted the factor $\mathrm{i}\,\Gamma_c/2$ in the denominator of the first term in the sum over intermediate states c to account for its finite decay width Γ_c. It also eliminates a singularity for $\hbar\omega_1 = \mathcal{E}_c - \mathcal{E}_a$. The second term in the sum over intermediate states is non-resonant since the denominator is always large.

The intermediate states $|c\rangle$ may be real or virtual, as illustrated in Fig. 13.1 for different cases. Equation (13.6) includes two "resonant" terms that are distinguished by the denominators. In practice, there are two situations determined by the relative size of $\hbar\omega_i$ and the average resonance energy $\mathcal{E}_0 \simeq \mathcal{E}_c - \mathcal{E}_a$, since the natural decay energy width $\Gamma_c/2$ is typically small. The most common situation corresponds to $\hbar\omega_i \approx \mathcal{E}_0$. The transition rate is then determined by the larger first term, since $\hbar\omega_1 - \mathcal{E}_0$ in the denominator is small while $\hbar\omega_2 + \mathcal{E}_0$ is large. The second resonant term can then be omitted, referred to as the *rotating wave approximation.*

The case $\hbar\omega_i \gg \mathcal{E}_0$ shown in Fig. 13.1a, labeled "Thomson scattering", corresponds to transitions through virtual states of much higher energy than any true resonant states. In this case both terms in (13.6) contribute. As discussed by Loudon [5], this corresponds to a description of Thomson scattering through a second order process where the usual Hamiltonian $e^2A^2/(2m_e)$ in (9.22) is replaced by the photoelectric Hamiltonian $e\,\mathbf{p}\cdot\mathbf{A}/m_e$. However, this description of Thomson scattering as a

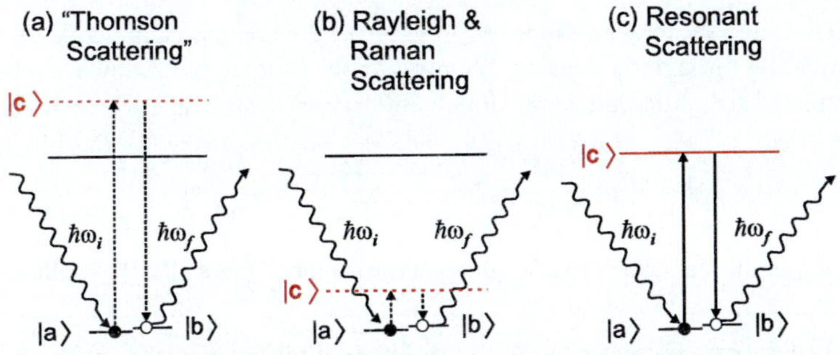

Fig. 13.1 Illustration of second order scattering processes in the optical and x-ray regime expressed by (13.6). The shown states $|a\rangle$ and $|b\rangle$ are the initial and final electronic states. The intermediate states $|c\rangle$ may be real or virtual as discussed in the text. As discussed in the text, the description of "Thomson scattering" through a second order process is somewhat problematic

second order process is problematic since it does not properly converge to the conventional Thomson scattering cross section derived in Sect. 12.5 at high x-ray energies.

Illustrated in Fig. 13.1b is the case of *Rayleigh* and *Raman* scattering in the *optical* range. Since air molecules have larger electronic energy differences than the energy of optical light, the scattering of sun light by the earth's atmosphere is dominated by virtual transitions through lower energy states. Rayleigh scattering is elastic $\mathcal{E}_b = \mathcal{E}_a$, while Raman scattering is inelastic. The Raman process allows observation of states $|b\rangle$ that cannot be reached directly by dipole transitions $|a\rangle \rightarrow |b\rangle$ because of selections rules.

In *resonant* scattering, illustrated in Fig. 13.1c, the intermediate states $|c\rangle$ are real electronic states through which the system passes to the final state $|b\rangle$. In the shown configuration picture (see discussion below in Sect. 13.2.2 and Fig. 13.2), the initial state $|a\rangle$ is the electronic ground state, and in the x-ray regime the intermediate state $|c\rangle$ consists of an excited electron in an empty valence state and a hole in a core state. The final state $|b\rangle$ is reached by an electronic decay, which may be envisioned by the electron decaying into the core hole or the core hole bubbling up to the filled valence shell. The final state $|b\rangle$ is an excited electronic state whose energy is higher than that of the ground state $|a\rangle$. In x-ray science $\mathcal{E}_b - \mathcal{E}_a$ ranges from meV vibrational excitations to electronic excitations within the valence shell up to a few eV.

By introducing the unit vector $\hat{\mathbf{r}}$ and factoring the radial matrix element $\mathcal{R}$, we obtain from (13.6) in the dipole and rotating wave approximations

$$\frac{\mathcal{W}^{\text{res}}_{a\to b}}{\mathrm{d}\Omega_2\,\mathrm{d}\hbar\omega_2} = \Phi_1\,\frac{4\pi^2\hbar\omega_1\hbar\omega_2\,\alpha_{\mathrm{f}}^2}{\lambda_2^2}\,\mathcal{R}^4\,\left(n_{p_2\mathbf{k}_2}+1\right)$$
$$\times\sum_b^{\omega_2\le\omega_1}\left|\sum_c\frac{\langle b|\hat{\mathbf{r}}\cdot\boldsymbol{\epsilon}^*_{p_2}|c\rangle\langle c|\hat{\mathbf{r}}\cdot\boldsymbol{\epsilon}_{p_1}|a\rangle}{\hbar\omega_1-(\mathcal{E}_c-\mathcal{E}_a)+\mathrm{i}\,\Gamma_c/2}\right|^2$$
$$\times\,\delta[(\mathcal{E}_b-\mathcal{E}_a)-(\hbar\omega_1-\hbar\omega_2)]\,. \tag{13.7}$$

Transition rates $\mathcal{W}$ may be expressed in terms of the corresponding cross sections σ by dividing through the incident photon flux Φ, yielding the following relations for the integrated, differential, and double differential scattering cross sections σ

$$\mathcal{W} = \Phi\,\sigma, \quad \frac{\mathcal{W}}{\mathrm{d}\Omega} = \Phi\,\frac{\mathrm{d}\sigma}{\mathrm{d}\Omega}, \quad \frac{\mathcal{W}}{\mathrm{d}\Omega\,\mathrm{d}\omega} = \Phi\,\frac{\mathrm{d}\sigma}{\mathrm{d}\Omega\,\mathrm{d}\omega}. \tag{13.8}$$

The differential scattering cross section is then obtained from (13.7) as follows.

The general expression for the double differential *resonant* cross section, which describes elastic (REXS) and inelastic (RIXS) x-ray scattering, is given in the dipole and rotating wave approximations by

$$\frac{\mathrm{d}\sigma^{\mathrm{res}}_{a\to b}}{\mathrm{d}\Omega_2\,\mathrm{d}\hbar\omega_2} = \frac{4\pi^2\hbar\omega_1\hbar\omega_2\,\alpha_{\mathrm{f}}^2}{\lambda_2^2}\,\mathcal{R}^4\,\left(n_{p_2\mathbf{k}_2}+1\right)$$
$$\times \sum_b^{\omega_2\le\omega_1} \left| \sum_c \frac{\langle b|\hat{\mathbf{r}}\cdot\boldsymbol{\epsilon}^*_{p_2}|c\rangle\langle c|\hat{\mathbf{r}}\cdot\boldsymbol{\epsilon}_{p_1}|a\rangle}{\hbar\omega_1 - (\mathcal{E}_c - \mathcal{E}_a) + \mathrm{i}\,\Gamma_c/2} \right|^2$$
$$\times\,\delta[(\mathcal{E}_b - \mathcal{E}_a) - (\hbar\omega_1 - \hbar\omega_2)]\,. \tag{13.9}$$

The states $|a\rangle$, $|b\rangle$, $|c\rangle$ are electronic states of the sample and may include their vibrationally split substates. The Dirac δ-function assures strict energy conservation between the initial state $|a\rangle$ and final state $|b\rangle$ that does *not* involve the intermediate states $|c\rangle$. It has the important consequence that the *intrinsic* (i.e. in the absence of instrumental contributions) REXS and RIXS linewidth of the scattered photons is *not* limited by the finite intermediate state widths Γ_c, as first pointed out by Weisskopf in 1931 [6].

13.2.2 One-Electron Versus Configuration Picture

Figure 13.2 illustrates different ways to picture elastic and inelastic resonant scattering for the case of $2p \leftrightarrow 3d$ scattering.

On the left we show how resonant scattering is pictured in a one-electron model by extension of the x-ray absorption and emission picture. This model follows the motion of the "active" electron from the $2p$ core to the $3d$ valence shell and back to the core. If the same electron is involved, we have elastic scattering, if an electron of different energy is involved in the de-excitation process, the scattering is inelastic. As illustrated in the middle, one can view resonant scattering also by considering the motion of a hole with the energy diagram turned upside down. On the right we show the proper *configuration* picture which corresponds to Fig. 13.1. It clearly shows that

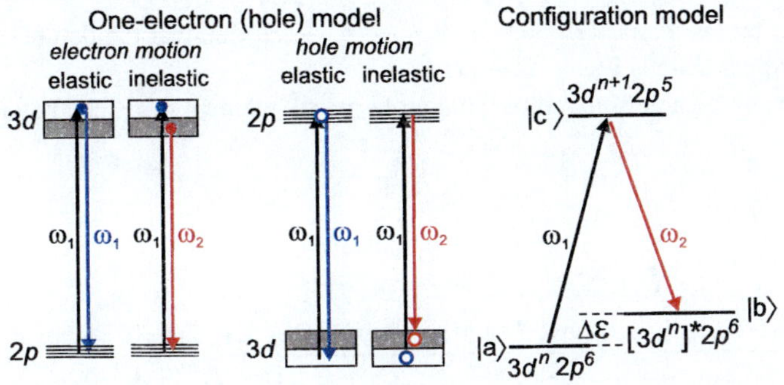

Fig. 13.2 Resonant elastic and inelastic x-ray scattering in different models. On the left we show how in a one-electron model we follow the motion of the "active" electron from the $2p$ core to the $3d$ valence shell and back to the core. If the same electron is involved, we have elastic scattering, and if in the de-excitation process an electron of different energy is involved, the scattering is inelastic. The same process can also be pictured by considering the motion of a hole as shown in the middle. The energy diagram is turned upside down. On the right we show the proper configuration picture where all three states are products of core and valence states

the core hole state is the intermediate excited state $|c\rangle$ through which the system passes from the initial state $|a\rangle$ to the final state $|b\rangle$.

There are two important messages in the configuration picture which will emerge from our later detailed treatments, but we shall already highlight them in a box.

> The core hole intermediate state $|c\rangle$ has no influence on energy conservation between the initial and final states. Therefore the core hole energy width does not limit the energy resolution of the resonant scattering process $|a\rangle \rightarrow |b\rangle$.
> The paths through the intermediate core hole state $|c\rangle$ allow final states $|b\rangle$ to be reached that are not accessible through direct dipole transitions $|a\rangle \rightarrow |b\rangle$. This is similar to optical Raman scattering [7].

13.2.3 Coherent Second Order Versus Consecutive First Order Processes

In the following we shall adopt the electronic configuration picture and label the ground state $|a\rangle$, the intermediate state $|c\rangle$, and final state $|b\rangle$, as shown on the right of Fig. 13.2. We consider two intermediate states $|c_1\rangle$ and $|c_2\rangle$ to illustrate their interference contribution. We assume that the two states are separated by $\Delta = \mathcal{E}_{c_2} - \mathcal{E}_{c_1} > 0$ and have the same natural decay width Γ. For convenience, we assume *linear polarization* so that the transition matrix elements are real and describe the dependence on the incident photon energy $\hbar\omega_1$ in terms of the detuning energy from the upper state is $|c_1\rangle$ as $\Omega_{c_1} = \Omega = \hbar\omega_1 - (\mathcal{E}_{c_1} - \mathcal{E}_{c_a})$ so that the detuning energy from the upper state is $\Omega_{c_2} = \Omega - \Delta$.

The absolute value squared of the double matrix element in (13.9) then takes the form

$$\left|\sum_{m=c_1,c_2} \frac{\langle b|\hat{\mathbf{r}}\cdot\boldsymbol{\epsilon}_{p_2}|m\rangle\langle m|\hat{\mathbf{r}}\cdot\boldsymbol{\epsilon}_{p_1}|a\rangle}{\hbar\omega_1 - (\mathcal{E}_m - \mathcal{E}_a) + \mathrm{i}\Gamma/2}\right|^2$$

$$= \underbrace{\frac{|\langle b|\hat{\mathbf{r}}\cdot\boldsymbol{\epsilon}_{p_2}|c_1\rangle|^2\,|\langle c_1|\hat{\mathbf{r}}\cdot\boldsymbol{\epsilon}_{p_1}|a\rangle|^2}{\Omega^2 + \Gamma^2/4} + \frac{|\langle b|\hat{\mathbf{r}}\cdot\boldsymbol{\epsilon}_{p_2}|c_2\rangle|^2\,|\langle c_2|\hat{\mathbf{r}}\cdot\boldsymbol{\epsilon}_{p_1}|a\rangle|^2}{(\Omega-\Delta)^2 + \Gamma^2/4}}_{\text{separate excitation and decay processes}}$$

$$+ \underbrace{2\frac{\langle b|\hat{\mathbf{r}}\cdot\boldsymbol{\epsilon}_{p_2}|c_1\rangle\langle c_1|\hat{\mathbf{r}}\cdot\boldsymbol{\epsilon}_{p_1}|a\rangle\langle b|\hat{\mathbf{r}}\cdot\boldsymbol{\epsilon}_{p_2}|c_2\rangle\langle c_2|\hat{\mathbf{r}}\cdot\boldsymbol{\epsilon}_{p_1}|a\rangle}{\Omega(\Omega-\Delta) + \frac{\Delta^2\Gamma^2/4}{\Omega(\Omega-\Delta)+\Gamma^2/4} + \Gamma^2/4}}_{\text{interference of the two processes}}. \quad (13.10)$$

The first underbrace identifies separate paths through the two intermediate states caused by vibrational or electronic spilitting. The second underbrace expresses interference of the paths. The existence of interference is a consequence of a fundamental rule of quantum mechanics, most clearly articulated and emphasized by Feynman [8, 9]. In a process between well-defined initial and final states the probability amplitudes of all alternative indistinguishable paths must first be summed before the total *probability* is calculated as the absolute value squared. In diffraction, for example, we have to sum (integrate) over all possible *spatial paths* of photons (or electrons), as was utilized in the quantum derivation of the van Cittert–Zernike theorem in Sect. 4.7. In REXS/RIXS we have to instead sum over all *energetic paths* of photons through electronic intermediate states.

Equation 13.10 expresses a fundamental rule of quantum mechanics, most clearly articulated and emphasized by Feynman [8, 9]. When considering a process between well-defined initial and final states, *probability amplitudes* of all alternative paths are first summed before the total *probability* is calculated as the absolute value squared. In diffraction, for example, we have to sum (integrate) over all possible *spatial paths* of photons (or electrons), as was utilized in the quantum derivation of the van Cittert–Zernike theorem in Sect. 4.7. In REXS/RIXS we have to sum over all *energetic paths* of photons through electronic intermediate states.

When interference between different intermediate state paths is neglected, the general case reduces as follows

$$\underbrace{\left|\sum_c \frac{\langle b|\hat{\mathbf{r}}\cdot\boldsymbol{\epsilon}^*_{p_2}|c\rangle\langle c|\hat{\mathbf{r}}\cdot\boldsymbol{\epsilon}_{p_1}|a\rangle}{\hbar\omega_1 - (\mathcal{E}_c - \mathcal{E}_a) + \mathrm{i}\Gamma/2}\right|^2}_{\text{includes interference}} \longrightarrow \underbrace{\sum_c \left|\frac{\langle b|\hat{\mathbf{r}}\cdot\boldsymbol{\epsilon}^*_{p_2}|c\rangle\langle c|\hat{\mathbf{r}}\cdot\boldsymbol{\epsilon}_{p_1}|a\rangle}{\hbar\omega_1 - (\mathcal{E}_c - \mathcal{E}_a) + \mathrm{i}\Gamma/2}\right|^2}_{\text{no interference}}. \quad (13.11)$$

The second order KHD formalism, which corresponds to an inseparable coherent excitation/decay process on the left, is approximated on the right by independent consecutive first order KHD processes. The presence of interference terms changes the spectral shape of RIXS spectra and interference of vibrational amplitudes also affects the quasi-elastic REXS lineshape. This will be illustrated by the REXS/RIXS spectra of N_2 and O_2 below.

13.2.4 REXS/RIXS Terminologies

The complicated general resonant term (13.10) has led to much discussion in the literature following the pioneering theoretical work of Gel'mukhanov et al. [10] and Åberg and Tulkki [3, 4, 11] around 1980. By now the theory has been applied to matter in various forms and compared to experiments.

The extensive work in the atomic and molecular (AMO) community has been reviewed by Gel'mukhanov et al. [2, 12]. It is interesting that it has taken three to four decades for synchrotron experiments to achieve similar resolution ($\sim$ 100 meV) as the early AMO XES studies by the Uppsala group in the 1970s, represented by the spectra of Ne in Fig. 12.6 and N_2 in Fig. 12.7. While early XES studies used electron excitation of gases and grating spectrographs operated in higher order, synchrotron-based photon in/out experiments long suffered by intensity/resolution trade-offs which were overcome only with third generation x-ray sources and large dedicated grating-based spectrometers [13–17]. At present, the best soft x-ray RIXS instruments have a resolution of about 30 meV at photon energies extending to about 1 keV [18]. In the hard x ray region around 10 keV, instrumental bandwidths of 4 meV have been achieved in RIXS spectrometers based on flat-quartz crystal optics [19].

Early AMO studies have been extended over the years to various forms of matter. By now RIXS has also been utilized for the study of polymers [20, 21], chemisorption systems [22], metal complexes [23–25], solids with weak and strong valence correlations [26–29], and solids under high pressure [30]. The rich landscape of RIXS and its utilization by different communities has also led to the use of different approximations of the general KHD formalism, resulting in terminologies that we will briefly outline below.

Today the generic terms "x-ray emission spectroscopy", XES, or "fluorescence spectroscopy" are mainly used when the radiative decay follows a non-specific creation of core holes, as by high energy or broad bandwidth x-rays or by ionizing electrons.

The modern use of resonant excitations has led to a distinction of the REXS/RIXS processes based on the role of the *intermediate states* through which the system passes from the ground to the final state. In particular, one considers whether the photon probability amplitude or wavefunction passes through the allowed intermediate core hole states with or without interference of the paths, as expressed by (13.11). This picture closely resembles Feynman's quantum formulation of the double-slit experiment (see Sect. 8.5.1), with the spatial (diffraction) paths through the slits replaced by energetic (spectroscopy) paths through two or more intermediate states. In RIXS, the inclusion of the interference of all energetic paths, which is a natural consequence of their indistinguishability in quantum mechanics, has been denoted by the terms "one-step RIXS" [31, 32] or "coherent inelastic scattering" [33, 34].

When interference between the paths is neglected, the "up" and "down" processes may be viewed as distinguishable, decoupled or incoherent. This approximation expressed by (13.11) has led to the terms "incoherent inelastic scattering" [33, 34] or "two-step RIXS" [12, 31, 32]. The term "resonantly excited XES" has also been used [22, 35]. Note, however, that this case still differs from XES and only merges into it in the limit that the incident photon energy has a broad bandwidth or is well above the emission energy so that the first excitation step is essentially non-resonant.

In the condensed matter community, where much emphasis is placed on the study of correlated materials, the names "direct RIXS" and "indirect RIXS" have been introduced [27, 29, 36, 37]. These terms are not linked to the one- versus two-step distinctions above but break down the intermediate state in the KHD formalism in a way that de-emphasizes ("direct") or emphasizes ("indirect") the role of the core hole. The intermediate state Hamiltonian is divided into that of the ground state Hamiltonian of the quantum system without a core hole, and a perturbative part that describes its change after the excitation step. The latter consists of the excitonic state created by the coupling of the core hole and the active electron and the interaction of this excitonic state with the other electrons in the system. The details of this division, which may be phrased in terms of Green's function propagators, and the relative size of the two terms then facilitate the description of correlated materials [27].

The *direct RIXS* process emphasizes the role of the ground state term, and dynamical aspects of the core hole in the intermediate state are neglected. In practice, this description applies when the RIXS process is dominated by strong resonant excitations $c \rightarrow V^*$ from a core state c to unoccupied valence states V^*, followed by reversed transitions $V \rightarrow c$ from occupied valence states V back into the core hole. This situation is encountered in K-shell studies of low-Z molecules, where the $1s \leftrightarrow 2p$ transitions involve molecular orbitals with large $2p$ contributions. Similarly, L-shell (and M-shell) studies of the important $3d$ transition metal compounds are dominated by $2p \leftrightarrow 3d$ ($3p \leftrightarrow 3d$) transitions. In both cases, "up" and "down" transitions are strong due the rather localized nature of the valence states involved. Since low-Z K-edges and transition metal L-edges lie in the spectral range < 1000 eV, direct RIXS typically dominates in the soft x-ray region.

The term *indirect RIXS* is used when the excitation process involves a "non-resonant" excitation to somewhat delocalized valence states that lie a few eV above the Fermi level E_F. The origin of the name "indirect" is best seen from the example of K-shell RIXS spectra of $3d$ transition metal compounds, involving hard x-ray energies. If there were no localized $3d$ states around E_F, the excitation of a $1s$ electron into delocalized $4p$ states above the Fermi level would predominantly lead to a REXS process where the active excited electron itself refills the core hole. From a direct RIXS point of view, one would expect to see no inelastic losses. The *presence* of localized $3d$ states around E_F, however, changes this picture. Then a RIXS signature may be observed which arises *indirectly* through an energy loss associated with an electron-hole excitation or "shake-up" within the d states.

In this chapter we will restrict our discussion to the simpler *direct RIXS* process and illustrate the difference in its one-step and two-step description. The discussion of RIXS is facilitated by starting with the somewhat simpler REXS process.

13.3 Quantum Formulation of REXS

The simplest case of (13.9) is resonant *elastic* x-ray scattering or REXS, defined as excitations from the electronic ground state through various intermediate core hole states back to the *electronic* ground state. Using the designation of electron states in the configuration model in Fig. 13.1, REXS consists of processes $|b\rangle = |a\rangle$. We consider only the resonant or second order term in the general KHD scattering expression (9.21) and ignore the elastic Thomson term and its possible interference with the resonant term.

If we include the splitting of electronic states through nuclear vibrations, we have to be more specific in our definition of REXS. The initial state of a sample consists of an electronic ground state that in the Born–Oppenheimer approximation may contain several nuclear vibrational states. The true ground state may be prepared by cooling the sample. In contrast, the intermediate core excited states are "hot" so that higher vibrational states are always occupied.

A true *elastic* scattering process that conserves energy in the photon system is a symmetric up and down process that leads back to the vibrational ground state. Processes that lead to higher vibrational final states or other low energy excited states are referred to as *quasi-elastic*. We will consider quasi-elastic processes as a specific case of resonant *inelastic* x-ray scattering or RIXS. In the following we shall first discuss the case of pure REXS, defined by a resonant scattering process from and back to a well-defined ground state.

13.3.1 The Fundamental REXS Cross Section

For the REXS case we have $|b\rangle = |a\rangle$ and with $\mathcal{E}_c - \mathcal{E}_a = \mathcal{E}_{ca}$, the general expression (13.9) reduces to

$$\frac{\mathrm{d}\sigma^{\mathrm{scat}}_{a\to a}}{\mathrm{d}\Omega_2\,\mathrm{d}\hbar\omega_2}=\frac{\mathrm{d}\sigma_{\mathrm{REXS}}}{\mathrm{d}\Omega_2\,\mathrm{d}\hbar\omega_2}=\frac{4\pi^2\hbar\omega_1\hbar\omega_2\,\alpha_{\mathrm{f}}^2}{\lambda_2^2}\,\mathcal{R}^4\,\left(n_{p_2\mathbf{k}_2}+1\right)$$
$$\times\left|\sum_c\frac{\langle a|\hat{\mathbf{r}}\cdot\boldsymbol{\epsilon}^*_{p_2}|c\rangle\langle c|\hat{\mathbf{r}}\cdot\boldsymbol{\epsilon}_{p_1}|a\rangle}{\hbar\omega_1-\mathcal{E}_{ca}+\mathrm{i}\Gamma_c/2}\right|^2\delta(\hbar\omega_1-\hbar\omega_2).\tag{13.12}$$

We have deliberately maintained the δ-function notation that enforces energy conservation. As mentioned earlier, this has the important consequence that the *intrinsic* (i.e. in the absence of instrumental contributions) REXS linewidth of the scattered photons is *not* limited by the intermediate state width Γ_c.

The multiplication with the δ-function in (13.12) effectively cuts out a slice of the intermediate state Lorentzian, resulting in an infinitely sharp peak. In practice, this corresponds to a measurement with δ-function-like instrumental bandwidths of the incident x-rays, determined by a monochromator, and of the scattered x-rays, determined by a spectrometer. The monochromator cuts out a narrow slice of the intermediate state Lorentzian, and this slice is then convoluted with the bandwidth of the emission spectrometer. This differs from the XES case, where the incident bandwidth does not enter and the emitted linewidth Γ_c is directly convoluted with the spectrometer resolution. For a δ-function-like spectrometer resolution, the measured XES linewidth will be Γ_c, since it is preserved upon convolution with a δ-function spectrometer width.

Under the assumption of infinitesimally small instrumental bandwidths, the REXS (and RIXS) linewidth is not limited by the intermediate state width Γ_c as a consequence of strict energy conservation expressed by the Dirac δ-function. In contrast, the intrinsic linewidth in XES (and XAS) is determined by the natural decay width Γ_c of the intermediate (final) states.

13.3.2 REXS with Finite Instrumental Resolution

In practice, the measured REXS cross section is obtained by considering the instrumental resolution functions of the monochromator that defines the incident bandwidth and of the emission spectrometer that analyzes the scattered photons. For grating-based RIXS instruments, the resolution functions are taken to be Gaussians. The theoretical cross section must then be convoluted with the Gaussians. It turns out that for REXS, the δ-function in (13.12) can be replaced by a normalized Gaussian that contains the monochromator FWHM, Γ_1, and spectrometer FWHM Γ_2 on equal footing with the following result.[2]

[2] I would like to thank Faris Gel'mukhanov for the mathematical derivation of this result.

$$\frac{d\sigma_{\text{REXS}}}{d\Omega_2\, d\hbar\omega_2} = \frac{4\pi^2 \hbar\omega_1 \hbar\omega_2\, \alpha_f^2}{\lambda_2^2} \mathcal{R}^4 \left(n_{p_2\mathbf{k}_2} + 1\right)$$
$$\times \left| \sum_c \frac{\langle a|\hat{\mathbf{r}} \cdot \boldsymbol{\epsilon}^*_{p_2}|c\rangle\langle c|\hat{\mathbf{r}} \cdot \boldsymbol{\epsilon}_{p_1}|a\rangle}{\hbar\omega_1 - \mathcal{E}_{ca} + \mathrm{i}\,\Gamma_c/2} \right|^2 G(\hbar\omega_1 - \hbar\omega_2), \qquad (13.13)$$

where

$$G(\hbar\omega_1 - \hbar\omega_2) = \frac{a}{\sqrt{2\pi}\sqrt{\Gamma_1^2 + \Gamma_2^2}} \exp\left[-\frac{a^2(\hbar\omega_1 - \hbar\omega_2)^2}{2\,(\Gamma_1^2 + \Gamma_2^2)}\right]. \qquad (13.14)$$

Here $a = 2\sqrt{\ln 4} = 2.355$ and the Gaussian is normalized to have unit integrated area (see Appendix A.2.2) like the δ-function it replaces in (13.12).[3]

The above formulation describes the case where the incident photon energy is defined by a monochromator of energy bandwidth (FWHM) Γ_1 around $\hbar\omega_1$ and the scattered radiation is analyzed by a spectrometer with bandwidth (FWHM) Γ_2 around $\hbar\omega_2$. Hence there is only a REXS signal when the energy distributions between the incident and scattered photons overlap. In practice, the Gaussian of combined width $\Gamma_G = \sqrt{\Gamma_1^2 + \Gamma_2^2}$ takes a slice out of the Lorentzian of width Γ_c so that the measured REXS linewidth is no longer determined by the intermediate state core hole width Γ_c but by that of the instrumental resolution function Γ_G.

13.4 Spontaneous and Stimulated REXS Versus XAS

In our discussion above, we have assumed that both the incident and scattered radiation by atoms are analyzed in energy. If we do not analyze the scattered photon energy but only measure the energy integrated scattered intensity that falls into a detector of angular acceptance $d\Omega_2$ around a certain scattering direction $\mathbf{k}_2$, the double differential cross section (13.12) turns into a single differential cross section. The REXS intensity then depends only on the tuning of the incident energy, assumed to have a very small bandwidth, with (13.12) integrated over the emitted bandwidth $d\hbar\omega_2$. By use of the sifting property of the δ-function and denoting $\hbar\omega = \hbar\omega_1 = \hbar\omega_2$, we obtain

$$\frac{d\sigma_{\text{REXS}}}{d\Omega_2} = \frac{4\pi^2 (\hbar\omega)^2\, \alpha_f^2}{\lambda^2} \mathcal{R}^4 \left(n_{p_2\mathbf{k}_2} + 1\right) \left| \sum_c \frac{\langle a|\hat{\mathbf{r}} \cdot \boldsymbol{\epsilon}^*_{p_2}|c\rangle\langle c|\hat{\mathbf{r}} \cdot \boldsymbol{\epsilon}_{p_1}|a\rangle}{\hbar\omega - \mathcal{E}_{ca} + \mathrm{i}\,\Gamma_c/2} \right|^2. \qquad (13.15)$$

This case is of interest since we can directly compare the REXS intensity to the XAS intensity, since both depend only on the tuning of the incident energy.

[3] When for a given incident energy $\hbar\omega_1$ the Gaussian is integrated over $\hbar\omega_2$, it reduces to unity.

For *spontaneous* REXS, we have $n_{p_2\mathbf{k}_2}+1=1$ and the photons are omitted in random directions as a consequence of being created out of the random zero-point quantum vacuum. Their polarization vectors $\boldsymbol{\epsilon}^*_{p_2}$ lie in the plane perpendicular to the emission direction $\mathbf{k}_2$ into the detector with acceptance angle $\mathrm{d}\Omega_2$.

For *stimulated* REXS, we have $n_{p_2\mathbf{k}_2}+1=n_{p_2\mathbf{k}_2}$. Now the scattered photons are clones of the incident ones $n_{p_2\mathbf{k}_2}$ that stimulate the decays, and the scattered photons propagate into the same direction $\mathbf{k}_2$ and with the same polarization p_2.

13.4.1 Spontaneous REXS Versus XAS

We first consider spontaneous REXS and utilize the fact that the differential elastic scattering cross section is directly linked with the x-ray scattering length introduced in Sect. 6.2.3 by the relation (6.19) which in our case reads

$$\frac{\mathrm{d}\sigma_{\mathrm{REXS}}}{\mathrm{d}\Omega_2}=|f(\boldsymbol{\epsilon}_{p_1},\boldsymbol{\epsilon}_{p_2})|^2. \tag{13.16}$$

From (13.15) we directly obtain the following expression for the *spontaneous* scattering length

$$\boxed{f(\boldsymbol{\epsilon}_{p_1},\boldsymbol{\epsilon}_{p_2})=\frac{2\pi\hbar\omega\,\alpha_{\mathrm{f}}}{\lambda}\,\mathcal{R}^2\sum_c\frac{\langle a|\hat{\mathbf{r}}\cdot\boldsymbol{\epsilon}^*_{p_2}|c\rangle\langle c|\hat{\mathbf{r}}\cdot\boldsymbol{\epsilon}_{p_1}|a\rangle}{\hbar\omega-\mathcal{E}_{ca}+\mathrm{i}\,\Gamma_c/2}.} \tag{13.17}$$

By assuming a single intermediate state $|c\rangle$ of decay width $\Gamma=\Gamma_c$, and denoting the resonance energy as $\mathcal{E}_0=\mathcal{E}_{ca}$ and the incident and scattered polarizations as $\boldsymbol{\epsilon}_p=\boldsymbol{\epsilon}_{p_1}$ and $\boldsymbol{\epsilon}_{p'}=\boldsymbol{\epsilon}_{p_2}$, we can write (13.17) in the form

$$\begin{aligned}f(\boldsymbol{\epsilon}_p,\boldsymbol{\epsilon}_{p'})=f'-\mathrm{i}f''&=\frac{2\pi\hbar\omega\,\alpha_{\mathrm{f}}}{\lambda}\,\mathcal{R}^2\langle a|\hat{\mathbf{r}}\cdot\boldsymbol{\epsilon}^*_{p'}|c\rangle\langle c|\hat{\mathbf{r}}\cdot\boldsymbol{\epsilon}_p|a\rangle\\&\times\left\{\frac{\hbar\omega-\mathcal{E}_0}{(\hbar\omega-\mathcal{E}_0)^2+\Gamma^2/4}-\mathrm{i}\,\frac{\Gamma/2}{(\hbar\omega-\mathcal{E}_0)^2+\Gamma^2/4}\right\}.\end{aligned} \tag{13.18}$$

For the *resonant* case, quantum theory directly provides the link between the real and imaginary parts of the scattering lengths[4]

$$f'=\frac{2(\hbar\omega-\mathcal{E}_0)}{\Gamma}\,f''. \tag{13.19}$$

The *spontaneous* REXS scattering cross section is given by

[4] This is the same as the classical relation between the corresponding scattering factors $F'=f'/r_0$ and $F''=f''/r_0$ given by (6.44). Note that in general, f' also contains a non-resonant Thomson contribution r_0Z according to (7.113).

$$\frac{\mathrm{d}\sigma_{\mathrm{REXS}}^{\mathrm{spon}}}{\mathrm{d}\Omega} = |f(\boldsymbol{\epsilon}_p, \boldsymbol{\epsilon}_{p'})|^2 = \frac{4\pi^2(\hbar\omega)^2\,\alpha_{\mathrm{f}}^2}{\lambda^2}\,\mathcal{R}^4\,\frac{|\langle a|\hat{\mathbf{r}}\cdot\boldsymbol{\epsilon}_{p'}^*|c\rangle|^2|\langle c|\hat{\mathbf{r}}\cdot\boldsymbol{\epsilon}_p|a\rangle|^2}{(\hbar\omega-\mathcal{E}_0)^2+\Gamma^2/4} \tag{13.20}$$

which may be rewritten as

$$\frac{\mathrm{d}\sigma_{\mathrm{REXS}}^{\mathrm{spon}}}{\mathrm{d}\Omega} = \frac{1}{4\pi\Gamma}\underbrace{\frac{8\pi^2\hbar\omega\,\alpha_{\mathrm{f}}}{\lambda^2}\,\mathcal{R}^2\,|\langle a|\hat{\mathbf{r}}\cdot\boldsymbol{\epsilon}_{p'}^*|c\rangle|^2}_{\Gamma_{ca}^{p'}}$$
$$\times\underbrace{\frac{8\pi\hbar\omega\,\alpha_{\mathrm{f}}}{\Gamma}\,\mathcal{R}^2|\langle c|\hat{\mathbf{r}}\cdot\boldsymbol{\epsilon}_p|a\rangle|^2\frac{\Gamma^2/4}{(\hbar\omega-\mathcal{E}_0)^2+\Gamma^2/4}}_{\sigma_{\mathrm{XAS}}^p=2\lambda f''^p}. \tag{13.21}$$

The imaginary part of the forward scattering length f''^p is linked to the x-ray absorption cross section, $\sigma_{\mathrm{XAS}}^p = 2\lambda f''^p$ by the optical theorem previously discussed in Sects. 7.6 and 7.7.2. The underbrackets correspond to our previous results given by (10.40) and (10.39). We can then write the spontaneous REXS cross section in the convenient short form

$$\boxed{\frac{\sigma_{\mathrm{REXS}}^{\mathrm{spon}}}{\mathrm{d}\Omega} = \frac{1}{4\pi}\frac{\Gamma_{ca}^{p'}}{\Gamma}\,\sigma_{\mathrm{XAS}}^p.} \tag{13.22}$$

Note the different polarizations p and p'. If we assume charge scattering and incident linear polarization, the integration of (13.21) over the solid angle of emission Ω will yield a factor of $8\pi/3$ as for elastic Thomson scattering (see (6.21)), and we obtain

$$\sigma_{\mathrm{REXS}}^{\mathrm{spon}} = \frac{2}{3}\frac{\Gamma_{ca}^{p'}}{\Gamma}\,\sigma_{\mathrm{XAS}}^p. \tag{13.23}$$

13.4.1.1 Decomposition of Spontaneous REXS

According to (13.18), the complex scattering length has the real part

$$f' = \frac{2\pi\hbar\omega\,\alpha_{\mathrm{f}}}{\lambda}\,\mathcal{R}^2\langle a|\hat{\mathbf{r}}\cdot\boldsymbol{\epsilon}_{p'}^*|c\rangle\langle c|\hat{\mathbf{r}}\cdot\boldsymbol{\epsilon}_p|a\rangle\,\frac{\hbar\omega-\mathcal{E}_0}{(\hbar\omega-\mathcal{E}_0)^2+\Gamma^2/4} \tag{13.24}$$

which depends on the incident polarization and the polarization of the randomly scattered photons. It is zero at resonance, $\hbar\omega = \mathcal{E}_0$, and for charge scattering with linear polarization when $\boldsymbol{\epsilon}_{p'} \perp \boldsymbol{\epsilon}_p$ (see (13.25) below).

The imaginary part is linked by the optical theorem to forward scattering, and if we consider charge scattering, the polarization is preserved, $p = p'$. We then have

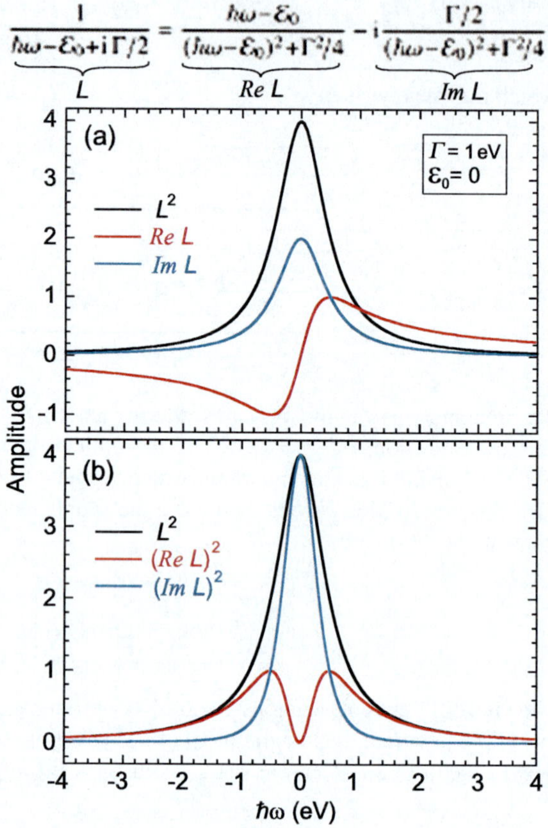

Fig. 13.3 Plot of the real and imaginary parts of the lineshape function defined by the curly brackets in (13.18), given at the top of the figure. The x-ray absorption cross section is proportional to Im L, while the resonant scattering cross section is proportional to L^2. The squared real (red) and imaginary (light blue) parts in **(b)** have the same integrated value of π

$$f'' = \frac{2\pi\hbar\omega\,\alpha_f}{\lambda}\,\mathcal{R}^2|\langle c|\hat{\mathbf{r}}\cdot\boldsymbol{\epsilon}_p|a\rangle|^2\,\frac{\Gamma/2}{(\hbar\omega-\mathcal{E}_0)^2+\Gamma^2/4}. \tag{13.25}$$

The real and imaginary parts of the Lorentzian lineshape functions of f' and f'' are listed at the top of Fig. 13.3, and their lineshape contributions are plotted underneath in Fig. 13.3a.

The absolute values squared of the real and imaginary parts of the lineshapes, which determine the scattered intensity, are illustrated in Fig. 13.3b. When we integrate the *squared* real (red) and imaginary (light blue) functions over photon energy, we find the same integrated value of π. The area under the total probability function, shown in black, therefore consists of equal probability contributions of the shown red (real) and blue (imaginary) parts to the total value of 2π. In accordance with the optical theorem, we may associate the imaginary part f'' with the spontaneously forward scattered photons and the real part f' with photons scattered out of the forward direction.

13.4.2 Stimulated REXS Versus XAS

In addition to spontaneous forward scattered REXS there can also be stimulated REXS which naturally preserves the direction and polarization, $\mathbf{k} = \mathbf{k}'$ and $p = p'$ of the photons driving the decays. The stimulated REXS cross section is obtained from (13.15) as

$$\frac{\sigma_{\text{REXS}}^{\text{stim}}}{\text{d}\Omega} = |f(\epsilon_p)|^2 = \frac{4\pi^2(\hbar\omega)^2\,\alpha_f^2}{\lambda^2}\,\mathcal{R}^4\,n_{p\mathbf{k}}\left|\sum_c \frac{|\langle c|\hat{\mathbf{r}}\cdot\boldsymbol{\epsilon}_p|a\rangle|^2}{\hbar\omega - \mathcal{E}_{ca} + \text{i}\,\Gamma_c/2}\right|^2 \tag{13.26}$$

For a single intermediate state the stimulated REXS cross section is related to the XAS cross section (10.39) and the two cross sections, written in the same notation, are linked according to

$$\frac{\sigma_{\text{REXS}}^{\text{stim}}}{\text{d}\Omega} = \frac{n_{p\mathbf{k}}}{4\pi\Gamma}\underbrace{\frac{8\pi^2\,\hbar\omega\,\alpha_f}{\lambda^2}\,\mathcal{R}^2\,|\langle a|\hat{\mathbf{r}}\cdot\boldsymbol{\epsilon}_p|c\rangle|^2}_{\Gamma_{ca}^p}$$
$$\times\underbrace{\frac{8\pi\,\hbar\omega\,\alpha_f}{\Gamma}\,\mathcal{R}^2|\langle c|\hat{\mathbf{r}}\cdot\boldsymbol{\epsilon}_p|a\rangle|^2\frac{\Gamma^2/4}{(\hbar\omega-\mathcal{E}_0)^2+\Gamma^2/4}}_{\sigma_{\text{XAS}}^p}, \tag{13.27}$$

where $\Gamma_{ca}^p = \Gamma_{ac}^p$ is the dipole transition width previously given by (10.40). By combining (13.27) with (13.23) we can therefore write the total REXS cross section in the short form

$$\boxed{\frac{\sigma_{\text{REXS}}}{\text{d}\Omega} = \frac{\sigma_{\text{REXS}}^{\text{spon}}}{\text{d}\Omega} + \frac{\sigma_{\text{REXS}}^{\text{stim}}}{\text{d}\Omega} = \frac{1}{4\pi}\left\{\frac{\Gamma_{ca}^{p'}}{\Gamma} + n_{p\mathbf{k}}\frac{\Gamma_{ca}^{p}}{\Gamma}\right\}\sigma_{\text{XAS}}^p.} \tag{13.28}$$

This expression for the combined spontaneous and stimulated contributions has an important consequence, referred to as the "*no-cloning*" theorem [38–42]. While stimulated decays expressed by the second term in (13.28) indeed produce a cloned twin of the stimulating photon, the presence of spontaneous scattering, expressed by the first term, also has a forward scattering component. This spontaneous contribution prevents perfect cloning of a single photon. The theorem breaks down, however, when at large $n_{p\mathbf{k}}$ stimulated scattering completely dominates.

13.4.3 Link to Semi-classical Results

Our KHD results may be compared to the semi-classical Thomson, XAS, and REXS cross sections summarized in Sect. 6.4.1. The classical resonant XAS and REXS cross sections are given by (6.46) and (6.47), which at resonance are given by

$$\text{Classical: } \sigma_{\text{XAS}} = 2\lambda r_0 \frac{\mathcal{E}_0}{\Gamma} \frac{(\Gamma/2)^2}{(\hbar\omega - \mathcal{E}_0)^2 + (\Gamma/2)^2}, \quad \sigma_{\text{REXS}}^{\text{spon}} = \frac{4\pi}{3} \frac{r_0}{\lambda} \frac{\mathcal{E}_0}{\Gamma} \sigma_{\text{XAS}}. \tag{13.29}$$

The corresponding KHD expressions for a given polarization are obtained from (10.38) and (13.21) as

$$\text{Quantum: } \sigma_{\text{XAS}} = \frac{\lambda^2}{\pi} \frac{\Gamma_{ac}}{\Gamma} \frac{(\Gamma/2)^2}{(\hbar\omega - \mathcal{E}_0)^2 + (\Gamma/2)^2}, \quad \sigma_{\text{REXS}}^{\text{spon}} = \frac{2}{3} \frac{\Gamma_{ca}}{\Gamma} \sigma_{\text{XAS}}, \tag{13.30}$$

where $\Gamma_{ac} = \Gamma_{ca}$ is given by (10.40) or (13.21). This yields the quantum expression for the classical electron radius

$$r_0 = \frac{\Gamma_{ac}}{\mathcal{E}_0} \frac{\lambda}{2\pi}. \tag{13.31}$$

13.5 Intermediate State Interference Effects in REXS

We have seen in Sect. 13.2.3 that a key question in the interpretation of experimental RIXS/REXS spectra is whether intermediate state interference effects may be neglected, which considerably simplifies the description. The role of such interference effects is best illustrated for the REXS case where the intermediate core hole state contains just *two* electronic or vibrational levels and REXS is described by (13.15). As discussed in Sect. 13.4, REXS is then directly related to XAS with the two intermediate REXS states corresponding to two final states in XAS. The XAS spectrum consists of just two peaks whose separation and relative intensities may be conveniently varied to explore their manifestations in the size of the REXS interference effects.

In REXS, the initial and final states are the same well-defined ground state of the total nuclear-electronic system and the system can pass through the two *energetically* separated intermediate states, and the interference of these paths is of interest. The process is analogous to the quantum description of the double-slit experiment where photons from a source point may pass through *spatially* separated intermediate states, represented by the two slits, to the detector plane.

In our treatment we assume that the two intermediate states decay with the same energy width Γ back to the ground state. For our model calculations we assume a core to valence transition energy of about 1 keV, which according to Table 10.2 corresponds to a core hole lifetime width of about $\Gamma = 0.5$ eV for both K- and L-shell excitations. In order to highlight the fundamental physics we assume incident radiation of negligible instrumental broadening and that the scattered intensity is measured at a certain momentum transfer with an energy integrating detector, as assumed in (13.15). Both REXS and XAS spectra are then only dependent on the tuning of the incident photon energy.

The XAS spectra are calculated with the same Γ value for the two peaks according to (10.55) and defining the resonant peak energies as $\mathcal{E}_1 = \mathcal{E}_{c_1} - \mathcal{E}_a$ and $\mathcal{E}_2 = \mathcal{E}_{c_2} - \mathcal{E}_a$ and the splitting energy $\Delta = \mathcal{E}_2 - \mathcal{E}_1$. We write the XAS cross section (10.55) for two Lorentzian peaks in a convenient short form by defining the detuning energy as $\Omega = \hbar\omega - \mathcal{E}_1$ and lumping all prefactors in the cross section into the following intensity form

$$I_{\mathrm{XAS}} = I_1 \frac{\Gamma^2/4}{\Omega^2 + \Gamma^2/4} + I_2 \frac{\Gamma^2/4}{(\Omega - \Delta)^2 + \Gamma^2/4}, \tag{13.32}$$

where the Lorentzian lineshape function reduces to unity at resonance.

By inspection of the XAS and REXS cross sections (10.55) and (13.15) we can rewrite the general REXS cross-section expression (13.15) which includes intermediate state interference in a corresponding intensity form

$$\boxed{\text{with interference}: \; I_{\mathrm{REXS}}^{\mathrm{1step}} = C \left| I_1 \frac{\Gamma/2}{\Omega + \mathrm{i}\Gamma/2} + I_2 \frac{\Gamma/2}{\Omega - \Delta + \mathrm{i}\Gamma/2} \right|^2}, \tag{13.33}$$

where C is a constant scaling factor and the squared XAS intensities I_i arise from the fact that the REXS expression contains an angular transition matrix element that is the square of the XAS one if we ignore polarization effects. The corresponding REXS expression that does *not include* intermediate state interference effects reads

$$\boxed{\text{no interference}: \; I_{\mathrm{REXS}}^{\mathrm{2step}} = C \left\{ I_1^2 \frac{\Gamma^2/4}{\Omega^2 + \Gamma^2/4} + I_2^2 \frac{\Gamma^2/4}{(\Omega - \Delta)^2 + \Gamma^2/4} \right\}}. \tag{13.34}$$

The size of the interference effect is given by the difference $\Delta I_{\mathrm{REXS}} = I_{\mathrm{REXS}}^{\mathrm{1step}} - I_{\mathrm{REXS}}^{\mathrm{2step}}$ given by

$$\Delta I_{\mathrm{REXS}} = C \left\{ 2 I_1 I_2 \frac{\Gamma^2/4}{\left[\Omega(\Omega - \Delta) + \frac{\Delta^2\Gamma^2/4}{\Omega(\Omega-\Delta)+\Gamma^2/4}\right] + \Gamma^2/4} \right\}. \tag{13.35}$$

For any two-peak absorption spectrum I_{XAS}, we can now calculate the corresponding one-step and two-step REXS cross sections and their difference ΔI_{REXS}. Since we know from (13.35) that the interference intensity scales *linearly* with the intensities, we pick $I_1 = 1$ and $I_2 = 0.5$. We then obtain the plots in Fig. 13.4 for the XAS and REXS intensities (scaled by setting $C = 1$) as a function of detuning energy Ω for four values of Δ/Ω.

In Fig. 13.4a we have plotted the reference XAS spectrum according to (13.32) for four values of Δ/Γ as a function of detuning energy Ω relative to the first

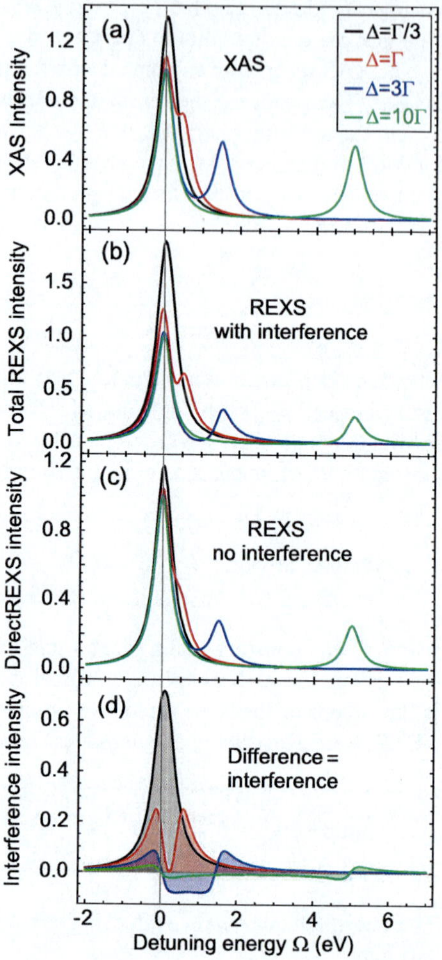

Fig. 13.4 **a** Reference XAS spectra according to (13.32) as a function of detuning energy Ω, assuming $I_1 = 1$ and $I_2 = 0.5$ for four values of Δ / Γ indicated in color. **b** one-step REXS spectra including interference according to (13.33) with $C = 1$. **c** two-step REXS spectra according to (13.34) with $C = 1$. **d** REXS intensity difference of (**b**) minus (**c**) given by (13.35)

XAS peak. The corresponding one-step REXS intensity (13.33) and two-step REXS intensity (13.34) are plotted in (b) and (c), and the interference intensity given by the difference of panels (b) and (c) and expressed by (13.35) is plotted in (d). We find that the interference contribution is constructive for $\Delta < \Gamma$, while for $\Delta > \Gamma$ the interference changes from constructive to destructive near the XAS resonance positions.

In particular, the expression identified by square brackets in (13.35) becomes Ω^2 for $\Delta = 0$ and Δ^2 for $\Omega = 0$, revealing similar roles of the detuning energy Ω and the splitting energy Δ, since they are both defined relative to the first peak energy $\mathcal{E}_1$. The interference intensity thus falls off from its maximum value with a *Lorentzian* dependence according to

$$\Delta I_{\text{REXS}} = I_1 I_2 \frac{1}{1 + 4x^2/\Gamma^2}, \tag{13.36}$$

where $x = \Omega$ or Δ. Because of the long energy tail of the Lorentzian, the interference effects linger on to relatively large values of x/Γ.

In REXS, interference of pathways through intermediate states of intrinsic decay width Γ, separated by Δ, modulates the elastic peak intensity as a function of the incident photon energy. The interference contribution is constructive for $\Delta \leq \Gamma$ and for $\Delta > \Gamma$ changes from constructive to destructive close to the XAS resonance positions. Interference effects decrease with increasing values of Δ and Γ.

13.5.1 REXS Interference Contour Map

In REXS/RIXS it is often convenient to view the intensity distribution in the form of a 2D contour map using the expression (13.12) with the axes given by $\hbar\omega_1$ and $\hbar\omega_2$ or with the $\hbar\omega_2$ axis replaced by the energy loss $\hbar\omega_2 - \hbar\omega_1$. Such a contour plot for the REXS *interference* intensity is shown in Fig. 13.5 where we have replaced (13.35) by

$$\Delta I_{\text{REXS}} = 2 I_1 I_2 \frac{\Gamma^2/4}{\left[\Omega(\Omega - \Delta) + \frac{\Delta^2\Gamma^2/4}{\Omega(\Omega-\Delta)+\Gamma^2/4}\right] + \Gamma^2/4} \exp\left[-\frac{a^2(\hbar\omega_1 - \hbar\omega_2)^2}{2\Gamma_G^2}\right] \tag{13.37}$$

which explicitly distinguishes $\hbar\omega_1$ from $\hbar\omega_2$ linked through a Gaussian distribution of FWHM Γ_G according to (13.13). For the plots in Fig. 13.5 we have used $I_1 = 100$, $I_2 = 50$, $\Gamma = 0.5$ eV, $\Delta = 3\Gamma$, and $\Gamma_G = 0.25$ eV. The line $\hbar\omega_1 = \hbar\omega_2$ where $\Gamma_G = 0$ correspond to the case shown in blue in Fig. 13.4d, with the intensity scaled by 100.

13.5.2 REXS Scattering Time

Our treatment in the energy domain also allows us to picture the REXS process in the time domain, treated in more detail in [2, 12, 43, 44]. As discussed in Sect. 12.2.3, the scattering process *at resonance* may be described by a $1/e$ intermediate state lifetime given by (12.18) or

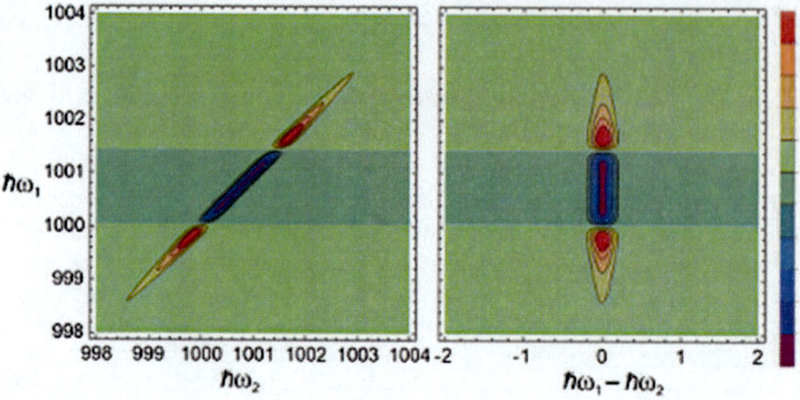

Fig. 13.5 REXS contour maps of the interference intensity ΔI_{REXS} given by (13.37) with $I_1 = 100$, $I_2 = 50$, $\Gamma = 0.5\,\text{eV}$, $\Delta = 3\Gamma$, and $\Gamma_G = 0.25\,\text{eV}$, as a function of the incident photon energy $\hbar\omega_1$ and scattered energy $\hbar\omega_2$ (left) and energy loss $\hbar\omega_1 - \hbar\omega_2$ (right)

$$\tau_\Gamma = \frac{\hbar}{\Gamma}. \tag{13.38}$$

When we detune the photon energy by Ω from the resonance energy $\mathcal{E}_1$ for a single intermediate state, the effective scattering time decreases according to

$$\tau_{\text{scat}} = \frac{\hbar}{\sqrt{\Gamma^2 + 4\Omega^2}}. \tag{13.39}$$

For a detuning energy of $\Omega = 5\Gamma$, the scattering time τ_{scat} reduces from the resonance duration τ_Γ by a factor of 10. For our value $\Gamma = 0.5\,\text{eV}$, for example, the resonance scattering time is $\tau_{\text{scat}} = \tau_\Gamma \simeq 1.3\,\text{fs}$, and it reduces to $\tau_{\text{scat}} \simeq 130\,\text{as}$ for a detuning of energy of $\Omega = 2.5\,\text{eV}$. The case $\tau_{\text{scat}} \ll \tau_\Gamma$ has been referred to as the *fast collision approximation* [45, 46].

For two degenerate intermediate states we have $\Delta = \mathcal{E}_2 - \mathcal{E}_1 = 0$ and the scattering time also follows (13.39), where Ω is the detuning energy from the first resonance at $\mathcal{E}_1$. When the interference intensity is measured at energy $\mathcal{E}_1$, i.e. $\Omega = \hbar\omega - \mathcal{E}_1 = 0$, (13.39) is replaced by

$$\tau_{\text{scat}} = \frac{\hbar}{\sqrt{\Gamma^2 + 4\Delta^2}}. \tag{13.40}$$

Now the energy separation Δ from the first resonance is equivalent to the detuning energy from the first resonance in (13.39).

13.5.3 REXS Interference in Molecular Spectra: N_2 and O_2

We now illustrate the REXS formulalism for the resonant K-shell spectra of the N_2 and O_2 molecules. These cases are particularly pedagogical because the K-shell XAS spectra shown in Figs. 10.11d and 10.12c exhibit vibrational fine structure that in REXS (and RIXS) leads to intermediate state interference effects.

13.5.3.1 N_2 REXS

We start with N_2 studied in detail by Kjellsson et al. [32, 47]. In our discussion we follow the configurational transition diagram shown in Fig. 13.6 which follows our electronic state designation on the right side of Fig. 13.2 with $|b\rangle = |a\rangle$.

The electronic ground state of N_2 is that previously shown in Fig. 10.11a, b and at room temperature only the lowest vibrational state of energy $\mathcal{E}_{\nu=0} = 0$ is occupied owing to the relatively large vibrational splitting $\mathcal{E}_{\nu=1} = 290\,\text{meV} \gg k_\text{B}T \simeq 25\,\text{meV}$. The XAS final state, which is the REXS intermediate state, is also vibrationally split as illustrated in Fig. 13.6, leading to the vibrational fine structure in the XAS spectrum. For N_2 we have the situation $\Delta \simeq 2\Gamma$. The relative intensities in the XAS spectrum are calculated within the Born–Oppenheimer approximation [49] by modifying the electronic matrix elements by vibrational mode specific Franck-Condon factors which depend on the overlap of the distance-dependent vibrational wavefunctions as shown for the XAS case in Fig. 10.11c.

For the REXS case, we can take a shortcut and simply deduce the relative size of the squared matrix elements $M_c = |\langle c|\hat{\mathbf{r}} \cdot \boldsymbol{\epsilon}|a\rangle|^2$, $c = 0, 1, 2...$ from the relative intensities in the measured XAS spectrum. The theoretical XAS spectrum is then calculated according to (10.55) by use of $\Gamma = 120\,\text{meV}$ for all vibrational components, with an intermediate vibrational state splitting of 230 meV, and transition matrix elements $M_0 = 1$, $M_1 = 0.94$, $M_2 = 0.49$, $M_3 = 0.25$, and $M_4 = 0.09$, obtained by matching the relative intensities of the calculated spectrum to those in the experimental spectrum in Fig. 10.11d [48]. The so calculated XAS spectrum is shown in black on the upper left in Fig. 13.6a.

This XAS parameterization was used to calculate the REXS spectra in Fig. 13.6b. The red REXS spectrum includes interference and was calculated with (13.13). The blue spectrum was calculated by neglect of interference through simplifying the double matrix elements in (13.13) according to (13.11). Their difference (red-blue), shown in Fig. 13.6c reveals the modulation caused by interference. The effect of interference is seen to be sizeable and as a function of energy may be constructive (red shading) or destructive (blue shading).

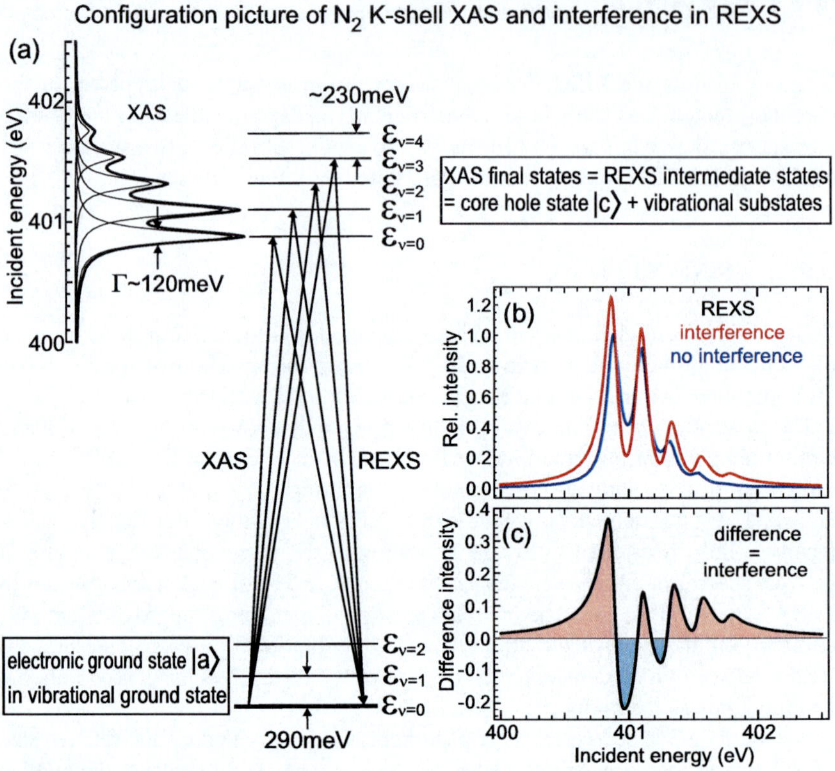

Fig. 13.6 **a** N_2 XAS and REXS K-shell transitions in an electronic configuration picture that includes vibrational splittings in the electronic ground state $|a\rangle$ and the core hole excited state $|c\rangle$ labeled as in the configuration model in Fig. 13.2. REXS consists of transitions from the lowest vibrational state of the electronic ground state $|a\rangle$ via all vibrational substates of the intermediate core hole state c back to the ground state. The theoretical XAS spectrum (black) is calculated according to (10.55), assuming the same core hole width $\Gamma = 120\,\text{meV}$ for all vibrational peaks and vibrational final state splittings of 230 meV. The relative size of the transition matrix elements was simply matched to the intensities of the experimental spectrum in Fig. 10.11d [48]. **b** The red REXS spectrum includes intermediate state interference and was calculated with (13.13), while the blue spectrum neglects interference by simplifying the double matrix element according to (13.11). In both cases the same parameters were used. **c** Difference spectrum of the red minus the blue spectrum in (**b**). We have shaded the modulation induced by constructive and destructive interference in red and blue, respectively

13.5.3.2 O_2 REXS

The corresponding theoretical XAS and REXS spectra of the O_2 molecule are shown in Fig. 13.7. The relevant transitions have previously been shown in Fig. 10.12. The spectra in Fig. 13.7 were calculated using the same procedure as for N_2 but using an electronic decay width of $\Gamma_c = \Gamma = 150\,\text{meV}$ and a vibrational final (intermediate)

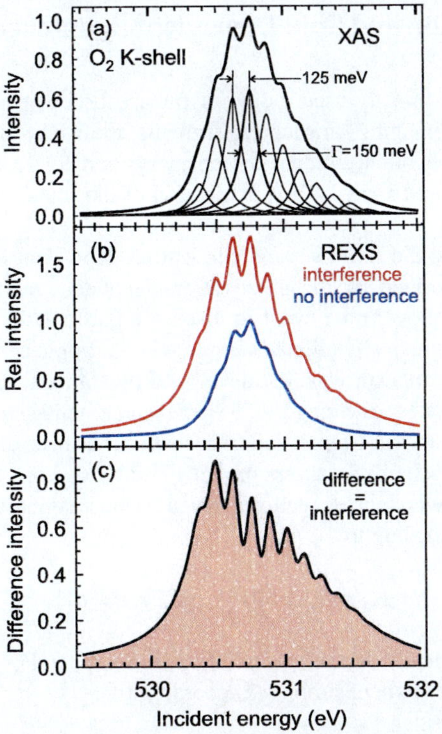

Fig. 13.7 Comparison of theoretical O_2 XAS and REXS K-shell spectra. **a** XAS spectrum calculated according to (10.55) by use of the values in Fig. 10.12, $\Gamma_a = \Gamma = 150$ meV, vibrational final state splittings of 125 meV and matrix elements obtained by matching the intensities of the calculated spectrum to those in the experimental spectrum in Fig. 10.12c [50]. **b** REXS spectra with interference, shown in red, calculated with (13.13) and without interference (blue) by simplifying the double matrix element (13.13) according to (13.11). We have used the same parameters as for the XAS spectrum. **c** Difference spectrum (red-blue) representing the modulation induced by interference which is constructive (red shading) at all energies

state splitting of 125 meV between the thirteen vibrational states indicated in the O_2 XAS spectrum in Fig. 13.7a. The relative sizes of the thirteen matrix elements were deduced by matching the theoretical XAS spectrum shown in Fig. 13.7a to the experimental XAS spectrum in Fig. 10.12c.[5]

The REXS spectra calculated by including interference (red) and without interference (blue) are shown in Fig. 13.7b. In contrast to the N_2 case their difference (red-blue) shown in (c) is positive at all energies, indicating constructive interference. Since for O_2 Δ is slightly less than Γ, the interference intensity is always positive similar to the cases for $\Delta \leq \Gamma$ shown in black and red in Fig. 13.4d.

[5] Our synthesized spectrum in Fig. 13.7a agrees well with the high-resolution O_2 XAS spectrum recorded by Strocov et al. [15].

13.6 Polarization and Spin Dependent Spontaneous REXS

In practice, REXS is mostly used today for polarization dependent scattering and diffractive coherent imaging. In such experiments, the REXS intensity is typically measured as a function of the incident photon energy and polarization with no energy analysis of the scattered x-rays. The polarization of the scattered photons may also be analyzed [51–54].

We have seen that the *spontaneous* polarization dependent REXS cross section (13.15) is fully described by the polarization dependent scattering length given by (13.17). In the following we want to discuss a particularly elegant solution of $f(\boldsymbol{\epsilon}_{p_1}, \boldsymbol{\epsilon}_{p_2})$ for a magnetically aligned sample with isotropic charge distribution.

The associated formalism was first developed by Hannon and Trammel in conjunction with Mössbauer scattering [46, 55] and later extended to its elegant modern form by Hannon, Trammel, Blume, and Gibbs which has been discussed in [45, 56–60]. The quantum mechanical expression (13.17) for $f(\boldsymbol{\epsilon}_{p_1}, \boldsymbol{\epsilon}_{p_2})$ not only describes REXS but also XAS whose cross section is given by the imaginary part of the *forward scattering* length according to

$$\sigma_{\mathrm{XAS}} = 2\lambda\, \mathrm{Im}\big[f(\boldsymbol{\epsilon}_{p_1}, \boldsymbol{\epsilon}_{p_2}), \mathbf{q}{=}0)\big]. \tag{13.41}$$

This reflects the optical theorem as discussed in Sect. 7.6. For the XAS case, the REXS double matrix element simplifies according to

$$\text{REXS:}\ \langle a|\hat{\mathbf{r}}\cdot\boldsymbol{\epsilon}^*_{p_2}|c\rangle\langle c|\hat{\mathbf{r}}\cdot\boldsymbol{\epsilon}_{p_1}|a\rangle \implies \text{XAS:}\ \big|\langle c|\hat{\mathbf{r}}\cdot\boldsymbol{\epsilon}_{p_1}|a\rangle\big|^2. \tag{13.42}$$

We have previously used the scattering length to phenomenologically describe the dichroic transmission through magnetic solids in Chap. 7 and dichroic diffraction in Chap. 8. All of our previous semi-classical derivations remain valid, but it is important to recognize the following key difference between the semi-empirical and quantum mechanical formulation.

Our previous use of the photon energy and polarization dependent scattering length $f^p(\hbar\omega)$ in Chaps. 7 and 8 was based on its *empirical* determination from *experimental data*. In this section we derive it quantum mechanically, which allows its calculation from *first principles*. The link between the scattering length f^p and optical parameters β^p and δ^p given in Sect. 7.11.1 remains the same, so that our quantum derivation of f^p may also be viewed as the quantum mechanical calculation of the energy and polarization dependent optical parameters.

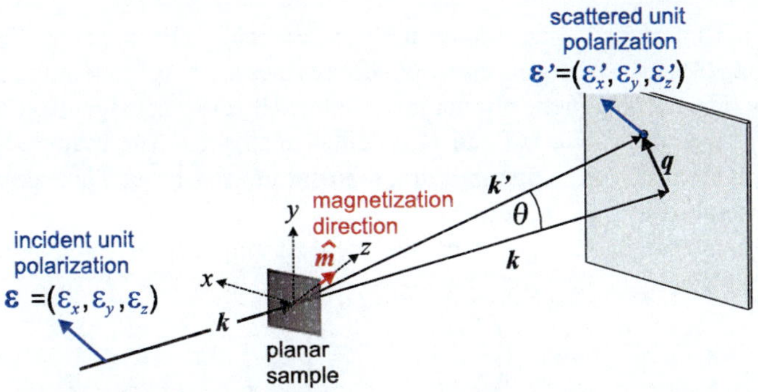

Fig. 13.8 Assumed geometry for the derivation of the polarization dependent x-ray scattering length for a magnetic sample. The photons are incident with wavevector **k**, and their *linear* polarization is described in the (x, y, z) coordinate system of the sample by a unit vector $\boldsymbol{\epsilon} = (\epsilon_x, \epsilon_y, \epsilon_z)$. The sample in the form of a film is assumed to have its magnetization $\mathbf{M} = M\hat{\mathbf{m}}$ aligned along the unit vector $\hat{\mathbf{m}}$ shown in red, with two possible orientations $\pm\hat{\mathbf{m}}$. The scattered photons propagate along $\mathbf{k}'$ with momentum change **q** and their polarization in the frame of the sample is $\boldsymbol{\epsilon}' = (\epsilon'_x, \epsilon'_y, \epsilon'_z)$

13.6.1 The Polarization Dependent Scattering Length

In the following we shall derive an expression for $f(\boldsymbol{\epsilon}_{p_1}, \boldsymbol{\epsilon}_{p_2})$ that describes polarization dependent magnetic scattering and absorption effects associated with x-ray magnetic *linear* dichroism (XMLD) and x-ray magnetic *circular* dichroism (XMCD). We choose the geometry shown in Fig. 13.8 and assume *linearly polarized* incident x-rays, so that the unit polarization vectors are *real*. Our use of linear polarization vectors is for convenience only, and they may be converted into circular polarization vectors as discussed in Sect. 13.6.1.1 below.

The following treatment is a simplified version of that in [45, 56–58] and covers the cases of linear and circular magnetic dichroism measured in transmission through a magnetic film, treated phenomenologically in Sect. 7.11. It also provides the basis for diffraction of a film, which is nothing but elastic scattering modified by interference effects as illustrated in Fig. 8.16.

In our treatment we will follow the notation in Fig. 13.8. We assume *linearly polarized* incident x-rays, so that the unit polarization vectors are *real* (see Sect. 3.2.6) and their components may readily be expressed in the Cartesian coordinate system of the sample. With $\boldsymbol{\epsilon} = \boldsymbol{\epsilon}_{p_1} = \boldsymbol{\epsilon}^*_{p_1}$ and $\boldsymbol{\epsilon}' = \boldsymbol{\epsilon}_{p_2} = \boldsymbol{\epsilon}^*_{p_2}$, the scattering length (13.17) becomes

$$f(\boldsymbol{\epsilon}, \boldsymbol{\epsilon}') = \frac{2\pi\hbar\omega\,\alpha_{\mathrm{f}}}{\lambda}\,\mathcal{R}^2 \sum_c \frac{\langle a|\hat{\mathbf{r}}\cdot\boldsymbol{\epsilon}'|c\rangle\langle c|\hat{\mathbf{r}}\cdot\boldsymbol{\epsilon}|a\rangle}{\hbar\omega - \mathcal{E}_{ca} + \mathrm{i}\,\Gamma_c/2}. \tag{13.43}$$

The photons are incident on our planar sample along their wavevector **k**, and their linear polarization vector in the (x, y, z) frame of the sample is given by $\boldsymbol{\epsilon} =$

$(\epsilon_x, \epsilon_y, \epsilon_z)$. The sample in the form of a film is assumed to be magnetically aligned with $\hat{\mathbf{m}}$ along the surface normal, corresponding to the case of perpendicular magnetic anisotropy. The ferromagnetic orientation is defined by the unit magnetization vector $\hat{\mathbf{m}}$ which is *unidirectional* and can point either along $\pm z$. The scattered photons propagating along $\mathbf{k}'$ have experienced a momentum transfer $\mathbf{q}$. Their polarization has the general form $\boldsymbol{\epsilon}' = (\epsilon'_x, \epsilon'_y, \epsilon'_z)$.

In our sample coordinate system, the incident and scattered unit polarization vectors are given by

$$\boldsymbol{\epsilon} = \begin{pmatrix} \epsilon_x \\ \epsilon_y \\ \epsilon_z \end{pmatrix}, \quad \boldsymbol{\epsilon}' = \begin{pmatrix} \epsilon'_x \\ \epsilon'_y \\ \epsilon'_z \end{pmatrix}, \tag{13.44}$$

where $\epsilon_x^2 + \epsilon_y^2 + \epsilon_z^2 = (\epsilon'_x)^2 + (\epsilon'_y)^2 + (\epsilon'_z)^2 = 1$. The unit position vector in the dipole operators $\boldsymbol{\epsilon} \cdot \hat{\mathbf{r}}$ and $\boldsymbol{\epsilon}' \cdot \hat{\mathbf{r}}$ in (13.43) may be expressed in terms of spherical tensors

$$\hat{\mathbf{r}} = \begin{pmatrix} e_x \\ e_y \\ e_z \end{pmatrix} = \begin{pmatrix} \frac{1}{\sqrt{2}}\left[C_{-1}^{(1)} - C_1^{(1)}\right] \\ \frac{\mathrm{i}}{\sqrt{2}}\left[C_{-1}^{(1)} + C_1^{(1)}\right] \\ C_0^{(1)} \end{pmatrix}. \tag{13.45}$$

This will allow the calculation of the angular part of the transition matrix elements according to Sect. 10.4.6.

For *elastic scattering* the double matrix element needs to connect the ground state $|a\rangle$ via an excited state $|c\rangle$ back to the ground state $|a\rangle$, and by considering the angular momenta of the states we see that the matrix element $\langle c|C_{-1}^{(1)}|a\rangle = \langle l'm'_l|C_{-1}^{(1)}|lm_l\rangle$ couples the states m_l and $m'_l = m_l - 1$, and the operator $C_1^{(1)}$ couples m and $m'_l = m_l + 1$. Therefore only the matrix elements of the form $\langle a|C_{\pm 1}^{(1)}|c\rangle\langle c|C_{\mp 1}^{(1)}|a\rangle$ and the matrix element $\langle a|C_0^{(1)}|c\rangle\langle c|C_0^{(1)}|a\rangle$ that preserve m_l in the up-down process are non-zero. The double matrix elements are rewritten keeping only the non-zero terms and use of the operator identities $\langle c|C_q^{(1)}|a\rangle = (-1)^q\langle c|(C_{-q}^{(1)})^*|a\rangle = (-1)^q\langle a|C_{-q}^{(1)}|c\rangle^*$ and by use of $\langle a|C_{-1}^{(1)}|c\rangle\langle c|C_{-1}^{(1)}|a\rangle = \langle a|C_1^{(1)}|c\rangle\langle c|C_1^{(1)}|a\rangle$ which is readily verified by inspection of Fig. 10.6. After some algebra one obtains

$$\begin{aligned} \langle a|\mathbf{r}\cdot\boldsymbol{\epsilon}'|c\rangle\langle c|\mathbf{r}\cdot\boldsymbol{\epsilon}|a\rangle = \frac{1}{2}\Bigg\{ & \underbrace{(\epsilon'_x\epsilon_x + \epsilon'_y\epsilon_y + \epsilon'_z\epsilon_z)}_{\boldsymbol{\epsilon}'\cdot\boldsymbol{\epsilon}}\left(|\langle c|C_{-1}^{(1)}|a\rangle|^2 + |\langle c|C_1^{(1)}|a\rangle|^2\right) \\ & + \underbrace{\epsilon'_z e_z\,\epsilon_z e_z}_{(\boldsymbol{\epsilon}'\cdot\hat{\mathbf{m}})(\boldsymbol{\epsilon}\cdot\hat{\mathbf{m}})}\left(2|\langle c|C_0^{(1)}|a\rangle|^2 - |\langle c|C_{-1}^{(1)}|a\rangle|^2 - |\langle c|C_1^{(1)}|a\rangle|^2\right) \\ & - \mathrm{i}\underbrace{(\epsilon'_x\epsilon_y - \epsilon'_y\epsilon_x)e_z}_{(\boldsymbol{\epsilon}'\times\boldsymbol{\epsilon})\cdot\hat{\mathbf{m}}}\left(|\langle c|C_1^{(1)}|a\rangle|^2 - |\langle c|C_{-1}^{(1)}|a\rangle|^2\right)\Bigg\}, \end{aligned} \tag{13.46}$$

where we have used the unit vector expression $\hat{\mathbf{m}} = (0, 0, e_z)$ in Fig. 13.8. Our result separates the polarization dependence, expressed in terms of the real linear polarization components from the associated transition matrix elements written in terms of spherical tensors. The non-vanishing polarization dependent angular matrix elements $\langle ML|C_q^{(1)}|ml\rangle$ for transitions between s, p, and d states are given in Fig. 10.6.

By inclusion of non-resonant Thomson *charge* scattering given by (12.36) and neglect of the much smaller Thomson *spin* scattering (see (6.23) and (9.11)), we obtain the full expression for the quantum mechanical scattering length in the compact form

$$\begin{aligned} f(\boldsymbol{\epsilon}, \boldsymbol{\epsilon}') = & \underbrace{r_0 F^0(\mathbf{q})\,(\boldsymbol{\epsilon}' \cdot \boldsymbol{\epsilon})}_{\substack{\text{non-resonant} \\ \text{Thomson charge}}} \\ & + \frac{\pi \hbar\omega\, \alpha_{\mathrm{f}}}{\lambda} \mathcal{R}^2 \Big\{ \underbrace{\mathbb{A}\,(\boldsymbol{\epsilon}' \cdot \boldsymbol{\epsilon})}_{\substack{\text{resonant} \\ \text{isotropic charge}}} \; \underbrace{+\mathbb{B}\,(\boldsymbol{\epsilon}' \cdot \hat{\mathbf{m}})(\boldsymbol{\epsilon} \cdot \hat{\mathbf{m}})}_{\substack{\text{uniaxial magnetic:} \\ \text{XMLD}}} - \underbrace{\mathrm{i}\,\mathbb{C}\,(\boldsymbol{\epsilon}' \times \boldsymbol{\epsilon}) \cdot \hat{\mathbf{m}}}_{\substack{\text{unidirectional} \\ \text{magnetic: XMCD}}} \Big\}. \end{aligned} \tag{13.47}$$

The absolute value squared $|f(\boldsymbol{\epsilon}, \boldsymbol{\epsilon}')|^2$ is the differential scattering cross section which is seen to contain interference between the Thomson and resonant terms.

The prefactors $\mathbb{A} - \mathbb{C}$ are resonant matrix elements given by

$$\mathbb{A} = \sum_c \frac{|\langle c|C_{-1}^{(1)}|a\rangle|^2 + |\langle c|C_1^{(1)}|a\rangle|^2}{\hbar\omega - \mathcal{E}_{ca} + \mathrm{i}\,\Gamma_c/2} \tag{13.48}$$

$$\mathbb{B} = \sum_c \frac{2|\langle c|C_0^{(1)}|a\rangle|^2 - |\langle c|C_{-1}^{(1)}|a\rangle|^2 - |\langle c|C_1^{(1)}|a\rangle|^2}{\hbar\omega - \mathcal{E}_{ca} + \mathrm{i}\,\Gamma_c/2} \tag{13.49}$$

$$\mathbb{C} = \sum_c \frac{|\langle c|C_1^{(1)}|a\rangle|^2 - |\langle c|C_{-1}^{(1)}|a\rangle|^2}{\hbar\omega - \mathcal{E}_{ca} + \mathrm{i}\,\Gamma_c/2}. \tag{13.50}$$

The *real* and *imaginary* parts in the expressions $\mathbb{A} - \mathbb{C}$ of the resonant scattering length are separated by use of

$$\frac{1}{\hbar\omega - \mathcal{E}_{ca} + \mathrm{i}\,\Gamma_c/2} = \underbrace{\frac{\hbar\omega - E_{ca}}{(\hbar\omega - \mathcal{E}_{ca})^2 + \Gamma_c^2/4}}_{\text{real part}} - \mathrm{i}\, \underbrace{\frac{2}{\Gamma_c} \frac{\Gamma_c^2/4}{(\hbar\omega - \mathcal{E}_{ca})^2 + \Gamma_c^2/4}}_{\text{imaginary part}}. \tag{13.51}$$

In (13.47) we have indicated by underbrackets the meaning of the various terms. The first term describes *non-resonant* elastic Thomson scattering by the atomic charge. The polarization factor $\boldsymbol{\epsilon}' \cdot \boldsymbol{\epsilon}$ is that previously derived classically for scattering by a single electron or atom in Sect. 6.2 and illustrated in Fig. 6.2. The polarization is preserved in the forward direction.

The other terms describe elastic *resonant* scattering. The second term (prefactor $\mathbb{A}$) describes *resonant* charge scattering by a spherically symmetric charge distribution and is the resonant analogue of non-resonant Thompson scattering with the same polarization dependence.

The third term with prefactor $\mathbb{B}$ represents a *uniaxial* magnetic effect since it scales with the square of $\hat{\mathbf{m}}$. It also exists if the *unidirectional* vector $\hat{\mathbf{m}}$ in a ferromagnet is replaced by a *uniaxial* magnetic symmetry in an antiferromagnet. It corresponds to the XMLD effect in XAS and arises from a spin-orbit coupling induced charge distortion along the magnetic axis. The charge distortion is revealed by rewriting the matrix element in the quadrupolar form along z given by $3z^2 - r^2$ according to

$$\begin{aligned} &2|\langle c|C_0^{(1)}|a\rangle|^2 - |\langle c|C_{-1}^{(1)}|a\rangle|^2 - |\langle c|C_1^{(1)}|a\rangle|^2 \\ &\quad = \underbrace{3|\langle c|C_0^{(1)}|a\rangle|^2}_{3z^2} - \underbrace{\left(\sum_c |\langle c|C_0^{(1)}|a\rangle|^2 + |\langle c|C_{-1}^{(1)}|a\rangle|^2 + |\langle c|C_{+1}^{(1)}|a\rangle|^2 \right)}_{\text{isotropic, } r^2} . \end{aligned} \tag{13.52}$$

The last term in (13.47), identified by the prefactor $\mathbb{C}$ depends on the existence of a *unidirectional* magnetization direction $\hat{\mathbf{m}}$, as present in a ferromagnet. Its sign depends on the direction of $\hat{\mathbf{m}}$ and corresponds to the XMCD effect in absorption.

13.6.1.1 Linear Versus Circular Polarization Vectors

The polarization dependence of the scattering process is completely determined by the three unit vectors $\boldsymbol{\epsilon}$, $\boldsymbol{\epsilon}'$, and $\hat{\mathbf{m}}$ that form the following irreducible components in Cartesian coordinates

$$\begin{aligned} \boldsymbol{\epsilon}' \cdot \boldsymbol{\epsilon} &= \varepsilon'_x \varepsilon_x + \varepsilon'_y \varepsilon_y + \varepsilon'_z \varepsilon_z \\ (\boldsymbol{\epsilon}' \cdot \hat{\mathbf{m}})(\boldsymbol{\epsilon} \cdot \hat{\mathbf{m}}) &= \varepsilon'_z e_z \varepsilon_z e_z \\ (\boldsymbol{\epsilon}' \times \boldsymbol{\epsilon}) \cdot \hat{\mathbf{m}} &= (\varepsilon'_x \varepsilon_y - \varepsilon'_y \varepsilon_x)\, e_z . \end{aligned} \tag{13.53}$$

The corresponding expressions in a circular basis system are more complicated as seen from the operator expressions for linear and circular light propagating along different Cartesian directions in Table 11.1. The general case simplifies for special geometries such as for forward scattering $\mathbf{q} = 0$ and $\mathbf{k} \parallel \hat{\mathbf{m}} \parallel z$. This is the XAS case discussed in Sect. 7.11, leading to the XMCD effect in a circular basis and a Faraday rotation in a linear basis. The expressions for x-ray transmission derived previously in Sects. 7.11.3 and 7.11.5 can be shown to follow from the above scattering formalism.

13.7 Spontaneous Versus Stimulated REXS by an Atomic Sheet

The "in-beam" forward scattered fields have a well-defined phase relative to the incident fields according to Sect. 7.5.4, and the scattered polarization remains in the plane perpendicular to the propagation direction but may be rotated for spin dependent scattering. Forward scattering has a *spontaneous* phase shifted contribution (see Sect. 7.5.4) and a *stimulated* in-phase and polarization conserving contribution due to the fact that photons clone themselves in the stimulation process.

The spontaneous versus stimulated response of individual atoms treated in Sect. 13.4 may be extended to that of a thin sheet, whose plane is oriented perpendicular to the direction of the incident photons. The sheet is envisioned to have a small enough thickness that the exponential Beer-Lambert attenuation law can be approximated to second order in the sheet thickness.

The intensity transmitted through an atomic thin sheet of thickness Δ and areal density $N_\mathrm{a}/A = \rho_\mathrm{a}\Delta$, where ρ_a is the bulk atomic number density, is given up to second order by the semi-classical expression (7.70)

$$I_\mathrm{trans} = I_0\Bigg\{1 - \underbrace{2\lambda f''\frac{N_\mathrm{a}}{A}}_{\substack{\text{1st order XAS}\\ \text{loss probability}}} \underbrace{-\frac{8\pi}{3}\left[(r_0 Z + f')^2 + (f'')^2\right]\frac{N_\mathrm{a}}{A}}_{\substack{\text{incoherent REXS loss}\\ \text{probability}}} + \underbrace{2\lambda^2 (f'')^2 \frac{N_\mathrm{a}^2}{A^2}}_{\substack{\text{coh. REXS gain=}\\ \text{2nd ord. XAS prob.}}}\Bigg\}. \tag{13.54}$$

This expression is valid in quantum theory under the assumption that all transition probabilities are *spontaneous* without inclusion of stimulated contributions. It holds to second order in the sheet thickness Δ or number of atoms N_a. The incident intensity is given by the quantum expression (10.9) as

$$I_0 = \frac{n_\mathrm{pk}\hbar\omega c}{V_\mathrm{pk}} = \frac{n_0\hbar\omega c}{A\Delta}. \tag{13.55}$$

The out-of-beam REXS probability is taken as an incoherent superposition of the atomic contributions ($\propto N_\mathrm{a}$), while the in-beam (forward) scattering contribution may be equivalently described as a second order XAS or coherent forward scattered REXS contribution ($\propto N_\mathrm{a}^2$). For a thin sheet we can distinguish contributions to the transmitted intensity in terms of their order in the sheet thickness Δ or number of atoms N_a.

In first order, the XAS loss is linear in Δ and N_a, and we also have a REXS loss contribution due to "incoherent" out-of-beam scattering contributions by the atoms. The designation "incoherent" reflects the classical picture that interference of the fields emitted by randomly positioned atoms averages out. In second order, we have the last term in (13.54), which may be interpreted either as a "coherent" (in phase) forward scattering REXS contribution or arising from expansion of the exponential

Beer-Lambert absorption law (7.32) to second order in the sheet thickness Δ. With increasing sheet thickness the coherently forward scattered REXS component will dominate.

Neglecting the incoherent REXS term, we may write (13.54) in our quantum mechanical notation as

$$I_{\rm trans} = I_0 \left\{ 1 - \underbrace{\frac{N_{\rm a}}{A}\frac{\lambda^2\Gamma_x}{\pi\Gamma}}_{\substack{\text{1st order XAS}\\ \text{loss probability}}} + \underbrace{\frac{1}{2}\left[\frac{N_{\rm a}}{A}\frac{\lambda^2\Gamma_x}{\pi\Gamma}\right]^2}_{\substack{\text{coh. REXS gain=}\\ \text{2nd ord. XAS prob.}}} \right\}. \tag{13.56}$$

The single atom spontaneous response discussed in Sect. 13.4 is replaced by the response of the entire sheet of $N_{\rm a}$ atoms contained in an illuminated lateral area A. Since the single atom cross section $\sigma_{\rm XAS}^{\rm spon}$ represents the fractional atomic area that corresponds to unit absorption probability, the absorption probability of all $N_{\rm a}$ atoms in the illuminated area A simply scales *linearly* with the number of atoms as

$$P_{\rm XAS}^{N_{\rm a}} = \frac{N_{\rm a}}{A}\frac{\lambda^2\Gamma_x}{\pi\Gamma} = \frac{N_{\rm a}}{A} 2\lambda f'' = \frac{N_{\rm a}}{A}\sigma_{\rm XAS}^{\rm spon}. \tag{13.57}$$

The spontaneous second order REXS contribution due to coherent forward scattering is equivalent to the second order XAS term. It scales with the number of atoms squared $N_{\rm a}^2$ as

$$P_{\rm REXS}^{N_{\rm a}} = 2\lambda^2 (f'')^2 \frac{N_{\rm a}^2}{A^2} = \frac{1}{2}\left[\frac{N_{\rm a}}{A}\frac{\lambda^2\Gamma_x}{\pi\Gamma}\right]^2 = \frac{1}{2}\frac{N_{\rm a}^2}{A^2}\left(\sigma_{\rm XAS}^{\rm spon}\right)^2. \tag{13.58}$$

Its positive sign leads to a transmission increase which is small for *spontaneous* forward scattering due to the small size of $f'' \simeq 10^{-4}$ nm (see Fig. 7.12) but may be enhanced through stimulation, not included in (13.54).

13.7.1 Forward Scattering by an Atomic Sheet

Upon stimulation by $n_{\boldsymbol{pk}}$ of energy $\hbar\omega$ within the bandwidth $\Delta_{\boldsymbol{pk}}$, linked to the incident intensity by (see (10.8))

$$n_{\boldsymbol{pk}} = I_0 \frac{\lambda^3}{c\Delta_{\boldsymbol{pk}}}, \tag{13.59}$$

the spontaneously transmitted intensity (13.56) becomes

$$\begin{aligned} I_{\text{trans}} &\simeq I_0\left\{1-\frac{N_{\text{a}}}{A}\frac{\lambda^2\Gamma_x}{\pi\Gamma}+\frac{1}{2}\left[\frac{N_{\text{a}}}{A}\frac{\lambda^2\Gamma_x}{\pi\Gamma}\right]^2(1+n_{p\mathbf{k}})\right\} \\ &\simeq I_0\left\{1-\frac{N_{\text{a}}}{A}2\lambda f_0''+2\frac{N_{\text{a}}^2}{A^2}\lambda^2(f_0'')^2(1+n_{p\mathbf{k}})\right\} \\ &\simeq I_0\left\{1-\rho_{\text{a}}\Delta\,\sigma_{\text{XAS}}^{\text{spon}}+\frac{1}{2}\rho_{\text{a}}^2\Delta^2\left(\sigma_{\text{XAS}}^{\text{spon}}\right)^2(1+n_{p\mathbf{k}})\right\}. \end{aligned} \tag{13.60}$$

The spontaneous part of this expression reflects expansion of the exponential *spontaneous* Beer-Lambert law (7.32) to second order in the sheet thickness Δ. It is implicity contained in the areal density $N_{\text{a}}/A=\rho_{\text{a}}\Delta$, where ρ_{a} is the bulk atomic number density (see Table 7.2). The stimulated part of (13.60) reduces to the spontaneous forward scattered REXS intensity for $n_{p\mathbf{k}}=1$, as expected.

In practice, film thicknesses exceed the validity range of our expression (13.60), and in Chap. 15 we will extend the KHD perturbation treatment of a thin sheet to cover higher incident intensities and films of arbitrary thickness.

13.8 Resonant Inelastic X-Ray Scattering: RIXS

We have previously derived the general expression (13.9) for the RIXS cross section in the dipole approximation. In our derivation we maintained the Dirac δ-function in the second order KHD formula to conserve energy according to $\delta[(\mathcal{E}_b-\mathcal{E}_a)-(\hbar\omega_1-\hbar\omega_2)]$. In contrast to the REXS case, however, the final RIXS state $|b\rangle$ will have a finite decay width Γ_b. We then need to replace the δ-function by a Lorentzian of energy FWHM Γ_b.[6] Like the δ-function, the Lorentzian has to be normalized to have unit integrated area and the differential RIXS cross section then has the following fundamental form.

The differential RIXS cross section, corresponding to a negligible *instrumental* linewidth contribution, is given by

$$\begin{aligned} \frac{\mathrm{d}\sigma_{\text{RIXS}}}{\mathrm{d}\Omega_2\,\mathrm{d}\hbar\omega_2} &= \frac{4\pi^2\hbar\omega_1\hbar\omega_2\,\alpha_{\text{f}}^2}{\lambda_2^2}\,\mathcal{R}^4\left(n_{p_2\mathbf{k}_2}+1\right) \\ &\times\sum_b^{\omega_2\le\omega_1}\left|\sum_c\frac{\langle b|\hat{\mathbf{r}}\cdot\boldsymbol{\epsilon}_{p_2}^*|c\rangle\langle c|\hat{\mathbf{r}}\cdot\boldsymbol{\epsilon}_{p_1}|a\rangle}{\hbar\omega_1-\mathcal{E}_{ca}+\mathrm{i}\,\Gamma_c/2}\right|^2\underbrace{L(\hbar\omega_1-\hbar\omega_2-\mathcal{E}_{ba})}_{\text{Raman term}}, \end{aligned} \tag{13.61}$$

where $\mathcal{E}_{ij}=\mathcal{E}_i-\mathcal{E}_j$ and

[6] The Lorentzian represents the final density of states of dimension [1/energy] according to (10.36).

$$L(\hbar\omega_1-\hbar\omega_2-\mathcal{E}_{ba}) = \frac{2}{\pi\Gamma_b}\frac{\Gamma_b^2/4}{(\hbar\omega_1-\hbar\omega_2-\mathcal{E}_{ba})^2+\Gamma_b^2/4} \tag{13.62}$$

is the final state Lorentzian of width Γ_b whose argument $\hbar\omega_1-\hbar\omega_2-\mathcal{E}_{ba}$ links it to the *Raman effect* in optical spectroscopy. Since it *multiplies* the intermediate state squared matrix element containing the larger core hole width Γ_c, the Lorentzian L effectively takes a slice out of the intermediate state width Γ_c in a RIXS map (see Fig. 13.5) that projects the incident and emitted energies $\hbar\omega_1$ and $\hbar\omega_2$ onto each other. The RIXS *spectral resolution* is therefore determined by the narrower width Γ_b.

The cross section (13.61) establishes the link between optical Stokes Raman spectroscopy and RIXS, which is seen to extend the capabilities of *optical Raman spectroscopy* into the x-ray region. RIXS has therefore also been called "resonant x-ray raman scattering" which provides access to electronic or spin dependent final states of energies that exceed typical vibrational energies.

The remarkable fact that the RIXS linewidth is narrower than the core hole width is a consequence of the second order KHD expression, where all paths through intermediate states do not enter in energy conservation. Second order RIXS and first order XES fundamentally differ in the achievable spectral resolution since according to (12.15) the XES linewidth Γ is the *sum* of the core hole and final state widths.

RIXS extends optical Raman scattering. It is element specific and allows the study of higher energy electronic final states that cannot be reached by direct dipole transitions. RIXS has a significant advantage over XES in terms of its *intrinsic* spectral resolution since the much larger core hole width which dominates XES is effectively eliminated in RIXS.

13.8.1 Two-Step RIXS

The *two-step* RIXS differential cross section is obtained from the general expression (13.61) for negligibly small instrumental bandwidths by taking the sum over intermediate states out of the squared matrix element so that

$$\frac{d\sigma_{\mathrm{RIXS}}}{d\Omega_2\, d\hbar\omega_2} = \frac{4\pi^2\hbar\omega_1\hbar\omega_2\,\alpha_f^2}{\lambda_2^2}\mathcal{R}^4\left(n_{p_2\mathbf{k}_2}+1\right) \times\sum_b\sum_c \frac{|\langle b|\hat{\mathbf{r}}\cdot\boldsymbol{\epsilon}_{p_2}^*|c\rangle|^2\,|\langle c|\hat{\mathbf{r}}\cdot\boldsymbol{\epsilon}_{p_1}|a\rangle|^2}{(\hbar\omega_1-(\mathcal{E}_c-\mathcal{E}_a))^2+\Gamma_c^2/4}L(\hbar\omega_1-\hbar\omega_2-\mathcal{E}_{ba}). \tag{13.63}$$

This can be rewritten in terms of the dipolar transition width introduced for XAS in (10.40) and XES in (12.17) in the three state RIXS form

$$
\begin{aligned}
\frac{\mathrm{d}\sigma_{\mathrm{RIXS}}}{\mathrm{d}\Omega_2\,\mathrm{d}\hbar\omega_2} &= \frac{\lambda_1^2}{2\pi^3}\left(n_{p_2\mathbf{k}_2}+1\right)\sum_b\sum_c \\
&\times \underbrace{\frac{8\pi^2\hbar\omega_1\alpha_{\mathrm{f}}}{\lambda_1^2}\mathcal{R}^2\left|\langle c|(\hat{\mathbf{r}}\cdot\boldsymbol{\epsilon}_{p_1}|a\rangle\right|^2}_{\text{absorption: }\Gamma_{ac}^{p_1}}\frac{1}{\Gamma_c^2}\frac{(\Gamma_c/2)^2}{[\hbar\omega_1-(\mathcal{E}_c-\mathcal{E}_a)]^2+(\Gamma_c/2)^2} \\
&\times \underbrace{\frac{8\pi^2\hbar\omega_2\alpha_{\mathrm{f}}}{\lambda_2^2}\mathcal{R}^2\left|\langle b|(\hat{\mathbf{r}}\cdot\boldsymbol{\epsilon}_{p_2}^*)|c\rangle\right|^2}_{\text{emission: }\Gamma_{cb}^{p_2}}\frac{1}{\Gamma_b}\frac{(\Gamma_b/2)^2}{[\hbar\omega_2-(\mathcal{E}_c-\mathcal{E}_b)]^2+(\Gamma_b/2)^2},
\end{aligned}
\tag{13.64}
$$

where the underbrackets define the dipolar transition energy widths of the excitation and de-excitation steps in close analogy to the expressions (10.40) and (12.17) for XAS and XES. The two-step RIXS cross section is seen to correspond to two consecutive first order absorption and emission processes.

13.8.1.1 The Integrated Double Differential RIXS Cross Section

The two-step double differential RIXS cross section (13.64) contains two decay lifetime widths, those of the core hole intermediate state Γ_c and final state Γ_b. We can eliminate the final state width by integrating over all emission energies $\hbar\omega_2$ and obtain

$$
\begin{aligned}
\frac{\mathrm{d}\sigma_{\mathrm{RIXS}}}{\mathrm{d}\Omega_2} &= \frac{\lambda_1^2}{4\pi^2}\left(n_{p_2\mathbf{k}_2}+1\right) \\
&\times\sum_c\frac{\Gamma_{ac}^{p_1}}{\Gamma_c^2}\frac{(\Gamma_c/2)^2}{(\hbar\omega_1-(\mathcal{E}_c-\mathcal{E}_a))^2+(\Gamma_c/2)^2} \\
&\times\sum_b\Gamma_{cb}^{p_2}\underbrace{\int_{-\infty}^{\infty}\frac{2}{\pi\Gamma_b}\frac{(\Gamma_b/2)^2}{(\hbar\omega_2-[\hbar\omega_1-(\mathcal{E}_b-\mathcal{E}_a)])^2+(\Gamma_b/2)^2}\mathrm{d}\hbar\omega_2}_{=1}
\end{aligned}
\tag{13.65}
$$

to obtain the compact expression

$$
\frac{\mathrm{d}\sigma_{\mathrm{RIXS}}}{\mathrm{d}\Omega_2}=\frac{\lambda_1^2}{4\pi^2}\left(n_{p_2\mathbf{k}_2}+1\right)\sum_c\sum_b\frac{\Gamma_{ac}^{p_1}\,\Gamma_{cb}^{p_2}}{\Gamma_c^2}\frac{(\Gamma_c/2)^2}{(\hbar\omega_1-(\mathcal{E}_c-\mathcal{E}_a))^2+(\Gamma_c/2)^2}. \tag{13.66}
$$

In a simple three state model and denoting the intermediate state width as $\Gamma_c=\Gamma=\Gamma^{\mathrm{X}}+\Gamma^{\mathrm{A}}$ (see (10.35)) we can regroup the terms and can link the two-step single differential RIXS with the XAS cross section according to

$$\frac{d\sigma_{\mathrm{RIXS}}}{d\Omega_2} = \frac{1}{4\pi}\left(1+n_{p\mathbf{k}}\right)\frac{\Gamma_{cb}^{p_2}}{\Gamma}\underbrace{\frac{\lambda_1^2}{\pi}\frac{\Gamma_{ac}^{p_1}}{\Gamma}\frac{(\Gamma/2)^2}{(\hbar\omega_1-(\mathcal{E}_c-\mathcal{E}_a))^2+(\Gamma/2)^2}}_{\sigma_{\mathrm{XAS}}^{p_1}}. \quad (13.67)$$

If we average the emission rate over polarization we obtain $\Gamma^{\mathrm{X}} = \langle\Gamma_{cb}^{p_2}\rangle = \frac{1}{3}\sum_{p_2}\Gamma_{cb}^{p_2}$ which is the radiative XES width that determines the fluorescence yield Y_{f} defined in Sect. 12.4, so that

$$\frac{d\sigma_{\mathrm{RIXS}}}{d\Omega_2} = \frac{1}{4\pi}\left(1+n_{p\mathbf{k}}\right)\underbrace{\frac{\Gamma^{\mathrm{X}}}{\Gamma}}_{Y_{\mathrm{f}}}\underbrace{\frac{\lambda_1^2}{\pi}\frac{\Gamma_{ac}^{p_1}}{\Gamma}\frac{(\Gamma/2)^2}{(\hbar\omega_1-(\mathcal{E}_c-\mathcal{E}_a))^2+(\Gamma/2)^2}}_{\sigma_{\mathrm{XAS}}^{p_1}}, \quad (13.68)$$

where Γ^{X} is given by (12.22) or

$$\Gamma^{\mathrm{X}} = \frac{8\pi^2\hbar\omega\,\alpha_{\mathrm{f}}}{\lambda^2}\mathcal{R}^2\frac{L(N_{\mathrm{e}}+1)}{3(2L+1)(2c+1)}. \quad (13.69)$$

If we additionally integrate over the 4π solid angle of inelastic emission which is isotropic, we obtain the RIXS cross section

$$\boxed{\sigma_{\mathrm{RIXS}} = \underbrace{\frac{\Gamma^{\mathrm{X}}}{\Gamma}}_{Y_{\mathrm{f}}}\left(n_{p_2\mathbf{k}_2}+1\right)\sigma_{\mathrm{XAS}}^{p_1}.} \quad (13.70)$$

We can summarize our results as follows.

> When the spontaneous ($n_{p_2\mathbf{k}_2} = 0$) RIXS cross section is integrated over emission energies and angles, it becomes the absorption cross section times the fluorescence yield, as required by energy conservation. In this case all intermediate state interference effects sum to zero and cancel.

13.9 RIXS with Finite Instrumental Resolution

The RIXS cross sections (13.61) and (13.64) have an *intrinsic* linewidth (i.e. for negligible instrumental bandwidths) determined by that of the final electronic state Γ_b. Even with state-of-the-art instrumentation, measured RIXS spectra typically still contain instrumental bandwidth contributions. The measured linewidth is then given by a convolution of (13.61) with the Gaussian monochromator and spectrometer resolution functions. In contrast to REXS, however, the two functions do not enter equivalently and one needs to carry out the convolutions separately. Also unlike REXS, where the convolution of the intrinsic linewidth is facilitated by the energy conserving δ-function in (13.12), the RIXS convolution includes not only the expression for the intermediate state matrix element but the final state Lorentzian $L(\hbar\omega_1 - \hbar\omega_2 - \mathcal{E}_{ba})$ given by (13.62). In this case there is unfortunately no general analytical solution for the convoluted lineshapes which have the form of Voigt functions.

13.9.1 The Case of Small Final State Width

In practice, the RIXS final state $|b\rangle$ typically lies above the ground state $|a\rangle$ only by a small energy of order $\mathcal{E}_b - \mathcal{E}_a \simeq 1\,\mathrm{eV}$, where the final state widths Γ_b are associated with excited valence electron or vibrational states. In these cases we have $\Gamma_b \ll \Gamma_c$ and one may approximate the final state Lorentzian by a Dirac δ-function. This significantly facilitates the convolution with the Gaussian monochromator transmission function of FWHM Γ_{mo} and yields an analytical expression.

In the convolution, the energy-dependent prefactor in the RIXS cross section (13.61) may be approximated by a constant C since $\hbar\omega_1 \simeq \hbar\omega_2$ and pulled out of the convolution integral so that we obtain

$$\left[\frac{\mathrm{d}\sigma_{\mathrm{RIXS}}}{\mathrm{d}\Omega_2\,\mathrm{d}\hbar\omega_2}\right]_{\mathrm{mo}} = C\int_{-\infty}^{\infty}\sum_b\left|\sum_c \frac{\langle b|\hat{\mathbf{r}}\cdot\boldsymbol{\epsilon}^*_{p_2}|c\rangle\langle c|\hat{\mathbf{r}}\cdot\boldsymbol{\epsilon}_{p_1}|a\rangle}{x-\mathcal{E}_{cb}+\mathrm{i}\Gamma_c/2}\right|^2$$

$$\times\,\delta(x-\hbar\omega_2)\,\frac{a}{\sqrt{2\pi}\,\Gamma_{\mathrm{mo}}}\exp\left[-\frac{a^2(\hbar\omega_1-x-\mathcal{E}_{ba})^2}{2\,\Gamma_{\mathrm{mo}}^2}\right]\mathrm{d}x$$

$$= \underbrace{C\sum_b\left|\sum_c \frac{\langle b|\hat{\mathbf{r}}\cdot\boldsymbol{\epsilon}^*_{p_2}|c\rangle\langle c|\hat{\mathbf{r}}\cdot\boldsymbol{\epsilon}_{p_1}|a\rangle}{\hbar\omega_2-\mathcal{E}_{cb}+\mathrm{i}\Gamma_c/2}\right|^2}_{\substack{\text{non-dispersive}\\ \text{“resonance” term}}}\;\underbrace{G(\hbar\omega_1-\hbar\omega_2-\mathcal{E}_{ba})}_{\substack{\text{dispersive}\\ \text{“Raman” term}}}, \tag{13.71}$$

where we have written the Gaussian in terms of the FWHM Γ_{mo}, where $a = 2\sqrt{\ln 4}$ see Appendix (A.2.2)

$$C \simeq \frac{4\pi^2\hbar\omega_1\hbar\omega_2\,\alpha_{\mathrm{f}}^2}{\lambda_2^2}\,\mathcal{R}^4\,\left(n_{p_2\mathbf{k}_2}+1\right) \tag{13.72}$$

and

$$G(\hbar\omega_1-\hbar\omega_2-\mathcal{E}_{ba}) = \frac{a}{\sqrt{2\pi}\,\Gamma_{\mathrm{mo}}}\exp\left[-\frac{a^2(\hbar\omega_1-\hbar\omega_2-\mathcal{E}_{ba})^2}{2\,\Gamma_{\mathrm{mo}}^2}\right]. \tag{13.73}$$

Hence for a small intrinsic final state width $\Gamma_b \ll \Gamma_c$, the intrinsic final state Lorentzian in (13.61) may be replaced by a Gaussian that accounts for the finite incident FWHM Γ_{mo} defined by a monochromator. Note that the monochromator function *multiplies* the intermediate state expression rather than convolutes it. This was pointed out by Tulkki and Åberg [11] in their discussion of early RIXS experiments [61, 62].

13.9.2 Reduction of RIXS to XES

In the limit of large incident bandwidth, the Gaussian monochromator function becomes a constant and the RIXS linewidth is determined by the intermediate state width Γ_c. Then the narrow linewidth advantage of RIXS over XES is lost. Furthermore, in the absence of intermediate state interference effects, the two-step RIXS process becomes equivalent to XES, as expected for broadband excitation.

13.10 Examples of RIXS Capabilities

We finish this chapter by showing characteristic K-shell RIXS spectra for the N_2 and O_2 molecules and L-shell RIXS spectra for transition metal atoms which are examples of our theoretical formalism.

13.10.1 K-Shell RIXS of N_2 and O_2

The K-shell RIXS spectra of the N_2 and O_2 molecules are particularly beautiful because they exhibit vibrational fine structure of both ground and excited electronic states. In their description, one needs to take into account all interference paths through the vibrational substates of the intermediate core hole state $|c\rangle$. Typically, the two-step RIXS description does not properly account for the intensities in the RIXS spectra although the measured peak positions are approximately predicted.

13.10.1.1 N_2 RIXS

In contrast to the REXS process for N_2, illustrated in Fig. 13.6, RIXS decays involve energy transfer to the electronic and vibrational degrees of freedom of the sample. The lowest energy losses involve higher vibrational states of the electronic ground state of energy $\mathcal{E}_{\nu=0} = 0$ shown in Fig. 13.6. In cases where these small energy losses remain unresolved, one speaks of *quasi-elastic* scattering. In addition, there are also RIXS transitions associated with decays to vibrational substates of *excited electronic* states $|b\rangle$. The REXS and RIXS processes for N_2 are schematically illustrated in Fig. 13.9.

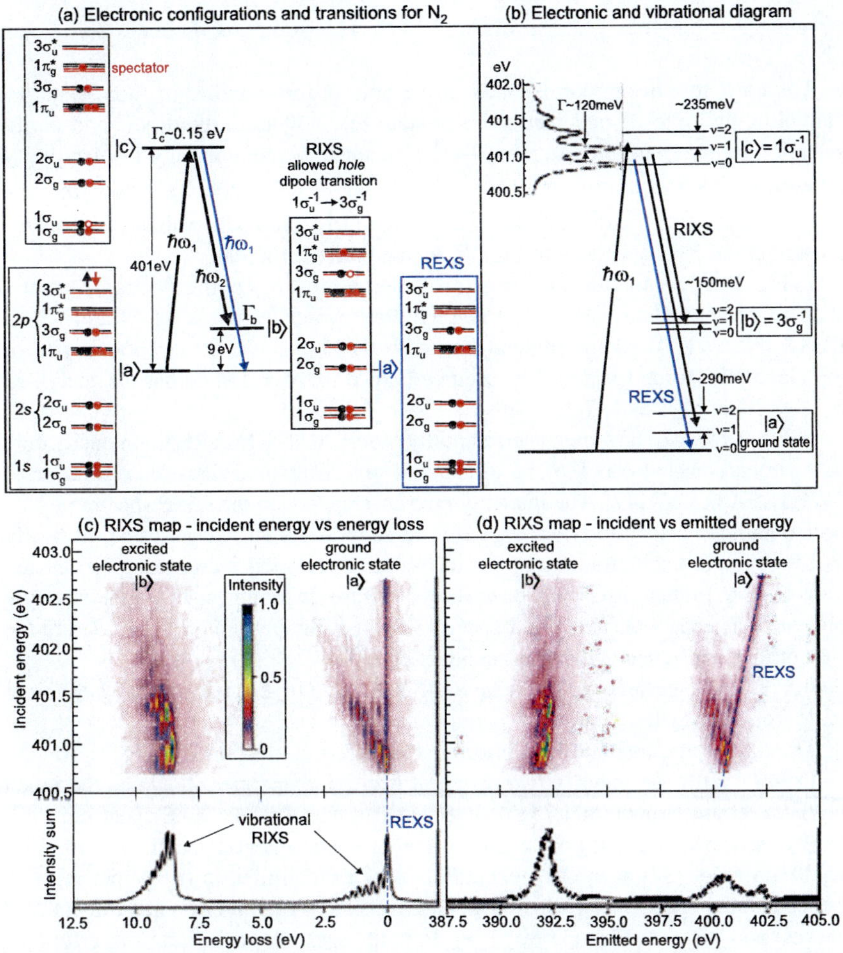

Fig. 13.9 **a** RIXS K-shell electronic configuration diagram and transitions for N_2. The electronic states are those previously shown in Fig. 10.11 with the core hole decay channel added. The RIXS decay consists of the dipole allowed transition from the core hole $1\sigma_u^{-1}$ state $|c\rangle$ to the valence hole $3\sigma_g^{-1}$ state $|b\rangle$, and the REXS decay leads back to the ground state $|a\rangle$. **b** Inclusion of the vibrational substates in the diagram in (**a**), with the XAS spectrum taken from [48, 50]. **c** Experimental N_2 RIXS/REXS intensity map, incident energy versus energy loss, recorded with an instrumental Gaussian resolution of FWHM$\simeq$150 meV [47]. The energy loss spectrum shown at the botton is the intensity map summed over incident energy. **d** Same as (c) as a function of emitted energy (L. Kjellsson, private commun.)

Figure 13.9a shows the electronic configurations and transitions for the REXS and RIXS processes in N_2. The initial intermediate and final states states are labeled according to the configuration diagram in Fig. 13.2. The intermediate state reached by the $1\sigma_u \rightarrow 1\pi_g^*$ transition is the same as the XAS final state whose vibrational fine structure was previously shown in Fig. 10.11b. The excitation produces core holes of $1\sigma_u$ symmetry.

The core hole decay corresponds to the hole in the intermediate core state $|c\rangle$ "bubbling up" into an occupied valence state $|b\rangle$, subject to the dipole and parity selection rules. Therefore after $1\sigma_u \rightarrow 1\pi_g^*$ resonant excitation only inverse parity allowed REXS hole transition $1\sigma_u^{-1} \rightarrow 1\pi_g^{-1}$ and the RIXS hole transition $1\sigma_u^{-1} \rightarrow 3\sigma_g^{-1}$ are allowed. In particular, decays leading to $1\pi_u^{-1}$ and $2\sigma_u^{-1}$ valence hole states, present in the XES spectrum in Fig. 12.7c, are parity forbidden.

In Fig. 13.9b we include the vibrational energy splittings of the electronic ground state $|a\rangle$, intermediate state $|c\rangle$, and final state which is $|b\rangle$ in RIXS and $|a\rangle$ in REXS [32, 47]. At room temperature, only the lowest $\nu = 0$ vibrational state of the electronic ground state $|a\rangle$ is occupied since the $\nu = 1$ state has an energy of $\mathcal{E}_{\nu=1} = 290\,\text{meV} >> k_\text{B}T \simeq 25\,\text{meV}$.

In Fig. 13.9c we show the experimentally observed N_2 RIXS/REXS intensity map of the transitions in (a) and (b), recorded with an instrumental Gaussian resolution of FWHM$\simeq$150 meV [47]. The intensity map corresponds to the RIXS spectra plotted horizontally as a function of *energy loss* relative to the REXS peak and vertically as a function of excitation energy. As the incident energy is increased, quasi-elastic transitions to higher vibrational levels of the electronic ground state are increasingly observed, as expected. The RIXS spectra shown at the bottom of Fig. 13.9c are the sum of those recorded at different incident energies.

In Fig. 13.9d we have replotted the RIXS spectra in (c) by changing the horizontal scale from energy loss to emitted photon energy.

The role of intermediate state interference effects is illustrated in Fig. 13.10.

In Fig. 13.10a we identify two different excitation energies shown as the green (400.75 eV) and brown (401.75 eV) lines superimposed on the XAS spectrum [48]. In (b) we show as black points and curves the experimental spectra recorded by Kjellsson et al. [32] at the two exciation energies identified in (a). Superimposed are calculated spectra by inclusion of intermediate states interference effects (red curves) and without (blue curves) [32]. In both cases, the inclusion of interference shows considerably better agreement.

13.10.1.2 O_2 RIXS

In Fig. 13.11 we illustrate the corresponding K-shell RIXS case for O_2. As for N_2, we illustrate in Fig. 13.11a the RIXS configuration diagram for O_2. In (b) we show the K-shell RIXS spectra recorded with FWHM$\simeq$50 meV instrumental resolution by Hennies et al. [63].

The *quasi-elastic* RIXS spectrum, extending to an energy loss of about 2 eV, beautifully shows the development of the vibrational series with increasing excitation energy. In comparison with the N_2 case, no vibrational peaks are missing due to intermediate state interference effects. This is due to the fact that for O_2 the vibrational splitting Δ is slightly less than the natural decay width Γ of the intermediate state, so

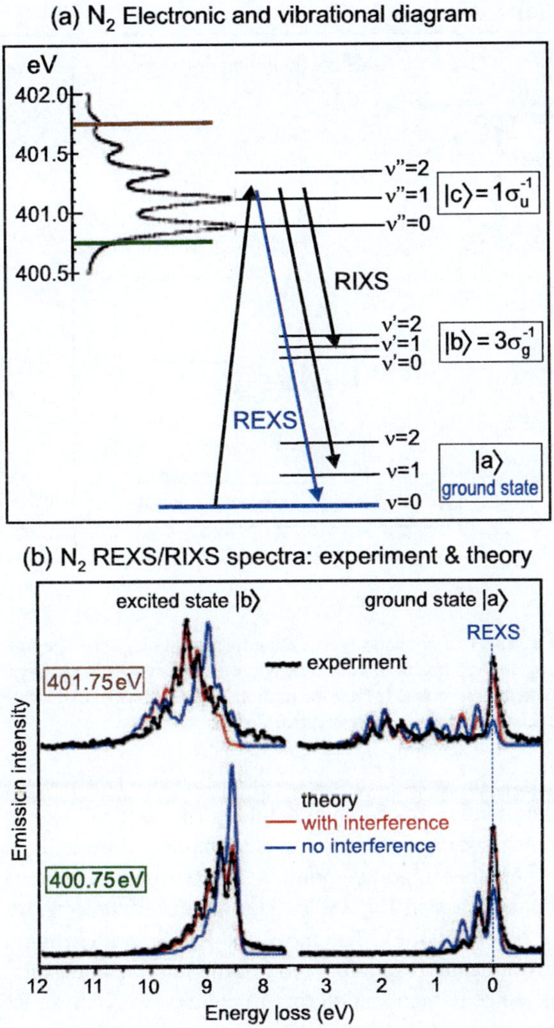

Fig. 13.10 **a** K-shell N_2 RIXS configuration diagram with two different excitation energies identified in the XAS spectrum [48] in green (400.75 eV) and brown (401.75 eV). **b** Experimental RIXS spectra (black points and curves) recorded for the green and brown excitation energies identified in (**a**), and their comparison to theoretical RIXS spectra, calculated by inclusion of interference (red curves) and without interference (blue curves) [32]

that the interference intensity is always positive as previously shown for the REXS case in Fig. 13.7c.

Since the O_2 XAS spectrum resembles an asymmetric Gaussian of about 0.7 eV FWHM, the collapse of the vibrational fine structure upon detuning of the incident energy from the peak (blue spectrum) to before the peak (pink spectrum) may be readily explained in a time-dependent picture as a reduction of the RIXS scattering time upon detuning. The coherently excited nuclear wavefunction develops until the core hole decays. As discussed in Sect. 13.5.2, the effective intermediate state

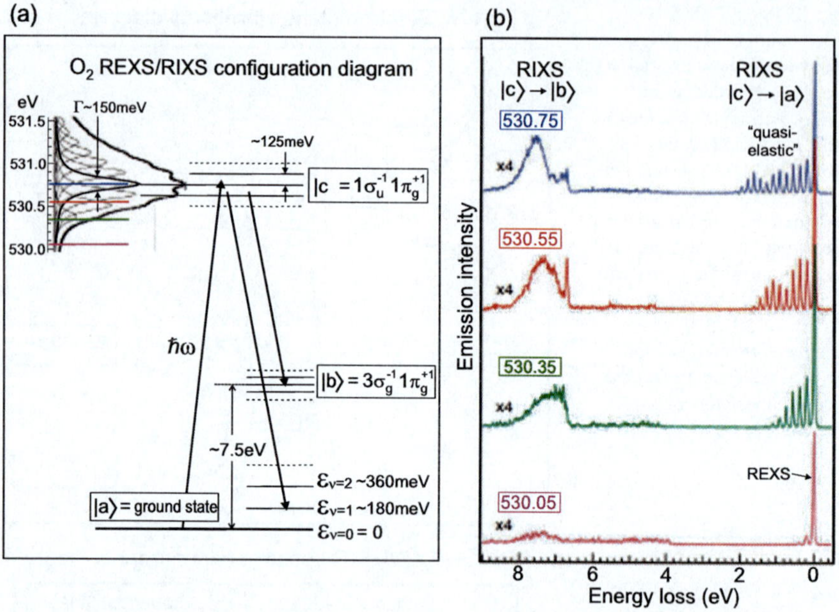

Fig. 13.11 **a** K-shell O_2 RIXS configuration diagram. The intermediate states $|c\rangle$ reached by the $1\sigma_u \rightarrow 1\pi_g^*$ transitions are shown as the XAS spectrum [50], with four different excitation energies indicated in color. **b** High-resolution (FWHM$\simeq$50 meV) experimental RIXS spectra [63] at the excitation energies indicated in color

duration is reduced upon detuning (the fast collision approximation), and the RIXS process samples only the nuclear motion close to the ground state [64].

Another important point is the extremely narrow width of the quasi-elastic vibrational peaks in Fig. 13.7b. The natural lifetime widths of the vibrational peaks is of order of 1 meV. The measured RIXS width is about 50 meV, limited only by the instrumental resolution and not the intermediate state width of $\Gamma = 150$ meV. This is a nice demonstration that in contrast to XAS, in RIXS the fundamental spectral resolution is not determined by the intermediate state lifetime Γ but rather by the lifetime of the final state.

The RIXS spectrum of the excited electronic state $|b\rangle = 3\sigma_g^{-1} 1\pi_g^{+1}$ is centered around a loss energy of 7.5 eV, and the vibrational structure is only partly visible in the 6.6–7.2 eV region because it is obscured by an underlying broad feature. It arises from the presence of another state in a multi-configuration picture that interacts with the $3\sigma_g^{-1} 1\pi_g^{+1}$ state creating the complicated RIXS band around 7.5 eV energy loss [65]. For details the reader is referred to [63, 65, 66].

13.10.2 L-edge RIXS of Transition Metal Oxides

The low energy excitations of the transition metal oxides are of great importance because of the many intriguing properties of such systems due to electron-electron correlation effects [67], such as low temperature superconductivity and colossal magnetoresistance, i.e. the dramatical change of the electrical resistance of manganese-based perovskite oxides in the presence of a magnetic field [68]. Transition metal compounds also play important roles in chemistry and biochemistry due to their ability to change their oxidation state [69], with bonds ranging from ionic in lower oxidation states to covalent in higher oxidations states. As such they exhibit calalytic activity which underlies biochemical transformations in nature and which are utilized in many industrial processes. Below we shall discuss the use of RIXS to explore the low lying excitations in La_2CuO_4 which for many years has served as a basis for the development of related low temperature superconductors.

13.10.2.1 Cu L_3 RIXS of La_2CuO_4

The electronic structure of La_2CuO_4 or LCO has been of keen interest because its central role in high temperature superconductivity. In transition metal compounds, the $3d$ orbitals are split by ligand field effects. Relative to the band-like $3d$ states in metals discussed in the following section, the orbitals can be assumed to be localized. For La_2CuO_4, the $2p_{3/2} \rightarrow 3d$ excitation step consists of promotion of a core electron into the half empty $d_{x^2-y^2}$ orbital. The importance of RIXS measurements arises from the fact that transitions between the d valence orbitals, which reveal the ligand field splitting, are optically forbidden by the dipole selection rule but may be observed by the L_3 ($2p_{3/2}$ core state) RIXS process because it proceeds via an intermediate core hole state as shown in Fig. 13.12.

In the ground state, Cu is in the $3d^9$ configuration with an electron missing in one spin state of the highest energy $d_{x^2-y^2}$ orbital. We have illustrated the ligand field splitting of the $3d$ shell in the electronic ground state $|a\rangle$ of La_2CuO_4 in the lower left corner of Fig. 13.12a and have indicated the two possible spin states by different colors.

The RIXS process illustrated for spin-conserving transitions in Fig. 13.12a involves the shown electron/hole configurations. In the intermediate state the active electron remains in the $d_{x^2-y^2}$ orbital. If the active electron itself fills the core hole, referred to a participator decay [1, 22], the scattering process is elastic. In the RIXS process there are three possible final states that can be reached from the intermediate state. They are created by bubbling up of the hole from the core to three possible valence hole states, as shown on the right side of Fig. 13.12a. The three final states lie above the ground state by different energies $\Delta\mathcal{E}_b$ with corresponding lifetimes Γ_b.

As shown in Fig. 13.12b, the Cu L_3 *XAS spectrum* of La_2CuO_4 is dominated by a pronounced resonance (black curve) near 933 eV due to a $2p_{3/2} \rightarrow 3d_{x^2-y^2}$ core to

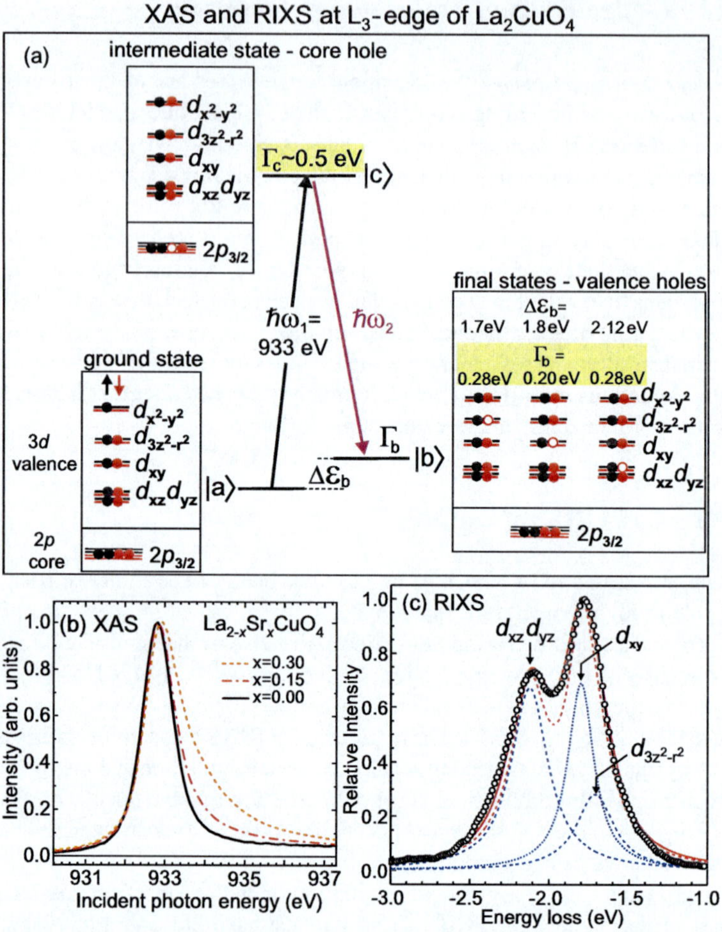

Fig. 13.12 **a** RIXS configurations for Cu L_3 excitation ($\simeq 932\,\mathrm{eV}$) in La_2CuO_4 [70]. Opposite spin states are shown in different colors, and we only show transitions that preserve the spin. There are three final states which lie above the ground state by the listed energies $\Delta\mathcal{E}_b$ and have the shown decay energy widths Γ_b. **b** XAS spectrum of $La_{2-x}Sr_xCuO_4$ (G. Ghiringhelli and N. B. Brooks, private communication) for different Sr doping. **c** RIXS spectrum revealing the three final state configurations in (**a**). The blue curves are Lorenzians at the positions $\Delta\mathcal{E}_b$ with associated FWHM Γ_b. The dashed red curves are the sum of the dashed blue curves, and the solid red curve is the dashed red curve convoluted with a Gaussian, representing the instrumental resolution of 130 meV FWHM

valence transition. As indicated, the resonance changes in shape upon replacing some of the La by Sr atoms (G. Ghiringhelli and N. B. Brooks, private communication). The XAS resonance is similar to the previously shown spectrum of $La_{1.85}Sr_{0.15}CuO_4$ in Fig. 7.18 [71] and that of Cu-phthalocyanine in Fig. 11.3b [72, 73].

The resulting *RIXS spectrum* is shown in Fig. 13.12c. It consists of three peaks whose positions and decay widths are obtained from fits with Lorentzians, shown in dashed blue. Their summed contributions are shown in dashed read. The final lineshape fits, shown in solid red were obtained by convolution of the dashed red curve with a Gaussian of 130 meV, representing the experimental resolution of the grating emission spectrometer.

13.10.2.2 Observation of Magnons and Their Wavevector Dispersion

For *elastic* charge scattering the phase of the scattered photons is shifted by 180° so that the scattered photons have exactly the opposite phase to the incident ones. For elastic magnetic scattering the phase is either rotated by +90° or −90°, depending on whether the incident wavevector **k** is parallel or antiparallel to the magnetization direction. For *inelastic* charge and magnetic scattering the polarization of the scattered photons is the same as for the respective elastic cases, but the scattering phase is random. Since the $2p_{3/2}$ states are spin-orbit coupled, they support both spin-conserving and non-conserving transitions, making RIXS spin dependent through polarization dependent selections rules.

Spin-flip transitions are only allowed for certain orientations of the antiferromagnetic spin axis relative to the lattice [14]. The polarization of the scattered radiation $\boldsymbol{\epsilon}^*_{p_2}$ is determined by that of the incident photons $\boldsymbol{\epsilon}_{p_1}$ according to $\boldsymbol{\epsilon}^*_{p_2} \cdot \boldsymbol{\epsilon}_{p_1}$ for charge and $\boldsymbol{\epsilon}^*_{p_2} \times \boldsymbol{\epsilon}_{p_1}$ for spin scattering [74]. One can therefore distinguish charge and spin dependent scattering through polarization dependent selections rules. If the core states are strongly spin-orbit coupled as for the $2p$ shell, they can support a change of the photon angular momentum and inelastic spin-flip scattering [75, 76].

As an example of spin-related features in RIXS and their energy dispersion with momentum transfer we show in Fig. 13.13 the Cu L_3 ($\simeq$932 eV) excitation spectrum of La_2CuO_4 [77]. The spectra shown in (a) as a function of momentum transfer $\mathbf{q}_{\|}$ (G. Ghiringhelli and N. B. Brooks, private communication) extend in energy to the region of the $d-d$ excitations shown in Fig. 13.12. At lower energy loss, a magnon peak, labeled B, is also observed. Its energy position disperses with $\mathbf{q}_{\|}$. Figure 13.13b shows the enlarged magnon region for a particular $\mathbf{q}_{\|}$, and the complete dispersion curve of peak B is shown as light blue data points in (c). The RIXS results are seen to be in excellent agreement with the dispersion curve measured by inelastic neutron scattering, which is superimposed as a dashed magenta curve.

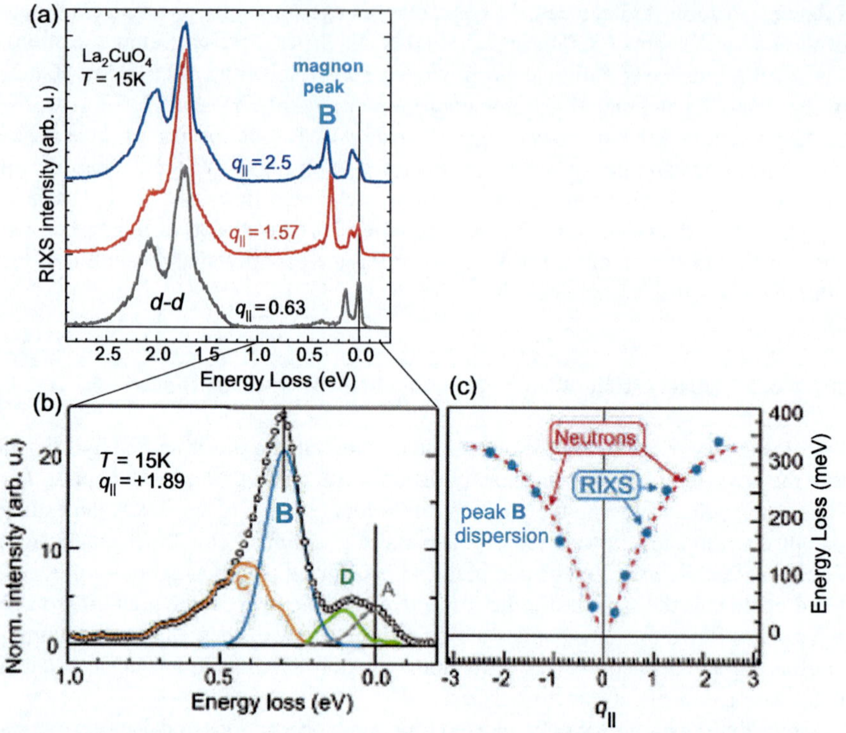

Fig. 13.13 **a** Extended Cu L$_3$ ($\simeq$932 eV) RIXS spectra of La$_2$CuO$_4$ as a function of momentum transfer $\mathbf{q}_{\|}$ [77] and G. Ghiringhelli and N. B. Brooks, private communication. The spectra show the higher energy $d - d$ excitations of Fig. 13.12 and a lower energy magnon loss peak labeled B. **b** Enlarged region of the magnon loss region. **c** Energy dispersion of the magnon loss peak as a function of $\mathbf{q}_{\|}$ along the $(-\pi, 0) \rightarrow (\pi, 0)$ direction in the Brillouin zone, shown as light blue data points. Superimposed as a dashed magenta curve is the dispersion measured with inelastic neutron scattering

13.10.3 L-Edge RIXS of Transition Metals

There is a distinct difference between the RIXS response of transition metal ions in *non-metallic compounds* such as La$_2$CuO$_4$ discussed above and that of the elementary 3*d transition metals*, whose RIXS spectra have been extensively studied [35, 78–81]. In the 3d transition metals, the XAS excitations from 2p core to 3d valence band states have been schematically illustrated in Fig. 10.14 and the corresponding XES decays in Fig. 12.9a. Their link in a configurational model is shown in Fig. 13.2.

In our discussion we assume that the 3d valence electrons of the transition metal are not magnetically aligned and their charge distribution is isotropic, which strictly holds for cubic lattice symmetry and approximately even for hexagonal metals. The polarization independent RIXS process may then be pictured as illustrated in

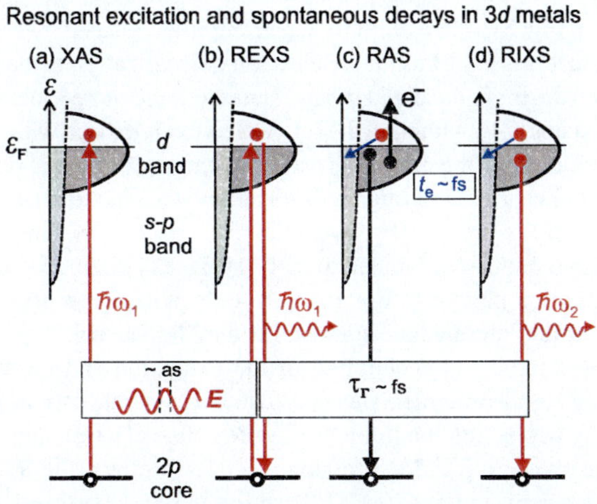

Fig. 13.14 Simple picture of excitation and de-excitation processes associated with RIXS at the L-edge of $3d$ transition metals. Ignoring magnetic alignment, the metal valence band consists of delocalized $s-p$ and localized d states, whose density of states is schematically illustrated. **a** Resonant excitation (XAS) of a $2p$ core electron to empty $3d$ valence states on the attosecond timescale. **b** Resonant elastic x-ray scattering (REXS), involving up and down motion of the "active electron". The emission occurs on the core hole clock timescale $\tau_\Gamma = \hbar/\Gamma$ of order of femtoseconds with a probability expressed by the fluorescence yield as illustrated in Fig. 12.4. **c** Resonant Auger spectroscopy (RAS) following non-radiative core hole decay on the femtosecond timescale τ_Γ. In metals, the excited electron delocalizes into the $s-p$ bands on a timescale $t_e \simeq \tau_\Gamma$. **d** Resonant inelastic x-ray scattering (RIXS) where the core hole is filled by radiative decays of electrons in the filled $3d$ band

Fig. 13.14. The figure distinguishes the densities of states of the more localized $3d$ valence band states and the more delocalized $s-p$ states.

In the excitation or XAS process, illustrated in Fig. 13.14a, a $2p$ core electron is excited into empty $3d$ valence states on the attosecond timescale. In the REXS process, shown in Fig. 13.14b, the excited electron decays back into the core hole and energy is conserved between the up and down motion of the "active electron", so that the energy contained in the incident photon is re-emitted. The spontaneous emission process is in random directions, while the stimulated one is directional. Although the radiative dipole process has lower probability corresponding to a decay width $\Gamma^X \sim 10^{-3}$ eV, the total decay width contains a considerably larger ghost contribution Γ^A due to the faster Auger process as discussed in Sect. 12.2.5 with the transition widths shown in Fig. 12.10.

After the excitation process, the intermediate state consists of a core hole with the active electron excited into empty $3d$ states. The core hole is effectively screened by the active electron in the intermediate state. There is little difference in the intermediate state created by "excitonic" active electron/core hole screening and collective core hole screening by metallic electrons. One may therefore assume that the active electron is absent due to rapid delocalization with extended metal states, as shown by the blue arrows in Fig. 13.14c, d, with its screening role replaced by collective metal states. The physical origin of this active electron delocalization is the hybridization (mixing) between the more localized d band states and the more itinerant or delocalized $s-p$ band states [74]. Although the active electron is excited with a larger dipole transition probability into the empty 3d valence states [82], it will quickly lose its angular momentum character through hybridization. This picture has emerged from extensive studies of charge transfer dynamics in core hole systems ranging from metals to atoms and molecules bonded to metal surfaces [1, 22, 35, 78, 79, 81, 83].

In the configuration picture of Fig. 13.2, the delocalization process of the active electron occurs in the intermediate state and it may be viewed to precede the core hole decay. This results in a decoupling of the excitation and decay steps, often called "dephasing" in the intermediate state. The intermediate core hole state is then well described by neglecting the influence of the active electron and RIXS may be described by the two-step RIXS expression (13.64). In effect, the RIXS process in transition metals is very similar to the XES process that follows threshold excitation. The advantage of threshold or resonant excitation to empty states just above the Fermi level is that it reduces multi-electron shake-up structure, as illustrated for Cu metal in Fig. 12.9d.

In a simple model, the empty and filled d band degeneracies are given by the number of holes and electrons in our *polarization averaged* dipolar XAS transition width (10.59) and the XES decay width (12.22). The two-step RIXS cross section is then given by (13.68) or

$$\boxed{\frac{\mathrm{d}\sigma_{\mathrm{RIXS}}}{\mathrm{d}\Omega_2}=\frac{1}{4\pi}\left(1+n_{\mathrm{pk}}\right)\underbrace{\frac{\Gamma^{\mathrm{X}}}{\Gamma}}_{Y_{\mathrm{f}}}\sigma_{\mathrm{XAS}}.} \tag{13.74}$$

Here the polarization averaged XAS cross section is given by

$$\sigma_{\mathrm{XAS}}=\frac{\lambda_1^2}{\pi}\frac{\langle\Gamma_{ac}^{p_1}\rangle}{\Gamma}\frac{(\Gamma/2)^2}{(\hbar\omega_1-(\mathcal{E}_c-\mathcal{E}_a))^2+(\Gamma/2)^2} \tag{13.75}$$

and the transition width for absorption is

$$\langle\Gamma_{ac}^{p_1}\rangle=\frac{8\pi^2\hbar\omega_1\,\alpha_{\mathrm{f}}}{\lambda_1^2}\mathcal{R}^2\frac{LN_{\mathrm{h}}}{3(2L+1)} \tag{13.76}$$

and for emission is given by (13.69).

If the emitted photon energy is analyzed by a spectrometer, the RIXS spectrum will then be representative of the density of the filled 3d band, as discussed in Sect. 12.3.4.

For metals, the RIXS process may simply be envisioned as resonant absorption followed by XES.

13.10.3.1 XAS, XES, REXS, and RIXS Cross Sections

We can now summarize resonant cross sections involving transitions between core states of angular momentum c and valences states of angular momentum L that are occupied by N_e electrons and N_h holes by the following *polarization averaged* cross sections and transition widths. In all cases we assume Lorentzian transition widths (FWHM) denoted by Γ_i. We note that in practice there is often a trade-off between the width of resonant transitions and their peak value. This is due to either superpositions of vibrational peaks as shown for the K-shell resonances of N_2 and O_2 in Figs. 10.11 and 10.12 or band structure broadening as shown in Fig. 10.17 for the L_3 XAS resonance in Fe, Co, and Ni.

Under the assumption of natural Lorentzian lineshapes, the XAS case is given by (10.60) or

$$\sigma_{\text{XAS}} = \frac{\lambda^2}{\pi}\frac{\Gamma_{\text{XAS}}}{\Gamma}\frac{(\Gamma/2)^2}{(\hbar\omega - \mathcal{E}_0)^2+(\Gamma/2)^2} \tag{13.77}$$

$$\Gamma_{\text{XAS}} = \frac{8\pi^2\hbar\omega\,\alpha_{\text{f}}}{\lambda^2}\,\mathcal{R}^2\frac{L N_{\text{h}}}{3(2L+1)}. \tag{13.78}$$

The XES case is given by (12.16) and (13.80) as

$$\sigma_{\text{XES}} = \frac{\lambda^2}{\pi}\frac{\Gamma_{\text{XES}}}{\Gamma}\frac{(\Gamma/2)^2}{(\hbar\omega - \mathcal{E}_0)^2+(\Gamma/2)^2}\left(1+n_{\mathbf{pk}}\right) \tag{13.79}$$

$$\Gamma_{\text{XES}} = \Gamma^{\text{X}} = \frac{8\pi^2\hbar\omega\,\alpha_{\text{f}}}{\lambda^2}\,\mathcal{R}^2\frac{L N_{\text{e}}}{3(2L+1)(2c+1)}. \tag{13.80}$$

The REXS scattering cross section is obtained from (13.28) as

$$\sigma_{\text{REXS}} = \left\{\frac{\Omega}{4\pi} + n_{\mathbf{pk}}\right\}\frac{\Gamma_{\text{REXS}}}{\Gamma}\,\sigma_{\text{XAS}}, \tag{13.81}$$

where for charge scattering $\Omega = 8\pi/3$ according to (13.23) and

$$\Gamma_{\text{REXS}} = \Gamma_{\text{XAS}}. \tag{13.82}$$

Finally, the two-step RIXS case is described by the angle integrated expression (13.74) or

$$\sigma_{\text{RIXS}} = \frac{\Gamma_{\text{RIXS}}}{\Gamma}\left(n_{\mathbf{pk}}+1\right)\sigma_{\text{XAS}} \tag{13.83}$$

and

$$\Gamma_{\rm RIXS} = \Gamma_{\rm XES} = \Gamma^{\rm X} = \frac{8\pi^2 \hbar\omega_2\, \alpha_{\rm f}}{\lambda_2^2} \mathcal{R}^2 \frac{L N_{\rm e}}{3(2L+1)(2c+1)}. \tag{13.84}$$

It is apparent that for a given c and L the cross sections and transition widths may differ by factors that involve the orbital degeneracies (not including spin) given by $2c+1$ and $2L+1$ and the valence shell population expressed by $N_{\rm e}$ electrons and $N_{\rm h}$ holes. For the 3d transition metals, Fe, Co, and Ni the number of valence holes $N_{\rm h}$ and electrons $N_{\rm e}$ are given in Table 10.5. In the following we shall discuss the Co case as an example.

13.10.3.2 L$_3$ RIXS of Co Metal

For the Co L-edge, we have $c = 1$, $L = 2$ and $Y_{\rm f} = 8 \times 10^{-3}$. The spontaneous XAS and XES cross sections happen to be the same because $N_{\rm h} = 2.5$ and $N_{\rm e} = 7.5$, so that

$$\sigma_{\rm XAS} = \sigma_{\rm XES} \gg \sigma_{\rm RIXS} = 8 \times 10^{-3} \sigma_{\rm XAS}. \tag{13.85}$$

The RIXS case for L$_3$ resonant excitation in the 3d metals is illustrated in Fig. 13.15a in a simple band model for the occupied and unoccupied 3d states previously used in Figs. 10.14 and 12.9a.

The model assumes that at a given energy relative to the Fermi level, $E_{\rm F}$, the band averaged DOS may be described by a mixture of all d orbitals. Also, in the 3d transition metals, the numbers of d electrons in filled states $N_{\rm e}$ and holes in empty states $N_{\rm h}$ are not integers [74], and only their sum is given by the total spin orbital degeneracy of $2(2L+1) = 10$. RIXS spectra of transition metals are best recorded by tuning the incident energy just above the L$_3$ threshold, determined by the Fermi level. The lack of excess incident energy avoids multi-electron shake-up structure [80], and the resonant excitation process increases the cross section.

The measured resonant XAS and the spontaneous RIXS spectra for the L$_3$ excitation in Co metal are shown in Fig. 13.15b [79]. Both consist of resonance like peaks associated with the filled and empty DOS, as expected. Although we have shown the XAS and RIXS spectra normalized to the same intensity, the RIXS intensity is considerably lower. According to (13.70) and (13.85), the angle integrated emission cross section is reduced relative to the absorption cross section because of the small fluorescence yield of $\simeq 10^{-2}$. In addition, only about one of $10^{-4} - 10^{5}$ emitted photons is scattered into the solid angle of a typical emission spectrometer [84, 85]. In practice, the relative XES intensity is therefore smaller by a factor of order $10^{-6} - 10^{-7}$. We will see in Chap. 15 that by stimulating the emission at high incident intensity, the RIXS intensity may become comparable to the XAS intensity.

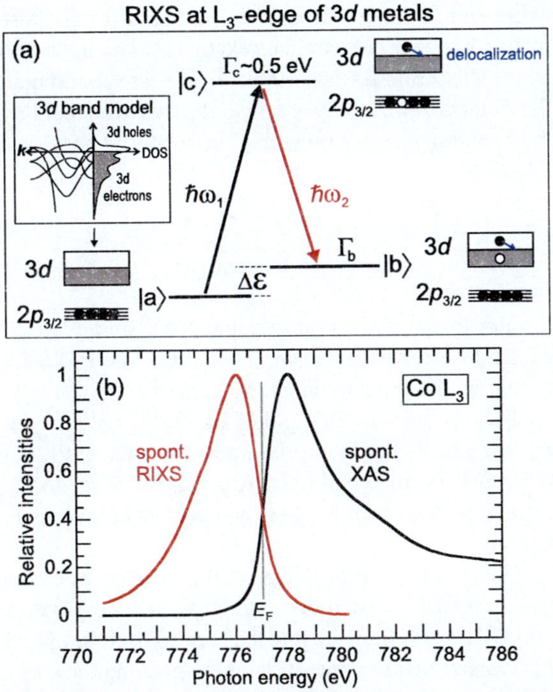

Fig. 13.15 **a** RIXS diagram for the L_3 excitation in $3d$ transition metals, involving $2p_{3/2}$ core and $3d$ valence electrons and holes. The intermediate state $|c\rangle$ and final state $|b\rangle$ are shown with the "active" electron in empty d states. In practice, it typically delocalizes as indicated by the blue arrow. **b** Measured spontaneous L_3-edge XAS spectrum for a Co/Pd multilayer [84] and RIXS spectrum for Co metal [79]. The Fermi energy is marked as E_F

13.10.4 RIXS of Chemisorbed Molecules: Polarization Dependence

The properties of chemisorbed molecules on metal surfaces have been extensively investigated with many techniques because their behavior underlies the broad and important field of heterogeneous catalysis [22, 86, 87]. From a RIXS perspective, there is a significant difference in the description of free and chemisorbed molecules. We have seen in Sect. 13.8 for the fundamental cases of N_2 and O_2 that the intermediate state exhibits vibrational substates leading to interfering channels in the RIXS cross section. Also, in these cases one may assume that during the RIXS process the "active electron" remains as a spectator in the x-ray emission process [88].

As reviewed by Nilsson and Pettersson [22], the chemisorption bond mixes molecular and metal states so that the valence electronic structure of chemisorbed molecules on metal surfaces is in many aspects metallic. This causes a screening response of electronic excitations on the molecules due to coupling to the continuum of the delocalized metal states. After the resonant absorption step, the intermediate state consists of a molecular core hole with the active electron in an unfilled molecular orbital which already screens the core hole [89]. As discussed for the transition metal case above, the intermediate core hole may then be viewed as a fully screened state and may be represented as a single state in the RIXS process. The two-step RIXS expression (13.64) may thus also be used for chemisorbed molecules. This simple picture has been supported by extensive experimental RIXS investigations pioneered by A. Nilsson in the late 1990s. It was also found that for weakly adsorbed species, the delocalization of the active electron may be slower so that the RIXS spectra may exhibit features reflecting "spectator" behavior of the active electron [22, 35, 83, 90, 91].

13.10.4.1 Polarization Dependence

Similar to polarization dependent XAS studies of chemisorbed molecules by use of the XNLD effect shown in Fig. 11.3, the RIXS intensity also strongly depends on the *incident* polarization. Studies of chemisorbed systems have not yet utilized analysis of the *scattered* polarization, but only the fact that the polarization vector $\boldsymbol{\epsilon}_{p_2}$ necessarily lies in the plane perpendicular to the emission direction $\mathbf{k}_2$ because of the transverse nature of light. Analysis of the scattered polarization has been utilized in studies of molecular gases and correlated materials as discussed in Sect. 13.10.5 below.

The dependence of RIXS on the incident polarization readily follows from the two-step RIXS description, which in the decay step restricts the dipole and parity allowed transitions back into the core hole. By simply detecting only photons emitted in symmetry chosen directions $\mathbf{k}_2$, e.g. along the surface normal or the two in-plane directions, the emitted polarization $\boldsymbol{\epsilon}_{p_2}$ is naturally restricted to the plane perpendicular to $\mathbf{k}_2$ [92]. This fact has been used extensively to determine the nature of the filled valence orbitals of oriented chemisorbed molecules [22]. An example was previously illustrated in Fig. 1.20 by polarization dependent RIXS spectra for benzene (C_6H_6) molecules bonded to a Cu(110) and Ni(100) surfaces [93, 94]. The difference between the spectra for the two cases nicely shows how the weaker chemisorption bond to Cu leaves the benzene molecular orbitals largely unperturbed, while the stronger bond to Ni perturbs them significantly.

13.10.4.2 Example: RIXS of Glycine on Cu(110)

A particularly beautiful XAS and RIXS example is the case of the low symmetry molecule glycine, the simplest amino acid, chemisorbed on the low symmetry Cu(110) surface [95, 96]. The chemisorption geometry is shown in Fig. 13.16a, which was deduced from polarization dependent XAS spectra shown in (b) and RIXS spectra in (c).

Glycine (NH_2CH_2COOH) consists of carboxyl (HC-OOH) and methylamine (H_2N-CH_3) groups that bond together by replacing two C-H with a single C-C bond, as shown on the left of Fig. 13.16a. Upon chemisorption, the molecular skeleton is preserved, but the acidic H atom is removed and the resultant glycate molecule bonds to the Cu surface atoms as shown on the right side of the figure.

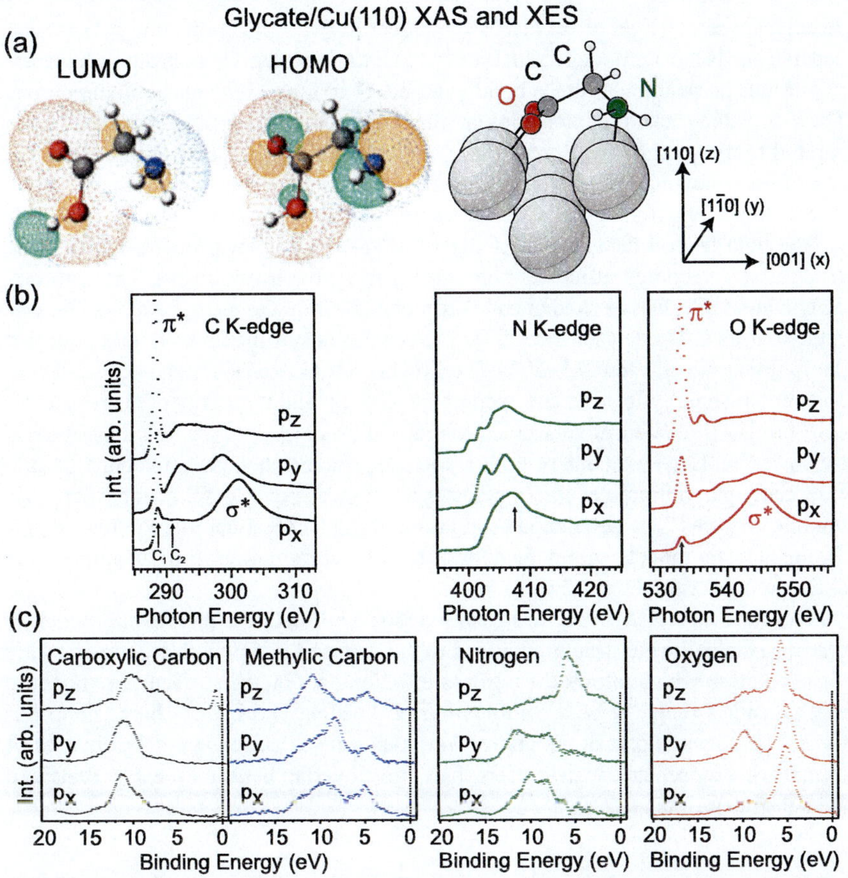

Fig. 13.16 **a** Atomic model of glycine, showing the lowest unoccupied (LUMO) and highest occupied (HOMO) molecular orbitals and the geometry of glycate bonded to Cu atoms on the (110) surface. **b** Polarization dependent XAS spectra. The arrows indicate the excitation energies of the RIXS spectra. **c** Polarization dependent RIXS spectra measured for excitation of the C atom in the carboxyl-COO (labeled C_1 in (b)) and in the methyl-CH_2 group (C_2), and for the N and O atoms excited at the arrow positions in (**b**) [95, 96]. The spectra are normalized to the same area, and the binding energy is referenced to the XPS 1*s* binding energy

The presence of C, N, and O atoms with directional covalent bonds provides an ideal case for demonstrating the power of polarized XAS and RIXS spectra. They provide atom specific and symmetry selective bonding information through spectra of unoccupied (XAS) and occupied (XES) orbitals. In all cases the electronic K-shell transitions involve spherically symmetric 1*s* core holes and orientation dependent $2p_x$, $2p_y$, $2p_z$ valence orbitals.

The polarization dependent *XAS spectra* shown in Fig. 13.16b were recorded with an instrumental resolution of 0.1 eV and provide direct orientational information through the search light effect discussed in Sect. 11.3.1. For glycine, the polarization dependence is most pronounced in the C and O XAS spectra. On both atoms there are π^* orbitals perpendicular to the bond axis and σ^* orbitals along the bond directions. The associated transition intensities are the largest when the **E** polarization vector is parallel to the spatial direction of the π^* and σ^* orbitals, respectively. The relative size of the resonances for $\mathbf{E} \parallel x, y, z$ in coordinate system shown in (a) then directly reveals the orientation of the bonds between C, N, and O atoms.

The polarization dependence of the *RIXS spectra* is shown in Fig. 13.16c. The spectra were recorded at incident energies marked by arrows in (b). The carboxyl carbon atom, C_1, was excited at its π^* resonance, while the methyl carbon, C_2, was excited at its C-H σ^* resonance [97]. The resolutions of the incident and scattered photons were set to about 0.5 eV. The spectra are normalized to the same area and put on a binding energy scale by subtracting the XPS 1s binding energy of the respective core levels. The valence electronic structure at binding energies less than about 6 eV corresponds to molecular orbitals responsible for the chemisorption bonds, while essentially pure intramolecular bonding orbitals give rise to RIXS features at higher binding energies. The detailed interpretation of the RIXS structure requires support through cluster models within the density functional framework [96]. It supports the chemisorption geometry shown in (a).

The XAS and RIXS spectra in conjunction with detailed theoretical modeling reveal a stable glycate structure bonded to the surface through both oxygens and the nitrogen atom which allows the adsorbate to lie down on the surface. The bonding mechanism is found to be a combination of covalent and image charge contributions. The covalent part of the interaction mainly takes place via a hybridization of orbitals of z-symmetry, which offers the largest overlap between the adsorbate and the substrate orbitals.

13.10.5 Utilization of the Scattered Polarization

In the soft x-ray region, the small fluorescence yield (see Fig. 12.10) has impeded analysis of the emitted polarization since it decreases the detected RIXS signal by more than an order of magnitude. It has only been recently implemented and utilized for a selected study at the Co L_3 resonance in a superconducting cuprate [98]. It clearly will be utilized more in the future for the study of correlated materials based on improved spectrometers [18].

At higher x-ray energies the fluorescence yield is larger and Bragg diffracting crystals can be used [99, 100]. This was utilized around 1990 for polarization analysis of RIXS spectra recorded at the Cl K-edge ($\sim$ 2800 eV) of chlorine containing hydrocarbon gases by Lindle et al. [101, 102]. One may think that the random molecular orientation in gases does not show a polarization dependent RIXS response, but this is not so. The basic origin of its existence is quite simple once one understands the polarization dependent response of oriented molecules as discussed in Sect. 13.10.4.

If a linearly polarized beam induces resonant transitions in randomly oriented molecules, the excitation probability of a molecule at a given incident energy will depend and its orientation relative to the incident **E**-vector. This is most easily understood by considering transition to unoccupied π^* or σ^* orbitals. For example, in diatomic molecules like N_2 or O_2 the π orbitals are perpendicular to the internuclear axis and the σ orbitals parallel to it. In ring-like molecules like benzene, the π orbitals are perpendicular to the ring plane and the σ orbitals are parallel to it. If the incident photon energy is chosen to correspond to a $1s \rightarrow \pi^*$ transition, the **E**-direction will preferentially create core holes on molecules that are oriented with their π system along **E**. The emitted RIXS intensity will therefore be dominated by emission from these oriented molecules. Hence the emitted radiation will not have random polarization but an emission direction dependent polarization, similar to that of oriented chemisorbed molecules.

The polarization dependence has been worked out in detail by Gel'mukhanov [12, 103] and may be written as

$$\sigma_{\text{RIXS}}(\boldsymbol{\epsilon}, \boldsymbol{\epsilon}') = \sigma_0 \left\{1 + R(3\cos^2\theta - 1)\right\}, \quad \theta = \angle(\boldsymbol{\epsilon}, \boldsymbol{\epsilon}')$$
$$R = \frac{1}{5}\left[3\left(\hat{\mathbf{d}}_{\hbar\omega'} \cdot \hat{\mathbf{d}}_{\hbar\omega}\right)^2 - 1\right] \tag{13.86}$$

where σ_0 is the isotropic cross section and $\hat{\mathbf{d}}_{\hbar\omega}$ is the dipole moment of the XAS transition and $\hat{\mathbf{d}}_{\hbar\omega'}$ that of the XES transition, which are determined by the orientation of the empty valence orbital in the up transition and filled orbital in the down transition.

13.11 RIXS and Reduced Linewidth XAS (HERFD)

In our configurational model in Fig. 13.2, the upper electronic core hole state $|c\rangle$ typically has a larger width than the lower final state $|b\rangle$. This is utilized in RIXS, where according to (13.61) and (13.62) the intrinsic linewidth is effectively determined by the width Γ_b of the lower valence hole state. This valuable property of RIXS may also be exploited to improve the energy resolution in XAS spectra by a technique that has been referred to as *lifetime reduced* or *high energy resolution fluorescence detection* XAS, or "HERFD" for short [28, 104].

XAS spectra may be recorded by scanning the incident photon energy and monitoring any secondary electron or photon signal generated in the process of filling the created core hole. These secondary signals are proportional to the probability that the core hole was created by x-ray absorption in the first place. In principle, one should integrate the emitted electron or photon signal over energy and angle, as given by (13.70), but it is usually sufficient to select a fraction of the signal as utilized in partial electron or fluorescence yield detection [105].

XAS spectra of width smaller than the natural core hole width were first recorded by Hämäläinen et al. [106] by use of a RIXS-like experimental arrangement. They tuned the incident energy across an absorption resonance and used a spectrometer to detect a slice of the x-ray emission signal that was narrower than the core hole life time width. Today, the HERFD technique is typically employed for absorption resonances in the hard x-ray region (around 10 keV or more) whose intrinsic core hole width is large (a few eV). In these cases, the deep core hole is filled by electrons from upper core shells rather directly from the valence shell. In a configuration picture the HERFD final state is then a hole in an outer *core* shell that has a smaller natural width than the original deep core hole. The HERFD process is thus nothing but a RIXS process, but in contrast to conventional RIXS where the final state is a valence hole, it is a core hole in HERFD.

The HERFD technique is based on reformulating the *two-step* RIXS expression (13.64), which for negligible instrumental linewidths has the form

$$\frac{d\sigma_{\mathrm{RIXS}}}{d\Omega_2\, d\hbar\omega_2} = \underbrace{\frac{\lambda_1^2}{2\pi^3}\left(n_{p_2\mathbf{k}_2}+1\right)\sum_c\sum_b \frac{\Gamma_{ac}^{p_1}}{\Gamma_c^2}\frac{\Gamma_{cb}^{p_2}}{\Gamma_b}}_{\text{determines resonance "intensity"}}$$
$$\times \underbrace{\frac{(\Gamma_c/2)^2}{(\hbar\omega_1-\mathcal{E}_{ca})^2+(\Gamma_c/2)^2}\frac{(\Gamma_b/2)^2}{(\hbar\omega_2-\mathcal{E}_{cb}])^2+(\Gamma_b/2)^2}}_{\text{determines resonance width}}. \tag{13.87}$$

As indicated by underbrackets, the resonant cross section consists of "intensity" and lineshape expressions. In the latter we typically have $\Gamma_b < \Gamma_c$, so that the multiplication of the two Lorentzians effectively takes a slice of width Γ_b out of the larger width Γ_c. The fact that the two Lorentzians are centered at different incident energies $\hbar\omega_1 = \mathcal{E}_{ca}$ and emitted energies $\hbar\omega_2 = \mathcal{E}_{cb}$ is taken care of by a projection of the two energy axes onto each other as clarified by the following example of HERFD.

13.11.1 HERFD XAS at the Pt L_3-Edge

As an example we discuss the sharpening of the L_3 threshold resonance in Pt metal. This case has received considerable attention owing to the importance of Pt in catalytic and electro-catalytic processes [107]. Figure 13.17 illustrates the key energy levels and electronic transitions of importance and the RIXS configuration diagram of the excitation and emission processes.

We have simplified the discussion by only considering the change of the threshold "white line" resonance which corresponds to transitions from $2p_{3/2}$ core states to empty $5d_{5/2}$ valence states. In practice, the resonance is superimposed on an edge-like step [108], and the general RIXS case has been discussed by de Groot et al. [24].

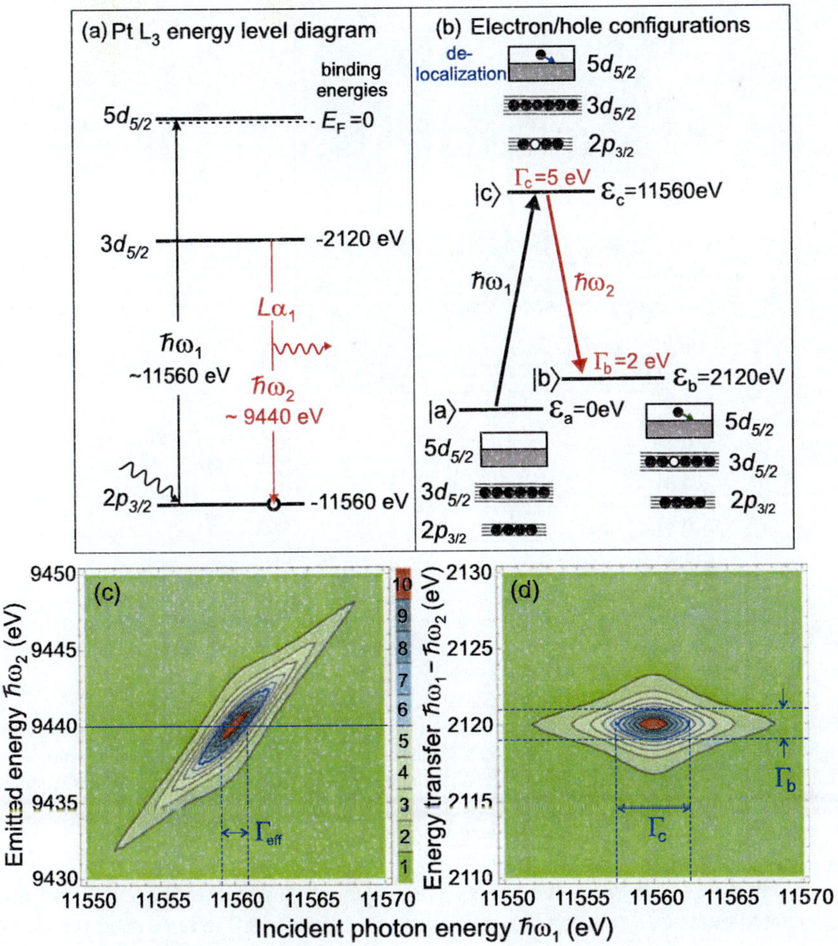

Fig. 13.17 **a** Binding energy diagram of key energy levels of Pt involved in an L_3 resonant transition from $2p_{3/2}$ core states to empty $5d_{5/2}$ valence states. We have assumed that the $5d_{5/2}$ state is located at the Fermi energy with binding energy $\mathcal{E}_{5d_{5/2}} = 0\,\text{eV}$, and the other binding energies are $\mathcal{E}_{3d_{5/2}} = -2,120\,\text{eV}$ and $\mathcal{E}_{2p_{3/2}} = -11,560\,\text{eV}$. **b** Electron/hole configuration diagrams relative to the ground state $|a\rangle$ of zero energy. The intermediate core hole state $|c\rangle$ of energy $\mathcal{E}_c = -\mathcal{E}_{2p_{3/2}}$ is shown in the fast collision approximation with the active electron in an empty d state and a decay energy width $\Gamma_c = 5\,\text{eV}$. The blue arrow reflects possible delocalization of the active electron before decay of the intermediate state. The final state $|b\rangle$ of energy $\mathcal{E}_b = -\mathcal{E}_{3d_{5/2}}$ and decay width $\Gamma_b = 2\,\text{eV}$ lies above the ground state by the energy transfer $\hbar\omega_1 - \hbar\omega_2$. **c** RIXS plane contour diagrams given by (13.87), representing the RIXS intensity distribution as a function of incident photon energy $\hbar\omega_1$ versus emitted photon energy $\hbar\omega_2$. The thicker blue contour corresponds to the FWHM of the intensity distribution. Also indicated are the effective width FWHM $\Gamma_{\text{eff}} = 1.86\,\text{eV}$ given by (13.88) and the widths Γ_c and Γ_b. **d** Replacement of the emitted energy $\hbar\omega_2$ scale in (c) by the photon energy transfer $\hbar\omega_1 - \hbar\omega_2$

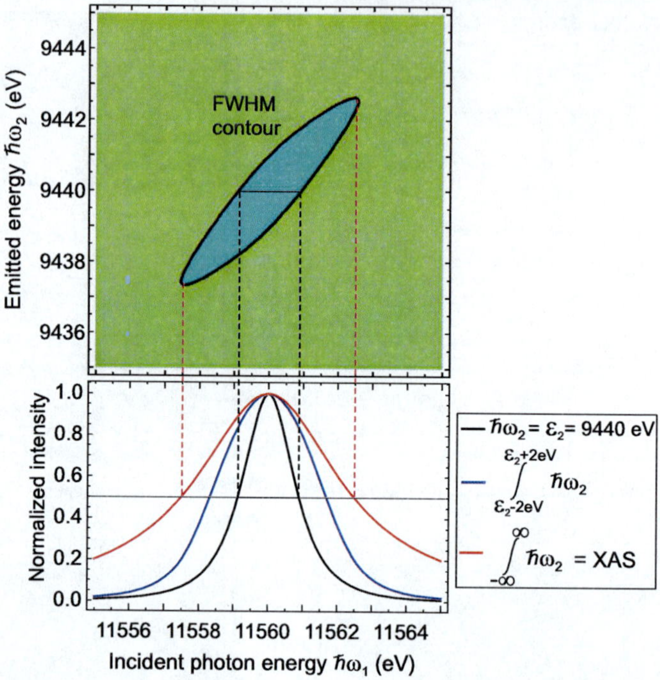

Fig. 13.18 Top: RIXS FWHM contour diagram according to (13.87) indicated by the rim of the light blue area. Bottom: Corresponding RIXS spectra measured with a flat-top detector window of different width, centered around the peak emission energy of 9440 eV. Black: infinitesimally small width, blue: 4 eV window width, red: very large width. The red curve represents the integration over all emission energies and is the conventional x-ray absorption resonance

In Fig. 13.17a we show the energy level diagram for the key Pt levels and the associated excitation (absorption) and emission transitions. We have used the binding energies $\mathcal{E}_{5d_{5/2}}=0\,\text{eV}$, $\mathcal{E}_{3d_{5/2}}=-2,120\,\text{eV}$, and $\mathcal{E}_{2p_{3/2}}=-11,560\,\text{eV}$ which are close to the exact values [109]. The corresponding RIXS diagram of the associated electron/hole configurations and the filling of the states is shown in Fig. 13.17b. The RIXS process involves the ground state $|a\rangle$ of zero energy, the intermediate core hole state $|c\rangle$ of energy $\mathcal{E}_c = 11,560\,\text{eV}$ and total decay energy width $\Gamma_c=5\,\text{eV}$, and the final state $|b\rangle$ of energy $\mathcal{E}_b=2,120\,\text{eV}$ and reduced decay width $\Gamma_b=2\,\text{eV}$. The assumed energy widths are close to the literature values [110].

In Fig. 13.17b the intermediate and final states still have the active electron as a spectator in the empty *d* orbital. As discussed above, we can assume for metals that the active electron will delocalize as indicated by a small blue arrow. The final 5*d* state then has a single hole in the occupied band and is the same as that in valence band photoemission.

In Fig. 13.17c we show the RIXS intensity contour map according to (13.87) as a function if incident ($\hbar\omega_1$) and scattered ($\hbar\omega_2$) photon energies, and in (d) the scattered energy has been replaced by the energy loss $\hbar\omega_1 - \hbar\omega_2$. For the simulations we have ignored polarization effects and set the transition matrix elements $\Gamma_{ac}^{p_1} = \Gamma_{cb}^{p_2} = 1$. The relative intensity is represented by colored contours in the figures, with the contour representing the FWHM of the intensity distribution shown as a thicker blue oval contour in (c). Its intersect with the blue horizontal emission energy line at $\hbar\omega_2 = 9,440\,\text{eV}$ defines the minimum achievable energy width of $\Gamma_{\text{eff}} = 1.86\,\text{eV}$, given approximately by

$$\Gamma_{\text{eff}} \simeq \sqrt{\frac{1}{1/\Gamma_c^2 + 1/\Gamma_b^2}}. \tag{13.88}$$

This HERFD-XAS width is narrower that the conventional XAS width of $\Gamma_c = 5\,\text{eV}$ and is approximately equal to $\Gamma_b = 2\,\text{eV}$ illustrated in (c). The RIXS width expands into the XAS width by integrating the RIXS cross section over all emitted photon energies according to (13.70).

The dependence of the HERFD-XAS width on the width of the detection window of the emitted photons, assumed to have a flat-top shape, is illustrated in more detail in Fig. 13.18. The point of the figure is to show how integration over an increasing window of the emitted intensity causes the HERFD-XAS spectrum to evolve into the conventional XAS one, as described by (13.70). We note that the reduction of RIXS to XAS spectra does not apply in general, but only for the two-step RIXS case where the excitation and emission processes can be decoupled [23, 25, 28, 104, 111].

References

1. P.A. Brühwiler, O. Karis, N. Mårtensson, Rev. Mod. Phys. **74**, 703 (2002)
2. F. Gel'mukhanov, M. Odelius, S.P. Polyutov, A. Föhlisch, V. Kimberg, Rev. Mod. Phys. **93**, 035 001 (2021)
3. T. Åberg, Phys. Scr. **21**, 495 (1980)
4. J. Tulkki, T. Åberg, J. Phys. B **13**, 3341 (1980)
5. R. Loudon, *The Quantum Theory of Light*, 3rd edn. (Clarendon Press, Oxford, 2000)
6. V. Weisskopf, Ann. Phys. Leipzig **9**, 23 (1931)
7. J. Michl, E.W. Thulstrup, *Spectroscopy with Polarized Light* (VCH Publishers Inc, New York, 1995)
8. R.P. Feynman, R.B. Leighton, M. Sands, *The Feynman Lectures on Physics*, Vol. III. (Addison-Wesley, Reading, 1965)
9. R.P. Feynman, A.R. Hibbs, *Quantum Mechanics and Path Integrals* (McGraw-Hill Inc, New York, 1965)
10. F. Gel'mukhanov, L.N. Mazalov, A. Kontratenko, Chem. Phys. Lett. **46**, 133 (1977)
11. J. Tulkki, T. Åberg, J. Phys. B **15**, L435 (1982)
12. F. Gel'mukhanov, H. Ågren, Phys. Rep. **312**, 87 (1999)
13. G. Ghiringhelli et al., Rev. Sci. Instrum. **77**, 113 108 (2006)
14. L.J.P. Ament, G. Ghiringhelli, M. Moretti Sala, L. Braicovich, J. van den Brink, Phys. Rev. Lett. **103**, 117 003 (2009)
15. V.N. Strocov, T. Schmitt, U. Flechsig, T. Schmidt, A. Imhof, Q. Chen, J. Raabe, R. Betemps, D. Zimoch, J. Krempasky, X. Wang, M. Grioni, A. Piazzalunga, L. Patthey, J. Synchrotron Rad. **17**, 631 (2010)
16. T. Schmitt, F.M.F. de Groot, J.E. Rubensson, J. Synchrotron Rad. **21**, 1065 (2014)

17. N.B. Brookes, G. Ghiringhelli, P. Glatzel, M. Moretti Sala, Synchr. Rad. News **31**, 26 (2018)
18. N.B. Brookes et al., Nucl. Instr. Meth. A **903**, 175 (2018)
19. J. Kim, D. Casa, A. Said, R. Krakora, B.J. Kim, E. Kasman, X. Huang, T. Gog, Sci. Rep. **8**, 1958 (2018)
20. F. Gel'mukhanov, H. Ågren, Phys. Rev. B **57**, 2780 (1998)
21. M. Magnuson, L. Yang, J.H. Guo, C. Såthe, A. Agui, J. Nordgren, Y. Luo, H. Ågren, N. Johansson, W. Salaneck, L. Horsburgh, A. Monkman, J. Elec. Spectrosc. Relat. Phenom. **101**, 573 (1999)
22. A. Nilsson, L.G.M. Pettersson, Surf. Sci. Rep. **55**, 49 (2004)
23. P. Carra, M. Fabrizio, B.T. Thole, Phys. Rev. Lett. **74**, 3700 (1995)
24. F.M.F. de Groot, M.H. Krisch, J. Vogel, Phys. Rev. B **66**, 195 112 (2002)
25. P. Glatzel, U. Bergmann, Coord. Chem. Rev. **249**, 65 (2005)
26. A. Kotani, S. Shin, Rev. Mod. Phys. **73**, 203 (2001)
27. L.J.P. Ament, M. van Veenendaal, T.P. Devereaux, J.P. Hill, J. van den Brink, Rev. Mod. Phys. **83**, 705 (2011)
28. J.J. Kas, J.J. Rehr, J.A. Soininen, P. Glatzel, Phys. Rev. B **83**, 235 114 (2011)
29. M. van Veenendaal, *Theory of Scattering and Absorption of X-Rays* (Cambridge University Press, Cambridge, 2015)
30. J.P. Rueff, A. Shukla, Rev. Mod. Phys. **82**, 847 (2010)
31. J.E. Rubensson, J. Elec. Spectrosc. Relat. Phenom. **110–111**, 135 (2000)
32. L. Kjellsson, V. Ekholm, M. Agåker, C. Såthe, A. Pietzsch, H.O. Karlsson, N. Jaouen, A. Nicolaou, M. Guarise, C. Hague, J. Lüning, S. Chiuzbăian, J.E. Rubensson, Phys. Rev. A **103**, 022 812 (2021)
33. W. Schülke, Nucl. Instrum. Methods A **280**, 338 (1989)
34. W. Schülke, *Electron Dynamics by Inelastic X-Ray Scattering* (Oxford University Press, Oxford, 2007)
35. A. Föhlisch, J. Hasselström, P. Bennich, N. Wassdahl, O. Karis, A. Nilsson, L. Triguero, M. Nyberg, L.G.M. Pettersson, Phys. Rev. B **61**, 16 229 (2000)
36. J. van den Brink, M. van Veenendaal, J. Phys. Chem. Solids **66**, 2145 (2005)
37. J. van den Brink, M. van Veenendaal, Europhys. Lett. **73**, 121 (2006)
38. D. Dieks, Phys. Lett. A **92** 271 (1982)
39. P.W. Milonni, M.L. Hardies, Phys. Lett. A **92**, 321 (1982)
40. W.K. Wootters, W.H. Zurek, Nature **299**, 802 (1982)
41. L. Mandel, Nature **304**, 188 (1983)
42. A. Lamas-Linares, C. Simon, J.C. Howell, D. Bouwmeester, Science **296**, 712 (2002)
43. F. Gel'mukhanov, T.P.P. Sałek, H. Ågren, Phys. Rev. A **59**, 380 (1999)
44. R. Feifel et al., Phys. Rev. A **69**, 022 707 (2004)
45. J. Luo, G.T. Trammell, J.P. Hannon, Phys. Rev. Lett. **71**, 287 (1993)
46. J.P. Hannon, G.T. Trammell, Hyperfine Int. **123/124**, 127 (1999)
47. L. Kjellsson, *X-ray Spectroscopy on Diatomic and Cationic Molecules*. Ph.D. Thesis, Department of Physics and Astronomy, Uppsala University (2021)
48. K. Prince et al., J. Synchrotron Rad. **5**, 565 (1998)
49. M. Born, J.R. Oppenheimer, Ann. Phys. **389**, 457 (1927)
50. K. Prince, M. Vondráček, J. Karvonen, M. Coreno, R. Camilloni, L. Avaldi, M. de Simone, J. Elec. Spectrosc. Relat. Phenom. **101–103**, 141 (1999)
51. D. Gibbs, D. Harshman, E. Isaacs, D.B. McWhan, D. Mills, C. Vettier, Phys. Rev. Lett. **61**, 1241 (1988)
52. C.C. Kao, J.B. Hastings, E.D. Johnson, D.P. Siddons, G.C. Smith, G.A. Prinz, Phys. Rev. Lett. **65**, 373 (1990)
53. C.C. Kao, C.T. Chen, E.D. Johnson, J.B. Hastings, H.J.L. Nad, G.H. Ho, G. Meigs, J.M. Brot, S.L. Hulpert, Y.U. Ydzerda, C. Vettier, Phys. Rev. B **50**, 9599 (1994)
54. L. Sève, N. Jaouen, J.M. Tonnerre, D. Raoux, F. Bartolomé et al., Phys. Rev. B **60**, 9662 (1999)
55. J.P. Hannon, G.T. Trammell, Phys. Rev. **169**, 315 (1968)

56. M. Blume, J. Appl. Phys. **57**, 3615 (1985)
57. J.P. Hannon, G.T. Trammell, M. Blume, D. Gibbs, Phys. Rev. Lett. **61**, 1245 (1988)
58. M. Blume, D. Gibbs, Phys. Rev. B **37**, 1779 (1988)
59. M. Altarelli, Resonant x-ray scattering: a theoretical introduction. in *Magnetism: A Synchrotron Radiation Approach, Springer Lecture Notes in Physics*, Vol. 697. ed. by E. Beaurepaire, H. Bulou, F. Scheurer, J.P. Kappler (Springer, Berlin, 2006), p. 201
60. M.W. Haverkort, Phys. Rev. Lett. **105**, 167 404 (2010)
61. P. Eisenberger, P.M. Platzman, H. Winick, Phys. Rev. Lett. **36**, 623 (1976)
62. J.P. Briand, D. Girard, V.O. Kostroun, P. Chevalier, K. Wohrer, J.P. Mossé, Phys. Rev. Lett. **41**, 1625 (1981)
63. F. Hennies et al., Phys. Rev. Lett. **104**, 193 002 (2010)
64. M. Neeb, J.E. Rubensson, M. Biermann, W. Eberhardt, J. Elec. Spectrosc. Relat. Phenom. **67**, 261 (1994)
65. P. Glans, K. Gunnelin, P. Skytt, J.H. Guo, N. Wassdahl, J. Nordgren, H. Ågren, F.K. Gel'mukhanov, T. Warwick, E. Rotenberg, Phys. Rev. Lett. **76**, 2448 (1996)
66. Y.P. Sun et al., Phys. Rev. Lett. **110**, 223 001 (2013)
67. Y. Tokura, Phys. Today **56**, 50 (2003)
68. A.P. Ramirez, J. Phys. Condens. Matter **9**, 8171 (1997)
69. G. Frenking, N. Fröhlich, Chem. Rev. **100**, 717 (2000)
70. M. Moretti Sala et al., New J. Phys. **13**, 043 026 (2011)
71. C.T. Chen, L.H. Tjeng, J. Kwo, H.L. Kao, P. Rudolf, F. Sette, R.M. Fleming, Phys. Rev. Lett. **68**, 2543 (1992)
72. S. Stepanow, A. Mugarza, G. Ceballos, P. Moras, J.C. Cezar, C. Carbone, P. Gambardella, Phys. Rev. B **82**, 014 405 (2010)
73. S. Stepanow, A.L. Rizzini, C. Krull, J. Kavich, J.C. Cezar, F. Yakhou-Harris, P.M. Sheverdyaeva, P. Moras, C. Carbone, G. Ceballos, A. Mugarza, P. Gambardella, J. Am. Chem. Soc. **136**, 5451 (2014)
74. J. Stöhr, H.C. Siegmann, *Magnetism: From Fundamentals to Nanoscale Dynamics* (Springer, Heidelberg, 2006)
75. F.M.F. de Groot, P. Kuiper, G.A. Sawatzky, Phys. Rev. B **57**, 14 584 (1998)
76. M. van Veenendaal, Phys. Rev. Lett. **96**, 117 404 (2006)
77. L. Braicovich, J. van den Brink, V. Bisogni, M. Moretti Sala, L.J.P. Ament, N.B. Brookes, G.M.D. Luca, M. Salluzzo, T. Schmitt, V.N. Strocov, G. Ghiringhelli, Phys. Rev. Lett. **104**, 077 002 (2010)
78. L.C. Duda, J. Stöhr, D.C. Mancini, A. Nilsson, N. Wassdahl, J. Nordgren, M.G. Samant, Phys. Rev. B **50**, 16 758 (1994)
79. A. Nilsson, J. Stöhr, T. Wiell, M. Aldén, P. Bennich, N. Wassdahl, M. Samant, S.S.P. Parkin, N. Mårtensson, J. Nordgren, B. Johansson, H.L. Skriver, Phys. Rev. B **54**, 2917 (1996)
80. M. Magnuson, N. Wassdahl, J. Nordgren, Phys. Rev. B **56**, 12 238 (1997)
81. M. Magnuson, J.E. Rubensson, A. Föhlisch, N. Wassdahl, A. Nilsson, N. Mårtensson, Phys. Rev. B **68**, 045 119 (2003)
82. H. Ebert, J. Stöhr, S.S.P. Parkin, M. Samant, A. Nilsson, Phys. Rev. B **53**, 16 067 (1996)
83. O. Karis, A. Nilsson, M. Weinelt, T. Wiell, C. Puglia, N. Wassdahl, N. Mårtensson, M. Samant, J. Stöhr, Phys. Rev. Lett. **76**, 1380 (1996)
84. D.J. Higley, Z. Chen, M. Beye, M. Hantschmann, A.H. Reid, V. Mehta, O. Hellwig, G.L. Dakovski et al., Commun. Phys. **5**, 83 (2022)
85. R. Qiao, Q. Li, Z. Zhuo, S. Sallis, O. Fuchs, M. Blum, L.Weinhardt, C. Heske, J. Pepper, M. Jones et al., Rev. Sci. Instrum. **88**, 033 106 (2017)
86. A. Nilsson, L.G.M. Pettersson, B. Hammer, T. Bligaard, C.H. Christenen, J.K. Nørskov, Catal. Lett. **100**, 111 (2005)
87. F. Zaera, Prog. Surf. Sci. **69**, 1 (2001)
88. J.E. Rubensson et al., Phys. Rev. Lett. **114**, 133 001 (2015)
89. J. Stöhr, R. Jaeger, Phys. Rev. B **26**, 4111 (1982)
90. A. Nilsson, N. Mårtensson, Phys. B **208/209**, 19 (1995)

91. A. Nilsson et al., Appl. Phys. A **65**, 147 (1997)
92. A. Nilsson, M. Weinelt, T. Wiell, P. Bennich, O. Karis, N. Wassdahl, J. Stöhr, M.G. Samant, Phys. Rev. Lett. **78**, 2847 (1997)
93. M. Weinelt, N. Wassdahl, T. Wiell, O. Karis, J. Hasselström, P. Bennich, A. Nilsson, J. Stöhr, M. Samant, Phys. Rev. B **58**, 7351 (1998)
94. L. Triguero, A. Föhlisch, P. Väterlein, J. Hasselström, M. Weinelt, L.G.M. Pettersson, Y. Luo, H. Ågren, A. Nilsson, J. Am. Chem. Soc. **122**, 12 310 (2000)
95. J. Hasselström, O. Karis, M. Nyberg, L.G.M. Pettersson, M. Weinelt, N. Wassdahl, A. Nilsson, J. Phys. Chem. B **104**, 11 480 (2000)
96. M. Nyberg, J. Hasselström, O. Karis, N. Wassdahl, M. Weinelt, A. Nilsson, L.G.M. Pettersson, J. Chem. Phys. **112**, 5420 (2000)
97. J. Hasselström, O. Karis, M. Weinelt, N. Wassdahl, A. Nilsson, M.G. Samant, J. Stöhr, Surf. Sci. **407**, 221 (1998)
98. L. Braicovich, M. Minola, G. Dellea, M.L. Tacon, M.M. Sala, C. Morawe, J.C. Peffen, R. Supruangnet, F. Yakhou, G. Ghiringhelli, N.B. Brookes, Rev. Sci. Instrum. **85**, 115 104 (2014)
99. Z. Hussain, E. Umbach, D.A. Shirley, J. Stöhr, J. Feldhaus, Nucl. Instrum. Methods **195**, 115 (1982)
100. S.H. Southworth, Nucl. Instrum. Methods **87**, 247 (1994)
101. D.W. Lindle, P.L. Cowan, R.E. LaVilla, T. Jach, R.D. Deslattes, B. Karlin, J.A. Sheehy, T.J. Gil, P.W. Langhoff, Phys. Rev. Lett. **60**, 1010 (1988)
102. D.W. Lindle, P.L. Cowan, T. Jach, R.E. LaVilla, R.D. Deslattes, R.C.C. Perera, Phys. Rev. A **43**, 2353 (1991)
103. F. Gel'mukhanov, H. Ågren, Phys. Rev. A **49**, 4378 (1994)
104. P. Glatzel, M. Sikora, M. Fernández-García, Eur. Phys. J. Spec. Top. **169**, 207 (2009)
105. J. Stöhr, *NEXAFS Spectroscopy* (Springer, Heidelberg, 1992)
106. K. Hämäläinen, D.P. Siddons, J.B. Hastings, L.E. Berman, Phys. Rev. Lett. **67**, 2850 (1991)
107. D. Friebel, H. Ogasawara, A. Nilsson, Operando x-ray and electron spectroscopy in oxygen electrocatalysis. in *Surface and Interface Science, Vol. 8: Interfacial Electrochemistry*, ed. by K. Wandelt (Wiley, New York, 2020)
108. B. Qi, I. Perez, P.H. Ansari, F. Lu, M. Croft, Phys. Rev. B **36**, 2972 (1987)
109. A.C. Thompson et al., *X-Ray Data Booklet*, (3rd edn, LBNL, Berkeley, 2009). Available at http://xdb.lbl.gov
110. J.L. Campbell, T. Papp, At. Data Nucl. Data Tables **77**(1), 1 (2001)
111. P.W. Loeffen, R.F. Pettifer, S. Müllender, M.A. van Veenendaal, J. Röhler, D.S. Sivia, Phys. Rev. B **54**, 14 877 (1996)

Part IV
Multi-photon Interaction Processes

Chapter 14
Resonant Non-linear X-Ray Processes in Atoms

14.1 Introduction

The *Kramers-Heisenberg-Dirac* (KHD) formalism has served the x-ray community well for the first 100+ years of x-ray science, and we shall see that it is valid even when the incident x-ray intensity is raised well above the level available at the brightest synchrotron radiation sources. The goal of this chapter is to explore when the KHD perturbation approach ceases to be a good description, and how the basic processes we refer to as x-ray absorption, x-ray scattering, and x-ray diffraction change in this largely unchartered x-ray territory that has been opened by the advent of XFELs.

The development of a "strong-field" theory goes back to a seminal paper by I. I. Rabi in 1937 [1] where he considered the manipulation of nuclear spins by means of the magnetic field component of microwaves. Because of the low frequency of microwaves, the spins in a sample can still follow the magnetic field component of the EM wave. Also the long wavelength of microwaves, ranging from centimeters to meters, rendered the radiation naturally coherent so that its interaction with matter could be described by assuming *classical EM waves* of variable field strength.

It was the development of powerful microwave sources during the Second World War for RADAR technology that furthermore paved the way for the development of nuclear magnetic resonance (NMR) in the 1940s. A key step in the development of NMR was the *time-dependent* description of the motion of the nuclear spin system in a strong EM field by Felix Bloch in 1946 [2], now known as the *Bloch equations*. In honor of the pioneering contributions of I. I. Rabi and F. Bloch to the development of the "strong-field" approach presented below, we shall refer to it as the *Bloch-Rabi* (BR) model. In optics, the theory is also referred to as the *Maxwell-Bloch* theory indicating its classical origin [3].

The powerful time-dependent BR model goes beyond the time-independent KHD perturbation treatment. At optical and higher frequencies the direct interaction of the magnetic fields of EM waves and the spins in a sample, treated by the original Bloch formalism, is negligibly small because the spins can no longer follow the high-frequency fields (see Sects. 7.3.3 and 7.3.4). Instead it is now the interaction of the

J. Stöhr, *The Nature of X-Rays and Their Interactions with Matter*, Springer Tracts in Modern Physics 288, https://doi.org/10.1007/978-3-031-20744-0_14

electric field of the EM waves and the electronic charge in a sample that dominates. In analogy to the two-state ("up" and "down") spin system, the electronic response of the sample can be cast into that of a two-state electronic system, as pointed out in 1957 by Feynman et al. [4]. With the invention of the laser in the 1960s, the BR formalism was readily extended, and the original spin-based Bloch equations were replaced by the "optical Bloch equations", resulting in the birth of the entire new field of non-linear optics.

Although the BR formalism has been extensively utilized in laser science, there was no need for x-ray scientists to use this formalism before the advent of XFELs since a simpler approach, the KHD perturbation theory, was available. The latter theory naturally lent itself to picturing x-ray interaction processes in a "photon" or particle picture, where a photon was lost in a sample through "absorption" or changed its direction through "scattering" or in a wave picture gave rise to "diffraction" phenomena. The need for a treatment beyond the KHD perturbation approach did not arise until the advent of x-ray free electron lasers in the new millennium.

In this chapter we will discuss the BR theory in the modern language of the density matrix formalism for the case of a single atom. The atom is treated as a simple two-level system, with electronic excitations of core electrons into valence holes, and de-excitations of valence electrons into core holes. The simple atomic case treated here nicely shows how the BR theory seamlessly evolves into the KHD theory at low x-ray intensities. The central goal of this chapter is to derive the equations governing the interplay of the populations of the lower and upper states of the resonantly driven two-level atom.

The optical Bloch equations can only be solved analytically in certain cases, and these will be discussed and the solutions elucidated by use of parameters that are characteristic for the resonant response near x-ray absorption edges. We will also compare the time-dependent BR theory to the time-independent KHD theory, emphasizing the connection between the temporal properties of the incident radiation that enters into the BR theory and the energy bandwidth that enters into the KHD theory. Another important aspect is the relationship between stimulated forward scattering and absorption, which will be shown to satisfy a sum rule.

We will also touch upon the famous Einstein rate equations, developed already in 1916/17 [5, 6], about 10 years before the development of quantum mechanics. As discussed earlier, Einstein's theory for the first time distinguished spontaneous and stimulated emission, which was required to balance the energy in Planck's famous blackbody radiation law formulated in 1900. We shall compare the results of the Einstein rate equation model, which assumes broad bandwidth radiation to the BR theory which specifically assumes narrow bandwidth radiation that is resonantly tuned.

At the end of this chapter we will discuss additional fundamental aspects of the two-level atom which in the optical region is referred to as "resonance fluorescence" [7–10]. This topic considers the spontaneous fluorescence response of a two-level atom in the strong-field limit, measured at right angles to the incident beam and its linear polarization vector. We will see that the resonant florescence spectrum directly exhibits an energy splitting whose quantum electrodynamics treatment is of fundamental importance because it can be compared to the semi-classical Bloch-Rabi treatment. Resonance fluorescence has also played a role in identifying so-called *antibunching* behavior of photons [11, 12], in contrast to the bunching behavior discussed in Sect. 5.2.2.

The atomic treatment presented here will be extended in Chap. 15 to solid samples, where the two-level atom is comprised of a collective multi-atom system.

14.2 X-Ray Induced Atomic Core to Valence Transitions

While the description of the temporal behavior of a two-level electronic system at *optical* frequencies can be found in many textbooks, e.g. in [7, 9, 13, 14], we will consider the *x-ray* case where the electronic system is described by two levels in an atom, as previously illustrated for the KHD case in Fig. 10.3. Up and down transitions in such a "two-level" atom are induced by the electric component of the EM field or in a quantum picture by loss and generation of photons, and we consider the transitions and populations of a two-level atom illustrated in Fig. 14.1.

In contrast to the KHD case, where the populations of the lower and upper states are assumed to be binary, either 0 or 1, our goal here is to formulate the transition probabilities between the two states in terms of time-dependent changes of the population distributions $\rho_{11}(t)$ and $\rho_{22}(t)$, which through the normalization condition $\rho_{11} + \rho_{22} = 1$ are temporally out of phase. In the x-ray case, the lower state is a core level and the upper state a valence level. We describe the core-to-valence *dipolar* transition width through a polarization average, which we define as equivalent "up"

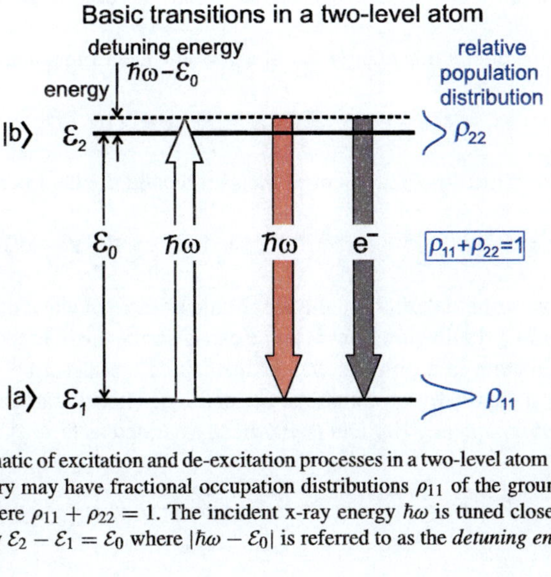

Fig. 14.1 Schematic of excitation and de-excitation processes in a two-level atom which in contrast to the KHD theory may have fractional occupation distributions ρ_{11} of the ground and ρ_{22} of the excited state, where $\rho_{11} + \rho_{22} = 1$. The incident x-ray energy $\hbar\omega$ is tuned close to the two-level resonance energy $\mathcal{E}_2 - \mathcal{E}_1 = \mathcal{E}_0$ where $|\hbar\omega - \mathcal{E}_0|$ is referred to as the *detuning energy*

and "down" transitions in accordance with (10.40) and (12.18)

$$\boxed{\Gamma_x = \frac{8\pi^2\alpha_f\hbar\omega}{\lambda^2}|\langle b|\mathbf{r}\cdot\boldsymbol{\epsilon}_p|a\rangle|^2 = \frac{8\pi^2\alpha_f\hbar\omega}{\lambda^2}|\langle a|\mathbf{r}\cdot\boldsymbol{\epsilon}_p|b\rangle|^2.} \tag{14.1}$$

For our two-level atom the transition width Γ_x is the same as the previously introduced phenomenological decay width Γ^X shown in Figs. 10.3 and 12.10.

In practice, the dipolar transition with Γ_x may be obtained from the peak value of the measured (spontaneous) absorption cross section σ_{XAS} linked by our previous expression (10.39) as

$$\Gamma_x = \frac{\pi\Gamma}{\lambda^2}\sigma_{XAS}. \tag{14.2}$$

14.2.1 *Interaction Energy and Hamiltonian: The Rabi Frequency*

Owing to the fact that the description of transitions between two spin or electronic states has a long history, the "two-level" case has been treated in great detail both classically and quantum mechanically [3, 15, 16]. The full quantum optics treatment of the coupling of a two-level atom to a single field mode is today known as the Jaynes-Cummings model [14, 16]. Before we present the treatment of the two-level case by the semi-classical optical Bloch equations, we need to briefly review the link of the classical field and quantum mechanical photon descriptions given in Chap. 3. This leads to a definition of the important Rabi frequency used in the BR formalism.

Classically, the interaction energy $\mathcal{E}_{int}$ of a periodic electromagnetic wave with an atom is given by

$$\mathcal{E}_{int} = -e\,\mathbf{r}\cdot\mathbf{E}(\mathbf{r},t) = \mathcal{D}\cdot\mathbf{E}(\mathbf{r},t), \tag{14.3}$$

where $\mathcal{D} = -e\mathbf{r}$ is the dipole moment of the sample and the field is given by (3.13) or

$$\mathbf{E}(\mathbf{r},t) = \mathbf{E}_0\left[e^{i(\mathbf{k}\cdot\mathbf{r}-\omega t)} + e^{-i(\mathbf{k}\cdot\mathbf{r}-\omega t)}\right] = 2\,\mathbf{E}_0\,\cos(\mathbf{k}\cdot\mathbf{r}-\omega t). \tag{14.4}$$

We note that for our field definition, the amplitude of the real wave field on the right side is given by $2E_0$. Following Rabi [1], one can describe the interaction of the field with a two-level atom in a time-dependent formalism that describes the frequency of the changes in the relative populations ρ_{11} and ρ_{22}. This characteristic frequency is called the *Rabi frequency* $\mathcal{W}_R$ and is classically defined as [7, 14]

$$\mathcal{W}_R = \frac{\mathcal{D}\cdot\mathbf{E}(\mathbf{r},t)}{2\hbar} = \frac{\mathcal{D}\cdot\mathbf{E}_0\,\cos(\mathbf{k}\cdot\mathbf{r}-\omega t)}{\hbar}. \tag{14.5}$$

Quantum mechanically, one uses an interaction Hamiltonian that is equivalent to the classical expression

$$\mathcal{H}_{\text{int}} = -e\,\mathbf{r}\cdot\mathbf{E}(\mathbf{r},t) = \mathcal{D}\cdot\mathbf{E}(\mathbf{r},t), \tag{14.6}$$

where the operator $\mathbf{E}(\mathbf{r}, t)$ is expressed by (3.61) and (3.68) as

$$\mathbf{E}(\mathbf{r},t) = \mathbf{E}_0\left[\boldsymbol{a}\,e^{\mathrm{i}(\mathbf{k}\cdot\mathbf{r}-\omega t)} + \boldsymbol{a}^\dagger\,e^{-\mathrm{i}(\mathbf{k}\cdot\mathbf{r}-\omega t)}\right] = \mathbf{E}^+(\mathbf{r},t) + \mathbf{E}^-(\mathbf{r},t). \tag{14.7}$$

Here the operators $\boldsymbol{a}$ and $\mathbf{E}^+(\mathbf{r}, t)$ destroy photons and $\boldsymbol{a}^\dagger$ and $\mathbf{E}^-(\mathbf{r}, t)$ create photons and

$$E_0 = \sqrt{\frac{\hbar\omega_{\mathbf{k}}}{2\epsilon_0 V}}. \tag{14.8}$$

The Rabi frequency may also be defined quantum mechanically within the *time-independent* KHD perturbation theory. In this formalism the light-atom interaction is described by (13.1), which may be written in the dipolar form

$$\mathcal{W}_{i\to f} = \frac{2\pi}{\hbar}\left|\sum_m \frac{\langle f|\mathcal{D}\cdot\mathbf{E}^-|m\rangle\langle m|\mathcal{D}\cdot\mathbf{E}^+|i\rangle}{\mathcal{E}_i - \mathcal{E}_m}\right|^2 \rho(\mathcal{E}_f)\,\delta(\mathcal{E}_f - \mathcal{E}_i), \tag{14.9}$$

In the evaluation of (14.9) for a two-level atom, one makes the rotating wave approximation as discussed in Sect. 13.2. This corresponds to use of only half of $\mathcal{W}_{i\to f}$, so that the quantum definition of the Rabi frequency in the KHD formalism is given by [7][1]

$$\boxed{\text{Quantum Rabi frequency:}\;\; \mathcal{W}_{\mathrm{R}} = 2\,\frac{\langle\,\mathcal{H}_{\text{int}}\,\rangle}{\hbar}.} \tag{14.10}$$

The importance of the Rabi frequency is that it is a fundamental building block in the optical Bloch equations which we will discuss below in Sect. 14.3.

14.2.2 *The Rabi Frequency in the X-Ray Regime*

In the x-ray region, it is convenient to express the *Rabi frequency* by $\mathcal{W}_{\mathrm{R}} = \Gamma_{\mathrm{R}}/\hbar$. The energy width Γ_{R}, which we shall refer to as the *Rabi energy*, is then given by the quantum mechanical expectation value of the interaction Hamiltonian according to (14.10) or

$$\boxed{\Gamma_{\mathrm{R}} = \hbar\mathcal{W}_{\mathrm{R}} = 2\langle\,\mathcal{H}_{\text{int}}\,\rangle.} \tag{14.11}$$

[1] We follow Loudon's notation of the optical Bloch equations and define the Rabi frequency (14.10) consistently with Loudon's derivation in Sect. 2.4 yielding his (2.4.14) which leads to his (4.10.11).

Its value depends on the incident intensity, and we shall see later in Sect. 14.11 that the situation where it becomes comparable with the total excited state decay width, i.e. $\Gamma_R \sim \Gamma$, is particularly interesting.

The expectation value of the Hamiltonian $\langle \mathcal{H}_{\text{int}} \rangle$ which describes the interaction of an electric field with a two-level atomic system, illustrated in Fig. 14.1, is evaluated as $\langle \mathcal{H}_{\text{int}} \rangle = \langle f| - e\mathbf{r} \cdot \mathbf{E}|i\rangle$, where $|i\rangle$ and $|f\rangle$ consist of products of photon and electron states as discussed in Sect. 10.2 for the resonant absorption case. We can therefore adopt the evaluation of the squared matrix element in that section and obtain in analogy to (10.3)

$$|\langle \mathcal{H}_{\text{int}} \rangle|^2 = |\langle f|e\mathbf{r} \cdot \mathbf{E}|i\rangle|^2 = \underbrace{\frac{n_{p\mathbf{k}}c}{V_{p\mathbf{k}}} 2\pi\hbar^2\omega}_{\text{photon part}} \underbrace{\alpha_f}_{\text{coupling}} \underbrace{|\langle b|\mathbf{r} \cdot \boldsymbol{\epsilon}|a\rangle|^2}_{\text{electron part}}. \tag{14.12}$$

This gives the simple expression

$$|\langle \mathcal{H}_{\text{int}} \rangle|^2 = \underbrace{\frac{n_{p\mathbf{k}}c}{V_{p\mathbf{k}}}}_{\Phi_0} \frac{\lambda^2\hbar}{4\pi} \Gamma_x. \tag{14.13}$$

It contains the photon flux Φ_0 linked by (10.9) to the incident number of photons per mode, $n_{p\mathbf{k}}$, and to the incident intensity I_0 as

$$\Phi_0 = \frac{n_{p\mathbf{k}}c}{V_{p\mathbf{k}}} = \frac{n_{p\mathbf{k}}c}{\lambda^3} \frac{\Delta_{p\mathbf{k}}}{\hbar\omega} = \frac{I_0}{\hbar\omega}, \tag{14.14}$$

where $V_{p\mathbf{k}} = \lambda^3\,\hbar\omega/\Delta_{p\mathbf{k}}$ according to (3.106) which reduces to $V_{p\mathbf{k}} = \lambda^3$ for the case of unit bandwidth. The photon intensity I_0 has the dimension [energy/(area$\times$ time].

The polarization averaged Rabi energy is obtained as[2]

$$\boxed{\Gamma_R^2 = 4\,|\langle \mathcal{H}_{\text{int}} \rangle|^2 = \frac{n_{p\mathbf{k}}c}{V_{p\mathbf{k}}} \frac{\lambda^2\hbar}{\pi} \Gamma_x.} \tag{14.15}$$

We obtain the following key relations between the key parameters entering into the KHD and BR descriptions of REXS.

[2] This is consistent with Loudon's (4.10.11) [7].

Coherent elastic scattering in the two-level scheme in Fig. 14.1 is defined by the incident radiation in terms of the number of incident photons n_{pk} in the energy bandwidth Δ_{pk} or the associated intensity I_0 and the atomic dipole transition width Γ_x. These quantities define the Rabi energy Γ_{R} according to

$$\Gamma_{\mathrm{R}}^2 = I_0 \frac{\lambda^3}{2\pi^2 c} \Gamma_x = n_{pk} \frac{\Delta_{pk}}{2\pi^2} \Gamma_x, \tag{14.16}$$

where $\Gamma_x/\hbar$ is the dipolar transition rate given by (14.1). This implies

$$n_{pk} = I_0 \frac{\lambda^3}{c\Delta_{pk}}. \tag{14.17}$$

In the *time-dependent* BR theory the two-level response is determined by the Rabi rate $\Gamma_{\mathrm{R}}/\hbar$ which describes the buildup of the upper state population. This buildup is neglected in the KHD theory where the occupations of the lower and upper states are binary, either 0 or 1.

In practice, the size of the non-linear electronic processes in Fig. 14.1 may be judged by the size of the *spontaneous* XAS cross section (see (10.55)) since it determines the initiation of the electronic processes. In the following will specifically denote it by a $\diamond$ superscript as

$$\sigma_{\mathrm{XAS}}^{\diamond} = \frac{\lambda^2}{\pi} \frac{\Gamma_x}{\Gamma} \frac{(\Gamma/2)^2}{(\hbar\omega - \mathcal{E}_0)^2 + (\Gamma/2)^2}, \tag{14.18}$$

where Γ_x is given by (14.1).

14.3 The Optical Bloch Equations

We consider the transitions between two levels illustrated in Fig. 14.1 based on the following assumptions.

- The two-level system consists of two electronic levels with energies $\mathcal{E}_1$ and $\mathcal{E}_2$, separated by an energy $\mathcal{E}_2 - \mathcal{E}_1 = \mathcal{E}_0$. After excitation the excited state decays with a natural rate $\Gamma/\hbar$, where Γ is FWHM of the Lorentzian shaped energy distribution of the emitted radiation.
- The energy bandwidth Δ of the incident radiation may vary between the highly monochromatic limit $\Delta \ll \Gamma$ and the broadband limit $\Delta = \hbar\omega$. In the monochromatic limit one can describe the resonant response as a function of "detuning" energy $\hbar\omega - \mathcal{E}_0$ from exact resonance excitation.

14.3.1 *Time-Dependent Transitions in a Two-Level System: Density Matrix Formulation*

We follow Fig. 14.1 and denote the fractional *populations* of the ground and excited states by real dimensionless quantities ρ_{11} and ρ_{22}, which are linked according to $\boxed{\rho_{22} = 1 - \rho_{11}}$. We also allow coupling between the two states, expressed as the complex electronic *coherences* ρ_{12} and ρ_{21}. They can be envisioned as the dynamic dipole moments induced by the oscillating electric field according to (14.3). The elements ρ_{ij} form a 2×2 *atomic density matrix* given by

$$\boldsymbol{\rho} = \begin{pmatrix} \rho_{11} & \rho_{12} \\ \rho_{21} & \rho_{22} \end{pmatrix}. \tag{14.19}$$

We have

$$\mathrm{Tr}\boldsymbol{\rho} = \rho_{11} + \rho_{22} = 1 \quad \text{and} \quad \rho_{12} = \rho_{21}^* \tag{14.20}$$

and

$$\mathrm{Tr}(\rho)^2 \begin{cases} = 1 \text{ for pure states} \\ < 1 \text{ for mixed states} \end{cases}. \tag{14.21}$$

The last expression reflects the fact that the diagonal matrix elements are 0 or 1 for pure states but lie between 0 and 1 for mixed states.

Let us consider a two-state electronic system described in the absence of an electromagnetic interaction by wavefunctions ψ_1 and ψ_2 of the two electronic states of energies $\mathcal{E}_1$ and $\mathcal{E}_2$. In the presence of an interaction with a periodic electric field, the wavefunction of the system may be expressed as

$$\Psi(\mathbf{r}, t) = a(t)\,\psi_1(\mathbf{r}, t) + b(t)\,\psi_2(\mathbf{r}, t). \tag{14.22}$$

Then the quantities

$$\rho_{11}(t) = |a(t)|^2 \quad \text{and} \quad \rho_{22}(t) = |b(t)|^2 \tag{14.23}$$

are just the probabilities of finding the system in the two states, respectively, and the coherences represent the interferences of the two states given by

$$\rho_{12}(t) = a(t)\,b^*(t) \quad \text{and} \quad \rho_{21}(t) = b(t)\,a^*(t) = \rho_{12}^*(t). \tag{14.24}$$

We have

$$\begin{pmatrix} \rho_{11} & \rho_{12} \\ \rho_{21} & \rho_{22} \end{pmatrix} = \begin{pmatrix} |a|^2 & ab^* \\ ba^* & |b|^2 \end{pmatrix}. \tag{14.25}$$

Using the density matrix approach one can derive the important *optical Bloch equations* for a two-level-system by describing the time evolution of its density operator $\rho(t)$ in the Schrödinger picture

$$\rho(t) = \mathrm{e}^{-\mathrm{i}\mathcal{H}_{\mathrm{int}}t/\hbar}\rho(0)\,\mathrm{e}^{\mathrm{i}\mathcal{H}_{\mathrm{int}}t/\hbar}, \tag{14.26}$$

where $\mathcal{H}_{\mathrm{int}}$ is *time independent*. The time evolution of the system is then entirely described by the density matrix according to the time-dependent Liouville-von Neumann equation

$$\frac{\mathrm{d}\rho(t)}{\mathrm{d}t} = -\frac{\mathrm{i}}{\hbar}\,[\mathcal{H}_{\mathrm{int}}, \rho(t)]\,. \tag{14.27}$$

The *Rabi frequency* enters through the relation (14.10) or

$$\langle\mathcal{H}_{\mathrm{int}}\rangle = \frac{\hbar\,\mathcal{W}_{\mathrm{R}}}{2} \tag{14.28}$$

The equations describe the time evolution of the coupled system. After making the *rotating wave approximation* which neglects terms of frequency much higher than the resonant frequency of the two-level system (see Sect. 13.2) and eliminating a phase factor by going to a rotating frame, we can state the Bloch equations as follows [7].

The *optical Bloch equations* describe the population changes in an upper electronic state, ρ_{22}, and lower state, ρ_{11}, separated by an energy $\mathcal{E}_0$ in response to an incident field of frequency ω under the condition $\rho_{22} = 1 - \rho_{11}$. Also included are interference contributions ρ_{12} and ρ_{21} between the states defined by (14.24). The incident photon energy $\hbar\omega$ may be slightly detuned by $|\hbar\omega - \mathcal{E}_0|$ from exact resonance $\hbar\omega_0 = \mathcal{E}_0$. The time-dependent process depends on the Rabi frequency $\mathcal{W}_{\mathrm{R}}$, defined through the interaction Hamiltonian (14.28) and the total excited state spontaneous decay rate, $\Gamma/\hbar$, which damps the oscillations of the populations. One may also include an additional dephasing rate $\Gamma_{\mathrm{deph}}/\hbar$, which enters as $\Gamma' = \Gamma + \Gamma_{\mathrm{deph}}$ into the off-diagonal coherences. The coupled Bloch equations then read [7]

$$\frac{\mathrm{d}\rho_{22}}{\mathrm{d}t} = -\frac{\mathrm{d}\rho_{11}}{\mathrm{d}t} = -\mathrm{i}\frac{\mathcal{W}_{\mathrm{R}}}{2}\,(\rho_{12} - \rho_{21}) - \frac{\Gamma}{\hbar}\rho_{22} \tag{14.29}$$

and

$$\frac{\mathrm{d}\rho_{12}}{\mathrm{d}t} = \frac{\mathrm{d}\rho_{21}^*}{\mathrm{d}t} = \mathrm{i}\frac{\mathcal{W}_{\mathrm{R}}}{2}\,(\rho_{11} - \rho_{22}) + \left[\mathrm{i}(\omega_0 - \omega) - \frac{\Gamma'}{2\hbar}\right]\rho_{12}\,. \tag{14.30}$$

Note that the rates $\Gamma/\hbar$ and $\Gamma'/\hbar$ enter differently by a factor of 2, which is discussed below.

The first equation (14.29) expresses the time dependence of the *populations* of the ground and excited states which change with the Rabi frequency $\mathcal{W}_{\mathrm{R}}$ and spontaneous "down" transitions with a rate $\Gamma/\hbar$ that damp the oscillations through energy loss from the field. For resonant core-valence transitions in the x-ray regime, Γ is the total spontaneous decay rate of the excited atom in the absence of any external stimulus. Hence $\Gamma = \Gamma^{X} + \Gamma^{A}$ is the sum of the spontaneous radiative decay rate and the spontaneous Auger decay rate as discussed in Sect. 12.4.

The last term $\Gamma\rho_{22}/\hbar$ in (14.29) causes a spontaneous decay of the population of the excited state. It may be thought of as a two-step process. First, in the excitation process energy is transferred from the field to the electronic system. However, the energy is not transferred back to the field but rather to a spontaneously emitted photon that is emitted into a random direction (hence mostly out of the beam) or to Auger electrons, with the Auger decay being 100 times more probable at soft x-ray energies. For a system of N_a atoms, the populations are given by the fraction of atoms that are in the ground and excited states, respectively

$$\rho_{11} = \frac{N_1}{N_a} \quad \text{and} \quad \rho_{22} = \frac{N_2}{N_a}. \tag{14.31}$$

The second Bloch equation (14.30) expresses the time evolution of the coherences given by (14.24) which are products of the complex coefficients of the ground and excited state *wavefunctions*. From (14.30) we see that the wavefunction interference depends on the population difference driven by the Rabi rate, the energy mismatch determined by $\hbar\omega - \hbar\omega_0$, and a total damping constant $\Gamma' = \Gamma + \Gamma_{deph}$ which includes the possible presence of a pure dephasing contribution Γ_{deph}, whose origin is discussed below.

14.3.2 Damping Constants: Longitudinal Versus Transverse Relaxation

Dating back to the original use of the Bloch equations for the description of NMR, one often distinguishes two kinds of processes by the introduction of Γ and Γ' in (14.29) and (14.30). In NMR, the two damping constants have specific names.

In general, a spin may be oriented along any direction in space. If a static magnetic field is applied as in NMR, say along z, a spin will precess about the quantization axis z. The spin wavefunction for an arbitrary spin orientation is then described by a coherent superposition of the binary spin-up and spin-down basis states, where the coefficients of the two basis states are complex. Their magnitude reflects the tilt angle from the z field direction, and their orientation in the (x, y) plane perpendicular to the field is described by the phases of the complex coefficients [17]. One then distinguishes changes of the spin system along z, i.e. spin flips, and refers to this as a *longitudinal* process. In contrast, any change of the spin orientation in the (x, y) plane is referred to as a *transverse* process. Changes in the spin system are driven by an externally applied microwave field in the (x, y) plane, with the *magnetic* part of the EM field acting on the *spin*.

Since in NMR the lower state is a pure spin-down basis state and the upper a pure spin-up basis state, a longitudinal process is a spin-flip process that changes the up and down *populations*. The characteristic time associated with population changes is called the *population* or *longitudinal* relaxation time (or *spin-flip* time in NMR), $T_1 = \hbar/\Gamma$. In contrast, any change in the orientation of the spin projection in the (x, y) plane is just a change in their phase angle in that plane and is called the *transverse* relaxation or *dephasing* time (or *spin dephasing* time in NMR), $T_2 = 2\hbar/\Gamma_{\mathrm{deph}}$. We have

$$\frac{\Gamma}{\hbar} = \frac{1}{T_1}, \quad \frac{\Gamma_{\mathrm{deph}}}{\hbar} = \frac{2}{T_2}, \quad \frac{\Gamma'}{\hbar} = \frac{1}{T_1} + \frac{2}{T_2}. \tag{14.32}$$

In laser science, the *optical Bloch equations*, also referred to as the *Maxwell-Bloch* equations, describe the change of the electronic population driven by the *electric* part of the EM field acting on the *charge*. One distinguishes two types of relaxation processes which affect the measured energy linewidth.

Homogeneous damping processes preserve the central resonance energy of the atoms in the sample and the lineshape is *Lorentzian*.
Inhomogeneous damping processes distribute the central resonance energy of the atoms in sample, and the lineshape is *Gaussian*.

The most important homogeneous damping processes are that of *spontaneous emission*, leading to a natural Lorentzian lifetime width Γ of the excited state. Another process is due to *elastic collisions* in gases. From an atom point of view, the electron oscillation is interrupted but continues with a different phase after the collision. Such processes may be included through Γ_{deph}, also referred to as *collision broadening* [7]. In an equivalent picture one may also describe this process by a finite coherence time of the incident field, whose phase takes a random jump after the coherence time equals to the time between collision. If dephasing processes are absent, we simply have $\Gamma = \Gamma'$ in (14.30).

Examples of *inhomogeneous* damping processes in gases are *inelastic collisions* that lead to a change in the excited state population or the *Doppler effect* which smears the central resonance frequency. In solids, important mechanisms are local inhomogeneities in the atomic environment or long range interactions between the valence electrons which either cause local variations in bonding energies or a smearing of the energy through band structure effects. In the soft x-ray region Γ is determined mostly by the Auger decay width. Measured soft x-ray resonances in solids are typically broader than this natural linewidth due to smearing of the valence states through bonding interactions or band structure effects and will be discussed in Sect. 14.4.1.

14.4 Definition of Transition Rates in the BR Theory

The BR transition rates for the two-level atom are naturally defined in terms of the spontaneous transition matrix element, expressed through the dipole transition width Γ_x given by (14.1), the intensity-dependent Rabi transition width Γ_{R} expressed by

(14.38), and the Bloch equation solutions for the populations $\rho_{11}(t)$ and $\rho_{22}(t)$. The BR transition rates are defined as follows.

$$\text{Absorption:} \quad \mathcal{W}_{\text{XAS}} = \frac{\Gamma_{\text{R}}^2}{\hbar\Gamma}\, \rho_{11}^{\infty} \tag{14.33}$$

$$\text{Spont. Auger:} \quad \mathcal{W}_{\text{A}} = \frac{\Gamma^{\text{A}}}{\hbar}\, \rho_{22}^{\infty} \tag{14.34}$$

$$\text{Spont. Scattering:} \quad \mathcal{W}_{\text{REXS}}^{\text{spon}} = \frac{\Gamma_x}{\hbar}\, \rho_{22}^{\infty} \tag{14.35}$$

$$\text{Stim. Scattering:} \quad \mathcal{W}_{\text{REXS}}^{\text{stim}} = \frac{\Gamma_{\text{R}}^2}{\hbar\Gamma}\, \rho_{22}^{\infty}. \tag{14.36}$$

The populations are seen to be key parameters that determine the various transition rates, and in the following sections we will discuss in which cases analytical solutions for the populations can be obtained. The solutions will be illustrated by model calculations.

14.4.1 X-Ray Interaction Parameters for Model Calculations

In the following sections we will take a look at the solutions of the optical Bloch equations by choosing realistic x-ray parameters. We have seen in Chap. 10 that strong resonances exist in atoms in the form of Rydberg resonances and in molecules in the form of transitions to unfilled molecular orbitals. In solids some of the strongest resonances are the 3*d* transition metal "white line" L-shell resonances shown in Fig. 10.16.

In this chapter we choose our interaction parameters to reflect the Co metal L_3 XAS resonance which has a peak cross section of $\sigma_{\text{XAS}} = 6.25\,\text{Mb}$, previously illustrated in Fig. 10.17 and shown again in Fig. 14.2.

Also shown in Fig. 14.2a is the considerably smaller spontaneous elastic scattering cross section, $\sigma_{\text{REXS}}^{\text{spon}}$ given by (13.23) which with the polarization averaged transition width (14.1) becomes[3]

$$\sigma_{\text{REXS}}^{\text{spon}} = \frac{2}{3}\frac{\Gamma_x}{\Gamma} \underbrace{\frac{\lambda^2}{\pi}\frac{\Gamma_x}{\Gamma}\frac{(\Gamma/2)^2}{(\hbar\omega - \mathcal{E}_0)^2 + (\Gamma/2)^2}}_{\sigma_{\text{XAS}}}. \tag{14.37}$$

The underbracket identifies the XAS cross section (14.18). The goal of the present chapter is to explore the increase of (14.37) through stimulation, as reflected by the arrow in Fig. 14.2a. We shall explore this through the additional term $n_{p_2\mathbf{k}_2}$ in the total KHD expression (13.28) and its link to the optical Bloch equations.

[3] The factor of 2/3 arises from the emitted intensity node along the polarization vector.

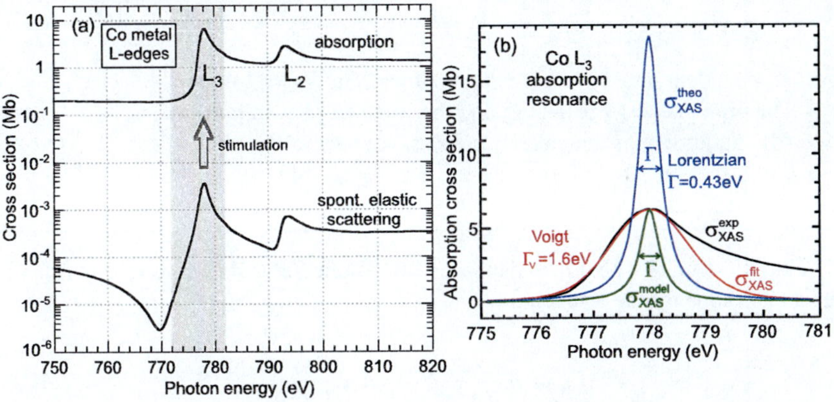

Fig. 14.2 **a** XAS cross section of the L-edge resonances in Co metal given by $\sigma_{\mathrm{XAS}} = 6.25\,\mathrm{Mb}$ and the spontaneous elastic scattering cross section $\sigma^{\mathrm{spon}}_{\mathrm{REXS}} = 3.5 \times 10^{-3}\,\mathrm{Mb}$. In this chapter we will discuss the increase of the scattering cross section through stimulation, as reflected by the arrow. **b** Enlarged Co L_3 resonance previously shown in Fig. 10.17b. We choose our modeling parameter for the BR treatment as indicated by the green curve of Lorentzian shape and FWHM $\Gamma = 0.43\,\mathrm{eV}$ and a peak cross section of $\sigma^{\mathrm{model}}_{\mathrm{XAS}} = 6.25\,\mathrm{Mb}$ that matches the experimental value for Co metal. It corresponds to a transition matrix element defined by (14.1) of $\Gamma_x = 0.34\,\mathrm{meV}$

To explore stimulation effects in REXS we shall use the model value $\sigma^{\mathrm{model}}_{\mathrm{XAS}} = 6.25\,\mathrm{Mb}$ corresponding to $\Gamma_x = 0.34\,\mathrm{meV}$ in conjunction with the natural Lorentzian linewidth $\Gamma = 430\,\mathrm{meV}$ from Table 10.2. Our chosen value of $\Gamma_x = 0.34\,\mathrm{meV}$ corresponds to the Lorentzian green curve in Fig. 14.2b, which has the same peak Co L_3 absorption cross section as shown in (a), also shown as a black curve in (b). Our model value of the transition matrix element $\Gamma_x = 0.34\,\mathrm{meV}$ is smaller than the value $\Gamma_x = 0.96\,\mathrm{meV}$ (see Table 10.5) which corresponds to the shown blue Lorentzian curve of the same width $\Gamma = 430\,\mathrm{meV}$ (also see Fig. 10.17b). We have chosen our model values to represent typical peak cross sections in the soft x-ray region, also reflected by the N_2 and O_2 XAS spectra shown in Figs. 10.11 and 10.12.

14.4.2 Practical Units and Beam Parameter Conversions

Our model calculations employ specific units for the beam parameters. We will express the incident intensity I_0 in units of [mJ/cm^2/fs]= [TW/cm^2], typically used in x-ray laser science.[4] The various physical quantities will be stated in the following units.

---

[4] In the conversion of our units it is convenient to use $1\,\mathrm{mJ/cm^2} = 6.25\times10^4\,\mathrm{meV/nm^2}$. Other conversions are accomplished by use of Appendix A.1.

- The incident intensity I_0 is in units of [mJ/cm²/fs]= [TW/cm²].
- The photon energy is in [eV] and the wavelength λ in [nm].
- All energy widths denoted Γ_i and Δ_{pk} are in [meV].
- Cross sections are in [nm²], where $1\,\mathrm{nm}^2 = 10^4\,\mathrm{Mb}$.

With these units where I_0 is measured in [mJ/cm²/fs] and Γ_{R} and Δ_{pk} in [meV], the expressions for the Rabi energy

$$\Gamma_{\mathrm{R}}^2 = I_0 \frac{\lambda^3}{2\pi^2 c} \Gamma_x = n_{pk} \frac{\Delta_{pk}}{2\pi^2} \Gamma_x \tag{14.38}$$

become for our Co L_3 resonance model

$$\boxed{\Gamma_{\mathrm{R}}[\mathrm{meV}] = 3.8\sqrt{I_0}} \tag{14.39}$$

$$\boxed{n_{pk} = 208.0 \frac{\lambda^3 I_0}{\Delta_{pk}} = 836.3 \frac{I_0}{\Delta_{pk}}.} \tag{14.40}$$

We will see in Sect. 14.6.3 that the mode-based description of the incident photons n_{pk} given by (14.17) may be replaced by the number of photons, denoted n_Γ, that are contained in the atomic core hole clock volume V_Γ under conservation of the photon density $n_{pk}/V_{pk} = n_\Gamma/V_\Gamma$. In this case we have $\Delta_{pk} = 2\pi^2\Gamma$ and (14.40) is replaced with Γ in [meV] by

$$\boxed{n_\Gamma = 10.55 \frac{\lambda^3 I_0}{\Gamma} = 42.4 \frac{I_0}{\Gamma}} \tag{14.41}$$

and the Rabi energy is given by

$$\boxed{\Gamma_{\mathrm{R}}^2 = \Gamma\Gamma_x n_\Gamma = 42.4\,\Gamma_x I_0.} \tag{14.42}$$

14.5 Analytical Solutions of the Bloch Equations

The key parameters determined by the Bloch equations are the temporal values of the lower state population $\rho_{11}(t)$ and the corresponding upper state population $\rho_{22}(t) = 1 - \rho_{11}(t)$. The absorption, emission, Auger, and elastic scattering rates are given by (14.33)–(14.36) in terms of the populations in conjunction with the dipolar energy widths Γ_x defined in (14.1), the atomic clock width Γ listed in Table 10.2, and the Rabi energy width Γ_R expressed by (14.38) in terms of the incident number of photons or intensity.

The Bloch equations have no general time-dependent solution and in general are solved numerically. However, a closed form solution for the populations $\rho_{11}(t)$ and $\rho_{22}(t)$ does exist for the following special cases.

- *Weak incident intensity*. Here the BR theory gives similar results to the KHD perturbation theory under certain assumptions, discussed below.
- *Exact resonance excitation.* This case corresponds to a small incident bandwidth or long coherence time of the incident photons, mimicking a long well-defined classical EM wavetrain. A time-dependent solution exists for any incident intensity.
- *Steady-state equilibrium.* This situation occurs when the incident beam can act on the two-level atom for a sufficiently long time, $t \to \infty$, that an equilibrium between the lower and upper state populations can be established.

We will discuss these three cases now.

14.5.1 Arbitrary Bandwidth: Low Incident Intensity

The first case where an analytical solution for the excited state population can be found is when the *incident intensity is low* so that $4\Gamma_R < \Gamma, \Gamma'$ (see (14.47) below). In this case the excited state population remains low as well. Under the initial conditions $\rho_{22}(t=0) = 0$ and $\rho_{12}(t=0) = 0$ one obtains [7, 18] by denoting the detuning energy $\Delta\mathcal{E} = \mathcal{E}_0 - \hbar\omega$ and $\Gamma' = \Gamma + \Gamma_{deph}$

$$\begin{aligned}\rho_{22}(t) = & \frac{(\Gamma_R/2)^2\,\Gamma'}{\Gamma\left[\Delta\mathcal{E}^2 + (\Gamma'/2)^2\right]} \\ & \times \left\{ 1 + \frac{2(\Gamma - \Gamma'/2)\left[\Delta\mathcal{E}^2 + (\Gamma'/2)^2\right]}{\Gamma'\left[\Delta\mathcal{E}^2 + (\Gamma - \Gamma'/2)^2\right]} \exp\left[-\frac{\Gamma t}{\hbar}\right] \right. \\ & - \frac{\Gamma[\Gamma'(\Gamma - \Gamma'/2) + \Delta\mathcal{E}^2]}{\Gamma'\left[\Delta\mathcal{E}^2 + (\Gamma - \Gamma'/2)^2\right]} \cos\left[\frac{\Delta\mathcal{E}\,t}{\hbar}\right] \exp\left[-\frac{\Gamma' t}{2\hbar}\right] \\ & \left. + \frac{2\Delta\mathcal{E}\,\Gamma(\Gamma - \Gamma')}{\Gamma'\left[\Delta\mathcal{E}^2 + (\Gamma - \Gamma'/2)^2\right]} \sin\left[\frac{\Delta\mathcal{E}\,t}{\hbar}\right] \exp\left[-\frac{\Gamma' t}{2\hbar}\right] \right\}. \end{aligned} \tag{14.43}$$

For $\Gamma = \Gamma'$, (14.43) reduces to

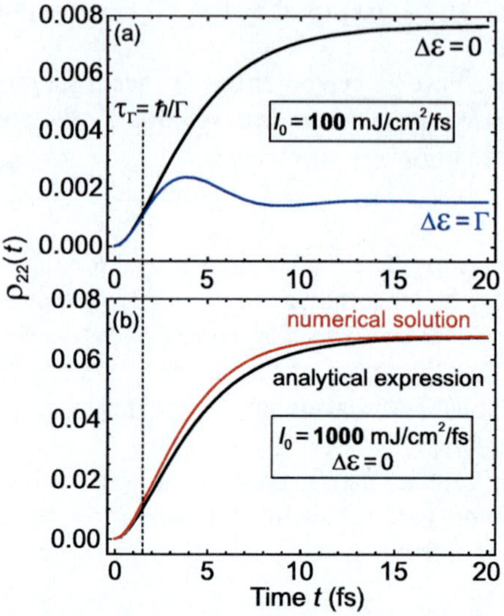

Fig. 14.3 **a** Dependence of upper state population $\rho_{22}(t)$ on the time t according to the low intensity approximation given by (14.44) for two values of $\Delta\mathcal{E} = 0$ and $\Delta\mathcal{E} = \Gamma = 430\,\text{meV}$. We have assumed $\Gamma_R[\text{meV}] = 3.8\sqrt{I_0}$ (see (14.39)), with $I_0 = 1000\,\text{mJ/cm}^2/\text{fs}$, $\Gamma = 430\,\text{meV}$ and $\lambda = 1.59\,\text{nm}$. The vertical black dashed line marks the core hole clock time τ_Γ. **b** Comparison of the low intensity analytical approximation (14.44) with the full numerical solution of the Bloch equations for an increased incident intensity of $I_0 = 1000\,\text{mJ/cm}^2/\text{fs}$. The onset of a discrepancy reveals the high intensity limit of the analytical solution (14.44)

$$\boxed{\rho_{22}(t) = \frac{\Gamma_R^2/4}{\Delta\mathcal{E}^2 + \Gamma^2/4}\left\{1 + \exp\left[-\frac{\Gamma t}{\hbar}\right] - 2\cos\left[\frac{\Delta\mathcal{E}}{\hbar}t\right]\exp\left[-\frac{\Gamma t}{2\hbar}\right]\right\}.} \tag{14.44}$$

The so-obtained excited state population is determined at a time t after the field is turned on instantaneously at $t = 0$. The off-diagonal density matrix elements or coherences $\rho_{12} = \rho_{21}^*$ for $\rho_{21}(0) = \rho_{12}(0) = 0$ are given by

$$\rho_{21}(t) = \frac{-\mathrm{i}\Gamma_R/2}{\mathrm{i}\Delta\mathcal{E} + \Gamma/2}\left\{1 - \exp\left[-\frac{(\mathrm{i}\Delta\mathcal{E} + \Gamma/2)t}{\hbar}\right]\right\}. \tag{14.45}$$

We have illustrated the dependence of the upper state population $\rho_{22}(t)$ on the time t for our model parameters in Fig. 14.3.

Figure 14.3a illustrates the excited state population $\rho_{22}(t)$ given by (14.44) at an incident intensity of $I_0 = 100\,\text{mJ/cm}^2/\text{fs}$ for our Co L_3 resonance parameters specified in Sect. 14.4.1, for two values of the detuning energy $\Delta\mathcal{E} = 0$ and $\Delta\mathcal{E} = \Gamma = 430\,\text{meV}$. As stated above, the analytical solution given by (14.44) is valid only if the incident intensity is not too strong so that the excited state population ρ_{22} remains relatively small. This is seen to be still the case for $I_0 = 100\,\text{mJ/cm}^2/\text{fs}$, where $\rho_{22}(t) \leq 8 \times 10^{-3}$. The analytical solutions shown in Fig. 14.3a are indistinguishable from the numerical solutions of the Bloch equations.

We also see that the black and blue curves for different detuning energies in Fig. 14.3a are identical for times that are shorter than the Co $2p$ core hole clock time $t < \tau_\Gamma = 1.53$ fs (see Fig. 12.4), marked as a vertical black dashed line. This is mathematically revealed by expanding the curly brackets in (14.44) up to second order in t, which yields

$$\rho_{22}(t) = \frac{\Gamma_R^2/4}{\Delta\mathcal{E}^2+\Gamma^2/4} \underbrace{\frac{\left(\Delta\mathcal{E}^2+\Gamma^2/4\right)t^2}{\hbar^2}}_{g(t)} = \frac{\Gamma_R^2\, t^2}{4\,\hbar^2}. \tag{14.46}$$

We see that at short times, $\rho_{22}(t)$ is independent of the detuning energy $\Delta\mathcal{E} = \mathcal{E}_0 - \hbar\omega$.

The validity range of (14.44) is demonstrated in Fig. 14.3b by comparing its predictions for a ten times higher incident intensity of $I_0 = 1000\,\text{mJ/cm}^2\text{/fs}$ and $\Delta\mathcal{E} = 0$, shown as a black curve, to that of the full numerical solution of the Bloch equations (red curve). Now the excited population has also increased by about a factor of ten and the full Bloch equation solution begins to deviate from the analytical solution (14.44).

The analytical solution (14.44) is valuable since it is valid for all cases of incident bandwidth. It is therefore possible to directly compare it to the results obtained with other formulations such as KHD perturbation theory and the famous Einstein rate equations which we shall do in Sects. 14.6 and 14.9 below.

14.5.2 Exact Resonance: Arbitrary Incident Intensity

Another case where an analytical solution of the Bloch equations exists is when highly monochromatic incident radiation is tuned to the resonance, i.e. $\Delta\mathcal{E} = 0$. At low incident intensity, this is just a special case of (14.44), but at high incident intensity there is no analytical solution for finite detuning energy $\Delta\mathcal{E}$ that reduces to the special case $\Delta\mathcal{E} = 0$.

For $\Delta\mathcal{E} = 0$ an analytical solution exists that is valid for *all incident intensities* at a time t after instantaneous turn on of the field at $t = 0$. In the absence of a dephasing contribution, i.e. $\Gamma' = \Gamma$ and defining the generalized Rabi frequency as

$$\omega_R = \frac{\sqrt{\Gamma_R^2 - \Gamma^2/16}}{\hbar} \tag{14.47}$$

and under the initial conditions $\rho_{21}(t=0) = \rho_{12}(t=0) = \rho_{22}(t=0) = 0$ the solution is given by [18]

$$\boxed{\rho_{22}(t) = \frac{\Gamma_R^2}{\Gamma^2 + 2\Gamma_R^2}\left\{1 - \left[\cosh(\omega_R t) + \frac{3\Gamma}{4\hbar\omega_R}\sinh(\omega_R t)\right] e^{-3\Gamma t/(4\hbar)}\right\}.} \tag{14.48}$$

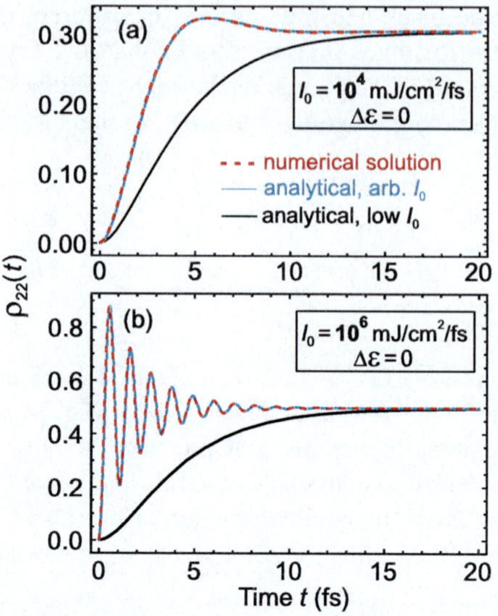

Fig. 14.4 **a** Comparison of the time dependent upper state population $\rho_{22}(t)$ for $\Delta\mathcal{E} = 0$ and $I_0 = 10^4$ mJ/cm^2/fs by use of the low intensity analytical expression (14.44) (black curve), the resonant arbitrary-intensity analytical expression (14.48) (light blue curve), and the numerical solution of the Bloch equations (dashed red). We have used $\Gamma_R[\text{meV}] = 3.8\sqrt{I_0}$, with $\Gamma = 430$ meV and $\lambda = 1.59$ nm (see (14.39)). **b** Same for $I_0 = 10^6$ mJ/cm^2/fs. As previously seen in Fig. 14.3b the low intensity approximation (14.44) (black curve) increasingly breaks down with increasing $I_0 \geq 10^3$ mJ/cm^2/fs although it still exhibits the proper steady state (long time) value

Again we have $\rho_{11}(t) = 1 - \rho_{22}(t)$. The excited state population $\rho_{22}(t)$ approaches the equilibrium value ($t \to \infty$) through a damped oscillation, indicated by the sinh and cosh terms as illustrated for our model parameters in Fig. 14.4. Expression (14.48) is valid for all incident intensities but is restricted to resonant excitation $\Delta\mathcal{E} = 0$. It can be shown to reduce to (14.44) with $\Delta\mathcal{E} = 0$ at low intensity. Equation (14.48) does *not* hold for broad band incident light as will be discussed in Sect. 14.9.

Figure 14.4a illustrates the case for $I_0 = 10^4$ mJ/cm^2/fs and (b) for $I_0 = 10^6$ mJ/cm^2/fs for $\Delta\mathcal{E} = 0$. Shown as black curves is the low intensity approximation (14.44) which according to Fig. 14.3b begins to break down around $I_0 = 10^3$ mJ/cm^2/fs, and for the shown I_0 values deviates significantly from the proper results, obtained either by the resonant analytical solution (14.48) (light blue curve) for arbitrary incident intensity or the numerical solution of the Bloch equations (dashed red). The increasing oscillations with time are the famous Rabi oscillations.

The dependence of $\rho_{22}(t)$ for resonant excitation, $\Delta\mathcal{E} = 0$, expressed by (14.48) is illustrated in more detail over a broad range of intensities in Fig. 14.5. We see that initially the excited state population changes strongly with time and incident intensity. By use of a strong incident beam of large degeneracy parameter, one may create an ultrafast population inversion where the excited state population exceeds 0.5. As shown by the green curve for an incident intensity of 10^7 mJ/cm^2/fs, the

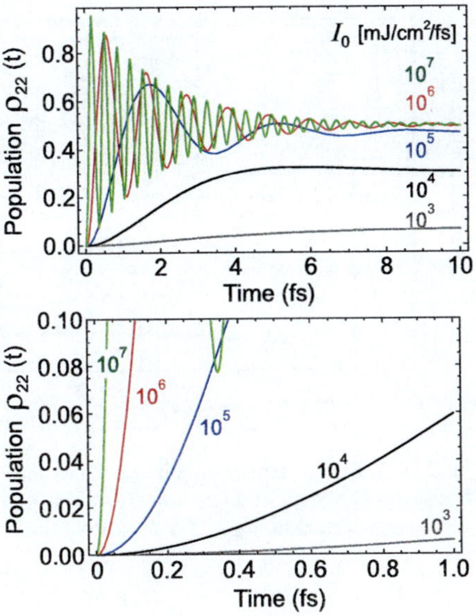

Fig. 14.5 Excited state population as a function of time after instantaneous turn on of the intensity according to (14.48), for resonant excitation, $\Delta\mathcal{E} = 0$, at various incident intensities I_0. We used $\Gamma_R = 3.8\sqrt{I_0}\,[\text{meV}]$ and $\Gamma = 430\,\text{meV}$

population inversion can happen on a subfemtosecond timescale that is shorter than the atomic clock decay time $\tau = \hbar/\Gamma$ of order of a few femtosconds (see Fig. 12.4). When a multi-atom system is excited that way, the first spontaneous decays of some atoms can stimulate decays on other atoms which can amplify as in a laser, a process called *amplified spontaneous emission* or ASE. This was first demonstrated by use of XFEL pulses by Rohringer et al. [19]. The amplification process resembles the SASE process in an XFEL, with the difference that SASE starts from noise.

As shown in Fig. 14.5a, the Rabi oscillations damp out in time and the excited state population stabilizes at an equilibrium value that depends on the incident intensity. We see that it takes about 5–10 fs at the highest intensities for the excited state population to reach a steady state. To create an excited state population of more than 10%, intensities higher than 10^3 mJ/cm^2/fs are needed. This means that the KHD perturbation theory which neglects population changes will break down.

The relationship between the excited state population buildup as a function of time for different incident intensities and as a function of intensity for different times is shown in more detail in Fig. 14.6. The colored points on the vertical line at 1 fs in (a) correspond to the points of the same color in (b).

It is instructive to consider (14.48) for the case of no damping, i.e. $\Gamma = 0$. We obtain

$$\rho_{22}(t) = \frac{1}{2}\left[1 - \cos(\mathcal{W}_R t)\right], \tag{14.49}$$

so that the occupation of the excited state oscillates about the value $1/2$ with the Rabi frequency $\mathcal{W}_R$. The occupation of the ground state undergoes the same oscillation with opposite phase since

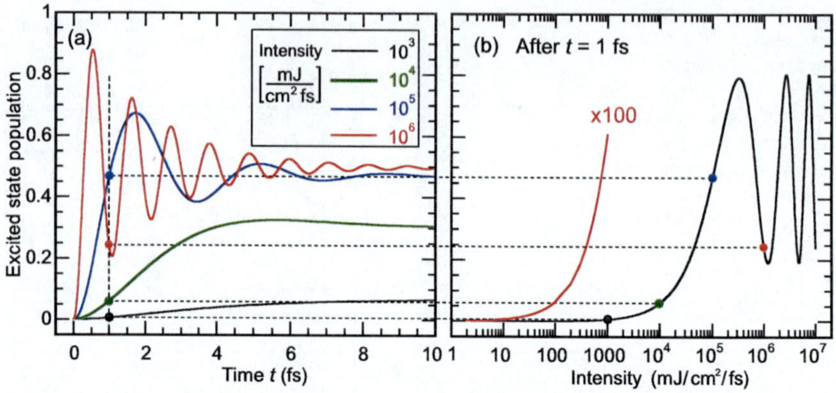

Fig. 14.6 **a** Excited state population as a function of time according to (14.48), for different incident intensities I_0. **b** Same as a function of intensity at different times. We have used parameters for the Co L_3-edge excitation, $\Gamma_R = 3.8\sqrt{I_0}$ [meV] and $\Gamma = 430$ meV

$$\rho_{11}(t) = 1 - \rho_{22}(t) = \frac{1}{2}\left[1 + \cos(\mathcal{W}_R t)\right]. \tag{14.50}$$

In retrospect, this justifies our earlier definition of the Rabi frequency in Sect. 14.2.

14.5.3 Excitations by Transform-Limited and SASE Pulses

The analytical solution (14.48) gives the excited state population under the assumption that an incident coherently oscillating EM wave has acted on the electronic system for a time t and the population is probed at t. This corresponds to the assumption of a flat-top (FT) transform-limited incident pulse of length t. In practice, pulses typically do not have a FT shape and one needs to convolute the FT distribution with the detailed pulse shape. As discussed in Sect. 4.4.1 the pulse shape affects the Time-bandwidth product which, for example, differs by a factor of 2 for FT and Gaussian pulses (see Table 4.2).

A more profound issue is that the internal structure of SASE XFEL pulses consists of coherent spikes of different widths and peak intensities [20]. In studies that depend on the temporal coherence, the different spikes act over different time periods t on the atoms causing dramatic spike-to-spike changes in the achieved upper state population (see Fig. 14.5). The action of SASE pulses then needs to be accounted for by a statistical average.

One may assume a distribution of coherent spike lengths t during a pulse of length τ_p, so that the average upper state population is given by an integration of (14.48) over a range of coherence times t up to $\tau \le \tau_p$ according to

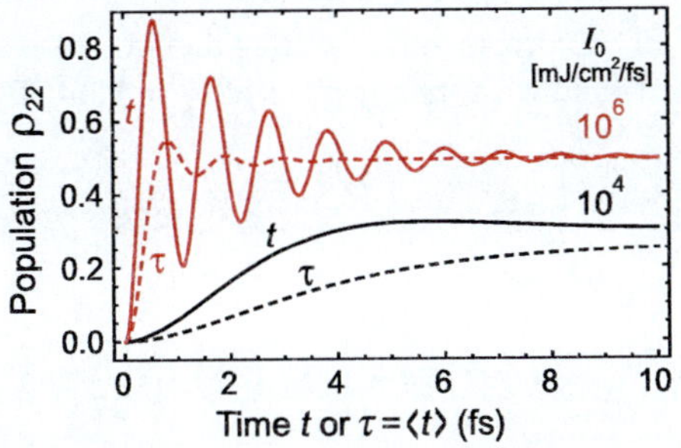

Fig. 14.7 Excited state population for two incident intensities I_0, distinguished by color, as a function of well-defined coherence time t (solid lines) after instantaneous pulse turn on according to (14.48) or averaged over a distribution of flat-top pulses of coherence time up to τ, expressed by (14.52) (dashed lines). We have used parameters for the Co L_3-edge excitation, $\Gamma_R = 3.8\sqrt{I_0}$ [meV] and $\Gamma = 430$ meV

$$\rho_{22}(\tau) = \frac{1}{\tau}\int_0^{\tau} \rho_{22}(t)\,\mathrm{d}t. \tag{14.51}$$

Assuming flat-top spikes of length up to τ, we obtain the analytical solution,

$$\rho_{22}(\tau) = \frac{\Gamma_R^2/4}{(\Gamma/2)^2 + \Gamma_R^2/2}\left\{1 - \frac{24\Gamma\hbar\omega_R\left[1-\cos(\omega_R\tau)\,\mathrm{e}^{-3\Gamma\tau/(4\hbar)}\right]}{\tau\,\omega_R\left[16(\hbar\omega_R)^2+9\Gamma^2\right]}\right.$$
$$\left. - \frac{\left[16(\hbar\omega_R)^2-9\Gamma^2\right]\sin(\omega_R\tau)\,\mathrm{e}^{-3\Gamma\tau/(4\hbar)}}{\tau\,\omega_R\left[16(\hbar\omega_R)^2+9\Gamma^2\right]}\right\}. \tag{14.52}$$

Figure 14.7 illustrates the expected behavior for two incident intensities. The integration over the pulse length (dashed curves) simply averages out the instantaneous-time Rabi oscillations and somewhat delays the onset of the population increase.

For maximum coherence times much longer than the core hole lifetime, $\tau \gg \tau_\Gamma = \hbar/\Gamma$, the term in wavy brackets in (14.52) approaches unity and $\rho_{22}(\tau) \rightarrow \rho_{22}(t)$. For large incident intensity, $\Gamma_R \gg \Gamma$ and long times t or τ we obtain the steady-state value $\rho_{22}(\infty) = \rho_{11}(\infty) = 1/2$.

14.5.4 Solution for the Steady-State or Long Time Limit

We have seen in Fig. 14.4a that the excited state population for times exceeding the fundamental core hole clock decay time settles down to an equilibrium value. The long time solution may be obtained by solving simplified Bloch equations. As previously discussed in Sect. 14.5.1 we allow dephasing by use of $\Gamma' = \Gamma + \Gamma_{\mathrm{deph}}$.

The steady-state condition corresponds to setting the time derivatives of the density matrix elements to zero, and we obtain from (14.29) and (14.30)

$$\frac{\Gamma}{\hbar}\rho_{22}^{\infty} = -\,\mathrm{i}\,\frac{\mathcal{W}_{\mathrm{R}}}{2}\left(\rho_{12}^{\infty} - \rho_{21}^{\infty}\right) \tag{14.53}$$

and,

$$\frac{\mathcal{W}_{\mathrm{R}}}{2}\left(\rho_{11}^{\infty} - \rho_{22}^{\infty}\right) = -\left[(\mathcal{E}_0 - \omega) + \mathrm{i}\frac{\Gamma'}{2\hbar}\right]\rho_{12}^{\infty}, \tag{14.54}$$

where we have denoted the steady-state values of the density matrix elements as $\rho_{ij}(t=\infty) = \rho_{ij}^{\infty}$. We have written the equations with the population terms on the left and the coherences on the right. The density matrix solution in the steady-state limit $t \to \infty$ for arbitrary detuning, $\hbar\omega - \mathcal{E}_0$, and in the presence of dephasing can be derived in closed form and is given by [7]

$$\rho_{22}^{\infty}(\hbar\omega) = \frac{\Gamma_{\mathrm{R}}^2\Gamma'/(4\Gamma)}{(\hbar\omega - \mathcal{E}_0)^2 + (\Gamma'/2)^2 + \Gamma_{\mathrm{R}}^2\Gamma'/(2\Gamma)} = 1 - \rho_{11}^{\infty}. \tag{14.55}$$

The solution is independent of the initial conditions. For the typical case $\Gamma' = \Gamma$ we obtain as follows.

In the steady-state limit $t \to \infty$ and arbitrary detuning, $\hbar\omega - \mathcal{E}_0$, the populations are given by

$$\rho_{22}^{\infty}(\hbar\omega) = \frac{\Gamma_{\mathrm{R}}^2}{\Gamma^2}\,\frac{(\Gamma/2)^2}{(\hbar\omega - \mathcal{E}_0)^2 + (\Gamma/2)^2\left[1 + 2\Gamma_{\mathrm{R}}^2/\Gamma^2\right]} \tag{14.56}$$

and

$$\rho_{11}^{\infty}(\hbar\omega) = 1 - \rho_{22}^{\infty}(\hbar\omega). \tag{14.57}$$

The steady-state behavior expressed by (14.56) is illustrated in Fig. 14.8a for zero detuning $\hbar\omega = \mathcal{E}_0$, (black lines), and finite detuning, $\hbar\omega - \mathcal{E}_0 = \Gamma/2$ (blue lines), while in (b) we illustrate the change of Γ_{R} with intensity. Detuning is seen to delay the buildup of the upper state population, as expected.

At the intersect $\Gamma_{\mathrm{R}} = \Gamma$, the upper state population in Fig. 14.8a becomes 1/3 and the lower state population is reduced to 2/3, as indicated by dashed lines. For larger values of Γ_{R} the populations equalize to $\rho_{22} = \rho_{11} = 0.5$. We will see that absorption and stimulated emission then become equal and the atoms become transparent.The gray shaded regions, corresponding to $\Gamma_{\mathrm{R}} \ll \Gamma$, indicate the validity of the KHD theory, which neglects the intensity dependence of the lower and upper state populations.

The off-diagonal elements are obtained as

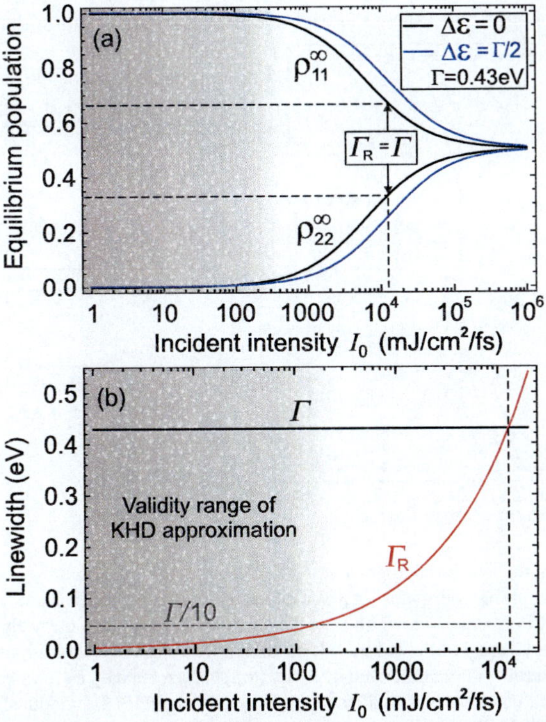

Fig. 14.8 **a** Steady-state solutions for the populations ρ_{22}^{∞} given by (14.56) and $\rho_{11}^{\infty} = 1 - \rho_{22}^{\infty}$ for resonant excitation $\Delta\mathcal{E} = \hbar\omega - \mathcal{E}_0 = 0$ (black) and off-resonance excitation $\Delta\mathcal{E} = \Gamma/2$ (blue), as a function of the incident intensity I_0. We have used the parameters for the Co L_3-edge excitation, $\Gamma_R[\text{eV}] = 3.8 \times 10^{-3}\sqrt{I_0}$ and $\Gamma = 0.43\,\text{eV}$. **b** Plot of the Rabi energy width Γ_R relative to the total decay width Γ as a function of I_0. The gray shaded regions identify the validity range of the KHD theory

$$
\begin{aligned}
\rho_{21}^{\infty}(\hbar\omega) = (\rho_{12}^{\infty})^* &= \frac{(\Gamma_R/2)(\hbar\omega - \mathcal{E}_0 - \mathrm{i}\Gamma'/2)}{(\hbar\omega - \mathcal{E}_0)^2 + (\Gamma'/2)^2 + \Gamma_R^2\Gamma'/(2\Gamma)} \\
&= \frac{2\Gamma}{\Gamma'\Gamma_R}\,\rho_{22}^{\infty}\,(\hbar\omega - \mathcal{E}_0 - \mathrm{i}\Gamma'/2).
\end{aligned}
\tag{14.58}
$$

At low incident intensity, $\Gamma_R \ll \Gamma, \Gamma'$, the last term in the denominator can be neglected and for $\Gamma = \Gamma'$ we obtain (14.45) in the limit $t \to \infty$. At resonance we obtain the simple relation

$$
\rho_{21}^{\infty} = -\mathrm{i}\frac{\Gamma}{\Gamma_R}\,\rho_{22}^{\infty} \tag{14.59}
$$

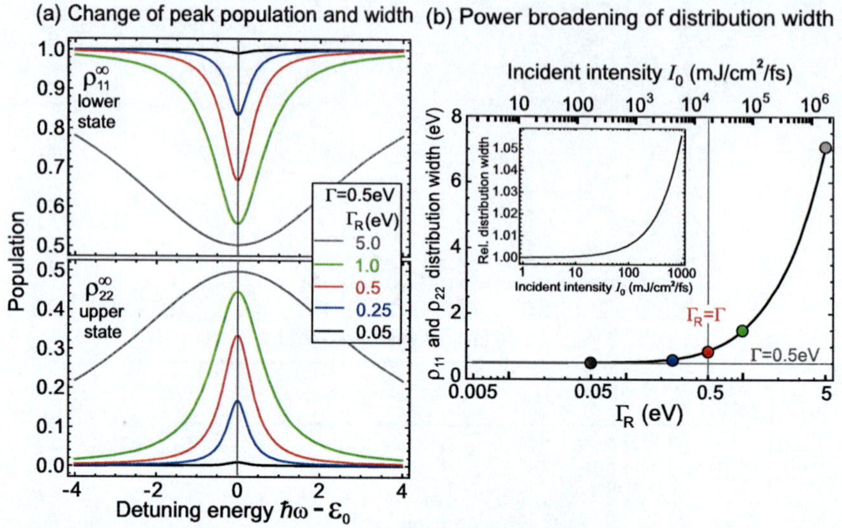

Fig. 14.9 **a** Plot of the lower and upper state population distribution functions (14.57) and (14.56) as a function of detuning energy$\hbar\omega - \mathcal{E}_0$ for different values of the Rabi energy Γ_R relative to the natural FWHM Lorentzian linewidth $\Gamma = 0.5\,\text{eV}$. Note the complementary changes of the peak population values and distribution widths. **b** Distribution widths given by (14.60) as a function of the Rabi energy Γ_R (bottom scale) or incident intensity (top scale) related by $\Gamma_R[\text{eV}] = 3.8 \times 10^{-3}\sqrt{I_0}$ (see (14.39)). On the curve we have indicated as colored filled circles the widths of the curves of the same color in (**a**). The inset shows the relatively small width increase at lower incident intensities

14.5.5 Power Broadening of the BR Linewidth

In Fig. 14.9a we show the change of the population distribution functions $\rho_{22}^{\infty}(\hbar\omega)$ and $\rho_{11}^{\infty}(\hbar\omega)$ given by (14.56) and (14.57) as a function of detuning energy $\hbar\omega - \mathcal{E}_0$ for different values of the Rabi energy Γ_R. For convenience we have assumed a natural lifetime width $\Gamma = 0.5\,\text{eV}$, instead of $\Gamma = 0.43\,\text{eV}$ in Fig. 14.8a. It is intuitively clear that the peak population values determine the peak transition rates and their energy distributions determine the detuning response. We will formally define the different transition rates and linewidths in terms of population distribution functions and transition matrix elements later, and at this stage the qualitative link suffices.

In Fig. 14.9b we plot the widths (FWHM) of the distributions ρ_{22}^{∞} and ρ_{11}^{∞} given by

$$FWHM = \sqrt{\Gamma^2 + 2\Gamma_R^2} \tag{14.60}$$

as a function of the Rabi energy Γ_R (bottom scale) and the incident intensity (top scale). The scales are related in the shown units by $\Gamma_R[\text{eV}] = 3.8 \times 10^{-3}\sqrt{I_0}$ (see (14.39)). The shown colored circles indicate the widths of the curves of the same color in (a). The distribution width increase with intensity is referred to as *power broadening*.

The population distributions as a function of detuning energy in Fig. 14.9a for the lower and upper populations are mirror images around the average value 1/2. The peak population values follow the black curves in Fig. 14.8a, except for a slight difference due to the value $\Gamma = 0.43\,\text{eV}$ used in that figure and $\Gamma = 0.5\,\text{eV}$ used in Fig. 14.9a.

At *low incident intensity*, $\Gamma_R \ll \Gamma$, the distributions (14.56) and (14.57) assume the Lorentzian form, e.g.

$$\rho_{22}^{\infty} = \frac{\Gamma_R^2}{\Gamma^2} \frac{\Gamma^2/4}{(\hbar\omega - \mathcal{E}_0)^2 + \Gamma^2/4}. \tag{14.61}$$

At high intensity $2\Gamma_R^2 \gg \Gamma^2$, the distributions become

$$\rho_{22}^{\infty} = \frac{\Gamma_R^2/4}{(\hbar\omega - \mathcal{E}_0)^2 + \Gamma_R^2/2}, \quad \text{and} \quad \rho_{11}^{\infty} = \frac{(\hbar\omega - \mathcal{E}_0)^2 + \Gamma_R^2/4}{(\hbar\omega - \mathcal{E}_0)^2 + \Gamma_R^2/2}, \tag{14.62}$$

and they assume a constant value of $1/2$ for $\Gamma_R \gg \hbar\omega - \mathcal{E}_0$.

The steady-state up and down transition rates between the two levels near the resonance value increases with incident intensity, but the relative increase off-resonance is larger than on resonance since there is more "headroom" for off-resonance growth. A nice early experimental study of power broadening in the optical regime can be found in [21].

14.6 Link of KHD and Low Intensity BR Rates

The time dependent BR and perturbative time-independent KHD theories can both describe the transition rates in a two-level atom in the low intensity limit where a closed form solution of the Bloch equations exists as discussed in Sect. 14.5.1. The question arises whether the two theories predict the same behavior and if, yes, how the formalisms can be related. We would expect that the two theories may be equivalent under certain assumptions since at low incident intensities the populations of the lower and upper states do not change significantly. We recall that the KHD perturbation approach treats the populations as binary with $\rho_{11} = 1$ and $\rho_{22} = 0$ for the excitation step and the reverse, $\rho_{22} = 1$ and $\rho_{22} = 1$, for the emission step. The binary populations in the KHD theory must therefore not deviate significantly from those predicted by the low intensity BR theory.

This is indeed confirmed by Fig. 14.3, where in the resonant excitation step the upper state population $\rho_{22}(t)$ is seen to remain quite small, $\rho_{22} < 10^{-2}$, for our model parameters and for an incident intensity of $I_0 \simeq 100\,\text{mJ/cm}^2\text{/fs}$. Considering that the intensity at state-of-the-art synchrotron sources is only of order $I_0 \simeq 10^{-6}\,\text{mJ/cm}^2\text{/fs}$

as discussed in Sect. 4.2.8, the intensity range covered by the KHD theory is seen to be quite remarkable. We can therefore directly compare the predictions of the BR and KHD theories up to about $I_0 \simeq 100\,\text{mJ/cm}^2\text{/fs}$. This is facilitated by the existence of an analytical solution of the Bloch equations given in Sect. 14.5.1 which can then be compared to the prediction of the KHD REXS theory derived in Sect. 13.4.

Before we formally compare the two theories, let us take a look at their explicit dependence on the incident beam and atom-specific properties. In both formulations we can describe the incident intensity as previously discussed in Sect. 10.2.1 in terms of the number of photons $n_{p\mathbf{k}}$ of bandwidth $\Delta_{p\mathbf{k}}$ in the mode volume $V_{p\mathbf{k}} = \lambda^3\,\hbar\omega/\Delta_{p\mathbf{k}}$ so that we can rewrite (14.14) in the intensity form

$$I_0 = \frac{n_{p\mathbf{k}}\hbar\omega\, c}{V_{p\mathbf{k}}} = \frac{n_{p\mathbf{k}}\Delta_{p\mathbf{k}}\, c}{\lambda^3}. \tag{14.63}$$

The response of the two-level atom in the KHD theory can be factored into a photon part described by (14.63) and an atomic part expressed in terms of the dipolar transition width Γ_x between the two states given by (14.1) and the atom-specific core hole clock width Γ listed in Table 10.2. Γ_x furthermore links the fundamental photon energy and wavelength to the dipolar atomic transition matrix elements through the fine structure constant, as elegantly expressed by (10.40) and (12.18).

In the BR formulation, the key parameter is the Rabi energy which according to (14.38) is a function of both the beam properties, expressed by (14.63), and the atomic transition width Γ_x given by (14.1). The key distinction of the two theories is the coherence time t during which the incident beam interacts with the atom. It does not explicitly appear in the KHD theory, and in the following we shall explore its hidden ramification.

14.6.1 The KHD Transition Rates

We start by reviewing the expressions for the KHD cross sections and associated transition rates for the case of a two-level atom. The spontaneous XAS cross section $\sigma_{\text{XAS}}^{\diamond}$ is that previously given by (14.18) or

$$\sigma_{\text{XAS}}^{\diamond} = \frac{\lambda^2}{\pi}\frac{\Gamma_x}{\Gamma}\,\frac{\Gamma^2/4}{\Delta\mathcal{E}^2 + \Gamma^2/4}, \tag{14.64}$$

where Γ_x is the polarization averaged transition width defined in (14.1) as

$$\Gamma_x = \frac{8\pi^2\alpha_{\text{f}}\hbar\omega}{\lambda^2}\,|\langle b|\mathbf{r}\cdot\boldsymbol{\epsilon}_p|a\rangle|^2. \tag{14.65}$$

The total spontaneous plus stimulated REXS cross section is obtained from (13.28) as

$$\boxed{\frac{\sigma_{\text{REXS}}}{\text{d}\Omega} = \frac{1}{4\pi}\frac{\Gamma_x}{\Gamma}\left\{1 + n_{p\mathbf{k}}\right\}\sigma_{\text{XAS}},} \tag{14.66}$$

where we have omitted the Dirac δ-function linking the incident and emitted photon energies in (13.12) which assures energy conservation and in practice does not limit the REXS linewidth to the XAS value Γ as discussed in Sect. 13.3.

The associated XAS and REXS transition rates $\mathcal{W}$ are just the cross sections times the incident photon flux. We obtain the following KHD transition rates $\mathcal{W}$ (dimension [1/time]), driven by $n_{p\mathbf{k}}$ incident photons in the mode volume $V_{p\mathbf{k}} = \lambda^3\,\hbar\omega/\Delta_{p\mathbf{k}}$, where $\Delta_{p\mathbf{k}}$ is the incident bandwidth

$$\mathcal{W}_{\text{XAS}} = \frac{n_{p\mathbf{k}}c}{V_{p\mathbf{k}}}\sigma_{\text{XAS}} = n_{p\mathbf{k}}\frac{\Gamma_x}{\hbar}\frac{\Delta_{p\mathbf{k}}}{2\pi^2\Gamma}\frac{\Gamma^2/4}{(\hbar\omega - \mathcal{E}_0)^2 + \Gamma^2/4} \tag{14.67}$$

and

$$\text{KHD:}\;\; \mathcal{W}_{\text{REXS}} = \frac{n_{p\mathbf{k}}c}{V_{p\mathbf{k}}}\sigma_{\text{REXS}} = \left\{1+n_{p\mathbf{k}}\right\}\frac{\Gamma_x}{\Gamma}\mathcal{W}_{\text{XAS}}. \tag{14.68}$$

> The REXS rate in the KHD theory is time *in*dependent and instead depends on the bandwidth $\Delta_{p\mathbf{k}}$ of the incident radiation. It is typically associated with a box-like mode volume according to $\Delta_{p\mathbf{k}} = \lambda^3\,\hbar\omega/V_{p\mathbf{k}}$, reflecting a flat-top energy distributions of width $\Delta_{p\mathbf{k}}$.

14.6.2 Link of Bandwidth in KHD and Time in BR Rates

In the BR formulation, the REXS rate is *time dependent* and has spontaneous and stimulated contributions according to (14.35) and (14.36) or

$$\text{BR:}\;\; \mathcal{W}_{\text{REXS}} = \mathcal{W}_{\text{REXS}}^{\text{spon}} + \mathcal{W}_{\text{REXS}}^{\text{stim}} = \frac{\Gamma_x}{\hbar}\rho_{22}(t) + \frac{\Gamma_{\text{R}}^2}{\hbar\Gamma}\rho_{22}(t), \tag{14.69}$$

where according to (14.44) we have at low intensity

$$\rho_{22}(t) = \frac{\Gamma_{\text{R}}^2}{\Gamma^2}\frac{\Gamma^2/4}{\Delta\mathcal{E}^2+\Gamma^2/4}\underbrace{\left\{1 + \exp\left[-\frac{\Gamma t}{\hbar}\right] - 2\cos\left[\frac{\Delta\mathcal{E}}{\hbar}t\right]\exp\left[-\frac{\Gamma t}{2\hbar}\right]\right\}}_{g(t)}. \tag{14.70}$$

By use of (14.38) for the Rabi energy, or

$$\Gamma_{\rm R}^2 = n_{p\mathbf{k}} \frac{\Delta_{p\mathbf{k}}}{2\pi^2} \Gamma_x \tag{14.71}$$

the total BR REXS rate (14.69) becomes

$$\text{BR:}\ \hbar\mathcal{W}_{\rm REXS} = \left\{1 + n_{p\mathbf{k}} \frac{\Delta_{p\mathbf{k}}}{2\pi^2\Gamma}\right\} n_{p\mathbf{k}} \frac{\Delta_{p\mathbf{k}}}{2\pi^2\Gamma} \frac{\Gamma_x^2}{\Gamma} \frac{\Gamma^2/4}{\Delta\mathcal{E}^2 + \Gamma^2/4}\, g(t) \tag{14.72}$$

which can be directly compared with the KHD expression (14.68) or

$$\text{KHD:}\ \hbar\mathcal{W}_{\rm REXS} = \left\{1 + n_{p\mathbf{k}}\right\} n_{p\mathbf{k}} \frac{\Delta_{p\mathbf{k}}}{2\pi^2\Gamma} \frac{\Gamma_x^2}{\Gamma} \frac{\Gamma^2/4}{\Delta\mathcal{E}^2 + \Gamma^2/4}. \tag{14.73}$$

The two expressions are identical for

$$\boxed{\Delta_{p\mathbf{k}} = \frac{2\pi^2\Gamma}{g(t)}.} \tag{14.74}$$

For short times, the expression $g(t)$ may be expanded up to second order to yield

$$\Delta_{p\mathbf{k}} = \frac{8\pi^2\hbar^2}{\Gamma\, t^2}. \tag{14.75}$$

The link between the mode bandwidth $\Delta_{p\mathbf{k}}$ in the KHD formation and the coherence time t in the BR theory, given by (14.74) is plotted for resonance excitation, $\Delta\mathcal{E} = 0$, as a red curve in Fig. 14.10. The short time limit (14.75), shown in gray, intersects the long time limit $\Delta_{p\mathbf{k}} = 2\pi^2\Gamma$, shown as a dashed gray horizontal line at the blue vertical demarkation time $t = 2\tau_\Gamma$, where $\tau_\Gamma = \hbar/\Gamma = 1.53$ fs is the core hole clock time (see Fig. 12.4).

From Fig. 14.10 it is apparent that in the long time or steady state limit, an interesting relation exists between the incident mode and atomic emission bandwidth in the form $\Delta_{p\mathbf{k}} = 2\pi^2\Gamma$. In the following we will take a closer look at this relationship involving the peculiar factor $2\pi^2$.

14.6.3 Mode-Based Versus Atom-Based Coherence Volumes

In the photon-atom interaction, the number of incident photons $n_{p\mathbf{k}}$ of polarization mode p, direction $\mathbf{k}$, and bandwidth $\Delta_{p\mathbf{k}}$ are assumed to be contained in a quantization box whose volume is given by (3.106) or

$$V_{p\mathbf{k}} = \lambda^3 \frac{\hbar\omega}{\Delta_{p\mathbf{k}}}. \tag{14.76}$$

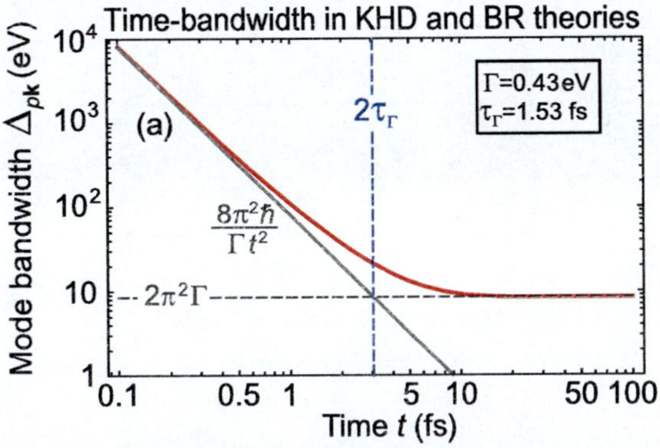

Fig. 14.10 Relationship between the mode bandwidth $\Delta_{p\mathbf{k}}$ of the incident photons in KHD theory and the instantaneous elapsed time t for a two-level atom in the BR theory. The red curve is given by (14.74) for zero detuning, $\Delta\mathcal{E} = 0$, and the vertical blue line at $t = 2\tau_\Gamma$ separates the short and long time behavior in the BR theory, with the short time response (14.75) shown as a gray line and the long time (equilibrium) response by the dashed gray horizontal line

The key quantity in the quantum mechanical description of the incident beam is the *photon density* or *energy density* which because of the constant speed of light, c, is reflected by the expression for the quantum mechanical intensity (10.9) or

$$I_0 = \frac{n_0 c\,\hbar\omega}{V} = \frac{n_{p\mathbf{k}} c\,\hbar\omega}{V_{p\mathbf{k}}}. \tag{14.77}$$

Here n_0 is the number of photons of energy $\hbar\omega$ in the "volume" V. This *normalization volume* is just an artificial construct, typically taken as a box, that falls out when physical observables are calculated. The intensity I_0 may therefore be expressed in different ways by preserving the density $n_0/V = n_{p\mathbf{k}}/V_{p\mathbf{k}}$.

It is often convenient to consider another specific "volume", namely that associated with an atom. Since each resonant electronic decay of energy $\hbar\omega$ has a definite linewidth Γ, we may write (14.77) also in terms the number of photons n_Γ of bandwidth Γ contained in the *atom-specific* volume V_Γ as

$$I_0 = \frac{n_{p\mathbf{k}} c\,\hbar\omega}{V_{p\mathbf{k}}} = \frac{n_\Gamma c\,\hbar\omega}{V_\Gamma}. \tag{14.78}$$

The photon density is conserved according to $n_{p\mathbf{k}}/V_{p\mathbf{k}} = n_\Gamma/V_\Gamma$. We obtain the following result [22].

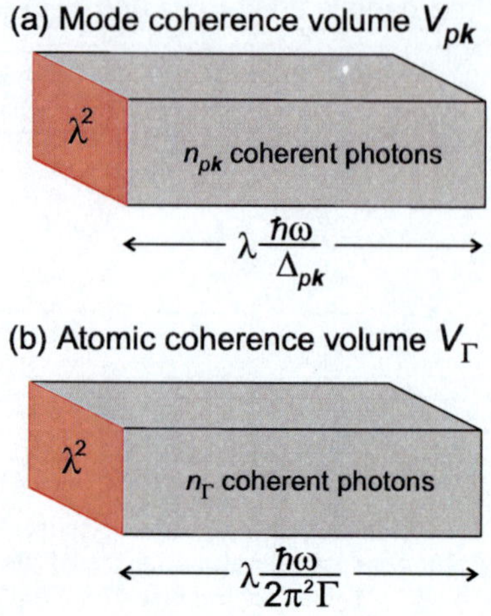

Fig. 14.11 Illustration of different longitudinal x-ray coherence volumes with the same lateral coherence area. **a** The *mode coherence volume* $V_{\mathbf{pk}}$ containing $n_{\mathbf{pk}}$ photons is the product of the lateral minimum coherence area $A_{\mathrm{coh}} = \lambda^2$ and the longitudinal coherence length $\ell_{\mathrm{coh}} = \lambda\, \hbar\omega/\Delta_{p\mathbf{k}}$. **b** In the *atomic clock coherence volume* V_Γ containing n_Γ photons, the longitudinal coherence length is given by $\ell_{\mathrm{coh}} = \lambda\, \hbar\omega/(2\pi^2\Gamma)$, where Γ is the atom-specific decay width

We can define an atom-specific coherence volume according to

$$V_\Gamma = \frac{\hbar\omega\,\lambda^3}{2\pi^2\Gamma}. \tag{14.79}$$

It contains the number of photons n_Γ of bandwidth Γ and is linked to the number of photon in mode $p\mathbf{k}$ of bandwidth $\Delta_{p\mathbf{k}}$ by

$$n_\Gamma = n_{p\mathbf{k}}\, \frac{\Delta_{p\mathbf{k}}}{2\pi^2\Gamma}. \tag{14.80}$$

This equation describes the density-conserving conversion of single mode photons of wavevector $\mathbf{k}$ and bandwidth $\Delta_{p\mathbf{k}}$ into continuous-mode photons of bandwidth Γ that may propagate into any $\mathbf{k}$ direction in the entire 4π solid angle.

The link of the numbers $n_{p\mathbf{k}}$ in volume $V_{p\mathbf{k}}$ and n_Γ in volume V_Γ may be envisioned as illustrated in Fig. 14.11. The associated photon fluxes may simply be pictured as the number of photons of bandwidth $\Delta_{\hbar\omega}$ and longitudinal coherence length $\ell = \lambda\, \hbar\omega/\Delta_{\hbar\omega}$ that flow through an area λ^2 with the speed of light.

We can summarize as follows.

For the description of x-ray interactions with atoms, we may use different normalization volumes that contain the incident photons, defined either by the illuminated sample volume, V, illustrated in Fig. 3.2, or the volumes illustrated in Fig. 14.11.

When at low incident intensity the interaction is treated in the *time-dependent* BR theory versus the *bandwidth-dependent* KHD theory, care has to be exercised since the time-bandwidth conversion between the two theories is given by (14.74), rather than by a simple linear time-bandwidth uncertainty product.

In summary, we have the following relations,

$$\text{Illuminated sample volume:}\ \ V = A\,d$$
$$\text{Mode coherence volume:}\ \ V_{pk} = \lambda^3 \frac{\hbar\omega}{\Delta_{pk}}$$
$$\text{Atomic coherence volume:}\ \ V_\Gamma = \lambda^3 \frac{\hbar\omega}{2\pi^2\Gamma}. \tag{14.81}$$

The photon intensity is then given by the number of photons of energy $\hbar\omega$ that traverse these volumes at the speed of light $c = \lambda\omega/(2\pi)$ as

$$\boxed{I_0 = \frac{n_0 c\,\hbar\omega}{V} = \frac{n_{pk} c\,\hbar\omega}{V_{pk}} = \frac{n_\Gamma c\,\hbar\omega}{V_\Gamma}.} \tag{14.82}$$

We can also rewrite the incident photon flux $\Phi_0 = I_0/\hbar\omega$ in (14.14) as

$$\Phi_0 = \frac{n_{pk} c}{\lambda^3} \frac{\Delta_{pk}}{\hbar\omega} = \frac{n_\Gamma c}{\lambda^3} \frac{2\pi^2\Gamma}{\hbar\omega}. \tag{14.83}$$

14.6.4 Zero-Point Field in the Bloch-Rabi Formalism

In conjunction with the KHD theory, we have discussed in Sect. 12.2.2.1 how spontaneous x-ray emission can be associated with the zero-point (ZP) field which can be expressed in terms of the mode volume according to (12.3) or

$$|E^k_{\rm ZP}|^2 = \frac{\hbar\omega}{2\epsilon_0 V_{pk}} = \frac{\Delta_{pk}}{2\epsilon_0\lambda^3}. \tag{14.84}$$

The same relationship also readily follows from the BR theory. Again, the zero-point field can be derived by comparing the spontaneous and stimulated scattering rates which will be equal if the incident field has the strength of the ZP field. By use of (14.69), the ZP field is determined from the condition

$$\frac{\mathcal{W}^{\mathrm{spon}}_{\mathrm{REXS}}}{\mathcal{W}^{\mathrm{stim}}_{\mathrm{REXS}}} = \frac{\Gamma_x \Gamma}{\Gamma_{\mathrm{R}}^2} = 1. \tag{14.85}$$

We can now insert the value of the Rabi energy from (14.88) to express the ZP field and obtain it either in the form (14.84) as the number of virtual photons in the volume $V_{p\mathbf{k}}$ or equivalently in terms of the number of virtual photons in the volume V_Γ and with (14.81) we obtain

$$|E^{\Gamma}_{\mathrm{ZP}}|^2 = \frac{\hbar\omega}{2\epsilon_0 V_\Gamma} = \frac{\pi^2 \Gamma}{\epsilon_0 \lambda^3}. \tag{14.86}$$

The *energy density* of the zero-point field per photon of energy $\hbar\omega$ is given by $\hbar\omega/V$, where V is either $V_{p\mathbf{k}}$ or V_Γ, so that

$$\text{ZP energy density:}\ \mathcal{E}_{\mathrm{ZP}} = \frac{\Delta_{p\mathbf{k}}}{\lambda^3} = \frac{2\pi^2\Gamma}{\lambda^3}. \tag{14.87}$$

In the following we will show that the bandwidth relation $\Delta_{p\mathbf{k}} = 2\pi^2\Gamma$ holds for all steady-state or long time absorption, emission and scattering rates in the BR and KHD theories up to incident intensities where saturation effects occur.

14.7 Link of BR Rates and KHD Rates in the Steady-State

We now consider the case of arbitrary incident intensity in the long time limit, where the two-level atom assumes an equilibrium population in the lower and upper states as discussed in Sect. 14.5.4 and illustrated in Fig. 14.8. For this case, the comparison of the BR and KHD theories is again facilitated by the existence of a closed form expression for the populations, expressed for ρ_{22}^{∞} by (14.56).

14.7.1 Steady-State Rate Expressions

In the long time limit or steady state, all relevant rate expressions may be summarized as follows.

The Rabi rate per atom is given by (14.38) and with the new expressions (14.77) and (14.78) we have

$$\Gamma_{\mathrm{R}}^2 = I_0 \frac{\lambda^3}{2\pi^2 c} \Gamma_x = n_{p\mathbf{k}} \frac{\Delta_{p\mathbf{k}}}{2\pi^2} \Gamma_x = n_\Gamma \Gamma \Gamma_x \tag{14.88}$$

where the dipolar transition width is given by (14.1) and linked to the peak spontaneous XAS cross section by (14.18). We have

$$\Gamma_x = \frac{8\pi^2 \alpha_f \hbar\omega}{\lambda^2} |\langle b|\mathbf{r}\cdot\boldsymbol{\epsilon}_p|a\rangle|^2 = \frac{\pi\Gamma}{\lambda^2}\,\sigma^{\diamond}_{\mathrm{XAS}} \tag{14.89}$$

In equilibrium (long time limit), the excited state population is given by (14.56) or

$$\rho_{22}^{\infty} = \frac{\Gamma_{\mathrm{R}}^2/4}{(\hbar\omega-\mathcal{E}_0)^2+\Gamma^2/4+\Gamma_{\mathrm{R}}^2/2} = \frac{n_\Gamma \frac{\Gamma_x}{\Gamma}\,\Gamma^2/4}{(\hbar\omega-\mathcal{E}_0)^2+\Gamma^2/4\left[1+2n_\Gamma\frac{\Gamma_x}{\Gamma}\right]} \tag{14.90}$$

and $\rho_{22}^{\infty} = 1-\rho_{11}^{\infty}$. This compares to the natural Lorentzian lineshape

$$\mathcal{L} = \frac{\Gamma^2/4}{(\hbar\omega-\mathcal{E}_0)^2+\Gamma^2/4}. \tag{14.91}$$

In all cases the quantities Γ_i represent transition energy widths (FWHM, dimension [energy]). In the following we denote the corresponding transition rates as $\mathcal{W}_i = \Gamma_i/\hbar$ (dimension [1/time]) and denote their low intensity limit for $\Gamma_{\mathrm{R}} \ll \Gamma$ by a $\diamond$ superscript. The various rates are given by

Absorption:

$$\mathcal{W}_{\mathrm{XAS}} = \frac{\Gamma_{\mathrm{R}}^2}{\hbar\Gamma}\,\rho_{11}^{\infty}, \quad \overset{\Gamma_{\mathrm{R}}\ll\Gamma}{\Longrightarrow} \quad \mathcal{W}^{\diamond}_{\mathrm{XAS}} = n_\Gamma\,\frac{\Gamma_x}{\hbar} \tag{14.92}$$

Spontaneous Auger:

$$\mathcal{W}_{\mathrm{A}} = \frac{\Gamma^{\mathrm{A}}}{\hbar}\,\rho_{22}^{\infty}, \quad \Longrightarrow \quad \mathcal{W}^{\diamond}_{\mathrm{A}} = n_\Gamma\,\frac{\Gamma_x\,\Gamma^{\mathrm{A}}}{\hbar\Gamma}\,\mathcal{L} \tag{14.93}$$

Spontaneous Scattering:

$$\mathcal{W}^{\mathrm{spon}}_{\mathrm{REXS}} = \frac{\Gamma_x}{\hbar}\,\rho_{22}^{\infty}, \quad \Longrightarrow \quad \mathcal{W}^{\diamond}_{\mathrm{REXS}} = n_\Gamma\,\frac{\Gamma_x^2}{\hbar\Gamma}\,\mathcal{L} \tag{14.94}$$

Stimulated Scattering:

$$\mathcal{W}^{\mathrm{stim}}_{\mathrm{REXS}} = \frac{\Gamma_{\mathrm{R}}^2}{\hbar\Gamma}\,\rho_{22}^{\infty}, \quad \Longrightarrow \quad \mathcal{W}^{\diamond\,\mathrm{stim}}_{\mathrm{REXS}} = n_\Gamma^2\,\frac{\Gamma_x^2}{\hbar\Gamma}\,\mathcal{L}. \tag{14.95}$$

The REXS expressions correspond to the case where all of the scattered radiation is collected, i.e. when an integration is performed over the emitted photon energy. Since at high intensity $\rho_{11}^{\infty} = \rho_{22}^{\infty} = 0.5$, we obtain $\mathcal{W}_{\rm XAS} = \mathcal{W}_{\rm REXS}^{\rm stim}$ so that the XAS and stimulated REXS rates become the same.

At high intensity, $\Gamma_{\rm R} \gg \Gamma$, all BR rates show a power-broadened behavior on *detuning energy* $\hbar\omega - \mathcal{E}_0$ through their dependence on ρ_{11}^{∞} or ρ_{22}^{∞} as illustrated in Fig. 14.9. The detuning response is determined by the *incident* bandwidth which is implicitly assumed to be negligibly small. For the absorption case, the detuning response also determines the measured XAS resonance width. For the REXS case, the linewidth of the emitted photons is *not* determined by the detuning response but by conservation of the energy bandwidth between incident and emitted photons. The expressions (14.94) and (14.95) correspond to the case where the energy of the emitted radiation is not resolved but integrated over. This corresponds to integration of the Dirac δ-function in (13.12) over $\hbar\omega_2$ for negligible instrumental resolution and similarly integration over the Gaussian in (13.13) for finite instrumental resolution. In both cases, the integration yields 1.

The BR rates reduce to the KHD rates at low intensity with the exception of the XAS lineshape. As shown by (14.92) the low intensity BR rate does not contain the Lorentzian lineshape factor present in the KHD expression (14.67), but the resonant values are the same. With increasing intensity, the BR XAS rate assumes a resonant lineshape because $\rho_{11}^{\infty}(\hbar\omega)$ begins to deviate from a constant unit value as shown in Fig. 14.9a.

14.7.2 Illustration of the BR Rates and Their Saturation

As an example, we have plotted in Fig. 14.12 the change of the various rates for the L_3 resonance excitation ($\hbar\omega = \mathcal{E}_0$) of Co with incident intensity I_0. The curves in Fig. 14.12 were calculated by assuming resonance $\hbar\omega = \mathcal{E}_0$ according to (14.92)–(14.95). We used the parameters for the Co L_3 resonance $\sigma_{\rm XAS}^{\rm model} = 6.25\,$Mb (see Fig. 14.2) or $\Gamma_{\rm R}\,[\text{eV}] = 3.8\times10^{-3}\sqrt{I_0}$ (see (14.39)) and $\Gamma = 0.43\,$eV. The steady-state populations as a function of intensity are those plotted previously in Fig. 14.8. The plotted transition linewidths $\hbar\mathcal{W}_i$ in units of [eV] correspond to rates $\mathcal{W}_i$ where $\hbar = 0.658\,$eV fs. The linewidths calculated by means of the BR theory are plotted as solid lines, those in the low intensity approximation as dashed lines, which also represent the KHD theory, whose validity is emphasized by gray shading. From our results shown in Figs. 14.8b and 14.12 we can state as follows.

For a two-level atom, the KHD theory remains valid as long as the Rabi energy $\Gamma_{\rm R}$, which according to (14.39) scales with the incident intensity as

$\sqrt{I_0}$, remains smaller than the natural decay width Γ of the excited state by $\Gamma_R \leq 0.1\Gamma$.

While the spontaneous rates are proportional to the incident intensity, the stimulated REXS rate is proportional to the square of the incident intensity. It becomes equal to the spontaneous rate around 10 mJ/cm^2/fs, meaning that the stimulating incident field is now as strong as the zero-point field, which corresponds to the condition $n_\Gamma = 1$, as indicated on top of Fig. 14.12.

At an incident intensity of > 1000 mJ/cm^2/fs, the BR widths $\hbar\mathcal{W}_i$ (solid lines) begin to deviate from the KHD widths $\hbar\mathcal{W}_i^\diamond$ (dashed). The stimulated REXS rate becomes equal to the Auger decay rate around 10 J/cm^2/fs, and at this point we have $\Gamma_R = \Gamma$. According to (14.56) and Fig. 14.8a, the upper state population has increased to $1/3$ and the lower state population is reduced to $2/3$. Both spontaneous rates now begin to saturate and even the stimulated rate increases at a reduced rate,

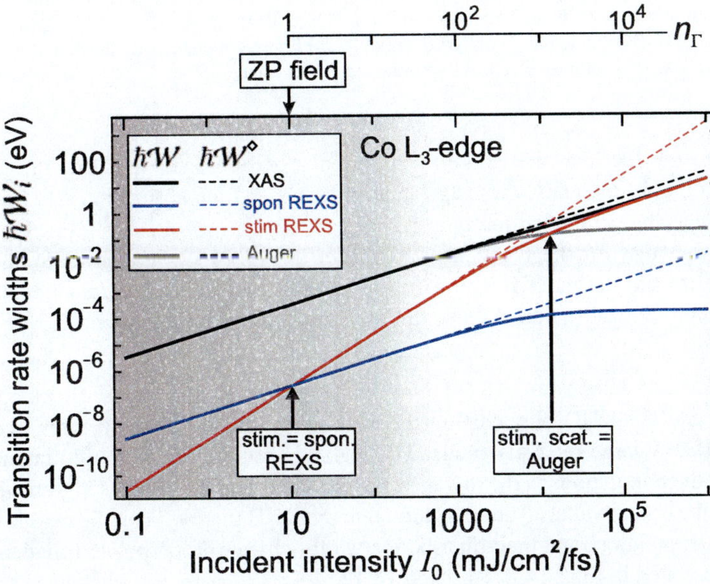

Fig. 14.12 Calculated BR widths $\hbar\mathcal{W}_i$ (solid lines) and low intensity approximation widths $\hbar\mathcal{W}_i^\diamond$ (dashed lines) for the *equilibrium* (long time) case versus incident intensity I_0. The widths were calculated according to (14.92)–(14.95). We have assumed resonance, $\hbar\omega = \mathcal{E}_0$, and parameters for the L$_3$-edge in Co metal, $\sigma_{\mathrm{XAS}}^{\mathrm{model}} = 6.25$ Mb, $\Gamma_R = 3.8 \times 10^{-3}\sqrt{I_0}$ [eV] and $\Gamma = 0.43$ eV. Note that in the low intensity case the XAS and Auger rates are the same. The stimulated and spontaneous REXS rates become the same at the left vertical line, corresponding to one stimulating photon ($n_\Gamma = 1$) in the atomic clock coherence volume. At the second vertical line, the stimulated REXS rate becomes as large as the Auger rate. At larger intensities, the BR XAS (solid black) and stimulated REXS (solid red) rates become the same. The gray shaded region identifies the validity range of the KHD theory

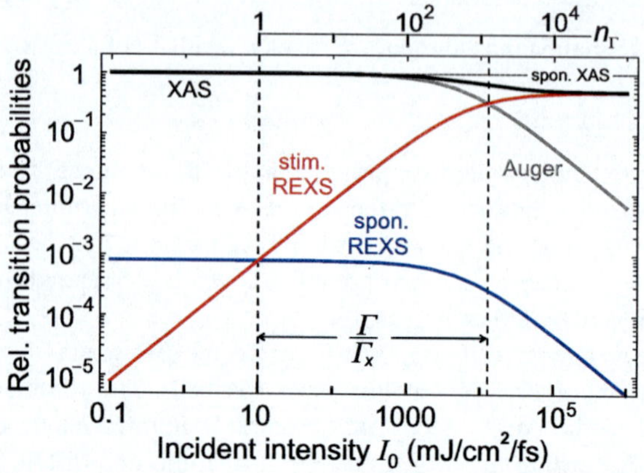

Fig. 14.13 Resonant, $\hbar\omega = \mathcal{E}_0$, relative BR transition probabilities as a function of incident intensity. We have used the same parameters as for Fig. 14.12. The shown transition probabilities are the respective BR rates, normalized to the absorption rate $\mathcal{W}^{\diamond}_{\mathrm{XAS}}$. Colors distinguish the relative XAS (black), Auger (gray), spontaneous REXS (blue), and stimulated REXS (red) probabilities. At the bottom we also indicate that it takes $n_\Gamma = \Gamma/\Gamma_x \simeq 10^3$ photons (see Fig. 14.12) to make the stimulated REXS equal to the Auger probability. At the highest intensities the XAS and stimulated REXS probabilities saturate at half the spontaneous XAS value

owing to the fact that the excited state population approaches that of the ground state. At higher intensity, the Auger rate saturates and the absorption rate becomes the same as the stimulated rate.

The fact that for the resonant case shown in Fig. 14.12 the solid BR lines are identical to the dashed KHD lines up to a remarkably high incident intensity of 1000 mJ/cm^2/fs, allows us to make the following important statement.

For a two-level atom excited at resonance, $\hbar\omega = \mathcal{E}_0$, the optical Bloch equations and the KHD perturbation theory give the same transition rates over a remarkably large intensity range. The KHD theory only fails at high intensity when the populations in the lower and upper states begin to deviate significantly from the binary values 0 and 1 assumed in the KHD theory. Then the peak value of the populations and their distribution widths change due to power broadening as illustrated in Figs. 14.8, 14.9 and 14.12.

Figure 14.12 shows that the spontaneous rate increase does not change at the point $\sim$10 mJ/cm^2/fs where the stimulated rate becomes the same. Only when the stimulated rate becomes as strong as the Auger rate do, we see a significant change in all rates. This is more clearly shown in Fig. 14.13 where all rates are normalized to the absorption rate $\mathcal{W}^{\diamond}_{\mathrm{XAS}}$. At the highest incident intensities the two dominant rates, the absorption and stimulated rates, contain half the power each, owing to the fact that the ground and excited states are equally populated.

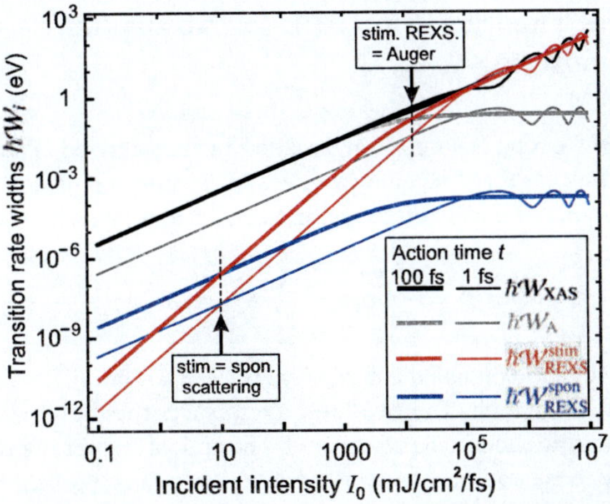

Fig. 14.14 Transition widths $\hbar W_i$ versus incident intensity for action times $t = 1$ fs (thin lines) and 100 fs (thick lines) according to (14.97). Parameters are the same as for Figs. 14.12 and 14.13. The thick $t = 100$ fs curves are indistinguishable from the BR curves of the same color in Fig. 14.12 which correspond to $t \to \infty$

14.7.3 Time Dependence of Rates at Resonance

For resonant excitation, we can follow the temporal evolution of all BR rates as a function of the "action time" t by use of (14.48) or

$$\rho_{22}(t) = \frac{\Gamma_R^2}{\Gamma^2 + 2\Gamma_R^2} \left| 1 - \left[\cosh(\omega_R t) + \frac{3\Gamma}{4\hbar\omega_R} \sinh(\omega_R t) \right] e^{-3\Gamma t/(4\hbar)} \right|, \quad (14.96)$$

where ω_R is given by (14.47) and $\rho_{11}(t) = 1 - \rho_{22}(t)$. The rates in (14.92)–(14.95) then assume the forms

$$\mathcal{W}_{XAS} = \frac{\Gamma_R^2}{\hbar\Gamma} \rho_{11}(t), \quad \mathcal{W}_A = \frac{\Gamma_A}{\hbar} \rho_{22}(t)$$
$$\mathcal{W}^{spon}_{REXS} = \frac{\Gamma_x}{\hbar} \rho_{22}(t), \quad \mathcal{W}^{stim}_{REXS} = \frac{\Gamma_R^2}{\hbar\Gamma} \rho_{22}(t). \quad (14.97)$$

In Fig. 14.14 we have plotted the XAS, Auger, and spontaneous and stimulated scattering rates as a function of incident intensity for two action times 1 fs and 100 fs. As the pulse length increases one needs higher intensity to achieve the same transition rate. The crossing of the rates with incident power is independent of time.

14.8 Optical Theorem: Sum Rule for Absorption and Scattering

In the BR theory the total absorption and scattering rates derived in Sect. 14.7 satisfy an important *sum rule* that links the two rates. This sum rule is a manifestation of the *optical theorem* previously discussed in Sect. 7.6 and given by (7.56) or

$$\mathrm{Im}\, f(\mathbf{Q}{=}0) = \frac{k}{4\pi}\,\sigma_{\mathrm{tot}} = \frac{1}{2\lambda}\,(\sigma_{\mathrm{abs}} + \sigma_{\mathrm{scat}}) \simeq \frac{\sigma_{\mathrm{abs}}}{2\lambda}. \tag{14.98}$$

Since the essence of the optical theorem is *energy conservation*, it must hold even in the presence of stimulation. The two-level atom is a particularly simple case since only elastic scattering needs to be considered. The optical theorem can then simply be stated by energy conservation between the different transition probabilities illustrated in Fig. 14.13. The balance between the spontaneous and stimulated contributions of the total XAS and REXS rates is more clearly revealed by the normalized probabilities plotted in Fig. 14.15.

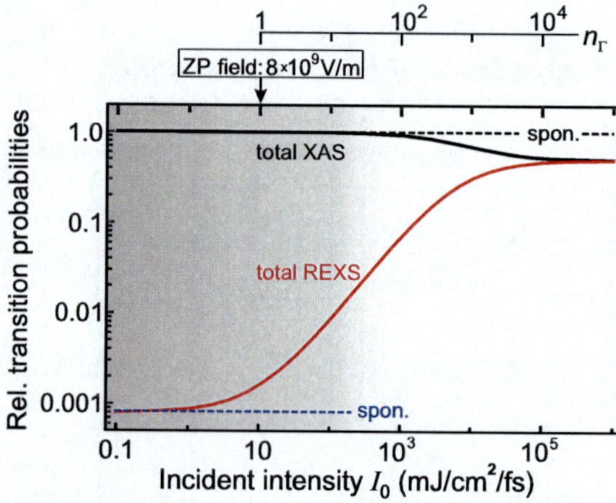

Fig. 14.15 Illustration of the sum rule given by (14.99) for resonant excitation, $\hbar\omega = \mathcal{E}_0$. Plotted are the relative BR XAS probability (black) and sum of the spontaneous and stimulated BR REXS (red) probabilities, given by the respective BR rates, normalized to the spontaneous (low intensity) XAS rate. We have used the same parameters as for Figs. 14.12 and 14.13 and again identified the validity range of the KHD theory by gray shading

14.8.1 XAS and REXS Cross-Section Sum Rule

The optical theorem (14.98) can be expressed in terms of the relative atomic BR transition probabilities in Fig. 14.15 which account for rebalancing the up and down rates due to population changes. By use of the labels of the black and red curves in the figure we have the probability balance equation

$$\underbrace{P_{\mathrm{XAS}}^{\mathrm{spon}} - P_{\mathrm{REXS}}^{\mathrm{stim}}}_{\text{total XAS}} + \underbrace{P_{\mathrm{REXS}}^{\mathrm{spon}} + P_{\mathrm{REXS}}^{\mathrm{stim}}}_{\text{total REXS}} \simeq P_{\mathrm{XAS}}^{\mathrm{spon}}. \tag{14.99}$$

With increasing incident intensity, the stimulated gain in the REXS probability, $P_{\mathrm{REXS}}^{\mathrm{stim}}$, is compensated by an equal loss in the XAS probability. Since the areas under the XAS and stimulated REXS probability distributions are also preserved at all incident intensities, as shown in Fig. 14.9a, we have energy conservation. Note that the Auger rate does not explicitly enter into the sum rule because at low incident intensity it is identical to the (spontaneous) absorption rate, and at high intensity it is negligible.

The sum rule holds over the entire intensity range in the BR formalism, which properly accounts for the population changes that lead to a trade-off between the absorption and stimulated channels. Since the sum rule of probabilities represents energy conservation, it can also be phrased in terms of cross sections and similar to (14.98) we have

$$\boxed{\text{Atomic sum rule:}\ \underbrace{\sigma_{\mathrm{XAS}}^{\mathrm{spon}} - \sigma_{\mathrm{REXS}}^{\mathrm{stim}}}_{\sigma_{\mathrm{XAS}}^{\mathrm{tot}}} + \underbrace{\sigma_{\mathrm{REXS}}^{\mathrm{spon}} + \sigma_{\mathrm{REXS}}^{\mathrm{stim}}}_{\sigma_{\mathrm{REXS}}^{\mathrm{tot}}} \simeq \sigma_{\mathrm{XAS}}^{\mathrm{spon}}.} \tag{14.100}$$

The sum rule holds since $\sigma_{\mathrm{XAS}}^{\mathrm{spon}} \gg \sigma_{\mathrm{REXS}}^{\mathrm{spon}}$. The spontaneous XAS cross section is obtained from (14.92) with (14.83) as

$$\sigma_{\mathrm{XAS}}^{\mathrm{spon}} = \frac{W_{\mathrm{XAS}}^{\diamond}}{\Phi_0} = \frac{\lambda^2 \Gamma_x}{\pi \Gamma}. \tag{14.101}$$

The spontaneous parts of the total REXS cross section are obtained from (14.94) with (14.83) as

$$\sigma_{\mathrm{REXS}}^{\mathrm{spon}} = \frac{W_{\mathrm{REXS}}^{\mathrm{spon}}}{\Phi_0} = \frac{\lambda^2 \Gamma_x}{\pi \Gamma} \frac{\rho_{22}^{\infty}}{n_\Gamma} \simeq \frac{\lambda^2 \Gamma_x}{\pi \Gamma} \frac{\Gamma_x}{\Gamma}. \tag{14.102}$$

On the far right we have neglected the decrease at high intensity and assumed the low intensity value of $\rho_{22}^{\infty}/n_\Gamma = \Gamma_x/\Gamma$ (see right side of (14.94)). It corresponds to the dashed blue line in Fig. 14.15. The stimulated REXS part is derived from (14.95) with (14.83) as

$$\sigma_{\mathrm{REXS}}^{\mathrm{stim}} = \frac{\mathcal{W}_{\mathrm{REXS}}^{\mathrm{stim}}}{\Phi_0} = \frac{\lambda^2 \Gamma_x}{\pi \Gamma} \rho_{22}^{\infty}. \tag{14.103}$$

14.8.2 Atom Transmission Sum Rule

The sum rule (14.99) links the total XAS and REXS transition probabilities without considering the directions of the absorbed and scattered photons. We can rewrite the sum rule in terms of the change due to loss and gain in the direction defined by the incident photons. In other words we want to derive a sum rule that reflects photon transmission by an atom in the presence of all linear and NL processes in the BR theory. Such a formulation will help us in the next chapter when we consider the transmission through a solid. There we will see that loss, conventionally attributed to "absorption", can indeed by compensated by stimulated forward scattering.

To see how transparency of an atom arises, we consider a hypothetical incident beam whose lateral size is that of the atomic area, which we will denote as A. We then consider what fraction of the photons going into the atomic area will come back out. We know from Sect. 13.4.1 that the *spontaneous REXS signal* emitted by the atom consist of two parts containing equal energy. Half the energy is emitted into 4π, the other into the forward direction. Both parts are, however, negligible relative to the spontaneous XAS loss at low incident intensity and stimulated REXS gain at high intensity. Hence we simply ignore it.

The transmission loss through the atom will then arise from photons that hit the atom in the fractional atomic area given by the atomic cross section

$$\sigma_{\mathrm{XAS}}^{\mathrm{spon}} = \underbrace{\frac{\lambda^2}{\pi}}_{\substack{\text{atom}\\ \text{area}}} \underbrace{\frac{\Gamma_x}{\Gamma}}_{\substack{\text{loss}\\ \text{probability}}} . \tag{14.104}$$

As indicated by underbrackets, in this formulation, one may consider the Breit–Wigner cross-sectional area λ^2/π (see Sect. 6.5.1) to represent the atomic area A, and the factor $\Gamma_x/\Gamma \leq 1$ is just the absorption probability of the incident photons within this area.[5] The transmission probability sum rule of a single atom can then be written by denoting loss by a minus sign and gain by a plus sign as

$$\begin{aligned} P_{\mathrm{trans}}^{\mathrm{atom}} &= 1 - \underbrace{\frac{\pi}{\lambda^2}\frac{\lambda^2 \Gamma_x}{\pi \Gamma}}_{\substack{\text{spont. loss}\\ \text{probability}}} + \underbrace{\frac{\pi}{\lambda^2}\frac{\lambda^2 \Gamma_x}{\pi \Gamma}\frac{\Gamma_x}{\Gamma}}_{\substack{\text{spon. gain}\\ \text{probability}}} + \underbrace{2\frac{\pi}{\lambda^2}\frac{\lambda^2 \Gamma_x}{\pi \Gamma}\rho_{22}^{\infty}}_{\substack{\text{stim. gain}\\ \text{probability}}} \\ &= 1 - \frac{\Gamma_x}{\Gamma} + \frac{\Gamma_x^2}{\Gamma^2} + 2\frac{\Gamma_x}{\Gamma}\rho_{22}^{\infty}. \end{aligned} \tag{14.105}$$

[5] In the optical region one has $\Gamma_x = \Gamma$ as discussed in Sect. 6.5.1.

The factor of 2 in front of the stimulated gain term arises from calling the NL reduction in total "absorption" in (14.100) a stimulated REXS gain in forward scattering. At low incident intensity we have $\rho_{22}^{\infty} \to 0$ and at high incident intensity $\rho_{22}^{\infty} \to 1/2$.

14.8.3 BR Versus KHD Stimulated Enhancement: Saturation

The establishment of an equilibrium between the excitation and stimulated decay rates shown in Fig. 14.15, known as saturation of the XAS channel, causes a difference between the KHD and BR theory. This is best seen by expressing the stimulated BR rate in the high intensity limit $n_\Gamma \to \infty$ in terms of the spontaneous rate. By use of (14.90), (14.94) and (14.95) we find the saturation resonance value

$$\mathcal{W}_{\mathrm{REXS}}^{\mathrm{stim}} = \mathcal{W}_{\mathrm{REXS}}^{\diamond} \underbrace{\frac{\Gamma}{2\Gamma_x}}_{\text{enhancement}} . \tag{14.106}$$

The maximum enhancement factor $\Gamma/(2\Gamma_x)$ is plotted in Fig. 14.16 for the K-edge of low-Z atoms and the L_3-edge of the 3d transition metals.

Since according to Fig. 14.15 the stimulated rate reaches half of the absorption rate at high intensity, one often finds the following expression for the dependence of the stimulated scattering rate on the spontaneous absorption rate and the incident and saturation intensities

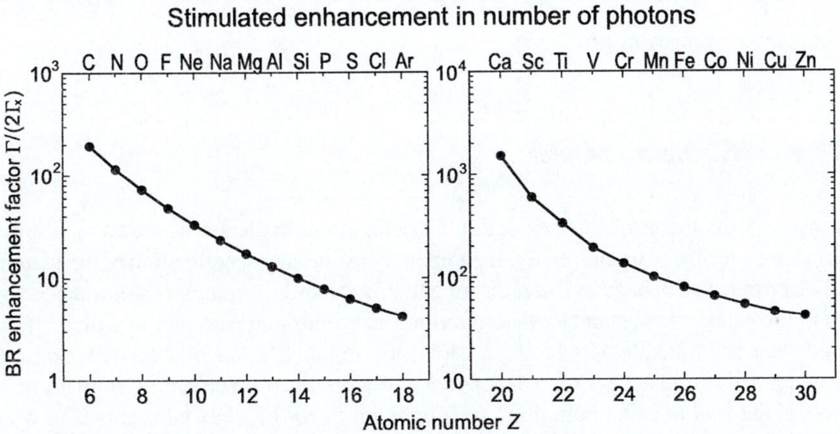

Fig. 14.16 Maximum enhancement factor at resonance of the stimulated REXS rate over the spontaneous rate in the BR theory according to (14.106)

$$\boxed{\text{Saturation: } \mathcal{W}^{\text{stim}}_{\text{REXS}} = \mathcal{W}^{\diamond}_{\text{XAS}} \frac{I_0/I_{\text{sat}}}{1 + I_0/I_{\text{sat}}} \longrightarrow \frac{\mathcal{W}^{\diamond}_{\text{XAS}}}{2},} \tag{14.107}$$

where at high intensity $n_\Gamma \rightarrow \Gamma/(2\Gamma_x)$ or

$$I_0 = n_\Gamma \frac{\pi^2 c\, 2\Gamma}{\lambda^3} \longrightarrow I_{\text{sat}} = \frac{\pi^2 c\, \Gamma^2}{\lambda^3 \Gamma_x}. \tag{14.108}$$

We note that the enhancement factors shown in Fig. 14.16 do not account for the (much larger) angular enhancement in photon density caused by the confinement of stimulated emission into the forward direction.

14.9 BR, KHD, and Einstein Treatment of a Two-Level System

In this section we will compare the BR and KHD treatments of a two-level atom to the famous phenomenological Einstein model put forward 1916/17 to describe the interaction of optical EM radiation with matter [5, 6]. As pointed out already in the Introduction chapter of our book (see Sect. 1.2.2), Einstein's work, clearly distinguished spontaneous and stimulated emission for the first time and provided a quantum derivation of Planck's empirical radiation law [23, 24] published 17 years earlier. Below, we start with a short review of the model and then show how it is a special case of the KHD and BR theories that were developed later in time. In fact, one of the early successes of Dirac's theory [25] was the derivation of Einstein's spontaneous emission rate.

14.9.1 Einstein's Model

Einstein's treatment of the interaction of radiation with electronic states is simply amazing because of its simplicity. It completely avoids any specification of the nature of electronic states involved, and through thermodynamic arguments alone succeeds in linking absorption, spontaneous, and stimulated emission rates of a so-called black body, a term coined by Kirchhoff in 1860 [26]. In practice, the blackbody spectrum was defined by the temperature dependent emission spectrum emerging from a small hole in the wall of a platinum (high melting point of 2041 K) box whose interior was originally blackened with iron oxide [27]. The practical implementation of a true blackbody is actually very difficult as reviewed in [28]. The blackbody spectrum covers an energy range that depends on the temperature T of the interior wall of the box.

Einstein assumed that the emitted radiation resulted from molecules inside the box which were in thermal equilibrium with the box. The molecules were assumed

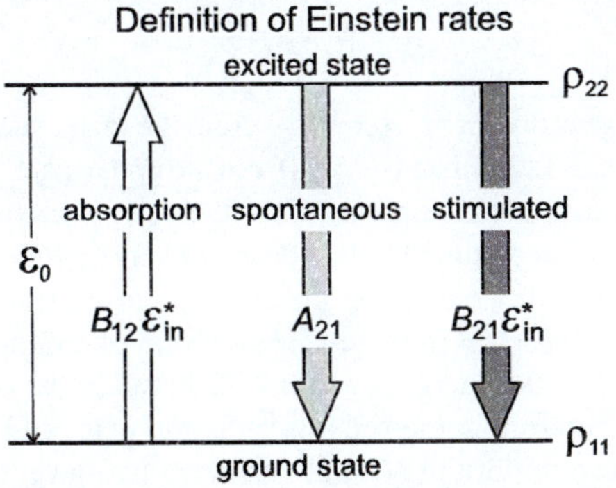

Fig. 14.17 Einstein model, balancing "up" and "down" transition rates for a two-level system with dimensionless fractional populations ρ_{11} and ρ_{22}, where $\rho_{11} + \rho_{22} = 1$. The rates are given by the Einstein coefficients A_{21}, and B_{12}, B_{21}, with the dimension [1/time], and the incident normalized energy $\mathcal{E}^*_{\text{in}} = \mathcal{E}_{\text{in}}/\mathcal{E}_{\text{ZP}}$, where $\mathcal{E}_{\text{in}}$ is the energy of the incident EM wave, and $\mathcal{E}_{\text{ZP}}$ is the zero-point energy which causes radiative spontaneous decays

to be two-level systems with a continuum of energy separations $\mathcal{E}_0$ and populations determined thermodynamically by a Boltzmann distribution. Einstein phenomenologically distinguished three different rates, absorption, spontaneous emission, and stimulated emission, in the two-level molecules that were determined by constant transition matrix elements but depended on temperature through the Boltzmann populations of the levels.

The work, which introduced the stimulated emission concept, preceded the invention of the maser and laser by more than 40 years. One may ask why Einstein did not already think of utilizing stimulated emission? In Einstein's model the maximum intensity or *saturation intensity* of a blackbody light source is approached when all radiation modes are filled with one photon, each. In practice, blackbody light sources have intensities that are about a factor of 10 less than the saturation intensity, and thus spontaneous emission dominates over stimulated emission. The concept of a laser requires the recognition that one needs to abandon the equilibrium concept and temporarily create a population inversion and extend the two-level concept to more levels with suitable relative decay times.

In the following we will briefly review the essence of Einstein's model by utilizing Fig. 14.17. The difference between the optical and x-ray cases is the prominent role of Auger decays which in the soft x-ray region completely dominate. However, we can simply account for Auger decays by adding its "ghost" linewidth contribution to the radiative spontaneous linewidth as discussed in Sect. 12.2.5.

As illustrated in Fig. 14.17, the Einstein model considers three types of first order transition rates in a two-level system. The response to an incident field is expressed as a time-dependent change in the populations of the lower and upper states, which are defined as unitless fractional populations that at any time obey the conservation law

$$\rho_{11}(t) + \rho_{22}(t) = 1. \tag{14.109}$$

The population change is driven by excitations from the lower state, i.e. x-ray absorption, and spontaneous and stimulated decays from the upper state. The Einstein model is based on two fundamental assumptions. First, the *energy density* of the light acting on the two-level system is assumed to be constant over the space of the system and constant in time. The latter means that the system can be assumed to be in equilibrium. Second, the *intensity distribution* or spectrum of the light is assumed to be smooth and broader than the energy separation $\mathcal{E}_0$ between the two levels. The two-level molecules are therefore immersed in a radiation field, and the Einstein model is designed to find the equilibrium populations in the two-level molecules.

The quantities A_{21}, B_{12}, and B_{21} in Fig. 14.17 are called *Einstein coefficients*. As reviewed by Hilborn [29, 30], they can be found with different dimensions in the literature. Following our formulation of the BR and KHD theories in terms of transition rates, we will also express the coefficients in terms of rates with dimension [1/time]. This is accomplished by defining the energy density of the photon bath by a normalized dimensionless quantity denoted $\mathcal{E}^*_{\rm in}$ in Fig. 14.17 according to

$$\mathcal{E}^*_{\rm in} = \frac{\mathcal{E}_{\rm in}}{\mathcal{E}_{\rm ZP}} = \frac{|E_{\rm in}|^2}{|E_{\rm ZP}|^2}. \tag{14.110}$$

It expresses the incident energy density $\mathcal{E}_{\rm in}$ in units of the zero-point energy density $\mathcal{E}_{\rm ZP}$ given by (14.87). The equilibrium transition rates are then determined by a rate equation for the population balance given by

$$\frac{d\rho_{11}}{dt} = -B_{12}\mathcal{E}^*_{\rm in}\,\rho_{11} + B_{21}\mathcal{E}^*_{\rm in}\,\rho_{22} + A_{21}\,\rho_{22} = -\frac{d\rho_{22}}{dt}. \tag{14.111}$$

The two-level system, assumed to consist of many molecules with different two-level separations, is immersed in a radiation bath where thermal equilibrium is accounted for by applying a Boltzmann distribution to the level populations. One then finds that the three Einstein coefficients are interrelated and the time-dependent solution of (14.111) can be obtained with $A = A_{21}$ and $B = B_{12} = B_{21}$. With the initial state condition $\rho_{11} = 1$ one obtains the solution

$$\boxed{\rho_{22}(t) = 1 - \rho_{11}(t) = \frac{B\mathcal{E}^*_{\rm in}}{A + 2B\mathcal{E}^*_{\rm in}}\,\{1 - \exp[-(A + 2B\mathcal{E}^*_{\rm in})\,t\,]\}.} \tag{14.112}$$

This reduces to the steady-state solution

$$\rho_{22}^{\infty} = 1 - \rho_{11}^{\infty} = \frac{B\mathcal{E}^*_{\rm in}}{A + 2B\mathcal{E}^*_{\rm in}}. \tag{14.113}$$

For $\mathcal{E}^*_{\rm in} = 0$ the excited state population evolves according to

$$\frac{d\rho_{22}}{dt} = -A\,\rho_{22} \tag{14.114}$$

which yield the result

$$\rho_{22}(t) = \rho_{22}(0)\,e^{-At}. \tag{14.115}$$

If the excited state decays exponentially with a natural line width $\Gamma = \Gamma^{A} + \Gamma_{x}$, we obtain the expected result,

$$\rho_{22}(t) = \rho_{22}(0)\,e^{-\Gamma t/\hbar}. \tag{14.116}$$

14.9.2 Reduction of the BR to the Einstein Theory

Since the Bloch equations cover various scenarios of incident light, ranging from narrow to broad bandwidth, one would expect that the Einstein theory should be recovered in the limit of low incident intensity and broad bandwidth. This is indeed the case.

At low intensity where $2\Gamma_{R}^{2} \ll \Gamma^{2}$ we can directly compare the analytical results of the BR theory given by (14.44) to the results of the broad bandwidth Einstein theory. If we integrate (14.44) given by

$$\rho_{22}(t)=\frac{\Gamma_{R}^{2}/4}{(\hbar\omega-\mathcal{E}_0)^2+(\Gamma/2)^2}\left\{1+e^{-\Gamma t/\hbar}-2\cos\left[\frac{\hbar\omega-\mathcal{E}_0}{\hbar}\,t\right]e^{-\Gamma t/(2\hbar)}\right\} \tag{14.117}$$

over the incident photon energy, corresponding to broadband incident radiation, we obtain

$$\int_{-\infty}^{\infty} \rho_{22}(t)\,d\hbar\omega = \frac{\pi\Gamma_{R}^{2}}{2\Gamma}\left\{1-e^{-\Gamma t/\hbar}\right\}. \tag{14.118}$$

We now need to take into account that our energy integration was over a Lorentzian lineshape. To obtain an equivalent expression to the broad flat-top energy distribution assumed in the Einstein theory, our integration result (14.118) needs to be multiplied by a factor $2/(\pi\Gamma)$ (see (10.36) and (10.37)), so that we obtain

$$\boxed{\text{Low intensity BR}:\quad \rho_{22}(t)=\frac{\Gamma_{R}^{2}}{\Gamma^{2}}\left\{1-e^{-\Gamma t/\hbar}\right\}.} \tag{14.119}$$

When we compare this result with the Einstein expression (14.112), which for low incident energy density or $2B\mathcal{E}_{in}^{*} \ll A$ reduces to

$$\boxed{\text{Low intensity Einstein}: \quad \rho_{22}(t) = \frac{B\mathcal{E}^*_{\text{in}}}{A}\left\{1 - \mathrm{e}^{-At}\right\}} \tag{14.120}$$

we find equivalence for

$$\boxed{A = \frac{\Gamma}{\hbar}, \quad \text{and} \quad B\mathcal{E}^*_{\text{in}} = \frac{\Gamma_{\text{R}}^2}{\hbar\Gamma} = \frac{n_\Gamma \Gamma_x}{\hbar}.} \tag{14.121}$$

This establishes the connection of the Einstein and BR theories for light of low intensity and broad bandwidth, encountered for conventional light sources.

14.9.2.1 Illustration of Difference of Einstein and BR Theories

The time-dependent solutions of the Bloch equations, integrated over a broad bandwidth window, do not in general merge into those of the Einstein rate equations. The reason is that the Bloch equations allow mixing of the lower and upper state wavefunctions through the coherences, which are left out of the Einstein rate equations. At low incident intensity the mixing contributions due to the coherences are negligible and the BR results merge into the Einstein results when integrated over a large enough bandwidth window as illustrated above.

At arbitrary intensities, the Einstein result is given by (14.112), which by use of (14.121) may be rewritten as

$$\boxed{\rho_{22}(t) = \frac{\Gamma_{\text{R}}^2}{\Gamma^2 + 2\Gamma_{\text{R}}^2}\left\{1 - \exp\left[-\frac{(\Gamma^2 + 2\Gamma_{\text{R}}^2)\,t}{\Gamma\,\hbar}\right]\right\}.} \tag{14.122}$$

This may be compared to the time dependent solution of the Bloch equation, $\rho_{22}(t)$, determined numerically and then integrated over bandwidth. For the Einstein and energy integrated BR result to have the same long time (equilibrium) value ρ_{22}^{∞}, the numerical BR result needs to be renormalized by multiplying by a factor of

$$f_{\text{norm}} = \frac{2}{\pi\Gamma}\frac{1}{\sqrt{1 + 2\Gamma_{\text{R}}^2/\Gamma^2}}. \tag{14.123}$$

This renormalization accounts for integration over the Lorentzian-like BR lineshape similar to our previous renormalization of (14.118).

The difference between the time dependence of the upper state populations $\rho_{22}(t)$ in the two theories is illustrated in Fig. 14.18, where the BR bandwidth integration effect is demonstrated for different incident intensities.

At the lowest intensities, $I_0 = 0.1\,\text{mJ/cm}^2\text{/fs}$ and $I_0 = 100\,\text{mJ/cm}^2\text{/fs}$, the BR theory is seen to seamlessly merge into the Einstein result with increasing bandwidth integration. This is no longer the case for $I_0 = 10^5\,\text{mJ/cm}^2\text{/fs}$ where the strong mixing of the lower and upper states by the coherences in the BR theory always causes a deviation from the Einstein theory due to remaining Rabi oscillations. In all cases, the threshold delay of the increase of $\rho_{22}(t)$ in the BR theory which is present for resonant excitation (gray line) relative to the Einstein theory reduces with increasing bandwidth integration and the equilibrium values at large t become the same.

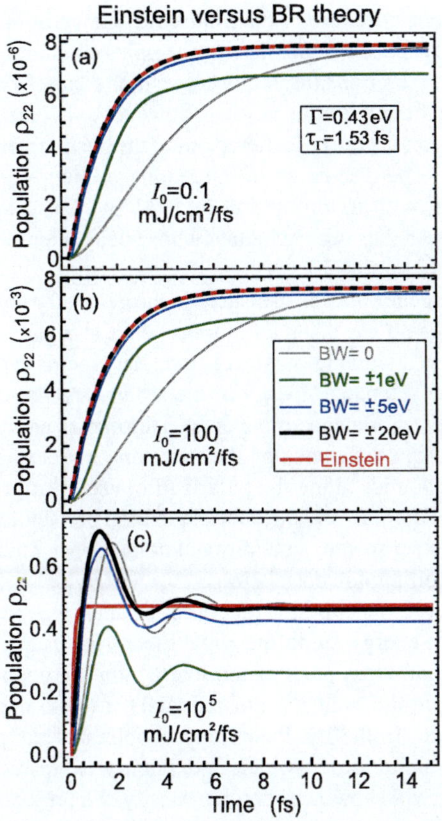

Fig. 14.18 Comparison of the time dependent populations $\rho_{22}(t)$ for three different incident intensities, **a** $I_0 = 0.1\,\text{mJ/cm}^2\text{/fs}$, **b** $I_0 = 100\,\text{mJ/cm}^2\text{/fs}$, and **c** $I_0 = 10^5\,\text{mJ/cm}^2\text{/fs}$. For each intensity the Einstein result, given by (14.122), shown in red is compared to the numerical solutions of the Bloch equations for the same parameters, $\Gamma_R = 3.8 \times 10^{-3}\sqrt{I_0}\,[\text{eV}]$ and $\Gamma = 0.43\,\text{eV}$. The resonant (no bandwidth integration) BR result is shown in gray and the bandwidth (BW) integrated and renormalized (multiplication by f_{norm} in (14.123)) results are shown for integration windows BW= $\pm 1\,\text{eV}$ in green, BW= $\pm 5\,\text{eV}$ in blue and BW= $\pm 20\,\text{eV}$ in black. In (**a**) and (**b**) the red and black curves are the same as indicated

14.10 Resonance Fluorescence

14.10.1 Introduction

The interplay of absorption, spontaneous scattering, and stimulated scattering near a resonance in a single two-level atom inserted into a beam is impossible to measure in a transmission geometry because of the miniscule change of intensity. We shall see in Chap. 15 that this will become possible for samples containing many atoms where the change in signal upon transmission becomes quite large. The dominance of the incident intensity for the single atom case can however be avoided by measurement of the intensity that is spontaneously scattered out of the beam, referred to as "resonance fluorescence" [7, 9]. The discussion of the resonance fluorescence spectrum has a long history, starting with its discussion in 1931 by Weisskopf [31]. This is the spectrum of the spontaneous decay channel whose peak intensity corresponds to the blue line in Figs. 14.12 and 14.13.

Resonance fluorescence may be measured in the geometry illustrated in Fig. 14.19 where the incident beam traverses an atomic beam at right angles, whose atoms have been prepared in a suitable two-level state. The incident photon energy $\hbar\omega$ is typically tuned to the resonance energy of the atoms, and the emitted fluorescence spectrum is measured with a spectrometer/interferometer and a photon detector as a function of the energy difference of the incoming and scattered photons. Since elastic scattering has a node along the polarization vector of the incident radiation, one often chooses the detector (spectrometer) position perpendicular to the **E**-vector, as shown. In some experiments, an additional coincidence circuit is used to detect time-dependent photon-photon correlations.

For a two-level atom one would expect that the fluorescence spectrum is just a peak at the resonance energy $\mathcal{E}_0$. In the KHD theory for example, this is indeed so because the photon and atom parts of the interaction are coupled only by a weak perturbation, so that in the evaluation of the KHD formula the photon and atomic parts can be separated. In the BR theory, the coupling of the photon and electron system is treated to higher order, which is explicitly expressed by the coherences $\rho_{12} = \rho_{21}^*$. Below we will show that at high intensity, the presence of the coherences, which were already shown to introduce Rabi oscillations of the populations, leads to a remarkable splitting effect of the spontaneously scattered spectrum, where the splitting itself is directly given by the Rabi energy Γ_R.

The existence of such a splitting was first pointed out and theoretically derived by Mollow [32] in a pioneering paper in 1969. The splitting occurs only at very high incident intensity, more specifically, when the Rabi interaction energy Γ_R of the photon-atom system becomes comparable to the total decay energy width Γ of the upper state. Semi-classically, the splitting arises from the coupling of a strong oscillating electric field to the electronic system, referred to as the dynamical or *AC Stark effect* also known as the *Autler–Townes effect* after Stanley Autler and Charles Townes who discovered the effect in 1955 [33].

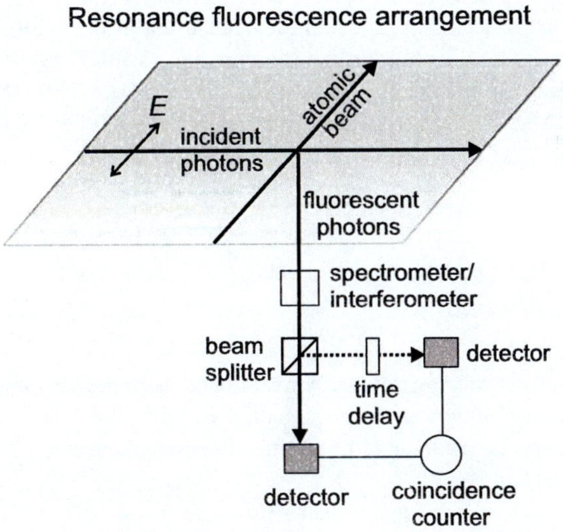

Fig. 14.19 Schematic of experimental geometry for resonance fluorescence measurements. The incident laser beam intercepts an atomic beam of low density at right angles. The atomic beam is prepared in a suitable electronic state beforehand, typically though interaction with another laser beam. The incident photon beam, shown to be linearly polarized, is typically tuned to exact resonance of the two-level atoms. The spontaneously scattered fluorescence spectrum is recorded as a function of the energy difference of the incident and scattered photons by use of a spectrometer or interferometer whose optical axis is perpendicular to the laser beam, its polarization axis, and the atomic beam. Also shown is a coincidence circuit that may be employed to investigate temporal correlations of two fluorescent photons as a function of delay time

Mollow's theory of resonance fluorescence was followed in the 1970s by theoretical [8, 15, 18, 34] and experimental work [11, 35–38]. In particular, Milonni [15] showed that the resonance fluorescence spectrum of a two-level atom may be described equivalently by the semi-classical Bloch equations and the quantum electrodynamics treatment of spontaneous emission. In a special version of the experiment, Kimble, Dagenais, and Mandel [11, 38] used the additional coincidence circuit shown in Fig. 14.19 to demonstrate the existence of so-called photon *antibunching* in the time-dependent photon-photon correlation signal in resonance fluorescence that cannot be explained classically.

Today, the study of strong light-matter interactions, first explored through single atom resonance fluorescence, has been extended by the use of EM radiation in the microwave [39] and hard x-ray regimes [40] and for the study of the collective properties of coupled matter as reviewed in [41]. The work in the microwave regime is particularly novel in that it is performed by use of so-called artificial atoms, a superconducting macroscopic two-level system on a chip, whose quantum behavior is of great importance for the development of quantum information science.

In the following we will discuss key aspects of the resonance fluorescence signal, such as the distinction of its elastic and inelastic components through the degrees of first and second order coherence and the calculation of the splitting. The theory is non-trivial, and we will only outline the derivation of the key results, referring the interested reader to the detailed derivations in references [7, 9, 18, 34, 42]. The x-ray case for SASE XFEL pulses has been theoretically studied by Cavaletto et al. [43].

14.10.2 Second Quantization of the p · A Interaction Hamiltonian

We begin by briefly reviewing how the conventional photoelectric interaction Hamiltonian in (9.16) may be written in the so-called *second quantization* form. We start with the interaction Hamiltonian (14.6) written in terms of the electric field operators according to (3.61) as

$$\mathcal{H}_{\text{int}} = -e\,\mathbf{r} \cdot (\mathbf{E}^{+} + \mathbf{E}^{-}). \tag{14.124}$$

In quantum optics it is common to express this Hamiltonian in *second quantization* form. To do so, we take the energy of the atomic ground state $|a\rangle$ as zero and consider the Hamiltonian for the atom, $\mathcal{H}_{\text{at}}$, in the excited state $|b\rangle$ of energy $\hbar\omega_{ba}$ plus the interaction Hamiltonian as

$$\mathcal{H} = \mathcal{H}_{\text{at}} + \mathcal{H}_{\text{int}} = \hbar\omega_{ba}|b\rangle\langle b| - e\,\mathbf{r}\cdot(\mathbf{E}^{+} + \mathbf{E}^{-}). \tag{14.125}$$

The dipole matrix elements of the form $\langle b|\mathbf{r}|a\rangle = \langle a|\mathbf{r}|b\rangle^{*}$ connect states of opposite parity. We now introduce operators $\hat{\pi}^{\dagger}$ that raise the atom from $a \rightarrow b$ and $\hat{\pi}$ that make it decay from $b \rightarrow a$ as

$$\hat{\pi}^{\dagger} = |b\rangle\langle a| \quad \text{and} \quad \hat{\pi} = |a\rangle\langle b|. \tag{14.126}$$

They have the properties

$$\hat{\pi}^{\dagger}|a\rangle = |b\rangle, \;\; \hat{\pi}|b\rangle = |a\rangle, \;\; \hat{\pi}^{\dagger}|b\rangle = \hat{\pi}|a\rangle = 0, \;\; \hat{\pi}^{\dagger}\hat{\pi} = |b\rangle\langle b|, \;\; \hat{\pi}\hat{\pi}^{\dagger} = |a\rangle\langle a|. \tag{14.127}$$

We also have $\hat{\pi}^{\dagger}\hat{\pi}^{\dagger} = \hat{\pi}\hat{\pi} = 0$ and the closure relation for the two-level system,

$$\hat{\pi}^{\dagger}\hat{\pi} + \hat{\pi}\hat{\pi}^{\dagger} = |a\rangle\langle a| + |b\rangle\langle b| = 1. \tag{14.128}$$

The interaction Hamiltonian in the dipole approximation can then be expressed as [42]

$$\mathcal{H}_{\text{int}} = -\langle b|e\,\mathbf{r}|a\rangle(\mathbf{E}^{+} + \mathbf{E}^{-})\,(\hat{\pi}^{\dagger} + \hat{\pi}), \tag{14.129}$$

where we have taken the matrix element to be real.[6] The connection with the BR treatment becomes apparent by realizing that (14.129) can be directly expressed in terms of the resonant Rabi energy, defined by (14.11), according to

$$\boxed{\mathcal{H}_{\text{int}} = |\langle f|e\,\mathbf{r}\cdot(\mathbf{E}^+ + \mathbf{E}^-)|i\rangle|\,(\hat{\pi}^\dagger + \hat{\pi}) = \frac{\Gamma_{\text{R}}}{2}\,(\hat{\pi}^\dagger + \hat{\pi}).} \tag{14.130}$$

We are now in a position to consider the coherence properties of the spontaneously emitted fluorescent intensity.

14.10.3 The Degree of First Order Temporal Coherence

As the name implies, the off-diagonal elements or "coherences" $\rho_{21}(t) = \rho_{12}^*(t)$ of the density matrix are indeed related to the temporal degree of first order coherence $g^{(1)}(t_1, t_2)$ of the scattered radiation, whose quantum formulation we have previously discussed in Sect. 4.5. It turns out that the resonance fluorescence spectrum can be obtained by a Fourier transform of the degree of first order coherence, as will be discussed below in Sect. 14.11. For this reason we shall derive it now.

The complex degree of first order temporal coherence is defined by (4.84) as the two-field correlation function at the same spatial point. For *quasi-stationary* light it may be written in the form (4.87) as

$$g^{(1)}(t) = \frac{\langle \boldsymbol{E}^-(t_0)\boldsymbol{E}^+(t_0+t)\rangle}{\langle \boldsymbol{E}^-(t_0)\boldsymbol{E}^+(t_0)\rangle} \tag{14.131}$$

For the case of resonant excitation one in interested in $g^{(1)}(t)$ of the *scattered fields*. The field operators in (14.131) are then replaced by those of the combined photon-atom system defined in Sect. 14.10.2, and one writes [7]

$$g^{(1)}(t) = \frac{\langle \hat{\pi}^\dagger(t_0)\hat{\pi}(t_0+t)\rangle}{\langle \hat{\pi}^\dagger(t_0)\hat{\pi}(t_0)\rangle}. \tag{14.132}$$

An analytical solution for the degree of first order temporal coherence $g^{(1)}(t)$ defined by (14.132) can only be obtained for the case of *exact resonance* and was worked out analytically by Carmichael and Walls [34]. The key part is the reduction of the two time correlation function in (14.132) to a single time correlation function by means of the quantum regression theorem that only depends on the time difference

[6] The relation (14.129) is typically approximated by considering the frequency dependence of the operators. While the electric fields oscillate with the frequency of the incident light, the transition operators $\hat{\pi}^\dagger$ and $\hat{\pi}$ are associated with the frequency ω_{ba} of the optical transition. Multiplying out the right side of (14.129), two terms have a slow frequency dependence, where the frequency difference is just the detuning energy $\hbar(\omega-\omega_{ba})$, while the other two terms have a high-frequency dependence corresponding to transitions that are far off-resonance. Omission of these terms corresponds to the *rotating wave approximation* with the result $\mathcal{H}_{\text{int}} = -\langle b|e\,\mathbf{r}|a\rangle\,(\mathbf{E}^+\hat{\pi}^\dagger + \mathbf{E}^-\hat{\pi})$.

t. At resonance, $g^{(1)}(t)$ may be expressed in the form [7]

$$\text{Exact resonance:} \quad g^{(1)}(t) = \frac{(\rho_{21}^{\infty})^*}{\rho_{22}^{\infty}} a_1(t) + a_2(t). \tag{14.133}$$

Here t is the correlation time of two fluorescence *fields* at the same spatial point, which in principle could be measured by replacing the *intensity* or photon-photon coincidence circuit in Fig. 14.19 with a Michelson-type interferometer. In practice, one measures the Fourier transform of $g^{(1)}(t)$ which according to the Wiener-Khintchine theorem is the normalized fluorescence spectrum or power spectral density. We will see in the following Sect. 14.11 that the measured resonance fluorescence spectrum contains coherent and incoherent contributions which directly yield the desired coherence information. Here we continue with the determination of the quantities in (14.133).

The long time limits ρ_{21}^{∞} and ρ_{22}^{∞} for $\Gamma' = \Gamma$ are obtained from (14.58) and (14.55) as

$$\rho_{21}^{\infty} = \frac{-\mathrm{i}\Gamma_{\mathrm{R}}\Gamma}{\Gamma^2 + 2\Gamma_{\mathrm{R}}^2}, \quad \text{and} \quad \rho_{22}^{\infty} = \frac{\Gamma_{\mathrm{R}}^2}{\Gamma^2 + 2\Gamma_{\mathrm{R}}^2}. \tag{14.134}$$

Using the following definition (note difference to (14.47)),

$$\hbar\omega_{\mathrm{R}} = \sqrt{\Gamma_{\mathrm{R}}^2 - \frac{\Gamma^2}{16}} \tag{14.135}$$

the terms $a_1(t)$ and $a_2(t)$ can be expressed analytically as[7]

[7] Our expressions for $a_1(t)$ and $a_2(t)$ correspond to the notation of Scully and Zubairy [9]. They are equivalent to Loudon's expressions [7] for $\alpha_1(t)$ and $\alpha_2(t)$

$$\alpha_1(t) = \frac{-\mathrm{i}\Gamma_{\mathrm{R}}\Gamma}{\Gamma^2+2\Gamma_{\mathrm{R}}^2}\left\{1 + \frac{\mathrm{i}\,\mathrm{e}^{-3\Gamma t/(4\hbar)}}{8\Gamma\hbar\omega_{\mathrm{R}}}\right. \times \left.\left[\left(4\left(\frac{\Gamma}{2}+\mathrm{i}\hbar\omega_{\mathrm{R}}\right)^2 - \frac{\Gamma^2}{4}\right)\mathrm{e}^{\mathrm{i}\omega_{\mathrm{R}}t} - \left(4\left(\frac{\Gamma}{2}-\mathrm{i}\hbar\omega_{\mathrm{R}}\right)^2 - \frac{\Gamma^2}{4}\right)\mathrm{e}^{-\mathrm{i}\omega_{\mathrm{R}}t}\right]\right\} \tag{14.136}$$

and

$$\alpha_2(t) = \frac{\mathrm{e}^{-\Gamma t/(2\hbar)}}{2} - \frac{\mathrm{i}\,\mathrm{e}^{-3\Gamma t/(4\hbar)}}{8\hbar\omega_{\mathrm{R}}}\left[\left(\frac{\Gamma}{2} + 2\mathrm{i}\hbar\omega_{\mathrm{R}}\right)\mathrm{e}^{\mathrm{i}\omega_{\mathrm{R}}t} - \left(\frac{\Gamma}{2} - 2\mathrm{i}\hbar\omega_{\mathrm{R}}\right)\mathrm{e}^{-\mathrm{i}\omega_{\mathrm{R}}t}\right] \tag{14.137}$$

.

$$a_1(t) = \rho_{21}^{\infty}\left\{1-\left[\cos(\omega_R\, t)-\frac{\Gamma_R^2-\Gamma^2/4}{\Gamma\hbar\omega_R}\sin(\omega_R\, t)\right]e^{-3\Gamma t/(4\hbar)}\right\} \tag{14.138}$$

and

$$a_2(t)=\frac{e^{-\Gamma t/(2\hbar)}}{2}+\frac{e^{-3\Gamma t/(4\hbar)}}{8\hbar\omega_R}[4\hbar\omega_R\cos(\omega_R\, t)+\Gamma\sin(\omega_R\, t)]\,. \tag{14.139}$$

Hence, at resonance, we obtain the first order correlation function as follows [7, 9].

At *exact resonance*, the degree of first order temporal coherence in resonance fluorescence is given by

$$\begin{aligned} g^{(1)}(t) = \exp[-i\omega t]\Bigg\{&\frac{\Gamma^2}{\Gamma^2+2\Gamma_R^2}+\frac{1}{2}\exp\left[-\frac{\Gamma t}{2\hbar}\right] \\ &+\frac{(\Gamma/2-2i\hbar\omega_R)^2}{8i\hbar\omega_R(3\Gamma/2+2i\hbar\omega_R)}\exp\left[-i\omega_R t-\frac{3\Gamma t}{4\hbar}\right] \\ &-\frac{(\Gamma/2+2i\hbar\omega_R)^2}{8i\hbar\omega_R(3\Gamma/2-2i\hbar\omega_R)}\exp\left[\,i\omega_R t-\frac{3\Gamma t}{4\hbar}\right]\Bigg\}, \end{aligned} \tag{14.140}$$

where

$$\hbar\omega_R = \sqrt{\Gamma_R^2-\frac{\Gamma^2}{16}}. \tag{14.141}$$

This expression is valid for both $\Gamma_R \geq \Gamma/4$ and $\Gamma_R \leq \Gamma/4$. We shall see that the first term in curly brackets corresponds to the coherent fraction and the last three terms exhibit chaotic light behavior.

We have plotted the time dependence of $|g^{(1)}(t)|$ in Fig. 14.20 and the intensity dependence which enters through Γ_R in Fig. 14.21, using parameters for the L_3 resonance of a Co atom.[8]

14.10.4 $g^{(1)}(t)$ From Numerical Solutions of Bloch Equations

The analytical expression for $g^{(1)}(t)$ given by (14.140), valid at resonance, depends on a single time. It can also be calculated numerically from the Bloch equations, but this involves two calculations with different initial conditions, reflecting that in

[8] The corresponding second order degree of temporal coherence is shown in Fig. 14.25.

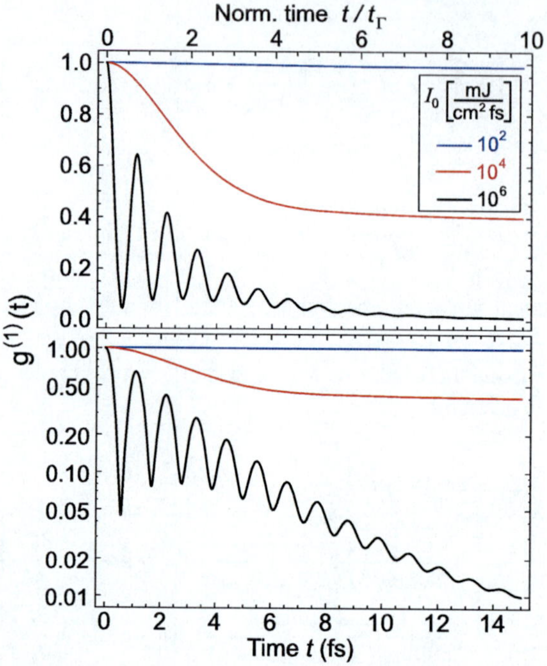

Fig. 14.20 Time dependence of the first order degree of temporal coherence $|g^{(1)}(t)|$ of the radiation scattered by a resonantly excited Co atom according to (14.140), for three incident intensities. Top on linear scale, bottom on logarithmic scale. We have assumed a two-level Co atom with upper state decay energy width $\Gamma = 0.43\,\text{eV}$ (decay time $t_\Gamma = \hbar/\Gamma = 1.53\,\text{fs}$), a resonance energy of $\mathcal{E}_0 = 778\,\text{eV}$ ($\lambda = 1.59\,\text{nm}$), and a resonant transition matrix element $\Gamma_x = 0.34\,\text{meV}$. The field is assumed to be instantaneously turned on at $t = 0$

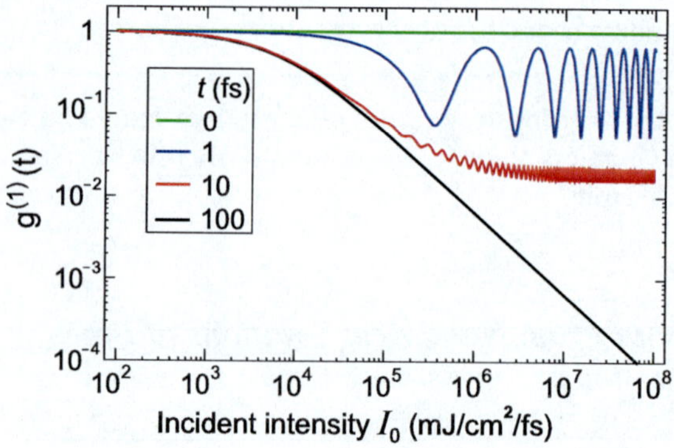

Fig. 14.21 Intensity dependence of the first order degree of temporal coherence $|g^{(1)}(t)|$ of the radiation scattered by a resonantly excited Co atom according to (14.140). The field is assumed to be instantaneously turned on at $t = 0$. We have assumed a two-level Co atom using the same parameters as for Fig. 14.20

essence $g^{(1)}(t)$ represents a two time correlation function. It can be shown that we can write (14.133) entirely in terms of two solutions of the numerical Bloch equations for two different initial conditions as follows

$$g^{(1)}(t) = \frac{1}{2}\left\{ 1 + \underbrace{\rho_{21}(t)}_{\rho_{21}(0)=1,\rho_{22}(0)=\mathrm{i}\Gamma/2\Gamma_R} + \underbrace{\left(\frac{2(\rho_{21}^{\infty})^*}{\rho_{22}^{\infty}} - 1 \right) \rho_{21}(t) - \frac{\rho_{22}(t)}{\rho_{22}^{\infty}}}_{\rho_{21}(0)=\rho_{22}(0)=0} \right\}. \tag{14.142}$$

14.11 The Resonant Fluorescence Spectrum

14.11.1 Form of the Spectrum

Before we formally derive the fluorescence spectrum recorded as a function of energy difference between the resonantly tuned incident photons, $\hbar\omega = \mathcal{E}_0$, and the scattered photons of energy $\hbar\omega_s$, we schematically illustrate in Fig. 14.22a the modification of the two-level scheme by the AC Stark effect and the resulting fluorescence spectrum. The key to the modification of the fluorescence spectrum with incident intensity is the dynamic Stark splitting of the upper and lower states, as schematically shown in the figure. The shown level splitting may also be expressed through dressed states in the Jaynes-Cummings model [14, 16].

The resonance fluorescence spectrum is recorded with the incident photon energy tuned to the resonance energy $\hbar\omega_0 = \mathcal{E}_0$ as a function of the difference between the scattered energy $\hbar\omega_s$ and incident energy $\hbar\omega_0$. At low incident intensity, the spectrum consists of a narrow δ-function-like peak reflecting strict energy conservation in accordance with the REXS expression (13.12). When the incident intensity I_0 becomes so large that the corresponding Rabi energy $\Gamma_R \propto \sqrt{I_0}$ (see (14.38)) becomes comparable to the natural linewidth Γ of the upper state, the dynamic Stark splitting of the upper and lower states causes the low intensity single line spectrum (red) to evolve into additional broader components shown in blue and green, known as the *Mollow triplet*, as schematically shown on the right of Fig. 14.22a. The splitting of the Mollow triplet is seen to be a direct measure of the Rabi energy.

In Fig. 14.22b, c we show a modern version of a Mollow triplet spectrum, recorded with microwaves scattered on an artificial atom in the form of a macroscopic superconducting loop interrupted by a Josephson junction, that is inductively coupled to a 1D transmission line [39]. The incident wave in the line is scattered by the artificial atom and can be detected in either the forward or backward direction.

In the following we will discuss the derivation of the resonant fluorescence spectrum.

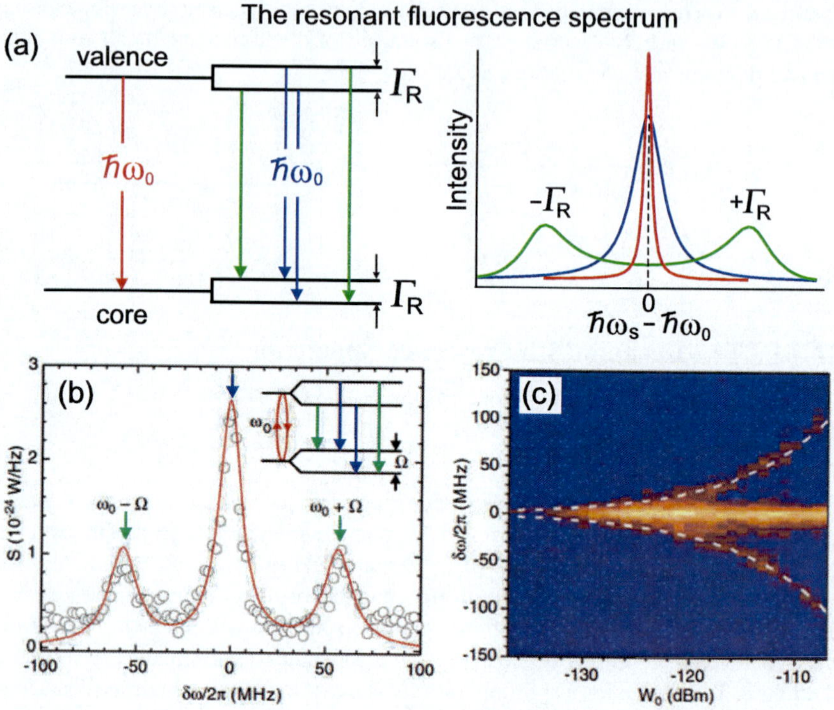

Fig. 14.22 **a** Schematic of the dynamic Stark effect splitting of a two-level atom in a strong incident field. The red component is the conventional spontaneous REXS spectrum, as a function of the energy difference $\hbar\omega_0 - \hbar\omega_s$ of the resonantly tuned incident light of energy $\hbar\omega_0 = \mathcal{E}_0$ and the spontaneously scattered light of energy $\hbar\omega_s$, which in principle is a Dirac δ-function but shown with finite width. The blue and green components arise from the Stark splitting by Γ_R. **b** A modern example of the resonant fluorescence spectrum of a strongly driven, so-called artificial atom, in a microwave field of Rabi frequency of 57 MHz corresponding to an incident microwave power of -112 dBm or 6.3×10^{-15} W [39]. Experimental data are shown by circles, and the red curve is a theoretical fit. **c** Splitting of the fluorescence emission spectrum in (**b**) as a function of the driving power [39]

14.11.2 *Fourier Transform of $g^{(1)}(t)$: The Fluorescence Spectrum*

The temporal dependence of $g^{(1)}(t)$ can be used to obtain the frequency dependence of the emitted radiation, i.e. the fluorescence spectrum, which is usually measured in the geometry illustrated in Fig. 14.19. The normalized spectrum, referred to as the *power spectral density*, can be obtained by use of the Wiener-Khintchine theorem as the Fourier transform of $g^{(1)}(t)$. The scattered spectrum of energy $\hbar\omega_s$ is symmetrical around the incident photon energy $\hbar\omega_0$ and given by

$$G(\hbar\omega_s) = \frac{1}{2\pi\hbar}\int_0^\infty g^{(1)}(t)\left[e^{-i\omega_s t/\hbar} + e^{i\omega_s t/\hbar}\right] dt. \tag{14.143}$$

It is possible to obtain an analytical form of $G(\hbar\omega_s)$, but one needs to deal with the zero value encountered in (14.141) which leads to a spectral singularity at $\Gamma_R = \Gamma/4$. We shall therefore separately give the solutions for the incident intensity ranges defined by $\Gamma_R < \Gamma/4$ and $\Gamma_R > \Gamma/4$ [7].

14.11.3 The Low Intensity Spectrum: $\Gamma_R < \Gamma/4$

In the *low intensity* range $\Gamma_R < \Gamma/4$ we define the real quantity

$$\hbar\omega_R = \sqrt{\frac{\Gamma^2}{16} - \Gamma_R^2} \tag{14.144}$$

and obtain by evaluation of (14.143) with (14.140) [7]

$$\begin{aligned} G(\hbar\omega_s) = &\underbrace{\frac{\Gamma^2}{\Gamma^2 + 2\Gamma_R^2}\,\delta(\hbar\omega_0 - \hbar\omega_s)}_{\text{elastic spectrum}} \\ &+ \underbrace{\frac{1}{\pi\Gamma}\frac{(\Gamma/2)^2}{(\hbar\omega_0 - \hbar\omega_s)^2 + (\Gamma/2)^2}}_{\text{inelastic spectrum}, S_1} \\ &- \underbrace{\frac{1}{16\pi\hbar\omega_R}\frac{(\Gamma/2 + 2\hbar\omega_R)^2}{(\hbar\omega_0 - \hbar\omega_s)^2 + (3\Gamma/4 - \hbar\omega_R)^2}}_{\text{inelastic spectrum}, S_{2a}} \\ &+ \underbrace{\frac{1}{16\pi\hbar\omega_R}\frac{(\Gamma/2 - 2\hbar\omega_R)^2}{(\hbar\omega_0 - \hbar\omega_s)^2 + (3\Gamma/4 + \hbar\omega_R)^2}}_{\text{inelastic spectrum}, S_{2b}} \end{aligned} \tag{14.145}$$

and the spectral power density is normalized as

$$\int_{-\infty}^{\infty} G(\hbar\omega_s)\, d(\hbar\omega_s) = 1. \tag{14.146}$$

In (14.145) we have indicated by underbrackets and labels the four components of the spectrum for later discussion. The designations of the underbracketed expressions as "elastic" and "inelastic" are a consequence of the fact that with increasing intensity the inelastic components dominate and it is seen from Figs. 14.20 and 14.21 that the properties of the fluorescent radiation change from completely coherent (elastic) at low to incoherent (inelastic) at high intensity.

14.11.4 High Intensity Spectrum: $\Gamma_R > \Gamma/4$

At *high intensity* in the range $\Gamma_R > \Gamma/4$ we use the definition of the real quantity

$$\hbar\omega_R = \sqrt{\Gamma_R^2 - \frac{\Gamma^2}{16}}, \tag{14.147}$$

and the normalized spectral density is given by [7]

$$\begin{aligned} G(\hbar\omega_s) = & \underbrace{\frac{\Gamma^2}{\Gamma^2 + 2\Gamma_R^2}\,\delta(\hbar\omega_0 - \hbar\omega_s)}_{\text{elastic spectrum}} \\ & + \underbrace{\frac{1}{\pi\Gamma}\frac{(\Gamma/2)^2}{(\hbar\omega_0 - \hbar\omega_s)^2 + (\Gamma/2)^2}}_{\text{inelastic central peak}, S_1} \\ & + \underbrace{\frac{\Gamma}{16\pi}\frac{3\hbar\omega_R(\Gamma_R^2 - \Gamma^2/2) + (5\Gamma_R^2 - \Gamma^2/2)(\hbar\omega_0 - \hbar\omega_s + \hbar\omega_R)}{\hbar\omega_R(\Gamma_R^2 + \Gamma^2/2)\left\{(\hbar\omega_0 - \hbar\omega_s + \hbar\omega_R)^2 + (3\Gamma/4)^2\right\}}}_{\text{inelastic side peak}, S_{2a}} \\ & + \underbrace{\frac{\Gamma}{16\pi}\frac{3\hbar\omega_R(\Gamma_R^2 - \Gamma^2/2) - (5\Gamma_R^2 - \Gamma^2/2)(\hbar\omega_0 - \hbar\omega_s - \hbar\omega_R)}{\hbar\omega_R(\Gamma_R^2 + \Gamma^2/2)\left\{(\hbar\omega_0 - \hbar\omega_s - \hbar\omega_R)^2 + (3\Gamma/4)^2\right\}}}_{\text{inelastic side peak}, S_{2b}} . \end{aligned} \tag{14.148}$$

The spectral power density is again normalized according to (14.146).

The first term in the low and high intensity spectra (14.145) and (14.148) contains a Dirac δ-function. It arises from the fact that the spectra are derived from an analytical $g^{(1)}(t)$ function given by (14.140) that is only valid at *strict resonance*, corresponding to a vanishingly small incident bandwidth. The spectral δ-function is a consequence of strict energy conservation between the incident and scattered energy in REXS, previously encountered in the KHD REXS cross section (13.12).

14.11.5 Calculated Fluorescence Spectra

In Fig. 14.23 we show spectra calculated with (14.145) and (14.148) assuming our Co L_3 model parameters for different incident intensities, which are specified through the values of Γ_R (see (14.39)). We replaced the δ-functions in (14.145) and (14.148) by finite width Lorentzians of FWHM $\Gamma_0 = 0.1\,\text{eV}$ with the same unit integrated area according to

$$\delta(\hbar\omega_0 - \hbar\omega_s) \leftrightarrow \frac{2}{\pi\Gamma_0}\frac{(\Gamma_0/2)^2}{(\hbar\omega_0 - \hbar\omega_s)^2 + (\Gamma_0/2)^2}. \qquad (14.149)$$

Our chosen Lorentzian lineshape of the elastically scattered component is arbitrary but serves to illustrate the change of the total elastic intensity given by the area under the curve.

At low intensity $\Gamma_R < \Gamma/4$ the spectrum is dominated by the coherent elastic peak, as illustrated in Fig. 14.23a, calculated with (14.145) at the upper incident intensity limit $\Gamma_R = \Gamma/4.01$. The curves shown in Fig. 14.23b correspond to $\Gamma_R = \Gamma/\sqrt{2}$ where the elastic (coherent) and inelastic (incoherent) integrated spectral intensities are equal. At higher incident intensity the elastic spectrum (red) further decreases as shown in Figs. 14.23c, d, and for $\Gamma_R = 3\Gamma$ three inelastic peaks have emerged that are often referred to as the Mollow triplet [7, 32, 42].

The elastic red peak in the panels of Fig. 14.23 contains the following fraction of the total intensity (a) 89%, (b) 50%, (c) 33%, (d) 5% and the intensities in [mJ/cm^2/fs] corresponding to the Γ_R values are (a) 814 , (b) 6.5×10^3, (c) 1.3×10^4, (d) 1.2×10^5 (also see Fig. 14.24b).

14.11.6 *Coherent and Incoherent Parts of the Equilibrium Spectrum*

The total spectra in Fig. 14.23 represented by the black curves may be separated into coherently and incoherently scattered components. The *coherent* components correspond to the red curves and the *incoherent* components to the sum of the blue and green curves. In the long time limit $t \to \infty$ where equilibrium is established one can derive simple expressions for the integrated coherent and incoherent intensities. Since the total integrated spectral densities (black curves) are normalized to unity according to (14.146), the total normalized scattered rate is

$$R_{\text{tot}} = \int_{-\infty}^{\infty} G(\hbar\omega_s)\, \mathrm{d}(\hbar\omega_s) = 1. \qquad (14.150)$$

The coherent contribution is the first term

$$R_{\text{coh}} = \frac{\Gamma^2}{\Gamma^2 + 2\Gamma_R^2} \underbrace{\int_{-\infty}^{\infty} \delta(\hbar\omega_0 - \hbar\omega_s)\, \mathrm{d}(\hbar\omega_s)}_{1} = \frac{\Gamma^2}{\Gamma^2 + 2\Gamma_R^2} \qquad (14.151)$$

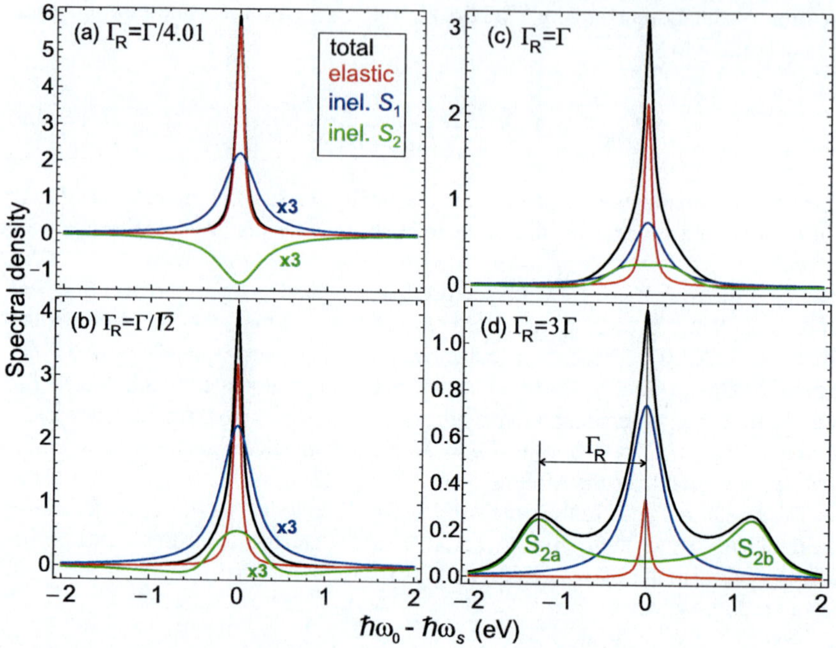

Fig. 14.23 Spectral density of the scattered radiation $G(\hbar\omega_s)$ given by (14.145) and (14.148) for low and high incident intensities, with the Dirac δ-function replaced by a Lorentzian of FWHM $\Gamma_0 = 0.1$ eV. The black curves represent the total scattered spectra with integrated unit normalization according to (14.146). The different incident intensities correspond to the four Γ_R values in units of $\Gamma = 0.43$ eV (Co L_3 resonance). We assumed an incident resonantly tuned photon energy of $\hbar\omega_0 = 778$ eV ($\lambda_0 = 1.59$ nm) and a resonant transition matrix element $\Gamma_x = 0.34$ meV. The red curves are the elastically scattered components. The blue curve is the central inelastically scattered component S_1 and the green curves the sum of the inelastically scattered components $S_2 = S_{2a} + S_{2b}$. **a** Calculated with (14.145) and $\Gamma_R = \Gamma/4.01$. **b–d** Calculated with (14.148) and $\Gamma_R = \Gamma/\sqrt{2}$, $\Gamma_R = \Gamma$, and $\Gamma_R = 3\Gamma$, respectively

and the incoherent contribution is just the difference

$$R_{\text{inc}} = R_{\text{tot}} - R_{\text{coh}} = 1 - \frac{\Gamma^2}{\Gamma^2 + 2\Gamma_R^2}. \tag{14.152}$$

We can now revisit the different rates plotted in Fig. 14.12, which correspond to resonant excitation and equilibrium ($t \to \infty$). The absorption rate shown in black and stimulated rate shown in red correspond to changes of the incident photon rate in the forward "in-beam" direction, and they remain the same as in Fig. 14.12.

The resonant fluorescence rate measured in the 90° "out-of-beam" direction corresponds to the *spontaneously* scattered intensity shown as a blue curve in Fig. 14.12. We are now able to quantitatively decompose this curve into coherent and incoherent spectral components shown in Fig. 14.23. The spontaneous scattering rate

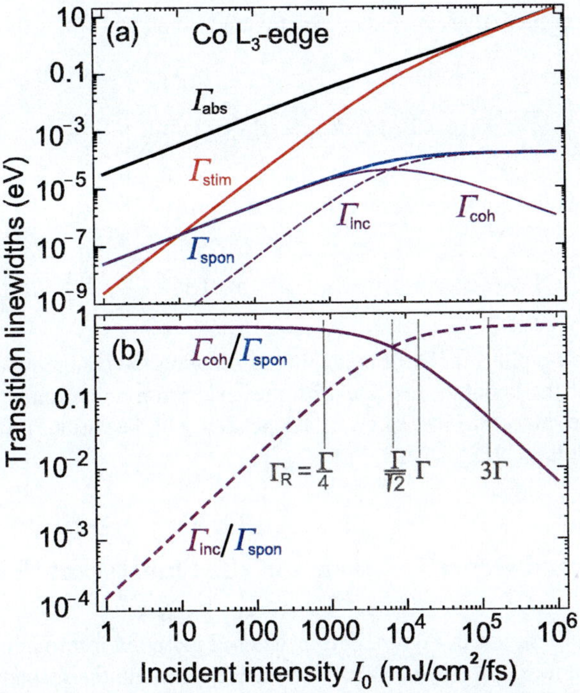

Fig. 14.24 **a** Intensity dependence of the various transition rates $\Gamma_i/\hbar$ as in Fig. 14.12. The transition energy width Γ_i was calculated for resonant excitation ($\hbar\omega = \mathcal{E}_0$) and equilibrium ($t \to \infty$). The absorption (black) and stimulated (red) rates are the same as in Fig. 14.12. The spontaneous scattering rate Γ_{spon} (blue) was calculated according to (14.154) and its coherent component (solid purple) from (14.155) and the incoherent component (dashed purple) from (14.156). We have assumed a two-level Co atom with upper state decay energy width $\Gamma = 0.43\,\text{eV}$ (decay time $t_\Gamma = \hbar/\Gamma = 1.53\,\text{fs}$), a resonance energy of $\mathcal{E}_0 = 778\,\text{eV}$ ($\lambda = 1.59\,\text{nm}$), and a resonant transition matrix element $\Gamma_x = 0.34\,\text{meV}$ **b** Relative fractions of the elastic and inelastic scattering rates contributing to the total rate, as indicated. In gray we have indicated by four vertical lines the specific incident intensities in Fig. 14.23 corresponding to Γ_{R} in units of Γ

$\mathcal{W}_{\text{REXS}}^{\text{spon}} = \Gamma_{\text{spon}}/\hbar$ is given by (14.94) or

$$\Gamma_{\text{spon}} = \Gamma_x \underbrace{\frac{\Gamma_{\text{R}}^2}{\Gamma^2 + 2\Gamma_{\text{R}}^2}}_{\rho_{22}^{\infty}} \tag{14.153}$$

where ρ_{22}^{∞} is the equilibrium resonant excited state population and Γ_x the dipolar transition width. The total spontaneous rate, $\Gamma_{\text{spon}}/\hbar$, is just the dimensionless unit-normalized total scattering rate R_{tot} times a the scaling factor $\Gamma_x \rho_{22}^{\infty}$, i.e.

$$\Gamma_{\text{spon}} = \Gamma_x\, \rho_{22}^{\infty}\, R_{\text{tot}}. \tag{14.154}$$

Similarly, the coherently scattered rate is obtained from (14.151) by the same scaling factor as

$$\Gamma_{\mathrm{coh}} = \Gamma_x\, \rho_{22}^{\infty}\, R_{\mathrm{coh}} = \Gamma_x \frac{\Gamma^2}{\Gamma_{\mathrm{R}}^2} \left(\rho_{22}^{\infty}\right)^2 \tag{14.155}$$

and the incoherently scattered rate is obtained from (14.152) as

$$\Gamma_{\mathrm{inc}}(\hbar\omega) = \Gamma_x\, \rho_{22}^{\infty}\, R_{\mathrm{inc}} = \Gamma_x \rho_{22}^{\infty} \left(1 - \frac{\Gamma^2}{\Gamma_{\mathrm{R}}^2} \rho_{22}^{\infty}\right). \tag{14.156}$$

We can now replot Fig. 14.12 by explicitly showing the coherent and incoherent component of the spontaneous rate with the absorption and stimulated scattering rates being the same as in Fig. 14.12. The dependence of the various rates on incident intensity is plotted in Fig. 14.24.

14.12 Second Order Coherence of the Fluorescent Photons

The coincidence circuit in Fig. 14.19 may be used to obtain information on photon-photon correlations in the resonance fluorescence signal. In the second quantization formulation (see Sect. 14.10.2), the degree of *second order coherence* in resonance fluorescence is given by [7]

$$g^{(2)}(t) = \frac{\langle \hat{\pi}^{\dagger}(t_0)\hat{\pi}^{\dagger}(t_0+t)\hat{\pi}(t_0+t)\hat{\pi}(t_0)\rangle}{\langle \hat{\pi}^{\dagger}(t_0)\hat{\pi}(t_0)\rangle^2}. \tag{14.157}$$

It can be shown that for a two-level system $g^{(2)}(t)$ is linked to the excited state population $\rho_{22}(t)$ by the general relationship [7, 12, 18, 42]

$$\boxed{g^{(2)}(t) = \frac{\rho_{22}(t)}{\rho_{22}^{\infty}}}. \tag{14.158}$$

While the general solution is obtained by numerical solution of the Bloch equation, it is instructive to consider cases where numerical solutions of the Bloch equations exist as discussed in Sect. 14.5. We shall now illustrate the evolution of $g^{(2)}(t)$ for three cases with analytical solutions.

14.12.1 Weak Incident Beam

We start with the solution of the Bloch equations for a weak beam given by (14.44). It is derived under the assumption that the incident photons are in the same mode $(p, \mathbf{k})$, i.e. have the same polarization, propagation direction $\mathbf{k}$, and same photon energy $\hbar\omega = \hbar c k$. In practice, the incident photons have a finite bandwidth and we start by assuming that it is very small relative to the natural atomic resonance width given by Γ, so that we can study the influence of detuning of the incident photon energy $\hbar\omega$ from the exact resonance energy $\mathcal{E}_0$. The second order degree of temporal coherence is then obtained from (14.44) as

$$g^{(2)}(t) = 1 + \mathrm{e}^{-\Gamma t/\hbar} - 2\cos\left[\frac{\hbar\omega - \mathcal{E}_0}{\hbar}\,t\right]\mathrm{e}^{-\Gamma t/(2\hbar)} \tag{14.159}$$

which at resonance, $\hbar\omega - \mathcal{E}_0 = 0$ assumes the simple form

$$g^{(2)}(t) = \left[1 - \mathrm{e}^{-\Gamma t/(2\hbar)}\right]^2. \tag{14.160}$$

We have also seen in Sect. 14.9.2 that (14.159), when integrated over photon energy, is equivalent to the Einstein result given by (14.119). In this case we simply have

$$g^{(2)}(t) = 1 - \mathrm{e}^{-\Gamma t/\hbar}. \tag{14.161}$$

In Fig. 14.25a we have plotted the temporal evolution of the degree of second order temporal coherence, $g^{(2)}(t)$ for weak incident intensities fulfilling the condition $4\Gamma_R < \Gamma = \Gamma'$ (see (14.47)).

The red curve is given by (14.159) assuming resonant, $\Delta\mathcal{E} = \hbar\omega - \mathcal{E}_0 = 0$, while the blue curve corresponds to detuning by $\Delta\mathcal{E} = 2\Gamma$. We also show in black the corresponding Einstein result (14.161), which represents the case of broad bandwidth (short coherence time) incident radiation. We have assumed the parameters given in the caption.

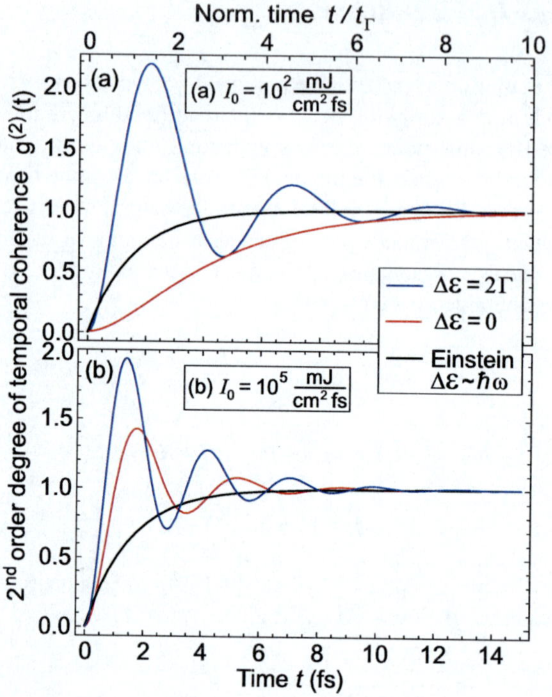

Fig. 14.25 **a** Second order degree of temporal coherence $g^{(2)}(t)$ for an incident intensity, $I_0 = 100\,\mathrm{mJ/cm^2/fs}$ as a function of relative arrival time (same pathlength) of transmitted and scattered photons by a driven atom. We have assumed a two-level Co atom with upper state decay energy width $\Gamma = 0.43\,\mathrm{eV}$ (decay time $t_\Gamma = \hbar/\Gamma = 1.53\,\mathrm{fs}$), a resonance energy of $\mathcal{E}_0 = 778\,\mathrm{eV}$ ($\lambda = 1.59\,\mathrm{nm}$) and a resonant transition matrix element $\Gamma_x = 0.34\,\mathrm{meV}$. The red curve is calculated with (14.159) assuming resonant excitation, $\Delta\mathcal{E} = \hbar\omega - \mathcal{E}_0 = 0$ and the blue curve with non-resonant excitation given by $\Delta\mathcal{E} = 2\Gamma$. The corresponding Einstein result shown in black is calculated with (14.161). **b** Same as in (**a**) with $I_0 = 10^5\,\mathrm{mJ/cm^2/fs}$. The on-resonance curve can be calculated analytically with (14.164), while the off-resonance curve was calculated by numerical solution of the Bloch equations

14.12.2 Large Dephasing: Einstein Result

More insight into the evolution of the second order degree of coherence from the narrow bandwidth BR to the broad bandwidth Einstein theory is provided by the weak field solution (14.43) in the presence of a dephasing contribution which adds to the natural linewidth Γ according to $\Gamma' = \Gamma + \Gamma_{\mathrm{deph}}$. In this case we have with $\Delta\mathcal{E} = \hbar\omega - \mathcal{E}_0$

$$\begin{aligned} g^{(2)}(t) = 1 &+ \frac{2(\Gamma - \Gamma'/2)\left[\Delta\mathcal{E}^2 + (\Gamma'/2)^2\right]}{\Gamma'\left[\Delta\mathcal{E}^2 + (\Gamma - \Gamma'/2)^2\right]}\, \mathrm{e}^{-\Gamma t/\hbar} \\ &- \frac{2\Gamma[\Gamma'(\Gamma - \Gamma'/2) + \Delta\mathcal{E}^2]/2}{\Gamma'\left[\Delta\mathcal{E}^2 + (\Gamma - \Gamma'/2)^2\right]} \cos\left[\frac{\Delta\mathcal{E}\, t}{\hbar}\right] \mathrm{e}^{-\Gamma' t/2\hbar} \\ &+ \frac{2\Delta\mathcal{E}\,\Gamma(\Gamma - \Gamma')}{\Gamma'\left[\Delta\mathcal{E}^2 + (\Gamma - \Gamma'/2)^2\right]} \sin\left[\frac{\Delta\mathcal{E}\, t}{\hbar}\right] \mathrm{e}^{-\Gamma' t/2\hbar}. \end{aligned} \tag{14.162}$$

Since the Einstein theory corresponds to broadband excitation, we assume that the dephasing term completely dominates and with $\Gamma' \gg \Gamma$, $\Delta\mathcal{E}$ we obtain

$$\begin{aligned} g^{(2)}(t) &= 1 - \mathrm{e}^{-\Gamma t/\hbar} - \left\{ \frac{2\Gamma}{\Gamma'} \cos\left[\frac{\Delta\mathcal{E}\, t}{\hbar}\right] + \frac{2\Delta\mathcal{E}\Gamma}{(\Gamma'/2)^2} \sin\left[\frac{\Delta\mathcal{E}\, t}{\hbar}\right] \right\} \mathrm{e}^{-\Gamma' t/2\hbar} \\ &\simeq 1 - \mathrm{e}^{-\Gamma t/\hbar} \end{aligned} \tag{14.163}$$

which merges into the Einstein result (14.161) for large dephasing. This can be readily envisioned as the change of the red and blue curves to the black curve in Fig. 14.25a.

14.12.3 Resonant Case of Arbitrary Intensity

While the weak field expression for $g^{(2)}(t)$ is independent of the incident intensity as long as $4\Gamma_{\mathrm{R}} < \Gamma = \Gamma'$ (see (14.47)), we now explore its changes at higher intensities. In this case an analytical solution to the Bloch equations only exists at exact resonance, but we will also show results obtained by numerical solution of the Bloch equations. We start with the analytical expression for exact resonance excitation given by (14.48), from which we obtain

$$g^{(2)}(t) = \left| 1 - \left[\cosh(\omega_{\mathrm{R}}\, t) + \frac{3\Gamma}{4\hbar\omega_{\mathrm{R}}} \sinh(\omega_{\mathrm{R}}\, t)\right] \mathrm{e}^{-3\Gamma t/(4\hbar)} \right|, \tag{14.164}$$

where the absolute value is required since the quantity

$$\omega_{\mathrm{R}} = \frac{1}{\hbar}\sqrt{\frac{\Gamma^2}{16} - \Gamma_{\mathrm{R}}^2} \tag{14.165}$$

may be imaginary.

The results for high incident intensity, $I_0 = 10^5$ mJ/cm^2/fs, and resonant excitation given by (14.164) are shown in red in Fig. 14.25b. They are compared to the Einstein result (black) and the numerical solution of the Bloch equation for a detuning energy of $\Delta\mathcal{E} = 2\Gamma$ (blue).

In practice, the temporal evolution of $g^{(2)}(t)$ measured with monochromatic light rather than the broadband light assumed in the Einstein model exhibits a rich oscillatory structure with a range of values from 0 to around 2, as shown by the red and blue curves in Fig. 14.25. The Rabi oscillations are due to the fact that the excited and ground state populations oscillate in time with opposite phases (since $\rho_{22} + \rho_{11} = 1$) before they become equal at long times.

At delay times $t \simeq 10\,t_\Gamma$ the second order degree of coherence settles down to unity. This does not mean that the fluorescence light is second order coherent, but rather that it becomes essentially uncorrelated and the two-photon coincidences are accidental. As discussed in Sect. 5.4, true second order coherence requires that both $g^{(1)}(t)$ and $g^{(2)}(t)$ are unity for the same time t.

In Fig. 14.26 we illustrate the rich structure of $g^{(2)}(t, \Delta\mathcal{E})$ in a 3D plot as a function of time and detuning energy for the same incident intensity of $I_0 = 10^5$ mJ/cm²/fs as in Fig. 14.25b.

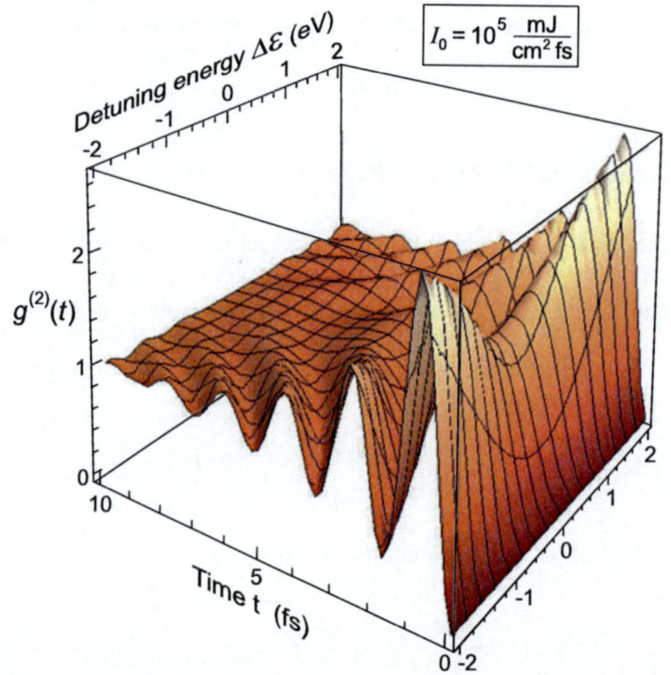

Fig. 14.26 Three-dimensional plot of the second order degree of temporal coherence $g^{(2)}(t)$ of a Co atom for $I_0 = 10^5$ mJ/cm²/fs as in Fig. 14.25b. The figure illustrates the rich structure of $g^{(2)}(t, \Delta\mathcal{E})$ as a function of time and detuning energy -2eV $\leq \Delta\mathcal{E} \leq 2$ eV ($\simeq 5\Gamma$), the latter dependence being symmetrical about $\Delta\mathcal{E} = 0$

14.12.4 *Photon Antibunching in Resonance Fluorescence*

Remarkably, the degree of second order temporal coherence of the elastically scattered radiation in the presence of stimulation evolves from zero, at $t = 0$ to unity at $t \simeq 10t_\Gamma$. The value of $g^{(2)}(t=0)=0$ has received a lot of attention since it violates the classically allowed range of values $1 \leq g^{(2)} \leq \infty$ [18]. This result was first experimentally demonstrated by Kimble et al. [11] using a very dilute atomic beam and more recently with single trapped $^{24}Mg^{+}$ ions by Höffges et al. [44]. In particular, the zero value of $g^{(2)}(t=0)$ is due to the inability of the single two-level atom to emit a pair of photons simultaneously. It was one of the first illustrations of states of light that can only be explained by quantum theory. During the emission of a photon, the atom decays into its ground state and needs to be re-excited before it can emit a second photon. The emitted photons are therefore spaced out in time and are said to be "antibunched" with an average separation of the excited state lifetime $\hbar/\Gamma$.

Figure 14.27 illustrates the opposite signature of "bunching" and "antibunching" in $g^{(2)}(t)$.

The red curve corresponds to the photon-photon coincidence count rate as a function of their relative photon arrival for light produced by a chaotic lightsource with a Lorentzian frequency distribution (exponential decay). At short times, photons arrive more likely together, i.e. are "bunched", than at longer timescales. This is the *photon bunching* effect observed by Hanbury Brown and Twiss [45–47]. The black curve is taken from Fig. 14.25 (Einstein model) and represents photon-photon coincidence of resonant fluorescence light. It shows the opposite behavior or "antibunching" at short times. The two curves are reflections of each other around the horizontal line $g^{(2)}(t)=1$ and the bunching property of chaotic light mirrors the antibunching property of light of an atom driven by a strong field.

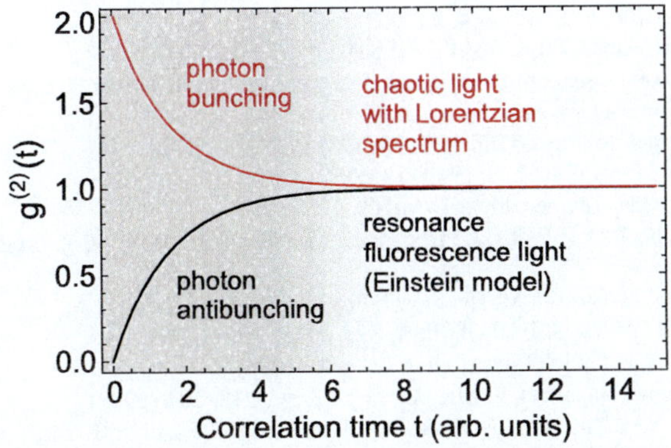

Fig. 14.27 Second order degree of temporal coherence $g^{(2)}(t)$ as a function of relative photon-photon arrival time for two different types of light. The red curve corresponds to chaotic light with a Lorentzian spectral distribution, while the black curve is taken from Fig. 14.25 and corresponds to resonant fluorescent light emitted by a strongly driven atom

References

1. I.I. Rabi, Phys. Rev. **51**, 652 (1937)
2. F. Bloch, Phys. Rev. **70**, 460 (1946)
3. L. Allen, J.H. Eberly, *Optical Resonance and Two Level Atoms* (Wiley, New York, 1975)
4. R.P. Feynman, F.L. Vernon, R.W. Hellwarth, J. Appl. Phys. **28**, 49 (1957)
5. A. Einstein, Verh. Deut. Phys. Ges. **18**, 318 (1916)
6. A. Einstein, Phys. Z. **18**, 121 (1917)
7. R. Loudon, *The Quantum Theory of Light*, 3rd edn. (Clarendon Press, Oxford, 2000)
8. H.J. Kimble, L. Mandel, Phys. Rev. A **13**, 2123 (1976)
9. M.O. Scully, M.S. Zubairy, *Quantum Optics* (Cambridge University Press, Cambridge, 1997)
10. L. Mandel, E. Wolf, *Optical Coherence and Quantum Optics* (Cambridge University Press, Cambridge, 1995)
11. H.J. Kimble, M. Dagenais, L. Mandel, Phys. Rev. Lett. **39**, 691 (1977)
12. H. Paul, Rev. Mod. Phys. **54**, 1061 (1982)
13. J.J. Sakurai, *Modern Quantum Mechanics, Revised* (Addison-Wesley, Reading, Mass., 1994)
14. C.C. Gerry, P.L. Knight, *Introductory Quantum Optics* (Cambridge University Press, Cambridge, 2005)
15. P.W. Milonni, Phys. Rep. **25C**, 1 (1976)
16. E.T. Jaynes, F.W. Cummings, Proc. IEEE **51**, 89 (1963)
17. J. Stöhr, H.C. Siegmann, *Magnetism: From Fundamentals to Nanoscale Dynamics* (Springer, Heidelberg, 2006)
18. R. Loudon, Rep. Prog. Phys. **43**, 913 (1980)
19. N. Rohringer et al., Nature **481**, 488 (2012)
20. N. Hartmann, G. Hartmann, R. Heider, M.S. Wagner, M. Ilchen et al., Nat. Photonics **12**, 215 (2018)
21. M.L. Citron, H.R. Gray, C.W. Gabel, J.C.R. Stroud, Phys. Rev. A **16**, 1507 (1977)
22. J. Stöhr, A. Scherz, Phys. Rev. Lett. **115**, 107 402 (2015)
23. M. Planck, Verh. Deut. Phys. Ges. **2**, 202 (1900)
24. M. Planck, Verh. Deut. Phys. Ges. **2**, 237 (1900)
25. P.A.M. Dirac, Proc. Roy. Soc. A **114**, 243 (1927)
26. G. Kirchhoff, Ann. Phys. Chem. **109**, 275 (1860)
27. O. Lummer, F. Kurlbaum, Verh. Deut. Phys. Ges. **17**, 106 (1898)
28. P.M.L. Robitaille, Prog. Phys. **4**, 3 (2009)
29. R.C. Hilborn, Am. J. Phys. **50**, 982 (1982)
30. Hilborn, A 2002 update of the original 1982 paper by Hilborn (Am. J. Phys. **50**, 982 (1982)) is available by the same author on Arxiv
31. V. Weisskopf, Ann. Phys. Leipzig **9**, 23 (1931)
32. B.R. Mollow, Phys. Rev. **188**, 1969 (1969)
33. S.H. Autler, C.H. Townes, Phys. Rev. **100**, 703 (1955)
34. H. J. Carmichael, D. F. Walls, J. Phys. B: At. Mol. Phys. **9**, 1199 (1976); erratum ibid. **9**, 2755 (1976)
35. F. Schuda, J.C.B. Stroud, M. Hercher, J. Phys. B **7**, L198 (1974)
36. R.E. Grove, F.Y. Wu, S. Ezekiel, Phys. Rev. A **15**, 227 (1977)
37. H.J. Kimble, L. Mandel, Phys. Rev. A **15**, 689 (1977)
38. H.J. Kimble, M. Dagenais, L. Mandel, Phys. Rev. A **18**, 201 (1978)
39. O. Astafiev, A.M. Zagoskin, A. A. Abdumalikov. Jr., Y.A. Pashkin, T. Yamamoto, K. Inomata, Y. Nakamura, J.S. Tsai, Science **327**, 840 (2010)
40. R. Röhlsberger, J. Evers, S. Shwartz, Quantum and nonlinear optics with hard x-rays. in *Synchrotron Light Sources and Free-Electron Lasers*, ed. by E. Jaeschke, S. Khan, J. Schneider, J. Hastings (Springer, Cham., Heidelberg, 2020), p. 1399
41. D.S. Dovzhenko, S.V. Ryabchuk, Y.P. Rakovich, I.R. Nabiev, Nanoscale **10**, 3589 (2018)
42. D. A. Steck, *Quantum and Atom Optics*. Available online at http://steck.us/teaching (2017)

43. S.M. Cavaletto, C. Buth, Z. Harman, E.P. Kanter, S.H. Southworth, L. Young, C.H. Keitel, Phys. Rev. A **86**(033), 402 (2012)
44. J.T. Höffges, H.W. Baldauf, T. Eichler, S.R. Helmfrid, H. Walther, Opt. Commun. **133**, 170 (1997)
45. R.H. Brown, R.Q. Twiss, Nature **178**, 1046 (1956)
46. R.H. Brown, R.Q. Twiss, Proc. Roy. Soc. Lond. A **242**, 300 (1957)
47. R.H. Brown, R.Q. Twiss, Proc. Roy. Soc. Lond. A **243**, 291 (1958)

Chapter 15
Non-linear Absorption and Scattering Processes in Solids

15.1 Introduction and Chapter Overview

15.1.1 *Brief Introduction*

In the optical regime, the advent of the laser required the consideration of non-linear interactions with matter beginning in the 1960s. Over the years they have led to the new field of non-linear optics [1, 2] or strong-field physics [3]. In the x-ray regime, non-linear phenomena naturally became important and have been explored with VUV and x-ray free electron lasers.

Of particular interest are spectroscopic x-ray studies of the modification of the electronic structure of solids [4–7], x-ray transparency in non-resonant absorption [8–11] and resonant x-ray absorption [7, 12–14], and the related effects of stimulated x-ray emission [15–20] and stimulated resonant inelastic scattering [21–23]. We also discuss the limits of resonant diffraction imaging induced by sample damage [24] and stimulated forward scattering [25, 26]. There are many other aspects of non-linear x-ray phenomena not covered here as outlined in Sect. 1.5.4.

15.1.2 *Chapter Overview*

The goal of the present chapter is to develop a description of the most fundamental non-linear response of a *solid*, namely the change in x-ray absorption with incident intensity. Since x-ray absorption has the largest interaction cross section, its understanding is fundamental for all XFEL experiments. X-ray absorption is also the simplest process to quantify since one simply has to measure the change of x-ray intensity upon transmission through a film.

In the past we have relied on the fact that even x-ray pulses of modern synchrotron radiation sources typically give information of the "as-prepared" state of a sample of interest. The "as-prepared" state may be any state of a *solid* that can be stabilized

J. Stöhr, *The Nature of X-Rays and Their Interactions with Matter*, Springer Tracts in Modern Physics 288, https://doi.org/10.1007/978-3-031-20744-0_15

over the effective measurement time. In conventional experiments it is the time the sample is exposed to the beam. In pump-probe experiments it is given by the individual 100 ps pulse length of storage rings. The conventional key paradigm of x-ray science has been that x-rays can be viewed as a weakly interacting probe that provides information of the as-prepared state of the sample before the measurement.

Relative to conventional experiments with synchrotron sources, there are two fundamental differences in XFEL pulse interactions with matter. The number of photons per pulse is increased and the pulse length shortened by similar factors of order $10^4 - 10^5$. At best, XFEL pulses allow us to take ultrafast snapshots on timescales faster than the picosecond atomic motion and comparable to the femtosecond time evolution of chemical bonds. At worst, the intense XFEL pulses will modify the electronic properties even during a single pulse, so that the pulse average already has an electronic damage component, and crystallographic damage will be observed when multiple consecutive pulses are used. It is therefore of fundamental importance to understand when the paradigm of measuring the "as-prepared" state of interest breaks down.

In the first part of this chapter we will explore the "damage" problem by discussing the temporal evolution of changes in solids triggered by energy transfer from the incident photons to atoms. This entails understanding how the largest interaction process, x-ray absorption, changes as a function of intensity and the temporal evolution of the energy transfer of the incident photons to the electronic system and the lattice in solids. We are particularly interested in exploring changes in the electronic structure which happens on the shortest time scales. We will show how such changes are reflected by changes of the x-ray absorption spectrum at relatively low incident intensities.

The central part of this chapter deals with the description of fundamental nonlinear effects that appear in x-ray absorption at higher incident intensities. We pay particular attention to the description of stimulated resonant forward scattering in solids, building on the previous chapter where the phenomenon was treated for atoms. Our treatment of solids also builds on our previous discussion of their semi-classical response in Chap. 7, and their quantum treatment by the KHD theory, discussed for XAS in Chap. 10, and REXS and RIXS in Chap. 13.

In the process we emphasize analytical solutions that allow us to readily treat various cases. Besides treating the change in absorption in the presence of stimulated REXS, we will also consider the effect of increased in-beam forward scattering on diffraction. Since diffraction is based on the out-of-beam scattered intensity, we will show that it is strongly affected by the enhancement of the in-beam forward scattered intensity and may vanish altogether. Besides conventional diffraction by charge domains we also consider the dichroic diffraction of magnetic domains which will be shown to exhibit particularly large non-linear effects.

Finally, we will consider the case of stimulated inelastic resonance scattering, stimulated RIXS. We will show that the main advantage of stimulated RIXS comes from the fact that stimulation is directional, so that the scattered photons produced by stimulation are driven into the small solid angle of the detector, which may lead to enhancements of the order of 10^5. In practice, however, the intensity required for stimulated RIXS also causes electronic changes even during ultrashort few-femtosecond pulses. At present, this appears to be a show stopper for stimulated RIXS in solids.

15.2 The Fundamental Damage Issue of XFEL Radiation

Historically, the use of x-rays for the study of the electronic and atomic structure of matter was based on the premise that x-rays provide information on the sample in the "*as-prepared*" state, which is either the electronic ground state at the temperature of interest or a state which has been prepared by application of an external stimulus like pressure, fields, or an electronic excitation by another source, e.g. an optical laser. Until the XFEL era, sample perturbations by the x-ray beam itself could typically be ignored and the interaction treated as a small perturbation, e.g. by the lowest order KHD perturbation theory as outlined in Chap. 9.

Early considerations of the effect of XFEL pulses on matter [27] already emphasized potential problems associated with changes of the state of the sample during or after the x-ray pulses. One of the key questions concerns the time scales associated with typical XFEL pulses (see Sect. 2.8) relative to the intrinsic timescales associated with charge, spin, and atomic motions in matter, illustrated in Fig. 1.12. If the pulses are much shorter than the temporal changes in matter, the pulse modifications can be either neglected since they have not materialized during the pulse or, provided that a time zero can be established, the temporal evolution of matter can be studied in detail. For comparable timescales of pulses and processes in matter, photons in the second half of the pulse will already produce a response that differs from the "as-prepared" state.

We know typical motion times of atoms, since they determine the speed of sound. More generally, the concept of *inertia* leads to the motto "the smaller (less mass), the faster", as illustrated in Fig. 1.12. This concept underlies the Born–Oppenheimer principle, stating that the lighter electrons move faster than the heavy nuclei. For the same reason, it was suggested already before XFELs came into existence [28] that ultrashort XFEL pulses of 1–10 fs length can traverse a sample fast enough that the recorded picture will show the "as-prepared" *atomic structure* before pulse-induced atomic motions can manifest themselves. In other words, one may beat the speed of sound with the speed of light (more precisely, with the speed of a femtosecond x-ray pulse). On the other hand, mass scaling predicts that this argument does not apply to the *electronic structure* of a sample. The valence charge distributions of the thousand times lighter electrons, that form the all-important bonding orbitals between atoms, may change on the same timescale as the fastest x-ray pulses. In the following we will consider typical intensities and XFEL pulse lengths to elucidate when "damage" in one form or another will be observed.

15.2.1 X-Ray Beam Parameters

We characterize the x-rays by three parameters, the photon density n_{ph}/V, the frequency $\omega = 2\pi c/\lambda$, and the pulse length τ, which we assume to be the same as the coherence time. Together with the constant speed of light c, these quantities define the cross-sectional area of the beam $A = V/(c\tau)$, and with the addition of Planck's constant $\hbar$, they define the quantum mechanical flux, fluence, and intensity (see Sect. 3.6.2), as

$$\text{Flux}: \quad \Phi = \frac{n_{\text{ph}}\, c}{V} \tag{15.1}$$

$$\text{Fluence}: \quad \mathcal{F} = \frac{n_{\text{ph}}\, \hbar\omega}{A} = \frac{n_{\text{ph}}\, \hbar\omega\, c\tau}{V} \tag{15.2}$$

$$\text{Intensity}: \quad I = \frac{n_{\text{ph}}\, \hbar\omega}{A\tau} = \frac{n_{\text{ph}}\, \hbar\omega\, c}{V}. \tag{15.3}$$

With our assumption that the pulses are temporally coherent (transform limited), the intensity is simply the fluence divided by the pulse length. We note that this is typically not fulfilled for SASE XFEL sources where the coherence time is determined by the length of coherent subpulses (spikes) within the total pulse (see Figs. 2.29 and 2.30). For coherently driven processes like stimulated REXS it is important to distinguish pulse length and coherence time (see Sect. 15.6.3).

In the following we will address the question up to which fluence or intensity one can expect to probe a largely unperturbed electronic state and lattice of a solid sample. Since the answer to this question depends on the type of material and the XAS cross sections of the atoms, we will consider a metal and assume an x-ray absorption cross section of about 5 Mb/atom, which as shown in Fig. 1.15 is characteristic of strong absorption.

15.2.2 Temporal Evolution of Matter after X-Ray Excitation

Simple arguments, already confirmed by ultrafast optical laser studies [29, 30], naturally explain the temporal evolution of matter after a severe instantaneous energetic impulse. The photon energy is first transferred to the more agile electronic system of the sample. The energy in the electronic system is then transferred to the atomic phonon system, heating up the lattice. The timescale of electronic and lattice modifications relative to the x-ray pulse length lies at the heart of all XFEL experiments since it directly determines when the conventional "weak perturbation" picture breaks down. The change in matter may manifest itself over a wide range from mild perturbation to complete destruction. An example of a mild change would be an increased electronic or lattice temperature. The other extreme is a transfer of energy of tens of eV to all atoms in the sample, resulting in a plasma like state and complete destruction of the sample. This case is itself interesting since the transiently created state is of interest to the field of high energy density (HED) science [31].

As an example of a response of a solid let us consider a metal. In the optical regime, the interactions of femtosecond laser pulses with metals such as Ag and Au have been thoroughly studied, as reviewed by Rethfeld et al. [30]. Energy is first transferred from the photon to the electron system on the timescale of the incident pulse. The hot electron system then cools and equilibrates on timescales of 100 fs – 1 ps by interaction with the phonons of the initially cold lattice. This process is typically described by the so-called *two-temperature model*, based on different temperatures of the electronic system and the lattice [32, 33]. With time, the temperatures in the two thermodynamic reservoirs equilibrate, leading to the establishment of a Fermi-Dirac electron distribution around the Fermi energy [34–36]. After deposition of large laser energy, the regular phonon modes of the lattice will become non-linear leading to severe atomic motion and diffusion that may culminate in melting.

More recently, the initial optical studies of electronic-to-lattice heat transfer have been extended to ultrafast spin-lattice dynamics through the study of magnetic solids [37–40]. The two-temperature model is then replaced by a *three-temperature model*, where the third temperature is that of the spin reservoir. Strong perturbations of the electronic and magnetic structure, in particular the loss of the magnetization on femtosecond timescales, have been observed that has led to the concept of ultrafast "all optical magnetic switching". Most of the optical studies were carried out at photon energies of ~1.5–2 eV, pulse lengths of about 50 fs, and fluences of the order of 10 mJ/cm^2. Since such intensities of less than 1 mJ/cm^2/fs are at the lower end of possible XFEL intensities that extend to about 10^8 mJ/cm^2/fs or 10^{20} W/cm^2 [41], it is clear that typical XFEL studies are bound to encounter strong non-linear effects of various types.

In the following we will follow the time evolution of the processes in matter after being triggered by a femtosecond XFEL pulse, starting with the initial x-ray excitation of atoms. Since *x-ray absorption* has the largest cross section, the temporal evolution of the sample is triggered by the quasi-instantaneous x-ray absorption process during the incident pulse. In the process, part of the incident energy $n_{\rm ph}\,\hbar\omega$ contained in the x-ray pulse of length τ is transferred to the number of atoms $N_{\rm a} = \rho_{\rm a} V$ in the illuminated sample volume $V = Ad$, where d is the effective x-ray penetration depth. The fraction of atoms that are excited is determined by the XAS cross section $\sigma_{\rm XAS}$ which determines the penetration depth of the x-rays into the sample. While $\sigma_{\rm XAS}$ can vary by orders of magnitude for different atoms and photon energies, the atomic number density of solids varies relatively little and typically lies between 50 and 100 atoms/nm^3. Hence the penetration depth, typically expressed as one x-ray absorption length ℓ_x defined as $\ell_x = 1/(\sigma_{\rm XAS}\rho_{\rm a})$, mainly depends on $\sigma_{\rm XAS}$.

The initial transfer of energy mainly involves excitations of core electrons as illustrated in Fig. 1.27 on the attosecond timescale. As illustrated in Fig. 9.5, one needs to distinguish between resonant transitions to bound states and excitations into higher lying continuum states. In bound-state transitions, the excited electron is *trapped* and is not emitted as a photoelectron. Ignoring direct x-ray excitations of valence electrons, which have a lower cross section, changes of the valence states occupation will occur as an aftereffect of filling the created core hole on a timescale set be the core hole clock $\tau_\Gamma = \hbar/\Gamma$ (see Fig. 12.4). Auger electrons created in the core hole decay are mainly responsible for electron reshuffling in the valence states through inelastic electron-electron scattering. The number of Auger electrons created in the core hole decay is typically larger than that of the spontaneously emitted x-rays, and *electron-electron* scattering cross section is larger than *photon-electron* scattering cross sections.

The kinetic energy of the Auger electrons of hundreds of eV is partially converted through inelastic electron-electron scattering and cascading into low energy electronic excitations of a few eV. While scattered electrons with kinetic energies above the surface potential barrier (work function) of $\simeq$5 eV can escape into vacuum, most electrons (>90%) cascade down in less than 10 fs to energies below the work function barrier and spread over nanometers [42, 43]. We note that the scattering dynamics of electrons with kinetic energies in the 5–100 eV range is particularly complicated, and their equilibration is a topic of ongoing research [44].

In x-ray excitations to *continuum states* above the XAS "edge", photoelectrons will be directly produced. They may scatter inelastically on their way out of a solid sample, leading to the well-known low energy electron tail in photoemission spectra. The photo electron scattering processes may therefore already lead to changes in the valence occupation before the core hole decays. The changes will be enhanced through the scattering of Auger electrons following the decay.

15.2.3 Energy Transfer to the Electronic System

For both resonant and continuum state excitations, the electron scattering cascade leads to a hot electron distribution around the Fermi energy E_F, caused by reshuffling electrons from below to above the Fermi level. The so-created electron/hole distribution around the Fermi level may significantly differ from the step-like low-temperature distribution, where all states below E_F are filled with electrons. This is illustrated in Fig. 15.1a where the filled states at low temperature are shown shaded in gray.

In the figure we have assumed that at an elevated electronic temperature $T_e = T$, the electron distribution is described by a Fermi-Dirac distribution function[1]

$$f(\mathcal{E}) = \frac{1}{e^{(\varepsilon - E_F)/k_B T_e} + 1}. \tag{15.4}$$

[1] We equate the Fermi level with the Fermi energy and denote them by E_F. This ignores the difference between the Fermi energy and the chemical potential μ of a thermodynamic system which are strictly the same only at $T_e = 0$. Since $E_F \gg k_B T_e$, we have $\mu = E_F$ to a very good approximation all the way to the melting point. This justifies the common practice of using the two quantities interchangeably for metals.

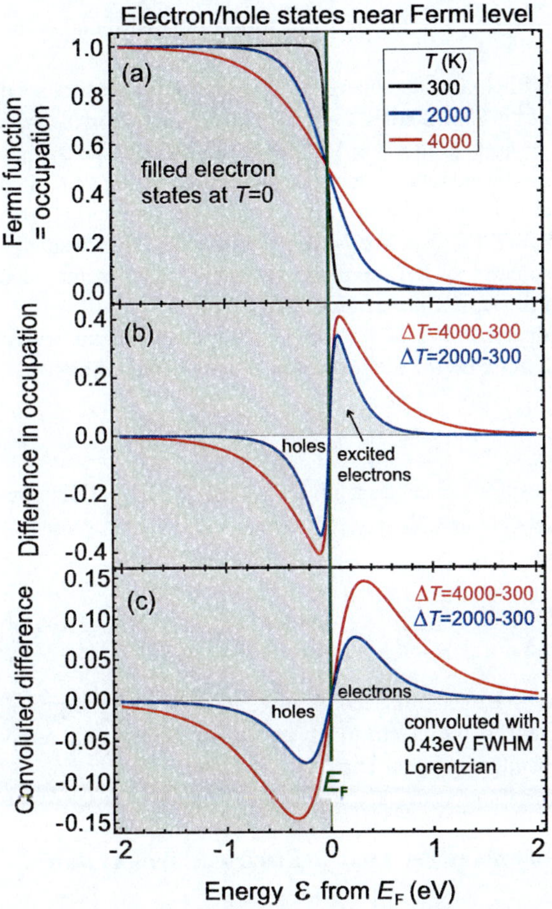

Fig. 15.1 **a** Electron distribution at $T = 0\,\mathrm{K}$ indicated by gray shading, and its change according to the Fermi-Dirac function (15.4) at elevated temperatures T, for three values, where T is the equilibrated temperature $T = T_e = T_p$ of the electron and phonon systems. **b** Difference between the $T = 2000\,\mathrm{K}$ and $T = 300\,\mathrm{K}$ Fermi-Dirac functions in (**a**) shown (blue), and between the $T = 4000\,\mathrm{K}$ and $T = 300\,\mathrm{K}$ functions (red), revealing the reshuffling of electrons to states above E_F and creation of holes below E_F due to electron conservation. **c** Curves in (**b**) convoluted with a Lorentzian of FWHM $\Gamma = 0.43\,\mathrm{eV}$ (Co L_3 value). The intensity difference due to electron/hole redistribution would be observed in an XAS experiment with spectral resolution limited by Γ

Here $\varepsilon - E_F$ is the energy relative to the Fermi level, taken as the zero of energy in Fig. 15.1, T_e is the temperature of the electronic system, and $k_B = 8.62 \times 10^{-5}\,\mathrm{eV/K}$ is the Boltzmann constant. In principle, the electronic temperature T_e can only be defined if the electronic system has equilibrated with itself through electron-electron collisions and has the form of (15.4). After a sudden excitation of the electronic system, equilibration involves many collisions and occurs on timescales of about

10–100 fs [30, 34, 35, 45]. Figure 15.1a shows the distributions for three different temperatures.

Figure 15.1b illustrates the change in distribution of electrons, corresponding to the difference of the $T = 2000$ K and $T = 300$ K function in (a), shown in blue, and between the $T = 4000$ K and $T = 300$ K functions (red). The change in temperature ΔT leads to electron excitations to states above E_F leaving behind hole states below E_F due to conservation of electrons. In (c) we show the curves in (b) convoluted with a Lorentzian of FWHM $\Gamma = 0.43$ eV (Co L_3 value). The shown change in distribution would be measured in an XAS experiment, whose instrumental resolution is much less than the intrinsic lifetime width Γ (see Fig. 15.3).

For an incident number of photons n_0, the absorbed number of photons n_{abs} changes with depth x below the surface of a sample according to

$$n_{abs}(x) = n_0 \left(1 - e^{-\sigma_{XAS} \rho_a x}\right). \tag{15.5}$$

The energy deposition hence has a large gradient. A film of thickness d, illuminated within an area A contains $N_a = \rho_a A d$ atoms, and the average energy deposited per atom is given by

$$U_{at} = \frac{n_{abs} \hbar\omega}{N_a} = \frac{n_0 \hbar\omega}{\rho_a A d} \left(1 - e^{-\sigma_{XAS} \rho_a d}\right) = \frac{\mathcal{F}}{\rho_a d} \left(1 - e^{-\sigma_{XAS} \rho_a d}\right). \tag{15.6}$$

During the short time of a femtosecond pulse, the deposited energy will be transferred almost exclusively to the electronic system whose electronic temperature is then defined by the change in Fermi-Dirac distribution.

15.2.3.1 From Energy Per Atom to Electronic Temperature

The Fermi-Dirac distribution function determines the temperature dependent occupation of electron and hole states around E_F. In order to calculate the change in electronic temperature one needs to take into account the electronic density of states of a solid $\mathcal{D}(\varepsilon)$, which is typically listed in the literature in units of [states/atom/eV)] [46]. The zero of energy $\varepsilon = 0$ is taken to be at the bottom of the free electron-like valence states, and at zero temperature the states are filled up to the Fermi level E_F which for metals is of order of 10 eV [46–48]. At $T_e = 0$, the Fermi function is a step-like function so that the electron distribution is given by

$$n(\varepsilon) = \begin{cases} 1 & \varepsilon < E_F \\ 0 & \varepsilon > E_F \end{cases} \tag{15.7}$$

and the energy of the electronic system $\mathcal{E}_0 = \mathcal{E}_e(T_e = 0)$ is

$$\mathcal{E}_0 = \int_0^\infty n(\varepsilon)\, \mathcal{D}(\varepsilon)\, d\varepsilon. \tag{15.8}$$

For a free electron metal this works out to be $\mathcal{E}_0 = 3N_e E_F/5$, where N_e is the total number of valence electrons per atom. We are interested in the *change of energy* of the electronic system *per atom* relative to its ground state value $\mathcal{E}_0$. This arises from the rearrangement of electrons from below to above the Fermi level at finite temperature, expressed by the Fermi-Dirac distribution function. The problem was solved by Sommerfeld in 1928 [49] using a perturbation approach, but we use a simpler derivation. The *change in energy per atom* at an electronic temperature T_e relative to $T_e = 0$ is given by

$$\mathcal{E}_e \simeq \mathcal{D}(E_F)\, 2 \int_{E_F}^\infty \frac{\varepsilon - E_F}{e^{(\varepsilon - E_F)/k_B T_e} + 1}\, d\varepsilon = \mathcal{D}(E_F)\, \frac{\pi^2}{6} k_B^2 T_e^2. \tag{15.9}$$

We have approximated the density of states over the range of the Fermi-Dirac function by its value at E_F and have used the fact that there is an equal number of electron and hole states which equally contribute to the energy change. This allows us to integrate from the Fermi level to infinity and simply multiply by 2. The so obtained value $\mathcal{E}_e$ is the energy associated with all valence electron excitations per atom. It is the temperature induced change of the ground state energy $\mathcal{E}_0 = 3N_e E_F/5$. Our result agrees with the different derivation given in [47].

The specific heat of the electronic system *per atom* is obtained from (15.9) as

$$c_e = \frac{\delta \mathcal{E}_e}{\delta T_e} = \frac{\pi^2}{3} \mathcal{D}(E_F)\, k_B^2 T_e. \tag{15.10}$$

The quadratic dependence on T_e of the electronic energy associated with *all excited electrons per atom* given by (15.9) arises from the fact that the number of excited valence electrons *per atom*, N_e^*, is given by

$$N_e^* \simeq \mathcal{D}(E_F)\, 2 \int_{E_F}^\infty \frac{1}{e^{(\varepsilon - E_F)/k_B T_e} + 1}\, d\varepsilon = \mathcal{D}(E_F) \ln(4) k_B T_e. \tag{15.11}$$

Since N_e^* itself scales with $k_B T_e$, the energy normalized to *each excited electron* at an electronic temperature T_e is given by

$$\frac{\mathcal{E}_e}{N_e^*} = \frac{\pi^2}{12 \ln(2)} k_B T_e \simeq 1.19 k_B T_e. \tag{15.12}$$

The small difference from the expected value $k_B T_e$ per excited electron supports our approximation of pulling the density of states out of the integrals in (15.9) and (15.11).

Equation (15.9) is derived by assuming a temperature-independent density of states and that the electronic temperature is not too high. It has recently been shown to hold remarkably well for transition metals up to temperatures close to the melting point [50]. It should be used with caution at larger electronic temperatures that can easily be created with XFEL x-rays. Nevertheless, by equating the deposited energy per atom with the change in electronic energy per atom, $U_{at} = \mathcal{E}_e$, we obtain the temperature of the electronic system under the assumption that no energy is lost as

$$\boxed{T_e = \frac{1}{k_B}\sqrt{\frac{6\,U_{at}}{\pi^2 \mathcal{D}(E_F)}}.} \tag{15.13}$$

15.2.4 From Electronic to Lattice Temperature

With increasing time, up to a picosecond or so, energy may be transferred between electrons and the lattice by electron-phonon collisions. Although electron-phonon and electron-electron collision times may be comparable [47], energy transfer from the hot electrons to the lattice will be slower due to the large mass difference of electrons and atoms. The electron-lattice equilibration is typically described phenomenologically by the so-called *two-temperature model* [32, 33]. It distinguishes the temperatures T_e and T_a and specific heats c_e and c_a of the electronic system and atomic lattice. The equilibration of the systems is described by a time-dependent rate equation that links the two reservoirs through a coupling constant. Due to the difference in specific heats of the electron gas and the phonons [30], the equilibration may take picoseconds [30, 51, 52].

Historically, it was puzzling why the heat capacity per atom in the lattice described by the empirical law of Dulong and Petit[2] could account for the specific heat of metals without needing to consider the contributions of the electrons per atom. Both atoms and electrons were expected to contribute equally based on the equipartition theorem. In its modern form, the Dulong-Petit law states that the heat capacity *per atom* has a constant value of

$$c_a \simeq 3k_B. \tag{15.14}$$

[2] The law was suggested by P. L. Dulong and A. T. Petit in 1819 [53] as, "The atoms of all simple bodies have exactly the same capacity for heat." They deduced from experimental data for 13 elemental solids a constant value of the heat capacity per weight multiplied by the relative atomic weight of the elements.

This relation can be derived in the Einstein and Debye models for large lattice temperatures, and it is a rather good approximation for transition metals at room temperature. The puzzle about the missing contribution of the electrons is naturally resolved by Fermi-Dirac statistics. It shows that only a small fraction of the electrons near E_F can be thermally excited, while most of the energy of the electronic system is buried in the *Fermi sea*, extending from the Fermi-level down by about 10 eV to the bottom of the valence band. The electronic contribution of the excited electrons to the specific heat of the solid is very small at room temperature. This is directly seen by comparing the atomic specific heat c_a *per atom* to the electronic specific heat c_e *per atom*. They can be linked by comparing (15.10) and (15.14) as

$$c_e \simeq \underbrace{3k_B}_{c_a} \mathcal{D}(E_F)\, k_B T_e. \tag{15.15}$$

Since in metals $\mathcal{D}(E_F) \simeq 1\,\text{atom}^{-1}\text{eV}^{-1}$ [46] and at room temperature $k_B T_e \simeq 0.025\,\text{eV}$, we have $\mathcal{D}(E_F)\, k_B T_e \simeq 0.025/\text{atom}$. For this reason one typically finds $c_a + c_e \simeq c_a$. This is not valid at low temperatures of order 10 K, where the Dulong-Petit law does not hold and one may indeed detect the temperature dependent electronic contribution.

After absorption of an intense x-ray pulse, the electronic temperature may be initially of order of $k_B T_e \simeq 1-10\,\text{eV}$ per atom. In the absence of heat loss, one may estimate the equilibration temperature with the lattice by using the constant Dulong-Petit value for the specific heat per atom. The electronic temperature then decreases by a factor of 3, and the lattice heats up to the same temperature. The increase ΔT in lattice temperature, which determines whether the sample melts, can then be related to the fluence by use of (15.9) according to

$$\boxed{\Delta T \simeq \frac{U_{at}}{c_a} = \frac{\mathcal{F}}{3k_B \rho_a d}\left(1 - e^{-\sigma_{XAS}\rho_a d}\right).} \tag{15.16}$$

15.2.5 Ablation Threshold

Ablation thresholds for XFEL radiation have been studied by Hau-Riege [54], Yumoto et al. [55], and Koyama et al. [56]. In Fig. 15.2 we show ablation patterns recorded as a function of x-ray fluence with 10 keV x-rays for 200 nm Pt layers deposited on SiO_2 substrates [56].

Pt has an atomic number density of $\rho_a = 65.1\,\text{atoms/nm}^3$, and at 10 keV an x-ray absorption cross section $\sigma_{XAS} = 0.035\,\text{Mb/atom} = 3.5 \times 10^{-6}\,\text{nm}^2/\text{atom}$ [57]. The corresponding x-ray absorption length of 4.4 μm is much larger than the 200 nm thick film, and at the Pt ablation threshold $\mathcal{F} = 0.023\,\mu\text{J}/\mu\text{m}^2$ (see Fig. 15.2g), the energy absorbed per atom is obtained with (15.6) as $U_{at} \simeq 0.5\,\text{eV/atom}$. According

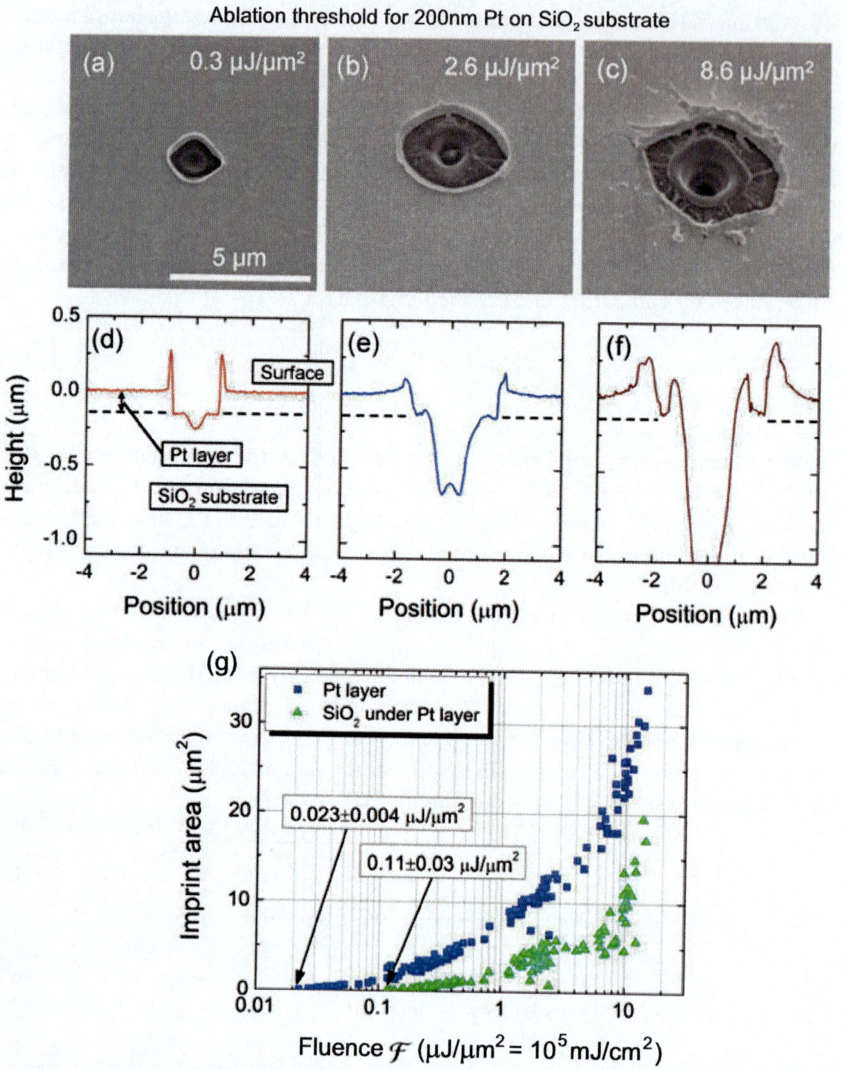

Fig. 15.2 Fluence-dependent ablation of Pt metal [56]. **a–c** SEM images of x-ray induced ablation craters for 200 nm of Pt on SiO_2 substrates. **d–f** Cross-sectional profiles of the craters. The dashed line indicates the position of the Pt/SiO_2 interface. **g** Imprint areas and ablation thresholds as a function of fluence for the Pt layer and SiO_2 underneath

to (15.16) the rise in lattice temperature is given by $\Delta T = 1930\,\text{K}$, and adding the initial room temparature of 300 K we obtain an ablation threshold temperature of $\simeq 2230\,\text{K}$ which compares to the melting temperature of 2041 K. These numbers are in reasonable agreement and support the simple theory outlined above.

15.2.6 *Summary*

We can summarize the evolution of the temperature in an x-ray excited solid as follows.

After interaction with an intense x-ray pulse, the energy absorbed from the pulse per atom first ends up in the electronic system. After it has equilibrated within itself over a period of 10–100 fs, one can define an electronic temperature T_e given by (15.13). The change of T_e from absolute zero is described by the Fermi-Dirac distribution function, which expresses the change in electron and hole population around the Fermi level, E_F. The population of electrons located in the deep Fermi sea well below E_F remains unchanged.

On the 100–1000 fs timescale the electronic system equilibrates with the lattice until they both have the same temperature. According to the Dulong-Petit law one finds as a rule of thumb an equilibrated temperature that is lower by a factor of 3 than the original temperature of the electronic system. This behavior is supported by experimental results for transition metals [34, 36].

15.3 Fluence-Dependent Changes of XAS Spectra

In this section we illustrate two fundamental non-linear effects that are observed with increase of the incident intensity.

15.3.1 *Redistribution of Valence Electrons*

As an example of a somewhat mild non-linear effect that appears at pulse fluences near the melting threshold we discuss the modification of the XAS L_3 resonance of Co in a Co/Pd multilayer. Such multilayers have been extensively studied because they can be prepared with a stable magnetization direction along the surface normal, and have therefore served as a prototype system for exploring both modification of the valence charge by use of linear polarization and the valence magnetization by use of circular polarization. Here we only consider changes in the valence charge by use of spectra recorded with incident linear polarization.

Intensity induced changes are illustrated in Fig. 15.3 for the 3*d* valence states of Co atoms in Co metal layers sandwiched between Pd [5, 6].

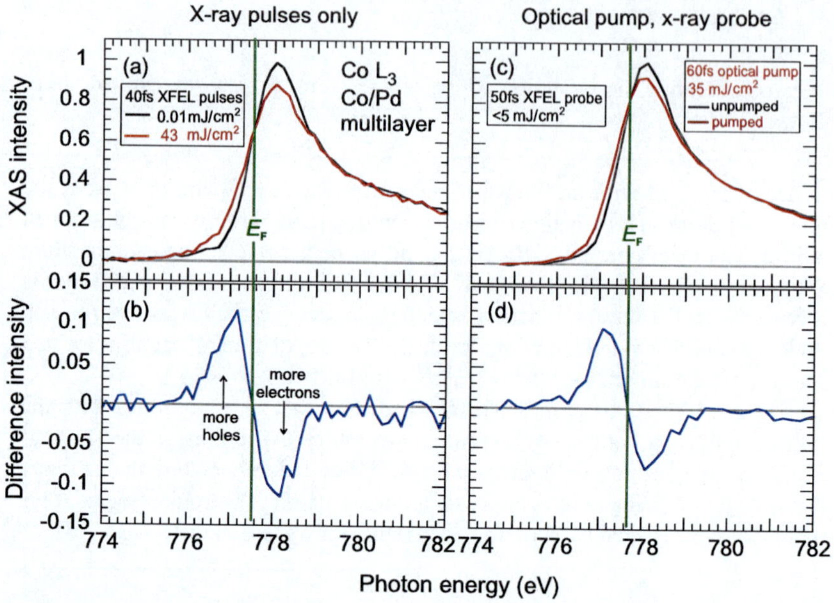

Fig. 15.3 Illustration of damage of the ground state valence band population, revealed as a decrease of the resonant Co L_3 XAS intensity in Co/Pd multilayers. **a, b** show changes in XAS when the x-ray fluence of 40 fs x-ray pulses is increased form 0.01 to 43 mJ/cm^2 [5] due to electron reshuffling from below to above the Fermi level, E_F, during the x-ray pulses. **c, d** show similar effects due to optical $\lambda \simeq 800$ nm ($\hbar\omega \simeq 1.6$ eV) excitations with 60 fs optical pump pulses of 35 mJ/cm^2 fluence that are coincident with weak 50 fs x-ray probe pulses [6]

In Fig. 15.3a, b we compare spectra recorded with 40 fs XFEL pulses of low fluence, $\mathcal{F}$= 0.01 mJ/cm^2 (black curve) and 43 mJ/cm^2 (red curve) and their difference spectrum shown in blue underneath. We note that the changes shown in Fig. 15.3a, b occur at relatively low incident *intensities* (fluence per pulse length) around 1 mJ/cm^2/fs. In this case the changes are entirely induced by the x-ray pulse itself [5]. They are observed due to the fact that the recorded spectrum is an average over the entire pulse length, and hence the photons near the end of the pulse already see the changes caused by the photons near the beginning of the pulse. We have also indicated the position of the Fermi level, which was inferred by the zero crossing in the difference curve in (b).

In Fig. 15.3c, d we show a similar effect that however was created in a pump-probe scenario using an optical 60 fs pump pulse of $\lambda \simeq 800$ nm ($\hbar\omega \simeq 1.6$ eV) of $\mathcal{F}$ =35 mJ/cm^2 fluence which was coincident with a weak x-ray probe pulse of 50 fs duration [6]. In this case, the changes of the x-ray spectra are entirely due to the optical pump pulses with an intensity of about 0.6 mJ/cm^2/fs. The changes in (b) and (d) show the same characteristic functional form, but are weaker in (d) due to the lower effective "pump" intensity.

The difference spectra in Fig. 15.3b, d are explained by Fig. 15.1c, which illustrated the expected change in the Fermi-Dirac distribution that would be observed in an XAS experiment with a lifetime-only spectral broadening. The experimental difference spectra shown in Fig. 15.3b, d closely resemble the theoretical ones in Fig. 15.1c because the Gaussian monochromator bandwidth of FWHM 250 meV was smaller than the Co L_3 FWHM 430 meV Lorentzian lifetime width.

The change in the XAS spectra therefore directly shows the electron redistribution of the $3d$ valence electrons around the Fermi level as a consequence of inelastic scattering of Auger electrons. We can estimate the initial energy deposition per Co atom by taking into account a 70% transmission loss due to Pd so that an effective fluence of $\mathcal{F} = 30\,\text{mJ/cm}^2$ was incident on the combined $d = 22\,\text{nm}$ of Co. This corresponds to an absorbed energy per atom of $U_{\text{at}} = 680\,\text{meV/atom}$ calculated with (15.6). The electronic temperature obtained with (15.13) is $k_B T_e = 370\,\text{meV/atom}$ or $T_e = 4300\,\text{K/atom}$. With the Dulong-Petit value for the specific heat per atom we would estimate a lattice temperature of 1430 K which is close to but below the Co melting temperature of 1768 K.

As mentioned earlier, the above estimates have to be considered with care but they seem reasonable. In the original reference, Higley et al. used a more detailed model and found that a deposited energy $U_{\text{at}} = 615\,\text{meV/atom}$ would melt the Co/Pd sample [5]. This would put our value of $U_{\text{at}} = 680\,\text{meV/atom}$ above rather than below the melting temperature. We can also fit of the difference curve distribution in Fig. 15.3b with a Fermi-Dirac function that is convoluted with a Voigt function that accounts for both the Co core hole width of FWHM 430 meV and an instrumental Gaussian width of FWHM of 250 meV. This yields an electronic temperature of 3300 K/atom which is about 25% lower than our estimated value above of $T_e = 4300\,\text{K/atom}$.

Considering the uncertainties of the above estimates, we can roughly associate an incident intensity of $I_0 \simeq 1\,\text{mJ/cm}^2\text{/fs}$ with an equilibrated sample temperature near the melting point around 1768 K. This agrees with the findings of Wang et al. [24] who used similar samples but studied the diffraction due to magnetic domains. They found that the magnetic domain pattern was altered after passage of 80 fs pulses and fluences $\mathcal{F} \geq 25\,\text{mJ/cm}^2$, or intensities of $I_0 \simeq 0.4\,\text{mJ/cm}^2\text{/fs}$. This was attributed to irreversible thermal damage caused by heating the sample above the Curie temperature of 750 K. Comparing the ratios of intensities and sample temperatures for the two cases, we find them to be approximately the same.

From Fig. 15.3 we can also conclude that after the electron cascade trickles down to energies of a few eV, x-ray excited samples reach states that are similar to those reached through optical excitation. The fluence-dependent XAS changes seen in Fig. 15.3c, d are reminiscent of those observed after optical excitation by Stamm et al. for nickel [58] and Cho et al. [59] for Cu.

For a given incident *intensity*, the valence electron redistribution observed after optical and x-ray excitation pulses are similar. Both are determined by electron/hole pairs created around the Fermi level by inelastic electron scattering.

15.3.2 X-Ray Transparency: An Introduction

At higher incident intensities a more dramatic and fascinating effect, *x-ray transparency*, can be observed as previously shown in Fig. 1.28. In that figure the change in non-resonant x-ray absorption above the $L_{3,2}$ edges of Al metal was illustrated. This *non-resonant* excitation of core electrons into continuum states, however, needs to be distinguished from *resonant* excitations into localized valence states as discussed in Chap. 10. The importance of x-ray absorption resonances was already pointed out in Sect. 1.4.2 which illustrated the particular importance of elemental, chemical, and spin dependent resonant transitions. It is therefore not surprising that the high-intensity or non-linear x-ray absorption behavior should be different for the two cases. This is in fact observed, as illustrated in Fig. 15.4 where we show the change in x-ray transmission (absorption) with incident intensity observed for two different cases.

As illustrated in Fig. 15.4a, the core electron may be excited to a partially empty *localized valence state* at the excitation threshold, giving rise to a strong absorption "resonance". In this case the final state is exciton like, with the excited electron remaining trapped and having insufficient energy to escape as a photoelectron. Alternatively, the core electron may be excited into higher energy free electron-like band states. The electron can then overcome the work function due to the surface potential barrier of the solid and escape as a photoelectron.

When the incident intensity is converted to the average number of photons per atom in the sample, we see that for *continuum* excitation above the Al L-edge, illustrated in Fig. 15.4b, the deviation of the transmission from the low intensity value starts when the number of incident photons becomes equal to the number of illuminated atoms in the film [8]. In contrast, for *resonant* excitation at the Co L_3-edge, shown in Fig. 15.4c, the non-linear transmission onset occurs much earlier, at less than 0.1 photon per Co atom. In both cases the transmission further increases with increasing intensity until the sample becomes transparent. We will later in Sect. 15.7 discuss in more detail the interesting physics behind the different behavior illustrated in Fig. 15.4, and here only give a simple explanation for the difference.

For *continuum* excitation, the non-linear transparency can be accounted for by considering when all illuminated atoms in the sample have a single core hole. The absorption will then stop, and the sample becomes transparent. This behavior is due to the fact that the *single photon* absorption cross section of a core shell σ_{XAS} completely dominates over the excitation cross sections of outer shells as shown in Fig. 1.27. Most importantly, once an electron from a given core shell has been excited, the photon excitation probability of the remaining core electrons in the same shell is negligibly small [60].

The whole saturation process for continuum excitation begins approximately when there is on average one photon per atom. This does not mean that there is also a core hole on each atom. Core hole creation increases probabilistically with the number of excitation attempts by photons that traverse the atomic volume. The increase of transparency beyond the vertical threshold line in Fig. 15.4b reflects the increasing probability of core hole production. It saturates only after on average more

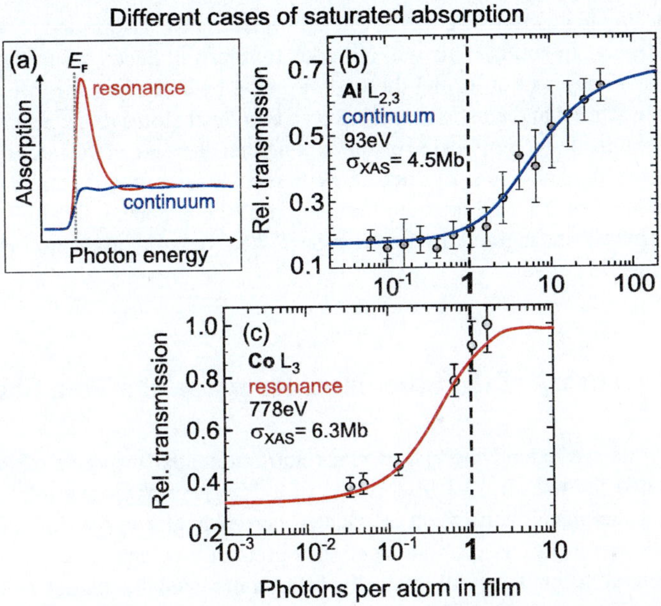

Fig. 15.4 Examples of the non-linear transmission of x-rays through films, leading to the disappearance of x-ray absorption or film transparency with increasing incident intensity. **a** We distinguish "resonance" (red) from "continuum" (blue) excitations in x-ray absorption, depending on whether the incident photon energy coincides with a strong XAS threshold resonance or excites core electrons into higher lying continuum states. **b** Continuum excitation case in (**a**). The transmission of 15 fs SASE pulses through a 53 nm thick Al film is shown as a function of incident intensity, expressed as the number of photons per illuminated Al atom. The incident photon energy was 93 eV or about 20 eV above the position of the $L_{2,3}$ edge (73 eV) [8], corresponding to a cross section $\sigma_{XAS} = 4.5$ Mb/atom. **c** Resonant excitation case in (**a**). We show the transmission of 25 fs SASE pulses whose energy was centered on the 778 eV L_3 resonance through $\simeq$20 nm of Co in a Co/Pd multilayer with a peak resonant cross section of $\sigma_{XAS} = 6.25$ Mb/atom [13]

than 100 photons have traversed the atomic volume and attempted an excitation. One may picture the required number of attempts for unit excitation probability as the ratio of the total atomic area relative to the smaller area given by the x-ray absorption cross section. Only when a photon hits within the small x-ray cross section area within the larger atomic area is there unit excitation probability.

In contrast to excitation into continuum states with a smooth density of states and cross section σ_{XAS}, *resonant* excitation is strongly photon energy and bandwidth dependent. Maximum absorption occurs when the photon energy is tuned to the resonance and the bandwidth is smaller than the resonance width. The bandwidth dependence means that the resonant process depends on the temporal coherence time of the incident photons. For a typical XFEL pulse, the transparency onset therefore depends on the coherent fraction of photons in the pulse rather than on the number of all photons in the pulse. In this case, the non-linear effect is not a "brute force" absorption effect but is enhanced by coherent stimulated elastic forward scattering.

In the following sections we will develop the quantitative description of non-linear x-ray absorption in solids. This will allow us to return in Sect. 15.7 to the topic of x-ray transparency with a formal theory. We begin by extending the BR theory of resonant absorption and elastic scattering by a two-level atoms to the case of a solid film. This extension is complicated by the exponential decrease of the intensity within a film of finite thickness. For this reason we first consider a film of atomic thickness which can be treated by a linear approximation of the exponential falloff and then use the Darwin-Prins method discussed in Sect. 7.8.2 to build up the non-linear film response sheet-by-sheet.

15.4 BR Theory of the Stimulated Response of a Thin Sheet

In this section we extend the spontaneous and stimulated response of individual atoms, treated in the Sect. 14.1, to the case of a thin sheet of atomic thickness. We assume that the plane of the sheet is oriented perpendicular to the direction of the incident photons as discussed from a classical point of view in Chap. 7.

Our general formalism will typically be accompanied by model calculations assuming L_3 resonance excitation of Co metal. Unless stated otherwise, our model parameters will be the polarization ($p = 0, \pm$) dependent parameters listed in Table 15.1. The linear polarization values are the same as those used in the atomic calculations in the previous chapter (see Sect. 14.4.1). The listed values for the optical parameters β^p are slightly smaller than those plotted for Fe and Co in Fig. 7.23 (also see Fig. 7.12 for Fe) since we have subtracted a small pre-edge background of about 0.2 Mb due to outer shell contributions.

Table 15.1 Parameters for the L_3 resonances of Fe, Co, and Ni metals

	ρ_a $\frac{\text{atoms}}{\text{nm}^3}$	$\mathcal{E}_0$ eV	λ_0 nm	σ^+_{XAS} Mb	σ^0_{XAS} Mb	σ^-_{XAS} Mb	Γ^+_x meV	Γ^0_x meV	Γ^-_x meV	β^+	β^0	β^-	f''^+ nm	f''^0 nm	f''^- nm
Fe	84.9	707	1.75	8.8	6.9	5.0	0.33	0.26	0.19	10.4†	8.2†	5.9†	0.25†	0.20†	0.14†
Co	90.9	778	1.59	7.9	6.25	4.65	0.43	0.34	0.25	9.1†	7.2†	5.4†	0.25†	0.20†	0.15†
Ni	91.4	853	1.45	5.1	4.35	3.7	0.37	0.32	0.27	5.4†	4.7†	3.9†	0.18†	0.15†	0.13†

†Listed values to be multiplied by 10^{-3}

Listed are the atomic number densities ρ_a, the resonance energies $\mathcal{E}_0$ and wavelengths λ_0, the polarization dependent (p=0, $\pm$) peak *experimental* cross sections $\sigma^p_{XAS} = [\sigma^\diamond_{XAS}]^p$ (1 Mb = 10^{-4} nm^2), assuming x-ray propagation along the magnetization direction, and the polarization-dependent dipole transition widths Γ^p_i, optical attenuation parameters β^p, and scattering lengths f''^p, derived from the listed σ^p_{XAS} values [see (7.30)]

15.4.1 Non-linear Response of an Atomic Sheet

We start with the transmission sum rule (14.105) for the single atom which links the spontaneous plus stimulated loss and gain probabilities. For the thin sheet transmission case, we similarly denote changes in transmission as "loss" (minus sign) and "gain" (plus sign). We include spontaneous and stimulated REXS forward scattering and, for now, neglect the spontaneous out-of-beam REXS contribution. The transmission sum rule for a thin sheet that balances absorption and forward scattering is then given by (13.56) modified by a stimulated enhancement as

$$P_{\text{trans}}^{\text{sheet}} \simeq 1 - \underbrace{\frac{N_{\text{a}}}{A}\frac{\lambda^2\Gamma_x}{\pi\Gamma}}_{\text{1st order XAS loss probability}} + \underbrace{\frac{1}{2}\left[\frac{N_{\text{a}}}{A}\frac{\lambda^2\Gamma_x}{\pi\Gamma}\right]^2}_{\text{coh. REXS gain = 2nd ord. XAS prob.}} + \underbrace{2\frac{N_{\text{a}}}{A}\frac{\lambda^2\Gamma_x}{\pi\Gamma}\widetilde{\rho}_{22}^{\infty}}_{\text{stim. REXS gain probability}} . \tag{15.17}$$

Here $N_{\text{a}}/A = \rho_{\text{a}}\Delta$ is the atomic areal density, given by the sheet thickness Δ and bulk atomic number density ρ_{a} (see Table 15.1). Δ is assumed to be of order of the diameter of an atom ($\simeq$0.2 nm) which allows us to write the transmission probability in the above linearized form. We have also introduced an excited state population factor $\widetilde{\rho}_{22}^{\infty}$, where the superscript ∞ reflects the long time or equilibrium value. It is distinguished by a tilde from the individual atom case discussed in the previous chapter. We will clarify its meaning below. This thin sheet expression reduces to the atomic version (14.105) with the replacement,[3]

$$\frac{N_{\text{a}}}{A} = \rho_{\text{a}}\Delta \longleftrightarrow \frac{\pi}{\lambda^2} \quad \text{and} \quad \widetilde{\rho}_{22}^{\infty} \Longleftrightarrow \rho_{22}^{\infty} \tag{15.18}$$

The area A containing N_{a} atoms reduces to the single atom area λ^2/π, and the collective excited state population of the illuminated N_{a} atoms in the sheet $\widetilde{\rho}_{22}^{\infty}$ turns into the single atom excited state population ρ_{22}^{∞}. The factor of 2 in the stimulated gain term in (15.17) arises from calling the non-linear (NL) reduction of absorption at large incident intensity a NL gain of the same size due to stimulated REXS forward scattering.

By multiplying the thin sheet transmission probabilities (14.105) by the incident intensity given by (14.82) as

$$I_0 = \frac{n_0 \hbar\omega\, c}{A\,\Delta}. \tag{15.19}$$

[3] Close comparison of (14.105) and (15.17) also reveals that the spontaneous REXS term for the thin sheet case is reduced by a factor of 1/2 which is absent for the atomic case. It has its origin in the second order expansion expressed by (7.66).

we can express (15.17) in the intensity form

$$I_{\mathrm{trans}} \simeq \underbrace{\frac{n_0 \hbar\omega c}{A\Delta}}_{I_0} \Bigg\{ 1 - \underbrace{\frac{N_{\mathrm{a}}}{A} \frac{\lambda^2 \Gamma_x}{\pi\Gamma}}_{\text{1st order XAS loss probability}} + \underbrace{\frac{1}{2}\left[\frac{N_{\mathrm{a}}}{A}\frac{\lambda^2\Gamma_x}{\pi\Gamma}\right]^2}_{\text{coh. REXS gain = 2nd ord. XAS prob.}} + \underbrace{2\frac{N_{\mathrm{a}}}{A}\frac{\lambda^2\Gamma_x}{\pi\Gamma}\widetilde{\rho}_{22}^{\infty}}_{\text{stim. REXS gain probability}} \Bigg\}. \tag{15.20}$$

The entire NL transmission response of the atomic sheet is expressed by the intensity-dependent factor $\widetilde{\rho}_{22}^{\infty}$ which we will discuss now.

15.4.2 *Effective Excited State Population and Enhancement Factor $\mathcal{G}_{\mathrm{coh}}$*

The N_{a}^2 dependence of the coherent forward scattering term in (15.20), which is also present in the classical formulation of spontaneous forward scattering given by (7.70), reveals that the second order forward scattered intensity involves the cooperative or coherent action of all atoms in the illuminated area. If all atoms contributed independently, the forward scattered intensity would scale linearly with N_{a} like the XAS term in (15.20) (or the incoherent scattering term in (7.70)).

This important fact shows that the stimulated REXS term in (15.20) must also scale with N_{a}^2, and hence $\widetilde{\rho}_{22}^{\infty}$ needs to scale with N_{a}/A. In addition, $\widetilde{\rho}_{22}^{\infty}$ needs to have the correct low and high intensity limits. At very high intensity, defined by an equilibrium of up and down transitions or saturation, we need to have $\widetilde{\rho}_{22}^{\infty} \rightarrow 1/2$. At an intensity corresponding to $n_\Gamma = 1$, the stimulated and spontaneous REXS intensities need to be the same and for $n_\Gamma \ll 1$ we need to have $\widetilde{\rho}_{22}^{\infty} \rightarrow 0$.

By consideration of these boundary conditions we write $\widetilde{\rho}_{22}^{\infty}$ in the same form as ρ_{22}^{∞} given by (14.90) with an additional unitless collective response factor $\mathcal{G}_{\mathrm{coh}}$

$$\widetilde{\rho}_{22}^{\infty} = \frac{\mathcal{G}_{\mathrm{coh}}\,\Gamma_{\mathrm{R}}^2/4}{(\hbar\omega - \mathcal{E}_0)^2 + (\Gamma/2)^2 + \mathcal{G}_{\mathrm{coh}}\,\Gamma_{\mathrm{R}}^2/2} = \frac{\mathcal{G}_{\mathrm{coh}}\, n_\Gamma \Gamma \Gamma_x/4}{(\hbar\omega - \mathcal{E}_0)^2 + (\Gamma/2)^2 + \mathcal{G}_{\mathrm{coh}}\, n_\Gamma \Gamma\Gamma_x/2}. \tag{15.21}$$

This expression fulfills the high intensity limit $\widetilde{\rho}_{22}^{\infty} \rightarrow 1/2$ for $n_\Gamma \rightarrow \infty$. At low intensity it reduces to

$$\widetilde{\rho}_{22}^{\infty} = \mathcal{G}_{\mathrm{coh}} \underbrace{\frac{n_\Gamma \Gamma_x}{\Gamma} \frac{\Gamma^2/4}{(\hbar\omega - \mathcal{E}_0)^2 + \Gamma^2/4}}_{\rho_{22}^{\infty}} \tag{15.22}$$

which is related to the low intensity atomic ρ_{22}^{∞} given by (14.61) as indicated. When inserted into (15.20) we find that for

$$\boxed{\mathcal{G}_{\rm coh} = \frac{\lambda^2 N_{\rm a}}{4\pi A} = \frac{\lambda^2 \rho_{\rm a} \Delta}{4\pi}} \tag{15.23}$$

all boundary conditions are fulfilled. For $n_\Gamma = 1$ the stimulated REXS term reduces to the spontaneous REXS term given by (13.56) and for $n_\Gamma \ll 1$ we find the proper limit $\widetilde{\rho}_{22}^{\infty} \to 0$.

We are now in a position to state the meaning of $\widetilde{\rho}_{22}^{\infty}$ which contains the collective (coherent) enhancement factor $\mathcal{G}_{\rm coh}$, first derived in [14].

For an *atomic sheet* oriented perpendicular to the x-ray propagation direction, all $N_{\rm a}$ atoms in the illuminated area A contribute collectively (coherently) to elastic scattering in the *forward* direction. This collective atomic action is reflected by $\mathcal{G}_{\rm coh}$ which also accounts for the fact that coherent forward scattering is confined to the solid angle cone $\mathrm{d}\Omega_{\rm coh} = \lambda^2/A$ of 4π [see (4.38)].

The enhancement factor $\mathcal{G}_{\rm coh}$ enters into the *effective* excited state population $\widetilde{\rho}_{22}^{\infty}$ according to (15.21). $\widetilde{\rho}_{22}^{\infty}$ is *not* the actual population of the upper states in the individual atoms in the sheet, ρ_{22}^{∞}, but is larger by $\mathcal{G}_{\rm coh}$ due to a collective atomic response.

15.5 Non-linear Transmission Through a Film of Finite Thickness

Our thin sheet NL transmission expression (15.20) is valid only up to second order in sheet thickness Δ. The limitation of the spontaneous absorption loss is illustrated in Fig. 15.5 by plotting the relative spontaneous transmission loss for L_3 resonance excitation of a Co thin film as a function of thickness Δ.

Shown in Fig. 15.5 as a black curve is the proper spontaneous attenuation according to the Beer-Lambert law (7.32) or

$$I_{\rm trans}^{\diamond} = I_0\,\mathrm{e}^{-\sigma_{\rm XAS}^{\diamond}\rho_{\rm a}\Delta} = I_0\,\mathrm{e}^{-2\lambda f''\rho_{\rm a}\Delta} = I_0\,\mathrm{e}^{-2\beta k\Delta} \tag{15.24}$$

where $\sigma_{\rm XAS}^{\diamond}$ is the polarization averaged spontaneous XAS cross section given by (14.18) as

$$\sigma_{\rm XAS}^{\diamond} = \frac{\lambda^2}{\pi}\frac{\Gamma_x}{\Gamma}\frac{(\Gamma/2)^2}{(\hbar\omega - \mathcal{E}_0)^2 + (\Gamma/2)^2} \tag{15.25}$$

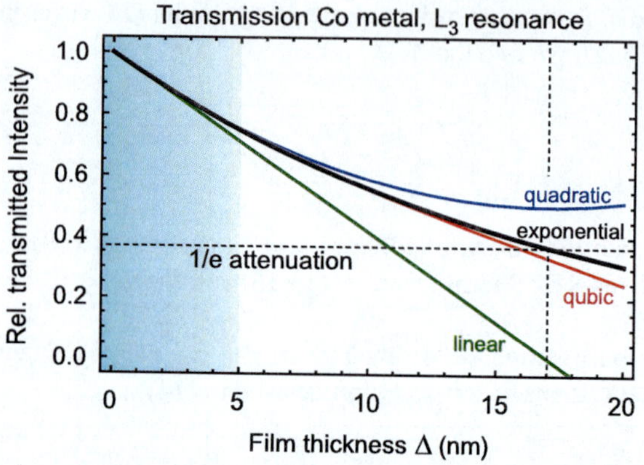

Fig. 15.5 Attenuation of the incident intensity upon transmission through a Co film of atomic density $\rho_a = 90.9$ atoms/nm^3 as a function of film thickness Δ assuming excitation at the L_3 resonance ($\lambda = 1.59$ nm, $\beta_0 = 7.5 \times 10^{-3}$ or $\sigma_{XAS}^{\diamond} = 6.5$ Mb). Shown is the exponential attenuation with Δ according to (15.24) (black curve), and approximations of the exponential by a power series in Δ given by (15.26), keeping only the linear term in Δ (green) and those up to second (blue) and third (red) order. We have indicated by blue shading the validity of the second order approximation. Also indicated by dashed black lines are the relative transmission value (1/e = 0.37) corresponding to one x-ray absorption length of 17 nm

and f'' and β are the conventional spontaneous response values defined in Sect. 7.5.5. In Fig. 15.5 we have used the optical parameter $\beta = 7.5 \times 10^{-3}$ from Table 7.2, corresponding to the peak value of the black curve in Fig. 7.24 and $\sigma_{XAS}^{\diamond} = 6.5$ Mb. In the figure we also show in different colors the approximations of (15.24) up to third order in Δ given in the β form by

$$I_{\text{trans}}^{\diamond} = I_0\, e^{-2\beta k \Delta} \simeq I_0 \left\{ 1 - 2\beta k \Delta + 2\beta^2 k^2 \Delta^2 - \frac{4}{3}\beta^3 k^3 \Delta^3 \right\}. \quad (15.26)$$

We see that our second order approximation of the spontaneous loss expressed by (15.20) and (13.60) remains valid up to a sheet thickness of $\Delta \simeq 7$ nm as indicated by blue shading.

Since in practice one typically uses larger film thicknesses, we need to consider the thickness correction of the spontaneous loss and NL gain. We start by rewriting the equilibrium excited state population of the sheet, $\widetilde{\rho}_{22}^{\infty}$, given by (15.21) which determines the NL response of our general second order transmission expression (15.20). We are interested in the dependence of $\widetilde{\rho}_{22}^{\infty}$ on the incident intensity I_0 and sheet thickness Δ. The I_0-dependence is obtained by use of (14.82) through the replacement

$$n_\Gamma = \frac{\lambda^3}{2\pi^2 \Gamma c} I_0 \quad (15.27)$$

Fig. 15.6 Dependence of $\tilde{\rho}_{22}^{\infty}$ on intensity I_0 and sheet thickness Δ according to (15.28) using the parameters for L_3 excitation in Co metal for linear polarization ($p = 0$) given in Table 15.1. Also shown for comparison is the plot of ρ_{22}^{∞} for a single atom (see Fig. 14.8a). The relative shift by $\mathcal{G}_{\rm coh}$ between the single atom case and the thin sheet case is illustrated for $\Delta = 2\,$nm

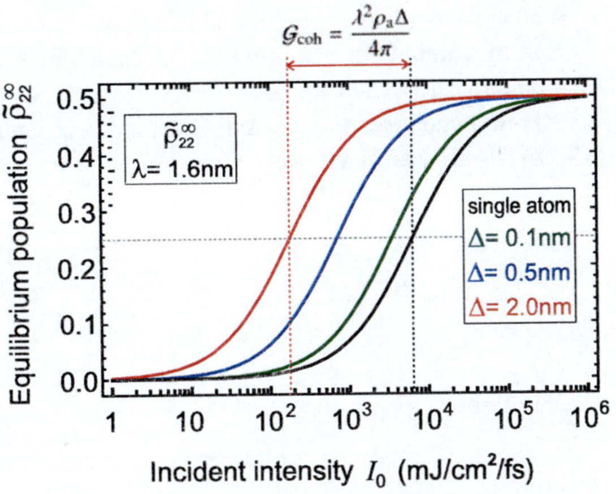

in (15.21). The Δ-dependence enters through the enhancement factor $\mathcal{G}_{\rm coh}$ given by (15.23). This yields

$$\boxed{\tilde{\rho}_{22}^{\infty} = \frac{I_0 \Delta\, \Gamma_x \rho_a \lambda^5/(32\pi^3 c)}{(\hbar\omega - \mathcal{E}_0)^2 + (\Gamma/2)^2 + I_0 \Delta\, \Gamma_x \rho_a \lambda^5/(16\pi^3 c)}}. \tag{15.28}$$

The dependence of this expression on I_0 for different values of Δ is illustrated in Fig. 15.6 for the example of L_3 resonance excitation in Co metal.

With increasing sheet thicknesses, the curves in Fig. 15.6 are simply shifted linearly by $\mathcal{G}_{\rm coh}$ to lower intensity relative to the single atom response, which corresponds to $\mathcal{G}_{\rm coh} = 1$. NL effects in sheet transmission therefore come into play at lower incident intensities when more atoms are in the beam. We will now derive the quantitative dependence of film transmission on thickness.

15.5.1 From Thin Sheet to Finite Thickness Film

In principle, x-ray propagation in a film of finite thickness could be incorporated into the differential optical Bloch equations which are then solved *numerically*. We have seen in Sect. 14.1 that in certain cases *analytical* solutions can be obtained that facilitate calculations and more clearly reveal the underlying physical processes. Our goal is to extend the analytical solution for the resonant excitation of individual atoms to include propagation in a film of finite thickness. To do so, we utilize the Darwin-Prins method developed in Sect. 7.8.2. We start with the thin sheet expression (15.20), which accounts for the competing resonant absorption and stimulation effects, and propagate the intensity through the N sheets of thickness Δ to the thickness $d = N\,\Delta$.

We begin with the expression (15.20) for a thin sheet and drop the second order XAS term which describes *spontaneous* REXS in the forward direction. We account for *stimulated* REXS by keeping the last term. This means we account for absorption in first order and resonant forward scattering in second order. By use of the expression (15.28) for $\widetilde{\rho}_{22}^{\infty}$ we then have

$$I_{\mathrm{trans}} \simeq I_0 \left\{ 1 - \rho_{\mathrm{a}} \Delta \frac{\lambda^2 \Gamma_x}{\pi \Gamma} + \rho_{\mathrm{a}} \Delta \frac{\lambda^2 \Gamma_x}{\pi \Gamma} \underbrace{\frac{I_0 \Delta\, \Gamma_x \rho_{\mathrm{a}} \lambda^5/(16\pi^3 c)}{(\hbar\omega - \mathcal{E}_0)^2 + (\Gamma/2)^2 + I_0 \Delta\, \Gamma_x \lambda^5 \rho_{\mathrm{a}}/(16\pi^3 c)}}_{2\widetilde{\rho}_{22}^{\infty}} \right\}. \tag{15.29}$$

By defining

$$\boxed{\mathbb{A} = \frac{\Gamma_x\, \rho_{\mathrm{a}} \lambda^5}{16\pi^3 c}, \quad \text{and} \quad \mathbb{B} = (\hbar\omega - \mathcal{E}_0)^2 + (\Gamma/2)^2} \tag{15.30}$$

we obtain the compact expression for the effective excited state population

$$2\widetilde{\rho}_{22}^{\infty}(\Delta) = \frac{I_0\, \Delta\, \mathbb{A}}{\mathbb{B} + I_0\, \Delta\, \mathbb{A}}, \tag{15.31}$$

so that we obtain by use of (15.25)

$$I_{\mathrm{trans}} \simeq I_0 \left\{ 1 - \sigma_{\mathrm{XAS}}^{\diamond}\, \rho_{\mathrm{a}} \Delta \left[1 - \underbrace{\frac{I_0\, \Delta\, \mathbb{A}}{\mathbb{B} + I_0\, \Delta\, \mathbb{A}}}_{2\,\widetilde{\rho}_{22}^{\infty}(\Delta)} \right] \right\}. \tag{15.32}$$

This expression nicely distinguishes the spontaneous XAS loss from the stimulated REXS gain which depends on $2\widetilde{\rho}_{22}^{\infty}(\Delta)$. For a thin film of thickness $d \gg \Delta$, it is instructive to consider two simple limits for the transmitted intensity. The proper, more complicated, sheet-to-sheet propagation of the fields must then give a result for the transmitted intensity that lies between these limits.

In one limit, we assume that *absorption* by a film of thickness d is properly accounted for by the exponential Beer-Lambert law but that the intensity I_0 in the stimulated gain term in (15.32) is taken as the incident intensity without sheet-by-sheet attenuation. This would *overestimate* the stimulated gain and yield

$$I_{\mathrm{trans}} = I_0 \exp\left[- \sigma_{\mathrm{XAS}}^{\diamond}\, \rho_{\mathrm{a}} d \left(1 - \underbrace{\frac{I_0\, d\, \mathbb{A}}{\mathbb{B} + I_0\, d\, \mathbb{A}}}_{2\,\widetilde{\rho}_{22}^{\infty}(d)} \right) \right]. \tag{15.33}$$

In the other limit I_0 in the stimulated gain term is taken as the attenuated intensity in the last sheet of the film. This would *underestimate* the gain according to

$$I_{\text{trans}} = I_0 \exp\Bigg[-\overset{\diamond}{\sigma}_{\text{XAS}}\, \rho_{\text{a}} d \Bigg(1 - \underbrace{\frac{I_0\, d\, \mathrm{e}^{-\overset{\diamond}{\sigma}_{\text{XAS}} \rho_{\text{a}} d}\, \mathbb{A}}{\mathbb{B} + I_0\, d\, \mathrm{e}^{-\overset{\diamond}{\sigma}_{\text{XAS}} \rho_{\text{a}} d}\, \mathbb{A}}}_{2\, \widetilde{\rho}_{22}^{\infty}(d\, \mathrm{e}^{-\overset{\diamond}{\sigma}_{\text{XAS}} \rho_{\text{a}} d})} \Bigg)\Bigg]. \tag{15.34}$$

The expression below the underbracket is meant to indicate that the functional dependence of the excited state population on d in (15.33) has been replaced by that on $d\, \mathrm{e}^{-\overset{\diamond}{\sigma}_{\text{XAS}} \rho_{\text{a}} d}$.

Owing to the fact that I_0 and d enter equivalently in $\widetilde{\rho}_{22}^{\infty}$, the two scenarios are described by shifted excited state population curves similar to Fig. 15.6. For example, for a Co film ($\rho_{\text{a}} = 90.9$ atoms/nm^3) of 20 nm thickness and a peak cross section $\overset{\diamond}{\sigma}_{\text{XAS}} = 6.25$ Mb we have $\mathrm{e}^{-\overset{\diamond}{\sigma}_{\text{XAS}} \rho_{\text{a}} d} \simeq 0.32$ and the two $\widetilde{\rho}_{22}^{\infty}$ curves would be shifted approximately linearly in I_0 by a factor of $\simeq 3$.

For the *spontaneous* transmission case, the x-ray absorption length into the film is intensity *independent*, and we can define an effective film thickness d_{eff} according to,

$$\boxed{d_{\text{eff}} = \int_0^d \mathrm{e}^{-\overset{\diamond}{\sigma}_{\text{XAS}}\, \rho_{\text{a}} x}\, \mathrm{d}x = \frac{1 - \mathrm{e}^{-\overset{\diamond}{\sigma}_{\text{XAS}}\, \rho_{\text{a}} d}}{\overset{\diamond}{\sigma}_{\text{XAS}}\, \rho_{\text{a}}}.} \tag{15.35}$$

This corresponds to projecting the exponential fall off of the intensity over the sample thickness d, given by the conventional Beer-Lambert law, onto a shorter distance d_{eff} over which the intensity is constant.

In the presence of stimulated forward scattering, the absorption in d_{eff} will be reduced and with increasing intensity the x-rays will penetrate deeper, until at $d_{\text{eff}} = d$ the entire film becomes transparent. This stimulated change may be approximated by use of the Darwin-Prins formalism to yield an effective thickness denoted by a tilde, given by

$$\tilde{d}_{\text{eff}} = \Delta \frac{1 - \exp\Big[-\overset{\diamond}{\sigma}_{\text{XAS}}\, \rho_{\text{a}} d \Big(1 - \frac{I_0\, \Delta\, \mathbb{A}}{\mathbb{B} + I_0\, \Delta\, \mathbb{A}}\Big)\Big]}{1 - \exp\Big[-\overset{\diamond}{\sigma}_{\text{XAS}}\, \rho_{\text{a}} \Delta \Big(1 - \frac{I_0\, \Delta\, \mathbb{A}}{\mathbb{B} + I_0\, \Delta\, \mathbb{A}}\Big)\Big]}. \tag{15.36}$$

Here Δ is chosen of the order of the atomic diameter so that the sheets consist of atomic monolayers with $\Delta \simeq 0.2$ nm. In the limit $\Delta \to 0$ we obtain $\tilde{d}_{\text{eff}} \to d_{\text{eff}}$.

The desired NL transmission expression for a film of arbitrary thickness d composed of $N = d/\Delta$ sheets of incremental thickness Δ is obtained as

$$I_{\text{trans}} = I_0 \exp\Bigg[-\overset{\diamond}{\sigma}_{\text{XAS}}\, \rho_{\text{a}} d \Bigg(1 - \underbrace{\frac{I_0\, \tilde{d}_{\text{eff}}\, \mathbb{A}}{\mathbb{B} + I_0\, \tilde{d}_{\text{eff}}\, \mathbb{A}}}_{2\, \widetilde{\rho}_{22}^{\infty}(\tilde{d}_{\text{eff}})} \Bigg)\Bigg]. \tag{15.37}$$

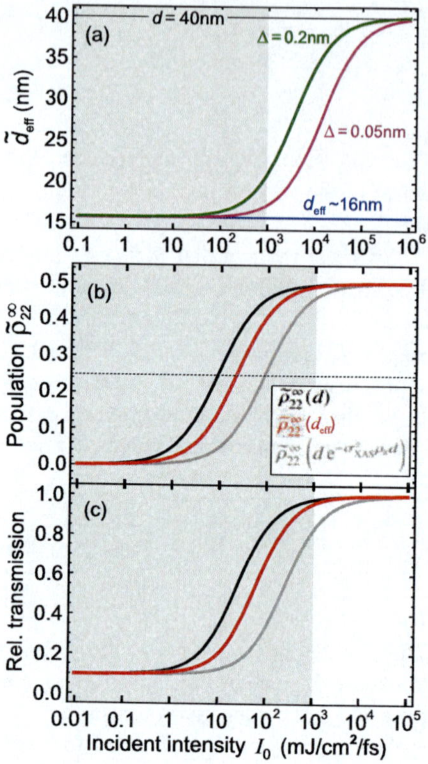

Fig. 15.7 **a** Dependence of $\tilde{d}_{\text{eff}}$ defined in (15.36) for resonant L_3 excitation of a 40 nm Co metal film consisting of $N = d/\Delta$ sheets on the incident intensity I_0. The green curve is for $\Delta = 0.2$ nm and the magenta curve for $\Delta = 0.05$ nm. The blue line corresponds to the spontaneous value $\tilde{d}_{\text{eff}} = d_{\text{eff}}$. **b** Upper state populations $\widetilde{\rho}_{22}^{\infty}$ for different thicknesses as a function of incident intensity I_0. The black curve is calculated with (15.33), the gray curve with (15.34), and the red curve with (15.37). **c** Corresponding transmission probability I_{trans}/I_0 according to (15.33) (black), (15.34) (gray) and (15.37) (red). The gray shading identifies the same intensity ranges in the three panels

In Fig. 15.7 we have illustrated the effect of propagation for a film of thickness d consisting of $N = d/\Delta$ sheets assuming resonant L_3 excitation ($\sigma_{\text{XAS}}^{\diamond} = 6.5$ Mb) of $d = 40$ nm Co metal film. This thickness is slightly larger than two x-ray absorption lengths ($\ell_x = 17$ nm).

Figure 15.7a shows the intensity dependence of $\tilde{d}_{\text{eff}}$ for $d = 40$ nm assuming $\Delta = 0.2$ nm (200 sheets) and $\Delta = 0.05$ nm (800 sheets). In both cases we have $\tilde{d}_{\text{eff}} \rightarrow d$ at high intensity, so that the entire film becomes transparent. In the limit $\Delta \rightarrow 0$ the curves shift to higher intensity. Then $\tilde{d}_{\text{eff}}$ remains at the spontaneous value $d_{\text{eff}} = 16$ nm indicated by a horizontal blue line over an increasingly large intensity range.

In Fig. 15.7b we illustrate the intensity-dependent effective excited state population, $\widetilde{\rho}_{22}^{\infty}$, averaged over all atoms in the film calculated for three assumptions. For the black curve the population is defined by (15.33) and the intensity that stimulates forward scattering is constant throughout the film at the value of the incident intensity I_0. The gray curve also assumes a constant stimulation throughout the film but with the population given by (15.34). The red curve corresponds to the proper inclusion of sheet-to-sheet stimulation in the presence of attenuation with the population defined by (15.37).

The intensity scales in (a) and (b) are aligned, emphasized by gray shading. We see that the $\tilde{d}_{\text{eff}}$ curves in (a) increase from their low intensity spontaneous value

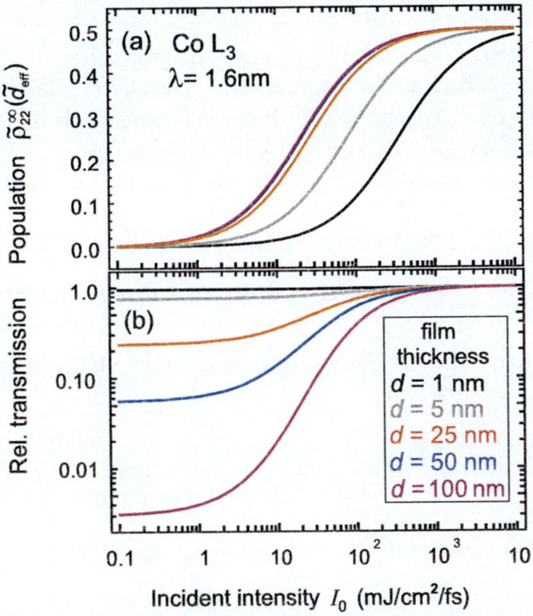

Fig. 15.8 **a** Effective upper state population $\tilde{\rho}_{22}^{\infty}(\tilde{d}_{\text{eff}})$ defined in (15.37) as a function of I_0 for different film thicknesses d composed of $N = d/\Delta$ sheets of thickness $\Delta = 0.1$ nm, using the parameters for L_3 excitation in Co metal. **b** Transmitted intensity normalized to the incident intensity I_0 according to (15.37) as a function of I_0 for different film thicknesses $d = 1, 5, 25, 50, 100$ nm

$\tilde{d}_{\text{eff}} = d_{\text{eff}}$ to their high intensity value $\tilde{d}_{\text{eff}} = d$ only at high intensity above about $I_0 \simeq 10^3$ mJ/cm^2/fs. Since the excited state populations in (b) have already saturated below this value, they may simply be calculated by use of $\tilde{d}_{\text{eff}} = d_{\text{eff}}$, indicated by the blue line in (a).

In Fig. 15.7c we have plotted the corresponding relative transmitted intensities I_{trans}/I_0 or transmission probability for the same three cases shown in (b), using the same color coding. The transmission probability expressed by (15.33) is shown as a black curve, that by (15.34) as a gray curve and by (15.37) as a red curve.

In Fig. 15.8 we illustrate the transmission behavior for different film thicknesses d as a function of incident intensity I_0. The curves are calculated using the parameters for L_3 excitation in Co metal according to (15.37), but the approximation $\tilde{d}_{\text{eff}} = d_{\text{eff}}$ gives nearly identical results. In (a) we show the effective upper state populations $\tilde{\rho}_{22}^{\infty}(d_{\text{eff}})$ defined in (15.37) for Co films of different thicknesses. In (b) we show the corresponding relative transmission intensities normalized to I_0.

The shown behavior of $\tilde{\rho}_{22}^{\infty}(\tilde{d}_{\text{eff}})$ in Fig. 15.8a extends the thickness range shown in Fig. 15.6.

For a film of thickness d, one may describe the propogation of the stimulated response by an effective film thickness $\tilde{d}_{\mathrm{eff}}$ given by (15.36), which can be approximated by d_{eff} given by (15.35). In the following we shall use this approximation $\tilde{d}_{\mathrm{eff}} = d_{\mathrm{eff}}$.

15.5.2 Summary: From Single Atom to Film Transmission

It is interesting to compare the forms of the excited population for the *single atom* case given by (14.90) or

$$[\rho_{22}^{\infty}]_{\mathrm{atom}} = \frac{\Gamma_{\mathrm{R}}^2/4}{(\hbar\omega - \mathcal{E}_0)^2 + \Gamma^2/4 + \Gamma_{\mathrm{R}}^2/2} \tag{15.38}$$

to the case of a *thin sheet* of thickness Δ, given by use of (15.21) and (15.23) as

$$[\widetilde{\rho}_{22}^{\infty}]_{\mathrm{sheet}} = \frac{\mathcal{G}_{\mathrm{coh}}\,\Gamma_{\mathrm{R}}^2/4}{(\hbar\omega - \mathcal{E}_0)^2 + (\Gamma/2)^2 + \mathcal{G}_{\mathrm{coh}}\,\Gamma_{\mathrm{R}}^2/2}, \quad \mathcal{G}_{\mathrm{coh}} = \frac{\lambda^2 \rho_{\mathrm{a}} \Delta}{4\pi}. \tag{15.39}$$

Finally, for a *film* of thickness d we have

$$\boxed{[\widetilde{\rho}_{22}^{\infty}]_{\mathrm{film}} \simeq \frac{\mathcal{G}_{\mathrm{eff}}\,\Gamma_{\mathrm{R}}^2/4}{(\hbar\omega - \mathcal{E}_0)^2 + (\Gamma/2)^2 + \mathcal{G}_{\mathrm{eff}}\,\Gamma_{\mathrm{R}}^2/2}, \quad \mathcal{G}_{\mathrm{eff}} \simeq \frac{\lambda^2 \rho_{\mathrm{a}} d_{\mathrm{eff}}}{4\pi}} \tag{15.40}$$

where we have used the approximation

$$\boxed{\tilde{d}_{\mathrm{eff}} \simeq d_{\mathrm{eff}} = \frac{1 - \mathrm{e}^{-\sigma_{\mathrm{XAS}}^{\diamond}\rho_{\mathrm{a}} d}}{\sigma_{\mathrm{XAS}}^{\diamond}\,\rho_{\mathrm{a}}}.} \tag{15.41}$$

The equations for the upper state population clearly show the evolution which can be summarized as follows.

For a resonantly excited *single atom*, the upper state population induced by a temporally coherent pulse of length τ_{coh} that is much longer than the excited state decay time τ_Γ (i.e. whose energy bandwidth $\Delta\hbar\omega$ is much narrower than the excited state decay width Γ) is given by (15.38).

For a thin *sheet of atoms* of atomic thickness Δ, the atomic response is changed by the *lateral coherence* of the incident pulse across the illuminated area. It causes a coherent enhancement, expressed by $\mathcal{G}_{\mathrm{coh}}$ in (15.39), that accounts for the collective action of the atoms in the sheet coupled by the *lateral coherence* of the incident beam. The N_a atoms in the atomic layer now contribute collectively, leading to a N_a^2 dependence of the coherent forward scattered signal (see (15.20) and Sect. 15.4.2).

For a *finite thickness film*, the *longitudinal coherence* of a laterally coherent pulse causes an additional longitudinal coupling of the N_a atoms in the film. It causes the collective response of an increased number of atoms contained in the effective film thickness $d_{\mathrm{eff}} > \Delta$. The atomic sheet enhancement factor $\mathcal{G}_{\mathrm{coh}}$ is then replaced by $\mathcal{G}_{\mathrm{eff}}$ as given by (15.40).

15.5.3 Sum Rule for Non-linear Film Transmission

Our result for the NL transmission through a film of finite thickness d given by (15.37) with (15.40) may also be stated in terms of a sum rule which is the film version of the single atom sum rule discussed in Sect. 14.8 and illustrated in Fig. 14.15. By lumping all NL contributions into a NL XAS cross section $\sigma_{\mathrm{XAS}}^{\mathrm{NL}}$, we can write (15.37) as

$$I_{\mathrm{trans}} = I_0 \exp\Big[- \Big(\sigma_{\mathrm{XAS}}^{\diamond} - \underbrace{2\,\sigma_{\mathrm{XAS}}^{\diamond}\widetilde{\rho}_{22}^{\infty}}_{\sigma_{\mathrm{XAS}}^{\mathrm{NL}}} \Big) \rho_a d \Big]. \tag{15.42}$$

As for the atomic case, the so-defined NL XAS cross-section $\sigma_{\mathrm{XAS}}^{\mathrm{NL}}$ corresponds to twice the NL REXS forward scattering cross section, i.e.

$$\boxed{\sigma_{\mathrm{XAS}}^{\mathrm{NL}} = 2\,\sigma_{\mathrm{REXS}}^{\mathrm{NL}}.} \tag{15.43}$$

This follows from the equivalence of the up and down rates in a two-level atom for $\rho_{22}^{\infty} = \rho_{11}^{\infty}$ expressed by (14.33) and (14.36).

As for the single atom case discussed in Sect. 14.8.1, we may therefore also divide the NL contribution into two equal parts according to

$$I_{\mathrm{trans}} = I_0 - I_{\mathrm{XAS}}^{\diamond} + 2I_{\mathrm{REXS}}^{\mathrm{stim}} = I_0 - \underbrace{(I_{\mathrm{XAS}}^{\diamond} - I_{\mathrm{REXS}}^{\mathrm{stim}})}_{I_{\mathrm{XAS}}} + \underbrace{I_{\mathrm{REXS}}^{\mathrm{stim}}}_{I_{\mathrm{REXS}}} \tag{15.44}$$

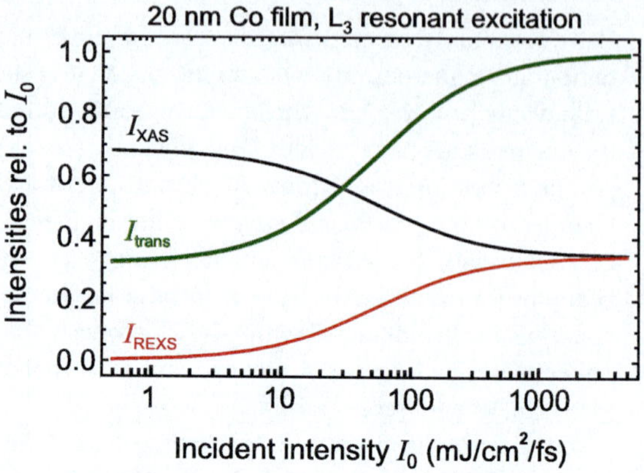

Fig. 15.9 Transmitted x-ray probability I_{trans}/I_0 (green), given by (15.47), through a $d = 20$-nm-thick Co film for L_3-edge excitation as a function of the incident intensity I_0. Also shown is its sum rule decomposition according to (15.45) into an XAS loss intensity, I_{XAS} (black), and stimulated REXS intensity I_{REXS} (red)

The so-defined sum rule contributions identified on the right side of (15.44) or

$$\boxed{I_{\text{trans}} = I_0 - I_{\text{XAS}} + I_{\text{REXS}}} \tag{15.45}$$

are illustrated in Fig. 15.9 for resonant L_3-edge excitation of a 20-nm-thick Co metal film.

The decomposition into XAS and REXS parts according to (15.45) is artificial, but it illustrates the existence of a sum rule in the BR theory due to the balance of up and down transitions at high intensity, leading to saturation. In the KHD perturbation theory, where saturation is not included, one typically assumes that the XAS contribution remains constant at the spontaneous value as a function of intensity and only stimulated forward scattering changes. The change in transmission would then only be half of that predicted by the BR theory.

In the following we shall for simplicity denote all NL contributions as a NL change of absorption, so that the total XAS cross section is defined as

$$\sigma^{\text{tot}}_{\text{XAS}} = \sigma^{\diamond}_{\text{XAS}} - \sigma^{\text{NL}}_{\text{XAS}}. \tag{15.46}$$

15.5.3.1 The Generalized Beer-Lambert Law in the Presence of Stimulation

In the presence of stimulation, the conventional Beer-Lambert law for a film of arbitrary thickness d given by (15.37) may also be written in the terms of modified scattering lengths or optical parameters. The *generalized Beer-Lambert law* which includes stimulated forward scattering is then given by the modified spontaneous version of (15.24) as

$$\boxed{I_{\text{trans}} = I_0\, \mathrm{e}^{-\left(\sigma_{\text{XAS}}^{\diamond} - \sigma_{\text{XAS}}^{\text{NL}}\right)\rho_{\text{a}} d} = I_0\, \mathrm{e}^{-2\lambda(f'' - f''_{\text{NL}})\rho_{\text{a}} d} = I_0\, \mathrm{e}^{-2(\beta - \beta_{\text{NL}})kd}}. \tag{15.47}$$

The first formulation on the right side of (15.47) expresses the transmitted intensity through the change in the spontaneous atomic cross section $\sigma_{\text{XAS}}^{\diamond}$ given by (14.18) as

$$\sigma_{\text{XAS}}^{\diamond} = \frac{\lambda^2}{\pi} \frac{\Gamma_x}{\Gamma} \frac{(\Gamma/2)^2}{(\hbar\omega - \mathcal{E}_0)^2 + (\Gamma/2)^2}. \tag{15.48}$$

The stimulated part is given by use of (15.40) as

$$\boxed{\sigma_{\text{XAS}}^{\text{NL}} = \sigma_{\text{XAS}}^{\diamond}\, 2[\widetilde{\rho}_{22}^{\infty}]_{\text{film}} \simeq \sigma_{\text{XAS}}^{\diamond} \frac{I_0\, d_{\text{eff}}\, \mathbb{A}}{\mathbb{B} + I_0\, d_{\text{eff}}\, \mathbb{A}}}, \tag{15.49}$$

where $\mathbb{A}$ and $\mathbb{B}$ are given by (15.30).

The second formulation in (15.47) involves the imaginary part of the *spontaneous* scattering length, given by the quantum mechanical expressions of Sect. 13.4.1 as

$$f'' = \frac{\lambda\, \Gamma_x}{2\pi\, \Gamma} \frac{\Gamma^2/4}{(\hbar\omega - \mathcal{E}_0)^2 + \Gamma^2/4}. \tag{15.50}$$

The NL contribution is given by

$$\boxed{f''_{\text{NL}} = f''\, 2[\widetilde{\rho}_{22}^{\infty}]_{\text{film}} \simeq f'' \frac{I_0\, d_{\text{eff}}\, \mathbb{A}}{\mathbb{B} + I_0\, d_{\text{eff}}\, \mathbb{A}}}. \tag{15.51}$$

For the resonant case, the corresponding real parts, f' and f'_{NL}, are linked to the imaginary parts f'' and f''_{NL} of the scattering length according to (13.19)

$$f' = \frac{2(\hbar\omega - \mathcal{E}_0)}{\Gamma} f'', \quad \text{and} \quad f'_{\text{NL}} = \frac{2(\hbar\omega - \mathcal{E}_0)}{\Gamma} f''_{\text{NL}}, \tag{15.52}$$

where the non-resonant Thomson contribution $r_0 Z$ to f' has been neglected.

The third expression on the right side of (15.47) utilizes the link between the atomic scattering lengths and optical parameters derived in Sect. 7.5.5. The spontaneous optical parameter β is given by the quantum expression

$$\beta = \frac{\lambda^2 \rho_{\text{a}}}{2\pi} \frac{\lambda \Gamma_x}{2\pi\Gamma} \frac{(\Gamma/2)^2}{(\hbar\omega - \mathcal{E}_0)^2 + (\Gamma/2)^2} = \frac{\lambda^2 \rho_{\text{a}}}{2\pi} f'', \tag{15.53}$$

and the NL part is

$$\boxed{\beta_{\text{NL}} = \beta\, 2[\widetilde{\rho}_{22}^{\infty}]_{\text{film}} \simeq \beta \frac{I_0\, d_{\text{eff}}\, \mathbb{A}}{\mathbb{B} + I_0\, d_{\text{eff}}\, \mathbb{A}}}. \tag{15.54}$$

The corresponding real parts, δ and $\delta_{\rm NL}$ are given by neglect of the Thomson contribution (see (7.52) and (7.53)) as

$$\delta = \frac{2(\hbar\omega - \mathcal{E}_0)}{\Gamma}\,\beta, \quad \text{and} \quad \delta_{\rm NL} = \frac{2(\hbar\omega - \mathcal{E}_0)}{\Gamma}\,\beta_{\rm NL}. \tag{15.55}$$

15.6 Polarization and Time-Dependent NL Transmission

Here we extend the polarization-averaged NL response in the long time limit given in Sect. 15.5.3.1 to include the polarization and time dependence.

15.6.1 The Polarization Dependent Generalized Beer-Lambert Law

The polarization dependent Beer-Lambert law in the presence of stimulation is given by adding the polarization label p to the polarization averaged expressions in Sect. 15.5.3.1. In the process we can follow the polarization dependent treatments previously done semi-classically in Sect. 7.11.1 and for the KHD theory in Chaps. 10–13.

The key polarization dependent parameter is the dipole transition matrix element, represented by the dipolar transition width Γ_x^p, where the polarization averaged expression for a transition between two state $|a\rangle \leftrightarrow |b\rangle$ given by (14.1) or

$$\Gamma_x = \frac{8\pi^2\alpha_{\rm f}\hbar\omega}{\lambda^2}\overline{|\langle b|\mathbf{r}\cdot\boldsymbol{\epsilon}_p|a\rangle|^2} \tag{15.56}$$

is replaced by (10.40) or[4]

$$\boxed{\Gamma_x^p = \frac{8\pi^2\alpha_{\rm f}\hbar\omega}{\lambda^2}|\langle b|\mathbf{r}\cdot\boldsymbol{\epsilon}_p|a\rangle|^2.} \tag{15.57}$$

The polarization averaged expression for $[\widetilde{\rho}_{22}^{\infty}]_{\rm film}$ given by (15.40) then includes a polarization dependent Rabi energy, defined as a polarization dependent interaction energy by generalization of (14.11). This leads to the polarization dependent version of (14.88)

$$\boxed{\left(\Gamma_{\rm R}^p\right)^2 = I_0^p\,\frac{\lambda^3}{2\pi^2 c}\,\Gamma_x^p = n_{p\mathbf{k}}\,\frac{\Delta_{p\mathbf{k}}}{2\pi^2}\,\Gamma_x^p = n_\Gamma^p\,\Gamma\,\Gamma_x^p.} \tag{15.58}$$

---

[4] Also see the REXS formulation expressed by (13.21).

For a film, $\mathcal{G}_{\text{eff}}$ also becomes polarization dependent through the effective thickness d_{eff} given by (15.35) which now depends on the polarization dependent spontaneous XAS cross section $[\sigma^{\diamond}_{\text{XAS}}]^p$ according to

$$\boxed{d^p_{\text{eff}} \simeq \frac{1-\mathrm{e}^{-[\sigma^{\diamond}_{\text{XAS}}]^p\,\rho_{\text{a}} d}}{[\sigma^{\diamond}_{\text{XAS}}]^p\,\rho_{\text{a}}},} \tag{15.59}$$

where (see (10.39))

$$[\sigma^{\diamond}_{\text{XAS}}]^p = \frac{\lambda^2}{\pi}\frac{\Gamma^p_x}{\Gamma}\,\frac{(\Gamma/2)^2}{(\hbar\omega-\mathcal{E}_0)^2+(\Gamma/2)^2}. \tag{15.60}$$

We then have

$$\mathcal{G}^p_{\text{eff}} = \frac{\lambda^2\rho_{\text{a}} d^p_{\text{eff}}}{4\pi}. \tag{15.61}$$

With (15.58) we obtain the effective upper state population for a film as

$$\begin{aligned}[\widetilde{\rho}^{\infty}_{22}]^p_{\text{film}} &\simeq \frac{\mathcal{G}^p_{\text{eff}}\,(\Gamma^p_{\text{R}})^2/4}{(\hbar\omega-\mathcal{E}_0)^2+\Gamma^2/4+\mathcal{G}^p_{\text{eff}}\,(\Gamma^p_{\text{R}})^2/2}\\ &= \frac{I^p_0\, d^p_{\text{eff}}\,\Gamma^p_x\,\rho_{\text{a}}\lambda^5/(32\pi^3 c)}{(\hbar\omega-\mathcal{E}_0)^2+\Gamma^2/4+I^p_0\, d^p_{\text{eff}}\,\Gamma^p_x\,\rho_{\text{a}}\lambda^5/(16\pi^3 c)}.\end{aligned} \tag{15.62}$$

The Beer-Lambert law given by (15.47) becomes the following polarization (p) dependent version

$$\boxed{I^p_{\text{trans}} = I^p_0\,\mathrm{e}^{-\left(\left[\sigma^{\diamond}_{\text{XAS}}\right]^p-\left[\sigma^{\text{NL}}_{\text{XAS}}\right]^p\right)\rho_{\text{a}} d} = I^p_0\,\mathrm{e}^{-2\lambda(f''^p-f''^p_{\text{NL}})\rho_{\text{a}} d} = I^p_0\,\mathrm{e}^{-2(\beta^p-\beta^p_{\text{NL}})kd}.} \tag{15.63}$$

This shows that all expressions in previous chapters that describe the polarization dependent *spontaneous* transmission or diffraction response of a film can be extended to include the NL response by adding the respective NL contributions with a minus sign. The opposite signs lead to a cancelation of the spontaneous and stimulated contributions at high intensity, called *self-induced x-ray transparency*.

The cross sections, scattering lengths, and optical parameters entering into (15.63) are explicitly given by the following expressions. The spontaneous *cross section* $[\sigma^{\diamond}_{\text{XAS}}]^p$ is given by (15.60) and the NL contribution by

$$\boxed{\left[\sigma^{\text{NL}}_{\text{XAS}}\right]^p = \left[\sigma^{\diamond}_{\text{XAS}}\right]^p\,2[\widetilde{\rho}^{\infty}_{22}]^p_{\text{film}} \simeq \left[\sigma^{\diamond}_{\text{XAS}}\right]^p\,\frac{I^p_0\, d^p_{\text{eff}}\,\mathbb{A}^p}{\mathbb{B}+I^p_0\, d^p_{\text{eff}}\,\mathbb{A}^p}.} \tag{15.64}$$

By generalization of (15.30) we have

$$\mathbb{A}^p = \frac{\Gamma_x^p \, \rho_a \lambda^5}{16\pi^3 c}, \quad \text{and} \quad \mathbb{B} = (\hbar\omega - \mathcal{E}_0)^2 + (\Gamma/2)^2. \tag{15.65}$$

The polarization dependent *scattering length* expressions are

$$f''^p = \frac{\lambda \, \Gamma_x^p}{2\pi \, \Gamma} \frac{\Gamma^2/4}{(\hbar\omega - \mathcal{E}_0)^2 + \Gamma^2/4} \tag{15.66}$$

and

$$f''^{\,p}_{\rm NL} = f''^p \, 2[\widetilde{\rho}_{22}^{\infty}]^p_{\rm film} \simeq f''^p \frac{I_0^p \, d_{\rm eff}^p \, \mathbb{A}^p}{\mathbb{B} + I_0^p \, d_{\rm eff}^p \, \mathbb{A}^p}. \tag{15.67}$$

Finally, the polarization dependent *optical parameters* are expressed through

$$\beta^p = \frac{\lambda^2 \rho_a}{2\pi} \frac{\lambda \Gamma_x^p}{2\pi \Gamma} \frac{(\Gamma/2)^2}{(\hbar\omega - \mathcal{E}_0)^2 + (\Gamma/2)^2} \tag{15.68}$$

and

$$\beta^p_{\rm NL} = \beta^p \, 2[\widetilde{\rho}_{22}^{\infty}]^p_{\rm film} \simeq \beta^p \frac{I_0^p \, d_{\rm eff}^p \, \mathbb{A}^p}{\mathbb{B} + I_0^p \, d_{\rm eff}^p \, \mathbb{A}^p}. \tag{15.69}$$

15.6.2 Non-linear Polarization Dependent Transmission by the Magnetic 3d Metals

As previously discussed in Sect. 10.5.5, XAS lineshapes for solids are often significantly broadened beyond the natural decay linewidths Γ listed in Table 10.2, and the cross sections are proportionately reduced as illustrated in Fig. 10.17. For convenience we have listed typical polarization dependent parameters for the L_3 resonances in Fe, Co, and Ni obtained from experiment in Table 15.1. The listed peak cross sections are those given in [14], and the linear polarization ones are the same as the polarization averaged ones plotted in Fig. 10.17 [see (11.29)]. In the following model calculations we will utilize the polarization dependent Co L_3 response parameters of Table 15.1.

In analogy to Fig. 10.17b, we show in Fig. 15.10 the polarization dependent theoretical and experimental Co L_3 cross sections $[\sigma^{\diamond}_{\rm XAS}]^p$ for a magnetically oriented sample and x-ray propagation along the magnetization direction. The theoretical cross sections in (a) shown in blue have the natural Lorentzian FWHM $\Gamma = 430$ meV, while the experimental cross sections in (b) (black lines) are compared to fits with Voigt profiles (red lines) consisting of a convolution of the natural Lorentzian lineshapes with a Gaussian of 1.4 eV FWHM. The latter accounts for band structure

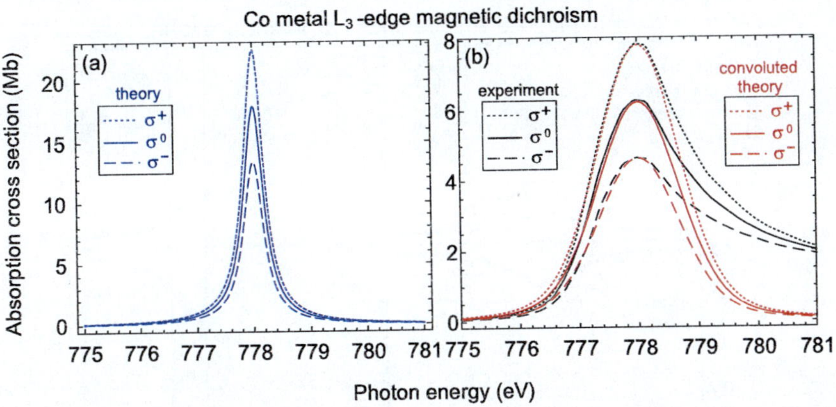

Fig. 15.10 **a** Theoretical polarization dependent spontaneous Co L3 XAS cross sections σ^p (blue lines) of Lorentzian lineshape as discussed in the text. **b** Comparison of the spontaneous experimental cross sections $\sigma^p = [\sigma^{\diamond}_{\rm XAS}]^p$ (black lines) and the theoretical cross sections in (**a**) convoluted with a Gaussian of 1.4 eV FWHM (red lines). The corresponding blue Lorentzian profiles in (**a**) and red Voigt ones in (**b**) have the same areas

broadened d valence states into which the $2p_{3/2}$ core electrons are excited. The areas under the blue curves in (a) and red curves in (b) are preserved.

We are now interested in the change of the spontaneous cross sections, given by the red Voigt profiles in Fig. 15.10b, upon stimulation. The total polarization dependent NL cross sections for a 20-nm-thick Co film in the Beer-Lambert law (15.63) is given in accordance with (15.46) by

$$[\sigma^{\rm tot}_{\rm XAS}]^p = \left[\sigma^{\diamond}_{\rm XAS}\right]^p - [\sigma^{\rm NL}_{\rm XAS}]^p \,. \tag{15.70}$$

The NL part given by (15.64) depends on the film thickness d which enters through the polarization dependent effective thickness $d^p_{\rm eff}$ given by (15.59). As an example we have listed in Table 15.2 the values for $d^p_{\rm eff}$ for a typical Co metal film thickness of $d = 20$ nm using the peak experimental cross sections $\sigma^p_{\rm XAS} = [\sigma^{\diamond}_{\rm XAS}]^p$ in Table 15.1. Also listed are the effective lateral coherent factors $\mathcal{G}^p_{\rm eff}$ given by (15.61) to illustrate the significant enhancement arising from coherent stimulated forward scattering.

The total polarization dependent NL cross sections (15.70) for a 20-nm-thick Co film are shown in Fig. 15.11a for three values of I_0. They were obtained by

Table 15.2 Polarization dependent effective thicknesses $d^p_{\rm eff}$ for a nominal Co metal film thickness of $d = 20$ nm, calculated according to (15.59)

$d^+_{\rm eff}$ (nm)	$d^0_{\rm eff}$ (nm)	$d^-_{\rm eff}$ [nm]	$\mathcal{G}^+_{\rm eff}$	$\mathcal{G}^0_{\rm eff}$	$\mathcal{G}^-_{\rm eff}$
10.7	12.0	13.5	195	219	247

Also given are the values for $\mathcal{G}^p_{\rm eff} = \lambda^2 \rho_a d^p_{\rm eff}/(4\pi)$ according to (15.61)

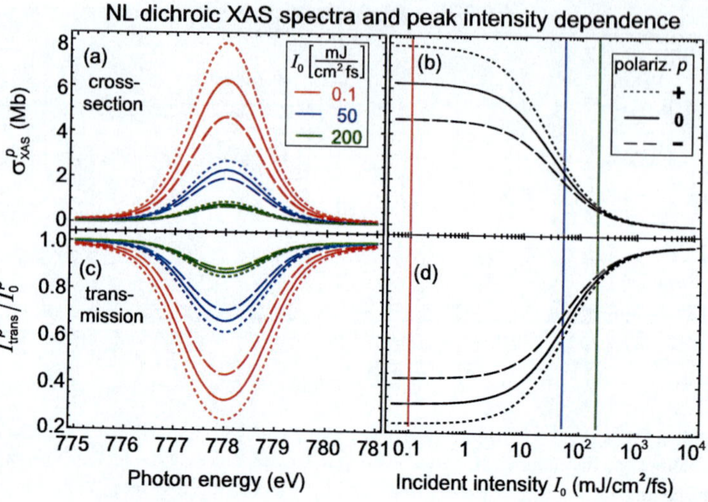

Fig. 15.11 **a** NL change of the polarization dependent XAS cross sections, indicated by dotted ($p = +$), solid ($p = 0$) and dashed ($p = -$) lines, for Co L_3 excitation for three incident intensities. The red curves for $I_0 = 0.1\,\text{mJ/cm}^2\text{/fs}$ are indistinguishable from the corresponding red curves in Fig. 15.10b, representing the spontaneous response. **b** Change of the resonant ($\hbar\omega = 778\,\text{eV}$) cross section values in (**a**) as a function of incident intensity. The color coded vertical lines in (**b**) correspond to the three I_0 values in (**a**). **c** NL transmission spectra I^p_{trans}/I^p_0 according to the NL Beer-Lambert law (15.63) and **d** the corresponding dependence of the peak transmission value on the incident intensity in analogy to (**a**) and (**b**)

assuming that the spontaneous Voigt spectral shapes are preserved upon stimulation. Figure 15.11b shows the change of the resonant ($\hbar\omega = 778\,\text{eV}$) cross-section values in (a) as a function of incident intensity, where the color coded vertical lines correspond to the three I_0 values in (a).

In analogy to Figs. 15.11a, b we show in (c) and (d) the corresponding NL transmission spectra I^p_{trans}/I^p_0 according to (15.63) and the dependence of the peak transmission value on the incident intensity. With increasing intensity the transmission becomes unity and the film becomes transparent. We note that our calculations were carried out with the assumption that the electronic populations have reached an equilibrium, i.e. $t \to \infty$, as expressed by $[\widetilde{\rho}^{\infty}_{22}]^p_{\text{film}}$ in (15.62). In the following we shall explore how the cross sections and dichroic transmission of the same film changes for pulses of finite coherence time width τ, typically encountered in experiments.

15.6.3 *Dependence on X-Ray Pulse Coherence Time*

We have seen in Sect. 14.1 that analytical solutions of the Bloch equation only exist in special cases, like the case of exact resonance and long time. In order to deal with time-dependent and non-resonant situations, the Bloch equations need to be solved numerically. For the film case, one then needs to account for the coherent enhancement factor $\mathcal{G}_{\text{eff}}$. This is done by redefining the atomic Rabi frequency $\mathcal{W}_{\text{R}}^{p} = \Gamma_{\text{R}}^{p}/\hbar$ in the Bloch equations (14.29) and (14.30) according to

$$[\mathcal{W}_{\text{R}}^{p}]_{\text{film}} = \mathcal{W}_{\text{R}}^{p}\sqrt{\mathcal{G}_{\text{eff}}^{p}} \simeq \mathcal{W}_{\text{R}}^{p}\sqrt{\frac{\lambda^2 \rho_{\text{a}} d_{\text{eff}}^{p}}{4\pi}}, \tag{15.71}$$

where the right side expression is obtained by use of (15.61). The analytical expression in the long time limit $[\widetilde{\rho}_{22}^{\infty}]_{\text{film}}^{p}$ is then replaced by the numerically calculated time-dependent excited state population $[\widetilde{\rho}_{22}^{t}]_{\text{film}}^{p}$. With these substitutions, our thin film expressions derived above are extended to include both the polarization and time dependence.

In the following, we consider how the effective upper state population in a film changes for a SASE pulse containing coherent spikes of flat-top duration times up to τ. The temporal evolution of the population in a film is similar to that for the single atom case and we can conveniently use the results of Sect. 14.5.3. In analogy to (14.52), an integration up to a coherence time τ gives the effective upper population in a film as

$$[\widetilde{\rho}_{22}^{\tau}]_{\text{film}}^{p} = \frac{\mathcal{G}_{\text{eff}}^{p}\,(\Gamma_{\text{R}}^{p})^2/4}{\Gamma^2/4 + \mathcal{G}_{\text{eff}}^{p}\,(\Gamma_{\text{R}}^{p})^2/2}\left\{1 - \frac{24\Gamma\hbar\omega_{\text{R}}^{p}\left[1-\cos(\omega_{\text{R}}^{p}\tau)\,\text{e}^{-3\Gamma\tau/(4\hbar)}\right]}{\tau\,\omega_{\text{R}}^{p}\left[16(\hbar\omega_{\text{R}}^{p})^2+9\Gamma^2\right]}\right.$$
$$\left. - \frac{\left[16(\hbar\omega_{\text{R}}^{p})^2-9\Gamma^2\right]\sin(\omega_{\text{R}}^{p}\tau)\,\text{e}^{-3\Gamma\tau/(4\hbar)}}{\tau\,\omega_{\text{R}}^{p}\left[16(\hbar\omega_{\text{R}}^{p})^2+9\Gamma^2\right]}\right\}, \tag{15.72}$$

where the generalized Rabi frequency is given by

$$\omega_{\text{R}}^{p} = \frac{\sqrt{\left(\Gamma_{\text{R}}^{p}\right)^2 - \Gamma^2/16}}{\hbar}. \tag{15.73}$$

With increasing coherence time $\tau \to \infty$, the terms in wavy brackets in (15.72) approach unity, and for large intensity ($\Gamma_{\text{R}}^{p} \gg \Gamma$) we then obtain $[\widetilde{\rho}_{22}^{\infty}]_{\text{film}}^{p} = [\widetilde{\rho}_{11}^{\infty}]_{\text{film}}^{p} = 1/2$.

As an example, we show in Fig. 15.12 the intensity-dependent increase in the excited state population (15.72) for L_3 resonant excitation of a 20 nm Co metal film for different values of τ. We assumed a peak XAS cross section (linear polarization) of Co atoms of $\sigma_{\text{XAS}}^{0} = 6.25$ Mb corresponding to $\Gamma_{x}^{0} = 0.34$ meV.

The increase of $[\widetilde{\rho}_{22}^{\tau}]_{\text{film}}^{0}$ with I_0^0 shown in Fig. 15.12 shows a shift with decreasing τ to higher intensity. This is expected because the driving field has less time to establish a stable upper state population. The shift resembles that previously shown in Fig. 15.8a with decreasing film thickness d. For short τ, the populations shown in Fig. 15.12 also exhibit Rabi oscillations that gradually damp out with increasing intensity.

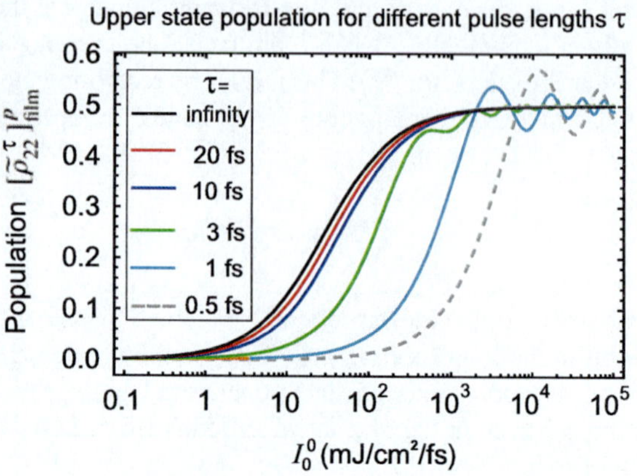

Fig. 15.12 Excited state population $[\tilde{\rho}_{22}^{\tau}]^{0}_{\rm film}$, given by (15.72), as a function of the linearly polarized incident intensity I_0^0 for different values of τ. We assumed L_3 resonant excitation $\hbar\omega = \mathcal{E}_0$ in a 20-nm-thick Co metal film as discussed in the text for linearly polarized light

Figure 15.13 illustrates the strong changes of the polarization dependent transmitted intensities for L_3 excitation of a 20 nm Co film with both the incident intensity $I_0 = I_0^p$ ($p = \pm, 0$) and length τ of incident coherent flat-top pulses. The plotted relative transmitted intensities are given by the NL polarization dependent Beer-Lambert law (15.63). In (a) we illustrate the transmitted dichroic intensities for $\tau = \infty$ as a function of incident intensity, previously shown in Fig. 15.11d. Figure 15.13b shows their dependence on the pulse coherence time τ for three values of I_0, chosen in size and color as those in Fig. 15.11.

In Fig. 15.13c, d we plot the dichroic difference between plus and minus angular momentum directions in (a) and (b). This corresponds to the transmitted XMCD signal in analogy to (7.123) for the spontaneous case. Again, we find a strong dependence on both I_0 and τ, with the largest changes relative to the spontaneous contrasts occurring at large intensities and when the coherence time τ becomes significantly longer than the Co L_3 core hole lifetime of $\tau_\Gamma = 1.5$ fs, indicated by a vertical gray line.

15.6.4 From Collective to Independent Atomic Response

In Sect. 15.4.2 we discussed the meaning of the coherent enhancement factor $\mathcal{G}_{\rm coh}$ for an atomic sheet which enters into the effective excited state population $\tilde{\rho}_{22}^{\infty}$. In later Sects. 15.5.1 and 15.6.1, the treatment was extended to films of finite thickness and inclusion of the polarization dependence, resulting in the expression (15.61) for $\mathcal{G}_{\rm eff}^{p}$ and the associated effective excited state population $[\tilde{\rho}_{22}^{\infty}]^{p}_{\rm film}$ given by (15.62).

The complete polarization dependent *effective* excited state population in a film reduces to that of a single atom through $\mathcal{G}_{\rm eff}^{p} \to 1$ according to (also see Sect. 15.5.2),

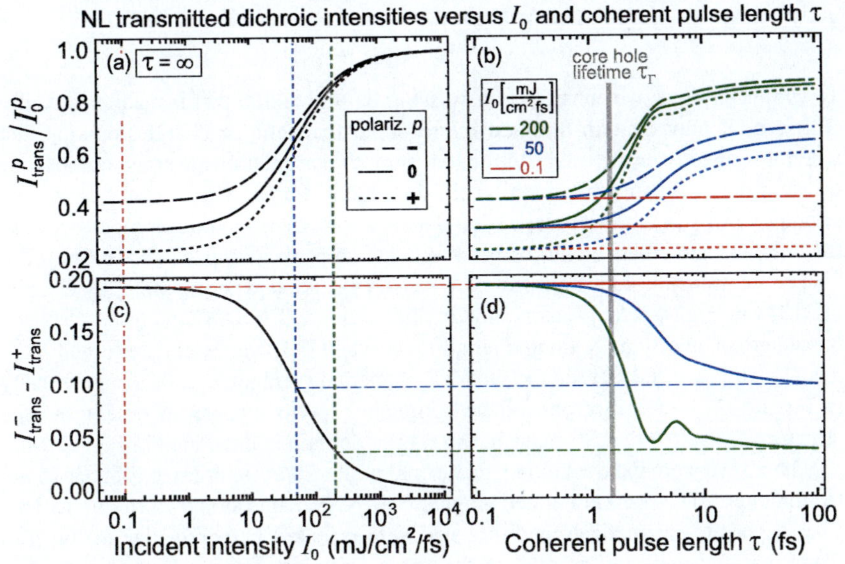

Fig. 15.13 **a** NL change of the polarization dependent transmitted intensities I^p_{trans}/I^p_0 according to (15.63), indicated by dashed ($p = -$), solid ($p = 0$) and dotted ($p = +$) curves, as a function of incident intensity. All curves are for $\tau = \infty$ and L_3 excitation in a 20 nm Co metal film. **b** NL change of I^p_{trans}/I^p_0 as a function of coherent flat-top pulse lengths τ for three color coded I_0 values, indicated by vertical lines in (**a**), corresponding to those in Fig. 15.11. The Co L_3 core hole lifetime of $\tau_\Gamma = 1.5$ fs is indicated by a vertical gray line. **c**, **d** show the dichroic difference or transmitted XMCD signal derived from (**a**) and (**b**)

$$\text{BR}: [\widetilde{\rho}^{\,\infty}_{22}]^p_{\text{film}} = \frac{\mathcal{G}^p_{\text{eff}}(\Gamma^p_{\text{R}})^2/4}{(\Delta\mathcal{E})^2+\Gamma^2/4+\mathcal{G}^p_{\text{eff}}(\Gamma^p_{\text{R}})^2/2} \longrightarrow [\rho^{\infty}_{22}]^p_{\text{atom}} = \frac{(\Gamma^p_{\text{R}})^2/4}{(\Delta\mathcal{E})^2+\Gamma^2/4+(\Gamma^p_{\text{R}})^2/2} \tag{15.74}$$

where $\Delta\mathcal{E} = \hbar\omega - \mathcal{E}_0$ is the detuning energy.

15.7 X-Ray Transparency

In this section we discuss the quantitative description of x-ray transparency, introduced in Sect. 15.3.2 for the two cases distinguished in Fig. 15.4a. We start with the description of transparency for *resonant excitations* which naturally follows from the two-level BR theory and compare it to experimental results.

15.7.1 Resonant Case: Co Metal

To simplify things, we consider transmission through a film that is macroscopically isotropic. It may contain oriented microscopic magnetic or charge domains, but when averaged over the total illuminated area, the orientations average out and the transmission in the forward direction is independent of polarization. The microscopic structure will still lead to polarization dependent transmission differences between the domains and create an out-of-beam diffraction pattern as discussed in Chap. 8.

The NL transmission and diffraction has been studied experimentally by Chen et al. [13] and Higley et al. [7] using linearly polarized XFEL SASE pulses of 25 fs and 5 fs length. The originally quoted length of 2.5 fs in [13] was later determined to be 5 fs [7]. The two studies used the same experimental arrangement, shown previously in Fig. 1.30. The relevant part of the complete experimental arrangement and the structure of the 25 fs SASE pulse in the energy domain is illustrated in Fig. 15.14.

The x-rays were incident along the normal of the Co/Pd multilayer films of 25 nm Co thickness. The films contained magnetic stripe domains as shown in Fig. 15.14a. Their opposite magnetization directions were oriented perpendicular to the film plane, parallel, and anti parallel to the beam direction. For incident linear polarization, there is no conventional XMCD dichoism effect in the forward direction ($\mathbf{q}=0$), of interest here. However, at finite momentum transfer $\mathbf{q} \neq 0$ there is a domain diffraction pattern whose interplay with the centrally transmitted intensity will be discussed in Sect. 15.9.

As shown in Fig. 15.14a, the pulses were split by a mirror with a sharp edge [61] into two similar intensity pulses that came to a focus at a Si chip containing 100-nm-thick silicon nitride membrane windows. Half of the pulses were transmitted through the Co/Pd magnetic multilayers deposited on top of the SiN, and their intensity was normalized by the other half transmitted through pure SiN. The images shown on the CCD correspond to horizontally dispersed spectra, with the upper one showing a central dip due to transmission loss by the Co/Pd film. The normalization is illustrated in Fig. 15.14b by comparison of the transmitted spectra of two halves of single pulses, recorded with the membranes, and Co/Pd multilayers removed from the x-ray paths.

The measured change of transmission through the film as a function of incident intensity of the 25 fs pulses is shown in Fig. 15.15 by data points shown as filled blue circles connected by a blue line. As mentioned earlier, the transmission for our linear polarization is independent of the magnetic orientation of the domains and is entirely a charge effect. Also shown as small blue open circles is a statistical simulation taken from [13]. The simulation was based on a statistical temporal description of the incident SASE pulses [62] which then served as input to the numerical solution of the time-dependent optical Bloch equations including the coherent enhancement factor.

Here we take a shortcut to account for the observed effects by utilizing the analytical expression for $[\widetilde{\rho}_{22}^{\tau}]_{\rm film}$, given by (15.72), where τ represents the upper limit of the width of the coherent SASE spikes. Inserting $\mathcal{G}_{\rm eff} = \lambda^2 \rho_{\rm a} d_{\rm eff}/(4\pi)$, given by (15.40), where $d_{\rm eff}$ has the approximated form (15.41), we obtain

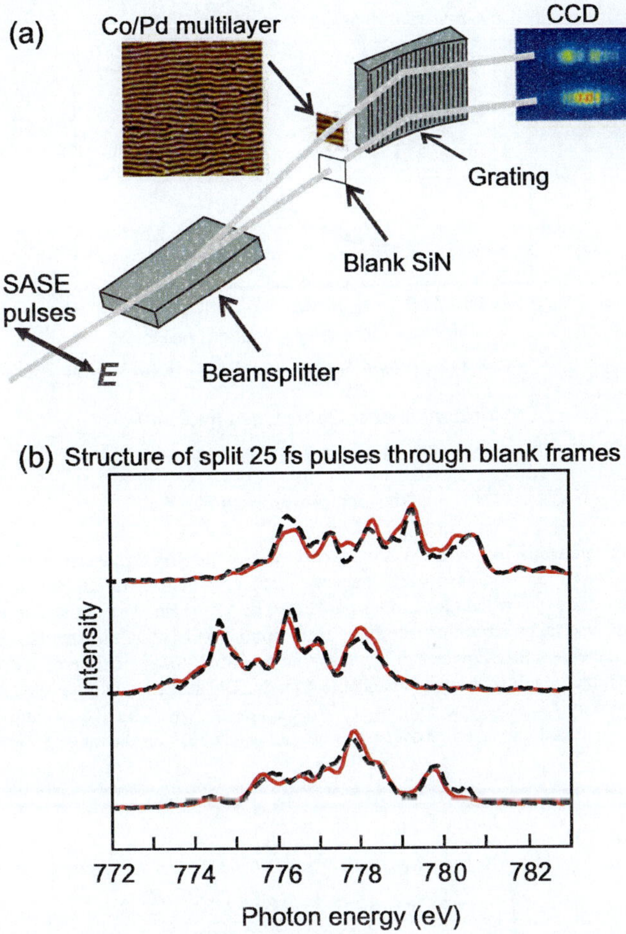

Fig. 15.14 **a** Simplified schematic of the experimental setup [7]. The x-ray beam is split into two components with an x-ray beam splitter. One of the beams passes through a blank SiN membrane, while the other passes through a membrane with a Co/Pd magnetic multilayer with stripe domains, as shown. The domains had opposite magnetic orientations perpendicular to the film plane. The beams emerging from the membranes in the forward direction are measured with a grating-based spectrometer. For linear polarization, there is no dichroism effect in transmission, and the existence of the domains is unimportant, as discussed in the text. **b** Examples of single-shot spectra for a 25 fs pulse, recorded when the membranes and Co/Pd multilayers were removed from the x-ray paths

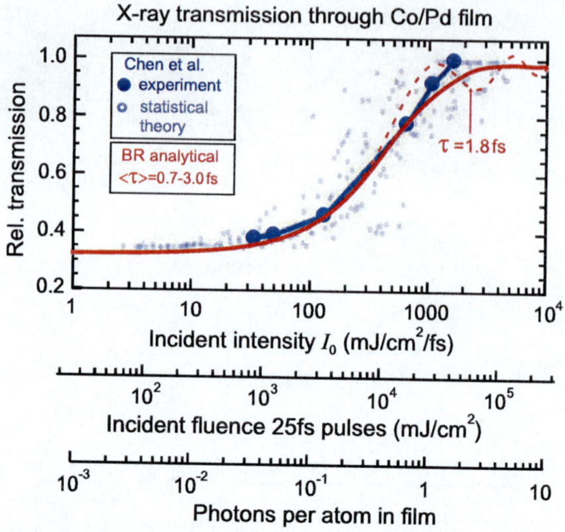

Fig. 15.15 X-ray transmission by a Co film, excited with 25 fs SASE pulses centered at the L_3 resonance, plotted as a function of incident intensity, fluence, or number of photons per atom. The experimental data shown as filled blue circles connected by a blue line were taken from [13]. The small blue points also taken from that reference are a simulation based on the statistical temporal description of the incident SASE pulses [62] and a numerical solution of the time-dependent optical Bloch equations including the coherent enhancement factor. The dashed red curve is the normalized transmission given by I_{trans}/I_0 according to (15.76) with $\widetilde{\rho}_{22}(\tau)$ given by (15.75) and $\tau = 1.8$ fs. The solid red curve includes an average over τ by integrating (15.75) over the range $1.7 \leq \tau \leq 3.0$ fs

$$\widetilde{\rho}_{22}(\tau) = \frac{I_0\, d_{\text{eff}}\, \Gamma_x \rho_{\text{a}} \lambda^5/(32\pi^3 c)}{(\Gamma/2)^2 + I_0\, d_{\text{eff}}\, \Gamma_x \rho_{\text{a}} \lambda^5/(16\pi^3 c)} \times \left\{ 1 - \frac{24\Gamma \hbar\omega_{\text{R}}^p \left[1 - \cos(\omega_{\text{R}}^p \tau)\, \mathrm{e}^{-3\Gamma\tau/(4\hbar)}\right]}{\tau\, \omega_{\text{R}}^p \left[16(\hbar\omega_{\text{R}}^p)^2 + 9\Gamma^2\right]} - \frac{\left[16(\hbar\omega_{\text{R}}^p)^2 - 9\Gamma^2\right] \sin(\omega_{\text{R}}^p \tau)\, \mathrm{e}^{-3\Gamma\tau/(4\hbar)}}{\tau\, \omega_{\text{R}}^p \left[16(\hbar\omega_{\text{R}}^p)^2 + 9\Gamma^2\right]} \right\}. \tag{15.75}$$

The transmitted intensity is given by the generalized Beer-Lambert expression

$$\boxed{I_{\text{trans}} = I_0\, \mathrm{e}^{-\overset{\diamond}{\sigma}_{\text{XAS}} \rho_{\text{a}} d [1 - 2\widetilde{\rho}_{22}(\tau)]}.} \tag{15.76}$$

The dashed red curve in Fig. 15.15 was calculated with (15.76) by use of (15.75) with $\tau = 1.8$ fs. To match the experimental spontaneous transmission value of 0.32 [13], we used $d = 20$ nm as the Co thickness instead of the nominal value of $d = 25$ nm. The statistical pulse description used in [13] instead yields a spread of the small blue circles in Fig. 15.15. The Rabi oscillations in the dashed red curve in Fig. 15.15 are then obscured. The solid red curve in Fig. 15.15 is obtained by smoothing out the

Rabi oscillations by integrating (15.75) over a range $1.7 \leq \tau \leq 3.0\,\mathrm{fs}$. It describes the measured data quite well.

We have now quantitatively accounted for the x-ray transparency effect previously shown in Fig. 15.4c. The early onset of transparency is a resonance effect where absorption is compensated by stimulated emission in the forward direction. The question arises how our *resonant* description of transparency developed above differs from that for *non-resonant* excitation, previously illustrated for Al in Figs. 1.28 and 15.4b. We now consider this case.

15.7.2 Resonant Versus Non-resonant X-Ray Transparency

One may understand the difference between induced transparency for resonant and continuum excitations in terms of different lifetimes of the final state in the XAS process. In both cases it consists of an excited electron and a core hole. In a simple picture, the total lifetime of the final state may be separated into a contribution from the core hole, τ_h, given by the core hole clock FWHM lifetime $\tau_\mathrm{h} = \tau_\Gamma$ (see (12.18)) and from the free electron-like band state denoted τ_e according to [63]

$$\frac{1}{\tau_\mathrm{tot}} = \frac{1}{\tau_\Gamma} + \frac{1}{\tau_\mathrm{e}}. \tag{15.77}$$

The corresponding FWHM transition widths are given by the uncertainty relations $\Gamma = \hbar/\tau_\Gamma$ and $\Gamma_\mathrm{e} = \hbar/\tau_\mathrm{e}$ so that (15.77) in energy space reads

$$\Gamma_\mathrm{tot} = \Gamma + \Gamma_\mathrm{e}. \tag{15.78}$$

In a *resonant* process, the excited electron remains bound for a time $\tau_\mathrm{e} > \tau_\Gamma$ and does not escape as a photoelectron. The total excited state lifetime in (15.77) is therefore determined by the shorter core hole clock lifetime and $\tau_\mathrm{tot} \simeq \tau_\Gamma$ (or $\Gamma_\mathrm{tot} \simeq \Gamma$ in (15.78)). The process is described by the optical Bloch equations that couple the lower and upper states by "coherences", as discussed in Sect. 14.3.1. The resonant process is determined by the stimulated increase in Rabi energy $\Gamma_\mathrm{R}^2 = n_\Gamma \Gamma_x \Gamma$. In an atom, saturation occurs when through stimulation $n_\Gamma \Gamma_x \to \Gamma$ and in a solid when $n_\Gamma \Gamma_x \mathcal{G}_\mathrm{eff} \to \Gamma$, as expressed by (15.74). The resonant process depends on the bandwidth of the incident photons, i.e. their coherence time. Transparency occurs when the coherent up-down process reaches an equilibrium balance between absorption and stimulated emission, so that $\rho_{22} = \rho_{11} = 0.5$.

In a *non-resonant* process, core electrons are excited into continuum states well above the Fermi level with a rather smooth density of states. The excited electron first resides in a free electron-like band state of the solid before it can escape into vacuum through the surface potential barrier as a photoelectron. The lifetime τ_e may be thought of as a band-specific "trapping" time τ_e. If this trapping time is shorter than the core hole lifetime, $\tau_e < \tau_\Gamma$, the total excited state lifetime in (15.77) is determined by $\tau_{tot} \simeq \tau_e$ (or $\Gamma_{tot} \simeq \Gamma_e$ in (15.78)). This is the opposite of the resonant case. For metals, τ_e has been found to be a few hundred attoseconds [64]. For example, for Cu free electron-like $s-p$ band states that lie about 20 eV above the Fermi level, it was found that $\tau_e \simeq 309 \pm 94$ as or $\Gamma_e = \hbar/\tau_e = 2.13 \pm 0.65$ eV [44].

The non-resonant XAS cross section is approximately independent of the incident photon bandwidth for typical values up to 10 eV or so. This means that it is insensitive to the coherence time of the incident radiation, and the coherent effects incorporated into the optical Bloch equations are absent. The up-down processes can simply be described by the Einstein rate equations discussed in Sect. 14.9. For solids, there is no coherent enhancement factor since the atoms respond independently.

In the following we give a simple description of x-ray transparency based on the solution of the Einstein rate equations, by accounting for the exponential attenuation of the incident intensity in the film. When the time-dependent upper state population in the Einstein theory given by (14.122) is integrated over the pulse length τ_p, we obtain

$$\rho_{22}(\tau_p) = \frac{\Gamma_R^2}{\Gamma_{tot}^2 + 2\Gamma_R^2}\left\{1 + \frac{\left[\exp\left(-\frac{(\Gamma_{tot}^2+2\Gamma_R^2)\tau_p}{\Gamma_{tot}\,\hbar}\right) - 1\right]\Gamma_{tot}\,\hbar}{(\Gamma_{tot}^2 + 2\Gamma_R^2)\,\tau_p}\right\}, \qquad (15.79)$$

where the effective width $\Gamma_{tot} \simeq \Gamma_e + \Gamma$ is determined by the larger of the two widths. In practice, for pulse lengths larger than a few femtoseconds, the term in curly brackets in (15.79) becomes unity.

The upper state population, $\rho_{22}(\tau_p)$, does not depend on the incident bandwidth or coherence time of the radiation. For the non-resonant excitation case, the pulses can therefore be incoherent "brute force" SASE pulses and both the photons and atomic excitation processes are uncorrelated. This is the key difference to the resonant case which requires coherent up-down processes, i.e. REXS.

By dropping the time-dependent term and expressing the Rabi energy in terms of the incident intensity by use of (14.88), we obtain (15.79) in the form

$$\rho_{22}^{NR} = \frac{I_0\Gamma_x\lambda^3/(2\pi^2 c)}{\Gamma_{tot}^2 + I_0\Gamma_x\lambda^3/(\pi^2 c)}. \qquad (15.80)$$

The normalized transmission $\mathcal{T} = I_{trans}/I_0$ for the *non-resonant* case is given by

$$\boxed{\text{Non-resonant}: \ \mathcal{T} = e^{-\sigma_{XAS}^{\diamond}\,\rho_a d\,(1-2\rho_{22}^{NR}\tilde{d}_{eff}/d)}}, \qquad (15.81)$$

where $\tilde{d}_{\text{eff}}$ is given by (15.41). This compares to the resonant expression (15.76) or

$$\boxed{\text{Resonant}: \quad \mathcal{T} = \mathrm{e}^{-\sigma_{\text{XAS}}^{\diamond}\,\rho_{\text{a}} d\,(1-2\widetilde{\rho}_{22}(\tau))}} \tag{15.82}$$

which depends on the *coherence time* τ through the excited state population $\widetilde{\rho}_{22}(\tau)$ given by (15.75).

The two transmission expressions differ in the form of the NL contribution, but both can lead to transparency, i.e. $\mathcal{T} = 1$. This requires that $2\rho_{22}^{\text{NR}} = 2\widetilde{\rho}_{22}(\tau) \rightarrow 1$ and $\tilde{d}_{\text{eff}} \rightarrow d$.

15.7.3 Non-resonant Transparency Above the Al L-Edge

We already illustrated the observed transparency for non-resonant excitation of Al above its $L_{3,2}$ threshold in Figs. 1.28a and 15.4b and gave a qualitative explanation. We now show that the experimental results can be quantitatively described by (15.81).

We start with the $L_{3,2}$ XAS spectrum of Al metal shown in Fig. 15.16a [65]. One of the issues with Al metal films is that they readily oxidize unless they are capped or kept in ultra-high ($\sim 10^{-10}$ Torr) vacuum [66]. The spectrum shown in Fig. 15.16a measured in transmission [65] likely contains small contributions from oxidized surface layers but is mostly representative of Al metal. Similarly, the films used in the transparency study by Nagler et al. [8] also contained oxidized layers. This complicates the analysis. We therefore consider the NL transmission for a pure Al metal film first.

The properties of Al metal and its x-ray response are relatively well understood. Al has an atomic density of $\rho_{\text{a}} = 60.3$ atoms/nm^3, and the XAS cross section at $\hbar\omega =$ 92 eV, about 20 eV above the L_3 (72.7 eV) and L_2 (73.2 eV) edges, is $\sigma_{\text{XAS}}^{\diamond} \simeq 4.5$ Mb (4.5×10^{-4} nm^2) as shown in Fig. 15.16a. This gives a $1/e$ x-ray absorption length of $\ell_x = 1/(\rho_{\text{a}}\sigma_{\text{XAS}}) = 36.9$ nm,. The calculated transmission value $\mathcal{T}_{\text{spon}} = 0.24$ is larger than the measured low intensity (spontaneous) value of $\mathcal{T}_{\text{spon}} \simeq 0.175$ shown in Fig. 15.16b. The latter would require $\sigma_{\text{XAS}}^{\diamond}\,\rho_{\text{a}} d$ in (15.81) to be 22% larger than expected for pure Al, most likely due to film oxidation.

The adopted value for the $L_{3,2}$-shell fluorescence yield is $Y_{\text{f}} = 7.5 \times 10^{-4}$, and the total core hole lifetime width is $\Gamma = 40$ meV [67, 68]. This corresponds to a dipole matrix element $\Gamma_x = Y_{\text{f}}\,\Gamma = 30\,\mu$eV. A theoretical study by Almbladh et al. [69] gave $\Gamma = 22$ meV and $\Gamma_x = 45\,\mu$eV. The other quantity of interest in non-resonant XAS is the total upper state width according to (15.78). One would expect a similar band trapping time of the photoelectron as for Cu, corresponding to $\Gamma_{\text{e}} = \hbar/\tau_{\text{e}} \sim 2$ eV ($\tau_{\text{e}} \sim 300$ as) [44]. Since Γ_{e} is considerably broader than the core hole width Γ, the upper state population (15.80) is described by $\Gamma_{\text{tot}} = \Gamma_{\text{e}}$.

Let us take a look at the temporal evolution of the transparency process. For non-resonant XAS, the temporal coherence time of the incident beam does not matter. Then the intensity and fluence (intensity $\times$ time) are simply related by the pulse

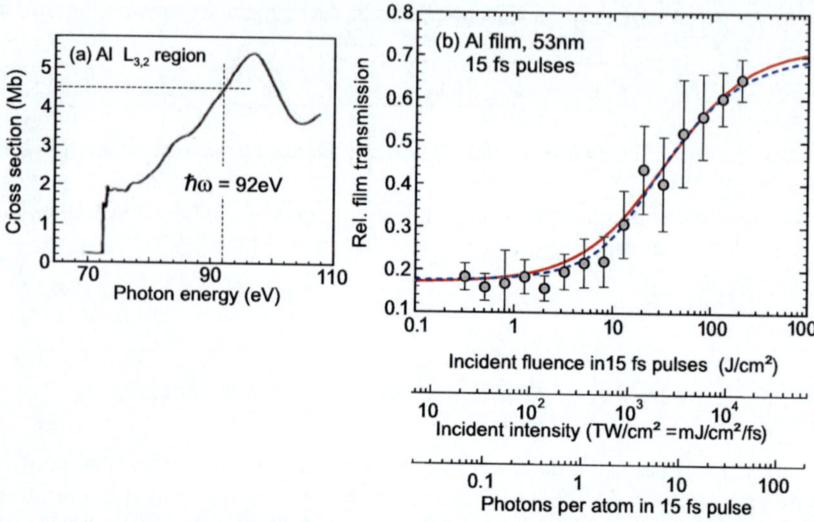

Fig. 15.16 **a** XAS spectrum above the Al L_3 (72.7 eV) and L_2 (73.2 eV) edges adopted from [65]. We have put the measured XAS spectrum on an approximate cross-section scale [57]. **b** Experimental data points for the NL increase of a 53-nm-thick Al film at $\hbar\omega = 92$ eV relative the spontaneous value [8], shown on different abscissa scales. The solid red curve is the theoretical prediction by (15.81) as discussed in the text. The dashed blue curve is taken from the original work [8]

length. At a certain central excitation energy, the XAS cross section into continuum states is approximately constant when averaged over a bandwidth up to about 10 eV or so. At first, the x-ray penetration depth is determined by the spontaneous value $\ell_x = 1/(\rho_a \sigma_{XAS})$, or the related effective spontaneous depth d_{eff} defined in (15.35). For a 53-nm-thick Al film we have $\ell_x = 37$ nm and $d_{eff} = 28$ nm. With increasing intensity, the core hole production extends deeper into the film since the atomic layers in the effective penetration depth become increasingly transparent. Transparency propagates layer by layer until the whole film becomes transparent, as reflected by $\tilde{d}_{eff} \to d$ according to (15.36).

As pointed out earlier, it is of great importance that the cross section for *multiple* L-shell core hole production is negligibly small relative to single core hole production. Hence transmission starts to increase approximately when there is on average one photon per atom, and it saturates when there are sufficient photons per atom that the excitation probability per atom becomes unity. If the atoms have a cross-sectional area A_{at}, saturation occurs when there are $n_{ph} \simeq A/\sigma_{XAS}$ photons per atom. The area of Al atoms seen by the incident photons is $A_{at} = 6.5 \times 10^{-2}$ nm^2 which compares to $\sigma_{XAS} \simeq 4.5 \times 10^{-4}$ nm^2. So it takes on average the attempts of $n_{ph} \simeq 150$ photons per atomic area to create a single L-shell core hole. This is in good accord with Fig. 15.16b (see the lowest scale).

The other crucial parameter is the refilling time of the created core holes, given by their lifetime. For Al $\tau_\Gamma = \hbar/\Gamma$ has the adopted value $\tau_\Gamma = 16$ fs [67, 68] which compares to a theoretical value of $\tau_\Gamma = 31$ fs [69]. For complete transparency to occur, the fluence in the short pulse length of $\tau_\mathrm{p} = 15$ fs would have to be large enough that the whole core hole production process is finished before it is partially compensated by core hole refilling, i.e. $\tau_\mathrm{p} \ll \tau_\Gamma$. Since this is marginally fulfilled for the Al L-edge case, complete transparency is not expected. This is indeed seen from the experimental data points in Fig. 15.16b which saturate below unity.

The solid red curve in Fig. 15.16b was calculated with (15.81) using the experimentally observed value of $\mathcal{T}_\mathrm{spon} = \mathrm{e}^{-\sigma^{\diamond}_\mathrm{XAS}\rho_\mathrm{a} d} = 0.175$, corresponding to a 22% increase of the product of the nominal values $\sigma^{\diamond}_\mathrm{XAS} \simeq 4.5$ Mb, $\rho_\mathrm{a} = 60.3$ atoms/nm^3 and $d = 53$ nm. The excited state population ρ_{22}^NR was evaluated according to (15.80) with $\lambda = 13.5$ nm, $\Gamma_\mathrm{tot} \simeq \Gamma_\mathrm{e} = 1.7$ eV ($\tau_\mathrm{e} = 390$ as), and the theoretical dipole matrix element $\Gamma_x = 45\,\mu$eV [69].

In the product $2\rho_{22}^\mathrm{NR}\tilde{d}_\mathrm{eff}/d$ in (15.81) we used $\tilde{d}_\mathrm{eff} = 43$ nm. This means that only the fraction $\tilde{d}_\mathrm{eff}/d = 0.81$ of the entire film becomes transparent. Since ρ_{22}^NR and $\tilde{d}_\mathrm{eff}$ enter equivalently in the theory, one may also attribute the incomplete transparency to an unsaturated excited state population $2\rho_{22}^\mathrm{NR} = 0.81$ averaged over all atoms in the entire thickness $\tilde{d}_\mathrm{eff} = d$. The physical origin of incomplete transparency is that the pulse length of $\tau_\mathrm{p} = 15$ fs is only slightly shorter than the core hole clock time $\tau_\Gamma \simeq 16$–31 fs [67, 69], so that the exited state population can already partially decay during the pulse. The dashed blue curve in Fig. 15.16b, taken from the original work [8], is nearly identical to our solid red curve. The measured transparency has also been treated by Hatada and Di Cicco [70, 71] in a rate equation model with similar results.

15.7.4 Non-linear Transparency Above the Fe K-Edge

A related investigation of x-ray induced transparency was carried out by Yoneda et al. using excitations to continuum states above the K-edge of Fe metal [10]. The XAS spectrum is shown in the inset of Fig. 15.17 as a gray curve [72] with the excitation energy of 7130 eV ($\lambda = 0.174$ nm) indicated by an arrow. The XAS cross section (right ordinate in the inset) has a value of $\sigma^{\diamond}_\mathrm{XAS} = 43$ Kb/atom which corresponds to the stated x-ray absorption coefficient by Yoneda et al..[5]

The experiment was carried out by monitoring the intensity-dependent transmission through a $d = 20$-μm-thick Fe foil (atomic density $\rho_\mathrm{a} = 85$ atoms/nm^3). Since the foil thickness greatly exceeded the spontaneous x-ray absorption length of $\ell_x = 1/(\rho_\mathrm{a}\sigma_\mathrm{XAS}) = 2.7\,\mu$m, the spontaneously transmitted intensity, $\mathcal{T}_\mathrm{spon} = \mathrm{e}^{-\sigma^{\diamond}_\mathrm{XAS}\rho_\mathrm{a} d} = 6.7 \times 10^{-4}$, was very small. The experiment involved 7 fs SASE pulses of large bandwidth (~40 eV) with remarkably high intensities up to 10^{20} W/cm^2, produced

[5] Note that the XAS cross section in Fig. 15.17 contains a background contribution of about 7 Kb/atom which was subtracted in Fig. 10.13.

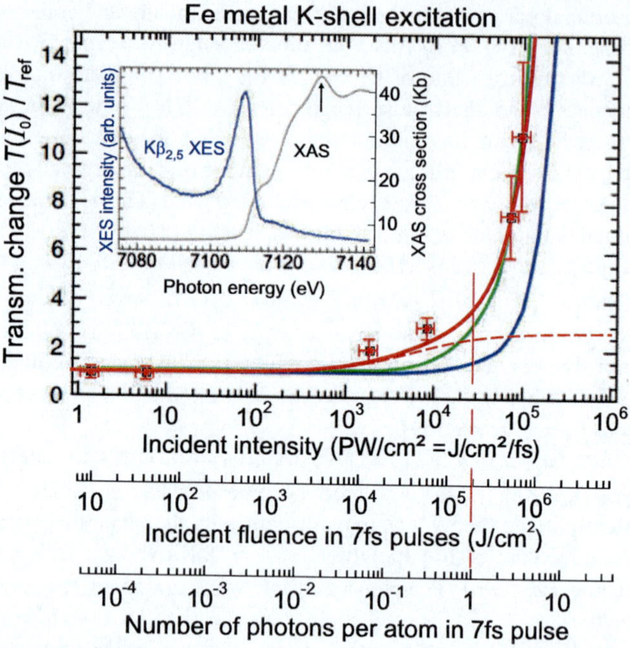

Fig. 15.17 Inset: Fe K-shell XAS (gray curve) and K$\beta_{2,5}$ (blue curve) XES spectra [72]. Main figure: Intensity-dependent NL transmission of 7 fs pulses, normalized the spontaneous transmission at low intensity, of a 20 μm Fe metal foil at a photon energy of 7130 eV, about 20 eV above the K-edge at 7112 eV [10]. The filled red squares and error bars are the experimental data points, and the solid blue and green curves are simulations taken from [10]. The red curve is calculated according to (15.81), with ρ_{22}^{NR} evaluated according to (15.80) with $\sigma_{\text{XAS}}^{\diamond} = 43$ Kb/atom, $\lambda = 0.174$ nm, $\Gamma_x = 4.5$ meV, and $\Gamma_{\text{tot}} = 1.33$ eV. The dashed red curve corresponds to $\tilde{d}_{\text{eff}} = d_{\text{eff}} = 2.7\,\mu$m in (15.81)

through exquisite focusing onto the sample. The high intensities were required due to the relatively small K-shell XAS cross section (see Fig. 7.14) and the short core hole lifetime of about 500 as (see below), which made it difficult to create core holes on a significant fraction of all atoms in the film during the 7 fs pulse. Experimental data points for the intensity-dependent transmission, normalized to the spontaneous transmission value at low intensity, are shown in Fig. 15.17 as filled red squares with associated error bars. The solid blue and green curves are simulations taken from the original work, corresponding to core hole lifetimes of $\tau_\Gamma = 500$ as ($\Gamma \simeq 1.3$ eV) (blue) and 2 fs ($\Gamma \simeq 0.33$ eV) (green).

The solid red curve in Fig. 15.17 was calculated according to (15.81) with the nominal values $\sigma_{\text{XAS}}^{\diamond} = 43$ Kb/atom, $\rho_{\text{a}} = 85$ atoms/nm^3, and $d = 20\,\mu$m. The excited state population ρ_{22}^{NR} given by (15.80) was calculated with $\lambda = 0.174$ nm and the following linewidths. For the Fe K-shell, the core hole width is given in the literature as $\Gamma = 1.25$ eV [68] or $\Gamma = 1.4$ eV [73]. Since these values are comparable to the expected excited state "band trapping" width, we used a value $\Gamma_{\text{tot}} = 1.33$ eV.

The dipole transition width Γ_x was determined from the partial fluorescence yield of the $\beta_{2,5}$ emission line, shown in blue in the inset of Fig. 15.17 [72]. This emission line corresponds to dipole transitions from Fe $4p$ valence states directly into the $1s$ core hole and its dipole matrix element Γ_x is the same as for the reverse $1s \rightarrow 4p$ core to valence excitation probed in XAS. According to theory[6] [72], the $\beta_{2,5}$ strength is almost exactly 1% of the total fluorescence yield $Y_f = 0.340$ [67]. This ratio is supported by the measured value for Ni metal [74]. The dipole matrix element is then given by $\Gamma_x \simeq 0.01\, Y_f\, \Gamma$, which by use of $\Gamma = 1.33$ eV yields $\Gamma_x = 4.5$ meV.

The red curve in Fig. 15.17 is seen to follow the experimental data points better than the published blue and green curves. In particular, the NL effect begins at lower intensity and initially follows the dashed red curve. It is calculated with $\tilde{d}_{\rm eff} = d_{\rm eff} = 2.7\,\mu$m in (15.81), which for the thick Fe foil is nearly the same [see 15.35)] as the spontaneous x-ray penetration depth $\ell_x = 2.7\mu$m. This shows that the early increase in transmission is due to the first stage of transparency induced in the spontaneous penetration depth. The transparency then progresses deeper into the foil until it increases more strongly when there is a photon for every atom in the film, marked by the vertical red line.

15.8 Polarization Dependent NL Transmission at Resonance

The following relatively short section finishes the treatment of the NL transmission through a film, excited at resonance, by considering the largest NL transmission effect and its polarization dependence for a magnetically aligned sample. It also serves as the basis for the next Sect. 15.9 which introduces the link between in-beam transmission ($\mathbf{q}=0$) and out-of-beam ($\mathbf{q}\neq 0$) diffraction.

15.8.1 The Maximum NL Transmission Effect

The largest NL effect in x-ray absorption occurs when the incident pulses are temporally coherent for a time considerably longer than the core hole clock time, as illustrated in Fig. 15.12, and are tuned to the resonance energy. For the example of resonant Co absorption with a core hole lifetime of $\tau_\Gamma = 1.5$ fs, the coherence time, needs to be about 20 fs or more. This situation is encountered either for a single transform-limited pulse of length longer than 20 fs, or for a longer SASE pulse that is put through a monochromator. In the latter case, the bandwidth of the monochromator filters out coherent fractions of the longer pulse, accompanied by a loss of intensity. A long pulse may then contain several coherent subpulses whose centers

[6] I would like to thank Craig Schwartz for the actual numbers.

are temporally shifted. In coherence dependent experiments, the subpulses act independently and just improve the statistics of the measured signal.

In the following we shall consider incident pulses of different coherence lengths. When divided by the coherence time, the pulses are assumed to have the same intensity. We will furthermore include the polarization dependence of transmission through a film, by considering that it is magnetically oriented along the x-ray propagation direction, chosen along the normal of the film. We know form Sect. 7.11.3 that incident circularly polarized x-rays will show a difference in transmission, the well-known XMCD effect, while the transmission for linearly polarized x-rays will be the average of the two circularly polarized cases. Of interest here is the intensity-dependent evolution of the XMCD effect.

15.8.2 Polarization Dependent Transmission

For a film that is uniformly magnetized in the direction $\hat{m}_z$, the transmitted intensity is given by the polarization dependent ($p = 0, \pm$) NL Beer-Lambert law given by (15.63) or

$$I^p_{\text{trans}} = I^p_0 \, \mathrm{e}^{-\left(\left[\sigma^{\diamond}_{\text{XAS}}\right]^p - \left[\sigma^{\text{NL}}_{\text{XAS}}\right]^p\right)\rho_{\text{a}} d} = I^p_0 \, \mathrm{e}^{-2\lambda\left(f''^p - f''^p_{\text{NL}}\right)\rho_{\text{a}} d} = I^p_0 \, \mathrm{e}^{-2\left(\beta^p - \beta^p_{\text{NL}}\right)kd}. \tag{15.83}$$

The cross sections, scattering lengths, and optical parameters in (15.83) are given in Sect. 15.6.1. In the following we will for convenience and consistency use the optical parameter formulation of Sects. 7.11 and 8.7.

In those sections the spontaneous isotropic charge response was expressed through the optical parameters β^0 and δ^0 ($p = 0$, linear polarization) given by (7.118), and the magnetic response was included through a change of these parameters denoted $\Delta\beta$ and $\Delta\delta$ as defined in (7.117). In order to directly see the additional NL contribution we write (15.83) in analogy to the spontaneous expression (7.120) in the total, spontaneous plus stimulated, form,

$$\boxed{I^{p,\hat{m}_z}_{\text{trans}} = I^p_0 \, \mathrm{e}^{-2[(\beta^0 - \beta^0_{\text{NL}}) + p\,\hat{m}_z(\Delta\beta - \Delta\beta_{\text{NL}})]kd}} \tag{15.84}$$

Here the spontaneous charge contribution, β^p ($p = 0, \pm$), and its NL part β^p_{NL}, which is in general time dependent, are given by (see (15.69) and (15.68))

$$\beta^p = \frac{\lambda^3 \rho_{\text{a}}}{4\pi^2} \frac{\Gamma^p_x}{\Gamma} \frac{(\Gamma/2)^2}{(\hbar\omega - \mathcal{E}_0)^2 + (\Gamma/2)^2}, \quad \text{and} \quad \beta^p_{\text{NL}}(\tau) = \beta^p \, 2\widetilde{\rho}^{\,p}_{22}(\tau), \tag{15.85}$$

where $\widetilde{\rho}_{22}(\tau)$ is expressed by (15.75). The magnetic corrections $\Delta\beta$ and $\Delta\beta_{\text{NL}}$ are given by

$$\Delta\beta = \frac{\beta^+ - \beta^-}{2}, \quad \text{and} \quad \Delta\beta_{\rm NL} = \frac{\beta^+_{\rm NL} - \beta^-_{\rm NL}}{2}. \tag{15.86}$$

The XMCD contrast is defined according to (7.122) and for the magnetization and polarization directions aligned according to $\hat{m}_z \epsilon_\pm = \pm 1$ we have

$$\Delta I_{\rm XMCD} = I^+_{\rm trans} - I^-_{\rm trans}. \tag{15.87}$$

The NL form of the spontaneous expression (7.123) then reads

$$\boxed{\Delta I_{\rm XMCD} = -2I_0\, {\rm e}^{-2(\beta^0 - \beta^0_{\rm NL})kd} \sinh[2(\Delta\beta - \Delta\beta_{\rm NL})kd].} \tag{15.88}$$

To illustrate the polarization and coherence time-dependent transmission, we reconsider the case shown in Fig. 15.15, corresponding to polarization independent transmission and a short coherence time of $\tau = 1.8$ fs that is comparable to the core hole clock time of $\tau_\Gamma = 1.5$ fs. We now compare it to the case of a magnetically aligned sample illuminated by a longer coherent pulse of $\tau = 20$ fs of the same intensity. This is done in Fig. 15.18.

In Fig. 15.18a we illustrate the effect of coherence time τ by plotting the total spontaneous plus stimulated transmission for a magnetically aligned sample according to (15.84) with $\hat{m}_z \epsilon_\pm = \pm 1$. The coherence time enters through the excited state population in (15.85). We see that the NL threshold for $\tau = 20$ fs (blue curves) occurs at a lower intensity than for $\tau = 1.8$ fs (red curves). In particular, the solid red curve for $p = 0$ and $\tau = 1.8$ fs is the same as the dashed red curve in Fig. 15.15.

The change in transmission due to stimulation relative to the spontaneous case is plotted in Fig. 15.18b. The solid curves reflect the change in relative transmission for linear polarization, $I^0_{\rm tot}/I^0_{\rm spon}$ for the two different coherence times. The dashed curves are the corresponding normalized dichroic XMCD ratios given by $(I^+_{\rm tot} - I^-_{\rm tot})/(I^+_{\rm spon} - I^-_{\rm spon})$. It is interesting that for $\tau = 20$ fs the change in transmission is nearly polarization independent as seen from the nearly identical solid and dot-dashed blue curves. At $\tau = 1.8$ fs the corresponding solid and dot-dashed red curves show differences caused by the Rabi oscillations.

In the following sections we will discuss the competition between the forward and out-of-beam scattered intensity in the presence of stimulation. We will show that an increase in transmission comes at the expense of the out-of-beam scattered or diffracted intensity, as a consequence of energy conservation.

15.9 Competition Between NL Transmission and Diffraction

In diffraction terminology, the transmitted intensity corresponds to zero momentum transfer, $\mathbf{q} = 0$. We are now interested in the intensity scattered out of the beam,

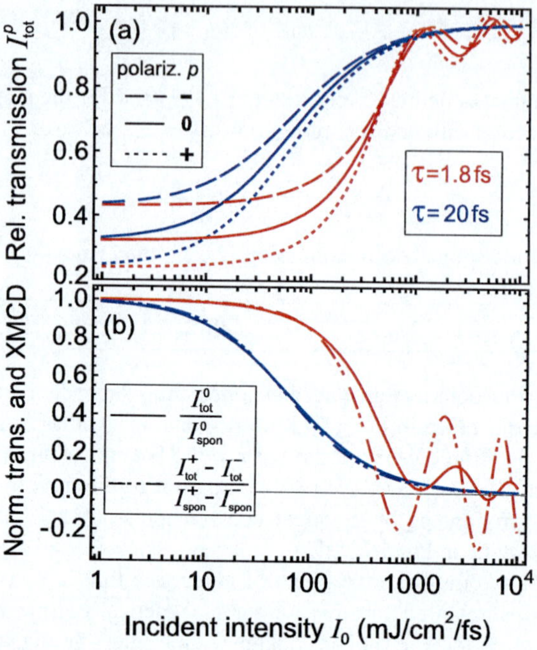

Fig. 15.18 **a** Polarization dependent ($p = 0, \pm$) total (spontaneous plus stimulated) transmission $I^p_{tot} = I^p_{trans}$ as a function of pulse intensity according to (15.84). The red curves are for SASE pulses with coherent regions up to $\tau = 1.8$ fs lengths, while the blue curves are for pulses of the same intensity with coherent regions up to $\tau = 20$ fs lengths. The solid red curve for $p = 0$ corresponds to the dashed red curve in Fig. 15.15. **b** Total intensities in (**a**), normalized by the spontaneous parts. The solid curves correspond to linear polarization, I^0_{tot}/I^0_{spon}, and the dashed curves are the dichroic XMCD ratios according to (15.88) as indicated

i.e. the diffracted intensity at $\mathbf{q} > 0$. One may first think that the weaker *diffracted* intensity remains unchanged from the spontaneous value since it is not stimulated. This is, however, *not* the case because of the fundamental law of conservation of energy between all available channels. An increase of intensity at $\mathbf{q} = 0$ therefore comes at the cost of that at $\mathbf{q} > 0$. We start by considering the NL change of the most fundamental diffraction pattern that of a homogeneous film with random bond and spin orientations in a circular aperture.

The spontaneous diffraction pattern for this case, which is closely related to the emission characteristics of a circular source, was derived in Sect. 4.4.3 and later used in Sect. 8.6.3 to demonstrate Babinet's amazing theorem. We now revisit it to demonstrate another fascinating phenomenon.

15.9.1 The NL Airy Pattern of a Film in a Circular Aperture

We have discussed multiple times the case of a circular aperture of area $A_s = \pi R^2$ that is coherently illuminated with a constant intensity across the aperture. In Sect. 4.4.3 it served to illustrate the Fourier optics definition of coherence, in Sect. 8.6.1 to define the diffraction limit, and in Sect. 5.5.5 to illustrate the change of the first (one photon) to second order (two photons) diffraction pattern. The conventional (first order) Airy diffraction pattern as a function of momentum transfer $q = k\rho/z_0$ is given by (4.71) or

$$\boxed{I^{(1)}(q, z_0) = \underbrace{I_0^{(1)} A_s}_{\mathcal{P}_s^{(1)}} \frac{A_s}{\lambda^2 z_0^2} \left[\frac{2J_1(qR)}{qR}\right]^2 = I_0^{(1)} \frac{A_s^2}{\lambda^2 z_0^2} \underbrace{\left[\frac{2J_1(qR)}{qR}\right]^2}_{\mathcal{A}_0(q)}.} \tag{15.89}$$

We have added a superscript "(1)" to specifically indicate that the intensity distribution corresponds to single photon counts. The first underbracket identifies the power $\mathcal{P}_s^{(1)}$ emitted by the source, expressed as the peak intensity times the source area. The power is preserved on propagation to the detector plane which is seen by substituting $q = k\rho/z_0$ and spatial integration over the detector plane in cylindrical coordinates according to

$$\mathcal{P}_d^{(1)} = \int_0^{2\pi}\int_0^{\infty} I^{(1)}(\rho, z_0)\, \rho\, d\rho = \mathcal{P}_s^{(1)}. \tag{15.90}$$

The second underbracket in (15.89) identifies the unitless normalized shape of the Airy pattern $\mathcal{A}_0(q)$. It may be written by expressing q in units of the momentum transfer $q_0 = 1.22\pi/R$ of the first Airy minimum, which defines the Rayleigh diffraction limit, according to

$$\mathcal{A}_0(q) = \left[\frac{2J_1(1.22\pi q/q_0)}{1.22\pi q/q_0}\right]^2. \tag{15.91}$$

It was previously plotted as a black curve in Fig. 8.12b.

In the presence of a film with random homogeneous bond and spin distributions, the corresponding spontaneous diffraction pattern is simply the Airy pattern of the hole, attenuated according to the conventional Beer-Lambert law, given by (8.25) or

$$\boxed{I_{\text{spon}}^{(1)}(q, z_0) = I_0^{(1)} \frac{A_s^2}{\lambda^2 z_0^2} \underbrace{\left[\frac{2J_1(qR)}{qR}\right]^2 e^{-2\beta kd}}_{\mathcal{A}_{\text{spon}}(q)}} \tag{15.92}$$

also previously shown in Fig. 8.12b as a blue curve, corresponding to $2\beta kd = 1$.

While diffraction by a hole alone or spontaneous diffraction by a film in a hole can be described by spherical wave emission from each point or atom in the aperture plane, the famous Huygens–Fresnel principle, this description breaks down in the presence of stimulated atomic scattering. In this case, the atoms in the film no longer spontaneously scatter photons into random directions, which on average is well modeled by the spherical waves assumed in the Huygens–Fresnel principle. Rather, the stimulating photons imprint their incident wavevector direction (mode) on the photon created in the stimulated decay. At maximum stimulation, the film becomes transparent as shown in Fig. 15.15.

15.9.1.1 Cloned Biphotons and the Two-Photon Airy Pattern

When the resonantly scattered (REXS) intensity is completely dominated by stimulation, one may envision all atoms as source points of two identical photons, which have been referred to as *cloned biphotons* [25, 26]. These biphotons are naturally coincident and lead to coincident two-photon counts at all points in the detector plane. The two-photon diffraction pattern is the *squared* one-photon Airy distribution given by (5.58) or

$$I^{(2)}(q, z_0) = \underbrace{I_0^{(2)} A_s}_{\mathcal{P}_s^{(2)}} \underbrace{\frac{1}{1-16/(3\pi^2)}}_{2.176} \frac{A_s}{\lambda^2 z_0^2} \left[\frac{2J_1(qR)}{qR}\right]^4 . \tag{15.93}$$

The power-conserving prefactor $1/(1-16/(3\pi^2)) = 2.176$ is slightly larger than 2 because the central peak of the Airy distribution also robs most of the power contained in the outer rings for the spontaneous case, which decreases from 16.2% to 0.2%. The total power is preserved on propagation to the detector plane which is seen by integrating over the detector plane in real space cylindrical coordinates according to

$$\mathcal{P}_d^{(2)} = \int_0^{2\pi}\int_0^{\infty} I^{(2)}(\rho, z_0)\, \rho\, d\rho = \mathcal{P}_s^{(2)}. \tag{15.94}$$

From a detection point of view, care has to be exercised with the cloned biphoton picture. As discussed in Sect. 5.6.1 one needs to distinguish a *collective two-photon state* given by (5.64) from its *binomial two-photon substate* identified in (5.66). The collective state contains an *average number* of two photons statistically distributed over a Poisson distribution shown in Fig. 5.11b, while the two-photon substate contains exactly two photons. These two-photon states differ in their degree of second order coherence as discussed in Sect. 5.6.3.1. The two-photon substate is *not* second order coherent, but it can be picked out by two-photon coincidence detection as demonstrated in [75]. In contrast, the collective two-photon state is second order

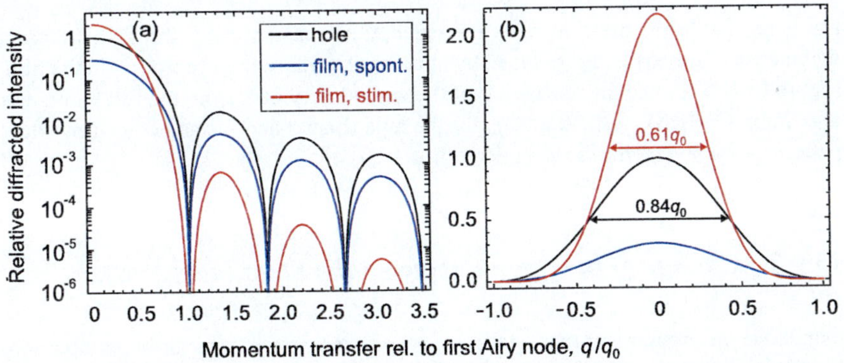

Fig. 15.19 **a** Airy diffraction patterns $\mathcal{A}_0(q)$ (black) according to (15.89), $\mathcal{A}_{\rm spon}(q)$ (blue) defined in (15.92) and $\mathcal{A}_{\rm stim}(q)$ given by (15.95) (red) as a function of normalized momentum transfer q/q_0, where $q_o = 1.22\pi/R$ is the first node of the Airy pattern. We have assumed a circular hole of radius $R = 725$ nm and $\boldsymbol{\lambda} = 1.6$ nm (Co L_3 resonance) and for the blue curve a film attenuation by a factor of $\simeq 3$. **b** Enlarged central part of the patterns on a linear scale, revealing the reduction of the width and increase of the intensity of the stimulated pattern

coherent and its diffraction pattern is naturally recorded by use of a photon number integrating detector such as a CCD. This is discussed in more detail in Chap. 16. In the following we will not worry about the detection details but are interested only in the difference between the one-photon and two-photon Airy patterns.

The two-photon diffraction pattern (15.93) corresponds to the completely stimulated pattern of a film emitting biphotons given by

$$I^{(2)}_{\rm stim}(q, z_0) = I^{(2)}_0 \frac{A_s^2}{\lambda^2 z_0^2} \underbrace{\frac{1}{1-16/(3\pi^2)} \left[\frac{2J_1(qR)}{qR}\right]^4}_{\mathcal{A}_{\rm stim}(q)} . \tag{15.95}$$

Note that the one-photon intensities given by (15.89) and (15.92) and the two-photon intensity in (15.95) are all normalized to the incident number of photons so that $I_0^{(2)} = 2I_0^{(1)}$. In Fig. 15.19 we have compared the normalized patterns per photon for a circular hole $\mathcal{A}_0(q)$ (black, see (15.89)), a Co film in the hole $\mathcal{A}_{\rm spon}(q)$ (blue, see (15.92)), and its completely stimulated pattern $\mathcal{A}_{\rm stim}(q)$ (red, see (15.95)). We used the parameters of the experiment reported in [297], $\lambda = 1.6$ nm (Co L_3 resonance) and a circular aperture hole of $R = 725$ nm. The patterns are plotted as a function of momentum transfer q/q_0, where $q_0 = 1.22\pi/R$ is the first node of the Airy pattern.

The conventional normalized Airy pattern of the hole shown as black curves in Fig. 15.19 contains 83.8% of the total power in the central disk. The stimulated pattern, given by (15.95) and shown in red, contains 99.8% of the power in the central disk, so that the outer Airy rings have negligible intensity. The spontaneous Airy pattern in the presence of a film given by (8.25) is shown in blue, assuming the presence of a $d = 20$-nm-thick Co film with optical constant $\beta = 0.0075$. In this case, the intensity and power are reduced by a factor of $\simeq 3$ through absorption in the film. Figure 15.19b reveals the narrowing of the central Airy disk from the full width half maximum (FWHM) value $0.844\, q_0$ for the hole (black) and spontaneous film (blue) patterns, to $0.606\, q_0$ for the stimulated film (red).

15.9.2 Change of the Spontaneous to the Stimulated Pattern

Our above discussion was restricted to the case of complete (saturation) stimulation. In order to derive the evolution with incident intensity from the blue spontaneous pattern in Fig. 15.19 to the red stimulated pattern, we describe the response of the film in terms of the spontaneous β and non-linear β_{NL} optical parameters (see Sect. 15.6.1). The development of the pattern is obtained by considering that the total (spontaneous and stimulated) q-dependent diffracted power $\mathcal{P}$ in the *detector* plane has to be equal to the transmitted power (at $q = 0$) in the *exit* plane of the sample, so that [12]

$$\underbrace{\mathcal{P}_{\mathrm{trans}}}_{\text{exit plane}} = \underbrace{F_{\mathrm{spon}}\, \mathcal{P}_{\mathrm{spon}}(q)}_{\text{spon. pattern}} + \underbrace{F_{\mathrm{stim}}\, \mathcal{P}_{\mathrm{stim}}(q)}_{\text{stim. pattern}} . \tag{15.96}$$

Here F_{spon} and F_{stim} with $F_{\mathrm{spon}} + F_{\mathrm{stim}} = 1$ are suitable intensity-dependent but q-independent coefficients that determine the relative contributions of the q-dependent spontaneous and stimulated patterns. We can determine the coefficients F_{spon} and F_{stim} by integrating the spontaneous and stimulated patterns in cylindrical coordinates over q, or equivalently over the real space distance $\rho = q z_0/k$ from the optical axis in the detector plane, and with $I_0 = I_0^{(1)} = I_0^{(2)}/2$ we obtain

$$\begin{aligned}
\mathcal{P}_{\mathrm{trans}} &= I_0 A_{\mathrm{s}}\, \mathrm{e}^{-2[\beta-\beta_{\mathrm{NL}}]\, kd} \\
\mathcal{P}_{\mathrm{spon}} &= \int_0^{2\pi}\!\!\int_0^{\infty} I^{(1)}_{\mathrm{spon}}(\rho, d)\, \rho\, \mathrm{d}\rho = I_0 A_{\mathrm{s}}\, \mathrm{e}^{-2\beta kd} \\
\mathcal{P}_{\mathrm{stim}} &= \int_0^{2\pi}\!\!\int_0^{\infty} I^{(1)}_{\mathrm{stim}}(\rho, d)\, \rho\, \mathrm{d}\rho = I_0 A_{\mathrm{s}},
\end{aligned} \tag{15.97}$$

where $I^{(1)}_{\mathrm{spon}}(q, z_0)$ is given by (15.92) and $I^{(1)}_{\mathrm{stim}}(q, z_0) = I^{(2)}_{\mathrm{stim}}(q, z_0)/2$ is the two-photon intensity (15.95) normalized per photon.

By inserting these expressions into (15.96) we can determine the coefficients and obtain the total power-normalized Airy diffraction pattern as

$$I_{\text{tot}}(q, z_0) = \underbrace{\frac{1 - e^{-2(\beta - \beta_{\text{NL}})kd}}{1 - e^{-2\beta kd}}}_{F_{\text{spon}}} I^{(1)}_{\text{spon}}(q, z_0) + \underbrace{\frac{e^{2\beta_{\text{NL}}kd} - 1}{e^{2\beta kd} - 1}}_{F_{\text{stim}}} I^{(1)}_{\text{stim}}(q, z_0), \qquad (15.98)$$

where

$$F_{\text{spon}} + F_{\text{stim}} = 1. \qquad (15.99)$$

The coefficient F_{spon} reflects the relative *decrease* of the spontaneous pattern, and F_{stim} the relative *increase* of the stimulated pattern under the condition of power conservation. The NL response arises through the intensity-dependent optical parameter β_{NL} given by (15.69) or

$$\beta_{\text{NL}} = \beta \, \frac{I_0 \, d_{\text{eff}} \, \mathbb{A}}{\mathbb{B} + I_0 \, d_{\text{eff}} \, \mathbb{A}}. \qquad (15.100)$$

The total pattern is seen to reduce to the spontaneous limit for $\beta_{\text{NL}} = 0$ and the stimulated limit for $\beta_{\text{NL}} = \beta$, as required.

In Fig. 15.20a we plot the intensity-dependent change of the Airy diffraction pattern $I_{\text{tot}}(q)$, given by (15.98) and normalized by $I_0 A_s^2/(\lambda^2 z_0^2)$, for a resonantly excited 20-nm-thick Co film in a circular aperture. We used the L_3 resonance value $\beta = 0.0072$, corresponding to $\sigma^{\diamond}_{\text{XAS}}$ =6.25 Mb.

The q-dependent patterns in Fig. 15.20a are shown for four incident intensities I_0, assuming laterally coherent x-rays with a temporal coherence length of 20 fs. The patterns contain different contributions of the normalized (intensity independent) spontaneous pattern $\mathcal{A}_{\text{spon}}(q)$, defined in (15.92), and its square, the stimulated pattern $\mathcal{A}_{\text{stim}}(q)$ defined in (15.95), weighted by the intensity-dependent coefficients F_{spon} and F_{stim}. They evolve from the blue pattern for $I_0 = 1$ mJ/cm^2/fs which is nearly identical to the blue spontaneous one in Fig. 15.19, to the orange pattern for $I_0 = 3 \times 10^3$ mJ/cm^2/fs, which approaches the red (completely) stimulated pattern in Fig. 15.19. In (b) we show the enlarged central part of the patterns in (a) on a linear scale to more clearly see the stimulated increase.

In Fig. 15.20c we show as a blue curve (left ordinate scale) the I_0-dependent decrease of the total *outer* Airy ring intensity integrated over all q. It is given by F_{spon} in (15.98). Superimposed as black-gray circles are the experimental data points by Wu et al. [12]. The intensity of the 50 fs pulses monochromatized SASE pulses used in the experiment was assumed to be contained in two coherent 20 fs subpulses to match experimental and theoretical coherence lengths. The experiment is discussed in more detail in Sect. 15.10.2. Also shown as a red curve (right ordinate) is the I_0-dependence of the cental $q = 0$ Airy intensity. It is given by $I_{\text{tot}}(q = 0)$ in (15.98), normalized by $I_0 A_s^2/(\lambda^2 z_0^2)$. Superimposed on the red curve as filled colored circles are the peak values of the four curves in (b) shown in corresponding colors.

15.9.3 X-Ray Soliton Model: Mode-Dependent Stimulation

The change of the diffraction pattern of a film with a homogeneous charge distribution in a circular aperture shown in Fig. 15.20 may also be described by a model based on the development of a spatial light soliton in the forward direction ($\mathbf{q} = 0$) through a non-linear propagation process through the thin film. In such a two-step model the circular aperture is first assumed to diffract the incident first order coherent beam into modes of different directions and intensities. In the second step the modes propagate non-linearly through the film to create a modified pattern as illustrated in Fig. 15.21.

In the optical regime, the propagation of spatial light solitons [76–78] is typically described semi-classically by an intensity-dependent change of the refractive

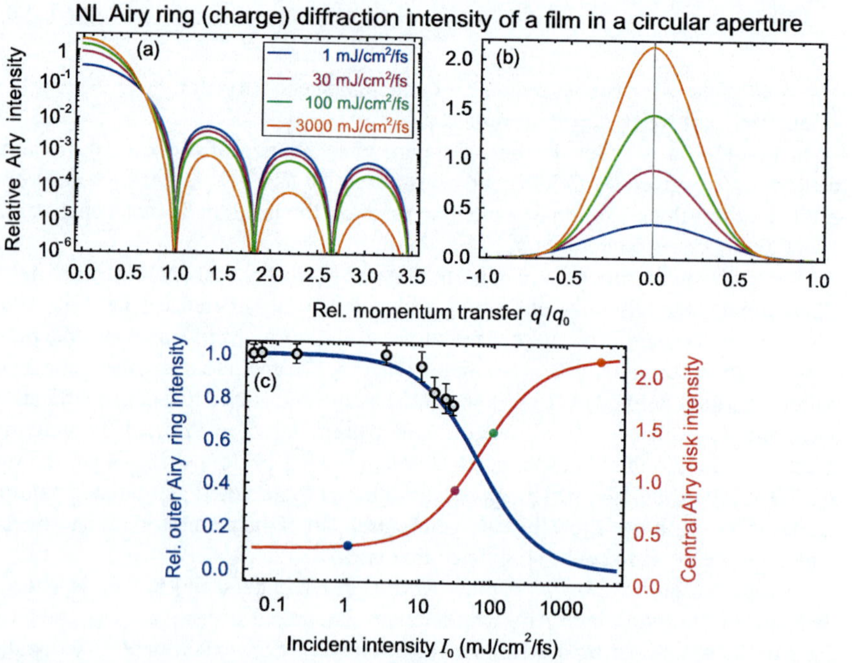

Fig. 15.20 **a** Airy patterns versus momentum transfer q in units of the first node of the Airy pattern at $q_0 = 1.22\pi/R$ for a 20 nm Co film in a circular aperture for four values of I_0. The patterns were calculated with (15.98) and normalized by $I_0 A_s^2/(\lambda^2 z_0^2)$. We assumed Co L_3 resonant excitation ($\lambda = 1.6$ nm) by laterally coherent x-rays with a longitudinal coherence time of 20 fs. **b** Central part of the pattern around $q = 0$ on a linear intensity scale calculated with the same parameters. (c) The blue curve (left ordinate scale) is the I_0-dependent change of the total *outer* Airy ring intensity given by F_{spon} in (15.98). It is compared to the data (black-gray circles) measured by Wu et al. [12] assuming the same 20 fs coherence length. The red curve (right ordinate) is the I_0-dependent change of the total Airy intensity $I_{\text{tot}}(q)$ at $q = 0$, given by (15.98) and normalized by $I_0 A_s^2/(\lambda^2 z_0^2)$. The peak values of the four curves in (**b**) are superimposed on the red curve as filled circles of corresponding colors

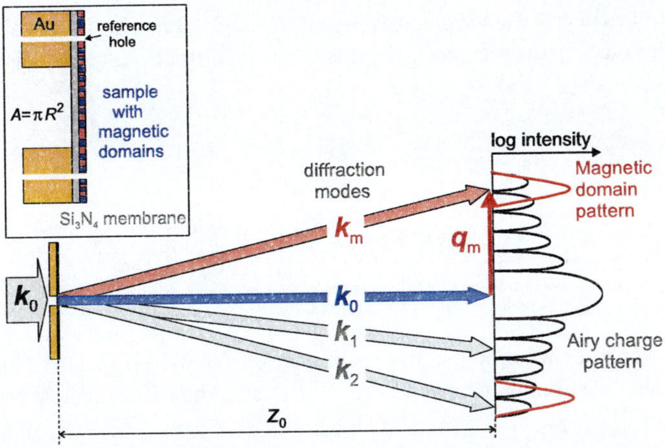

Fig. 15.21 Typical sample structure used for coherent imaging experiments. The beam is transmitted and diffracted by a thin film containing domains with the beam size defined by a cylindrical Au aperture of area $A = \pi R^2$. The sample structure is shown in detail in the inset. The diffracted beam is detected by a position sensitive detector at a distance z_0 from the sample. The Airy pattern due to the sample aperture and a schematic magnetic diffraction pattern determined by the average size of the domains is shown on the right, plotted on a logarithmic intensity scale that emphasizes the small outer modes. One may distinguish different diffraction modes by their wavevectors $\mathbf{k}_i$, as shown. The transmitted mode in the forward direction $\mathbf{k}_0$ has by far the largest intensity and is preferentially stimulated

index through the optical or AC Kerr effect [79]. The dependence of the refractive index of the medium on the intensity creates a refractive index change that mimics the transverse intensity pattern of the beam. This causes the natural linear beam expansion to become compensated by the non-linear self-focusing. The self-focusing effect depends on the shape of the intensity distribution in the material which modifies the refractive index. This may be expressed as a non-linear contribution that is proportional to the third order non-linear susceptibility [1].

For the spontaneous response case, the Airy diffraction pattern from the circular aperture, shown in black in Fig. 15.21 on a logarithmic scale, is attenuated by absorption in the film relative to an aperture without a film as previously illustrated in Fig. 15.19. At low incident intensities, the spontaneous charge response is independent of the location of the film which simply acts as an attenuator. For the stimulated case, the location of the film behind the aperture matters since the intensity decreases with distance from the aperture. The strongest non-linear effect will occur when the film is located just behind the circular aperture, and it is weakest when it is inserted right before the distant detector due to the spatial expansion of the beam.

In general, the q-dependent stimulation process consists of two contributions. First, at a given q (mode) the intensity change is dependent on the intensity in this mode $I(q)$. Second, there is intensity exchange between different q directions (modes), driven by optimization of total power conservation.

In the following we want to quantify the soliton model. Ignoring polarization labels to simplify notation, the incident and spontaneously diffracted intensity distributions are given by

$$\text{Flat–top}: \; I_{\text{in}}(r) = \begin{cases} I_0 & -R < r < +R \\ 0 & \text{otherwise} \end{cases} \tag{15.101}$$

$$I_{\text{diff}}(q) = I_0 \underbrace{\frac{\pi^2 R^4}{z_0^2 \lambda^2}}_{\text{propagation}} \left[\frac{2J_1(qR)}{qR} \right]^2 . \tag{15.102}$$

For convenience we drop the propagation term from the exit plane of the sample to the distant detection plane given by $\pi^2 R^4/(z_0^2\lambda^2)$. The stimulated pattern (15.95) per photon $I^{(1)}_{\text{stim}}(q, z_0) = I^{(2)}_{\text{stim}}(q, z_0)/2$ translated back to the exit plane of the sample is then given by

$$I^{(1)}_{\text{stim}}(q) = I_0 \frac{1}{1 - 16/(3\pi^2)} \left[\frac{2J_1(qR)}{qR} \right]^4 . \tag{15.103}$$

The transition from the spontaneous to the stimulated pattern may be expressed through the introduction of a non-linear optical parameter $\beta_{\text{NL}}(q)$, which in contrast to the $q = 0$ transmission case is now q-dependent. In this formulation we have

$$I^{(1)}_{\text{stim}}(q) = I_0 \frac{1}{1 - 16/(3\pi^2)} \left[\frac{2J_1(qR)}{qR} \right]^2 \mathrm{e}^{-2[\beta - \beta_{\text{NL}}(q)]kd} . \tag{15.104}$$

We now consider how (15.104) evolves into (15.103) through a change in the q-dependent non-linear absorption parameter $\beta_{\text{NL}}(q)$. Comparison of the two expressions yields the following form of $\beta_{\text{NL}}(q)$

$$\beta_{\text{NL}}(q) = \beta \left(1 + \ln \left[\left(\frac{2J_1(Rq)}{Rq} \right)^2 \right] \frac{1}{2\beta kd} \right) . \tag{15.105}$$

The non-linear absorption parameter is seen to contain a q-independent and q-dependent term. The q-independent term reflects the change in the central intensity at $q = 0$. The q-dependent term arises from the difference in stimulation at q relative to $q = 0$. The full intensity and polarization dependent expression for $\beta^p_{\text{NL}}(q)$ is obtained from (15.54) as

$$\beta^p_{\text{NL}}(q) = \beta^p \frac{I^p_0 \, d^p_{\text{eff}} \, \mathbb{A}^p}{\mathbb{B} + I^p_0 \, d^p_{\text{eff}} \, \mathbb{A}^p} \left(1 + \ln \left[\left(\frac{2J_1(Rq)}{Rq} \right)^2 \right] \frac{1}{2\beta^p kd} \right) , \tag{15.106}$$

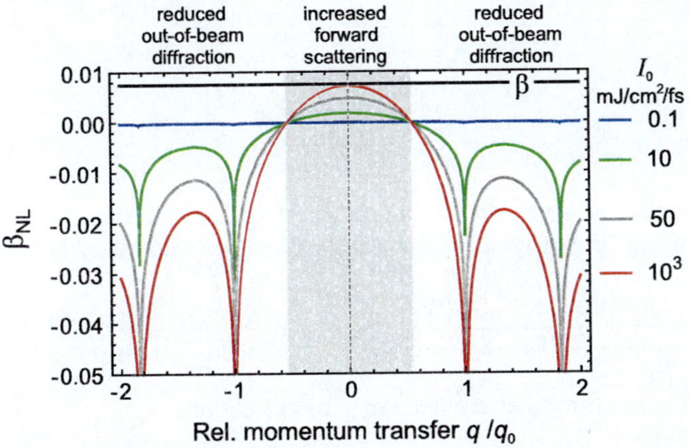

Fig. 15.22 Non-linear parameter $\beta^0_{\rm NL}(q)$ (linear polarization) given by (15.106), plotted for different incident intensities I^0_0 as a function of q/q_0 where $q_0 = 1.22\pi/R$ is the first node of the Airy pattern. The central region $|q| \lesssim 0.5q_0$ highlighted in gray corresponds to *increased* transmission due to stimulated forward scattering relative to the spontaneous case. It is accompanied by a *reduction* of the diffracted intensity in the region $|q| \gtrsim 0.5q_0$

where $\mathbb{A}^p$ and $\mathbb{B}$ are defined in (15.65) and $d^p_{\rm eff}$ by (15.59).

In Fig. 15.22 we have plotted $\beta^0_{\rm NL}(q)$ (linear polarization, $p = 0$) for resonant excitation of a 20-nm-thick Co film in the circular aperture. For simplicity we have taken $d^p_{\rm eff} = d = 20\,$nm. At low intensity we have $\beta_{\rm NL}(q) = 0$ (blue line) and with increasing intensity a q-dependent structure emerges that resembles the Airy pattern but changes sign as a function of q. It is seen to be positive for $|q| \lesssim 0.5q_0$ and negative for $|q| \gtrsim 0.5q_0$. The increase of the central part of $\beta_{\rm NL}(q)$ around $q = 0$ is compensated by a decrease at $|q| \gtrsim 0.5q_0$, resulting in increased central transmission and decreased out-of-beam diffraction. This leads to the evolution of the diffracted Airy pattern shown in Fig. 15.20.

15.10 Polarization Dependent NL Diffraction

For the description of polarization dependent diffraction in the presence of stimulation, we can conveniently utilize the expressions for conventional (spontaneous) diffraction derived in Sect. 8.7. We only need to make the following replacements in the formulas for the spontaneous diffraction patterns

$$\begin{aligned}\delta^p &\rightarrow \delta^p - \delta^p_{\rm NL}\\ \beta^p &\rightarrow \beta^p - \beta^p_{\rm NL}\\ \Delta\delta &\rightarrow \Delta\delta - \Delta\delta_{\rm NL}\\ \Delta\beta &\rightarrow \Delta\beta - \Delta\beta_{\rm NL}.\end{aligned} \tag{15.107}$$

The polarization dependent NL optical parameters are in general also energy and time dependent. The imaginary parts are given by (see (15.68)) and (15.69) or,

$$\beta^p = \frac{\lambda^3 \rho_{\rm a}}{4\pi^2}\frac{\Gamma^p_x}{\Gamma}\frac{(\Gamma/2)^2}{(\hbar\omega-\mathcal{E}_0)^2+(\Gamma/2)^2}, \quad \text{and} \quad \beta^p_{\rm NL}(\tau) = \beta^p\, 2\widetilde{\rho}^{\,p}_{22} \tag{15.108}$$

and the dichroic corrections $\Delta\beta$ and $\Delta\beta_{\rm NL}$ by (15.86) or

$$\Delta\beta = \frac{\beta^+ - \beta^-}{2}, \quad \text{and} \quad \Delta\beta_{\rm NL} = \frac{\beta^+_{\rm NL} - \beta^-_{\rm NL}}{2}. \tag{15.109}$$

By neglect of the non-resonant Thomson contribution, the real parts are related to the corresponding imaginary parts by (15.55) or

$$\delta^p = \frac{2(\hbar\omega-\mathcal{E}_0)}{\Gamma}\beta^p, \quad \text{and} \quad \delta^p_{\rm NL} = \frac{2(\hbar\omega-\mathcal{E}_0)}{\Gamma}\beta^p_{\rm NL} \tag{15.110}$$

and

$$\Delta\delta = \frac{\delta^+ - \delta^-}{2}, \quad \text{and} \quad \Delta\delta_{\rm NL} = \frac{\delta^+_{\rm NL} - \delta^-_{\rm NL}}{2}. \tag{15.111}$$

15.10.1 NL Diffraction by Magnetic Domains

We now consider the presence of magnetic domains as shown in Fig. 8.18 and indicated in Fig. 15.21 and assume that the charge distribution in the film is homogeneous. For simplicity we assume a random magnetic domain pattern with a single characteristic correlation length ℓ which gives rise to a magnetic diffraction ring at a momentum transfer $\mathbf{q}_m = 4\pi/\ell$, schematically indicated in red in Fig. 15.21.

We first consider the case where the beam defining aperture is absent. In this case the pattern in the distant detector plane will consist of a central bright spot due to the transmitted beam and an out-of-beam magnetic diffraction ring. For randomly oriented domains, the centrally transmitted XAS intensity is polarization independent, and the XMCD contrast, given by (15.88), vanishes due to an equal number of domains with $\hat{m}_z = \pm 1$. The centrally transmitted intensity given by (15.83) will increasingly become equal to the incident intensity I_0^p with increasing stimulation.

A magnetic diffraction pattern will be observed for both linear and circular incident polarizations as discussed in Sect. 8.7.3. The patterns are given by (8.52) for *circular polarization* and (8.59) for linear polarization. In the *absence of an aperture* the magnetic patterns are the same, independent of polarization ($p = 0, \pm$) as pointed out in Sect. 8.7.4. By dropping the aperture term $|D_{A_s}(\mathbf{q})|^2$ the spontaneous magnetic domain diffraction pattern is given by

$$I_{\text{diff}}(\mathbf{q}) = \frac{I_0}{\lambda^2 z_0^2}\, \mathrm{e}^{-2\beta k d}\, |D_{\hat{m}_z}(\mathbf{q})|^2 \left[(\Delta\delta)^2 + (\Delta\beta)^2\right] k^2 d^2. \tag{15.112}$$

With the substitutions (15.107) we readily obtain the NL pattern as

$$I_{\text{diff}}(\mathbf{q}) = \frac{I_0}{\lambda^2 z_0^2}\, \mathrm{e}^{-2[\beta-\beta_{\text{NL}}(\mathbf{q})]\, kd}\, |D_{\hat{m}_z}(\mathbf{q})|^2 \times \underbrace{\left\{\left[\Delta\beta - \Delta\beta_{\text{NL}}(\mathbf{q})\right]^2 + \left[\Delta\delta - \Delta\delta_{\text{NL}}(\mathbf{q})\right]^2\right\} k^2 d^2}_{C_{\text{NL}}(\mathbf{q})}. \tag{15.113}$$

Here we have specifically indicated the $\mathbf{q}$-dependence of the NL parameters, as shown in Fig. 15.22 for the Airy pattern. The vector notion of $\mathbf{q}$ allows for the magnetic diffraction pattern to deviate from cylindrical symmetry as shown in Fig. 8.18 due to a preferential domain orientation.

The magnetic domains are illuminated within a "field of view" area A_{eff} given by

$$A_{\text{eff}} = \frac{2\pi}{I_0} \int_0^{\infty} I_0(r)\, r\, \mathrm{d}r, \tag{15.114}$$

so that the total incident power is given by $\mathcal{P}_0 = I_0 A_{\text{eff}}$. For a flat-top incident intensity distribution we simply have $A_{\text{eff}} = \pi R^2$. By redefining $D_{\hat{m}_z}(\mathbf{q})$ as the Fourier transform of the dimensionless magnetization pattern alone, taken as $\hat{m}_z(x, y) = \pm 1$ in the $x - y$ sample plane according to

$$D_{\hat{m}_z}(\mathbf{q}) = \frac{1}{A_{\text{eff}}} \iint_{-\infty}^{\infty} \hat{m}_z(x, y)\, \mathrm{e}^{-\mathrm{i}(q_x x + q_y y)}\, \mathrm{d}x\, \mathrm{d}y, \tag{15.115}$$

we can rewrite (15.113) in the convenient form

$$\boxed{I_{\text{diff}}(\mathbf{q}) = I_0 \frac{A_{\text{eff}}^2}{\lambda^2 z_0^2}\, \mathrm{e}^{-2[\beta-\beta_{\text{NL}}(\mathbf{q})]\, kd}\, C_{\text{NL}}(\mathbf{q})\, |D_{\hat{m}_z}(\mathbf{q})|^2.} \tag{15.116}$$

We can simplify the $\mathbf{q}$-dependent formalism by distinguishing in-beam forward scattering around $q \simeq 0$ and out-of-beam diffraction $q > 0$. We know that the centrally transmitted intensity is enhanced by stimulation at the expense of the diffracted intensity, under the constraint of power conservation. We can therefore divide the total intensity in the detector plane into these two contributions. We furthermore replace the direction dependent momentum transfer $\mathbf{q} > 0$ by performing a cylindrical average around the central beam, so that

$$I_{\text{tot}}(q) = a\, I_{\text{trans}}(q \simeq 0) + b\, I_{\text{diff}}(q > 0), \tag{15.117}$$

where the coefficients a and b depend on the incident intensity and have to satisfy power conservation. The situation is similar to the division of the Airy pattern of an aperture into the central Airy disk intensity and an average over the outer ring intensity, discussed in Sect. 15.9.2. Because of $F_{\text{stim}} + F_{\text{spon}} = 1$ we can associate the coefficients a and b in (15.117) with $a = F_{\text{stim}}$ and $b = F_{\text{spon}}$ to obtain,

$$I_{\text{tot}}(q) = F_{\text{stim}}\, I_{\text{trans}}(q \simeq 0) + F_{\text{spon}}\, I_{\text{diff}}(q > 0). \tag{15.118}$$

The transmitted intensity is given by (15.84) which for linear polarization $p = 0$ is independent of $\hat{m}_z$ and given by

$$I_{\text{trans}}(q \simeq 0) = I_0\, \mathrm{e}^{-2(\beta - \beta_{\text{NL}})kd}. \tag{15.119}$$

The diffracted intensity due to the magnetic domains is given by (15.116)

$$I_{\text{diff}}(q > 0) = I_0\, \frac{A_{\text{eff}}^2}{\lambda^2 z_0^2}\, \mathrm{e}^{-2[\beta - \beta_{\text{NL}}]kd}\, C_{\text{NL}}\, |D_{\hat{m}_z}(q)|^2, \tag{15.120}$$

where all the q-dependent information is contained in $D_{\hat{m}_z}(q)$, which for a given $\mathbf{q}$ has been cylindrically averaged around the centrally transmitted intensity.

We can then divide the total intensity in the detector plane into in-beam transmitted and out-of-beam diffracted components, and by use of (15.116) and demanding power conservation we obtain

$$\boxed{I_{\text{tot}}(q, \hbar\omega) = I_0\, \mathrm{e}^{-2(\beta - \beta_{\text{NL}})kd} \Big\{ \underbrace{F_{\text{stim}}(1 - C_{\text{NL}})}_{\text{central transmission}} + \underbrace{F_{\text{spon}}\, C_{\text{NL}}\, \frac{A_{\text{eff}}^2}{\lambda^2 z_0^2}\, |D_{\hat{m}_z}(q)|^2}_{\text{out-of-beam diffraction}} \Big\}.} \tag{15.121}$$

Through the explicit photon energy dependent notation $I_{\text{tot}}(q, \hbar\omega)$ we indicate that the optical constants implicitly are strongly photon energy dependent. This is particularly important for diffractive imaging since $\beta(\hbar\omega)$ and $\delta(\hbar\omega)$, and their dichroic and NL contributions have their maximum values at different energies as illustrated in Fig. 8.30.

We first discuss the terms in (15.121) by assuming that the incident x-rays of narrow bandwidth are tuned to the XAS maximum, i.e. the maximum of $\beta(\hbar\omega)$. We will come back to the photon energy dependence in Sect. 15.10.3 when we discuss the case of broad bandwidth incident x-rays.

The NL transmission through the film is described by the modified exponential Beer-Lambert law by inclusion of

$$\beta_{\mathrm{NL}} = \beta \, \frac{I_0\, d_{\mathrm{eff}}\, \mathbb{A}}{\mathbb{B} + I_0\, d_{\mathrm{eff}}\, \mathbb{A}}. \tag{15.122}$$

We have $\beta_{\mathrm{NL}} \to 0$ for small I_0 and $\beta_{\mathrm{NL}} \to \beta$ for large I_0 . The intensity-dependent coefficients balance the transmitted and diffracted contributions through their dependence on β_{NL} according to the limits

$$F_{\mathrm{spon}} = \frac{1-\mathrm{e}^{-2(\beta-\beta_{\mathrm{NL}})kd}}{1-\mathrm{e}^{-2\beta kd}}, \quad \overset{\beta_{\mathrm{NL}}\to 0}{\Longrightarrow} 1, \text{ and } \overset{\beta_{\mathrm{NL}}\to \beta}{\Longrightarrow} 0 \tag{15.123}$$

and

$$F_{\mathrm{stim}} = \frac{\mathrm{e}^{2\beta_{\mathrm{NL}}kd}-1}{\mathrm{e}^{2\beta kd}-1}, \quad \overset{\beta_{\mathrm{NL}}\to 0}{\Longrightarrow} 0, \text{ and } \overset{\beta_{\mathrm{NL}}\to \beta}{\Longrightarrow} 1. \tag{15.124}$$

The diffracted pattern defined by the second underbracket in (15.121) is given by the Fourier transform of the binary domain pattern $|D_{\hat{m}_z}(q)|^2$ (see (15.115)). The factor $A_{\mathrm{eff}}^2/(\lambda^2 z_0^2)$ describes the propagation from the illuminated sample area A_{eff} to the detector plane at distance z_0. The strength (intensity) of the pattern depends on the incident intensity I_0 through the product $F_{\mathrm{spon}}\, C_{\mathrm{NL}}$. The magnetic domain response is expressed by C_{NL} which depends on the dichroic spontaneous and NL parameters taken up to second order according to,[7]

$$\begin{aligned} C_{\mathrm{NL}} &= \left\{\lfloor \Delta\beta - \Delta\beta_{\mathrm{NL}} \rfloor^2 + \lfloor \Delta\delta - \Delta\delta_{\mathrm{NL}} \rfloor^2\right\} k^2 d^2 \\ &\overset{\Delta\beta_{\mathrm{NL}}\to 0}{\Longrightarrow} C_{\mathrm{spon}} = [(\Delta\beta)^2 + (\Delta\delta)^2] k^2 d^2, \text{ and } \overset{\Delta\beta_{\mathrm{NL}}\to \Delta\beta}{\Longrightarrow} 0, \end{aligned} \tag{15.126}$$

where we have indicated the limits. Note that C_{spon} represents the strength of the spontaneous magnetic diffraction pattern which is small relative to the centrally transmitted intensity (of order 10%).[8] We obtain the following spontaneous and completely stimulated limits of the intensity in the detector plane

---

[7] The expression given for C_{NL} in [12] differs in the definition of the sign of the NL term from that in [13] which is the same as used here. It reduces in second order of the optical constants to our expression according to

$$\begin{aligned} C_{\mathrm{NL}} &= \cosh\left[2(\Delta\beta - \Delta\beta_{\mathrm{NL}})kd\right] - \cos\left[2(\Delta\delta - \Delta\delta_{\mathrm{NL}})kd\right] \\ &\simeq 2\left[(\Delta\beta - \Delta\beta_{\mathrm{NL}})^2 + (\Delta\delta - \Delta\delta_{\mathrm{NL}})^2\right] k^2 d^2. \end{aligned} \tag{15.125}$$

[8] For example, in the Airy diffraction pattern, the outer rings contain 16.2% and the central disk 83.8% of the total intensity (see Sect. 15.9.1.1).

$$I_{\mathrm{tot}}^{\mathrm{spon}}(q) = \begin{cases} I_{\mathrm{diff}}(q>0) = I_0\,\mathrm{e}^{-2\beta kd}\,\frac{A_{\mathrm{eff}}^2}{\lambda^2 z_0^2}\,[(\Delta\beta)^2+(\Delta\delta)^2]k^2d^2\,|D_{\hat{m}_z}(q)|^2 \\ I_{\mathrm{trans}}(q=0) = I_0\,\mathrm{e}^{-2\beta kd}\left\{1-[(\Delta\beta)^2+(\Delta\delta)^2]k^2d^2\right\} \simeq I_0\,\mathrm{e}^{-2\beta kd} \end{cases} \tag{15.127}$$

$$I_{\mathrm{tot}}^{\mathrm{stim}}(q) = \begin{cases} I_{\mathrm{diff}}(q>0) = 0 \\ I_{\mathrm{trans}}(q=0) = I_0 \end{cases}. \tag{15.128}$$

In the following we will consider two distinct NL diffraction cases for a sample with ferromagnetic domains, extending the spontaneous diffraction case treated in Sect. 8.7.3. Our theory developed above will be compared to experimental data obtained with pulses of coherence times longer than the core hole life time by Wu et al. [12] and pulses with comparable coherence and core hole life times by Chen et al. [13].

15.10.2 Narrow Bandwidth Resonant Case

We first consider an incident beam of *narrow bandwidth and long coherence time*, tuned to the peak of an XAS resonance. This case was studied experimentally by Wu et al. [12] as illustrated in Fig. 15.23a. The same geometry was previously shown in Fig. 8.18 to illustrate the separation of magnetic and charge diffraction from magnetic domains in a Co/Pd film, excited at the Co L_3 resonance.

The experiment was conducted at LCLS using linearly polarized laterally coherent SASE pulses of nominal 50 fs length. The pulses were sent through a monochromator which defined the photon energy of 778 ± 0.1 eV, matching the Co L_3 transition. They were then focused to 10 μm, overfilling the 1.45 μm circular Au aperture located before the samples. As shown in Fig. 15.23a the Au apertures contained five 100 nm reference holes that were completely drilled through and were used to measure the pulse fluences $\mathcal{F}$ incident onto the sample by use of a holography trick [80] (see Sect. 8.9.2).

Simulations of the SASE pulses transmitted through the Gaussian monochromator transmission function of 0.2 eV FWHM yielded total output pulses containing on average 1.5 temporally coherent regions of about 20 fs width [12]. Since each coherent 20 fs subpulse acts independently in a coherent REXS diffraction process, the average *experimental* intensity per total pulse was divided by 1.5 in order to compare it to the *theoretical* coherent intensity I_0 per pulse in units of [mJ/cm²/20 fs]. The so-determined coherent intensity per 20 fs is then simply divided by 20 to obtain the coherent intensity in units of [mJ/cm²/fs] that will be used below.

The sample was a Co/Pd multilayer, containing 20 nm of Co, with a nanoscale magnetic worm domain structure as shown in Fig. 15.23a. The diffraction pattern was recorded with a position sensitive CCD detector, with the total accumulated charge per pixel forming the diffraction pattern. No temporal gating of the detector and no electronic coincidence circuit were employed.

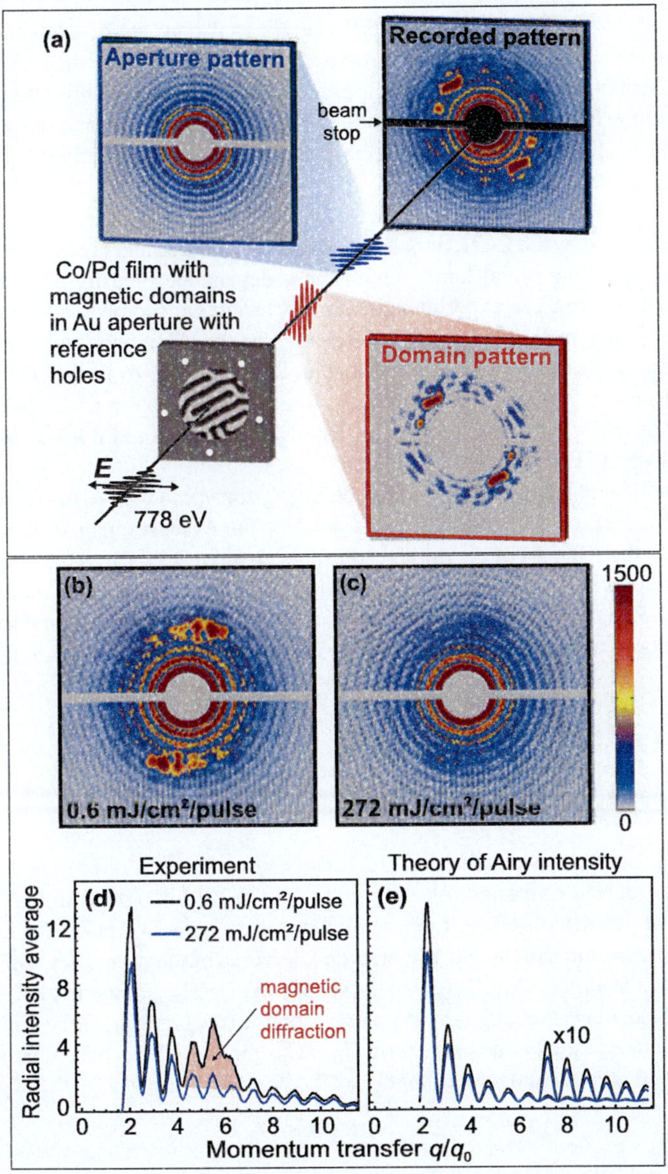

Fig. 15.23 **a** Coherent imaging experiment of a magnetic Co/Pd film with magnetic worm domains in a circular Au aperture illuminated by laterally coherent monochromatized XFEL pulses (FWHM 0.2 eV bandpass) tuned to the Co L_3 resonance at 778 eV [12]. A beam stop in front of the CCD detector eliminated the intense central part of the diffraction pattern. The linearly polarized incident radiation allows separation of the diffraction pattern from the Au aperture and the magnetic domains due to the polarization dependence of charge and magnetic scattering as shown. **b** Experimentally recorded diffraction pattern by averaging over many low-fluence shots (0.6 mJ/cm^2/pulse). **c** Pattern for a single high-fluence shot (272 mJ/cm^2/pulse). **d** Radial intensity average of the patterns in (**b**) and (**c**) for the two fluence values. **e** Simulation of the change in the Airy pattern of the aperture, assuming the presence of stimulated resonant scattering [25, 26]

As indicated in Fig. 15.23a, the diffracted signal consists of a magnetic part, determined by the domains and a charge part due to the aperture and the homogeneous charge distribution in the film, which according to Fig. 8.12b only attenuates the intensity of the first order diffraction pattern. For the employed linear polarization, the magnetic and charge patterns are separable [81] as indicated in the figure, previously discussed in Sect. 8.7.4.

In Fig. 15.23b, c we show the change of the diffraction pattern with increasing fluence, recorded by the CCD, of both the magnetic domains and the circular aperture [12]. Corresponding radial line scans of the q-dependent intensity distribution are shown in (d) and (e). The experimental intensity profiles are compared to simulations using the theory in Sect. 15.9.2 for the change of the Airy rings, whose decrease with intensity was previously shown in Fig. 15.20. Owing to the used beamstop, indicated in Fig. 15.23a, which prevented the saturation of the CCD detector, the expected intensity enhancement in the forward direction could not be observed in this experiment [12].

The incident pulses act on the atoms coherently for a time $\tau \simeq 20\,\mathrm{fs}$ which is much longer than the core hole decay time $\tau_\Gamma = 1.5\,\mathrm{fs}$. For exact resonance excitation we can neglect the real part of the optical constants given in Sect. 15.10 (see Fig. 7.25a). Furthermore, we can use the analytical solution of the Bloch equations for $\widetilde{\rho}^{\,p}_{22}(\tau)$ given by (15.75) to obtain $\beta^{p}_{\mathrm{NL}}(\tau)$. The out-of-beam diffraction pattern in the presence of stimulation, normalized to the spontaneous diffraction pattern, is then given by

$$\boxed{\left[\frac{I^{\mathrm{stim}}_{\mathrm{diff}}}{I^{\mathrm{spon}}_{\mathrm{diff}}}\right]_{q>0} = \underbrace{\frac{1-\mathrm{e}^{-2(\beta-\beta_{\mathrm{NL}})kd}}{1-\mathrm{e}^{-2\beta kd}}}_{F_{\mathrm{spon}}}\;\underbrace{\frac{(\Delta\beta-\Delta\beta_{\mathrm{NL}})^2}{(\Delta\beta)^2}}_{C_{\mathrm{NL}}/C_{\mathrm{spon}}},} \tag{15.129}$$

where C_{NL} and C_{spon} are defined in (15.126).

The normalized diffracted intensity given by (15.129) is plotted in Fig. 15.24 as a *red* curve superimposed on the experimental data points of [12]. We have also included in *blue* the simultaneously recorded decrease of the *outer Airy rings*, previously shown in Fig. 15.20c. The blue curve is given by F_{spon} alone in (15.129). The difference between the *magnetic* domain response (red) and the Airy ring *charge* response (blue) is solely due to the term $C_{\mathrm{NL}}/C_{\mathrm{spon}}$ in (15.129). There is good agreement between theory and experiment.

15.10.3 Broad Bandwidth Case

In order to demonstrate the effect of *temporal coherence* in resonant magnetic diffraction, we have also included in Fig. 15.24 in *green* color the diffraction part of the study by Chen et al. [13], whose experimental geometry and principal results were previously shown in Fig. 1.30. In this case, the sample was not enclosed by a circular aperture and hence the diffraction pattern was entirely due to the magnetic domains

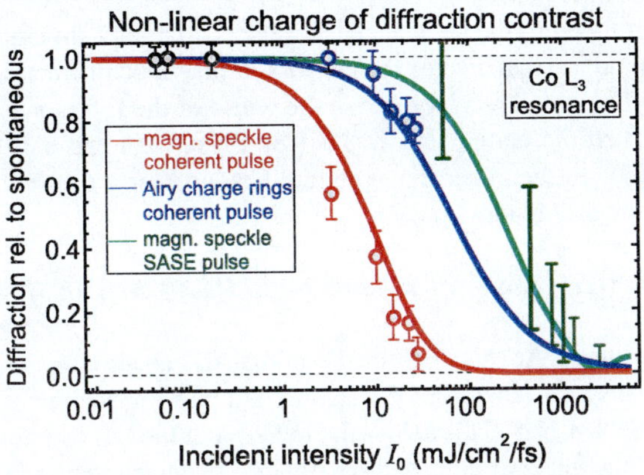

Fig. 15.24 Change of the diffracted magnetic speckle (red) and Airy ring charge (blue) contrast as a function of incident intensity for a Co/Pd multilayer containing 20 nm Co in a circular aperture, resonantly excited with monochromatized 50 fs SASE pulses of linear polarization [12]. The plotted data points and error bars are from [12], representing the ratio of the total NL contrast normalized by the spontaneous contrast. The solid *blue* curve is taken from Fig. 15.20c and corresponds to the F_{spon} term alone in (15.129). The *red* curve is calculated with the same parameters according to the full expression (15.129). In both cases the coherence time was taken to be $\tau_{\text{coh}} = 20$ fs. The *green* data points represent averaged magnetic diffraction intensities recorded for similar Co/Pd samples with 25 fs SASE pulses [13]. The simultaneously recorded central transmission increase was previously shown in Fig. 15.15. The green curve is a simulation discussed in the text

in a film that was uniform from a charge distribution point of view. The intensity-dependent transmission of the central beam was already discussed in Sect. 15.7.1, and the results were shown in Fig. 15.15.

In contrast to the monochromatic pulses used for the data shown in red (magnetic diffraction) and blue (Airy ring or charge diffraction), the experimental green data bars in Fig. 15.24 correspond to 25 fs SASE pulses, whose internal structure in the energy domain was shown in Fig. 15.14b.[9]

In particular, the red and green curves and data points in Fig. 15.24 both represent the change in *magnetic diffraction*. The red curve represents the change in diffraction for an incident pulse of narrow monochromatized bandwidth (0.2 eV) tuned to the peak of the L_3 resonance whose experimental width is about 1.6 eV as shown in Fig. 10.17b. Hence it optimizes the *pure resonance* response since all incident photons are tuned to the resonance. The green curve is shifted by a factor of about 30 to higher intensity relative to the red curve. This shift is larger than that expected from the ratio of the temporal coherence lengths alone, $20/1.8 \simeq 11$.

In contrast to the narrow monochromatized bandwidth of 0.2 eV used for the red data, the green data were recorded with SASE pulses of large 4 eV energy width (see Fig. 15.14b). The additional shift is due to the fact that most of the incident

[9] The experimental arrangement shown in Fig. 15.14a was actually part of a more complete arrangement shown previously in Fig. 1.30.

photons are not tuned to the resonance, so it takes a larger I_0 to create a significant upper state population that creates the NL loss of magnetic diffraction. In particular, when the SASE bandwidth is larger than the width of the L_3 resonance, the slightly off-resonance dichroic component $(\Delta\delta)^2$ (see Figs. 7.25a and 8.30) also enters in addition to $(\Delta\beta)^2$ in the general magnetic diffraction term C_{NL} given by (15.126) which more explicitly reads

$$C_{NL}(\hbar\omega) = \left\{[\Delta\beta(\hbar\omega) - \Delta\beta_{NL}(\hbar\omega)]^2 + [\Delta\delta(\hbar\omega) - \Delta\delta_{NL}(\hbar\omega)]^2\right\} k^2 d^2. \quad (15.130)$$

The peak position of $\Delta\delta(\hbar\omega)$ lies in the tail of the absorption resonance (i.e. of the β components), and therefore the sample is more transparent at that energy as pointed out in Sect. 8.11.5. This delays the increase of the NL component $\Delta\delta_{NL}$ with I_0, causing the spontaneous contribution $\Delta\delta(\hbar\omega)$ to persist to higher intensities and delay the disappearance of magnetic diffraction.

For the Co L_3 case, the 4 eV energy width also covers both the position of the XAS peak at 778 eV and the position of the XES peak at 776 eV (see Fig. 13.15). The incident SASE pulse can therefore stimulate both REXS and RIXS decays. The green curve in Fig. 15.24 was calculated by accounting for the $\Delta\delta(\hbar\omega)$ diffraction contribution, and taking into account that for Co metal the stimulated REXS channel is reduced by a factor of 2 because of an equally strong stimulated RIXS channel. This will be shown in the following section.

15.11 Stimulated Resonant Inelastic X-Ray Scattering

In the following sections we will discuss the intensity-dependent evolution of the RIXS process, which combines the XAS "up" and XES "down" processes as previously discussed in Sect. 13.10.3.2. We will begin by outlining the concept of stimulated RIXS and review its experimental observation in Co metal by Higley et al. [7]. We then use a simple extension of the BR theory to describe the REXS and RIXS channels and their saturation. In the end, we will compare our theoretical formulation to the Co L_3 data and their interpretation in the Kramers-Heisenberg-Dirac (KHD) formulation given in [7].

15.11.1 Stimulated L_3 REXS and RIXS for Co Metal

The REXS and RIXS processes associated with L_3 excitation in Co metal are illustrated in a configuration picture (see Sect. 13.2.2) in Fig. 15.25a, and the dipole matrix elements for the up/down $2p \leftrightarrow 3d$ transitions are shown in (b) in a one-electron picture.

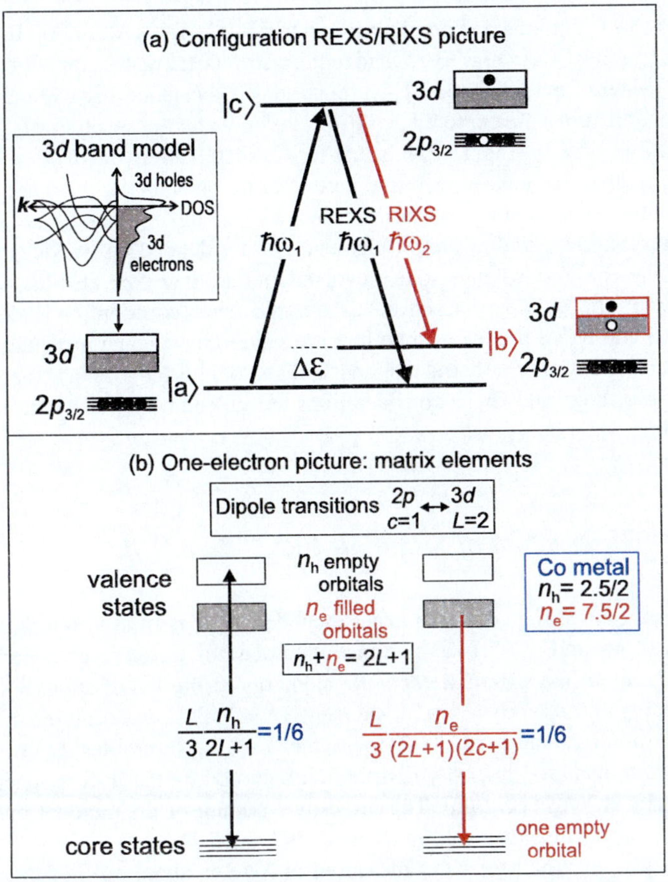

Fig. 15.25 **a** Configuration-based transitions for L-shell $2p \leftrightarrow 3d$ transitions between core and valence states for transition metals. Empty valence states are shown as empty boxes and filled valence states as filled boxes. Here $|a\rangle$ is the electronic ground state, $|c\rangle$ the intermediate core-excited state, and the final states are either the ground state $|a\rangle$ (REXS, black) or an exited valence state $|b\rangle$ of energy $\Delta\mathcal{E}$ above the ground state (RIXS, red). **b** Orbital degeneracies (without spin) of the electronic states involved in the transitions, and values of the angular transition matrix elements that take into account the degeneracies. For Co metal, there are $2n_{\rm h} = 2.5$ empty $3d$ spin orbitals and $2n_{\rm e} = 7.5$ filled $3d$ spin orbitals, as indicated [38]. This leads to the same up/down angular dipole matrix elements. The radial $2p \leftrightarrow 3d$ dipole matrix elements are also the same

Figure 15.25a summarizes previous diagrams, shown for XAS in Fig. 10.14, XES in Fig. 12.9, and RIXS in Fig. 13.15, and require no detailed discussion. For the Co L_3 case, the center of the XAS and REXS transitions is located at 778 eV, and the XES and RIXS transitions are centered at 776 eV as shown by the spectra in Fig. 13.15b.

The key message of Fig. 15.25b is that for Co metal, all up/down $2p \leftrightarrow 3d$ total transition matrix elements work out to be the same. The angular parts are the same because of the compensating degeneracies of the lower and upper states, as illustrated in the figure, and the radial matrix elements do not depend on the degeneracy or filling of the core and valence states involved and are therefore also the same (see Sect. 12.4.1). The L_3 ($2p_{3/2}$) fraction of the total L-shell contribution is simply 2/3. The REXS and RIXS processes therefore are seen to have equal probabilities. The two channels are furthermore independent and do not interfere because the states $|a\rangle$ and $|b\rangle$ are orthogonal. These considerations are indeed born out by experimental results which we will discuss now.

15.11.2 *Observation of Stimulated RIXS in a Solid*

The relevant part of the complete experimental arrangement used by Higley et al. [7] is that shown in Fig. 15.14. The transmitted photons were energy separated and detected by an *in-line spectrometer* with an energy resolution of about 0.8 eV. The photon energy of the about 4 eV wide incident SASE pulses was centered around the Co L_3 absorption resonance of 778 eV to produce Co $2p_{3/2}$ core holes. At low incident fluence, the transmitted intensity distribution recorded by the in-line spectrometer is the dispersed XAS spectrum. The intensity structure of the incident pulses was normalized out by the pulse splitting scheme in Fig. 15.14.

A typical transmitted intensity measured in XAS is about 30% of the incident intensity. In contrast, the *spontaneous* REXS and RIXS intensities are not only much weaker, but they are also emitted into a 4π solid angle. An upper estimate of the spontaneous RIXS intensity seen by a spectromter with finite acceptance angle is obtained by the product of the fluorescence yield Y_f times the solid angle of acceptance of the RIXS spectrometer, which is of order 10^{-5} of 4π steradian [82–86].

In our case, the Co LVV x-ray emission has a fluorescence yield of $Y_f = 8 \times 10^{-3}$ so that the spontaneous RIXS signal is only of order $10^{-7} - 10^{-6}$ of the transmitted signal.[10] This reveals the near impossibility of measuring the tiny *spontaneous* RIXS component of the total transmitted intensity. In practice, spontaneous RIXS is therefore measured which a spectrometer whose acceptance cone is at an angle with the direct beam. On the other hand, the transmission geometry is ideally suited to demonstrate the orders of magnitude enhancement of RIXS upon stimulation by the incident beam itself. The stimulated enhancement is similar to that of REXS, and the stimulated REXS and RIXS signals emitted in the forward direction can be

[10] The L_3 contribution is given by 2/3 of this value, which is the core state degeneracy ratio $2p_{3/2}/(2p_{3/2} + 2p_{1/2}) = 4/6$.

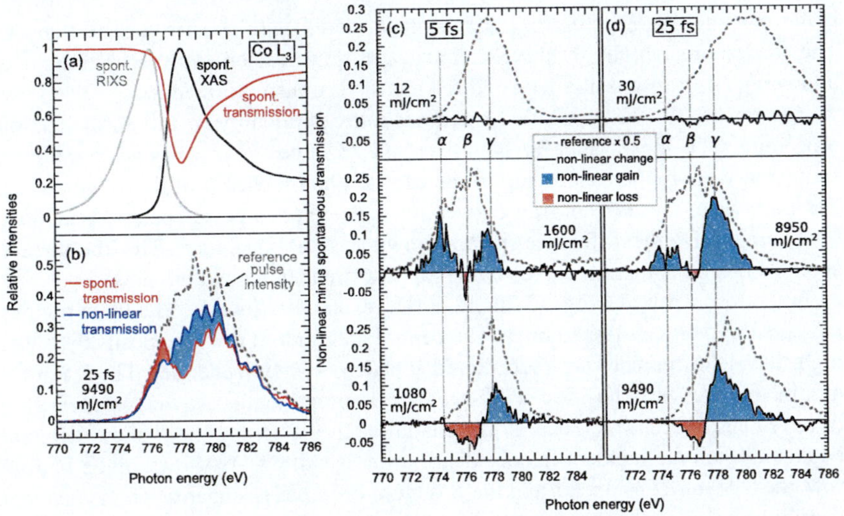

Fig. 15.26 **a** Comparison of the scaled spontaneous Co L_3 RIXS spectrum (gray) for Co metal [87] and the Co/Pd XAS spectrum (black), both recorded at synchrotron light sources (low-fluence limit). The red spectrum is the transmission version of the XAS spectrum. **b** Example of data extraction and normalization. The dashed gray line is the reference spectrum of a 25 fs pulse of 9490 mJ/cm^2 fluence transmitted through the SiN window, multiplied by 0.55 to account for the constant non-resonant absorption of the Co/Pd sample. The blue curve is the measured transmission spectrum through the Co/Pd sample at the stated high fluence. The red curve is the spontaneous (low fluence) transmission spectrum, obtained by multiplying the red spectrum in (**a**) by the dashed gray reference spectrum. Light blue shaded areas indicate non-linear gain and red areas non-linear loss. Panels (**c**) and (**d**) show as dashed lines the reference spectra transmitted through the SiN for 5 fs and 25 fs pulses for different pulse shapes and fluences. The associated *transmission difference spectra* are shown as solid black lines. They were obtained by subtraction of the spontaneous low-fluence spectra from the non-linear high-fluence spectra for the respective transmission curves. The shading of areas corresponds to the procedure (blue minus red curves) illustrated in (**b**). Each spectrum is an average of many shots as discussed in the text. The centers of three regions with non-linear response are denoted by dashed vertical lines and labeled α, β, and γ

directly compared by use of an in-line spectrometer capable of separating the photon energies (2 eV for Co) of the two signals.

Experimental results of [7] are shown in Fig. 15.26. Panel (a) shows the spontaneous RIXS (gray) and XAS (black) spectra recorded with synchrotron radiation taken from Fig. 13.15b. The shown spectra are arbitrarily scaled to the same unit peak value. Also shown as a red curve is the corresponding spontaneous transmission spectrum with a 32% transmission at the L_3 resonance.

Panel (b) illustrates the extraction of the transmission difference spectra to obtain the NL relative to the spontaneous response. The dashed gray line is the reference transmission spectrum of a 25 fs pulse. The red curve is the calculated spontaneous (low fluence) transmission spectrum through the Co sample, obtained by multiplying the red curve in (a) by the reference pulse transmission spectrum. The blue curve is the transmission spectrum measured for the indicated pulse length and fluence. The shaded areas highlight the NL changes in transmission, with light blue areas indicating NL transmission *gain* and red areas NL transmission *loss*.

The top row in Figs. 15.26c, d demonstrates the quality of the normalization procedure. The shown data for both 5 fs and 25 fs pulses correspond to multiple shots that were binned according to the central photon energy of the x-ray pulses. For each case, the dashed gray curves are the reference pulse spectra, scaled by 0.5 to emphasize the difference spectra shown as solid black lines. The low-fluence spontaneous transmission curves in the top row are identical within noise.

In the lower rows of Fig. 15.26c, d, a NL response is found in the three regions indicated by vertical dashed lines. The central and bottom rows, respectively, show high-fluence difference spectra, centered at different photon energies. The feature γ around the XAS peak position at 778 eV appears prominently as a *transmission gain* (blue) in all difference spectra. In contrast, the NL features α and β show different behavior when the incident fluence distribution is shifted. As shown in the bottom row, the α feature disappears and the β feature becomes stronger when the incident distribution shifts to higher energy.

Figure 15.27 shows a simple modeling of the observed results by assuming three types of non-linear effects. Stimulated REXS and RIXS are accompanied by a contribution due to electron redistribution previously shown in Figs. 15.3a, b.

In Fig. 15.27a we show the assumed spontaneous RIXS (gray), XAS (black), and transmission (red) spectra. The RIXS spectrum has been arbitrarily scaled to unit peak height. Also shown are the relative energy distributions and sizes of the three NL contributions, assumed to represent those at the highest incident fluences in Fig. 15.26. We assume that the stimulated RIXS (magenta) and REXS (blue) contributions have the shape of the spontaneous spectra in (a) and have the *same size* of 18% of the resonant XAS peak value and area. The electron redistribution (green) is modeled by the difference of two Fermi-Dirac distributions as illustrated in Fig. 15.1c. It has a peak value of 20% of the resonant XAS value and 8% of the integrated XAS area.

Depending on the energy distribution of the incident pulse, the three NL contributions contribute with different shapes and intensities as shown in Figs. 15.27b, c, modeled to reflect the two 25 fs high-fluence cases on the right side of Fig. 15.26. Finally, we show in Fig. 15.27d the change of the spontaneous transmission spectrum in (a), shown again in red, by adding to it the three NL contributions in Fig. 15.27b. The total NL transmission spectrum (black) exhibits strong NL effects whose spectral distortions are indicated by arrows.

The close quantitative agreement of our simulations with experimental results for corresponding incident pulse distributions is underscored by their direct comparison on the same vertical scales in Fig. 15.28. Of particular importance is that the stimulated RIXS spectrum, although, partly obscured by electron redistribution, appears on the same intensity scale as the change in elastic transmission. This arises from a

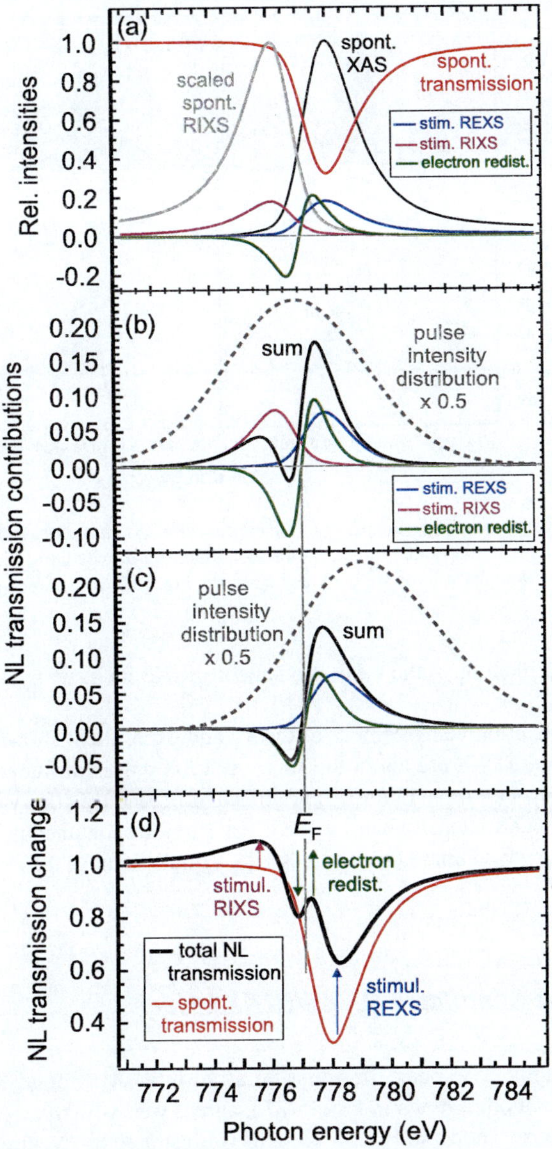

Fig. 15.27 **a** Model Co L_3 spontaneous RIXS (gray), XAS (black) and transmission (red) spectra, and three assumed non-linear (NL) contributions, stimulated RIXS (magenta), electron redistribution (green), and stimulated REXS (blue). The sizes of the NL contributions are referenced to the unit value of the XAS peak, and their shapes are discussed in the text. **b** Assumed incident pulse reference distribution (dashed), which allows all three NL channels to contribute to the sum shown in black. **c** Shifted pulse reference distribution (dashed) which eliminates the stimulated RIXS contribution. The remaining two add up the sum shown in black. **d** Change of the spontaneous transmission spectrum (red) taken from (**a**), to the total NL transmission one (black) for a wide incident energy distribution. The black curve is the red curve plus the the sum of all three NL contributions. Colored arrows indicate NL transmission gain (up arrows) and loss (down arrows) caused by the respective NL channels

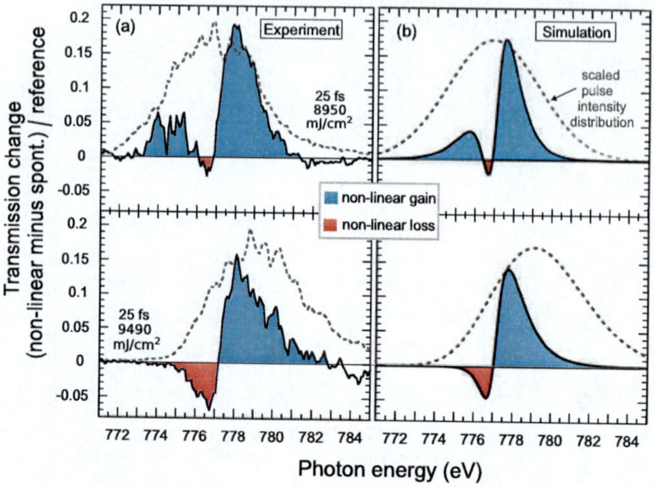

Fig. 15.28 **a** Experimental high-fluence non-linear effects for 25 fs pulses taken from Fig. 15.26d. **b** Simulated results with our model for similar incident intensity distributions, plotted on the same scale. The shown simulated curves are the black curves in Fig. 15.27b, c

stimulated amplification of the weak spontaneous RIXS intensity by a factor of order 10^6 (see below).

The good quantitative agreement of theory and experiment shows that the stimulated RIXS and REXS are about the *same* for 25 fs pulses of fluence in the range $9 - 9.5 \times 10^3$ mJ/cm^2 or an intensity around 350 mJ/cm^2/fs. At this value, the size of the stimulated RIXS intensity has a value of 18% of the spontaneous XAS intensity. In the following we compare these values with calculations for the stimulated RIXS rate.

15.11.3 The Stimulated REXS/RIXS Model

The approximately 65% peak transmission at an intensity around 350 mJ/cm^2/fs, predicted by our model shown in Fig. 15.27d, agrees well with the value at the same intensity previously shown in Fig. 15.15. This is illustrated by the direct comparison in Fig. 15.29.

Saturation occurs when at high incident intensity the population of the upper state, which is the intermediate state $|c\rangle$ in Fig. 15.25a, reaches 1/2. When the incident photons have an energy distribution wider than $\Delta\mathcal{E}$ in Fig. 15.25a, they can stimulate both REXS and RIXS decays. In the forward direction, the "impulsive" stimulation by the incident beam itself can lead to equally strong REXS and RIXS intensities. The respective cross sections [see (15.43) and (15.46)] are then given by

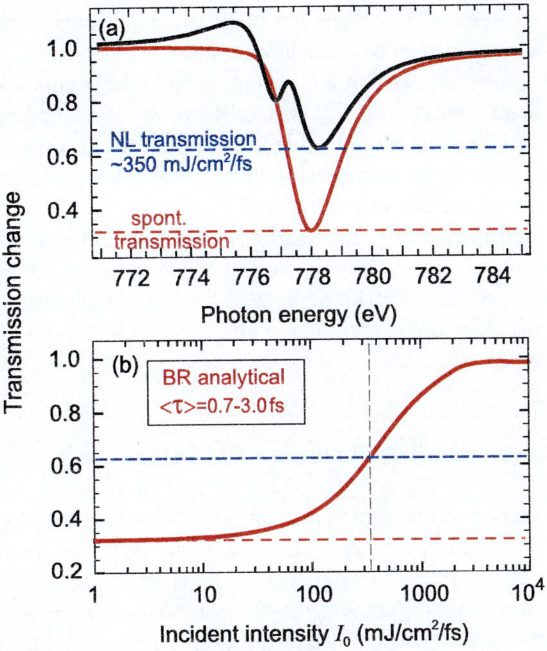

Fig. 15.29 **a** Spontaneous and NL transmission spectra taken from Fig. 15.27d, with horizontal dashed red and blue lines indicating the change in peak transmission values. **b** Transmission curve replicating that in Fig. 15.15 in the same color

$$\sigma_{\text{XAS}}^{\text{tot}} = \sigma_{\text{XAS}}^{\diamond} - \sigma_{\text{XAS}}^{\text{NL}} \tag{15.131}$$

$$\sigma_{\text{XAS}}^{\text{NL}} = \sigma_{\text{RIXS}}^{\text{NL}} + \sigma_{\text{REXS}}^{\text{NL}} \tag{15.132}$$

$$\sigma_{\text{RIXS}}^{\text{NL}} = \sigma_{\text{REXS}}^{\text{NL}} = \sigma_{\text{XAS}}^{\diamond} \widetilde{\rho}_{22}(\tau). \tag{15.133}$$

The distinction between the REXS and RIXS photons whose energy only differs by $\Delta\mathcal{E}$ was not considered in the study by Chen et al. [13], so that the change in transmission illustrated in Fig. 15.15 was attributed to stimulated REXS alone. As revealed by a more detailed data analysis by Higley et al. [7], the stimulated REXS and RIXS channels contribute equally as shown in Fig. 15.27a with an additional contribution from electron rearrangement around the Fermi level that is included in Fig. 15.29a.

15.11.3.1 Photon Stimulation Versus Electron Reshuffling in RIXS

The BR theory shown in Fig. 15.29b properly accounts for the peak transmission shown in (a), although it does not explicitly include electron-electron reshuffling. Our BR formulation of stimulated REXS/RIXS scattering corresponds to the *direct RIXS* model (see Sect. 13.2.4) for the intermediate state. In this formalism the atomic response is effectively described in a one-electron picture as the transition of a single active electron between atomic states.

One may envision the electron reshuffling contribution to arise from an *indirect RIXS* process [88–91], where the decay is accompanied by an intra-valence band shake-up that reshuffles electrons around the Fermi level. The treatment of the indirect RIXS case is considerably more complicated, similar to the description of inelastic electron-electron scattering of Auger electrons. It is conceivable that the two pictures are linked in higher order QED. The lower intensity results of Sect. 15.3.1 together with our present higher intensity results suggest that resonant NL photon-solid interactions may first be dominated by valence electron redistribution (indirect RIXS), with the photon stimulation effect (direct RIXS) increasing with photon degeneracy parameter.

15.11.3.2 The Stimulated RIXS Gain

We can now quantitatively compare the stimulated RIXS gain in the forward direction measured by Higley et al. [7] with our theoretical model. This is done in Fig. 15.30, where we show as magenta (5 fs) and green (25 fs) filled circles the RIXS probabilities taken from [7] at the highest measured intensity of about 350 mJ/cm^2/fs. The plotted probabilities are defined by the ratio of the RIXS and the spontaneous XAS intensities. The spontaneously XAS intensity, which served as a reference, is shown on top as a black horizontal line of unit value.

The lowest horizontal line in Fig. 15.30 indicates the L_3 fraction of the fluorescence yield (FY) given by $Y_f(2p_{3/2}) = \frac{2}{3}Y_f \simeq 5.3 \times 10^{-3}$, which corresponds to

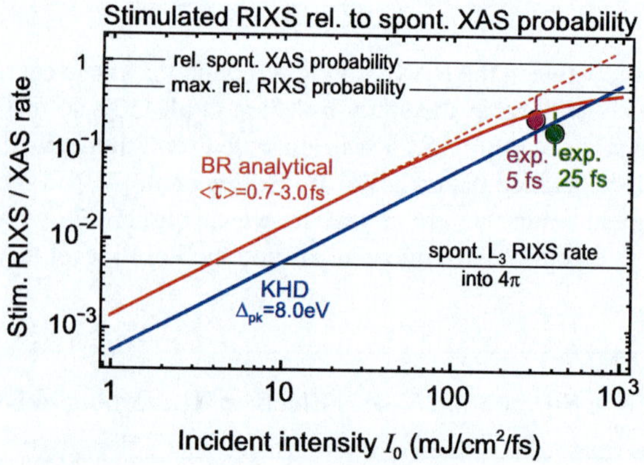

Fig. 15.30 Comparison of the I_0-dependent experimental and theoretical RIXS probabilities, defined as $\sigma^{\mathrm{NL}}_{\mathrm{RIXS}}/\sigma^{\mathrm{spon}}_{\mathrm{XAS}}$. Experimental data points around $I_0 = 350$ mJ/cm^2/fs, taken from [7], are shown as magenta (5 fs pulses) and green (25 fs) data points. The *red* theory curve is calculated with (15.134), the *dashed red* curve with (15.136) and the *blue* curve with (15.138), with parameters discussed in the text

the spontaneous L_3 RIXS signal emitted into 4π. In practice, only a tiny fraction, $\sim 10^{-5}$ of 4π, of this intensity falls into the detector acceptance cone. The main gain of the RIXS intensity seen by the detector is due to the fact that the stimulated RIXS intensity remains within the direction of the incident photons. It is therefore directly driven into the detector. The maximum stimulated RIXS intensity is only 1/2 of the spontaneous XAS value, as indicated by another horizontal line. Stimulated RIXS and REXS each contribute half of the total NL intensity. By comparison to the maximum value of 1/2, indicated by a horizontal line in Fig. 15.30, we see that at $I_0 = 350\,\mathrm{mJ/cm^2/fs}$ the calculated stimulated RIXS intensity shown as a *solid red* curve has nearly saturated.

The solid red curve in Fig. 15.30 was calculated assuming the Co metal value $\sigma^{\diamond}_{\mathrm{XAS}} = 6.25\,\mathrm{Mb}$ or transition matrix element $\Gamma_x = 0.34\,\mathrm{meV}$ with the cross section given by,

$$\text{BR theory}: \sigma^{\mathrm{NL}}_{\mathrm{RIXS}} = \sigma^{\diamond}_{\mathrm{XAS}}\, \widetilde{\rho}_{22}(\tau) = \sigma^{\diamond}_{\mathrm{XAS}} \frac{\mathcal{G}_{\mathrm{eff}}\,(\Gamma_{\mathrm{R}})^2/4}{\Gamma^2/4 + \mathcal{G}_{\mathrm{eff}}\,(\Gamma_{\mathrm{R}})^2/2}\, g(\tau) \tag{15.134}$$

The excited population $\widetilde{\rho}_{22}(\tau)$ is given by (15.72) with $p = 0$ or (15.75), with the time-dependent part expressed by

$$g(\tau) = 1 - \frac{24\Gamma\hbar\omega_{\mathrm{R}}\left[1 - \cos(\omega_{\mathrm{R}}\tau)\,\mathrm{e}^{-3\Gamma\tau/(4\hbar)}\right] - \left[16(\hbar\omega_{\mathrm{R}})^2 - 9\Gamma^2\right]\sin(\omega_{\mathrm{R}}\tau)\,\mathrm{e}^{-3\Gamma\tau/(4\hbar)}}{\omega_{\mathrm{R}}\tau\left[16(\hbar\omega_{\mathrm{R}})^2 + 9\Gamma^2\right]}. \tag{15.135}$$

As for the curves in Figs. 15.15 and 15.29b, the $g(\tau)$ expression was integrated over the range $0.7 \le \tau \le 3.0\,\mathrm{fs}$, resulting in the solid red curve in Fig. 15.30.

The dashed red curve in Fig. 15.30 was calculated in the low intensity approximation $\Gamma_{\mathrm{R}} \ll \Gamma$ and therefore does not exhibit saturation. As shown for the atomic case by (14.44), one can obtain a time-dependent analytical solution at low intensity which contains the detuning energy $\Delta\mathcal{E} = \hbar\omega - \mathcal{E}_0$. By use of $\Gamma_{\mathrm{R}}^2 = n_\Gamma \Gamma_x \Gamma$ (see (15.58)) and integrating (14.44) over time we obtain the low intensity expression of (15.134) in the same time notation τ as

$$\text{BR theory for } \Gamma_{\mathrm{R}} \ll \Gamma:\ \sigma^{\mathrm{NL}}_{\mathrm{RIXS}} = \sigma^{\diamond}_{\mathrm{XAS}}\, \mathcal{G}_{\mathrm{eff}}\, \underbrace{n_\Gamma \frac{\Gamma_x}{\Gamma} \frac{\Gamma^2/4}{(\hbar\omega - \mathcal{E}_0)^2 + \Gamma^2/4}\, h(\tau)}_{[\rho_{22}(\tau)]_{\mathrm{atom}}}, \tag{15.136}$$

where we have identified by an underbracket the low intensity single atom population $[\rho_{22}(\tau)]_{\mathrm{atom}}$, with the curly brackets in (14.44) integrated up to a time τ given by

$$h(\tau) = 1 + \frac{\hbar\left(1 - \mathrm{e}^{-\Gamma\tau/\hbar}\right)}{\tau\Gamma} - \frac{\hbar\Gamma\,\mathrm{e}^{-\Gamma\tau/(2\hbar)}}{\tau\left[\Delta\mathcal{E}^2 + \Gamma^2/4\right]} \left\{ \mathrm{e}^{\Gamma\tau/(2\hbar)} - \cos\left[\frac{\tau\Delta\mathcal{E}}{\hbar}\right] + \frac{2\Delta\mathcal{E}}{\Gamma} \sin\left[\frac{\tau\Delta\mathcal{E}}{\hbar}\right] \right\} \tag{15.137}$$

The factor $\mathcal{G}_{\text{eff}}$ in (15.136), which for Co L_3 excitation is $\mathcal{G}_{\text{eff}} = 219$ (see Table 15.2), is seen to simply multiply the *actual* population of the individual atoms in the film. For solids, both REXS and RIXS are enhanced in the forward direction by the coherent enhancement factor $\mathcal{G}_{\text{eff}} \simeq \lambda^2 \rho_a d_{\text{eff}}/(4\pi)$ in (15.40). The dashed red line in Fig. 15.30 was calculated with (15.136) for $\Delta\mathcal{E} = 0$ with the same integration over τ as for the red curve. By comparison of the two curves we see that the validity range of the low intensity BR theory is limited to the overlap region $I_0 < 100\,\text{mJ/cm}^2/\text{fs}$) of the red and dashed red curves in Fig. 15.30.

Finally we show as a *blue* line in Fig. 15.30 the result of the simple KHD direct RIXS formulation (13.83) used in [7] given by

$$\boxed{\text{KHD theory:}\quad \sigma^{\text{NL}}_{\text{RIXS}} = n_{p\mathbf{k}} \frac{\Gamma_x}{\Gamma} \sigma^{\text{at}}_{\text{XAS}} = \frac{\lambda^3}{c\,\Delta_{p\mathbf{k}}} I_0\, Y_{\text{f}}\, \sigma^{\text{at}}_{\text{XAS}},} \tag{15.138}$$

where on the right side we have used $n_{p\mathbf{k}} = \lambda^3 I_0/(c\Delta_{p\mathbf{k}})$ (see (15.58)) and identified the fluorescence yield Y_{f} with Γ_x/Γ.

In [7], the stimulated response was calculated with (15.138) by use of a Lorentzian *atomic cross section* to avoid the complications of solid-state effects, which enter through changes of the resonance width and intensity (see Fig. 15.10), exponential intensity decay through the film and a collective enhancement factor $\mathcal{G}_{\text{eff}}$. The blue curve is calculated with (15.138) and the parameters of [7], $\sigma^{\text{at}}_{\text{XAS}} = 41\,\text{Mb}$ corresponding to $\Gamma_x = 2.2\,\text{meV}$, and a bandwidth of $\Delta_{p\mathbf{k}}$=8 eV, except we used the value $Y_{\text{f}}(2p_{3/2}) = \frac{2}{3} Y_{\text{f}} \simeq 5.3 \times 10^{-3}$. Similar to the red dashed curve, the blue curve does not saturate and somewhat fortuitously predicts the magenta and green experimental values.

It is interesting to compare the low intensity BR expression (15.136) for a solid with the atomic KHD expression (15.138). By use of the relationship $n_{p\mathbf{k}} = n_\Gamma \frac{2\pi^2\Gamma}{\Delta_{p\mathbf{k}}}$ given by (14.80) which links the number of photons n_Γ of bandwidth Γ in the atomic coherence volume with the number of photons $n_{p\mathbf{k}}$ of bandwidth $\Delta_{p\mathbf{k}}$ in mode volume, we obtain at resonance

$$\begin{aligned} [\sigma^{\text{NL}}_{\text{RIXS}}]_{\text{BR}} &= n_\Gamma \frac{\Gamma_x}{\Gamma} h(\tau)\, \mathcal{G}_{\text{eff}}\, \sigma^{\diamond}_{\text{XAS}} \\ [\sigma^{\text{NL}}_{\text{RIXS}}]_{\text{KHD}} &= n_\Gamma \frac{\Gamma_x}{\Gamma} \frac{2\pi^2\Gamma}{\Delta_{p\mathbf{k}}} \sigma^{\text{at}}_{\text{XAS}}. \end{aligned} \tag{15.139}$$

By comparison we obtain

$$\underbrace{h(\tau)\, \mathcal{G}_{\text{eff}}\, \sigma^{\diamond}_{\text{XAS}}}_{\substack{\text{time-dep.BR theory} \\ \text{coll. atom (solid) response}}} = \underbrace{\frac{2\pi^2\Gamma}{\Delta_{p\mathbf{k}}} \sigma^{\text{at}}_{\text{XAS}}}_{\substack{\text{bandwidth-dep. KHD theory} \\ \text{indiv. atom response}}} \tag{15.140}$$

With the earlier given parameters and $\mathcal{G}_{\text{eff}} = 219$ (see Table 15.2) the left side of (15.140) yields the dashed red line in Fig. 15.30 and the right side the blue line. As has been discussed in Sect. 14.6.2 for the atomic REXS case, one may consider the stimulated RIXS cross sections in (15.139) as linking the time-dependent low intensity BR theory with the bandwidth-dependent KHD perturbation theory.

We conclude this chapter with the following statement.

> When atoms in a solid are exposed to a laterally and temporally coherent beam tuned to an atomic resonance, the *effective* excited state population is larger by a factor of $\mathcal{G}_{\text{eff}}$ than the true excited state population calculated for individual (independent) atoms. The enhancement factor $\mathcal{G}_{\text{eff}}$ accounts for the collective response of the coherently illuminated atoms.

References

1. R.W. Boyd, *Nonlinear Optics* (Academic Press, Cambridge, MA, 2008)
2. S. Mukamel, *Principles of Nonlinear Optical Spectroscopy* (Oxford University Press, Oxford, 1999)
3. T. Brabec, *Strong Field Laser Physics* (Springer, New York, 2008)
4. S. Schreck, M. Beye, J.A. Sellberg, T. McQueen, H. Laksmono, B. Kennedy, S. Eckert, D. Schlesinger, D. Nordlund et al., Phys. Rev. Lett. **113**, 153002 (2014)
5. D.J. Higley, A.H. Reid, Z. Chen, L.L. Guyader, O. Hellwig, A.A. Lutman, T. Liu, P. Shafer, T. Chase, G.L. Dakovski, A. Mitra, E. Yuan, J. Schlappa, H.A. Dürr, W.F. Schlotter, J. Stöhr, Nat. Comm. **10**, 5289 (2019)
6. L.L. Guyader, D.J. Higley, M. Pancaldi, T. Liu, Z. Chen, T. Chase, P.W. Granitzka, G. Coslovich et al., Appl. Phys. Lett. **120**, 032401 (2022)
7. D.J. Higley, Z. Chen, M. Beye, M. Hantschmann, A.H. Reid, V. Mehta, O. Hellwig, G.L. Dakovski et al., Commun. Phys. **5**, 83 (2022)
8. B. Nagler et al., Nat. Phys. **5**, 693 (2009)
9. S.M. Vinko, O. Ciricosta, B.I. Cho, K. Engelhorn, H.K. Chung, C.R.D. Brown, T. Burian, J. Chalupský, R.W. Falcone, C. Graves et al., Nature **482**, 59 (2012)
10. H. Yoneda, Y. Inubushi, M. Yabashi, T. Katayama, T. Ishikawa, H. Ohashi, H. Yumoto, K. Yamauchi, H. Mimura, H. Kitamura, Nat. Comm. **5**, 5080 (2014)
11. O.S. Humphries, R.S. Marjoribanks, Q.Y. van den Berg, E.C. Galtier, M.F. Kasim, H.J. Lee, A.J.F. Miscampbell, B. Nagler, R. Royle, J.S. Wark, S.M. Vinko, Phys. Rev. Lett. **125**, 195001 (2020)
12. B. Wu, T. Wang, C.E. Graves, D. Zhu, W.F. Schlotter, J.J. Turner, O. Hellwig, Z. Chen, H.A. Dürr, A. Scherz, J. Stöhr, Phys. Rev. Lett. **117**, 027401 (2016)
13. Z. Chen, D.J. Higley, M. Beye, M. Hantschmann, V. Mehta, O. Hellwig et al., Phys. Rev. Lett. **121**, 137403 (2018)
14. J. Stöhr, A. Scherz, Phys. Rev. Lett. **115**, 107402 (2015)
15. N. Rohringer et al., Nature **481**, 488 (2012)
16. M. Beye, S. Schreck, F. Sorgenfrei, C. Trabant, N. Pontius, C. Schüßler-Langeheine, W. Wurth, A. Föhlisch, Nature **501**, 191 (2013)
17. S. Schreck, M. Beye, A. Föhlisch, J. Mod. Opt. **62**, S34 (2015)
18. H. Yoneda, Y. Inubushi, K. Nagamine, Y. Michine, H. Ohashi, H. Yumoto, K. Yamauchi, H. Mimura et al., Nature **524**, 446 (2015)

19. T. Kroll et al., Phys. Rev. Lett. **120**, 133203 (2018)
20. T. Kroll, C. Weninger, F.D. Fuller, M.W. Guetg, A. Benediktovitch, Y. Zhang, A. Marinelli, R. Alonso-Mori, A. Aquila, M. Liang et al., Phys. Rev. Lett. **125**, 037404 (2020)
21. C. Weninger et al., Phys. Rev. Lett. **111**, 233902 (2013)
22. V. Kimberg, A. Sanchez-Gonzalez, L. Mercadier, C. Weninger, A. Lutman, D. Ratner, R. Coffee, M. Bucher, M. Mucke, M. Agåker et al., Faraday Discuss. **194**, 305 (2016)
23. N. Rohringer, Philos. Trans. R. Soc. A **377**, 20170471 (2019)
24. T. Wang et al., Phys. Rev. Lett. **108**, 267403 (2012)
25. J. Stöhr, Phys. Rev. Lett. **118**, 024801 (2017)
26. J. Stöhr, Adv. Opt. Photon. **11**, 215 (2019)
27. G. K. Shenoy and J. Stöhr (eds.), *LCLS-The First Experiments* (SLAC Publication SLAC-R-611, 2000)
28. R. Neutze, R. Wouts, D. van der Spoel, E. Weckert, J. Hajdu, Nature **406**, 752 (2000)
29. E.K.H. Bennemann, *Nonlinear Optics in Metals* (Oxford University Press, Oxford, 1998)
30. B. Rethfeld, A. Kaiser, M. Vicanek, G. Simon, Phys. Rev. B **65**, 214303 (2002)
31. P.W. Hatfield, J.A. Gaffney, G.J. Anderson et al., Nature **593**, 351 (2021)
32. S.I. Anisimov, A.M. Bonch-Bruevich, M.A. El'yashevich, Y.A. Imas, N.A. Pavlenko, G.S. Romanov, Sov. Phys. Tech. Phys. **11**, 945 (1967)
33. S.I. Anisimov, B.L. Kapeliovich, T.L. Perel'man, Sov. Phys. JETP **39**, 375 (1974)
34. R. Rhie, H. Dürr, W. Eberhardt, Phys. Rev. Lett. **90**, 247201 (2003)
35. H.A. Dürr, Nucl. Instr. Meth. A **601**, 132 (2009)
36. P. Tengdin, W. You, C. Chen, X. Shi, D. Zusin, Y. Zhang, C. Gentry, A. Blonsky, M. Keller, P.M. Oppeneer, H.C. Kapteyn, Z. Tao, M.M. Murnane, Science Advances **4**, eaap9744 (2018)
37. E. Beaurepaire, J.C. Merle, A. Daunois, J.Y. Bigot, Phys. Rev. Lett. **76**, 4250 (1996)
38. J. Stöhr, H.C. Siegmann, *Magnetism: From Fundamentals to Nanoscale Dynamics* (Springer, Heidelberg, 2006)
39. A. Kirilyuk, A.V. Kimel, T. Rasing, Rev. Mod. Phys. **82**, 2731 (2010)
40. A. Eschenlohr, U. Bovensiepen (eds.), Special issue on ultrafast magnetism. J. Phys.: Condens. Matter **29** (2017)
41. H. Mimura et al., Nat. Comm. **5** (2014)
42. W.S.M. Werner, Surf. Interface Anal. **31**, 141 (2001)
43. B. Ziaja, R.A. London, J. Hajdu, J. Appl. Phys. **97**, 064905 (2005)
44. C. Chen, Z. Tao, A. Carr, P. Matyba, T. Szilvási, S. Emmerich et al., Proc. Natl. Acad. Sci. U.S.A. **114**, E5300 (2017)
45. R. Knorren, K.H. Bennemann, R. Burgermeister, M. Aeschlimann, Phys. Rev. B **61**, 9427 (2000)
46. D.A. Papaconstantopoulos, *Handbook of the Band Structure of Elemental Solids* (Plenum, New York, 1986)
47. N.W. Ashcroft, N.D. Mermin, *Solid State Physics* (Holt, Rinehart and Winston, Philadelphia, 1976)
48. H. Ibach, H. Lüth, *Solid State Physics: An Introduction to Principles of Materials Science* (Springer, Berlin, 2003)
49. A. Sommerfeld, Z. Phys. **47**, 1 (1928)
50. X. Zhang, B. Grabowski, F. Körmann, C. Freysoldt, J. Neugebauer, Phys. Rev. B **95**, 165126 (2017)
51. P.B. Corkum, F. Brunel, N.K. Sherman, T. Srinivasan-Rao, Phys. Rev. Lett. **61**, 2886 (1988)
52. P. Maldonado, K. Carva, M. Flammer, P.M. Oppeneer, Phys. Rev. B **96**, 174439 (2017)
53. A.T. Petit, P.L. Dulong, Annales de Chimie et de Physique **10**, 395 (1819)
54. S.P. Hau-Riege, *High-Intensity X-ray-Interaction with Matter* (Wiley, Weinheim, Germany, 2011)
55. H. Yumoto, H. Mimura, T. Koyama, S. Matsuyama, K. Tono et al., Nature Phot. **7**, 43 (2012)
56. T. Koyama, H. Yumoto, Y. Senba, K. Tono, T. Sato et al., Opt. Express **21**, 15382 (2013)
57. B.L. Henke, E.M. Gullikson, J.C. Davis, At. Data Nucl. Data Tables **54**, 181 (1993)

58. C. Stamm, T. Kachel, N. Pontius, R. Mitzner, T. Quast, K. Holldack, S. Khan, C. Lupulescu, E.F. Aziz, M. Wietstruk et al., Nat. Mater. **6**, 740 (2007)
59. B.I. Cho, K. Engelhorn, A.A. Correa, T. Ogitsu, C.P. Weber et al., Phys. Rev. Lett. **106**, 167601 (2011)
60. G. Doumy et al., Phys. Rev. Lett. **106**, 083002 (2011)
61. N. Berrah, L. Fang, B.F. Murphy, E. Kukk, T.Y. Osipov, R. Coffee, K.R. Ferguson, H. Xiong, J.C. Castagna, V.S. Petrovic et al., Opt. Express **24**, 11768 (2016)
62. E.L. Saldin, E.A. Schneidmiller, M.V. Yurkov, New J. Phys. **12**, 035010 (2010)
63. J. Stöhr, *NEXAFS Spectroscopy* (Springer, Heidelberg, 1992)
64. Z. Tao, C. Chen, T. Szilvási, M. Keller, M. Mavrikakis, H. Kapteyn, M. Murnane, Science **353**, 62 (2016)
65. C. Gähwiller, F.C. Brown, Phys. Rev. B **2**, 1918 (1970)
66. A. Bianconi, R.Z. Bachrach, S.B.M. Hagstrom, S.A. Flodström, Phys. Rev. B **19**, 2837 (1979)
67. M.O. Krause, J. Phys. Chem. Ref. Data **8**, 307 (1979)
68. M.O. Krause, J.H. Oliver, J. Phys. Chem. Ref. Data **8**, 329 (1979)
69. C.O. Almbladh, A.L. Morales, G. Grossmann, Phys. Rev. B **39**, 3489 (1989)
70. K. Hatada, A.D. Cicco, J. Elec. Spectrosc. Relat. Phenom. **196**, 177 (2014)
71. K. Hatada, A.D. Cicco, Appl. Sci. **7**, 814 (2017)
72. D.R. Mortensen, G.T. Seidler, J.J. Kas, N. Govind, C.P. Schwartz, S. Pemmaraju, D.G. Prendergast, Phys. Rev. B **96**, 125136 (2017)
73. J.L. Campbell, T. Papp, At. Data Nucl. Data Tables **77**(1), 1 (2001)
74. H. Sorum, J. Bremer, J. Phys. F: Met. Phys. **12**, 2721 (1982)
75. R. Shimizu, K. Edamatsu, T. Itoh, Phys. Rev. A **74**, 013801 (2006)
76. M. Segev, G. Stegeman, Phys. Today **51**, 42 (1998)
77. G.I. Stegeman, M. Segev, Science **286**, 1518 (1999)
78. Y. Bromberg, Y. Lahini, E. Small, Y. Silberberg, Nat. Photon. **4**, 721 (2010)
79. E. Hecht: *Optics*, 4th ed. (Addison-Wesley, Reading, 2001)
80. W.F. Schlotter et al., Appl. Phys. Lett. **89**, 163112 (2006)
81. S. Eisebitt, M. Lörgen, W. Eberhardt, J. Lüning, J. Stöhr, C.T. Rettner, O. Hellwig, E.E. Fullerton, G. Denbeaux, Phys. Rev. B **68**, 104419 (2003)
82. G. Ghiringhelli et al., Rev. Sci. Instrum. **77**, 113108 (2006)
83. V.N. Strocov, T. Schmitt, U. Flechsig, T. Schmidt, A. Imhof, Q. Chen, J. Raabe, R. Betemps, D. Zimoch, J. Krempasky, X. Wang, M. Grioni, A. Piazzalunga, L. Patthey, J. Synchrotron Rad. **17**, 631 (2010)
84. T. Schmitt, F.M.F. de Groot, J.E. Rubensson, J. Synchrotron Rad. **21**, 1065 (2014)
85. R. Qiao, Q. Li, Z. Zhuo, S. Sallis, O. Fuchs, M. Blum, L. Weinhardt, C. Heske, J. Pepper, M. Jones et al., Rev. Sci. Instrum. **88**, 033106 (2017)
86. N.B. Brookes, G. Ghiringhelli, P. Glatzel, M. Moretti Sala: Synchr. Rad. News **31**, 26 (2018)
87. A. Nilsson, J. Stöhr, T. Wiell, M. Aldén, P. Bennich, N. Wassdahl, M. Samant, S.S.P. Parkin, N. Mårtensson, J. Nordgren, B. Johansson, H.L. Skriver, Phys. Rev. B **54**, 2917 (1996)
88. J. van den Brink, M. van Veenendaal, J. Phys. Chem. Solids **66**, 2145 (2005)
89. J. van den Brink, M. van Veenendaal, Europhys. Lett. **73**, 121 (2006)
90. L.J.P. Ament, M. van Veenendaal, T.P. Devereaux, J.P. Hill, J. van den Brink, Rev. Mod. Phys. **83**, 705 (2011)
91. M. van Veenendaal, *Theory of Scattering and Absorption of X-Rays* (Cambridge University Press, Cambridge, 2015)

Chapter 16
Quantum Diffraction: Emergence of the Quantum Substructure of Light

16.1 Introduction

In this chapter we discuss an entirely quantum mechanical description of diffraction[1]. We will introduce the remarkably simple new paradigm that diffraction patterns, long associated with wave interference, are instead direct signatures of the quantum states of light. In first order QED (corresponding to conventional quantum mechanics) this link has remained hidden leading to the wave-photon ambiguity. The ambiguity is shown to disappear in second order QED, where only the photon-based description can account for the observed diffraction patterns.

Historically, the description of diffraction has been based on the magical Huygens–Fresnel principle, discussed in Sect. 8.3.3. Its acceptance is mostly based on the fact that it gives the right result for conventional diffraction experiments. It therefore embodies a pragmatic rather than fundamental approach to describe physical processes. The principle postulates the mysterious emission of spherical waves from "points" even when there are no real atoms that can either emit light or scatter it! Another fascinating aspect of this description is that the detection of the "intensity", calculated as the absolute value squared of the interfering fields, is not specified. This ignores the well-known fact that, in practice, the pattern is recorded as a mosaic of local "bright spots" as shown in Fig. 8.10, where a photon energy packet $\hbar\omega$ is taken from the total wave field.

The simplest quantum formulation of diffraction is due to Feynman [1] who explained Young's double-slit experiment by use of his space-time probability amplitude formulation of quantum mechanics (QM) [2]. His treatment of diffraction may be viewed more generally as a demonstration of the inherent wave-particle duality underlying various formulations of quantum mechanics [3], first expressed through de Broglie's hypothesis that all matter has wave properties [4]. Relative to other formulation of QM, Feynman's formulation (see Sect. 4.8) is particularly appealing since single photon probability amplitudes closely resemble classical wave fields [5,

[1] It is based on a paper published on March 28, 2020 on the physics archive arXiv:2003.14217.

J. Stöhr, *The Nature of X-Rays and Their Interactions with Matter*, Springer Tracts in Modern Physics 288, https://doi.org/10.1007/978-3-031-20744-0_16

6]. Rather than resolving the wave-particle conflict, Feynman's treatment effectively consolidated it.

All formulations of diffraction within the confines of conventional QM are limited, however, by its well-known linearity [7]. In particular, QM may be viewed as a *first order* perturbation within the complete theory of light and matter, quantum electrodynamics (QED), which extends to infinite order. Of the three different formulations of QED by Tomonaga [8], Schwinger [9], and Feynman [10], it is again Feynman's formulation that is most appealing and of practical utility, as pointed out by Dyson [11] in showing their equivalence in 1949.

In particular, the concept of space-time probability amplitudes of single independent photons or electrons in QM may be extended to higher perturbative orders through the construction of probability amplitudes of an increasing number of particles. Remarkably, in first order the elementary building blocks of QED, photons and electrons, are described by the *same* probability amplitudes. This is the deeper reason why photons and electrons give the same conventional diffraction patterns [1, 12]. It is only in *second order* that the description of multi-particles states becomes different for Bosons (photons) and fermions (electrons), reflected for fermions by increasingly complex Feynman diagrams [11, 12].

The difference between the interference behavior of two Bosons and two fermions in second order QED is most beautifully demonstrated by a Hanbury Brown–Twiss type interference experiment carried out with two different charge-neutral isotopes of He atoms, Bosonic ^{4}He and fermionic ^{3}He [13]. Owing to their symmetric versus antisymmetric two-particle wave functions the Bosonic bunching peak is replaced by a fermionic antibunching tip.

In Feynman's formulation, the probability amplitudes of an increasing number of photons, corresponding to increasing orders in QED, are constructed by addition and multiplication of those of individual photons [5, 6]. The formulation for different cases is augmented by rules regarding the addition versus multiplication of single particle amplitudes [6, 12]. These rules become increasing complex for more than two photons. It is then advantageous to use the operator-based method based on the concepts of modern quantum optics [14, 15], introduced for the one-photon case in Sect. 4.7 and extended to the two-photon case in Sect. 5.5. This method pioneered by Glauber describes light in terms of orders of coherence $O = 1, 2, 3 \ldots \infty$ [16] which are equivalent to the orders of perturbation in Feynman's formulation of QED [6, 12].

In Glauber's formulation, the construction of probability amplitudes for increasing number of photons is facilitated by the use of products of an increasing number of photon birth and destruction (detection) operators that can create and destroy quantum states containing increasing numbers of photons. This leads to the link of quantum states and diffraction patterns, the central theme of the present chapter.

The existence of quantum states of light containing specific number of photons N was first experimentally verified by experiments in the late 1980s where individual photons ($N = 1$) [17] or photon pairs ($N = 2$) [18] were sent through a lossless beam splitter and their emergence from different output ports was examined by coincidence detection, as discussed in Sect. 5.2. In the process it became clear that Dirac's famous statement that photons do not interfere with each other [19] holds only in first order QED. This was more directly revealed by diffraction experiments in the late 1990s where the conventional illumination of Young's double slits was replaced by use of entangled photon pairs [20, 21], produced by parametric down conversion [22–24].

In this chapter, we utilize the concept of quantum diffraction to illustrate the new paradigm that within QED, diffraction images are direct signatures of different quantum states of light. This becomes apparent when the results of modern versions of Young's double-slit experiment, performed by illumination with differently modified laser light and photon-based detection, are compared to the patterns predicted by the formulation of diffraction within QED.

This direct link has remained hidden in the past because the treatment of diffraction by conventional quantum mechanics results in an accidental degeneracy of diffraction patterns for different quantum states. Diffraction has therefore continued to be explained by the *ad hoc* concepts of "*coherent*" and "*incoherent*" superposition of waves. Here we show that the degeneracy of patterns for different quantum states in first order is lifted upon extension of QED to second order. This evolution is shown to be particularly important since the wave-particle equivalence breaks down and the true photon-based nature of light emerges in the diffraction patterns.

As a consequence, the wave theory of diffraction can in principle be abandoned altogether today, and the framework of statistical optics [25, 26] may be replaced by the more fundamental quantum formulation of light [15, 16]. In particular, the broad and difficult concept of "partial coherence" in wave optics can now be succinctly defined through the degrees of coherence of specific quantum states in different orders of QED and their diffraction patterns.

16.2 Generation of Different States of Light

The advent of the laser has allowed the creation of different quantum states of light which are described by quantum optics [14, 15, 21, 27, 28]. In Fig. 16.1 we present different schemes that have been used to prepare double-slit-like sources. The reason for the shown order will become clear later when the cases will be linked to the evolution of their respective diffraction patterns from first to second order.

Figure 16.1a shows the cases of "coherent" illumination of the double slits by a conventional source that has been made (first order) "coherent" by use of a monochromator and pinhole or by a laser which is higher order coherent. The first and second order diffraction patterns have been studied by Shimizu et al. [29] using either a monochromatized halogen lamp or a Ti:sapphire laser to illuminate the slits.

On the right of the same figure we show the particular case where a strong incident near transform-limited pulse, whose width is approximately equal to the photon coherence time, is tuned to a well-defined atomic resonance in a thin film [30]. In the so-prepared source, absorption is compensated by stimulated emission [31], producing a second order coherent source [6].

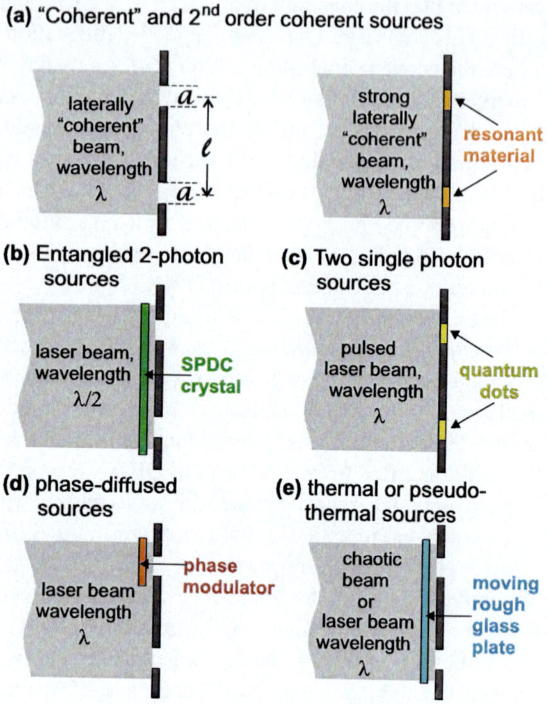

Fig. 16.1 Experimental schemes of preparing different double-slit-like sources representing different quantum states, as discussed in the text. The double-slit dimensions are labeled as previously in Fig. 8.10

In Fig. 16.1b, a laser is used to illuminate a suitable thin crystal that through spontaneous parametric down conversion produces two spatially entangled photons, each with half the incident photon energy [32]. The "entangled biphoton" diffraction case has been extensively studied in the literature [5, 6, 21].

In Fig. 16.1c two *single photons* are simultaneously emitted from two quantum sources. This case can be implemented by use of a laser pulse that triggers two quantum dots or two trapped atoms or ions to simultaneously emit single photons [33–36].

In Fig. 16.1d the slits are illuminated by *phase-diffused light*, implemented by Liu et al. [5] by splitting a coherent laser beam and modulating the phase of one of the beams with a vibrating mirror. The intensity falling onto the two slits is kept *constant* so that the photons with random phases still obey *Poisson* counting statistics.

Finally, as shown in Fig. 16.1e, the slits may be illuminated by *chaotic* light produced by a thermal source [37]. In practice, it is convenient to use higher intensity "pseudo-thermal" light generated by shining a laser on a rotating ground glass plate [38]. Such light exhibits both phase and intensity fluctuations with *Bose–Einstein* counting statistics [39]. It has been employed for double-slit diffraction by several groups [40–42].

The cases shown in Fig. 16.1 represent experimental schemes that generate specific quantum states of light whose characteristic first and second order diffraction patterns have been reported in the cited literature. In the following we shall illustrate how the observed diffraction patterns follow from the photon-based formulation of diffraction within QED, which directly links quantum states with their characteristic encoded diffraction signatures.

The presented formulation of quantum diffraction is general up to second order in QED and covers the infinite number of quantum states of light within QED. These quantum states give rise to the complex light behavior typically described by the broad and difficult concept of "partial coherence" in statistical optics [25, 26]. The quantum formulation allows the specification of this concept by directly defining coherence and diffraction through that of individual quantum states. These states span the entire range between the limiting cases of *coherent states* represented by a Poisson distribution and *chaotic states* associated with a Bose–Einstein distribution.

16.3 The Formulation of Quantum Diffraction

The experimental geometry for Young's double-slit experiment is illustrated in Fig. 16.2a with identification of the relevant coordinates. The photons emerging from the two source points A and B, which may be expanded into slits, are detected at a single point or two points in a distant detector plane.

The diffraction pattern is determined by the *spatial* variations in the number of detected quasi-monochromatic photons. The assumption of monochromaticity allows the reduction of the general space-time description of photons in QED to the spatial domain only, since the width of the photon energy distribution defines a time interval of photon arrival, given by the *coherence time* τ of the photons. This is taken into account in the design of diffraction experiments, as illustrated in Fig. 16.2b, c. The first order diffraction pattern is determined by the arrival probability of photons within the time increment τ at a given detection point. In first order, the classical and quantum patterns are the same and for the double slit are given by (8.20) which with $q = k\rho/z_0$ reads

$$\boxed{I(\rho) = 4I_0 \frac{a^2}{\lambda z_0} \mathrm{sinc}^2\left[\frac{ka}{2z_0}\rho\right] \cos^2\left[\frac{k\ell}{2z_0}\rho\right],} \tag{16.1}$$

where z_0 is the distance between the source and detector planes and ρ is the lateral distance from the center of the detector plane. The second order pattern is determined by the coincident arrival probability of two photons within τ at a single or two detection points.

The first and second order diffraction cases previously discussed in Sects. 4.7 and 5.5 correspond to the lowest orders, $m = 1, 2$, of perturbation in QED. For our treatment below we write the diffraction probability in the form (5.69) or

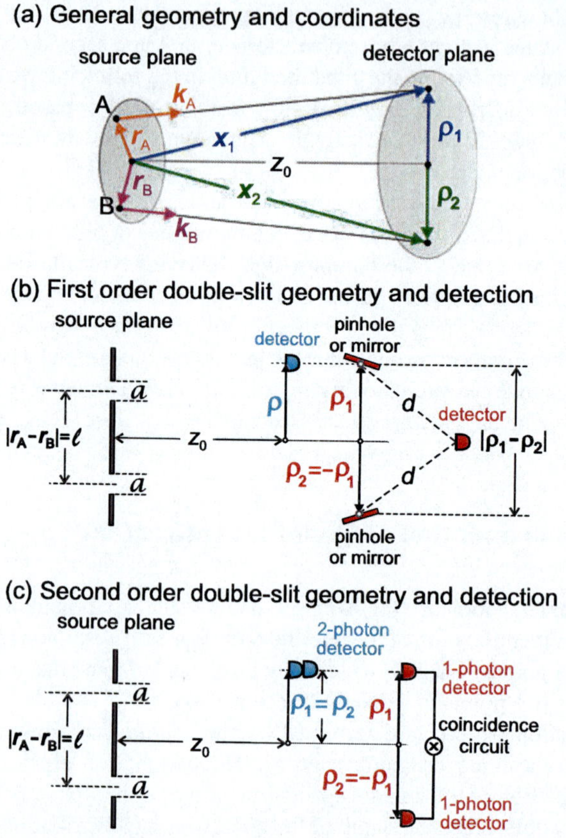

Fig. 16.2 **a** Assumed geometry and coordinates of photon propagation from points $\mathbf{r}_A$ and $\mathbf{r}_B$ in the source plane to distant points with coordinates $\boldsymbol{\rho}_i$ or $\mathbf{x}_i$ $(i = 1, 2)$ in the detector plane. **b** In first order, photons emitted from either slit are detected at a *single* point using the scenarios shown in blue and red. This corresponds to the classical diffraction case discussed in Sect. 8.5.1. A photon detector is either scanned as a function of its separation ρ from the optical axis or a detector on the optical axis detects the probability that photons have taken paths through points $\boldsymbol{\rho}_2 = -\boldsymbol{\rho}_1$, defined by scannable mirrors or pinholes. The detection probability is given by (5.69) with $m = 1$. **c** In second order, one measures the coincident arrival of two photons at a single or two points by use of the blue and red detection schemes. The detection probability is given by (5.69) with $m = 2$

$$\big\langle \Phi_s | \mathbf{P}^{(m)}(\boldsymbol{\rho}_1, \boldsymbol{\rho}_2) | \Phi_s \big\rangle = P_m \, \mathcal{G}^{(m)}(\boldsymbol{\rho}_1, \boldsymbol{\rho}_2). \tag{16.2}$$

Here $\mathbf{P}^{(m)}(\boldsymbol{\rho}_1, \boldsymbol{\rho}_2)$ is the detection probability operator, given by the conventional quantum optical correlation function of order m and the triangular brackets denote the quantum mechanical expectation value, evaluated for a given quantum state $|\Phi_s\rangle$ created at the source. As indicated on the right of (16.2), the diffraction pattern can be conveniently written in terms of a position-dependent shape function $\mathcal{G}^{(m)}(\boldsymbol{\rho}_1, \boldsymbol{\rho}_2)$ and an overall scaling factor P_m that is determined by conservation of emitted and detected photons. Before we evaluate the diffraction patterns for the double-slit case we briefly review our previous first order results derived in Sect. 4.7 and second order ones from Sect. 5.5.

16.3.1 First Order Diffraction Formulation

As derived in Sect. 4.7 the first order probability operator $\mathbf{P}^{(1)}(\boldsymbol{\rho}_1, \boldsymbol{\rho}_2)$ in (16.2) may be expressed either in the coordinates of the detection points $\boldsymbol{\rho}_i$ or $\mathbf{x}_i$ defined in Fig. 16.2a. For the case where the double slits are just two points, the detection probability is obtained as the expectation value of (4.124), or

$$\left\langle \Phi_s \left| \mathbf{P}^{(1)}(\boldsymbol{\rho}_1, \boldsymbol{\rho}_2) \right| \Phi_s \right\rangle = \frac{1}{4} \left\langle \Phi_s \middle| \mathbf{X} + \mathbf{Y} \middle| \Phi_s \right\rangle \tag{16.3}$$

where $|\Phi_s\rangle$ is a given quantum state created in the source. The operators $\mathbf{X}$ and $\mathbf{Y}$ are given by (4.125) and (4.126).

For the double-slit geometry in Fig. 16.2a the slit centers are separated by ℓ and each has a finite width a. The detection probability is calculated according to (4.139) by integration over the slits according to

$$\left\langle \mathbf{P}^{(1)}(\boldsymbol{\rho}_1, \boldsymbol{\rho}_2) \right\rangle = \frac{1}{4a^2} \sum_{\text{U,V slits}} \iint \left\langle \mathbf{a}^{\dagger}_{\mathbf{k}_U} \mathbf{a}_{\mathbf{k}_V} \right\rangle e^{i\frac{k}{z_0}(\mathbf{r}_U \cdot \boldsymbol{\rho}_1 - \mathbf{r}_V \cdot \boldsymbol{\rho}_2)} d\mathbf{r}_U \, d\mathbf{r}_V . \tag{16.4}$$

For a given quantum state, only certain matrix elements $\langle \mathbf{a}^{\dagger}_{\mathbf{k}_U} \mathbf{a}_{\mathbf{k}_V} \rangle$ will have non-zero values and they will determine the integration and hence the diffraction pattern (see Tables 16.2 and 16.3).

16.3.2 Second Order Diffraction Formulation

For two source points, the *second order* detection *probability* is given by (5.14) or

$$\left\langle \Phi_s \left| \mathbf{P}^{(2)}(\mathbf{x}_1, \mathbf{x}_2) \right| \Phi_s \right\rangle = \frac{1}{16} \left\langle \Phi_s \middle| \mathbf{A} + \mathbf{B} + \mathbf{C} + \mathbf{D} \middle| \Phi_s \right\rangle \tag{16.5}$$

with the four operators $\mathbf{A}, \mathbf{B}, \mathbf{C}, \mathbf{D}$ given by (5.15)–(5.18) and $|\Phi_s\rangle$ being the state created in the source.

For finite-sized slits our previous expression (5.42) takes the form

$$\left\langle \mathbf{P}^{(2)}(\boldsymbol{\rho}_1, \boldsymbol{\rho}_2)\right\rangle = \frac{1}{16a^4} \sum_{\mathrm{U,V,X,Y}} \iiiint_{\text{slits}} \left\langle \mathbf{a}^{\dagger}_{\mathbf{k}_{\mathrm{U}}} \mathbf{a}^{\dagger}_{\mathbf{k}_{\mathrm{V}}} \mathbf{a}_{\mathbf{k}_{\mathrm{X}}} \mathbf{a}_{\mathbf{k}_{\mathrm{Y}}} \right\rangle \mathrm{e}^{\mathrm{i}\frac{k}{z_0}(\mathbf{r}_{\mathrm{U}}-\mathbf{r}_{\mathrm{X}})\cdot\boldsymbol{\rho}_1} \mathrm{e}^{\mathrm{i}\frac{k}{z_0}(\mathbf{r}_{\mathrm{V}}-\mathbf{r}_{\mathrm{Y}})\cdot\boldsymbol{\rho}_2}$$
$$\mathrm{d}\mathbf{r}_{\mathrm{U}}\mathrm{d}\mathbf{r}_{\mathrm{V}}\mathrm{d}\mathbf{r}_{\mathrm{X}}\mathrm{d}\mathbf{r}_{\mathrm{Y}}. \tag{16.6}$$

For a given quantum state, only certain matrix elements $\langle \mathbf{a}^{\dagger}_{\mathbf{k}_{\mathrm{U}}} \mathbf{a}^{\dagger}_{\mathbf{k}_{\mathrm{V}}} \mathbf{a}_{\mathbf{k}_{\mathrm{X}}} \mathbf{a}_{\mathbf{k}_{\mathrm{Y}}} \rangle$ will have non-zero values and they will determine the integration and hence the diffraction pattern (see Tables 16.4 and 16.5).

16.3.3 *Order-Dependent Degree of Coherence*

In quantum optics, coherence is characterized by a *degree of spatial coherence* $g^{(m)}(\mathbf{x}_1, \mathbf{x}_2)$ which depends on the order m of perturbation in QED. The previously defined degrees of first ($m=1$) order (4.83) and second ($m=2$) order (5.4) spatial coherence may be written in the form

$$g^{(m)}(\mathbf{x}_1, \mathbf{x}_2) = \frac{\left\langle \mathbf{P}^{(m)}(\mathbf{x}_1, \mathbf{x}_2)\right\rangle}{\left[\left\langle \mathbf{P}^{(1)}(\mathbf{x}_1, \mathbf{x}_1)\right\rangle \left\langle \mathbf{P}^{(1)}(\mathbf{x}_2, \mathbf{x}_2)\right\rangle\right]^{m/2}}. \tag{16.7}$$

For the *coherent cases*, the numerators in $g^{(m)}(\mathbf{x}_1, \mathbf{x}_2)$ factor into the denominators, so that the diffraction fine structure contained in both the numerators and denominators is normalized out, yielding a constant. It is then convenient to plot the normalized diffraction pattern $\mathcal{G}^{(m)}(\mathbf{x}_1, \mathbf{x}_2)$ defined through (16.2) which preserves the characteristic diffraction structure. We shall utilize both complementary formulations in the present chapter.

16.4 The Quantum States of Light

The first and second order diffraction patterns, defined by (5.69) with $m = 1, 2$, are determined by quantum states involving two wavevector modes $\mathbf{k}_{\mathrm{A}}$ and $\mathbf{k}_{\mathrm{B}}$ as defined in Fig. 16.2a. In the following we will switch to the shorter and more convenient notation $\mathbf{k} = \mathbf{k}_{\mathrm{A}}$ and $\mathbf{k}' = \mathbf{k}_{\mathrm{B}}$.

The two-mode quantum states produced in the cases shown in Fig. 16.1 involve different numbers of photons. In general, we distinguish *collective states* which contain an *average* number of photons per mode from states that contain a *specific* number of photons per mode. We first discuss the two-mode multi-photon collective states associated with Fig. 16.1a, d and e and their decomposition into probability distributions of substates. Their specific two-photon substates are then linked to the central two cases in Fig. 16.1b, c, which involve only two photons.

16.4.1 Two-Mode Collective Quantum States

The two-mode collective *coherent state* produced in Fig. 16.1a and the *phase-diffused coherent state* in Fig. 16.1d are constructed from two single mode coherent states of the form (3.78) or

$$|\alpha\rangle_k = \sum_{m=0}^{\infty} \frac{\alpha_k^m}{e^{|\alpha_k|^2/2}\sqrt{m!}} |m\rangle_k. \tag{16.8}$$

Here α_k is a complex number and the coherent state contains an *average number of photons* $\langle n\rangle_k = |\alpha_k|^2$ in the mode k, distributed in a Poisson distribution around the average value $|\alpha_k|^2$. In general, the two modes may contain different numbers of photons and have different phases. In the following we shall assume that both modes contain the same *average* number of photons per mode, i.e. $|\alpha|^2 = \langle n\rangle = \langle n\rangle_k = \langle n\rangle_{k'}$. We then obtain with $\alpha_k = |\alpha|e^{i\phi_k}$ and $\alpha_{k'} = |\alpha|e^{i\phi_{k'}}$ the following general expression for a two-mode collective "coherent" state

$$|\alpha\rangle_k|\alpha\rangle_{k'} = \frac{1}{e^{|\alpha|^2}} \sum_{n=0}^{\infty}\sum_{m=0}^{\infty} \frac{|\alpha|^{n+m}\, e^{i(n\phi_k+m\phi_{k'})}}{\sqrt{n!\,m!}} |n\rangle_k\, |m\rangle_{k'}. \tag{16.9}$$

The *two-mode chaotic state* associated with Fig. 16.1e is constructed from two single mode chaotic states $|\beta\rangle_k$ given by (3.88) or

$$|\beta\rangle_k = \sum_{m=0}^{\infty} \sqrt{\frac{\langle n\rangle^m}{(1+\langle n\rangle)^{1+m}}}\, |m\rangle_k. \tag{16.10}$$

It contains an average number of $\langle n\rangle$ photons per mode in the form of a Bose–Einstein distribution. The *two-mode chaotic state* containing an *average number* $\langle n\rangle$ of photons in each mode is given by

$$|\beta\rangle_k|\beta\rangle_{k'} = \sum_{\ell=0}^{\infty}\sum_{m=0}^{\infty} e^{i(\phi_\ell+\phi_m)} \sqrt{\frac{\langle n\rangle^\ell \langle n\rangle^m}{(1+\langle n\rangle)^{\ell+m+2}}}\, |\ell\rangle_k|m\rangle_{k'}. \tag{16.11}$$

The states (16.9) and (16.11) are *collective* two-mode quantum states which contain the same *average* number of photons per mode, $\langle n\rangle = |\alpha|^2$. We now show that they may be written as a linear combination of two-mode *substates* that contain *specific numbers* of photons $N = 0, 1, 2, 3 \ldots \infty$ with probability distributions around the mean value $2\langle n\rangle = N$, denoting the total number of photons in the two modes.

16.4.2 The Collective Coherent State and Its Substates

The *two-mode coherent state* describes the case shown in Fig. 16.1a where the two slits are illuminated by the same average number of photons in both modes and the two modes have the same phases. We then have $\alpha_k = \alpha_{k'} = |\alpha| e^{i\phi}$ and the general expression (16.9) can be written in the form

$$|\Phi_{\mathrm{coh}}\rangle_{k,k'} = \frac{1}{e^{|\alpha|^2}} \sum_{N=0}^{\infty} |\alpha|^N e^{iN\phi} \sum_{m=0}^{N} \frac{1}{\sqrt{m!\,(N-m)!}} |m\rangle_k |N-m\rangle_{k'}. \tag{16.12}$$

The state is composed of *binomial substates* which contain *specific numbers N* of photons that are distributed according to a *Poisson* probability distribution around the average value $2\langle n \rangle$. We can write (see Sect. 5.6.1)

$$|\Phi_{\mathrm{coh}}\rangle_{k,k'} = \sum_{N=0}^{\infty} \underbrace{\frac{2^{N/2}\alpha^N}{\sqrt{N!}\, e^{|\alpha|^2}}}_{c_N^\alpha} |\phi_{\mathrm{cohN}}\rangle_{k,k'}. \tag{16.13}$$

The complex coefficients c_N^α fulfill the normalization $\sum_{N=0}^{\infty} |c_N^\alpha|^2 = 1$ and weigh the contributions of the *binomial substates* $|\phi_{\mathrm{cohN}}\rangle_{k,k'}$ which contain N-photons and are given by

$$|\phi_{\mathrm{cohN}}\rangle_{k,k'} = \frac{1}{2^{N/2}} \sum_{m=0}^{N} \sqrt{\frac{N!}{m!\,(N-m)!}} |m\rangle_k |N-m\rangle_{k'}. \tag{16.14}$$

The states fulfill the normalization $\langle \Phi_{\mathrm{coh}} | \Phi_{\mathrm{coh}} \rangle = \langle \phi_{\mathrm{cohN}} | \phi_{\mathrm{cohN}} \rangle = 1$.

16.4.3 The Collective Phase-Diffused Coherent State and Its Substates

The *phase-diffused laser light* encountered for the case in Fig. 16.1d corresponds to random phases between the two modes in (16.9). Since only the relative phase between the two modes is important, we may set $\phi_k = 0$ and denote the relative phase shift as $\varphi = \phi_{k'}$ to obtain

$$|\Phi_{\mathrm{dif}}\rangle_{k,k'} = \frac{1}{e^{|\alpha|^2}} \sum_{N=0}^{\infty} |\alpha|^N \sum_{m=0}^{N} \frac{e^{i(N-m)\varphi}}{\sqrt{m!\,(N-m)!}} |m\rangle_k |N-m\rangle_{k'}. \tag{16.15}$$

This can be written in terms of substates containing a specific number of N-photons with probabilities in form of a *Poisson* distribution around the average value $2\langle n\rangle$ according to

$$|\Phi_{\mathrm{dif}}\rangle_{k,k'} = \sum_{N=0}^{\infty} \underbrace{\frac{2^{N/2}\,\alpha^N}{\sqrt{N!}\,\mathrm{e}^{|\alpha|^2}}}_{c_N^{\alpha}} |\phi_{\mathrm{difN}}\rangle_{k,k'}, \tag{16.16}$$

where

$$|\phi_{\mathrm{difN}}\rangle_{k,k'} = \frac{1}{2^{N/2}} \sum_{m=0}^{N} \mathrm{e}^{\mathrm{i}(N-m)\varphi} \sqrt{\frac{N!}{m!\,(N-m)!}}\, |m\rangle_k\, |N-m\rangle_{k'}. \tag{16.17}$$

The states fulfill the normalization $\langle\Phi_{\mathrm{dif}}|\Phi_{\mathrm{dif}}\rangle = \langle\phi_{\mathrm{difN}}|\phi_{\mathrm{difN}}\rangle = 1$.

16.4.4 The Collective Chaotic State and Its Substates

The two-mode collective chaotic state and its substates, describing the case in Fig. 16.1e, are given by the general form (16.11) which may be rewritten as

$$|\Phi_{\mathrm{cha}}\rangle_{k,k'} = \sum_{N=0}^{\infty} \sqrt{\frac{\langle n\rangle^N}{(1+\langle n\rangle)^{N+2}}} \sum_{m=0}^{N} \mathrm{e}^{\mathrm{i}(\phi_m+\phi_{N-m})}\, |m\rangle_k |N-m\rangle_{k'}. \tag{16.18}$$

The phase factors account for the relative phase difference between number states $|m\rangle_k$ and $|N-m\rangle_{k'}$ in the two modes. It also contains substates with a specific number of N-photons. Their probability is distributed in the form of a *Bose–Einstein* distribution around the average value $2\langle n\rangle$. The state (16.18) may be written as

$$|\Phi_{\mathrm{cha}}\rangle_{k,k'} = \sum_{N=0}^{\infty} \underbrace{\sqrt{\frac{(N+1)\langle n\rangle^N}{(1+\langle n\rangle)^{N+2}}}}_{c_N^{\beta}} |\phi_{\mathrm{chaN}}\rangle_{k,k'}, \tag{16.19}$$

where $\sum_{N=0}^{\infty} |c_N^{\beta}|^2 = 1$ and the substates are

$$|\phi_{\mathrm{chaN}}\rangle_{k,k'} = \frac{1}{\sqrt{N+1}} \sum_{m=0}^{N} \mathrm{e}^{\mathrm{i}\phi_{m,N-m}} |m\rangle_k |N-m\rangle_{k'}. \tag{16.20}$$

The states fulfill the normalization $\langle\Phi_{\mathrm{cha}}|\Phi_{\mathrm{cha}}\rangle = \langle\phi_{\mathrm{chaN}}|\phi_{\mathrm{chaN}}\rangle = 1$.

16.4.5 Plots of the Substate Distributions

The substructure of the two-mode collective coherent, phase-diffused coherent, and chaotic states is illustrated in Fig. 16.3 for the cases of different orders of coherence O, defined by an average number of photons per mode $O = \langle n \rangle = 1, 2, 4, 9$.

The cases where $2\langle n \rangle = N$ are shown in enhanced colors. They represent the total number of photons in the two modes. The probabilities of all distributions sum to unity according to $\sum_{N=0}^{\infty} |c_N^{\alpha}|^2 = \sum_{N=0}^{\infty} |c_N^{\beta}|^2 = 1$.

> We distinguish the total *average number of photons* contained in the two-mode collective states $2\langle n \rangle$ from the total *specific number of photons* N contained in the substates.

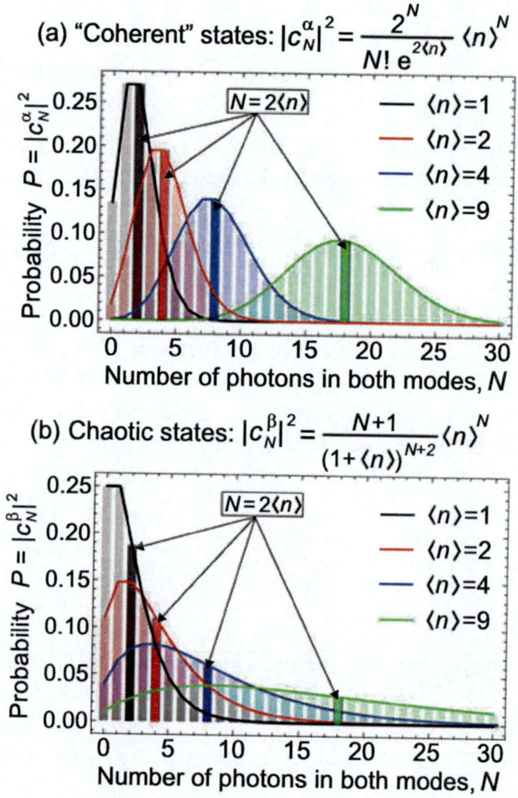

Fig. 16.3 Probability distributions of the substates of **a** the two-mode collective coherent state (16.13) and phase-diffused coherent state (16.16) and **b** the collective chaotic state (16.19). For the shown cases, the collective states contain different average number of photons per mode $\langle n \rangle = 1, 2, 4, 9$, while their substates contain different total numbers of N-photons in both modes. The distributions peak approximately at $N = 2\langle n \rangle$, indicated by enhanced colors, corresponding to the total number of photons in both modes

16.4.6 *Other Fundamental Quantum States*

16.4.6.1 N-Photon Entangled or NOON State

A particularly important state in quantum information science is the N-photon entangled state [43–46] given by

$$|\phi_{\text{entN}}\rangle_{k,k'} = \frac{1}{\sqrt{2}}\Big[|N\rangle_k|0\rangle_{k'} + \mathrm{e}^{\mathrm{i}\phi}|0\rangle_k|N\rangle_{k'}\Big] \tag{16.21}$$

which we have written in a form that reflects why it is also called a NOON state. It corresponds to N-photons being emitted into the *same* mode and none into the other. The specific state $N = 2$ given by

$$|\phi_{\text{ent2}}\rangle_{k,k'} = \frac{1}{\sqrt{2}}\Big[|2\rangle_k|0\rangle_{k'} + \mathrm{e}^{\mathrm{i}\varphi}|0\rangle_k|2\rangle_{k'}\Big] \tag{16.22}$$

describes the case in Fig. 16.2b.

16.4.6.2 N-Photon Number State

In the complementary case, an equal number of $N/2$ photons are emitted into separate modes, described by the two-mode number state

$$|\phi_{\text{numN}}\rangle_{k,k'} = \left|\frac{N}{2}\right\rangle_k \left|\frac{N}{2}\right\rangle_{k'} . \tag{16.23}$$

Owing to the indivisibility of photons the state exists only for $N \geq 2$. The specific state with $N = 2$

$$|\phi_{\text{num2}}\rangle_{k,k'} = |1\rangle_k|1\rangle_{k'} \tag{16.24}$$

describes the case in Fig. 16.2c.

16.4.7 *Summary of Key Multi-photon Quantum States*

The normalized two-mode multi-photon states representing the cases with corresponding names in Fig. 16.1 are summarized in Table 16.1.

Table 16.1 Quantum states of light associated with Fig. 16.1 [a]

Coherent states and substates:

$$|\Phi_{\mathrm{coh}}\rangle_{k,k'} = \sum_{N=0}^{\infty} c_N^{\alpha}\, |\phi_{\mathrm{cohN}}\rangle_{k,k'} = \sum_{N=0}^{\infty} \frac{2^{N/2}\,\alpha^N}{\sqrt{N!}\,\mathrm{e}^{|\alpha|^2}}\, |\phi_{\mathrm{cohN}}\rangle_{k,k'}$$

$$|\phi_{\mathrm{cohN}}\rangle_{k,k'} = \frac{1}{2^{N/2}} \sum_{m=0}^{N} \sqrt{\frac{N!}{m!\,(N-m)!}}\, |m\rangle_k\, |N-m\rangle_{k'}$$

Phase-diffused coherent states and substates:

$$|\Phi_{\mathrm{dif}}\rangle_{k,k'} = \sum_{N=0}^{\infty} c_N^{\alpha}\, |\phi_{\mathrm{difN}}\rangle_{k,k'} = \sum_{N=0}^{\infty} \frac{2^{N/2}\,\alpha^N}{\sqrt{N!}\,\mathrm{e}^{|\alpha|^2}}\, |\phi_{\mathrm{difN}}\rangle_{k,k'}$$

$$|\phi_{\mathrm{difN}}\rangle_{k,k'} = \frac{1}{2^{N/2}} \sum_{m=0}^{N} \mathrm{e}^{\mathrm{i}(N-m)\varphi} \sqrt{\frac{N!}{m!\,(N-m)!}}\, |m\rangle_k\, |N-m\rangle_{k'}$$

Chaotic states and substates:

$$|\Phi_{\mathrm{cha}}\rangle_{k,k'} = \sum_{N=0}^{\infty} c_N^{\beta}\, |\phi_{\mathrm{chaN}}\rangle_{k,k'} = \sum_{N=0}^{\infty} \sqrt{\frac{(N+1)\langle n\rangle^N}{(1+\langle n\rangle)^{N+2}}}\, |\phi_{\mathrm{chaN}}\rangle_{k,k'}$$

$$|\phi_{\mathrm{chaN}}\rangle_{k,k'} = \frac{1}{\sqrt{N+1}} \sum_{m=0}^{N} \mathrm{e}^{\mathrm{i}\phi_{m,N-m}}\, |m\rangle_k |N-m\rangle_{k'}$$

N-photon entangled (NOON) state:

$$|\phi_{\mathrm{entN}}\rangle_{k,k'} = \frac{1}{\sqrt{2}} \left(|N\rangle_k |0\rangle_{k'} + \mathrm{e}^{\mathrm{i}\varphi} |0\rangle_k |N\rangle_{k'} \right)$$

N-photon number state ($N > 1$):

$$|\phi_{\mathrm{numN}}\rangle_{k,k'} = |N/2\rangle_k |N/2\rangle_{k'}$$

[a]The collective states $|\Phi_i\rangle_{k,k'}$ have $\langle n\rangle$ photons in each mode, and for the coherent states we have $|\alpha|^2 = \langle n\rangle$. The substates $|\phi_j\rangle_{k,k'}$ have N-photons in both modes. All states have unit normalization and $\sum_{N=0}^{\infty} |c_N^{\alpha}|^2 = \sum_{N=0}^{\infty} |c_N^{\beta}|^2 = 1$

We also have the important sum rules in first order: $\sum_{N=0}^{\infty} \frac{N}{2} |c_N^{\alpha}|^2 = \sum_{N=0}^{\infty} \frac{N}{2} |c_N^{\beta}|^2 = \langle n\rangle$ and in second order: $\sum_{N=0}^{\infty} \frac{N(N-1)}{4} |c_N^{\alpha}|^2 = \sum_{N=0}^{\infty} \frac{N(N-1)}{6} |c_N^{\beta}|^2 = \langle n\rangle^2$.

16.5 First Order Double-Slit Diffraction Patterns

The key part of the calculation of the first order patterns $\langle \mathbf{P}^{(1)}(\boldsymbol{\rho}_1, \boldsymbol{\rho}_2) \rangle$ given by (16.4) is the evaluation of the matrix elements or expectation values of the operators $\mathbf{X}$ and $\mathbf{Y}$ given by (4.125) and (4.126) with the quantum states in Table 16.1. The lengthy calculations of the matrix elements involve application of the well-known quantum mechanical rules for the action of creation and annihilation operators on the number substates [15, 19] and utilize sum rules like those listed at the bottom of Table 16.1. The matrix elements are derived in Appendix A.9 and for convenience are listed in Table 16.2 for the collective states containing an average of $\langle n\rangle$ photons per mode and in Table 16.3 for the states containing a total of N-photons in both modes.

16.5.1 *Calculation of the First Order Patterns*

For the double-slit geometry in Fig. 16.2a, we have $\mathbf{r}_{\mathrm{A}} \parallel \mathbf{r}_{\mathrm{B}} \parallel \boldsymbol{\rho}_1 \parallel \boldsymbol{\rho}_2$ and the two slits of width a have a center-to-center separation of $\ell = |\mathbf{r}_{\mathrm{A}} - \mathbf{r}_{\mathrm{B}}| = 2r$. The integration in (16.4) over the slits consists of four integrals over source points $\mathbf{r}_U$ and $\mathbf{r}_V$ that

Table 16.2 Matrix elements $\langle \phi_X | \mathbf{O}^{(1)} | \phi_X \rangle = \langle \mathbf{O}^{(1)} \rangle_X$ of the four first order coherence operators $\mathbf{O}^{(1)}$ with the *collective* quantum states of Table 16.1 containing an average number of $\langle n \rangle$ photons *per mode* or a total of $2\langle n \rangle$ photons

Operator $\mathbf{O}^{(1)}$	$\langle \mathbf{O}^{(1)} \rangle_{\mathrm{coh}}$	$\langle \mathbf{O}^{(1)} \rangle_{\mathrm{dif}}$	$\langle \mathbf{O}^{(1)} \rangle_{\mathrm{cha}}$
$\mathbf{a}^{\dagger}_{\mathbf{k}_A}\mathbf{a}_{\mathbf{k}_A}$	$\langle n \rangle$	$\langle n \rangle$	$\langle n \rangle$
$\mathbf{a}^{\dagger}_{\mathbf{k}_B}\mathbf{a}_{\mathbf{k}_B}$	$\langle n \rangle$	$\langle n \rangle$	$\langle n \rangle$
$\mathbf{a}^{\dagger}_{\mathbf{k}_A}\mathbf{a}_{\mathbf{k}_B}$	$\langle n \rangle$	0	0
$\mathbf{a}^{\dagger}_{\mathbf{k}_B}\mathbf{a}_{\mathbf{k}_A}$	$\langle n \rangle$	0	0

Table 16.3 Matrix elements of the first order coherence operators $\mathbf{O}^{(1)}$ with quantum states containing a total of N-photons in two modes

Operator $\mathbf{O}^{(1)}$	$\langle \mathbf{O}^{(1)} \rangle_{\mathrm{cohN}}$	$\langle \mathbf{O}^{(1)} \rangle_{\mathrm{entN}}$	$\langle \mathbf{O}^{(1)} \rangle_{\mathrm{numN}}$	$\langle \mathbf{O}^{(1)} \rangle_{\mathrm{difN}}$	$\langle \mathbf{O}^{(1)} \rangle_{\mathrm{chaN}}$
$\mathbf{a}^{\dagger}_{\mathbf{k}_A}\mathbf{a}_{\mathbf{k}_A}$	$\frac{N}{2}$	$\frac{N}{2}$	$\frac{N}{2}$	$\frac{N}{2}$	$\frac{N}{2}$
$\mathbf{a}^{\dagger}_{\mathbf{k}_B}\mathbf{a}_{\mathbf{k}_B}$	$\frac{N}{2}$	$\frac{N}{2}$	$\frac{N}{2}$	$\frac{N}{2}$	$\frac{N}{2}$
$\mathbf{a}^{\dagger}_{\mathbf{k}_A}\mathbf{a}_{\mathbf{k}_B}$	$\frac{N}{2}$	0	0	0	0
$\mathbf{a}^{\dagger}_{\mathbf{k}_B}\mathbf{a}_{\mathbf{k}_A}$	$\frac{N}{2}$	0	0	0	0

may lie either in the same slit or in different slits. This accounts for correlations between photons born within the same or in different slits. To demonstrate how the patterns for two source points in changed upon integration over slits of finite size, we derive the results for the *collective coherent state* (16.12) and the collective chaotic state (16.18).

16.5.2 Coherent State

For the collective coherent state, $|\Phi_{\mathrm{coh}}\rangle$, all matrix elements $\langle \mathbf{a}^{\dagger}_{\mathbf{k}_X}\mathbf{a}_{\mathbf{k}_Y}\rangle$ for any two points X and Y are the same as seen from Table 16.2 and 16.3. They can hence be pulled out of the integral, and we obtain

$$
\begin{aligned}
\left\langle \mathbf{P}^{(1)}(\boldsymbol{\rho}_1,\boldsymbol{\rho}_2)\right\rangle_{\mathrm{coh}} = \frac{\langle n\rangle}{4a^2}\Bigg\{ & \underbrace{\int\limits_{\ell/2-a/2}^{\ell/2+a/2} \mathrm{e}^{\mathrm{i}\frac{k}{z_0} r_U \rho_1}\,\mathrm{d}r_U \int\limits_{\ell/2-a/2}^{\ell/2+a/2} \mathrm{e}^{-\mathrm{i}\frac{k}{z_0} r_V \rho_2}\,\mathrm{d}r_V}_{\text{upper slit}} \\
& + \underbrace{\int\limits_{-\ell/2-a/2}^{-\ell/2+a/2} \mathrm{e}^{\mathrm{i}\frac{k}{z_0} r_U \rho_1}\,\mathrm{d}r_U \int\limits_{-\ell/2-a/2}^{-\ell/2+a/2} \mathrm{e}^{-\mathrm{i}\frac{k}{z_0} r_V \rho_2}\,\mathrm{d}r_V}_{\text{lower slit}} \\
& + \underbrace{\int\limits_{\ell/2-a/2}^{\ell/2+a/2} \mathrm{e}^{\mathrm{i}\frac{k}{z_0} r_U \rho_1}\,\mathrm{d}r_U}_{\text{upper slit}} \underbrace{\int\limits_{-\ell/2-a/2}^{-\ell/2+a/2} \mathrm{e}^{-\mathrm{i}\frac{k}{z_0} r_V \rho_2}\,\mathrm{d}r_V}_{\text{lower slit}} \\
& + \underbrace{\int\limits_{-\ell/2-a/2}^{-\ell/2+a/2} \mathrm{e}^{\mathrm{i}\frac{k}{z_0} r_U \rho_1}\,\mathrm{d}r_U}_{\text{lower slit}} \underbrace{\int\limits_{\ell/2-a/2}^{\ell/2+a/2} \mathrm{e}^{-\mathrm{i}\frac{k}{z_0} r_V \rho_2}\,\mathrm{d}r_V}_{\text{upper slit}} \Bigg\}.
\end{aligned}
\tag{16.25}
$$

The integrals over the lower and upper slits are evaluated by use of

$$
\text{lower slit}: \quad \frac{1}{a}\int\limits_{-\ell/2-a/2}^{-\ell/2+a/2} \mathrm{e}^{\pm\mathrm{i}\frac{k}{z_0} r\,\rho}\,\mathrm{d}r = \mathrm{e}^{\mp\mathrm{i}\frac{k\ell}{2z_0}\rho}\,\mathrm{sinc}\left[\frac{k\,a}{2z_0}\,\rho\right] \tag{16.26}
$$

and

$$
\text{upper slit}: \quad \frac{1}{a}\int\limits_{\ell/2-a/2}^{\ell/2+a/2} \mathrm{e}^{\pm\mathrm{i}\frac{k}{z_0} r\,\rho}\,\mathrm{d}r = \mathrm{e}^{\pm\mathrm{i}\frac{k\ell}{2z_0}\rho}\,\mathrm{sinc}\left[\frac{k\,a}{2z_0}\,\rho\right]. \tag{16.27}
$$

The sinc functions factor and the exponential terms combine to cos-type modulations according to

$$
\left\langle \mathbf{P}^{(1)}(\boldsymbol{\rho}_1,\boldsymbol{\rho}_2)\right\rangle_{\mathrm{coh}} = \langle n\rangle \underbrace{\cos\left[\frac{k\,\ell}{2z_0}\,\rho_1\right]\cos\left[\frac{k\,\ell}{2z_0}\,\rho_2\right]}_{\text{point like slits}} \underbrace{\mathrm{sinc}\left[\frac{k\,a}{2z_0}\,\rho_1\right]\mathrm{sinc}\left[\frac{k\,a}{2z_0}\,\rho_2\right]}_{\text{finite slit width}}. \tag{16.28}
$$

The integrations over the finite slit widths are seen to give an additional sinc-type function due to the finite slit widths a that represents an envelope function of the finer cos-type modulation of point-like slits. For $\langle n\rangle = 1/2$ *per mode* or a single photon in the two modes and point-like sources we obtain our previous result shown in Fig. 4.17b.

16.5.2.1 Chaotic Case

For the collective chaotic state, only the matrix elements $\langle\Phi_{\text{cha}}|\mathbf{a}^{\dagger}_{\mathbf{k}_\text{I}}\mathbf{a}_{\mathbf{k}_\text{I}}|\Phi_{\text{cha}}\rangle = \langle n\rangle$ with I = A, B listed in Table 16.2 are non-zero . There is no interference of photon probability amplitudes from different slits. In addition, the integral over each slit involves a correlation of the form $\delta(\mathbf{r}_\text{U} - \mathbf{r}_\text{V})$ between different points, since according to Table 16.2 only probability amplitudes from the same points within the slits can interfere. In this case (16.4) reduces to

$$\left\langle \mathbf{P}^{(1)}(\boldsymbol{\rho}_1, \boldsymbol{\rho}_2)\right\rangle_{\text{cha}} = \frac{\langle n\rangle}{4a}\Bigg\{ \underbrace{\int\limits_{-\ell/2-a/2}^{-\ell/2+a/2} \mathrm{e}^{\mathrm{i}\frac{k}{z_0}r|\boldsymbol{\rho}_1-\boldsymbol{\rho}_2|}\,\mathrm{d}r}_{\text{lower slit}} + \underbrace{\int\limits_{\ell/2-a/2}^{\ell/2+a/2} \mathrm{e}^{\mathrm{i}\frac{k}{z_0}r|\boldsymbol{\rho}_1-\boldsymbol{\rho}_2|}\,\mathrm{d}r}_{\text{upper slit}} \Bigg\}. \tag{16.29}$$

The cross-slit interference terms are absent, and by use of (16.26) and (16.27) we obtain

$$\left\langle \mathbf{P}^{(1)}(\boldsymbol{\rho}_1, \boldsymbol{\rho}_2)\right\rangle_{\text{cha}} = \frac{\langle n\rangle}{2}\underbrace{\cos\left[\frac{k\,\ell}{2z_0}|\boldsymbol{\rho}_1 - \boldsymbol{\rho}_2|\right]}_{\text{point like slits}}\underbrace{\text{sinc}\left[\frac{k\,a}{2z_0}|\boldsymbol{\rho}_1 - \boldsymbol{\rho}_2|\right]}_{\text{finite slit width}}. \tag{16.30}$$

For $\langle n\rangle = 1/2$ *per mode* or a single photon in the two modes and $a = 0$ we obtain the pattern previously shown in Fig. 4.17c.

When we write the above patterns as $\langle\mathbf{P}^{(1)}(\boldsymbol{\rho}_1, \boldsymbol{\rho}_2)\rangle = P_1\,\mathcal{G}^{(1)}(\boldsymbol{\rho}_1, \boldsymbol{\rho}_2)$ according to (5.69), we find that in both (16.28) and (16.30) the prefactor P_1 scales with the average number $\langle n\rangle$ of photons/mode emitted by the source, but the shape of the patterns $\mathcal{G}^{(1)}(\boldsymbol{\rho}_1, \boldsymbol{\rho}_2)$ is independent of $\langle n\rangle$. We can summarize as follows.

> The diffraction patterns for two *finite-sized* slits are those of two *point-sized* slits embedded in sinc-type envelope functions. The patterns scale with the average number of photons $\langle n\rangle$ emitted by the slits, but their *shape* is the same for a given quantum state, independent how many photons it contains.

The patterns for the other quantum states in Table 16.1 are derived in Appendix A.9.

16.5.3 *Plots of the First Order Patterns*

The calculated shapes $\mathcal{G}^{(1)}(\boldsymbol{\rho}_1, \boldsymbol{\rho}_2)$ of all *first order* patterns for the quantum states in Table 16.1 (see Appendix A.9) are plotted in Fig. 16.4. For convenience, both the shapes and scaling factors P_1 for the states are listed on the right. Remarkably, the different quantum states result in only two kinds of characteristic patterns, revealing a degeneracy of the patterns for several of the states. The patterns are identical to those derived by wave or statistic optics for the limiting cases of "coherent" and "incoherent" light [6, 26].

The *coherent states* and their substates produce the same characteristic pattern, shown as dashed thick blue and red curves in Fig. 16.4a for $a = \ell/4$. The patterns are the same for the blue and red detection geometries in Fig. 16.2b. This is due to the fact that the detection probability $\langle \mathbf{P}^{(1)}(\boldsymbol{\rho}_1, \boldsymbol{\rho}_2)\rangle$ *factors* in the coordinates $\boldsymbol{\rho}_1$ and $\boldsymbol{\rho}_2$ [16]. For convenience we also show the two limits of point-like slits of width $a \ll \ell$ as a gray curve, and a single slit of width a, represented by the thin red-blue sinc^2 envelope function.

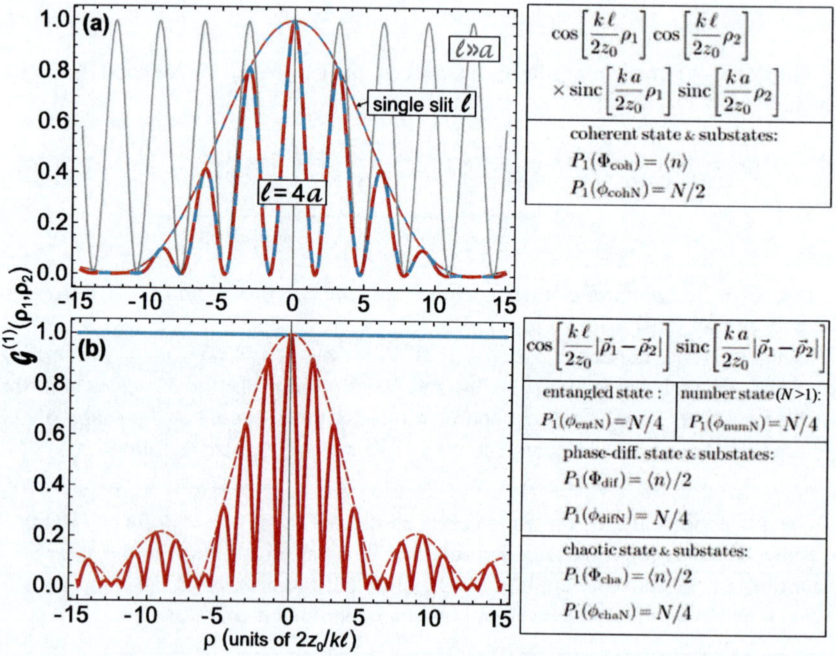

Fig. 16.4 First order double-slit diffraction patterns for the quantum states of light in Table 16.1, with the blue and red colors representing the two detection schemes in Fig. 16.2b. The shape of the patterns $\mathcal{G}^{(1)}(\boldsymbol{\rho}_1, \boldsymbol{\rho}_2)$ and scaling factors P_1 are given on the right. **a** Patterns $\mathcal{G}^{(1)}(\boldsymbol{\rho}_1, \boldsymbol{\rho}_2)$ for the coherent state and its substates representing the cases in Fig. 16.1a. The gray curve assumes point-like slits of width $a \ll \ell$, the thick red-blue pattern is that for $a = \ell/4$, and the thin red-blue line represents the sinc^2 envelope function which remains if the two slits are joined into a single slit of width a. **b** Patterns for all other quantum states and cases in Fig. 16.1b–e

All other states in Table 16.1 form the patterns shown in Fig. 16.4b. The detection probability does not factor in the coordinates $\boldsymbol{\rho}_1$ and $\boldsymbol{\rho}_2$, and the source is "incoherent" in first order. This results in *constant* photon counts for the blue detection geometry in Fig. 16.2b with a diffraction structure observed only for the red detection geometry.

In all cases, the *shapes* of the patterns $\mathcal{G}^{(1)}(\boldsymbol{\rho}_1, \boldsymbol{\rho}_2)$ in Fig. 16.4 are the same for the three *collective states* containing an average of $2\langle n \rangle$ photons and their respective *substates* containing N-photons. The shapes are even independent of the number of photons, N, contained in the substates. Emission of more than single photons, $N > 1$, only increases the scaling factor P_1, i.e. the overall detection probability. This explains the experimental fact that the first order pattern for the case shown in Fig. 16.1a is independent of whether the slits are illuminated by a thermal source that has been made first order coherent by energy (monochromator) and spatial (aperture) filtering or by a laser, although the light is coherent to different orders in QED for the two cases.

This means that the nature of the collective states remains preserved in the coefficients and phases of their respective substates, independent of the number of photons, N, they contain. The patterns in Fig. 16.4 may therefore be recorded with one-photon detectors that pick out the $N = 1$ state or charge integrating detectors which detect arbitrary numbers of arriving photons. Note that even the collective states with $\langle n \rangle = 1$ contain different number of photons $N = 1, 2, 3 \ldots$ as shown in Fig. 16.3. By use of charge or photon number integrating detectors the diffraction patterns therefore accumulate through the arrival of photons whose number statistically varies between $N = 1$ and about $N = 6$.

This shows that the first order patterns do not depend on the *photon degeneracy parameter* or number of emitted photons per mode, $\langle n \rangle$ or $N/2$. This agrees with Dirac's statement that photons do not interfere with each other [19]. The subtlety of this statement, however, is revealed by the fact that photons in the same mode are in principle *indistinguishable* as emphasized by Ou et al. [27, 28], and one would therefore expect them to interfere. The dilemma is resolved by the fact that a meaningful diffraction pattern always corresponds to the accumulation of a large number of detection events. Any interference structure that may be present when only few photons arrive in coincidence is increasingly averaged out upon appearance of a statistically meaningful pattern [47, 48].

In x-ray science, the *independence* of the Bragg diffraction pattern on the degeneracy parameter has the important consequence that all diffraction patterns observed for the last 100 years with weak sources such as Röntgen tubes remain the same when recorded in a single shot with an x-ray free electron laser, despite an increase of the degeneracy parameter by about 25 orders of magnitude [6]. This means that it does not matter whether the pattern assembles one-photon-at-a-time or in a "single shot". With increasing degeneracy parameter the statistics of the pattern is simply improved. The pattern may be recorded much faster, even in a single few-femtosecond shot [49]. Hence no new x-ray diffraction theory is needed to describe first order patterns.

16.5.4 Degree of First Order Coherence

The first order diffraction probabilities $\langle \mathbf{P}^{(1)}(\boldsymbol{\rho}_1, \boldsymbol{\rho}_2) \rangle$ for the cases in Fig. 16.4 are readily converted by means of (16.7) into the corresponding degrees of first order coherence. For the double-slit geometry we have

$$g^{(1)}(\boldsymbol{\rho}, -\boldsymbol{\rho}) = \frac{\langle \mathbf{P}^{(1)}(\boldsymbol{\rho}, -\boldsymbol{\rho}) \rangle}{\langle \mathbf{P}^{(1)}(\boldsymbol{\rho}, \boldsymbol{\rho}) \rangle}. \tag{16.31}$$

For the coherent case in Fig. 16.4a we obtain with (A.51) the expected result

$$g^{(1)}(\boldsymbol{\rho}, -\boldsymbol{\rho}) = 1. \tag{16.32}$$

The "incoherent" cases in Fig. 16.4b all yield the same expression

$$g^{(1)}(\boldsymbol{\rho}, -\boldsymbol{\rho}) = \cos\left[\frac{k\,\ell}{z_0}\rho\right] \operatorname{sinc}\left[\frac{k\,a}{z_0}\rho\right]. \tag{16.33}$$

Hence we have $g^{(1)}(\boldsymbol{\rho}, -\boldsymbol{\rho}) = \mathcal{G}^{(1)}(\boldsymbol{\rho}, -\boldsymbol{\rho})$ for these cases, represented by the red pattern Fig. 16.4b.

16.5.5 Reduction of First-Order Quantum to Wave Formalism

The *first order* quantum formulation of diffraction gives the same patterns as the classical or statistical optics descriptions. In the quantum formulation, however, different diffraction patterns may directly be linked to different quantum states created in the source.

The conventional *wave formalism* of diffraction emerges from the quantum descriptions by approximating the average over all photon emission directions by the concept of a spherical wave. One then makes the *ad hoc* assumption that all waves are created with the same birth phase. The distinction between coherence and incoherence arises from geometry alone. Spherical waves emitted from different source points are said to be coherent at a distant observation point, if one may approximate them by *plane waves* within a finite solid angle "observation cone" extending backward from the observation point to the source points. The picture may also be turned around by defining a coherent *emission cone* rather than a coherent *observation cone*. This simple geometrical concept means that within the solid angle coherence cone the curvature of the spherical wave is negligible on the scale of the wavelength. Classically one simply calculates the "intensity" at different points in the observation plane as the absolute value squared of interfering wave fields. The measurement process of the intensity remains unspecified.

One might have expected that an incoherent or chaotic source, reflected by the *chaotic state* in quantum optics, does not give rise to a diffraction pattern at all. The reason for its existence, revealed by the red pattern in Fig. 16.4b, lies in the fact that for the spatial phenomenon of diffraction one assumes space-time separability and temporal coherence, i.e. that the photons have the same well-defined energy or wavelength λ. If the source was chaotic in both space and time, there would indeed be no position-dependent diffraction structure.

In wave or statistical optics, the pattern in Fig. 16.4b for chaotic quantum states produced in the source is predicted by the powerful van Cittert–Zernike theorem [6, 26, 50–52]. It picks out the coherent fraction of light emitted by a source, even if the source is chaotic. In classical electromagnetism, radiation (i.e. acceleration fields [53]) can only separate from the charge if it is defined at least over the dimension of its average wavelength. Hence any light-emitting source contains a coherent fraction that arises from waves emitted from the minimum coherence area of order λ^2 [54, 55]. In classical optics, the 1D pattern in Fig. 16.4b arises from waves emitted from regions of lateral size $\simeq \lambda$ within the two slits that interfere at detection points.

The more fundamental photon nature of light, described by QED, just happens to reduce in first order to the conventional wave formalism. One has to realize, however, that the classical theory is based on *ad hoc* assumptions or postulates which make it work. The fundamental origin of these assumptions, like the perceived existence of spherical waves and the validity of the Fresnel–Huygens principle, emerges in lowest order QED as a consequence of the interference of single photon probability amplitudes associated with all possible photon paths to a given detection point.

The independence of the first order quantum pattern of the total number of emitted photons N, or the average number of photons $2\langle n\rangle$ (which reflect the photon degeneracy parameter of the source) also reveals why the detection process did not have to be specified in the classical formulation. The diffracted "intensity" is simply calculated as the absolute value squared of the "wave field", and the quality or statistics of the pattern improves linearly with "intensity". Historically, this allowed the use of wave concept long before the true photon nature of light was known.

In the following sections we will show that only QED can account for more sophisticated diffraction experiments carried out by changing the detection process, revealing the intrinsic limitation of the wave theory of light.

16.6 Second Order Double-Slit Diffraction Patterns

For the quantum states in Table 16.3, the *second order* diffraction patterns $\langle \mathbf{P}^{(2)}(\boldsymbol{\rho}_1, \boldsymbol{\rho}_2)\rangle$, given by (16.5), are calculated by use of the matrix elements of the four second order terms (5.15)–(5.18). The matrix elements, derived in Appendix A.10, are given in Table 16.4 for the collective states containing an average of $\langle n\rangle$ photons *per mode* (or a total of $2\langle n\rangle$) and in Table 16.5 for the states containing a total of N-photons in both modes.

By use of these matrix elements the integrations in (16.6) over the finite slit widths involve the integrals (16.26) and (16.27). As for the first order cases they will give sinc-type functions that account for the finite slit widths and can be factored. The

Table 16.4 Matrix elements of the second order normally ordered operator products, listed in the same sequence as in (5.15)–(5.18) with the *collective* quantum states of Table 16.1

Term	Operator $\mathbf{O}^{a}$	$\langle\mathbf{O}\rangle_{coh}$	$\langle\mathbf{O}\rangle_{dif}$	$\langle\mathbf{O}\rangle_{cha}$
A	$\mathbf{a}^\dagger_{\mathbf{k}_A}\mathbf{a}^\dagger_{\mathbf{k}_B}\mathbf{a}_{\mathbf{k}_A}\mathbf{a}_{\mathbf{k}_B}$	$\langle n\rangle^2$	$\langle n\rangle^2$	$\langle n\rangle^2$
B	$\mathbf{a}^\dagger_{\mathbf{k}_A}\mathbf{a}^\dagger_{\mathbf{k}_A}\mathbf{a}_{\mathbf{k}_A}\mathbf{a}_{\mathbf{k}_A}$	$\langle n\rangle^2$	$\langle n\rangle^2$	$2\langle n\rangle^2$
	$\mathbf{a}^\dagger_{\mathbf{k}_B}\mathbf{a}^\dagger_{\mathbf{k}_B}\mathbf{a}_{\mathbf{k}_B}\mathbf{a}_{\mathbf{k}_B}$	$\langle n\rangle^2$	$\langle n\rangle^2$	$2\langle n\rangle^2$
	$\mathbf{a}^\dagger_{\mathbf{k}_A}\mathbf{a}^\dagger_{\mathbf{k}_A}\mathbf{a}_{\mathbf{k}_B}\mathbf{a}_{\mathbf{k}_B}$	$\langle n\rangle^2$	0	0
	$\mathbf{a}^\dagger_{\mathbf{k}_B}\mathbf{a}^\dagger_{\mathbf{k}_B}\mathbf{a}_{\mathbf{k}_A}\mathbf{a}_{\mathbf{k}_A}$	$\langle n\rangle^2$	0	0
C, D	$\mathbf{a}^\dagger_{\mathbf{k}_A}\mathbf{a}^\dagger_{\mathbf{k}_A}\mathbf{a}_{\mathbf{k}_A}\mathbf{a}_{\mathbf{k}_B}$	$\langle n\rangle^2$	0	0
	$\mathbf{a}^\dagger_{\mathbf{k}_A}\mathbf{a}^\dagger_{\mathbf{k}_B}\mathbf{a}_{\mathbf{k}_B}\mathbf{a}_{\mathbf{k}_B}$	$\langle n\rangle^2$	0	0
	$\mathbf{a}^\dagger_{\mathbf{k}_A}\mathbf{a}^\dagger_{\mathbf{k}_B}\mathbf{a}_{\mathbf{k}_A}\mathbf{a}_{\mathbf{k}_A}$	$\langle n\rangle^2$	0	0
	$\mathbf{a}^\dagger_{\mathbf{k}_B}\mathbf{a}^\dagger_{\mathbf{k}_B}\mathbf{a}_{\mathbf{k}_A}\mathbf{a}_{\mathbf{k}_B}$	$\langle n\rangle^2$	0	0

[a]Note that $\mathbf{a}^\dagger_{\mathbf{k}_A}\mathbf{a}^\dagger_{\mathbf{k}_B}=\mathbf{a}^\dagger_{\mathbf{k}_B}\mathbf{a}^\dagger_{\mathbf{k}_A}$ and $\mathbf{a}_{\mathbf{k}_A}\mathbf{a}_{\mathbf{k}_B}=\mathbf{a}_{\mathbf{k}_B}\mathbf{a}_{\mathbf{k}_A}$. The order of two adjacent creation or destruction operators may be switched, and we have listed the operator with $\mathbf{k}_A$ first. The states contain an *average number* of $\langle n\rangle$ photons *per mode* or a total number of $2\langle n\rangle$ photons.

Table 16.5 Matrix elements as in Table 16.4 but for the states containing a *total number* of N-photons in the two modes

Term	Operator $\mathbf{O}$	$\langle\mathbf{O}\rangle_{cohN}$	$\langle\mathbf{O}\rangle_{entN}$	$\langle\mathbf{O}\rangle_{numN}$	$\langle\mathbf{O}\rangle_{difN}$	$\langle\mathbf{O}\rangle_{chaN}$
A	$\mathbf{a}^\dagger_{\mathbf{k}_A}\mathbf{a}^\dagger_{\mathbf{k}_B}\mathbf{a}_{\mathbf{k}_A}\mathbf{a}_{\mathbf{k}_B}$	$\frac{1}{4}N(N-1)$	0	$\frac{1}{4}N^2$	$\frac{1}{4}N(N-1)$	$\frac{1}{6}N(N-1)$
B	$\mathbf{a}^\dagger_{\mathbf{k}_A}\mathbf{a}^\dagger_{\mathbf{k}_A}\mathbf{a}_{\mathbf{k}_A}\mathbf{a}_{\mathbf{k}_A}$	$\frac{1}{4}N(N-1)$	$\frac{1}{2}N(N-1)$	$\frac{1}{4}N(N-2)$	$\frac{1}{4}N(N-1)$	$\frac{1}{3}N(N-1)$
	$\mathbf{a}^\dagger_{\mathbf{k}_B}\mathbf{a}^\dagger_{\mathbf{k}_B}\mathbf{a}_{\mathbf{k}_B}\mathbf{a}_{\mathbf{k}_B}$	$\frac{1}{4}N(N-1)$	$\frac{1}{2}N(N-1)$	$\frac{1}{4}N(N-2)$	$\frac{1}{4}N(N-1)$	$\frac{1}{3}N(N-1)$
	$\mathbf{a}^\dagger_{\mathbf{k}_A}\mathbf{a}^\dagger_{\mathbf{k}_A}\mathbf{a}_{\mathbf{k}_B}\mathbf{a}_{\mathbf{k}_B}$	$\frac{1}{4}N(N-1)$	$\delta(N,2)$	0	0	0
	$\mathbf{a}^\dagger_{\mathbf{k}_B}\mathbf{a}^\dagger_{\mathbf{k}_B}\mathbf{a}_{\mathbf{k}_A}\mathbf{a}_{\mathbf{k}_A}$	$\frac{1}{4}N(N-1)$	$\delta(N,2)$	0	0	0
C, D	$\mathbf{a}^\dagger_{\mathbf{k}_A}\mathbf{a}^\dagger_{\mathbf{k}_A}\mathbf{a}_{\mathbf{k}_A}\mathbf{a}_{\mathbf{k}_B}$	$\frac{1}{4}N(N-1)$	0	0	0	0
	$\mathbf{a}^\dagger_{\mathbf{k}_A}\mathbf{a}^\dagger_{\mathbf{k}_B}\mathbf{a}_{\mathbf{k}_B}\mathbf{a}_{\mathbf{k}_B}$	$\frac{1}{4}N(N-1)$	0	0	0	0
	$\mathbf{a}^\dagger_{\mathbf{k}_A}\mathbf{a}^\dagger_{\mathbf{k}_B}\mathbf{a}_{\mathbf{k}_A}\mathbf{a}_{\mathbf{k}_A}$	$\frac{1}{4}N(N-1)$	0	0	0	0
	$\mathbf{a}^\dagger_{\mathbf{k}_B}\mathbf{a}^\dagger_{\mathbf{k}_B}\mathbf{a}_{\mathbf{k}_A}\mathbf{a}_{\mathbf{k}_B}$	$\frac{1}{4}N(N-1)$	0	0	0	0

The symbol $\delta(N,2)$ means that we obtain 1 only for $N=2$ and 0 otherwise.

remaining sum of exponentials reduce to cos-type expressions, representing the case of point-like slits. As an example we shall discuss below the calculation of the pattern for the collective coherent state.

16.6.1 Coherent State

For the collective coherent state all matrix elements in Table 16.4 have the same value, independent of the permutations X,Y,U,V=A,B, given by $\langle\Phi_{coh}|\mathbf{a}^\dagger_{\mathbf{k}_U}\mathbf{a}^\dagger_{\mathbf{k}_V}\mathbf{a}_{\mathbf{k}_X}\mathbf{a}_{\mathbf{k}_Y}|\Phi_{coh}\rangle=\langle n\rangle^2$. They can therefore be factored and (16.6) is evaluated by use of (16.26) and (16.27) as

$$\left\langle\mathbf{P}^{(2)}(\rho_1,\rho_2)\right\rangle_{coh}=\langle n\rangle^2\,\mathrm{sinc}^2\left[\frac{k\,a}{2z_0}\rho_1\right]\mathrm{sinc}^2\left[\frac{k\,a}{2z_0}\rho_2\right]\sum Z. \tag{16.34}$$

We are left with a sum denoted $\sum Z$ consisting of sixteen exponential terms, each of the form

$$\mathrm{e}^{\pm\mathrm{i}\frac{k\ell}{2z_0}\rho_1}\,\mathrm{e}^{\pm\mathrm{i}\frac{k\ell}{2z_0}\rho_1}\mathrm{e}^{\pm\mathrm{i}\frac{k\ell}{2z_0}\rho_2}\,\mathrm{e}^{\pm\mathrm{i}\frac{k\ell}{2z_0}\rho_2}. \tag{16.35}$$

The two possible signs in each term correspond to parallel or antiparallel orientations of the source coordinates $\mathbf{r}_\mathrm{I}$ relative to the detector coordinates $\boldsymbol{\rho}_i$ when the dot products in (16.6) are evaluated. The sixteen terms are those in (5.15)–(5.18) with unit expectation values of all operators. The terms combine to cos-type terms representing the case where the two slits are points, previously treated in Sect. 5.5.2. After some algebra we obtain the final result

$$\left\langle \mathbf{P}^{(2)}(\boldsymbol{\rho}_1,\boldsymbol{\rho}_2)\right\rangle_{\mathrm{coh}} = \langle n\rangle^2 \underbrace{\cos^2\left[\frac{k\,\ell}{2z_0}\rho_1\right]\cos^2\left[\frac{k\,\ell}{2z_0}\rho_2\right]}_{\text{point like slits}} \underbrace{\mathrm{sinc}^2\left[\frac{k\,a}{2z_0}\rho_1\right]\mathrm{sinc}^2\left[\frac{k\,a}{2z_0}\rho_2\right]}_{\text{finite slit width}}. \tag{16.36}$$

The *second order* patterns for the other quantum states in Table 16.1 are derived in Appendix A.10.

16.6.2 Plots of the Second Order Patterns

The calculated shapes $\mathcal{G}^{(2)}(\boldsymbol{\rho}_1,\boldsymbol{\rho}_2)$ are plotted in Fig. 16.5 in the order of the cases in Fig. 16.1. Again the shapes and scaling factors P_2 are given on the right for convenience. For the same quantum states, the degeneracy still present in the first order patterns in Fig. 16.4b is now lifted. The diffraction patterns have become characteristic signatures of the different quantum states of light, revealing the new paradigm.

The pattern in Fig. 16.5a is characteristic of *coherent states* emitted by the source. The shape is independent of the total average number of photons, $2\langle n\rangle$, and the specific photon number, N, in the substates which only determine the scaling factor P_2. The coherent states and their substates give the same patterns for the two detection schemes in Fig. 16.2c.

For the specific source scheme on the right in Fig. 16.1a, the source is described by a second order collective coherent state that contains an average number of $\langle n\rangle = 2$ photons per mode. The collective state consists of a Poisson substate distribution extending up to about $N = 10$ as shown in Fig. 16.3a. The collective state may be imaged in two ways. The two-photon coincidence detection scheme in Fig. 16.2c will pick out its representative $N = 2$ substate defined by the arrival of photon pairs. On the other hand, a CCD detector will integrate over all arriving numbers of photons and thus record the entire collective state, as previously conjectured [56, 57]. This is discussed in more detail in conjunction with the degree of second order coherence of the collective coherent state and its substates in the following section.

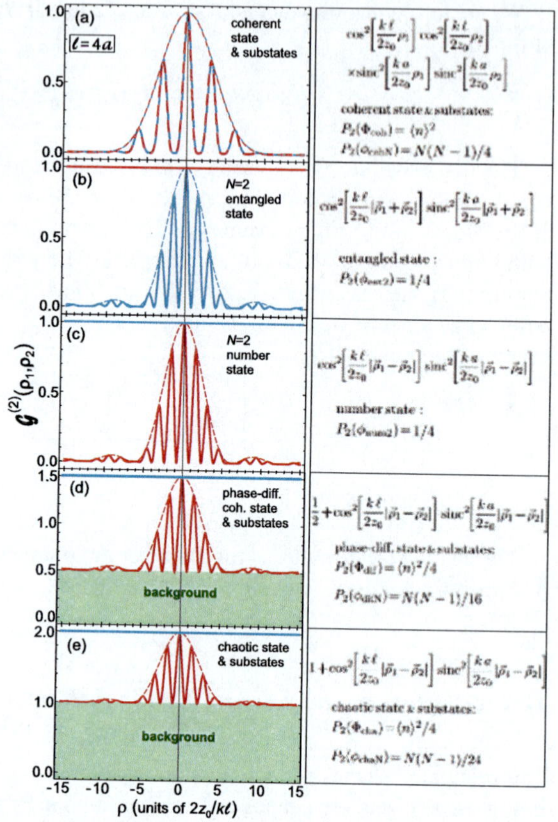

Fig. 16.5 Second order double-slit diffraction patterns for the indicated two-mode quantum states and their substates for $\ell = 4a$. Blue and red colors represent patterns for the two detection schemes in Fig. 16.2c. **a** Pattern $\mathcal{G}^{(2)}(\boldsymbol{\rho}_1, \boldsymbol{\rho}_2)$ for the coherent state and its binomial substates. **b** Pattern for the $N = 2$ entangled state, and **c** for the $N=2$ number state. **d** Pattern for the phase-averaged coherent state and its substates, and **e** for the chaotic state and its substates

The patterns of the $N=2$ entangled state $|\phi_{\text{ent2}}\rangle$ given by (16.22) and number state $|\phi_{\text{num2}}\rangle$ expressed by (16.24) are shown in Fig. 16.5b, c. They are complete opposites, corresponding to the exchange of the red and blue detection schemes in Fig. 16.2c. In particular, the well-studied entangled state $|\phi_{\text{ent2}}\rangle$ [21, 27, 28] incorporates the essence of quantum behavior since it is maximally entangled, and it plays a prominent role in quantum information science [43–46]. The complementary behavior of the two states holds a central position in quantum optics since their diffraction patterns cannot be explained by the wave formalism. The same behavior was already encountered for a circular flat-top source as shown in Fig. 5.8.

The key role of the two-photon entangled and number states is furthermore revealed by the fact that the two-photon case of all five source implementations in Fig. 16.1 may be represented as a linear combination of the two states, written in the general form

$$|\phi_{N=2}\rangle_{k,k'} = a_{11}\mathrm{e}^{\mathrm{i}\phi_{11}}|1\rangle_k\,|1\rangle_{k'} + a_{02}\mathrm{e}^{\mathrm{i}\phi_{02}}|0\rangle_k|2\rangle_{k'} + a_{20}\mathrm{e}^{\mathrm{i}\phi_{20}}|2\rangle_k|0\rangle_{k'}, \qquad (16.37)$$

where the coefficients a_{ij} are real and for equal occupation of the two modes we have $a_{02} = a_{20}$. The normalization condition furthermore links the coefficients by $a_{11}^2 + a_{20}^2 + a_{02}^2 = 1$.

In particular, the $N=2$ coherent substate (16.14), given by

$$|\phi_{\mathrm{coh2}}\rangle_{k,k'} = \frac{1}{2}\left\{\sqrt{2}|1\rangle_k|1\rangle_{k'} + \Big[|0\rangle_k|2\rangle_{k'} + |2\rangle_k|0\rangle_{k'}\Big]\right\} \tag{16.38}$$

corresponds to an *in-phase* addition of the $N = 2$ state $|\phi_{\mathrm{num2}}\rangle$ and the state $|\phi_{\mathrm{ent2}}\rangle$, identified by square brackets. Thus the coherent two-photon pattern in Fig. 16.5a arises from the interference of the number and entangled states, expressed by (A.73).

The $N = 2$ entangled and number states are also responsible for the evolution of the patterns in Figs. 16.5c–e. The number state $|\phi_{\mathrm{num2}}\rangle = |1\rangle_k|1\rangle_{k'}$ creates the diffraction fine structure shown in (c) which is superposed on a background, shown in green, in the patterns of the collective phase-diffused and chaotic states and their substates in (d) and (e). The size of the background in these patterns is determined by different relative contributions of the entangled and number states, which combine with random relative phases in the general two-photon state (16.37). The key difference of the $N = 2$ phase-diffused coherent substate (16.17) and the chaotic substate (16.20) is revealed by writing them respectively as

$$|\phi_{\mathrm{dif2}}\rangle_{k,k'} = \frac{1}{\sqrt{2}}\left\{\mathrm{e}^{\mathrm{i}\varphi}|1\rangle_k|1\rangle_{k'} + \frac{1}{\sqrt{2}}\Big[\mathrm{e}^{2\mathrm{i}\varphi}|0\rangle_k|2\rangle_{k'} + |2\rangle_k|0\rangle_{k'}\Big]\right\} \tag{16.39}$$

and

$$|\phi_{\mathrm{cha2}}\rangle_{k,k'} = \frac{1}{\sqrt{3}}\left\{\mathrm{e}^{\mathrm{i}\varphi_1}|1\rangle_k|1\rangle_{k'} + \Big[\mathrm{e}^{\mathrm{i}\varphi_2}|0\rangle_k|2\rangle_{k'} + |2\rangle_k|0\rangle_{k'}\Big]\right\}. \tag{16.40}$$

In the last expression we have rewritten (16.20) by eliminating an unimportant overall phase factor through the choice $\phi_{20} = 0$, $\phi_{02} = \varphi_2$, and $\phi_{11} = \varphi_1$.

The different contributions of the entangled substate in (16.39) and (16.40), identified by rectangular brackets, relative to the number state is the reason for the change in the green background in the patterns in Fig. 16.5c–e. The background quantitatively scales with the square of the coefficients expressing the number and entangled state contributions to these states.

The entangled state reflects the simultaneous birth of two photons within a given slit. This situation is encountered in practice for incoherent or chaotic sources since the creation of only single photons per slit, reflected by a pure $|1\rangle_k|1\rangle_{k'}$ state (pattern in Fig. 16.5c), requires special source preparation [33–36].

The dashed red envelope function of the chaotic state pattern in Fig. 16.5e is the 1D manifestation of the famous Hanbury Brown–Twiss (HBT) result [58–62], where a circular or rectangular 2D source is replaced by a 1D slit of width a. The HBT effect was first derived quantum mechanically by Fano [63] using Feynman's concepts of probability amplitudes and played an important role in Glauber's development of quantum optics as recalled in his Nobel lecture [64].

In QED, the HBT arises naturally because of the structure of the chaotic quantum states. This is best revealed by the two-photon chaotic substate (16.40) which contains an entangled part of 50%. It is therefore not surprising that many quantum optics experiments, first performed with entangled biphotons, can also be performed with chaotic light. A prominent example is "ghost imaging" [21, 65] which is also possible with chaotic sources [66], albeit with reduced contrast [67].

The HBT effect may also be explained classically as arising from "intensity fluctuations" [52, 68] which of course are nothing but the fluctuations in the created number of photons per unit time and area. In the semi-classical picture, the constant background is typically attributed to two photons that "accidentally" arrive at the two detectors in coincidence. This arrival condition is assured for entangled biphotons created in parametric down conversion by their simultaneous birth at the same place. It may occur for a chaotic source when two photons born at different times and positions "accidentally" arrive at the same time because the difference in birth time is compensated by the difference in travel time (distance) [6]. From a practical or detection point of view, the entangled and "accidental" scenarios are indistinguishable. The formal statistical optics derivation utilizes the so-called Reed theorem [69] or complex Gaussian moment theorem [26, 70], which veils the underlying fundamental quantum processes.

16.6.3 Degree of Second Order Coherence

The difference of the second order patterns in Fig. 16.5 is also reflected by the degree of second order coherence (16.7) of the respective quantum states, which for the double-slit geometry is expressed by

$$g^{(2)}(\boldsymbol{\rho}, -\boldsymbol{\rho}) = \frac{\langle \mathbf{P}^{(2)}(\boldsymbol{\rho}, -\boldsymbol{\rho}) \rangle}{\langle \mathbf{P}^{(1)}(\boldsymbol{\rho}, \boldsymbol{\rho}) \rangle^2}. \tag{16.41}$$

It is evaluated in Appendix A.11 for the different cases in Fig. 16.5.

For the *collective* coherent state $|\Phi_{\text{coh}}\rangle$, we have $g^{(2)}_{\text{coh}}(\boldsymbol{\rho}, -\boldsymbol{\rho}) = 1$ which together with $g^{(1)}_{\text{coh}}(\boldsymbol{\rho}, -\boldsymbol{\rho}) = 1$ given by (16.32) is the signature of a second order coherent state [16, 71]. It is remarkable that the coherent *substates* $|\phi_{\text{cohN}}\rangle$ are *not* second order coherent since $g^{(2)}_{\text{cohN}}(\boldsymbol{\rho}, -\boldsymbol{\rho}) = 1 - \frac{1}{N}$, which approaches the coherent value of unity only in the limit of a large number of photons in the substates (see Sect. 5.6.3.1).

In particular, we obtain $g^{(2)}_{\text{coh2}}(\boldsymbol{\rho}, -\boldsymbol{\rho}) = 1/2$ for the two-photon coherent substate $|\phi_{\text{coh2}}\rangle$ given by (16.38). This state, describing a *coherent biphoton* [6, 56], is created when a single photon clones itself in a stimulated decay process. Its lack of second order coherence, which was previously not recognized [57], arises from the fact that in the presence of a single photon, an atom may also decay with equal probability via spontaneous photon emission. This is expressed by the well-known factor $1 + n$, where 1 is the relative probability that an excited electronic state decays *spontaneously* in the absence of other photons, and n is the relative probability that the decay is *stimulated* by the presence of n photons in the same mode.

The lack of second order coherence of the $N = 2$ substate $|\phi_{\mathrm{coh2}}\rangle$, produced through single photon ($n = 1$) stimulation, is reflected by the so-called no cloning theorem [72–75]. Second order coherence and perfect cloning are only reached when many photons in the same mode cooperate to completely control excitation and de-excitation of an atom [31]. This corresponds to the case shown on the right in Fig. 16.1a where an incident temporally coherent pulse with high degeneracy parameter drives the atoms in the film to a collective second order coherent state with equal populations in the ground and excited states. The coupled atom-photon system then becomes second order coherent, and the "no cloning" theorem no longer applies.

The two-photon entangled state $|\phi_{\mathrm{ent2}}\rangle$, given by (16.22), yields the surprising result $g^{(2)}_{\mathrm{ent2}}(\boldsymbol{\rho}, -\boldsymbol{\rho}) = 1$. A state is second order coherent, however, only if both $g^{(2)}$ and $g^{(1)}$ are unity. This is not the case because $g^{(1)}$ is not unity according to (16.33). Instead, the state $|\phi_{\mathrm{ent2}}\rangle$ causes the constant background in Fig. 16.5d, e as discussed in the previous section. The unit value of the background is therefore deceiving since it does not reflect second order coherence, which in the semi-classical explanation has led to its description as "accidental coincidences".

For the two-photon number state $|\phi_{\mathrm{num2}}\rangle$, the collective phase-diffused coherent state $|\Phi_{\mathrm{dif}}\rangle$ and the collective chaotic state $|\Phi_{\mathrm{dif}}\rangle$, we find $g^{(2)}(\boldsymbol{\rho}, -\boldsymbol{\rho}) = \mathcal{G}^{(2)}(\boldsymbol{\rho}, -\boldsymbol{\rho})$. The red patterns in Fig. 16.5c–e therefore represent the degree of second order coherence of these states. For the substates $|\phi_{\mathrm{difN}}\rangle$ and $|\phi_{\mathrm{chaN}}\rangle$ the expressions for $g^{(2)}$, given by (A.108) and (A.110), are similar but contain additional prefactors.

The values of the degree of second order coherence $g^{(2)}(\boldsymbol{\rho}_1, \boldsymbol{\rho}_2)$ of the different quantum states complement the information revealed by the shapes of their diffraction patterns $\mathcal{G}^{(2)}(\boldsymbol{\rho}_1, \boldsymbol{\rho}_2)$ in Fig. 16.5. While the patterns of the collective states $\langle\Phi_{\mathrm{i}}|\mathbf{P}^{(2)}(\boldsymbol{\rho}_1, \boldsymbol{\rho}_2)|\Phi_{\mathrm{i}}\rangle$ and their substates $\langle\phi_{\mathrm{iN}}|\mathbf{P}^{(2)}(\boldsymbol{\rho}_1, \boldsymbol{\rho}_2)|\phi_{\mathrm{iN}}\rangle$ can in principle be distinguished through their scaling factors P_2, their difference is directly revealed by the normalized degree of second order coherence $g^{(2)}(\boldsymbol{\rho}_1, \boldsymbol{\rho}_2)$. Examples are the different values of $g^{(2)}_{\mathrm{coh}} = 1$ for the collective coherent state and $g^{(2)}_{\mathrm{coh2}} = 1/2$ for its two-photon substate which exhibit the same diffraction shapes.

16.6.4 The Evolution from First to Second Order

The behavior of independent photons in first order QED may also be accounted for by the wave theory, augmented by certain *ad hoc* recipes like the Huygens–Fresnel principle which make it work. In second order QED, phenomena arise that simply cannot be explained by the wave theory of light, clearly revealing its incompleteness. The hallmark of second order QED is the existence of correlations between photons, the most heralded phenomenon being photon *entanglement* over large distances [45, 46]. More precisely, the second order case covers phenomena associated with a number of $N \geq 2$ photons that arise from the correlations between their probability amplitudes. These correlations are absent in first order QED or conventional quantum mechanics, as expressed by Dirac's famous statement.

Comparison of the first order patterns in Fig. 16.4 with the second order ones in Fig. 16.5 directly reveals how the remaining degeneracy in the first order patterns is lifted in second order. In particular, the evolution leads to distinct patterns for the fundamental two-mode entangled and number states, and their central role becomes apparent.

In all cases, the shapes of the second order diffraction profiles (apart from any constant background) are seen to be the *square* of the corresponding first order patterns. For the coherent states the *effective width* of the first order pattern for $\ell \geq 2a$ is given by [6]

$$\frac{ka}{\pi z_0} \int_{-\infty}^{\infty} \cos^2\left[\frac{k\,\ell}{2z_0}\rho\right] \operatorname{sinc}^2\left[\frac{k\,a}{2z_0}\rho\right] \mathrm{d}\rho = 1 \tag{16.42}$$

while that of the second order coherent pattern is given by

$$\frac{ka}{\pi z_0} \int_{-\infty}^{\infty} \cos^4\left[\frac{k\,\ell}{2z_0}\rho\right] \operatorname{sinc}^4\left[\frac{k\,a}{2z_0}\rho\right] \mathrm{d}\rho = \frac{1}{2}. \tag{16.43}$$

Photon conservation then requires that the reduction in effective width of the second order pattern by a factor of 2 is compensated by a factor of 2 larger peak value. This illustrates that with increasing order of coherence the pattern is increasingly centered around the forward direction. When extended to higher order, this leads to the remarkable result that, in principle, an nth order coherent state no longer diffracts and the collective photon state propagates on particle-like trajectories [6].

16.7 Summary

The key message of this chapter is that diffraction patterns can simply be viewed as encoded signatures of different quantum states of light. This is revealed by a quantum formulation of diffraction that goes beyond the first order description in conventional quantum mechanics. The theoretical link between quantum states and their characteristic diffraction images is revealed by modern versions of Young's double-slit diffraction experiment, summarized in Fig. 16.1. Ironically, the very experiment that 200 years ago led to the notion that light is a wave, can therefore be used to disprove this hypothesis.

We note that the true photon nature of what we call electromagnetic (EM) radiation is not restricted to the short wavelength range extending from the optical to the x-ray regime. It is a universal feature of EM radiation despite the power of Maxwell's classical theory of electromagnetic waves. Owing to the lower energy of photons below the visible range, it just becomes increasingly difficult to detect them since typical detectors at room temperature have a thermal background noise corresponding to about 25 meV. Today, even microwave photons may be detected by use of photon counters based on Josephson junctions [77].

In principle, the wave theory can be abandoned altogether today. In practice, it has served us well and may continue to be used with the understanding of its limitations. The broad and difficult concept of "partial coherence" in statistical optics can be better defined in quantum optics, which is anchored in the fundamental theory of light and matter, QED. The definition of quantum coherence is more specific since it is directly linked to different quantum states of light whose interference and diffraction properties are quantified through an order-dependent degree of coherence. The quantum theory furthermore differentiates between the behavior of collective quantum states defined through statistical distributions of photons and states containing specific numbers of photons.

In *first order* QED, the fundamental photon nature of light just happens to be describable by the *ad hoc* wave theory, based on spherical light waves, the magical Huygens–Fresnel principle, and the assumption that the absolute value squared of the wave field gives the diffracted intensity. The limitations and non-fundamental nature of the wave theory become apparent only in *second order* QED, where the concept of spherical waves needs to be replaced by an average over photon emission directions, the Huygens–Fresnel principle becomes the quantum interference of photon probability amplitudes for different source-detector paths, and the "intensity" concept is replaced by the probability of photon detection.

References

1. R.P. Feynman, R.B. Leighton, M. Sands, *The Feynman Lectures on Physics*, vol. III (Addison-Wesley, Reading, 1965)
2. R.P. Feynman, Rev. Mod. Phys. **20**, 367 (1948)
3. D.F. Styer et al., Am. J. Phys. **70**, 288 (2002)
4. L. de Broglie, Ann. Phys. (Paris) **3**, 22 (1925)
5. J. Liu, G. Zhang, Phys. Rev. A **82**, 013822 (2010)
6. J. Stöhr, Adv. Optics and Photonics **11**, 215 (2019)
7. X.F. Pang, Y.P. Feng, *Quantum Mechanics in Nonlinear Systems* (World Scientific, New Jersey, 2005)
8. S. Tomonaga, Prog. Theor. Phys. **1**, 27 (1946)
9. J. Schwinger, Phys. Rev. **74**, 1439 (1948)
10. R.P. Feynman, Phys. Rev. **76**, 769 (1949)
11. F. Dyson, Phys. Rev. **75**, 486 (1949)
12. R.P. Feynman, *QED: The Strange Theory of Light and Matter* (Princeton University Press, Princeton NJ, 1985)
13. T. Jeltes et al., Nature **445**, 402 (2007)
14. R. Loudon, *The Quantum Theory of Light*, 3rd edn. (Clarendon Press, Oxford, 2000)
15. M.O. Scully, M.S. Zubairy, *Quantum Optics* (Cambridge University Press, Cambridge, 1997)
16. R. Glauber, Optical coherence and photon statistics, in *Quantum Optics and Electronics*, ed. by A.B.C. deWitt, C. Cohen-Tannoudji (Gordon and Breach, New York, 1965)

17. P. Grangier, G. Roger, A. Aspect, Europhys. Lett **1**, 173 (1986)
18. C.K. Hong, Z.Y. Ou, L. Mandel, Phys. Rev. Lett. **59**, 2044 (1987)
19. P.A.M. Dirac, *Quantum Mechanics*, 4th edn. (Oxford University Press, Oxford, 1958)
20. T.B. Pittman, Y.H. Shih, D.V. Strekalov, A.V. Sergienko, Phys. Rev. A **52**, R3429 (1995)
21. Y. Shih, *An Introduction to Quantum Optics: Photon and Biphoton Physics* (CRC Press, Boca Raton, Florida, 2011)
22. D.N. Klyshko, LJETP **6**, 490 (1967)
23. S.E. Harris, M.K. Oshman, R.L. Byer, Phys. Rev. Lett. **18**, 732 (1967)
24. D. Magde, H. Mahr, Phys. Rev. Lett. **18**, 905 (1967)
25. M. Born, E. Wolf, *Principles of Optics: Electromagnetic Theory of Propagation, Interference, and Diffraction of Light* (Cambridge University Press, Cambridge, 1999)
26. J.W. Goodman, *Statistical Optics*, 2nd edn. (Wiley, New York, 2015)
27. Z.Y.J. Ou, *Multi-photon Quantum Interference* (Springer, Heidelberg, New York, 2007)
28. Z.Y.J. Ou, *Quantum Optics for Experimentalists* (World Scientific, Singapore, 2017)
29. R. Shimizu, K. Edamatsu, T. Itoh, Phys. Rev. A **74**, 013801 (2006)
30. B. Wu, T. Wang, C.E. Graves, D. Zhu, W.F. Schlotter, J.J. Turner, O. Hellwig, Z. Chen, H.A. Dürr, A. Scherz, J. Stöhr, Phys. Rev. Lett. **117**, 027401 (2016)
31. J. Stöhr, A. Scherz: Phys. Rev. Lett. **115**, 107 402 (2015)
32. D.N. Klyshko, *Photons and Nonlinear Optics* (Gordon and Breach, New York, 1988)
33. C. Santori, D. Fattal, J. Vučković, G.S. Solomon, Y. Yamamoto, Nature **419**, 594 (2002)
34. R.B. Patel, A.J. Bennett, I. Farrer, C.A. Nicoll, D.A. Ritchie, A.J. Shields, Nat. Photon. **4**, 632 (2010)
35. E.B. Flagg, A. Muller, S.V. Polyakov, A. Ling, A. Migdall, G.S. Solomon, Phys. Rev. Lett. **104**, 137401 (2010)
36. A. Neuzner, M. Körber, O. Morin, S. Ritter, G. Rempe, Nat. Photon. **10**, 303 (2016)
37. Y.H. Zhai, X.H. Chen, D. Zhang, L.A. Wu, Phys. Rev. A **72**, 043805 (2005)
38. W. Martienssen, E. Spiller, Am. J. Phys. **32**, 919 (1964)
39. F.T. Arecchi, Phys. Rev. Lett. **15**, 912 (1965)
40. G. Scarcelli, A. Valencia, Y. Shih, Phys. Rev. A **70**, 051802(R) (2004)
41. J. Xiong, D.Z. Cao, F. Huang, H.G. Li, X.J. Sun, K. Wang, Phys. Rev. Lett. **94**, 173601 (2005)
42. Y. Zhou, F.L. Li, B. Bai, H. Chen, J. Liu, Z. Xu, H. Zheng, Phys. Rev. A **95**, 053809 (2017)
43. D. Bouwmeester, A.K. Ekert, A. Zeilinger (eds.), *The Physics of Quantum Information: Quantum Cryptography, Quantum Teleportation, Quantum Computation* (Springer, Heidelberg, 2000)
44. J.M. Raimond, M. Brune, S. Haroche, Rev. Mod. Phys. **73**, 565 (2001)
45. R. Horodecki, P. Horodecki, M. Horodecki, K. Horodecki, Rev. Mod. Phys. **81**, 865 (2009)
46. J.W. Pan, Z.B. Chen, C.Y. Lu, H. Weinfurter, A. Zeilinger, M. Żukowski, Rev. Mod. Phys. **84**, 777 (2012)
47. J. Javanainen, S.M. Yoo, Phys. Rev. Lett. **76**, 161 (1996)
48. L. Mandel, Rev. Mod. Phys. **71**, S274 (1999)
49. T. Wang et al., Phys. Rev. Lett. **108**, 267403 (2012)
50. P.H. van Cittert, Phys. Rev. B **1**, 201 (1934)
51. F. Zernike, Physica **5**, 785 (1938)
52. E. Wolf, *Introduction to the Theory of Coherence and Polarization of Light* (Cambridge University Press, Cambridge, 2007)
53. J.D. Jackson, *Classical Electrodynamics* (Wiley, New York, 1999)
54. M.J. Beran, G.B. Parrent, *Theory of Partial Coherence* (Prentice-Hall, Englewood Cliffs, New Jersey, 1964)
55. I.H. Deutsch, J.C. Garrison, Phys. Rev. A **43**, 2498 (1991)
56. J. Stöhr: Phys. Rev. Lett. **118**, 024 801 (2017)
57. In Refs. [6,55] it was assumed that the 2-photon coherent or cloned-biphoton state is second order coherent, which the present analysis reveals to be incorrect. Second order coherence only develops when the stimulation process involves many photons
58. R.H. Brown, R.Q. Twiss, Nature **177**, 27 (1956)

59. R.H. Brown, R.Q. Twiss, Nature **178**, 1046 (1956)
60. R.H. Brown, R.Q. Twiss: Proc. Roy. Soc. (London) A **242**, 300 (1957)
61. R.H. Brown, R.Q. Twiss: Proc. Roy. Soc. (London) A **243**, 291 (1958)
62. R.H. Brown, R.Q. Twiss: Proc. Roy. Soc. (London) A **248**, 222 (1958)
63. U. Fano, Am. J. Phys. **29**, 539 (1961)
64. R.J. Glauber, Rev. Mod. Phys. **78**, 1267 (2006)
65. D.V. Strekalov, A.V. Sergienko, D.N. Klyshko, Y.H. Shih, Phys. Rev. Lett. **74**, 3600 (1995)
66. R.S. Bennink, S.J. Bentley, R.W. Boyd: Phys. Rev. Lett. **89**, 113 601 (2002)
67. A. Valencia, G. Scarcelli, M. D'Angelo, Y. Shih: Phys. Rev. Lett. **94**, 063 601 (2005)
68. E.M. Purcell, Nature **178**, 1449 (1956)
69. I.S. Reed, I.R.E. Trans, Inf. Theory **8**, 194 (1962)
70. L. Mandel, E. Wolf, Rev. Mod. Phys. **37**, 231 (1965)
71. U.M. Titulaer, R.J. Glauber, Phys. Rev. **140**, B676 (1965)
72. D. Dieks, Phys. Lett. A **92**, 271 (1982)
73. L. Mandel, Nature **304**, 188 (1983)
74. A. Lamas-Linares, C. Simon, J.C. Howell, D. Bouwmeester, Science **296**, 712 (2002)
75. P.W. Milonni, M.L. Hardies, Phys. Lett. A **92**, 321 (1982)
76. W.K. Wootters, W.H. Zurek, Nature **299**, 802 (1982)
77. Y.F. Chen, D. Hover, S. Sendelbach, L. Maurer, S.T. Merkel, E.J. Pritchett, F.K. Wilhelm, R. McDermott: Phys. Rev. Lett. **107**, 217 401 (2011)

Appendix

A.1 The International System of Units (SI)

Table A.1 Physical quantities, symbols, and their SI units

Symbol	Physical quantity	Equivalent	Numerical value and units
$\mathbf{A}$	Vector potential		$\mathrm{V\,s\,m^{-1}}$
$\mathbf{E}$	Electric field		$\mathrm{V\,m^{-1}}$
$\mathbf{D}$	Displacement	$\epsilon\epsilon_0\mathbf{E}$	$\mathrm{A\,s\,m^{-2}}$
$\mathbf{H}$	Magnetic field		$\mathrm{A\,m^{-1}}$
$\mathbf{B}$	Magnetic Induction	$\mu\mu_0\mathbf{H}$	$\mathrm{V\,s\,m^{-2}}$ ($\equiv$ T)
$\mathbf{p}$	Electric dipole moment		$\mathrm{A\,s\,m}$
$\mathbf{m}$	Magnetic dipole moment		$\mathrm{V\,s\,m}$
$\mathbf{M}$	Magnetization	$\mathbf{m}/V$	$\mathrm{V\,s\,m^{-2}}$ ($\equiv$ T)
c	Speed of light	$1/\sqrt{\epsilon_0\mu_0}$	$2.998\times10^{8}\ \mathrm{m\,s^{-1}}$
ϵ_0	Dielectric constant	$1/\mu_0 c^2$	$8.854\times10^{-12}\ \mathrm{A\,s\,V^{-1}m^{-1}}$
μ_0	Permeability	$1/\epsilon_0 c^2$	$4\pi\times10^{-7}\ \mathrm{V\,s\,A^{-1}m^{-1}}$
$1/4\pi\epsilon_0$	SI units prefactor		$8.99\times10^{9}\ \mathrm{V\,m\,A^{-1}s^{-1}}$
μ_B	Bohr magneton	$e\hbar\mu_0/2m_e$	$1.165\times10^{-29}\ \mathrm{V\,m\,s}$
m_e	Electron mass		$9.109\times10^{-31}\ \mathrm{V\,A\,s^3\,m^{-2}}$
h	Planck's constant		$6.626\times10^{-34}\ \mathrm{V\,A\,s^2} = 4.136\,\mathrm{eV\,fs}$
$\hbar$	Planck's constant	$h/2\pi$	$1.055\times10^{-34}\ \mathrm{V\,A\,s^2} = 0.6582\,\mathrm{eV\,fs}$
k_B	Boltzmann's constant		$1.381\times10^{-23}\ \mathrm{V\,A\,s\,K^{-1}} =$ $8.617\times10^{-5}\ \mathrm{eV\,K^{-1}}$
N_A	Avogadro's number		6.02214×10^{23} atoms/mol
R_g	Universal gas constant	$k_B N_A$	$8.31446\ \mathrm{V\,A\,s\,K^{-1}\,mol^{-1}} =$ $5.188\times10^{19}\ \mathrm{eV\,K^{-1}mol^{-1}}$
$q=-e$	Electron charge		$-1.602\times10^{-19}\ \mathrm{A\,s}$
e/m_e	Electron charge/mass	ω/B	$1.759\times10^{11}\ \mathrm{rad\,s^{-1}\,T^{-1}}$

(continued)

J. Stöhr, *The Nature of X-Rays and Their Interactions with Matter*, Springer Tracts in Modern Physics 288, https://doi.org/10.1007/978-3-031-20744-0

Table A.1 (continued)

$m_e c^2$	Electron rest energy		0.819×10^{-13} V A s = 0.511 MeV
E_R	Rydberg energy	$m_e e^4/2\hbar^2(4\pi\epsilon_0)^2$	13.606 eV
a_0	Bohr radius	$4\pi\epsilon_0\hbar^2/m_e e^2$	0.529×10^{-10} m
r_0	Classical electron radius	$e^2/4\pi\epsilon_0 m_e c^2$	2.818×10^{-15} m
σ_e	Thomson cross section	$8\pi r_0^2/3$	0.665×10^{-28} m^2 = 0.665 barn
α_f	Fine structure constant	$e^2/4\pi\epsilon_0 \hbar c$	= 1/137.04
$\lambda\!\!\!^{-}_C$	Compton scattering length	$\hbar/m_e c$	3.862×10^{-13} m

Conversions of units

1 Oersted (Oe) = $(1000/4\pi)$ A m^{-1} = 79.59 A m^{-1}
1 Tesla (T) = 1 N A^{-1}m^{-1} = 1 V s m^{-2} (i.e. 1 T corresponds to 10^4 Oe)
1 Ohm (Ω) = 1 V A^{-1}
1 Coulomb (C) = 1 A s
1 Newton (N) = 1 V A s m^{-1}
1 Kilogram (kg) = 1 V A s^3 m^{-2}
1 Farad (F) = 1 A s V^{-1}
1 Joule (J) = 1 N m = 1 V A s
1 Watt (W) = 1 V A = 1 J s^{-1}
1 eV = 1.602×10^{-19} V A s
1 eV$/k_B$ = 1.1605×10^4 K (energy to temperature)
1 eV$/h$ = 2.418×10^{14} Hz (energy to cycle frequency)
1 eV$/hc$ = 8066 cm^{-1} (energy to wavenumber, also 1 cm^{-1} = 1 Kayser)
$h\nu$[eV] = $1239.852/\lambda$ [nm] (photon energy to wavelength and vice versa)
1 μ_B/μ_0 = 0.578×10^{-4} eV T^{-1}
1 barn (b) = 1×10^{-28} m^2
1 deg (°) = $\pi/180$ rad = 17.45 mrad
1 arcmin = 1/60° = 290.9 μrad

A.2 Resonance Lineshapes

A.2.1 *Lorentzian Lineshape and Integral*

The *Lorentzian lineshape* function centered at x_0 is

$$L(x) = A\,\frac{(\Gamma/2)^2}{(x-x_0)^2 + (\Gamma/2)^2}, \tag{A.1}$$

where A is the peak value and Γ is the full width at half maximum (FWHM). For $A = 2/(\pi\Gamma)$ the Lorentzian has unit area.

Its integral is a step function given by

$$I(x) = \int_{-\infty}^{x} L(u)\,\mathrm{d}u = H\left[\frac{1}{2} + \frac{1}{\pi}\arctan\left(\frac{x - x_0}{\Gamma/2}\right)\right], \tag{A.2}$$

where $H = A\,\Gamma\,\pi/2$ is the step height.

A.2.2 *Gaussian Lineshape and Integral*

The *Gaussian lineshape* function centered at x_0 is given by

$$G(x) = A\,\exp\left(-\frac{(x - x_0)^2}{2\,\sigma^2}\right) = A\,\exp\left(-\frac{a^2(x - x_0)^2}{2\,\Gamma_\mathrm{G}^2}\right). \tag{A.3}$$

Here A is the peak height, and σ is the rms (one standard deviation) width. The full width at half maximum (FWHM) Γ_G is related to the rms width according to $\Gamma_\mathrm{G} = a\,\sigma$, where $a = 2\sqrt{\ln 4} = 2.355$. For $A = 1/(\sigma\sqrt{2\pi}) = a/(\Gamma_\mathrm{G}\sqrt{2\pi})$ the Gaussian has unit area.

Its integral is a step function given by

$$I(x) = \int_{-\infty}^{x} G(u)\,\mathrm{d}u = H\left[\frac{1}{2} + \frac{1}{2}\,\mathrm{erf}\left(\frac{x - x_0}{\sqrt{2}\,\sigma}\right)\right], \tag{A.4}$$

where $H = A\,\sigma\sqrt{2\pi}$ is the step height and erf is the error function.

A.2.3 *Voigt Lineshape*

The Voigt function is a convolution of the Lorentzian of natural line FWHM Γ and a Gaussian of FWHM Γ_G.

With the same notations as for the Lorentzian and Gauss functions above, the *Voigt lineshape* function is given by

$$V(x) = A\,\frac{\mathrm{Re}\left\{\exp\left[-\frac{a^2(x - x_0 - \mathrm{i}\Gamma/2)^2}{2\Gamma_\mathrm{G}^2}\right]\,\mathrm{erfc}\left[\frac{a(\mathrm{i}(x - x_0) + \Gamma/2)}{\sqrt{2}\,\Gamma_\mathrm{G}}\right]\right\}}{\exp\left[\frac{a^2\Gamma^2}{8\Gamma_\mathrm{G}^2}\right]\,\mathrm{erfc}\left[\frac{a\Gamma}{2\sqrt{2}\,\Gamma_\mathrm{G}}\right]}. \tag{A.5}$$

Here $\mathrm{erfc}[x] = 1 - \mathrm{erf}[x]$ and A is the peak height. The profile has unit area for

$$A = \frac{a}{\sqrt{2\pi}\,\Gamma_G} \exp\left[\frac{a^2\Gamma^2}{8\Gamma_G^2}\right] \mathrm{erfc}\left[\frac{a\Gamma}{2\sqrt{2}\,\Gamma_G}\right], \tag{A.6}$$

and the FWHM linewidth is given by

$$\Gamma_V = 0.535\,\Gamma + \sqrt{0.217\,\Gamma^2 + \Gamma_G^2}\,. \tag{A.7}$$

A.3 Dirac δ-Function

The Dirac δ-function is zero for all $\mathbf{x}$ except at one point $\mathbf{x} = \mathbf{a}$ where it is infinite

$$\delta(\mathbf{x} - \mathbf{a}) = \begin{cases} 0 & \mathbf{x} \neq \mathbf{a} \\ \infty & \mathbf{x} = \mathbf{a}\,. \end{cases} \tag{A.8}$$

It is defined only through its integrated property, according to

$$\int_{-\infty}^{\infty} \delta(\mathbf{x} - \mathbf{a})\,d\mathbf{x} = 1 \tag{A.9}$$

and has the *sifting property*

$$\int_{-\infty}^{\infty} f(\mathbf{x})\,\delta(\mathbf{x} - \mathbf{a})\,d\mathbf{x} = f(\mathbf{a})\,. \tag{A.10}$$

It is commonly assumed that the dimension of $\mathbf{x}$ in the argument defines the dimension of the δ function, which may also be explicitly denoted as $\delta^{(n)}$.

A.4 Fourier Transforms and Parseval's Theorem

The Fourier transform formalism introduced below is based on J. W. Goodman's book "Introduction to Fourier Transforms" [1]. More recently, the same author has written a wonderful little tutorial on the practical implementation of continuous and discrete Fourier transforms using Mathematica [2].

A.4.1 1D Fourier Transform

The (forward) Fourier transform of a *one-dimensional* function $f(x)$ is defined as

$$F(p) = \int_{-\infty}^{\infty} f(x)\, \mathrm{e}^{-2\pi \mathrm{i} p x}\, \mathrm{d}x\,, \tag{A.11}$$

and the inverse transform is

$$f(x) = \int_{-\infty}^{\infty} F(p)\, \mathrm{e}^{2\pi \mathrm{i} p x}\, \mathrm{d}p. \tag{A.12}$$

The Fourier relations are accompanied by a powerful theorem involving the absolute values squared of the transform pair, called *Parseval's theorem*,[1] sometimes called Rayleigh's theorem or Plancherel's theorem

$$\int_{-\infty}^{\infty} |f(x)|^2\, \mathrm{d}x = \int_{-\infty}^{\infty} |F(p)|^2\, \mathrm{d}p\,. \tag{A.13}$$

In practice, one often encounters conjugate variables that are linked to the original variables by inclusion of a factor of 2π. Examples are the relations between time and the *angular* frequency, $\omega = 2\pi\nu = 2\pi/t$, or between wavelength and wavevector $k = 2\pi/\lambda$. By substituting $q = 2\pi p$ into (A.11) and (A.12) we obtain the following equivalent transform formulation, which we shall use in this book.[2] In general, the transforms $F(q)$ and $f(x)$ have different dimensions.

In this book we shall use the following formulation of the 1D Fourier transform and Parseval's theorem

$$F(q) = \int_{-\infty}^{\infty} f(x)\, \mathrm{e}^{-\mathrm{i} q x}\, \mathrm{d}x\,. \tag{A.16}$$

---

[1] It is named after Marc-Antoine Parseval des Chênes and dates back to 1799 when he presented it to the Académie des Sciences in Paris.

[2] Another definition often used in physics (and e.g. in Mathematica) is to split the factor of 2π between the forward and inverse transforms to make them symmetrical,

$$F(q) = \frac{1}{\sqrt{2\pi}} \int_{-\infty}^{\infty} f(x)\, \mathrm{e}^{-\mathrm{i} q x}\, \mathrm{d}x \tag{A.14}$$

and

$$f(x) = \frac{1}{\sqrt{2\pi}} \int_{-\infty}^{\infty} F(q)\, \mathrm{e}^{\mathrm{i} q x}\, \mathrm{d}q. \tag{A.15}$$

$$f(x) = \frac{1}{2\pi} \int_{-\infty}^{\infty} F(q)\, \mathrm{e}^{\mathrm{i}qx}\, \mathrm{d}q. \tag{A.17}$$

$$\int_{-\infty}^{\infty} |f(x)|^2\, \mathrm{d}x = \frac{1}{2\pi} \int_{-\infty}^{\infty} |F(q)|^2\, \mathrm{d}q \tag{A.18}$$

Parseval's theorem (A.18) assures energy conservation upon Fourier transformation (see below).

A.4.2 1D Transformation Under Preservation of Dimension

It is convenient to define Fourier transformed fields in the time t and angular frequency $\omega = 2\pi\nu$ (or energy $\hbar\omega$) domain with the same dimension, which for electric fields is expressed by [V/m] in the SI system. This requires introduction of a dimension-preserving constant C of dimension [time]. We write the temporal electric field distribution as

$$E(t) = E_0\, f(t)\,, \tag{A.19}$$

where $f(t)$ is a *dimensionless* temporal distribution function of unit peak value, so that $E(t=0) = E_0$. The fields in the time and (reciprocal) frequency domains can then be linked by a 1D Fourier transform according to

$$E(\omega) = E_0 \underbrace{\frac{1}{C} \int_{-\infty}^{\infty} f(t)\, \mathrm{e}^{-\mathrm{i}\omega t}\, \mathrm{d}t}_{F(\omega)}\,. \tag{A.20}$$

The frequency distribution function $F(\omega)$ is also *dimensionless* of unit peak value, so that $E(\omega=0) = E_0$, and its inverse Fourier transform reproduces $E(t)$ according to

$$E(t) = E_0 \underbrace{\frac{C}{2\pi} \int_{-\infty}^{\infty} F(\omega)\, \mathrm{e}^{\mathrm{i}\omega t}\, \mathrm{d}\omega}_{f(t)}\,. \tag{A.21}$$

Then Parseval's theorem links the associated intensities according to

$$\boxed{\int_{-\infty}^{\infty} |E(t)|^2\, \mathrm{d}t = \frac{C^2}{2\pi} \int_{-\infty}^{\infty} |E(\omega)|^2\, \mathrm{d}\omega} \tag{A.22}$$

and therefore assures intensity or energy conservation upon Fourier transforming. The constant C needs to satisfy the Fourier relations as well as Parseval's theorem. It may simply be determined by setting $\omega = 0$ in (A.20) or alternatively at $t = 0$ from (A.21), which for $E(t=0) = E(\omega=0) = E_0$ yields

$$C = \frac{\int_{-\infty}^{\infty} E(t)\, \mathrm{d}t}{E_0} = \frac{2\pi E_0}{\int_{-\infty}^{\infty} E(\omega)\, \mathrm{d}\omega}. \tag{A.23}$$

A.4.3 *Effective 1D Distribution Widths and Transform Limit*

Since Parseval's relation (A.22) assures intensity conservation, we may use it to define effective intensity distribution widths. The idea is to define the integrated intensity as its peak value E_0^2 times *effective widths* of the distribution functions in the time and frequency domains according to

$$\text{Effective widths: } \Delta_t = \frac{\int_{-\infty}^{\infty} |E(t)|^2\, \mathrm{d}t}{E_0^2} \quad \text{and} \quad \Delta_\omega = \frac{\int_{-\infty}^{\infty} |E(\omega)|^2\, \mathrm{d}\omega}{E_0^2}. \tag{A.24}$$

The so-defined integrated intensities $E_0^2 \Delta_t$ and $E_0^2 \Delta_\omega$ correspond to those under flat-top intensity distributions of peak values E_0^2 and widths Δ_t and Δ_ω containing the same photon energy. The product of the effective widths constitutes a time-bandwidth product that determines the transform limit. It is obtained as

$$\Delta_t \Delta_\omega = \frac{\int_{-\infty}^{\infty} |E(t)|^2\, \mathrm{d}t}{E_0^2} \frac{\int_{-\infty}^{\infty} |E(\omega)|^2\, \mathrm{d}\omega}{E_0^2} = \frac{2\pi}{C^2 E_0^4} \left| \int_{-\infty}^{\infty} |E(t)|^2\, \mathrm{d}t \right|^2, \tag{A.25}$$

where on the right we have used (A.22). Inserting the expression for C given by the temporal expression in (A.23) we obtain

$$\boxed{\Delta_t \Delta_\omega = 2\pi \underbrace{\frac{\left|\int_{-\infty}^{\infty} |E(t)|^2\, \mathrm{d}t\right|^2}{E_0^2 \left|\int_{-\infty}^{\infty} E(t)\, \mathrm{d}t\right|^2}}_{K}} \tag{A.26}$$

which expresses the *transform limit* or *time-bandwidth product* in the form (4.20) by assuming effective distribution widths and K given by (4.24) for a classical field.

A.4.4 2D Fourier Transform

The 2D transform is defined in analogy to (A.16) and (A.17)

$$F(q_x, q_y) = \iint_{-\infty}^{\infty} f(x, y)\, \mathrm{e}^{-\mathrm{i}(q_x x + q_y y)}\, \mathrm{d}x\, \mathrm{d}y \tag{A.27}$$

$$f(x, y) = \frac{1}{4\pi^2} \iint_{-\infty}^{\infty} F(q_x, q_y)\, \mathrm{e}^{\mathrm{i}(q_x x + q_y y)}\, \mathrm{d}q_x\, \mathrm{d}q_y\,. \tag{A.28}$$

Parseval's theorem for the 2D case reads

$$\iint_{-\infty}^{\infty} |f(x, y)|^2\, \mathrm{d}x\, \mathrm{d}y = \frac{1}{4\pi^2} \iint_{-\infty}^{\infty} |F(q_x, q_y)|^2\, \mathrm{d}q_x\, \mathrm{d}q_y\,. \tag{A.29}$$

A.4.5 2D Transformation Under Preservation of Dimension

In the following we shall consider the 2D Fourier transform under preservation of dimension similar to the 1D case treated in Sect. A.4.2. While the transformation step between conjugate 1D time—energy scalars is very similar to that between 2D space-momentum vectors, in the latter case one is typically also interested in including the propagation from the source to the detector plane, which contains the measured diffraction pattern.

Because of the linearity of the Fourier transform, however, the desired transformation $E(\mathbf{r}, 0) \leftrightarrow E(\mathbf{q}, z_0)$ can be broken down into a pure transform step $E(\mathbf{r}) \leftrightarrow E(\mathbf{q})$ times a propagation step $z{=}0 \leftrightarrow z{=}z_0$. It is therefore possible to consider the transform step alone, without propagation, as we shall do below. The intensity propagation can simply be included after the transformation step by a *propagation factor*. For a source of area A_s and a detector plane at large distance z_0 the intensity propagation is given by a multiplicative factor $A_s^2/(\lambda^2 z_0^2)$ as identified, for example, in the far-field diffraction pattern (8.25).

As in Sect. A.4.2 we describe the real space coherent field with coordinates $\mathbf{r} = (x, y)$ as

$$E(x, y) = E_0 f(x, y)\,, \tag{A.30}$$

where $f(x, y)$ is dimensionless distribution function of unit peak value. Its Fourier transform is then given in coordinates $\mathbf{q} = (q_x, q_y)$ as

$$E(q_x, q_y) = E_0 \underbrace{\frac{1}{C} \iint\limits_{-\infty}^{\infty} f(x, y)\, \mathrm{e}^{-\mathrm{i}(q_x x + q_y y)}\, \mathrm{d}x\, \mathrm{d}y}_{F(q_x, q_y)} . \tag{A.31}$$

Here $F(q_x, q_y)$ is a dimensionless reciprocal space distribution function of unit peak value and C of dimension [length^2] assures preservation of the field dimension. The back transformation reproduces the real space field according to

$$E(x, y) = E_0 \underbrace{\frac{C}{4\pi^2} \iint\limits_{-\infty}^{\infty} F(q_x, q_y)\, \mathrm{e}^{\mathrm{i}(q_x x + q_y y)}\, \mathrm{d}q_x\, \mathrm{d}q_y}_{f(x,y)} . \tag{A.32}$$

Parseval's theorem links the associated intensities according to

$$\iint\limits_{-\infty}^{\infty} |E(x, y)|^2\, \mathrm{d}x\, \mathrm{d}y = \frac{|C|^2}{4\pi^2} \iint\limits_{-\infty}^{\infty} |E(q_x, q_y)|^2\, \mathrm{d}q_x\, \mathrm{d}q_y . \tag{A.33}$$

The constant C needs to satisfy the Fourier relations as well as Parseval's theorem. It may simply be determined by setting $\mathbf{q}=0$ in (A.31) or alternatively at $\mathbf{r}=0$ in (A.32), and with $E(\mathbf{r}=0) = E(\mathbf{q}=0) = E_0$ we obtain

$$C = \frac{\iint_{-\infty}^{\infty} E(x, y)\, \mathrm{d}x\, \mathrm{d}y}{E_0} = \frac{4\pi^2 E_0}{\iint_{-\infty}^{\infty} E(q_x, q_y)\, \mathrm{d}q_x\, \mathrm{d}q_y} . \tag{A.34}$$

A.4.6 *Effective 2D Distribution Areas and Diffraction Limit*

Since Parseval's theorem assures energy conservation upon field transformation, we can now define *effective* real source areas and associated reciprocal space "areas" that determine the solid angle of emission. The effective source area of an arbitrary intensity distribution function is defined as

$$\text{Effective source area:}\ A_{\text{eff}} = \frac{\iint_{-\infty}^{\infty} |E(x, y)|^2\, \mathrm{d}x\, \mathrm{d}y}{E_0^2} . \tag{A.35}$$

The effective solid angle associated with the *reciprocal space* momentum transfer **q** is similarly defined as

$$\text{Effective solid angle:}\ \mathrm{d}\Omega_{\text{eff}} = \frac{\iint_{-\infty}^{\infty} |E(q_x, q_y)|^2\, dq_x\, dq_y}{k^2\, E_0^2}\,, \tag{A.36}$$

where $k = 2\pi/\lambda$. The product $A_{\text{eff}}\,\mathrm{d}\Omega_{\text{eff}}$ then defines the diffraction limit (see Fig. 4.1) and by use of (A.33) and then inserting C from (A.34) we obtain

$$\boxed{A_{\text{eff}}\,\mathrm{d}\Omega_{\text{eff}} = \underbrace{\frac{\left|\iint_{-\infty}^{\infty} |E(x, y)|^2\, dx\, dy\right|^2}{E_0^2 \left|\iint_{-\infty}^{\infty} E(x, y)\, dx\, dy\right|^2}}_{H}\, \lambda^2}\,. \tag{A.37}$$

This expresses the *diffraction limit* in the form (4.19) for effective source areas and solid emission angles with H given by (4.25) for a classical field.

A.5 Spherical Harmonics and Tensors

Following Racah, we define spherical tensors $C_m^{(l)}$ for $l \le 4$, in terms of the spherical harmonics as follows

$$C_m^{(l)} = \sqrt{\frac{4\pi}{2l+1}}\; Y_{l,m}(\theta, \phi)\,. \tag{A.38}$$

Table A.2 gives expressions for $C_m^{(l)}$ for $l = 0, 1, 2, 3, 4$.

Table A.2 Racah tensor operators $C_m^{(l)}$ for $l = 0, 1, 2, 3, 4$ defined in (A.38), expressed in Cartesian and spherical coordinates, according to $x = r \sin\theta \cos\phi$, $y = r \sin\theta \sin\phi$, $z = r\cos\theta$

$C_0^{(0)} = 1$	
$C_0^{(1)} = \cos\theta$	$= \frac{z}{r}$
$C_{\pm1}^{(1)} = \mp\frac{1}{\sqrt{2}} \sin\theta\, e^{\pm i\phi}$	$= \mp\frac{1}{\sqrt{2}} \frac{(x \pm iy)}{r}$
$C_0^{(2)} = \frac{1}{2}(3\cos^2\theta - 1)$	$= \frac{1}{2}\frac{(3z^2 - r^2)}{r^2}$
$C_{\pm1}^{(2)} = \mp\sqrt{\frac{3}{2}} \cos\theta \sin\theta\, e^{\pm i\phi}$	$= \mp\sqrt{\frac{3}{2}} \frac{(x \pm iy)z}{r^2}$
$C_{\pm2}^{(2)} = \sqrt{\frac{3}{8}} \sin^2\theta\, e^{\pm 2i\phi}$	$= \sqrt{\frac{3}{8}} \frac{(x \pm iy)^2}{r^2}$
$C_0^{(3)} = \frac{1}{2}(5\cos^3\theta - 3\cos\theta)$	$= \frac{1}{2}\frac{(5z^2 - 3r^2)z}{r^3}$

(continued)

Table A.2 (continued)

$C^{(3)}_{\pm 1} = \mp\sqrt{\frac{3}{16}}\,\sin\theta\,(5\cos^2\theta - 1)\,e^{\pm i\phi}$	$= \mp\sqrt{\frac{3}{16}}\,\frac{(x\pm iy)(5z^2-r^2)}{r^3}$
$C^{(3)}_{\pm 2} = \sqrt{\frac{15}{8}}\,\cos\theta\,\sin^2\theta\,e^{\pm 2i\phi}$	$= \sqrt{\frac{15}{8}}\,\frac{z(x\pm iy)^2}{r^3}$
$C^{(3)}_{\pm 3} = \mp\sqrt{\frac{5}{16}}\,\sin^3\theta\,e^{\pm 3i\phi}$	$= \mp\sqrt{\frac{5}{16}}\,\frac{(x\pm iy)^3}{r^3}$
$C^{(4)}_{0} = \frac{1}{8}\,(35\cos^4\theta - 30\cos^2\theta + 3)$	$= \frac{1}{8}\,\frac{(35z^4-30z^2r^2+3r^4)}{r^4}$
$C^{(4)}_{\pm 1} = \mp\sqrt{\frac{5}{16}}\,\sin\theta\,(7\cos^3\theta - 3\cos\theta)\,e^{\pm i\phi}$	$= \mp\sqrt{\frac{5}{16}}\,\frac{(x\pm iy)(7z^3-3zr^2)}{r^4}$
$C^{(4)}_{\pm 2} = \sqrt{\frac{5}{32}}\,\sin^2\theta\,(7\cos^2\theta - 1)\,e^{\pm 2i\phi}$	$= \sqrt{\frac{5}{32}}\,\frac{(x\pm iy)^2(7z^2-r^2)}{r^4}$
$C^{(4)}_{\pm 3} = \mp\sqrt{\frac{35}{16}}\,\cos\theta\,\sin^3\theta\,e^{\pm 3i\phi}$	$= \mp\sqrt{\frac{35}{16}}\,\frac{z(x\pm iy)^3}{r^4}$
$C^{(4)}_{\pm 4} = \sqrt{\frac{35}{128}}\,\sin^4\theta\,e^{\pm 4i\phi}$	$= \sqrt{\frac{35}{128}}\,\frac{(x\pm iy)^4}{r^4}$

The operators have the complex conjugate property $(C^{(l)}_m)^* = (-1)^m C^{(l)}_{-m}$

A.5.1 *Relations between First and Second Order Tensors $C^{(1)}_m$ and $C^{(2)}_m$*

Table A.3 Relationships of Racah tensor operators $C^{(1)}_m$ and $C^{(2)}_m$ derived from the expressions listed in Table A.2

$C^{(2)}_0 = C^{(1)}_0\,C^{(1)}_0 + C^{(1)}_{-1}\,C^{(1)}_1$
$C^{(2)}_{\pm 1} = \sqrt{3}\,C^{(1)}_0\,C^{(1)}_{\pm 1}$
$C^{(2)}_{\pm 2} = \sqrt{\frac{3}{2}}\,C^{(1)}_{\pm 1}\,C^{(1)}_{\pm 1}$

The operators have the complex conjugate property $(C^{(l)}_m)^* = (-1)^m C^{(l)}_{-m}$, and they commute $C^{(l)}_m\,C^{(k)}_q = C^{(k)}_q\,C^{(l)}_m$.

A.6 *s*, *p*, and *d* Orbitals

Table A.4 gives the often used *real s*, *p*, and *d* orbitals in terms of linear combinations of the (complex) spherical harmonics.

Table A.4 Mathematical description of s, p, and d orbitals in terms of spherical harmonics $Y_{l,m_l} = |l, m\rangle$

s	$=\frac{1}{\sqrt{4\pi}}$	$= Y_{0,0}$
p_x	$=\sqrt{\frac{3}{4\pi}}\frac{x}{r}$	$= \frac{1}{\sqrt{2}}(Y_{1,-1} - Y_{1,+1})$
p_y	$=\sqrt{\frac{3}{4\pi}}\frac{y}{r}$	$=\frac{i}{\sqrt{2}}(Y_{1,-1} + Y_{1,+1})$
p_z	$=\sqrt{\frac{3}{4\pi}}\frac{z}{r}$	$= Y_{1,0}$
d_{xy}	$=\sqrt{\frac{15}{4\pi}}\frac{xy}{r^2}$	$=\frac{i}{\sqrt{2}}(Y_{2,-2} - Y_{2,+2})$
d_{xz}	$=\sqrt{\frac{15}{4\pi}}\frac{xz}{r^2}$	$=\frac{1}{\sqrt{2}}(Y_{2,-1} - Y_{2,+1})$
d_{yz}	$=\sqrt{\frac{15}{4\pi}}\frac{yz}{r^2}$	$=\frac{i}{\sqrt{2}}(Y_{2,-1} + Y_{2,+1})$
$d_{x^2-y^2}$	$=\sqrt{\frac{15}{16\pi}}\frac{(x^2-y^2)}{r^2}$	$=\frac{1}{\sqrt{2}}(Y_{2,-2} + Y_{2,+2})$
$d_{3z^2-r^2}$	$=\sqrt{\frac{5}{16\pi}}\frac{(3z^2-r^2)}{r^2}$	$=Y_{2,0}$

The orbitals o_i are real and normalized according to $\int |o_i|^2 d\Omega = \int_0^{2\pi}\int_0^{\pi} |o_i|^2 \sin\theta \, d\theta \, d\phi = \langle o_i|o_i\rangle =$ 1.

A.7 Spin-Orbit Basis Functions and Matrix Elements

Table A.5 lists one-electron spin-orbit coupled functions $|l, s, j, m_j\rangle$ for s, p, and d electrons in terms of the uncoupled basis functions $|l, s, m_l, m_s\rangle$. The given transformation is a specific example of the more general transformation between functions in different coupling schemes given by the Clebsch–Gordon coefficients [3–5]. In Tables A.6 and A.7 we list the action of the spin and angular momentum operators in a Cartesian coordinate system.

Table A.5 Spin-orbit wave functions for s, p and d orbitals. The electron spin wavefunctions are $|s, 1/2\rangle = \uparrow$ (spin-up) and $= |s, -1/2\rangle = \downarrow$ (spin-down)

One-elec. label	Config. label	$\|l, s, j, m_j\rangle$ basis		$\|l, m_l, s, m_s\rangle$ basis
l_j	$^{2S+1}L_J$	j	m_j	Y_{l,m_l} ↑ or ↓
$s_{\frac{1}{2}}$	$^2S_{\frac{1}{2}}$	$\frac{1}{2}$	$+\frac{1}{2}$	$Y_{0,0}\uparrow$
			$-\frac{1}{2}$	$Y_{0,0}\downarrow$
$p_{\frac{1}{2}}$	$^2P_{\frac{1}{2}}$	$\frac{1}{2}$	$+\frac{1}{2}$	$\frac{1}{\sqrt{3}}(-Y_{1,0}\uparrow + \sqrt{2}\,Y_{1,+1}\downarrow)$
			$-\frac{1}{2}$	$\frac{1}{\sqrt{3}}(-\sqrt{2}\,Y_{1,-1}\uparrow + Y_{1,0}\downarrow)$
$p_{\frac{3}{2}}$	$^2P_{\frac{3}{2}}$	$\frac{3}{2}$	$+\frac{3}{2}$	$Y_{1,+1}\uparrow$
			$+\frac{1}{2}$	$\frac{1}{\sqrt{3}}(\sqrt{2}\,Y_{1,0}\uparrow + Y_{1,+1}\downarrow)$
			$-\frac{1}{2}$	$\frac{1}{\sqrt{3}}(Y_{1,-1}\uparrow + \sqrt{2}\,Y_{1,0}\downarrow)$
			$-\frac{3}{2}$	$Y_{1,-1}\downarrow$
$d_{\frac{3}{2}}$	$^2D_{\frac{3}{2}}$	$\frac{3}{2}$	$+\frac{3}{2}$	$\frac{1}{\sqrt{5}}(-Y_{2,+1}\uparrow + 2\,Y_{2,+2}\downarrow)$
			$+\frac{1}{2}$	$\frac{1}{\sqrt{5}}(-\sqrt{2}\,Y_{2,0}\uparrow + \sqrt{3}\,Y_{2,+1}\downarrow)$
			$-\frac{1}{2}$	$\frac{1}{\sqrt{5}}(-\sqrt{3}\,Y_{2,-1}\uparrow + \sqrt{2}\,Y_{2,0}\downarrow)$
			$-\frac{3}{2}$	$\frac{1}{\sqrt{5}}(-2\,Y_{2,-2}\uparrow + Y_{2,-1}\downarrow)$
$d_{\frac{5}{2}}$	$^2D_{\frac{5}{2}}$	$\frac{5}{2}$	$+\frac{5}{2}$	$Y_{2,+2}\uparrow$
			$+\frac{3}{2}$	$\frac{1}{\sqrt{5}}(2\,Y_{2,+1}\uparrow + Y_{2,+2}\downarrow)$
			$+\frac{1}{2}$	$\frac{1}{\sqrt{5}}(\sqrt{3}\,Y_{2,0}\uparrow + \sqrt{2}\,Y_{2,+1}\downarrow)$
			$-\frac{1}{2}$	$\frac{1}{\sqrt{5}}(\sqrt{2}\,Y_{2,-1}\uparrow + \sqrt{3}\,Y_{2,0}\downarrow)$
			$-\frac{3}{2}$	$\frac{1}{\sqrt{5}}(Y_{2,-2}\uparrow + 2\,Y_{2,-1}\downarrow)$
			$-\frac{5}{2}$	$Y_{2,-2}\downarrow$

The configuration label is either for one-electron or one-hole occupation in the particular subshell.

Table A.6 Spin operators $\mathbf{s}_\alpha$ ($\alpha = x, y, z$) acting on spin states $s_z = +1/2 = \uparrow$ and $s_z = -1/2 = \downarrow$, where we have assumed the quantization axis z

$\mathbf{s}_x\|\uparrow\rangle = \frac{1}{2}\|\downarrow\rangle$	$\mathbf{s}_y\|\uparrow\rangle = \frac{i}{2}\|\downarrow\rangle$	$\mathbf{s}_z\|\uparrow\rangle = \frac{1}{2}\|\uparrow\rangle$
$\mathbf{s}_x\|\downarrow\rangle = \frac{1}{2}\|\uparrow\rangle$	$\mathbf{s}_y\|\downarrow\rangle = -\frac{i}{2}\|\uparrow\rangle$	$\mathbf{s}_z\|\downarrow\rangle = -\frac{1}{2}\|\downarrow\rangle$

Eigenvalues are in units of $\hbar$.

Table A.7 Angular momentum operators $\mathbf{L}_\alpha$ ($\alpha = x, y,$ or z) acting on d orbitals, taken from Ballhausen [6]

$\mathbf{L}_x d_{xz} = -\mathrm{i}\,d_{xy}$	$\mathbf{L}_y d_{xz} = \mathrm{i}\,d_{x^2-y^2} - \mathrm{i}\sqrt{3}\,d_{3z^2-r^2}$	$\mathbf{L}_z d_{xz} = \mathrm{i}\,d_{yz}$
$\mathbf{L}_x d_{yz} = \mathrm{i}\sqrt{3}\,d_{3z^2-r^2} + \mathrm{i}\,d_{x^2-y^2}$	$\mathbf{L}_y d_{yz} = \mathrm{i}\,d_{xy}$	$\mathbf{L}_z d_{yz} = -\mathrm{i}\,d_{xz}$
$\mathbf{L}_x d_{xy} = \mathrm{i}\,d_{xz}$	$\mathbf{L}_y d_{xy} = -\mathrm{i}\,d_{yz}$	$\mathbf{L}_z d_{xy} = -\mathrm{i}\,2\,d_{x^2-y^2}$
$\mathbf{L}_x d_{x^2-y^2} = -\mathrm{i}\,d_{yz}$	$\mathbf{L}_y d_{x^2-y^2} = -\mathrm{i}\,d_{xz}$	$\mathbf{L}_z d_{x^2-y^2} = \mathrm{i}\,2\,d_{xy}$
$\mathbf{L}_x d_{3z^2-r^2} = -\mathrm{i}\sqrt{3}\,d_{yz}$	$\mathbf{L}_y d_{3z^2-r^2} = \mathrm{i}\sqrt{3}\,d_{xz}$	$\mathbf{L}_z d_{3z^2-r^2} = 0$

Because all matrix elements between orbitals o_i are purely imaginary we have $\langle o_i|\mathbf{L}_\alpha|o_n\rangle = -\langle o_i|\mathbf{L}_\alpha|o_n\rangle^* = -\langle o_n|\mathbf{L}_\alpha^\dagger|o_i\rangle = -\langle o_n|\mathbf{L}_\alpha|o_i\rangle$. Eigenvalues are in units of $\hbar$.

A.8 Matrix Elements of Spherical Tensors

Below we give tables of selected matrix elements of Racah's spherical tensors. The matrix elements obey the general relationship

$$\langle L, M|C_q^{(k)}|l, m\rangle^* = \langle L, M|C_q^{(k)}|l, m\rangle = (-1)^q \langle l, m|C_{-q}^{(k)}|L, M\rangle\,. \tag{A.39}$$

Matrix elements of spherical tensor operators can be conveniently obtained by use of the Wigner-Eckart theorem [5, 7, 8]. One can either use the analytical form of the Clebsch–Gordon coefficients given by Condon and Shortley [3] or Slater [4], or use modern computer programs such as *Mathematica*. By use of the notation of Slater's Table 20-5 for the CG coefficients we have

$$\begin{aligned}\langle J, M|C_{M_S}^{(S)}|L, M_L\rangle &= \frac{\langle J||C^{(S)}||L\rangle}{\sqrt{2J+1}}\,(L\,S\,M_L\,M_S\,|L\,S\,J\,M)\\ &= (-1)^{M_S}\,\langle L, M_L|C_{-M_S}^{(S)}|J, M\rangle\,.\end{aligned} \tag{A.40}$$

In Table A.8 we tabulate some important reduced matrix elements $\langle L'||\,C^{(k)}\,||L\rangle$. A complete listing is given by Cowan [7]. Matrix elements $\langle L, M|C_q^{(1)}|l, m\rangle$ are listed in Table 10.1.

Table A.8 Selected values for the reduced matrix elements $\langle L'||\,C^{(k)}\,||L\rangle$ for p, d, and f electrons

$\langle p\|\|C^{(0)}\|\|p\rangle = \sqrt{3}$	$\langle d\|\|C^{(0)}\|\|d\rangle = \sqrt{5}$	$\langle f\|\|C^{(0)}\|\|f\rangle = \sqrt{7}$
$\langle p\|\|C^{(1)}\|\|s\rangle = 1$	$\langle d\|\|C^{(1)}\|\|p\rangle = \sqrt{2}$	$\langle f\|\|C^{(1)}\|\|d\rangle = \sqrt{3}$
$\langle d\|\|C^{(2)}\|\|s\rangle = 1$	$\langle f\|\|C^{(2)}\|\|p\rangle = \frac{3}{\sqrt{5}}$	
$\langle p\|\|C^{(2)}\|\|p\rangle = -\sqrt{\frac{6}{5}}$	$\langle d\|\|C^{(2)}\|\|d\rangle = -\sqrt{\frac{10}{7}}$	$\langle f\|\|C^{(2)}\|\|f\rangle = -\sqrt{\frac{28}{15}}$

For s electrons we have $\langle s||\,C^{(0)}\,||s\rangle = 1$, and we have $\langle L'||C^{(k)}||L\rangle = (-1)^k\langle L||C^{(k)}||L'\rangle$ [7]. In particular, we have $\langle L+1||C^{(1)}||L\rangle = -\langle L||C^{(1)}||L+1\rangle = \sqrt{L+1}$ [5].

A.8.1 Polarization Dependent $p \to d$ Transition Probabilies

In Fig. A.8 we have assumed that the degeneracy in m_j of the s–o split $p_{3/2}$ and $p_{1/2}$ core states is lifted by application of an external **H** field along z. **H** acts on both the spin and orbital parts causing an inversion of the sign of the Zeeman components m_j of the $2p_{3/2}$ and $2p_{1/2}$ states. For an exchange field which only acts on the spin, the m_j substates would have the same order of signs [10].

A.8.2 Sum Rules For Matrix Elements of $C_m^{(1)}$ and $C_m^{(2)}$

Let ϕ_L denote one of the p ($L = 1$) or d ($L = 2$) functions listed in Table A.4. We are interested in converting the XNLD cross section (11.2), illustrated in a molecular orbital picture in Fig. 11.2, into the quadrupolar form (11.8), illustrated in Fig. 11.5. This requires expressing the transition matrix elements given by the first order spherical tensors $C^{(1)}$ in terms of those of $C^{(2)}$ by use of Table A.3 and then to those of the

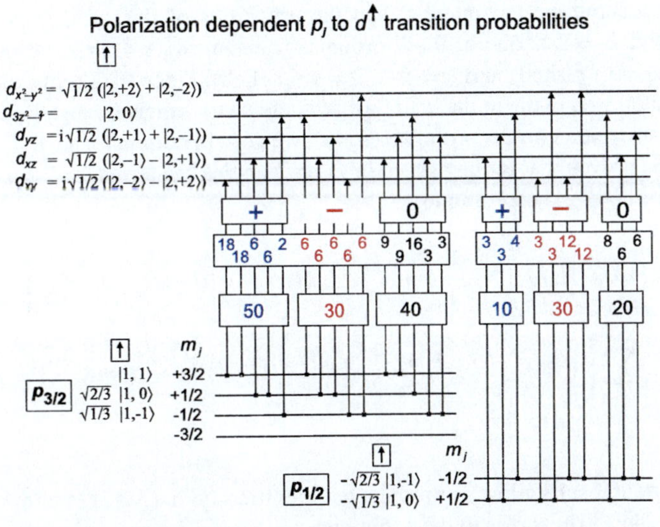

Fig. A.8 Polarization dependent transition probabilities, all multiplied by a factor of 90, from spin-orbit and exchange split p core states $|j, m_j\rangle$ to *spin-up* ($m_s = +1/2$) d valence orbitals. We have chosen the magnetic quantization axis $\mathbf{m} \parallel z$ and assumed $\mathbf{k} \parallel z$ for circular polarization ($p = \pm$) and $\mathbf{E} \parallel z$ for linear polarization ($p = 0$). The dipole transition operators are given according to Table 11.1 by $P_z^{+}/r = C_{+1}^{(1)}$ (labeled +) and $P_z^{-}/r = C_{-1}^{(1)}$ (labeled −) and $P_z^{0}/r = C_0^{(1)}$ (labeled 0). We have assumed that the degeneracy in m_j of the s–o split $p_{3/2}$ and $p_{1/2}$ core states is lifted by application of an external **H** field along z. We have also given the spin-up components $|1, m\rangle$ ($m = 0, \pm 1$) of the m_j core states according to Table A.5. The transition probabilities to the *spin-down* ($m_s = -1/2$) d states are obtained by exchanging the circular polarization labels + and −

quadrupole operators by use of (11.7). By also utilizing the sum rules in Sect. 10.4.7, we here give the key steps in doing so.

The matrix elements of the first order spherical tensors obey the conjugation symmetry property, operators

$$\langle LM|C_q^{(1)}|lm\rangle^* = \langle LM|C_q^{(1)}|lm\rangle = (-1)^q \langle lm|C_{-q}^{(1)}|LM\rangle \,. \tag{A.41}$$

One key trick in linking the matrix elements of the tensors $C^{(2)}$ and $C^{(1)}$ is the insertion of a complete set of states $\sum_{lm} |lm\rangle\langle lm| = 1$, called a *closure relation*. This leads to the link

$$\langle \phi_L|C_0^{(2)}|\phi_L\rangle = \sum_{l,m} |\langle lm|C_0^{(1)}|\phi_L\rangle|^2 - \frac{1}{2}|\langle lm|C_1^{(1)}|\phi_L\rangle|^2 - \frac{1}{2}|\langle lm|C_{-1}^{(1)}|\phi_L\rangle|^2 \,. \tag{A.42}$$

One then simplifies the sum over final states $|l,m\rangle$ in (A.42). Although the matrix elements involve a complete set of states $|lm\rangle$, in practice, all matrix elements are zero except for the $l = L \pm 1$ final states connected by the dipole operators $C_q^{(1)}$. One can further simplify by expressing the terms for $l = L + 1$ by those for $l = L - 1$ through the matrix elements in Table 10.1.

Let us assume we are interested in the properties of the $3d$ valence shell so that ϕ_L with $L = 2$ is one of the d orbitals. The sum in (A.42) then extends over $l = L + 1 = 3$ (f shell) and $l = L - 1 = 1$ (p-shell). Since for L-edge studies we are only interested in the in the $d \to p$ matrix elements which are just the conjugate $p \to d$ matrix elements according to (A.41), we need to eliminate the f shell terms.

It can be proven that for any state $|LM\rangle$ one can link the terms for $l = L + 1$ and $l = L - 1$ in (A.42) according to

$$\sum_{\substack{l=L+1\\ m}} \left\{ |\langle l\, m|C_0^{(1)}|L\, M\rangle|^2 - \frac{1}{2}|\langle l\, m|C_1^{(1)}|L\, M\rangle|^2 - \frac{1}{2}|\langle l\, m|C_{-1}^{(1)}|L\, M\rangle|^2 \right\}$$
$$= \frac{2L-1}{2L+3} \sum_{\substack{l=L-1\\ m}} \left\{ |\langle l\, m|C_0^{(1)}|L\, M\rangle|^2 - \frac{1}{2}|\langle l\, m|C_1^{(1)}|L\, M\rangle|^2 - \frac{1}{2}|\langle l\, m|C_{-1}^{(1)}|L\, M\rangle|^2 \right\} . \tag{A.43}$$

This important results allow us to evaluate the entire sum in (A.42) in terms of states $l = L - 1$, only. This leads to the expression

$$\sum_{\substack{l=L-1\\ m}} |\langle lm|C_0^{(1)}|\phi_L\rangle|^2 = \frac{|\langle L-1||C^{(1)}||L\rangle|^2}{3(2L+1)} + \frac{2L+3}{6L+3} \langle \phi_L|C_0^{(2)}|\phi_L\rangle \,. \tag{A.44}$$

By use of the identity

$$\begin{aligned}&\langle L, M-1|C^{(1)}_{-1}|L+1, M\rangle\langle L, M+1|C^{(1)}_{1}|L+1, M\rangle\\&= \tfrac{2L-1}{2L+3}\langle L, M-1|C^{(1)}_{-1}|L-1, M\rangle\langle L, M+1|C^{(1)}_{1}|L-1, M\rangle \qquad \text{(A.45)}\end{aligned}$$

we furthermore have

$$\begin{aligned}&\sum_{\substack{l=L-1\\ m}} \frac{1}{2}|\langle lm\,|C^{(1)}_{-1} \mp C^{(1)}_{1}|\phi_L\rangle|^2\\&= \frac{L}{6L+3} - \frac{2L+3}{2(6L+3)}\left\langle\phi_L\middle|C^{(2)}_0 \mp \sqrt{\frac{3}{2}}\left(C^{(2)}_2 + C^{(2)}_{-2}\right)\middle|\phi_L\right\rangle. \qquad \text{(A.46)}\end{aligned}$$

A.9 Quantum States and Diffraction Patterns

Here we give details of the calculations of the matrix elements leading to the first order diffraction patterns in Sect. 16.5. We here use the shorter and more convenient notation $\mathbf{k} = \mathbf{k}_A$ and $\mathbf{k}' = \mathbf{k}_B$.

A.9.1 Coherent State

For the *collective coherent state* (16.13) the non-vanishing matrix elements are given by

$$\begin{aligned}&\langle\Phi_{coh}|\mathbf{a}^\dagger_{\mathbf{k}}\,\mathbf{a}_{\mathbf{k}}|\Phi_{coh}\rangle = \langle\Phi_{coh}|\mathbf{a}^\dagger_{\mathbf{k}'}\,\mathbf{a}_{\mathbf{k}'}|\Phi_{coh}\rangle\\&= \langle\Phi_{coh}|\mathbf{a}^\dagger_{\mathbf{k}}\,\mathbf{a}_{\mathbf{k}'}|\Phi_{coh}\rangle = \langle\Phi_{coh}|\mathbf{a}^\dagger_{\mathbf{k}'}\,\mathbf{a}_{\mathbf{k}}|\Phi_{coh}\rangle\\&= \sum_{N=0}^{\infty}\frac{|\alpha|^{2N}}{e^{2|\alpha|^2}}\sum_{m=0}^{N}\frac{m}{m!(N-m)!} = |\alpha|^2 . \qquad \text{(A.47)}\end{aligned}$$

The one-photon detection probability is obtained as

$$\begin{aligned}\left\langle\mathbf{P}^{(1)}(\mathbf{x}_1,\mathbf{x}_2)\right\rangle_{coh} &= \left\langle\Phi_{coh}\left|\mathbf{P}^{(1)}(\mathbf{x}_1,\mathbf{x}_2)\right|\Phi_{coh}\right\rangle\\&= \frac{1}{4}\Bigg\{\underbrace{|\alpha|^2 e^{i\mathbf{k}\cdot(\mathbf{x}_2-\mathbf{x}_1)} + |\alpha|^2 e^{i\mathbf{k}'\cdot(\mathbf{x}_2-\mathbf{x}_1)}}_{\langle\Phi_{coh}|\mathbf{X}|\Phi_{coh}\rangle}\\&\quad + \underbrace{|\alpha|^2\, e^{i(\mathbf{k}'\cdot\mathbf{x}_2-\mathbf{k}\cdot\mathbf{x}_1)} + |\alpha|^2\, e^{i(\mathbf{k}\cdot\mathbf{x}_2-\mathbf{k}'\cdot\mathbf{x}_1)}}_{\langle\Phi_{coh}|\mathbf{Y}|\Phi_{coh}\rangle}\Bigg\}. \qquad \text{(A.48)}\end{aligned}$$

We can now express the coordinates $(\mathbf{k}, \mathbf{x})$ in terms of $(\mathbf{r}, \boldsymbol{\rho})$ in Fig. 16.2a and by assuming that the detector plane is at a large distance z_0 from the source we have the relations (4.127) or

$$\mathbf{k}_X \cdot \mathbf{x}_j \simeq k z_0 - \frac{k}{z_0} \mathbf{r}_X \cdot \boldsymbol{\rho}_j , \tag{A.49}$$

where $X = A, B$ and $\mathbf{k}_A = \mathbf{k}$ and $\mathbf{k}_B = \mathbf{k}'$ and $j = 1, 2$. For point sources located at $\mathbf{r}_A = -\mathbf{r}_B$, separated by $\ell = |\mathbf{r}_A - \mathbf{r}_B|$, and $\boldsymbol{\rho}_2 \parallel \boldsymbol{\rho}_1$ we then obtain

$$\left\langle \mathbf{P}^{(1)}(\boldsymbol{\rho}_1, \boldsymbol{\rho}_2) \right\rangle_{\text{coh}} = |\alpha|^2 \cos\left[\frac{k\ell}{2z_0}\rho_1\right] \cos\left[\frac{k\ell}{2z_0}\rho_2\right] . \tag{A.50}$$

The case of *finite-size* double-slit sources of widths a is evaluated by integrating over all points in the slits. This yields an additional envelope function and with $|\alpha|^2 = \langle n \rangle$ we obtain

$$\begin{aligned} \left\langle \mathbf{P}^{(1)}(\boldsymbol{\rho}_1, \boldsymbol{\rho}_2) \right\rangle_{\text{coh}} = & \langle n \rangle \cos\left[\frac{k\,\ell}{2z_0}\rho_1\right] \cos\left[\frac{k\,\ell}{2z_0}\rho_2\right] \\ & \times \operatorname{sinc}\left[\frac{k\,a}{2z_0}\rho_1\right] \operatorname{sinc}\left[\frac{k\,a}{2z_0}\rho_2\right] . \end{aligned} \tag{A.51}$$

A.9.2 N-Photon Substate of Coherent State

The diffraction pattern of the two-mode *coherent substate* (16.14) is calculated by use of the matrix elements

$$\begin{aligned} \langle \phi_{\text{cohN}} | \mathbf{a}_{\mathbf{k}}^\dagger \, \mathbf{a}_{\mathbf{k}'} | \phi_{\text{cohN}} \rangle &= \langle \phi_{\text{cohN}} | \mathbf{a}_{\mathbf{k}'}^\dagger \, \mathbf{a}_{\mathbf{k}} | \phi_{\text{cohN}} \rangle \\ &= \langle \phi_{\text{cohN}} | \mathbf{a}_{\mathbf{k}'}^\dagger \, \mathbf{a}_{\mathbf{k}'} | \phi_{\text{cohN}} \rangle = \langle \phi_{\text{cohN}} | \mathbf{a}_{\mathbf{k}}^\dagger \, \mathbf{a}_{\mathbf{k}} | \phi_{\text{cohN}} \rangle \\ &= \frac{1}{2^N} \sum_{n=0}^{N} \frac{n\, N!}{n!\,(N-n)!} = \frac{N}{2} . \end{aligned} \tag{A.52}$$

The diffraction pattern $\left\langle \phi_{\text{cohN}} | \mathbf{P}^{(1)}(\boldsymbol{\rho}_1, \boldsymbol{\rho}_2) | \phi_{\text{cohN}} \right\rangle$ is hence the collective coherent state result (A.51) with $\langle n \rangle$ replaced by $N/2$, i.e.

$$\begin{aligned} \left\langle \mathbf{P}^{(1)}(\boldsymbol{\rho}_1, \boldsymbol{\rho}_2) \right\rangle_{\text{cohN}} = & \frac{N}{2} \cos\left[\frac{k\,\ell}{2z_0}\rho_1\right] \cos\left[\frac{k\,\ell}{2z_0}\rho_2\right] \\ & \times \operatorname{sinc}\left[\frac{k\,a}{2z_0}\rho_1\right] \operatorname{sinc}\left[\frac{k\,a}{2z_0}\rho_2\right] . \end{aligned} \tag{A.53}$$

When the pattern is summed over N with the proper weight coefficients, we see from the relation

$$\sum_{N=0}^{\infty} \frac{N}{2}\,|c_N^{\alpha}|^2 = \langle n \rangle \tag{A.54}$$

that the coherent result (A.51) is obtained.

A.9.3 Phase-Diffused Coherent State

For the *phase-diffused coherent state* $|\Phi_{\rm dif}\rangle$ the non-vanishing matrix elements are given by

$$\begin{aligned}\langle \Phi_{\rm dif}|\mathbf{a}_{\mathbf{k}}^{\dagger}\,\mathbf{a}_{\mathbf{k}}|\Phi_{\rm dif}\rangle &= \langle \Phi_{\rm dif}|\mathbf{a}_{\mathbf{k}'}^{\dagger}\,\mathbf{a}_{\mathbf{k}'}|\Phi_{\rm dif}\rangle \\ &= \sum_{N=0}^{\infty} \frac{|\alpha|^{2N}}{e^{2|\alpha|^2}} \sum_{m=0}^{N} \frac{m}{m!(N-m)!} = |\alpha|^2\end{aligned} \tag{A.55}$$

and

$$\begin{aligned}\langle \Phi_{\rm dif}|\mathbf{a}_{\mathbf{k}}^{\dagger}\,\mathbf{a}_{\mathbf{k}'}|\Phi_{\rm dif}\rangle &= |\alpha|^2\, e^{i\varphi} \\ \langle \Phi_{\rm dif}|\mathbf{a}_{\mathbf{k}'}^{\dagger}\,\mathbf{a}_{\mathbf{k}}|\Phi_{\rm dif}\rangle &= |\alpha|^2\, e^{-i\varphi}\,.\end{aligned} \tag{A.56}$$

The $\mathbf{Y}$-term vanishes for a phase average over φ. For our assumed geometry in Fig. 1 (b) we obtain in the $(\mathbf{r}, \boldsymbol{\rho})$ coordinates and by integration over the slit width a and with $\langle n \rangle = |\alpha|^2$

$$\left\langle \mathbf{P}^{(1)}(\boldsymbol{\rho}_1, \boldsymbol{\rho}_2)\right\rangle_{\rm dif} = \frac{\langle n \rangle}{2}\cos\left[\frac{k\,\ell}{2z_0}|\boldsymbol{\rho}_1 - \boldsymbol{\rho}_2|\right]\operatorname{sinc}\left[\frac{k\,a}{2z_0}|\boldsymbol{\rho}_1 - \boldsymbol{\rho}_2|\right]. \tag{A.57}$$

A.9.4 N-Photon Substate of Phase-Diffused Coherent State

For the substate $|\phi_{\rm difN}\rangle$ of the phase-diffused coherent state, the matrix elements are given by (A.55) and (A.56) with $|\alpha|^2$ replaced by $N/2$. The pattern is obtained as

$$\left\langle \mathbf{P}^{(1)}(\boldsymbol{\rho}_1, \boldsymbol{\rho}_2)\right\rangle_{\rm difN} = \frac{N}{4}\cos\left[\frac{k\,\ell}{2z_0}|\boldsymbol{\rho}_1 - \boldsymbol{\rho}_2|\right]\operatorname{sinc}\left[\frac{k\,a}{2z_0}|\boldsymbol{\rho}_1 - \boldsymbol{\rho}_2|\right]. \tag{A.58}$$

When summed over N with the weight factor $|c_N^{\alpha}|^2$, it is seen from the relation

$$\sum_{N=0}^{\infty} |c_N^{\alpha}|^2 \frac{N}{4} = |\alpha|^2 = \frac{\langle n \rangle}{2} \tag{A.59}$$

that (A.58) becomes the pattern of the phase-diffused state given by (A.57), as required.

A.9.5 Chaotic State

The non-vanishing matrix elements of the *two-mode chaotic state* $|\Phi_{\text{cha}}\rangle$ are given by

$$\langle \Phi_{\text{cha}}|\mathbf{a}_k^\dagger \mathbf{a}_k|\Phi_{\text{cha}}\rangle = \langle \Phi_{\text{cha}}|\mathbf{a}_{k'}^\dagger \mathbf{a}_{k'}|\Phi_{\text{cha}}\rangle = \sum_{N=0}^{\infty} \frac{N\,\langle n\rangle^N}{(1+\langle n\rangle)^{N+1}} = \langle n\rangle \tag{A.60}$$

since all other matrix element contain phase factors that average to zero. The detection probability (1) is therefore the same as for the phase-diffused coherent state, and we have

$$\left\langle \mathbf{P}^{(1)}(\rho_1,\rho_2)\right\rangle_{\text{cha}} = \frac{\langle n\rangle}{2} \cos\left[\frac{k\,\ell}{2z_0}|\rho_1-\rho_2|\right] \text{sinc}\left[\frac{k\,a}{2z_0}|\rho_1-\rho_2|\right]. \tag{A.61}$$

A.9.6 N-Photon Substate of Chaotic State

For the chaotic substate $|\phi_{\text{chaN}}\rangle$ the matrix elements are given by (A.60) with $\langle n\rangle$ replaced by $N/2$. The pattern is obtained as

$$\left\langle \mathbf{P}^{(1)}(\rho_1,\rho_2)\right\rangle_{\text{chaN}} = \frac{N}{4} \cos\left[\frac{k\,\ell}{2z_0}|\rho_1-\rho_2|\right] \text{sinc}\left[\frac{k\,a}{2z_0}|\rho_1-\rho_2|\right]. \tag{A.62}$$

If we sum over N with the weight factors $|c_N^\beta|^2$, we obtain by use of

$$\sum_{N=0}^{\infty} |c_N^\beta|^2 \frac{N}{4} = \frac{\langle n\rangle}{2} \tag{A.63}$$

the chaotic result (A.61).

A.9.7 N-Photon Entangled (NOON) State

For the *two-mode N-photon entangled state* $|\phi_{\text{entN}}\rangle$, the matrix elements are

$$\begin{aligned} \langle\phi_{\rm entN}|\mathbf{a}^\dagger_{\mathbf{k}}\,\mathbf{a}_{\mathbf{k}'}|\phi_{\rm entN}\rangle &= \langle\phi_{\rm entN}|\mathbf{a}^\dagger_{\mathbf{k}'}\,\mathbf{a}_{\mathbf{k}}|\phi_{\rm entN}\rangle = 0 \\ \langle\phi_{\rm entN}|\mathbf{a}^\dagger_{\mathbf{k}}\,\mathbf{a}_{\mathbf{k}}|\phi_{\rm entN}\rangle &= \langle\phi_{\rm entN}|\mathbf{a}^\dagger_{\mathbf{k}'}\,\mathbf{a}_{\mathbf{k}'}|\phi_{\rm entN}\rangle = \frac{N}{2}\,. \end{aligned} \tag{A.64}$$

We obtain

$$\left\langle\mathbf{P}^{(1)}(\boldsymbol{\rho}_1,\boldsymbol{\rho}_2)\right\rangle_{\rm entN} = \frac{N}{4}\cos\left[\frac{k\,\ell}{2z_0}|\boldsymbol{\rho}_1-\boldsymbol{\rho}_2|\right]\operatorname{sinc}\left[\frac{k\,a}{2z_0}|\boldsymbol{\rho}_1-\boldsymbol{\rho}_2|\right]. \tag{A.65}$$

A.9.8 N-Photon Number State

For the *two-mode N-photon number state* $|\phi_{\rm numN}\rangle$ with $N \geq 2$ the matrix elements are evaluated as

$$\begin{aligned} \langle\phi_{\rm numN}|\mathbf{a}^\dagger_{\mathbf{k}}\,\mathbf{a}_{\mathbf{k}'}|\phi_{\rm numN}\rangle &= \langle\phi_{\rm numN}|\mathbf{a}^\dagger_{\mathbf{k}'}\,\mathbf{a}_{\mathbf{k}}|\phi_{\rm numN}\rangle = 0 \\ \langle\phi_{\rm numN}|\mathbf{a}^\dagger_{\mathbf{k}}\,\mathbf{a}_{\mathbf{k}}|\phi_{\rm numN}\rangle &= \langle\phi_{\rm numN}|\mathbf{a}^\dagger_{\mathbf{k}'}\,\mathbf{a}_{\mathbf{k}'}|\phi_{\rm numN}\rangle = \frac{N}{2} \end{aligned} \tag{A.66}$$

which is the same as for the entangled state (A.65).

A.10 Matrix Element of Second Order Coherence Operators

Here we give details of the calculations of the matrix elements leading to the second order diffraction patterns in Sect. 16.6. Again we use the notation $\mathbf{k} = \mathbf{k}_{\rm A}$ and $\mathbf{k}' = \mathbf{k}_{\rm B}$.

A.10.1 Coherent State

For the collective coherent state (16.13), the relevant matrix elements for term $\langle\mathbf{A}\rangle$ are given by

$$\begin{aligned} \langle\Phi_{\rm coh}|\mathbf{a}^\dagger_{\mathbf{k}}\,\mathbf{a}^\dagger_{\mathbf{k}'}\,\mathbf{a}_{\mathbf{k}}\,\mathbf{a}_{\mathbf{k}'}|\Phi_{\rm coh}\rangle &= \langle\Phi_{\rm coh}|\mathbf{a}^\dagger_{\mathbf{k}'}\,\mathbf{a}^\dagger_{\mathbf{k}}\,\mathbf{a}_{\mathbf{k}'}\,\mathbf{a}_{\mathbf{k}}|\Phi_{\rm coh}\rangle \\ &= \mathrm{e}^{-2|\alpha|^2}\sum_{N=0}^{\infty}|\alpha|^{2N}\sum_{n=0}^{N}\frac{n(N-n)}{n!\,(N-n)!} = |\alpha|^4\,. \end{aligned} \tag{A.67}$$

For term $\langle\mathbf{B}\rangle$ we have

$$\langle \Phi_{\rm coh}|\mathbf{a}_{\mathbf{k}}^{\dagger}\,\mathbf{a}_{\mathbf{k}}^{\dagger}\,\mathbf{a}_{\mathbf{k}}\,\mathbf{a}_{\mathbf{k}}|\Phi_{\rm coh}\rangle = \langle \Phi_{\rm coh}|\mathbf{a}_{\mathbf{k}'}^{\dagger}\,\mathbf{a}_{\mathbf{k}'}^{\dagger}\,\mathbf{a}_{\mathbf{k}'}\,\mathbf{a}_{\mathbf{k}'}|\Phi_{\rm coh}\rangle$$
$$= \mathrm{e}^{-2|\alpha|^2}\sum_{N=0}^{\infty}|\alpha|^{2N}\sum_{n=0}^{N}\frac{n(n-1)}{n!\,(N-n)!} = |\alpha|^4 \tag{A.68}$$

and

$$\langle \Phi_{\rm coh}|\mathbf{a}_{\mathbf{k}}^{\dagger}\,\mathbf{a}_{\mathbf{k}}^{\dagger}\,\mathbf{a}_{\mathbf{k}'}\,\mathbf{a}_{\mathbf{k}'}|\Phi_{\rm coh}\rangle = \langle \Phi_{\rm coh}|\mathbf{a}_{\mathbf{k}'}^{\dagger}\,\mathbf{a}_{\mathbf{k}'}^{\dagger}\,\mathbf{a}_{\mathbf{k}}\,\mathbf{a}_{\mathbf{k}}|\Phi_{\rm coh}\rangle = |\alpha|^4\,. \tag{A.69}$$

Similarly we obtain the matrix elements for terms $\langle\mathbf{C}\rangle$ and $\langle\mathbf{D}\rangle$

$$\langle \Phi_{\rm coh}|\mathbf{a}_{\mathbf{k}}^{\dagger}\,\mathbf{a}_{\mathbf{k}}^{\dagger}\,\mathbf{a}_{\mathbf{k}}\,\mathbf{a}_{\mathbf{k}'}|\Phi_{\rm coh}\rangle = \langle \Phi_{\rm coh}|\mathbf{a}_{\mathbf{k}}^{\dagger}\,\mathbf{a}_{\mathbf{k}'}^{\dagger}\,\mathbf{a}_{\mathbf{k}'}\,\mathbf{a}_{\mathbf{k}'}|\Phi_{\rm coh}\rangle$$
$$\langle \Phi_{\rm coh}|\mathbf{a}_{\mathbf{k}'}^{\dagger}\,\mathbf{a}_{\mathbf{k}}^{\dagger}\,\mathbf{a}_{\mathbf{k}}\,\mathbf{a}_{\mathbf{k}}|\Phi_{\rm coh}\rangle = \langle \Phi_{\rm coh}|\mathbf{a}_{\mathbf{k}'}^{\dagger}\,\mathbf{a}_{\mathbf{k}'}^{\dagger}\,\mathbf{a}_{\mathbf{k}'}\,\mathbf{a}_{\mathbf{k}}|\Phi_{\rm coh}\rangle = |\alpha|^4\,. \tag{A.70}$$

The second order detection probability becomes

$$\begin{aligned}\big\langle\mathbf{P}^{(2)}(\mathbf{x}_1,\mathbf{x}_2)\big\rangle_{\rm coh} = \frac{1}{4}\Bigg\{ & \underbrace{\frac{|\alpha|^4}{2}+\frac{|\alpha|^4}{2}\cos[(\mathbf{k}-\mathbf{k}')\cdot(\mathbf{x}_1-\mathbf{x}_2)]}_{\langle\Phi_{\rm coh}|\mathbf{A}|\Phi_{\rm coh}\rangle/4} \\ & \underbrace{+\frac{|\alpha|^4}{2}+\frac{|\alpha|^4}{4}\,\mathrm{e}^{-\mathrm{i}(\mathbf{k}-\mathbf{k}')\cdot(\mathbf{x}_1+\mathbf{x}_2)}+\frac{|\alpha|^4}{4}\,\mathrm{e}^{\mathrm{i}(\mathbf{k}-\mathbf{k}')\cdot(\mathbf{x}_1+\mathbf{x}_2)}}_{\langle\Phi_{\rm coh}|\mathbf{B}|\Phi_{\rm coh}\rangle/4} \\ & \underbrace{+\frac{|\alpha|^4}{2}\,\mathrm{e}^{-\mathrm{i}(\mathbf{k}-\mathbf{k}')\cdot\mathbf{x}_1}+\frac{|\alpha|^4}{2}\,\mathrm{e}^{\mathrm{i}(\mathbf{k}-\mathbf{k}')\cdot\mathbf{x}_1}}_{\langle\Phi_{\rm coh}|\mathbf{C}|\Phi_{\rm coh}\rangle/4} \\ & \underbrace{+\frac{|\alpha|^4}{2}\,\mathrm{e}^{-\mathrm{i}(\mathbf{k}-\mathbf{k}')\cdot\mathbf{x}_2}+\frac{|\alpha|^4}{2}\,\mathrm{e}^{\mathrm{i}(\mathbf{k}-\mathbf{k}')\cdot\mathbf{x}_2}}_{\langle\Phi_{\rm coh}|\mathbf{D}|\Phi_{\rm coh}\rangle/4}\Bigg\}\,,\end{aligned} \tag{A.71}$$

where we have identified the origin of the four terms by underbrackets. By use of $|\alpha|^2=\langle n\rangle$ and the identities $1+\cos[x]=2\cos^2[x/2]$ and $\cos a+\cos b=2\cos\left[\frac{a-b}{2}\right]\cos\left[\frac{a+b}{2}\right]$ this becomes

$$\begin{aligned}\big\langle\mathbf{P}^{(2)}(\mathbf{x}_1,\mathbf{x}_2)\big\rangle_{\rm coh} = \frac{\langle n\rangle^2}{4}\Bigg\{&\cos^2\left[\frac{1}{2}(\mathbf{k}-\mathbf{k}')\cdot(\mathbf{x}_1-\mathbf{x}_2)\right]+\cos^2\left[\frac{1}{2}(\mathbf{k}-\mathbf{k}')\cdot(\mathbf{x}_1+\mathbf{x}_2)\right] \\ &+2\cos\left[\frac{1}{2}(\mathbf{k}-\mathbf{k}')\cdot(\mathbf{x}_1-\mathbf{x}_2)\right]\cos\left[\frac{1}{2}(\mathbf{k}-\mathbf{k}')\cdot(\mathbf{x}_1+\mathbf{x}_2)\right]\Bigg\}\,.\end{aligned} \tag{A.72}$$

The pattern is seen to become symmetrical and the last term is just the interference term of the amplitudes associated with the first two terms according to

$$\langle\mathbf{C}\rangle+\langle\mathbf{D}\rangle = 2\sqrt{\langle\mathbf{A}\rangle\langle\mathbf{B}\rangle} \tag{A.73}$$

Converting to the $(\mathbf{r}, \boldsymbol{\rho})$ coordinates, we obtain for the slit separation $\ell = |\mathbf{r}_A - \mathbf{r}_B|$ and detector positions $\boldsymbol{\rho}_2 \parallel \boldsymbol{\rho}_1$

$$\langle \mathbf{P}^{(2)}(\boldsymbol{\rho}_1, \boldsymbol{\rho}_2) \rangle_{coh} = \langle n \rangle^2 \cos^2\left[\frac{k\,\ell}{2z_0}\rho_1\right] \cos^2\left[\frac{k\,\ell}{2z_0}\rho_2\right] . \tag{A.74}$$

For the case of two slits of width a we obtain

$$\langle \mathbf{P}^{(2)}(\boldsymbol{\rho}_1, \boldsymbol{\rho}_2) \rangle_{coh} = \langle n \rangle^2 \cos^2\left[\frac{k\,\ell}{2z_0}\rho_1\right] \cos^2\left[\frac{k\,\ell}{2z_0}\rho_2\right] \mathrm{sinc}^2\left[\frac{k\,a}{2z_0}\rho_1\right] \mathrm{sinc}^2\left[\frac{k\,a}{2z_0}\rho_2\right] . \tag{A.75}$$

The detection probability now factors into separate symmetric contributions from points $\boldsymbol{\rho}_1$ and $\boldsymbol{\rho}_2$, and the same pattern is observed for both detection schemes.

A.10.2 N-Photon Substate of Coherent State

Similarly, the coherent substate (16.14) gives equal contributions from all terms **A** – **D** expressed by the matrix elements

$$\begin{aligned} &\langle \phi_{cohN} | \mathbf{a}_k^\dagger \, \mathbf{a}_{k'}^\dagger \mathbf{a}_k \, \mathbf{a}_{k'} | \phi_{cohN} \rangle = \langle \phi_{cohN} | \mathbf{a}_k^\dagger \, \mathbf{a}_k^\dagger \mathbf{a}_k \, \mathbf{a}_k | \phi_{cohN} \rangle \\ &= \langle \phi_{cohN} | \mathbf{a}_k^\dagger \, \mathbf{a}_k^\dagger \mathbf{a}_{k'} \, \mathbf{a}_{k'} | \phi_{cohN} \rangle = \langle \phi_{cohN} | \mathbf{a}_k^\dagger \, \mathbf{a}_k^\dagger \, \mathbf{a}_k \, \mathbf{a}_{k'} | \phi_{cohN} \rangle \\ &= \frac{1}{2^N} \sum_{n=0}^{N} n(n-1) \frac{N!}{n!\,(N-n)!} = \frac{1}{4} N(N-1) \end{aligned} \tag{A.76}$$

and the same when exchanging k and k'. The pattern therefore is the same as for the coherent state with the substitution $|\alpha|^4 = \langle n \rangle^2$ by $N(N-1)/4$, i.e.

$$\begin{aligned} \langle \mathbf{P}^{(2)}(\boldsymbol{\rho}_1, \boldsymbol{\rho}_2) \rangle_{coh2} = \frac{N(N-1)}{4} &\cos^2\left[\frac{k\,\ell}{2z_0}\rho_1\right] \cos^2\left[\frac{k\,\ell}{2z_0}\rho_2\right] \\ &\times \mathrm{sinc}^2\left[\frac{k\,a}{2z_0}\rho_1\right] \mathrm{sinc}^2\left[\frac{k\,a}{2z_0}\rho_2\right] . \end{aligned} \tag{A.77}$$

When the pattern is summed over N with the proper weight coefficients, we see from the relation

$$\sum_{N=0}^{\infty} \frac{N(N-1)}{4} |c_N^\alpha|^2 = \langle n \rangle^2 \tag{A.78}$$

that the coherent result (A.75) is obtained.

A.10.3 Phase-Diffused Coherent State

For the *phase-diffused coherent state* $|\Phi_{\text{dif}}\rangle$ the matrix elements for term $\langle \mathbf{A}\rangle$ are obtained as

$$\begin{aligned}\langle \Phi_{\text{dif}}|\mathbf{a}_{\mathbf{k}}^{\dagger}\,\mathbf{a}_{\mathbf{k}'}^{\dagger}\,\mathbf{a}_{\mathbf{k}}\,\mathbf{a}_{\mathbf{k}'}|\Phi_{\text{dif}}\rangle &= \langle \Phi_{\text{dif}}|\mathbf{a}_{\mathbf{k}'}^{\dagger}\,\mathbf{a}_{\mathbf{k}}^{\dagger}\,\mathbf{a}_{\mathbf{k}'}\,\mathbf{a}_{\mathbf{k}}|\Phi_{\text{dif}}\rangle \\ = \sum_{N=0}^{\infty}\frac{|\alpha|^{2N}}{\mathrm{e}^{2|\alpha|^2}}\sum_{m=0}^{N}\frac{m(m-1)}{m!(N-m)!} &= |\alpha|^4 .\end{aligned} \tag{A.79}$$

For the first two terms in $\langle \mathbf{B}\rangle$ we have

$$\begin{aligned}\langle \Phi_{\text{dif}}|\mathbf{a}_{\mathbf{k}}^{\dagger}\,\mathbf{a}_{\mathbf{k}}^{\dagger}\,\mathbf{a}_{\mathbf{k}}\,\mathbf{a}_{\mathbf{k}}|\Phi_{\text{dif}}\rangle &= \langle \Phi_{\text{dif}}|\mathbf{a}_{\mathbf{k}'}^{\dagger}\,\mathbf{a}_{\mathbf{k}'}^{\dagger}\,\mathbf{a}_{\mathbf{k}'}\,\mathbf{a}_{\mathbf{k}'}|\Phi_{\text{dif}}\rangle \\ &= |\alpha|^4 ,\end{aligned} \tag{A.80}$$

while for the last two terms in $\langle \mathbf{B}\rangle$ we obtain

$$\begin{aligned}\langle \Phi_{\text{dif}}|\mathbf{a}_{\mathbf{k}}^{\dagger}\,\mathbf{a}_{\mathbf{k}}^{\dagger}\,\mathbf{a}_{\mathbf{k}'}\,\mathbf{a}_{\mathbf{k}'}|\Phi_{\text{dif}}\rangle &= \mathrm{e}^{2\mathrm{i}\varphi}\,|\alpha|^4 \\ \langle \Phi_{\text{dif}}|\mathbf{a}_{\mathbf{k}'}^{\dagger}\,\mathbf{a}_{\mathbf{k}'}^{\dagger}\,\mathbf{a}_{\mathbf{k}}\,\mathbf{a}_{\mathbf{k}}|\Phi_{\text{dif}}\rangle &= \mathrm{e}^{-2\mathrm{i}\varphi}\,|\alpha|^4 .\end{aligned} \tag{A.81}$$

The terms $\langle \mathbf{C}\rangle$ and $\langle \mathbf{B}\rangle$ are evaluated as

$$\begin{aligned}\langle \Phi_{\text{dif}}|\mathbf{a}_{\mathbf{k}}^{\dagger}\,\mathbf{a}_{\mathbf{k}}^{\dagger}\,\mathbf{a}_{\mathbf{k}}\,\mathbf{a}_{\mathbf{k}'}|\Phi_{\text{dif}}\rangle &= \langle \Phi_{\text{dif}}|\mathbf{a}_{\mathbf{k}}^{\dagger}\,\mathbf{a}_{\mathbf{k}'}^{\dagger}\,\mathbf{a}_{\mathbf{k}'}\,\mathbf{a}_{\mathbf{k}'}|\Phi_{\text{dif}}\rangle \\ &= \mathrm{e}^{\mathrm{i}\varphi}\,|\alpha|^4 \\ \langle \Phi_{\text{dif}}|\mathbf{a}_{\mathbf{k}'}^{\dagger}\,\mathbf{a}_{\mathbf{k}}^{\dagger}\,\mathbf{a}_{\mathbf{k}}\,\mathbf{a}_{\mathbf{k}}|\Phi_{\text{dif}}\rangle &= \langle \Phi_{\text{dif}}|\mathbf{a}_{\mathbf{k}'}^{\dagger}\,\mathbf{a}_{\mathbf{k}'}^{\dagger}\,\mathbf{a}_{\mathbf{k}'}\,\mathbf{a}_{\mathbf{k}}|\Phi_{\text{dif}}\rangle \\ &= \mathrm{e}^{-\mathrm{i}\varphi}\,|\alpha|^4 .\end{aligned} \tag{A.82}$$

All terms containing the phase φ vanish upon phase averaging, and we obtain with $|\alpha|^2 = \langle n\rangle$

$$\left\langle \mathbf{P}^{(2)}(\mathbf{x}_1,\mathbf{x}_2)\right\rangle_{\text{dif}} = \frac{\langle n\rangle^2}{4}\left\{\frac{1}{2}+\cos^2\left[\frac{1}{2}(\mathbf{k}-\mathbf{k}')\cdot(\mathbf{x}_1-\mathbf{x}_2)\right]\right\} . \tag{A.83}$$

When expressed in the coordinates $(\mathbf{r},\boldsymbol{\rho})$ and integrated over the slit width a, we obtain

$$\left\langle \mathbf{P}^{(2)}(\boldsymbol{\rho}_1,\boldsymbol{\rho}_2)\right\rangle_{\text{dif}} = \frac{\langle n\rangle^2}{4}\left\{\frac{1}{2}+\cos^2\left[\frac{k\ell}{2z_0}|\boldsymbol{\rho}_1-\boldsymbol{\rho}_2|\right]\operatorname{sinc}^2\left[\frac{ka}{2z_0}|\boldsymbol{\rho}_1-\boldsymbol{\rho}_2|\right]\right\} . \tag{A.84}$$

The pattern is constant, $\left\langle \mathbf{P}^{(2)}(\boldsymbol{\rho}_1,\boldsymbol{\rho}_2)\right\rangle_{\text{dif}} = 3\langle n\rangle^2/8$ for the detection geometry $\boldsymbol{\rho}_1 = \boldsymbol{\rho}_2$, and for $\boldsymbol{\rho}_1 = -\boldsymbol{\rho}_2$ the diffraction fine structure sits on a constant background $\langle n\rangle^2/8$.

A.10.4 N-Photon Substate of Phase-Diffused Coherent State

For the substate $|\phi_{\mathrm{difN}}\rangle$ the matrix elements are the same as those of the collective parent state with $|\alpha|^4$ replaced by $N(N-1)/4$ so that the pattern is

$$\left\langle \mathbf{P}^{(2)}(\boldsymbol{\rho}_1, \boldsymbol{\rho}_2)\right\rangle_{\mathrm{difN}} = \frac{N(N-1)}{16}\left\{\frac{1}{2} + \cos^2\left[\frac{k\ell}{2z_0}|\boldsymbol{\rho}_1 - \boldsymbol{\rho}_2|\right] \mathrm{sinc}^2\left[\frac{ka}{2z_0}|\boldsymbol{\rho}_1 - \boldsymbol{\rho}_2|\right]\right\}. \tag{A.85}$$

When summed over N with the weight factors $|c_N^{\alpha}|^2$, it is seen from the relation

$$\sum_{N=0}^{\infty} \frac{N(N-1)}{16} |c_N^{\alpha}|^2 = \frac{\langle n\rangle^2}{4} \tag{A.86}$$

that (A.85) becomes the pattern of the phase-diffused state given by (A.84), as required.

A.10.5 Chaotic State

For the *two-mode chaotic state* $|\Phi_{\mathrm{cha}}\rangle$ the terms in $\langle \mathbf{A}\rangle$ are evaluated as

$$\begin{aligned}\langle \Phi_{\mathrm{cha}}|\mathbf{a}_{\mathbf{k}}^{\dagger}\, \mathbf{a}_{\mathbf{k}'}^{\dagger}\, \mathbf{a}_{\mathbf{k}}\, \mathbf{a}_{\mathbf{k}'}|\Phi_{\mathrm{cha}}\rangle &= \langle \Phi_{\mathrm{cha}}|\mathbf{a}_{\mathbf{k}'}^{\dagger}\, \mathbf{a}_{\mathbf{k}}^{\dagger}\, \mathbf{a}_{\mathbf{k}'}\, \mathbf{a}_{\mathbf{k}}|\Phi_{\mathrm{cha}}\rangle \\ &= \left[\sum_{N=0}^{\infty} N \frac{\langle n\rangle^N}{(1+\langle n\rangle)^{N+1}}\right]^2 \\ &= \sum_{N=0}^{\infty}\sum_{m=0}^{N} m(N-m)\, \frac{\langle n\rangle^N}{(1+\langle n\rangle)^{N+2}} = \langle n\rangle^2 .\end{aligned} \tag{A.87}$$

Similarly, the contributions of the first two terms of $\langle \mathbf{B}\rangle$ are obtained as

$$\begin{aligned}\langle \Phi_{\mathrm{cha}}|\mathbf{a}_{\mathbf{k}}^{\dagger}\, \mathbf{a}_{\mathbf{k}}^{\dagger}\, \mathbf{a}_{\mathbf{k}}\, \mathbf{a}_{\mathbf{k}}|\Phi_{\mathrm{cha}}\rangle &= \langle \Phi_{\mathrm{cha}}|\mathbf{a}_{\mathbf{k}'}^{\dagger}\, \mathbf{a}_{\mathbf{k}'}^{\dagger}\, \mathbf{a}_{\mathbf{k}'}\, \mathbf{a}_{\mathbf{k}'}|\Phi_{\mathrm{cha}}\rangle \\ &= \sum_{N=0}^{\infty}\sum_{m=0}^{N} m(m-1) \frac{\langle n\rangle^N}{(1+\langle n\rangle)^{N+2}} = 2\langle n\rangle^2\end{aligned} \tag{A.88}$$

while the contributions from the other two terms in $\langle \mathbf{B}\rangle$ average to zero, i.e.

$$\langle \Phi_{\mathrm{cha}}|\mathbf{a}_{\mathbf{k}}^{\dagger}\mathbf{a}_{\mathbf{k}}^{\dagger}\mathbf{a}_{\mathbf{k}'}\mathbf{a}_{\mathbf{k}'}|\Phi_{\mathrm{cha}}\rangle = \langle \Phi_{\mathrm{cha}}|\mathbf{a}_{\mathbf{k}'}^{\dagger}\mathbf{a}_{\mathbf{k}'}^{\dagger}\mathbf{a}_{\mathbf{k}}\mathbf{a}_{\mathbf{k}}|\Phi_{\mathrm{cha}}\rangle = 0 . \tag{A.89}$$

The contribution by the term $\langle \mathbf{B} \rangle$ is therefore only a constant resulting in a background. The matrix elements associated with terms $\langle \mathbf{C} \rangle$ and $\langle \mathbf{D} \rangle$ are zero owing to the fact that they contain unpaired raising and lowering operators, i.e. $\langle \mathbf{C} + \mathbf{D} \rangle = 0$. Evaluation of detection probability and phase averaging yields

$$\left\langle \mathbf{P}^{(2)}(\mathbf{x}_1, \mathbf{x}_2) \right\rangle_{\text{cha}} = \frac{\langle n \rangle^2}{4} \left\{ 1 + \cos^2\left[\frac{1}{2} (\mathbf{k} - \mathbf{k}') \cdot (\mathbf{x}_1 - \mathbf{x}_2) \right] \right\}. \tag{A.90}$$

In the coordinates $(\mathbf{r}, \boldsymbol{\rho})$ and integration over the slit width a we obtain

$$\left\langle \mathbf{P}^{(2)}(\boldsymbol{\rho}_1, \boldsymbol{\rho}_2) \right\rangle_{\text{cha}} = \frac{\langle n \rangle^2}{4} \left\{ 1 + \cos^2\left[\frac{k\,\ell}{2z_0} |\boldsymbol{\rho}_1 - \boldsymbol{\rho}_2| \right] \operatorname{sinc}^2\left[\frac{k\,a}{2z_0} |\boldsymbol{\rho}_1 - \boldsymbol{\rho}_2| \right] \right\}. \tag{A.91}$$

A.10.6 N-Photon Substate of Chaotic State

For the chaotic substate $|\phi_{\text{chaN}}\rangle$ the matrix elements are those of the collective parent state with $\langle n \rangle^2$ replaced by $N(N-1)/6$ and we obtain

$$\left\langle \mathbf{P}^{(2)}(\boldsymbol{\rho}_1, \boldsymbol{\rho}_2) \right\rangle_{\text{chaN}} = \frac{N(N-1)}{24} \left\{ 1 + \cos^2\left[\frac{k\,\ell}{2z_0} |\boldsymbol{\rho}_1 - \boldsymbol{\rho}_2| \right] \operatorname{sinc}^2\left[\frac{k\,a}{2z_0} |\boldsymbol{\rho}_1 - \boldsymbol{\rho}_2| \right] \right\}. \tag{A.92}$$

By summing (A.92) over N with the weight factors $|c_N^{\beta}|^2$ we obtain by use of the relation

$$\sum_{N=0}^{\infty} |c_N^{\beta}|^2 \frac{N(N-1)}{24} = \frac{\langle n \rangle^2}{4} \tag{A.93}$$

the chaotic result (A.91), as required.

A.10.7 N-Photon Entangled (NOON) State

For the *two-modeN-photon entangled state* $|\phi_{\text{entN}}\rangle$ the matrix elements for $\mathbf{A}$ vanish

$$\langle \phi_{\text{entN}} | \mathbf{a}_{\mathbf{k}}^{\dagger} \mathbf{a}_{\mathbf{k}'}^{\dagger} \mathbf{a}_{\mathbf{k}'} \mathbf{a}_{\mathbf{k}} | \phi_{\text{entN}} \rangle = 0\,. \tag{A.94}$$

The matrix elements for the first two terms in $\mathbf{B}$ are

$$\langle \phi_{\text{entN}} | \mathbf{a}_{\mathbf{k}}^{\dagger} \, \mathbf{a}_{\mathbf{k}}^{\dagger} \mathbf{a}_{\mathbf{k}} \, \mathbf{a}_{\mathbf{k}} | \phi_{\text{entN}} \rangle = \langle \phi_{\text{entN}} | \mathbf{a}_{\mathbf{k}'}^{\dagger} \, \mathbf{a}_{\mathbf{k}'}^{\dagger} \mathbf{a}_{\mathbf{k}'} \, \mathbf{a}_{\mathbf{k}'} | \phi_{\text{entN}} \rangle = \frac{N(N-1)}{2} \tag{A.95}$$

while those for the second two terms in **B** are non-zero only for $N = 2$,

$$\langle \phi_{\text{entN}} | \mathbf{a}_{\mathbf{k}}^{\dagger} \, \mathbf{a}_{\mathbf{k}}^{\dagger} \mathbf{a}_{\mathbf{k}'} \, \mathbf{a}_{\mathbf{k}'} | \phi_{\text{entN}} \rangle = \langle \phi_{\text{entN}} | \mathbf{a}_{\mathbf{k}'}^{\dagger} \, \mathbf{a}_{\mathbf{k}'}^{\dagger} \mathbf{a}_{\mathbf{k}} \, \mathbf{a}_{\mathbf{k}} | \phi_{\text{entN}} \rangle = \delta(N, 2) \tag{A.96}$$

where $\delta(N, 2) = 1$ for $N = 2$ and zero otherwise.

The diffraction pattern is a constant for $N > 2$, given by $\langle \mathbf{P}^{(2)}(\mathbf{x}_1, \mathbf{x}_2) \rangle_{\text{entN}} = \frac{N(N-1)}{16}$. For the specific $N = 2$ case we obtain

$$\langle \mathbf{P}^{(2)}(\mathbf{x}_1, \mathbf{x}_2) \rangle_{\text{ent2}} = \frac{1}{16} \langle \phi_{\text{ent2}} | \mathbf{B} | \phi_{\text{ent2}} \rangle = \frac{1}{4} \cos^2 \left[\frac{1}{2} (\mathbf{k} - \mathbf{k}') \cdot (\mathbf{x}_1 + \mathbf{x}_2) \right]. \tag{A.97}$$

In the coordinates $(\mathbf{r}, \boldsymbol{\rho})$ and integrated over the slit width a the pattern becomes

$$\langle \mathbf{P}^{(2)}(\boldsymbol{\rho}_1, \boldsymbol{\rho}_2) \rangle_{\text{ent2}} = \frac{1}{4} \cos^2 \left[\frac{k \ell}{2 z_0} |\boldsymbol{\rho}_1 + \boldsymbol{\rho}_2| \right] \text{sinc}^2 \left[\frac{k a}{2 z_0} |\boldsymbol{\rho}_1 + \boldsymbol{\rho}_2| \right]. \tag{A.98}$$

A.10.8 N-Photon Number State

For the *two-mode N-photon number state* $|\phi_{\text{numN}}\rangle$ the matrix elements of **A** are given by

$$\langle \phi_{\text{numN}} | \mathbf{a}_{\mathbf{k}}^{\dagger} \, \mathbf{a}_{\mathbf{k}'}^{\dagger} \mathbf{a}_{\mathbf{k}'} \, \mathbf{a}_{\mathbf{k}} | \phi_{\text{numN}} \rangle = \frac{N^2}{4} \tag{A.99}$$

and for the first two terms in **B** we have

$$\begin{aligned} &\langle \phi_{\text{numN}} | \mathbf{a}_{\mathbf{k}}^{\dagger} \, \mathbf{a}_{\mathbf{k}}^{\dagger} \mathbf{a}_{\mathbf{k}} \, \mathbf{a}_{\mathbf{k}} | \phi_{\text{numN}} \rangle \\ &= \langle \phi_{\text{numN}} | \mathbf{a}_{\mathbf{k}'}^{\dagger} \, \mathbf{a}_{\mathbf{k}'}^{\dagger} \mathbf{a}_{\mathbf{k}'} \, \mathbf{a}_{\mathbf{k}'} | \phi_{\text{numN}} \rangle \\ &= \frac{N(N-2)}{4}. \end{aligned} \tag{A.100}$$

Those for the second two terms in **B** and those of **C** and **D** vanish. We obtain in general

$$
\begin{aligned}
\left\langle \mathbf{P}^{(2)}(\mathbf{x}_1, \mathbf{x}_2) \right\rangle_{\mathrm{numN}} &= \frac{1}{16} \langle \phi_{\mathrm{num2}} | \mathbf{A} + \mathbf{B} | \phi_{\mathrm{num2}} \rangle \\
&= \frac{N}{32} \left\{ 2N \cos^2\left[\frac{1}{2} (\mathbf{k} - \mathbf{k}') \cdot (\mathbf{x}_1 - \mathbf{x}_2) \right] + (N-2) \right\} . \quad \text{(A.101)}
\end{aligned}
$$

For the case $N = 2$ double-slit case we obtain

$$
\left\langle \mathbf{P}^{(2)}(\boldsymbol{\rho}_1, \boldsymbol{\rho}_2) \right\rangle_{\mathrm{num2}} = \frac{1}{4} \cos^2\left[\frac{k\,\ell}{2z_0} |\rho_1 - \rho_2| \right] \mathrm{sinc}^2\left[\frac{k\,a}{2z_0} |\rho_1 - \rho_2| \right] . \quad \text{(A.102)}
$$

A.11 Evaluation of the Degree of Second Order Coherence

The degree of second order coherence may be obtained by use of the first order matrix elements in Tables 16.2 and 16.3 and the second order matrix elements in Tables 16.4 and 16.5 or directly from the diffractions patterns.

For the *collective* coherent state $|\Phi_{\mathrm{coh}}\rangle$, the numerator in (16.41) factors into the denominator so that

$$
g^{(2)}_{\mathrm{coh}}(\boldsymbol{\rho}, -\boldsymbol{\rho}) = 1 . \quad \text{(A.103)}
$$

For the coherent *substates* $|\phi_{\mathrm{cohN}}\rangle$, we obtain from (A.53) and (A.77)

$$
g^{(2)}_{\mathrm{cohN}}(\boldsymbol{\rho}, -\boldsymbol{\rho}) = 1 - \frac{1}{N} . \quad \text{(A.104)}
$$

For the two-photon entangled state $|\phi_{\mathrm{ent2}}\rangle$ given by (16.22), we obtain from (A.65) and (A.98)

$$
g^{(2)}_{\mathrm{ent2}}(\boldsymbol{\rho}, -\boldsymbol{\rho}) = 1 . \quad \text{(A.105)}
$$

For the two-photon number state $|\phi_{\mathrm{num2}}\rangle$ given by (16.24), we obtain from (A.102) and $\left\langle \mathbf{P}^{(1)}(\boldsymbol{\rho}, \boldsymbol{\rho}) \right\rangle_{\mathrm{num2}} = \left\langle \mathbf{P}^{(1)}(\boldsymbol{\rho}, \boldsymbol{\rho}) \right\rangle_{\mathrm{ent2}}$

$$
g^{(2)}_{\mathrm{num2}}(\boldsymbol{\rho}, -\boldsymbol{\rho}) = \cos^2\left[\frac{k\,\ell}{z_0} \rho \right] \mathrm{sinc}^2\left[\frac{k\,a}{z_0} \rho \right] . \quad \text{(A.106)}
$$

For the collective phase-diffused coherent state $|\Phi_{\mathrm{dif}}\rangle$, we obtain

$$
g^{(2)}_{\mathrm{dif}}(\boldsymbol{\rho}, -\boldsymbol{\rho}) = \frac{1}{2} + \cos^2\left[\frac{k\,\ell}{z_0} \rho \right] \mathrm{sinc}^2\left[\frac{k\,a}{z_0} \rho \right] \quad \text{(A.107)}
$$

and for the substates $|\phi_{\mathrm{difN}}\rangle$ we have

$$g_{\mathrm{difN}}^{(2)}(\boldsymbol{\rho}, -\boldsymbol{\rho}) = \left[1 - \frac{1}{N}\right]\left\{\frac{1}{2} + \cos^2\left[\frac{k\,\ell}{z_0}\rho\right] \operatorname{sinc}^2\left[\frac{k\,a}{z_0}\rho\right]\right\}. \tag{A.108}$$

For the collective chaotic state $|\Phi_{\mathrm{cha}}\rangle$, we obtain

$$g_{\mathrm{cha}}^{(2)}(\boldsymbol{\rho}, -\boldsymbol{\rho}) = 1 + \cos^2\left[\frac{k\,\ell}{z_0}\rho\right] \operatorname{sinc}^2\left[\frac{k\,a}{z_0}\rho\right] \tag{A.109}$$

and for the substates $|\phi_{\mathrm{chaN}}\rangle$ we have

$$g_{\mathrm{chaN}}^{(2)}(\boldsymbol{\rho}, -\boldsymbol{\rho}) = \frac{2}{3}\left[1 - \frac{1}{N}\right]\left\{1 + \cos^2\left[\frac{k\,\ell}{z_0}\rho\right] \operatorname{sinc}^2\left[\frac{k\,a}{z_0}\rho\right]\right\}. \tag{A.110}$$

References

1. J.W. Goodman, *Introduction to Fourier Optics*, 4th edn. (W. H. Freeman, New York, 2017)
2. J.W. Goodman, *Fourier Transforms Using Mathematica* (SPIE Press, Bellingham, WA, 2020)
3. E.U. Condon, G.H. Shortley, *The Theory of Atomic Spectra* (Cambridge University Press, Cambridge, 1963)
4. J.C. Slater, *Quantum Theory of Atomic Structure*, vol. II (McGraw-Hill, New York, 1960)
5. B.W. Shore, D.H. Menzel, *Principles of Atomic Spectra* (Wiley, New York, 1968)
6. C.J. Ballhausen, *Molecular Electronic Structures of Transition Metal Complexes* (McGraw-Hill, New York, 1979)
7. R.D. Cowan, *The Theory of Atomic Structure and Spectra* (University of California Press, Berkeley, 1981)
8. J.J. Sakurai, *Modern Quantum Mechanics, Revised* (Addison-Wesley, Reading, Mass, 1994)
9. See Ref. [3] page 178 and Ref. [4], Sections 22.2, 25.1 and Appendix 20
10. J. Stöhr, H.C. Siegmann, *Magnetism: From Fundamentals to Nanoscale Dynamics* (Springer, Heidelberg, 2006)

Index

J. Stöhr, *The Nature of X-Rays and Their Interactions with Matter*, Springer Tracts in Modern Physics 288, https://doi.org/10.1007/978-3-031-20744-0

Y

Z